Fundamentals of Nonparametric Bayesian Inference

Explosive growth in computing power has made Bayesian methods for infinite-dimensional models – Bayesian nonparametrics – a nearly universal framework for inference, finding practical use in numerous subject areas. Written by leading researchers, this authoritative text draws on theoretical advances of the past 20 years to synthesize all aspects of Bayesian nonparametrics, from prior construction to computation and large sample behavior of posteriors. Because understanding the behavior of posteriors is critical to selecting priors that work, the large sample theory is developed systematically, illustrated by various examples of model and prior combinations. Precise sufficient conditions are given, with complete proofs, that ensure desirable posterior properties and behavior. Each chapter ends with historical notes and numerous exercises to deepen and consolidate the reader's understanding, making the book valuable for graduate students and researchers alike in statistics and machine learning, as well as application areas such as econometrics and biostatistics.

SUBHASHIS GHOSAL is Professor of Statistics at North Carolina State University.

AAD VAN DER VAART is Professor of Stochastics at Leiden University.

CAMBRIDGE SERIES IN STATISTICAL AND PROBABILISTIC MATHEMATICS

This series of high-quality upper-division textbooks and expository monographs covers all aspects of stochastic applicable mathematics. The topics range from pure and applied statistics to probability theory, operations research, optimization, and mathematical programming. The books contain clear presentations of new developments in the field and also of the state of the art in classical methods. While emphasizing rigorous treatment of theoretical methods, the books also contain applications and discussions of new techniques made possible by advances in computational practice.

A complete list of books in the series can be found at www.cambridge.org/statistics. Recent titles include the following:

18. *Statistical Mechanics of Disordered Systems,* by Anton Bovier
19. *The Coordinate-Free Approach to Linear Models,* by Michael J. Wichura
20. *Random Graph Dynamics,* by Rick Durrett
21. *Networks,* by Peter Whittle
22. *Saddlepoint Approximations with Applications,* by Ronald W. Butler
23. *Applied Asymptotics,* by A. R. Brazzale, A. C. Davison and N. Reid
24. *Random Networks for Communication,* by Massimo Franceschetti and Ronald Meester
25. *Design of Comparative Experiments,* by R. A. Bailey
26. *Symmetry Studies,* by Marlos A. G. Viana
27. *Model Selection and Model Averaging,* by Gerda Claeskens and Nils Lid Hjort
28. *Bayesian Nonparametrics,* edited by Nils Lid Hjort *et al.*
29. *From Finite Sample to Asymptotic Methods in Statistics,* by Pranab K. Sen, Julio M. Singer and Antonio C. Pedrosa de Lima
30. *Brownian Motion,* by Peter Mörters and Yuval Peres
31. *Probability (Fourth Edition),* by Rick Durrett
33. *Stochastic Processes,* by Richard F. Bass
34. *Regression for Categorical Data,* by Gerhard Tutz
35. *Exercises in Probability (Second Edition),* by Loïc Chaumont and Marc Yor
36. *Statistical Principles for the Design of Experiments,* by R. Mead, S. G. Gilmour and A. Mead
37. *Quantum Stochastics,* by Mou-Hsiung Chang
38. *Nonparametric Estimation under Shape Constraints,* by Piet Groeneboom and Geurt Jongbloed
39. *Large Sample Covariance Matrices and High-Dimensional Data Analysis,* by Jianfeng Yao, Shurong Zheng and Zhidong Bai
40. *Mathematical Foundations of Infinite-Dimensional Statistical Models,* by Evarist Giné and Richard Nickl
41. *Confidence, Likelihood, Probability,* by Tore Schweder and Nils Lid Hjort
42. *Probability on Trees and Networks (Volume 1),* by Russell Lyons and Yuval Peres
43. *Random Graphs and Complex Networks,* by Remco van der Hofstad
44. *Fundamentals of Nonparametric Bayesian Inference,* by Subhashis Ghosal and Aad van der Vaart

Fundamentals of Nonparametric Bayesian Inference

Subhashis Ghosal
North Carolina State University

Aad van der Vaart
Leiden University

CAMBRIDGE
UNIVERSITY PRESS

CAMBRIDGE
UNIVERSITY PRESS

University Printing House, Cambridge CB2 8BS, United Kingdom

One Liberty Plaza, 20th Floor, New York, NY 10006, USA

477 Williamstown Road, Port Melbourne, VIC 3207, Australia

314-321, 3rd Floor, Plot 3, Splendor Forum, Jasola District Centre, New Delhi - 110025, India

79 Anson Road, #06-04/06, Singapore 079906

Cambridge University Press is part of the University of Cambridge.

It furthers the University's mission by disseminating knowledge in the pursuit of education, learning and research at the highest international levels of excellence.

www.cambridge.org
Information on this title: www.cambridge.org/9780521878265

First published 2017
Reprinted 2017

A catalogue record for this publication is available from the British Library

Library of Congress Cataloging in Publication data
Names: Ghosal, Subhashis. | Vaart, A. W. van der.
Title: Fundamentals of nonparametric Bayesian inference / Subhashis Ghosal, North Carolina State University, Aad van der Vaart, Leiden University.
Description: Cambridge : Cambridge University Press, 2016. |
Series: Cambridge series in statistical and probabilistic mathematics |
Includes bibliographical references and indexes.
Identifiers: LCCN 2016056429 | ISBN 9780521878265
Subjects: LCSH: Nonparametric statistics. | Bayesian statistical decision theory.
Classification: LCC QA278.8 .G46 2016 | DDC 519.5/42–dc23
LC record available at https://lccn.loc.gov/2016056429

ISBN 978-0-521-87826-5 Hardback

To Mahasweta and Meghdoot. –S. G.

Contents

Expanded Contents

Glossary of Symbols

i.i.d.	independent and identically distributed
c.d.f.	cumulative distribution function
p.d.f.	probability density function
c.h.f.	cumulative hazard function
a.s.	almost surely
a.e.	almost everywhere
$\mathbb{1}$	indicator function
card(A), $\#A$	cardinality, number of elements of a set A
$\lfloor x \rfloor$	greatest integer less than or equal to x
$\lceil x \rceil$	smallest integer greater than or equal to x
$\langle x \rangle$	fractional part of x
f', $\dot{f}$	derivative of f
$f^{(k)}$	kth derivative of f
$f \circ g$	f composed with g
$\oplus$	direct sum of subspaces
$\times$	product of numbers, Cartesian product, product of measures
$\otimes$	Kronecker product of matrices, product of σ-fields
$\setminus$	difference of sets
$\propto$	proportional to
$a{:=}b$	a is equal to b by definition
$a_n \lesssim b_n$, $a_n \gtrsim b_n$	inequalities up to irrelevant multiplicative constant
$a_n \asymp b_n$	equality of order of magnitude
$a_n \sim b_n$	$a_n/b_n \to 1$
$a_n = O(b_n)$	a_n/b_n bounded
$a_n = o(b_n)$	$a_n/b_n \to 0$
$\gg$	bounded away from
$\to_p$	convergence in probability
$\rightsquigarrow$	convergence in law/distribution, weak convergence

$\leftarrow$	left side replaced by right side
$\leftrightarrow$	one-to-one correspondence
$o_p(a_n)$	sequence of random variables tending to zero in probability
$O_p(a_n)$	sequence of random variables bounded in probability
$=_d$	equality in distribution
$x^+ = \max(x, 0), x^- = \max(-x, 0)$	
$a \vee b, a \wedge b$	maximum and minimum of a and b respectively
f^+, f^-	positive and negative parts of a function
$\log_+, \log_-,$	positive and negative parts of logarithm
$\log^k x = (\log x)^k, k > 0$	
$a^{[k]} = a(a+1)\cdots(a+k-1)$	ascending factorial
$\binom{n}{k}$	number of k-combinations out of n
$((a_{ij}))$	matrix with (i, j)th entry a_{ij}
$A^\top$	transpose of matrix A
$\det(A)$	determinant of A
$\mathrm{tr}(A)$	trace of A
$\mathrm{eig}_j(A)$	jth smallest eigenvalue of A

Sets

$\mathbb{N}$	set of natural numbers $\{1, 2, \ldots\}$
$\mathbb{N}_n = \{1, \ldots, n\}, n \in \mathbb{N}$	
$\mathbb{N}_0 = \{0, 1, 2, \ldots\}$	
$\mathbb{Z}$	set of all integers
$\mathbb{R}$	real line
$\mathbb{Q}$	set of rational numbers
$\mathbb{R}^+$	set of positive real numbers
$\mathbb{Q}^+$	set of positive rational numbers
$\mathbb{C}$	set of complex numbers
$\mathbb{R}^k$	k-dimensional Euclidean space
$\mathbb{R}^\infty$	set of all sequences of real numbers
$\mathbb{S}_k$	k-dimensional unit simplex $\{(x_1, \ldots, x_k) \in \mathbb{R}^k : x_i \geq 0, \forall i, \sum_{i=1}^k x_i = 1\}$
$\mathbb{S}_\infty$	infinite-dimensional unit simplex $\{(x_1, x_2, \ldots) \in \mathbb{R}^\infty : x_i \geq 0, \forall i, \sum_{i=1}^\infty x_i = 1\}$
$\mathscr{R}, \mathscr{R}^k, \mathscr{R}^\infty$	Borel σ-fields on $\mathbb{R}$, $\mathbb{R}^k$, $\mathbb{R}^\infty$ respectively
$\mathrm{Re}(z)$	real part of complex number z
$\mathrm{Im}(z)$	imaginary part of complex number z
$\mathrm{vol}(A)$	(hyper)volume of A (in any dimension)
$\bar{A}$	closure of set A
$\mathrm{int}(A)$	interior of A
∂A	topological boundary of A
$\mathrm{lin}(B)$	linear span of B
$\mathrm{conv}(B)$	convex hull of B

$\overline{\text{lin}}(B)$	closed linear span of B
$\overline{\text{conv}}(B)$	closed convex hull of B
$\sigma\langle\mathscr{F}\rangle$	σ-field generated by $\mathscr{F}$

Spaces, Norms and Distances

$\mathbb{L}_p(\mathfrak{X}, \mathscr{X}, \mu)$	space of p-integrable functions on measure space $(\mathfrak{X}, \mathscr{X}, \mu)$, $0 < p < \infty$
$\mathbb{L}_\infty(\mathfrak{X}, \mathscr{X}, \mu)$	space of essentially bounded functions on measure space $(\mathfrak{X}, \mathscr{X}, \mu)$
$\mathbb{L}_p(\mathbb{R})$, $\mathbb{L}_p(\mathbb{R}^k)$	Lebesgue $\mathbb{L}_p$-spaces
$\mathfrak{L}_\infty(T)$	space of bounded functions on T
ℓ_p	p-summable real sequences
ℓ_∞	bounded real sequences
$\mathfrak{W}^\alpha$	real sequences of Sobolev smoothness α: (w_j) with $(j^\alpha w_j) \in \ell_2$
$\mathfrak{M}(\mathfrak{X})$	space of probability measures on a sample space $\mathfrak{X}$
$\mathfrak{M}_\infty(\mathfrak{X})$	space of (positive) measures on a sample space $\mathfrak{X}$
$\mathfrak{C}(\mathfrak{X})$	space of continuous functions on $\mathfrak{X}$
$\mathfrak{C}_b(\mathfrak{X})$	space of bounded continuous functions on $\mathfrak{X}$
$\mathfrak{C}^\alpha(\mathfrak{X})$	Hölder space of functions of smoothness α on $\mathfrak{X}$
$\mathfrak{UC}(\mathfrak{X})$	space of uniformly continuous functions on $\mathfrak{X}$
$\mathfrak{W}^\alpha(\mathfrak{X})$	Sobolev space of functions of smoothness α on $\mathfrak{X}$
$\mathfrak{B}^\alpha_{p,q}(\mathfrak{X})$	Besov space of functions of smoothness α on $\mathfrak{X}$
$\mathfrak{D}(\mathfrak{X})$	Skorohod space of cadlag functions on $\mathfrak{X}$
$\|f\|_p$	ℓ_p-norm or $\mathbb{L}_p$-norm of f, $1 \le p \le \infty$
$\|f\|_{p,G}$	$\mathbb{L}_p$-norm of f with respect to measure G, $1 \le p \le \infty$
$\|f\|_{\text{Lip}}$	Lipschitz norm of f, smallest number such that $\lvert f(x) - f(y)\rvert \le \|f\|_{\text{Lip}}\|x - y\|$
$\|f\|_{\mathfrak{C}^\alpha}$	Hölder α-norm of f
$\|f\|_{p,q,\alpha}$	Besov norm of f
$\|f\|_{2,2,\alpha}$	Sobolev norm of order α of f
$\|f\|_\infty$	Uniform (or supremum) norm
d_{KS}	Kolmogorov–Smirnov distance
d_{TV}	total variation distance
d_H	Hellinger distance
d_L	Lévy distance
d_{BL}	bounded Lipschitz distance
$K(P; Q)$	Kullback–Leibler (KL) divergence
$V_k(P; Q)$, $V_{k,0}(P; Q)$	KL variation of order k, centered
$K^+(P; Q)$, $K^-(P; Q)$	signed KL divergences
$V^+(P; Q)$, $V^-(P; Q)$	signed KL variations
$\rho_{1/2}$	affinity
ρ_α	Hellinger transform

D_α	α-divergence
R_α	Renyi divergence
$D(\epsilon, S, d)$	ϵ-packing number of B with respect to distance d
$N(\epsilon, S, d)$	ϵ-covering number of B with respect to distance d
$N_{[\,]}(\epsilon, S, d)$	ϵ-bracketing number of B with respect to distance d
$N_{]}(\epsilon, S, d)$	ϵ-upper bracketing number of B with respect to distance d
$\mathrm{supp}(P)$	topological support of a (probability) measure
$\mathrm{KL}(\Pi)$	Kullback–Leibler support of a prior Π

Random Variables and Distributions

$\mathrm{E}(X)$	expectation of X
$\mu f = \mu(f) = \int f\, d\mu$	
$\mathrm{var}(X)$	variance of X
$\mathrm{sd}(X)$	standard deviation of X
$\mathcal{L}(X)$	law (distribution) of X
$\mathrm{cov}(X, Y)$	covariance of X and Y
$\mathrm{Cov}(X)$	covariance matrix of random vector X
$X \perp\!\!\!\perp Y$	random variables X and Y are independent
$X =_d Y$	X and Y have the same distribution/law
$X \sim P$	X has distribution P
$X_i \stackrel{\mathrm{iid}}{\sim} P$	$X_1, X_2, \ldots$ are i.i.d. and have distribution P
$X_i \stackrel{\mathrm{ind}}{\sim} P_i$	$X_1, X_2, \ldots$ are independent and $X_i \sim P_i$
$\mathbb{P}_n$	empirical measure
$\mathbb{G}_n$	empirical process
$\mathrm{Nor}(\mu, \sigma^2)$	normal distribution with mean μ and variance σ^2
$\mathrm{Nor}_k(\mu, \Sigma)$	k-variate normal distribution with mean vector μ and dispersion matrix Σ
$\phi_{\mu,\Sigma}$	normal density with mean (vector) μ and dispersion (matrix) Σ
$\mathrm{Bin}(n, p)$	binomial distribution with parameters n and p
$\mathrm{Poi}(\lambda)$	Poisson distribution with mean λ
$\mathrm{Unif}(a, b)$	uniform distribution over (a, b)
$\mathrm{Exp}(\lambda)$	exponential distribution with mean $1/\lambda$
$\mathrm{Ga}(a, b)$	gamma distribution with shape a and scale b
$\mathrm{ga}(x; a, b)$	gamma density $x \mapsto (b^a/\Gamma(a))e^{-bx}x^{a-1}$
$\mathrm{Be}(a, b)$	beta distribution with parameter a and b
$\mathrm{be}(x; a, b)$	beta density $x \mapsto (\Gamma(a+b)/\Gamma(a)\Gamma(b))\, x^{a-1}(1-x)^{b-1}$
$\mathrm{Wei}(\alpha, \lambda)$	Weibull distribution with shape α and scale λ
Rad	Rademacher distribution, ± 1 with probability $1/2$
$\mathrm{MN}_k(n; p_1, \ldots, p_k)$	k-variate multinomial distribution with n trials
$\mathrm{Dir}(k; \alpha_1, \ldots, \alpha_k)$	k-dimensional Dirichlet distribution
$\mathrm{IGau}(\alpha, \gamma)$	inverse-Gaussian distribution
$\mathrm{NIGau}(k, \alpha)$	(multivariate) normalized inverse-Gaussian distribution on the unit simplex

$\mathrm{DP}(\alpha)$, DP_α	Dirichlet process and measure with base measure α
$\mathrm{IDP}(\alpha)$	invariant Dirichlet process with parameter α
$\mathrm{PT}(\mathcal{T}_m, \mathcal{A})$	Pólya tree process with partitions $\{\mathcal{T}_m\}$ and parameters $\mathcal{A}$
$\mathrm{PT}^*(\alpha, a_m)$	canonical Pólya tree process with mean α and parameters $\{a_m\}$

Preface

This book is our attempt to synthesize theoretical, methodological and computational aspects of Bayesian methods for infinite-dimensional models, popularly known as Bayesian nonparametrics. While Bayesian nonparametrics existed in an informal fashion as far back in time as Henri Poincaré, it became a serious methodology only after the introduction of the Dirichlet process by Thomas Ferguson in 1973. The development of computing algorithms especially suited for Bayesian analysis in the 1990s together with the exponential growth of computing resources enabled Bayesian nonparametrics to go beyond the simplest problems and made it a universally applicable paradigm for inference. Unfortunately, examples showed that seemingly natural priors can lead to uncomfortable properties of the posterior distribution, making a systematic study of the behavior of posterior distributions of infinite-dimensional parameters essential. The framework for such a study is frequentist. It assumes that the data are generated according to some "true" distribution, and the question is whether and how well the posterior distribution can recover this data-generating mechanism. This is typically studied when the sample size increases to infinity, although bounds for finite sample sizes are obtained implicitly within the analysis. Many significant results on such frequentist behavior of the posterior distribution have been obtained in the past 15–20 years building upon the seminal work of Lorraine Schwartz published in 1965. A fairly complete theory of the large sample behavior of the posterior distribution for dominated experiments is now presentable, covering posterior consistency, contraction rates, adaptation, and distributional approximations. A primary focus of this book is to describe this theory systematically, using examples of model and prior combinations, and to convey a general understanding of what type of priors make a Bayesian nonparametric method work. It is manifest that we emphasize positive aspects of Bayesian methods for infinite-dimensional models. We state and prove theorems that give precise conditions that ensure desirable properties of Bayesian nonparametric methods. Proofs are almost always given completely, often generalized or simplified compared with the existing literature. Historical notes at the end of each chapter point out the original sources. Some results improve on the literature or appear here for the first time.

The book is intended to give a treatment of all aspects of Bayesian nonparametrics: prior construction, computation and large sample behavior of the posterior distribution. It is important to consider these aspects together, as a prior is useful only if the posterior can be computed efficiently and has desirable behavior. These two properties of posteriors can thus throw light on how to construct useful priors and what separates a good prior from a not-so-good one. An introductory chapter describes the goals, advantages and difficulties of nonparametric Bayesian inference. Chapters 2 and 3 deal with methods of prior constructions on functions and distributions. In Chapter 4 the Dirichlet process is introduced and and studied. This continues in Chapter 5 with kernel mixtures with the Dirichlet process prior on the mixing distribution. This chapter also describes tools for computing the corresponding posterior distribution, such as Markov chain Monte Carlo techniques and variational methods. Posterior consistency theory is developed in Chapter 6 and applications are discussed in Chapter 7. Chapter 8 refines the theory to contraction rates, and applications are presented in Chapter 9. In Chapter 10 on Bayesian adaptation it is shown that an optimal contraction rate is often obtainable without knowing the complexity of the true parameter, by simply putting a good prior on this complexity. Related results on the behavior of Bayes factors are also presented in this chapter. Chapter 11 deals with properties of Gaussian process priors and the resulting posteriors for various inference problems, and gives an overview of methods for posterior computation. Bernstein–von Mises-type results for infinite-dimensional models are studied in Chapter 12. Priors based on independent increment processes, applied to survival models, are the subject of Chapter 13. As the corresponding models are not dominated, Bayes's theorem does not apply directly, so posterior theory and large sample theory is developed based on conjugacy. Finally Chapter 14 studies properties of discrete random distributions and the partition structures that arise when sampling from such distributions. Dependencies between the chapters are shown in Figure 1.

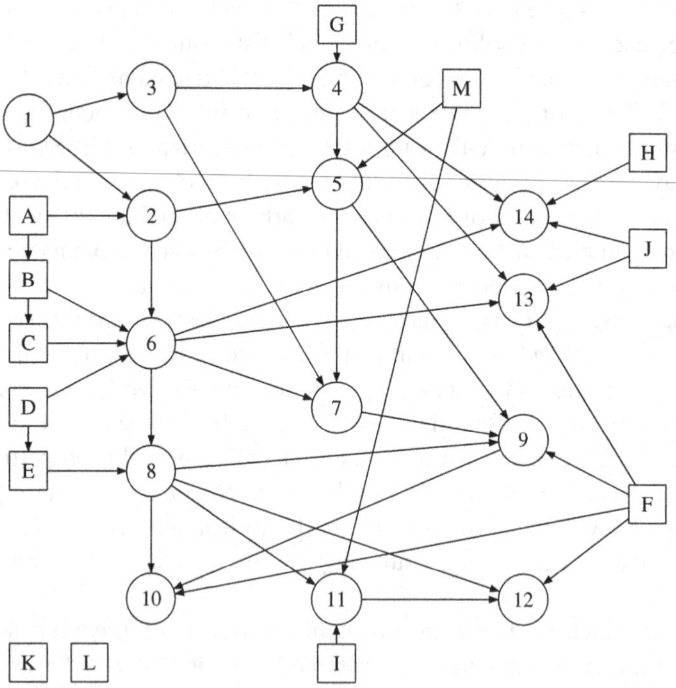

Figure 1 Dependencies between chapters. A chapter at the origin of an arrow must be read before the chapter at its end. Chapters from the main body of the book (given by numbers) are represented by circles, whereas chapters from the appendix (with letters) are given by squares.

Although the asymptotic behavior of posterior distributions is a main focus of the book, a considerable part of the book concerns properties of priors, posterior updating, and computation. This is true in particular for Chapters 1–5 and the biggest part of Chapters 13–14. Thus even those who (surprisingly) are not interested in large sample studies may find much useful material. The only aspect we do not cover is the application of the methods on specific data sets. There are several excellent books dedicated to applied Bayesian nonparametrics, for instance Dey et al. (1998) and the recent publication Müller et al. (2015).

Bayesian nonparametrics continues to develop, so that this book cannot be comprehensive. Recent developments that are not covereed or are barely covered are posterior contraction rates under stronger norms, Bayesian analysis of inverse problems, including deconvolution problems, the performance of empirical Bayes procedures and nonparametric Bernstein–von Mises results. Perhaps the biggest omission regards the coverage of (adaptive) credible regions, which is central to justifying (or not) the use of the full posterior distribution in a truly Bayesian sense. All these topics are still actively developing and their theory is incomplete at the present. Even without these topics the book has swelled considerably in size, more than we imagined in the beginning. Notwithstanding this growth, many interesting results within the book's scope had to be omitted because of space (and time) constraints. Problem sections at the end of each chapter describe many of the omitted results, with an original source cited. While this hardly does justice to the omitted results, readers can at least access the statements without going to the original sources. Most of these problems should be interpreted as pointers to supplementary reading rather than simple exercises. At the end of each chapter we also give a brief account of the historical development of the material, to the extent known to us. We apologize for (unintended) omissions, and any failures to give proper credit.

Reading the book requires familiarity with graduate-level mathematics, although some chapters and sections are more demanding in this respect than others. Probability theory and real analysis are needed almost throughout the book, and familiarity with measure theory is assumed, at least to the extent of understanding abstract integration. Measurability issues are addressed properly, but readers not concerned about measurability can typically skip these portions of the text and follow the rest of the material. Functional analysis beyond the conception of function spaces is not heavily used, except in the chapters on Gaussian processes and the Bernstein–von Mises theorem, and in the section on misspecification. Mathematical rigor is maintained, and formal proofs of almost all results are given. For the reader's convenience, we collected in 13 appendices many results, typically with proofs, which are needed in the main body of the book but not directly part of Bayesian nonparametrics. This concerns mathematical and probabilistic tools such as distances on measures, entropy, function approximation theory, Gaussian processes and completely random measures, as well as statistical tools from the theory of testing. Often prior reading of the relevant section of the appendix is advisable for better understanding a chapter or section in the main body of the book. The diagram in Figure 1 shows which sections of the appendix are used in which chapters in the main body.

One goal for writing this book was to provide a text appropriate for teaching a one-semester course on Bayesian nonparametrics. In our experience, Sections 1.2, 2.1–2.5, 3.2–3.5, 3.7, 4.1, 4.3, 4.5, 4.7, 5.1–5.3, 5.5, 6.1, 6.4, 6.7.1, 7.1.1, 7.1.2, 7.2.1, 7.4.1, 8.1, 8.2, 8.3.1, 9.1, 9.4 (without proof), 10.1, 10.3, 10.4.1, 10.5, 10.6.2, 11.1–11.3, 13.1, 13.3, 13.4, 13.4.1, 14.1, 14.2, 14.4, along with the relevant prerequisites from the appendix provide a good basis for a primarily theoretical course with some material omitted based on judgement. For a more methodology-oriented course, asymptotic properties other than the most basic consistency results may be omitted, and the material may be supplemented by the additional computational techniques in Section 11.7 and data applications from a good book on applied Bayesian nonparametrics.

The first draft of the book goes back nearly a decade, and parts of the book were written even earlier. It goes without saying that the landscape of Bayesian nonparametrics changed continually. The writing was nourished by opportunities to deliver graduate courses and short courses. The first author taught two courses at North Carolina State University as well as short courses at the Institut de Mathématiques de Luminy, France (2006), at the "Nonparametric Bayesian Regression" meeting in Cambridge, UK (2007), at Bilkent University, Ankara, Turkey (2007), and at Eurandom, Eindhoven, The Netherlands (2011), based on the drafts of the book available at the corresponding time. The second author used parts of the manuscript for master courses in the Netherlands and several short courses in Germany.

This book would not have materialized without contributions from various people. A large chunk of the presented material comes from our work with co-authors (including student co-authors): Eduard Belitser, Ismaël Castillo, Jayanta Ghosh, Jiezhun Gu, Bas Kleijn, Bartek Knapik, Willem Kruijer, Juri Lember, R.V. Ramamoorthi, Judith Rousseau, Anindya Roy, Weining Shen, Botond Szabó, Yongqiang Tang, Surya Tokdar, William Weimin Yoo, Yuefeng Wu, Harry van Zanten. Working with them has always been enjoyable and furthered our understanding of Bayesian nonparametrics. Igor Prünster was kind enough to read first drafts of the last two chapters of the manuscript. He provided us with many important references, answered many questions and supplied several missing proofs, especially in the last chapter. It would be difficult to imagine that chapter taking shape without his tireless help. S. M. Srivastava helped us to clarify matters related to descriptive set theory. Ismaël Castillo, Kolyan Ray and Botond Szabó read portions of the manuscript and provided useful resources. William Weimin Yoo and Julyan Arbel carefully read parts of the manuscript and pointed out many typos and small mistakes. Discussions with Peter Müller, Surya Tokdar, Yongdai Kim, Antonio Lijoi and Sasha Gnedin were instrumental in resolving many questions. The book would not have been possible without face-to-face meeting between us through occasional visits. Leiden University, VU Amsterdam, North Carolina State University and Eurandom hosted these visits. Funding from the Netherlands Organisation for Scientific Research (NWO), the Royal Netherlands Academy of Sciences (KNAW), the European Research Council (ERC Grant Agreement 320637), and the US National Science Foundation (NSF) were instrumental in supporting some of these visits, and also contributed to the research in general. We are indebted to the staff of Cambridge University Press, particularly Diana Gillooly, for their help, patience and constant encouragement throughout the long process.

Subhashis Ghosal
Aad van der Vaart

1

Introduction

Why adopt the nonparametric Bayesian approach for inference? The answer lies in the simultaneous preference for nonparametric modeling and desire to follow a Bayesian procedure. Nonparametric (and semiparametric) models can allow one to avoid the arbitrary and possibly unverifiable assumptions inherent in parametric models. Bayesian procedures may be desirable because of the conceptual simplicity of the Bayesian paradigm, philosophical reasons.

1.1 Motivation

Bayesian nonparametrics concerns Bayesian inference methods for nonparametric and semiparametric models. In the Bayesian nonparametric paradigm, a prior distribution is assigned to all relevant unknown quantities, whether finite or infinite dimensional. The *posterior distribution* is the conditional distribution of these quantities, given the data, and is the basis for all inference. This is the same as in any Bayesian inference, except that the unknown quantities or parameters may be infinite dimensional. A *model* completely specifies the conditional distribution of all observed, given all unobserved quantities, or parameters, while a *prior distribution* specifies the distribution of all unobservables. The posterior distribution involves an inversion of the order of conditioning and gives the distribution of the unobservables, given the observables. Existence of a regular version of the posterior distribution is guaranteed under mild conditions on the relevant spaces (see Section 1.3). From the Bayesian point of view, random effects and latent variables are unobservables and are treated in the same way as the unknown parameters used to describe the model. Distributions of these quantities, often considered as part of the model itself from the classical point of view, are part of the prior.

1.1.1 Classical versus Bayesian Nonparametrics

Nonparametric and semiparametric statistical models are increasingly replacing parametric models as a way to gain the flexibility necessary to address a wide variety of data. A nonparametric or semiparametric model involves at least one infinite-dimensional parameter and hence may also be referred to as an "infinite-dimensional model." Indeed, the nomenclature "nonparametric" is misleading in that it gives the impression that there are no parameters in the model, while in reality there are infinitely many unknown quantities. However, the term nonparametric is so popular that it makes little sense not to use it. The infinite-dimensional parameter is usually a function or measure. In the canonical example of a nonparametric model, the data are a random sample from a completely unknown distribution or density. More generally, models may be structured and data non-i.i.d., and functions of interest include the cumulative distribution function, density function, regression function, hazard rate, transition density of a Markov process, spectral density of a time series, response probability of a binary variable as a function of covariates, false discovery rate as a function of nominal level in multiple testing and receiver operating characteristic function between distributions. Non-Bayesian methods for the estimation of many of these functions are well understood, and widely accepted statistical procedures are available. Bayesian estimation methods for nonparametric problems started receiving attention over the past four decades.

Different people may like Bayesian methods for different reasons. To some the appeal is philosophical. Certain *axioms of rational behavior* lead to the conclusion that one ought to follow a Bayesian approach in order not to be irrational (see Bernardo and Smith 1994). Although the axioms themselves can be questioned, the impression that Bayesian methods are logically more consistent than non-Bayesian methods is widespread. In particular, expressing uncertainty in probabilities is more satisfying than using criteria that involve integration over the sample space – that is, bother about samples that could have, but have not, realized. Others justify the Bayesian paradigm by appealing to exchangeability and de Finetti's theorem (de Finetti 1937). This celebrated theorem concludes the existence of a "random parameter" instead of a "fixed parameter" based on a "concrete" set of observations and a relatively weak assumption on distributional invariance (see Schervish 1995). However, this argument leads to subjective specification of a prior, which is regarded as difficult. Decision theorists may be Bayesians because of the complete class theorem, which asserts that for any procedure there is a better Bayesian procedure, and that only the latter procedures are admissible, or essentially so (see Ferguson 1967). While this could be a strong reason for a frequentist to take the Bayesian route, there are difficulties in that the complete class theorem holds only when the parameter space is compact (and the loss function is convex) and that the argument does not say which prior to choose from among the class of all priors. People who believe in asymptotic theory might find Bayesian methods attractive for their large-sample optimality. However, many non-Bayesian procedures (most notably, the maximum likelihood estimator, or MLE) are also asymptotically optimal, hence the argument is not compelling.

Although the specification of a prior distribution may be challenging, the Bayesian approach is extremely straightforward, in principle – the full inference is based on the posterior distribution only. All inference tools are produced in one stroke, and one need not start afresh when the focus of attention changes from one quantity to another. In particular, the same analysis produces an estimate as well as an assessment of its accuracy (in terms of variability or a probable interval for the location of the parameter value). The Bayesian approach produces a "real probability" on the unknown parameter as a quantification of the uncertainty about its value, which may be used to construct a "credible interval" or test with a clear interpretation. The Bayesian approach also eliminates the problem of nuisance parameters by integrating them out, while classical procedures must often find ingenious ways to tackle them separately for each inference problem. Finally, prediction problems, which are often the primary objective of statistical analysis, are solved most naturally if one follows the Bayesian approach.

These conveniences come at a price, however. The Bayesian principle is also restrictive in nature, allowing no freedom beyond the choice of prior. This limitation can put Bayesian methods at a disadvantage vis-à-vis non-Bayesian methods, particularly when performance is evaluated by frequentist principles. For instance, even if only a part of the unknown parameter is of interest, a Bayesian must still specify a prior on the whole parameter, compute the posterior distribution, and integrate out the irrelevant part, whereas a classical procedure may be able to target the part of interest. Another problem is that no corrective measures are allowed in a Bayesian framework once the prior has been specified. In contrast, the MLE is known to be nonexistent or inconsistent in many infinite-dimensional problems, such as density estimation, but it can be modified by penalization, sieves (see Grenander 1981), partial likelihood (Cox 1972) or other devices (Murphy and van der Vaart 2000). An honest Bayesian cannot change the likelihood, change the prior by looking at the data or even change the prior with increasing sample size.

1.1.2 Parametric versus Nonparametric Bayes

Parametric models make restrictive assumptions about the data-generating mechanism, which may cause serious bias in inference. In the Bayesian framework, a parametric model assumption can be viewed as an extremely strong prior opinion. Indeed, a parametric model specification $X|\,\theta \sim p_\theta$, for $\theta \in \Theta \subset \mathbb{R}^d$, with a prior specification $\theta \sim \pi$, may be written as $X|\,p \sim p$ for $p \sim \Pi$ and a prior distribution Π on the set of all possible densities with the property that $\Pi(\{p_\theta\colon \theta \in \Theta\}) = 1$. Thus parametric modeling is equivalent to insisting on a prior that assigns probability one to a thin subset of all densities. This is a very strong prior opinion indeed, which is replaced by an open-minded view when following the nonparametric Bayesian approach.

To some extent, the nonparametric Bayesian approach also solves the problem of partial specification. Often a model is specified incompletely, without describing every detail of the data-generating mechanism. A familiar example is the Gauss-Markov setup of a linear model, where errors are assumed to be uncorrelated, mean-zero variables with constant variance, but no further distributional assumptions are imposed. Lacking a likelihood, a parametric Bayesian approach cannot proceed further. However, a nonparametric Bayesian approach can use a prior on the space of densities with mean zero as a model for the error distribution. More generally, incomplete model assumptions may be complemented by general assumptions involving infinite-dimensional parameters in order to build a complete model, which a nonparametric Bayesian approach can equip with infinite-dimensional priors.

1.2 Challenges of Bayesian Nonparametrics

This section describes some conceptual and practical difficulties that arise in Bayesian nonparametrics, along with possible remedies.

1.2.1 Prior Construction

A Bayesian analysis cannot proceed without a prior distribution on all parameters. A prior ideally expresses a quantification of knowledge from past experience and subjective feelings.

A prior on a function requires knowledge of many aspects of the function, including infinitesimal details, and the ability to quantify the information in the form of a probability measure. This poses an apparent conceptual contradiction: a nonparametric Bayesian approach is pursued to minimize restrictive parametric assumptions, but at the same time it requires specification of the minute details of a prior distribution for an infinite-dimensional parameter.

There seems to be overall agreement that subjective specification of a prior cannot be expected in complex statistical problems. Instead, inference must be based on a *default prior*. This is vaguely understood as a prior distribution that is proposed by some automatic mechanism that is not biased toward any particular parameter values and has low information content compared to the data.

Some of the earliest statistical analyses in history used the idea of *inverse probability* and came down to a default Bayesian analysis with respect to a uniform prior. Later uniform priors were strongly criticized for lacking invariance, which led to a decline in the popularity of Bayesian analysis until more invariance-friendly methods such as *reference analysis* or *probability matching* emerged. However, most of these ideas are restricted to finite-dimensional parametric problems.

A default prior need not be noninformative, but should be spread over the whole parameter space. Some key hyperparameters regulating the prior may be chosen by the user, whereas other details must be constructed by the default mechanism. Unlike in parametric situations, where noninformative priors are often improper, default priors considered in nonparametric Bayesian inference are almost always proper. Large support of the prior means that the prior is not too concentrated in some particular region. This situation generally ensures that the information contained in the prior is subdued gradually by the data if the sample size increases, so that eventually the data override the prior.

The following chapters discuss methods of prior construction for various problems of interest. Although a default prior is not unique in any sense, it is expected that over the years, based on theoretical results and practical experience, a handful of suitable priors will be short-listed and cataloged for consensus use in each inference problem.

1.2.2 Computation

The property of conjugacy played an important role in parametric Bayesian analysis, as it enabled the derivation of posterior distributions at a time when computing resources were lacking. Later sampling-based methods such as the Metropolis-Hastings algorithm and Gibbs sampling gave Bayesian analysis a tremendous boost. Without modern computing, nonparametric Bayesian analysis would hardly be practical.

However, we cannot simulate directly from the posterior distribution of a function unless it is parameterized by finitely many parameters. The function of interest must be broken up into more elementary finite-dimensional objects whose posterior distributions are accessible. For this reason, the structure of the prior is important. Useful structure may come from conjugacy or approximation. Often, a computational method combines analytic derivation and Markov chain Monte Carlo (MCMC) algorithms, and is based on innovative ideas. For instance, density estimation with a Dirichlet mixture prior, discussed in Chapter 4, uses an equivalent hierarchical mixture model involving a latent variable for each observation and integrates out the infinite-dimensional parameter, given the latent variables (see 5.3). Thus the problem of infinite dimension has been reduced to one of finite dimension. In another instance, in a binary response model with a Gaussian process prior, introducing normal latent variables brings in conjugacy (see Section 11.7.3).

1.2.3 Asymptotic Behavior

Putting a prior on a large parameter space makes it easy to be grossly wrong. Therefore "robustness" is important in Bayesian nonparametrics: the choice of prior should not influence the posterior distribution too much. This is difficult to study in a general framework. A more manageable task is the study of asymptotic properties of posterior distributions, as the information in the data increases indefinitely. For example, "posterior consistency" may be considered an asymptotic form of robustness. Loosely speaking, posterior consistency means that the posterior probability eventually concentrates in a (any) small neighborhood of the actual value of the parameter. This is a weak property shared by many prior distributions. Finer properties, such as the rate of contraction or a (functional) limit theorem, give more insight into the performance of different priors.

The study of asymptotic properties is more complex in the nonparametric than in the parametric context. In the parametric setting, good properties are guaranteed under mild conditions. Under some basic regularity conditions, it suffices that the true value of the parameter belongs to the support of the prior. In the infinite-dimensional context this is not enough. Consistency may fail for natural priors satisfying the support condition, meaning that even an infinite amount of data may not overcome the pull of a prior in a wrong direction. Consistent priors may differ strongly in accuracy, depending on their fine details, as can be measured through their rates of contraction. Unlike in the parametric setting, many priors do not "wash out," as the information in the data increases indefinitely.

Thus it makes sense to impose posterior consistency and a good rate of contraction as requirements on a "default prior." Several chapters in this book are devoted to the study of asymptotic behavior of the posterior distribution and other related quantities. Chapter 10 is devoted to combining priors hierarchically into an overall prior so as to make the posterior "adapt" to a large class of underlying true parameters.

1.3 Priors, Posteriors and Bayes's Rule

In the Bayesian framework, the data X follows a distribution determined by a parameter θ, which is itself considered to be generated from a *prior distribution* Π. The corresponding *posterior distribution* is the conditional distribution of θ, given X. This framework is identical in parametric and nonparametric Bayesian statistics, the only difference being the dimension of the parameter. Because the proper definitions of priors and (conditional) distributions require (more) care in the nonparametric case, we review the basics of conditioning and Bayes's rule in this section.

If the parameter set Θ is equipped with a σ-field $\mathscr{B}$, the prior distribution Π is a probability measure on the measurable space $(\Theta, \mathscr{B})$, and the distribution P_θ of X, given θ, is a regular conditional distribution on the sample space $(\mathfrak{X}, \mathscr{X})$ of the data,[1] then the pair (X, θ) has a well-defined joint distribution on the product space $(\mathfrak{X} \times \Theta, \mathscr{X} \otimes \mathscr{B})$, given by

$$\mathrm{P}(X \in A, \theta \in B) = \int_B P_\theta(A)\, d\Pi(\theta).$$

This gives rise to the *marginal distribution* of X, defined by

$$\mathrm{P}(X \in A) = \int P_\theta(A)\, d\Pi(\theta), \qquad A \in \mathscr{X},$$

and the conditional distribution, called *posterior distribution*, given by

$$\Pi(B|\, X) = \mathrm{P}(\theta \in B|\, X), \qquad B \in \mathscr{B}.$$

By Kolmogorov's definition, the latter conditional probabilities are always well defined, for every given $B \in \mathscr{B}$, as measurable functions of X such that $\mathrm{E}[\Pi(B|\, X)\mathbb{1}\{A\}(X)] = \mathrm{P}(X \in A, \theta \in B)$, for every $A \in \mathscr{X}$. If the measurable space $(\Theta, \mathscr{B})$ is not too big, then there also exists a *regular version* of the conditional distribution: a Markov kernel from $(\mathfrak{X}, \mathscr{X})$ into $(\Theta, \mathscr{B})$. We shall consider the existence of a regular version to be necessary in order to speak of a true posterior distribution. A sufficient condition is that Θ is a Polish space[2] and $\mathscr{B}$ its Borel σ-field.[3]

Even though the posterior distribution can usually be thus defined, some further care may be needed. It is inherent in the definition that the conditional probabilities $\mathrm{P}(\theta \in B|\, X)$ are unique only up to null sets under the marginal distribution of X. Using a regular version (on a standard Borel space) limits these null sets to a single null set that works for every measurable set B, but does not eliminate them altogether. This is hardly a concern if the full Bayesian setup is adopted, as this defines the marginal distribution of X as the appropriate data distribution. However, if the Bayesian framework is viewed as a method for inference only and it is allowed that the data X is generated according to some "true" distribution P_0 different from the marginal distribution of X in the Bayesian setup, then the exceptional "null sets" may well have nonzero mass under this "true" distribution, and it is impossible to speak of *the* posterior distribution. (To distinguish P_0 from the marginal distribution of X, we shall refer to the latter sometimes as the "Bayesian marginal distribution.")

Obviously, this indefiniteness can only happen under serious "misspecification" of the prior. In particular, no problem arises if

$$P_0 \ll \int P_\theta\, d\Pi(\theta),$$

which is guaranteed for instance if P_0 is dominated by P_θ for θ in a set of positive prior probability. In parametric situations the latter condition is very reasonable, but the nonparametric case can be more subtle, particularly if the set of all P_θ is not dominated. Then there may be a "natural" way of defining the posterior distribution consistently for all X, but it must

[1] I.e. a *Markov kernel* from $(\Theta, \mathscr{B})$ into $(\mathfrak{X}, \mathscr{X})$: the map $A \mapsto P_\theta(A)$ is a probability measure for every $\theta \in \Theta$ and the map $\theta \mapsto P_\theta(A)$ is measurable for every $A \in \mathscr{X}$.

[2] A topological space that is a complete separable metric space relative to some metric that generates the topology.

[3] More generally, it is sufficient that $(\Theta, \mathscr{B})$ is a *standard Borel space*: a measurable space admitting a bijective, bimeasurable correspondence with a Borel subset of a Polish space.

be kept in mind that this is not dictated by Bayes's rule alone. An important example of this situation arises with the nonparametric Dirichlet prior (see Chapter 4), where the marginal distribution may or may not dominate the distribution of the data.

For a dominated collection of measures P_θ, it is generally possible to select densities p_θ relative to some σ-finite dominating measure μ such that the map $(x, \theta) \mapsto p_\theta(x)$ is jointly measurable. Then a version of the posterior distribution is given by *Bayes's formula*

$$\Pi(B|\, X) = \frac{\int_B p_\theta(X)\, d\Pi(\theta)}{\int p_\theta(X)\, d\Pi(\theta)}. \tag{1.1}$$

Of course, this expression is defined only if the denominator $\int p_\theta(X)\, d\Pi(\theta)$, which is the (Bayesian) *marginal density* of X, is positive. Definitional problems arise (only) if this is *not* the case under the true distribution of the data. Incidentally, the formula also shows that a Polish assumption on $(\Theta, \mathscr{B})$ is sufficient, but not necessary, for existence of the posterior distribution: (1.1) defines a Markov kernel as soon as it is well defined.

In a vague sense, the *support* of a measure is a smallest set that contains all its mass. A precise definition is possible only under assumptions on the measurable space. We limit ourselves to Polish spaces, for which the following definition of support can be shown to be well posed.

Definition 1.1 (Support) The *support* of a probability measure on the Borel sets of a Polish space is the smallest closed set of probability one. Equivalently, it is the set of all elements of the space for which every open neighborhood has positive probability.

It is clear that a posterior distribution will not recover a "nonparametric" set of true distributions unless the prior has a large support. In Chapters 6 and 8 this requirement will be made precise in terms of posterior consistency (at a rate), which of course depends both on the prior and on the way the data distribution P_θ depends on the parameter θ. As preparation, when discussing priors in the following chapters, we pay special attention to their supports.

1.3.1 Absolute Continuity

Bayes's formula (1.1) is available if the model $(P_\theta: \theta \in \Theta)$ is dominated. This is common in parametric modeling, but may fail naturally in nonparametric situations. As a consequence, sometimes we perform Bayesian analysis without Bayes. Mathematically, this is connected to absolute continuity of prior and posterior distributions.

It seems natural that a prior distribution supported on a certain set yields a posterior distribution supported inside the same set. Indeed, the equality $\Pi(B) = \mathrm{EP}(\theta \in B|\, X)$ immediately gives the implication: if $\Pi(B) = 1$, then $\mathrm{P}(\theta \in B|\, X) = 1$, almost surely. The exceptional null set is again relative to the marginal distribution of X, and it may depend on the set B. The latter dependence can be quite serious. In particular, the valid complementary implication: if $\Pi(B) = 0$, then $\mathrm{P}(\theta \in B|\, X) = 0$ almost surely, should not be taken as proof that the posterior is always absolutely continuous with respect to the prior. The nonparametric Dirichlet prior exemplifies this, as the posterior is typically orthogonal to the prior (see Section 4.3.4).

Such issues do not arise in the case that the collection of distributions P_θ is dominated. Formula (1.1) immediately shows that the posterior is absolutely continuous relative to the prior in this case (where it is assumed that the formula is well posed). This can also be reversed. In the following lemma we assume that the posterior distribution is a regular conditional distribution, whence it is unique up to a null set.

Lemma 1.2 *If both $(\mathfrak{X}, \mathscr{X})$ and $(\Theta, \mathscr{B})$ are standard Borel spaces, then the set of posterior distributions $\mathrm{P}(\theta \in B \mid X = x)$, where $x \in \mathfrak{X}_0$ for a measurable set $\mathfrak{X}_0 \subset \mathfrak{X}$ of marginal probability one, is dominated by a σ-finite measure if and only if the collection $\{P_\theta : \theta \in \Theta_0\}$ is dominated by a σ-finite measure, for some measurable set $\Theta_0 \subset \Theta$ with $\Pi(\Theta_0) = 1$. In this case, the posterior distributions are dominated by the prior.*

Proof A collection of probability measures $\{P_\theta : \theta \in \Theta\}$ on a standard Borel space is dominated if and only if it is separable relative to the total variation distance, and in this case the measures permit densities $x \mapsto p_\theta(x)$ that are jointly measurable in (x, θ) (e.g. Strasser 1985, Lemmas 4.6 and 4.1). Formula (1.1) then gives a version of the posterior distribution, which is dominated by the prior. Any other version differs from this version by at most a null set $\mathfrak{X}_0^c$.

The converse follows by interchanging the roles of x and θ. If the set of posterior distributions is dominated by a σ-finite measure, then they can be represented by conditional densities $\pi(\theta \mid x)$ relative to the dominating measure, measurable in (x, θ), and we can reconstruct a regular version of the conditional distribution of x given θ by (1.1) with the roles of θ and x interchanged, which is dominated. By assumption, the original distributions P_θ give another regular version of this conditional distribution and hence agree with the dominated version on a set of probability one. □

1.4 Historical Notes

Laplace and Bayes pioneered the idea of inverse probability. Fisher was a strong critic of this approach because of invariance-related paradoxes. Jeffreys revived the idea of inverse probability by replacing the uniform prior with his famous prior, the square root of the determinant of Fisher's information. The theory of objective priors has now evolved extensively; see Bernardo and Smith (1994) for different approaches and references. A Bayesian nonparametric idea seems to have been first used by Poincaré for numerical interpolation. He considered the unknown function as a random series and function values as observed up to measurement error. He computed Bayes estimates by assigning priors on the coefficients in the expansion. Freedman (1963) considered tail-free priors, a general class that can avoid inconsistency problems. Computational issues were still not considered very seriously. A breakthrough came in the seminal paper in which Ferguson (1973) introduced the Dirichlet process, which initiated modern Bayesian nonparametrics. The Dirichlet process has only two easily interpretable hyperparameters and leads to an analytically tractable posterior distribution. This advance solved the problem of estimating a cumulative distribution by a Bayesian method and reduced the gap with the classical approach, which already offered solutions to density estimation, nonparametric regression and other curve estimation problems using kernel and other smoothing methods. Within a decade, a Bayesian solution to density estimation by Dirichlet mixture processes was available, although computational challenges remained daunting. A little later, Gaussian processes and Pólya tree processes were proposed as alternative solutions. The early nineties also witnessed the development of priors for survival analysis, including the beta process prior of Hjort (1990) and independent increment processes. The first half of the nineties saw the astonishing development of computational methods for Bayesian nonparametrics, ostensibly because of the advent of MCMC techniques and the availability of fast computers. Major initial contributions came from Escobar, West, MacEachern, Müller, Neal and others, fueled by the progress on MCMC ideas in general by Gelfand, Smith, Green, Chib and the development of the *WinBUGS* software by Gilks et al. (1994). Recently, the R-package *DP package* has been developed to solve Bayesian nonparametric computational problems; see Jara (2007). The earliest results on posterior consistency for nonparametric problems were obtained by Doob (1949), but those results give no clue about consistency at a given true value of the parameter. Freedman (1963, 1965) constructed examples of posterior inconsistency and showed, in general, that a posterior distribution can asymptotically misbehave for most priors (here, "most" is used in a topological sense). Freedman also introduced the concept of tail-freeness, a key property that can eliminate inconsistency problems. The most significant early result on consistency was due to Schwartz (1965), whose ideas led to a reasonably complete general theory of consistency. Schwartz showed that the support of the prior in a Kullback-Leibler sense is the key factor in determining consistency, together with a restriction on the size of the model imposed through a testing condition. The original Schwartz theorem is, however, incapable of yielding consistency for density estimation. Subsequent ideas were developed by Barron, Ghosh, Ghosal, Wasserman, Ramamoorthi, Walker and others. Diaconis and Freedman (1986b,a) and Doss (1985a,b) pointed out serious problems of inconsistency with the Dirichlet process prior if used for the error distribution in the location problem. This is an important phenomenon, as such a prior was considered natural for the location problem in the eighties. It emphasized that posterior consistency cannot be taken lightly in nonparametric problems. A more striking example of posterior inconsistency was recently constructed by Kim and Lee (2001) in the context of survival analysis. The study of rates of contraction in a general framework was started at the beginning of the previous decade by Ghosal, van der Vaart, Shen, Wasserman and others and a relatively complete theory is now available, even extending to observations that may not be independent or identically distributed. Rates of contraction have been computed for several priors such as Dirichlet process mixtures, priors based on splines and Gaussian processes. Results on rates of contraction that adapt to the underlying function class and related problems about model selection have been studied recently. The *Bernstein–von Mises theorem* for regular parametric models implies that the posterior distribution of the parameter centered at the MLE converges to the same normal distribution as that of the limit of the normalized MLE. Thus, asymptotically, Bayesian and sampling probabilities agree, so confidence regions of approximate frequentist validity may be generated from the posterior distribution. Cox, Freedman and others showed that such a result should not be expected for curve estimation problems. Some positive results have been obtained by Lo, Kim, Lee, Shen, Castillo, Nickl, Leahu, Bickel, Kleijn and others. The study of nonparametric Bayesian uncertainty quantification for curve estimation problems was started only recently, with first results and discussion by Szabó et al. (2015).

2

Priors on Function Spaces

A functional parameter in nonparametric statistics may be restricted only by qualitative assumptions, such as smoothness (as in standard nonparametric regression), but may also satisfy additional constraints, such as positivity (density, Poisson regression), integration to unity (probability density), monotonicity (distribution, or link function), etc. A prior distribution for such a parameter must take the restrictions into account. Given only qualitative restrictions, a probability measure on a general function space is a appropriate, whereas a special construction or a transformation is necessary to adhere to further restrictions. In this chapter we start with examples of priors on general function spaces, and next discuss priors satisfying constraints. Examples of the first type are random series and stochastic processes.

2.1 Random Basis Expansion

Given a set of "basis functions" $\psi_j: T \to \mathbb{R}$ on some domain $T \subset \mathfrak{X}$, indexed by j in some set J, a prior on functions $f: T \to \mathbb{R}$ can be constructed by writing

$$f = \sum_{j \in J} \beta_j \psi_j, \tag{2.1}$$

and putting priors on the coefficients β_j in this representation. There are many choices of bases: polynomials, trigonometric functions, wavelets, splines, spherical harmonics, etc. Each basis is suited to represent a certain type of function, where the quality is typically measured by the number of basis functions needed to attain a good approximation. The approximation properties of the basis together with the prior on the coefficients determine the suitability of the prior for a given application. If the coefficients are unrestricted by each other (the usual situation), then independent priors on the coefficients are reasonable. Normal priors are commonly used if the coefficients can take on every real value, typically with zero mean and variances chosen to control the "energy" at the different levels of the expansion.

Given infinitely many basis functions, the convergence of the *random series* (2.1) is not guaranteed. When the target functions form a (subset of a) Hilbert space, it is natural and convenient to use an orthonormal basis, and convergence is easy to establish.

Lemma 2.1 (Hilbert space) *If the functions* $(\psi_j: j \in \mathbb{N})$ *form an orthonormal basis of a Hilbert space* $\mathbb{H}$, *then the series* (2.1) *converges in* $\mathbb{H}$ *a.s. if and only if* $\sum_j \beta_j^2 < \infty$ *a.s. If the* β_j *are independent and for every* k *the vector* $(\beta_1, \ldots, \beta_k)$ *has full support* $\mathbb{R}^k$, *then the corresponding prior has support* $\mathbb{H}$. *In particular, for* $\beta_j \sim \mathrm{Nor}(\mu_j, \tau_j^2)$ *these conditions* *are verified if* $\sum_{j=1}^{\infty} \mu_j^2 < \infty$ *and* $\sum_{j=1}^{\infty} \tau_j^2 < \infty$. *If* $\beta_1, \beta_2, \dots$ *are also independent, then these conditions are necessary.*

Proof The first assertion is immediate from the fact that a deterministic series $\sum_{j=1}^{\infty} \beta_j \psi_j$ converges in $\mathbb{H}$ if and only if $\sum_j \beta_j^2 < \infty$. That $\{\psi_j : j \in \mathbb{N}\}$ forms an orthonormal basis means that any element $f_0 \in \mathbb{H}$ can be written $f_0 = \sum_j \beta_{0j} \psi_j$, where $\sum_{j>k} \beta_{j0}^2 \to 0$ as $k \to \infty$. That f_0 is in the support of the prior therefore follows from the fact that the event $\{(\beta_1, \dots, \beta_k): \sum_{j \le k} |\beta_j - \beta_{j0}|^2 < \epsilon, \sum_{j>k} \beta_j^2 < \epsilon\}$ has positive probability under the prior, for every $\epsilon > 0$ and sufficiently large k.

For a Gaussian sequence $(\beta_1, \beta_2, \dots)$, we have $\mathrm{E} \sum_j \beta_j^2 = \sum_j (\mu_j^2 + \tau_j^2)$, and hence $\sum_j \beta_j^2$ converges in mean if (and only if) the given series converges. The series $\sum_j \beta_j^2$ then also converges almost surely. For the necessity of the condition if the variables β_j are independent, we first note that $\beta_j \to 0$ as $j \to \infty$ and normality imply that $\mu_j \to 0$ and $\tau_j \to 0$. Furthermore, by the three-series theorem, convergence of the series $\sum_j \mathrm{E}(\beta_j^2 \mathbb{1}\{|\beta_j| < 1\})$ is necessary for almost sure convergence, and this can be translated into convergence of $\sum_j (\tau_j^2 \kappa_{j,2} + 2\tau_j \mu_j \kappa_{j,1} + \mu_j^2 \kappa_{j,0})$, where $\kappa_{j,k} = \mathrm{E}(Z^k \mathbb{1}\{A_j\})$ for a standard normal variable Z and $A_j = \{|\tau_j Z + \mu_j| < 1\}$. Since the events A_j increase to the full space as $\mu_j \to 0$ and $\tau_j \to 0$, it follows that $\kappa_{j,k} \to 1$, for $k = 0, 2$, and $\kappa_{j,1} \to 0$. Therefore $\sum_j (\mu_j^2 + \tau_j^2) < \infty$. □

For convergence of an infinite series (2.1) relative to other norms, there exist many precise results in the literature, but they depend on the basis functions, on the prior on the coefficients and on the desired type of convergence. The following lemma gives a simple condition for convergence relative to the *supremum norm* $\|f\|_\infty = \sup_{t \in T} |f(t)|$.

Lemma 2.2 (Uniform norm) *If* $\sum_{j=1}^{\infty} \|\psi_j\|_\infty \mathrm{E}|\beta_j| < \infty$, *then the series* (2.1) *converges uniformly, in mean. If* $\beta_1, \beta_2, \dots$ *are independent, then the uniform convergence is also in the almost sure sense. If any finite set of coefficients* $(\beta_1, \dots, \beta_k)$ *has full support* $\mathbb{R}^k$, *then the corresponding prior has support equal to the closure of the linear span of the functions* ψ_j *in the set of bounded functions relative to the uniform norm.*

Proof By the monotone convergence theorem, $\mathrm{E} \sum_{j=1}^{\infty} |\beta_j| \|\psi_j\|_\infty = \sum_{j=1}^{\infty} \mathrm{E}|\beta_j| \|\psi_j\|_\infty < \infty$. This implies the almost sure pointwise absolute convergence of the series $\sum_{j=1}^{\infty} |\beta_j| \psi_j$ and hence its pointwise convergence, almost surely. The uniform convergence in mean of the series is next immediate from the inequality $\mathrm{E}\|\sum_{j>n} \beta_j \psi_j\|_\infty \le \sum_{j>n} |\beta_j| \mathrm{E}\|\psi_j\|_\infty$. That the series converges almost surely if the variables are independent follows by the Ito-Nisio theorem, which says that a series of independent random elements in a separable Banach space converges almost surely if and only if it converges in mean. By Markov's inequality, $\mathrm{P}(\|\sum_{j>k} \beta_j \psi_j\|_\infty < \epsilon) \to 1$, as $k \to \infty$, and hence it is certainly positive for sufficiently large k. If $(\beta_1, \dots, \beta_k)$ has full support, then $\mathrm{P}(\|\sum_{j \le k} \beta_j \psi_j - f\|_\infty < \epsilon) > 0$, for every $\epsilon > 0$ and every f in the linear span of $\psi_1, \dots, \psi_k$. Combined this gives the last assertion. □

For a finite set of basis functions, the sum (2.1) is always well defined, but the prior has a finite-dimensional support and hence is not truly nonparametric. This may be remedied by using several dimensions simultaneously, combined with a prior on the dimension.

Basis functions are often constructed so that for certain target functions f of interest, there exist coefficients $f_1, f_2, \dots$ such that for every $J \in \mathbb{N}$ and k the dimension of the domain of the functions,

$$\Big\| f - \sum_{j=1}^{J} f_j \psi_j \Big\| \lesssim \Big(\frac{1}{J}\Big)^{\alpha/k} \|f\|_\alpha^*. \tag{2.2}$$

Then truncating the basis at J terms allows us to approximate the function, in a given norm $\|\cdot\|$, with error bounded by a power of $1/J$. The power is typically determined by a *regularity parameter* α (e.g. the number of times f is differentiable) divided by the dimension k of the domain. The approximation is moderated by a quantitative measure $\|f\|_\alpha^*$ of the regularity, for instance the norm of the αth derivative. Results of this nature can be found in approximation theory for every of the standard types of bases, including wavelets, polynomials, trigonometric functions and splines, where the norms used in the inequality depend on the type of basis (see Appendix E for a brief review).

A rough statistical message is as follows. The truncated series $\sum_{j=1}^{J} f_j \psi_j$ gives a J-dimensional parametric model, and hence ought to be estimable with mean square error of the order J/n, for n expressing the informativeness of the data (e.g. n independent replicates), i.e. a mean square error "per parameter" of the order $1/n$. Under (2.2) the bias incurred by estimating the truncation rather than f is of the order $(1/J)^{\alpha/k}$. Equating square bias and mean square error gives

$$\Big(\frac{1}{J}\Big)^{2\alpha/k} \asymp \frac{J}{n}.$$

The solution $J \asymp n^{k/(2\alpha+k)}$ gives an estimation error of the order $n^{-\alpha/(2\alpha+k)}$. This is the typical optimal nonparametric rate of estimation for functional parameters on a k-dimensional domain that are regular of order α. It is attained by standard procedures such as kernel estimation, truncated series, penalized splines, etc.

In the Bayesian setting we put a prior on the coefficients f_j and hope that the posterior distribution will come out right. The informal calculation suggests that if the unknown function is a priori thought to be regular of order α, then a series prior that mostly charges the first $J \asymp n^{k/(2\alpha+k)}$ coefficients might do a good job. For instance, we might put noninformative priors on the first $n^{k/(2\alpha+k)}$ coefficients and put the remaining coefficients equal to zero, or we might make the priors shrink to 0 at an appropriate speed as $j \to \infty$ (e.g. Gaussian priors $\mathrm{Nor}(0, \tau_j^2)$ with τ_j tending to zero). A problem is that the truncation point, or rate of decay, must depend on the regularity α, which we may only have guessed. A natural Bayesian solution is to put a prior on α, or perhaps more simply on the number J of terms in the truncated series, or on the scale of the variance.

We analyze constructions of this type in Chapters 9, 10 and 11.

2.2 Stochastic Processes

A *stochastic process* indexed by a set T is a collection $W = (W(t): t \in T)$ of random variables defined on a common probability space. A *sample path* $t \mapsto W(t)$ of W is a "random function," and hence the law of W, the "law of the set of sample paths," is a prior on a space of functions $f: T \to \mathbb{R}$.

For a precise definition, W must be defined as a measurable map $W: (\Omega, \mathcal{U}, \mathrm{P}) \mapsto (\mathfrak{F}, \mathscr{F})$ from a probability space into a function space $\mathfrak{F}$, equipped with a σ-field $\mathscr{F}$. This is not always easy, but there are many ready examples of stochastic processes.

2.2.1 Gaussian Processes

A *Gaussian process* W is a stochastic process such that finite-dimensional distributions of $(W(t_1), \ldots, W(t_m))$, for all $t_1, \ldots, t_m \in T$, $m \in \mathbb{N}$, are multivariate normally distributed. The finite-dimensional distributions are completely specified by the *mean function* $\mu: T \to \mathbb{R}$ and *covariance kernel* $K: T \times T \to \mathbb{R}$, given by

$$\mu(t) = \mathrm{E}[W(t)], \qquad K(s,t) = \mathrm{cov}\,(W(s), W(t)).$$

The mean function is a deterministic shift of the sample paths, and in prior construction the Gaussian process is often taken with zero mean and a desired offset is made part of the statistical model, possibly with its own prior.

The finite-dimensional distributions determine the process as a measurable map in $\mathbb{R}^T$, but for uncountable domains T, they do not determine the sample paths. However, there may exist a version with special sample paths, often regular of some type, and the process may be seen as a map into the space of functions of this type. For instance, a Brownian motion is a Gaussian process with $T = [0, \infty)$ with mean zero and covariance function $\min(s,t)$, and can be realized to have continuous sample paths (which are then even regular of order 1/2), so that it can be viewed as a map in the space of continuous functions.

A function $K: T \times T \to \mathbb{R}$ is called *nonnegative-definite* if all matrices of the type $((K(t_i, t_j)))$, for $t_1, \ldots, t_m \in T$ and $m \in \mathbb{N}$, are nonnegative. Any such nonnegative-definite function is the covariance function of some Gaussian process. Thus there is a great variety of such processes. For some applications with $T \subset \mathbb{R}^k$ covariance functions of the type $K(s,t) = \psi(s-t)$ are of special interest as they generate *stationary processes*: the corresponding (mean zero) process is invariant in law under translations: the distribution of $(W(t+h): t \in T)$ is the same for every $h \in \mathbb{R}^k$.

Many Gaussian processes have "full" support, for instance in a space of continuous functions, and therefore are suitable to model a function with unrestricted shape taking values in $\mathbb{R}$. However, their performance as a prior strongly depends on the fine properties of the covariance function and sample paths. This performance may also be influenced by changing the *length scale* of the process: using $t \mapsto W(\lambda t)$ as a prior process, for some positive number λ, possibly chosen itself from a prior distribution.

A random series (2.1) with independent normal variables β_j as coefficients, is an example of a Gaussian process prior. Conversely, every Gaussian process can be represented as a random series with independent normal coefficients (see Theorem I.25). The special choice with the basis functions ψ_j equal to the eigenfunctions of the covariance kernel is called the *Karhunen-Loève expansion*.

Gaussian priors are conjugate in the Gaussian nonparametric regression model. Outside this linear Gaussian setting, computation of the posterior distribution is considerably more complicated. We discuss computational schemes and many results on Gaussian priors in Chapter 11.

2.2.2 Increasing Processes

Because Gaussian variables can assume every real value, Gaussian processes are unsuitable as priors for monotone functions $f:\mathbb{R} \to \mathbb{R}$ or $f:[0,\infty) \to \mathbb{R}$, such as cumulative distribution and hazard functions, or for monotone link functions in generalized linear models. Independent increment processes with nonnegative increments are the most popular examples of priors for increasing functions. Gamma processes and, more generally, Lévy processes with nonnegative increments (*subordinators*, see Appendix J) are in this class. These processes have the remarkable property that they increase by jumps only. Their sample paths either converge to a finite positive random variable or diverge to infinity, a.s.

Given an increasing process $S = (S(t): t \geq 0)$, we may define a random cumulative distribution function (c.d.f.) F on $[0, \tau]$ by

$$F(t) = S(t)/S(\tau),$$

for any $\tau < \infty$, and also for $\tau = \infty$ if the sample paths of S are bounded. For S the gamma process, this leads to the Dirichlet process, discussed in Chapter 4.

An increasing process with unbounded sample paths may be used as a random cumulative hazard function, which will subsequently induce a prior on distributions. This approach is fruitful in the context of Bayesian survival analysis, discussed in Chapter 13. A similar but technically different idea is to consider a random cumulative distribution function defined by $F(t) = 1 - e^{-S(t)}$. If S is not continuous, then it is not the cumulative hazard function of F, but it can be close to it.

Increasing processes, and hence priors on sets of increasing functions, can be constructed without the measure-theoretic problems associated with the construction of (general) random measures. We focus on increasing functions with domain $\mathbb{R}$. The idea is to specify the desired distribution of every finite-dimensional marginal $(F(t_1), \ldots, F(t_k))$, with $t_1, \ldots, t_k$ belonging to a given countable, dense subset of $\mathbb{R}$ – for instance, the set of rational numbers $\mathbb{Q}$. Provided these distributions are consistent, Kolmogorov's consistency theorem allows construction of a full process. The following proposition shows that this will automatically have right-continuous, nondecreasing sample paths if the bivariate and univariate marginal distributions satisfy certain conditions.

Proposition 2.3 *For every $k \in \mathbb{N}$ and every $t_1, \ldots, t_k$ in a countable dense set $Q \subset \mathbb{R}$ let $\mu_{t_1,\ldots,t_k}$ be a probability distribution on $\mathbb{R}^k$ such that*

(i) *if $T' \subset T$, then $\mu_{T'}$ is the marginal distribution obtained from μ_T by integrating out the coordinates $T \setminus T'$;*

(ii) $t_1 < t_2$ *implies* $\mu_{t_1,t_2}\{(x_1,x_2): x_1 \le x_2\} = 1$;
(iii) $t_n \downarrow t$ *implies* $\mu_{t_n} \rightsquigarrow \mu_t$.

Then there exists a stochastic process $(F(t): t \in \mathbb{R})$ *with nondecreasing, right-continuous sample paths such that* $(F(t_1), \ldots, F(t_k))$ *has distribution* $\mu_{t_1,\ldots,t_k}$ *for every* $t_1, \ldots, t_k \in Q$ *and* $k \in \mathbb{N}$. *If, moreover,*

(iv) $t \downarrow -\infty$ *implies* $\mu_t \rightsquigarrow \delta_0$ *and* $t \uparrow \infty$ *implies* $\mu_t \rightsquigarrow \delta_1$, *for* δ_x *the degenerate measure at* x,

then the sample paths of the stochastic process F *are cumulative distribution functions on* $\mathbb{R}$.

Proof Because the marginal distributions are consistent by assumption (i), Kolmogorov's consistency theorem implies existence of a stochastic process $(F(t): t \in Q)$ with the given marginal distributions. Because only countably many exceptional null sets arise in relation (ii), the sample paths of F are nondecreasing a.s. To show a.s. right continuity on Q, fix $t \in Q$ and let $t_k \downarrow t$, $t_k \in Q$. Monotonicity gives that $F^*(t) := \lim_{k\to\infty} F(t_k) \ge F(t)$ a.s., while property (iii) implies that $F(t_k) \rightsquigarrow F(t)$. This implies that $F^*(t) = F(t)$ a.s. (see Problem L.1). Again the exceptional null sets can be pulled together to a single null set, which implies a.s. right continuity simultaneously at every point.

Any right-continuous, nondecreasing function $F: Q \to \mathbb{R}$ can be uniquely extended to a function $F: \mathbb{R} \to \mathbb{R}$ with the same properties by simply taking a right limit at every $t \in \mathbb{R} \setminus Q$.

This function is a cumulative distribution function if and only if its limits at $-\infty$ and ∞ are 0 and 1, respectively. Under the additional assumption (iv), this can be established a.s. $[\mu]$ by a similar argument. □

Example 2.4 (Subordinators) As an example, consider the construction of *subordinators*, with given marginal distributions μ_t. Assume that $(\mu_t: t \ge 0)$ are a continuous semigroup (i.e. $\mu_{s+t} = \mu_s * \mu_t$ for all $s, t \ge 0$ and $\mu_t \rightsquigarrow \delta_0$ as $t \to 0$) of probability measures concentrated on $(0, \infty)$, and define $\mu_{t_1,\ldots,t_k}$ for rational numbers $0 < t_1 < t_2 < \cdots < t_k$ as the joint distribution of the cumulative sums of k independent variables with distributions $\mu_{t_1}, \mu_{t_2-t_1}, \ldots, \mu_{t_k-t_{k-1}}$, respectively. The semigroup properties make the marginal distributions consistent and ensure property (iii), whereas (ii) is immediate from the nonnegativity of the support of μ_t. The random distribution function F has independent increments with $F(t) - F(s) \sim \mu_{t-s}$.

2.3 Probability Densities

Probability densities are functions that are nonnegative and integrate to 1. In this section we discuss priors that observe these constraints. A prior induced by a prior on distributions that charges absolutely continuous distributions only (e.g. certain Pólya tree processes, Section 3.7) would be an alternative.

Natural metrics on the set of all densities relative to a given dominating measure μ are the *total variation distance* and the *Hellinger distance*

$$d_{TV}(p,q) = \tfrac{1}{2}\int |p-q|\,d\mu,$$

$$d_H(p,q) = \sqrt{\int (\sqrt{p}-\sqrt{q})^2\,d\mu}.$$

For a standard Borel sample space and a σ-finite dominating measure, the set of μ-densities is Polish under these metrics, and hence the support is well defined. Because the two distances induce the same topology, the total variation and Hellinger supports are the same.

It will be seen in Chapter 6 that support relative to the Kullback-Leibler divergence is crucial for posterior consistency. The *Kullback Leibler divergence* between two densities p and q is defined as

$$K(p;q) = \int \Big(\log \frac{p}{q}\Big)\,p\,d\mu.$$

Because it is not a metric, the notion of a "support" does not follow from general principles. We say that a density p_0 is in the *Kullback-Leibler support* of a prior Π if, for all $\epsilon > 0$,

$$\Pi\Big(p\colon K(p_0;p) < \epsilon\Big) > 0.$$

Because $d_H^2(p,q) \le K(p;q)$, for any pair of probability densities p, q, the Kullback-Leibler support is always smaller than the Hellinger support. See Appendix B for further details on these distances and discrepancies.

2.3.1 Exponential Link Function

Given a measurable function $f\colon \mathfrak{X} \to \mathbb{R}$ with a μ-integrable exponential e^f, we can define a probability density p_f by

$$p_f(x) = e^{f(x)-c(f)}, \tag{2.3}$$

where $c(f) = \log \int e^f\,d\mu$ is the norming constant. We can construct a prior on densities by putting a prior on the function f.

If the prior on f is a random series, then the induced prior (2.3) is an *exponential family*

$$\bar{p}_\beta(x) = \exp\Big(\sum_j \beta_j \psi_j(x) - c(\beta)\Big).$$

The basis functions ψ_j are the sufficient statistics of the family. If the series is infinite, then the exponential family is also of infinite dimension.

The prior (2.3) will have full support relative to the Kullback-Leibler divergence as soon as the prior on f has full support relative to the uniform norm. This follows from the following lemma, which even shows that the Hellinger distance and Kullback-Leibler divergence between two densities of the form (2.3) are essentially bounded by the uniform norm between the corresponding exponents. The discrepancy $V_2(p,q)$ is the square Kullback-Leibler variation, defined in (B.2).

Lemma 2.5 *For any measurable functions $f, g\colon \mathfrak{X} \to \mathbb{R}$,*

(i) $d_H(p_f, p_g) \le \|f - g\|_\infty e^{\|f-g\|_\infty/2}$,
(ii) $K(p_f; p_g) \lesssim \|f - g\|_\infty^2 e^{\|f-g\|_\infty}(1 + \|f - g\|_\infty)$,
(iii) $V_2(p_f; p_g) \lesssim \|f - g\|_\infty^2 e^{\|f-g\|_\infty}(1 + \|f - g\|_\infty)^2$.

Furthermore, if $f(x_0) = g(x_0)$ for some $x_0 \in \mathfrak{X}$, then

(iv) $\|f - g\|_\infty \le 2D(f, g)\|p_f - p_g\|_\infty$, *for* $D(f, g) = (\min_x(p_f(x) \wedge p_g(x))^{-1}$.

Proof The triangle inequality and simple algebra give

$$d_H(p_f, p_g) = \left\| \frac{e^{f/2}}{\|e^{f/2}\|_2} - \frac{e^{g/2}}{\|e^{g/2}\|_2} \right\|_2 \le 2\frac{\|e^{f/2} - e^{g/2}\|_2}{\|e^{g/2}\|_2}.$$

Because $|e^{f/2} - e^{g/2}| \le e^{g/2}e^{|f-g|/2}|f - g|/2$ for any $f, g \in \mathbb{R}$, the square of the right side is bounded by

$$\frac{\int e^g e^{|f-g|}|f - g|^2\, d\nu}{\int e^g\, d\nu} \le e^{\|f-g\|_\infty}\|f - g\|_\infty^2.$$

This proves assertion (i) of the lemma.

Next assertions (ii) and (iii) follow from (i) and the equivalence of K, V_2 and the squared Hellinger distance if the quotient of the densities is uniformly bounded (see Lemma B.2). From $g - \|f - g\|_\infty \le f \le g + \|f - g\|_\infty$ it follows that $c(g) - \|f - g\|_\infty \le c(f) \le c(g) + \|f - g\|_\infty$. Therefore $\|\log(p_f/p_g)\|_\infty = \|f - g - c(f) + c(g)\|_\infty \le 2\|f - g\|_\infty$. Assertions (ii) and (iii) now follow by Lemma B.2.

For the final assertion we note first that $|f - g - c(f) + c(g)| = |\log p_f - \log p_g|$ is bounded above by $D(f, g)|p_f - p_g|$. Evaluating this as $x = x_0$, we see that $|c(f) - c(g)| \le D(f, g)|p_f(x_0) - p_g(x_0)|$. The result follows by reinserting this in the initial inequality. □

2.3.2 Construction through Binning

Consider a partition of $\mathbb{R}$ into intervals of length $h > 0$, and randomly distribute probabilities to the intervals, for instance by the techniques to be discussed in Section 3.3, and distribute the mass uniformly inside the intervals. This yields a random histogram of bandwidth h. Finally, give h a nonsingular prior that allows arbitrary small values of h with positive probability.

Since any integrable function can be approximated in $\mathbb{L}_1$-distance by a histogram of sufficiently small bandwidth, the resulting prior has full $\mathbb{L}_1$-support if the weights themselves are chosen to have full support equal to the countable unit simplex.

The idea is easily extendable to $\mathbb{R}^k$ or regularly shaped subsets of $\mathbb{R}^k$.

2.3.3 Mixtures

A powerful technique to construct a prior on densities via one on probabilities is through mixtures. For given probability density functions $x \mapsto \psi(x, \theta)$ indexed by a parameter θ and measurable in its two arguments, and a probability distribution F, let

$$p_F(x) = \int \psi(x, \theta)\, dF(\theta).$$

Then a prior on F induces a prior on densities. The function ψ is referred to as the *kernel function* and the measure F as the *mixing distribution.*

The choice of a kernel will depend on the sample space. For the entire Euclidean space, a location-scale kernel, such as the normal kernel, is often appropriate. For the unit interval, beta distributions form a flexible two-parameter family. On the positive half line, mixtures of gamma, Weibull or lognormal distributions may be used. The use of a uniform kernel leads to random histograms as discussed in the method of binning in Section 2.3.2.

Mixtures can form a rich family. For instance, location-scale mixtures, with a kernel of the form $x \mapsto \sigma^{-1}\psi((x-\mu)/\sigma)$ for some fixed density ψ and $\theta = (\mu, \sigma)$, may approximate any density in the $\mathbb{L}_1$-sense if σ is allowed to approach 0. This is because

$$\int \frac{1}{\sigma}\psi\left(\frac{\cdot - \mu}{\sigma}\right) f(\mu)\, d\mu \to f(\cdot), \qquad \text{as } \sigma \to 0, \tag{2.4}$$

in $\mathbb{L}_1$ by *Fejér's theorem* (cf. Helson (1983)). Thus, a prior on densities may be induced by putting a prior on the mixing distribution F and a prior on σ, which supports zero. Similarly, one can consider location-scale mixture $\int \sigma^{-1}\psi((x-\mu)/\sigma)\, dF(\mu, \sigma)$, where F is a bivariate distribution on $\mathbb{R} \times (0, \infty)$.

The mixing distribution F may be given a prior following any of the methods discussed in Chapter 3. A Dirichlet process prior is particularly attractive, as it leads to efficient computational algorithms (cf. Chapter 5) and a rich theory of convergence properties (cf. Chapter 7 and 9).

2.3.4 Feller Approximation

In the previous section several different choices of kernels were mentioned, mostly motivated by the underlying sample space (full line, half line, interval, etc.). The choice of kernel is also intimately connected to the approximation scheme it generates. For instance, motivation for a location-scale kernel comes from (2.4), whereas beta mixtures can be based on the theory for Bernstein polynomials, as reviewed in Section E.1. Feller (1968) described a general constructive approximation scheme for continuous functions on an interval, which may be used to propose a kernel in a canonical way.

A *Feller random sampling scheme* on an interval I is a collection $\{Z_{k,x}: k \in \mathbb{N}, x \in I\}$ of random variables such that $\mathrm{E}(Z_{k,x}) = x$ and $V_k(x) := \mathrm{var}(Z_{k,x}) \to 0$ as $k \to \infty$. Then $Z_{k,x} \rightsquigarrow \delta_x$, and hence, for any bounded, continuous real-valued function F,

$$A(x; k, F) := \mathrm{E}(F(Z_{k,x})) \to F(x), \qquad k \to \infty.$$

Thus $A(\cdot; k, F)$ is an approximation of the function F. If $Z_{k,x}$ possesses a density $z \mapsto g_k(z; x)$ relative to some dominating measure ν_k, then we can write the approximation in the form

$$A(x; k, F) = \int F(t) g_k(t; x)\, d\nu_k(t) = \iint_{[z,\infty)} g_k(t; x)\, d\nu_k(t)\, dF(z).$$

Thus a (continuous) distribution function F can be approximated using mixtures of the kernel $\bar{G}_k(z; x) := \int_{[z,\infty)} g_k(t; x)\, d\nu_k(t)$. Under suitable conditions mixtures of the derivative $(\partial/\partial x)\bar{G}_k(z; x)$ of this kernel will approximate a corresponding density function.

An example of a Feller scheme is $Z_{k,x} = k^{-1}\sum_{i=1}^k Y_{i,x}$, for i.i.d. random variables $Y_{i,x}$ with mean x and finite variance. Consider specializing the distribution of $Y_{i,x}$ to a natural exponential family, given by a density of the form $p_\theta(y) = \exp(\theta y - \psi(\theta))$ relative to some σ-finite dominating measure independent of θ. The density $z \mapsto g_k(z; x)$ of $Z_{k,x}$ relative to some (other) dominating measure ν_k can then be written in the form $g_k(z; x) = \exp(k\theta z - k\psi(\theta))$, and the usual identities for exponential families show that $\mathrm{E}(Z_{k,x}) = \mathrm{E}Y_{i,x} = \psi'(\theta)$ and $\mathrm{var}(Y_{i,x}) = \psi''(\theta)$. The first implies that the parameter $\theta(x)$ of the family must be chosen to solve $\psi'(\theta(x)) = x$. Differentiating across this identity with respect to x and using the second identity we see that $\theta'(x) = 1/\psi''(\theta(x)) = 1/V(x)$ for $V(x) = \mathrm{var}(Y_{i,x})$. This permits us to express $(\partial/\partial x)g_k(z; x)$ as $kg_k(z; x)(z - x)/V(x)$, leading to the approximation scheme for the density of the cumulative distribution function F given by

$$a(x; k, F) = \int h_k(x; z)\, dF(z), \tag{2.5}$$

for

$$h_k(x; z) = \frac{k}{V(x)}\int_{[z,\infty)} (t - x) g_k(t; x)\, d\nu_k(t).$$

The following proposition shows that this mixture indeed approaches $f = F'$ as $k \to \infty$. It is implicitly assumed that the equation $\psi'(\theta(x)) = x$ is solvable: the approximation scheme only works on the interval of possible mean values of the given exponential family (the image of its natural parameter set under the increasing map ψ').

Proposition 2.6 *If F has bounded and continuous density f, then $a(x; k, F) \to f(x)$ as $k \to \infty$.*

Proof Without loss of generality set $x = 0$; otherwise redefine $Z_{k,x}$ to $Z_{k,x} - x$. Abbreviate $Z_{k,0}$ to Z_k, $V(0)$ to V, and $h_k(0; z)$ to $h_k(z)$. The latter function can be written in the form $h_k(z) = k\mathrm{E}\mathbb{1}\{Z_k \geq z\}Z_k/V$. It is nonnegative for every z, as is immediate for $z \geq 0$, and follows for $z \leq 0$ by rewriting it as $-k\mathrm{E}\mathbb{1}\{Z_k < z\}Z_k/V$, using the fact that $\mathrm{E}Z_k = 0$. Furthermore, by Fubini's theorem, for every z,

$$\bar{H}_k(z) := \int_z^\infty h_k(t)\, dt = \mathrm{E}kZ_k^2\mathbb{1}\{Z_k \geq z\}/V - z\mathrm{E}kZ_k\mathbb{1}\{Z_k \geq z\}/V.$$

For fixed k the first term tends to $\mathrm{E}kZ_k^2/V = 1$ and to 0 as z tends to $-\infty$ or ∞, respectively. Because $\mathrm{E}Z_k = 0$, the second term can also be written $-z\mathrm{E}kZ_k\mathbb{1}\{Z_k < z\}/V$. Therefore its absolute value is bounded above by $\mathrm{E}kZ_k^2\mathbb{1}\{Z_k < z\}/V$ as $z < 0$ and by $\mathrm{E}kZ_k^2\mathbb{1}\{Z_k \geq z\}/V$ as $z > 0$ and hence it tends to zero both as $z \to -\infty$ and as $z \to \infty$. We conclude that $\bar{H}_k$ is a survival function, for every fixed k.

By the law of large numbers $\mathrm{P}(Z_k > z) \to 0$ as $k \to \infty$ for $z > 0$. Because $\sqrt{k}Z_k$ is uniformly square integrable in k (it has variance V and tends to a Nor(0, V)-variable), it follows for $z > 0$ by the same bounds as previously that $\bar{H}_k(z) > 2\mathrm{E}kZ_k^2\mathbb{1}\{Z_k \geq z\}/V \to 0$, as $k \to \infty$. Similarly we see that $\bar{H}_k(z) \to 1$ for any $z < 0$. We conclude that $H_k \rightsquigarrow \delta_0$, whence $a(0; k, F) = \int f(z)\, dH_k(z) \to f(0)$, by the assumed continuity and boundedness of f. □

2.4 Nonparametric Normal Regression

Consider a continuous response variable Y which depends on the value of a covariate X through the regression equation $Y = f(X) + \varepsilon$, for a function $f: \mathfrak{X} \to \mathbb{R}$ on a measurable space $\mathfrak{X}$, and a stochastic error variable ε. If $\varepsilon | X \sim \mathrm{Nor}(0, \sigma^2)$, then the conditional density of Y given $X = x$ is

$$p_{f,\sigma}(y|\, x) = \frac{1}{\sigma\sqrt{2\pi}} e^{-(y-f(x))^2/(2\sigma^2)}.$$

In the *random design* model the covariate X is a random variable. We denote its marginal distribution by G and write $p_{f,\sigma}$ for the joint density of a single observation (X, Y), relative to the product of G and Lebesgue measure. In the *fixed design* model the covariates are deterministic values $x_1, \ldots, x_n$. We denote their empirical distribution by G, and use the same notation $p_{f,\sigma}$ for the density of a vector of $(Y_1, \ldots, Y_n)$ of independent responses at these covariate values.

A prior on (f, σ) induces a prior on $p_{f,\sigma}$. The following lemma shows that a prior on f with large support relative to the $\mathbb{L}_2(G)$-norm combined with a suitable independent prior on σ induces a prior with a large Hellinger or Kullback-Leibler support.

Lemma 2.7 *For any measurable functions $f, g: \mathfrak{X} \to \mathbb{R}$ and any $\sigma, \tau > 0$,*

(i) $d_H(p_{f,\sigma}, p_{g,\tau}) \le \|f - g\|_{2,G}/(\sigma + \tau) + 2|\sigma - \tau|/(\sigma + \tau)$.
(ii) $K(p_{f,\sigma}; p_{g,\tau}) = \frac{1}{2}\|f - g\|_{2,G}^2/\tau^2 + \frac{1}{2}\big[\log(\tau^2/\sigma^2) + (\sigma^2/\tau^2) - 1\big]$.
(iii) $V_{2,0}(p_{f,\sigma}; p_{g,\tau}) = (\sigma/\tau)^2\|f - g\|_{2,G}^2/\tau^2 + \frac{1}{2}(\sigma^2 - \tau^2)^2/\tau^4$.

Proof For (i) we first find by direct calculation using the Gaussian integral, that the affinity between two univariate normal distributions is given by

$$\rho_{1/2}\big(\mathrm{Nor}(f, \sigma^2), \mathrm{Nor}(g, \tau^2)\big) = \sqrt{1 - (\sigma - \tau)^2/(\sigma^2 + \tau^2)} e^{-|f-g|^2/(4\sigma^2+4\tau^2)}.$$

Next we integrate with respect to G to find the affinity between $p_{f,\sigma}$ and $p_{g,\tau}$, use the relationship between the square Hellinger distance and affinity (see Lemma B.5(i)), and finally use the inequalities $|1 - \sqrt{1-s}| \le s$, for $s \in [0, 1]$, and $1 - e^{-t} \le t$, for $t \ge 0$. Equalities (ii) and (iii) similarly can first be established when G is a Dirac measure by calculations on the normal distribution, and next be integrated for a general G. □

2.5 Nonparametric Binary Regression

In the binary regression model we try to predict a binary outcome (zero-one response, membership in one of two classes, success or failure, etc.) by the value of a covariate x. If we model the outcome as a Bernoulli variable with success probability given as a function $x \mapsto p(x)$ of the covariate, then it is natural to predict (or classify) a new example x according to whether $p(x)$ exceeds a certain threshold $c \in [0, 1]$ or not (e.g. $c = 1/2$).

To estimate the function p from "training data" we put a prior on the set of functions $p: \mathfrak{X} \to [0, 1]$ on the covariate space. Because the prior should observe that the range is the unit interval, a general prior on functions is unsuitable. This may be remedied by using a (inverse) *link function* $H: \mathbb{R} \to [0, 1]$ that maps the real line to the unit interval with derivative $h = H'$. We may then use a general prior on functions f and model the probability response function as

$$p(x) = H(f(x)).$$

Monotone link functions (in particular cumulative distribution functions) are customary. In particular, the *logistic link function* $H(t) = (1 + e^{-t})^{-1}$ and *Gaussian link function* $H(t) = \Phi(t)$, the cumulative distribution function of the standard normal distribution, are popular.

The conditional density of a Bernoulli variable $Y|\, x \sim \text{Bin}(1, H(f(x))$ is

$$p_f(y|\, x) = H(f(x))^y (1 - H(f(x)))^{1-y}, \qquad y \in \{0, 1\}. \tag{2.6}$$

For sampling instances of (X, Y), this is the relevant density (relative to the marginal distribution of the covariate X and counting measure on $\{0, 1\}$), if we assume that the covariate distribution is not informative about the response.

The following lemma allows us to derive the support of the induced prior on p_f from the support of the prior on f. For the logistic link function, the induced prior has full Kullback-Leibler support as soon as the prior on f has full support relative to the $\mathbb{L}_2(G)$-metric, for G the marginal distribution of the covariate. The Gaussian link is slightly less well behaved, but a similar result holds with the uniform norm if f_0 is bounded.

Lemma 2.8 *For any measurable functions $f, g: \mathfrak{X} \to \mathbb{R}$ and any $r \geq 1$,*

(i) $\|p_f - p_g\|_r = 2^{1/r}\|H(f) - H(g)\|_{r,G} \leq 2^{1/r}\|h\|_\infty \|f - g\|_{2,G}$.
(ii) $d_H(p_f, p_g) \leq \sqrt{I(h)}\|f - g\|_{2,G}$, *for* $I(h) = \int (h'/h)^2\, dH$.
(iii) $K(p_f; p_g) \lesssim \|(f - g)S(|f| \vee |g|)^{1/2}\|_{2,G}^2$.
(iv) $V_2(p_f; p_g) \lesssim \|(f - g)S(|f| \vee |g|)\|_{2,G}^2$.

Here $S = 1$ for H the logistic link and $S(u) = |u| \vee 1$ for the Gaussian link. Furthermore, the inequalities are true with $S = 1$ for any strictly monotone, continuously differentiable link function if the functions f, g take values in a fixed compact interval.

Proof Assertion (i) follows, because, for any x,

$$|p_f(0|\, x) - p_g(0|\, x)| = |p_f(1|\, x) - p_g(1|\, x)| = |H \circ f(x) - H \circ g(x)|.$$

The second inequality follows by bounding the right side by the mean value theorem. For (ii) we argue similarly, but must bound the differences $|\sqrt{H \circ f} - \sqrt{H \circ g}|$ and $|\sqrt{1 - H \circ f} - \sqrt{1 - H \circ g}|$. The appropriate derivatives are now $\frac{1}{2}h/\sqrt{H}$ and $\frac{1}{2}h/\sqrt{1 - H}$, which are both bounded by $\frac{1}{2}\sqrt{I(h)}$, since $h(x)^2 = |\int_{-\infty}^x h'(s)\, ds|^2 \leq I(h)H(x)$, in view of the Cauchy-Schwarz inequality, and the analogous inequality in the right tail.

To derive (iii) we express the Kullback-Leibler divergence as $K(p_f; p_g) = \int \psi_f(g)\, dG$, where

$$\psi_u(v) = H(u) \log \frac{H(u)}{H(v)} + (1 - H(u)) \log \frac{1 - H(u)}{1 - H(v)}.$$

The function ψ_u possesses derivative $\psi_u'(v) = (h/(H(1-H)))(v)(H(v)-H(u))$. Here the function $h/(H(1-H))$ is identically equal to 1 for the logistic link and is asymptotic to the identity function, as its argument tends to $\pm\infty$ for the Gaussian link; hence it is bounded in absolute value by the function S as given. Because the function ψ_u vanishes at u, the mean value theorem gives that $|\psi_f(g)| \leq S(|f| \vee |g|)|f-g|$, and assertion (iii) follows.

For the proof of (iv) we write $V_2(p_f, p_g) = \int \phi_f(g)\,dG$, where the function ϕ_f is as ψ_f, but with the logarithmic factors squared. The result follows because both $\log(H(g)/H(f))$ and $\log[(1-H(g))/(1-H(f)]$ are bounded in absolute value by $S(|f| \vee |g|)|f-g|$. □

2.6 Nonparametric Poisson Regression

In the Poisson regression model, a response variable is Poisson distributed with mean parameter $\lambda(x)$ depending on a covariate x. In the nonparametric situation, $x \mapsto \lambda(x)$ is an unknown function $\lambda: \mathfrak{X} \to (0,\infty)$. A link function $H: \mathbb{R} \to (0,\infty)$ maps a general function $f: \mathbb{R} \to \mathbb{R}$ to the positive half line through

$$\lambda(x) = H(f(x)).$$

A prior on f induces a prior on λ. The exponential link function $H(f) = e^f$ is a canonical choice, but functions that increase less rapidly may have advantages, as seen in Example 2.11, below.

The conditional density of a Poisson variable $Y|\,x \sim \text{Poi}(H(f(x))$ is given by

$$p_f(y|\,x) = \frac{1}{y!}e^{-H(f(x))}H(f(x))^y, \qquad y = 0, 1, 2, \ldots.$$

The following lemma relates the relevant distances on p_f to the $\mathbb{L}_2(G)$-distance on f, for G, the marginal distribution of the variable x.

Lemma 2.9 *For any measurable functions $f, g: \mathfrak{X} \to (a,b) \subset \mathbb{R}$, and $\|K\|_\infty = \sup\{|K(s)|: a < s < b\}$, the uniform norm of a function with domain (a,b),*

(i) $d_H(p_f, p_g) \leq \|H'/\sqrt{H}\|_\infty \|f-g\|_{2,G}$;
(ii) $K(p_f; p_g) \leq (C_f\|H'/H\|_\infty^2 + C_f\|H''/H\|_\infty + \|H''\|_\infty)\|f-g\|_{2,G}^2$, *for* $C_f = \sup_x |H(f(x))|$;
(iii) $V_{2,0}(p_f; p_g) \leq C_f\|H'/H\|_\infty^2 \|f-g\|_{2,G}^2$.

Proof The three distances between Poisson distributions Q_λ with mean λ are given by

$$d_H^2(Q_\lambda, Q_\mu) = 2(1 - e^{-(\sqrt{\lambda}-\sqrt{\mu})^2/2}) \leq (\sqrt{\lambda}-\sqrt{\mu})^2,$$
$$K(Q_\lambda; Q_\mu) = \mu - \lambda + \lambda\log\frac{\lambda}{\mu},$$
$$V_{2,0}(Q_\lambda; Q_\mu) = \lambda\Big(\log\frac{\lambda}{\mu}\Big)^2.$$

The quantities in the lemma are obtained by replacing λ and μ in the right sides by $H(f(x))$ and $H(g(x))$, and integrating over x with respect to G. For (i) and (iii) we use that the functions $g \mapsto \sqrt{H(g)} - \sqrt{H(f)}$ and $g \mapsto \sqrt{H(f)} \log H(f)/H(g)$ have first derivatives $\frac{1}{2}H'/\sqrt{H}$ and $-\sqrt{H(f)}(H'/H)$, respectively. For (ii) we use that the first derivative of the function $g \mapsto H(g) - H(f) + H(f) \log H(f)/H(g)$ vanishes at $g = f$ and that the second derivative of the function takes the form $H'' - H(f)(H''/H) + H(f)(H'/H)^2$. ☐

Example 2.10 (Exponential link) For the exponential link function $H(x) = e^x$ we may take the interval (a, b) equal to $(-\infty, M)$, for some constant M. Then the constants $\|H'/\sqrt{H}\|_\infty$ and $\|H''\|_\infty$ in the lemma are bounded by e^M, while the constants $\|H'/H\|_\infty$ and $\|H''/H\|_\infty$ are equal to 1. If f takes its values in $(-\infty, M]$, then $C_f \leq e^M$, and all multiplicative constants in the lemma are bounded by a multiple of e^M.

Example 2.11 (Quadratically increasing link) The function $H: \mathbb{R} \to (0, \infty)$ defined by $H(x) = e^x$, for $x \leq 1$, and $H(x) = e + e(x-1) + e(x-1)^2/2$, for $x \geq 1$, is twice continuously differentiable, and such that the constants $\|H'/\sqrt{H}\|_\infty$, $\|H''\|_\infty$, $\|H'/H\|_\infty$ and $\|H''/H\|_\infty$ are all finite when computed for the full real line $(a, b) = \mathbb{R}$. Thus, for this link function, the Hellinger distance on the densities p_f is bounded by the $\mathbb{L}_2(G)$-distance on the functions f. If f is bounded above, then $C_f < \infty$, and the Kullback-Leibler discrepancies between p_f and any other density p_g will be bounded by the square $\mathbb{L}_2(G)$-distance between f and g.

2.7 Historical Notes

Random basis expansions have been in use for a long time, beginning with the work of Poincaré. The term "infinite-dimensional exponential family" seems to be first mentioned by Verdinelli and Wasserman (1998) and Barron et al. (1999). Gaussian processes as priors for density estimation were first used by Leonard (1978) and later by Lenk (1988). Nonparametric mixtures of kernels with a Dirichlet process as mixing distribution appeared in Ghorai and Rubin (1982), Ferguson (1983) and Lo (1984). Feller approximation schemes to construct priors were considered by Petrone and Veronese (2010). Lemmas 2.5 and 2.8 come from van der Vaart and van Zanten (2008a).

Problems

2.1 (Hjort 1996) For density estimation on $\mathbb{R}$, use finite Hermite polynomial expansion

$$\sum_{j=1}^{m} \phi_\sigma(x-\mu)\Big[1 + \sum_{j=3}^{m} \frac{\kappa_j}{j!} H_j((x-\mu)/\sigma)\Big]$$

for large m, where κ_js are cumulants and the Hermite polynomials H_js are given by $(d^j/dx^j)\phi(x) = (-1)^j H_j(x)\phi(x)$, to construct a prior on a density.

The above method has the drawback that cumulants need not be finite for many densities. Show that the alternative expansion

$$\sum_{j=1}^{m} \phi_\sigma(x-\mu)\Big[\sum_{j=0}^{m} \frac{\delta_j}{\sqrt{j!}} H_j(\sqrt{2}(x-\mu)/\sigma)\Big]$$

leads to finite values of δ_js for any density and hence can be used for constructing a prior on a probability density.

2.2 (Petrone and Veronese 2010) Show that $a(x; k, F)$ in (2.5) is indeed the density of $A(x; k, F)$.

2.3 (Petrone and Veronese 2010) Show that in a Feller sampling scheme:

(a) if the sampling scheme is based on $Y \sim \text{Nor}(x, \sigma^2)$, then the kernel $h_k(x; z)$ is normal;
(b) if the sampling scheme is based on $Y \sim \text{Bin}(1, x)$, then the kernel $h_k(x; z)$ is a Bernstein polynomial;
(c) if the sampling scheme is based on $Y \sim \text{Poi}(x)$, then the kernel $h_k(x; z)$ is gamma;
(d) if the sampling scheme is based on $Y \sim \text{Ga}(x)$, then the kernel $h_k(x; z)$ is inverse-gamma.

2.4 (Log Lipschitz link function) Let $\Psi: \mathbb{R} \to (0, \infty)$ be a monotone function whose logarithm $\log \Psi$ is uniformly Lipschitz with constant L. For a function $f: \mathfrak{X} \to \mathbb{R}$ on a measurable space $\mathfrak{X}$ such that $c(f) := \int \Psi(f)\, d\mu < \infty$, define the probability density $p_f(x) = \Psi(f(x))/c(f)$. Show that $d_H(p_f, p_g) \lesssim L\|f - g\|_\infty e^{L\|f-g\|_\infty}$ and $(K + V_{2,0})(p_f; p_g) \lesssim L^2\|f - g\|_\infty e^{4L\|f-g\|_\infty}$. This generalizes Lemma 2.5 (i) to (iii), which consider the special case that Ψ is the exponential function. [Hint: Derive first that $|\Psi(f)^a - \Psi(g)^a| \leq \Psi(f)^a e^{aL\epsilon}$ if $\|f - g\|_\infty = \epsilon$, for $a = 1, 2$. This yields the bound on the Hellinger distance as in the proof of Lemma 2.5. Furthermore, it shows that $c(g) \leq c(f)(1 + L\epsilon e^{L\epsilon})$, which together with the Lipschitz property gives a uniform bound on $\log(p_f/p_g)$.]

2.5 (Lipschitz link function) For given measurable functions $\Psi: \mathbb{R} \to (0, \infty)$ and $f: \mathfrak{X} \to \mathbb{R}$ such that $c(f) := \int \Psi(f)\, d\mu < \infty$, define the probability density $p_f(x) = \Psi(f(x))/c(f)$. For $r \geq 1$, show that $\|p_f^{1/r} - p_g^{1/r}\|_r \leq 2\|\Psi(f)^{1/r} - \Psi(g)^{1/r}\|_r/c(f)^{1/r}$ and $|c(f)^{1/r} - c(g)^{1/r}| \leq \|\Psi(f)^{1/r} - \Psi(g)^{1/r}\|_r$, and also that $\|p_f - p_g\|_\infty \leq (1 + \mu(\mathfrak{X})\|p_g\|_\infty)\|\Psi(f) - \Psi(g)\|_\infty/c(f)$. Conclude that if Ψ is uniformly Lipschitz on an interval that contains the ranges of f and g, then $\|p_f - p_g\|_1 \lesssim \mu(\mathfrak{X})\|f - g\|_\infty/c(f)$ and $|c(f) - c(g)| \lesssim \mu(\mathfrak{X})\|\|f - g\|_\infty$, and also that $\|p_f - p_g\|_\infty \leq (1 + \mu(\mathfrak{X})\|p_g\|_\infty)\|f - g\|_\infty/c(f)$. Furthermore, conclude that if $\sqrt{\Psi}$ is Lipschitz on an interval that contains the ranges of f and g, then $d_H(p_f, p_g) \lesssim \mu(\mathfrak{X})\|f - g\|_\infty/c(f)$ and $|c(f)^{1/2} - c(g)^{1/2}| \lesssim \mu(\mathfrak{X})\, \|f - g\|_\infty$.

2.6 (van der Vaart and van Zanten 2008a) In the setting of Lemma 2.8, let H be the cumulative distribution function Φ of the standard normal distribution. Show that $\max\{K(p_w, p_{w_0}), V_2(p_w, p_{w_0})\} \lesssim \|w - w_0\|_{2,G_0}^2 + \|w - w_0\|_{4,G}^4$, where $dG_0 = (w_0^2 \vee 1)\, dG$.

3

Priors on Spaces of Probability Measures

In the nonparametric setting it is natural to place a prior distribution directly on the law of the data. After presenting a general background on priors on spaces of measures, in this chapter we introduce several methods of constructing priors: stick breaking, successive partitioning, random distribution functions, etc. In particular, we discuss the rich class of tail-free processes and their properties which includes the important special case of a Pólya tree process.

3.1 Random Measures

Constructing a prior distribution on a space of probability measures comes with some technical complications. To limit these as much as possible, we assume that the sample space $(\mathfrak{X}, \mathscr{X})$ is a Polish space, and consider priors on the collection $\mathfrak{M} = \mathfrak{M}(\mathfrak{X})$ of all probability measures on $(\mathfrak{X}, \mathscr{X})$.

A prior Π on $\mathfrak{M}$ can be viewed as the law of a *random measure* P (a map from some probability space into $\mathfrak{M}$), and can be identified with the collection of "random probabilities" $P(A)$ of sets $A \in \mathscr{X}$. It is natural to choose the measurability structure on $\mathfrak{M}$ so that at least each $P(A)$ is a random variable; in other words, $(P(A): A \in \mathscr{X})$ is a *stochastic process* on the underlying probability space. In this chapter we choose the σ-field $\mathscr{M}$ on $\mathfrak{M}$, the minimal one to make this true: we set $\mathscr{M}$ equal to the smallest σ-field that makes all maps $M \mapsto M(A)$ from $\mathfrak{M}$ to $\mathbb{R}$ measurable, for $A \in \mathscr{X}$, and consider priors Π that are measures on $(\mathfrak{M}, \mathscr{M})$. Although other measurability structures are possible, the σ-field $\mathscr{M}$ is attractive for two reasons.

First, it is identical to the Borel σ-field for the weak topology on $\mathfrak{M}$ (the topology of convergence in distribution in this space, see Proposition A.5). As $\mathfrak{M}$ is Polish under the weak topology (see Theorem A.3), this means that $(\mathfrak{M}, \mathscr{M})$ is a standard Borel space. As is noted in Section 1.3, this is desirable for the definition of posterior distributions and also permits us to speak of the support of a prior, called the *weak support* in this situation. Furthermore, the parameter θ from Section 1.3 that indexes the statistical model $(P_\theta: \theta \in \Theta)$, can be taken equal to the distribution P itself, with $\mathfrak{M}$ (or a subset) as the parameter set, giving a model of the form $(P: P \in \mathfrak{M})$. With respect to the σ-field $\mathscr{M}$ on the parameter set $\mathfrak{M}$, the data distributions are trivially "regular conditional probabilities":

(i) $P \mapsto P(A)$ is $\mathscr{M}$-measurable for every $A \in \mathscr{X}$,
(ii) $A \mapsto P(A)$ is a probability measure for every realization of P.

This mathematically justifies speaking of "drawing a measure P from the prior Π and next sampling observations X from P."

Second, the fact that $\mathscr{M}$ is generated by all maps $M \mapsto M(A)$, for $A \in \mathscr{X}$, implies that a map $P: (\Omega, \mathscr{U}, \mathrm{P}) \to (\mathfrak{M}, \mathscr{M})$ defined on some probability space is measurable precisely if the induced measure $P(A)$ of every set $A \in \mathscr{X}$ is a random variable. Thus, as far as measurability goes, a random probability measure can be identified with a random element $(P(A): A \in \mathscr{X})$ in the product space $\mathbb{R}^{\mathscr{X}}$ (or $[0, 1]^{\mathscr{X}}$).

3.1.1 Other Topologies

In the preceding we work with a measurability structure that is linked to the weak topology (or topology of convergence in distribution). Other natural topologies lead to different measurability structures. Two topologies are common.

A stronger topology is induced by the *total variation distance*

$$d_{TV}(P, Q) = \sup_{A \in \mathscr{X}} |P(A) - Q(A)|.$$

In general, the σ-field $\mathscr{M}$ is smaller than the Borel σ-field arising from the total variation distance on $\mathfrak{M}$. This difference disappears if the set of measures of interest is dominated: the traces of the two σ-fields on a dominated subset $\mathfrak{M}_0 \subset \mathfrak{M}$ are equal (see Proposition A.10; $\mathscr{M}$ is always equal to the ball σ-field). However, this equivalence does not extend to the topologies and the supports of priors: in general the weak support is bigger than the support relative to the total variation distance, even on dominated sets of measures.

The topology of uniform convergence of cumulative distribution functions, in the case that $\mathfrak{X} = \mathbb{R}^k$, generated by the *Kolmogorov-Smirnov distance* defined in (A.5), is intermediate between the weak and total variation topologies. Thus the Kolmogorov-Smirnov support is also smaller than the weak support. However distributions with a *continuous* distribution function are in the Kolmogorov-Smirnov support as soon as they are in the weak support. This is a consequence of Pólya's theorem (see Proposition A.11), which asserts that weak convergence to a continuous distribution function implies uniform convergence.

Appendices A and B contain extended discussions of these issues.

3.2 Construction through a Stochastic Process

One general method of constructing a random measure is to start with the stochastic process $(P(A): A \in \mathscr{X})$, constructed using Kolmogorov's consistency theorem, and next show that this process can be realized within $\mathfrak{M}$, viewed as a subset of $\mathbb{R}^{\mathscr{X}}$. As measures have much richer properties than can be described by the finite-dimensional distributions involved in Kolmogorov's theorem, this approach is nontrivial, but it can be pushed through by standard arguments. The details are as follows.

For every finite collection $A_1, \ldots, A_k$ of Borel sets in $\mathfrak{X}$, the vector $(P(A_1), \ldots, P(A_k))$ of probabilities obtained from a random measure P is an ordinary random vector in $\mathbb{R}^k$. The construction of P may start with the specification of the distributions of all vectors of this type. A simple, important example would be to specify these as Dirichlet distributions with parameter vector $(\alpha(A_1), \ldots, \alpha(A_k))$, for a given Borel measure α. For any *consistent* specification of the distributions, Kolmogorov's theorem allows us to construct on a suitable probability space $(\Omega, \mathscr{U}, \mathrm{P})$ a stochastic process $(P(A): A \in \mathscr{X})$ with the given finite-dimensional distributions. If the marginal distributions correspond to those of a random measure, then it will be true that

(i) $P(\varnothing) = 0$, $P(\mathfrak{X}) = 1$, a.s.
(ii) $P(A_1 \cup A_2) = P(A_1) + P(A_2)$, a.s., for any disjoint A_1, A_2.

Assertion (i) follows, because the distributions of $P(\varnothing)$ and $P(\mathfrak{X})$ will be specified to be degenerate at 0 and 1, respectively, while (ii) can be read off from the degeneracy of the joint distribution of the three variables $P(A_1)$, $P(A_2)$ and $P(A_1 \cup A_2)$. Thus the process $(P(A): A \in \mathscr{X})$ will automatically define a *finitely additive* measure on $(\mathfrak{X}, \mathscr{X})$.

A problem is that the exceptional null sets in (ii) might depend on the pair (A_1, A_2). If restricted to a countable subcollection $\mathscr{X}_0 \subset \mathscr{X}$, there would only be countably many pairs and the null sets could be gathered in a single null set. Then still when extending (ii) to σ-additivity, which is typically possible by similar distributional arguments, there would be uncountably many sequences of sets. This problem can be overcome through existence of a *mean measure*

$$\mu(A) = \mathrm{E}[P(A)].$$

For a valid random measure P, this necessarily defines a Borel measure on $\mathfrak{X}$. Existence of a mean measure is also sufficient for existence of a version of $(P(A): A \in \mathscr{X})$ that is a measure on $(\mathfrak{X}, \mathscr{X})$.

Theorem 3.1 (Random measure) *Suppose that $(P(A): A \in \mathscr{X})$ is a stochastic process that satisfies (i) and (ii) and whose mean $A \mapsto \mathrm{E}[P(A)]$ is a Borel measure on $\mathfrak{X}$. Then there exists a version of P that is a random measure on $(\mathfrak{X}, \mathscr{X})$. More precisely, if P is defined on the complete probability space $(\Omega, \mathscr{U}, \mathrm{P})$, then there exists a measurable map $\tilde{P}: (\Omega, \mathscr{U}, \mathrm{P}) \to (\mathfrak{M}, \mathscr{M})$ such that $P(A) = \tilde{P}(A)$ almost surely, for every $A \in \mathscr{X}$.*

Proof Let $\mathscr{X}_0$ be a countable field that generates the Borel σ-field $\mathscr{X}$, enumerated arbitrarily as $A_1, A_2, \ldots$. Because the mean measure $\mu(A) := \mathrm{E}[P(A)]$ is regular, there exists for every $i, m \in \mathbb{N}$ a compact set $K_{i,m} \subset A_i$ with $\mu(A_i \setminus K_{i,m}) < 2^{-2i-2m}$. By Markov's inequality

$$\mathrm{P}(P(A_i \setminus K_{i,m}) > 2^{-i-m}) \le 2^{i+m}\mathrm{E}P(A_i \setminus K_{i,m}) \le 2^{-i-m}.$$

Consequently, the event $\Omega_m = \cap_i \{P(A_i \setminus K_{i,m}) \le 2^{-i-m}\}$ possesses probability at least $1 - 2^{-m}$, and $\liminf \Omega_m$ possesses probability 1, by the Borel-Cantelli lemma.

Because $\mathscr{X}_0$ is countable, the null sets involved in (i) and (ii) with $A_1, A_2 \in \mathscr{X}_0$ can be aggregated into a single null set N. For every $\omega \notin N$, the process P is a finitely additive measure on $\mathscr{X}_0$, with the resulting usual properties of monotonicity and subadditivity. By increasing N, if necessary, we can also ensure that it is subadditive on all finite unions of sets $A_i \setminus K_{i,m}$.

Let $A_{i_1} \supset A_{i_2} \supset \cdots$ be an arbitrary decreasing sequence of sets in $\mathscr{X}_0$ with empty intersection. Then, for every fixed m, the corresponding compacts $K_{i_j,m}$ also possess empty intersection, whence there exists a finite J_m such that $\cap_{j \le J_m} K_{i_j,m} = \varnothing$. This implies that

$$A_{i_{J_m}} = \cap_{j=1}^{J_m} A_{i_j} \setminus \cap_{j=1}^{J_m} K_{i_j,m} \subset \cup_{j=1}^{J_m} (A_{i_j} \setminus K_{i_j,m}).$$

Consequently, on the event $\Omega_m \setminus N$,

$$\limsup_j P(A_{i_j}) \le P(A_{i_{J_m}}) \le \sum_{j=1}^{J_m} P(A_{i_j} \setminus K_{i_j,m}) \le 2^{-m}.$$

Thus on the event $\Omega_0 := \liminf \Omega_m \setminus N$, the limit is zero. We conclude that for every $\omega \in \Omega_0$, the restriction of $A \mapsto P(A)$ to $\mathscr{X}_0$ is countably additive. By Carathéodory's theorem, it extends to a measure $\tilde{P}$ on $\mathscr{X}$.

By construction $\tilde{P}(A) = P(A)$, almost surely, for every $A \in \mathscr{X}_0$. In particular, $\mathrm{E}[\tilde{P}(A)] = \mathrm{E}[P(A)] = \mu(A)$, for every A in the field $\mathscr{X}_0$, whence by uniqueness of extension the mean measure of $\tilde{P}$ coincides with the original mean measure μ on $\mathscr{X}$. For every $A \in \mathscr{X}$, there exists a sequence $\{A_m\} \subset \mathscr{X}_0$ such that $\mu(A \,\Delta\, A_m) \to 0$. Then both $P(A_m \,\Delta\, A)$ and $\tilde{P}(A_m \,\Delta\, A)$ tend to zero in mean. Finite-additivity of P gives that $|P(A_m) - P(A)| \le P(A_m \,\Delta\, A)$, almost surely, and by σ-additivity the same is true for $\tilde{P}$. This shows that $\tilde{P}(A) = P(A)$, almost surely, for every $A \in \mathscr{X}$.

This also proves that $\tilde{P}(A)$ is a random variable for every $A \in \mathscr{X}$, whence $\tilde{P}$ is a measurable map in $(\mathfrak{M}, \mathscr{M})$. □

Rather than starting from the process $(P(A): A \in \mathscr{X})$ indexed by all Borel sets, we may wish to start from a smaller set $(P(A): A \in \mathscr{X}_0)$ of variables, for some $\mathscr{X}_0 \subset \mathscr{X}$. As shown in the proof of the preceding theorem, a countable collection $\mathscr{X}_0$ suffices, but compact sets play a special role.

Theorem 3.2 (Random measure) *Suppose that $(P(A): A \in \mathscr{X}_0)$ is a stochastic process that satisfies (i) and (ii) for a countable field $\mathscr{X}_0$ that generates $\mathscr{X}$ and is such that for every $A \in \mathscr{X}_0$ and $\epsilon > 0$ there exists a compact $K_\epsilon \subset \mathfrak{X}$ and $A_\epsilon \in \mathscr{X}_0$ such that $A_\epsilon \subset K_\epsilon \subset A$ and $\mu(A \setminus A_\epsilon) < \epsilon$, where μ is the mean $\mu(A) = \mathrm{E}[P(A)]$. Then there exists a random measure that extends P to $\mathscr{X}$.*

The proof of the theorem follows the same lines, except that, if the compacts K_ϵ are not elements of $\mathscr{X}_0$, the bigger sets $A \setminus A_\epsilon$ must be substituted for $A \setminus K_\epsilon$ when bounding the P-measure of this set.

For instance, for $\mathfrak{X} = \mathbb{R}^k$ we can choose $\mathscr{X}_0$ equal to the finite unions of cells $(a, b]$, with the compacts and A_ϵ equal to the corresponding finite unions of the intervals $[a_\epsilon, b]$ and $(a_\epsilon, b]$ for a_ϵ descending to a. By restricting to rational endpoints we obtain a countable collection.

3.3 Countable Sample Spaces

A probability distribution on a countable sample space (equipped with the σ-field of all its subsets) can be represented as an infinite-length probability vector $s = (s_1, s_2, \ldots)$. A prior on the set $\mathfrak{M}$ of all probability measures on a countable sample space can therefore be identified with the distribution of a random element with values in the countable-dimensional unit simplex

$$\mathbb{S}_\infty = \Big\{ s = (s_1, s_2, \ldots) : s_j \geq 0, j \in \mathbb{N}, \sum_{j=1}^{\infty} s_j = 1 \Big\}.$$

The usual σ-field $\mathscr{M}$ on $\mathbb{S}_\infty \equiv \mathfrak{M}$ can be characterized in various ways, the simplest being that it is generated by the coordinate maps $s \mapsto s_i$, for $i \in \mathbb{N}$. This shows that a map p from some probability space into $\mathbb{S}_\infty$ is a random element if and only if every coordinate variable p_i is a random variable. Hence a prior simply corresponds to an infinite sequence of nonnegative random variables $p_1, p_2, \ldots$ that add up to 1.

We can also embed $\mathbb{S}_\infty$ in $\mathbb{R}^\infty$, from which it then inherits its measurable structure (the projection σ-field) and a topology: the topology of coordinatewise convergence. By Scheffé's theorem the restriction of this topology to $\mathbb{S}_\infty$ is equivalent to the topology derived from the norm $\|s\|_1 = \sum_{j=1}^{\infty} |s_j|$ of the space $\ell_1 := \{s = (s_1, s_2, \ldots) \in \mathbb{R}^\infty : \sum_{j=1}^{\infty} |s_j| < \infty\} \supset \mathbb{S}_\infty$. The weak topology on $\mathbb{S}_\infty$, viewed as probability measures, coincides with these topologies as well. Thus the projection σ-field $\mathscr{M}$ is also the Borel σ-field for these topologies and $(\mathfrak{M}, \mathscr{M})$ is a Polish space.

The general approach of Section 3.2 of constructing priors using Kolmogorov's theorem applies, but can be simplified by ordering the coordinates: it suffices to construct consistent marginal distributions for $(p_1, \ldots, p_k)$, for every $k = 1, 2, \ldots$ In the next sections, we also present two structural methods.

The support of a prior Π relative to the ℓ_1-norm is the set of all p_0, such that $\Pi(p: \|p - p_0\|_1 < \epsilon) > 0$, for every $\epsilon > 0$. Since the norm topology coincides with the topology of coordinatewise convergence (and hence finite intersections of sets of the form $\{p: |p_j - p_{0j}| < \epsilon\}$ form a base of the topology), it can also be described as the set of all p_0 such that $\Pi(p: |p_j - p_{0j}| < \epsilon, 1 \leq j \leq k) > 0$, for every $\epsilon > 0$ and $k \in \mathbb{N}$. This reduction to the finite-dimensional marginal distributions makes it easy to construct priors with large support.

3.3.1 Construction through Normalization

Given nonnegative random variables $Y_1, Y_2, \ldots$ such that $\sum_{j=1}^{\infty} Y_j$ is positive and converges a.s., we can define a prior on $\mathbb{S}_\infty$ by putting

$$p_k = \frac{Y_k}{\sum_{j=1}^{\infty} Y_j}, \qquad k \in \mathbb{N}. \tag{3.1}$$

A simple, sufficient condition for the convergence of the random series is that $\sum_{j=1}^{\infty} \mathrm{E}(Y_j) < \infty$. It is convenient to use independent random variables.

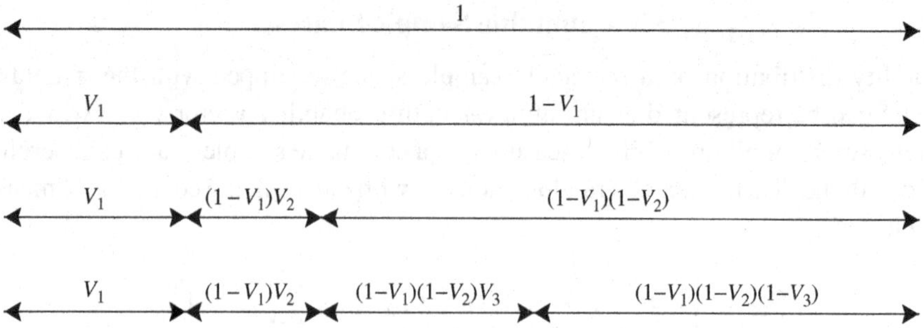

Figure 3.1 Stick breaking. A stick of length 1 is sequentially broken in smaller sticks in proportions given by a random sequence $V_1, V_2, \ldots$

Lemma 3.3 *If $Y_1, Y_2, \ldots$ are independent, positive random variables with $\sum_{j=1}^{\infty} Y_j < \infty$ a.s. and marginal densities that are positive everywhere in $(0, \infty)$, then the support of the prior defined by (3.1) is the full space $\mathbb{S}_\infty$.*

Proof Fix an arbitrary point $p_0 \in \mathbb{S}$, $\epsilon > 0$ and $k \in \mathbb{N}$. Since $\sum_{j=1}^{\infty} Y_j$ converges a.s., there exists $K \geq k$ such that the event $\{\sum_{j=K+1}^{\infty} Y_j < \epsilon/2\}$ has positive probability. By choosing K larger, if necessary, we can also ensure that $\sum_{j=K+1}^{\infty} p_{0j} < \epsilon/2$. By continuity of the maps $(y_1, \ldots, y_K) \mapsto y_j/(\sum_{j=1}^{K} y_j + c)$ from $\mathbb{R}^K \to \mathbb{R}$, the set

$$\mathcal{N}_K = \left\{ (y_1, \ldots, y_K) \colon \max_{1 \leq j \leq k} \left| \frac{y_j}{\sum_{l=1}^{K} y_l} - p_{0j} \right| \vee \left| \frac{y_j}{\sum_{l=1}^{K} y_l + \epsilon/2} - p_{0j} \right| < \epsilon \right\}$$

is open in $(0, \infty)^K$. Also $\mathcal{N}_K$ is nonempty, since $(p_{01}, \ldots, p_{0K}) \in \mathcal{N}_K$, as is easy to verify. Thus the event $\{(Y_1, \ldots, Y_K) \in \mathcal{N}_K\}$ has positive probability by the assumption of positive densities.

On the intersection of the events $\{\sum_{j=K+1}^{\infty} Y_j < \epsilon/2\}$ and $\{(Y_1, \ldots, Y_K) \in \mathcal{N}_K\}$, we have that $|Y_j/(\sum_{l=1}^{\infty} Y_l) - p_{0j}| < \epsilon$ for all $1 \leq j \leq k$. Because the events are independent and both have positive probability, the intersection has positive probability. □

3.3.2 Construction through Stick Breaking

Stick-breaking is a technique to construct a prior directly on $\mathbb{S}_\infty$. The problem at hand is to distribute the total mass 1, which we identify with a stick of length 1, randomly to each element of $\mathbb{N}$. We first break the stick at a point given by the realization of a random variable $0 \leq V_1 \leq 1$ and assign mass V_1 to $1 \in \mathbb{N}$. We think of the remaining mass $1 - V_1$ as a new stick, and break it into two pieces of relative lengths V_2 and $1 - V_2$ according to the realized value of another random variable $0 \leq V_2 \leq 1$. We assign mass $(1 - V_1)V_2$ to the point 2, and are left with a new stick of length $(1 - V_1)(1 - V_2)$. Continuing in this way, we assign mass to the point j equal to

$$p_j = \Big(\prod_{l=1}^{j-1} (1 - V_l) \Big) V_j. \tag{3.2}$$

Clearly, by continuing to infinity, this scheme will attach a random subprobability distribution to $\mathbb{N}$ for any sequence of random variables $V_1, V_2, \ldots$ with values in $[0, 1]$. Under mild conditions, the probabilities p_j will sum to one.

Lemma 3.4 (Stick-breaking) *The random subprobability distribution* $(p_1, p_2, \ldots)$ *lies in* $\mathbb{S}_\infty$ *almost surely if and only if* $\mathrm{E}[\prod_{l=1}^j (1 - V_l)] \to 0$ *as* $j \to \infty$. *For independent variables* $V_1, V_2, \ldots$, *this condition is equivalent to* $\sum_{l=1}^\infty \log \mathrm{E}(1 - V_l) = -\infty$. *In particular, for i.i.d. variables* $V_1, V_2, \ldots$, *it suffices that* $\mathrm{P}(V_1 > 0) > 0$. *If for every* $k \in \mathbb{N}$ *the support of* $(V_1, \ldots, V_k)$ *is* $[0, 1]^k$, *then the support of* $(p_1, p_2, \ldots)$ *is the whole space* $\mathbb{S}_\infty$.

Proof By induction, it easily follows that the leftover mass at stage j is equal to $1 - \sum_{l=1}^j p_l = \prod_{l=1}^j (1 - V_l)$. Hence the random subprobability distribution will lie in $\mathbb{S}_\infty$ a.s. if and only if $\prod_{l=1}^j (1 - V_l) \to 0$ a.s. Since the leftover sequence is decreasing, nonnegative and bounded by 1, the almost sure convergence is equivalent to convergence in mean. If the V_js are independent, then this condition becomes $\prod_{l=1}^j (1 - \mathrm{E}(V_l)) \to 0$ as $j \to \infty$, which is equivalent to the condition $\sum_{l=1}^\infty \log \mathrm{E}(1 - V_l) = -\infty$.

The last assertion follows, because the probability vector $(p_1, \ldots, p_k)$ is a continuous function of $(V_1, \ldots, V_k)$, for every k. ☐

In the stick-breaking construction, if X is a random variable distributed according to probability mass function $p = (p_1, p_2, \ldots)$,

$$V_j = \frac{p_j}{1 - \sum_{l=1}^{j-1} p_l} = \mathrm{P}(X = j \mid X \geq j). \tag{3.3}$$

This is known as the *discrete hazard rate* for X. Thus the construction may be interpreted in terms of hazard rates, or equivalently in terms of the negative log-hazard $h_j = -\log V_j$. Any consistent system of joint distributions of h_js lying in $[0, \infty)$ (in particular, independent) gives rise to a possibly defective prior on p. For independent h_js, the necessary and sufficient condition for this to be proper prior is given by $\sum_{j=1}^\infty \log(1 - \mathrm{E}(e^{-h_j})) = -\infty$.

3.3.3 Countable Dirichlet Process

The *countable Dirichlet distribution* can be obtained both through normalization and stick breaking.

In the construction by normalization, we choose the variables $Y_1, Y_2, \ldots$ to be independent with $Y_j \sim \mathrm{Ga}(\alpha_j, 1)$, for $j \in \mathbb{N}$. Then, by Proposition G.2, the vector $(p_1, p_2, \ldots)$ defined in (3.1) satisfies, for any $k \in \mathbb{N}$,

$$\Big(p_1, \ldots, p_k, 1 - \sum_{j=1}^k p_j\Big) \sim \mathrm{Dir}\Big(k + 1; \alpha_1, \ldots, \alpha_k, \sum_{j=k+1}^\infty \alpha_j\Big). \tag{3.4}$$

In particular, every p_j is $\mathrm{Be}(\alpha_j, \sum_{l \neq j} \alpha_l)$-distributed.

In the construction by stick breaking, we choose the variables $V_1, V_2, \ldots$ independent with $V_j \sim \mathrm{Be}(\alpha_j, \sum_{l=j+1}^\infty \alpha_l)$ and define $(p_1, p_2, \ldots)$ by (3.2). To see that this yields the same distribution, it suffices to check that the joint distribution of $(p_1, \ldots, p_k)$

is the same in the two cases, for every $k \in \mathbb{N}$. Equivalently, it suffices to show that $V_1, V_2, \ldots$, defined by (3.3) from $p_1, p_2, \ldots$ satisfying (3.4), are independent variables with $V_j \sim \text{Be}(\alpha_j, \sum_{l=j+1}^{\infty} \alpha_l)$. This is a consequence of the aggregation properties of the finite-dimensional Dirichlet distribution described in Proposition G.3, as noted in Corollary G.5.

Explicit calculation of the posterior distribution given a sample of i.i.d. observations $X_1, \ldots, X_n$ from p is possible in this case. The likelihood can be written as $p \mapsto \prod_j p_j^{N_j}$, for N_j the number of observations equal to j. The vector $N = (N_1, N_2, \ldots)$ is a sufficient statistic, and hence the posterior given N is the same as the posterior given the original observations. For any $l \in \mathbb{N}$,

$$\Big(N_1, \ldots, N_l, n - \sum_{j=1}^{l} N_j\Big) \sim \text{MN}_{l+1}\Big(n; p_1, \ldots, p_l, 1 - \sum_{j=1}^{l} p_j\Big).$$

Therefore, by (3.4) with l replacing k and the conjugacy of the finite Dirichlet distribution with the multinomial likelihood, expressed in Proposition G.8, it follows that the posterior density of $(p_1, \ldots, p_l, 1 - \sum_{j=1}^{l} p_j)$ given $N_1, \ldots, N_l$ is given by

$$\text{Dir}\Big(l + 1; \alpha_1 + N_1, \ldots, \alpha_l + N_l, \sum_{j=l+1}^{\infty} \alpha_j + n - \sum_{j=1}^{l} N_j\Big). \tag{3.5}$$

Marginalizing to the first $k \leq l$ cells, it follows from Proposition G.3 that the posterior density of $(p_1, \ldots, p_k, 1 - \sum_{j=l}^{k} p_j)$ given $N_1, \ldots, N_l$ is given by

$$\text{Dir}\Big(k + 1; \alpha_1 + N_1, \ldots, \alpha_k + N_k, \sum_{j=k+1}^{\infty} \alpha_j + n - \sum_{j=1}^{k} N_j\Big). \tag{3.6}$$

Because this depends only on $(N_1, \ldots, N_k)$, the posterior density of $(p_1, \ldots, p_k, 1 - \sum_{j=l}^{k} p_j)$ given $N_1, \ldots, N_l$ is the same for every $l \geq k$. Because $\sigma\langle N_1, \ldots, N_l\rangle$ increases to $\sigma\langle N_1, N_2, \ldots\rangle$ as $l \to \infty$, it follows by the martingale convergence theorem that the preceding display also gives the posterior density of $(p_1, \ldots, p_k, 1 - \sum_{j=l}^{k} p_j)$ given N.

The distribution (3.6) has the same form as the distribution in the right side of (3.4), whence the posterior distribution is also a countable Dirichlet distribution. The posterior parameters depend on the prior parameters and the cell counts and can be found by the straightforward updating rule $\alpha \mapsto \alpha + N$.

In analogy with the finite dimensional Dirichlet distribution, it is natural to call the prior the *Dirichlet process* on $\mathbb{N}$ or the *countable Dirichlet process*. We shall write $(p_1, p_2, \ldots) \sim \text{DP}((\alpha_1, \alpha_2, \ldots))$. This Dirichlet process will be generalized to arbitrary spaces in the next chapter and will be seen to admit similar explicit expressions. It is a central object in Bayesian nonparametrics.

From (3.6) and the properties of the finite-dimensional Dirichlet distribution, the posterior mean, variance and covariances are seen to be

$$\text{E}(p_j | X_1, \ldots, X_n) = \frac{\alpha_j + N_j}{\sum_{l=1}^{\infty} \alpha_l + n}, \tag{3.7}$$

$$\operatorname{var}(p_j | X_1, \ldots, X_n) = \frac{(\alpha_j + N_j)(\sum_{l \neq j} \alpha_l + n - N_j)}{(\sum_{l=1}^{\infty} \alpha_l + n)^2 (\sum_{l=1}^{\infty} \alpha_l + n + 1)}, \tag{3.8}$$

$$\operatorname{cov}(p_j, p_{j'} | X_1, \ldots, X_n) = -\frac{(\alpha_j + N_j)(\alpha_{j'} + N_{j'})}{(\sum_{l=1}^{\infty} \alpha_l + n)^2 (\sum_{l=1}^{\infty} \alpha_l + n + 1)}. \tag{3.9}$$

3.4 Construction through Structural Definitions

In this section we collect priors on measures on a general Polish space that are defined explicitly or through an iterative algorithm.

3.4.1 Construction through a Distribution on a Dense Subset

An easy method to obtain a prior with full support is to assign positive prior mass to every point in a given countable dense subset. Such a prior distribution can be considered a default prior if the point masses are constructed by a default mechanism. The set $\mathfrak{M}$ of Borel measures on a Polish space $\mathfrak{X}$ is also Polish by Theorem A.3, and hence has a countable dense subset to which this construction can be applied.

Often it is meaningful to construct the prior using a sequence of finite subsets that gradually improve the approximation. For instance, at stage m we choose a finite subset $\mathfrak{M}_m$ which approximates the elements of $\mathfrak{M}$ within a distance ϵ_m, and put the discrete uniform distribution on $\mathfrak{M}_m$. A convex combination over m (or the sequence of discrete priors) may be regarded as a default prior. If $\epsilon_m \downarrow 0$, then the weak support of a prior of this type is the whole of $\mathfrak{M}$. It will be seen in Chapters 6 and 8 that such a prior, with carefully chosen support points guided by covering numbers, results in posterior distributions with good large-sample properties.

Computation of the resulting posterior distribution may be difficult to carry out.

3.4.2 Construction through a Randomly Selected Discrete Set

Given an integer $N \in \mathbb{N} \cup \{\infty\}$, nonnegative random variables $W_{1,N}, \ldots, W_{N,N}$ with $\sum_{i=1}^{N} W_{i,N} = 1$ and random variables $\theta_{1,N}, \ldots, \theta_{N,N}$ taking their values in $(\mathfrak{X}, \mathscr{X})$, we can define a random probability measure by

$$P = \sum_{i=1}^{N} W_{i,N} \delta_{\theta_{i,N}}.$$

The realizations of this prior are discrete with finitely or countably many support points, which may be different for each realization. Given the number N of support points, their "weights" $W_{1,N}, \ldots, W_{N,N}$ and "locations" $\theta_{1,N}, \ldots, \theta_{N,N}$ are often chosen independent. Assume that $\mathfrak{X}$ is a separable metric space with a metric d.

Lemma 3.5 *If the support of N is unbounded and given $N = n$, the weights and locations are independent with full supports $\mathbb{S}_n$ and $\mathfrak{X}^n$, respectively, for every n, then P has full support $\mathfrak{M}$.*

Proof Because the finitely discrete distributions are weakly dense in $\mathfrak{M}$, it suffices to show that P gives positive probability to any weak neighborhood of a distribution $P^* = \sum_{i=1}^k w_i^* \delta_{\theta_i^*}$ with finite support. All distributions $P' := \sum_{i=1}^k w_i \delta_{\theta_i}$ with $(w_1, \ldots, w_k)$ and $(\theta_1, \ldots, \theta_k)$ sufficiently close to $(w_1^*, \ldots, w_k^*)$ and $(\theta_1^*, \ldots, \theta_k^*)$ are in such a weak neighborhood. So are the measures $P' = \sum_{i=1}^\infty w_i \delta_{\theta_i}$ with $\sum_{i>k} w_i$ sufficiently small and $(w_1, \ldots, w_k)$ and $(\theta_1, \ldots, \theta_k)$ sufficiently close to their targets, as before.

If $\mathrm{P}(N = \infty) > 0$, then the assertion follows upon considering the events $\{\sum_{i>k} W_{i,\infty} < \epsilon, \max_{i \le k} |W_{i,k'} - w_i^*| \vee d(\theta_{i,k}, \theta_i^*) < \epsilon\}$. These events have positive probability, as they refer to an open subset of $\mathbb{S}_\infty \times \mathfrak{X}^k$.

If N is finite almost surely, then the assertion follows from the assumed positive probability of the events $\{N = k', \max_{i \le k'} |W_{i,k'} - w_i^*| \vee d(\theta_{i,k'}, \theta_i^*) < \epsilon\}$ for every $\epsilon > 0$ and some $k' > k$, where we define $w_i^* = 0$ and θ_i^* arbitrarily for $k < i \le k'$. □

The prior is computationally more tractable if N is finite and bounded, but such a prior does not have full support on an infinite sample space. To achieve reasonable large sample properties, N must either depend on the sample size n, or be given a prior with infinite support.

An important special case is obtained by choosing $N \equiv \infty$, yielding a prior of the form

$$P = \sum_{i=1}^\infty W_i \delta_{\theta_i}. \tag{3.10}$$

Further specializations are to choose $\theta_1, \theta_2, \ldots$ an i.i.d. sequence in $\mathfrak{X}$, and independently to choose the weights $W_1, W_2, \ldots$ by the stick-breaking algorithm of Section 3.3.2. If the common distribution of the θ_i has support equal to the full space $\mathfrak{X}$ and the stick-breaking weights are as in Lemma 3.4, then this prior has full support. The assumed independence of locations and weights gives that the mean measure of P is equal to the distribution of the θ_i:

$$\mathrm{E}P(A) = \mathrm{P}(\theta_i \in A), \qquad A \in \mathscr{X}.$$

Because the realizations $\{\theta_1, \theta_2, \ldots\}$ will be dense in $\mathfrak{X}$ a.s., it is then also true that $P(U) = \sum_{i:\theta_i \in U} W_i > 0$ a.s. for every open U, as soon as the stick-breaking weights are strictly positive. In Section 4.2.5 the Dirichlet process prior will be seen to take this form.

Lemma 3.6 *If* $(W_1, W_2, \ldots)$ *are stick-breaking weights based on stick lengths* $V_i \overset{iid}{\sim} H$ *for a fully supported measure* H *on* $[0, 1]$*, independent of* $\theta_i \overset{iid}{\sim} G$ *with full support* $\mathfrak{X}$*, then the random measure* (3.10) *has full support* $\mathfrak{X}$*. Furthermore, the mean measure of* P *is* G*.*

3.4.3 Construction through Random Rectangular Partitions

Consider the sample space $\mathfrak{X} = [0, 1]$ (other domains like $\mathbb{R}$ may be handled through appropriate transformations), and fix a probability measure μ on the unit square $[0, 1]^2$, not entirely concentrated on the boundary. Draw a point randomly from the unit square according to μ. This divides the unit square into four rectangles: lower left (LL), upper left (UL), lower right (LR) and upper right (UR). Keep LL and UR, including their boundaries, and

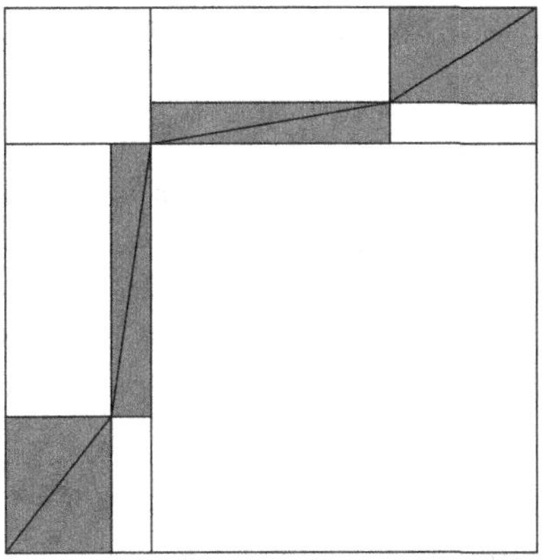

Figure 3.2 Random rectangular partitions. The shaded rectangles are the ones kept after two rounds of choosing splitting points. The curve following their diagonals is the stage 2 approximation of the random distribution function.

discard the other two. Construct the probability measures induced by μ on LL and UR by affinely mapping the unit square onto these rectangles. Draw a point randomly from LL and discard its UL and LR. Similarly, draw a point randomly from UR and discard its UL and LR. Continue this process indefinitely (see Figure 3.2).

The union of the rectangles kept by this algorithm forms an infinite sequence of decreasing compact sets. Their intersection is the graph of a continuous increasing function from $[0, 1]$ onto $[0, 1]$ and hence defines a random cumulative distribution function on $[0, 1]$. To verify this claim, observe that each vertical cross-section of the compact sets forms a sequence of nested closed *intervals* whose lengths converge to zero a.s., and hence the intersection is a singleton a.s. Alternatively, it can be seen that the function is given by the limit of the sequence of functions obtained by joining the split points at a given stage (see Figure 3.2). This is a sequence of increasing continuous functions converging uniformly a.s., and hence the limit is also increasing and continuous a.s.

This construction is similar to the one of Section 3.5 with the difference that presently both probabilities and partitions are generated randomly. Some examples of measures μ of particular interest are the uniform distributions on the sets:

(a) the vertical line segment $x = 1/2, 0 < y < 1$;
(b) the horizontal line segment $0 < x < 1, y = 1/2$;
(c) the unit square.

Case (a) is equivalent to each time breaking the horizontal line segment evenly into two subintervals and assigning probabilities according to the uniform distribution; it is the special case of the tree construction of Section 3.5, where the $V_{\varepsilon 0}$s are i.i.d. uniformly distributed. Case (b) is equivalent to breaking the horizontal axis randomly into two and assigning probability half to each segment; this generates binary quantiles of a distribution. Case (c) seems to be a natural default choice.

If μ is concentrated on the diagonal, then the procedure always leads to the uniform distribution, and hence the prior is degenerate.

3.4.4 Construction through Moments

If the domain is a bounded interval in $\mathbb{R}$, then the sequence of moments uniquely determines the probability measure. Hence a prior on the space of probability measures can be induced from one on the sequence of moments. One may control the location, scale, skewness and kurtosis of the random probability by using subjective priors on the first four moments. Priors for the higher-order moments are difficult to elicit, and some default method should be used. Maintaining the necessary constraints in the prior specification linking various moments is difficult; hence, the approach may be hard to implement.

3.4.5 Construction through Quantiles

A prior for quantiles is much easier to elicit than for moments. One may put priors on all dyadic quantiles honoring the order restrictions. Conceptually, this operation is the opposite of specifying a tree-based prior as considered in Section 3.5. For quantile priors the masses are predetermined and the partitions are random. In practice, one may put priors only for a finite number of quantiles, and then distribute the remaining masses uniformly over the corresponding interval.

3.4.6 Construction by Normalization

A prior distribution on a probability measure P needs to honor not only the countable additivity of P, but also the normalization condition $P(\mathfrak{X}) = 1$. The additional restriction typically rules out assignments such as independence of $P(A)$ and $P(B)$, when A and B are disjoint. One possible approach is to disregard this restriction and construct a prior distribution Π_∞ on $\mathfrak{M}_\infty(\mathfrak{X})$, the space of positive finite measures, and apply a renormalization step $\mu \mapsto \mu/\mu(\mathfrak{X})$ in the end. If the prior Π_∞ has full support $\mathfrak{M}_\infty(\mathfrak{X})$ with respect to the weak topology, then the prior Π on $\mathfrak{X}$ obtained by renormalization will also have full support $\mathfrak{M}(\mathfrak{X})$ with respect to the weak topology, as the following result shows.

Proposition 3.7 *Let Π_∞ be a probability distribution on $\mathfrak{M}_\infty(\mathfrak{X})$ and $\mu_0 \in \mathfrak{M}_\infty(\mathfrak{X})$ be such that, for any collection of bounded continuous functions $g_1, \dots, g_k$ and every $\epsilon > 0$,*

$$\Pi_\infty\Big(\mu{:}\,\Big|\int g_j\,d\mu - \int g_j\,d\mu_0\Big| < \epsilon\Big) > 0.$$

Then the probability measure $\mu_0/\mu_0(\mathfrak{X})$ belongs to the weak support of the probability measure Π on $\mathfrak{M}(\mathfrak{X})$ induced by the map $\mu \mapsto \mu/\mu(\mathfrak{X})$ on $\mathfrak{M}_\infty(\mathfrak{X})$ equipped with Π_∞.

Proof A typical neighborhood for the weak topology takes the form $\{P{:}\,|\int g_j dP - \int g_j dP_0| < \epsilon\}$, for given bounded continuous functions $g_1, \dots, g_k$ and $\epsilon > 0$. If $|\int g_j\,d\mu - \int g_j\,d\mu_0| < \delta$ and $|\mu(\mathfrak{X}) - \mu_0(\mathfrak{X})| = |\int 1\,d\mu - \int 1\,d\mu_0| < \delta$, then

$$\begin{aligned}\left|\frac{\int g_j\,d\mu}{\int g_j\,d\mu_0} - \frac{\int g_j\,d\mu}{\mu(\mathfrak{X})}\right| &\le \frac{1}{\mu(\mathfrak{X})}\Big|\int g_j\,d\mu - \int g_j\,d\mu_0\Big| + \int |g_j|\,d\mu\frac{|\mu(\mathfrak{X}) - \mu_0(\mathfrak{X})|}{\mu(\mathfrak{X})\mu_0(\mathfrak{X})}\\ &\le \frac{1}{\mu_0(\mathfrak{X}) - \delta}\Big(\delta + \frac{\max_j \|g_j\|_\infty \delta}{\mu_0(\mathfrak{X})}\Big) < \epsilon.\end{aligned}$$

Thus, for sufficiently small δ, the measure $\mu/\mu(\mathfrak{X})$ belongs to the given neighborhood. ☐

A random finite measure μ on $\mathfrak{M}_\infty(\mathfrak{X})$ such that the variables $\mu(A)$ and $\mu(B)$ are independent, for every disjoint sets A and B, is called a *completely random measure*, as discussed in Appendix J. The probability measure obtained by normalization of such a measure is a *normalized completely random measure*, and is studied in Section 14.7.

3.5 Construction through a Tree

Consider a sequence $\mathcal{T}_0 = \{\mathfrak{X}\}$, $\mathcal{T}_1 = \{A_0, A_1\}$, $\mathcal{T}_2 = \{A_{00}, A_{01}, A_{10}, A_{11}\}$, and so on, of measurable partitions of the sample space $\mathfrak{X}$, obtained by splitting every set in the preceding partition into two new sets. With $\mathcal{E} = \{0, 1\}$ and $\mathcal{E}^* = \cup_{m=0}^{\infty} \mathcal{E}^m$, the set of all finite strings $\varepsilon_1 \cdots \varepsilon_m$ of 0s and 1s, we can index the 2^m sets in the mth partition $\mathcal{T}_m$ by $\varepsilon \in \mathcal{E}^m$, in such a way that $A_\varepsilon = A_{\varepsilon 0} \cup A_{\varepsilon 1}$ for every $\varepsilon \in \mathcal{E}^*$. Here $\varepsilon 0$ and $\varepsilon 1$ are the extensions of the string ε with a single symbol 0 or 1; the empty string indexes $\mathcal{T}_0$ (see Figure 3.3). Let $|\varepsilon|$ stand for the length of a string ε and let $\varepsilon\delta$ be the concatenation of two strings $\varepsilon, \delta \in \mathcal{E}^*$. The set of all finite unions of sets A_ε, for $\varepsilon \in \mathcal{E}^*$, forms a subfield of the Borel sets. We assume throughout that the splits are chosen rich enough that this generates the Borel σ-field.

Because the probability of any A_ε must be distributed to its "offspring" $A_{\varepsilon 0}$ and $A_{\varepsilon 1}$, a probability measure P must satisfy the *tree additivity* requirement $P(A_\varepsilon) = P(A_{\varepsilon 0}) + P(A_{\varepsilon 1})$. The relative weights of the offspring sets are the conditional probabilities

$$V_{\varepsilon 0} = \mathrm{P}(A_{\varepsilon 0} | \, A_\varepsilon), \qquad \text{and} \qquad V_{\varepsilon 1} = \mathrm{P}(A_{\varepsilon 1} | \, A_\varepsilon). \tag{3.11}$$

This motivates to define, for a given specification of a set $(V_\varepsilon : \varepsilon \in \mathcal{E}^*)$ of $[0, 1]$-valued random variables,

$$P(A_{\varepsilon_1 \cdots \varepsilon_m}) = V_{\varepsilon_1} V_{\varepsilon_1 \varepsilon_2} \cdots V_{\varepsilon_1 \cdots \varepsilon_m}, \qquad \varepsilon = \varepsilon_1 \cdots \varepsilon_m \in \mathcal{E}^m. \tag{3.12}$$

If $V_{\varepsilon 0} + V_{\varepsilon 1} = 1$ for every ε, then the stochastic process $(P(A_\varepsilon) : \varepsilon \in \mathcal{E}^*)$ will satisfy the tree-additivity condition and define a finitely additive measure on the field of all finite unions of sets A_ε, for $\varepsilon \in \mathcal{E}^*$.

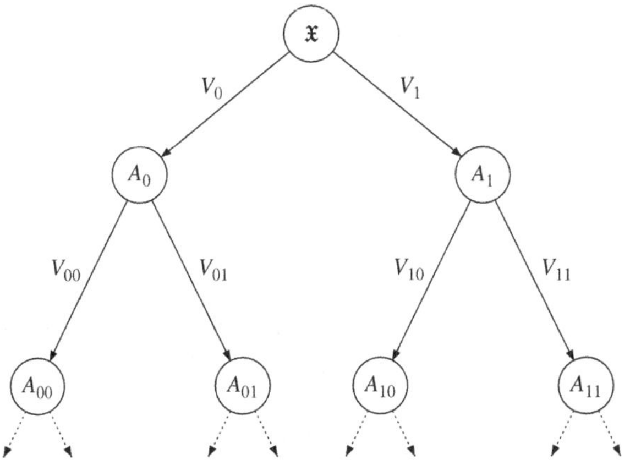

Figure 3.3 Tree diagram showing the distribution of mass over the first two partitions $\mathfrak{X} = A_0 \cup A_1 = (A_{00} \cup A_{01}) \cup (A_{10} \cup A_{11})$ of the sample space. Mass at a given node is distributed to its two children proportionally to the weights on the arrows. Every pair of Vs on arrows originating from the same node add to 1.

Countable additivity is not immediate, but may be established using a mean measure by the approach of Theorem 3.2.

Theorem 3.8 *Consider a sequence of partitions $\mathcal{T}_m = \{A_\varepsilon : \varepsilon \in \mathcal{E}^m\}$ that generates the Borel sets in $(\mathfrak{X}, \mathscr{X})$ and is such that every A_ε is the union of all $A_{\varepsilon\delta}$ whose closure is compact and satisfies $\overline{A}_{\varepsilon\delta} \subset A_\varepsilon$, where $\delta \in \mathcal{E}^*$. If $(V_\varepsilon : \varepsilon \in \mathcal{E}^*)$ is a stochastic process with $0 \le V_\varepsilon \le 1$ and $V_{\varepsilon 0} + V_{\varepsilon 1} = 1$ for all $\varepsilon \in \mathcal{E}^*$, and there exists a Borel measure μ such that $\mu(A_\varepsilon) = \mathrm{E}[V_{\varepsilon_1} V_{\varepsilon_1\varepsilon_2} \cdots V_{\varepsilon_1\cdots\varepsilon_m}]$, for every $\varepsilon = \varepsilon_1 \cdots \varepsilon_m \in \mathcal{E}^*$, then there exists a random Borel measure P satisfying* (3.12).

Proof For fixed $\varepsilon \in \mathcal{E}^*$ there are at most countably many $A_{\varepsilon\delta}$ as stated, and their union is A_ε. Thus for any given $\eta > 0$ there exists a finite subcollection whose union $B_{\varepsilon,\eta}$ satisfies $\mu(A_\varepsilon \setminus B_{\varepsilon,\eta}) < \eta$. The corresponding union $K_{\varepsilon,\eta}$ of the closures $\bar{A}_{\varepsilon\delta}$ is compact and satisfies $B_{\varepsilon,\eta} \subset K_{\varepsilon,\eta} \subset A_\varepsilon$. Thus we are in the situation of Theorem 3.2, with P defined by (3.12) as a finitely additive measure on the field consisting of all finite unions of A_ε. □

In the case of $\mathfrak{X} = \mathbb{R}$ splits in intervals are natural. The condition of the preceding theorem is then met as soon as a mean measure exists. Alternatively, an explicit condition for countable additivity can be given as follows.

Suppose that we use left-open and right-closed cells $(a, b]$, except for a single cell of the form (a, ∞) that is unbounded to the right at every level, and choose the index such that $A_{\varepsilon 0}$ lies below $A_{\varepsilon 1}$. (Thus the sets A_ε with ε a string of only 0s or 1s are unbounded to the left and right, respectively; the other sets are bounded.) Furthermore, suppose that the split points are dense in $\mathbb{R}$, so that the partitions generate the Borel σ-field.

To check in this special case that P defined by (3.12) extends to a countably additive random measure on the Borel sets, it suffices to check that its induced distribution function is right continuous at every of the boundary points of the partitions and tends to 0 and 1 as $t \to -\infty$ or $+\infty$. This can be easily translated in the conditional probabilities V_ε.

Theorem 3.9 *Consider a sequence of partitions $\mathcal{T}_m = \{A_\varepsilon : \varepsilon \in \mathcal{E}^m\}$ of $\mathbb{R}$ such that $A_{\varepsilon 0} = \{x \in A_\varepsilon : x \le a_\varepsilon\}$ and $A_{\varepsilon 1} = \{x \in A_\varepsilon : x > a_\varepsilon\}$, where $a_\varepsilon \in \operatorname{int} A_\varepsilon$ and $\{a_\varepsilon : \varepsilon \in \mathcal{E}^*\}$ is dense in $\mathbb{R}$. If $(V_\varepsilon : \varepsilon \in \mathcal{E}^*)$ is a stochastic process with $0 \le V_\varepsilon \le 1$ and $V_{\varepsilon 0} + V_{\varepsilon 1} = 1$ for all $\varepsilon \in \mathcal{E}^*$, then P defined by* (3.12) *extends to a random measure on the Borel sets in $\mathbb{R}$ a.s. if and only if*

$$\mathrm{E}[V_\varepsilon V_{\varepsilon 0} V_{\varepsilon 00} \cdots] = 0 \text{ for all } \varepsilon \in \mathcal{E}^*, \quad \text{and } \mathrm{E}[V_1 V_{11} V_{111} \cdots] = 0. \tag{3.13}$$

Proof The probabilities (3.12) define a "distribution function" on the collection $A = \{a_\varepsilon : \varepsilon \in \mathcal{E}^*\}$ of endpoints of the partitions. If this function is right continuous and has the appropriate limits at $\pm\infty$, then it is a valid distribution function on A, which extends (uniquely) to a distribution function on $\mathbb{R}$.

The left endpoint of A_ε (for $\varepsilon \in \mathcal{E}^*$, $\varepsilon \ne 000\cdots$) is also the left endpoint of the cells $A_{\varepsilon 0}, A_{\varepsilon 00}, \ldots$ Therefore right continuity of the distribution function of P at this point holds if and only if $V_{\varepsilon 0} V_{\varepsilon 00 \ldots} = 0$. Similarly convergence to 0 and 1 in the tails correspond to $V_0 V_{00} \cdots = 0$ and $V_1 V_{11} \cdots = 0$. By assumption (3.13) these sequences tend to zero in mean. Because they are monotone and bounded, the convergence is also a.s. The exceptional null sets add to a single null set, as $\mathcal{E}^*$ has only countably many elements. □

Rather than in two subsets, the sets may be split in different (even varying, including infinite) numbers of subsets, as long as the V-variables that regulate the mass distribution satisfy the appropriate restriction. In particular, in the case that $\mathfrak{X} = \mathbb{R}^k$, splitting in 2^k sets would be natural. (These splits could also be incorporated in a longer, binary tree with sequential splits along the coordinate axes.) The preceding theorem can be extended to this case, although the condition for countable additivity (3.13) will become more complicated.

To study properties of tree-based priors, we consider a given partitioning tree $\mathcal{T}_1, \mathcal{T}_2, \ldots$ and a random measure P on the Borel sets. The measure may have been constructed through the partitioning tree, but this is not assumed in the following. Given a tree and a random measure P, we define *splitting variables* $(V_\varepsilon, \varepsilon \in \mathcal{E}^*)$ through (3.11) and then the representation (3.12) holds.

Tree-based priors have full weak support under weak conditions. Call a collection of sets $\mathscr{X}_0 \subset \mathscr{X}$ *convergence-forcing* for weak convergence if $P_m(A) \to P(A)$ for every $A \in \mathscr{X}_0$, for probability measures $\{P_m\}$ and P, implies that $P_m \rightsquigarrow P$. An example is the set of all cells with endpoints in a dense subset of $\mathbb{R}^k$.

Theorem 3.10 (Weak support) *Let $\mathcal{T}_m = \{A_\varepsilon: \varepsilon \in \mathcal{E}^m)$ be a sequence of successive binary partitions that generates the Borel sets and such that $\cup_m \mathcal{T}_m$ is convergence-forcing for weak convergence. If P is a random probability measure with splitting variables $(V_{\varepsilon 0}: \varepsilon \in \mathcal{E}^*)$ of which every finite-dimensional subvector has full support (a unit cube), then the weak support of P is the full space $\mathfrak{M}$.*

Proof Because $\cup_m \mathcal{T}_m$ is convergence-forcing for weak convergence, the topology generated by the metric $d(P, Q) = \sum_{\varepsilon \in \mathcal{E}^*} |P(A_\varepsilon) - Q(A_\varepsilon)| 2^{-2|\varepsilon|}$, is at least as strong as the weak topology. Therefore, it suffices to show that every open d-ball receives positive prior mass. For given $\eta > 0$, we can find m such that $\sum_{j>m} 2^{-j} < \eta$. Hence all probability measures P with

$$\sum_{\varepsilon \in \mathcal{E}^m} |P(A_\varepsilon) - P_0(A_\varepsilon)| < \eta$$

are in a d-ball of radius 2η around a given P_0. The latter event can be described in terms of a continuous function of the vectors $(V_\varepsilon: \varepsilon \in \mathcal{E}^j)$, for $j = 1, \ldots, m$. The inverse image of the set under this continuous map is open and has positive mass under the joint distribution of these vectors, as this has full support by assumption. □

3.6 Tail-Free Processes

It is simple and appealing to choose the splitting variables in a tree-based prior independent across the levels of the hierarchy. This leads to a "tail-free" random measure. In this section we study this property in general.

Consider a given partitioning tree $\mathcal{T}_1, \mathcal{T}_2, \ldots$ and a random measure P on the Borel sets, and define the splitting variables $(V_\varepsilon, \varepsilon \in \mathcal{E}^*)$ as in (3.11).

Definition 3.11 (Tail-free) The random measure P is a *tail-free process* with respect to the sequence of partitions $\mathcal{T}_m$ if $\{V_0\} \perp\!\!\!\perp \{V_{00}, V_{10}\} \perp\!\!\!\perp \cdots \perp\!\!\!\perp \{V_{\varepsilon 0}: \varepsilon \in \mathcal{E}^m\} \perp\!\!\!\perp \cdots$.

A degenerate prior is certainly tail-free according to this definition (with respect to any sequence of partitions), since all its V-variables are degenerate at appropriate values. Nontrivial and important examples are the Pólya-tree and Dirichlet processes, which are discussed respectively in Section 3.7 and Chapter 4.

Tail-free processes enjoy an obvious *equivariance* property under transformations: if P is a tail-free process with respect to a given sequence of partitions $\mathcal{T}_m = \{A_\varepsilon: \varepsilon \in \mathcal{E}^m\}$ and g is a measurable isomorphism, then the induced random measure $P \circ g^{-1}$ is tail-free with respect to the partitions $\{g^{-1}(A_\varepsilon): \varepsilon \in \mathcal{E}^m\}$. In the case that $\mathfrak{X} = \mathbb{R}$, where ordered partitions in intervals, as in Theorem 3.9, are natural, the transformed partitions will again consist of ordered intervals if g is monotonically increasing.

On the other hand, the tail-free property is not preserved under the formation of mixtures. For instance, a mixture of a nondegenerate and a degenerate tail-free process fails to be tail-free. In statistical applications this means that if the partitions and/or the V-variables involve hyperparameters, which are given a nondegenerate distribution, then the resulting mixture prior will not be tail-free, except in trivial cases. In Chapter 6 this will be seen to have important negative implications for posterior consistency.

The factorization (3.12) of the probabilities $P(A_\varepsilon)$ of a tail-free process and independence of the V-variables make it easy to compute means and variances of probabilities and log-probabilities of the sets in the partitions. Part (iv) of the following proposition shows that the probabilities of sets appearing later in the hierarchy have more variation in the logarithmic scale.

Proposition 3.12 (Moments) *For every tail-free process P and $\varepsilon = \varepsilon_1\varepsilon_2\cdots \in \mathcal{E}^*$,*

(i) $\mathrm{E}(P(A_\varepsilon)) = \prod_{j=1}^{|\varepsilon|} \mathrm{E}(V_{\varepsilon_1\cdots\varepsilon_j})$.
(ii) $\mathrm{var}(P(A_\varepsilon)) = \prod_{j=1}^{|\varepsilon|} \mathrm{E}(V^2_{\varepsilon_1\cdots\varepsilon_j}) - (\mathrm{E}(P(A_\varepsilon)))^2$.
(iii) $\mathrm{E}(\log P(A_\varepsilon)) = \sum_{j=1}^{|\varepsilon|} \mathrm{E}(\log V_{\varepsilon_1\cdots\varepsilon_j})$.
(iv) $\mathrm{var}(\log P(A_\varepsilon)) = \sum_{j=1}^{|\varepsilon|} \mathrm{var}(\log V_{\varepsilon_1\cdots\varepsilon_j})$.

Consequently, if $\sup_{\varepsilon\in\mathcal{E}^*} \mathrm{E}(V_\varepsilon^2) < 1/2$, *then* $\max\{P(A_\varepsilon): \varepsilon \in \mathcal{E}^m\} \to 0$ *a.s. at an exponential rate.*

Proof Parts (i)–(iv) are immediate from (3.12) and the independence of the splitting variables. For the final assertion we bound the maximum by the sum and apply (i) and (ii) to see that, if $r < 1$ is a uniform upper bound on $2\mathrm{E}(V_\varepsilon^2)$,

$$\mathrm{E}[\max_{\varepsilon\in\mathcal{E}^m} P(A_\varepsilon)]^2 \le \sum_{\varepsilon\in\mathcal{E}^m} \prod_{j=1}^{m} \mathrm{E}(V^2_{\varepsilon_1\cdots\varepsilon_j}) \le 2^m (r/2)^m = r^m. \tag{3.14}$$

The result is now immediate from the Borel-Cantelli lemma. □

In applications, we may wish to choose the partitions and the V-variables such that the tail-free process possesses a given mean measure μ. Part (i) of the preceding proposition shows that this can always be achieved by combining equal probability splits at every level (i.e. $\mu(A_{\varepsilon 0}) = \mu(A_{\varepsilon 1})$ for every $\varepsilon \in \mathcal{E}^*$) with V-variables with mean $\mathrm{E}(V_\varepsilon) = 1/2$. For a continuous measure μ on $\mathfrak{X} = \mathbb{R}$ this is always possible: the first split is at the median of μ, the second at the quartiles, the third at the octiles, etc.; the mth partition will have the $(k/2^m)$th quantiles of μ as its boundary points and consist of sets of μ-probability 2^{-m}. We refer to a partition with $\mu(A_\varepsilon) = 2^{-|\varepsilon|}$, for every $\varepsilon \in \mathcal{E}^*$, as a *canonical partition* relative to μ. (On $\mathbb{R}$ with interval splits this is uniquely determined by μ, whereas on a general space μ is determined by the partition, but not the other way around.)

The uniform distribution over the unit interval may seem a natural default choice for meeting the condition $\mathrm{E}(V_\varepsilon) = 1/2$. However, the resulting prior is peculiar in that its realizations are continuous, but a.s. singular with respect to the Lebesgue measure (see Lemma 3.17).

The mass $P(A_\varepsilon)$ of a partitioning set at level m can be expressed in the V-variables up to level m (see (3.12)), while, by their definition (3.11), the V-variables at higher levels control conditional probabilities. Therefore, tail-freeness makes the distribution of mass *within* every partitioning set in $\mathcal{T}_m$ independent of the distribution of the total mass one *among* the sets in $\mathcal{T}_m$. Definition 3.11 refers only to masses of partitioning sets, but under the assumption that the partitions generate the Borel sets, the independence extends to all Borel sets.

Lemma 3.13 *If P is a random measure that is tail-free relative to a sequence of partitions $\mathcal{T}_m = \{A_\varepsilon : \varepsilon \in \mathcal{E}^m\}$ that generates the Borel sets $\mathscr{X}$ in $\mathfrak{X}$, then for every $m \in \mathbb{N}$ the process $(P(A|\, A_\varepsilon): A \in \mathscr{X}, \varepsilon \in \mathcal{E}^m)$ is independent of the random vector $(P(A_\varepsilon): \varepsilon \in \mathcal{E}^m)$.*

Proof Because P is a random measure, its mean measure $\mu(A) = \mathrm{E}P(A)$ is a well defined Borel probability measure. As $\mathcal{T} := \cup_m \mathcal{T}_m$ is a field, which generates the Borel σ-field by assumption, there exists for every $A \in \mathscr{X}$ a sequence A_n in $\mathcal{T}$ such that $\mu(A_n \,\Delta\, A) \to 0$. Because P is a random measure, $P(A_n|\, A_\varepsilon) \to P(A|\, A_\varepsilon)$ in mean and hence a.s. along a subsequence. It follows that the random variable $P(A|\, A_\varepsilon)$ is measurable relative to the completion of the σ-field generated by the variables $P(C|\, A_\varepsilon)$, for $C \in \mathcal{T}$. Every of these conditional probabilities is a finite sum of probabilities of the form $P(A_{\varepsilon\delta}|\, A_\varepsilon) = V_{\varepsilon\delta_1} \cdots V_{\varepsilon\delta_1\cdots\delta_k}$, for $\delta = \delta_1 \cdots \delta_k \in \mathcal{E}^k$ and $k \in \mathbb{N}$. Therefore, by tail-freeness this σ-field is independent of the σ-field generated by the variables $P(A_\varepsilon) = V_{\varepsilon_1} \cdots V_{\varepsilon_1\cdots\varepsilon_m}$, for $\varepsilon = \varepsilon_1 \cdots \varepsilon_m \in \mathcal{E}^m$. □

Relative to the σ-field $\mathscr{M}$ generated by all maps $M \mapsto M(A)$, the process $(P(A|\, A_\varepsilon): A \in \mathscr{X})$ contains all information about the conditional random measure $P(\cdot|\, A_\varepsilon)$. Thus the preceding theorem truly expresses that the "conditional measure within partitioning sets is independent of the distribution of mass among them."

Suppose that the data consist of an i.i.d. sample $X_1, \ldots, X_n$ from a distribution P, which is a priori modeled as a tail-free process. For each $\varepsilon \in \mathcal{E}^*$, denote the number of observations falling in A_ε by

$$N_\varepsilon := \#\{1 \le i \le n : X_i \in A_\varepsilon\}. \tag{3.15}$$

For each m the vector $(N_\varepsilon : \varepsilon \in \mathcal{E}^m)$ collects the counts of all partitioning sets at level m. The following theorem shows that this vector contains all information (in the Bayesian sense) about the probabilities $(P(A_\varepsilon): \varepsilon \in \mathcal{E}^m)$ of these sets: the additional information about the precise positions of the X_i within the partitioning sets is irrelevant.

Theorem 3.14 *A random measure P is tail-free relative to a given sequence of partitions $\mathcal{T}_m = \{A_\varepsilon: \varepsilon \in \mathcal{E}^m\}$ that generates the Borel sets if and only if for every m and n the posterior distribution of $(P(A_\varepsilon): \varepsilon \in \mathcal{E}^m)$ given an i.i.d. sample $X_1, \ldots, X_n$ from P is the same as the posterior distribution of this vector given $(N_\varepsilon: \varepsilon \in \mathcal{E}^m)$ defined in* (3.15), *a.s.*

Proof Fix m. Given P the data $X = (X_1, \ldots, X_n)$ can be generated in two steps. First we generate a multinomial vector $N = (N_\varepsilon: \varepsilon \in \mathcal{E}^m)$ with parameters n and $(P(A_\varepsilon): \varepsilon \in \mathcal{E}^m)$. Next, given N, we generate an i.i.d. sample of size N_ε from the measure $P(\cdot|\, A_\varepsilon)$, independently for every $\varepsilon \in \mathcal{E}^m$, and randomly order the n values so obtained.

The random measure P can also be generated in two steps. First we generate the vector $\theta := (P(A_\varepsilon): \varepsilon \in \mathcal{E}^m)$, and second the process $\eta := (P(A|\, A_\varepsilon): A \in \mathscr{X}, \varepsilon \in \mathcal{E}^m)$. For a tail-free measure P these steps are independent.

The first step of the generation of the data depends on θ only, and the second only on (N, η). The set of (conditional) independencies $\eta \perp\!\!\!\perp (N, \theta)$, if P is tail-free, and $X \perp\!\!\!\perp \theta|\, (N, \eta)$ implies that $\theta \perp\!\!\!\perp X|\, N$. Indeed, these assumption imply that $\mathrm{E}(f(X)g(\theta)|\, N, \eta) = \mathrm{E}(f(X)|\, N, \eta)\mathrm{E}(g(\theta)|\, N)$, for any bounded, measurable functions f and g, from which the assertion follows by taking the conditional expectation given N left and right. Now $\theta \perp\!\!\!\perp X|\, N$ is equivalent to the "only if" part of the theorem for this special representation of prior and data. Because the assertion depends on the joint distribution of (P, X, N) only, it is true in general.

For the proof that dependence on the cell counts only implies tail-freeness, let $N' = (N_\varepsilon: \varepsilon \in \mathcal{E}^{m+1})$. First note that $\theta \perp\!\!\!\perp X|\, N$ implies $\theta \perp\!\!\!\perp N'|\, N$, which can also be represented as $p(\theta|\, N') = p(\theta|\, N)$. Thus

$$\mathrm{P}(\theta \in C, N' = y')\mathrm{P}(N = y) = \mathrm{P}(\theta \in C, N = y)\mathrm{P}(N' = y'),$$

for every measurable set C and every pair $y' = (y'_\varepsilon: \varepsilon \in \mathcal{E}^{m+1})$ and $y = (y_\varepsilon: \varepsilon \in \mathcal{E}^m)$ that are compatible (i.e. $y_\varepsilon = \sum_{\delta \in \mathcal{E}} y'_{\varepsilon\delta}$ for all ε). Because given P, the vectors N' and N possess multinomial distributions and θ is a function of P, this can be further rewritten as

$$\mathrm{E}\mathbb{1}_{\theta \in C}\binom{n}{y'} \prod_{\varepsilon \in \mathcal{E}^{m+1}} P(A_\varepsilon)^{y'_\varepsilon} \mathrm{P}(N = y) = \mathrm{E}\mathbb{1}_{\theta \in C}\binom{n}{y} \prod_{\varepsilon \in \mathcal{E}^m} P(A_\varepsilon)^{y_\varepsilon} \mathrm{P}(N' = y').$$

In view of the definition of θ we conclude that, provided that $\mathrm{P}(N = y) > 0$,

$$\mathrm{E}\Big(\prod_{\varepsilon \in \mathcal{E}^{m+1}} P(A_\varepsilon)^{y'_\varepsilon} |\, \theta\Big) = \prod_{\varepsilon \in \mathcal{E}^m} P(A_\varepsilon)^{y_\varepsilon}\, c(y'),$$

for $c(y') = \binom{n}{y}\mathrm{P}(N' = y')/\mathrm{P}(N = y)/\binom{n}{y'}$. If $\mathrm{P}(N = y) = \mathrm{E}\binom{n}{y}\prod_{\varepsilon \in \mathcal{E}^m} P(A_\varepsilon)^{y_\varepsilon} = 0$, then the variable in the left side is zero and the identity is true with $c(y') = 0$. In other words, the conditional mixed moment of degree y' of the vector $V = (P(A_{\varepsilon\delta})/P(A_\varepsilon): \varepsilon \in \mathcal{E}^m, \delta \in \mathcal{E})$ given $(P(A_\varepsilon): \varepsilon \in \mathcal{E}^m)$ is equal to a deterministic number $c(y')$. This being true for every y' in n times the unit simplex, for every n, implies independence of V and $(P(A_\varepsilon): \varepsilon \in \mathcal{E}^m)$. ∎

Given P the vector $(N_\varepsilon : \varepsilon \in \mathcal{E}^m)$ possesses a multinomial distribution with parameters n and $(P(A_\varepsilon): \varepsilon \in \mathcal{E}^m)$. Finding the posterior distribution of the latter vector of cell probabilities therefore reduces to the finite dimensional problem of multinomial probabilities. This not only makes computations easy, but also means that asymptotic properties of the posterior distribution follow those of parametric problems, for instance easily leading to consistency in an appropriate sense, as discussed in Chapter 6. The result also justifies the term "tail-free" in that posterior computation can be carried out without looking at the tail of the prior.

Tail-free processes form a conjugate class of priors, in the sense that the posterior process is again tail-free.

Theorem 3.15 (Conjugacy) *The posterior process corresponding to observing an i.i.d. sample $X_1, \ldots, X_n$ from a distribution P that is a priori modeled by a tail-free process is tail-free with respect to the same sequence of partitions as in the definition of the prior.*

Proof We must show that the vectors $(V_{\varepsilon 0}: \varepsilon \in \mathcal{E}^m)$ defined in (3.11) are mutually conditionally independent across levels m, given the data. It suffices to show sequentially for every m that this vector is conditionally independent of the vectors corresponding to lower levels. Because the vectors $(V_\varepsilon: \varepsilon \in \cup_{k \leq m} \mathcal{E}^k)$ and $(P(A_\varepsilon): \varepsilon \in \mathcal{E}^m)$ generate the same σ-field, it suffices to show that $(V_{\varepsilon 0}: \varepsilon \in \mathcal{E}^m)$ is conditionally independent of $(P(A_\varepsilon): \varepsilon \in \mathcal{E}^m)$, for every fixed m.

Together these vectors are equivalent to the vector $(P(A_\varepsilon): \varepsilon \in \mathcal{E}^{m+1})$. Therefore, by Theorem 3.14 the joint posterior distribution of the latter vectors depends only on the cell counts, $N = (N_{\varepsilon\delta}: \varepsilon \in \mathcal{E}^m, \delta \in \mathcal{E})$, and "conditionally given the data" can be interpreted as "given this vector N." Writing $V = (V_{\varepsilon\delta}: \varepsilon \in \mathcal{E}^m, \delta \in \mathcal{E})$ and $\theta = (\theta_\varepsilon: \varepsilon \in \mathcal{E}^m)$, for $\theta_\varepsilon = P(A_\varepsilon)$, we can write the likelihood for (V, θ, N) as

$$\binom{n}{N} \prod_{\varepsilon \in \mathcal{E}^m, \delta \in \mathcal{E}} (\theta_\varepsilon V_{\varepsilon\delta})^{N_{\varepsilon\delta}} \, d\Pi_1(V) \, d\Pi_2(\theta).$$

Here Π_1 and Π_2 are the marginal (prior) distributions of V and θ, and we have used that these vectors are independent under the assumption that P is tail-free. Clearly the likelihood factorizes in parts involving (V, N) and involving (θ, N). This shows that V and θ are conditionally independent given N. □

The atoms in the σ-field generated by $\mathcal{T}_0, \mathcal{T}_1, \ldots$ are the sets $\cap_{m=1}^\infty A_{\varepsilon_1 \cdots \varepsilon_m}$, for $\varepsilon_1 \varepsilon_2 \cdots$ an infinite string of 0s and 1s. If the sequence of partitions generates the Borel sets, then these atoms are single points (or empty). Their probabilities under a tree-based random measure are $\prod_{m=1}^\infty V_{\varepsilon_1 \cdots \varepsilon_m}$, and hence $P\{x\} = 0$ a.s. for every x as soon as these infinite products are zero. For tail-free random measures this is typical, as the means $\mathrm{E}[V_\varepsilon]$ will usually be smaller and bounded away from 1.

However, this does not imply that P is a.s. continuous (atomless), but only says that P has no *fixed atoms*. Because there are uncountably many possible atoms (or infinitely long sequences of 0s and 1s), every realization of P may well have some atom. In fact, every realization of P can be discrete, as is illustrated by the Dirichlet process priors discussed in Chapter 4.

In practice, an atomless, or even absolutely continuous, random measure may be preferable. This may be constructed by choosing splitting variables that are "close" to the splits of a deterministic absolutely continuous measure. For instance, to make the random measure resemble the canonical measure for the partition, we choose the splitting variables close to 1/2. In the other case, when the mass is frequently divided unevenly between two offspring sets in the splitting tree, then the realizations of the resulting random measure will tend to concentrate more around some points and fail to have a density. The following theorem makes "close to $1/2$" precise.

Theorem 3.16 (Absolute continuity) *Let $\mathcal{T}_m = \{A_\varepsilon : \varepsilon \in \mathcal{E}^m)$ be a sequence of successive binary partitions that generates the Borel sets. If P is a random measure with splitting variables $(V_\varepsilon : \varepsilon \in \mathcal{E}^*)$ that satisfy, for an arbitrary probability measure μ,*

$$\sup_{m\in\mathbb{N}} \max_{\varepsilon\in\mathcal{E}^m} \frac{\mathrm{E}(\prod_{j=1}^m V_{\varepsilon_1\cdots\varepsilon_j}^2)}{\mu^2(A_{\varepsilon_1\cdots\varepsilon_m})} < \infty, \tag{3.16}$$

then almost all realizations of P are absolutely continuous with respect to μ, i.e. $\Pi(P \ll \mu) = 1$. In particular, this condition is satisfied if P is tail-free, the partitions are canonical relative to μ and the following two conditions hold:

$$\sum_{m=1}^\infty \max_{\varepsilon\in\mathcal{E}^m} \left|\mathrm{E}(V_\varepsilon) - \frac{1}{2}\right| < \infty, \qquad \sum_{m=1}^\infty \max_{\varepsilon\in\mathcal{E}^m} \mathrm{var}(V_\varepsilon) < \infty. \tag{3.17}$$

In this case a density process $x \mapsto p(x)$ is given by, for $x \in \cap_{m=1}^\infty A_{x_1\cdots x_m}$[1]*,*

$$p(x) = \prod_{j=1}^\infty (2V_{x_1x_2\cdots x_j}). \tag{3.18}$$

For the canonical partitions relative to the Lebesgue measure on $\mathfrak{X} = (0,1)$ and fully supported and independent variables $V_{\varepsilon 0}$, for $\varepsilon \in \mathcal{E}^*$, any version of this process is discontinuous almost surely at every boundary point of the partitions.

Proof The σ-fields $\sigma(\mathcal{T}_1) \subset \sigma(\mathcal{T}_2) \subset \cdots$ are increasing and generate the Borel σ-field. If the restrictions P_m and μ_m of two fixed (nonrandom) Borel probability measures P and μ to $\sigma(\mathcal{T}_m)$ satisfy $P_m \ll \mu_m$, then by Lemma L.7, the sequence of densities $p_m = dP_m/d\mu_m$ tends μ almost surely to the density of the absolutely continuous part of P with respect to μ. In particular, if p_m tends to p, and $\int p\,d\mu = 1$, then P is absolutely continuous with respect to μ, with density p. We shall apply this to the present *random* measures.

Condition (3.16) implies that $\mu(A_\varepsilon) > 0$, for every $\varepsilon \in \mathcal{E}^m$, and hence μ_m dominates any measure on $\sigma(\mathcal{T}_m)$. The density of P_m is given by

$$p_m = \sum_{\varepsilon\in\mathcal{E}^m} \frac{P(A_\varepsilon)}{\mu(A_\varepsilon)} \mathbb{1}_{A_\varepsilon}, \tag{3.19}$$

[1] This condition defines a binary coding $x_1x_2\cdots$ for any $x \in \mathfrak{X}$. In the case of the canonical partitions of the unit interval in cells, it coincides with the infinite binary expansion of x (not ending in $000\cdots$).

and satisfies, for Π the law of P,

$$\iint p_m^2\,d\mu\,d\Pi(p) = \sum_{\varepsilon\in\mathcal{E}^m} \frac{\mathrm{E}(\prod_{j=1}^m V_{\varepsilon_1\cdots\varepsilon_j}^2)}{\mu(A_{\varepsilon_1\cdots\varepsilon_m})} \le \max_{\varepsilon\in\mathcal{E}^m} \frac{\mathrm{E}(\prod_{j=1}^m V_{\varepsilon_1\cdots\varepsilon_j}^2)}{\mu^2(A_{\varepsilon_1\cdots\varepsilon_m})}.$$

By assumption, the right side is bounded in m, which implies that the sequence p_m is uniformly $(\mu\times\Pi)$-integrable. It follows that $\iint p\,d\mu\,d\Pi = \lim_{m\to\infty}\iint p_m\,d\mu\,d\Pi = 1$, which implies that $\int p\,d\mu = 1$ a.s. $[\Pi]$.

For a canonical partition the denominator of the quotient in (3.16) is 2^{-2m} and the quotient can be written in the form $\mathrm{E}\prod_{j=1}^m(2V_{\varepsilon_1\cdots\varepsilon_j})^2 = \prod_{j=1}^m \mathrm{E}(2V_{\varepsilon_1\cdots\varepsilon_j})^2$, if P is tail-free. If $|\mathrm{E}(V_\varepsilon)-1/2|\le\delta_j$ and $var(V_\varepsilon)\le\gamma_j$ for $\varepsilon\in\mathcal{E}^j$, then

$$\mathrm{E}(V_{\varepsilon_1\cdots\varepsilon_j}^2) = var(V_{\varepsilon_1\cdots\varepsilon_j}) + (\mathrm{E}(V_{\varepsilon_1\cdots\varepsilon_j}))^2 \le (1+4\gamma_j+4\delta_j+4\delta_j^2)/4.$$

Therefore, the expression in (3.16) is bounded by $\prod_{j=1}^\infty(1+4\gamma_j+4\delta_j+4\delta_j^2)$, which is finite if $\sum_j(\delta_j+\gamma_j)<\infty$.

If $x\in A_{x_1\cdots x_m}$ and $\mu(A_\varepsilon)=2^{-|\varepsilon|}$, then the right-hand side of (3.19) evaluated at x reduces to $2^{-m}P(A_{x_1\cdots x_m})$. Combined with (3.12) this gives formula (3.18).

For the proof of the last assertion consider an arbitrary point $x = x_1x_2\cdots\in(0,1)$ in its infinite binary expansion (so not ending in $000\cdots$), so that $A_{x_1\cdots x_m} = (x_m^-, x_m^+]$, for $x_m^- = x_1\cdots x_m000\cdots < x \le x_m^+ = x_m^- + 2^{-m} = x_1\cdots x_m111\cdots$. The finite binary expansion of x_m^+ is given by $x_m^+ = x_1\cdots x_{j_m-1}1\cdots1000\cdots$, where j_m is the biggest integer $j\le m$ with $x_j=0$ and the zeros start at coordinate $m+1$; hence, the partitioning set immediately to the right of $A_{x_1\cdots x_m}$ is $A_{x_1\cdots x_{j_m-1}1\cdots1}$. If F is the cumulative distribution function of P, then

$$F(x_m^-+2^{-m}) - F(x_m^-) = P(A_{x_1\cdots x_m}) = \prod_{j=1}^m V_{x_1\cdots x_j},$$

$$F(x_m^++2^{-m}) - F(x_m^+) = P(A_{x_1\cdots x_{j_m-1}1\cdots1}) = \prod_{j=1}^{j_m-1} V_{x_1\cdots x_j}\prod_{j=j_m}^m V_{x_1\cdots x_{j_m-1}1\cdots1}.$$

If F is absolutely continuous with density f that is continuous at x, then $2^m(F(y_m+2^{-m}) - F(x_m)) = 2^m\int_{y_m}^{y_m+2^{-m}} f(s)\,ds \to f(x)$ for any $y_m\to x$. In particular the quotient of the right sides of the preceding display tends to 1 as $m\to\infty$.

If x is a boundary point of the (m_0+1)th partition, then $x = x_1\cdots x_{m_0}0111\cdots$ for some m_0, and $j_m = m_0+1$ for $m>m_0$ and hence the quotient tends to the infinite product $\prod_{j>m_0}(V_{x_1\cdots x_j}/V_{x_1\cdots x_{m_0}1\cdots1})$. The logarithm of this expression is a sum of independent, continuous variables and hence possesses a continuous (atomless) distribution; in particular the probability that it is $\log 1 = 0$ is zero. ◻

It follows that the realizations of a tail-free measure with splitting variables such that $\mathrm{E}|V_\varepsilon - 1/2|^2 \to 0$ as $|\varepsilon|\to\infty$ at a fast enough speed possess densities relative to the canonical measure for the partition. On the other hand, fixed (for instance, uniform) splitting variables give widely unbalanced divisions with appreciable probabilities and lead to singular measures.

Lemma 3.17 *Let P be a tail-free process with respect to a canonical partition relative to a probability measure μ and with i.i.d. splitting variables $(V_{\varepsilon 0}: \varepsilon \in \mathcal{E}^*)$.*

(a) *If* $\mathrm{P}(V_{\varepsilon 0} \neq 1/2) > 0$*, then P is a.s. singular with respect to μ.*
(b) *If* $\mathrm{E}(V_{\varepsilon 0}) = 1/2$ *and* $\mathrm{P}(0 < V_{\varepsilon 0} < 1) > 0$*, then P is a.s. nonatomic.*

Proof As in the proof of Theorem 3.16, the density of the absolutely continuous part of P with respect to μ can be computed as the limit of the sequence p_m given in (3.19), where presently we substitute $\mu(A_\varepsilon) = 2^{-|\varepsilon|}$. For (a) it suffices to prove that this limit is zero a.s.

Define a map from $\mathfrak{X}$ to $\{0,1\}^\infty$ by $x \mapsto \varepsilon_1\varepsilon_2\cdots$ if $x \in \cap_{j=1}^\infty A_{\varepsilon_1\varepsilon_2\cdots\varepsilon_j}$. Write the map as $x \mapsto x_1x_2\cdots$, so that $x \in A_{x_1\cdots x_m}$, for every m. Formula (3.19) can then be written in the form

$$\frac{1}{m}\log p_m(x) = \frac{1}{m}\log\left(2^m P(A_{x_1\cdots x_m})\right) = \frac{1}{m}\sum_{j=1}^m \log\left(2V_{x_1\cdots x_j}\right).$$

If X is a random element in $\mathfrak{X}$ with law μ and $X_1X_2\ldots$ is its image under the map $x \mapsto x_1x_2\cdots$ defined previously, then $\mathrm{P}(X_1 = \varepsilon_1, \ldots, X_m = \varepsilon_m) = \mu(A_\varepsilon) = 2^{-m}$, for every $\varepsilon = \varepsilon_1\cdots\varepsilon_m \in \mathcal{E}^m$, and hence $X_1, X_2, \ldots$ are i.i.d. Bernoulli variables.

The variables $V_{x_1\cdots x_j}$ are independent and distributed as V if $x_j = 0$ and as $1 - V$ if $x_j = 1$, for V a variable with a given distribution with $\mathrm{P}(V \neq 1/2) > 0$. By the strong law of large numbers μ-almost every sequence $X_1, X_2, \ldots$ has equal limiting frequencies of 0s and 1s. For any such realization $x_1x_2\cdots$, another application of the strong law of large numbers gives that

$$\frac{1}{m}\log p_m(x) \to \frac{1}{2}\mathrm{E}(\log(2V)) + \frac{1}{2}\mathrm{E}(\log(2(1-V))), \qquad \text{a.s.}$$

Because $4v(1-v) < 1$ for all $v \neq 1/2$, under assumption (a) the right side of this display is negative, which implies that $p_m(x) \to 0$.

We have shown that the event $\{\lim p_m(x) = 0\}$ has probability 1 for every x in a set of μ-measure 1. Using Fubini's theorem we can also formulate this in the form that $\lim p_m(x) = 0$ for a. e. x $[\mu]$, with probability one.

To prove (b), observe that the probability of any atom is bounded by $\max_{\varepsilon\in\mathcal{E}^m} P(A_\varepsilon)$ for every m, and hence is 0 a.s., in view of Proposition 3.12, since $\mathrm{E}(V^2) < \mathrm{E}(V)$ whenever $\mathrm{P}(0 < V < 1) > 0$, and both $V_{\varepsilon 0}$ and $V_{\varepsilon 1} = 1 - V_{\varepsilon 0}$ have mean 1/2 by assumption. □

The last two theorems concern the event that a random measure P is singular or absolutely continuous. From Proposition A.7 it follows that the sets of singular probability measures and absolutely continuous probability measures are indeed measurable subsets of $\mathfrak{M}$, as are the sets of nonatomic and discrete probability measures.

The following theorem shows that it is no coincidence that these events were trivial in all cases: tail-free processes satisfy the following zero-one law.

Theorem 3.18 (Zero-one law) *A random measure that is tail-free with respect to a sequence of partitions that generates the Borel σ-field and with splitting variables such that $0 < V_\varepsilon < 1$ for all $\varepsilon \in \mathcal{E}^*$ is, with probability zero or one,*

(i) *absolutely continuous with respect to a given measure,*

(ii) *mutually singular with respect to a given measure,*
(iii) *atomless,*
(iv) *discrete.*

Proof We prove only (i); the proofs of the other parts are virtually identical. Fix $m \in \mathbb{N}$, and decompose P as $P(A) = \sum_{\varepsilon \in \mathcal{E}^m} P(A|\, A_\varepsilon)P(A_\varepsilon)$. Because $P(A_\varepsilon) > 0$ for all $\varepsilon \in \mathcal{E}^m$, the measure P is absolutely continuous with respect to a given measure λ if and only if $P(\cdot|\, A_\varepsilon)$ is so for all $\varepsilon \in \mathcal{E}^m$. Since every measure $P(\cdot|\, A_\varepsilon)$ is describable in terms of (only) $(V_{\varepsilon\delta}, \delta \in \mathcal{E}^*)$, so is the event that $P(\cdot|\, A_\varepsilon)$ is absolutely continuous with respect to λ. In other words, this event is tail-measurable for the sequence of independent random vectors $(V_\varepsilon\colon \varepsilon \in \mathcal{E}^1), (V_\varepsilon\colon \varepsilon \in \mathcal{E}^2), \ldots$ By Kolmogorov's zero-one law, it has probability zero or one. ☐

Under the sufficient condition for absolute continuity given in Theorem 3.16, a tail-free process automatically has all absolutely continuous measures in its total variation support.

Theorem 3.19 (Strong support) *Let P be a random measure that is tail-free relative to a μ-canonical sequence of partitions that generates the Borel σ-field. If the splitting variables V_ε satisfy* (3.17)*, and the random vectors $(V_{\varepsilon 0}\colon \varepsilon \in \mathcal{E}^m)$ have full support* $[0,1]^{2^m}$*, then the total variation support of P consists of all probability measures that are absolutely continuous relative to μ.*

Proof By Theorem 3.16 P is μ-absolutely continuous with density p given by (3.18). As the partitions generate the Borel sets, the probability densities p_0 that are constant on every set in some partition $\mathcal{T}_m$, for some m, are dense in the set of all μ-probability densities relative to $\mathbb{L}_1(\mu)$-metric. For $p_m(x) = \prod_{j\le m}(2V_{x_1x_2\cdots x_j})$,

$$\int |p - p_0|\, d\mu \le \int \left|\frac{p}{p_m} - 1\right| d\mu \|p_m\|_\infty + \|p_m - p_0\|_\infty.$$

Because under tail-freeness the process $x \mapsto (p/p_m)(x) = \prod_{j>m}(2V_{x_1\cdots x_j})$ is independent of the process $x \mapsto p_m(x)$, the prior probability that the left side is smaller than ϵ is bigger than

$$\Pi\Big(\int \left|\frac{p}{p_m} - 1\right| d\mu < \frac{\epsilon}{2\|p_0\|_\infty + \epsilon}\Big)\Pi(\|p_m - p_0\|_\infty < \epsilon/2). \tag{3.20}$$

The second probability refers to the event that the vector $(V_{\varepsilon 0}\colon \varepsilon \in \cup_{j\le m}\mathcal{E}^j)$ belongs to a certain nonempty open set, and hence is positive by assumption. The assumptions on the means and variances of the variables V_ε and the Cauchy-Schwarz inequality imply that $\sum_{j=1}^\infty |2V_{x_1\cdots x_j} - 1| < \infty$ a.s. and hence, for every x and $m \to \infty$,

$$\frac{p}{p_m}(x) = \prod_{j>m}(2V_{x_1\cdots x_j}) \sim e^{\sum_{j>m}(2V_{x_1\cdots x_j}-1)} \to 1, \qquad \text{a.s.}$$

Using Fubini's theorem we conclude that $p(x)/p_m(x) \to 1$ almost surely under the measure $\mu \times \Pi$, as $m \to \infty$. Furthermore, for $n > m$,

$$\iint \Big(\frac{p_n}{p_m}\Big)^2 \, d\mu \, d\Pi(p) = \sum_{\varepsilon \in \mathcal{E}^n} \mathrm{E} \prod_{m<j\le n} (2V_{x_1\cdots x_j})^2 2^{-n} \le \max_{\varepsilon \in \mathcal{E}^n} \prod_{m<j\le n} \mathrm{E}(2V_{x_1\cdots x_j})^2.$$

With notation as in the proof of Theorem 3.16, the right-hand side can be bounded above by $\prod_{m<j\le n}(1+4\gamma_j+4\delta_j+4\delta_j^2)$, which remains finite in the limit $n \to \infty$. By Fatou's lemma $\iint (p/p_m)^2 \, d\mu \, d\Pi(p)$ is smaller than the limit and hence is bounded in m. This shows that p/p_m is uniformly integrable relative to $\mu \times \Pi$, and we can conclude that $\mathrm{E}\int |(p/p_m) - 1| \, d\mu \to 0$, as $m \to \infty$. This implies that the first probability in (3.20) tends to one, as $m \to \infty$, for any ϵ. Given a piecewise constant p_0 and $\epsilon > 0$, we choose m sufficiently large that the latter probability is positive and such that p_0 is piecewise constant on $\mathcal{T}_m$ to finish the proof. □

3.7 Pólya Tree Processes

Let $\mathcal{T}_m = \{A_\varepsilon : \varepsilon \in \mathcal{E}^m\}$ be a sequence of successive, binary, measurable partitions of the sample space, as before, and for a random measure P on $(\mathfrak{X}, \mathscr{X})$, define random variables $V_{\varepsilon 0}$ as in (3.11).

Definition 3.20 (Pólya tree) A random probability measure P is said to follow a *Pólya tree process* with parameters $(\alpha_\varepsilon : \varepsilon \in \mathcal{E}^*)$ with respect to the sequence $\{\mathcal{T}_m\}$ of partitions if the variables $V_{\varepsilon 0}$, for $\varepsilon \in \mathcal{E}^* \cup \{\varnothing\}$, are mutually independent and $V_{\varepsilon 0} \sim \mathrm{Be}(\alpha_{\varepsilon 0}, \alpha_{\varepsilon 1})$.

Thus a Pólya tree process is a tail-free process with splitting variables $V_{\varepsilon 0}$ that are mutually independent not only between, but also *within* the levels, and possess beta-distributions. The parameters α_ε of the beta distributions must be nonnegative; we allow the value 0 provided that $\alpha_{\varepsilon 0} + \alpha_{\varepsilon 1} > 0$, for every $\varepsilon \in \mathcal{E}^*$, with the beta-distributions $\mathrm{Be}(0, \alpha)$ and $\mathrm{Be}(\alpha, 0)$ interpreted as the degenerate distributions at 0 and 1, respectively, for any $\alpha > 0$.

We denote this process by $\mathrm{PT}(\mathcal{T}_m, \alpha_\varepsilon)$. If the partitions are canonical relative to a measure μ, then we also write $\mathrm{PT}(\mu, \alpha_\varepsilon)$. The choice $\mathrm{E}(V_{\varepsilon 0}) = 1/2$, which then renders the mean measure equal to μ, corresponds to setting the two parameters of each beta distribution equal: $\alpha_{\varepsilon 0} = \alpha_{\varepsilon 1}$. The parameter vector (α_ε), then, still has many degrees of freedom. A reasonable default choice is to use a single parameter per level, that is $\alpha_\varepsilon = a_m$ for all $\varepsilon \in \mathcal{E}^m$ and a sequence of numbers a_m. We refer to the resulting priors as a *canonical Pólya tree process* with center measure μ and rate sequence a_m, and denote it by $\mathrm{PT}^*(\mu, a_m)$. In practice it may be desirable to choose different parameters at the first few levels, since then the probability of choosing a density close to some target distribution may be raised without affecting the essential path properties of the random distribution.

Here are some easy facts or consequences of the preceding:

(i) The image $P \circ g^{-1}$ of a Pólya tree process under a measurable isomorphism g on the sample space is a Pólya tree process with respect to the transformed partitions $\{g(A_\varepsilon) : \varepsilon \in \mathcal{E}^m\}$, with the same set of V-variables.

(ii) The Pólya tree exists as a random measure on $\mathbb{R}$ if, for all $\varepsilon \in \mathcal{E}^*$ (see (3.13) and use that a Be(α, β)-variable has mean $\alpha/(\alpha+\beta)$),

$$\frac{\alpha_{\varepsilon 0}}{\alpha_{\varepsilon 0}+\alpha_{\varepsilon 1}} \times \frac{\alpha_{\varepsilon 00}}{\alpha_{\varepsilon 00}+\alpha_{\varepsilon 01}} \times \cdots = 0, \qquad \frac{\alpha_{1}}{\alpha_{0}+\alpha_{1}} \times \frac{\alpha_{11}}{\alpha_{10}+\alpha_{11}} \times \cdots = 0.$$

(iii) The first two moments of probabilities of partitioning sets under a Pólya tree process distribution are (see Proposition 3.12)

$$\mathrm{E}(P(A_{\varepsilon_1\cdots\varepsilon_m})) = \prod_{j=1}^{m} \frac{\alpha_{\varepsilon_1\cdots\varepsilon_j}}{\alpha_{\varepsilon_1\cdots\varepsilon_{j-1}0}+\alpha_{\varepsilon_1\cdots\varepsilon_{j-1}1}}, \tag{3.21}$$

$$\mathrm{E}(P(A_{\varepsilon_1\cdots\varepsilon_m})^2) = \prod_{j=1}^{m} \frac{\alpha_{\varepsilon_1\cdots\varepsilon_j}(\alpha_{\varepsilon_1\cdots\varepsilon_j}+1)}{(\alpha_{\varepsilon_1\cdots\varepsilon_{j-1}0}+\alpha_{\varepsilon_1\cdots\varepsilon_{j-1}1})(\alpha_{\varepsilon_1\cdots\varepsilon_{j-1}0}+\alpha_{\varepsilon_1\cdots\varepsilon_{j-1}1}+1)}.$$

(iv) A Pólya tree process with $\alpha_\varepsilon > 0$, for every $\varepsilon \in \mathcal{E}^*$, has full weak support as soon as the sequence of partitions is convergence-forcing (see Theorem 3.10).

(v) A Pólya tree process with parameters such that $|\alpha_\varepsilon - a_m| \le K$ for every $\varepsilon \in \mathcal{E}^m$ and some constant K and numbers a_m with $\sum_{m=1}^\infty a_m^{-1} < \infty$, is absolutely continuous with respect to a canonical measure μ for the partition, with density $x \mapsto p(x)$ satisfying, for $x \in \cap_{m=1}^\infty A_{x_1\cdots x_m}$,

$$\mathrm{E}(p(x)) = \prod_{j=1}^{\infty} \frac{2\alpha_{x_1\cdots x_j}}{\alpha_{x_1\cdots x_{j-1}0}+\alpha_{x_1\cdots x_{j-1}1}}, \tag{3.22}$$

$$\mathrm{E}(p(x)^2) = \prod_{j=1}^{\infty} \frac{4\alpha_{x_1\cdots x_j}(\alpha_{x_1\cdots x_j}+1)}{(\alpha_{x_1\cdots x_{j-1}0}+\alpha_{x_1\cdots x_{j-1}1})(\alpha_{x_1\cdots x_{j-1}0}+\alpha_{x_1\cdots x_{j-1}1}+1)}.$$

Furthermore, its total variation support consists of all probability measures that are absolutely continuous relative to μ. (Observe that the assumption implies $|\mathrm{E}(V_\varepsilon) - 1/2| \le c/a_m$ and $var V_\varepsilon \le c/a_m$, for every $\varepsilon \in \mathcal{E}^m$, and some constant c, and apply Theorem 3.16.)

Like the tail-free class, the Pólya tree class of priors enjoys the conjugacy property.

Theorem 3.21 (Conjugacy) *The posterior process corresponding to observing an i.i.d. sample $X_1, \ldots, X_n$ from a distribution P that is a priori modeled as a Pólya tree process prior* $\mathrm{PT}(\mathcal{T}_m, \alpha_\varepsilon)$ *is a Pólya tree process* $\mathrm{PT}(\mathcal{T}_m, \alpha_\varepsilon^*)$, *where $\alpha_\varepsilon^* := \alpha_\varepsilon + \sum_{i=1}^n \mathbb{1}(X_i \in A_\varepsilon)$, for every $\varepsilon \in \mathcal{E}^*$.*

Proof Because the posterior process is tail-free by Theorem 3.15, it suffices to show that under the posterior distribution the variables $V_{\varepsilon 0} = P(A_{\varepsilon 0}|\, A_\varepsilon)$ *within* every given level are independent beta variables with the given parameters $(\alpha_{\varepsilon 0}^*, \alpha_{\varepsilon 1}^*)$. Fix $m \in \mathbb{N}$ and set $V = (V_\varepsilon : \varepsilon \in \cup_{k \le m} \mathcal{E}^k)$, where $V_{\varepsilon 1} = 1 - V_{\varepsilon 0}$. By Theorem 3.14 the (joint posterior) distribution of V given $X_1, \ldots, X_n$ is the same as their conditional distribution given the vector of cell counts $N = (N_\varepsilon : \varepsilon \in \mathcal{E}^m)$. The marginal likelihood of $V_{\varepsilon 0}$ is proportional to $V_{\varepsilon 0}^{\alpha_{\varepsilon 0}}(1 - V_{\varepsilon 0})^{\alpha_{\varepsilon 1}} = V_{\varepsilon 0}^{\alpha_{\varepsilon 0}} V_{\varepsilon 1}^{\alpha_{\varepsilon 1}}$, and these variables are independent. The conditional likelihood of N

given V is multinomial, with the probabilities $P(A_\varepsilon)$ determined by (3.12). Therefore, the joint likelihood of (N, V) is proportional to

$$\prod_{|\varepsilon|=m}(V_{\varepsilon_1}V_{\varepsilon_2\varepsilon_2}\cdots V_{\varepsilon_1\cdots\varepsilon_m})^{N_\varepsilon}\prod_{|\varepsilon|\le m}V_\varepsilon^{\alpha_\varepsilon}=\prod_{|\varepsilon|\le m}V_\varepsilon^{N_\varepsilon+\alpha_\varepsilon},$$

where $N_\varepsilon = \sum_{\delta\in\mathcal{E}^{m-|\varepsilon|}} N_{\varepsilon\delta}$ denotes the number of observations falling in A_ε, also for ε with $|\varepsilon| < m$. We recognize the right side as a product of beta-likelihoods with parameters α_ε^*. □

In conjunction with equations (3.21) and (3.22), the preceding result gives expressions for the posterior means and variances of probabilities and densities, simply by replacing α_ε with α_ε^*. For instance, for the canonical Pólya tree process, formula (3.22) shows that the mean posterior density, with $N_{x_1\cdots x_j}$ the number of observations falling in the set in the jth partition that contains x (i.e. $x \in \cap_{j=1}^\infty A_{x_1\cdots x_j}$ and N defined as in (3.15)),

$$\mathrm{E}(p(x)|\,X_1,\ldots,X_n)=\prod_{j=1}^{\infty}\frac{2a_j+2N_{x_1\cdots x_j}}{2a_j+N_{x_1\cdots x_{j-1}}}. \tag{3.23}$$

If no observation falls in $A_{x_1\cdots x_j}$, then $N_{x_1\cdots x_j} = 0$, and the jth factor in the product is 1, as are all terms for higher j, as the sets are decreasing. The infinite product then reduces to a finite product. This happens for every $x \notin \{X_1, \ldots, X_n\}$ for sufficiently large $j = j(x)$. Furthermore, on every partitioning set A_ε that does not contain observations, the (in)finite product is constant, as all x in such a set share their membership in the partitioning sets at the coarser levels. This implies that the mean posterior density is very regular, in contrast to the sample paths of the posterior density, which are discontinuous at every boundary point by the last assertion of Theorem 3.16.

The condition $\sum_{m=1}^\infty a_m^{-1} < \infty$ ensures the absolute continuity of the realizations of a Pólya prior, and hence effectively reduces the model to a dominated set of measures. The posterior is then absolutely continuous with respect to the prior, by Bayes's rule, as noted in Lemma 1.2. It turns out that prior and posterior are mutually singular if the series diverges.

Theorem 3.22 *Consider the posterior distribution corresponding to observing an i.i.d. sample of finite size from a distribution P that is a priori modeled as a canonical Pólya tree process* $\mathrm{PT}^*(\lambda, a_m)$. *Then prior and posterior are mutually absolutely continuous or mutually singular, depending on whether $\sum_{m=1}^\infty a_m^{-1}$ converges or diverges.*

Proof The map $\phi: \mathfrak{M} \to [0, 1]^{\mathcal{E}^*}$ given by $P \mapsto (P(A_\varepsilon), \varepsilon \in \mathcal{E}^*)$ is a measurable isomorphism, relative to $\mathscr{M}$ and the projection σ-field, and so is the further map into the vector $(V_{\varepsilon 0}: \varepsilon \in \mathcal{E}^*)$ of conditional probabilities, also viewed as subset of $[0, 1]^{\mathcal{E}^*}$. Therefore, prior and posterior are absolutely continuous or singular whenever the corresponding joint laws of $(V_{\varepsilon 0}: \varepsilon \in \mathcal{E}^*)$ are so. Prior and posterior are both Pólya tree processes; hence, the variables $V_{\varepsilon 0}$ are independent beta variables. Therefore, by Kakutani's theorem (see Corollary B.9), the joint laws are absolutely continuous or singular depending on whether the product of the affinities of the marginal distributions of the $V_{\varepsilon 0}$ under prior and posterior is positive or zero.

The affinity of two beta distributions with parameters (α, β) and (α^*, β^*) is equal to

$$\frac{B((\alpha+\alpha^*)/2, (\beta+\beta^*)/2)}{\sqrt{B(\alpha, \beta)B(\alpha^*, \beta^*)}},$$

where B stands for the beta function $B(a,b) = \int_0^1 x^{a-1}(1-x)^{b-1}dx = \Gamma(a)\Gamma(b)/\Gamma(a+b)$. The beta distributions of the variables $V_{\varepsilon 0}$ at a given level $m = |\varepsilon| + 1$ have parameters (a_m, a_m) under prior and posterior, except for the indices ε of the partitioning cells A_ε that contain observations, where the parameters are $(a_m + N_{\varepsilon 0}, a_m + N_{\varepsilon 1})$ for the posterior. The product of the affinities for all variables $V_{\varepsilon 0}$ is therefore given by

$$\prod_{m=1}^{\infty} \prod_{\varepsilon \in \mathcal{E}^{m-1}} \frac{B(a_m + N_{\varepsilon 0}/2, a_m + N_{\varepsilon 1}/2)}{\sqrt{B(a_m, a_m)B(a_m + N_{\varepsilon 0}, a_m + N_{\varepsilon 1})}}.$$

With some effort, using Stirling's formula, it can be shown that, for fixed numbers $c, d \geq 0$, and $a \to \infty$,

$$B(a+c, a+d) = \sqrt{\frac{2\pi}{a}}\, 2^{-2a-c-d}\Big[1 + \frac{1}{a}\Big(c^2 + d^2 - \frac{c+d}{4} + \frac{1}{12}\Big) + O\Big(\frac{1}{a^2}\Big)\Big],$$

where the second order term is uniform in bounded c, d. Therefore, if $a_m \to \infty$, the product of affinities in the second last display can, again with some effort, be seen to be equal to

$$\prod_{m=1}^{\infty} \prod_{\varepsilon \in \mathcal{E}^{m-1}} \Big[1 - \frac{1}{a_m}\Big(\frac{N_{\varepsilon 0}^2 + N_{\varepsilon 1}^2}{4}\Big) + O\Big(\frac{1}{a_m^2}\Big)\Big].$$

Since $N_{\varepsilon 0}^2 + N_{\varepsilon 1}^2 > 0$ for at least one and at most n indices ε, for every m, the product is zero if $\sum_m a_m^{-1} = \infty$, and positive otherwise.

If a_m is bounded by a given constant for infinitely many m, then the corresponding terms of the double product are uniformly bounded by a constant strictly less than 1, and hence the product is zero. □

3.7.1 Relation with the Pólya Urn Scheme

In a *Pólya urn model* one repeatedly draws a random ball from an urn containing balls of finitely many types, replacing it with given numbers of balls of the various varieties before each following draw. In our situation there are two types of balls, marked "0" or "1," and every ball drawn from the urn is replaced by two balls of the same type (the chosen ball is replaced and an extra ball of the same type is added). There is an interesting link between Pólya tree processes and the Pólya urn scheme, which can be helpful for certain calculations.

The link employs infinitely many urns, arranged in a tree structure (see Figure 3.4). In the (binary) *Pólya tree urn model* there is a Pólya urn with balls marked "0" or "1" at every node of the (binary) tree, and balls are drawn sequentially from the urns that are situated on a random path of nodes through the tree. The path is determined, from the root down, by choosing at every node the left or right branch depending on the type of ball drawn from the urn at the node. After each draw the urn is updated, just as for an ordinary Pólya urn, thus creating a new tree of urns. Urns that are not on the path remain untouched and are copied unaltered to the new tree. The new tree is visited along a new, random path, starting

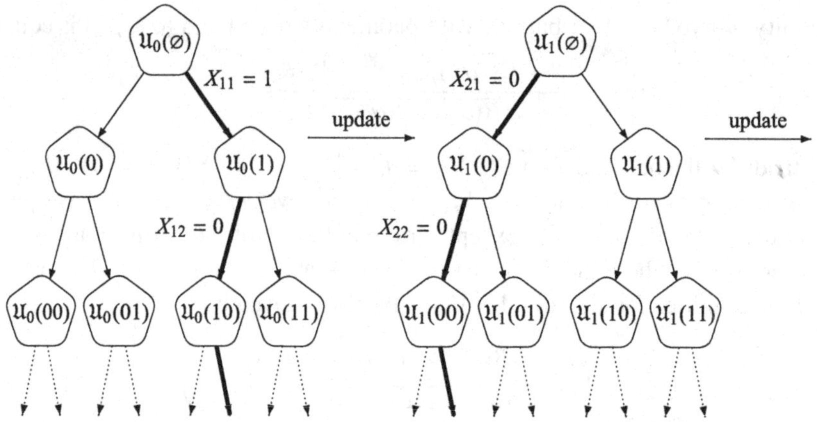

Figure 3.4 Pólya tree urn model. Shown are the urns at the first three levels of the initial tree (left) and its first update (right). Realizations of the paths chosen through the trees are indicated by bold arrows, the balls drawn from the urns on the paths are indicated on the outgoing arrows.

again at the root, and proceeding as before, albeit with different transition probabilities, as the contents of some urns have been updated.

We index the tree, as usual, by the elements of $\mathcal{E}^* \cup \varnothing$, with the empty string $\varnothing$ corresponding to the root and the strings of length m to the 2^m nodes at the mth level. The process starts with an initial tree of urns, which we denote by $\mathfrak{U}_0 = (\mathfrak{U}_0(\varepsilon): \varepsilon \in \mathcal{E}^* \cup \varnothing)$, and this is sequentially replaced by urn trees $\mathfrak{U}_1, \mathfrak{U}_2, \ldots$ Every urn $\mathfrak{U}_m(\varepsilon)$ corresponds to a pair of non-negative real numbers (r_0, r_1), which "physically" stand for the numbers of balls of colors 0 and 1. The numbers of balls in the initial urns $\mathfrak{U}_0(\varepsilon)$ are given; in the later urns they are random. Non-integer numbers (r_0, r_1) are permitted, in the understanding that a ball drawn from the urn is "0" with probability $r_0/(r_0 + r_1)$ and "1" with probability $r_1/(r_0 + r_1)$.

Every urn tree $\mathfrak{U}_i$ leads to an infinite sequence of draws, whose realizations we record as $X_i = X_{i1}, X_{i2}, X_{i3}, \ldots$, an infinite string of 0s and 1s. For instance, X_{i1} is the color of the ball drawn from urn $\mathfrak{U}_i(\varnothing)$, X_{i2} is the ball drawn from $\mathfrak{U}_i(X_{i1})$, X_{i3} is drawn from $\mathfrak{U}_i(X_{i1}X_{i2})$, etc. With this notation, the updating scheme can be described as $\mathfrak{U}_{i+1}(\varnothing) = \mathfrak{U}_i(\varnothing) \cup \{X_{i1}\}, \mathfrak{U}_{i+1}(X_{i1}) = \mathfrak{U}_i(X_{i1}) \cup \{X_{i2}\}$, etc.

As is well known, the consecutive draws from a Pólya urn, for instance the sequence $X_{11}, X_{21}, X_{31}, \ldots$ in our scheme, is exchangeable (see Feller 1968, page 120). It can be verified that, similarly, the sequence $X_1, X_2, \ldots$ is exchangeable, in the state space $\{0, 1\}^\infty$. Consequently $Y_1, Y_2, \ldots$, defined through the binary expansions $Y_i = \sum_{j=1}^\infty X_{ij} 2^{-j}$ based on X_i, are exchangeable random variables in $[0, 1]$. By de Finetti's theorem (e.g. Theorem 1.49 of Schervish 1995), there exists a unique measure Π on $\mathfrak{M} = \mathfrak{M}[0, 1]$ such that, for every Borel sets $A_1, \ldots, A_n$,

$$\mathrm{P}(Y_1 \in A_1, \ldots, Y_n \in A_n) = \int \prod_{i=1}^n P(A_i)\, d\Pi(P).$$

The de Finetti measure Π acts as a prior for P: it happens to be a Pólya tree process. For a precise proof of the following theorem, see Mauldin et al. (1992).

Theorem 3.23 *The de Finetti measure corresponding to the Pólya urn tree model is the Pólya tree process* $\mathrm{PT}(\lambda, \alpha_\varepsilon)$, *where* λ *is the Lebesgue measure on* $[0, 1]$ *and* $(\alpha_{\varepsilon 0}, \alpha_{\varepsilon 1}) = \mathfrak{U}(\varepsilon)$ *is the configuration of the initial urn at node* $\varepsilon \in \mathcal{E}^* \cup \varnothing$.

Its connection with Pólya urn schemes justifies the name Pólya tree. Many properties of Pólya tree processes can be elegantly derived from the urn scheme formulations; see Problems 3.6 and 3.7.

3.7.2 Mixtures of Pólya Tree Processes

Although the support of a Pólya tree can be reasonably large (and it will be seen in Chapter 6 that the posterior distribution is consistent under certain conditions), practical implementation requires some meaningful elicitation of the mean measure. Often a class of target measures can be identified, like the normal family, but it is hard to single out one member from the family. Therefore it is natural to consider a mean measure that contains finitely many unspecified parameters and put a further prior on this set of hyperparameters. The resulting hierarchical prior is a mixture of Pólya tree processes.

Let the observations be a random sample $X_1, \ldots, X_n$ of real variables. Suppose that G_θ and g_θ are the distribution function and density of the elicited mean measure, and that the hyperparameter θ is given the prior density ρ. Given θ, let P follow the canonical Pólya tree process $\mathrm{PT}^*(G_\theta, a_m)$. Then the prior mean measure is $A \mapsto \mathrm{E}(P(A)) = \int G_\theta(A)\rho(\theta)\, d\theta$. Assume that $\sum_{m=1}^\infty a_m^{-1} < \infty$, so that the random measure P admits a density p a.s.

The canonical partitions can be constructed for all θ simultaneously as $(G_\theta^{-1}(A_\varepsilon)\colon \varepsilon \in \mathcal{E}^*)$, for $(A_\varepsilon\colon \varepsilon \in \mathcal{E}^*)$, the canonical partition of the unit interval relative to the uniform measure. Given θ, the prior and posterior are ordinary Pólya tree processes, with the updating given in Theorem 3.21. In particular, (3.23) gives, with $N_{G_\theta(x)_1 \cdots G_\theta(x)_j}$ the number of transformed observations $G_\theta(X_i)$ falling in the same jth level partitioning set as $G_\theta(x)$,

$$\mathrm{E}(p(x)|\, \theta, X_1, \ldots, X_n) = g_\theta(x) \prod_{j=1}^{\infty} \frac{2a_j + 2N_{G_\theta(x)_1 \cdots G_\theta(x)_j}}{2a_j + N_{G_\theta(x)_1 \cdots G_\theta(x)_j}}. \tag{3.24}$$

The posterior density is obtained by integrating this relative to the posterior density of θ. By Bayes's theorem, the latter is given by

$$\rho(\theta|\, X_1, \ldots, X_n) \propto p(X_1, \ldots, X_n|\, \theta)\rho(\theta) = \int \prod_{i=1}^{n} p(X_i)\, d\Pi(P|\, \theta)\, \rho(\theta),$$

where $d\Pi(P|\, \theta)$ is the Pólya tree process $\mathrm{PT}^*(G_\theta, a_m)$. For a single observation X_1 this evaluates to $\rho(\theta)$ times the right side of (3.24) evaluated at $x = X_1$. The general formula can also be written explicitly, but depends on the pattern of joint memberships of the observations to the partitioning sets (at lower levels). The resulting expression can be evaluated numerically, and gives a method for density estimation.

This often has a desirable smoothness property. For instance, in the case that θ is a location parameter, the partitions defined by G_θ shift continuously as θ varies, resulting in a posterior expected mixture density that is continuous everywhere. The case that θ is a scale parameter is similar, except for a singularity at the point zero.

Theorem 3.24 *Let e be the function* $x \mapsto \mathrm{E}(p(x)|\, X_1, \ldots, X_n)$ *of pointwise posterior mean of* $p(x)$*, given a random sample* $X_1, \ldots, X_n$ *from a distribution* P *drawn from a canonical Pólya tree process* $\mathrm{PT}^*(G_\theta, a_m)$ *with* $\sum_m a_m^{-1} < \infty$ *with partitions as described. Let* G_θ *have Lebesgue density* g_θ*, and let* g *be a fixed bounded, continuous probability density with respect to Lebesgue measure on* $\mathbb{R}$*.*

(i) *If* $g_\theta(x) = g(x-\theta)$*, then e is continuous everywhere, a.s.*
(ii) *If* $g_\theta(x) = \theta^{-1} g(x/\theta)$ *and* $\int_0^\infty \theta^{-1} \rho(\theta)\, d\theta < \infty$*, then e is continuous everywhere except at zero, a.s.*

Proof The posterior mean density is the integral of (3.24) relative to the posterior distribution of θ. The latter is absolutely continuous and does not depend on the argument x. The former is an ordinary Pólya-tree mean density, and it is uniformly bounded, and continuous at every x that is not a boundary point of the sequence of partitions. To see this, note that the tail of the infinite product becomes uniformly in x close to 1, as $\sum_j a_j^{-1} < \infty$ and

$$\prod_{j>m}\Bigl(1-\frac{n}{2a_j}\Bigr) \le \prod_{j>m} \frac{2a_j + 2N_{G_\theta(x)_1\cdots G_\theta(x)_j}}{2a_j + N_{G_\theta(x)_1\cdots G_\theta(x)_j}} \le \prod_{j>m}\Bigl(1+\frac{n}{a_j}\Bigr). \tag{3.25}$$

Furthermore, for every given m, a non-boundary point x shares its mth level partitioning set with an open neighborhood surrounding it, rendering the first m terms of the product identical for every point in the neighborhood.

If D are the boundary points of the canonical uniform partition, then $G_\theta^{-1}(D)$ are the boundary points for the Pólya tree given θ. In the location case, this is the set $G^{-1}(D) + \theta$, whereas in the scale case this is $\theta G^{-1}(D)$.

It follows that in the location case the function (3.24) is continuous at x if $x \notin G^{-1}(D)+\theta$, or equivalently if $\theta \notin G^{-1}(D) - x$, which is a countable set of θ and hence a null set for the posterior distribution of θ. Because the function (3.24) is uniformly bounded in θ by $\|g\|_\infty$ times the right side of (3.25) for $m = 0$, the assertion follows from the dominated convergence theorem.

In the scale case the function (3.24) is continuous at x if $x \notin \theta G^{-1}(D)$; equivalently, $1/\theta \notin G^{-1}(D)/x$ if $x \neq 0$, which again is a countable set of θ. (If $x = 0$ and $0 \in G^{-1}(D)$, then $x \in \theta G^{-1}(D)$ for all θ.) In this case, (3.24) is bounded above by $\theta^{-1}\|g\|_\infty$ times the right side of (3.25) for $m = 0$, which is integrable relative to the posterior distribution of θ for almost every $X_1, \ldots, X_n$ as $\mathrm{E} \int \theta^{-1} \rho(\theta|\, X_1, \ldots, X_n)\, d\theta = \mathrm{E}\theta^{-1} < \infty$ by assumption. The assertion follows again by the dominated convergence theorem. □

An alternative way of mixing Pólya trees keeps the partitions intact, but varies the parameters α_ε with θ. A desired prior mean density g_θ can then be obtained from solving the relation

$$g_\theta(x) = \prod_{j=1}^{\infty} \frac{2\alpha_{x_1\cdots x_j}(\theta)}{\alpha_{x_1\cdots x_{j-1}0}(\theta) + \alpha_{x_1\cdots x_{j-1}1}(\theta)}. \tag{3.26}$$

We shall refer to this type of mixtures as *Pólya tree mixtures of the second kind* and to the former mixtures as *Pólya tree mixtures of the first kind*. As the partitions do not vary, the density of a mixture of second kind is discontinuous at all boundary points of the partition just like a usual Pólya tree. However, since the counts in the partitioning intervals are free of θ, the posterior expected density has a simpler expression than for a mixture of the first kind. A Pólya tree mixture of the second kind is useful in the context of testing a parametric family against a nonparametric alternative using Bayes factors.

3.7.3 Partially Specified Pólya Tree

Rather than continuing the splitting tree indefinitely, one may stop at a certain level m, and distribute the remaining probability masses uniformly over the partitioning set at the terminating level. On $[0, 1]$ with the standard binary intervals, this yields a prior on all regular histograms of width 2^{-m}. If the splitting variables are independent and possess beta distributions, then the resulting "partially specified Pólya tree" can formally be viewed as a regular Pólya tree with $\alpha_\varepsilon = \infty$ for all ε with $|\varepsilon| > m$ and $\mathrm{Be}(\infty, \infty)$ interpreted as the Dirac measure at $1/2$.

Computing the posterior distribution reduces to a multinomial problem where only the counts of the sets A_ε, for $\varepsilon \in \mathcal{E}^m$ are observed (consult Theorem 3.14 in this context). Furthermore, the density estimate given by (3.23) contains only m factors and is differentiable (with zero derivative) on the interior of all mth level partitioning sets.

One may similarly consider a mixture of these partially specified Pólya trees. Then, by a slight modification of the arguments given in the proof of Theorem 3.24, it follows that if g is bounded and continuous, then for a location mixture, the density estimate is differentiable everywhere except at the observations. For a scale mixture, if $\int \theta^{-1}\rho(\theta)\,d\theta < \infty$, the density estimate is differentiable everywhere except at zero and at the observations. The result can be extended to higher-order derivatives by appropriately strengthening the integrability condition.

Notwithstanding the differentiability of the density estimate, a partially specified Pólya tree process can support only histograms of some specified bandwidth, and hence not all densities can be consistently estimated by it as a prior. In practice, one considers a sequence of priors and works with a level fine enough for the accuracy associated with a given sample size, that is, one lets $m_n \to \infty$. Such a sequence of priors can maintain both consistency and differentiability properties.

3.7.4 Evenly Split Pólya Tree

A Pólya tree has full weak support, but rough sample paths. In contrast, a partially specified Pólya tree gives piecewise constant sample paths, but has limited support. The partial specification in a Pólya tree corresponds to half-half splitting of probabilities from a stage onwards, rather than continuing to split randomly, as in a Pólya tree. By introducing a random choice between even and random splitting, this makes it possible to design a process that combines the Pólya and partially specified Pólya trees and enjoys both large weak support and piecewise constant sample paths.

Suppose that the splitting variables are independent and distributed according to the mixture distribution, for some choice of numbers $\alpha_\varepsilon > 0$ and $\beta_\varepsilon \in [0, 1]$, for $\varepsilon \in \mathcal{E}^*$,

$$V_{\varepsilon_1\cdots\varepsilon_m 0} \sim (1 - \beta_{\varepsilon_1\cdots\varepsilon_m})\delta_{1/2} + \beta_{\varepsilon_1\cdots\varepsilon_m}\mathrm{Be}(\alpha_{\varepsilon_1\cdots\varepsilon_m 0}, \alpha_{\varepsilon_1\cdots\varepsilon_m 1}).$$

If $0 < \beta_\varepsilon < 1$ for all $\varepsilon \in \mathcal{E}^*$ and $\sum_{\varepsilon \in \mathcal{E}^*} \beta_\varepsilon < \infty$, then only finitely many splits can be random, almost surely, by the Borel-Cantelli lemma, and the other splits will be even. This ensures that the realizations of this process are almost surely piecewise constant functions with only finitely many points of discontinuity. As the nature of the splits is not predetermined, it is clear from Theorem 3.10 that the process has full weak support. In fact, Theorem 3.19 implies that every density is in the total variation support of the process. The process is not a Pólya tree, but it can be considered a mixture of partially specified Pólya trees. The mixing takes place over all splitting regimes that give even splits at all, but finitely many, occasions.

The parameters α_ε and β_ε may be chosen to depend only on the string lengths, i.e. $\alpha_\varepsilon = a_m$ and $\beta_\varepsilon = b_m$, for all $\varepsilon \in \mathcal{E}^m$. If $\sum_{m=1}^\infty 2^m b_m < \infty$, then for any choice of $a_m > 0$ the process is fully supported in both the weak and the total variation sense. In particular the parameters a_m need not grow with m. This is in sharp contrast with the usual Pólya tree, which requires $\sum_{m=1}^\infty a_m^{-1} < \infty$ already for the existence of a density.

3.8 Historical Notes

Priors for a probability distribution on a countable sample space were extensively discussed in Freedman (1963). Dubins and Freedman (1963) introduced the method of random rectangular partitions, discussed the singularity property and characterized the prior mean. Tail-free processes were studied by Freedman (1963, 1965) and Fabius (1964). The concept also appeared in Doksum (1974), who considered a very related concept, neutral to the right process, for priors on survival function. Theorem 3.16 is due to Kraft (1964). Theorem 3.14 was exploited by Freedman (1965) to show posterior consistency for tail-free process priors; its converse statement is due to Fabius (1964). Theorems 3.15 and 3.10 were obtained by Freedman (1965). Pólya trees were studied extensively by Lavine (1992, 1994), who showed conjugacy and derived formulas for density estimates, but many of the ideas actually appeared implicitly in the works of Freedman (1965) and Ferguson (1974). The connection with Pólya urn schemes was uncovered by Mauldin et al. (1992), based on earlier ideas of Blackwell and MacQueen (1973). Theorem 3.22 is due to Drăghici and Ramamoorthi (2000). Pólya tree mixtures of the first kind were studied by Hanson and Johnson (2002), who also gave a version of Theorem 3.24. Pólya tree mixtures of the second kind appeared in Berger and Guglielmi (2001) in the context of testing a parametric family against a nonparametric alternative.

Problems

3.1 Consider a stick-breaking scheme $p_k = V_k \prod_{j=1}^{k-1}(1 - V_j)$, where the V_ks are independent with $\mathrm{P}(V_k = 0) = 1 - k^{-2}$ and $\mathrm{P}(V_k = 1 - e^{-k}) = k^{-2}$. Show that $\sum_{k=1}^\infty \mathrm{E}[\log(1 - V_k)] = -\infty$, yet $\sum_{k=1}^\infty p_k < 1$ a.s. [This contradicts the claim made in Ishwaran and James (2001) that the condition $\sum_{l=1}^\infty \mathrm{E}[\log(1 - V_l)] = -\infty$ is necessary and sufficient to obtain a probability measure.]

3.2 Suppose we want to construct a prior on the space of discrete distributions with decreasing probability mass function. Construct a random element ξ on $\mathbb{S}_\infty$ and put $p_k = \sum_{j=k}^{\infty} \xi_j/j$. Show that every prior must be the law of a random vector $(p_1, p_2, \ldots,)$ of this form.

3.3 Construct a prior on each of the following spaces: (a) space of finite nonnegative measures on $\mathbb{N}$, (b) ℓ_1, (c) ℓ_2.

3.4 (Ghosh and Ramamoorthi 2003) Suppose that for every measurable partition $\{B_1, \ldots, B_k\}$ of a Polish space $(\mathfrak{X}, \mathscr{X})$, a probability $\Pi_{B_1,\ldots,B_k}$ has been specified with the property that if $\{A_1, \ldots, A_l\}$ is a coarser partition, i.e. $A_j = \cup_{i:B_i \subset A_j} B_i$ for all j, then the distribution of

$$\Big(\sum_{i:B_i \subset A_1} P(B_i), \ldots, \sum_{i:B_i \subset A_l} P(B_i) \Big),$$

where $(P(B_1), \ldots, P(B_k)) \sim \Pi_{B_1,\ldots,B_k}$, is $\Pi_{A_1,\ldots,A_l}$. If further $\Pi_{B_n} \rightsquigarrow \delta_0$ for any $B_n \downarrow \varnothing$ and $\Pi_{\mathfrak{X}} = \delta_1$, then there exists a unique probability measure Π on $\mathfrak{M}(\mathfrak{X})$ such that the induced distribution of $(Q(B_1), \ldots, Q(B_k))$, when $Q \sim \Pi$, is equal to $\Pi_{B_1,\ldots,B_k}$ for every measurable partition $\{B_1, \ldots, B_k\}$ of $\mathfrak{X}$.

3.5 Let P and Q be random probability measures on a Polish space $(\mathfrak{X}, \mathscr{X})$. Let $\mathscr{C}$ be a generator for the Borel σ-field on $\mathfrak{X}$. Suppose that for every finite collection $C_1, \ldots, C_n \in \mathscr{C}$,

$$(P(C_1), \ldots, P(C_n)) =_d (Q(C_1), \ldots, Q(C_n)).$$

Show that this identity then holds for every finite collection of Borel sets $C_1, \ldots, C_n$.

3.6 (Mauldin et al. 1992) Derive the posterior updating formula for a Pólya tree process prior through the urn scheme representation. Derive the marginal distribution, and the conditions for its continuity from the urn scheme representation.

3.7 (Mauldin et al. 1992) Generalize the Pólya urn scheme to k types of balls, and hence construct a Pólya tree process where the mass of any set at any level is distributed to k children of the set following a k-dimensional Dirichlet distribution. The construction is useful in defining Pólya tree processes on $\mathbb{R}^d$ with $k = 2^d$.

3.8 (Dubins and Freedman 1963) Consider the prior of Subsection 3.4.3. The mean cumulative distribution function $\bar{\mu}$ can be characterized as the unique fixed point of the map $T_\mu: \mathfrak{M}[0, 1] \to \mathfrak{M}[0, 1]$ where

$$(T_\mu F)(x) = \int_0^1 \int_x^1 \beta F(x/\alpha)\, \mu(d\alpha, d\beta) + \int_0^1 \int_0^x \big[\beta + (1-\beta) F((x-\alpha)/(1-\alpha))\big]\, \mu(d\alpha, d\beta).$$

Show that if $\mu = \delta_r \times \nu$, where ν has mean w, then $\bar{\mu}$ satisfies the equation

$$F(x) = \begin{cases} wF(x/r), & 0 \le x \le r, \\ w + (1-w)F((x-r)/(1-r)), & r \le x \le 1. \end{cases}$$

In particular, when μ is the uniform distribution on the vertical line $x = 1/2$, as in case (a) of Subsection 3.4.3, then $\bar{\mu}$ is the uniform distribution on $[0, 1]$. Show that $\bar{\mu}(x) = (2/\pi) \sin^{-1} \sqrt{x}$ for cases (b) and (c).

3.9 Consider a canonical Pólya tree process $P \sim \mathrm{PT}^*(\lambda, cr_m)$ and observations $X_1, \ldots, X_n \mid P \overset{\text{iid}}{\sim} P$. Show that P under the posterior converges weakly to δ_λ as $c \to$

∞ and converges weakly to the random measure $\sum_{i=1}^{n} p_i \delta_{X_i}$, where $(p_1, \ldots, p_n) \sim \mathrm{Dir}(n; 1, \ldots, 1)$, as $c \to 0$.

3.10 Derive conditions for the existence of higher-order derivatives of the posterior expected density everywhere for a location mixture of Pólya tree process.

3.11 Derive conditions for the existence of higher-order derivatives of the posterior expected density everywhere except at 0 for scale mixtures of Pólya tree process.

4

Dirichlet Processes

The Dirichlet process is of fundamental importance in Bayesian nonparametrics. It is a default prior on spaces of probability measures, and a building block for priors on other structures. In this chapter we define and construct the Dirichlet process by several methods, and derive many properties of interest, both simple and involved: expressions for moments, support, convergence, discreteness, self-similarity, thickness of tails, distribution of functionals, characterizations. We also study posterior conjugacy, joint and predictive distribution of observations and mixtures of Dirichlet processes.

4.1 Definition and Basic Properties

Throughout the chapter $\mathfrak{M}$ denotes the set of probability measures on a Polish space $(\mathfrak{X}, \mathscr{X})$. Unless stated otherwise, it is equipped with the weak topology and the corresponding Borel σ-field $\mathscr{M}$. The phrase *random measure on* $(\mathfrak{X}, \mathscr{X})$ means a probability measure on $(\mathfrak{M}, \mathscr{M})$.

Definition 4.1 (Dirichlet process) A random measure P on $(\mathfrak{X}, \mathscr{X})$ is said to possess a *Dirichlet process* distribution $\mathrm{DP}(\alpha)$ with *base measure* α, if for every finite measurable partition $A_1, \ldots, A_k$ of $\mathfrak{X}$,

$$(P(A_1), \ldots, P(A_k)) \sim \mathrm{Dir}(k; \alpha(A_1), \ldots, \alpha(A_k)). \tag{4.1}$$

In this definition α is a given finite positive Borel measure on $(\mathfrak{X}, \mathscr{X})$. We write $|\alpha| = \alpha(\mathfrak{X})$ its total mass (called *prior precision*), and $\bar{\alpha} = \alpha/|\alpha|$ the probability measure obtained by normalizing α (called *center measure*), and use the notation $P \sim \mathrm{DP}(\alpha)$ to say that P has a Dirichlet process distribution with base measure α; and also $P \sim \mathrm{DP}(MG)$ in the case that $\mathfrak{X} = \mathbb{R}^k$, $M = |\alpha|$ and $\bar{\alpha} = \alpha/M$ has cumulative distribution function G[1].

Existence of the Dirichlet process is not obvious, but a proof is deferred to Section 4.2. The finite-dimensional Dirichlet distribution in the right side of (4.1) is reviewed in Appendix G.

It is immediate from the definition that for any measurable function $\psi: \mathfrak{X} \to \mathfrak{Y}$ between standard Borel spaces, the random measure $P \circ \psi^{-1}$ on $\mathfrak{Y}$ obtained from a Dirichlet process distribution $P \sim \mathrm{DP}(\alpha)$ possesses the $\mathrm{DP}(\alpha \circ \psi^{-1})$-distribution.

[1] The definition could be extended to *finitely additive Dirichlet processes P on a subfield $\mathscr{X}_0 \subset \mathscr{X}$ by requiring that* (4.1) *hold for all partitions whose elements belong to $\mathscr{X}_0$. Then it suffices that α is finitely additive.*

By applying (4.1) to the partition $\{A, A^c\}$ for a Borel set A, the vector $(P(A), P(A^c))$ possesses the $\mathrm{Dir}(2; \alpha(A), \alpha(A^c))$-distribution; equivalently, the variable $P(A)$ is $\mathrm{Be}(\alpha(A), \alpha(A^c))$-distributed.

In particular $P(A) > 0$ almost surely for every measurable set A with $\alpha(A) > 0$. Because the exceptional null set depends on A, this does not imply that P and α are mutually absolutely continuous. In fact, in Theorem 4.14 they are shown to be almost surely singular unless α is atomic.

4.1.1 Expectations, Variances and Co-Variances

The center measure is the mean measure of the Dirichlet prior.

Proposition 4.2 (Moments) *If $P \sim \mathrm{DP}(\alpha)$, then, for any measurable sets A and B,*

$$\mathrm{E}(P(A)) = \bar{\alpha}(A), \tag{4.2}$$

$$\mathrm{var}\,(P(A)) = \frac{\bar{\alpha}(A)\bar{\alpha}(A^c)}{1+|\alpha|}, \tag{4.3}$$

$$\mathrm{cov}\,(P(A), P(B)) = \frac{\bar{\alpha}(A\cap B) - \bar{\alpha}(A)\bar{\alpha}(B)}{1+|\alpha|}, \tag{4.4}$$

Proof The first two results are immediate from the properties of the beta distribution of $P(A)$. To prove the third relation, first assume that $A \cap B = \varnothing$. Then $\{A, B, A^c \cap B^c\}$ is a partition of $\mathfrak{X}$, whence $(P(A), P(B), P(A^c \cap B^c)) \sim \mathrm{Dir}(3; \alpha(A), \alpha(B), \alpha(A^c \cap B^c))$. By the expressions for the moments of finite Dirichlet distribution given by Corollary G.4,

$$\mathrm{cov}\,(P(A), P(B)) = -\frac{\alpha(A)\alpha(B)}{|\alpha|^2(1+|\alpha|)},$$

which is the same as (4.4) in the special case $A \cap B = \varnothing$. For the proof of the general case we decompose $A = (A\cap B)\cup(A\cap B^c)$ and $B = (A\cap B)\cup(A^c\cap B)$, and write the left-hand side of (4.4) as

$$\mathrm{cov}\,(P(A\cap B) + P(A\cap B^c), P(A\cap B) + P(A^c\cap B)).$$

We write this as a sum of four terms and use both (4.3) and the special case of (4.4) twice to reduce this to the right side of (4.4). ☐

The first assertion of the proposition can also be rewritten as $\int P(A)\,d\mathrm{DP}_\alpha(P) = \bar{\alpha}(A)$. Interpreting P in this equation as the law of an observation X from P, we see

$$\text{if } P \sim \mathrm{DP}(\alpha) \text{ and } X\,|\,P \sim P, \text{ then } X \sim \bar{\alpha}. \tag{4.5}$$

Thus $\bar{\alpha}$ is the marginal law of "an observation from the Dirichlet process," as well as the prior mean. It is also called the *center measure* of the Dirichlet process.

By (4.4) the total mass $|\alpha|$ controls the variability of any $P(A)$ around its prior mean, and hence is appropriately called the *precision parameter*. For instance, if one expects the $\mathrm{Nor}(0, 1)$ model to hold, but is not absolutely confident about it, one may propose a Dirichlet process prior with center measure $\mathrm{Nor}(0, 1)$ and precision $|\alpha|$ reflecting the degree of confidence in the prior guess. A more precise assessment of $|\alpha|$ can be based on the posterior distribution, which we obtain later on. However, we shall see later that the "precision parameter" also has another role and the rule "small $|\alpha|$ means less prior information" should be interpreted cautiously.

Proposition 4.2 implies that different choices of α generally lead to different Dirichlet processes: if $\alpha_1 \neq \alpha_2$, then $\mathrm{DP}(\alpha_1) \neq \mathrm{DP}(\alpha_2)$, unless $\bar{\alpha}_1 = \bar{\alpha}_2 = \delta_x$ for some x. This is obvious if $\bar{\alpha}_1 \neq \bar{\alpha}_2$, since the two random measures then have different means by (4.2). In case $\bar{\alpha}_1 = \bar{\alpha}_2$, but $|\alpha_1| \neq |\alpha_2|$, the variances of $P(A)$ under the two priors are different by (4.3), for every A with $0 < \bar{\alpha}_1(A) = \bar{\alpha}_2(A) < 1$; provided the base measures are nondegenerate, there is at least one such A.

Proposition 4.2 can be generalized to integrals of functions with respect to P.

Proposition 4.3 (Moments) *If $P \sim \mathrm{DP}(\alpha)$, then for any measurable functions ψ and ϕ for which the expression on the right-hand side is meaningful,*

$$\mathrm{E}(P\psi) = \int \psi \, d\bar{\alpha}, \tag{4.6}$$

$$\mathrm{var}(P\psi) = \frac{\int (\psi - \int \psi \, d\bar{\alpha})^2 \, d\bar{\alpha}}{1 + |\alpha|}, \tag{4.7}$$

$$\mathrm{cov}(P\psi, P\phi) = \frac{\int \psi\phi \, d\bar{\alpha} - \int \psi \, d\bar{\alpha} \int \phi \, d\bar{\alpha}}{1 + |\alpha|}. \tag{4.8}$$

Proof For indicator functions $\psi = \mathbb{1}_A$ and $\phi = \mathbb{1}_B$ the assertions reduce to those of Proposition 4.2. They extend by linearity (for the first) or bilinearity (for the second and third) to simple functions ψ and ϕ, and next by splitting in positive and negative parts and monotone convergence (for the first) or continuity in the $\mathbb{L}_2$-sense (for the second and third) to general measurable functions. ☐

Remark 4.4 In particular $\int |\psi| \, d\alpha < \infty$ implies that $P|\psi| < \infty$ a.s. $[\mathrm{DP}(\alpha)]$. We shall see in Sections 4.3.5 and 4.3.7 that the converse is false.

4.1.2 Self-Similarity

The Dirichlet process is tail-free in the sense of Definition 3.11, for any sequence of successive partitions of the sample space. In fact, it possesses much stronger conditional independence properties.

For a measure P and measurable set B, let $P|_B$ stand for the restriction measure $P|_B(A) = P(A \cap B)$, and P_B for the conditional measure $P_B(A) = P(A|\, B)$, for B with $P(B) > 0$.

Theorem 4.5 (Self-similarity) *If $P \sim \mathrm{DP}(\alpha)$, then $P_B \sim \mathrm{DP}(\alpha|_B)$, and the variable and processes $P(B)$, $(P_B(A): A \in \mathscr{X})$ and $(P_{B^c}(A): A \in \mathscr{X})$ are mutually independent, for any $B \in \mathscr{X}$ such that $\alpha(B) > 0$.*

Proof Because $P(B) \sim \mathrm{Be}(\alpha(B), \alpha(B^c))$, the condition that $\alpha(B) > 0$ implies that $P(B) > 0$ a.s., so that the conditional probabilities given B are well defined.

For given partitions $A_1, \ldots, A_r$ of B and $C_1, \ldots, C_s$ of B^c, the vector

$$X:= (P(A_1), \ldots, P(A_r), P(C_1), \ldots, P(C_s))$$

possesses a Dirichlet distribution $\mathrm{Dir}(r+s; \alpha(A_1), \ldots, \alpha(A_r), \alpha(C_1), \ldots, \alpha(C_s))$. By Proposition G.3 the four variables or vectors

$$Z_1:= \sum_{i=1}^{r} X_i, \qquad Z_2:= \sum_{i=r+1}^{r+s} X_i, \qquad \Big(\frac{X_1}{Z_1}, \ldots, \frac{X_r}{Z_1}\Big), \qquad \Big(\frac{X_{r+1}}{Z_2}, \ldots, \frac{X_{r+s}}{Z_2}\Big)$$

are mutually independent, and the two vectors have Dirichlet distributions with the restrictions of the original parameters. These are precisely the variables $P(B)$, $P(B^c)$ and vectors with coordinates $P_B(A_i)$ and $P_{B^c}(C_i)$. □

Theorem 4.5 shows that the Dirichlet process "localized" by conditioning to a set B is again a Dirichlet process, with base measure the restriction of the original base measure. Furthermore, processes at disjoint localities are independent of each other, and also independent of the "macro level" variable $P(B)$. Within any given locality, mass is further distributed according to a Dirichlet process, independent of what happens to the "outside world." This property may be expressed by saying that locally a Dirichlet process is like itself; in other words it is *self similar*.

Theorem 4.5 also shows that the Dirichlet process is a special Pólya tree process: for any sequence of successively refined binary partitions $\{A_0, A_1\}, \{A_{00}, A_{01}, A_{10}, A_{11}\}, \ldots$, the splitting variables $V_{\varepsilon 0} = P(A_{\varepsilon 0}|\, A_\varepsilon)$ satisfy

$$V_{\varepsilon 0} \sim \mathrm{Be}(\alpha(A_{\varepsilon 0}), \alpha(A_{\varepsilon 1})), \tag{4.9}$$

and all variables $V_{\varepsilon 0}$, for $\varepsilon \in \mathcal{E}^*$, are mutually independent. A special property of the Dirichlet process is that it is a Pólya tree for *any* sequence of partitions.

Another special property is that the parameters of the beta-distributions are additive in the sense that $\alpha(A_{\varepsilon 0}) + \alpha(A_{\varepsilon 1}) = \alpha(A_\varepsilon)$, for every $\varepsilon \in \mathcal{E}^*$. Generally in a Pólya tree, the probabilities $P(A_\varepsilon)$ are products of independent beta variables (cf. (3.12)). In a Dirichlet process these probabilities are themselves beta-distributed. These properties are closely related (see Problem G.4). The additivity of the parameters is actually a distinguishing property of the Dirichlet process within the class of Pólya trees, as shown in Section 4.4.

4.1.3 Conjugacy

One of the most remarkable properties of the Dirichlet process prior is that the posterior distribution is again a Dirichlet process. Consider observations $X_1, X_2, \ldots, X_n$ sampled independently from a distribution P that was drawn from a Dirichlet prior distribution. By an abuse of language, which we shall follow, such observations are often termed a *sample from the Dirichlet process*.

Theorem 4.6 (Conjugacy) *The* $\mathrm{DP}(\alpha + \sum_{i=1}^n \delta_{X_i})$*-process is a version of the posterior distribution given an i.i.d. sample* $X_1, \ldots, X_n$ *from the* $\mathrm{DP}(\alpha)$*-process.*

Proof Because the Dirichlet process is tail-free by Theorem 4.5, and a given measurable partition $\{A_1, \ldots, A_k\}$ of $\mathfrak{X}$ can be viewed as part of a sequence of successive binary partitions, the posterior distribution of the vector $(P(A_1), \ldots, P(A_k))$ given $X_1, \ldots, X_n$ is the same as the posterior distribution of this vector given the vector $N = (N_1, \ldots, N_k)$ of cell counts, defined by $N_j = \#(1 \le i \le n: X_i \in A_j)$. Given P the vector N possesses a multinomial distribution with parameter $(P(A_1), \ldots, P(A_k))$, which has a $\text{Dir}(k; \alpha(A_1), \ldots, \alpha(A_k))$ prior distribution. The posterior distribution can be obtained using Bayes's rule applied to these finite-dimensional vectors, as in Proposition G.8. □

Theorem 4.6 can be remembered as the updating rule $\alpha \mapsto \alpha + \sum_{i=1}^n \delta_{X_i}$ for the base measure of the Dirichlet distribution. In terms of the parameterization $\alpha \leftrightarrow (M = |\alpha|, \bar{\alpha})$ of the base measure, this rule takes the form

$$M \mapsto M + n \quad \text{and} \quad \bar{\alpha} \mapsto \frac{M}{M+n}\bar{\alpha} + \frac{n}{M+n}\mathbb{P}_n, \tag{4.10}$$

where $\mathbb{P}_n = n^{-1}\sum_{i=1}^n \delta_{X_i}$ is the empirical distribution of $X_1, \ldots, X_n$. Combining the theorem with Proposition 4.2, we see that

$$\text{E}(P(A)|\, X_1, \ldots, X_n) = \frac{|\alpha|}{|\alpha| + n}\bar{\alpha}(A) + \frac{n}{|\alpha| + n}\mathbb{P}_n(A). \tag{4.11}$$

Thus the posterior mean (the "Bayes estimator" of P) is a convex combination of the prior mean $\bar{\alpha}$ and the empirical distribution, with weights $M/(M+n)$ and $n/(M+n)$, respectively. For a given sample it is close to the prior mean if M is large, and close to the empirical distribution (which is based only on the data) if M is small relative to n. Thus M determines the extent to which the prior controls the posterior mean — a Dirichlet process prior with precision M contributes information equivalent to a sample of size M (although M is not restricted to integer values). This invites to view M as the *prior sample size*, or the "number of pre-experiment samples." In this interpretation the sum $M + n$ is the *posterior sample size*.

For a fixed prior (i.e. fixed M), the posterior mean (4.11) behaves asymptotically as $n \to \infty$ like the empirical distribution $\mathbb{P}_n$ to the order $O(n^{-1})$, a.s. Thus it possesses the same asymptotic properties regarding consistency (Glivenko-Cantelli theorem) and asymptotic normality (Donsker theorem) as the empirical distribution (see Appendix F). In particular, if $X_1, X_2, \ldots$ are sampled from a "true distribution" P_0, then the posterior mean will tend a.s. to P_0.

In addition the full posterior distribution will contract to its mean, whenever the posterior sample size tends to infinity. Indeed, by combining Theorem 4.6 and (4.3), we see, for $\tilde{\mathbb{P}}_n$ the posterior mean (4.11),

$$\text{var}\,(P(A)|\, X_1, \ldots, X_n) = \frac{\tilde{\mathbb{P}}_n(A)\tilde{\mathbb{P}}_n(A^c)}{1 + M + n} \le \frac{1}{4(1 + M + n)}. \tag{4.12}$$

In Sections 4.3.2 and 12.2 we shall strengthen this "setwise contraction" to contraction of the posterior distribution as a whole.

Consequently, if the data are sampled from a true distribution P_0, then the posterior distribution of P converges weakly to the measure degenerate at P_0. It is remarkable that this is true for *any* base measure α: the prior is not even required to "support the true value" P_0, in the sense of assigning positive probabilities to (weak) neighborhoods of P_0. (It should be noted that "the" posterior distribution is uniquely defined only up to the marginal distribution of the observations. The remarkable consistency is true for the particular version given in Theorem 4.6, not for *every* version; cf. Example 6.5.)

4.1.4 Marginal and Conditional Distributions

The joint distribution of a sample $X_1, X_2, \ldots$ from a Dirichlet process (i.e. $P \sim \mathrm{DP}(\alpha)$ and $X_1, X_2, \ldots \mid P \stackrel{\text{iid}}{\sim} P$) has a complicated structure, but can be conveniently described by its sequence of *predictive distributions*: the laws of X_1, $X_2 \mid X_1$, $X_3 \mid X_1, X_2$, etc.

Because $\mathrm{P}(X_1 \in A) = \mathrm{E}P(X_1 \in A \mid P) = \mathrm{E}P(A) = \bar\alpha(A)$, the marginal distribution of X_1 is $\bar\alpha$.

Because $X_2 \mid (P, X_1) \sim P$ and $P \mid X_1 \sim \mathrm{DP}(\alpha + \delta_{X_1})$, we can apply the same reasoning again, but now conditionally given X_1, to see that $X_2 \mid X_1$ follows the normalization of $\alpha + \delta_{X_1}$. This is a mixture of $\bar\alpha$ and δ_{X_1} with weights $|\alpha|/(|\alpha|+1)$ and $1/(|\alpha|+1)$, respectively.

Repeating this argument, using that $P \mid X_1, \ldots, X_{i-1} \sim \mathrm{DP}(\alpha + \sum_{j=1}^{i-1} \delta_{X_j})$, we find that

$$X_i \mid X_1, \ldots, X_{i-1} \sim \begin{cases} \delta_{X_1}, & \text{with probability } 1/(|\alpha| + i - 1), \\ \vdots & \qquad \vdots \\ \delta_{X_{i-1}}, & \text{with probability } 1/(|\alpha| + i - 1), \\ \bar\alpha, & \text{with probability } |\alpha|/(|\alpha| + i - 1). \end{cases} \tag{4.13}$$

Being a mixture of a product of identical distributions, the joint distribution of $X_1, X_2, \ldots$ is exchangeable, so re-labeling does not affect the structure of (4.13).

The recipe (4.13) is called the *generalized Pólya urn scheme*, and can be viewed as a continuous analog of the familiar Pólya urn scheme. Consider balls which can carry a continuum $\mathfrak{X}$ of "colors." Initially the "number of balls" is $M = |\alpha|$, which may be any positive number, and the colors are distributed according to $\bar\alpha$. We draw a ball from the collection, observe its color X_1, and return it to the urn along with an additional ball of the same color. The total number of balls is now $M + 1$, and the colors are distributed according to $(M\bar\alpha + \delta_{X_1})/(M + 1)$. We draw a ball from this updated urn, observe its color X_2, and return it to the urn along with an additional ball of the same color. The probability of picking up the ball that was added after the first draw is $1/(M+1)$, in which case $X_2 = X_1$; otherwise, with probability $M/(M+1)$, we make a fresh draw from the original urn. This process continues indefinitely, leading to the conditional distributions in (4.13).

Clearly the scheme (4.13) will produce ties in the values $X_1, \ldots, X_n$ with positive probability. The joint distribution of $(X_1, \ldots, X_n)$ is a mixture of distributions with components described by the collection of partitions $\{S_1, \ldots, S_k\}$ of the set of indices $\{1, 2, \ldots, n\}$. The coordinates X_i with index i belonging to the same partitioning set S_j are identical, with their common value distributed as $\bar\alpha$, independent of the other common values. For instance, if $n = 3$, then the partition $\{\{1, 2\}, \{3\}\}$ corresponds to a component that is supported on the "diagonal" $D_{12;3} = \{(x_1, x_2, x_3) \in \mathbb{R}^3 : x_1 = x_2\}$, which can be described as the distribution of a vector (Y_1, Y_2, Y_3) with $Y_1 = Y_2 \sim \bar\alpha$ independent of $Y_3 \sim \bar\alpha$.

The induced random partition of $\{1, 2, \ldots, n\}$ is of interest by itself, and known as the *Chinese restaurant process*. It is discussed in the next section and in Chapter 14.

The following result gives an expression for the mean of certain product functions. Let $\mathscr{S}_n$ stand for the collection of all partitions of $\{1, \ldots, n\}$, and write a partition as $\mathcal{S} = \{S_1, \ldots, S_N\}$, for $N = \#\mathcal{S}$ the number of partitioning sets; and let $s^{[n]} = s(s+1)\cdots(s+n-1)$ stand for the *ascending factorial*.

Proposition 4.7 *If $X_1, \ldots, X_n$ are a random sample from a Dirichlet process with base measure α, or equivalently arise from the generalized Pólya urn scheme* (4.13)*, then for any measurable functions $g_i\colon \mathscr{X} \to \mathbb{R}$ such that the right side is defined,*

$$\mathrm{E}\Big(\prod_{i=1}^n g_i(X_i)\Big) = \frac{1}{|\alpha|^{[n]}} \sum_{\mathcal{S}\in\mathscr{S}_n} \prod_{j=1}^{\#\mathcal{S}} \Big[(\#S_j - 1)! \int \prod_{l\in S_j} g_l(x)\, d\alpha(x)\Big]. \tag{4.14}$$

Proof By factorizing the joint distribution of $(X_1, \ldots, X_n)$ as the product of its predictive distributions, we can write the left side as

$$\int \cdots \int \prod_{i=1}^n g_i(x_i) \prod_{j=1}^n \frac{d(\alpha + \delta_{x_{j-1}} + \cdots + \delta_{x_1})(x_j)}{|\alpha| + j - 1}.$$

We next expand the second product, thus obtaining a sum of $n!$ terms, corresponding to choosing α, or one of $\delta_{x_{j-1}}, \ldots \delta_{x_1}$ in the jth term, for $j = 1, \ldots, n$. Every choice of a Dirac measure sets two coordinates equal, while every choice of α "opens" a partitioning set S_j. Thus the $n!$ terms of the sum can be grouped according to the partitions of $\{1, 2, \ldots, n\}$. That every partition appears as many times as claimed in the proposition is perhaps best proved by induction.

For $n = 1$ the proposition is true. By (4.13) the left side of the proposition for $n+1$ can be written as

$$\mathrm{E}\Big[\prod_{i=1}^n g_i(X_i) \int g_{n+1}(x) \frac{d(\alpha + \sum_{j=1}^n \delta_{X_j})(x)}{|\alpha| + n}\Big]$$
$$= \frac{1}{|\alpha| + n}\Big[\mathrm{E}\Big(\prod_{i=1}^n g_i(X_i)\Big) \int g_{n+1}\, d\alpha + \sum_{j=1}^n \mathrm{E}\Big(\prod_{i\neq j} g_i(X_i)\,(g_j g_{n+1})(X_j)\Big)\Big].$$

By the proposition for n we can replace both expectations by a sum over $\mathscr{S}_n$. The resulting expression can be rewritten as a sum over $\mathscr{S}_{n+1}$, where the first term delivers all partitions of $\{1, 2, \ldots, n+1\}$ with the one-point set $\{n+1\}$ as one of the partitioning sets and the second term the remaining partitions, where $n+1$ is joined to the partitioning set of the partition of $\{1, \ldots, n\}$ to which j belongs. ☐

4.1.5 Number of Distinct Values

In this section we investigate the number of distinct values in a random sample $X_1, \ldots, X_n$ from the Dirichlet process, and other aspects of the induced partitioning of the set $\{1, \ldots, n\}$. For simplicity we assume throughout that the base measure α is atomless, so that with probability one the ith value X_i in the Pólya scheme (4.13) is distinct from the previous $X_1, \ldots, X_{i-1}$ precisely if it is drawn from $\bar{\alpha}$.

For $i \in \mathbb{N}$ define $D_i = 1$ if the ith observation X_i is a "new value," i.e. if $X_i \notin \{X_1, \ldots, X_{i-1}\}$, and set $D_i = 0$ otherwise. Then $K_n = \sum_{i=1}^n D_i$ is the number of distinct values among the first n observations. Let $\mathcal{L}(K_n)$ stands for its law (distribution).

Proposition 4.8 (Distinct values) *If the base measure α is atomless with total mass $|\alpha| = M$, then the variables $D_1, D_2, \ldots$ are independent Bernoulli variables with success probabilities* $\mathrm{P}(D_i = 1) = M/(M + i - 1)$. *Consequently, for fixed M, as $n \to \infty$,*

(i) $\mathrm{E}(K_n) \sim M \log n \sim \mathrm{var}(K_n)$,
(ii) $K_n / \log n \to M$, a.s.,
(iii) $(K_n - \mathrm{E}K_n)/sd(K_n) \rightsquigarrow \mathrm{Nor}(0, 1)$,
(iv) $d_{TV}(\mathcal{L}(K_n), \mathrm{Poi}(\mathrm{E}(K_n))) = O(1/\log n)$.

Proof The first assertion follows because given $X_1, \ldots, X_{i-1}$ the variable X_i is "new" if and only if it is drawn from $\bar{\alpha}$, which happens with probability $M/(M + i - 1)$. Then assertion (i) can be derived from the exact formulas

$$\mathrm{E}(K_n) = \sum_{i=1}^n \frac{M}{M + i - 1}, \qquad \mathrm{var}(K_n) = \sum_{i=1}^n \frac{M(i-1)}{(M + i - 1)^2}.$$

Furthermore, assertion (ii) follows from Kolmogorov's strong law of large numbers for independent variables, since

$$\sum_{i=1}^\infty \frac{\mathrm{var}(D_i)}{(\log i)^2} = \sum_{i=1}^\infty \frac{M(i-1)}{(M + i - 1)^2 (\log i)^2} < \infty.$$

Next (iii) is a consequence of the Lindeberg central limit theorem. Finally for (iv) we apply the Chen-Stein approximation (see Chen 1975, Arratia et al. 1990, Chen et al. 2011) to the distribution of a sum of independent Bernoulli variables to see that

$$d_{TV}(\mathcal{L}(K_n), \mathrm{Poi}(\mathrm{E}(K_n))) \leq 2 \sum_{i=1}^n (\mathrm{E}(D_i))^2 \frac{1 - e^{-\mathrm{E}(K_n)}}{\mathrm{E}(K_n)}. \tag{4.15}$$

Here $\sum_{i=1}^n (\mathrm{E}(D_i))^2 = M^2 \sum_{i=1}^n (M + i - 1)^{-2}$ is bounded in n, whereas $\mathrm{E}(K_n)$ tends to infinity at order $\log n$, by (i). □

Thus the number of distinct values in a (large) sample from a distribution taken from a fixed Dirichlet prior is logarithmic in the sample size. Furthermore, the fluctuations of this number around its mean are of the order $\sqrt{\log n}$, and we might approximate its distribution by a Poisson.

The distribution of the number of distinct values depends on the base measure, but only through its prior strength $M = |\alpha|$. From the formula for the exact mean and variance, given in the preceding proof, we can derive that, for all M and n,

$$1 \vee M \log\Big(1 + \frac{n}{M}\Big) \leq \mathrm{E}(K_n) \leq 1 + M \log\Big(1 + \frac{n}{M}\Big),$$
$$\mathrm{var}(K_n) \leq M \log\Big(1 + \frac{n}{M}\Big).$$

Thus small M allows few distinct observations. As $M \to 0$, the mean number tends to one and the variance to zero, and there can be only one distinct observation. (See also Theorem 4.16 in this context.) That the posterior distribution of P is a $(M,n)/(M+n)$-mixture of the base measure and the empirical distribution motivated us earlier to think of M as the *prior strength*. However, that all observations must be identical is a strong prior opinion, and hence this interpretation may be inappropriate as $M \to 0$. Especially in density estimation and clustering using Dirichlet mixtures, the limit as $M \to 0$ should not be seen as rendering a "noninformative prior."

The following proposition gives the exact distribution of K_n. The sum in the formula for $C_n(k)$ involves $\binom{n-1}{k-1}$ terms, and hence computation could be demanding for large n.

Proposition 4.9 (Distinct values) *If the base measure α is atomless with total mass M, then*

$$\mathrm{P}(K_n = k) = C_n(k)\, n!\, M^k \frac{\Gamma(M)}{\Gamma(M+n)}, \tag{4.16}$$

where, for $k = 1, 2, \ldots,$

$$C_n(k) = \frac{1}{n!} \sum_{S\subset\{1,\ldots,n-1\}:|S|=n-k} \prod_{j\in S} j. \tag{4.17}$$

Proof The event $K_n = k$ can happen in two ways: $K_{n-1} = k-1$ and the nth observation is new, or $K_{n-1} = k$ and the nth observation is a previous observation. This gives the following recursion relation for $p_n(k, M) := \mathrm{P}(K_n = k|\, M)$, for $1 \le k \le n$:

$$p_n(k, M) = \frac{M}{M+n-1} p_{n-1}(k-1, M) + \frac{n-1}{M+n-1} p_{n-1}(k, M). \tag{4.18}$$

For $C_n(k) := p_n(k, 1)$ this recurrence relation (with $M = 1$) implies

$$C_n(k) = \frac{1}{n} C_{n-1}(k-1) + \frac{n-1}{n} C_{n-1}(k). \tag{4.19}$$

We shall first verify (4.16) with $C_n(k) = p_n(k, 1)$, and next derive (4.17).

We prove (4.16) (for every k) by induction on n. For $n = 1$ this equation reduces to $p_1(1, M) = C_1(1)$ for $k = 1$, which is true by definition, and to $0 = 0$ for other k. Assuming the result for $n-1$, we can apply (4.18) to express $p_n(k, M)$ into $p_{n-1}(k-1, M)$ and $p_{n-1}(k, M)$, and next (4.16) for $n-1$ to express the latter two in $C_{n-1}(k-1)$ and $C_{n-1}(k)$. With some algebra and (4.19) this can next be reduced to the right side of (4.16) for n.

In terms of the generating function $A_n(s) = \sum_{k=1}^{\infty} C_n(k) s^k$ of the sequence $C_n(k)$ (which is actually a polynomial of finite degree), the recurrence (4.19) can be written $A_n(s) = n^{-1}(sA_{n-1}(s) + (n-1)A_{n-1}(s)) = ((s+n-1)/n)\, A_{n-1}(s)$. Repeatedly applying this and using the fact that $A_1(s) = s$, we see that

$$A_n(s) = A_1(s) \prod_{j=2}^{n} \frac{s+j-1}{j} = \frac{1}{n!} \prod_{j=0}^{n-1} (s+j).$$

Expanding the product and collecting the coefficients of s^k, we obtain (4.17). ∎

We say that $X_1, \dots, X_n$ possesses *multiplicity class* $C(m_1, \dots, m_n)$ if in the set $\{X_1, \dots, X_n\}$, m_1 values appear exactly once, m_2 (different) values appear twice, and so on. Here $m_1, \dots, m_n$ are nonnegative integers with $m_1 + 2m_2 + \cdots + nm_n = n$ (and many entries of $m_1, \dots, m_n$ will be zero). Let $s^{[n]} = s(s+1)\cdots(s+n-1)$ stand for the ascending factorial. *Ewens's sampling formula* gives the probability that a sample from the Dirichlet process has a given multiplicity class.

Proposition 4.10 (Ewens's sampling formula) *A random sample $X_1, \dots, X_n$ from a Dirichlet process with nonatomic base measure of strength $|\alpha| = M$ possesses multiplicity class $C(m_1, \dots, m_n)$ with probability equal to*

$$\frac{n!}{M^{[n]}} \prod_{i=1}^{n} \frac{M^{m_i}}{i^{m_i} m_i!}.$$

Proof Consider the event that $C(m_1, \dots, m_n)$ occurs with the m_1 values of multiplicity 1 appearing first (so $X_1, \dots, X_{m_1}$ are distinct and occur once in the full sample), with the m_2 pairs of multiplicity 2 appearing next (so $X_{m_1+1} = X_{m_1+2} \neq \cdots \neq X_{m_1+2m_2-1} = X_{m_1+2m_2}$), followed by the m_3 triples of multiplicity 3 etc. From the Pólya scheme (4.13) it is straightforward to see that the probability of this event is given by

$$\frac{\prod_{i=1}^{n} ((i-1)!)^{m_i} M^{\sum_{i=1}^{n} m_i}}{M^{[n]}}.$$

Because the vector $(X_1, \dots, X_n)$ is exchangeable, the probability that it possesses multiplicity class $C(m_1, \dots, m_n)$ is the probability in the display multiplied by the number of configurations of distinct and equal values that give this multiplicity class. Here, "configuration" refers to the pattern of equal and distinct values, without taking the values into account.

We can represent the m_1 values of multiplicity 1 by the symbols $1^1, 1^2, \dots, 1^{m_1}$, the m_2 pairs of multiplicity 2 by $2^1, 2^1, 2^2, 2^2, \dots, 2^{m_2}, 2^{m_2}$, the m_3 triplets of multiplicity 3 by $3^1, 3^1, 3^1, 3^2, \dots, 3^{m_3}, 3^{m_3}, 3^{m_3}$, etc. The configurations can then be identified with the permutations of these n symbols, where permutations that are identical after swapping superscripts within a multiplicity level are considered equivalent. There are $n!$ possible orderings if the n symbols are all considered different, but for every i the m_i groups of multiplicity i (given by different superscripts) can be rearranged in $m_i!$ orders, and every group of multiplicity i (of identical symbols, e.g. the group $3^1, 3^1, 3^1$) can be internally permuted in $i!$ ways. Consequently the total number of configurations is

$$\frac{n!}{\prod_{i=1}^{n}(m_i!(i!)^{m_i})}.$$

Multiplying the two preceding displays and using the relation $(i-1)!/i! = 1/i$, we obtain the result. ☐

A random sample $X_1, \dots, X_n$ from the Dirichlet process also induces a nontrivial random partition of $\{1, 2, \dots, n\}$, corresponding to the pattern of ties. (More precisely, we consider the equivalence classes under the relation $i \equiv j$ if and only if $X_i = X_j$.) In the following proposition we obtain its distribution. In Chapter 14 we return to this topic and discuss the distribution of partitions induced by more general processes.

Proposition 4.11 (Dirichlet partition) *A random sample $X_1, \dots, X_n$ from a Dirichlet process with atomless base measure of strength $|\alpha| = M$ induces a given partition of $\{1, 2, \dots, n\}$ into k sets of sizes $n_1, \dots, n_k$ with probability equal to*

$$\frac{M^k \Gamma(M) \prod_{j=1}^k \Gamma(n_j)}{\Gamma(M+n)}. \tag{4.20}$$

Proof By exchangeability the probability depends on the sizes of the partitioning sets only. The probability that the partitioning set of size n_1 consists of the first n_1 variables, the one of size n_2 of the next n_2 variables, etc. can be obtained by multiplying the appropriate conditional probabilities for the consecutive draws in the Pólya urn scheme in their natural order of occurrence. For $\bar{n}_j = \sum_{l=1}^j n_l$, it is given by

$$\frac{M}{M}\frac{1}{M+1}\cdots\frac{n_1-1}{M+n_1-1} \times \cdots \times \frac{M}{M+\bar{n}_{k-1}}\frac{1}{M+\bar{n}_{k-1}+1}\cdots\frac{n_k-1}{M+\bar{n}_{k-1}+n_k-1}.$$

This can be rewritten as in the proposition. □

4.2 Constructions

We describe several methods for constructing the Dirichlet process. The first construction, via a stochastic process, is sufficient to prove existence, but the other methods give additional characterizations.

4.2.1 Construction via a Stochastic Process

Definition 4.1 specifies the joint distribution of the vector $(P(A_1), \dots, P(A_k))$, for any measurable partition $\{A_1, \dots, A_k\}$ of the sample space. In particular, it specifies the distribution of $P(A)$, for every measurable set A, and hence the *mean measure* $A \mapsto \mathrm{E}(P(A))$. By Proposition 4.2, this is the center measure $\bar{\alpha}$, which is a valid Borel measure by assumption. Therefore Theorem 3.1 implies existence of the Dirichlet process $\mathrm{DP}(\alpha)$, provided the specification of distributions can be consistently extended to any vector of the type $(P(A_1), \dots, P(A_k))$, for arbitrary measurable sets and not just partitions, in such a way that it gives a finitely-additive measure.

An arbitrary collection $A_1, \dots, A_k$ of measurable sets defines a collection of 2^k atoms of the form $A_1^* \cap A_2^* \cap \cdots \cap A_k^*$, where A^* stands for A or A^c. These atoms $\{B_j : j = 1, \dots, 2^k\}$ (some of which may be empty) form a partition of the sample space, and hence the joint distribution of $(P(B_j): j = 1, \dots, 2^k)$ is defined by Definition 4.1. Every A_i can be written as a union of atoms, and $P(A_i)$ can be defined accordingly as the sum of the corresponding $P(B_j)$s. This defines the distribution of the vector $(P(A_1), \dots, P(A_k))$.

To prove the existence of a stochastic process $(P(A): A \in \mathscr{X})$ that possesses these marginal distributions, it suffices to verify that this collection of marginal distributions is consistent in the sense of Kolmogorov's extension theorem. Consider the distribution of the vector $(P(A_1), \ldots, P(A_{k-1}))$. This has been defined using the coarser partitioning in the 2^{k-1} sets of the form $A_1^* \cap A_2^* \cap \cdots \cap A_{k-1}^*$. Every set in this coarser partition is a union of two sets in the finer partition used previously to define the distribution of $(P(A_1), \ldots, P(A_k))$. Therefore, consistency pertains if the distributions specified by Definition 4.1 for two partitions, where one is finer than the other, are consistent.

Let $\{A_1, \ldots, A_k\}$ be a measurable partition and let $\{A_{i1}, A_{i2}\}$ be a further measurable partition of A_i, for $i = 1, \ldots, k$. Then Definition 4.1 specifies that

$$(P(A_{11}), P(A_{12}), P(A_{21}), \ldots, P(A_{k1}), P(A_{k2})) \\ \sim \mathrm{Dir}(2k; \alpha(A_{11}), \alpha(A_{12}), \alpha(A_{21}), \ldots, \alpha(A_{k1}), \alpha(A_{k2})).$$

In view of the group additivity of finite dimensional Dirichlet distributions given by Proposition G.3, this implies

$$\Big(\sum_{j=1}^{2} P(A_{1j}), \ldots, \sum_{j=1}^{2} P(A_{kj})\Big) \sim \mathrm{Dir}\Big(k; \sum_{j=1}^{2} \alpha(A_{1j}), \ldots, \sum_{j=1}^{2} \alpha(A_{kj})\Big).$$

Consistency follows as the right side is $\mathrm{Dir}(k; \alpha(A_1), \ldots, \alpha(A_k))$, since α is a measure.

That $P(\varnothing) = 0$ and $P(\mathfrak{X}) = 1$ almost surely follow from the fact that $\{\varnothing, \mathfrak{X}\}$ is an eligible partition in Definition 4.1, whence $(P(\varnothing), P(\mathfrak{X})) \sim \mathrm{Dir}(2; 0, |\alpha|)$ by (4.1). That $P(A_1 \cup A_2) = P(A_1) + P(A_2)$ almost surely for every disjoint pair of measurable sets A_1, A_2, follows similarly from consideration of the distributions of the vectors $(P(A_1), P(A_2), P(A_1^c \cap A_2^c))$ and $(P(A_1 \cup A_2), P(A_1^c \cap A_2^c))$, whose three and two components both add up to 1.

We have proved the existence of the Dirichlet process distribution $\mathrm{DP}(\alpha)$ for every Polish sample space and every base measure α.

4.2.2 Construction through Distribution Function

When $\mathfrak{X} = \mathbb{R}$ we can also construct the Dirichlet process through its cumulative distribution function, following the scheme of Proposition 2.3.

For any finite set $-\infty < t_1 < t_2 < \cdots < t_k = \infty$ of rational numbers, let

$$(F(t_1), F(t_2) - F(t_1), \ldots, 1 - F(t_{k-1})) \sim \mathrm{Dir}(k + 1, MG(t_1), \\ MG(t_2) - MG(t_1), \ldots, M - MG(t_{k-1})),$$

in agreement with (4.1). The resulting distributions of $(F(t_1), F(t_2), \ldots, F(t_k))$ can be checked to form a consistent system in the sense of (i) of Proposition 2.3, and (ii) of the proposition is satisfied, since the increments, which are Dirichlet distributed, are nonnegative.

For given $s_n \downarrow s$, the distribution of $F(s_n)$ is $\mathrm{Be}(MG(s_n), M - MG(s_n))$ and converges weakly to $\mathrm{Be}(MG(s), M - MG(s))$, which is the distribution of $F(s)$. Finally to verify (iv) of Proposition 2.3, we note that $0 \leq \mathrm{E}(F(s)) = G(s) \to 0$ as $s \downarrow -\infty$, while $1 \geq \mathrm{E}(F(s)) = G(s) \to 1$ as $s \uparrow \infty$.

4.2.3 Construction through a Gamma Process

The *gamma process* (see Appendix J) is a random process $(S(u): u \geq 0)$ with nondecreasing, right continuous sample paths $u \mapsto S(u)$ and independent increments satisfying $S(u_2) - S(u_1) \sim \mathrm{Ga}(u_2 - u_1, 1)$, for $u_1 < u_2$. Thus, for given $M > 0$ and cumulative distribution function G on $\mathfrak{X} = \mathbb{R}$, we can define a random distribution function by $F(t) = S(MG(t))/S(M)$.

The corresponding distribution is the DP(MG)-distribution. Indeed, for any partition $-\infty = t_0 < t_1 < \cdots < t_k = \infty$, we have $S(MG(t_i)) - S(MG(t_{i-1})) \overset{\text{ind}}{\sim} \mathrm{Ga}(MG(t_i) - MG(t_{i-1}), 1)$, whence $(F(t_1) - F(t_0), \ldots, F(t_k) - F(t_{k-1})) \sim \mathrm{Dir}(k; MG(t_1) - MG(t_0), \ldots, MG(t_k) - MG(t_{k-1}))$-distribution, by Proposition G.2. It follows that (4.1) holds for any partition in intervals. As intervals form a measure-determining class, the corresponding measure is the Dirichlet process (cf. Proposition A.5).

Rather than only the distribution function, we can also construct the full measure directly by viewing the gamma process as a *completely random measure*, as described in Appendix J. This is a random measure $(U(A): A \in \mathscr{X})$ such that $U(A_i) \overset{\text{ind}}{\sim} \mathrm{Ga}(\alpha(A_i), 1)$, for every measurable partition $\{A_1, \ldots, A_k\}$. Then $P(A) = U(A)/U(\mathfrak{X})$ is a random probability measure that satisfies (4.1), in view of Proposition G.2.

The construction using a completely random measure works not only for the positive real line, but for every Polish sample space $\mathfrak{X}$.

4.2.4 Construction through Pólya Urn Scheme

A *Pólya sequence* with parameter α is a sequence of random variables $X_1, X_2, \ldots$ whose joint distribution satisfies

$$X_1 \sim \bar{\alpha}, \qquad X_{n+1} | X_1, \ldots, X_n \sim \frac{\alpha + \sum_{i=1}^n \delta_{X_i}}{|\alpha| + n}, \qquad n \geq 1.$$

In Section 4.1.4 a random sample from a distribution generated from a Dirichlet process with base measure α is shown to be a Pólya sequence. Such a sequence can of course also be constructed without reference to the Dirichlet process prior. In this section we reverse the construction, and obtain the Dirichlet process from the Pólya urn sequence, invoking *de Finetti's theorem* (see Schervish 1995, Theorem 1.49).

It can be verified that a Pólya sequence $X_1, X_2, \ldots$ is exchangeable. Therefore, by de Finetti's theorem, there exists a random probability measure P such that $X_i | P \overset{\text{iid}}{\sim} P$, $i = 1, 2, \ldots$. We claim that (the law of) P is the DP(α)-process.

In fact, given existence of the Dirichlet process, we can refer to Section 4.1.4 to see that the Dirichlet process generates a Pólya sequence. Because the random measure given by de Finetti's theorem is unique, there is nothing more to prove.

A proof that does not appeal to an existence result needs further argumentation. It suffices to show that the joint distribution of $(P(A_1), \ldots, P(A_k))$ for any measurable partition $\{A_1, \ldots, A_k\}$, where P is the de Finetti measure, satisfies (4.1). These are the almost sure limits of the sequences of the empirical marginals $(\mathbb{P}_n(A_1), \ldots, \mathbb{P}_n(A_k))$. Define a sequence $Y_1, Y_2, \ldots$ of variables that register the partitioning sets containing $X_1, X_2, \ldots$, by $Y_i = j$ if $X_i \in A_j$. These variables take their values in the finite set $\{1, \ldots, k\}$ and can be seen to form a Pólya sequence themselves, with parameter β given by $\beta\{j\} = \alpha(A_j)$. Because the existence of the finite-dimensional Dirichlet process $\mathrm{DP}(\beta)$ is clear, we can use the argument of the preceding paragraph to see that the de Finetti measure Q of the exchangeable sequence $Y_1, Y_2, \ldots$ satisfies $Q \sim \mathrm{DP}(\beta)$. Now $\mathbb{P}_n(A_j) = \mathbb{Q}_n\{j\} \to Q\{j\}$, almost surely, and hence $(P(A_1), \ldots, P(A_k)) = (Q\{1\}, \ldots, Q\{k\})$ possesses the correct Dirichlet distribution.

4.2.5 Stick-Breaking Representation

The *stick-breaking representation* of a Dirichlet process expresses it as a random discrete measure of the type discussed in Section 3.4.2, with stick-breaking weights, as in Section 3.3.2, based on the beta-distribution. The random support points are generated from the center measure.

The representation gives an easy method to simulate a Dirichlet process, at least approximately.

Even though the representation is explicit and constructive, the proof given below that it gives the Dirichlet process presumes existence of the Dirichlet process.

Theorem 4.12 (Sethuraman) *If* $\theta_1, \theta_2, \ldots \stackrel{iid}{\sim} \bar{\alpha}$ *and* $V_1, V_2, \ldots \stackrel{iid}{\sim} \mathrm{Be}(1, M)$ *are independent random variables and* $W_j = V_j \prod_{l=1}^{j-1}(1 - V_l)$, *then* $\sum_{j=1}^{\infty} W_j \delta_{\theta_j} \sim \mathrm{DP}(M\bar{\alpha})$.

Proof Because $\mathrm{E}(\prod_{l=1}^{j}(1 - V_l)) = (M/(M+1))^j \to 0$, the stick-breaking weights W_j form a probability vector a.s. (c.f. Lemma 3.4), so that P is a probability measure a.s.

For $j \geq 1$ define $W'_j = V_{j+1} \prod_{l=2}^{j}(1 - V_l)$ and $\theta'_j = \theta_{j+1}$. Then $W_{j+1} = (1 - V_1) W'_j$ for every $j \geq 1$ and hence

$$P := W_1 \delta_{\theta_1} + \sum_{j=2}^{\infty} W_j \delta_{\theta_j} = V_1 \delta_{\theta_1} + (1 - V_1) \sum_{j=1}^{\infty} W'_j \delta_{\theta'_j}.$$

The random measure $P' := \sum_{j=1}^{\infty} W'_j \delta_{\theta'_j}$ has exactly the same structure as P, and hence possesses the same distribution. Furthermore, it is independent of (V_1, θ_1).

We conclude that P satisfies the distributional equation (4.21) given below, and the theorem follows from Lemma 4.13. □

The distributional equation for the Dirichlet process used in the preceding proof is of independent interest, and will be a valuable tool later on. For independent random variables $V \sim \mathrm{Be}(1, |\alpha|)$ and $\theta \sim \bar{\alpha}$, consider the equation

$$P =_d V \delta_\theta + (1 - V) P. \tag{4.21}$$

We say that a random measure P that is independent of (V, θ) is a solution to equation (4.21) if for every measurable partition $\{A_1, \ldots, A_k\}$ of the sample space the random vectors obtained by evaluating the random measures on its left and right sides are equal in distribution in $\mathbb{R}^k$[2].

[2] By Proposition A.5 this is equivalent to the random measures on either side being equal in distribution as random elements in $\mathfrak{M}$, justifying the notation $=_d$.

Lemma 4.13 *For given independent $\theta \sim \bar{\alpha}$ and $V \sim \mathrm{Be}(1, |\alpha|)$, the Dirichlet process* $\mathrm{DP}(\alpha)$ *is the unique solution of the distributional equation* (4.21).

Proof That the Dirichlet process is a solution is immediate from (4.1) and Proposition G.11. Uniqueness is a consequence of Lemma L.2. □

4.3 Further Properties

4.3.1 Discreteness and Support

A realization from the Dirichlet process is discrete with probability one, also when the base measure is absolutely continuous. This is perhaps disappointing, especially if the intention is to model absolutely continuous probability measures. In particular, the Dirichlet process cannot be used as a prior for density estimation.

Theorem 4.14 (Discreteness) *Almost every realization from the* $\mathrm{DP}(\alpha)$ *is a discrete measure:* $\mathrm{DP}_\alpha(P \in \mathfrak{M}\colon P \text{ is discrete}) = 1$.

Proof The event in the theorem is a measurable subset of $\mathfrak{M}$, by Proposition A.7.

The stick-breaking representation exhibits the Dirichlet distribution explicitly as a random discrete measure. The representation through the gamma process also gives a direct proof of the theorem, as the gamma process increases only by jumps (see Appendix J).

We present a third proof based on the posterior updating formula for the Dirichlet process. A measure P is discrete if and only if $P\{x\colon P\{x\} > 0\} = 1$. If $P \sim \mathrm{DP}(\alpha)$ and $X|\, P \sim P$, then

$$\mathrm{P}(P\{X\} > 0) = \int P\{x\colon P\{x\} > 0\}\, d\mathrm{DP}_\alpha(P).$$

Hence $\mathrm{DP}(\alpha)$ gives probability one to the discrete measures if and only if $\mathrm{P}(P\{X\} > 0) = 1$. By conditioning in the other direction, we see that this is (certainly) the case if $\mathrm{P}(P\{X\} > 0|\, X = x) = 1$ for every x. Because the posterior distribution of P given $X = x$ is $\mathrm{DP}(\alpha + \delta_x)$, given $X = x$ the variable $P\{X\}$ possesses a $\mathrm{Be}(\alpha\{x\} + 1, \alpha(\{x\}^c))$-distribution, and hence is a.s. positive. □

Notwithstanding its discreteness, the support of the Dirichlet process relative to the weak topology is typically very large. It is the whole of $\mathfrak{M}$ as soon as the base measure is fully supported in $\mathfrak{X}$.

Theorem 4.15 (Support) *The weak support of* $\mathrm{DP}(\alpha)$ *is the set* $\{P \in \mathfrak{M}\colon \mathrm{supp}(P) \subset \mathrm{supp}(\alpha)\}$. *Further if $\mathfrak{X}$ is a Euclidean space and P_0 is atomless and belongs to the weak support of* $\mathrm{DP}(\alpha)$, *then it also belongs to the Kolmogorov-Smirnov support.*

Proof The set $\mathcal{H} := \{P\colon \mathrm{supp}(P) \subset \mathrm{supp}(\alpha)\}$ is weakly closed. Indeed, if $P_n \rightsquigarrow P$, then $P(F) \geq \limsup_{n\to\infty} P_n(F)$, for every closed set, by the Portmanteau theorem. For $F = \mathrm{supp}(\alpha)$ and $P_n \in \mathcal{H}$, we have $P_n(F) = 1$ and hence $P(F) = 1$. This shows that $\mathrm{supp}(P) \subset F$, and hence $P \in \mathcal{H}$.

We now first show that supp(DP(α)) $\subset \mathcal{H}$. For F = supp(α) we have $\alpha(F^c) = 0$, and hence a random measure P with $P \sim$ DP(α) satisfies $P(F^c) \sim \text{Be}(0, \alpha(F))$, whence $P(F^c) = 0$ a.s. This shows that $P \in \mathcal{H}$ a.s., so that $\text{DP}_\alpha(\mathcal{H}) = 1$, and hence supp(DP($\alpha$)) $\subset \mathcal{H}$.

To prove the converse inclusion $\mathcal{H} \subset$ supp(DP(α)), we must show that $\text{DP}_\alpha(\mathcal{U}) > 0$ for every weakly open neighborhood $\mathcal{U}$ of any $P_0 \in \mathcal{H}$. Let $\mathscr{E}$ be the collection of all finite unions of sets from a countable base for the open sets in $\mathfrak{X}$. If $\mathscr{E}$ is enumerated arbitrarily as $\mathscr{E} = \{E_1, E_2, \ldots\}$, then $d(P, Q) = \sum_i 2^{-i}\big|P(E_i) - Q(E_i)\big|$ is a well-defined metric, which generates a topology that is stronger than the weak topology. Indeed, suppose that $P_n(E) \to P(E)$, for every $E \in \mathscr{E}$. For every open set G in $\mathfrak{X}$, there exists a sequence E^m in $\mathscr{E}$ with $E^m \uparrow G$, and hence $\liminf P_n(G) \geq \liminf P_n(E^m) = P(E^m)$, for every m. By the Portmanteau theorem, this implies that $P_n \rightsquigarrow P$. Because from all balls around a point, at most countably many can be *dis*continuity sets for a given measure P_0, we can also ensure that all sets in the collection $\mathcal{E}$ are P_0-continuity sets.

Thus it suffices to show that $\text{DP}_\alpha(P\colon d(P, P_0) < \delta) > 0$ for every $\delta > 0$, or equivalently that $\text{DP}_\alpha(P\colon \cap_{i=1}^k |P(E_i) - P_0(E_i)| < \delta) > 0$ for every finite collection $E_1, \ldots, E_k \in \mathscr{E}$ and $\delta > 0$. This certainly follows if the corresponding statement is true for the set of all intersections $F_1 \cap \cdots \cap F_k$, where $F_i \in \{E_i, E_i^c\}$, whence we may assume that $E_1, \ldots, E_k$ form a partition of $\mathfrak{X}$, so that $(P(E_1), \ldots, P(E_k)) \sim \text{Dir}(k; \alpha(E_1), \ldots, \alpha(E_k))$. If the vector $(\bar\alpha(E_1), \ldots, \bar\alpha(E_k))$ is in the interior of the simplex, then the corresponding Dirichlet distribution has full support, and hence it charges any neighborhood of $(P_0(E_1), \ldots, P_0(E_k))$. The cases that $\alpha(E_i) = 0$ or $\alpha(E_i^c) = 0$ for some i, when $P(E_i) = 0$ or $P(E_i) = 1$ a.s. under DP(α), must be considered separately.

If $\alpha(E) = 0$, then certainly $\alpha(E^o) = 0$, for E^o the interior of E, and hence $E^o \subset \text{supp}(\alpha)^c \subset \{P_0\}^c$, if $P_0 \in \mathcal{H}$. Thus $0 = P_0(E^o) = P_0(E)$, if E is a P_0 continuity set, so that $P(E) = P_0(E)$ almost surely under DP(α), and hence $\text{DP}_\alpha(|P(E) - P_0(E)| < \delta) = 1 > 0$. We may apply this with both E_i and E_i^c to remove those partitioning sets with α-measure 0. □

4.3.2 Convergence

The Dirichlet process depends continuously on its parameter relative to the weak topology. The following theorem also gives weak limits if the total mass of the base measure tends to 0 or ∞. In the latter case the "prior precision" increases indefinitely and the Dirichet process collapses to a degenerate measure at its mean measure.

Theorem 4.16 *Let α_m be a sequence of finite measures on $\mathfrak{X}$ such that $\bar\alpha_m \rightsquigarrow \bar\alpha$ for some Borel probability measure $\bar\alpha$ on $\mathfrak{X}$.*

(i) *If $|\alpha_m| \to 0$, then* $\text{DP}(\alpha_m) \rightsquigarrow \delta_X$, *for $X \sim \bar\alpha$.*
(ii) *If $|\alpha_m| \to M \in (0, \infty)$, then* $\text{DP}(\alpha_m) \rightsquigarrow \text{DP}(\alpha)$, *where $\alpha = M\bar\alpha$.*
(iii) *If $|\alpha_m| \to \infty$, then* $\text{DP}(\alpha_m) \rightsquigarrow \delta_{\bar\alpha}$, *the distribution degenerate at $\bar\alpha$.*

If, moreover, $\bar\alpha_m(A) \to \bar\alpha(A)$ for every Borel set A, then $\int \psi\, dP_m \rightsquigarrow \int \psi\, dP$ for every Borel measurable function $\psi\colon \mathfrak{X} \to \mathbb{R}$ that is uniformly integrable relative to α_m, where

$P_m \sim \mathrm{DP}(\alpha_m)$ and P is a random measure distributed according to the limit distribution given in each case (i)–(iii).

Proof By Theorem A.6 a sequence of random measures on $(\mathfrak{M}, \mathscr{M})$ is tight if and only if the corresponding sequence of mean measures on $(\mathfrak{X}, \mathscr{X})$ is tight. Since the sequence of mean measures $\bar{\alpha}_m$ of the given Dirichlet processes is weakly convergent by assumption, it is tight by Prohorov's theorem. Hence, by Prohorov's theorem in the other direction, it suffices to identify the weak limit point(s) of the sequence $\mathrm{DP}(\alpha_m)$ as the one given, in each case (i)–(iii).

If the random measure Q is distributed as a weak limit point, then by the continuous mapping theorem $\int \psi\, dQ$ is a weak limit of the sequence $\int \psi\, dP_m$, for $P_m \sim \mathrm{DP}(\alpha_m)$ and every continuous $\psi: \mathfrak{X} \to [0, 1]$. We shall show in each of the cases (i)–(iii) that $\int \psi\, dQ =_d \int \psi\, dP$, for P a random measure with the limit distribution claimed in (i)–(iii). By Proposition A.5 this implies that P and Q are equal in distribution, and the proof is complete.

A given continuous $\psi: \mathfrak{X} \to [0, 1]$ can be approximated from below by simple functions $-1 \leq \psi_k \uparrow \psi$. Without loss of generality these can be chosen of the form $\sum_{i=1}^k a_i \mathbb{1}\{A_i\}$, for constants $a_i \geq -1$ and $\{A_1, \ldots, A_k\}$ a measurable partition of $\mathfrak{X}$ into α-continuity sets. (For instance $\sum_i \xi_{i-1,k} \mathbb{1}\{\xi_{i-1,k} < \psi \leq \xi_{i,k}\}$, for $-1 < \xi_{1,k} < \xi_{2,k} < \cdots < \xi_{k.k} < 1$ a grid with meshwidth tending to zero and such that every set $\{x: \psi(x) = \xi_{i,k}\}$ is an α-continuity set. By the continuity of ψ the boundary of a set $\{x: a < \psi(x) \leq b\}$ is contained in the set $\{x: \psi(x) = a\} \cup \{x: \psi(x) = b\}$. Because sets of the latter form are disjoint for different a or b, at most countably many b can fail to give a continuity set, and hence a grid $\{\xi_{i,k}\}$ exists.) Suppose that we can show that, as $m \to \infty$, for every partition in α-continuity sets,

$$(P_m(A_1), \ldots, P_m(A_k)) \rightsquigarrow (P(A_1), \ldots, P(A_k)). \tag{4.22}$$

Then $\int \psi_k\, dP_m \rightsquigarrow \int \psi_k\, dP$, by the continuous mapping theorem. Because $\int \psi\, dQ$ is the weak limit of $\int \psi\, dP_m \geq \int \psi_k\, dP_m$, it follows that $\int \psi\, dQ \geq_s \int \psi_k\, dP$ in the sense of stochastic ordering (e.g. by ordering of cumulative distribution functions), for every k. Letting $k \to \infty$, we conclude that $\int \psi\, dQ$ is stochastically larger than $\int \psi\, dP$. By applying the argument also to $1 - \psi$, we see that the two variables are equal in distribution. It remains to prove (4.22) in each of the three cases.

Proof of (ii). The vector on the left of (4.22) is $\mathrm{Dir}(k; \alpha_m(A_1), \ldots, \alpha_m(A_k))$-distributed. Because $\alpha_m(A_j) \to \alpha(A_j)$ for every j by assumption and construction, these distributions tend weakly to $\mathrm{Dir}(k; \alpha(A_1), \ldots, \alpha(A_k))$ by Proposition G.6, which is the law of the right side of (4.22), if $P \sim \mathrm{DP}(\alpha)$.

Proof of (i) and (iii). We prove (4.22) by verifying the convergence of all (mixed) moments. By Corollary G.4, for any nonnegative integers $r_1, \ldots, r_k$ and $u^{[r]}$ the ascending factorial,

$$\mathrm{E}[P_m(A_1)^{r_1} \times \cdots \times P_m(A_k)^{r_k}] = \frac{\alpha_m(A_1)^{[r_1]} \times \cdots \times \alpha_m(A_k)^{[r_k]}}{|\alpha_m|^{[\sum_i r_i]}}.$$

If $\bar{\alpha}_m(A_j) \to \bar{\alpha}(A_j)$ and $|\alpha_m| \to \infty$ as in (iii), then this tends to $\bar{\alpha}(A_1)^{r_1} \cdots \bar{\alpha}(A_k)^{r_k}$, which is the corresponding mixed moment when P is identically $\bar{\alpha}$. If $|\alpha_m| \to 0$ as in (i), then there are two cases. If exactly one $r_j \neq 0$, then the moment tends to $\bar{\alpha}(A_j)$, which is equal to $E(P(A_j)^{r_j}) = E[\delta_X(A_j)^{r_j}] = \bar\alpha(A_j)$, if $X \sim \bar\alpha$. On the other hand, if at least two of $r_1, \ldots, r_k$ are nonzero, then the mixed moment is zero under the limit measure δ_X, as $P(A_j) = \delta_X(A_j)$ is nonzero for exactly one j, and the mixed moments in the preceding display also tend to zero, as at least two factors in the numerator tend to zero, while only $|\alpha_m|$ does so in the denominator and is bigger; all other factors tend to positive integers.

For the proof of the final assertion consider first a bounded, measurable function ψ, and let L be a weak limit point of the sequence $\int \psi\, dP_m$. The function ψ can be approximated from below by a sequence of indicator functions $0 \le \psi_k \uparrow \psi$. Under the additional assumption of convergence of $P_m(A)$ for every Borel set, the convergence (4.22) is true for any measurable partition $\{A_1, \ldots, A_k\}$. Consequently $\int \psi_k\, dP_m \rightsquigarrow \int \psi_k\, dP$, for every k. Arguing as before, we find that $L \ge_s \int \psi\, dP$. Applying the argument also to $1-\psi$, we see that $L =_d \int \psi\, dP$, and the proof for a bounded ψ is complete. A general measurable ψ is the sum of the bounded function $\psi \mathbb{1}\{|\psi| \le M\}$ and the function $\psi \mathbb{1}\{|\psi| > M\}$. Here $E \int |\psi| \mathbb{1}\{|\psi| > M\}\, dP_m = \int |\psi| \mathbb{1}\{|\psi| > M\}\, d\alpha_m$ can be made arbitrarily small by choosing large M, uniformly in m. □

As a corollary, the above theorem implies limiting behavior of the posterior of a Dirichlet process in two different asymptotic scenarios – as the sample size goes to infinity and in the noninformative limit as the prior sample size goes to zero. Recall that the posterior distribution based on a random sample $X_1, \ldots, X_n$ from a distribution drawn from a DP(α)-prior is DP($\alpha + \sum_{i=1}^n \delta_{X_i}$).

Corollary 4.17 *Suppose $X_1, X_2, \ldots$ are an i.i.d. sample from P_0.*

(i) *If $n \to \infty$, then* DP($\alpha + \sum_{i=1}^n \delta_{X_i}$) $\rightsquigarrow \delta_{P_0}$, *a.s.,*
(ii) *If $|\alpha| \to 0$, then* DP($\alpha + \sum_{i=1}^n \delta_{X_i}$) $\rightsquigarrow$ DP($\sum_{i=1}^n \delta_{X_i}$), *a.s.*

Proof The total mass of the base measure $\alpha + \sum_{i=1}^n \delta_{X_i}$ tends to infinity in case (i) and to n in case (ii). As the empirical distribution tends weakly to P_0 almost surely as $n \to \infty$, by Proposition F.3, the center measure tends to P_0 and $n^{-1} \sum_{i=1}^n \delta_{X_i}$, respectively. Thus the assertions follow from Theorem 4.16 (iii) and (ii). □

Assertion (i) means that the posterior is "consistent"; this is discussed more precisely in Chapter 6. Assertion (ii) gives the *noninformative limit* as the prior sample size goes to zero, so that the effect of the prior is eliminated from the posterior. The limiting posterior process is known as the *Bayesian bootstrap*. The result is one of the few concrete noninformative analyses in Bayesian nonparametrics. The Bayesian bootstrap generates random probability measures that are supported only at the observed points, and gives rise to a resampling scheme similar to the usual empirical (or Efron's) bootstrap. It is discussed further in Section 4.7.

4.3.3 Approximations

In this section we discuss two approximations to the Dirichlet process, both taking the form of a random discrete distribution, as discussed in Section 3.4.2, with finite support.

In the first, the weights are assigned by a finite Dirichlet distribution. Because this has only finitely many parameters, it is convenient for approximate inference on a computer.

Definition 4.18 (Dirichlet-multinomial) The *Dirichlet-multinomial process* of order N and parameters G and $(\alpha_1, \ldots, \alpha_N)$ is the random probability measure $\sum_{k=1}^{N} W_k \delta_{\theta_k}$, where $(W_1, \ldots, W_N) \sim \mathrm{Dir}(N; \alpha_1, \ldots, \alpha_N)$ and $\theta_1, \ldots, \theta_N \stackrel{\text{iid}}{\sim} G$ are independent.

The name "Dirichlet-multinomial process" stems from the fact that a random sample $X_1, \ldots, X_n$ generated from a distribution sampled from such a process can be represented as $X_i = \theta_{s_i}$, and $(R_1, \ldots, R_N)| W \sim \mathrm{MN}_N(n; W_1, \ldots, W_N)$, where $R_j = \#\{i: s_i = j\}$. This representation shows, in particular, that the sample $X_1, \ldots, X_n$ will tend to have ties even when n is much smaller than N.

The Dirichlet-multinomial assigns probability $\sum_{k:\theta_k \in A} W_k$ to a given set A. Because aggregation of Dirichlet weights to a smaller number of cells gives a Dirichlet distribution, by Proposition G.3, given the vector $\theta = (\theta_1, \ldots, \theta_N)$ the Dirichlet-multinomial process is a Dirichlet process with base measure $\sum_{k=1}^{N} \alpha_k \delta_{\theta_k}$.

In particular, this shows that $\mathrm{E}(\int \psi \, dP | \theta) = \int \psi \, d(\sum_k \bar{\alpha}_k \delta_{\theta_k}) = \sum_k \bar{\alpha}_j \psi(\theta_k)$, for any measurable function $\psi \in \mathbb{L}_1(G)$, for $\bar{\alpha}_k = \alpha_k / \sum_k \alpha_k$. In particular, the expectation $\mathrm{E}(\int \psi \, dP) = \int \psi \, dG$ does not depend on the parameters $\alpha_1, \ldots, \alpha_N$.

As it is finitely supported, a sample from the Dirichlet-multinomial is easily generated. If the parameters are chosen appropriately, this will be almost as good as generating from a Dirichlet process, as part (ii) of the following theorem shows.

Theorem 4.19 *Let P_N be a Dirichlet-multinomial process of order $N \to \infty$ with parameters $(\alpha_{1,N}, \ldots, \alpha_{N,N})$ and G, satisfying $\max_{1 \le k \le N} \alpha_{k,N}/\alpha_{.N} \to 0$, for $\alpha_{.N} = \sum_{k=1}^{N} \alpha_{k,N}$.*

- (i) *If $\alpha_{.N} \to 0$, then $\int \psi \, dP_N \rightsquigarrow \psi(\theta)$, where $\theta \sim G$, for any bounded continuous ψ. In particular, this is true if $\alpha_{k,N} = \lambda_N$ and $N\lambda_N \to 0$.*
- (ii) *If $\alpha_{.N} \to M$, then $\int \psi \, dP_N \rightsquigarrow \int \psi \, dP$, where $P \sim \mathrm{DP}(MG)$, for any $\psi \in \mathbb{L}_1(G)$.*
- (iii) *If $\alpha_{.N} \to \infty$, then $\int \psi \, dP_N \to_p \int \psi \, dG$, for any $\psi \in \mathbb{L}_1(G)$. In particular, this is true if $\alpha_{k,N} = \lambda_N$ and $N\lambda_N \to \infty$.*

Furthermore, if $\theta_i \stackrel{iid}{\sim} G$ and the weights are given by $W_{k,N} = \Gamma_k / \sum_{j=1}^{N} \Gamma_j$, where $\Gamma_k \stackrel{ind}{\sim} \mathrm{Ga}(\lambda_k, 1)$ with $\sum_{k=1}^{\infty} \lambda_k^2 / k^2 < \infty$ and $N^{-1} \sum_{k=1}^{N} \lambda_k \to \lambda > 0$ (hence including the case $\alpha_{k,N} = \lambda_k$), then the convergence in (iii) is also almost surely, for any $\psi \in \mathbb{L}_2(G)$.

Proof Given $\theta_1, \ldots, \theta_N$ the Dirichlet-multinomial process is a Dirichlet process with base measure $\sum_k \alpha_{k,N} \delta_{\theta_k}$. The integral of a function ψ relative to the center measure is equal to $\sum_k \bar{\alpha}_{k,N} \psi(\theta_k)$, for $\bar{\alpha}_{k,N} = \alpha_{k,N}/\alpha_{.N}$, and has expectation and variance

$$\mathrm{E}\Big(\sum_k \bar{\alpha}_{k,N} \psi(\theta_k)\Big) = \int \psi \, dG,$$

$$\mathrm{var}\Big(\sum_k \bar{\alpha}_{k,N} \psi(\theta_k)\Big) = \sum_k \bar{\alpha}_{k,N}^2 \, \mathrm{var}\, \psi(\theta_1) \le \max_k \bar{\alpha}_{k,N} \int \psi^2 \, dG.$$

In particular, $\int \psi \, d(\sum_k \bar{\alpha}_{k,N} \delta_{\theta_k})$ tends in probability to $\int \psi \, dG$, for every $\psi \in \mathbb{L}_2(G)$, in particular for every bounded, continuous ψ. It follows that conditionally on $\theta_1, \theta_2, \ldots$ the sequence of center measures converges weakly to G, in probability. The total mass of the base measure is $\alpha_{.N}$ and converges to 0, M and ∞ in the cases (i)–(iii). Therefore, Theorem 4.16 gives that the Dirichlet-multinomial $P_N = \sum_{k=1}^N W_{k,N}\delta_{\theta_k}$ tends in distribution in $\mathfrak{M}$ to δ_θ for $\theta \sim G$, DP(MG), and δ_G, in the three cases (i)–(iii), conditionally on $\theta_1, \theta_2, \ldots$ in probability, and then also unconditionally, as the three limits do not depend on the sequence $\theta_1, \theta_2, \ldots$ Moreover, the last convergence holds in probability since G is a fixed probability measure. Because the map $P \mapsto \int \psi\, dP$ from $\mathfrak{M}$ to $\mathbb{R}$ is continuous, for every bounded, continuous function $\psi: \mathfrak{X} \to \mathbb{R}$, the continuous mapping theorem shows that assertions (i)–(iii) of the theorem are true for every bounded continuous function ψ.

It suffices to improve the conclusion to $\psi \in \mathbb{L}_1(G)$ in cases (ii) and (iii), and to almost sure convergence in the special case of (iii) mentioned in the final assertion of the theorem. For the upgrade to integrable ψ we apply the final assertion of Theorem 4.16. Because $\mathrm{E}\int |\psi|\mathbb{1}\{|\psi| > M\}\, d(\sum_k \bar\alpha_{k,N}\delta_{\theta_k}) = \int |\psi|\mathbb{1}\{|\psi| > M\}\, dG$, any function $\psi \in \mathbb{L}_1(G)$ satisfies the uniform integrability assumption, in mean. Secondly, the argument given previously already showed that given $\theta_1, \theta_2, \ldots$ the center measures converge to G not only weakly, but also setwise.

To prove the final statement of the theorem we apply Kolmogorov's strong law. Because $N^{-1}\sum_k \mathrm{E}(\Gamma_k) \to \lambda$ and $\sum_k \mathrm{var}(\Gamma_k)/k^2 = \sum_k \lambda_k/k^2 < \infty$ by assumption, this gives first that $N^{-1}\sum_{k=1}^N \Gamma_k \to \lambda$ a.s. Second, $N^{-1}\sum_{k=1}^N \Gamma_k\psi(\theta_k) \to \lambda\int\psi\, dG$ a.s., since $\sum_k \mathrm{var}\,(\Gamma_k\psi(\theta_k))/k^2 \le \sum_k(\lambda_k + \lambda_k^2)\int\psi^2\, dG/k^2 < \infty$. ☐

The tempting "default choice" $\alpha_{1,N} = \cdots = \alpha_{N,N} = 1$ belongs to case (iii) of Theorem 4.19, and hence yields a sequence of Dirichlet-multinomial priors that converges to the prior for P degenerate at G. It is inappropriate for inference (unless G is equal to the data-generating measure). The corresponding Dirichlet-multinomial distribution with $N = n$ and weights at the observations is the same as the Bayesian bootstrap given in Corollary 4.17. This converges to the correct distribution as $n \to \infty$, but is a posterior distribution, with $\theta_1, \theta_2, \ldots$ observations, and not a prior.

A second approximation to the Dirichlet process is obtained by truncating the stick-breaking representation. To guarantee an error smaller than a predetermined level, the truncation point can be chosen random. For stick-breaking weights $W_i = \prod_{j=1}^{i-1}(1 - V_j)V_i$, for $V_1, V_2, \ldots \stackrel{\text{iid}}{\sim} \mathrm{Be}(1, M)$, and independent $\theta_0, \theta_1, \ldots \stackrel{\text{iid}}{\sim} G$, consider for a given, small $\epsilon > 0$,

$$P_\epsilon = \sum_{i=1}^{N_\epsilon} W_i\delta_{\theta_i} + \Bigl(1 - \sum_{i=1}^{N_\epsilon} W_i\Bigr)\delta_{\theta_0}, \qquad N_\epsilon = \inf\Bigl\{N \ge 1: \sum_{i=1}^{N} W_i > 1 - \epsilon\Bigr\}. \tag{4.23}$$

We shall call the distribution of P_ϵ an *ϵ-Dirichlet process*, and denote it by DP(α, ϵ)[3].

Proposition 4.20 (ϵ-Dirichlet process) *For every $\epsilon > 0$ the total variation distance between the ϵ-Dirichlet process P_ϵ defined in* (4.23) *and the Dirichlet process $P = \sum_{i=1}^\infty W_i\delta_{\theta_i}$ satisfies $d_{TV}(P_\epsilon, P) \le \epsilon$ a.s. Consequently, the Lévy distance defined in* (A.1) *between the distributions* DP(α, ϵ) *and* DP(α) *of these random measures on* $(\mathfrak{M}, \mathscr{M})$,

[3] An alternative construction with the same properties is $P_\epsilon = \sum_{i\le N_\epsilon} W_i\delta_{\theta_i}/\sum_{i\le N_\epsilon} W_i$.

equipped with the total variation distance, satisfies $d_L(\mathrm{DP}(\alpha,\epsilon),\mathrm{DP}(\alpha)) \le \epsilon$. *Furthermore, as* $\epsilon \to 0$ *we have* $\int \psi\, dP_\epsilon \to \int \psi\, dP$ *a.s., for any measurable function* ψ *that is* P*-integrable a.s. Finally, the number* $N_\epsilon + 1$ *of support points of* P_ϵ *satisfies* $N_\epsilon - 1 \sim \mathrm{Poi}(M \log_- \epsilon)$.

Proof The first assertion is clear from the fact that $|P(A) - P_\epsilon(A)| \le \sum_{i>N_\epsilon} W_i < \epsilon$, for any measurable set $A \subset \mathfrak{X}$. We conclude that $P \in \mathcal{A}$ for an arbitrary $\mathcal{A} \in \mathscr{M}$ implies that $P_\epsilon \in \{Q: d_{TV}(Q,\mathcal{A}) < \epsilon\}$ and hence also $P_\epsilon \in \{Q: d_L(Q,\mathcal{A}) < \epsilon\}$, because the Lévy distance d_L (on $\mathfrak{M} = \mathfrak{M}(\mathfrak{X})$) satisfies $d_L \le d_{TV}$. Consequently $\mathrm{DP}_{\alpha,\epsilon}(\mathcal{A}) = \mathrm{P}(P_\epsilon \in \mathcal{A}) \le \mathrm{P}(P \in \mathcal{A}^\epsilon) = \mathrm{DP}_\alpha(\mathcal{A}^\epsilon)$. Similarly $\mathrm{DP}_\alpha(\mathcal{A}) \le \mathrm{DP}_{\alpha,\epsilon}(\mathcal{A}^\epsilon)$, and the assertion follows from the definition of the Lévy distance d_L (on $\mathfrak{M}(\mathfrak{M})$ this time).

The almost sure convergence of $\int \psi\, dP_\epsilon = \sum_{i=1}^{N_\epsilon} W_i \psi(\theta_i) + \bar{W}_\epsilon \psi(\theta_0)$, for $\bar{W}_\epsilon = 1 - \sum_{i=1}^{N_\epsilon} \le \epsilon$, to $\sum_{i=1}^\infty W_i\psi(\theta_i) = \int \psi\, dP$ is clear from the facts that $N_\epsilon \to \infty$ and $\bar{W}_\epsilon \to 0$.

Because $\sum_{i\le N} W_i = 1 - \prod_{i\le N}(1 - V_i)$, we have $N_\epsilon = \inf\{N: -\sum_{i\le N}\log(1-V_i) > \log_- \epsilon\}$ by the definition of N_ϵ. Because $-\log(1-V_i) \overset{\text{iid}}{\sim} \mathrm{Exp}(M)$, the variable $N_\epsilon - 1$ can be identified with the number of events occurring in the time interval $[0, \log_- \epsilon]$ in a homogeneous Poisson process with rate M. This implies the final assertion of the proposition. □

The last assertion of Proposition 4.20 implies that on average $\mathrm{E}(N_\epsilon + 1) = 2 + M \log_- \epsilon$ terms are needed to attain a level of accuracy ϵ.

In the case that $\mathfrak{X} = \mathbb{R}$, we can also consider the Kolmogorov-Smirnov distance. Since this is bounded by the total variation distance, it follows that $d_{KS}(F_\epsilon, F) \le \epsilon$ almost surely, for F_ϵ and F the cumulative distribution functions of P_ϵ and P. This next implies that continuous functionals $\psi(F)$ can be approximated by the corresponding functional $\psi(F_\epsilon)$. Examples include the KS-distance itself, and quantiles of the distribution.

4.3.4 Mutual Singularity of Dirichlet Processes

The prior and posterior distributions given a random sample from a Dirichlet process with a atomless base measure are mutually singular. This follows from Theorem 3.22, since the Dirichlet process is a special case of a Pólya tree process (where we can choose a canonical binary partition with $\alpha_\varepsilon = \alpha(A_\varepsilon) = 2^{-|\varepsilon|}$, so that $\sum_{m=1}^\infty 2^m = \infty$). In this subsection, we show more generally, without referring to Pólya tree processes, that two Dirichlet processes with different base measures are typically mutually singular.

The mutual singularity of prior and posterior can arise, because the Dirichlet prior is not concentrated on a dominated set of measures (see Lemma 1.2). The following general theorem also implies the mutual singularity of the posterior distributions based on different Dirichlet priors. That (even slightly) different priors give quite different behavior is somewhat unusual.

A given base measure α can be decomposed as the sum $\alpha = \alpha_a + \alpha_c$ of a discrete measure α_a (i.e. $\alpha_a(D^c) = 0$ for a countable set D) and a measure α_c without atoms (i.e. $\alpha_c\{x\} = 0$ for every x). We refer to the two measures as the *atomic part* and *continuous part* of α, respectively.

Theorem 4.21 *If the continuous parts of α_1 and α_2 are unequal, or their atomic parts have different supports, then* $\mathrm{DP}(\alpha_1)$ *and* $\mathrm{DP}(\alpha_2)$ *are mutually singular.*

Proof Let $\alpha_1 = \alpha_{1,c} + \alpha_{1,a}$ and $\alpha_2 = \alpha_{2,c} + \alpha_{2,a}$ be the decompositions of the two base measures, S_1 and S_2 the supports of their atomic parts, and $\mathfrak{X}' = \mathfrak{X} \setminus (S_1 \cup S_2)$.

The Dirichlet measures $\mathrm{DP}(\alpha_1)$ and $\mathrm{DP}(\alpha_2)$ are probability measures on the space of measures $(\mathfrak{M}, \mathscr{M})$ on the sample space $(\mathfrak{X}, \mathscr{X})$. We shall show that there exist disjoint measurable subsets C_1, C_2 of $\mathfrak{X}^\infty$ such that $\mathrm{DP}_{\alpha_i}(P \in \mathfrak{M}: P^\infty(C_i) = 1) = 1$, for $i = 1, 2$. The theorem then follows as $\{P \in \mathfrak{M}: P^\infty(C_1) = 1\} \cap \{P \in \mathfrak{M}: P^\infty(C_2) = 1\} = \varnothing$.

For simplicity of notation consider P^∞ as the law of a random sequence $X_1, X_2, \ldots$ drawn from $P \sim \mathrm{DP}(\alpha_i)$. Let $Y_1, Y_2, \ldots$ be the distinct values in $X_1, X_2, \ldots$ that do not belong to $S_1 \cup S_2$, and let K_n be the number of these values among $X_1, \ldots, X_n$. (More precisely, define $D_i = 1$ if $X_i \notin \{X_1, \ldots, X_{i-1}\} \cup S_1 \cup S_2$; let $K_n = \sum_{i=1}^n D_i$; and put $Y_j = X_{\tau_j}$, for $\tau_1 = 1$ and $\tau_j = \min\{n: K_n = j\}$ for $j \in \mathbb{N}$.) Then $Y_1, Y_2, \ldots$ can be seen to form an i.i.d. sequence from $\bar{\alpha}_{i,c}$, the normalized continuous part of the base measure (cf. Problem 4.21). By the strong law of large numbers $n^{-1} \sum_{j=1}^n \mathbb{1}\{Y_j \in A\} \to \bar{\alpha}_{i,c}(A)$, for every measurable set A. Furthermore $K_n / \log n \to |\alpha_{i,c}|$ almost surely, by a slight extension of the result of Proposition 4.8(ii).

If $\alpha_{1,c} \neq \alpha_{2,c}$, then either $\bar{\alpha}_{1,c} \neq \bar{\alpha}_{2,c}$ or $|\alpha_{1,c}| \neq |\alpha_{2,c}|$, or both. In the first case there exists a measurable set A with $\bar{\alpha}_{1,c}(A) \neq \bar{\alpha}_{2,c}(A)$, and then the two events $\{n^{-1} \sum_{j=1}^n \mathbb{1}\{Y_j \in A\} \to \bar{\alpha}_{i,c}(A)\}$, for $i = 1, 2$, are disjoint, up to a null set. In the second case the two events $\{K_n / \log n \to |\alpha_{i,c}|\}$, for $i = 1, 2$, are disjoint up to a null set.

Finally if $S_1 \neq S_2$, then there exists x that is in exactly one S_i, say $x \in S_2 \setminus S_1$. Then the event $\cup_{j \geq 1}\{X_j = x\}$ has probability one under $P \sim \mathrm{DP}(\alpha_2)$ and probability zero if $P \sim \mathrm{DP}(\alpha_1)$. □

The conditions in the theorem cannot be relaxed further. If $\alpha_{1,c} = \alpha_{2,c}$ and $\mathrm{supp}(\alpha_{1,a}) = \mathrm{supp}(\alpha_{2,a})$, then $\mathrm{DP}(\alpha_1)$ and $\mathrm{DP}(\alpha_2)$ need not be mutually singular.

4.3.5 Tails of a Dirichlet Process

By Theorem 4.15 the support of a $\mathrm{DP}(MG)$-process on $\mathbb{R}$ is equal to the support of the center measure G. In this section we study the tails of the Dirichlet process, i.e. the behavior of $F(x)$ as $G(x) \downarrow 0$, or $1 - F(x)$ as $G(x) \uparrow 1$, for F the cumulative distribution function of the Dirichlet process. In view of the representation of $\mathrm{DP}(MG)$ given in Section 4.2.3, this is the same as that of a gamma process.

Theorem 4.22 *If F is the cumulative distribution function of the* $\mathrm{DP}(MG)$*-process, then*

$$\liminf_{G(x)\downarrow 0} F(x) \exp\Big(\frac{r \log|\log MG(x)|}{MG(x)}\Big) = \begin{cases} 0, & \text{if } r < 1, \\ \infty, & \text{if } r > 1, \end{cases} \quad \text{a.s.},$$

$$\limsup_{G(x)\downarrow 0} F(x) \exp\Big(\frac{1}{MG(x)|\log MG(x)|^r}\Big) = \begin{cases} 0, & \text{if } r > 1, \\ \infty, & \text{if } r \leq 1, \end{cases} \quad \text{a.s.}$$

The same results hold if F and G are replaced by $1 - F$ and $1 - G$.

Proof We can represent $F(x)$ as $\gamma(MG(x))/\gamma(M)$ for a standard gamma process γ (see Section 4.2.3 and Appendix J). The assertion on the lim inf is then immediate from Theorem J.19(ii).

According to Theorem J.19(iii), for a given convex, increasing function h, the variable $\limsup_{t\downarrow 0}\gamma(t)/h(t)$ is equal to 0 or ∞ if the integral $\int_0^1\int_{h(t)}^{\infty} x^{-1}e^{-x}\,dx\,dt$ converges, or diverges, respectively. Combining the inequalities $e^{-1}\log(1/t) \leq \int_t^1 x^{-1}e^{-x}\,dx \leq \log(1/t)$, for $0 < t < 1$, and $0 \leq \int_1^{\infty} x^{-1}e^{-x}\,dx \leq e^{-1}$, we see

$$e^{-1}\log\frac{1}{h(t)} \leq \int_{h(t)}^{\infty} x^{-1}e^{-x}\,dx \leq e^{-1} + \log\frac{1}{h(t)}.$$

Thus the relevant integral converges or diverges if and only if $\int_0^1 \log_- h(t)\,dt$ converges or diverges. For $\log h(t) = t^{-1}|\log t|^{-r}$ this is the case if $r < 1$ or $r \geq 1$, respectively.

The result for the upper tail can be obtained by the same method, after noting that the stochastic process $s \mapsto \gamma(1)-\gamma(s)$ is equal in distribution to the process $s \mapsto \gamma(1-s)$. □

The theorem shows that the lim inf and lim sup of the tails have different orders. This is caused by the irregularity of the sample paths of the Dirichlet (or gamma) process, which increase by infinitely many (small) jumps on a countable dense set. This also causes that there is no deterministic function that gives the exact order of the lim inf, in the sense of existence of $\liminf_{G(x)\to 0} F(x)/h(x)$ for some deterministic function h as a number in $(0,\infty)$, rather than 0 or ∞ (cf. Theorem J.19(i)).

The message of the theorem can be summarized (imprecisely) in the bounds, for $r > 1$, a.s. eventually as $G(x)$ is small enough,

$$\exp\Bigl(-\frac{r\log|\log MG(x)|}{MG(x)}\Bigr) \leq F(x) \leq \exp\Bigl(-\frac{1}{MG(x)|\log MG(x)|^r}\Bigr). \tag{4.24}$$

Similar bounds are valid for the upper tail. These inequalities show that the tails of F are much thinner ("exponentially much") than the tails of the base measure G. This may supplement assertion (4.6), which shows that these tails are equal "on average." In particular, the variable $\int|\psi|\,dP$ may well be finite a.s. even if $\int|\psi|\,d\alpha = \infty$ (and the former variable is not integrable).

Example 4.23 (Normal distribution) The upper tail of the standard normal distribution G satisfies $1 - G(x) \sim x^{-1}(2\pi)^{-1/2}e^{-x^2/2}$. As $r > 1$ can be arbitrarily close to 1, we can absorb any constant in the power of x. Inequalities (4.24) imply, for fixed M as $x \to \infty$,

$$\exp\bigl[-2r(2\pi)^{1/2}M^{-1}e^{x^2/2}\log x\bigr] \leq 1 - F(x) \leq \exp\bigl[-e^{x^2/2}x^{-2r}\bigr].$$

Thus the tails are (much) thinner than the tails of the extreme value distribution.

Example 4.24 (Cauchy distribution) The upper tail of the standard Cauchy distribution G satisfies $1 - G(x) \sim \pi^{-1}x^{-1}$, for $x \to \infty$. Again we can absorb any constant in powers of $\log x$. Inequalities (4.24) imply, for fixed M as $x \to \infty$,

$$\exp\big[-r\pi M^{-1}x\log\log x\big] \le 1 - F(x) \le \exp\big[-x/|\log x|^r\big].$$

Thus the tails of F are almost exponential, even though G has no mean.

4.3.6 Distribution of Median

The *median* m_F of a DP(MG)-process on $\mathbb{R}$ is defined as any random variable m_F such that $F(m_F-) \le \frac{1}{2} \le F(m_F)$, for F the (random) cumulative distribution function of the DP(MG)-process. The following theorem gives an exact expression for its distribution, which is known as the *median-Dirichlet distribution* with parameters M and G. The theorem also shows that "any median of the random variable m_F is a median of the base measure G," a property that may be considered the analog of the equation $\mathrm{E}(\int \psi\, dF) = \int \psi\, dG$ for means.

Theorem 4.25 *Any median m_F of $F \sim$ DP(MG) has cumulative distribution function H given by*

$$H(x) = \int_{1/2}^{1} \frac{\Gamma(M)}{\Gamma(MG(x))\Gamma(M(1-G(x)))} u^{MG(x)-1}(1-u)^{M(1-G(x))-1}\, du.$$

The cumulative distribution function H is continuous if G is so, and has the same (set of) medians as G.

Proof If $m_F \le x$, then $F(x) \ge F(m_F) \ge 1/2$, by the definition of m_F; furthermore, $m_F > x$ implies $F(x) \le F(m_F-) \le 1/2$. This shows that $\{F(x) > 1/2\} \subset \{m_F \le x\} \subset \{F(x) \ge 1/2\}$. As $F(x)$ has a $\mathrm{Be}(MG(x), M\bar{G}(x))$-distribution, the probabilities of the events on left and right are equal and hence $\mathrm{P}(m_F \le x) = \mathrm{P}(F(x) \ge 1/2)$. This gives the formula for H by employing the formula for the beta distribution. The continuity of H is clear from the continuity of the beta distribution in its parameter.

The beta distributions $B_\theta := \mathrm{Be}(M\theta, M(1-\theta))$ form a one-dimensional exponential family in the parameter $0 < \theta < 1$, which possesses a strictly monotone likelihood ratio. Furthermore, $B_{1/2}$ is symmetric about $1/2$. Therefore, the preceding derivation gives $\mathrm{P}(m_F \le x) = B_{G(x)}([1/2, 1]) \ge 1/2$ if and only if $G(x) \ge 1/2$. By a similar derivation $\mathrm{P}(m_F > x) = \mathrm{P}(F(x) \le 1/2) = B_{G(x)}([0, 1/2])$. Replacing x by $x - \epsilon$ and taking the limit as $\epsilon \downarrow 0$, we find that $\mathrm{P}(m_F \ge x) = B_{G(x-)}([0, 1/2])$, whence $\mathrm{P}(m_F < x) \le 1/2$ if and only if $G(x-) \le 1/2$. Combining the preceding, we see that x is a median of m_F if and only if $G(x-) \le 1/2 \le G(x)$. □

4.3.7 Distribution of Mean

Consider the mean functional $\int \psi\, dP$, for a given measurable function $\psi: \mathfrak{X} \to \mathbb{R}$ and $P \sim \mathrm{DP}(\alpha)$.

Let $\mathrm{Re}\, z$ and $\mathrm{Im}\, z$ stand for the real and imaginary parts of a complex number z, and let $\log z = \log|z| + i \arg z$ denote the principal branch of the complex logarithm (with $\arg z \in [-\pi, \pi)$).

Theorem 4.26 *The mean* $\int \psi\, dP$ *for* $P \sim \mathrm{DP}(\alpha)$ *is defined as an a.s. finite random variable for any measurable function* $\psi\colon \mathfrak{X} \to \mathbb{R}$ *such that* $\int \log(1+|\psi|)\, d\alpha < \infty$. *Provided that* $\beta := \alpha \circ \psi^{-1}$ *is not degenerate its distribution is absolutely continuous with Lebesgue density* h *and cumulative distribution function* H *given by*

$$h(s) = \frac{1}{\pi}\int_0^\infty \mathrm{Re}\Big[\exp\Big(-\int \log(1+it(s-x))\, d\beta(x)\Big)\int \frac{d\beta(x)}{1+it(s-x)}\Big]\, dt,$$

$$H(s) = \frac{1}{2} - \frac{1}{\pi}\int_0^\infty \frac{1}{t}\,\mathrm{Im}\Big[\exp\Big(-\int \log(1+it(s-x))\, d\beta(x)\Big)\Big]\, dt.$$

Proof Since $\int \psi\, dP = \int x\, dP \circ \psi^{-1}(x)$ and $P \circ \psi^{-1} \sim \mathrm{DP}(\alpha \circ \psi^{-1})$, it suffices to consider the special case where $\psi(x) = x$ on $\mathfrak{X} = \mathbb{R}$ and $P \sim \mathrm{DP}(\beta)$. We can then represent the cumulative distribution function of P as $\gamma(\beta(x))/\gamma(M)$, for β also denoting the cumulative distribution function of β and $M = |\alpha| = \beta(\infty)$.

The a.s. finiteness of $\int x\, dP(x)$ follows from Lemma 4.27.

Furthermore, the distribution function of the mean $\int x\, d\gamma(\beta(x))/\gamma(M)$ can be written $H(s) = \mathrm{P}(\int \psi\, d\gamma \le 0)$ for the function ψ defined by $\psi(x) = x - s$. Lemma 4.27 gives the characteristic function of the random variable $\int \psi\, d\gamma$, and an inversion formula (see Problem L.2) gives that

$$H(s) + H(s-) = 1 - \frac{2}{\pi}\lim_{\epsilon\downarrow 0, T\uparrow\infty}\int_\epsilon^T \frac{1}{t}\,\mathrm{Im}\Big[e^{-\int \log(1+it(s-x))\, d\beta(x)}\Big]\, dt. \tag{4.25}$$

To complete the proof we show that the integrand is integrable near 0 and ∞, so that the limit exists as an ordinary integral, and justify differentiation under the integral, which indeed gives the formula for the derivative h.

Because $\log z = \log|z| + i \arg z$, the integrand in the right side is given by

$$t^{-1}\exp\Big[-\int \log(1+t^2(s-x)^2)\, d\beta(x)/2\Big]\sin\Big(\int \tan^{-1}(t(s-x))\, d\beta(x)\Big).$$

Both the exponential and the sine functions are bounded by 1, and hence integrability is an issue only for t near 0 and ∞. Because $\int \log(1+t^2(s-x)^2)\, d\beta(x) \ge c_1 \log(1+c_2t^2(s-c_3)^2)$, for some positive constants c_1, c_2 and constant c_3 (that depend on α), the expression is bounded above by

$$t^{-1}\exp\Big[-c_1\log(1+c_2t^2(s-c_3)^2)\Big] = \frac{1}{t(1+c_2t^2(s-c_3)^2)^{c_1}}.$$

The integral over $t > T$ converges and hence tends to zero as $T \uparrow \infty$. Because $|\tan^{-1} x| \lesssim |x|/(1+|x|)$ and $|\sin x| \le |x|$, the expression is bounded by a multiple of $\int |s-x|/(1+t|s-x|)\, d\alpha(x)$, whence

$$\int_0^\epsilon \int \frac{|s-x|}{(1+t|s-x|)}\, d\beta(x)\, dt = \int \log(1+\epsilon|s-x|)\, d\beta(x).$$

This tends to zero as $\epsilon \downarrow 0$. ∎

Lemma 4.27 *If γ is a gamma process with intensity measure α, then $\int |\psi|\, d\gamma < \infty$ a.s. for any measurable function $\psi: \mathbb{R} \to \mathbb{R}$ with $\int \log(1 + |\psi|)\, d\alpha < \infty$*[4]*. Furthermore, the characteristic function of this variable is given by*

$$\mathrm{E}\Big[\exp\Big(it \int \psi\, d\gamma\Big)\Big] = \exp\Big[-\int \log(1 - it\psi)\, d\alpha\Big]. \tag{4.26}$$

Proof First consider a function of the type $\psi = \sum_{i=1}^{k} a_i \mathbb{1}_{A_i}$, for constants $a_1, \ldots, a_k$ and disjoint measurable sets $A_1, \ldots, A_k$. As $\gamma(A_i) \overset{\text{ind}}{\sim} \mathrm{Ga}(\alpha(A_i), 1)$, the left side of (4.26) is then equal to

$$\prod_{i=1}^{k} \Big(\frac{1}{1 - ita_i}\Big)^{\alpha(A_i)} = \exp\Big[-\sum_{i=1}^{k} \alpha(A_i) \log(1 - ita_i)\Big].$$

This is identical to the right side of (4.26), and hence this equation is true for every simple function ψ of this type.

Given a general measurable function ψ satisfying the integrability condition of the lemma, let ψ_n be a sequence of simple, measurable functions with $0 \le \psi_n^+ \uparrow \psi^+$ and $0 \le \psi_n^- \uparrow \psi^-$. Because $|\log(1 - it)| \le \log(1 + |t|) + 2\pi$, we have $\int \log(1 - it\psi_n)\, d\alpha \to \int \log(1 - it\psi)\, d\alpha$, by the dominated convergence theorem, and hence also

$$\log \mathrm{E}\Big[\exp\Big(it \int \psi_n\, d\gamma\Big)\Big] = -\int \log(1 - it\psi_n)\, d\alpha \to -\int \log(1 - it\psi)\, d\alpha.$$

The limit on the right side is continuous at $t = 0$, and hence by Lévy's continuity theorem the sequence $\int \psi_n\, d\gamma$ converges weakly to a proper random variable whose characteristic function is given by the right side.

Because every realization of γ is a finite measure, we also have $\int \psi_n^+\, d\gamma \to \int \psi^+\, d\gamma$ a.s., by the monotone convergence theorem, and similarly for the negative parts. By the argument of the preceding paragraph applied to the positive and negative parts instead of the ψ_n, the two limit variables are proper, and hence we can take the difference to see that $\int \psi_n\, d\gamma \to \int \psi\, d\gamma$ a.s. Thus the weak limit found in the preceding paragraph is $\int \psi\, d\gamma$. We conclude that this variable possesses the right side of the preceding display as its log characteristic function. □

4.4 Characterizations

It was noted in Section 4.1.2 that the Dirichlet process is tail-free. It is also neutral in the sense defined below, and has the property that the posterior distribution of cell probabilities depends on the observations only through their cell counts. Each of these properties distinguishes the Dirichlet process from other random measures, except for the following three trivial types of random measures P:

(T1) $P = \rho$ a.s., for a deterministic probability measure ρ;
(T2) $P = \delta_X$, for a random variable $X \sim \rho$;

[4] This condition is also known to be necessary.

(T3) $P = W\delta_a + (1 - W)\delta_b$, for deterministic $a, b \in \mathfrak{X}$ and an arbitrary random variable W with values in $[0, 1]$.

The random measures (T1) and (T2) are limits of Dirichlet processes, as the precision parameter goes to infinity and zero, respectively. The third (T3) arises because random probability vectors of length two cannot be characterized by conditional independence properties.

The concept of a random measure that is tail-free relative to a given sequence of measurable (binary) partitions of the sample space (which are successive refinements) is introduced in Definition 3.11. In the following proposition we call a random measure *universally tail-free* if it is tail-free relative to *any* sequence of measurable binary partitions.

Whereas tail-freeness entails conditional independence across successive layers of ever finer partitions, "neutrality" refers to partitions at a given level. A random measure P is called *neutral* if for every finite measurable partition $\{A_1, \ldots, A_k\}$ the variable $P(A_1)$ is conditionally independent of the vector $(P(A_2)/(1 - P(A_1)), \ldots, P(A_k)/(1 - P(A_1)))$ given $P(A_1) < 1$[5]. As tail-freeness, neutrality has a constructive interpretation: after fixing the probability $P(A_1)$ of the first set, the *relative* sizes of the probabilities (or conditional probabilities) of the remaining sets of the partition are generated independently of $P(A_1)$. The neutrality principle can be applied a second time, to the coarsened partition $\{A_1 \cup A_2, A_3, \ldots, A_k\}$, to show that given the probability assigned to the first two sets the conditional probabilities of $A_3, \ldots, A_k$ are again generated independently, etc.

An alternative property is *complete neutrality*, which entails that the random variables $P(A_1), P(A_2|\, A_1^c), P(A_3|\, A_1^c \cup A_2^c), \ldots$ are independent (after properly providing for conditioning on events of probability zero). For a given partition in a given ordering, this appears to be slightly stronger, but when required for any ordering of the partitioning sets (or any partition) these concepts are equivalent (see Proposition G.7).

Theorem 4.28 *The following three assertions about a random measure P are equivalent:*

(i) *P is universally tail-free.*
(ii) *P is neutral.*
(iii) *P is a Dirichlet process, or one of the random measures (T1), (T2) or (T3).*

Proof That a Dirichlet process is universally tail-free is noted in Section 4.1.2; that it is also neutral follows most easily from the representation of the finite-dimensional Dirichlet distribution through gamma variables, given in Proposition G.2(i). That random measures of types (T1) and (T2) are universally tail-free and neutral follows by taking limits of Dirichlet processes. That a random measure (T3) is universally tail-free and neutral can be checked directly. Thus (iii) implies both (i) and (ii).

We now prove that (i) implies (ii). An arbitrary partition $\{A_1, \ldots, A_k\}$ can be built into a sequence of successively refined partitions $\{A_\varepsilon : \varepsilon \in \mathcal{E}^*\}$ as $B_0 = A_1, B_{10} = A_2, B_{110} = A_3$, etc. By (3.12) the probabilities of these sets satisfy $P(B_0) = V_0$, $P(B_{10}) = (1 - V_0)V_{10}$, $P(B_{110}) = (1 - V_0)(1 - V_{10})V_{110}$, etc. Under tail-freeness (i) the variables V_ε corresponding to ε of different lengths are independent. This implies that the variable $P(A_1)$ is neutral in the vector $(P(A_1), \ldots, P(A_k))$, whence P is neutral.

[5] See Section 13.4 for a related concept "neutral to the right."

It now suffices to prove that (ii) implies (iii). Under neutrality the variables $X_i = P(A_i)$ satisfy condition (i) of Proposition G.7, for any measurable partition $\{A_1, \ldots, A_k\}$. Consider two cases: there is, or there is not, a partition such that at least three variables $P(A_i)$ do not vanish with probability one. In the case that there is not, the support of almost every realization of P must consist of at most two given points, and then P is of type (T3). For the remainder of the proof assume that there is a partition $\{A_1, \ldots, A_k\}$ such that at least three variables $P(A_i)$ do not vanish with probability one. We may leave off the sets A_i such that $X_i = P(A_i)$ vanishes identically, and next apply Proposition G.7 (iv) to the vector of remaining variables. If this set consists of k variables we can conclude that this possesses a $\overline{\text{Dir}}(k; M, \rho_1, \ldots, \rho_k)$-distribution, as defined preceding Proposition G.7. The vector $(\rho_1, \ldots, \rho_k)$ is the mean of this distribution; hence, ρ_i is equal to $\rho(A_i)$, for ρ the mean measure $\rho(A) = \mathrm{E}P(A)$ of P.

The parameter M is either positive and finite or takes one of the values 0 or ∞. This division in two cases is independent of the partition. Indeed, given another partition we can consider its common refinement with $\{A_1, \ldots, A_k\}$, whose probability vector must then also possess an extended Dirichlet distribution $\overline{\text{Dir}}$. If this has M-parameter equal to 0 or ∞, then this distribution is discrete. This is possible if and only if the distribution of the aggregate vector $(P(A_1), \ldots, P(A_k))$ is also discrete.

The same argument combined with Proposition G.3 shows that the value of M in case $M \in (0, \infty)$ is the same for every partition. The random measure P is $\text{DP}(M\rho)$ in this case.

If $M = \infty$, then the distribution of $(P(A_1), \ldots, P(A_k))$ is concentrated at the single point $(\rho(A_1), \ldots, \rho(A_k))$. If this is true for every partition, then P is of type (T1). If $M = 0$ for some partition, then the distribution of $(P(A_1), \ldots, P(A_k))$ is concentrated on the vertices of the k-dimensional unit simplex, and we can represent it as the distribution of $(\delta_X(A_1), \ldots, \delta_X(A_k))$, for a random variable X with $\mathrm{P}(X \in A_i) = \rho_i$, giving a random measure P of type (T2). □

For a tail-free process, the posterior distribution of a set depends on a random sample of observations only through the count of that set. By Theorem 3.14, this actually characterizes tail-freeness for a particular sequence of partitions. Combined with the fact that a Dirichlet process is tail-free for any sequence of partitions, we obtain the following characterization. The property may be considered a drawback of a Dirichlet process, since it indicates that the posterior distribution is insensitive to the location of the observations. In particular, the Dirichlet process provides no smoothing.

Corollary 4.29 *If for every n and measurable partition $\{A_1, \ldots, A_k\}$ of the sample space the posterior distribution of $(P(A_1), \ldots, P(A_k))$ given a random sample $X_1, \ldots, X_n$ from P depends only on $(N_1, \ldots, N_k)$, for $N_j = \sum_{i=1}^n \mathbb{1}\{X_i \in A_j\}$, then P is a Dirichlet process or one of the trivial processes (T1), (T2) and (T3).*

It was seen in Section 4.1.2 that a Dirichlet process is a Pólya tree. The following result shows that its distinguishing property within the class of Pólya tree processes is the additivity of the parameters of the beta-splitting variables across partitions.

Theorem 4.30 *A Pólya tree prior* $\mathrm{PT}(\mathcal{T}_m, \alpha_\varepsilon)$ *on a sequence of partitions* $\mathcal{T}_m$ *that generates the Borel sets is a Dirichlet process if and only if* $\alpha_\varepsilon = \alpha_{\varepsilon 0} + \alpha_{\varepsilon 1}$ *for all* $\varepsilon \in \mathcal{E}^*$.

Proof That a Dirichlet process is a Pólya tree is already noted in Section 4.1.2, with the parameters given by $\alpha_\varepsilon = \alpha(A_\varepsilon)$ by (4.1.2), which are clearly additive.

Conversely, given a $\mathrm{PT}(\mathcal{T}_m, \alpha_\varepsilon)$-process P, define a set function α on the partitioning sets $\{A_\varepsilon: \varepsilon \in \mathcal{E}^*\}$ by $\alpha(A_\varepsilon) = \alpha_\varepsilon$. If the parameters α_ε are additive as in the theorem, then α is additive and extends to a finitely additive positive measure on the field generated by all partitions. By assumption the splitting variables $V_{\varepsilon 0} = P(A_{\varepsilon 0} | A_\varepsilon)$ of the Pólya tree process are independent and possess $\mathrm{Be}(\alpha_{\varepsilon 0}, \alpha_{\varepsilon 1})$-distributions. In particular $P(A_0) = V_0 \sim \mathrm{Be}(\alpha_0, \alpha_1)$, whence $(V_0, 1 - V_0) \sim \mathrm{Dir}(2; \alpha_0, \alpha_1)$. By (3.12)

$$\begin{aligned}&(P(A_{00}), P(A_{01}), P(A_{10}), P(A_{11}))\\ &\quad = (V_0 V_{00}, V_0(1 - V_{00}), (1 - V_0)V_{10}, (1 - V_0)(1 - V_{10})).\end{aligned}$$

This possesses a $\mathrm{Dir}(4; \alpha_{00}, \alpha_{01}, \alpha_{10}, \alpha_{11})$-distribution by the converse part of Proposition G.3, applied to the partition of $\{1, 2, 3, 4\}$ into the sets $\{1, 2\}$ and $\{3, 4\}$, since the required conditions (i)–(iii) are satisfied. In a similar manner $(P(A_\varepsilon): \varepsilon \in \mathcal{E}^m)$ possesses a $\mathrm{Dir}(2^m; (\alpha_\varepsilon: \varepsilon \in \mathcal{E}^m))$-distribution, for any given m.

Thus the random measure P satisfies (4.1) for every of the partitions $\mathcal{T}_m$. Since this sequence of partitions is assumed to generate the Borel σ-field, this determines P as a Dirichlet process, by Proposition A.5. □

4.5 Mixtures of Dirichlet Processes

Application of the Dirichlet process prior requires a choice of a base measure α. It is often reasonable to choose the center measure $\bar{\alpha}$ from a specific family such as the normal family, but then the parameters of the family must still be specified. It is natural to give these a further prior. Similarly, one may put a prior on the precision parameter $|\alpha|$.

For a base measure α_ξ that depends on a parameter ξ, the Bayesian model then consists of the hierarchy

$$X_1, \ldots, X_n | P, \xi \overset{\text{iid}}{\sim} P, \qquad P | \xi \sim \mathrm{DP}(\alpha_\xi), \qquad \xi \sim \pi. \tag{4.27}$$

We denote the induced (marginal) prior on P by $\mathrm{MDP}(\alpha_\xi, \xi \sim \pi)$. Many properties of this *mixture of Dirichlet processes* (MDP) prior follow immediately from those of a Dirichlet process. For instance, any P following an MDP is almost surely discrete. However, unlike a Dirichlet process, an MDP is not tail-free.

Expressions for prior mean and variance are readily obtained from the corresponding ones for the Dirichlet process (see Proposition 4.2) and conditioning:

$$\begin{aligned}\mathrm{E}(P(A)) &= \int \bar{\alpha}_\xi(A)\, d\pi(\xi) =: \bar{\alpha}_\pi(A),\\ \mathrm{var}\,(P(A)) &= \int \frac{\bar{\alpha}_\xi(A)\bar{\alpha}_\xi(A^c)}{1 + |\alpha_\xi|}\, d\pi(\xi) + \int (\bar{\alpha}_\xi(A) - \bar{\alpha}_\pi(A))^2\, d\pi(\xi).\end{aligned}$$

The expressions for means and variances of functions $\int \psi\, dP$ generalize likewise.

Example 4.31 (Normal location) In a "robustified" location problem $X = \xi + \varepsilon$ we put priors both on the location parameter and the error distribution. The specification $\varepsilon|\, H \sim H$ and $H \sim \mathrm{DP}(\mathrm{Nor}(0, 1))$ is equivalent to specifying $X|\, P, \xi \sim P$ and $P|\, \xi \sim \mathrm{DP}(\mathrm{Nor}(\xi, 1))$. If $\xi \sim \mathrm{Nor}(0, 1)$, then after observing n i.i.d. observations $X_1, \ldots, X_n$,

$$\mathrm{E}(P(A)|\, X_1, \ldots, X_n) = \frac{1}{n+1}\int \Phi(A-\xi)\phi(\xi)\, d\xi + \frac{n}{n+1}\mathbb{P}_n(A),$$

where Φ is the cumulative distribution function of the standard normal distribution and ϕ is its density.

The posterior distribution of P in the scheme (4.27) turns out to be another MDP. Given ξ we can use the posterior updating rule for the ordinary Dirichlet process, and obtain that

$$P|\, \xi, X_1, \ldots, X_n \sim \mathrm{DP}(\alpha_\xi + n\mathbb{P}_n). \tag{4.28}$$

To obtain the posterior distribution of P given $X_1, \ldots, X_n$, we need to mix this over ξ relative to its posterior distribution given $X_1, \ldots, X_n$. By Bayes's theorem the latter has density proportional to

$$\xi \mapsto \pi(\xi)\, p(X_1, \ldots, X_n|\, \xi).$$

Here the marginal density of $X_1, \ldots, X_n$ given ξ (the second factor) is described by the generalized Pólya urn scheme (4.13) with α_ξ instead of α. In general, this has a somewhat complicated structure due to the ties between the observations. However, for a posterior calculation we condition on the observed data $X_1, \ldots, X_n$, and know the partition that they generate. Given this information the density takes a simple form. In particular, if the observations are distinct (which happens with probability one if the observations actually follow a continuous distribution), then the Pólya urn scheme must have simply generated a random sample from the center measure $\bar{\alpha}_\xi$, in which case the preceding display becomes

$$\pi(\xi)\prod_{i=1}^n d\alpha_\xi(X_i)\prod_{i=1}^n \frac{1}{|\alpha_\xi| + i - 1},$$

for $d\alpha_\xi$ a density of α_ξ. This is somewhat awkward in a nonparametric model, and it will be seen later that it has negative implications towards consistency. Further calculations depend on the specific family and its parameterization.

The posterior mean and variance of $P(A)$ for a set A are those of an MDP process with parameters as in (4.28), and given by, for $\alpha_{\xi,n} = \alpha_\xi + n\mathbb{P}_n$ the base measure,

$$\mathrm{E}(P(A)|\, X_1, \ldots, X_n) = \int \bar{\alpha}_{\xi,n}(A)\, d\pi(\xi|\, X_1, \ldots, X_n) =: \tilde{P}_n(A), \tag{4.29}$$

$$\begin{aligned}\mathrm{var}\,(P(A)|\, X_1, \ldots, X_n) &= \int \frac{\bar{\alpha}_{\xi,n}(A)\bar{\alpha}_{\xi,n}(A^c)}{1 + |\alpha_\xi| + n}\, d\pi(\xi|\, X_1, \ldots, X_n)\\ &\quad + \int (\bar{\alpha}_{\xi,n}(A) - \tilde{\mathbb{P}}_n(A))^2\, d\pi(\xi|\, X_1, \ldots, X_n).\end{aligned}$$

If the precision $|\alpha_\xi|$ is bounded above uniformly in ξ, then $\bar{\alpha}_{\xi,n}$ and hence the posterior mean $\tilde{P}_n$ is equivalent to the empirical measure $\mathbb{P}_n$ up to order n^{-1} as $n \to \infty$, just as the posterior mean of the ordinary Dirichlet. Furthermore, the posterior variance is bounded above by $5/(4(n+1)) \to 0$. Thus the posterior distribution of $P(A)$ contracts to a Dirac measure at the true probability of A whenever $|\alpha_\xi|$ is bounded above. (The latter condition can be relaxed substantially; see Problem 4.35.)

Typically the precision parameter M and center measure G in $\alpha = MG$ will be modeled as independent under the prior. The posterior calculation then factorizes in these two parameters. To see this, consider the following scheme to generate the parameters and observations:

(i) Generate M from its prior.
(ii) Given M generate a random partition $S = \{S_1, \ldots, S_{K_n}\}$ according to the distribution given in Proposition 4.11.
(iii) Generate G from its prior, independently of (M, S).
(iv) Given (S, G) generate a random sample of size K_n from G, independently of M, and set X_i with $i \in S_j$ equal to the jth value in this sample.

By the description of the Pólya urn scheme this indeed gives a sample $X_1, \ldots, X_n$ from the mixture of Dirichlet processes MDP($MG, M \sim \pi, G \sim \pi$). We may now formally write the density of $(M, S, G, X_1, \ldots, X_n)$ in the form, with π abusively denoting prior densities for both M and G and p conditional densities of observed quantities,

$$\pi(M)\, p(S|\, M)\, \pi(G)\, p(X_1, \ldots, X_n|\, G, S).$$

Since this factorizes in terms involving M and G, these parameters are also independent under the posterior distribution, and the computation of their posterior distributions can be separated.

The term involving M depends on the data through K_n only (the latter variable is *sufficient* for M). Indeed, by Proposition 4.11 it is proportional to

$$M \mapsto \pi(M)\frac{M^{K_n}\Gamma(M)}{\Gamma(M+n)} \propto \pi(M)M^{K_n}\int_0^1 \eta^{M-1}(1-\eta)^{n-1}\,d\eta.$$

Rather than by (numerically) integrating this expression, the posterior density is typically computed by simulation. Suppose that $M \sim \mathrm{Ga}(a,b)$ a priori, and consider a fictitious random vector (M, η) with $0 \le \eta \le 1$ and joint (Lebesgue) density proportional to

$$\pi(M)M^{K_n}\eta^{M-1}(1-\eta)^{n-1} \propto M^{a+K_n-1}e^{-M(b-\log\eta)}\eta^{-1}(1-\eta)^{n-1}.$$

Then by the preceding display the marginal density of M is equal to its posterior density (given K_n, which is fixed for the calculation). Thus simulating from the distribution of (M, η) and dropping η simulates M from its posterior distribution. The conditional distributions are given by

$$M|\,\eta, K_n \sim \mathrm{Ga}(a+K_n, b-\log\eta), \qquad \eta|\,M, K_n \sim \mathrm{Be}(M, n). \tag{4.30}$$

We can use these in a *Gibbs sampling scheme*: given an arbitrary starting value η_0, we generate a sequence $M_1, \eta_1, M_2, \eta_2, M_3, \ldots$, by repeatedly generating M from its conditional distribution given (η, K_n) and η from its conditional distribution given (M, K_n), each time setting the conditioning variable (η or M) equal to its last value.

4.6 Modifications

In many applications the underlying probability measure is required to satisfy specific constraints, but still be nonparametric in nature. For instance, natural constraints on the error distribution in a location or regression problem are symmetry, zero mean or zero median, leading to the identifiability of the regression function. The first two subsections discuss methods of imposing constraints on a distribution obtained from a Dirichlet process. The third subsection discusses a modification of the Dirichlet process that can regulate correlation between probabilities of sets.

4.6.1 Invariant Dirichlet Process

Two methods to construct a nonparametric prior supported on the space of all probability measures symmetric about zero are to symmetrize a Dirichlet process $Q \sim \mathrm{DP}(\alpha)$ on $\mathbb{R}$ to $A \mapsto (Q(A) + Q(-A))/2$, where $-A = \{x: -x \in A\}$, and to "unfold" a Dirichlet process Q on $[0, \infty)$ to $A \mapsto (Q(A \cap [0, \infty)) + Q(-A \cap [0, \infty)))/2$. These methods can be generalized to constructing random probability measures that are invariant under a group of transformations of the sample space. Symmetry is the special case of the group $\{1, -1\}$ on $\mathbb{R}$ consisting of the identity and the reflection in zero.

Let $\mathfrak{G}$ be a compact metrizable group acting continuously on the Polish sample space $\mathfrak{X}$.[6] For a Borel measure Q and $g \in \mathfrak{G}$ the measure $A \mapsto \int \mathbb{1}\{g(x) \in A\}\, dQ(x)$ is well defined and measurable in g by Fubini's theorem; we denote it (as usual) by $Q \circ g^{-1}$. A measure P on $\mathfrak{X}$ is called *invariant* under the group $\mathfrak{G}$ if $P \circ g^{-1} = P$ for all $g \in \mathfrak{G}$. There are two basic methods of constructing invariant measures. Both utilize the (unique, both left and right) Haar probability measure on $\mathfrak{G}$, which we denote by μ.

The first method is to start with an arbitrary Borel measure Q on the sample space, and consider the measure

$$A \mapsto \int Q \circ g^{-1}(A)\, d\mu(g). \tag{4.31}$$

From the fact that $g \circ h \sim \mu$ for every $h \in \mathfrak{G}$ if g is distributed according to the Haar measure μ, it can be seen that this measure is invariant.

The second construction of an invariant random measure follows the idea of unfolding. The *orbit* of a point $x \in \mathfrak{X}$ is the set $\mathfrak{O}(x) := \{g(x): g \in \mathfrak{G}\}$. The orbits are the equivalence classes under the relationship $x \equiv y$ if and only if $g(x) = y$ for some $g \in \mathfrak{G}$, and are invariant under the action of $\mathfrak{G}$. The set of all orbits is denoted by $\mathfrak{X}/\mathfrak{G}$ and is a measurable space relative to the *quotient σ-field*: the largest σ-field making the *quotient map* $x \mapsto \mathfrak{O}(x)$ measurable. If $\mathfrak{X}/\mathfrak{G}$ is a standard Borel space, then *Burgess's theorem* (cf. Theorem 5.6.1 of Srivastava 1998) allows us to choose a representative from each orbit to form a Borel measurable subset $\mathfrak{R}$ of $\mathfrak{X}$ that is isomorphic to $\mathfrak{X}/\mathfrak{G}$.[7] We may now start with a Borel probability measure Q on $\mathfrak{R}$ and "unfold" it to the full space as

[6] Every $g \in \mathfrak{G}$ is a map $g: \mathfrak{X} \mapsto \mathfrak{X}$, the group operation of $\mathfrak{G}$ is composition of maps $\circ$, and the *group action* $(g, x) \mapsto g(x)$ is continuous.

[7] Since $\mathfrak{G}$ is compact in our case, the full power of Burgess's theorem is not necessary. The same result can be derived from the simpler Effros theorem (cf. Theorem 5.4.2 of Srivastava 1998).

$$A \mapsto \int Q(g^{-1}(A) \cap \mathfrak{R})\, d\mu(g). \tag{4.32}$$

The formula has the same structure as (4.31), and hence the measure is invariant.

The second construction is the special case of the first where the measure Q is concentrated on the subset $\mathfrak{R} \subset \mathfrak{X}$. If the action of $\mathfrak{G}$ is *free* ($g(x) \neq x$ for any $x \in \mathfrak{X}$ and any $g \in \mathfrak{G}$ not equal to the identity), then $\mathfrak{X}$ can be identified with the product space $\mathfrak{R} \times \mathfrak{G}$ through the map $g(r) \leftrightarrow (r, g)$, and any invariant probability measure P on $\mathfrak{X}$ can be represented as $Q \times \mu$, giving $P(A) = Q \times \mu((r, g){:}\, g(r) \in A) = \int Q(g^{-1}(A))\, d\mu(g)$.[8] Thus every invariant measure can be constructed using either method. This may be extended to the situation that a single point $x \in \mathfrak{X}$ is invariant under *every* $g \in \mathfrak{G}$, as is the case for the symmetry group on $\mathbb{R}$, or the group of rotations.

The preceding constructions apply equally well to *random* measures. Thus we may start with a Dirichlet process $P \sim \mathrm{DP}(\alpha)$ on the full space $\mathfrak{X}$ and make it invariant using (4.31), or take a Dirichlet process Q on $\mathfrak{R} = \mathfrak{X}/\mathfrak{G}$ and apply (4.32). Both constructions lead to an "invariant Dirichlet process," as given in the following definition. Say that a set $A \subset \mathfrak{X}$ is *invariant* if $g(A) = A$ for every $g \in \mathfrak{G}$, and call a random measure *invariant* if all its realizations are invariant.

Definition 4.32 (Invariant Dirichlet process) An invariant random measure P is said to follow an *invariant Dirichlet process* $\mathrm{IDP}(\alpha, \mathfrak{G})$ with (invariant) base measure α, if for every measurable partition of $\mathfrak{X}$ into $\mathfrak{G}$-invariant sets $A_1, \ldots, A_k$,

$$(P(A_1), \ldots, P(A_k)) \sim \mathrm{Dir}(k; \alpha(A_1), \ldots, \alpha(A_k)).$$

The base measure α in this definition is a finite positive Borel measure on $\mathfrak{X}$, which without loss of generality is taken to be invariant. Since by definition an invariant set A satisfies $g^{-1}(A) = A$ for any $g \in \mathfrak{G}$, it is obvious that the invariant version of Q in (4.31) coincides with Q on invariant sets. Hence if $Q \sim \mathrm{DP}(\alpha)$ for an invariant base measure, then the invariant version (4.31) is an $\mathrm{IDP}(\alpha, \mathfrak{G})$-process. Alternatively we may use the second construction (4.32), starting from a $\mathrm{DP}(\tilde{\alpha})$-process Q on the quotient space $\mathfrak{R}$, with arbitrary base measure $\tilde{\alpha}$ on $\mathfrak{R}$. For an invariant set A the integral (4.32) is simply $Q(A \cap \mathfrak{R})$ and hence the marginal distributions of the process $P(A)$ in this construction are $\mathrm{Dir}(k; \tilde{\alpha}(A_1 \cap \mathfrak{R}), \ldots, \tilde{\alpha}(A_k \cap \mathfrak{R}))$. By unfolding $\tilde{\alpha}$ these can also be written as in Definition 4.32 for an invariant measure α on $\mathfrak{X}$.[9]

It should be noted that an invariant Dirichlet process is itself not a Dirichlet process (as the name might suggest), but is in fact a mixture of Dirichlet processes. A Dirichlet process $P \sim \mathrm{DP}(\alpha)$ for an invariant base measure α possesses some distributional invariance properties, but is not invariant. For instance, in the symmetry case a Dirichlet process will split the mass to the negative and positive half lines by a $\mathrm{Be}(1/2, 1/2)$-variable, whereas the invariant Dirichlet process (4.31) makes a deterministic $(1/2, 1/2)$ split.

[8] Every $\mathfrak{G}$-invariant set A can be represented as $\tilde{A} \times \mathfrak{G}$, where $\tilde{A} = \{x \in \mathfrak{R}{:}\, \mathfrak{O}(x) \cap A \neq \varnothing\}$.

[9] Both constructions can be extended to any distribution on $\mathfrak{M}$, including the Pólya tree process or the Dirichlet mixture process. The correspondence between the two constructions is then not as neat as in the Dirichlet case, where the same invariant random measure is obtained by making the base measures α and $\tilde{\alpha}$ correspond.

Many of the properties of an invariant Dirichlet process are similar to those of a Dirichlet process. Suppose $P \sim \mathrm{IDP}(\alpha, \mathfrak{G})$ for an invariant measure α.

(i) For any invariant α-integrable or nonnegative measurable function ψ we have $\mathrm{E}(\int \psi\, dP) = \int \psi\, d\bar{\alpha}$. In particular, the marginal distribution of $X|\, P \sim P$ is $\bar{\alpha}$.
(ii) Any $\mathfrak{G}$-invariant probability measure P_0 with $\mathrm{supp}(P_0) \subset \mathrm{supp}(\alpha)$ belongs to the weak support of $\mathrm{IDP}(\alpha, \mathfrak{G})$.
(iii) If $X_1, \ldots, X_n|\, P \stackrel{\text{iid}}{\sim} P$ and $P \sim \mathrm{IDP}(\alpha, \mathfrak{G})$, then

$$P|\, X_1, \ldots, X_n \sim \mathrm{IDP}\Big(\alpha + \sum_{i=1}^{n} \int \delta_{g(X_i)}\, d\mu(g), \mathfrak{G}\Big). \tag{4.33}$$

Unlike the Dirichlet process, the invariant Dirichlet process is not tail-free: the partitions in a tail-free representation get mixed up by the action of the group $\mathfrak{G}$.

Property (i) follows easily from the representation $P = \int Q \circ g^{-1}\, d\mu(g)$ for $Q \sim \mathrm{DP}(\alpha)$, while property (ii) can be proved along the lines of Theorem 4.15. To prove (iii), observe that by the first representation of the IDP, the model may be described hierarchically as a mixture of Dirichlet processes:

$$X_i|\, Q, g_1, \ldots, g_n \stackrel{\text{ind}}{\sim} Q \circ g_i, \qquad Q \sim \mathrm{DP}(\alpha), \qquad g_i \stackrel{\text{iid}}{\sim} \mu, \quad i = 1, \ldots, n.$$

It follows that $(g_1(X_1), \ldots, g_n(X_n))|\, (Q, g_1, \ldots, g_n) \stackrel{\text{iid}}{\sim} Q$ and hence $Q|\, (X_1, \ldots, X_n, g_1, \ldots, g_n) \sim \mathrm{DP}(\alpha + \sum_{i=1}^n \delta_{g_i(X_i)})$ by Theorem 4.6 applied conditionally given $g_1, \ldots, g_n$. The law P of the observations is the deterministic transformation $P = \int Q \circ g^{-1}\, d\mu(g)$ of Q and hence its posterior is obtained by substituting the posterior of Q in this integral. This posterior is certainly invariant, as it has the form (4.31). It remains to show that its marginal distribution on an invariant partition is a Dirichlet with base measure as given. Because $P(A) = Q(A)$ for every invariant set A, we may concentrate on the posterior law of $(Q(A_1), \ldots, Q(A_k))$ for an invariant partition. By the preceding, it is a mixture of $\mathrm{Dir}((\alpha + \sum_{i=1}^n \delta_{g_i(X_i)})(A_1), \ldots, (\alpha + \sum_{i=1}^n \delta_{g_i(X_i)})(A_k))$-distributions relative to the posterior distribution of $(g_1, \ldots, g_n)$. Since $\delta_{g(x)}(A) = \delta_x(A)$ for invariant set A and every g, the components of this mixture are actually identical and the mixture is a Dirichlet distribution with parameters $((\alpha + \sum_{i=1}^n \delta_{X_i})(A_1), \ldots, (\alpha + \sum_{i=1}^n \delta_{X_i})(A_1))$. For a representation through an invariant base measure we replace each term $\delta_{X_i}(A_j)$ by its invariant version $\int \delta_{g(X_i)}(A_j)\, d\mu(g)$, and we arrive at the form (4.33).

The probability measure $\int \delta_{g(x)}\, d\mu(g)$, which appears in the base measure of the posterior distribution in (4.33), is restricted to the orbit $\mathfrak{O}(x)$ of x. If the action of $\mathfrak{G}$ on $\mathfrak{X}$ is free, then it can be regarded as the "uniform distribution on this orbit".

Example 4.33 (Symmetry) Distributions on $\mathbb{R}$ that are symmetric about 0 are invariant under the reflection group $\mathfrak{G} = \{1, -1\}$ consisting of the map $x \mapsto g \cdot x$. The $\mathrm{IDP}(\alpha, \{-1, 1\})$-process is the *symmetrized Dirichlet process* considered at the beginning of the section. The posterior mean of the cumulative distribution function F based on a random sample of size n is given by, for $\alpha(x) = \alpha(-\infty, x]$,

$$\mathrm{E}(F(x)|\, X_1, \dots, X_n) = \frac{\alpha(x) + \frac{1}{2}\sum_{i=1}^n (\mathbb{1}(X_i \le x) + \mathbb{1}(X_i \ge -x))}{|\alpha| + n}. \tag{4.34}$$

This is the natural Bayes estimator for F.

Example 4.34 (Rotational invariance) A rotationally invariant random measure on $\mathbb{R}^k$ may be constructed as an invariant Dirichlet process under the action of the group of orthogonal $k \times k$ matrices. The action of the orthogonal group is free if the origin is removed and hence the posterior base measure is the sum of the base measure and the uniform distributions on the orbits of $X_1, \dots, X_n$. This is one of the natural examples of a group of infinite cardinality.

Example 4.35 (Exchangeability) The exchangeable measures on $\mathbb{R}^k$ are invariant measures under the action of the permutation group Σ_k of order k on the order of the coordinates of the vectors. Therefore an invariant Dirichlet process under the action of the permutation group is a suitable prior on the exchangeable measures. The permutation group action is free. The posterior expectation of the cumulative distribution function given a random sample of observations $X_1 = (X_{1,1}, \dots, X_{1,k}), \dots, X_n = (X_{n,1}, \dots, X_{n,k})$ from an IDP(α, Σ_k)-prior, is given by

$$\begin{aligned}&\mathrm{E}(F(y_1, \dots, y_k)|\, X_1, \dots, X_n)\\ &\quad= \frac{\alpha(y_1, \dots, y_k) + \frac{1}{k!}\sum_{i=1}^n \sum_{\sigma\in\Sigma_k} \mathbb{1}\{X_{i,\sigma(1)} \le y_1, \dots, X_{i,\sigma(k)} \le y_k\}}{|\alpha| + n}.\end{aligned}$$

Example 4.36 (Orthant symmetry) The group consisting of the coordinatewise multiplications $(x_1, \dots, x_k) \mapsto (g_1x_1, \dots, g_kx_k)$ of vectors in $\mathbb{R}^k$ by elements $g \in \{-1, 1\}^k$ generates measures that are invariant under relocation of orthants. It generalizes the notion of symmetry on the line. The posterior expectation of the cumulative distribution function of an IDP(α, $\{-1, 1\}^k$)-random measure, given observations $X_1, \dots, X_n$ as in the preceding example, is given by

$$\begin{aligned}&\mathrm{E}(F(y_1, \dots, y_k)|\, X_1, \dots, X_n)\\ &\quad= \frac{\alpha(y_1, \dots, y_k) + \frac{1}{k!}\sum_{i=1}^n \sum_{g\in\{-1,1\}^k} \mathbb{1}\{g_1X_{i,1} \le y_1, \dots, g_kX_{i,k} \le y_k\}}{|\alpha| + n}.\end{aligned}$$

4.6.2 Constrained Dirichlet Process

Given a finite measurable partition $\{A_1, \dots, A_k\}$ of the sample space (called *control sets*), the conditional distribution of a DP(α)-process P given $(P(A_1), \dots, P(A_k)) = w$ for a given vector $w = (w_1, \dots, w_k)$ in the k-dimensional unit simplex is called a *constrained Dirichlet process.*

Such a process can actually be constructed as a finite mixture $P = \sum_{j=1}^k w_j P_j$ of independent DP($w_j\alpha|_{A_j}$)-processes P_j (with orthogonal supports). This follows, since by the self-similarity of Dirichlet processes (see Theorem 4.5),

$$(P(A_1), \dots, P(A_k)) \perp\!\!\!\perp P_{A_1} \perp\!\!\!\perp \cdots \perp\!\!\!\perp P_{A_k},$$

where each P_A is a DP$(\alpha|_A)$-process. The posterior distribution can be obtained as for general mixtures of Dirichlet processes (see Section 4.5). In particular the posterior mean follows (4.29).

The particular case of a binary partition $A_1 = (-\infty, 0)$, $A_2 = [0, \infty)$ of $\mathbb{R}$ with $w_1 = w_2 = 1/2$ yields a random distribution with median 0. Constraints on multiple quantiles can be placed in a similar manner.

Restriction on moments are possible as well. For instance, a random measure P with mean 0 can be generated as $P(A) = Q(A - \int x\, dQ(x))$ for $Q \sim \mathrm{DP}(\alpha)$. However, such a scheme does not lead to a simple representation in terms of an MDP, because the change of location is random and dependent on Q.

4.6.3 Penalized Dirichlet Process

That the Dirichlet process selects discrete distributions with probability one (see Theorem 4.14) may be construed as an absence of smoothing. This lack of smoothing leads to anomalous relationships between the random probabilities of collections of sets.

Given a measurable set A and an observation $X = x$ with $x \notin A$, we have under the DP(α) prior on the distribution of X, since $\delta_x(A) = 0$,

$$\mathrm{E}(P(A)|\, X = x) = \frac{\alpha(A)}{|\alpha| + 1} < \frac{\alpha(A)}{|\alpha|} = \mathrm{E}(P(A)). \tag{4.35}$$

By itself it is natural that the expectation of $P(A)$ decreases, since $x \notin A$ makes A "less likely." However, the posterior expectation is smaller than the prior expectation no matter the proximity of x to A: the Dirichlet process does not respect closeness. This is counter-intuitive if one is used to continuity, as one tends to believe that an observation in a locality enhances the probability of the locality.

Another anomalous property is the negative correlation between probabilities of disjoint sets, as is evident from (4.4), even when the sets are adjacent to each other. This blanket attachment of negative correlation again indicates a complete lack of smoothing. This may for instance be considered undesirable for random histograms. The histograms obtained from a continuous density should have similar heights across neighboring cells, and a random histogram should comply with this criterion by assigning positive correlation to probability ordinates of neighboring cells. The Dirichlet process is too flexible for this purpose. Although binning takes care of smoothing within a cell, no smoothing across bins is offered, which will result in a less efficient density estimator.

In order to rectify this problem we may introduce positive correlation among the probabilities of neighboring cells, by penalizing too much variation between neighboring cells.

Consider density estimation on a bounded interval by random histograms. Because the binning at any stage gives rise to only finitely many cell probabilities, it is enough to modify the finite-dimensional Dirichlet distribution. Instead of the finite-dimensional Dirichlet density, consider the density proportional to $(p_1, \ldots, p_k) \mapsto p_1^{\alpha_1 - 1} \cdots p_k^{\alpha_k - 1} e^{-\lambda \Delta(p)}$, where $\Delta(p)$ is a penalty term for roughness. The choice $\lambda = 0$ returns the ordinary Dirichlet distribution. Meaningful choices of the penalty $\Delta(p)$ are $\sum_{j=1}^{k-1} (p_{j+1} - p_j)^2$, which helps control variation of successive cell probabilities, $\sum_{j=2}^{k-1} (p_{j+1} - 2p_j + p_{j-1})^2$, which helps control the second-order differences of cell probabilities, and $\sum_{j=1}^{k-1}(\log p_{j+1} - \log p_j)^2$, which helps control the ratios of successive cell probabilities. Interestingly, the resulting posterior distribution based on a random sample $X_1, \ldots, X_n |\, p \overset{\text{iid}}{\sim} p$ has the same form as the prior, with α_i updated to $\alpha_i + \sum_{j=1}^{n} \mathbb{1}\{X_j = i\}$. The posterior mean, if desired, could be obtained by numerical integration or by the Metropolis-Hastings algorithm.

4.7 Bayesian Bootstrap

The weak limit $\mathrm{DP}(\sum_{i=1}^n \delta_{X_i})$ of the posterior distribution $\mathrm{DP}(\alpha + \sum_{i=1}^n \delta_{X_i})$ in the non-informative limit as $|\alpha| \to 0$ is called the *Bayesian bootstrap* (BB) distribution. Its center measure is the empirical measure $\mathbb{P}_n = n^{-1}\sum_{i=1}^n \delta_{X_i}$ and a random probability generated from it is necessarily supported on the observation points. In this sense it compares to Efron's bootstrap, justifying the name. In fact, both the Bayesian and Efron's bootstrap can be represented as $\sum_{i=1}^n W_i \delta_{X_i}$, where $(W_1, \ldots, W_n) \sim \mathrm{Dir}(n; 1, \ldots, 1)$ for the Bayesian bootstrap (the uniform distribution on the unit simplex) and $n(W_1, \ldots, W_n) \sim \mathrm{MN}_n(n; n^{-1}, \ldots, n^{-1})$ for Efron's bootstrap. Thus the Bayesian bootstrap assigns continuous weights (and in particular, puts positive weight to every observation on every realization), and leads to smoother empirical distributions of resampled variables than Efron's bootstrap, which typically assigns zero values to some observations.

When n is moderate or large, the empirical part in the Dirichlet posterior $\mathrm{DP}(\alpha + \sum_{i=1}^n \delta_{X_i})$ will dominate, and hence the posterior distribution of a linear or non-linear function will be close to that under the Bayesian bootstrap. The biggest advantage of the Bayesian bootstrap relative to the true posterior is that samples are easily generated from the Bayesian bootstrap, allowing the approximation of the posterior distribution of any quantity of interest by simple Monte Carlo methods. The representation $W_i = Y_i / \sum_{j=1}^n Y_j$, where $Y_i \overset{\text{iid}}{\sim} \mathrm{Exp}(1)$, for the weights of the Bayesian bootstrap (cf. Proposition G.2) is convenient here. The distributions of resampled variables may be used to compute both estimates and credible intervals.

By Theorem 4.16, the Bayesian bootstrap based on a random sample $X_1, \ldots, X_n$ from a distribution P_0 converges weakly to the degenerate measure at P_0, as $n \to \infty$. This suggests that the Bayesian bootstrap-based estimators and credible intervals have frequentist validity like Efron's bootstrap. This has indeed been shown for the more general class of *exchangeable bootstrap* processes (see van der Vaart and Wellner 1996, Chapter 3.6). We discuss a large sample property of the Bayesian bootstrap in Chapter 12.

The density of a mean functional $\mu = \int \psi\, dP$ under the Bayesian bootstrap may also be obtained analytically from Theorem 4.26. With $Y_i = \psi(X_i)$, it leads to the following formula

$$p(\mu |\, X_1, \ldots, X_n) = (n-1) \sum_{i=1}^n \frac{(Y_i - \mu)_+^{n-2}}{\prod_{j \neq i}(Y_i - Y_j)},$$

provided that $Y_i \neq Y_j$ a.s. for all $i \neq j$. The density is a spline function of order $(n-2)$ with knots at $Y_1, \ldots, Y_n$, and has $(n-3)$ continuous derivatives.

The multidimensional mean functional $\mu = (\int \psi_1\, dP, \ldots, \int \psi_s\, dP)^\top$, can similarly be shown to be an $(n-s-2)$-times continuously differentiable if $Y_i := (\psi_j(X_i);\, j =$

$1, \dots, s)^\top$, $i = 1, \dots, n$, do not lie on any lower-dimensional hyperplane. The resulting density function is called the s-variate B-spline with knots at $Y_1, \dots, Y_n$. Some recursive formula permit writing an s-variate B-spline in terms of $(s-1)$-variate ones; see Micchelli (1980). Nevertheless, the resulting density is very complicated, so it is easier to adopt a simulation based approach. Choudhuri (1998) discussed a simulation technique exploiting the log-concavity of the density of multivariate mean functional of the Bayesian bootstrap, and also showed that the sampling based approach with N resamples estimates a credible set within $O(N^{-1/(s+2)} \log N)$ in the metric defined by the Lebesgue measure of symmetric difference of sets.

The Bayesian bootstrap may be smoothed by convolving it with a kernel giving a substitute for posterior of density function which is computable without any MCMC technique. However, the bandwidth parameter must be supplied depending on the sample size.

The weights $W_i = Y_i / \sum_{j=1}^n Y_j$ in the Bayesian bootstrap may be generalized to Y_is other than i.i.d. standard exponential, leading to what is called the *Bayesian bootstrap clone*. The particular choice Ga(4, 1) for the distribution of the Y_i leads to $(W_1, \dots, W_n) \sim \mathrm{Dir}(n; 4, \dots, 4)$. This is sometimes used to achieve higher-order agreement between credibility and confidence of a Bayesian bootstrap interval. The Bayesian bootstrap clone in turn is a special case of the *exchangeable bootstrap*, which only assumes that the weights are exchangeable.

4.8 Historical Notes

Ferguson (1973) introduced the Dirichlet process, which put Bayesian nonparametrics on firm ground by exhibiting practical feasibility of posterior computations. The importance of the Dirichlet process in Bayesian nonparametrics is comparable to that of the normal distribution in probability and general statistics. Ferguson (1973) also obtained prior moments, showed posterior conjugacy, and discussed the naive construction as well as the construction through a gamma process, based on an earlier idea of Ferguson and Klass (1972). Blackwell and MacQueen (1973) noticed the connection between the Pólya urn scheme and the Dirichlet process and presented the construction of a Dirichlet process through the Pólya urn scheme. The stick-breaking representation first appeared in McCloskey (1965), and was rediscovered in the statistical context by Sethuraman (1994). It has turned out to be very helpful in dealing with complicated functionals of a Dirichlet process. The ϵ-Dirichlet process approximation was studied by Muliere and Tardella (1998) and Gelfand and Kottas (2002). The discreteness property of the Dirichlet process was inferred by Ferguson (1973) from the pure jump nature of the gamma process, and by Blackwell and MacQueen (1973) as a consequence of the Pólya urn scheme. The proof presented here is possibly due to Savage. The weak support of the Dirichlet process was characterized by Ferguson (1973). Weak convergence of a sequence of Dirichlet processes were discussed in Sethuraman and Tiwari (1982). Ganesh and O'Connell (2000) observed that the posterior distribution of a Dirichlet process satisfies a large deviation principle under the weak topology (see Problem 4.14). Finer moderate deviation properties of a random probability measure "exchangeable with respect to a sequence of partitions," centered at the mean and scaled by an appropriately growing sequence, were studied by Eichelsbacher and Ganesh (2002). Such random processes include the Dirichlet process as well as Pólya tree processes. Convergence of the multinomial-Dirichlet process to the general Dirichlet process was shown by Ishwaran and Zarepour (2002), and is especially useful for computation using BUGS; see Gilks et al. (1994). The marginal distribution of a Dirichlet sample and the distribution of the number of distinct observations were discussed by Antoniak (1974). Identifiability of the Dirichlet base measure based on infinitely many random samples and mutual singularity under non-atomicity were shown by Korwar and Hollander (1973). Tails of a Dirichlet process were studied by Doss and Sellke (1982). Formulas for the distribution of the mean functional of a Dirichlet process, including numerical approximations to the analytic formulas, were obtained by Cifarelli and Regazzini (1990), Regazzini et al. (2002) using the method to study ratios developed in Gurland (1948), and generalized to normalized independent increment processes by Regazzini et al. (2003). Several characterizations of the Dirichlet process were stated by Ferguson (1974) without proof. Antoniak (1974) studied mixtures of Dirichlet processes. Dalal (1979) studied symmetrized Dirichlet processes. The Dirichlet process conditioned to have median zero has been used in many contexts, such as in Doss (1985a,b) in modeling of error distributions in a location problem. The penalization idea to the Dirichlet distribution and its use in histogram smoothing was discussed by Hjort (1996). The Bayesian bootstrap was introduced by Rubin (1981) and later studied by many authors. Gasparini (1996) and Choudhuri (1998) studied the multidimensional mean functional in detail. Large sample properties of the Bayesian bootstrap were studied in Lo (1987) and Præstgaard and Wellner (1993); also see van der Vaart and Wellner (1996), Chapter 3.7.

Problems

4.1 Show that if $P \sim \mathrm{DP}(\alpha)$ on $\mathfrak{X}$ and $\psi: \mathfrak{X} \to \mathfrak{Y}$ is measurable, then $P \circ \psi^{-1} \sim \mathrm{DP}(\alpha \circ \psi^{-1})$ on $\mathfrak{Y}$.

4.2 Find an expression for $\mathrm{E}[\prod_{i=1}^{k} \int \psi_i \, dP]$ for $k = 3, 4, \ldots$

4.3 Let $P \sim \mathrm{DP}(\alpha)$ on a product space $\mathfrak{X} \times \mathfrak{Y}$, and let P_1 and $P_{2|1}$, and $\bar{\alpha}_1$ and $\bar{\alpha}_{2|1}$, the marginal distributions on the first coordinate and the conditional distributions of the second given the first coordinate of P and $\bar{\alpha}$.

(a) Show that $P_1 \sim \mathrm{DP}(|\alpha|\bar{\alpha}_1)$. [Hint: Apply Problem 4.1 with the projection function or the stick-breaking representation of a Dirichlet process.]

(b) Show that almost surely $P_{2|1}$ is given by $\mathrm{DP}(\alpha(\{X\} \times \mathfrak{Y})\bar{\alpha}_{2|1})$ if $\alpha(\{X\} \times \mathfrak{Y}) > 0$, and is given by δ_Y for $Y \sim \bar{\alpha}_{2|1}$ otherwise, and is distributed independently of P_1. [Hint: For the first case, use the fact that P conditioned on $\{X\} \times \mathfrak{Y}$ is a Dirichlet process in view of the self-similarity property. Use the stick-breaking representation in the second case and observe that there can be at most one support point in this case.]

Note that if $P_1(\{x\}) = 0$, then $P_{2|1}(\cdot| x)$ can be defined arbitrarily. Furthermore, for $\alpha_1(\{x\}) = 0$, the measure δ_Y with $Y \sim \bar{\alpha}_{2|1}(\cdot|X)$ can be viewed as a limiting Dirichlet process with precision parameter $\alpha_1(\{x\}) = 0$ and center measure $\bar{\alpha}_{2|1}(\cdot|X)$. Hence the two different forms of the conditional distribution can be unified with a broader notation.

4.4 (Two sample testing, Ferguson 1973) Let $X \sim F$ and $Y \sim G$ independently, and suppose that we want to estimate $\mathrm{P}(X \leq Y)$ with m independent X observations and

n independent Y observations. Put independent Dirichlet priors on F and G. Find an expression for the Bayes estimate of $\mathrm{P}(X \le Y)$ and show that it reduces to the classical Mann-Whitney U-statistic under a noninformative limit.

4.5 (Tolerance interval, Ferguson 1973) Let F be an unknown c.d.f. on $\mathbb{R}$ and $0 < q < 1$ be given. Consider the problem of estimating the qth quantile $t_q(F)$ of F with the loss function $L(a, t_q(F)) = pF(a) + q\mathbb{1}\{a < t_q(F)\}$, where $0 < p < 1$ is also given, indicating relatively high penalty for underestimating $t_q(F)$. Let $F \sim \mathrm{DP}(MG)$.

Show that the Bayes risk of the no-data estimation problem is given by the rth quantile of G, where r is the unique minimizer of

$$u \mapsto pu + q \int_0^q \frac{\Gamma(M)}{\Gamma(Mu)\Gamma(M(1-u))} z^{Mu-1}(1-z)^{M(1-u)-1}\,dz,$$

to be denoted by $r(p, q, M)$.

Now if a sample $X_1, \ldots, X_n \overset{\text{iid}}{\sim} F$ is available, obtain the Bayes estimator of $t_q(F)$.

4.6 Prove Theorem 4.6 by directly checking that

$$\mathrm{E}(\mathrm{DP}_{\alpha+\delta_X}(C)\mathbb{1}\{X \in A\}) = \mathrm{P}\{P \in C, X \in A\},$$

where $C \in \mathscr{M}$, $A \in \mathscr{X}$, X has marginal distribution $\bar{\alpha}$ on the left-hand side, and (X, P) follows the Dirichlet model $X|\, P \sim P$ and $P \sim \mathrm{DP}(\alpha)$. [Hint: Consider C a finite dimensional projection set and then show that the two sides lead to the same sets of moments.]

4.7 Theorem 4.5 can be generalized to more than two localities. If $P \sim \mathrm{DP}(\alpha)$ and $A_1, \ldots, A_k \in \mathscr{X}$ are disjoint measurable sets such that $\alpha(A_i) > 0$, then $\{P(A_i) \colon i = 1, \ldots, k\} \perp\!\!\!\perp P_{A_i} \perp\!\!\!\perp \cdots \perp\!\!\!\perp P_{A_k}$ and $P_{A_i} \sim \mathrm{DP}(\alpha_{|A_i})$, for $i = 1, \ldots, k$.

4.8 Using the distributional equation (4.21), construct a Markov chain on $\mathfrak{M}$ which has stationary distribution $\mathrm{DP}(\alpha)$.

4.9 Prove Proposition 4.3 using (4.21).

4.10 Prove the posterior updating formula using the stick-breaking representation.

4.11 Let $E_1, E_2, \ldots \overset{\text{iid}}{\sim} \mathrm{Exp}(1)$. Define $\Gamma_0 = 0$, $\Gamma_k = \sum_{j=1}^k E_j$, $k \in \mathbb{N}$. Let $\theta_1, \theta_2, \ldots \overset{\text{iid}}{\sim} G$. Show that $P := \sum_{k=1}^\infty (e^{-\Gamma_{k-1}/M} - e^{-\Gamma_k/M})\delta_{\theta_k}$ is a random probability measure having distribution $\mathrm{DP}(MG)$.

4.12 (Muliere and Tardella 1998) Obtain sharp probability inequalities for the number of terms present in an ϵ-Dirichlet process defined in Subsection 4.3.3.

4.13 Consider a stick-breaking process identical to the Dirichlet process except that the stick-breaking distribution is generalized to $\mathrm{Be}(a, b)$. Find expressions for prior expectation and variance.

4.14 (Ganesh and O'Connell 2000) Large deviation principle: If $X_i \overset{\text{iid}}{\sim} P_0$, then for all Borel subsets $\mathcal{B}$ of $\mathfrak{M}$,

$$\begin{aligned}
-\inf_{P \in \mathcal{B}^o} K(P_0; P) &\le \liminf_{n\to\infty} \frac{1}{n} \log \mathrm{DP}_{\alpha+\sum_{i=1}^n \delta_{X_i}}(\mathcal{B}) \\
&\le \limsup_{n\to\infty} \frac{1}{n} \log \mathrm{DP}_{\alpha+\sum_{i=1}^n \delta_{X_i}}(\mathcal{B}) \le -\inf_{P \in \bar{\mathcal{B}}} K(P_0; P), \quad a.s.[P_0^\infty].
\end{aligned}$$

4.15 Write down the density of the marginal distribution of a sample of size $n = 3$ from a Dirichlet process explicitly. Develop a systematic notation for the description of the density of the joint distribution for a general n.

4.16 Obtain (4.14) using mathematical induction.

4.17 (Bell number) Compute the number of terms in Proposition 4.7; (this is known as a *Bell number* or *Bell's exponential number*).

4.18 Derive an asymptotic formula for $\mathrm{var}(K_n)$ when n and M both can vary (cf. Proposition 4.8(i)).

4.19 Show Part (iii) of Proposition 4.8 by computing moment-generating functions.

4.20 For $M = 1, 10$, and $n = 30, 100$, numerically evaluate the exact distribution of K_n, the number of distinct observations in a sample of size n from a Dirichlet process using Proposition 4.9.

Evaluate the same expression through a simulation scheme involving independent Bernoulli variables.

Compare these with the normal and the Poisson approximations described in Subsection 4.1.5.

4.21 Show that the *distinct* values in a random sequence $X_1, X_2, \ldots$ sampled from a distribution generated from a $\mathrm{DP}(\alpha)$-prior with atomless base measure form an i.i.d. sequence from $\bar{\alpha}$. (Thus define $\tau_1 = 1$ and $\tau_j = \min\{n: \sum_{i=1}^n D_i = j\}$ for $D_i = 1$ if $X_i \notin \{X_1, \ldots, X_{i-1}\}$, and put $Y_j = X_{\tau_j}$.)

4.22 While sampling from a Dirichlet process with precision parameter M, show that the maximum likelihood estimator for M based on data $K_n = k$ is unique, consistent and asymptotically equivalent to $K_n/\log n$.

4.23 For the center measure $G = \mathrm{Unif}(0, 1)$, show that the rth raw moment of m_F is given by $\int_0^1 B(\frac{1}{2}; Mx^{1/r}, (1 - x^{1/r}))\,dx$, where $B(\cdot; a, b)$ stands for the incomplete beta function.

Numerically evaluate and plot the mean, variance, skewness and kurtosis of this distribution as functions of M.

For $M = 2, 20$, numerically evaluate and plot the density function of m_F.

4.24 Obtain the distribution of the pth quantile of a random $F \sim \mathrm{DP}(MG)$.

4.25 In Theorem 4.26, show that if G is symmetric, so will be H about the same point.

4.26 (Diaconis and Kemperman 1996) If $P \sim \mathrm{DP}(G)$, where G is $\mathrm{Unif}(0, 1)$, then show that the density of $\int x\,dP$ is given by $e\pi^{-1}\sin(\pi y)y^{-y}(1 - y)^{y-1}\mathbb{1}\{0 < y < 1\}$.

4.27 (Regazzini et al. 2002) If $P \sim \mathrm{DP}(G)$, where G is $\mathrm{Exp}(\lambda)$, then show that the density of $\int x\,dP$ is given by $\pi^{-1}y^{-1}\sin(\pi - \pi e^{-\lambda y})\exp[e^{-\lambda y}\mathrm{PV}(\int_{-\infty}^{\lambda y} t^{-1}e^t\,dt)]$, where PV stands for the principal value of the integral.

4.28 (Yamato 1984, Cifarelli and Regazzini 1990, Lijoi and Regazzini 2004) Let $P \sim \mathrm{DP}(G)$, where G has Cauchy density given by $\sigma\pi^{-1}[1 + \sigma^2(x - \mu)^2]^{-1}$. Using the identity $\int \log|x - y|dG(x) = \frac{1}{2}\log[\{1 + \sigma^2(y - \mu)^2\}/\sigma^2]$, show that the distribution of the Dirichlet mean $\int x\,dP$ is again G. Further, the property characterizes the Cauchy distribution.

4.29 Let $\alpha = \theta_0\delta_0 + \theta_1\delta_1$. If $P \sim \mathrm{DP}(\alpha)$, then $\int x\,dP = P(\{1\}) \sim \mathrm{Be}(\theta_1, \theta_0)$. Derive the same result from the general formula for the distribution of a Dirichlet mean.

4.30 Let $P \sim \mathrm{DP}(MG)$ and $L = \int x\,dP(x)$. In view of (4.21), it follows that $L =_d V\theta + (1-V)L$, where $\theta \sim G$ and $V \sim \mathrm{Be}(1, M)$. Use this relation to simulate a Markov chain whose stationary distribution is the distribution of L. For G chosen as normal, Cauchy, exponential or uniform, and $M = 1, 10$, obtain the distribution of L by the MCMC method, and compare with their theoretical distribution given by Theorem 4.26.

4.31 (Lo 1991) Let $X_1, X_2, \ldots \mid P \stackrel{\text{iid}}{\sim} P$, where P has an arbitrary prior distribution. Suppose that there exists a nonstochastic sequence $a_n > 0$ and nonstochastic probability measures P_n such that $\mathrm{E}(P \mid X_1, \ldots, X_n) = (1-a_n)P_n + a_n n^{-1}\sum_{i=1}^n \delta_{X_i}$ for all n. Show that $a_n = n/(a_1^{-1} + n - 1)$, $P_n = \mathrm{E}(P)$ and $P \sim \mathrm{DP}(a_1\mathrm{E}(P))$.

4.32 For a location mixture of a Dirichlet process, obtain expressions for covariances of probabilities of sets.

4.33 Using conditioning arguments, show that

$$\mathrm{MDP}\Big(M+1, \frac{MG+\delta_\theta}{M+1}, \theta \sim G\Big) \equiv \mathrm{DP}(MG) \equiv \mathrm{MDP}(M, G, \theta \sim \delta_0).$$

Thus unlike the DP, the parameters of MDP are not necessarily identifiable.

4.34 Using the monotonicity of distribution functions, show that if M_θ is bounded in a mixture of Dirichlet process model, then the posterior distribution of P is consistent at any true distribution P_0 in the Kolmogorov-Smirnov sense.

4.35 Consider samples generated from a atomless true distribution P_0. Let the modeled P be given a mixture of Dirichlet process distribution $\mathrm{MDP}(MG_\xi, (M, \xi) \sim \pi_M \times \pi_\xi)$. Show that, with probability one, the posterior distribution of M is always equal to its prior π_M, and the relative weight $M/(M+n)$ of the first term in the posterior base measure $\bar{\alpha}_{\xi,n}$ in (4.29) always converges to zero. Conclude that the posterior mean is a consistent estimator.

Under the same condition, show that the posterior variance converges to zero. Thus the posterior distribution of any set, as well as the posterior distribution of P in the Kolmogorov-Smirnov sense, is consistent. [Hint: For the last part use a uniformization argument as in the proof of Glivenko-Cantelli theorem.]

4.36 Obtain the posterior distribution for a constrained Dirichlet process prior.

4.37 Describe a stick-breaking representation of an invariant Dirichlet process.

4.38 Let $X_1, \ldots, X_n \mid P \stackrel{\text{iid}}{\sim} P$ and $P \sim \mathrm{DP}(\sum_{i=1}^N \delta_{\theta_i})$, the Bayesian bootstrap distribution based on the sample $(\theta_1, \ldots, \theta_N)$. Assume that $\theta_1, \ldots, \theta_n$ are all distinct, and let K_n be the number of distinct elements in $\{X_1, \ldots, X_n\}$. Show that K_n follows the *Bose-Einstein distribution*:

$$\mathrm{P}(K_n = k) = \frac{\binom{N}{k}\binom{n-1}{k-1}}{\binom{N+n-1}{N-1}}, \quad k = 1, \ldots, \min(n, N).$$

4.39 (Ishwaran and James 2001) Consider a truncated stick-breaking process prior Π_N defined by $P_N = \sum_{k=1}^N V_k \delta_{\theta_k}$, where $V_k = [\prod_{j=1}^{k-1}(1-Y_j)]Y_k$, $Y_1, \ldots, Y_{N-1} \stackrel{\text{iid}}{\sim} \mathrm{Be}(1, M)$, $Y_N = 1$ and $\theta_1, \ldots, \theta_N \stackrel{\text{iid}}{\sim} G$. Let $\psi(\cdot \mid \theta)$ be a parametric family of probability densities, $X_i \mid Z_i \sim \psi(\cdot \mid Z_i)$, where $Z_1, \ldots, Z_n \stackrel{\text{iid}}{\sim} P_N$ and $P_N \sim \Pi_N$. Let $m_N^{(n)}$ stand for the marginal distribution of $(X_1, \ldots, X_n)$. Let $m_\infty^{(n)}$ be the marginal distribution of $(X_1, \ldots, X_n)$ when $X_i | Z_i \sim \psi(\cdot | Z_i)$, $Z_1, \ldots, Z_n \stackrel{\text{iid}}{\sim} P$ and $P \sim \Pi := \text{DP}(MG)$. Show that $d_{TV}(m_N^{(n)}, m_\infty^{(n)}) \approx 4ne^{-(N-1)/M}$.

4.40 Let $U_1, \ldots, U_{n-1} \stackrel{\text{iid}}{\sim} \text{Unif}(0, 1)$. Define the order statistics $U_{0:n-1} = 0$, $U_{n:n-1} = 1$ and $U_{j:n-1}$ the jth smallest of $\{U_1, \ldots, U_{n-1}\}$. If $W_j = U_{j:n-1} - U_{j-1:n-1}$, $j = 1, \ldots, n$, show that $(W_1, \ldots, W_n) \sim \text{Dir}(n; 1, \ldots, 1)$, the weights of the Bayesian bootstrap distribution.

5

Dirichlet Process Mixtures

The discreteness of the Dirichlet process makes it useless as a prior for estimating a density. This can be remedied by convolving it with a kernel. Although it is rarely possible to characterize the resulting posterior distribution analytically, a variety of efficient and elegant algorithms allow us to approximate it numerically. These algorithms exploit the special properties of the Dirichlet process derived in the preceding chapter. The present chapter starts with a general introduction to Dirichlet process mixtures and next discusses a number of computational strategies.

5.1 Dirichlet Process Mixtures

For each θ in a parameter set Θ let $x \mapsto \psi(x, \theta)$ be a probability density function, measurable in its two arguments. Densities $p_F(x) = \int \psi(x, \theta)\, dF(\theta)$, with F equipped with a Dirichlet process prior, are known as *Dirichlet process mixtures*. If the kernel also depends on an additional parameter $\varphi \in \Phi$, giving mixtures $p_{F,\varphi}(x) = \int \psi(x, \theta, \varphi)\, dF(\theta)$, it is more appropriate to call the result a "mixture of Dirichlet process mixture," but the name Dirichlet process mixture even for this case seems more convenient.

In this section we discuss a method of posterior computation for these mixtures. To include also other problems such as regression, spectral density estimation and so on, we allow the observations to be non-i.i.d. For $x \mapsto \psi_i(x; \theta, \varphi)$ probability density functions (relative to a given σ-finite dominating measure ν), consider

$$X_i \overset{\text{ind}}{\sim} p_{i,F,\varphi}(x) = \int \psi_i(x; \theta, \varphi)\, dF(\theta), \qquad i = 1, \dots, n. \tag{5.1}$$

We equip F and φ with independent priors $F \sim \mathrm{DP}(\alpha)$ and $\varphi \sim \pi$. (Independence of the observations is not crucial; later we also consider mixture models for estimating the transition density of a Markov chain.) The resulting model can be equivalently written in terms of n latent variables $\theta_1, \dots, \theta_n$ as

$$X_i \,|\, \theta_i, \varphi, F \overset{\text{ind}}{\sim} \psi_i(\cdot; \theta_i, \varphi), \qquad \theta_i \,|\, F, \varphi \overset{\text{iid}}{\sim} F, \qquad F \sim \mathrm{DP}(\alpha), \qquad \varphi \sim \pi. \tag{5.2}$$

The posterior distribution of any object of interest can be described in terms of the posterior distribution of (F, φ) given $X_1, \dots, X_n$. The latent variables $\theta_1, \dots, \theta_n$ help to make the description simpler, since $F \,|\, \theta_1, \dots, \theta_n \sim \mathrm{DP}(\alpha + \sum_{i=1}^n \delta_{\theta_i})$, and given $\theta_1, \dots, \theta_n$, the observations $X_1, \dots, X_n$ are ancillary with respect to (F, φ), whence the conditional distribution of F given $\theta_1, \ldots, \theta_n, X_1, \ldots, X_n$ is free of the observations. In particular, for any measurable function ψ,

$$\mathrm{E}\Big(\int \psi\, dF \,|\, \varphi, \theta_1, \ldots, \theta_n, X_1, \ldots, X_n\Big) = \frac{1}{|\alpha| + n}\Big[\int \psi\, d\alpha + \sum_{j=1}^{n} \psi(\theta_j)\Big]. \tag{5.3}$$

The advantage of this representation is that the infinite-dimensional parameter F has been eliminated. To compute the posterior expectation, it now suffices to average out the right-hand side with respect to the posterior distribution of $(\theta_1, \ldots, \theta_n)$, and that of φ. Proposition 5.2 and Theorem 5.3 below give two methods for the first, an analytic one and one based on simulation, where the second one is the more practical one.

Example 5.1 (Density estimation) The choice $\psi(\theta) = \psi_i(x, \theta, \varphi)$ in (5.3) gives the Bayes estimate of the density $p_{i,F,\varphi}(x)$. This consists of a part attributable to the prior and a part due to the observations. We might ignore the prior part and view the conditional expectation

$$\frac{1}{n}\sum_{j=1}^{n} \int \psi_i(x, \theta_j, \varphi)\, d\Pi_n(\theta_j \,|\, X_1, \ldots, X_n)$$

of the second part as a partial Bayesian estimator, which has the influence of the prior guess reduced.

Let $\mathscr{S}_n$ stand for the class of all partitions of the set $\{1, \ldots, n\}$, and let W be the probability measure on $\mathscr{S}_n$ satisfying

$$W(\mathcal{S}) \propto \prod_{j=1}^{\#\mathcal{S}} (\#S_j - 1)! \int \prod_{l \in S_j} \psi_l(X_l; \theta)\, d\alpha(\theta), \qquad \mathcal{S} = \{S_1, \ldots, S_k\}.$$

Proposition 5.2 *In the model* (5.2), *for any nonnegative or integrable function* ψ,

$$\mathrm{E}\Big(\int \psi\, dF \,|\, \varphi, X_1, \ldots, X_n\Big)$$
$$= \frac{1}{|\alpha| + n}\Big[\int \psi\, d\alpha + \sum_{\mathcal{S} \in \mathscr{S}_n} W(\mathcal{S}) \sum_{j=1}^{\#\mathcal{S}} \#S_j \frac{\int \psi(\theta) \prod_{l \in S_j} \psi_l(X_l; \theta, \varphi)\, d\alpha(\theta)}{\int \prod_{l \in S_j} \psi_l(X_l; \theta, \varphi)\, d\alpha(\theta)}\Big].$$

Proof It suffices to show that the conditional expectation of the second term on the right of (5.3) given $X_1, \ldots, X_n$, agrees with the second term of the formula. As φ is fixed throughout, we drop it from the notation.

Because $\theta_1, \ldots, \theta_n$ are a random sample from a $\mathrm{DP}(\alpha)$-distribution, their (marginal) prior distribution Π is described by Proposition 4.7. By Bayes's rule the posterior density of $(\theta_1, \ldots, \theta_n)$ is proportional to $\prod_{i=1}^{n} \psi_i(X_i; \theta_i) d\Pi(\theta_1, \ldots, \theta_n)$. Hence, for any measurable functions g_i,

$$\mathrm{E}\Big(\prod_{i=1}^n g_i(\theta_i)|\, X_1,\ldots,X_n\Big) = \frac{\int\cdots\int \prod_{i=1}^n g_i(\theta_i)\prod_{i=1}^n \psi_i(X_i;\theta_i)\,d\Pi(\theta_1,\ldots,\theta_n)}{\int\cdots\int \prod_{i=1}^n \psi_i(X_i;\theta_i)\,d\Pi(\theta_1,\ldots,\theta_n)}$$
$$= \frac{\sum_{\mathcal{S}\in\mathscr{S}_n}\prod_{j=1}^{\#\mathcal{S}}(\#S_j-1)!\int\prod_{l\in S_j}\big(g_l(\theta_l)\psi_l(X_l;\theta)\big)\,d\alpha(\theta)}{\sum_{\mathcal{S}\in\mathscr{S}_n}\prod_{j=1}^{\#\mathcal{S}}(\#S_j-1)!\int\prod_{l\in S_j}\psi_l(X_l;\theta)\,d\alpha(\theta)},$$

by Proposition 4.7. For $g_i = \psi$ and $g_j = 1$ for $j \neq i$, we can rewrite this in the form

$$\mathrm{E}(\psi(\theta_i)|\, X_1,\ldots,X_n) = \sum_{\mathcal{S}\in\mathscr{S}_n} W(\mathcal{S})\frac{\int\prod_{l\in S(i)}\psi(\theta)\psi_l(X_l;\theta)\,d\alpha(\theta)}{\int\prod_{l\in S(i)}\psi_l(X_l;\theta)\,d\alpha(\theta)},$$

where $S(i)$ is the partitioning set in $\mathcal{S} = \{S_1,\ldots,S_k\}$ that contains i. When summing this over $i = 1,\ldots,n$, we obtain $\#\mathcal{S}$ different values, with multiplicities $\#S_1,\ldots,\#S_k$. □

Unfortunately, the analytic formula of Proposition 5.2 is of limited practical importance because it involves a sum with a large number of terms (cf. Problem 4.17). In practice, we use a simulation technique: we draw repeatedly from the posterior distribution of $(\theta_1,\ldots,\theta_n,\varphi)$, evaluate the right-hand side of (5.3) for each realization, and average out over many realizations.

The next theorem explains a Gibbs sampling scheme to simulate from the posterior distribution of $(\theta_1,\ldots,\theta_n)$, based on a weighted generalized Pólya urn scheme. Inclusion of a possible parameter φ and other hyperparameters is tackled in the next section.

We use the subscript $-i$ to denote every index $j \neq i$, and $\theta_{-i} = (\theta_j\colon j \neq i)$.

Theorem 5.3 (Gibbs sampler) *In the model* (5.2) *the conditional posterior distribution of* θ_i *is given by*

$$\theta_i|\,\theta_{-i},\varphi,X_1,\ldots,X_n \sim \sum_{j\neq i} q_{i,j}\delta_{\theta_j} + q_{i,0}G_{b,i}, \tag{5.4}$$

where $(q_{i,j}\colon j \in \{0,1,\ldots,n\}\setminus\{i\})$ *is the probability vector satisfying*

$$q_{i,j} \propto \begin{cases}\psi_i(X_i;\theta_j,\varphi), & j\neq i,\ j\geq 1,\\ \int\psi_i(X_i;\theta,\varphi)\,d\alpha(\theta), & j=0,\end{cases} \tag{5.5}$$

and $G_{b,i}$ *is the* baseline posterior measure *given by*

$$dG_{b,i}(\theta|\,\varphi,X_i) \propto \psi_i(X_i;\theta,\varphi)\,d\alpha(\theta). \tag{5.6}$$

Proof Since the parameter φ is fixed throughout, we suppress it from the notation. For measurable sets A and B,

$$\mathrm{E}(\mathbb{1}_A(X_i)\mathbb{1}_B(\theta_i)|\,\theta_{-i},X_{-i}) = \mathrm{E}\Big(\mathrm{E}(\mathbb{1}_A(X_i)\mathbb{1}_B(\theta_i)|\,F,\theta_{-i},X_{-i})|\,\theta_{-i},X_{-i}\Big).$$

Because (θ_i, X_i) is conditionally independent of (θ_{-i}, X_{-i}) given F, the inner conditional expectation is equal to $\mathrm{E}(\mathbb{1}_A(X_i)\mathbb{1}_B(\theta_i)|\,F) = \int\int \mathbb{1}_A(x)\mathbb{1}_B(\theta)\psi_i(x;\theta)\,d\mu(x)\,dF(\theta)$. Upon inserting this expression, in which F is the only variable, in the display we see that in the outer layer of conditioning the variables X_{-i} are superfluous, by the conditional independence of F and X_{-i} given θ_{-i}. Therefore, by Proposition 4.3 the preceding display is equal to

$$\frac{1}{|\alpha|+n-1}\iint \mathbb{1}_A(x)\mathbb{1}_B(\theta)\psi_i(x;\theta)\,d\mu(x)\,d\Big(\alpha+\sum_{j\neq i}\delta_{\theta_j}\Big)(\theta).$$

This determines the joint conditional distribution of (X_i,θ_i) given (θ_{-i},X_{-i}). By Bayes's rule (applied to this joint law conditionally given (θ_{-i},X_{-i})) we infer that

$$\mathrm{P}(\theta_i\in B|\,X_i,\theta_{-i},X_{-i})=\frac{\int_B\psi_i(X_i;\theta)\,d(\alpha+\sum_{j\neq i}\delta_{\theta_j})(\theta)}{\int\psi_i(X_i;\theta)\,d(\alpha+\sum_{j\neq i}\delta_{\theta_j})(\theta)}.$$

This in turn is equivalent to the assertion of the theorem. □

Remark 5.4 The posterior distribution of F is actually a mixture of Dirichlet processes: for Π^* described in (5.4),

$$F|\,X_1,\ldots,X_n,\varphi,\alpha\sim\mathrm{MDP}\Big(\alpha+\sum_{i=1}^n\delta_{\theta_i},(\theta_1,\ldots,\theta_n)\sim\Pi^*\Big).$$

5.2 MCMC Methods

In this section we present five algorithms to simulate from the posterior distribution in the Dirichlet process mixture model:

$$X_i|\,\theta_i,\varphi,M,\xi,F\overset{\text{ind}}{\sim}\psi_i(\cdot;\theta_i,\varphi),\qquad\theta_i|\,F,\varphi,M,\xi\overset{\text{iid}}{\sim}F,\qquad F|\,M,\xi\sim\mathrm{DP}(MG_\xi),$$

where φ, M and ξ are independently generated hyperparameters. The basic algorithm uses the Gibbs sampling scheme of Theorem 5.3 to generate $\theta_1,\ldots,\theta_n$ given $X_1,\ldots,X_n$ in combination with the Gibbs sampler for the posterior distribution of M given in Section 4.5, and/or additional Gibbs steps. The prior densities of the hyperparameters are denoted by a generic π.

Algorithm 1 Generate samples by sequentially executing steps (i)–(iv) below:

(i) Given the observations and φ, M and ξ, update each θ_i sequentially using (5.4) inside a loop $i=1,\ldots,n$.
(ii) Update $\varphi\sim p(\varphi|\,\theta_1,\ldots,\theta_n,X_1,\ldots,X_n)\propto\pi(\varphi)\prod_{i=1}^n\psi_i(X_i;\theta_i,\varphi)$.
(iii) Update $\xi\sim p(\xi|\,\theta_1,\ldots,\theta_n)\propto\pi(\xi)p(\theta_1,\ldots,\theta_n|\,\xi)$, where the marginal distribution of $(\theta_1,\ldots,\theta_n)$ is as in the Pólya scheme (4.13).
(iv) Update M and next the auxiliary variable η using (4.30), for K_n the number of distinct values in $\{\theta_1,\ldots,\theta_n\}$.

Because sampling from the generalized Pólya urn will rarely yield a new value (cf. Theorem 4.8), the θ-values will involve many ties and can be stored and updated more efficiently. Let $\{\mu_1,\ldots,\mu_k\}$ be the distinct values in $\{\theta_1,\ldots,\theta_n\}$, let $N_1,\ldots,N_k$ be their multiplicities, and let $(s_1, \ldots, s_n)$ be pointers such that $s_i = s$ if $\theta_i = \mu_s$. Then $(\theta_1, \ldots, \theta_n)$ is in one-to-one correspondence with $(k, \mu_1, \ldots, \mu_k, s_1, \ldots, s_n)$. The pointers $(s_1, \ldots, s_n)$ determine the partition of $\{1, \ldots, n\}$ corresponding to the ties in $\theta_1, \ldots, \theta_n$. In the Pólya urn scheme for generating $\theta_1, \ldots, \theta_n$ the values μ_s of the ties are drawn independently from the center measure $\bar{\alpha}$. Given the pointers $s_1, \ldots, s_n$ they completely determine $\theta_1, \ldots, \theta_n$. Given the pointers $s_1, \ldots, s_n$, the observations $X_1, \ldots, X_n$ can be viewed as next generated as k independent samples $(X_i\colon s_i = s)$ of sizes $N_s = \#(i\colon s_i = s)$ from the densities $(x_i\colon s_i = s) \mapsto \prod_{i:s_i=s} \psi_i(x_i; \mu_s, \varphi)$. The conditional density of the tie values given the observations and still given the pointers follows by Bayes's rule. This leads to the following algorithm, which alternates between updating the pointers $(s_1, \ldots, s_n)$ and the values $\mu_1, \ldots, \mu_k$ of the ties.

The parameters φ, ξ and M are treated as in Algorithm 1 and therefore are dropped from the notation.

Algorithm 2 Perform the two steps:

(i) Update the full configuration vector $(s_1, \ldots, s_n)$ in a Gibbs loop according to $\mathrm{P}(s_i = s \mid s_{-i}, \text{rest}) = q^*_{i,s}$, for, with $N_{-i,s} = N_s - \mathbb{1}\{s = s_i\}$,

$$q^*_{i,s} \propto \begin{cases} N_{-i,s}\psi_i(X_i; \mu_s), & \text{if } s = s_j,\ j \neq i, \\ \int \psi_i(X_i; \theta)\, d\alpha(\theta), & \text{if } s = s_{i,\text{new}}. \end{cases}$$

Here $s_{i,\text{new}}$ is a value that does not appear in the current vector s_{-i}. If a new value $s = s_{i,\text{new}}$ is selected in the ith step, generate a corresponding μ_s from $G_{b,i}$ given in (5.6). Update k as the number of different elements in $s_1, \ldots, s_n$.

(ii) Update the full vector $(\mu_1, \ldots, \mu_k)$ by drawing the coordinates independently from the densities proportional to $\mu_s \mapsto \prod_{i:s_i=s} \psi_i(X_i; \mu_s)\, d\alpha(\mu_s)$.

In some applications, such as clustering problems, the values $\{\mu_1, \ldots, \mu_k\}$ may not be of interest. The following algorithm "integrates them out" and simulates only values of the configuration vector $(s_1, \ldots, s_n)$. A disadvantage of this algorithm is that it cannot update hyperparameters, but these could be estimated using an empirical Bayes approach.

Algorithm 3 Update the configuration vector $(s_1, \ldots, s_n)$ by

$$\mathrm{P}(s_i = s \mid s_{-i}, \text{rest}) \propto \frac{\int N_{-i,s}\psi_i(X_i; \theta) \prod_{j:s_j=s} \psi_j(X_j; \theta)\, d\alpha(\theta)}{\int \prod_{j \neq i: s_j = s} \psi_j(X_j; \theta)\, d\alpha(\theta)},$$

where an empty product means 1.

All three algorithms are feasible, even for large n, but only if the integrals in (5.6) and (5.5) are analytically computable and drawing from the baseline posterior distribution $G_{b,i}$ is straightforward. This virtually restricts to the case where the base measure α is conjugate to the kernels ψ_i. In many applications conjugacy arises naturally and is not too restrictive, especially if α is modeled flexibly using some hyperparameters. Some common cases are discussed in Section 5.5.

In the case where a conjugate base measure does not exist none of the three algorithms is suitable. Several algorithms have been developed for this situation. We present one such method, the *no gaps algorithm*. The basis of reasoning in this algorithm is that the configuration indices $s_1, \ldots, s_n$ ought to form a consecutive sequence without a gap, and this constraint should be respected in all moves of the MCMC. With the same notations as before, the algorithm has the following steps:

Algorithm 4 ("No gaps" algorithm) For k_{-i} the number of distinct elements in the set $\{s_1, \ldots, s_{i-1}, s_{i+1}, \ldots, s_n\}$:

(a) For $i = 1, \ldots, n$, repeat the following steps.
  (i) If $s_i \neq s_j$ for all $j \neq i$, leave s_i unchanged with probability $k_{-i}/(k_{-i}+1)$ and with probability $1/(k_{-i}+1)$ assign it to the new label $k_{-i}+1$. Proceed to step (iii).
  (ii) If $s_i = s_j$ for some $j \neq i$, draw a value for $\mu_{k_{-i}+1} \sim G$ if this is not available in the previous iteration.
  (iii) Label s_i to be $s \in \{1, \ldots, k_{-i}+1\}$ according to the probabilities

$$\mathrm{P}(s_i = s \,|\, s_{-i}, \text{rest}) \propto \begin{cases} N_{-i,s}\psi_i(X_i;\, \mu_s), & 1 \leq s \leq k_{-i}, \\ \frac{M}{k_{-i}+1}\psi_i(X_i;\, \mu_s), & s = k_{-i}+1. \end{cases}$$

(b) For all $s \in \{1, \ldots, k\}$, draw a new value $(\mu_s |\, X_i{:}\, s_i = s)$ by a one-step Metropolis-Hastings algorithm, or some other method of sampling for which the target distribution is invariant.

The "no gaps" algorithm deliberately uses a unidentifiable model to improve mixing. We outline why this algorithm works in the no-data case. The basic reasoning is the same in general.

The probability of obtaining a fixed configuration $(s_1, \ldots, s_n)$ is given by (4.20) as

$$P(s_1, \ldots, s_n) = \frac{M^k \Gamma(M) \prod_{j=1}^{k} \Gamma(N_j)}{\Gamma(M+n)}.$$

If gaps are to be avoided, there are $k!$ labeling possible. Thus

$$P(s_1, \ldots, s_n, \text{labels}) = \frac{M^k \Gamma(M) \prod_{j=1}^{k} \Gamma(N_j)}{k!\, \Gamma(M+n)}.$$

An observation cannot move if it would empty a cluster, unless it is in cluster k. The probabilities for index i to move to the jth cluster are proportional to

$$\begin{aligned} &\frac{\Gamma(M) M^{k_{-i}} N_{-i,j} \prod_{l=1}^{k_{-i}} \Gamma(N_{-i,l})}{\Gamma(M+n)(k_{-i})!}, && j = 1, \ldots, k_{-i}, \\ &\frac{\Gamma(M) M^{k_{-i}+1} \prod_{l=1}^{k_{-i}} \Gamma(N_{-i,l})}{\Gamma(M+n)(k_{-i}+1)!}, && j = k_{-i}+1. \end{aligned}$$

This is equivalent to saying that the probabilities are proportional to $N_{-i,j}$ for $j = 1, \ldots, k_{-i}$ and $M/(k_{-i}+1)$ for $j = k_{-i}+1$.

The permutation step immediately before each transition says that a cluster i of size 1 is kept the same with probability $k_{-i}/(k_{-i}+1)$ and moved with probability $1/(k_{-i}+1)$. Given a move, the destination is j with probability proportional to $N_{-i,j}$, for $j = 1, \ldots, k_{-i}$, and is $k_{-i}+1$ (that is, a new cluster is opened) with probability proportional to $M/(k_{-i}+1)$.

The next algorithm is based on the stick-breaking representation $F = \sum_{j=1}^\infty W_j \delta_{\theta_j}$ of the mixing distribution and "slicing." The stick-breaking weights $W_j = V_j \prod_{l<j}(1 - V_l)$ in this representation are in one-to-one correspondence with the relative cuts V_j of the sticks, whence we can use the V_j and W_j interchangeably in the description of the algorithm. In terms of the stick-breaking representation the mixed density (5.1) (with the parameter φ omitted) becomes

$$p_{i,F}(x) = \sum_j W_j \psi_i(x; \theta_j).$$

We shall write this presently as $p_i(x|\, W, \theta)$, for $W = (W_1, W_2, \ldots)$ and $\theta = (\theta_1, \theta_2, \ldots)$ the vectors of all latent variables. We use similar notation to collect all coordinates of other variables.

It is helpful to rewrite the sampling scheme (5.2) in terms of the W_j and θ_j, and also to augment it with latent variables $s_1, \ldots, s_n$ that denote the components (or θ_j) from which $X_1, \ldots, X_n$ are drawn. (These variables have another meaning than in Algorithm 2; also the present θ_j have another interpretation than the variables θ_i in (5.2).) This gives the equivalent description

$$X_i|\, \theta, s, W \overset{\text{ind}}{\sim} \psi_i(\cdot; \theta_{s_i}), \quad s_i|\, V, \theta \overset{\text{iid}}{\sim} W, \quad V_j \overset{\text{iid}}{\sim} \text{Be}(1, M), \quad \theta_j \overset{\text{iid}}{\sim} \bar{\alpha}. \tag{5.7}$$

It follows that the conditional likelihood of $X_1, \ldots, X_n, s_1, \ldots, s_n$ takes the form

$$p(X_1, \ldots, X_n, s_1, \ldots, s_n|\, \theta, W) = \prod_{i=1}^n \psi_i(X_i; \theta_{s_i}) W_{s_i}.$$

The next step is to augment the scheme with further latent variables $U_1, \ldots, U_n$, the *slicing variables*, such that the joint (conditional) density is given by

$$p(X_1, \ldots, X_n, s_1, \ldots, s_n, U_1, \ldots, U_n|\, \theta, W) = \prod_{i=1}^n \psi_i(X_i; \theta_{s_i}) \mathbb{1}\{U_i < W_{s_i}\}.$$

Given a uniform dominating measure this indeed integrates out relative to $U_1, \ldots, U_n$ to the preceding display, and hence provides a valid augmentation.

The next algorithm is a Gibbs sampler for the posterior density obtained by multiplying the preceding display by prior densities of θ and W (or equivalently V, which is a product of the beta densities $M(1 - v_j)^{M-1}$).

Algorithm 5 (Slicing algorithm) Sequentially update the vectors U, θ, V, s using the conditional densities in the given order:

(i) $p(U_i|\, U_{-i}, \text{rest}) \propto \mathbb{1}\{U_i < W_{s_i}\}, \qquad i = 1, \ldots, n;$
(ii) $p(\theta_j|\, \theta_{-j}, \text{rest}) \propto \prod_{i:s_i=j} \psi_i(X_i; \theta_j)\, d\alpha(\theta_j), \qquad j = 1, 2 \ldots;$
(iii) $p(V_j|\, W_{-j}, \text{rest}) \propto M(1 - V_j)^{M-1} \prod_{i=1}^n \mathbb{1}\{U_i < W_{s_i}\}, \qquad j = 1, 2 \ldots;$
(iv) $p(s_i|\, s_{-i}, \text{rest}) \propto \psi_i(X_i; \theta_{s_i}) \mathbb{1}\{U_i < W_{s_i}\}, \qquad i = 1, \ldots, n.$

The products in (ii) or (iii) can be vacuous, in which case they are interpreted as 1. As in the other algorithms described above, conjugacy makes step (ii) easier, but is not essential.

Steps (ii) and (iii) are written as updates of the infinite sequences θ and W, but in fact only their coordinates θ_{s_i} and W_{s_i} appear in the algorithm, and $\max(s_1, \ldots, s_n)$ will typically remain small, so that in practice truncation to a small set is feasible. Furthermore, the conditional distributions in steps (ii) and (iii) are identical to the prior for $j > \max(s_1, \ldots, s_n)$; no update is needed for such θ_j. The set $\{s_1, \ldots, s_n\}$ varies over the cycles of the Gibbs sampler. Because the conditional density of s_i vanishes at k if $W_k \leq U_i$, all s_i will be smaller than N if $\max_{k>N} W_k \leq \min_i U_i$, which is implied by $\sum_{j=1}^N W_j \geq 1 - \min_i U_i$. In view of Proposition 4.20 the minimal N is distributed as one plus a Poisson random variable with parameter $M \log_-(\min_i U_i)$.

The density for updating V_j (or W_j) in (iii) is a beta-density restricted to the set $\cap_i \{U_i < W_{s_i}\}$. Since $W_j = \prod_{l<j}(1 - V_l)V_j$, this set can by some algebra be rewritten in the form $a_j < V_j < b_j$, for

$$a_j = \frac{\max_{i:s_i=j} U_i}{\prod_{l<j}(1 - V_l)}, \qquad b_j = \min_{i:s_i>j}\Bigl(1 - \frac{U_i}{V_{s_i} \prod_{l<s_i, l\neq j}(1 - V_l)}\Bigr).$$

Sampling from this truncated distribution can be simply implemented through the quantile method, as the cumulative distribution function of the $\text{Be}(1, M)$-distribution has the explicit form $v \mapsto (1 - v)^M$. In fact, the algorithm can be easily extended to more general stick-breaking distributions as long as the truncated cumulative distribution function has a form suitable for inversion.

An easy way of computing with Dirichlet mixtures is replacing $F \sim \text{DP}(\alpha)$ by the process F_N following the Dirichlet-multinomial distribution described by Theorem 4.19. This reduces the parameter of the problem from infinite-dimensional to finite-dimensional, allowing a treatment by common MCMC methods, such as the Metropolis-Hastings algorithm. However, the right choice of N is unclear. Thumb rules such as $N = cn$ for small n and $N = \sqrt{n}$ for large n have been proposed, but a full theoretical justification is unavailable. (However, see Problem 4.39.)

5.3 Variational Algorithm

A *variational algorithm* is an iterative, deterministic method to determine an approximation to the posterior density of a given predefined form. Given a flexible but tractable class $\mathcal{Q}$ of probability densities on the parameter space Θ it seeks to find $q \in \mathcal{Q}$ that minimizes the Kullback-Leibler divergence

$$K(q; \pi(\cdot| X)) = -\int q(\theta) \log\Bigl(\frac{p(X| \theta)\pi(\theta)}{q(\theta)}\Bigr)\, d\nu(\theta) + \log p(X) \tag{5.8}$$

between q and the posterior density $\theta \mapsto \pi(\theta| X) = p(X| \theta)\pi(\theta)/p(X)$. Typically $\mathcal{Q}$ is taken to be a high-dimensional parametric family, and maximization is carried out by sequentially optimizing over a single parameter, keeping the other parameters fixed, iterating "until convergence." The method is similar in spirit to the EM algorithm in that it involves computing and maximizing integrals, and shares the coordinatewise iterations with Gibbs sampling.

Because the second term on the right of (5.8), the log-marginal density of X, does not depend on q, the minimization can be carried out by maximizing the integral. If the posterior density is contained in $\mathcal{Q}$, then the left side will be minimized to zero, and hence the maximum value of the integral will equal $\log p(X)$. Therefore the value of the maximization problem is an estimate of $\log p(X)$. This by-product of the algorithm is useful, e.g. for model selection.

Variational methods were originally developed for exponential families with incomplete data and conjugate priors. In that situation the integral of the log-density reduces to an expression involving the mean of the variable, whence the approach is commonly known as a *mean-field variational method*. Here we discuss an implementation of the algorithm for approximating the posterior density in Dirichlet process mixture models.

We consider the model (5.7), with the difference that presently we truncate the vectors θ and W to vectors of length N. This corresponds to using a truncated stick-breaking approximation $F_N = \sum_{j=1}^N W_j \delta_{\theta_j}$ for the mixing distribution F (which follows a Dirichlet process in the "ideal" model). The weights in this approximation are for $j = 1, \ldots, N-1$, given as usual by $W_j = V_j \prod_{l=1}^{j-1}(1 - V_l)$, for $V_j \stackrel{\text{iid}}{\sim} \mathrm{Be}(1, M)$, but presently we set $V_N = 1$ to ensure that $\sum_{j=1}^N W_j = 1$. (This corresponds to taking $\theta_0 = \theta_{N_\epsilon}$ in the ϵ-Dirichlet process (4.23).) The posterior density of the set of latent variables now satisfies

$$\begin{aligned} p(s_1, \ldots, s_n, \theta_1, \ldots, \theta_N, V_1 \ldots, V_{N-1} | X_1, \ldots, X_n) \\ \propto \prod_{i=1}^n \psi_i(X_i; \theta_{s_i}) \prod_{i=1}^n W_{s_i} \prod_{j=1}^N g(\theta_j) \prod_{j=1}^{N-1} \mathrm{be}(V_j; 1, M). \end{aligned}$$

We approximate this by a density q under which all the latent variables s_i, θ_j, V_j are independent, every s_i has a completely unknown distribution on $\{1, \ldots, N\}$ given by a probability vector $(\pi_{i,1} \ldots, \pi_{i,N})$ that is allowed to be different for different coordinates, $\theta_j \sim g_{\xi_j}$ for some parameterized family of densities g_ξ, and $V_j \stackrel{\text{ind}}{\sim} \mathrm{Be}(a_j, b_j)$ for parameters a_j, b_j. Thus

$$q(s_1, \ldots, s_n, \theta_1, \ldots, \theta_N, V_1, \ldots, V_{N-1}) = \prod_{i=1}^n \pi_{i,s_i} \prod_{j=1}^N g_{\xi_j}(\theta_j) \prod_{j=1}^{N-1} \mathrm{be}(V_j; a_j, b_j).$$

Next we determine the parameters $\pi_{i,j}, \xi_j, a_j, b_j$ by the variational method.

The integral in (5.8) is the expectation under q of

$$\sum_{i=1}^n \Bigl(\log \psi_i(X_i; \theta_{s_i}) + \log \frac{W_{s_i}}{\pi_{i,s_i}}\Bigr) + \sum_{j=1}^N \log \frac{g}{g_{\xi_j}}(\theta_j) + \sum_{j=1}^{N-1} \log \frac{\mathrm{be}(V_j; 1, M)}{\mathrm{be}(V_j; a_j, b_j)}.$$

After substituting $W_j = V_j \prod_{l<j}(1 - V_l)$ and the expressions for the beta densities, we can evaluate this expectation to

$$\sum_{i=1}^{n}\sum_{j=1}^{N}\pi_{i,j}\Big[\mathrm{E}_q(\log\psi_i(X_i;\theta_j))+\mathrm{E}_q\log V_j+\sum_{l<j}\mathrm{E}_q\log(1-V_l)-\log\pi_{i,j}\Big]$$
$$-\sum_{j=1}^{N}K(g_{\xi_j};g)$$
$$+\sum_{j=1}^{N-1}\Big[\log\frac{M\Gamma(a_j)\Gamma(b_j)}{\Gamma(a_j+b_j)}-(a_j-1)\mathrm{E}_q\log V_j+(M-b_j)\mathrm{E}_q\log(1-V_j)\Big].$$

A beta variable $V\sim\mathrm{Be}(a,b)$ has logarithmic moment $\mathrm{E}(\log V)=\Psi(a)-\Psi(a+b)$ and hence $\mathrm{E}(\log(1-V))=\Psi(b)-\Psi(a+b)$, for the *digamma function* $\Psi(x)=\frac{d}{dx}\log\Gamma(x)$. This allows to replace $\mathrm{E}_q\log V_j$ and $\mathrm{E}_q\log(1-V_j)$ by concrete expressions involving (a_j,b_j). The resulting expression must be maximized with respect to the variational parameters $\xi_j,\pi_{i,j},a_j,b_j$, for $i=1,\ldots,n$ and $j=1,\ldots,N$. A typical method is "coordinatewise ascent": the expression is repeatedly maximized with respect to one parameter at a time, with the other parameters fixed. The updating formulas are known as *variational updates*.

Example 5.5 We illustrate the method for normal mixtures, with conjugate choices for the densities constituting q. (We fix the standard deviation σ of the mixing distribution; extensions to varying σ are straightforward.) Specifically, we let $\psi_i(\cdot;\theta)$ be the density of the $\mathrm{Nor}(\theta,\sigma^2)$-distribution, the center measure G of the Dirichlet prior a $\mathrm{Nor}(\mu,\tau^2)$-distribution, and replace g_ξ by the density of a $\mathrm{Nor}(\xi,\eta^2)$-distribution (and ξ by (ξ,η)). The integral in (5.8) then becomes

$$-\sum_{i=1}^{n}\sum_{j=1}^{N}\pi_{i,j}\Big[\log(\sigma\sqrt{2\pi})+\frac{(X_i-\xi_j)^2+\eta_j^2}{2\sigma^2}\Big]-\sum_{i=1}^{n}\sum_{j=1}^{N}\pi_{i,j}\log\pi_{i,j}$$
$$+\sum_{i=1}^{n}\sum_{j=1}^{N-1}\pi_{i,j}(\Psi(a_j)-\Psi(a_j+b_j))+\sum_{i=1}^{n}\sum_{j=1}^{N}\pi_{i,j}\sum_{l<j}(\Psi(b_l)-\Psi(a_l+b_l))$$
$$-\sum_{j=1}^{N}\Big[\log\frac{\tau}{\eta_j}-\frac{\tau^2-\eta_j^2}{2\tau^2}+\frac{(\mu-\xi_j)^2}{2\tau^2}\Big]+\sum_{j=1}^{N-1}\log\frac{M\Gamma(a_j)\Gamma(b_j)}{\Gamma(a_j+b_j)}$$
$$-\sum_{j=1}^{N-1}\Big[(a_j-1)(\Psi(a_j)-\Psi(a_j+b_j))-(M-b_j)(\Psi(b_j)-\Psi(a_j+b_j))\Big].$$

Maximizing this with respect to one parameter at a time leads to the following variational updates, which must be executed cyclically until convergence occurs:

(i) $\xi_j\rightarrow\Big(\frac{\sum_{i=1}^n\pi_{i,j}X_i}{\sigma^2}+\frac{\mu}{\tau^2}\Big)\Big(\frac{\sum_{i=1}^n\pi_{i,j}}{\sigma^2}+\frac{1}{\tau^2}\Big)^{-1}$;

(ii) $\eta_j^2\rightarrow\Big(\frac{\sum_{i=1}^n\pi_{i,j}}{\sigma^2}+\frac{1}{\tau^2}\Big)^{-1}$;

(iii) update $\pi_{i,j}$ for $j = 1, \dots, N-1$, to $Kc_{i,j}/(1+K\sum c_{i,l})$, and $\pi_{i,N}$ to $1/(1+K\sum c_{i,l})$, where $K = \exp\big[-((X_i - \xi_N)^2 + \eta_N^2)/(2\sigma^2)\big]$ and

$$c_{i,j} = \exp\Big(-\frac{(X_i - \xi_j)^2 + \eta_j^2}{2\sigma^2} - \sum_{k=1}^{j} \Psi(a_k + b_k) + \Psi(a_j) + \sum_{k=1}^{j-1} \Psi(b_k)\Big).$$

(iv) a_j is updated to the solution a of the equation

$$\Psi'(a)\Big(a - 1 + \sum_{i=1}^{n} \pi_{i,j}\Big) - \Psi'(a + b_j)\Big(\sum_{i=1}^{n}\sum_{l=j}^{N} \pi_{i,l} + M - b_j - a\Big) = 0.$$

(v) b_j is updated to the solution b of of the equation

$$\Psi'(b)\Big(\sum_{i=1}^{n}\sum_{l=j}^{N} \pi_{i,l} + M - b\Big) - \Psi'(a_j + b)\Big(\sum_{i=1}^{n}\sum_{l=j}^{N} \pi_{i,l} + M + 1 - a_j - b\Big) = 0.$$

The updates of a_j and b_j are easy to implement, since the *trigamma function* $\Psi'(x) = \frac{d^2}{dx^2} \log \Gamma(x)$ is available in the library of most mathematical software packages.

5.4 Predictive Recursion Deconvolution Algorithm

Consider the mixture model $p_F(x) = \int \psi(x; \theta)\, dF(\theta)$ as in (5.1), but with a kernel $\psi(\cdot; \theta)$ that does not depend on i or an additional nuisance parameter. Given $F \sim \mathrm{DP}(MG)$ and a single observation X_1 from the mixture density p_F, the posterior mean of the mixing distribution satisfies, by Proposition 5.2 with $n = 1$ (or (5.3) in combination with Theorem 5.3 for the conditional distribution of θ_1 given X_1),

$$\mathrm{E}(F(A)|\, X_1) = \frac{M}{M+1} G(A) + \frac{1}{M+1} \frac{\int_A \psi(X_1; \theta)\, dG(\theta)}{\int \psi(X_1; \theta)\, dG(\theta)}.$$

Given a center measure G with a density g this suggests as estimator for a density f of the mixing distribution F:

$$\hat{f}_1(\theta) = \frac{M}{M+1} g(\theta) + \frac{1}{M+1} \frac{\psi(X_1; \theta) g(\theta)}{\int \psi(X_1; t)\, dG(t)}.$$

This is a convex combination of the prior density g and the "baseline posterior density," which treats g as the prior for θ and the kernel $\psi(X_1; \theta)$ as the likelihood. This formula can be forwarded into a sequential updating algorithm, which incorporates the observations one by one. The initial step starts from a true prior density $g = \hat{f}_0$, but each following step uses the output $\hat{f}_{i-1}$ of the previous step as the prior. Generalizing also the weights of the convex combination to given numbers $w_1, w_2, \ldots \in (0, 1)$, we are led to the recursive algorithm, often called *Newton's algorithm*:

$$\hat{f}_i(\theta) = (1 - w_i) \hat{f}_{i-1}(\theta) + w_i \frac{\psi(X_i; \theta)\, \hat{f}_{i-1}(\theta)}{\int \psi(X_i; t)\, d\hat{F}_{i-1}(t)}. \tag{5.9}$$

The corresponding estimates of the mixture densities are given by $\hat{p}_i(x) := \int \psi(x;\theta)\, d\hat{F}_i(\theta)$.

Attractive features of this algorithm are simplicity, making for fast computations, and the built-in absolute continuity of the estimate of the mixing distribution. An unappealing aspect is that the final estimate $\hat{f}_n$ depends on the ordering of the observations. In particular, it is not a Bayes estimator,[1] even if the initial motivation came from the posterior mean of a Dirichlet process mixture prior. The dependence on the ordering can be removed (or alleviated) by averaging over all (or many random) permutations of the observations. We can hope that the resulting estimator will be a reasonable approximation to a proper Bayesian estimator.

On the theoretical side, the estimator's properties are difficult to study, since it is neither a Bayes estimator nor an optimizer of a criterion function, nor does it have a closed form. The next theorem shows consistency; its proof uses martingale convergence techniques.

The weights are assumed to satisfy $\sum_{i=1}^{\infty} w_i = \infty$ and $\sum_{i=1}^{\infty} w_i^2 < \infty$. The first can be motivated by the fact that the weight $\prod_{i=1}^{n}(1-w_i)$ of the initial estimate $\hat{f}_0$ in $\hat{f}_n$ converges to zero if and only if $\sum_{i=1}^{\infty} w_i = \infty$. The second ensures that the weights decay to zero fast enough that the final estimate is not too much affected by the later terms, but information provided by additional observations accumulates properly. The theorem makes the following technical assumption:

$$\sup_{\theta_1,\theta_2,\theta_3\in\Theta} \int \frac{\psi^2(x;\theta_1)}{\psi^2(x;\theta_2)}\, \psi(x;\theta_3)\, d\nu(x) < \infty. \tag{5.10}$$

For consistency of $\hat{f}_n$ (in a weak sense) the mixing distribution is assumed identifiable. Most common kernels, including $\mathrm{Nor}(\theta,\sigma^2)$, σ fixed, $\mathrm{Ga}(\alpha,\theta)$, α fixed and $\mathrm{Poi}(\theta)$, satisfy this condition.

Let f_0 denote the density of the true mixing measure F_0 and let $p_0(x) = \int \psi(x;\theta)\, dF_0(\theta)$ be the true density of the observations.

Theorem 5.6 *If $\sum_{i=1}^{\infty} w_i = \infty$ and $\sum_{i=1}^{\infty} w_i^2 < \infty$, $K(p_0; p_G) < \infty$, and (5.10) holds, then $K(p_0;\hat{p}_n) \to 0$ a.s. If, moreover, the map $F \mapsto p_F$ is one-to-one, the kernel $\psi(x;\theta)$ is bounded and continuous in θ for every $x \in \mathfrak{X}$, and for every $\epsilon > 0$ and compact subset $\mathfrak{X}_0 \subset \mathfrak{X}$, there exists a compact subset $\Theta_0 \subset \Theta$ such that $\int_{\mathfrak{X}_0} \psi(x;\theta)\, d\nu(x) < \epsilon$ for all $\theta \notin \Theta_0$, then $\hat{F}_n \rightsquigarrow F_0$ a.s.*

Proof Because convergence in Kullback-Leibler divergence is stronger than $\mathbb{L}_1$- and weak convergence, the last assertion of the theorem follows from the first and Lemma K.11.

For the proof of the first assertion we first show that $K(p_0;\hat{p}_n)$ converges a.s. to a finite limit. We can write

$$\log\frac{\hat{p}_n}{\hat{p}_0}(x) = \sum_{i=1}^{n} \log\frac{\hat{p}_i}{\hat{p}_{i-1}}(x) = \sum_{i=1}^{n} \log\left(1 + w_i Z_i(x)\right),$$

[1] The only exception is $n = 1$, when $\hat{f}_1$ coincides with the Bayes estimator for the Dirichlet mixture prior where $M = (w_1^{-1} - 1)$ and G has density $\hat{f}_0$.

for the stochastic processes

$$Z_i(x) = \int \frac{\psi(x;\theta)\psi(X_i;\theta)}{\hat{p}_{i-1}(x)\hat{p}_{i-1}(X_i)}\, d\hat{F}_{i-1}(\theta) - 1.$$

For R defined by the relation $\log(1+x) = x - x^2R(x)$, this gives that

$$K(p_0;\hat{p}_0) - K(p_0;\hat{p}_n) = \sum_{i=1}^{n}\int w_i Z_i\, dP_0 - \sum_{i=1}^{n}\int w_i^2 Z_i^2 R(w_i Z_i)\, dP_0. \qquad (5.11)$$

By Taylor's theorem there exists a constant $\xi \in [0,1]$ such that $R(x) = \frac{1}{2}/(1+\xi x)^2$. Therefore $R(x) \le \frac{1}{2}/(1-w)^2$ for $x \ge -w$, whence the second term on the right is bounded above by $\frac{1}{2}\sum_{i=1}^{n} w_i^2/(1-w_i)^2 \int Z_i^2\, dP_0$. For $\mathscr{F}_i = \sigma(X_1,\ldots,X_i)$,

$$\begin{aligned}\mathrm{E}\Big(\int Z_i\, dP_0 \,|\, \mathscr{F}_{i-1}\Big) &= \int \Big(\int \frac{\psi(x;\theta)}{\hat{p}_{i-1}(x)}\, dP_0(x)\Big)^2 d\hat{F}_{i-1}(\theta) - 1 \\ &\ge \Big(\int\int \frac{\psi(x;\theta)}{\hat{p}_{i-1}(x)}\, dP_0(x)\, d\hat{F}_{i-1}(\theta)\Big)^2 - 1 = \Big(\int dP_0(x)\Big)^2 - 1 = 0.\end{aligned}$$

Thus $S_n := \sum_{i=1}^{n} w_i \int Z_i\, dP_0$ is a submartingale. By the inequality $(a+b)^2 \le 2a^2 + 2b^2$ and the Cauchy-Schwarz inequality $Z_i^2(x) \le 2k_{i-1}(x)k_{i-1}(X_i) + 2$, for the function defined by $k_{i-1}(x) = \int \psi^2(x;\theta)/\hat{p}_{i-1}^2(x)\, d\hat{F}_{i-1}(\theta)$. Consequently,

$$\begin{aligned}\tfrac{1}{2}\mathrm{E}\Big(\int Z_i^2\, dP_0 \,|\, \mathscr{F}_{i-1}\Big) &\le \Big(\int\int \frac{\psi^2(x;\theta)}{\hat{p}_{i-1}^2(x)}\, d\hat{F}_{i-1}(\theta)\, dP_0(x)\Big)^2 + 1 \\ &\le \Big(\int\int\int\int \frac{\psi^2(x;\theta_1)\psi(x;\theta_3)}{\psi^2(x;\theta_2)}\, d\mu(x)\, d\hat{F}_{i-1}(\theta_1) d\hat{F}_{i-1}(\theta_2)\, dF_0(\theta_3)\Big)^2 + 1.\end{aligned}$$

In the last step we apply Jensen's inequality to $1/\hat{p}_{i-1}^2(x)$, using the convexity of $x \mapsto x^{-2}$ on $(0,\infty)$. The right is bounded by 1 plus the square of the constant given in assumption (5.10). Because $R(x) \ge 0$ it follows that $Q := \frac{1}{2}\sum_{i=1}^{\infty} w_i^2/(1-w_i)^2 \int Z_i^2\, dP_0$ is nonnegative and integrable. Because $S_n \le K(p_0;\hat{p}_0) - K(p_0;\hat{p}_n) + Q \le K(p_0;\hat{p}_0) + Q$ by (5.11), where $\hat{p}_0 = p_G$ is fixed, it follows that S_n is a submartingale that is bounded above by an integrable random variable. By a martingale convergence theorem it converges a.s. (and in $\mathbb{L}_1$) to a finite limit.

If the limit were strictly positive, then the sequence $\sum_{i=1}^{n} w_i K(p_0;\hat{p}_{i-1})$ would increase to infinity, as the series $\sum_i w_i$ diverges by assumption. Thus to prove that the limit is zero, it suffices to show that the sequence $\sum_{i=1}^{n} w_i K(p_0;\hat{p}_{i-1})$ is almost surely bounded. By the inequality $\log x \le x - 1$, for $x > 0$, this sequence is bounded above by $\sum_{i=1}^{n} w_i \int (p_0/\hat{p}_{i-1} - 1)\, dP_0$. We show that the latter sequence tends to a limit by a similar argument as the preceding, but now at the level of f rather that p_F.

Writing $\log(\hat{f}_n/\hat{f}_0) = \sum_{i=1}^{n} \log(\hat{f}_i/\hat{f}_{i-1}) = \sum_{i=1}^{n} \log(1 + w_iY_i)$, we obtain in analogy with (5.11),

$$K(f_0;\hat{f}_0) - K(f_0;\hat{f}_n) = \sum_{i=1}^{n}\int w_i Y_i\, dF_0 - \sum_{i=1}^{n}\int w_i^2 Y_i^2 R(w_i Y_i)\, dF_0,$$

for the stochastic processes $Y_i(\theta) = \psi(X_i;\theta)/\hat{p}_{i-1}(X_i) - 1$. The second term on the right is bounded above by $Q' := \frac{1}{2}\sum_{i=1}^\infty w_i^2/(1-w_i)^2 \int Y_i^2\,dF_0$, which can be seen to be an integrable random variable by similar arguments as before. Since $\int Y_i\,dF_0 = (p_0/\hat{p}_{i-1})(X_i) - 1$, the increments of the first term satisfy

$$\mathrm{E}\Big(\int Y_i\,dF_0|\,\mathscr{F}_{i-1}\Big) = \int\Big(\frac{p_0}{\hat{p}_{i-1}} - 1\Big)\,dP_0 \geq \int \log\frac{p_0}{\hat{p}_{i-1}}\,dP_0 \geq 0.$$

Thus $S_n' := \sum_{i=1}^n w_i \int Y_i\,dF_0$ is a submartingale. It is bounded above by $K(f_0; g) + Q'$ and hence converges a.s. to a finite limit by a martingale convergence theorem.

The increments of the martingale $M_n' := S_n' - \sum_{i=1}^n \mathrm{E}(\int w_i Y_i\,dF_0|\,\mathscr{F}_{i-1})$ have conditional second moments bounded by $\mathrm{E}(\int w_i^2 Y_i^2\,dF_0|\,\mathscr{F}_{i-1})$, whose sum converges in $\mathbb{L}_1$. Therefore, the martingale M_n' converges a.s., and hence the compensator $\sum_{i=1}^n \mathrm{E}(\int w_i Y_i\,dF_0|\,\mathscr{F}_{i-1})$ converges, which is what we wanted to prove. □

5.5 Examples of Kernels

Example 5.7 (Density estimation on $\mathbb{R}$ using normal mixtures) For estimating a density function supported on the whole of $\mathbb{R}$ the density $x \mapsto \phi_\sigma(x-\mu)$ of the normal $\mathrm{Nor}(\mu, \sigma^2)$-distribution is a natural kernel. One may consider either a location-only mixture or a location-scale mixture.

For a location-only mixture, we equip $\theta = \mu$ with a mixing distribution F and consider $\varphi = \sigma$ an additional parameter of the kernel. The conjugate center measure G for the Dirichlet process prior on F is a normal distribution $\mathrm{Nor}(m, \eta^2)$. Given σ, the updating rule given by the generalized Pólya urn scheme of Theorem 5.3 becomes

$$q_{i,j} \propto \begin{cases} \sigma^{-1}e^{-(X_i-\mu_j)^2/(2\sigma^2)}, & j \neq i, j \geq 1, \\ M(\sigma^2+\eta^2)^{-1/2}e^{-(x-m)^2/(2(\sigma^2+\eta^2))}, & j = 0, \end{cases}$$

with the baseline posterior described by

$$dG_b(\mu|\,\sigma, X_i) \sim \mathrm{Nor}\Big(\frac{\sigma^{-2}X_i + \eta^{-2}m}{\sigma^{-2}+\eta^{-2}}, \frac{1}{\sigma^{-2}+\eta^{-2}}\Big).$$

The bandwidth parameter σ may be kept constant (but dependent on the sample size), or, more objectively, be equipped with a prior. The inverse-gamma prior $\sigma^{-2} \sim \mathrm{Ga}(s, \beta)$ is conjugate for the model given $(X_1, \ldots, X_n, \mu_1, \ldots, \mu_n)$, which is essentially a parametric model $X_i - \mu_i \stackrel{\text{iid}}{\sim} \mathrm{Nor}(0, \sigma^2)$. By elementary calculations,

$$\sigma^{-2}|\,X_1, \ldots, X_n, \mu_1, \ldots, \mu_n \sim \mathrm{Ga}\Big(s + n/2, \beta + \tfrac{1}{2}\sum_{i=1}^n (X_i - \mu_i)^2\Big).$$

For location-scale mixtures, we equip the parameter $\theta = (\mu, \sigma)$ with a mixing distribution, which in turn is modeled by a Dirichlet process prior. The normal-inverse-gamma distribution (that is, $\sigma^{-2} \sim \mathrm{Ga}(s, \beta)$ and $\mu|\,\sigma \sim \mathrm{Nor}(m, \sigma^2/a)$) is conjugate in this case. A simple computation shows that

$$q_{i,j} \propto \begin{cases} \dfrac{M\sqrt{a}\Gamma(s+1/2)\beta^s}{\sqrt{1+a}\,\Gamma(s)\{\beta+a(X_i-m)^2/(2(1+a))\}^{s+1/2}}, & j=0,\\ \sigma_j^{-1}e^{-(X_i-\mu_j)^2/(2\sigma_j^2)}, & j\neq i,\ j\geq 1,\end{cases}$$

and the baseline posterior is described by

$$\mu|\,\sigma, X_i \sim \mathrm{Nor}((X_i+am)/(1+a), \sigma^2/(1+a)),$$
$$\sigma^{-2}|\,X_i \sim \mathrm{Ga}(s+1/2, \beta+a(X_i-m)^2/(2(1+a))).$$

Example 5.8 (Uniform scale mixtures) Any nonincreasing probability density p on $[0,\infty)$ can be represented as a mixture of the form (see Williamson 1956)

$$p(x) = \int_0^\infty \theta^{-1}\mathbb{1}\{0\leq x\leq \theta\}\,dF(\theta).$$

This lets us put a prior on the class of *decreasing densities* by putting a Dirichlet process prior on F, with center measure G supported on $(0,\infty)$. In this case, the posterior updating formulas of Theorem 5.3 hold with

$$q_{i,j} \propto \begin{cases} \theta_j^{-1}\mathbb{1}\{X_i<\theta_j\}, & j\neq 0,\\ M\int_{X_i}^\infty \theta^{-1}\,dG(\theta), & j=0,\end{cases}$$
$$dG_b(\theta|\,X_i) \propto \theta^{-1}\mathbb{1}\{\theta > X_i\}\,dG(\theta).$$

Symmetrization of decreasing densities yield *symmetric, unimodal densities*. Such densities are representable as mixtures $p(x) = \int_0^\infty (2\theta)^{-1}\mathbb{1}\{-\theta\leq x\leq\theta\}\,dF(\theta)$ and can be equipped with a prior by, as before, putting a Dirichlet process prior on F. It is left as an exercise to derive computational formulas for the posterior.

Possibly *asymmetric unimodal densities* can be written as two-parameter mixtures of the form

$$p(x) = \int (\theta_2+\theta_1)^{-1}\mathbb{1}\{-\theta_1\leq x\leq\theta_2\}\,dF(\theta_1,\theta_2).$$

We may put a Dirichlet process prior on F, with center measure G supported on $(0,\infty)\times(0,\infty)$, or alternatively model F a priori as $F=F_1\times F_2$, for independent $F_1\sim \mathrm{DP}(M_1G_1)$ and $F_2\sim\mathrm{DP}(M_2G_2)$. In the second case it is convenient to index by a pair of integers (i,j), and write the updating formulas as $q_{i_1j_1:i_2j_2} = q^{(1)}_{i_1,j_1}q^{(2)}_{i_2,j_2}$, where

$$q^{(1)}_{i_1j_1} \propto \begin{cases} \theta_{1j_1}^{-1}\mathbb{1}\{X_i > -\theta_{1j_1}\}, & j_1\neq 0,\\ M_1\int_{X_i^-}^\infty \theta_1^{-1}\mathbb{1}\{X_i>-\theta_1\}\,dG_1(\theta_1), & j_1=0,\end{cases}$$

$$q^{(2)}_{i_2j_2} \propto \begin{cases} \theta_{2j_2}^{-1}\mathbb{1}\{X_i < \theta_{2j_2}\}, & j_2\neq 0,\\ M_2\int_{X_i^+}^\infty \theta_2^{-1}\mathbb{1}\{X_i<\theta_2\}\,dG_2(\theta_2), & j_2=0,\end{cases}$$

and

$$dG_{1b}(\theta_1|\,X_i) \propto \int_{\theta_2\in(X_i^+,\infty)} (\theta_2+\theta_1)^{-1}\,dG_2(\theta_2)\mathbb{1}\{\theta_1>-X_i\}\,dG_1(\theta_1),$$

$$dG_{2b}(\theta_2|\,X_i) \propto \int_{\theta_1\in(X_i^-,\infty)} (\theta_2+\theta_1)^{-1}\,dG_1(\theta_1)\mathbb{1}\{\theta_2<X_i\}\,dG_2(\theta_2).$$

Example 5.9 (Mixtures on the half line) Densities on the half line occur in applications in survival analysis and reliability. Also, the error distribution in quantile regression, which has zero as a fixed quantile, may be modeled by a combination of two densities on the half line.

Mixtures of exponential distributions are reasonable models for a decreasing, convex density on the positive half line. More generally, mixtures of gamma densities may be considered. Other possible choices of kernels include the inverse-gamma, Weibull and lognormal. Multi-dimensional parameters may be mixed using a joint mixing distribution or separately modeled by one-dimensional mixing distributions. Weibull and lognormal mixtures are dense in the space of densities on the positive half line relative to the $\mathbb{L}_1$-distance, provided that the shape parameter can assume arbitrarily large (for the Weibull) or small (for the lognormal) values. This follows because these two kernels form location-scale families in the log-scale, so that (2.4) applies.

We discuss computation for Weibull mixtures, with the prior on the mixing distribution equal to the Dirichlet-multinomial. The latter was seen to approximate the Dirichlet process in Theorem 4.19. Other examples can be treated similarly.

The Dirichlet-multinomial distribution is particularly convenient for application using BUGS. It suffices to describe the model and the prior, since the appropriate algorithm for drawing from conditional distributions is obtained by the program itself. Let $\text{Wei}(\alpha, \lambda)$ stand for the Weibull distribution with shape parameter α and rate parameter λ; that is, the density of X is given by $\alpha\lambda^\alpha x^{\alpha-1}e^{-\lambda^\alpha x^\alpha}$, $x > 0$. Let g stand for the density of the center measure. Then the model and the prior can be described through the following hierarchical scheme:

(i) $X_i \overset{\text{ind}}{\sim} \text{Wei}(\alpha, \lambda_{s_i})$,
(ii) $(s_1, \ldots, s_N)$ i.i.d. with $\text{P}(s_i = j) = w_j$, $j = 1, \ldots, N$,
(iii) $(w_1, \ldots, w_N) \sim \text{Dir}(N; M/N, \ldots, M/N)$,
(iv) $\lambda_i \overset{\text{iid}}{\sim} g$,
(v) $\alpha \sim \pi$.

Example 5.10 (Mixtures of beta and Bernstein polynomials) Beta distributions form a flexible two-parameter family of distributions on the unit interval, and generate a rich class of mixtures. Particular shapes of mixtures may be promoted by restricting the values of the beta parameters. For instance, a (vertically reflected) J-shaped density may be modeled by mixtures of $\text{Be}(a, b)$, $a < 1 \leq b$. This type of density is used to model the density of p-values under the alternative in multiple hypothesis testing or to model service time distributions.

We may consider a two-dimensional mixing distribution F on the parameters (a, b) of the beta distributions or mix a and b separately using two univariate mixing distributions, F_1 and F_2. In both cases we can employ Dirichlet process priors, where in the second case it is natural to model F_1 and F_2 as a priori independent. To obtain the reflected J-shape, it suffices to restrict the support of the base measure of F within $(0, 1) \times [1, \infty)$, or that of F_1 and F_2 within $(0, 1)$ and $[1, \infty)$, respectively.

In fact, mixtures of beta densities with *integer* parameters are sufficient to approximate any probability density. These are closely related to *Bernstein polynomials*, as discussed in Section E.1. The kth Bernstein polynomial associated to a given cumulative distribution function F on $(0, 1]$ is

$$B(x; k, F) := \sum_{j=0}^{k} F\Big(\frac{j}{k}\Big)\binom{k}{j} x^j (1-x)^{k-j}.$$

This is in fact itself a cumulative distribution function on [0, 1], which approximates F as $k \to \infty$. The corresponding density is

$$b(x; k, F) := \sum_{j=1}^{k} F\Big(\frac{j-1}{k}, \frac{j}{k}\Big] \mathrm{be}(x; j, k-j+1). \tag{5.12}$$

If F has a continuous density f, then $b(\cdot; k, F)$ approximates f uniformly as $k \to \infty$. The approximation rate is k^{-1} if f has bounded second derivative (see Proposition E.3), which is lower than the rate k^{-2} obtainable with other k-dimensional approximation schemes (such as general polynomials, splines or Gaussian mixtures). This slower approximation property will lead to slower contraction of a posterior distribution. However, the constructive nature of the approximation and preservation of nonnegativity and monotonicity are desirable properties of Bernstein polynomials.

The Bernstein density (5.12) is a mixture of beta densities. Modeling F as a Dirichlet process $\mathrm{DP}(MG)$ and independently, the degree k by a prior ρ on $\mathbb{N}$ yields the *Bernstein-Dirichlet process*. The beta densities $\mathrm{be}(x; j, k-j+1)$ may be thought of as smooth analogs of the uniform kernels $k\mathbb{1}\{(j-1)/k < x \leq j/k\}$, and thus the Bernstein-Dirichlet process can be considered a smoothing of the Dirichlet process. The tuning parameter k works like the reciprocal of a bandwidth parameter.

For computing the posterior distribution based on a random sample $X_1, \ldots, X_n$ from the Bernstein-Dirichlet process, it is convenient to introduce auxiliary latent variables $Y_1, \ldots, Y_n$ that control the label of the beta component from where $X_1, \ldots, X_n$ are sampled: if $(j-1)/k < Y_i \leq j/k$, then $X_i \sim \mathrm{Be}(\cdot; j, k-j+1)$. The prior then consists of the hierarchy

(i) $k \sim \rho$;
(ii) $F \sim \mathrm{DP}(MG)$;
(iii) $Y_1, \ldots, Y_n | \, k, F \stackrel{\text{iid}}{\sim} F$;
(iv) $X_1, \ldots, X_n | \, k, F, Y_1, \ldots, Y_n \stackrel{\text{ind}}{\sim} p(\cdot | \, k, Y_i)$, where

$$p(x | \, k, Y_i) = \sum_{j=1}^{k} \mathrm{be}(x; j, k-j+1)\mathbb{1}\{(j-1)/k < Y_i \leq j/k\}.$$

The posterior distribution can be computed by Gibbs sampling, as follows, with $z(y, k) := j$ if $(j-1)/k < y \leq j/k$:

(i) $k | \, X_1, \ldots, X_n, Y_1, \ldots, Y_n \sim \rho(\cdot | \, \mathrm{X}, \mathrm{Y})$, for

$$\rho(k | \, \mathrm{X}, \mathrm{Y}) \propto \rho(k) \prod_{i=1}^{n} \mathrm{be}(X_i; z(Y_i, k), k - z(Y_i, k) + 1);$$

(ii) $Y_i|\, k, X_1, \ldots, X_n, Y_{-i} \sim \sum_{j\neq i} q_{i,j}\delta_{Y_j} + q_{i,0}G_{b,i}$, for

$$q_{i,j} \propto \begin{cases} Mb(X_i; k, G), & j = 0, \\ \mathrm{be}(X_i; z(Y_j, k), k - z(Y_j, k) + 1), & j \neq i, 1 \leq j \leq k, \end{cases}$$

$$dG_{b,i}(y|\, k, Y_i) \propto g(y)\mathrm{be}(X_i; z(y, k), k - z(y, k) + 1).$$

Example 5.11 (Random histograms) Random histograms were briefly discussed in Section 2.3.2 as a method for prior construction via binning. Here we consider the particular situation where the corresponding probability distribution is given a Dirichlet process prior. The sample space, an interval in the real line, is first partitioned into intervals $\{I_j(h): j \in J\}$ of (approximately) length h, which is chosen from a prior. Next, a probability density is formed by distributing mass 1 first to the intervals according to the Dirichlet process and next uniformly within every interval.

The resulting prior can be cast in the form of a Dirichlet mixture, with the kernel

$$\psi(x, \theta, h) = \frac{1}{h}\sum_{j\in J} \mathbb{1}\{x \in I_j(h)\}\mathbb{1}\{\theta \in I_j(h)\}.$$

The corresponding mixture with mixing distribution F is

$$p_{h,F}(x) = \frac{1}{h}\sum_{j\in J} F(I_j(h))\mathbb{1}\{x \in I_j(h)\}.$$

We now equip h with a prior π and, independently, F with a DP(α) prior.

Analytic computation of the posterior mean of $p_{h,F}(x)$, given a random sample $X_1, \ldots, X_n$ from this density, is possible, assisted by a single one-dimensional numerical integration, without using simulation. As a first step, given the bandwidth h,

$$\mathrm{E}(p_{h,F}(x)|\, h, X_1, \ldots, X_n) = \frac{1}{h}\sum_{j\in J} \frac{\alpha(I_j(h)) + N_j(h)}{|\alpha| + n}\mathbb{1}\{x \in I_j(h)\}, \tag{5.13}$$

for $N_j(h) = \#\{i: X_i \in I_j(h)\} = n\mathbb{P}_n(I_j(h))$ the number of observations falling in the jth partitioning set. It remains to integrate out the right side with respect to the posterior distribution of h.

Because the prior enters the likelihood only through the probabilities $F(I_j(h))$ of the partitioning sets $I_j(h)$, the posterior distribution is only dependent on the vector of cell counts $(N_j(h): j \in J)$, by Theorem 3.14. The likelihood given (h, F) is proportional to $\prod_j F(I_j(h))^{N_j(h)}$, and hence the likelihood given only h is proportional to the expectation of this expression with respect to F, that is

$$\mathrm{E}\Big(\prod_{j\in J} F(I_j(h))^{N_j(h)}\Big) = \prod_{j\in J} (\alpha(I_j(h)))^{[N_j(h)]}/|\alpha|^{[n]},$$

by Corollary G.4; here $a^{[n]} = a(a+1)\cdots(a+n-1)$ is the ascending factorial. (Note that the product is a finite product, since $N_j(h) = 0$ except for finitely many j.) Thus, by Bayes's theorem, the posterior density satisfies

$$\pi(h|\, X_1, \dots, X_n) \propto \pi(h) \prod_{j \in J} \alpha(I_j(h))^{[N_j(h)]}. \tag{5.14}$$

Thus the posterior distribution of $p_{h,F}$ is conjugate, where the updating rule is given by $\alpha \mapsto \alpha + \sum_{i=1}^n \delta_{X_i}$ and $\pi \mapsto \pi(\cdot|\, X_1, \dots, X_n)$.

Finally, the posterior mean $\mathrm{E}(p_{h,F}(x)|\, X_1, \dots, X_n)$ can be computed as the posterior expectation of (5.13), and is given by

$$\frac{\int h^{-1} \sum_{j \in J} \frac{\alpha(I_j(h)) + N_j(h)}{|\alpha| + n} \mathbb{1}\{x \in I_j(h)\} \pi(h) \prod_{k \in J} \alpha(I_k(h))^{[N_k(h)]}\, dh}{\int \pi(h) \prod_{k \in J} \alpha(I_k(h))^{[N_k(h)]}\, dh}. \tag{5.15}$$

Example 5.12 (Hierarchical Dirichlet process) This example is qualitatively different from the previous examples in that a nonparametric mixture of nonparametric distributions is considered. Suppose

$$X_{i,1}, \dots, X_{i,n_j}|\, F_i, \bar\alpha \overset{\text{iid}}{\sim} F_i, \quad F_1, \dots, F_k|\, \bar\alpha \overset{\text{iid}}{\sim} \mathrm{DP}(M\bar\alpha), \quad \bar\alpha \sim \mathrm{DP}(\alpha_0).$$

Thus, k distributions are independently generated from a Dirichlet process whose base measure is itself randomly drawn from a source Dirichlet process, and a sample of observations is drawn from every of these k distributions. The discreteness of the Dirichlet process forces $\bar\alpha$ to be discrete, and hence the F_i will share support points. Such "Dirichlet mixtures of Dirichlet processes" are useful for hierarchical clustering. For further discussion, see Section 14.1.1.

Example 5.13 (Feller priors) The *Feller random sampling scheme* introduced in Section 2.3.4 approximates a given density f on an interval $I \subset \mathbb{R}$ by a mixture

$$a(x; k, F) = \int h_k(x; z)\, dF(z),$$

with the kernel h_k derived from the density $g_k(\cdot; x)$ of a Feller random sampling scheme $Z_{k,x}$. Specifically, if $Z_{i,k,x} = k^{-1} \sum_{i=1}^k Y_{i,x}$ with the $Y_{i,x}$ i.i.d. variables following an exponential family with parameters so that $\mathrm{E}Y_{i,x} = x$, the relationship between h_k and g_k can be made explicit and the scheme shown to be consistent in the sense that $a(\cdot; k, F)$ converges pointwise to f, at least if f is bounded and continuous.

We can form a prior on densities by putting a Dirichlet process on the mixing distribution F in (2.5). In view of Proposition 2.6 and Scheffe's theorem, the resulting Feller prior will have full total variation support if the Dirichlet process prior for F has full weak support.

The Feller prior for the Nor$(\theta, 1)$ family is a Dirichlet process mixture of normal location densities with $\sigma = k^{-1}$, while a Bernoulli sampling scheme gives the Bernstein polynomial prior. Another interesting choice is the Poisson family, which leads to a discrete gamma Dirichlet mixture $a(x; k, F) = \sum_{j=1}^\infty F((j-1)/k, j/k]\mathrm{ga}(x; j, k)$. This is thus seen to have every continuous density on the half line in its $\mathbb{L}_1$-support. Finally, the gamma family leads to a mixture of inverse-gamma distributions $\int (kz)^k x^{-(k+1)} e^{-kz/x}\, dF(z)/\Gamma(k)$.

5.6 Historical Notes

Bayesian density estimation was first considered by Ghorai and Rubin (1982). Dirichlet process mixtures were introduced by Ferguson (1983) and Lo (1984). Computational formulas were discussed by Kuo (1986) and Lo (1984). The Gibbs sampler based on the weighted generalized Pólya urn scheme was contributed by Escobar (1994), Escobar and West (1995) and MacEachern (1994), and refined by MacEachern and Müller (1998), Dey et al. (1998) and Neal (2000). These papers also discuss extensive applications of Dirichlet process mixtures of normals. The slicing algorithm is due to Walker (2007). The variational algorithm has its roots in the statistical physics literature. Beal and Ghahramani (2003) developed the ideas in the context of parametric exponential families. Blei and Jordan (2006) adapted the ideas to suit posterior computations in Dirichlet process mixture models. The predictive recursion deconvolution algorithm appeared in the papers Newton et al. (1998) and Newton and Zhang (1999). An initial proof of convergence appeared in Newton (2002), and was corrected by Ghosh and Tokdar (2006) and Tokdar et al. (2009). Brunner and Lo (1989) and Brunner (1992) considered uniform scale mixtures. Bernstein polynomials priors were studied by Petrone (1999b,a). Gasparini (1996) studied random histograms. Feller approximation priors were introduced by Petrone and Veronese (2010).

Problems

5.1 Generalize (5.14) and (5.15) when α is allowed to depend on h.

5.2 Obtain variational updates in a Dirichlet process mixture of normal model when σ^2 is unknown and is given an inverse-gamma prior distribution.

5.3 (Ghosal et al. 2008) Let a density p be a mixture of $\mathrm{Be}(a, 1)$, $0 < a \leq 1$. Show that $H(y) := \int_0^{e^{-y}} p(x)\, dx$ is a completely monotone function of y on $[0, \infty)$, that is, $(-1)^k H^{(k)}(y) \geq 0$ for all $k \in \mathbb{N}$, $y \geq 0$. Conversely, if $H(y)$ is a completely monotone function of y on $[0, \infty)$, then p is a mixture of $\mathrm{Be}(a, 1)$, $0 < a < \infty$.

Similarly, if p is a mixture of $\mathrm{Be}(1, b)$, $b \geq 1$, then $H(y) := \int_{e^{-y}}^{\infty} p(x)\, dx$ is a completely monotone function of y on $[0, \infty)$. Conversely, if $H(y)$ is a completely monotone function of y, then p is a mixture of $\mathrm{Be}(1, b)$, $0 < b < \infty$.

5.4 (Petrone 1999b) If for all k, $\rho(k) > 0$ and w_k has full support on $\mathbb{S}_k$, then show that every distribution on (0,1] is in the weak support of the Bernstein polynomial prior, and every continuous distribution is in the support for the topology induced by the Kolmogorov-Smirnov distance d_{KS}.

5.5 Verify that $b(x; k, F)$ is the density of $B(x; k, F)$ in Example 5.10.

5.6 (Petrone 1999a) Consider a Bernstein polynomial prior $p(x) = b(x; k, F)$, where $F \sim \mathrm{DP}(MG)$ and $k \sim \rho$. Let $\xi_{jk} = G((j-1)/k, j/k])$ and let $p^*(x)$ be the Bayesian density estimate given observations $X_1, \ldots, X_n$. Put $N_{jk} = \#\{i : X_i \in ((j-1)/k, j/k]\}$. Show that

(a) $p^*(x) \to \sum_{k=1}^{\infty} b(x; k, G)\rho_\infty(k|\, X_1, \ldots, X_n)$ as $M \to \infty$, for the probability mass function

$$\rho_\infty(k|\, X_1, \ldots, X_n) \propto \rho(k) \sum_{(z_{1k},\ldots,z_{kk})\in\{1,\ldots,k\}^k} \prod_{j=1}^{k} \xi_{jk}^{N_{jk}} \prod_{i=1}^{n} \mathrm{be}(X_i;\, z_{ik}, k - z_{ik} + 1);$$

(b) $p^*(x) \to \sum_{k=1}^{\infty} b(x; k, w_k)\rho_\infty(k|\, X_1, \ldots, X_n)$ as $M \to 0$, where the jth component of $w_k \in \mathbb{S}_k$ is proportional to $\xi_{jk} \prod_{i=1}^{n} \mathrm{be}(X_i;\, j, k - j + 1)$ and

$$\rho_\infty(k|\, X_1, \ldots, X_n) \propto \rho(k) \sum_{j=1}^{k} \xi_{jk} \prod_{i=1}^{n} \mathrm{be}(X_i;\, j, k - j + 1).$$

6

Consistency: General Theory

Posterior consistency concerns the convergence of the posterior distribution to the true value of the parameter of the distribution of the data when the amount of data increases indefinitely. Schwartz's theorem concerning consistency in models with replicated observations is a core result. In this chapter we extend it in many ways, also to general observations, such as Markov processes. Other results include Doob's theorem on almost everywhere consistency, consistency of tail-free processes and examples of inconsistency. We conclude with alternative approaches to proving consistency of posterior distributions and posterior-like processes.

6.1 Consistency and Its Implications

For every $n \in \mathbb{N}$, let $X^{(n)}$ be an observation in a sample space $(\mathfrak{X}^{(n)}, \mathscr{X}^{(n)})$ with distribution $P_\theta^{(n)}$ indexed by a parameter θ belonging to a first countable topological space Θ.[1] For instance, $X^{(n)}$ may be sample of size n from a given distribution P_θ, and $(\mathfrak{X}^{(n)}, \mathscr{X}^{(n)}, P_\theta^{(n)})$ the corresponding product probability space. Given a prior Π on the Borel sets of Θ, let $\Pi_n(\cdot|\, X^{(n)})$ be a version of the posterior distribution: a given choice of a regular conditional distribution of θ given $X^{(n)}$.

Definition 6.1 (Posterior consistency) The posterior distribution $\Pi_n(\cdot|\, X^{(n)})$ is said to be *(weakly) consistent* at $\theta_0 \in \Theta$ if $\Pi_n(U^c|\, X^{(n)}) \to 0$ in $P_{\theta_0}^{(n)}$-probability, as $n \to \infty$, for every neighborhood U of θ_0. The posterior distribution is said to be *strongly consistent* at $\theta_0 \in \Theta$ if this convergence is in the almost-sure sense.

Both forms of consistency are of interest. Naturally, strong consistency is more appealing as it is stronger, but it may require more assumptions. To begin with, it presumes that the observations $X^{(n)}$ are defined on a common underlying probability space (with, for each n, the measure $P_\theta^{(n)}$ equal to the image $P_\theta^{(\infty)} \circ (X^{(n)})^{-1}$ of the probability measure $P_\theta^{(\infty)}$ on this space), or at least that their joint distribution is defined, whereas weak consistency makes perfect sense without any relation between the observations across n.

The definition of strong consistency allows the exceptional null set where convergence does not occur to depend on the neighborhood U. However, because we assume throughout that Θ is first countable, consideration of countably many neighborhoods suffices, and a single null set always works.

[1] A topological space is *first countable* if for every point there is a countable base for the neighborhoods.

In the case that the parameter set Θ is a metric space, the neighborhoods U in the definition can be restricted to balls around θ_0, and consistency is equivalent to the convergence $\Pi_n(\theta\colon d(\theta,\theta_0) > \epsilon \mid X^{(n)}) \to 0$ as $n \to \infty$ in probability or almost surely, for every $\epsilon > 0$. Thus, the posterior distribution contracts to within arbitrarily small distance ϵ to θ_0. This can also be described as weak convergence of the posterior distribution to the Dirac measure at θ_0.

Proposition 6.2 (Consistency by convergence to Dirac) *On a metric space Θ the posterior distribution $\Pi_n(\cdot\mid X^{(n)})$ is consistent (or strongly consistent, respectively) at θ_0 if and only if $\Pi_n(\cdot\mid X^{(n)}) \rightsquigarrow \delta_{\theta_0}$ in $P_{\theta_0}^{(n)}$-probability (or almost surely $[P_{\theta_0}^{(\infty)}]$, respectively), as $n \to \infty$.*

Proof With the help of the Portmanteau theorem, Theorem A.2, it can be verified that a sequence of deterministic probability measures Π_n satisfies $\Pi_n \rightsquigarrow \delta_{\theta_0}$ if and only if $\Pi_n(\theta\colon d(\theta,\theta_0) > \epsilon) \to 0$ as $n \to \infty$ for every $\epsilon > 0$. The set of $\epsilon > 0$ in this statement can also be restricted to a countable set. This immediately gives the proposition for the case of strong consistency. This can be extended to weak consistency by using the fact that convergence in probability is equivalent to every subsequence having an almost surely converging subsequence (after first defining the observations on a suitable common probability space). □

Remark 6.3 Weak convergence of random measures "in probability" is best defined by metrizing the topology of weak convergence. A more precise statement of the preceding proposition is that consistency of the posterior distribution is equivalent to convergence of the random variables

$$d_W(\Pi_n(\cdot\mid X^{(n)}), \delta_{\theta_0})$$

to zero in probability (or almost surely) under the measures $P_{\theta_0}^{(n)}$ (or $P_{\theta_0}^{(\infty)}$), for d_W a metric on $\mathfrak{M}(\Theta)$ that metrizes the topology of weak convergence. For a separable metric space Θ such a metric exists, while for a general metric space it is still possible to metrize convergence in distribution to limits with separable support (in particular when the limit is a Dirac measure) (see e.g. van der Vaart and Wellner 1996, Chapter 1.12). Convergence in distribution of measures on a general topological space requires care, which makes defining consistency through convergence to a Dirac measure unattractive in that case.

Rather than through the posterior mass of abstract neighborhoods U, posterior consistency can be characterized through the posterior distributions of a set of test functions. Suppose that Ψ is a collection of real functions $\psi\colon \Theta \to \mathbb{R}$ such that, for any net $\{\theta_m\} \subset \Theta$,

$$\theta_m \to \theta_0, \qquad \text{if} \qquad \psi(\theta_m) \to \psi(\theta_0), \text{ for every } \psi \in \Psi. \tag{6.1}$$

For a countable set Ψ, it will be sufficient to limit this requirement to *sequences* θ_m.

For each ψ there is an induced posterior distribution $\Pi_n(\theta\colon \psi(\theta) \in \cdot \mid X^{(n)})$ on $\mathbb{R}$, which by definition is consistent at $\psi(\theta_0)$ if it converges in probability (or almost surely) in distribution to the Dirac measure at $\psi(\theta_0)$. Consistency of all these induced posterior distributions is equivalent to the posterior consistency for θ.

Lemma 6.4 (Consistency by functionals) *If Ψ is a set of measurable real functions on Θ so that* (6.1) *holds, then the posterior distribution is (strongly) consistent at θ_0 if for each $\psi(\theta)$ the induced posterior is (strongly) consistent at $\psi(\theta_0)$. If the functions ψ are uniformly bounded, then the latter is equivalent to the pair of conditions* $\mathrm{E}(\psi(\theta)|\, X^{(n)}) \to \psi(\theta_0)$ *and* $\mathrm{var}\,(\psi(\theta)|\, X^{(n)}) \to 0$*, in probability (or almost surely).*

Proof We claim that for any open neighborhood U of θ_0, there exists a finite set $\psi_1, \ldots, \psi_k$ and $\epsilon > 0$ such that $\cap_{i=1}^k \{\theta\colon |\psi_i(\theta) - \psi_i(\theta_0)| < \epsilon\} \subset U$. In this case $\Pi_n(U^c|\, X^{(n)})$ can be bounded by the sum of the probabilities $\Pi_n(|\psi_i(\theta) - \psi_i(\theta_0)| \geq \epsilon|\, X^{(n)})$, which tends to zero for fixed k, and the theorem follows.

If the claim were false, there would exist a neighborhood U of θ_0 such that for every finite subset m of Ψ there is some $\theta_m \notin U$ such that $|\psi(\theta_m) - \psi(\theta_0)| \leq (\#m)^{-1}$, for every $\psi \in m$. Then the net θ_m with the finite subsets of Ψ partially ordered by inclusion would satisfy $\psi(\theta_m) \to \psi(\theta_0)$ as $m \to \infty$, for every fixed ψ. Hence $\theta_m \to \theta_0$ by (6.1), which contradicts that $\theta_m \notin U$ for every m.

If Ψ is countable, we may order it arbitrarily as $\psi_1, \psi_2, \ldots$ and use the special net with m the finite set $\{\psi_1, \ldots, \psi_m\}$, which is a sequence.

A sequence of deterministic distributions on a compact subset of $\mathbb{R}$ tends weakly to a Dirac measure at a point if and only if the expectations tend to this point and the variances tend to zero. This extends trivially to almost sure weak convergence of random distributions (covering strong consistency), and to weak convergence in probability by arguing along subsequences. □

The following example shows that consistency depends not only on the true parameter and the prior, but also on the version of the posterior distribution. This discrepancy can arise because the posterior distribution is defined only up to a null set under the Bayesian marginal distribution of $X^{(n)}$, and this may not dominate this variable's distribution $P_{\theta_0}^{(n)}$ under a fixed parameter (see Section 1.3).

Example 6.5 (Dependence of consistency on version) Consider the model with observations $X_1, \ldots, X_n|\, P \stackrel{\text{iid}}{\sim} P$ and a Dirichlet prior $P \sim \mathrm{DP}(\alpha)$, for some base measure α. By Theorem 4.6 the $\mathrm{DP}(\alpha + \sum_{i=1}^n \delta_{X_i})$-distribution is a version of the posterior distribution, and this is consistent at any distribution P_0, by Corollary 4.17. Now, given a measurable set B with $\alpha(B) = 0$ and some probability measure Q, consider the random measure

$$\mathcal{P}^*_{X_1,\ldots,X_n} = \begin{cases} Q, & \text{if } X_i \in B\ \forall i = 1, \ldots, n, \\ \mathrm{DP}(\alpha + \sum_{i=1}^n \delta_{X_i}), & \text{otherwise.} \end{cases}$$

The Pólya urn scheme given in Section 4.1.4 describes the marginal distribution of the observation $(X_1, \ldots, X_n)$ as a mixture of distributions of vectors of the type $(Y_1, \ldots, Y_n)$, where $Y_i = Y_j$ if i and j belong to the same set in a given partition of $\{1, 2, \ldots, n\}$ and the tied values are generated i.i.d. from $\bar{\alpha}$. This shows that it gives probability zero to the set $B^n = B \times \cdots \times B$. As a posterior distribution is unique only up to null sets in the marginal distribution of the data, it follows that $\mathcal{P}^*_{X_1,\ldots,X_n}$ is another version of the posterior distribution.

If $P_0^n(B^n) \to 0$, then $\mathcal{P}^*_{X_1,\ldots,X_n}$ differs from the consistent sequence $\mathrm{DP}(\Phi + \sum_{i=1}^n \delta_{X_i})$ only on events with probability tending to zero, and hence it is (also) consistent at P_0. However, in the other case the new version is inconsistent whenever $Q \neq P_0$. For instance, this situation arises if P_0 and α are orthogonal and the set B is chosen so that $P_0(B) = 1$ and $\alpha(B) = 0$. As another example, if $B = \{0\}$, then $P_0^n(B^n) \to 0$ for every $P_0 \neq \delta_0$ and hence consistency pertains, but the posterior is inconsistent at $P_0 = \delta_0$ unless $Q = \delta_0$.

Consistency is clearly dependent on the topology on Θ under consideration. If Θ is a class of distributions, common choices are the weak topology, the topology induced by the Kolmogorov-Smirnov distance (if the sample space is Euclidean) and the topology induced by the total variation distance. The weak topology gives the weakest form of consistency of these three, but can be appropriate if only the probability measure needs to be estimated. The total variation distance is appropriate in density estimation, as it is one-half times the $\mathbb{L}_1$-distance on the densities, and is also equivalent to the Hellinger distance (in that they generate the same topology).

Consistency is a frequentist concept, but should be of interest also to a subjective Bayesian, by the following reasoning. A prior or posterior distribution expresses our knowledge about the unknown parameter, and a degenerate prior models perfect knowledge. Thus, by its definition, consistency entails convergence toward perfect knowledge. This is desirable, for instance, if n is the sample size and increases indefinitely, and also in every other setting where information accumulates without bound as $n \to \infty$.

From the point of view of an objective Bayesian (who believes in an unknown true model), consistency can be motivated by a "what if" argument. Suppose that an experimenter generates observations from a given distribution and presents the resulting data to a Bayesian without revealing the source distribution. It would be embarrassing if, even with large sample size, the Bayesian failed to come close to finding the mechanism used by the experimenter. Thus, consistency can be thought of as a validation of the Bayesian method: it ensures that an infinite amount of data overrides the prior opinion.

Consistency can also be linked to robustness with respect to the choice of a prior distribution: a Bayesian procedure should change little when the prior is only slightly modified. Now for any prior distribution Π_0, the posterior distribution $\Pi_{n,1}(\cdot|\, X^{(n)})$ based on a random sample $X^{(n)}$ of observations and the "contamination" prior $\Pi_1 = (1-\epsilon)\Pi_0 + \epsilon\delta_{p_0}$ is consistent at p_0 (see Problem 6.7). Therefore, if the posterior $\Pi_{n,0}(\cdot|\, X^{(n)})$ based on the prior Π_0 were *in*consistent at p_0, then for a metric d_W for the weak convergence the distance $d_W(\Pi_0(\cdot|\, X^{(n)}), \Pi_1(\cdot|\, X^{(n)}))$ between the two sequences of posterior distributions would not go to zero. Because this would be true even for small $\epsilon > 0$, robustness would be violated.

From a subjective Bayesian's point of view no true parameter exists, whence consistency may appear a useless concept at first. However, the true parameter may be eliminated and consistency be equivalently defined in terms of *merging* of predictive distributions of future observations, fully within the Bayesian paradigm. Roughly speaking, the calculations of different Bayesians using different priors will tend to agree (in the sense of the weak topology) if and only if their posteriors are consistent. In other words, two Bayesians with different priors will have their opinions "merging" if and only if consistency holds.

This is formalized in the following theorem, whose proof we omit (see Diaconis and Freedman 1986b, Appendix A). Consider i.i.d. observations $X_1, X_2, \ldots$ from a distribution P_θ on a sample space $\mathfrak{X}$, and denote the (posterior) predictive distribution of the future observations $(X_{n+1}, X_{n+2}, \ldots)$, given $X^{(n)} = (X_1, \ldots, X_n)$ when using prior Π by $Q_{\Pi,n} = \int P_\theta^\infty \, d\Pi_n(\theta|\, X^{(n)})$.

Theorem 6.6 (Merging of opinions) *Let Θ be a Borel subset of a Polish space and $\mathfrak{X}$ a standard Borel space, and assume that the map $\theta \mapsto P_\theta$ is one-to-one and measurable from Θ into $\mathfrak{M}(\Theta)$, equipped with the weak topology. Then, for any prior probability distribution Π on Θ equivalent are*

(i) *The posterior distribution $\Pi_n(\cdot|\, X^{(n)})$ relative to Π is consistent at every $\theta \in \Theta$.*
(ii) *The posterior distributions $\Pi_n(\cdot|\, X^{(n)})$ and $\Gamma_n(\cdot|\, X^{(n)})$ relative to Π and Γ merge in the sense that $d_W(\Pi_n(\cdot|\, X^{(n)}), \Gamma_n(\cdot|\, X^{(n)})) \to 0$ as $n \to \infty$ a.s. $[\int P_\theta^\infty \, d\Gamma(\theta)]$, for any prior probability distribution Γ.*

If, furthermore, the map $\theta \mapsto P_\theta$ is continuous and has a continuous inverse, then these statements are also equivalent to

(iii) *The predictive distributions relative to Π and Γ merge in that $d_W(Q_{\Pi,n}, Q_{\Gamma,n}) \to 0$ as $n \to \infty$ a.s. $[\int P_\theta^\infty \, d\Gamma(\theta)]$, for any prior probability distribution Γ.*

Posterior consistency entails the contraction of the full posterior distribution to the true value of the parameter. Naturally, an appropriate summary of its location should provide a point estimator that is consistent, in the usual sense of consistency of estimators. The following proposition gives a summary that works without further conditions. (The value $1/2$ can be replaced by any other number strictly between 0 and 1; possible problems with selecting a measurable version can be alleviated by restricting the estimator to a countable dense subset of Θ.)

Proposition 6.7 (Point estimator) *Suppose that the posterior distribution $\Pi_n(\cdot|\, X^{(n)})$ is consistent (or strongly consistent) at θ_0 relative to the metric d on Θ. Then $\hat{\theta}_n$, defined as the center of a (nearly) smallest ball that contains posterior mass at least $1/2$ satisfies $d(\hat{\theta}_n, \theta_0) \to 0$ in $P_{\theta_0}^{(n)}$-probability (or almost surely $[P_{\theta_0}^{(\infty)}]$, respectively).*

Proof For $B(\theta, r) = \{s \in \Theta\colon d(s, \theta) \le r\}$ the closed ball of radius r around $\theta \in \Theta$, let $\hat{r}_n(\theta) = \inf\{r\colon \Pi_n(B(\theta, r)|\, X^{(n)}) \ge 1/2\}$, where the infimum over the empty set is infinity. Taking the balls closed ensures that $\Pi_n(B(\theta, \hat{r}_n(\theta))|\, X^{(n)}) \ge 1/2$, for every θ. Let $\hat{\theta}_n$ be a near minimizer of $\theta \mapsto \hat{r}_n(\theta)$ in the sense that $\hat{r}_n(\hat{\theta}_n) \le \inf_\theta \hat{r}_n(\theta) + 1/n$.

By consistency $\Pi_n(B(\theta_0, \epsilon)|\, X^{(n)}) \to 1$ in probability or almost surely, for every $\epsilon > 0$. As a first consequence $\hat{r}_n(\theta_0) \le \epsilon$ with probability tending to one, or eventually almost surely, and hence $\hat{r}_n(\hat{\theta}_n) \le \hat{r}_n(\theta_0) + 1/n$ is bounded by $\epsilon + 1/n$ with probability tending to one, or eventually almost surely. As a second consequence, the balls $B(\theta_0, \epsilon)$ and $B(\hat{\theta}_n, \hat{r}_n(\hat{\theta}_n))$ cannot be disjoint, as their union would contain mass nearly $1 + 1/2$. This shows that $d(\theta_0, \hat{\theta}_n) \le \epsilon + \hat{r}_n(\hat{\theta}_n)$ with probability tending to one, or eventually almost surely, which is further bounded by $2\epsilon + 1/n$. □

An alternative point estimator is the *posterior mean* $\int \theta \, d\Pi_n(\theta | X^{(n)})$. This is attractive for computational reasons, as it can be approximated by the average of the output of a simulation run. Usually the posterior mean is also consistent, but in general this may require additional assumptions. For instance, for the parameter set equal to a Euclidean space, convergence of the posterior distributions relative to the weak topology does not imply convergence of its moments.

For the posterior mean to make sense, the parameter set Θ must be a subset of a vector space and some other conditions must be fulfilled. If the parameter θ is a probability measure, e.g. describing the law of a single observation, then the posterior mean is well defined as the *mean measure* of the posterior distribution by Fubini's theorem. If the parameter is an element of a separable normed space, then the posterior mean can be defined as a Pettis integral (see Section I.4), provided that $\int \|\theta\| \, d\Pi_n(\theta | X^{(n)}) < \infty$. The geometric version of *Jensen's inequality* says that the mean $\int \theta \, d\Pi(\theta)$ of a probability measure Π that concentrates on a closed convex set is contained in this set. Even when the metric on Θ is not induced by a norm, it is often true that the balls around points are convex. In the following theorem, only assume that Θ is a metric space where notions of expectation and convex combinations are defined.

Theorem 6.8 (Posterior mean) *Assume that the balls of the metric space (Θ, d) are convex and the geometric version of Jensen's inequality holds on closed balls. Suppose that $d(\theta_{n,1}, (1-\lambda_n)\theta_{n,1}+\lambda_n\theta_{n,2}) \to 0$ as $\lambda_n \to 0$, for any sequences $\theta_{n,1}$ and $\theta_{n,2}$. Then (strong) consistency of the posterior distributions $\Pi_n(\cdot | X^{(n)})$ at θ_0 implies (strong) consistency of the sequence of posterior means $\int \theta \, d\Pi_n(\theta | X^{(n)})$ at θ_0.*

Proof For fixed $\epsilon > 0$ we can decompose the posterior mean $\bar\theta_n = \int \theta \, d\Pi_n(\theta | X^{(n)})$ as $\bar\theta_n = (1-\lambda_n)\bar\theta_{n,1} + \lambda_n\bar\theta_{n,2}$, where $1-\lambda_n = \Pi_n(\theta\colon d(\theta,\theta_0) \le \epsilon | X^{(n)})$, and $\bar\theta_{n,1}$ and $\bar\theta_{n,2}$ are the means of the probability measures obtained by restricting and renormalizing the posterior distribution to the sets $\{\theta\colon d(\theta,\theta_0) \le \epsilon\}$ and $\{\theta\colon d(\theta,\theta_0) > \epsilon\}$, respectively. By the geometric form of Jensen's inequality and the assumed convexity of balls, $d(\bar\theta_{n,1},\theta_0) \le \epsilon$. By the consistency of the posterior distribution, the constants λ_n tend to zero. The result now follows with the help of the triangle inequality $d(\theta_0,\bar\theta_n) \le d(\theta_0,\bar\theta_{n,1}) + d(\bar\theta_{n,1}, (1-\lambda_n)\bar\theta_{n,1} + \lambda_n\bar\theta_{n,2})$ and the condition on d. □

If the metric d is convex in its arguments and uniformly bounded, then its balls are convex, and the condition on the metric d holds, as $d(\theta_{n,1}, (1-\lambda_n)\theta_{n,1} + \lambda_n\theta_{n,2}) \le 0 + \lambda_n d(\theta_{n,1},\theta_{n,2}) \to 0$, as $\lambda_n \to 0$. This shows that the theorem applies to the total variation norm, the Kolmogorov-Smirnov distance, the $\mathbb{L}_r$-norms for $r \ge 1$ and also to the weak topology on probability measures, metrized by the bounded Lipshitz distance (see Section A.2).

The Hellinger distance is not convex, but its square $P \mapsto d_H^2(P_0, P)$ is convex, which allows us to reach the same conclusion.

Provided that consistency is defined also for more general "divergence measures" like the Kullback-Leibler divergence, these could also be included if convex.

6.2 Doob's Theorem

Doob's theorem basically says that for any fixed prior, the posterior distribution is consistent at every θ except those in a "bad set" that is "small" when seen from the prior point of view. We first present the theorem and next argue that the message is not as positive as it may first seem.

The result requires that the experiments $(\mathfrak{X}^{(n)}, \mathscr{X}^{(n)}, P_\theta^{(n)}: \theta \in \Theta)$ are suitably linked and that Θ is a separable metric space with Borel σ-field $\mathscr{B}$. We assume that the observations $X^{(n)}$ can be defined as measurable functions $X^{(n)}: \mathfrak{X}^{(\infty)} \to \mathfrak{X}^{(n)}$ in a single experiment $(\mathfrak{X}^{(\infty)}, \mathscr{X}^{(\infty)}, P_\theta^{(\infty)}: \theta \in \Theta)$, with corresponding induced laws $P_\theta^{(n)} = P_\theta^{(\infty)} \circ (X^{(n)})^{-1}$. Furthermore, we assume that the informativeness increases with n in the sense that the induced σ-fields $\sigma\langle X^{(n)}\rangle$ are increasing in n (i.e. form a filtration on $(\mathfrak{X}^{(\infty)}, \mathscr{X}^{(\infty)})$. Given a prior Π, we next define a distribution Q on the product space $(\mathfrak{X}^{(\infty)} \times \Theta, \mathscr{X}^{(\infty)} \otimes \mathscr{B})$ by

$$Q(A \times B) = \int_B P_\theta^{(\infty)}(A)\, d\Pi(\theta).$$

The variable $\vartheta: \mathfrak{X}^{(\infty)} \times \Theta \to \Theta$ defined by $\vartheta(x, \theta) = \theta$ possesses the prior distribution Π as its law under Q. With abuse of notation we view $X^{(n)}$ also as maps defined on the product space (depending only on the first coordinate), with the induced σ-fields $\sigma\langle X^{(n)}\rangle$ contained in the product σ-field $\mathscr{X}^{(\infty)} \otimes \mathscr{B}$ (but being of product form). In this setting, the conditional distribution of $X^{(n)}$ given $\vartheta = \theta$ is $P_\theta^{(n)}$ and the posterior law $\Pi_n(\cdot|\, X^{(n)})$ is the conditional law of ϑ given $X^{(n)}$.

Theorem 6.9 (Doob) *Let Π be an arbitrary prior on the Borel sets of a separable metric space Θ. If ϑ is measurable relative to the Q-completion of $\sigma\langle X^{(1)}, X^{(2)}, \ldots\rangle$, then the posterior distribution $\Pi_n(\cdot|\, X^{(n)})$ is strongly consistent at θ, for Π-almost every θ. In fact $\int f(\theta')\, d\Pi_n(\theta'|\, X^{(n)}) \to f(\theta)$, almost surely $[P_\theta^{(\infty)}]$, for Π-almost every θ and every $f \in \mathbb{L}_1(\Pi)$.*

Proof For a given bounded measurable real-valued function f on Θ, the martingale convergence theorem gives that $\mu_n(X^{(n)}) := \mathrm{E}(f(\vartheta)|\, X^{(n)}) \to \mathrm{E}(f(\vartheta)|\, \sigma\langle X^{(1)}, X^{(2)}, \ldots\rangle)$, almost surely under Q. By assumption, the limit is equal to $f(\vartheta)$, almost surely. It follows that

$$1 = Q\Big(\lim_{n\to\infty} \mu_n(X^{(n)}) = f(\vartheta)\Big) = \int P_\theta^{(\infty)}\Big(\lim_{n\to\infty} \mu_n(X^{(n)}) = f(\theta)\Big)\, d\Pi(\theta).$$

The integrand must be equal to 1 for all θ except those in a Π-null set N. For every $\theta \notin N$ we have $\int f(\theta)\, d\Pi_n(\theta|\, X^{(n)}) = \mu_n(X^{(n)}) \to f(\theta)$.

For every bounded measurable function f there exists such a null set. Because Θ is a separable metric space, there exists a *countable* collection of such functions that determines convergence in distribution (see e.g. Theorem 1.12.2 in van der Vaart and Wellner 1996). For every θ not in the union of these countably many null sets, we have that $\Pi_n(\cdot|\, X^{(n)}) \rightsquigarrow \delta_\theta$. ◻

The assumption that $\vartheta \in \sigma\langle X^{(1)}, X^{(2)}, \ldots\rangle$ can be paraphrased as saying that, in the Bayesian model, with unlimited observations, it is possible to find the parameter exactly. The condition holds, for instance, if θ, or even a bimeasurable transformation $h(\theta)$ of the parameter, is consistently estimable from the observations $X^{(n)}$. Indeed, if $T_n(X^{(n)}) \to h(\theta)$ almost surely $[P_\theta^{(\infty)}]$ for every $\theta \in \Theta$, then

$$Q\Big(\lim_{n\to\infty} T_n(X^{(n)}) = h(\vartheta)\Big) = \int P_\theta^{(\infty)}\Big(\lim_{n\to\infty} T_n(X^{(n)}) = h(\theta)\Big)\, d\Pi(\theta) = 1,$$

by the dominated convergence theorem. Thus $h(\vartheta)$ inherits its measurability (up to a Q-null set) from $\lim_{n\to\infty} T_n(X^{(n)})$. Various conditions ensuring existence of consistent estimators are available in the literature (e.g. Ibragimov and Has'minskiĭ 1981, Bickel et al. 1998, or van der Vaart (1998), Chapters 5, 10). Alternatively, the assumption may be verified directly. In the i.i.d. setup, with $\mathfrak{X}^{(\infty)} = \mathfrak{X}^\infty$ and $X^{(n)}$ the projection on the first n coordinates, some basic regularity conditions on the model suffice.

Proposition 6.10 *Let $(\mathfrak{X}, \mathscr{X}, P_\theta: \theta \in \Theta)$ be experiments with $(\mathfrak{X}, \mathscr{X})$ a standard Borel space and Θ a Borel subset of a Polish space. If $\theta \mapsto P_\theta(A)$ is Borel measurable for every $A \in \mathscr{X}$ and the model $\{P_\theta: \theta \in \Theta\}$ is identifiable, then there exists a Borel measurable function $f: \mathfrak{X}^\infty \to \Theta$ such that $f(x_1, x_2, \ldots) = \theta$ a.s. $[P_\theta^\infty]$, for every $\theta \in \Theta$.*

Proof By Proposition A.5 and the measurability of $\theta \mapsto P_\theta(A)$ for every Borel set A, the map $P: \theta \mapsto P_\theta$ can be viewed as a Borel measurable map from Θ into the space $\mathfrak{M}(\mathfrak{X})$ equipped with the weak topology (with metric d_W). Likewise, the empirical distribution $\mathbb{P}_n$ of n i.i.d. observations $X_1, \ldots, X_n$ from P_θ can be viewed as a measurable map from $\mathfrak{X}^\infty$ into $\mathfrak{M}(\mathfrak{X})$. By Proposition F.3, $d_W(\mathbb{P}_n, P_\theta) \to 0$ almost surely $[P_\theta^\infty]$. We can also view both $\mathbb{P}_n$ and P as maps on the product space $\mathfrak{X}^\infty \times \Theta$ (depending only on the first and second coordinates, respectively), and then by Fubini's theorem have that $d_W(\mathbb{P}_n, P) \to 0$ almost surely under the measure Q on $\mathfrak{X}^\infty \times \Theta$, defined as previously by $Q(A \times B) = \int_B P_\theta^\infty(A)\, d\Pi(\theta)$. Hence the random variable P taking values in the Polish space $(\mathfrak{M}(\mathfrak{X}), d_W)$ is the a.s. limit of the sequence of the random variables $\mathbb{P}_n$, taking values in the same space. As $\lim \mathbb{P}_n$ is measurable relative to the σ-field $\sigma\langle X_1, X_2, \ldots\rangle$ on $\mathfrak{X}^\infty \times \Theta$ and $\mathfrak{X}^\infty$ is Polish by assumption, it follows that $P_\theta = \tilde{f}(X_1, X_2, \ldots)$, Q-almost surely for some measurable function $\tilde{f}: \mathfrak{X}^\infty \to \mathfrak{M}(\mathfrak{X})$ (see Dudley 2002, Theorem 4.2.8).

Since $\theta \mapsto P_\theta$ is one-to-one, its range is a measurable subset of $\mathfrak{M}(\mathfrak{X})$ and its inverse $\jmath: P_\theta \mapsto \theta$ is measurable by Kuratowski's theorem (Parthasarathy 2005; or Srivastava 1998, Theorems 4.5.1 and 4.5.4). We define f as the composition $f = \jmath \circ \tilde{f}$. □

Doob's theorem is remarkable in many respects. Virtually no condition is imposed on the model regarding dependence or distribution, or about the parameter space. As posterior consistency implies existence of a consistent estimator (see Proposition 6.7), the only condition of the theorem is also necessary. Also of interest is that the result applies to any version of the posterior distribution.

The implication of Doob's theorem goes far beyond consistency because the posterior measure converges to the degenerate measure at the true θ not just weakly, but in a much stronger sense where the posterior expectation of every integrable function f approaches the function value $f(\theta)$. Weak convergence requires that the latter hold only for bounded, continuous functions. In particular, this means that the posterior probability of every set A

converges to the indicator $\mathbb{1}\{\theta \in A\}$ under $P_\theta^{(\infty)}$ a.s. θ $[\Pi]$, and hence if $\Pi(\{\theta_0\}) > 0$, then $\Pi_n(\{\theta_0\}|X^{(n)}) \to 1$ a.s. under $P_{\theta_0}^\infty$.

The theorem implies that a Bayesian will "almost always" have consistency, as long as she is certain of her prior. Since null sets are "negligibly small," a troublesome value of the parameter "will not obtain."

However, such a view is very dogmatic. No one in practice can be certain of the prior, and troublesome values of the parameter may really obtain. In fact, the Π-null set could be very large if *not* judged from the point of view of the prior. To see an extreme example, consider a prior that assigns all its mass to some fixed point θ_0. The posterior then also assigns mass one to θ_0 and hence is *in*consistent at every $\theta \neq \theta_0$. Doob's theorem is still true, of course; the point is that the set $\{\theta: \theta \neq \theta_0\}$ is a null set under the present prior. Thus Doob's theorem should not create a false sense of satisfaction about Bayesian procedures in general. It is important to know, for a given "reasonable" prior, at which parameter values consistency holds. Consistency at every parameter in a set of prior probability one is not enough. This explains that in Section 6.3 we reach a very different conclusion when measuring "smallness" of exceptional sets in a different way.

An exception is the case that the parameter set Θ is countable. Then Doob's theorem shows that consistency holds at θ as long as Π assigns positive mass to it. More generally, consistency holds at any atom of a prior. In fact, if θ_0 is an atom of Π, then $\Pi(\theta = \theta_0|X^{(n)}) \to 1$ a.s. $[P_{\theta_0}^{(\infty)}]$. If the parameter space is Euclidean and the null sets of the prior distribution are also Lebesgue null sets, then the conclusion from Doob's theorem may be still attractive even though this does not give consistency at all points. In a semiparametric model, if the parametric part is only of interest, the method may give rise to posterior consistency except on a Lebesgue null set for the parametric part. Even in these cases the theorem is of "asymptopia" type only, in that at best it gives convergence without quantification of the approximation error, or uniformity in the parameter.

6.3 Inconsistency

If $\Pi(U) = 0$ for some set U, then $\Pi_n(U|\,X^{(n)}) = 0$ almost surely under the Bayesian marginal distribution $\int P_\theta^{(n)}\,d\Pi(\theta)$ of the observation $X^{(n)}$, and hence also under its "true" law $P_{\theta_0}^{(n)}$ if this is absolutely continuous with respect to its marginal law. (The latter is necessary to ensure that the posterior distribution is almost surely uniquely defined under the true law.) Therefore, in this situation it is necessary for consistency that the prior charge every neighborhood of the true parameter θ_0. In most finite-dimensional problems, this necessary condition is also sufficient, but not so in infinite dimensions.

Consider the simplest infinite-dimensional problem, that of estimating a discrete distribution on $\mathbb{N}$. The parameter set $\Theta = \mathfrak{M}(\mathbb{N})$ can be identified with the infinite-dimensional unit simplex $\mathbb{S}_\infty$, and the weak and total variation topologies are equivalent. One might expect that the posterior is consistent at every point if the prior assigns positive probability to every of its neighborhoods, but this is not the case.

Example 6.11 (Freedman 1963) Let θ_0 be the geometric distribution with parameter $1/4$. There exists a prior Π on $\mathbb{S}_\infty$ that gives positive mass to every neighborhood of

θ_0, but the posterior concentrates in the neighborhoods of the geometric distribution with parameter $3/4$.

This counter-example is generic in a topological sense. A subset A of a topological space is considered to be topologically small, and said to be *meager* (or of the *first category*), if it can be written as a countable union of *nowhere dense* sets (sets whose closures have no interior points). The following result shows that the misbehavior of the posterior described in the preceding example is not just a pathological case, but a common phenomenon.

Theorem 6.12 *Under the product of the usual topology on $\Theta = \mathfrak{M}(\mathbb{N})$ and the weak topology on $\mathfrak{M}(\Theta)$, the set of all pairs $(\theta, \Pi) \in \Theta \times \mathfrak{M}(\Theta)$ such that the posterior is consistent at θ is meager in $\Theta \times \mathfrak{M}(\Theta)$. Furthermore, the set of $(\theta, \Pi) \in \Theta \times \mathfrak{M}(\Theta)$ such that*

$$\limsup_{n\to\infty} \mathrm{E}_\theta \Pi_n(U|\, X_1, \ldots, X_n) = 1, \text{ for all open } U \neq \varnothing, \tag{6.2}$$

is the complement of a meager set.

Proof The first assertion is a consequence of the second. For the proof of the second let $\Theta^+ = \{\theta = (\theta_1, \theta_2, \ldots) \colon \theta_i > 0, \forall i\}$ be the interior of the infinite-dimensional simplex, and let $\mathfrak{M}^+(\Theta^+) = \{\Pi \colon \Pi(\Theta^+) > 0\}$ be the set of priors that assign some mass to this, i.e. do not concentrate all their mass on the "faces" $\{\theta \colon \theta_i = 0, \text{ some } i\}$ of Θ.

The set $\mathfrak{M}^+(\Theta^+)$ is *residual*, i.e. its complement $\{\Pi \colon \Pi(\Theta^+) = 0\}$ is meager within $\mathfrak{M}(\Theta)$. Indeed, as Θ^+ is open, this complement is closed by the Portmanteau theorem; it also has empty interior, since $(1 - 1/m)\Pi + (1/m)\delta_\theta \rightsquigarrow \Pi$ as $m \to \infty$ for any θ, and is clearly not in the set if $\theta \in \Theta^+$. As a consequence the set $\Theta \times \mathfrak{M}^+(\Theta^+)$ is residual within $\Theta \times \mathfrak{M}(\Theta)$. The remainder of the proof may therefore focus on showing that the set of $(\theta, \Pi) \in \Theta \times \mathfrak{M}^+(\Theta^+)$ that satisfy (6.2) is residual within $\Theta \times \mathfrak{M}^+(\Theta^+)$.

The likelihood for an i.i.d. sample $X_1, \ldots, X_n \stackrel{\text{iid}}{\sim} \theta$ relative to counting measure on $\mathfrak{X}^n$ is the function $\theta \mapsto \prod_{i=1}^n \theta_{X_i}$, and the posterior distribution satisfies $d\Pi_n(\theta|\, X_1, \ldots, X_n) \propto \prod_{i=1}^n \theta_{X_i}\, d\Pi(\theta)$. For $\Pi \in \mathfrak{M}^+(\Theta^+)$, the norming constant is positive and the posterior is uniquely defined. Furthermore, when restricted to the domain $\Theta \times \mathfrak{M}^+(\Theta^+)$, the map

$$(\theta_0, \Pi) \mapsto \mathrm{E}_{\theta_0} \int \chi(\theta)\, d\Pi_n(\theta|\, X_1, \ldots, X_n) = \mathrm{E}_{\theta_0} \frac{\int \chi(\theta) \prod_{i=1}^n \theta_{X_i}\, d\Pi(\theta)}{\int \prod_{i=1}^n \theta_{X_i}\, d\Pi(\theta)}$$

is continuous for any fixed, bounded, continuous function $\chi \colon \Theta \to \mathbb{R}$.

For $\theta^+ \in \Theta^+$ let $\mathfrak{M}^f(\Theta, \theta^+)$ be the set of priors Π that are finitely supported with exactly one support point θ^+ belonging to Θ^+ (and hence the finitely many other support points belonging to one of the faces $\{\theta \colon \theta_i = 0\}$ of the simplex). From the facts that any prior Π can be approximated arbitrarily closely by a finitely supported measure, the Dirac measure δ_θ can be approximated by Dirac measures $\delta_{(\theta_1, \ldots, \theta_N, 0, 0, \ldots)}$ and θ^+ can be given arbitrarily small weight, it follows that $\mathfrak{M}^f(\Theta, \theta^+)$ is dense in $\mathfrak{M}^+(\Theta^+)$.

For $\Pi \in \mathfrak{M}^f(\Theta, \theta^+)$ and every $\theta_0 \in \Theta^+$,

$$\Pi_n(\cdot|\, X_1, \ldots, X_n) \rightsquigarrow \delta_{\theta^+}, \quad \text{a.s. } [P_{\theta_0}^\infty].$$

In other words, the posterior distribution for data resulting from a fully supported θ_0 eventually chooses for the (unique) fully supported point θ^+ in the support of the prior, also if $\theta^+ \neq \theta_0$. The reason is that for every $\theta \neq \theta^+$ in the support of Π eventually some observation will be equal to a value $x \in \mathfrak{X}$ such that $\theta_x = 0$, whence the posterior distribution $\Pi_n(\{\theta\}| X_1 \ldots, X_n) \propto \prod_{i=1}^n \theta_{X_i} \, \Pi(\{\theta\})$ attaches zero mass to that point.

For $\theta^+ \in \Theta^+$, let $\theta \mapsto \chi_k(\theta; \theta^+)$ be a sequence of bounded continuous functions with $1 \geq \chi_k(\cdot; \theta^+) \downarrow \mathbb{1}_{\{\theta^+\}}$, as $k \to \infty$. By the preceding display, for every k,

$$\limsup_{n\to\infty} \mathrm{E}_{\theta_0} \int \chi_k(\theta; \theta^+) \, d\Pi_n(\theta| X_1, \ldots, X_n) = 1. \tag{6.3}$$

This is true for any prior $\Pi \in \mathfrak{M}^f(\Theta, \theta^+)$, and any $\theta_0 \in \Theta^+$.

For $\theta^+ \in \Theta^+$ and $k, j, m \in \mathbb{N}$, define sets $F_{\theta^+,k,j,m}$ as the intersections

$$\bigcap_{n\geq m} \Big\{ (\theta_0, \Pi) \in \Theta \times \mathfrak{M}^+(\Theta^+) \colon \mathrm{E}_{\theta_0} \int \chi_k(\theta; \theta^+) \, d\Pi_n(\theta| X_1, \ldots, X_n) \leq 1 - j^{-1} \Big\}.$$

These sets are closed in $\Theta \times \mathfrak{M}^+(\Theta^+)$ by the continuity of the maps in their definition. They have empty interior, because any pair (θ_0, Π) can be arbitrarily closely approximated by an element of $\Theta^+ \times \mathfrak{M}^f(\Theta, \theta^+)$, but no such pair is contained in $F_{\theta^+,k,j,m}$ by (6.3). Thus $F_{\theta^+,k,j,m}$ is nowhere dense.

Finally the set of $(\theta, \Pi) \in \Theta \times \mathfrak{M}^+(\Theta^+)$ such that (6.2) fails is the union of all $F_{\theta^+,k,j,m}$ over θ^+ in a countable, dense subset of Θ^+, $k \in \mathbb{N}$, $j \in \mathbb{N}$, $m \in \mathbb{N}$. This follows since every open U contains θ^+ with $\mathbb{1}_{\{\theta^+\}} \leq \chi_k(\cdot; \theta^+) \leq \mathbb{1}_U$. □

Thus, except for a relatively small collection of pairs of (θ, Π), the posterior distribution is inconsistent. The second assertion of the theorem even shows that the posterior distribution visits every open set in the parameter set infinitely often, thus "wanders aimlessly around the parameter space." While this result cautions about the naive use of Bayesian methods, it does not mean that Bayesian methods are useless. Indeed, a pragmatic Bayesian's only aim may be to find some prior that is consistent at various parameter values (and complies with his subjective belief, if available). Plenty of such priors may exist, even though "many more" are not appropriate. This phenomenon may be compared with the role of differentiable functions within the class of all functions. Functions that are differentiable at some point are a small minority in the topological sense, and nowhere differentiable functions are abundant. Nevertheless, the functions which we generally work with are differentiable at nearly all places.

Dirichlet process priors can be seen to be consistent by inspection of the (Dirichlet) posterior (see Chapter 4), and this extends to mixtures of Dirichlet processes, provided the prior precision parameters of the Dirichlet components are uniformly bounded (see (4.29) and the subsequent discussion in Section 4.5). The following example shows that boundedness of the prior precision cannot be missed. It also exemplifies that a convex mixture of consistent priors may itself be inconsistent.

Example 6.13 (Mixtures of Dirichlet processes) Consider again estimating a discrete distribution on $\mathbb{N}$ based on a random sample $X_1, \ldots, X_n$ from a distribution $\theta \in \Theta = \mathfrak{M}(\mathbb{N}) =$

$\mathbb{S}_\infty$. Consider a prior of the form $\Pi = \frac{1}{2}\mathrm{DP}(\alpha) + \frac{1}{2}\delta_\phi$, for α and ϕ the elements of Θ satisfying

$$\alpha_i = 2^{-i}, \qquad \phi_i \propto \frac{1}{i(\log i)^2}.$$

Then the posterior distribution in the model with observations $X_1, \ldots, X_n | \theta \sim \theta$ and prior $\theta \sim \Pi$ satisfies $\Pi_n(\cdot | X_1, \ldots, X_n) \rightsquigarrow \delta_\phi$, θ_0^∞-almost surely, for every $\theta_0 \in \Theta$ such that $\theta_{0i} = \phi_i$ for some $i \geq 10$ and the Kullback-Leibler divergence $K(\theta_0; \phi) < \infty$. In other words, the posterior distribution is strongly *in*consistent whenever $\theta_0 \neq \phi$. This is true in spite of the fact that the prior has full support, through its $\mathrm{DP}(\alpha)$-component.

If the Dirac measure δ_ϕ is viewed as a Dirichlet process with center measure ϕ and infinite precision, then the prior Π is a mixture of two Dirichlet processes. Alternatively, the same inconsistency can be created with infinite mixtures of Dirichlet processes of finite precision. The idea is to let the components have precision tending to infinity, so that in the limit they approximate the degenerate Dirichlet δ_ϕ: as alternate prior, consider $\Pi = \sum_{j=1}^\infty 2^{-j}\mathrm{DP}(\alpha_j)$, for $\alpha_j \in \Theta$ is defined by

$$\bar\alpha_{j,i} \propto \begin{cases} \phi_i, & \text{if } i \leq j, \\ 2^{-i}, & \text{otherwise}, \end{cases} \qquad |\alpha_j| \to \infty.$$

In this setting the posterior distribution behaves the same as before.

The intuition is that "thick-tailed" distributions θ (with $\theta_i \to 0$ slowly) are unexpected according to every Dirichlet prior $\mathrm{DP}(\alpha_j)$, but less so as $j \to \infty$. A thick-tailed true distribution will produce large observations, which makes the posterior distribution choose for the heavy-tailed components of the prior.

The proofs of these results can be based on the general Theorem 6.17 ahead. For the two-component mixture this is particularly straightforward. The one-point set $\{\phi\}$ has Kullback-Leibler divergence $K(\theta_0; \phi)$ equal to a finite constant c, and its prior mass is bigger than $\frac{1}{2}$, both by construction. Thus the posterior distributions contract to $\{\phi\}$ (and hence are inconsistent) if for every $\epsilon > 0$ there exist tests such that $P_{\theta_0}^n \psi_n \to 0$ and $\int_{d(\theta,\phi)>\epsilon} P_\theta^n(1 - \psi_n) \leq e^{-nC}$ for some constant $C > 0$. Now for any $c_n > 0$ and DP_α the Dirichlet prior,

$$P_{\theta_0}^n(X_{(n)} \leq c_n) = \Big(1 - \sum_{i>c_n} \theta_{0i}\Big)^n \leq \Big(1 - \frac{1}{\log c_n}\Big)^n,$$

$$\int_{\theta\neq\phi} P_\theta^n(X_{(n)} > c_n)\, d\Pi(\theta) \leq \tfrac{1}{2} n \int P_\theta(X_1 > c_n)\, d\mathrm{DP}_\alpha(\theta) \leq \tfrac{1}{2} n \sum_{i>c_n} \alpha_i \leq 2^{-c_n}.$$

Therefore, the tests $\psi_n = \mathbb{1}\{X_{(n)} \leq c_n\}$ for $c_n = e^{\sqrt{n}}$ satisfy the requirements, and contraction to ϕ follows.

For the mixture prior we first note that $\mathrm{DP}(\alpha_j) \rightsquigarrow \delta_\phi$ as $j \to \infty$, by Theorem 4.16 as $\bar\alpha_j \rightsquigarrow \phi$ and $|\alpha_j| \to \infty$. Therefore $\liminf_{j\to\infty} \mathrm{DP}_{\alpha_j}(\theta\colon K(\theta_0; \theta) < c)$ is bounded below by $\delta_\phi(\theta\colon K(\theta_0; \theta) < c) = 1$, by the Portmanteau theorem, and hence $\Pi(\theta\colon K(\theta_0; \theta) < c) > 0$. Next we use the same tests ψ_n, but replace the estimate of the errors of the second kind by

$$\int_{\theta\neq\phi} P_\theta^n(X_{(n)} > c_n)\, d\Pi(\theta) \le \sum_{j<c_n} 2^{-j}\tfrac{1}{2}n \sum_{i>c_n} \alpha_{j,i} + \sum_{j\ge c_n} 2^{-j}.$$

This is readily seen to be exponentially small.

The inconsistency in the preceding example is caused by having too little prior mass near the true value of the parameter. More drastic examples of inconsistency occur in semiparametric problems. In the following example we consider the problem of estimating a location parameter θ based on observations $X_i = \theta + e_i$, with a (symmetrized) Dirichlet process prior on the distribution of the errors. Consistency of the posterior distribution for θ then depends strongly on the combination of shapes of the base measure and the error distribution. This is a bit surprising, as the parameter of interest is only one-dimensional. In the example, the location parameter is identified by the symmetry of the error distribution, but similar results hold if the median of the errors is zero and the Dirichlet prior is conditioned to have median zero.

Example 6.14 (Symmetric location) Consider a sample of observations $X_i = \theta + e_i$, where θ is an unknown constant and the errors $e_1, \ldots, e_n$ are an i.i.d. sample from a distribution H that is symmetric about 0. Thus θ is identified as the point of symmetry of the distribution of the observations.

We equip the error distribution H with a Dirichlet prior, and independently the parameter θ with a Nor(0, 1)-prior. As H is symmetric, a symmetrized Dirichlet is natural, but for simplicity consider first an ordinary (true) Dirichlet with a symmetric absolutely continuous base measure $\alpha := MC$, $M > 0$, C with density c. The resulting model can be written hierarchically as

$$X_i = Y_i + \theta, \qquad Y_1, \ldots, Y_n \,|\, H \overset{\text{iid}}{\sim} H, \qquad H \sim \text{DP}(MC), \qquad \theta \sim \text{Nor}(0, 1).$$

The marginal distribution of the variables $Y_1, \ldots, Y_n$ is given by the Pólya urn with base measure MC, described in Section 4.1.4. Given θ, the observations $X_1, \ldots, X_n$ are simple shifts of these variables and hence follow a Pólya urn with base measure $MC(\cdot - \theta)$. Conditioned on the event F that the outputs of the urn are all distinct, the outputs are simply a random sample from the base measure; in other words:

$$p(X_1, \ldots, X_n \,|\, \theta, F) = \prod_{i=1}^n c(X_i - \theta). \tag{6.4}$$

If the true (non-Bayesian) distribution of the observations is continuous, then the event F *will* occur with probability one, and it is unnecessary to consider its complement. By Bayes's rule the posterior density for θ is then proportional to the expression in the preceding display times the prior density of θ. We may analyze its performance by methods for parametric models: the infinite-dimensional parameter has been removed. However, although the expression in the display is a parametric likelihood, it is not the likelihood of the data, and the analysis must proceed by treating the expression as a misspecified likelihood. Under regularity conditions the posterior will asymptotically concentrate on the set of θ that minimizes the Kullback-Leibler divergence between the true distribution of $(X_1, \ldots, X_n)$ and the misspecified likelihood (see Theorem 8.36 ahead).

We do not pursue the full details of this analysis, but point out some striking results (found by Diaconis and Freedman 1986a). If the base measure C is standard Cauchy and the true error distribution H_0 has two point masses of size $1/2$ at the points $\pm a$, where $a > 1$, then there are two points of minimum Kullback-Leibler divergence: $\gamma_{+,-} := \theta_0 \pm \sqrt{a^2 - 1}$, for θ_0 the true location. If the true error distribution is continuous, but with a density that puts mass nearly $1/2$ close to $\pm a$, then this situation is approximated, and can be matched exactly by fine-tuning the distribution. Then the posterior distribution contracts to the points $\{\gamma_+, \gamma_-\}$ rather than to θ_0. It is even true that the posterior distribution will choose between the two points based on the fractions of observations near $\pm a$ and alternate between the γ_+ and γ_-, with the result that, for any $\epsilon > 0$, for *both* $\gamma = \gamma_+$ and $\gamma = \gamma_-$,

$$\limsup_{n\to\infty} \Pi_n(|\theta - \gamma| < \epsilon | X_1, \ldots, X_n) = 1, \qquad \text{a.s.}$$

Similar results are true if the base measure is a general t-distribution.

On the positive side, if the true error distribution is symmetric and strongly unimodal, then the true value θ_0 is the unique point of minimum of the Kullback-Leibler divergence, and consistency will obtain. Furthermore, if the base measure is log-concave (like the normal or double-exponential distribution), then consistency obtains at any symmetric density.

Regrettably, the negative performance of the posterior distribution does not disappear if the prior for the error distribution H is symmetrized. The observations for a symmetrized Dirichlet process prior (see Section 4.6.1) on H, and an independent Nor(0, 1)-prior on θ can be written $X_i = S_i Y_i + \theta$, for the hierarchy

$$\theta \sim \text{Nor}(0, 1), \quad S_1, \ldots, S_n \stackrel{\text{iid}}{\sim} \text{Rad}, \quad Y_1, \ldots, Y_n | H \stackrel{\text{iid}}{\sim} H, \quad H \sim \text{DP}(C).$$

(The S_i are *Rademacher variables*: $\text{P}(S_i = 1) = \text{P}(S_i = -1) = 1/2$.) As before, the marginal distribution of $(Y_1, \ldots, Y_n)$ follows the Pólya urn scheme (4.13). Presently we need to consider this distribution, both given the event F that there are no ties among the Y_i and given the events $G_{k,l}$ that $Y_k = Y_l$ is the only tie (for $1 \le k < l \le n$). Given F the variables $Y_1, \ldots, Y_n$ are a random sample from $C(\cdot - \theta)$, as before, and taking account of the symmetry of C, we again find formula (6.4) for the likelihood (given F). The posterior distribution of θ, given F, is continuous with density proportional to (6.4) times $\phi(\theta)$.

In the present situation there is no one-to-one relationship between ties among the X_i and the Y_i, whence F is *not* the event that the X_i are distinct. In fact, on the event $\{S_k = S_l\}$ we have $Y_k = Y_l$ if and only if $X_k = X_l$, but on the event $\{S_k = -S_l\}$ we have $Y_k = Y_l$ if and only if $X_k + X_l = 2\theta$. Because the second possibility will arise in the data (for some θ), we also consider the likelihood given $G_{k,l}$. Given $G_{k,l}$, the Pólya scheme produces $n - 1$ i.i.d. variables Y_i for $i \neq l$ from the base measure C and sets $Y_l = Y_k$. The resulting X_i consist of a random sample $(X_i : i \neq l)$ from C and have either $X_l = X_k$ (when $S_k = S_l$) or $X_k + X_l = 2\theta$ (when $S_k = -S_l$), both with probability $1/2$. The conditional likelihood given the intersection $G'_{k,l} = G_{k,l} \cap \{S_k = -S_l\}$ can be written symbolically in the form

$$p(X_1, \ldots, X_n | \theta, G'_{k,l}) = \prod_{\substack{i=1\\ i\neq l}}^{n} c(X_i - \theta)\delta_{2\theta}(X_k + X_l).$$

This implies that the posterior distribution of θ given $X_1, \ldots, X_n, G'_{k,l}$ is a Dirac measure at the point $\theta_{k,l} := (X_k + X_l)/2$.

The posterior distribution, given the union $E := F \cup \cup_{k<l} G'_{k,l}$, is the convex combination of the posterior distributions given the events in this union, with weights equal to $\mathrm{P}(F \mid X_1, \ldots, X_n, E)$ and $\mathrm{P}(G'_{k,l} \mid X_1, \ldots, X_n, E)$. From the Pólya urn scheme it is easy to see that the unconditional probabilities of F and a single $G_{k,l}$ are equal to $M^n/M^{[n]}$ and $M^{n-1}/M^{[n]}$, respectively, whence the prior weights of these events satisfy $\mathrm{P}(F \mid E) = M\mathrm{P}(G'_{k,l} \mid E)$. For the posterior weights we compute

$$p(X_1, \ldots, X_n \mid F) = \int \prod_{i=1}^{n} c(X_i - \theta)\, \phi(\theta)\, d\theta,$$

$$p(X_1, \ldots, X_n \mid G'_{k,l}) = \int \prod_{\substack{i=1\\ i \neq l}}^{n} c(X_i - \theta) \delta_{2\theta}(X_k + X_l)\, \phi(\theta)\, d\theta.$$

The second marginal conditional distribution shifts the $(n-1)$-dimensional, singular distribution of $X_1, \ldots, X_n \mid \theta, G'_{k,l}$ over the full space $\mathbb{R}^n$, and is equivalent to Lebesgue measure (if C has full support). Therefore, the posterior weights of the events F and $G'_{k,l}$ satisfy

$$\mathrm{P}(F \mid X_1, \ldots, X_n, E) \propto \int \prod_{i=1}^{n} c(X_i - \theta)\, \phi(\theta)\, d\theta M,$$

$$\mathrm{P}(G'_{k,l} \mid X_1, \ldots, X_n, E) \propto \int \prod_{\substack{i=1\\ i \neq l}}^{n} c(X_i - \theta) \delta_{2\theta}(X_k + X_l)\, \phi(\theta)\, d\theta,$$

where the inverse proportionality factor is the sum of the right sides of the first equation and of $n(n-1)/2$ second equations (for $1 \leq k < l \leq n$). Thus we find that the posterior distribution of θ given E (and the observations) is the convex combination of a continuous distribution and a discrete distribution on the points $\theta_{k,l}$.

If the true distribution of the observations is continuous, then all observations X_i will be distinct with probability one, and so will all pairs $X_k + X_l$. Since this sure event can be seen to be contained in $E := F \cup \cup_{k<l} G'_{k,l}$, it is unnecessary to consider the posterior distribution outside E.

The remainder of the analysis is again to study the posterior of a misspecified likelihood. The analysis is more involved, as both the absolutely continuous and the discrete parts of the posterior distribution may dominate, depending on the true density of the observations (in particular the induced density of the pairs $(X_k + X_l)/2$). However, the final result turns out to be the same (see Diaconis and Freedman 1986a for details).

6.4 Schwartz's Theorem

In this section we take the parameter equal to a probability density p, belonging to a parameter set $\mathcal{P}$, a class of probability densities relative to a given dominating measure ν on a sample space $(\mathfrak{X}, \mathscr{X})$. The parameter set $\mathcal{P}$ is equipped with an appropriate topology and σ-field, and the prior is a probability measure Π on $\mathcal{P}$. We consider estimating p based on a random sample $X_1, \dots, X_n$ of observations, with p_0 denoting the true density. As notational convention we use the corresponding uppercase letter P to denote the probability measure specified by a density p.

Throughout the section the prior is considered to be fixed. Priors that change with n are considered in Section 6.7.

A key condition for posterior consistency is that the prior assigns positive probability to any Kullback-Leibler (or KL) neighborhood of the true density. Write $K(p_0; p) = \int p_0 \log(p_0/p)\, d\nu$ for the Kullback-Leibler divergence, and for a set $\mathcal{P}_0$ of densities let $K(p_0; \mathcal{P}_0) = \inf_{p \in \mathcal{P}_0} K(p_0; p)$ be the minimal divergence of p_0 to an element of $\mathcal{P}_0$.

Definition 6.15 (Kullback-Leibler property) A density p_0 is said possess the *Kullback-Leibler property* relative to a prior Π if $\Pi(p: K(p_0; p) < \epsilon) > 0$ for every $\epsilon > 0$. This is denoted $p_0 \in \mathrm{KL}(\Pi)$, and we also say that p_0 belongs to the *Kullback-Leibler support* of Π.

Restricted to a class of densities on which pointwise evaluation is continuous, the Kullback-Leibler divergence is a measurable function of its second argument relative to the Borel σ-field of the total variation metric, and hence Kullback-Leibler neighborhoods are measurable in the space of densities equipped with this Borel σ-field (see Problem B.5). However, for our purpose it is sufficient to interpret the Kullback-Leibler property in the sense of inner probability: it suffices that there exist measurable sets $\mathcal{B} \subset \{p: K(p_0; p) < \epsilon\}$ with $\Pi(\mathcal{B}) > 0$.

Schwartz's theorem is the basic result on posterior consistency for dominated models. It has two conditions: the true density p_0 should belong to the Kullback-Leibler support of the prior, and the hypothesis $p = p_0$ should be testable against complements of neighborhoods of p_0. The first is clearly a Bayesian condition, but the second may be considered a condition to enable recovery of p_0 by any statistical method. Although in its original form the theorem has limited applicability, extensions go deeper, and lead to a rich theory of posterior consistency. Also the theory of contraction rates, developed in Chapter 8, uses similar ideas.

In the present context *tests* ϕ_n are understood to refer both to measurable mappings $\phi_n: \mathfrak{X}^n \to [0, 1]$ and to the corresponding statistics $\phi_n(X_1, \dots, X_n)$. We write $P^n \phi_n$ for the power $\mathrm{E}_p \phi_n(X_1, \dots, X_n) = \int \phi_n \, dP^n$ of the tests. The topology in the following theorem is assumed to have a countable local basis (e.g. metrizable), but is otherwise arbitrary.

Theorem 6.16 (Schwartz) *If $p_0 \in \mathrm{KL}(\Pi)$ and for every neighborhood $\mathcal{U}$ of p_0 there exist tests ϕ_n such that $P_0^n \phi_n \to 0$ and $\sup_{p \in \mathcal{U}^c} P^n(1 - \phi_n) \to 0$, then the posterior distribution $\Pi_n(\cdot \mid X_1, \dots, X_n)$ in the model $X_1, \dots, X_n \mid p \overset{iid}{\sim} p$ and $p \sim \Pi$ is strongly consistent at p_0.*

Proof It must be shown that $\Pi_n(\mathcal{U}^c \mid X_1, \dots, X_n) \to 0$ almost surely, for every given neighborhood $\mathcal{U}$ of p_0. By Lemma D.11 it is not a loss of generality to assume that the tests ϕ_n as in the theorem have exponentially small error probabilities in the sense that $P_0^n \phi_n \le e^{-Cn}$ and $\sup_{p \in \mathcal{U}^c} P^n(1 - \phi_n) \le e^{-Cn}$, for some positive constant C. By the Borel-Cantelli lemma $\phi_n \to 0$ almost surely under P_0^∞. Then the theorem follows from an application of Theorem 6.17(a), with $\mathcal{P}_0 = \{p: K(p_0; p) < \epsilon\}$ a Kullback-Leibler neighborhood of size $c = \epsilon < C$ and $\mathcal{P}_n = \mathcal{U}^c$. ◻

Part (a) of the following theorem contains the crux of Schwartz's theorem. It has a much wider applicability. For instance, it was used in Example 6.13 to prove the *in*consistency of certain priors. Part (b) of the result shows that the posterior does not concentrate on sets of exponentially small mass. Although it is a special case of (a) it is often applied explicitly to extend the applicability of Schwartz's theorem.

Theorem 6.17 (Extended Schwartz) *If there exist a set $\mathcal{P}_0 \subset \mathcal{P}$ and number c with $\Pi(\mathcal{P}_0) > 0$ and $K(p_0; \mathcal{P}_0) \le c$, then $\Pi_n(\mathcal{P}_n | X_1, \ldots, X_n) \to 0$ a.s. $[P_0^\infty]$ for any sets $\mathcal{P}_n \subset \mathcal{P}$ such that either* (a) *or* (b) *holds for some constant $C > c$:*

(a) *there exist tests ϕ_n such that $\phi_n \to 0$ a.s. $[P_0^\infty]$ and $\int_{\mathcal{P}_n} P^n(1 - \phi_n)\, d\Pi(p) \le e^{-Cn}$.*
(b) $\Pi(\mathcal{P}_n) \le e^{-Cn}$.

In particular, the posterior distribution $\Pi_n(\cdot | X_1, \ldots, X_n)$ in the model $X_1 \ldots, X_n | p \overset{iid}{\sim} p$ and $p \sim \Pi$ is strongly consistent at p_0 if $p_0 \in \mathrm{KL}(\Pi)$ and for every neighborhood $\mathcal{U}$ of p_0 there exists a constant $C > 0$, measurable partitions $\mathcal{P}_n = \mathcal{P}_{n,1} \cup \mathcal{P}_{n,2}$, and tests ϕ_n such that

(i) $P_0^n \phi_n < e^{-Cn}$ *and* $\sup_{p \in \mathcal{P}_{n,1} \cap \mathcal{U}^c} P^n(1 - \phi_n) < e^{-Cn}$,
(ii) $\Pi(\mathcal{P}_{n,2}) < e^{-Cn}$.

Proof The second part of the theorem follows by applying, for every given neighborhood $\mathcal{U}$ of p_0 of square radius smaller than C, the first part (a) with the choices $\mathcal{P}_0 = \mathcal{U}$ and $\mathcal{P}_n = \mathcal{U}^c \cap \mathcal{P}_{n,1}$, and the first part (b) with the choices $\mathcal{P}_0 = \mathcal{U}$ and $\mathcal{P}_n = \mathcal{P}_{n,2}$.

For the proof of the first part we first show that, for any $c' > c$, eventually a.s. $[P_0^\infty]$:

$$\int \prod_{i=1}^n \frac{p}{p_0}(X_i)\, d\Pi(p) \ge \Pi(\mathcal{P}_0) e^{-c'n}. \tag{6.5}$$

The integral in the left side is bounded below by $\Pi(\mathcal{P}_0) \int \prod_{i=1}^n (p/p_0)(X_i)\, d\Pi_0(p)$, for Π_0 the renormalized restriction of Π to $\mathcal{P}_0$. By Jensen's inequality its logarithm is bounded below by $\log \Pi(\mathcal{P}_0) + \int \log \prod_{i=1}^n (p/p_0)(X_i)\, d\Pi_0(p)$. The second term times n^{-1} coincides with the average $n^{-1} \sum_{i=1}^n \int \log(p/p_0)(X_i)\, d\Pi_0(p)$, and tends almost surely to its expectation, by the strong law of large numbers. By Fubini's theorem the expectation is $-\int K(p_0; p)\, d\Pi_0(p) \ge -c$, by the assumption and the definition of Π_0. This concludes the proof of (6.5).

For the proof of (a) we note that, in view of Bayes's rule (1.1), with probability one under P_0^∞,

$$\Pi_n(\mathcal{P}_n | X_1, \ldots, X_n) \le \phi_n + \frac{(1 - \phi_n) \int_{\mathcal{P}_n} \prod_{i=1}^n (p/p_0)(X_i)\, d\Pi(p)}{\int \prod_{i=1}^n (p/p_0)(X_i)\, d\Pi(p)}.$$

The first term on the right tends to zero by assumption. By (6.5), the denominator of the second term is bounded below by $\Pi(\mathcal{P}_0) e^{-c'n}$ eventually a.s., for every $c' > c$. The proof is complete once it is shown that $e^{c'n}$ times the numerator tends to zero almost surely, for some $c' > c$. By Fubini's theorem,

$$P_0^n\Big[(1-\phi_n)\int_{\mathcal{P}_n}\prod_{i=1}^n \frac{p}{p_0}(X_i)\,d\Pi(p)\Big] = \int_{\mathcal{P}_n} P_0^n\Big[(1-\phi_n)\prod_{i=1}^n \frac{p}{p_0}(X_i)\Big]\,d\Pi(p)$$
$$\leq \int_{\mathcal{P}_n} P^n(1-\phi_n)\,d\Pi(p) \leq e^{-Cn}.$$

Since $\sum_n e^{c'n}e^{-Cn} < \infty$ if $c < c' < C$, it follows by Markov's inequality that $e^{c'n}$ times the numerator tends to zero almost surely.

Assertion (b) is the special case of (a) with the tests chosen equal to $\phi_n = 0$, □

Remark 6.18 If $p_0 \in \mathrm{KL}(\Pi)$, then the integral $I_n = \int \prod_{i=1}^n (p/p_0)(X_i)\,d\Pi(p)$ satisfies $n^{-1}\log I_n \to 0$, almost surely $[P_0^\infty]$. Indeed, (6.5) applied with $\mathcal{P}_0$ a Kullback-Leibler neighborhood of radius ϵ gives $\liminf n^{-1}\log I_n \geq -\epsilon'$, for every $\epsilon' > \epsilon > 0$. Conversely, by Markov's inequality $P_0^n(n^{-1}\log I_n > \epsilon) \leq e^{-n\epsilon}P_0^n I_n \leq e^{-n\epsilon}$, for every n, and therefore we have $\limsup n^{-1}\log I_n \leq 0$, by the Borel-Cantelli lemma.

The convergence $n^{-1}\log I_n \to 0$ shows that the lower bound $e^{-c'n}$ in (6.5) is pessimistic: the constant c' should actually be (close to) zero. We shall exploit more accurate bounds with $c' = c'_n \to 0$ only when discussing rates of convergence in Chapter 8. Then we shall take the *amount* of prior mass near p_0 into account; exploitation of only its positiveness (the Kullback-Leibler property) makes Schwartz's theory unnecessarily weak.

Example 6.19 (Finite-dimensional models) If the model is smoothly parameterized by a finite-dimensional parameter that varies over a bounded set, then consistent tests as required in Theorem 6.16 exist under mere regularity conditions. For unbounded Euclidean sets, some conditions are needed. (See e.g. Le Cam 1986, Chapter 16, or van der Vaart 1998, Lemmas 10.4 and 10.6.)

Example 6.20 (Weak topology) For the weak topology on the set of probability measures (corresponding to the densities in $\mathcal{P}$) consistent tests as in Theorem 6.16, always exist. Therefore, the posterior distribution is consistent for the weak topology at any density p_0 that has the Kullback-Leibler property for the prior.

To construct the tests, observe that sets of the type $\mathcal{U} = \{p: P\psi < P_0\psi + \epsilon\}$, for continuous functions $\psi: \mathfrak{X} \to [0,1]$ and $\epsilon > 0$, form a subbase for the weak neighborhood system at a probability measure P_0. By Hoeffding's inequality, Theorem K.1, the test $\phi_n = \mathbb{1}\{n^{-1}\sum_{i=1}^n \psi(X_i) > P_0\psi + \epsilon/2\}$ satisfies $P_0^n\phi_n \leq e^{-n\epsilon^2/2}$ and $P^n(1-\phi_n) \leq e^{-n\epsilon^2/2}$, for any $P \in \mathcal{U}^c$. A general neighborhood contains a finite intersection of neighborhoods from the subbase, and can be tested by the maximum of the tests attached to every of the subbasic neighborhoods. The errors of the first and second kind of this aggregate test are bounded by $Ne^{-n\epsilon^2/2}$ and $e^{-n\epsilon^2/2}$, respectively, for N the number of subbasic neighborhoods in the union.

Example 6.21 (Countable sample spaces) All the usual topologies (including total variation) on $\mathfrak{M}(\mathfrak{X})$ for a countable, discrete sample space $\mathfrak{X}$ coincide with the weak topology. The preceding example shows that the posterior distribution is strongly consistent at every parameter θ_0 such that

$$\Pi\Big(\theta = (\theta_1, \theta_2, \ldots)\colon \sum_{j=1}^{\infty} \theta_{0,j} \log \frac{\theta_{0,j}}{\theta_j} < \delta\Big) > 0. \tag{6.6}$$

In particular, for a prior with full support consistency pertains at any θ_0 that is essentially finite dimensional in the sense that $\{j \in \mathbb{N}\colon \theta_{0,j} > 0\}$ is finite.

Example 6.11 shows that consistency may not hold for arbitrary θ_0, and Theorem 6.12 suggests that (6.6) fails for "most" combinations (Π, θ_0). However, the condition is reasonable and many combinations seem to work.

In its original form Schwartz's theorem requires that the complement of every neighborhood of p_0 can be "tested away." For strong metrics, such as the $\mathbb{L}_1$-distance, such tests may not exist, but the posterior distribution may still be consistent. Theorem 6.17(b) can be very helpful here, as it shows that the posterior distribution will give negligible mass to a set of very small prior mass; such a set need not be tested. The following proposition shows that this possibility of splitting of the model in sets that are testable or of small prior mass is also necessary for posterior consistency (at an exponential rate).

Proposition 6.22 (Necessity of testing) *If $p_0 \in \mathrm{KL}(\Pi)$, then the following are equivalent for any sequence of sets $\mathcal{P}_n \subset \mathcal{P}$:*

(i) *There exists $C > 0$ such that $\Pi_n(\mathcal{P}_n | X_1, \ldots, X_n) \le e^{-nC}$, eventually a.s. $[P_0^\infty]$.*
(ii) *There exist partitions $\mathcal{P}_n = \mathcal{P}_{n,1} \cup \mathcal{P}_{n,2}$, test functions ϕ_n, and $C_1, C_2 > 0$ with:*

  (a) *$\phi_n \to 0$ a.s. $[P_0^\infty]$ and $\sup_{p \in \mathcal{P}_{n,1}} P^n(1 - \phi_n) \le e^{-C_1 n}$;*
  (b) *$\Pi(\mathcal{P}_{n,2}) \le e^{-C_2 n}$, eventually.*

Proof That (ii) implies that $\Pi_n(\mathcal{P}_n | X_1, \ldots, X_n) \to 0$ follows from Theorem 6.17. That this convergence happens at exponential speed can be seen by inspection of the proof.

To show that (i) implies (ii) let S_n be the event $\{\Pi_n(\mathcal{P}_n | X_1, \ldots, X_n) > e^{-Cn}\}$, and define tests $\phi_n = \mathbb{1}\{S_n\}$ and sets $\mathcal{P}_{n,2} = \{p \in \mathcal{P}_n\colon P^n(S_n^c) \ge e^{-Cn/2}\}$ and $\mathcal{P}_{n,1} = \mathcal{P}_n \setminus \mathcal{P}_{n,2}$. Then $\phi_n = 0$ for all sufficiently large n a.s. $[P_0^\infty]$ by (i), and $P^n(1 - \phi_n) = P(S_n^c) < e^{-Cn/2}$, for every $p \in \mathcal{P}_{n,1}$, by the definition of the set $\mathcal{P}_{n,1}$, which establishes (a) with $C_1 = C/2$. Because $e^{Cn/2} P^n(S_n^c) \ge 1$ for $p \in \mathcal{P}_{n,2}$,

$$e^{-Cn/2}\Pi(\mathcal{P}_{n,2}) \le \int_{\mathcal{P}_{n,2}} P^n(S_n^c)\, d\Pi(p) = \mathrm{E}\Pi_n(\mathcal{P}_{n,2} | X_1, \ldots, X_n)\mathbb{1}\{S_n^c\},$$

by Bayes's theorem. This is bounded above by e^{-Cn} by the definition of S_n and the fact that $\mathcal{P}_{n,2} \subset \mathcal{P}_n$, which yields (b) with $C_2 = C/2$. □

Appendix D discusses the construction of appropriate tests. It turns out that tests for *convex* alternatives exist as soon as these have a minimal, positive $\mathbb{L}_1$-distance from the null hypothesis P_0 (where a larger distance gives a bigger constant in the exponent). Unfortunately, the alternatives $\mathcal{U}^c$ that must be tested in the application of Schwartz's theorem are complements of balls around p_0 and hence are not convex. In general a positive distance to p_0 without convexity is not enough, and consistent tests may not exist.

Now tests for nonconvex alternatives can be constructed by covering them with convex sets, and aggregating the separate tests for each of the sets in the cover, simply by taking the maximum of the test functions: given nonrandomized tests the null hypothesis is rejected as soon as one of the tests rejects it. The probability of an error of the first kind of such an aggregated test depends on the number of sets in the cover. Typically we cover by balls for some metric, and then the number of tests is bounded by the *covering number* of the alternative. The ϵ-covering number of a set $\mathcal{P}$ is defined as the minimal number of d-balls of radius ϵ needed to cover $\mathcal{P}$, and denoted by $N(\epsilon, \mathcal{P}, d)$. Appendix C discusses these numbers in detail.

If combined with a partition $\mathcal{U}^c = \mathcal{P}_{n,1} \cup \mathcal{P}_{n,2}$ in sets $\mathcal{P}_{n,1}$ that are tested and sets $\mathcal{P}_{n,2}$ of negligible prior mass, the appropriate cover refers to sets $\mathcal{P}_{n,1}$ that change with n. The covering numbers then depend on n. As shown in the following theorem they can be allowed to increase exponentially.

The theorem yields consistency relative to any distance d that is bounded above by (a multiple of) the Hellinger distance, for instance the $\mathbb{L}_1$-distance.

Theorem 6.23 (Consistency by entropy) *Given a distance d that generates convex balls and satisfies $d(p_0, p) \leq d_H(p_0, p)$ for every p, suppose that for every $\epsilon > 0$ there exist partitions $\mathcal{P} = \mathcal{P}_{n,1} \cup \mathcal{P}_{n,2}$ and a constant $C > 0$, such that, for sufficiently large n,*

(i) $\log N(\epsilon, \mathcal{P}_{n,1}, d) \leq n\epsilon^2$.
(ii) $\Pi(\mathcal{P}_{n,2}) \leq e^{-Cn}$.

Then the posterior distribution in the model $X_1, \ldots, X_n \mid p \overset{iid}{\sim} p$ and $p \sim \Pi$ is strongly consistent relative to d at every $p_0 \in \mathrm{KL}(\Pi)$.

Proof For given $\epsilon > 0$, cover the set $\mathcal{P}_{n,1}$ with a (minimal) collection of $N(\epsilon, \mathcal{P}_{n,1}, d)$ balls of radius ϵ. From every ball that intersects the set $\mathcal{U}^c := \{p: d(p, p_0) \geq 4\epsilon\}$, take an arbitrary point p_1 in $\mathcal{U}^c$. Then the balls of radius 2ϵ around the points p_1 cover $\mathcal{P}_{n,1} \cap \mathcal{U}^c$ and have minimal distance at least 2ϵ to p_0. By Proposition D.8 for every p_1 there exists a test ψ_n such that $P_0^n \psi_n \leq e^{-2n\epsilon^2}$ and $P^n(1-\psi_n) \leq e^{-2n\epsilon^2}$, for every p with $d(p, p_1) < 2\epsilon$. Define a test ϕ_n as the maximum of all tests ψ_n attached to some p_1 in this way. Then ϕ_n has power bigger than any of the individual tests, and hence satisfies $P^n(1 - \phi_n) \leq e^{-2n\epsilon^2}$, for every $p \in \mathcal{P}_{n,1} \cap \mathcal{U}^c$. Furthermore, because ϕ_n is smaller than the sum of the tests ψ_n, its size is bounded by $(\#p_1)e^{-2n\epsilon^2} \leq e^{n\epsilon^2} e^{-2n\epsilon^2}$.

The assertion of the theorem is now a consequence of Theorem 6.17. ☐

Example 6.24 (Prior partition) If $p_0 \in \mathrm{KL}(\Pi)$ and for every $\epsilon > 0$ there exists a partition $\mathcal{P} = \cup_j \mathcal{P}_j$ in sets of Hellinger diameter smaller than ϵ and $\alpha < 1$ such that $\sum_j \Pi(\mathcal{P}_j)^\alpha < \infty$, then the posterior distribution is strongly Hellinger consistent at p_0.

This sufficient condition is attractive in that it refers to the prior only, and seemingly avoids a testing or entropy condition, but in fact it implies the conditions of Theorem 6.23, for $d = d_H$. To see this, given $\epsilon, c > 0$ let $J = \{j: \Pi(\mathcal{P}_j) \geq e^{-cn}\}$, and consider the partition of $\mathcal{P}$ into the two sets $\mathcal{P}_{n,1} = \cup_{j \in J} \mathcal{P}_j$ and $\mathcal{P}_{n,2} = \cup_{j \notin J} \mathcal{P}_j$. Because $\sum_j \Pi(\mathcal{P}_j) \leq 1$ the cardinality of J is at most e^{cn}, whence $\log N(\epsilon, \mathcal{P}_{n,1}, d_H) \leq \log \#J \leq cn$, as every partitioning set fits into a single ball of radius ϵ. Furthermore $\Pi(\mathcal{P}_{n,2}) \leq \sum_{j\notin J} \Pi(\mathcal{P}_j) \leq \sum_j \Pi(\mathcal{P}_j)^\alpha (e^{-cn})^{1-\alpha}$, which is exponentially small.

Sequence of priors

In the preceding, the prior is assumed not to depend on the number of observations. Schwartz's theorem extends to a sequence Π_n of prior distributions, but with weak rather than strong consistency as conclusion. We say that a density p_0 satisfies the *Kullback-Leibler property* relative to a sequence of priors Π_n if $\liminf_{n\to\infty} \Pi_n(p\colon K(p_0; p) < \epsilon) > 0$, for every $\epsilon > 0$. Weak consistency may be derived under this assumption in the setting of Theorem 6.17, and hence also in the setting of Theorem 6.23.

Theorem 6.25 (Schwartz, sequence of priors) *If p_0 satisfies the Kullback-Leibler property relative to the sequence of priors Π_n and for every neighborhood $\mathcal{U}$ of p_0 there exists a constant $C > 0$, a measurable partition $\mathcal{P}_n = \mathcal{P}_{n,1} \cup \mathcal{P}_{n,2}$, and tests ϕ_n such that $P_0^n \phi_n < e^{-Cn}$ and $\sup_{p\in\mathcal{P}_{n,1}\cap\mathcal{U}^c} P^n(1-\phi_n) < e^{-Cn}$ and $\Pi_n(\mathcal{P}_{n,2}) < e^{-Cn}$, then the posterior distribution $\Pi_n(\cdot|\, X_1, \ldots, X_n)$ in the model $X_1 \ldots, X_n|\, p \overset{iid}{\sim} p$ and $p \sim \Pi_n$ is weakly consistent at p_0.*

The theorem can be proved by proceeding as in the proof of Theorem 6.17, except that the a.s. lower bound (6.5) on the denominator of the posterior distribution is replaced by the lower bound given in the following lemma, for a sufficiently large constant D and ϵ such that $2D\epsilon < C$.

Lemma 6.26 (Evidence lower bound) *For any probability measure Π on $\mathcal{P}$, and any constants $D > 1$ and $\epsilon \geq n^{-1}$, with P_0^n-probabiity at least $1 - D^{-1}$,*

$$\int \prod_{i=1}^n \frac{p(X_i)}{p_0(X_i)}\, d\Pi(p) \geq \Pi(p\colon K(p_0; p) < \epsilon) e^{-2Dn\epsilon}. \tag{6.7}$$

Proof The integral becomes smaller by restricting it to the set $B := \{p\colon K(p_0; p) < \epsilon\}$. By next dividing the two sides of the equation by $\Pi(B)$, we can write the inequality in terms of the prior restricted and renormalized to a probability measure on B. Thus without loss of generality we may assume that Π is concentrated on B. By Jensen's inequality applied to the logarithm,

$$\log \int \prod_{i=1}^n \frac{p(X_i)}{p_0(X_i)}\, d\Pi(p) \geq \int \log \prod_{i=1}^n \frac{p(X_i)}{p_0(X_i)}\, d\Pi(p) =: Z.$$

Hence, the probability of the complement of the event in (6.7) is bounded above by $\mathrm{P}(Z < -2Dn\epsilon) \leq \mathrm{E}(-Z)^+/(2Dn\epsilon)$, by Markov's inequality. By another application of Jensen's inequality and since $(-\log x)^+ = \log_+(1/x)$,

$$\mathrm{E}(-Z)^+ \leq \mathrm{E}\Big[\int \log_+ \prod_{i=1}^n \frac{p_0(X_i)}{p(X_i)}\, d\Pi(p)\Big] = \int K^+(p_0^n; p^n)\, d\Pi(p).$$

Because $K^+ \leq K + \sqrt{K/2}$, by Lemma B.13, and the integral is over B by construction, the right is bounded above by $n\epsilon + \sqrt{n\epsilon/2} \leq 2n\epsilon$, whence $\mathrm{P}(Z < -2Dn\epsilon) \leq 1/D$. □

6.5 Tail-Free Priors

The supports of Dirichlet and Pólya tree priors are (often) not dominated, which precludes a direct application of Schwartz's theorem. However, they were among the earliest examples of consistent priors on measures. The key to their (weak) consistency is that the posterior distribution of the masses of the sets in a partition depends only on the vector of counts of the sets (see Lemma 3.14), which reduces the setting to one involving a finite-dimensional multinomial vector.

Theorem 6.27 *Let the prior distribution Π be tail-free with respect to a sequence $\mathcal{T}_m = \{A_\varepsilon : \varepsilon \in \mathcal{E}^m)$ of successive binary partitions whose union generates the Borel σ-field and is convergence-forcing for weak convergence, and with splitting variables $(V_{\varepsilon 0}: \varepsilon \in \mathcal{E}^*)$ of which every finite-dimensional subvector has full support (a unit cube). Then the posterior distribution $\Pi_n(\cdot| X_1, \ldots, X_n)$ in the model $X_1, X_2, \ldots | P \overset{iid}{\sim} P$ and $P \sim \Pi$ is strongly consistent with respect to the weak topology, at any probability measure P_0.*

Proof Because $\cup_m \mathcal{T}_m$ is convergence-forcing for weak convergence, the topology generated by the metric $d(P, Q) = \sum_{\varepsilon\in\mathcal{E}^*} |P(A_\varepsilon) - Q(A_\varepsilon)| 2^{-2|\varepsilon|}$ is at least as strong as the weak topology. Therefore, it is certainly sufficient to prove consistency relative to d. Since every d-ball of P_0 contains a set of the form $\cap_{A\in\mathcal{T}_m} \{P: |P(A) - P_0(A)| < \delta\}$ (for sufficiently large m and small δ), it suffices to show that the posterior distribution of $(P(A): A \in \mathcal{T}_m)$ is consistent (relative to the Euclidean metric on $\mathbb{R}^{2^m}$).

By Lemma 3.14, this posterior distribution is the same as the posterior distribution of $(P(A): A \in \mathcal{T}_m)$, given the vector of counts of the cells in $\mathcal{T}_m$. Thus the problem reduces to a finite-dimensional multinomial problem, with parameter $\theta = (P(A): A \in \mathcal{T}_m)$, with induced prior distribution θ and with true parameter $\theta_0 = (P_0(A): A \in \mathcal{T}_m)$. Because $\Pi(P: |P(A) - P_0(A)| < \epsilon, A \in \mathcal{T}_m) > 0$, for any $\epsilon > 0$ by the condition of full support of the splitting variables (cf. the proof of Theorem 3.10), the true vector θ_0 is in the support of the prior for θ. The proof can now be completed by an application of Example 6.21. □

The above theorem implies in particular that the posterior based on a Pólya tree prior is consistent with respect to the weak topology if all of its parameters α_ε, for $\varepsilon \in \mathcal{E}^*$, are positive. For the special case of a Dirichlet process this conclusion was reached in Corollary 4.17 even without the restriction on the support.

6.6 Permanence of the Kullback-Leibler Property

The Kullback-Leibler property is useful in many consistency studies, in particular in semiparametric problems, also beyond density estimation, because it is stable under various operations, unlike the tail-free property. In this section we show that the property is preserved under forming mixtures, symmetrization, averaging and conditioning.

Proposition 6.28 (Mixture priors) *In the model $p|\xi \sim \Pi_\xi$ and $\xi \sim \rho$ we have $p_0 \in \mathrm{KL}(\int \Pi_\xi d\rho(\xi))$ for any p_0 such that $\rho\{\xi: p_0 \in \mathrm{KL}(\Pi_\xi)\} > 0$.*

Proof The mixture measure $\Pi = \int \Pi_\xi d\rho(\xi)$ satisfies, for any set B,

$$\Pi\{p: K(p_0; p) < \epsilon\} \geq \int_B \Pi_\xi(p: K(p_0; p) < \epsilon)\, d\rho(\xi).$$

For B the set of ξ such that $p_0 \in \mathrm{KL}(\Pi_\xi)$ the integrand is positive. This set has positive ρ-probability by assumption. □

Proposition 6.29 (Products) *If $p \sim \Pi_1$, $q \sim \Pi_2$, then $p_0 \times q_0 \in \mathrm{KL}(\Pi_1 \times \Pi_2)$ whenever $p_0 \in \mathrm{KL}(\Pi_1)$ and $q_0 \in \mathrm{KL}(\Pi_2)$.*

Proof This is a consequence of the additivity of Kullback-Leibler divergence: $K(p_0 \times q_0; p \times q) = K(p_0; p) + K(q_0; q)$. □

Many preservation properties follow from the fact that the Kullback-Leibler divergence decreases as the information in an experiment decreases. Specifically, if Y is an observation with density f, then the induced density p_f of a given measurable transformation $X = T(Y, U)$, where U is a random variable with a fixed distribution independent of Y, satisfies $K(p_f; p_g) \leq K(f; g)$, for any pair of densities f, g (see Lemma B.11). Equivalently, this is true if X is a randomization of Y through a Markov kernel, in the sense that p_f is a density of a variable X drawn from a conditional distribution $\Psi(\cdot| Y)$, and $Y \sim f$. (These two forms of randomization are in fact equivalent if $\mathfrak{X}$ is Polish.)

Now if $K(p_f; p_g) \leq K(f; g)$, then $\{g: K(f; g) < \epsilon\} \subset \{g: K(p_f; p_g) < \epsilon\}$, and hence $f \in \mathrm{KL}(\Pi)$ for a prior on f implies that $p_f \in \mathrm{KL}(\tilde{\Pi})$, for $\tilde{\Pi}$ the induced prior on p_f.

Proposition 6.30 (Transformation) *If $p \sim \Pi$ and $\tilde{\Pi}$ is the induced prior on $p \circ T^{-1}$ for a given measurable transformation, then $p_0 \in \mathrm{KL}(\Pi)$ implies $p_0 \circ T^{-1} \in \mathrm{KL}(\tilde{\Pi})$.*

This follows from the above discussion as a special case where $X = T(Y)$, not depending on the auxiliary random variable U.

Proposition 6.31 (Mixtures) *If $f \sim \Pi$ and $\tilde{\Pi}$ is the induced prior on the mixture density $p_f(x) = \int \psi(x; \theta) f(\theta)\, d\nu(\theta)$, for a given probability density kernel ψ, then $f_0 \in \mathrm{KL}(\Pi)$ implies $p_{f_0} \in \mathrm{KL}(\tilde{\Pi})$.*

This follows from the above discussion with the aid of Example B.12.

Proposition 6.32 (Symmetrization) *If $p \sim \Pi$ is a density on $\mathbb{R}$ and $\bar{\Pi}$ is the induced prior on its symmetrization $\bar{p}(x) = (p(x) + p(-x))/2$, then $p_0 \in \mathrm{KL}(\Pi)$ implies that $\bar{p}_0 \in \mathrm{KL}(\bar{\Pi})$. The same result is true for the symmetrization $\bar{p}(x) = \bar{p}(-x) = \frac{1}{2}p(x)$, for $x \geq 0$, of a density p on $[0, \infty)$.*

The conclusions follow from the above discussion by considering the transformation $T(X, U) = |X|\mathrm{sign}(U - 1/2)$.

Proposition 6.33 (Invariance) *If $p \sim \Pi$ is a density on $\mathfrak{X}$ and $\bar{p}$ is a density of the invariant measure $\bar{P}(A) = \int P(gA)\,d\mu(g)$ induced by p under the action of a compact metrizable group $\mathfrak{G}$ on $\mathfrak{X}$ having normalized Haar measure μ, then $p_0 \in \mathrm{KL}(\Pi)$ implies that $\bar{p}_0 \in \mathrm{KL}(\bar{\Pi})$, for $\bar{\Pi}$ the induced prior.*[2] *If the action of the group $\mathfrak{G}$ is free and $\mathfrak{X}$ is Borel isomorphic to $\mathfrak{R} \times \mathfrak{G}$, where $\mathfrak{R}$ is a subset of $\mathfrak{X}$ isomorphic to the space of orbits as in Subsection 4.6.1, then the same is true for the invariant densities $\bar{p}$ induced by densities p on $\mathfrak{R}$ with respect to an arbitrary σ-finite measure ν through $\bar{p}(x) = p(y)$, where y is the orbit of x.*

This result clearly follows by considering the transformation $T(X, g) = gX$ with g having the known distribution ν, the normalized Haar measure on $\mathfrak{G}$.

Proposition 6.34 (Domination) *If for every $C > 1$ there exists a density $p_1 \in \mathrm{KL}(\Pi)$ with $p_0 \le Cp_1$ a.e., then $p_0 \in \mathrm{KL}(\Pi)$.*

Proof By Lemma B.14 the Kullback-Leibler divergence $K(p_0; p)$ is bounded above by $\log C + C(\delta + \sqrt{\delta/2})$ if $K(p_1; p) < \delta$. We can make this arbitrarily small by first choosing C close to 1 and next δ close to zero. □

Corollary 6.35 *If $\nu(\mathfrak{X}) < \infty$ and every strictly positive continuous probability density belongs to the Kullback-Leibler support of Π, then $p_0 \in \mathrm{KL}(\Pi)$ for every continuous p_0.*

Proof For given $\eta > 0$ the probability density $p_1 = C^{-1}(p_0 \vee \eta)$, for $C = \int (p_0 \vee \eta)\,d\nu$, is well defined, strictly positive and continuous, and $p_0 \le Cp_1$. Furthermore, $C \downarrow 1$ as $\eta \downarrow 0$ by the dominated convergence theorem. Since $p_1 \in \mathrm{KL}(\Pi)$, the result follows by Proposition 6.34. □

In the following proposition we write p_θ for the shifted density $p(\cdot - \theta)$ (with $p_{0,\theta}$ the shifted version of a density p_0), and $\bar{p}$ for the symmetrized density $x \mapsto \frac{1}{2}(p(x) + p(-x))$. Combined these notations yield the density $\overline{p_\theta}$ given by $x \mapsto \frac{1}{2}(p(x - \theta) + p(-x - \theta))$, which is symmetric about 0 (and not θ). If p is itself symmetric, then the latter density can also be written $\frac{1}{2}(p_\theta + p_{-\theta})$.

Proposition 6.36 *Given priors Π on symmetric densities on $\mathbb{R}$ and μ on $\mathbb{R}$, let $\bar{\Pi}$ denote the induced prior on $p_\theta(\cdot) = p(\cdot - \theta)$ when $(p, \theta) \sim \Pi \times \mu$. If p_0 is symmetric and $\overline{p_{0,\theta}} \in \mathrm{KL}(\Pi)$ for every sufficiently small $\theta > 0$, θ_0 is in the support of μ and $K(p_0; p_{0,\theta}) \to 0$ as $\theta \to 0$, then $p_{0,\theta_0} \in \mathrm{KL}(\bar{\Pi})$.*

Proof Since $K(p_{0,\theta_0}; p_\theta) = K(p_0; p_{\theta-\theta_0})$, we can assume that $\theta_0 = 0$. It can be verified that $K(p_0; p_\theta) = K(p_{0,\theta}; p)$ for symmetric densities p_0 and p. The right side can be

[2] If the dominating measure is invariant then $\bar{p}(x) = \int p(g(x))\,d\mu(g)$.

decomposed as $K(p_{0,\theta}; \overline{p_{0,\theta}}) + P_{\theta_0} \log(\overline{p_{0,\theta}}/p) = K(p_{0,\theta}; \overline{p_{0,\theta}}) + K(\overline{p_{0,\theta}}; p)$, because the function $\log(\overline{p_{0,\theta}}/p)$ is symmetric. Because $\overline{p_{0,\theta}} = \frac{1}{2}(p_{0,\theta} + p_{0,-\theta})$, the first term on the right is bounded above by $\frac{1}{2}0 + \frac{1}{2}K(p_{0,\theta}; p_{0,-\theta})$. Combining all inequalities we find $K(p_0; p_\theta) \leq \frac{1}{2}K(p_{0,\theta}; p_{0,-\theta}) + K(\overline{p_{0,\theta}}; p)$. The first term on the right is $\frac{1}{2}K(p_0; p_{0,-2\theta})$ and is arbitrarily small as θ is close to $\theta_0 = 0$; the second term is small with positive Π probability by the assumption that $\overline{p_{0,\theta}} \in \mathrm{KL}(\Pi)$. □

The Kullback-Leibler property is also preserved when one passes from the prior to the posterior given an observation, and uses the posterior as a new prior in combination with future data. An interesting aspect is that the prior can be improper, as long as the posterior is proper. (The Kullback-Leibler property for an improper prior is defined in the obvious way.)

Let $X_1, \ldots, X_m$ be i.i.d. with density p following a possibly improper prior distribution Π.

Proposition 6.37 *If* $p_0 \in \mathrm{KL}(\Pi)$, *then* $p_0 \in \mathrm{KL}(\Pi_n(\cdot|\, X_1, \ldots, X_m))$ *a.s.* $[P_0^m]$ *on the event* $\{\int \prod_{i=1}^m p(X_i)\, d\Pi(p) < \infty\}$.

Proof For any $\epsilon > 0$, on

$$E_\epsilon = \{\Pi_n(p: K(p_0; p) < \epsilon |\, X_1, \ldots, X_m) = 0\} \cap \{\int \prod_{i=1}^m p(X_i)\, d\Pi(p) < \infty\}$$

we have that

$$0 = \Pi_n(p: K(p_0; p) < \epsilon |\, X_1, \ldots, X_m) = \frac{\int_{K(p_0;p)<\epsilon} \prod_{i=1}^m p(X_i)\, d\Pi(p)}{\int \prod_{i=1}^m p(X_i) d\Pi(p)}.$$

Thus the numerator vanishes a.s. on E_ϵ. Integrating it over this event and applying Fubini's theorem, we find that $\int_{K(p_0;p)<\epsilon} P^m(E_\epsilon)\, d\Pi(p) = 0$. This implies that there exists $\tilde{p}$ with $K(p_0; \tilde{p}) < \epsilon$ such that $\tilde{P}^m(E_\epsilon) = 0$. Since, by the definition of Kullback-Leibler divergence, $K(p_0; \tilde{p}) < \epsilon < \infty$ is possible only if $P_0 \ll \tilde{P}^*$, it follows that $P_0^m(E_\epsilon) = 0$.

Finally let ϵ run through the positive rational numbers, and accumulate the corresponding null sets E_ϵ in a single null set. □

Remark 6.38 The preceding result implies that Schwartz's consistency theorem, Theorem 6.16, generalizes to improper priors Π for which the posterior distribution is proper for some m a.s. $[P_0^\infty]$. It suffices to apply the theorem conditionally on the first m observations, for m an index which gives a proper posterior, which can be viewed as the prior for the model with the remaining observations.

6.7 General Observations

In this section we return to the general set-up of the introduction, and consider a sequence of statistical experiments $(\mathfrak{X}^{(n)}, \mathscr{X}^{(n)}, P_\theta^{(n)}: \theta \in \Theta_n)$, with parameter sets Θ_n, and observations $X^{(n)}$. Let Π_n stand for a sequence of prior distributions for θ. We assume that every $P_\theta^{(n)}$ admits a density $p_\theta^{(n)}$ with respect to a σ-finite measure $\nu^{(n)}$ on $\mathfrak{X}^{(n)}$, which is jointly measurable in the parameter and the observation, so that (a version of) the posterior distribution $\Pi_n(\cdot|\, X^{(n)})$ is given by Bayes's formula (1.1).

All entities may depend on n, including the prior Π_n and the "true" values $\theta_{n,0}$ of the parameter.

For probability densities p and q, let $K(p;q) = \int p\log(/q)$ and $V_{k,0}^+(p;q) = \int p((\log(p/q) - K(p;q))^+)^k$, respectively, the Kullback-Leibler divergence and kth order positive Kullback-Leibler variation between two probability densities p and q, as defined in (B.2).

Theorem 6.39 *If for some $k \geq 2$ and $c > 0$ there exist measurable sets $B_n \subset \Theta_n$ with $\liminf \Pi_n(B_n) > 0$ and*

$$\sup_{\theta\in B_n} \frac{1}{n} K(p_{\theta_{n,0}}^{(n)}; p_\theta^{(n)}) \leq c, \qquad \sup_{\theta\in B_n} \frac{1}{n^k} V_{k,0}^+(p_{\theta_{n,0}}^{(n)}; p_\theta^{(n)}) \to 0, \tag{6.8}$$

then $\Pi_n(\tilde\Theta_n|\, X^{(n)}) \to 0$ in $P_{\theta_{n,0}}^{(n)}$-probability for any sets $\tilde\Theta_n \subset \Theta_n$ such that either (a) *or* (b) *holds for a constant $C > c$,*

(a) *there exist tests ϕ_n with $P_{\theta_{n,0}}^{(n)}\phi_n \to 0$ and $\int_{\tilde\Theta_n} P_\theta^{(n)}(1-\phi_n)\, d\Pi_n(\theta) \leq e^{-Cn}$;*

(b) $\Pi_n(\tilde\Theta_n) \leq e^{-Cn}$.

The convergence is also in the almost sure sense if ϕ_n in (a) and $\mathbb{1}\{A_n^c\}$ for A_n defined in (6.10) *below tend to zero almost surely under $P_{\theta_{n,0}}^{(n)}$.*

Proof The proof is similar to that of Theorem 6.17. Let $\tilde\Pi_n$ be the renormalized restriction of Π_n to the set B_n. Then $\Pi_n \geq \Pi_n(B_n)\tilde\Pi_n$, and hence, by Jensen's inequality,

$$\begin{aligned}
&\frac{1}{n}\log\Big[e^{c'n}\int \frac{p_\theta^{(n)}}{p_{\theta_{n,0}}^{(n)}}(X^{(n)})\, d\Pi_n(\theta)\Big] \\
&\qquad \geq c' + \int \frac{1}{n}\log\frac{p_\theta^{(n)}}{p_{\theta_{n,0}}^{(n)}}(X^{(n)})\, d\tilde\Pi_n(\theta) + \frac{1}{n}\log \Pi_n(B_n).
\end{aligned} \tag{6.9}$$

Because the integrand has expectation $-n^{-1}K(p_{\theta_{n,0}}^{(n)}; p_\theta^{(n)}) \geq -c$, for $\theta \in B_n$, the right-hand side is bigger than $c' - c - \epsilon + o(1)$, on the event

$$A_n := \Big\{\frac{1}{n}\int \Big(\log\frac{p_\theta^{(n)}}{p_{\theta_{n,0}}^{(n)}}(X^{(n)}) - P_{\theta_{n,0}}^{(n)}\log\frac{p_\theta^{(n)}}{p_{\theta_{n,0}}^{(n)}}\Big)\, d\tilde\Pi_n(\theta) > -\epsilon\Big\}. \tag{6.10}$$

By Markov's inequality followed by Jensen's inequality, we obtain

$$\begin{aligned}
P_{\theta_{n,0}}^{(n)}(A_n^c) &\leq \frac{1}{(n\epsilon)^k}\mathrm{E}_{\theta_{n,0}}\Big|\int \Big(\log\frac{p_{\theta_{n,0}}^{(n)}}{p_\theta^{(n)}}(X^{(n)}) - P_{\theta_{n,0}}^{(n)}\log\frac{p_{\theta_{n,0}}^{(n)}}{p_\theta^{(n)}}\Big)^+ d\tilde\Pi_n(\theta)\Big|^k \\
&\leq \frac{1}{(n\epsilon)^k}\int V_{k,0}^+(p_{\theta_{n,0}}^{(n)}; p_\theta^{(n)})\, d\tilde\Pi_n(\theta).
\end{aligned}$$

The right side tends to zero by assumption, for any $\epsilon > 0$.

If $c' > c$, then we can choose $\epsilon > 0$ so that $c' - c - \epsilon > 0$, and then the left side of (6.9) is bounded below by a positive number on the event A_n. Then the multiple

$$e^{c'n} \int (p_\theta^{(n)} / p_{\theta_{n,0}})^{(n)} (X^{(n)})\, d\Pi_n(\theta)$$

of the denominator in Bayes's formula tends to infinity on A_n, and hence, eventually,

$$P_{\theta_{n,0}}^{(n)} \Pi_n(\tilde{\Theta}_n | X^{(n)})(1 - \phi_n)\mathbb{1}\{A_n\} \leq e^{c'n} \int_{\tilde{\Theta}_n} P_{\theta_{n,0}}^{(n)} \Big[(1 - \phi_n) \frac{p_\theta^{(n)}}{p_{\theta_{n,0}}^{(n)}}\Big]\, d\Pi_n(\theta).$$

This can be further bounded above by both $e^{c'n} \int_{\tilde{\Theta}_n} P_\theta^{(n)}(1 - \phi_n)\, d\Pi_n(\theta)$ and $e^{c'n}\Pi_n(\tilde{\Theta}_n)$. These upper bounds tend to zero under (a) and (b), respectively, if c' is chosen such that $c < c' < C$.

Thus $\Pi_n(\tilde{\Theta}_n | X^{(n)})(1 - \phi_n)\mathbb{1}\{A_n\}$ tends to zero in probability. We combine this with the convergence to zero of $\Pi_n(\tilde{\Theta}_n | X^{(n)})(\phi_n + \mathbb{1}\{A_n^c\}) \leq \phi_n + \mathbb{1}\{A_n^c\}$, which is true by assumption and by construction.

The strengthening to almost sure convergence is valid as the convergence in probability is exponentially fast. □

Typically the preceding theorem will be applied with B_n equal to a neighborhood of the true parameter, chosen small so that c can be small, and $\tilde{\Theta}_n$ the complement of another neighborhood, which should be testable versus the true parameter or possess small prior mass in order to satisfy (a) or (b).

The Kullback-Leibler variations $V_{k,0}$ help to control the variability of the log-likelihood around its mean, so that the probability of the events A_n defined in (6.10) tends to one. Theorems 6.16 and 6.17 illustrate that consideration of only the Kullback-Leibler divergence itself may suffice. These results are limited to a fixed prior $\Pi_n = \Pi$ and based on the law of large numbers for i.i.d. variables, which is valid under existence of first moments only. A similar simplification is possible for other special cases, the crux being to ensure that the events A_n are asymptotically of probability one.

By a slightly more elaborate proof we also have the following variation of the theorem. It is restricted to a fixed prior and parameter set, and assumes that the experiments are linked across n, with the observation $X^{(n)}$ defined on a single underlying probability space and its law $P_\theta^{(n)}$ the image of a fixed measure $P_\theta^{(\infty)}$ on this space.

Theorem 6.40 (Fixed prior) *If there exist $c > 0$ and a measurable set $B \subset \Theta$ with $\Pi(B) > 0$ and, for every $\theta \in B$,*

$$\frac{1}{n} K(p_{\theta_0}^{(n)}; p_\theta^{(n)}) \leq c, \qquad \frac{1}{n}\Big(\log \frac{p_\theta^{(n)}}{p_{\theta_0}^{(n)}}(X^{(n)}) - P_{\theta_0}^{(n)} \log \frac{p_\theta^{(n)}}{p_{\theta_0}^{(n)}}\Big) \to 0, \text{ a.s. } [P_{\theta_0}^{(\infty)}], \tag{6.11}$$

then $\Pi_n(\tilde{\Theta}_n | X^{(n)}) \to 0$ almost surely $[P_{\theta_0}^{(\infty)}]$ for any sets $\tilde{\Theta}_n \subset \Theta_n$ such that either (a) with $\phi_n \to 0$ a.s. $[P_{\theta_0}^{(\infty)}]$ or (b) of Theorem 6.39 holds for some constant $C > c$.

Proof For given $\epsilon > 0$, let $\hat{\Pi}_n$ be the renormalized restriction of Π to the set

$$\hat{B}_n = \Big\{\theta \in B: \inf_{k\geq n} \frac{1}{k}\Big(\log \frac{p_\theta^{(k)}}{p_{\theta_0}^{(k)}}(X^{(k)}) - P_{\theta_0}^{(k)} \log \frac{p_\theta^{(k)}}{p_{\theta_0}^{(k)}}\Big) \geq -\epsilon\Big\}.$$

Although the set $\hat{B}_n \subset \Theta$ and measure $\hat{\Pi}_n$ are presently random, the argument leading to (6.9) in the proof of Theorem 6.39 still applies, and yields that the left side of (6.9) is bounded below by

$$c' - c - \epsilon + \frac{1}{n} \log \Pi(\hat{B}_n).$$

On the event $E_n = \{\Pi(\hat{B}_n) \geq \frac{1}{2}\Pi(B)\}$ this is further bounded below by $c' - c - \epsilon + o(1)$, and the remainder of the proof of Theorem 6.39 can be copied to show that the posterior mass tends to zero on the event $\liminf E_n$. It suffices to prove that $P_{\theta_0}^{(\infty)}(\limsup E_n^c) = 0$. Because $E_1^c \supset E_2^c \supset \cdots$ this is equivalent to the probability of the events E_n^c tending to zero.

If we write $\hat{B}_n$ as $\{\theta \in B: \inf_{k\geq n} Z_k(\theta, X^{(k)}) \geq -\epsilon\}$, then

$$\Pi(\hat{B}_n) = \Pi(B) - \Pi\Big(\theta \in B: \inf_{k\geq n} Z_k(\theta, X^{(k)}) < -\epsilon\Big),$$

and the event E_n^c can be written in the form $\{\Pi(\theta \in B: \inf_{k\geq n} Z_k(\theta, X^{(k)}) < -\epsilon) > \frac{1}{2}\Pi(B)\}$. By Markov's inequality followed by Fubini's theorem,

$$P_{\theta_0}^{(\infty)}(E_n^c) \leq \frac{2}{\Pi(B)}(\Pi \times P_{\theta_0}^{(\infty)})\Big(\theta \in B, \inf_{k\geq n} Z_k(\theta, X^{(k)}) < -\epsilon\Big).$$

By assumption the sequence $Z_n(\theta, X^{(n)})$ tends to zero almost surely $[P_{\theta_0}^{(\infty)}]$, for every fixed $\theta \in B$. Therefore $P_{\theta_0}^{(\infty)}(\inf_{k\geq n} Z_k(\theta, X^{(k)}) < -\epsilon) \to 0$ for each such θ, whence the right side tends to zero, by Fubini's theorem. □

In Section 8.3 we evaluate the Kullback-Leibler divergence and variation for several statistical settings, in the framework of obtaining rates of convergence. Here we consider only independent observations and ergodic Markov processes.

6.7.1 Independent Observations

For $X^{(n)}$ consisting of independent observations the Kullback-Leibler divergence and variation can be expressed or bounded in the same quantities of the individual observations. Furthermore, tests exist automatically relative to the *root average square Hellinger metric*

$$d_{n,H}(\theta_1, \theta_2) = \sqrt{\frac{1}{n}\sum_{i=1}^n d_H^2(P_{n,\theta_1,i}, P_{n,\theta_2,i})}.$$

This yields the following corollary of Theorem 6.39, and extension of Theorem 6.23.

Theorem 6.41 *Let $P_\theta^{(n)}$ be the product of measures $P_{n,\theta,1},\ldots,P_{n,\theta,n}$. If for every $\epsilon>0$ there exist measurable sets $B_n\subset\Theta_n$ with $\liminf\Pi_n(B_n)>0$ and*

$$\sup_{\theta\in B_n}\frac{1}{n}\sum_{i=1}^n K(p_{n,\theta_{n,0},i};\,p_{n,\theta,i})\le\epsilon,\qquad \sup_{\theta\in B_n}\frac{1}{n^2}\sum_{i=1}^n V_{2,0}(p_{n,\theta_{n,0},i};\,p_{n,\theta,i})\to 0,$$

then $\Pi_n(\tilde\Theta_n|\,X^{(n)})\to 0$ in $P_{\theta_{n,0}}^{(n)}$-probability for any sets $\tilde\Theta_n\subset\Theta_n$ such that either (a) or (b) of Theorem 6.39 holds for a constant $C>0$. In particular, if for every $\epsilon>0$ there exists a partition $\Theta_n=\Theta_{n,1}\cup\Theta_{n,2}$ and $C>0$ such that

(i) $\log N(\epsilon,\Theta_{n,1},d_{n,H})\le 3n\epsilon^2$,
(ii) $\Pi(\Theta_{n,2})\le e^{-Cn}$,

then $\Pi_n(d_{n,H}(\theta,\theta_{n,0})>\epsilon|\,X^{(n)})\to 0$ in $P_{\theta_{n,0}}^{(n)}$-probability, for every $\epsilon>0$.

Proof The first assertion is an immediate corollary of Theorem 6.39. For the proof of the second we construct the tests for $\Theta_{n,1}=\{\theta\colon d_{n,H}(\theta,\theta_{n,0})>\epsilon\}$ by the same arguments as in the proof of Theorem 6.23, where we obtain the basic tests from Proposition D.9. □

The preceding theorem applies to general triangular arrays of independent observations. For sequences $X^{(n)}=(X_1,\ldots,X_n)$ of independent variables with $X_i\sim p_{\theta,i}$ and a fixed prior Π, we may apply Theorem 6.40, with the condition of almost sure convergence of the log-likelihood ratios verified by Kolmogorov's theorem: for $\theta\in B$,

$$\sum_{i=1}^\infty\frac{1}{i^2}V_{2,0}(p_{\theta_0,i};\,p_{\theta,i})<\infty.$$

Other versions of the law of large numbers may be brought in as well.

6.7.2 Markov Processes

Consider $X^{(n)}=(X_1,\ldots,X_n)$ for a Markov process $\{X_n;\,n\ge 0\}$ with stationary transition density $(x,y)\mapsto p_\theta(y|\,x)$ relative to a dominating measure ν on the state space $(\mathfrak{X},\mathscr{X})$. For Q_θ the invariant distribution corresponding to p_θ (assumed to exist and to be unique) define the *Kullback-Leibler divergence between two transition densities* as

$$K(p_{\theta_0};\,p_\theta)=\int\log\frac{p_{\theta_0}}{p_\theta}(y|\,x)\,p_{\theta_0}(y|\,x)\,d\nu(y)\,dQ_{\theta_0}(x).$$

Say that a prior Π on the parameter set Θ possesses the *KL property* at θ_0, and written $p_{\theta_0}\in\mathrm{KL}(\Pi)$, if $\Pi(\theta\colon K(p_{\theta_0};\,p_\theta)<\epsilon)>0$ for every $\epsilon>0$.

Theorem 6.42 (Markov chain, fixed prior) *Let $P_\theta^{(n)}$ be the distribution of $(X_1,\ldots,X_n)$ for a Markov chain $(X_t\colon t\ge 0)$. If the chain is positive Harris recurrent under θ_0 and $p_{\theta_0}\in\mathrm{KL}(\Pi)$, then $\Pi_n(\tilde\Theta_n|\,X^{(n)})\to 0$ in $P_{\theta_0}^{(n)}$-probability for any sets $\tilde\Theta_n\subset\Theta_n$ such that either (a) or (b) of Theorem 6.39 holds for a constant $C>0$.*

Proof The condition that the Markov chain is Harris recurrent is equivalent to validity of the strong law for all functions that are Q_{θ_0}-integrable, and also equivalent to triviality of the invariant σ-field relative to any initial condition x for the chain (see Theorem 17.1.7 of Meyn and Tweedie 1993). The last shows that the property is inherited by the chain of pairs (X_{t-1}, X_t), and hence

$$\frac{1}{n}\sum_{i=1}^{n} \log \frac{p_{\theta_0}}{p_\theta}(X_i \mid X_{i-1}) \to K(p_{\theta_0}; p_\theta). \qquad \text{a.s. } [P_{\theta_0}^{(\infty)}].$$

Thus, Theorem 6.40 applies. □

6.8 Alternative Approaches

In this section we discuss alternative approaches to formulate and prove posterior consistency. These complement Schwartz's theory as expounded in Sections 6.4 and 6.7, but actually never give weaker conditions.

6.8.1 Separation

The existence of exponentially consistent tests, used in Section 6.4, is equivalent to *separation* of measures (see Le Cam 1986). Schwartz's original proof of her theorem implicitly used this concept.

For a probability density p write p^k for its k-fold product.

Definition 6.43 A probability density p_0 and a set $\mathcal{V}$ of probability densities are *strongly δ-separated* at stage $k \in \mathbb{N}$ if $\rho_{1/2}(p_0^k, \int p^k \, d\mu(p)) < \delta$, for every probability measure μ on $\mathcal{V}$.

For example, the ball $\mathcal{V} = \{p \in \mathcal{P}: \|p - p_1\|_1 < \delta\}$ of radius δ around a point p_1 with $\|p_1 - p_0\|_1 > 2\delta$ is strongly δ-separated at stage 1. Strong separation for some k automatically entails exponential separation of product measures. The following lemma is essentially a restatement of Lemma D.6, and allows a variation of Schwartz's theorem in terms of strong δ-separation.

Lemma 6.44 *If p_0 and $\mathcal{V}$ are strongly δ-separated at stage k, then $\rho_{1/2}(p_0^n, \int p^n \, d\mu(p)) < e^{-\lfloor n/k \rfloor \log_- \delta}$, for any probability measure μ on $\mathcal{V}$.*

Theorem 6.45 *If $p_0 \in \mathrm{KL}(\Pi)$, then $\Pi(\mathcal{V} \mid X_1, \ldots, X_n) \to 0$ a.s. $[P_0^\infty]$, for any set $\mathcal{V}$ that is strongly δ separated from p_0 at stage k, for some δ and k.*

Proof Inequality (6.5) in the proof of Theorem 6.17, applied with $\mathcal{P}_0$ an ϵ-Kullback-Leibler neighborhood of p_0, shows that $\int \prod_{i=1}^n (p/p_0)(X_i) \, d\Pi(p)$ is bounded below by $e^{-\epsilon n}$, for any $\epsilon > 0$, eventually almost surely. It suffices that $e^{\epsilon n} L_n$, for $L_n = \int_{\mathcal{V}} \prod_{i=1}^n (p/p_0)(X_i) \, d\Pi(p)$, tends to zero almost surely for some $\epsilon > 0$. Now $P_0^n \sqrt{L_n} = \rho_{1/2}(p_0^n, \int_{\mathcal{V}} p^n \, d\Pi(p)) \le e^{-cn}$, for some $c > 0$, by Lemma 6.44. □

6.8.2 Le Cam's Inequality

Le Cam's inequality allows to verify posterior consistency by a testing argument, in the spirit of Schwartz's theorem, but it substitutes the total variation distance for the Kullback-Leibler divergence. Thus, avoiding the Kullback-Leibler property, it is applicable to models that are not dominated.

We state the inequality in abstract form, for the posterior distribution based on a single observation X, in the Bayesian setting where a probability distribution P is drawn from a prior Π on the collection $\mathfrak{M}$ of probability measures on the sample space, and next X is drawn according to P. The posterior distribution $\Pi(\cdot|\, X)$ is the conditional distribution of P given X. It is not assumed that Π concentrates on a dominated set of measures, and hence Bayes's formula may not be available.

We write $P_{\mathcal{U}}$ for the average $P_{\mathcal{U}} = \int_{\mathcal{U}} P\, d\Pi(P)/\Pi(\mathcal{U})$ of the measures P in a measurable subset $\mathcal{U} \subset \mathfrak{M}$.

Lemma 6.46 (Le Cam) *For any pair of measurable sets $\mathcal{U}, \mathcal{V} \subset \mathfrak{M}$, test ϕ, and measure $P_0 \in \mathfrak{M}$,*

$$P_0\Pi(\mathcal{V}|\, X) \leq d_{TV}(P_0, P_{\mathcal{U}}) + P_0\phi + \frac{1}{\Pi(\mathcal{U})}\int_{\mathcal{V}} P(1-\phi)\, d\Pi(P).$$

Proof Since both $0 \leq \phi \leq 1$ and $0 \leq \Pi(\mathcal{V}|\, X) \leq 1$, for any probability measure Q,

$$P_0\Pi(\mathcal{V}|\, X) \leq P_0\phi + P_0[\Pi(\mathcal{V}|\, X)(1-\phi)] \leq P_0\phi + d_{TV}(P_0, Q) + Q[\Pi(\mathcal{V}|\, X)(1-\phi)].$$

The measure $Q = P_{\mathcal{U}}$ is bounded above by $1/\Pi(\mathcal{U})$ times $\int P\, d\Pi(P)$. The latter is the marginal distribution of X in the model $P \sim \Pi$ and $X|\, P \sim P$. Therefore, $\Pi(\mathcal{U})$ times the third term on the right is bounded above by $\mathrm{E}[\Pi(\mathcal{V}|\, X)(1-\phi)(X)] = \mathrm{E}[\mathbb{1}\{P \in \mathcal{V}\}(1-\phi)(X)]$, by the orthogonality (projection property) of conditional expectations. Since $\mathrm{E}[(1-\phi)(X)|\, P] = P(1-\phi)$, this can be further rewritten as $\int_{\mathcal{V}} P(1-\phi)\, d\Pi(P)$, leading to the third term in the bound of the lemma. □

In a consistency proof the set $\mathcal{U}$ in Le Cam's inequality is taken equal to a small neighborhood of P_0, to ensure that the first term on the right $d_{TV}(P_0, P_{\mathcal{U}})$ is small. A useful bound is then obtained if there exists a test of P_0 versus $\mathcal{V}$ with integrated error probability that is significantly smaller than $\Pi(\mathcal{U})$.

In applications where $X = (X_1, \ldots, X_n)$ is a sample of n i.i.d. observations, the measure P_0 in the inequality, which is the true distribution of X, must be replaced with a product measure P_0^n, and $\mathcal{U}$ will consist of product measures P^n as well. By Lemmas B.1 and B.8 the total variation distance $d_{TV}(P_0^n, P^n)$ between product measures is bounded above by $2\sqrt{n}d_H(P_0, P)$, and hence a good choice for $\mathcal{U}$ is the set of products of measures in a Hellinger ball of radius a small multiple of $1/\sqrt{n}$. This gives the following theorem, which applies to every topology on P for any sequence of priors Π_n.

Theorem 6.47 *If for every neighborhood $\mathcal{U}$ of P_0 and every $\epsilon > 0$ there exist tests ϕ_n and $C > 0$ such that $P_0^n\phi_n \leq e^{-Cn}$ and*

$$\int_{\mathcal{U}^c} P^n(1-\phi_n)\,d\Pi_n(P) \le e^{-Cn}\Pi_n\Big(P: d_H(P, P_0) < \frac{\epsilon}{\sqrt{n}}\Big),$$

then the posterior distribution $\Pi_n(\cdot|\, X_1,\ldots,X_n)$ *in the model* $X_1,\ldots,X_n|\, P \stackrel{iid}{\sim} P$ *and* $P \sim \Pi_n$ *is strongly consistent at* P_0.

Similar as in Schwartz's theorem, the requirements are existence of good tests and sufficient prior mass in a neighborhood of the true measure. Presently the neighborhood is relative to the Hellinger distance rather than the Kullback-Leibler divergence, and it shrinks with n. If

$$\frac{1}{n}\log\frac{1}{\Pi_n(P: d_H(P, P_0) < \epsilon/\sqrt{n})} \to 0, \tag{6.12}$$

then $e^{-Cn}\Pi_n(P: d_H(P, P_0) < \epsilon/\sqrt{n})$ is of the order $e^{-Cn+o(n)}$, and the testing condition takes a similar form as in Section 6.4.

Some insight in the size of $\Pi_n(P: d_H(P, P_0) < \epsilon/\sqrt{n})$ is obtained by considering a fixed prior Π that distributes its mass "uniformly." If we could place $N(\epsilon/\sqrt{n})$ disjoint balls of radius $\epsilon/\sqrt{n}$ in the support $\mathcal{P}$ of the prior, then every ball would obtain prior mass $1/N(\epsilon/\sqrt{n})$. The number of balls ought to be comparable to the covering number $N(\epsilon/\sqrt{n}, \mathcal{P}, d_H)$ of the model, and then (6.12) would be equivalent to $\log N(\epsilon/\sqrt{n}, \mathcal{P}, d_H) = o(n)$. This is true if the entropy $\log N(\epsilon, \mathcal{P}, d_H)$ is of smaller order than $(1/\epsilon)^2$, meaning that the model could be big, but not too big.

For further illustration, we derive posterior consistency for a *default prior* constructed by mixing finite uniform distributions on ϵ_m-nets, for a sequence of approximation levels $\epsilon_m \downarrow 0$. Consider a target model $\mathcal{P}$ that is totally bounded relative to the Hellinger distance, let Π_m be the uniform discrete distribution on a minimal ϵ_m-net over $\mathcal{P}$ for the Hellinger distance, and let $\Pi = \sum_{m=1}^{\infty}\lambda_m\Pi_m$ be a fixed overall prior on $\mathcal{P}$, for a given infinite probability vector $(\lambda_1, \lambda_2, \ldots)$.

Theorem 6.48 *For a set* $\mathcal{P}$ *of probability measures with* $\epsilon^2\log N(\epsilon, \mathcal{P}, d_H) \to 0$ *as* $\epsilon \to 0$, *a sequence* $\epsilon_m \downarrow 0$ *such that* $\epsilon_{m-1}/\epsilon_m = O(1)$ *and weights such that* $\epsilon_m^2\log(1/\lambda_m) \to 0$, *construct a prior* Π *as indicated. Then the posterior distribution* $\Pi_n(\cdot|\, X_1,\ldots,X_n)$ *in the model* $X_1,\ldots,X_n|\, P \stackrel{iid}{\sim} P$ *and* $P \sim \Pi$ *is strongly consistent relative to the Hellinger distance at any* $P_0 \in \mathcal{P}$.

Proof Given $\epsilon > 0$ and m such that $\epsilon_m < \epsilon/\sqrt{n} \le \epsilon_{m-1}$ a Hellinger ball of radius $\epsilon/\sqrt{n}$ contains at least one support of Π_m, whence $\Pi(P: d_H(P, P_0) < \epsilon/\sqrt{n}) \ge \lambda_m/N(\epsilon_m, \mathcal{P}, d_H)$. It follows that (6.12) is satisfied if both $n^{-1}\log N(\epsilon_m, \mathcal{P}, d_H)$ and $n^{-1}\log(1/\lambda_m)$ tend to zero. As $n^{-1} \le (\epsilon_{m-1}/\epsilon)^2 \lesssim \epsilon_m^2$, this follows from the assumptions.

By the compactness of $\mathcal{P}$ there exist tests of P_0 versus the complement of a Hellinger ball of positive radius with error probabilities bounded by e^{-Cn}, for some $C > 0$, by combination of Lemma D.3 and Proposition D.8, or by Example 6.20 and the fact that the weak and strong topologies coincide on strongly compact sets. □

Example 6.49 Consider the set $\mathcal{P}$ of all probability measures on $[0,1]^d$ that possess a Lebesgue density p such that $\sqrt{p}$ possesses Hölder norm of order α (see Definition C.4) bounded by a fixed constant. By Proposition C.5 this has Hellinger entropy of the order $(1/\epsilon)^{d/\alpha}$, whence the entropy condition is satisfied for $\alpha > d/2$. Thus, consistency pertains unless λ_m decays too rapidly relative to ϵ_m. For instance, if $\epsilon_m = m^{-1}$, then λ_m with tails thicker than any distribution proportional to e^{-am^2} suffices. The choice $\epsilon_m = 2^{-m}$ allows even double-exponential tails for λ_m.

6.8.3 Predictive Consistency

Given a sequence of priors Π_n for a density p and i.i.d. observations $X_1, X_2, \ldots$, the *predictive density* $\hat{p}_i$ for X_i is the posterior mean based on $X_1, \ldots, X_{i-1}$:

$$\hat{p}_i(x) = \mathrm{E}(p(x)|\, X_1, \ldots, X_{i-1}) = \frac{\int p(x) \prod_{j=1}^{i-1} p(X_j)\, d\Pi_n(p)}{\int \prod_{j=1}^{i-1} p(X_j)\, d\Pi_n(p)}. \tag{6.13}$$

The conditional mean in the central expression is in the Bayesian setup with $p \sim \Pi_n$ and $X_1, \ldots, X_n|\, p \overset{\text{iid}}{\sim} p$.

The posterior mean of the usual posterior distribution is $\hat{p}_{n+1}$, and ordinarily we would be interested in its consistency (and the contraction of the full posterior distribution). We have seen that this pertains under a combination of the Kullback-Leibler property and a testing (or entropy) condition, where the latter cannot be omitted in general (see Example 7.12), and is necessary for consistency at an exponential rate (see Proposition 6.22). Interestingly, the Kullback-Leibler property alone is sufficient for consistency of the Cesàro averages of the predictive densities, also for strong metrics.

Theorem 6.50 *If $n^{-1} \log \Pi_n(p\colon K(p_0; p) < \epsilon) \to 0$ for every $\epsilon > 0$, then it holds that $K(p_0; n^{-1} \sum_{i=1}^n \hat{p}_i) \le n^{-1} \sum_{i=1}^n K(p_0; \hat{p}_i) \to 0$ in mean under P_0^n. In particular, this is true for a fixed prior $\Pi_n = \Pi$ with $p_0 \in \mathrm{KL}(\Pi)$.*

We shall prove a more general version of the theorem, within a setup with arbitrary, not necessarily i.i.d. observations. Let Π_n be a prior on densities on the sample space $(\mathfrak{X}^\infty, \mathscr{X}^\infty)$, and consider an infinite vector of observations in the Bayesian model

$$p_\infty \sim \Pi_n, \qquad (X_1, X_2, \ldots)|\, p_\infty \sim p_\infty.$$

The *predictive density* at stage i is defined in this setting as the conditional density of X_i given $X_1, \ldots, X_{i-1}$. As before, we denote it by $\hat{p}_i$. In terms of the density p_i of $(X_1, \ldots, X_i)$, it can be written

$$\hat{p}_i(x) = p_i(x|\, X_1, \ldots, X_{i-1}) = \frac{\int p_i(X_1, \ldots, X_{i-1}, x)\, d\Pi_n(p)}{\int p_{i-1}(X_1, \ldots, X_{i-1})\, d\Pi_n(p)}. \tag{6.14}$$

A true density $p_{0,\infty}$ on $(\mathfrak{X}^\infty, \mathscr{X}^\infty)$ similarly defines densities $p_{0,i}$ for $(X_1, \ldots, X_i)$ and conditional densities $\hat{p}_{0,i}$ of X_i, given $X_1, \ldots, X_{i-1}$. For i.i.d. observations $\hat{p}_{0,i} = p_0$, for every i, and the preceding display reduces to (6.13).

Theorem 6.51 *If $n^{-1} \log \Pi_n(p_\infty\colon n^{-1} K(p_{0,n}; p_n) < \epsilon) \to 0$ for every $\epsilon > 0$, then it holds that $K(n^{-1} \sum_{i=1}^n \hat{p}_{0,i}; n^{-1} \sum_{i=1}^n \hat{p}_i) \le n^{-1} \sum_{i=1}^n K(\hat{p}_{0,i}; \hat{p}_i) \to 0$ in mean under $P_{0,\infty}$.*

Proof The first inequality is a consequence of Jensen's inequality and the convexity of the function $(u, v) \mapsto u \log(u/v)$. To prove the convergence to zero, observe that

$$\frac{\hat{p}_i(X_i)}{\hat{p}_{0,i}(X_i)} = \frac{\int (p_i/p_{0,i})(X_1, \ldots, X_i)\, d\Pi_n(p)}{\int (p_{i-1}/p_{0,i-1})(X_1, \ldots, X_{i-1})\, d\Pi_n(p)} = \frac{I_{i,n}}{I_{i-1,n}},$$

for $I_{i,n} = \int (p_i/p_{0,i})(X_1, \ldots, X_i)\, d\Pi_n(p)$ for $i = 1, \ldots, n$, and $I_{0,n} = 1$. It follows that

$$\frac{1}{n}\sum_{i=1}^{n} \log \frac{\hat{p}_{0,i}(X_i)}{\hat{p}_i(X_i)} = -\frac{1}{n} \log I_{n,n}. \tag{6.15}$$

The expectation of the left side under $P_{0,\infty}$ is also the expectation of $n^{-1}\sum_{i=1}^{n} K(\hat{p}_{0,i}; \hat{p}_i)$. It is clearly nonnegative, whence it suffices to show that the expectation of the right side is asymptotically bounded above by ϵ, for every $\epsilon > 0$. If $\Pi_{n,\epsilon}$ is the renormalized restriction of Π_n to $B_{n,\epsilon} := \{p\colon n^{-1}K(p_{0,n}; p_n) < \epsilon\}$, then

$$I_{n,n} \geq \Pi_n(B_{n,\epsilon}) \int \frac{p_n}{p_{0,n}}(X_1, \ldots, X_n)\, d\Pi_{n,\epsilon}(p).$$

By Jensen's inequality,

$$\begin{aligned} P_{0,\infty}\Big[-\frac{1}{n}\log I_{n,n}\Big] &\leq -\frac{1}{n}\log \Pi_n(B_{n,\epsilon}) - P_{0,n}\frac{1}{n}\int \log \frac{p_n}{p_{0,n}}(X_1, \ldots, X_n)\, d\Pi_{n,\epsilon}(p) \\ &= -\frac{1}{n}\log \Pi_n(B_{n,\epsilon}) + \int \frac{1}{n} K(p_{0,n}; p_n)\, d\Pi_{\epsilon,n}(p) \leq -\frac{1}{n}\log \Pi_n(B_{n,\epsilon}) + \epsilon. \end{aligned}$$

The right side tends to ϵ as $n \to \infty$, by assumption. □

As the Kullback-Leibler divergence is larger than the squared Hellinger distance (see Lemma B.1), it also follows that, in P_0^n-probability,

$$d_H^2\Big(\frac{1}{n}\sum_{i=1}^{n} \hat{p}_{0,i}; \frac{1}{n}\sum_{i=1}^{n} \hat{p}_i\Big) \to 0.$$

In other words, the distance between the Cesàro averages of the predictive and true conditional densities tends to zero. In particular, for i.i.d. observations the Cesàro mean of the predictive densities is a weakly consistent density estimator at the true density p_0 whenever $p_0 \in \mathrm{KL}(\Pi)$, or more generally if a sequence of priors satisfies the condition of the theorem. (In Lemma 6.52 below the consistency is shown to be in the almost sure sense for a fixed prior.)

Besides that its consistency can be proved under weak conditions, there appears to be no good motivation for the Cesàro estimator $n^{-1}\sum_{i=1}^{n} \hat{p}_i$. Its asymmetric, sequential use of i.i.d. observations can be remedied by applying the procedure of Rao-Blackwell, which leads to the estimator

$$\frac{1}{n}\sum_{i=1}^{n}\frac{1}{\binom{n}{i}}\sum_{1\le j_1<\cdots<j_i\le n}\mathrm{E}(p|\,X_{j_1},\ldots,X_{j_i}). \tag{6.16}$$

This estimator contains $2^n - 1$ terms, and hence is computationally prohibitive, but it inherits the Hellinger consistency, by the convexity of d_H^2.

6.8.4 Martingale Approach

In Section 6.8.3 the predictive density $\hat{p}_i$, defined in (6.13) for i.i.d. observations and in (6.14) in general, was seen to approximate the true conditional density $\hat{p}_{0,i}$ (or p_0 in the i.i.d. case), as $n \to \infty$. When using a fixed prior Π, for a given function $\ell\colon [0,\infty) \to \mathbb{R}$ we may study this further through the variables

$$M_n = \sum_{i=1}^{n}\Big[\ell\Big(\frac{\hat{p}_i}{\hat{p}_{0,i}}(X_i)\Big) - \mathrm{E}_0\Big[\ell\Big(\frac{\hat{p}_i}{\hat{p}_{0,i}}(X_i)\Big)|\,X_1,\ldots,X_{i-1}\Big]\Big].$$

Note here that the hats in $\hat{p}_i$ and $\hat{p}_{0,i}$ refer to the conditioning on the "past" observations $X_1,\ldots,X_{i-1}$.

The variables M_n are sums of variables with zero conditional means given the past and hence form a martingale. We take the conditional expectations E_0 relative to the "frequentist" model $(X_1, X_2, \ldots) \sim p_{0,\infty}$, whence the martingale property is also relative to this model. The leading term of M_n can be viewed as a cumulative sum of discrepancies $\ell((\hat{p}_i/\hat{p}_{0,i})(X_i))$ between the predictive and true density, and the "compensator"

$$\mathrm{E}_0\big[\ell((\hat{p}_i/\hat{p}_{0,i})(X_i))|\,X_1,\ldots,X_{i-1}\big]$$

as the integrated version of this discrepancy. For the special cases $\ell(x) = \log x$ and $\ell(x) = 2(\sqrt{x}-1)$ this compensator is

$$\mathrm{E}_0\Big(\log\frac{\hat{p}_i}{\hat{p}_{0,i}}(X_i)|\,X_1,\ldots,X_{i-1}\Big) = -K(\hat{p}_{0,i};\hat{p}_i), \tag{6.17}$$

$$\mathrm{E}_0\Big(2\Big(\sqrt{\frac{\hat{p}_i}{\hat{p}_{0,i}}}(X_i) - 1\Big)|\,X_1,\ldots,X_{i-1}\Big) = -d_H^2(\hat{p}_{0,i};\hat{p}_i). \tag{6.18}$$

Martingale convergence theory gives a handle on these objects.

Lemma 6.52 *If $\sum_{n=1}^{\infty} n^{-2}\,\mathrm{var}_0\,\ell((\hat{p}_n/\hat{p}_{0,n})(X_n)) < \infty$, then $n^{-1}M_n \to 0$ almost surely. This condition is automatic for $\ell(x) = 2(\sqrt{x}-1)$. For i.i.d. observations, assuming the Kullback-Leibler property of the prior Π at p_0, this implies that $n^{-1}\sum_{i=1}^{n} d_H^2(\hat{p}_i, p_0) \to 0$, almost surely $[P_0^\infty]$.*

Proof The martingale differences $\Delta M_n = M_n - M_{n-1}$ have conditional mean zero and conditional variances satisfying $\mathrm{E}_0\,\mathrm{var}_0\,(\Delta M_n|\,X_1,\ldots,X_{n-1}) \le \mathrm{var}_0\,\ell((\hat{p}_n/\hat{p}_{0,n})(X_n))$. Hence the convergence of $n^{-1}M_n$ follows from the strong law for martingale differences (which follows from the martingale convergence theorem and Kronecker's lemma).

The variance is bounded above by the second moment $\mathrm{E}_0\ell^2((\hat{p}_n/\hat{p}_{0,n})(X_n))$, which in turn is bounded above by $4\mathrm{E}_{p_0}d_H^2(\hat{p}_n;\hat{p}_{0,n}) \le 8$ for $\ell(x) = 2(\sqrt{x}-1)$.

In view of (6.18), $n^{-1}M_n \to 0$ implies that $\limsup n^{-1}\sum_{i=1}^n d_H^2(\hat p_i, \hat p_{0,i})$ is bounded above by

$$\limsup_{n\to\infty} -\frac{2}{n}\sum_{i=1}^n\Bigl(\sqrt{\frac{\hat p_i}{\hat p_{0,i}}(X_i)}-1\Bigr) \le \limsup_{n\to\infty} -\frac{1}{n}\sum_{i=1}^n \log\frac{\hat p_i}{\hat p_{0,i}}(X_i),$$

since $2(\sqrt{x}-1)\ge \log x$, for every $x \ge 0$. By (6.15) the variable on the right is equal to $-n^{-1}\log I_n$, where in the case of i.i.d. observations for $I_n = \int\prod_{i=1}^n(p/p_0)(X_i)\,d\Pi(p)$. This tends to zero almost surely by Remark 6.18. □

The denominator $\int p_n(X_1,\ldots,X_n)\,d\Pi(p)$ of the posterior distribution is the marginal density of the observations in the Bayesian model. It can be factorized as the product of the predictive densities, as $p_1(X_1)p_2(X_2|\,X_1)\cdots p_n(X_n|\,X_1,\ldots,X_{n-1}) = \prod_{i=1}^n \hat p_i(X_i)$. The posterior probability of a set $\mathcal{U}$ can be factorized similarly. If $\Pi_{\mathcal{U}}$ is the renormalized restriction of Π to a set $\mathcal{U}$, then the predictive density of X_i in the model $p_\infty \sim \Pi_{\mathcal{U}}$ and $(X_1, X_2,\ldots)|\,p_\infty \sim p_\infty$ is

$$\hat p_i(X_i|\,\mathcal{U}) = p_i(X_i|\,X_1,\ldots,X_{i-1},\mathcal{U}) = \frac{\int p_i(X_1,\ldots,X_i)\,d\Pi_{\mathcal{U}}(p)}{\int p_{i-1}(X_1,\ldots,X_{i-1})\,d\Pi_{\mathcal{U}}(p)}.$$

With this notation we can write

$$\Pi_n(\mathcal{U}|\,X_1,\ldots,X_n) = \frac{\prod_{i=1}^n \hat p_i(X_i|\,\mathcal{U})\,\Pi(\mathcal{U})}{\prod_{i=1}^n \hat p_i(X_i)} = \frac{1}{I_n}\prod_{i=1}^n\frac{\hat p_i(X_i|\,\mathcal{U})}{\hat p_{0,i}(X_i)}\Pi(\mathcal{U}),$$

for $I_n = \int (p_n/p_{0,n})(X_1,\ldots,X_n)\,d\Pi(p)$. An intuitive interpretation of the center formula is that the posterior mass of a set $\mathcal{U}$ will tend to one if and only if in the limit the priors $\Pi_{\mathcal{U}}$ and Π lead to the same predictive densities.

The formula on the far right invites an alternative consistency proof: the posterior mass of a set $\mathcal{U}$ such that $\hat p_i(\cdot|\,\mathcal{U})$ is unlike $\hat p_{0,i}$ will tend to zero. For i.i.d. observations this idea can be made precise by noting that I_n is bounded below by $e^{-\epsilon n}$, for any $\epsilon > 0$, if $p_0 \in \mathrm{KL}(p_0)$, by (6.5) or Remark 6.18. Thus the far right expression tends to zero if the product is exponentially small. Now, for any $\alpha \in (0,1)$ and ρ_α the Hellinger transform,

$$\mathrm{E}_0\biggl(\Bigl(\frac{\hat p_i(X_i|\,\mathcal{U})}{\hat p_{0,i}(X_i)}\Bigr)^\alpha\,|\,X_1,\ldots,X_i\biggr) = \rho_\alpha(\hat p_i(\cdot|\,\mathcal{U}), \hat p_{0,i}).$$

If $\mathcal{U}$ is convex, then $\hat p_i(\cdot|\,\mathcal{U})$ is contained in $\mathcal{U}$ by Jensen's inequality, and hence the Hellinger transform is bounded by $\rho_\alpha(\mathcal{U}, \hat p_{0,i}) = \sup_{p\in\mathcal{U}}\rho_\alpha(p, \hat p_{0,i})$. By applying this argument for $i = n, n-1, \ldots, 1$ and with $\hat p_{0,i} = p_0$, we can "peel off" the observations one by one and find

$$\mathrm{E}_0\Bigl(\prod_{i=1}^n \frac{\hat p_i(X_i|\,\mathcal{U})}{\hat p_{0,i}(X_i)}\Pi(\mathcal{U})\Bigr)^\alpha \le \rho_\alpha(\mathcal{U}, p_0)^n\,\Pi(\mathcal{U})^\alpha.$$

Thus the numerator in the posterior is exponentially small for any convex $\mathcal{U}$ such that $\rho_\alpha(\mathcal{U}, p_0) < 1$. A slight extension of this argument allows to recover the result of Example 6.24 (see Problem 6.17), which was obtained by Schwartz's testing approach. Of course, these methods are closely related, as the Hellinger affinities provide bounds on the testing rates (see Appendix D).

6.8.5 α-Posterior

Another way to obtain Hellinger consistency under only the Kullback-Leibler property is to change the definition of the posterior distribution. For $\alpha \geq 0$ the *α-posterior distribution* based observations $X_1, \ldots, X_n$ from a density p following a prior Π is defined by

$$\Pi_n^{(\alpha)}(p \in B \mid X_1, \ldots, X_n) = \frac{\int_B \prod_{i=1}^n p^\alpha(X_i)\, d\Pi_n(p)}{\int \prod_{i=1}^n p^\alpha(X_i)\, d\Pi_n(p)}. \tag{6.19}$$

The α-posterior can be interpreted as the usual posterior distribution with respect to the data-dependent prior Π^* defined by

$$d\Pi^*(p) \propto \prod_{i=1}^n p^{\alpha-1}(X_i)\, d\Pi(p). \tag{6.20}$$

Example 6.53 For p the density of the normal Nor(θ, σ^2)-distribution with unknown mean θ and known variance σ^2, and a Nor(μ, τ^2) prior, the α-posterior distribution is the normal distribution with mean and variance

$$\frac{(\alpha n\bar{X}/\sigma^2) + (\mu/\tau^2)}{(\alpha n/\sigma^2) + (1/\tau^2)}, \qquad \frac{1}{(\alpha n/\sigma^2) + (1/\tau^2)}.$$

This leads to the interpretation that either the variance is misspecified as σ^2/α, or the sample size misspecified as $n\alpha$. The α-posterior mean is equal to the usual posterior mean with respect to the Nor$(\mu, \alpha\tau^2)$-prior, but the corresponding posterior variances are different. The latter implies that a credible region of the α-posterior does not have asymptotically correct frequentist coverage.

Theorem 6.54 *If $p_0 \in \mathrm{KL}(\Pi)$ for a fixed prior Π, then for any $0 < \alpha < 1$ the α-posterior distribution $\Pi_n^{(\alpha)}(\cdot \mid X_1, \ldots, X_n)$ in the model $X_1, \ldots, X_n \mid p \stackrel{iid}{\sim} p$ and $p \sim \Pi$ is strongly consistent at p_0. In particular, the α-posterior mean is strongly Hellinger consistent at p_0.*

Proof By the same argument as in the proof of Schwartz's theorem (see (6.5)) we can derive that $e^{\epsilon n} \int \prod_{i=1}^n (p/p_0)^\alpha(X_i)\, d\Pi(p) \to \infty$ almost surely $[P_0^\infty]$, for any $\epsilon > 0$. Furthermore, since $P_0(p/p_0)^\alpha = \rho_\alpha(p, p_0)$ is the Hellinger transform, for any neighborhood $\mathcal{U}$ of p_0,

$$P_0^n\Big[\int_{\mathcal{U}^c} \prod_{i=1}^n \frac{p^\alpha}{p_0^\alpha}(X_i)\, d\Pi(p)\Big] = \int_{\mathcal{U}^c} \rho_\alpha(p, p_0)^n\, d\Pi(p).$$

For $\mathcal{U}$ a Hellinger ball of radius ϵ, the integrand is bounded above by e^{-Cn}, for $C = \epsilon^2 \min(\alpha, 1-\alpha)$, by Lemma B.5. The theorem follows upon choosing $\epsilon < C$. □

In the above result, we can even allow $\alpha = \alpha_n$ to increase to 1 slowly; see Problem 6.15. A generalization to non-identically distributed observations is described in Problem 6.16.

Example 6.55 (Pólya tree) Let $X_1, X_2, \ldots \mid P \stackrel{\text{iid}}{\sim} P$, where $P \sim \mathrm{PT}(\mathcal{T}_m, \alpha_\varepsilon)$. Assume that the condition (3.16) holds, so that the prior is absolutely continuous almost surely, and the

α-posterior is well defined. We claim that this is again a Pólya tree process with parameters updated as

$$\alpha_\varepsilon \mapsto \alpha_\varepsilon + \alpha \sum_{i=1}^n \mathbb{1}\{X_i \in B_\varepsilon\}, \qquad \varepsilon \in \mathcal{E}^*. \tag{6.21}$$

The result will follow from the posterior updating rule for the ordinary Pólya tree process prior and posterior (the display with $\alpha = 1$) if the data-dependent prior Π^* in (6.20) is a Pólya tree process with updated parameters $\alpha_\varepsilon + (\alpha - 1)n \sum_{i=1}^n \mathbb{1}\{X_i \in B_\varepsilon\}$. To see this, it is convenient to describe the Pólya tree process in terms of the splitting variables $V = (V_{\varepsilon 0})_{\varepsilon \in \mathcal{E}^*}$, which are independent $\mathrm{Be}(\alpha_{\varepsilon 0}, \alpha_{\varepsilon 1})$-variables. The density of the Pólya tree process Π can be expressed as $p = \phi(V)$ for some measurable map $\phi\colon [0,1]^{\mathcal{E}^*} \to \mathcal{P}$, and hence for every measurable function h and λ the uniform measure on $[0,1]^{\mathcal{E}^*}$,

$$\int h(p)\, d\Pi(p) = \int_{[0,1]^{\mathcal{E}^*}} h \circ \phi(v) \prod_{\varepsilon \in \mathcal{E}^*} \frac{v_{\varepsilon 0}^{\alpha_{\varepsilon 0}-1}(1 - v_{\varepsilon 0})^{\alpha_{\varepsilon 1}-1}}{B(\alpha_{\varepsilon 0}, \alpha_{\varepsilon 1})}\, d\lambda(v).$$

The density $\prod_{i=1}^n p(X_i)^{\alpha-1}$ of the prior Π^* relative to Π is expressed in the splitting variables through (3.18), and it follows that

$$\int h(p)\, d\Pi^*(p) = \frac{\int h(p) \prod_{i=1}^n p(X_i)^{\alpha-1}\, d\Pi(p)}{\int \prod_{i=1}^n p(X_i)^{\alpha-1}\, d\Pi(p)}$$
$$= \frac{1}{c} \int_{[0,1]^{\mathcal{E}^*}} h \circ \phi(v) \prod_{i=1}^n \prod_{j=1}^\infty (2 v_{X_{i,1}\cdots X_{i,j}})^{\alpha-1} \prod_{\varepsilon \in \mathcal{E}^*} \frac{v_{\varepsilon 0}^{\alpha_{\varepsilon 0}}(1 - v_{\varepsilon 0})^{\alpha_{\varepsilon 1}}}{B(\alpha_{\varepsilon 0}, \alpha_{\varepsilon 1})}\, d\lambda(v).$$

Here $X_{i,1}X_{i,2}, \cdots$ is the binary expansion of X_i over the partitioning tree, and c is the normalizing constant (which depends on $X_1, \ldots, X_n$). Upon comparing the preceding displays we see that Π^* is indeed a Pólya tree process as indicated.

The canonical Pólya tree $\mathrm{PT}^*(\lambda, a_m)$ with $\sum_{m=1}^\infty a_m^{-1} < \infty$ possesses the Kullback-Leibler property at every p_0 with $K(p_0; \lambda) < \infty$, by Theorem 7.1. Consequently the α-posterior described by (6.21) and the resulting α-posterior mean are Hellinger consistent by Theorem 6.54. As in (3.23), the α-posterior mean of $p(x)$ is given by

$$\prod_{j=1}^\infty \frac{2a_j + 2\alpha \sum_{i=1}^n \mathbb{1}\{X_i \in B_{\epsilon_1\cdots\epsilon_j}\}}{2a_j + \alpha \sum_{i=1}^n \mathbb{1}\{X_i \in B_{\epsilon_1\cdots\epsilon_{j-1}}\}}. \tag{6.22}$$

Interestingly, this coincides with the expression for the usual posterior mean (3.23) with parameters αa_m. We can conclude that the ordinary posterior mean for $\mathrm{PT}^*(\lambda, a_m)$ is also consistent with respect to d_H.

6.9 Historical Notes

Consistency, as the most basic asymptotic property, has been long studied in the frequentist literature. Doob (1949) and Le Cam (1953) were apparently the first to formalize the concept of posterior consistency. Le Cam (1953) used uniform strong laws of large numbers to address posterior consistency in abstract spaces, similar to Wald's method for consistency of the maximum likelihood estimator. Freedman (1963, 1965) first addressed posterior consistency issues in concrete infinite-dimensional spaces. The importance of selecting an appropriate version of the posterior was emphasized by Diaconis and Freedman (1986b). Theorem 6.6 about merging in the weak*-sense, due to Diaconis and Freedman (1986b), is based on a similar idea of merging in the variation distance by Blackwell and Dubins (1962) (see Problem 6.1). Theorem 6.9 and Proposition 6.10 are due to Doob (1949). The example of inconsistency in Problem 6.11 for infinite-dimensional spaces with priors having the true distribution in their weak support is due to Freedman (1963), who was the first to construct such examples. Theorem 6.12 appeared in Freedman (1965), and Example 6.21 in Freedman (1963). Example 6.14 on inconsistency in the location model was constructed by Diaconis and Freedman (1986b,a). The inconsistency is related to the inconsistency of an M-estimator shown in Freedman and Diaconis (1982). Parallel results when the error distribution is constrained to median zero instead of being symmetric were obtained by Doss (1985a,b). The concept of Kullback-Leibler support and the role of exponentially consistent tests were first pointed out by Schwartz (1965). Her theorem was partly inspired by Freedman (1963)'s results for discrete sample spaces. The weak consistency result of Example 6.20 is also due to Schwartz (1965). The observation in Theorem 6.17 that the test can be restricted to a part of the parameter space of overwhelming prior mass appeared in an unpublished technical report by Barron in 1988. Theorem 6.23 is a strengthening of a result due to Ghosal et al. (1999b), who extended the scope of Schwartz's theory to address density estimation problems. A similar result was obtained by Barron et al. (1999) under stronger conditions involving bracketing entropy numbers and a direct proof bounding likelihood ratios. The first part of Example 6.24 is due to Walker (2004). Stability of the Kullback-Leibler property seems to have first been noticed by Ghosal et al. (1999a), especially for symmetrization and small location shift. Proposition 6.37 is due to Choi and Ramamoorthi (2008). Consistency for independent, non-identically distributed observations was addressed by Amewou-Atisso et al. (2003), who treated linear regression models without the provision of a sieve. Subsequently, many other authors addressed the issue, including Choudhuri et al. (2004a), Choi and Schervish (2007) and Wu and Ghosal (2008b). Dependent cases were addressed by Tang and Ghosal (2007b,a) for Markov processes (Theorem 6.42) and for general models by Roy et al. (2009). The results as presented here borrow in form from the rate results in Ghosal and van der Vaart (2007a). The idea of separation goes back to Schwartz (1965), and was later considered in a paper by Choi and Ramamoorthi (2008). Barron (1999) seems to have first observed that just the Kullback-Leibler property is sufficient to imply the convergence of the Cesàro Kullback-Leibler risk. Theorem 6.54 (with $\alpha = \frac{1}{2}$) as well as Example 6.55 are due to Walker and Hjort (2001). The martingale approach to consistency was introduced by Walker (2003, 2004); an extension to ergodic Markov processes can be found in Ghosal and Tang (2006). The prior constructed in Theorem 6.48 was suggested as an automatic prior for density estimation by Ghosal et al. (1997), who showed consistency of the posterior.

Problems

6.1 (Blackwell and Dubins 1962) In the setup of Theorem 6.6, assume in addition that two priors Π and Γ are such that P_Π^∞ and P_Γ^∞ are mutually absolutely continuous. Show that $\|Q_{\Pi,n} - Q_{\Gamma,n}\|_{TV} \to 0$ a.s. $[P_\Pi^\infty]$.

6.2 (Breiman et al. 1964) Consider real-valued i.i.d. observations from P_θ, $\theta \in \Theta = [0, 1]$. Show that there exists $\tilde{f}(X_1, X_2, \ldots)$ such that $P_\theta\{\tilde{f}(X_1, X_2, \ldots) = \theta\} = 1$ for all $\theta \in \Theta$ if and only if $\{\theta \mapsto P_\theta(X \le x): x \in \mathfrak{X}\}$ generates the Borel σ-field.

6.3 (Ghosh et al. 1994) Consider a sequence $\{(\mathfrak{X}^{(n)}, \mathscr{X}^{(n)}, p^{(n)}(x^{(n)}, \theta)); \theta \in \Theta\}$ of dominated statistical experiments, where Θ is a finite-dimensional Euclidean space and the true value θ_0 is an interior point of Θ. Consider prior densities π_1 and π_2 which are positive and continuous at θ_0. Suppose that the posterior distributions π_{1n} and π_{2n} are strongly consistent (respectively, weakly consistent) at θ_0. Show that the two posterior distributions merge in total variation sense, i.e., $d_{TV}(\pi_{1n}, \pi_{2n}) \to 0$ a.s. (respectively, in probability) under the true distribution.

6.4 (Breiman et al. 1964) Let $0 < C < 1$ and $1 = a_0 > a_1 > a_2 > \cdots$ be defined by $\int_{a_k}^{a_{k-1}} (e^{1/x^2} - C)\, dx = 1 - C$. Consider the parameter space $\Theta = \mathbb{N}$ and for every $k \in \mathbb{N}$, let p_k be a probability density on $[0, 1]$ defined by $p_k(x) = e^{x^{-2}}$ for $a_k < x < a_{k-1}$, and $p_k(x) = C$ otherwise. Consider i.i.d. observations $X_1, X_2, \ldots$ from this model.

(a) Show that $a_k \to 0$ as $k \to \infty$.
(b) Show that the likelihood function is maximized on $\{k: a_k > \min(X_1, \ldots, X_n)\}$ and hence the MLE exists.
(c) Let $I_n = I_n(X_1, \ldots, X_n)$ be the unique k such that $\min(X_1, \ldots, X_n) \in (a_k, a_{k-1})$. Show that for any $j \in \mathbb{N}$, $\sum_{i=1}^n \log(p_{I_n}/p_j) \to \infty$ in P_j probability, and hence the MLE tends to infinity. In particular, the MLE is inconsistent everywhere. (Note that the posterior is consistent everywhere by Doob's theorem.)

6.5 (Doksum and Lo 1990) Consider the setup of Example 6.14 without any symmetrization. Show that the "partial posterior" of θ given the median is consistent.

6.6 (Barron 1986) Let p_0 be the true density of i.i.d. observations and let $q(x_1, \ldots, x_n | \mathcal{N}) = \int_{\mathcal{N}} \prod_{i=1}^n p(x_i; \theta)\, d\Pi_n(p)/\Pi_n(\mathcal{N})$, where Π_n is the prior for p and $\mathcal{N}$ is a neighborhood of p_0 with $\Pi_n(\mathcal{N}) > 0$. We say that the *conditional mixture likelihood* $q(X_1, \ldots, X_n | \mathcal{N})$ matches with the true likelihood $\prod_{i=1}^n p_0(X_i)$ if for every $\epsilon > 0$,

$$e^{-n\epsilon} q(X_1, \ldots, X_n | \mathcal{N}) \le \prod_{i=1}^n p_0(X_i) \le e^{n\epsilon} q(X_1, \ldots, X_n | \mathcal{N}) \tag{6.23}$$

for all sufficiently large n a.s. $[P_0^\infty]$. Show that the second condition in (6.23) is always satisfied, and first holds if $p_0 \in \mathrm{KL}_n(\Pi_n)$ and $\mathcal{N} \supset \{p: K(p_0; p) < \epsilon\}$ for sufficiently small $\epsilon > 0$.

6.7 Let $X_1, \ldots, X_n \overset{\text{iid}}{\sim} p$. Show that a prior for p of the form $(1-\epsilon)\Pi_0 + \epsilon\delta_{p_0}$ is consistent at p_0 with respect to d_H or the $\mathbb{L}_1$-distance.

6.8 Consider a model where a $[0, 1]$-values observation has density $p := \pi u_0 + (1 - \pi) f$, where u_0 is a fixed density on $[0, 1]$ positive everywhere and f is an arbitrary density on $[0, 1]$. Let $\pi \sim \mu$ and $f \sim \Pi$ be a priori independent. Suppose that the true value of π is π_0, $0 < \pi_0 < 1$ and $\pi_0 \in \mathrm{supp}(\mu)$. Let the true value of f be f_0, where f_0 belongs to the $\mathbb{L}_1$-support of Π. Show that $p_0 := \pi_0 u_0 + (1 - \pi_0) f_0$ is in the Kullback-Leibler support of the prior induced on p.

6.9 (Salinetti 2003) Let a family of densities be parameterized by $\theta \in \Theta$. If the Kullback-Leibler property holds at the true density p_{θ_0}, then consistency holds if

$$\limsup_{n\to\infty} \sup_{\theta\in\mathcal{U}^c} n^{-1} \sum_{i=1}^{n} \log \frac{p(X_i;\theta)}{p(X_i;\theta_0)} < 0 \tag{6.24}$$

for every neighborhood $\mathcal{U}$ of θ_0. Show that (6.24) may be derived from a weaker form of uniform strong law of large numbers such as *hypo-convergence* or Wald's method of deriving consistency of MLE.

6.10 (Ghosal and van der Vaart 2003) Show that (6.24) implies existence of uniformly consistent tests in Theorem 6.16.

6.11 (Choi and Ramamoorthi 2008) Often, especially in the i.i.d. case, one derives a stronger conclusion than consistency, namely: for any fixed neighborhood $\mathcal{U}$ of the true density p_0, the posterior probability $\Pi_n(\mathcal{U}^c | X_1, \dots, X_n) \to 0$ exponentially fast a.s. $[P_0]$. Call this property strong *exponential consistency*. By the following example, show that the assertion is strictly stronger than consistency:

Let Π be a prior such that the posterior is inconsistent at f_0 with respect to the Hellinger (equivalently, $\mathbb{L}_1$) distance, and consider the prior $\Pi^* = \frac{1}{2}\Pi + \frac{1}{2}\delta_{p_0}$. Show that the posterior based on Π^* is strongly Hellinger consistent, but is not strongly exponentially consistent.

6.12 The Kullback-Leibler property is not necessary for consistency, even when the family is dominated. Consider the family Unif$[\theta - 1, \theta + 1]$, $\theta \in \mathbb{R}$, i.e., density given by $p_\theta(x) = \frac{1}{2}\mathbb{1}\{\theta - 1 \le x \le \theta + 1\}$, and the prior $\theta \sim \mathrm{Nor}(0, 1)$. Show that $K(p_\theta; p_{\theta'}) = \infty$ whenever $\theta \ne \theta'$, so the Kullback-Leibler support of the prior (or any other prior) is empty; yet the posterior is consistent at any value of θ.

6.13 Prove the following generalization of Lemma 6.26 for general observations: if $p_\theta^{(n)}$ is the joint density of $X^{(n)}$, $\theta \in \Theta$, θ_0 is the true parameter and Π any prior on θ, then show that

$$P_{\theta_0}^{(n)} \left\{ \int \frac{p_\theta^{(n)}}{p_{\theta_0}^{(n)}} d\Pi(\theta) < \Pi(\theta\colon K(p_{\theta_0}^{(n)}; p_\theta^{(n)}) < n\delta) e^{-2bn\delta} \right\} \le b^{-1}.$$

Hence show that in Theorem 6.39, the second condition on Kullback-Leibler variations can be dropped at the expense of strengthening the second assumption for arbitrary $C > 0$. State a posterior consistency result based on the revised assertions.

6.14 (Choi and Ramamoorthi 2008) Derive Example 6.24 using Theorem 6.45.

6.15 Show that Theorem 6.54 holds for $\alpha = \alpha_n \to 1$ sufficiently slowly.

6.16 For any sequence of independent random observations, define the α-posterior in an analogous way. Show that Theorem 6.54 holds with the squared Hellinger distance d_H^2 replaced by the average squared Hellinger distance $n^{-1}\sum_{i=1}^n d_H^2(p_{0i}, p_i)$ under the assumptions that there exists a sequence of sets B_n, such that $\liminf \Pi_n(B_n) > 0$ and (6.8) holds on B_n.

6.17 Use the argument at the end of Section 6.8.4 to show: If $p_0 \in \mathrm{KL}(\Pi)$ and for every $\epsilon > 0$ there exists a partition $\mathcal{P} = \cup_j \mathcal{P}_j$ and $\alpha < 1$ such that $\sum_j \Pi(\mathcal{P}_j)^\alpha < \infty$, then the posterior distribution is strongly Hellinger consistent at p_0. [Hint: Given a Hellinger ball $\mathcal{U}$ of radius $2\epsilon > 0$ around p_0 let $\mathcal{P} = \cup_j \mathcal{P}_j$ be a partition for ϵ as given, and let J be the set of indices of the partitioning sets that intersect $\mathcal{U}^c$. Because the diameters are smaller than ϵ, the convex hull of each of the latter sets has distance at least ϵ to p_0, whence

$$\rho_\alpha(\text{conv}(\mathcal{P}_j), p_0) \leq 1 - \min(\alpha, 1-\alpha)\epsilon^2 \leq e^{-Cn}.$$

Finally, $\Pi_n(\cup_{j\in J} \mathcal{P}_j| X_1, \ldots, X_n) \leq \sum_{j\in J} \Pi_n(\mathcal{P}_j| X_1, \ldots, X_n)^\alpha$, and we can apply the preceding to each term of the sum to find a bound of the form $e^{c'n} \sum_j e^{-Cn} \Pi(\mathcal{P}_j)^\alpha$.]

6.18 (Walker 2004, Ghosal and Tang 2006) Consider an ergodic Markov chain with transition density f given a prior Π. Let the Kullback-Leibler property hold at the true value f_0. Assume that $\mathcal{A}_n \supset \mathcal{A}_{n+1}$ are random sets and $\mathbb{1}\{f \in \mathcal{A}_n\}$ is an $\mathscr{X}_n$-measurable random variable for all f and n. Show that $\Pi(\mathcal{A}_n| X_0, \ldots, X_n) \to 0$ a.s. if

$$\liminf_{N\to\infty} N^{-1} \sum_{n=0}^{N-1} d_H^2(f_0(\cdot| X_n), f_{n,\mathcal{A}_n}(\cdot| X_n)| X_n) > 0 \text{ a.s.,} \tag{6.25}$$

where

$$f_{n,\mathcal{A}_n}(y|X_n) = \frac{\int_{\mathcal{A}_n} f(y|X_n) \prod_{i=1}^n f(X_i|X_{i-1})\, d\Pi(f)}{\int_{\mathcal{A}_n} \prod_{i=1}^n f(X_i|X_{i-1})\, d\Pi(f)}$$

and $d_H^2(f(\cdot|x), g(\cdot|x)|x) = \int (\sqrt{f(y|x)} - \sqrt{g(y|x)})^2 d\nu(y)$.

Specialize the result to density estimation based on i.i.d. variables.

6.19 (Walker 2004) Show that for i.i.d. observations the Kullback-Leibler property implies consistency with respect to the weak topology using the i.i.d. case of Problem 6.18. [Hint: Consider a subbasic neighborhood for the weak topology $\mathcal{U} = \{p: P\psi < P_0\psi + \epsilon\}$, where ψ is a bounded continuous function. Observe that its complement is also convex!].

7

Consistency: Examples

In this chapter we apply the result of the preceding chapter in a number of statistical settings: density estimation with a variety of priors (Pólya trees, Dirichlet process mixtures, Gaussian processes), nonparametric and semiparametric regression problems, and time series models. The Kullback-Leibler property of priors plays a central role, and is discussed extensively.

7.1 Priors with the Kullback-Leibler Property

In this section we prove that common priors such as the Pólya tree process and Dirichlet process mixtures and for density estimation possess the Kullback-Leibler property under appropriate conditions. Another important class of priors, namely those based on Gaussian processes, also posses the same property under appropriate conditions. However, we shall prove much stronger results in Chapter 11 under essentially the same conditions, so we refrain from presenting separate results on Kullback-Leibler property of Gaussian processes in this section.

7.1.1 Pólya Trees

The canonical Pólya tree $\mathrm{PT}^*(\lambda, a_m)$ is defined as a prior on probability measures in Definition 3.20 (and the subsequent discussion). Its parameters are a probability measure λ and a sequence of constants a_m. When $\sum_{m=0}^{\infty} a_m^{-1} < \infty$ the realizations from this prior are a.s. absolutely continuous relative to λ, by Theorem 3.16, and the prior can be viewed as a prior on λ-densities. The following theorem shows that it automatically possesses the Kullback-Leibler property.

Theorem 7.1 *Let P follow a canonical Pólya tree prior* $\mathrm{PT}^*(\lambda, a_m)$ *and such that* $\sum_{m=1}^{\infty} a_m^{-1} < \infty$*. Then P possesses a density p with respect to λ a.s. and the prior* $\mathrm{PT}^*(\lambda, a_m)$ *possesses the Kullback-Leibler property at any density p_0 such that $K(p_0; \lambda) < \infty$. Furthermore, if $V_2(p_0; \lambda) < \infty$, then $V_2(p_0; p) < \infty$ a.s.* $[\mathrm{PT}^*(\lambda, a_m)]$.

Proof As shown in Theorem 3.16, the density p of $P \sim \mathrm{PT}^*(\lambda, a_m)$ exists a.s. and can be represented as the infinite product (3.18). In the present situation, the variables V_ε possess $\mathrm{Be}(a_{|\varepsilon|}, a_{|\varepsilon|})$-distributions, all variables $V_{\varepsilon 0}$ are independent, and $V_{\varepsilon 1} = 1 - V_{\varepsilon 0}$, for $\varepsilon \in \mathcal{E}^*$. The discretization of p_0 to the partition at stage m has density $p_{0,m} = 2^m \sum_{\varepsilon \in \mathcal{E}^m} P(A_\varepsilon) \mathbb{1}\{A_\varepsilon\}$ relative to λ (which has $\lambda(A_\varepsilon) = 2^{-|\varepsilon|}$). For $\varepsilon = \varepsilon_1 \cdots \varepsilon_m \in \mathcal{E}^m$ we can factorize $P(A_\varepsilon) = \prod_{j=1}^{m} v_{\varepsilon_1 \cdots \varepsilon_j}$, for $v_{\varepsilon\delta} = P(A_{\varepsilon\delta} | A_\varepsilon)$. As $p_{0,m} = \mathrm{E}_\lambda(p_0 | \mathscr{T}_m)$, where

$\mathcal{T}_m = \sigma\langle T_m \rangle$, the martingale convergence theorem gives that $p_{0,m} \to p_0$ λ-a.e. as $m \to \infty$. This leads to the representations

$$p_{0,m}(x) = \prod_{j=1}^{m}(2v_{x_1\cdots x_j}), \qquad p_0(x) = \prod_{j=1}^{\infty}(2v_{x_1\cdots x_j}). \tag{7.1}$$

The quotient p_0/p can also be written as an infinite product, and the Kullback-Leibler divergence $K(p_0; p) = \int p_0 \log(p_0/p)\, d\lambda$ can be decomposed as, for any $m \geq 1$,

$$\int \Big[\sum_{j=1}^{m} \log \frac{v_{x_1\cdots x_j}}{V_{x_1\cdots x_j}} + \sum_{j=m+1}^{\infty} \log(2v_{x_1\cdots x_j}) - \sum_{j=m+1}^{\infty} \log(2V_{x_1\cdots x_j})\Big]\, p_0(x)\, d\lambda(x).$$

We split this in the three terms indicated in square brackets. For given m, the event that the first term is smaller than some given $\delta > 0$ can be written as the event that a finite set of independent beta-variables $V_{x_1\cdots x_j}$ falls into a nonempty open set of the appropriate unit cube (it is nonempty as $(v_{x_1\cdots x_j}\colon j = 1, \ldots, m,\ (x_1, \ldots, x_m) \in \mathcal{E}^m)$ belongs to it). This has positive probability for any $\delta > 0$. The second term is small for large m by the last assertion of Lemma B.10. The third term is independent of the first. The expectation and variance of the integrand are both bounded by $\sum_{j=m+1}^{\infty} a_j^{-1}$, uniformly in x, by Lemma G.12. It follows that the third term tends to zero in probability, and hence is certainly smaller than a given $\delta > 0$ with positive probability.

This concludes the proof of the Kullback-Leibler property.

Since $V_2(p_0; p) \leq 2P_0(\log p_0)^2 + 2P_0(\log p)^2$, for the final assertion it is enough to show that $P_0(\log p)^2 < \infty$, a.s. In view of (3.18) the expectation of the latter random variable is bounded above by

$$\mathrm{E} \int \Big[\sum_{j=1}^{\infty} \log(2V_{x_1\cdots x_j})\Big]^2 p_0(x)\, d\lambda(x).$$

The second moment of the series is the sum of its variance and the square of its expectation, which were already noted to be bounded by a multiple of $\sum_{j=1}^{\infty} a_j^{-1} < \infty$, uniformly in x, by Lemma G.12. ◻

7.1.2 Kernel Mixtures

By equipping the parameters F and φ in mixtures of the type $p_{F,\varphi}(x) = \int \psi(x; \theta, \varphi)\, dF(\theta)$ with priors, we obtain a prior on densities. In this section we derive conditions on the kernel $\psi(\cdot; \theta, \varphi)$, a given family of probability density functions, so that the resulting prior possesses the Kullback-Leibler property as soon as the priors on F and φ have full (weak) support.

We assume that the sample space $\mathfrak{X}$ and the parameter spaces Θ and Φ of θ and φ are Polish spaces, equipped with their Borel σ-fields. The parameter φ may be vacuous, in which case the conditions simplify.

The true density p_0 need not itself possess mixture form, but must be approximable by mixtures. For given F_ϵ, φ_ϵ we may decompose

$$K(p_0; p_{F,\varphi}) = K(p_0; p_{F_\epsilon,\varphi_\epsilon}) + P_0 \log \frac{p_{F_\epsilon,\varphi_\epsilon}}{p_{F,\varphi}}. \tag{7.2}$$

We conclude that p_0 possesses the Kullback-Leibler property if, for every $\epsilon > 0$, the pair $(F_\epsilon, \varphi_\epsilon)$ can be chosen such that $K(p_0; p_{F_\epsilon,\varphi_\epsilon}) < \epsilon$ and such that the set of $(F_\epsilon, \varphi_\epsilon)$ so that the second term is bounded by ϵ and has positive prior probability. The second is certainly true if $(F, \varphi) \mapsto P_0 \log(p_{F_\epsilon,\varphi_\epsilon}/p_{F,\varphi})$ is continuous at $(F_\epsilon, \varphi_\epsilon)$, and $(F_\epsilon, \varphi_\epsilon)$ is in the support of its prior. In most examples these conditions are easiest to verify if F_ϵ is chosen compactly supported.

The following lemma gives explicit assumptions. For $D \subset \Theta$ and $N \subset \Phi$ define

$$\underline{\psi}(x; D, N) = \inf_{\theta\in D,\varphi\in N} \psi(x; \theta, \varphi), \qquad \overline{\psi}(x; \theta, N) = \sup_{\varphi\in N} \psi(x; \theta, \varphi).$$

Theorem 7.2 (Mixtures) *Suppose that for every $\epsilon > 0$ there exist $F_\epsilon \in \mathrm{supp}(\Pi)$ and $\varphi_\epsilon \in \mathrm{supp}(\mu)$, and open sets $D \supset \mathrm{supp}(F_\epsilon)$ and $N \ni \varphi_\epsilon$ with $K(p_0; p_{F_\epsilon,\varphi_\epsilon}) < \epsilon$ and that the following conditions hold.*

(A1) $\{\theta \mapsto \psi(x; \theta, \varphi), \varphi \in N\}$ *is uniformly bounded and equicontinuous on D, for any x.*
(A2) $\int \log\big(p_{F_\epsilon,\varphi_\epsilon}(x)/\underline{\psi}(x; D, N)\big)\, dP_0(x) < \infty$.
(A3) $\varphi \mapsto \psi(x; \theta, \varphi)$ *is continuous at φ_ϵ, for any x and $\theta \in D$.*
(A4) $\int \overline{\psi}(x; \theta, N)\, dF_\epsilon(\theta) < \infty$, *for every x.*

Then $p_0 \in \mathrm{KL}(\Pi^*)$ for the prior Π^* on $p_{F,\varphi}$ induced by the prior $\Pi \times \mu$ on (F, φ).

Proof Finiteness of $K(p_0; p_{F_\epsilon,\varphi_\epsilon})$ implies that $p_{F_\epsilon,\varphi_\epsilon}(x) > 0$ for P_0-a.e. x. Therefore the decomposition (7.2) is valid. It suffices to show that the second term on the right is bounded by a multiple of ϵ with positive prior probability. Because F_ϵ and φ_ϵ are contained in the supports of their priors by assumption, this is true if the limit superior of this term as $F \rightsquigarrow F_\epsilon$ and $\varphi \to \varphi_\epsilon$ is nonpositive. If F_D is the restriction of F to D, then $p_{F,\varphi} \geq p_{F_D,\varphi}$, and hence the second term on the right increases if F is replaced by F_D. Because D contains the support of F_ϵ in its interior, so that $F_\epsilon(\partial D) = 0$, we have $F_D \rightsquigarrow F_\epsilon$ if $F \rightsquigarrow F_\epsilon$. Therefore, it suffices to show the limit of this term is nonpositive for *sub*probability measures $F \rightsquigarrow F_\epsilon$ supported within D.

By condition (A1), we have that $p_{F,\varphi}(x) \to p_{F_\epsilon,\varphi}(x)$ uniformly in $\varphi \in N$, for every x, as $F \rightsquigarrow F_\epsilon$ (and all are supported on D). By (A3) and (A4) and the dominated convergence theorem we also have that $p_{F_\epsilon,\varphi}(x) \to p_{F_\epsilon,\varphi_\epsilon}(x)$ as $\varphi \to \varphi_\epsilon$, for every x. Combined with the last assertion this shows that $p_{F,\varphi}(x) \to p_{F_\epsilon,\varphi_\epsilon}(x)$ as $F \rightsquigarrow F_\epsilon$ and $\varphi \to \varphi_\epsilon$. Also for F supported on D,

$$\log \frac{p_{F_\epsilon,\varphi_\epsilon}}{p_{F,\varphi}}(x) \leq \log \frac{p_{F_\epsilon,\varphi_\epsilon}(x)}{\underline{\psi}(x; D, N)}.$$

By (A2) the functions on the right are dominated by a P_0-integrable function. It follows that $\limsup P_0 \log(p_{F_\epsilon,\varphi_\epsilon}/p_{F,\varphi}) \leq P_0 \log 1 = 0$, by the dominated convergence theorem. ☐

Next, specialize the kernel function to be of location-scale type

$$\psi(x; \mu, h) = \frac{1}{h^d}\chi\Big(\frac{x-\mu}{h}\Big),$$

for a given probability density function χ defined on $\mathfrak{X} = \mathbb{R}^d$. We first consider location-scale mixtures obtained by putting a distribution on $\theta := (\mu, h)$, and next location mixtures created by mixing only over $\theta := \mu$ and considering $\varphi := h$ as an additional parameter.

Theorem 7.3 (Location-scale mixtures) *Assume that*

(B1) *χ is bounded, continuous and positive everywhere,*
(B2) $\int p_0(x) \log p_0(x)\, dx < \infty$,
(B3) $-\int p_0(x) \log \inf_{\|y\|<\delta} p_0(x-y)\, dx < \infty$, *for some* $\delta > 0$,
(B4) $\inf_{\|y\|<\|x\|^\eta} \chi(x-y) \geq \underline{\chi}(x)$ *for large* $\|x\|$ *and a function* $\underline{\chi}$ *that is decreasing as its argument moves away from zero and satisfies* $-\int p_0(x) \log \underline{\chi}(2x|x|^\eta)\, dx < \infty$, *for some* $\eta \in (0, 1)$.

Then $p_0 \in \mathrm{KL}(\Pi)$ *for the prior* Π *on* $p_F = \int h^{-d} \chi((\cdot - \mu)/h)\, dF(\mu, h)$ *induced by a prior on* F *with full support* $\mathfrak{M}(\mathbb{R}^d \times (0, \infty))$; *and also for the prior on* $p_{F,h} = \int h^{-d} \chi((\cdot - \mu)/h)\, dF(\mu)$ *induced by a product prior on* (F, h) *with full support on* $\mathfrak{M}(\mathbb{R}^d) \times (0, \infty)$.

Proof Since there is no parameter φ, it suffices to verify conditions (A1)–(A2) of Theorem 7.2. We choose the measure F_ϵ on $\mathbb{R}^d \times (0, \infty)$ of the form $F_m := P_m \times \delta_{h_m}$, where P_m is P_0 restricted and renormalized to $\{\mu: \|\mu\| \leq m\}$, and δ_h is the Dirac measure at $h \in (0, \infty)$. We shall show that $K(p_0; p_{F_m}) \to 0$ as $m \to \infty$, for $h_m = m^{-\eta}$ and fixed $\eta > 0$, so that $K(p_0; p_{F_m}) < \epsilon$ for large enough m. The measure F_m is supported within the open set $D = \{\mu: \|\mu\| < m'\} \times (a, b)$, for $m' > m$ and $0 < a < h_m < b$. Boundedness and continuity of the map $(\mu, h) \mapsto \chi_h(x - \mu) := \chi((x - \mu)/h)/h^d$ on D as in condition (A1) is clear from the continuity of χ. Only Condition (A2) remains to be verified.

As is known from the theory of kernel approximation, $p_0 * \chi_h \to p_0$ in $\mathbb{L}_1$, as $h \to 0$. Since $\|p_m * \chi_h - p_0 * \chi_h\|_1 \leq \|p_m - p_0\|_1 \to 0$, as $m \to \infty$, we see by the triangle inequality that $p_{F_m} = p_m * \chi_{h_m} \to p_0$ in $\mathbb{L}_1$. To show that $K(p_0; p_{F_m}) \to 0$ it suffices to show that $\log(p_0/p_{F_m})$ is uniformly integrable in $\mathbb{L}_1(p_0)$. As $-\log x \geq 1 - x$ for $x > 0$, this function is bounded from below by $1 - p_{F_m}/p_0$, which is uniformly integrable relative to p_0, as $p_{F_m} \to p_0$ in $\mathbb{L}_1$. It suffices to show that $\log(p_0/p_{F_m})$ is bounded above by a p_0-integrable function.

Letting $c_m^{-1} = \int_{\|x\|<m} p_0(x)\, dx$, we can write

$$p_{F_m}(x) = c_m \int_{\|y\|<m} p_0(y) \chi_{h_m}(x - y)\, dy.$$

This immediately gives the estimate, for $\|x\| \leq m/2$ so that $\|y\| < m$ if $\|x - y\| < h_m$,

$$p_{F_m}(x) \geq c_m \inf_{\|y\|<h_m} p_0(x - y) \chi_{h_m}(y: \|y\| < h_m).$$

For $\|x\| > m/2$ and $\|y\| < C$, we have (easily, as $h_m = m^\eta$) for sufficiently large m that $\|y/h_m\| \leq \|x/h_m\|^\eta$ and hence $\chi_{h_m}(x - y) \geq \underline{\chi}(x/h_m)$, by the definition of $\underline{\chi}$. Since $\|x/h_m\| = \|x\| m^\eta$ is closer to the origin than $2x\|x\|^\eta$, if $\|x\| > m/2$, the assumed monotonicity of $\underline{\chi}$ yields $\chi_{h_m}(x - y) \geq \underline{\chi}(2x\|x\|^\eta)$, so that, for $\|x\| > m/2$,

$$p_{F_m}(x) \geq c_m P_0(y: \|y\| < C) \underline{\chi}(2x\|x\|^\eta).$$

Combining the preceding displays and noting that $c_m \to 1$, we find, for sufficiently large m,

$$p_{F_m}(x) \gtrsim \inf_{\|y\|<\epsilon} p_0(x-y) \wedge \underline{\chi}(2x\|x\|^\eta).$$

Hence $\log(p_0/p_{F_m})$ is bounded above by $\log p_0$ minus the negative logarithm of the function on the right side, which is integrable from above by assumptions (B2), (B3) and (B4).

For $D = \{\mu: \|\mu\| < m'\} \times (a, b)$ we have for $\|\mu/h\| \le \|x/h\|^\eta$

$$\inf_{(\mu,h)\in D} \frac{1}{h}\chi\Big(\frac{x-\mu}{h}\Big) \ge \frac{1}{a}\underline{\chi}\Big(\frac{x}{h}\Big).$$

In particular, this is valid for $\|x\|^\eta \ge K/a^{1-\eta}$, and then the right side is further bounded below by a multiple of $\underline{\chi}(x\|x\|^\eta)$. For $\|x\| < C$ and any given $C > 0$ the function on the left is bounded away from zero by the continuity and positivity of χ. It follows that minus the logarithm of the left side is integrable above. Because p_{F_m} is uniformly bounded, for fixed m, by the boundedness of χ, it follows that (A2) is satisfied.

The proof of the second part of the assertion, for location mixtures with a prior on scale, proceeds by similar arguments. □

Example 7.4 (Skew-normal kernel) The *skew-normal kernel* with skewness parameter λ is given by

$$\chi(x) = \frac{1}{\pi} e^{-x^2/2} \int_{-\infty}^{\lambda x} e^{-t^2/2} dt.$$

The value $\lambda = 0$ corresponds to the normal kernel. For this kernel the Kullback-Leibler property holds at every continuous density p_0 that has a finite moment of order $2 + \eta > 2$ and satisfies (B2)–(B3).

Because the skew-normal density is monotone at its two extremes, for (B4) it suffices to verify finiteness of $\int p_0(x) \log \chi(2x|x|^\eta)\, dx$, which is clear from the moment condition.

Example 7.5 (Multivariate normal kernel) For the standard multivariate normal kernel $\chi(x) = (2\pi)^{-d/2} e^{-\|x\|^2/2}$, the Kullback-Leibler property holds for every p_0 that satisfies conditions (B2)–(B3) and has $\int \|x\|^{2+\eta} p_0(x)\, dx < \infty$, for some $\eta > 0$.

As the kernel decreases monotonically with increasing $\|x\|$, condition (B4) can be checked by a monotonicity argument, as in the one-dimensional case.

Example 7.6 (Histograms) For given $\theta \in (0, 1]$ and $m \in \mathbb{N}$ let $\psi(\cdot; \theta, m)$, the density of the uniform distribution on the interval $((i-1)/m, i/m]$ if $(i-1)/m < \theta \le i/m$ (i.e. $\psi(x; , \theta, m) = m$ if both x and θ belong to $((i-1)/m, i/m]$ and is 0 otherwise). Then the mixture $\int \psi(\cdot; \theta, m)\, dF(\theta)$, for a measure F on $(0, 1]$ and fixed m, is a histogram with bins $((i-1)/m, i/m]$. Alternatively, the mixture $\int \psi(\cdot; \theta, m)\, dF(\theta, m)$, for a measure F on $(0, 1] \times \mathbb{N}$ is a histogram with variable bins, which may be used to estimate a density p_0 on the unit interval. We may construct a prior on histograms by putting a prior on F and/or m. In both cases any continuous density p_0 is contained in Kullback-Leibler support of the prior, as soon as the prior on F has full weak support, and in the case of fixed bins, the number m of bins is given a prior with infinite support in $\mathbb{N}$.

If p_0 is bounded away from zero, then this is easy to derive from Theorem 7.2, and the fact that any continuous function can be uniformly approximated by a histogram. If

p_0 is bounded away from zero this automatically also gives approximation relative to the Kullback-Leibler divergence. Extension to any continuous density next follows with the help of Corollary 6.35.

Example 7.7 (Bernstein polynomial kernel) The Bernstein polynomial prior on densities on $[0, 1]$ described in Example 5.10, with $(w_0, \ldots, w_k)|\, k \sim \Pi_k$ and $k \sim \mu$, for a prior Π_k with full support $\mathbb{S}_k$ and μ with infinite support in $\mathbb{N}$ contains every continuous density p_0 in its Kullback-Leibler support.

As every continuous p_0 is uniformly approximated by its associated Bernstein polynomial, this can be proved similarly as in Example 7.6.

Example 7.8 (Lognormal kernel) The two parameter *lognormal kernel*

$$\psi(x; \mu, \varphi) = (2\pi)^{-1/2} x^{-1} \varphi^{-1} e^{-(\log x - \mu)^2/(2\varphi^2)}$$

is supported on $(0, \infty)$ rather than on the full line. We can investigate whether a density is in the Kullback-Leibler support of lognormal mixtures by applying Theorem 7.3 after mapping the positive half line to $\mathbb{R}$ by the logarithmic function. Since this map is a bimeasurable bijection, the Kullback-Leibler divergence between the induced distributions is the same as between the original distributions, whence the Kullback-Leibler property is preserved.

Under the map $y = \log x$ the density p_0 transforms into $y \mapsto q_0(y) = e^y p_0(e^y)$, and the lognormal mixtures into normal mixtures. The conditions (B2)–(B3) on q_0 translate back into the conditions on the original density p_0 that the two integrals $\int_0^\infty p_0(x) \log(x p_0(x))\, dx$ and $-\int_0^\infty p_0(x) \log \inf_{\|u-1\|<\delta} (xu\, p_0(xu))\, dx$ are finite, while the condition that q_0 has a finite $2+\eta$ moment translates into convergence of the integral $\int_0^\infty p_0(x) |\log x|^{2+\eta}\, dx$. Any p_0 with these properties is in the Kullback-Leibler support of the prior.

Example 7.9 (Exponential kernel) A mixture p_F of the exponential scale family on $\mathfrak{X} = (0, \infty)$ satisfies

$$p_F(x) = \int \lambda e^{-\lambda x}\, dF(\lambda) = -\frac{d}{dx} \int e^{-\lambda x}\, dF(\lambda).$$

Consequently, the survival function $\int_x^\infty p_F(y)\, dy$ is the Laplace transform of a measure F. The set of Laplace transforms of measures on $(0, \infty)$ is exactly the set of *completely monotone functions*: functions G such that $(-1)^n d^n G(x)/dx^n \geq 0$, for all $n, x > 0$ (Feller 1971, Chapter XIII).

If F is given a prior with full support $\mathfrak{M}(0, \infty)$, then any density p_0 with a completely monotone survival function, finite first absolute moment, and $\int p_0(x) \log p_0(x)\, dx < \infty$ belongs to the Kullback-Leibler support of the induced prior on p_F.

This may be proved using Theorem 7.2, by taking F_ϵ the distribution P_0 truncated to a sufficiently large compact set in $(0, \infty)$.

7.1.3 Exponential Densities

In Section 2.3.1 it is seen that the Kullback-Leibler divergence between densities of the form

$$p_f(x) = e^{f(x) - c(f)}, \qquad c(f) = \log \int e^f\, d\nu, \tag{7.3}$$

is bounded above by a function of the uniform distance between the corresponding functions f. This readily gives the Kullback-Leibler property for the prior induced on these densities by a prior on f.

Theorem 7.10 *If f_0 is bounded and belongs to the uniform support of the prior Π, then $p_{f_0} \in \mathrm{KL}(\Pi^*)$ for Π^* the prior induced on p_f by the prior Π on f.*

Proof If $\|f - f_0\|_\infty < \epsilon$, then $\|f - f_0\|_\infty \le \|f_0\|_\infty + \epsilon$, and hence $K(p_{f_0}; p_f) < \epsilon^2 e^\epsilon (1+\epsilon)$, by Lemma 2.5. Then we have $\Pi^*(p_f: K(p_{f_0}; \bar{p}_f) < 6\epsilon^2) \ge \Pi(f: \|f - f_0\|_\infty < \epsilon)$, for $\epsilon < 1$, which is positive by assumption. ☐

Example 7.11 If $\mathfrak{X}$ is a compact subset of $\mathbb{R}^d$ and p_0 is continuous and strictly positive, then $f_0 = \log p_0$ is well defined and continuous, and hence belongs to the support of any fully supported Borel prior on $\mathfrak{C}(\mathfrak{X})$. Since $p_0 = p_{f_0}$ it follows that p_0 is in the Kullback-Leibler support of the induced exponential prior. By Corollary 6.35 this extends to *any* continuous density.

Concrete examples are fully supported Gaussian process priors, and series priors as in Lemma 2.2, which lead to infinite-dimensional exponential families.

7.2 Density Estimation

In Section 7.1 the Kullback-Leibler property was seen to hold under mild conditions for common priors, including Pólya tree processes, Dirichlet process mixtures, and random series. The Kullback-Leibler property implies consistency with respect to the weak topology (see Example 6.20), but this topology is too weak to be really useful for density estimation. Theorem 6.23 shows that the Kullback-Leibler property together with a complexity bound gives consistency for the total variation distance (or equivalently the Hellinger distance). The following counterexample confirms that the Kullback-Leibler property alone is not sufficient for consistency relative to this stronger topology.

In the remainder of this section we next investigate consistency in the total variation distance for normal mixtures, general Dirichlet process mixtures, Pólya trees and series priors. Other examples are treated in later chapters in the context of rates of contraction; in particular, priors based on Gaussian processes in Chapter 11.

Example 7.12 (Inconsistency) We shall construct a prior on the set of probability densities on $[0, 1]$ that has the Kullback-Leibler property relative to the uniform density p_0, but yields a posterior distribution such that $\limsup_{n\to\infty} \Pi_n(\|p - p_0\|_1 = 1 | X_1, \ldots, X_n) = 1$, almost surely, if $X_1, \ldots, X_n \stackrel{\text{iid}}{\sim} p_0$.

The prior puts mass $1/2$ on a model $\mathcal{P}_0$ and masses $3/(\pi N)^2$ on models $\mathcal{P}_N$, for $N = 1, 2, \ldots$ Here $\mathcal{P}_0 = \{p_\theta: 0 \le \theta \le 1\}$ is the parametric model consisting of the probability densities $p_\theta(x) = e^{-\theta + \sqrt{2\theta}\Phi^{-1}(x)}$, for Φ the standard normal cumulative distribution function, and the prior distributes the mass within the model according to the prior $\pi(\theta) \propto e^{-1/\theta}$. For every $N \ge 1$ the model $\mathcal{P}_N$ is the collection of all histograms on the partition $((j-1)/(2N^2), j/(2N^2)]$, for $j = 1, \ldots, 2N^2$, with N^2 ordinate values 2 and

N^2 ordinate values 0; the prior spreads the mass uniformly over the $\binom{2N^2}{N^2}$ elements of the model. The values 0 and 2 are chosen so that $\|p - p_0\|_1 = 1$, for all $p \in \mathcal{P}_N$ and all $N \geq 1$. Thus the claim on the posterior distribution follows if $\limsup \Pi_n(\mathcal{P}_0|\, X_1, \ldots, X_n) = 0$, almost surely. The prior possesses the Kullback-Leibler property in virtue of the fact that p_θ is equal to the uniform density if $\theta = 0$ and $\theta \mapsto p_\theta$ is smooth; in fact $K(p_0; p_\theta) = \theta$, for all $\theta \in [0, 1]$.

Intuitively the inconsistency is caused by the overly large number of densities at distance one from p_0. Because $\int \psi(x) p_N(x)\, dx \to \int \psi(x) p_0(x)\, dx$ as $N \to \infty$, for any bounded continuous function ψ and any $p_N \in \mathcal{P}_N$, the distance of $\mathcal{P}_N$ to p_0 for the weak topology actually tends to zero. This explains that the posterior can be consistent relative to the weak topology, as it should be since p_0 possesses the Kullback-Leibler property.

It remains to be shown that $\limsup \Pi_n(\mathcal{P}_0|\, X_1, \ldots, X_n) = 0$, almost surely. For $N^2 \geq n$ and any given observed values $X_1, \ldots, X_n$, there are at least $\binom{2N^2-n}{N^2}$ densities $p \in \mathcal{P}_N$ such that $\prod_{i=1}^n p(X_i) = 2^n$. Indeed, this value of the likelihood pertains if all observations fall in a bin with ordinate value 2; since the n observations fall in maximally n different bins, this leaves at least $2N^2 - n$ bins from which to choose the N^2 bins with value 0; each choice gives a density in $\mathcal{P}_N$. It follows that

$$\int_{\cup_{N>0}\mathcal{P}_N} \prod_{i=1}^n p(X_i)\, d\Pi(p) \geq 2^n \sum_{N \geq \sqrt{n}} \frac{1}{2} \frac{6}{\pi^2 N^2} \binom{2N^2 - n}{N^2} \Big/ \binom{2N^2}{N^2}$$

$$\geq 2^n \frac{3}{\pi^2} \sum_{N \geq \sqrt{n}} \frac{1}{N^2} \frac{1}{2^n} \gtrsim \frac{1}{\sqrt{n}}.$$

On the other hand, for c the norming constant of the prior,

$$\int_{\mathcal{P}_0} \prod_{i=1}^n p(X_i)\, d\Pi(p) = c \int_0^1 e^{-n\theta + \sqrt{2\theta} \sum_{i=1}^n \Phi^{-1}(X_i)} e^{-1/\theta}\, d\theta.$$

As $-n\theta - 1/\theta \leq -2\sqrt{n}$ for all n and $0 \leq \theta \leq 1$, and $\sum_{i=1}^n \Phi^{-1}(X_i) < 0$ infinitely often a.s. $[P_0^\infty]$ by the law of the iterated logarithm, the right-hand side of the display is at most $e^{-2\sqrt{n}}$ infinitely often a.s. $[P_0^\infty]$. The quotient of the two integrals on the left gives the relative mass that the posterior assigns to $\cup_{N>0}\mathcal{P}_N$ and $\mathcal{P}_0$, respectively. The claim follows, as the lim sup of the quotient is ∞ almost surely.

7.2.1 Normal Mixtures

Let $p_{F,\sigma} = \phi_\sigma * F$, for ϕ_σ the density of the normal distribution with mean 0 and variance σ^2, and consider the prior on densities induced by equipping σ and F with independent priors $F \sim \Pi$ and $\sigma \sim \mu$. Theorem 7.3 gives conditions for the Kullback-Leibler property.

Theorem 7.13 *If for every $\delta, \epsilon > 0$ there exists sequences a_n and $\sigma_n < S_n$ and a constant $C > 0$ such that $n^{-1} \log\log(S_n/\sigma_n) \to 0$ and*

$$\frac{a_n}{\sigma_n} < n\delta, \qquad \Pi(F\colon F[-a_n, a_n]^c > \epsilon) < e^{-Cn}, \qquad \mu([\sigma_n, S_n]^c) \leq e^{-Cn},$$

then the posterior distribution $\Pi_n(\cdot| X_1, \ldots, X_n)$ *for* $p_{F,\sigma}$ *in the model* $X_1, \ldots, X_n | (F, \sigma) \stackrel{iid}{\sim} p_{F,\sigma}$, *for* $(F, \sigma) \sim \Pi \times \mu$, *is strongly consistent relative to the total variation norm at every* p_0 *in the Kullback-Leibler support of the prior of* $p_{F,\sigma}$. *In particular, for* $\Pi = \mathrm{DP}(\alpha)$ *and* μ *a distribution with finite moment consistency holds if one of the following conditions holds:*

(i) α *has compact support and* $\mu(\sigma < h) \le e^{-c/h}$, *for* $h \downarrow 0$.
(ii) α *has sub-Gaussian tails and* $\mu(\sigma < h) \le e^{-c/h^2}$, *for* $h \downarrow 0$.

Proof Given $\epsilon' > 0$, let a_n, σ_n, S_n and C be as in the assumption for $\epsilon = \epsilon'/6$ and $\delta = (\epsilon')^3/(100 \log(60/\epsilon'))$. We apply Theorem 6.23 with $\mathcal{P}_{n,1}$ the set of mixtures $\phi_\sigma * F$ when $\sigma \in [\sigma_n, S_n]$ and $F[-a_n, a_n] \ge 1 - \epsilon$, and define $\mathcal{P}_{n,2}$ the complementary set of mixtures. Clearly

$$\Pi(\mathcal{P}_{n,2}) \le \mu([\sigma_n, S_n]^c) + \Pi(F\colon F[-a_n, a_n]^c > \epsilon) \le 2e^{-Cn}.$$

We shall show that the $\mathbb{L}_1$-entropy of $\mathcal{P}_{n,1}$ at ϵ' is bounded above by $n(\epsilon')^2/4$.

Fix an integer $N \sim a_n/(\sigma_n \epsilon)$, and a regular partition of $[-a_n, a_n]$ into N adjacent intervals $E_1, \ldots, E_N$ of equal length, with midpoints $\theta_1, \ldots, \theta_N$. Construct a set S^* of points in $(\sigma_n, S_n]$ by choosing an $\epsilon 2^{k-1}$-net in $(\sigma_n 2^{k-1}, \sigma_n 2^k]$, for $k = 1, 2, \ldots, \lceil \log_2(S_n/\sigma_n) \rceil$. Let $\mathcal{P}^*$ be the set of all densities of the form

$$\phi_{\sigma^*} * \sum_{i=1}^N p_i^* \delta_{\theta_i},$$

when σ^* ranges over S^*, and $p^* = (p_1^*, \ldots, p_N^*)$ ranges over an ϵ-net P^* over the N-dimensional unit simplex equipped with the ℓ_1-norm. The number of points in $\mathcal{P}^*$ is equal to $\#S^* \#P^*$, and hence $\log \#\mathcal{P}^* \le \log \log_2(S_n/\sigma_n) + \log(3/\epsilon) + N \log(5/\epsilon)$, in view of Proposition C.1. The parameter δ has been defined so that $N \log(5/\epsilon)$ is bounded above by $n(\epsilon')^2/8$, under the condition $a_n/\sigma_n < n\delta$. Furthermore, $\log(2/\epsilon\sigma_n) + \log(2S_n/\epsilon) \ll n$ by assumption, as $n \to \infty$.

For any $\sigma \in [\sigma_n, S_n]$ and arbitrary F, there exists $\sigma^* \in S^*$ and $p^* \in P^*$ such that $|\sigma - \sigma^*| < \epsilon \sigma_n 2^{k-1}$ if $\sigma \in [\sigma_n 2^{k-1}, \sigma_n 2^k]$, and $\sum_{i=1}^N |p_i^* - F(E_i)/F[-a_n, a_n]| < \epsilon$. If $F[-a_n, a_n] \ge 1 - \epsilon$, then also $\sum_{i=1}^N |p_i^* - F(E_i)| < 2\epsilon$. Because $\|\phi_{\sigma_1} - \phi_{\sigma_2}\|_1 \le 2|\sigma_1 - \sigma_2|/(\sigma_1 \wedge \sigma_2)$ and $\|\phi_\sigma * \delta_{\theta_1} - \phi_\sigma * \delta_{\theta_2}\|_1 \le |\theta_1 - \theta_2|/\sigma$, for any $\sigma_1, \sigma_2, \sigma, \theta_1, \theta_2$, we have

$$\|\phi_\sigma * F - \phi_{\sigma^*} * F\|_1 \le \frac{2|\sigma - \sigma^*|}{\sigma \wedge \sigma^*},$$

$$\Big\|\phi_{\sigma^*} * F - \phi_{\sigma^*} * \sum_{i=1}^N F(E_i)\delta_{\theta_i}\Big\|_1 \le F[-a_n, a_n]^c + \sum_{i=1}^N \int_{E_i} \|\phi_{\sigma^*} * \delta_\theta - \phi_{\sigma^*} * \delta_{\theta_i}\|_1 \, dF(\theta),$$

$$\Big\|\phi_{\sigma^*} * \sum_{i=1}^N F(E_i)\delta_{\theta_i} - \phi_{\sigma^*} * \sum_{i=1}^N p_i^* \delta_{\theta_i}\Big\|_1 \le \sum_{i=1}^N |F(E_i) - p_i^*|.$$

By the triangle inequality it follows that every element of $\mathcal{P}_{n,1}$ is approximated within $\mathbb{L}_1$-distance $6\epsilon < \epsilon'$ by some element of $\mathcal{P}^*$.

By Markov's inequality and by Proposition 4.2,

$$\mathrm{DP}_\alpha(F\colon F[-a_n, a_n]^c > \epsilon) \leq \epsilon^{-1}\bar\alpha([-a_n, a_n]^c).$$

If μ has a finite first moment, then $\mu([S_n, \infty)) \leq e^{-Cn}$ for $S_n \gtrsim e^{Cn}$, and $\log\log S_n = o(n)$. If α has compact support, then a_n can be chosen equal to a fixed, large constant, and $a_n/\sigma_n < n\delta$ if $\sigma_n \gtrsim 1/n$. The bound $\mu([0, \sigma_n)) \leq e^{-Cn}$ is then implied by $\mu([0, h)) \leq e^{-c/h}$. If α has sub-Gaussian tails, then a_n can be chosen of the order $\sqrt{n}$, and σ_n of the order $1/\sqrt{n}$, and a similar argument yields the stated bound. □

Example 7.14 (Inverse-gamma) The conjugate *inverse-gamma* prior $\sigma^{-2} \sim \mathrm{Ga}(a, b)$ satisfies $\mathrm{E}\sigma < \infty$ for $a > 1$, and $\mathrm{P}(\sigma < h) \sim e^{-b/h^2} h^{-2a+2}$. Thus both conditions (i) and (ii) are satisfied if $a > 1$.

For a multivariate generalization of the result, with matrix-valued bandwidth, see Problem 7.10. For Dirichlet mixtures, see also Section 7.2.2 and Section 9.4.

7.2.2 Dirichlet Process Mixtures of a General Kernel

Let $p_{F,\varphi}(x) = \int \psi(x; \theta, \varphi)\, dF(\theta)$ for a given family of probability densities $x \mapsto \psi(x; \theta, \varphi)$, indexed by Euclidean parameters $\theta \in \Theta \subset \mathbb{R}^k$ and $\varphi \in \Phi \subset \mathbb{R}^l$. Equip F with the Dirichlet process prior and φ by some other prior. The following theorem shows that consistency relative to the $\mathbb{L}_1$-norm holds under mild conditions as soon as the true density is in the Kullback-Leibler support of the prior. Section 7.1.2 gives examples of the latter.

Theorem 7.15 *If for any given $\epsilon > 0$ and n, there exist subsets $\Theta_n \subset \mathbb{R}^k$ and $\Phi_n \subset \mathbb{R}^l$ and constants $a_n, A_n, b_n, B_n > 0$ such that*

(i) $\|\psi(\cdot; \theta, \varphi) - \psi(\cdot; \theta', \varphi')\|_1 \leq a_n\|\theta - \theta'\| + b_n\|\varphi - \varphi'\|$, *for all* $\theta, \theta' \in \Theta_n$ *and* $\varphi, \varphi' \in \Phi_n$,
(ii) $\mathrm{diam}(\Theta_n) \leq A_n$ *and* $\mathrm{diam}(\Phi_n) \leq B_n$,
(iii) $\log(a_n A_n) \leq C \log n$ *for some* $C > 0$, *and* $\log(b_n B_n) \leq n\epsilon^2/(8l)$,
(iv) $\max(\bar\alpha(\Theta_n^c), \pi(\Phi_n^c)) \leq e^{-Cn}$, *for some* $C > 0$,

then the posterior distribution $\Pi_n(\cdot| X_1, \ldots, X_n)$ for $p_{F,\varphi}$ in the model $X_1, \ldots, X_n| (F, \varphi) \stackrel{iid}{\sim} p_{F,\varphi}$, for $(F, \varphi) \sim \mathrm{DP}(\alpha) \times \pi$, is strongly consistent relative to the total variation norm at every p_0 in the Kullback-Leibler support of the prior of $p_{F,\varphi}$.

Proof We apply Theorem 6.23 with d equal to the $\mathbb{L}_1$-distance divided by 2. For given $\epsilon > 0$ we choose $\mathcal{P}_{n,1}$ equal to, with $N \sim n\delta/\log n$ and δ sufficiently small, to be determined,

$$\mathcal{P}_{n,1} = \Big\{\sum_{j=1}^{\infty} w_j \psi(\cdot; \theta_j, \varphi)\colon (w_j) \in \mathbb{S}_\infty, \sum_{j>N} w_j < \frac{\epsilon}{8}, \theta_1, \ldots, \theta_N \in \Theta_n, \varphi \in \Phi_n\Big\}.$$

By the series representation $F = \sum_j W_j \delta_{\theta_j}$ of the Dirichlet process, given in Theorem 4.12, the prior density $p_{F,\varphi}$ is contained in $\mathcal{P}_{n,1}$, unless the weights of the representation satisfy

$\sum_{j>N} W_j \geq \epsilon/8$, or at least one of $\theta_1, \ldots, \theta_N \stackrel{\text{iid}}{\sim} \bar{\alpha}$ falls outside Θ_n, or $\varphi \notin \Phi_n$. It follows that

$$\Pi(\mathcal{P}_{n,1}^c) \leq \mathrm{P}\Big(\sum_{j>N} W_j \geq \frac{\epsilon}{8}\Big) + N\bar{\alpha}(\Theta_n^c) + \pi(\Phi_n^c).$$

The last two terms are exponentially small by assumption and the choice of N. The stick-breaking weights in the first term satisfy $W_j = V_j \prod_{l=1}^{j-1}(1 - V_l)$, for $V_l \stackrel{\text{iid}}{\sim} \mathrm{Be}(1, |\alpha|)$ and $\sum_{j>N} W_j = \prod_{j=1}^{N}(1 - V_j)$. Since $-\log(1 - V_l)$ possesses an exponential distribution, $R_N := -\log \sum_{j>N} W_j$ is gamma distributed with parameters N and $|\alpha|$, and hence $\mathrm{P}(R_N < r) \leq (|\alpha| r)^N/N! \leq (e|\alpha| r/N)^N$. Therefore, the first term is bounded above by $(e|\alpha|\log(8/\epsilon)/N)^N$, which is also exponentially small, by choice of N, for any $\delta > 0$

The functions of the form $\sum_{j=1}^{N} w_j \psi(\cdot; \theta_j, \varphi)$ with $(w_1, \ldots, w_N) \in \mathbb{S}_N$ form an $\epsilon/4$-net over $\mathcal{P}_{n,1}$ for the $\mathbb{L}_1$-norm. To construct an $3\epsilon/4$-net over these finite sums we restrict $(w_1, \ldots, w_N)$ to an $\epsilon/4$-net over $\mathbb{S}_N$, restrict $(\theta_1, \ldots, \theta_N)$ to an $\epsilon/(4a_n)$-net over Θ_n and φ to an $\epsilon/(4b_n)$-net over Φ_n. The cardinality of such a net is bounded above by

$$\Big(\frac{20}{\epsilon}\Big)^N \times \Big(\frac{12A_n a_n}{\epsilon}\Big)^{kN} \times \Big(\frac{12B_n b_n}{\epsilon}\Big)^l.$$

By the triangle inequality, it follows that $\log N(\epsilon, \mathcal{P}_{n,1}, \|\cdot\|_1) \leq n\epsilon^2$, provided δ is small enough. ☐

Conditions of the theorem typically hold for a location kernel with an additional scale parameter if the prior density of the scale parameter has a thin tail at zero and the base measure of the Dirichlet process has a sufficiently thin tail. In particular, consistency for Dirichlet process mixtures of a normal kernel can be derived from this theorem. The result can however go much beyond location kernels.

7.2.3 Pólya Tree Process

For parameters satisfying $\sum_{m=1}^{\infty} a_m^{-1} < \infty$, the canonical Pólya tree process $\mathrm{PT}^*(\lambda, a_m)$ possesses the Kullback-Leibler property at densities with finite Kullback-Leibler divergence relative to λ (see Section 7.1.1). Under a much more restrictive condition the posterior is consistent with respect to the total variation distance.

Theorem 7.16 *If $\sum_{m>k} a_m^{-1} \leq C2^{-k}$ for some constant C and every large k, then the posterior distribution $\Pi_n(\cdot|\, X_1, \ldots, X_n)$ in the model $X_1, \ldots, X_n|\, p \stackrel{iid}{\sim} p$ and $p \sim \mathrm{PT}^*(\lambda, a_m)$ is strongly consistent with respect to the total variation distance at every p_0 with $K(\lambda; p_0) < \infty$.*

Proof Since $p_0 \in \mathrm{KL}(\mathrm{PT}^*(\lambda, a_m))$ by Theorem 7.1, it suffices to verify conditions (i) and (ii) of Theorem 6.23. The random density p generated by the Pólya tree process can be represented as in (3.18), where $x_1 x_2 \cdots$ is the expansion of x relative to the partition, the $V_{\varepsilon 0}$ are independent $\mathrm{Be}(a_{|\varepsilon|}, a_{|\varepsilon|})$ variables, and $V_{\varepsilon 1} = 1 - V_{\varepsilon 0}$, for $\varepsilon \in \mathcal{E}^*$. Write p_k for the corresponding finite products $p_k(x) = \prod_{j=1}^{k}(2V_{x_1 \cdots x_j})$. For given $k = k_n \in \mathbb{N}$, consider the partition of the model in the sets $\mathcal{P}_{n,1} = \{p\colon \|\log(p/p_k)\|_\infty \leq \epsilon^2/8\}$ and their complements $\mathcal{P}_{n,2}$.

Because p_k is constant within every kth level partitioning set, we have

$$\big|\log p(x) - \log p(y)\big| \le 2\big\|\log(p/p_k)\big\|_\infty,$$

whenever x and y belong to the same partitioning set. Thus the modulus $\Delta \log p$ of every $p \in \mathcal{P}_{n,1}$ relative to the kth level partition is bounded above by $\epsilon^2/4$. Because $\epsilon^2/4 + \sqrt{2\epsilon^2/4} < \epsilon$, we conclude by Corollary C.3 that $\log N(\epsilon, \mathcal{P}_{n,1}, \|\cdot\|_1) \le 2^k C_\epsilon$ for some positive number C_ϵ not depending on k. Since the squared Hellinger distance is bounded by the $\mathbb{L}_1$-distance, the entropy relative to the Hellinger distance satisfies the same inequality, with a different constant C_ϵ. Condition (i) of Theorem 6.23 is satisfied provided k is chosen to satisfy $2^k \le D_\epsilon n$, for a specific constant D_ϵ that depends on ϵ only.

By Lemma G.12 the variables $\log(2V_{x_1,\ldots,x_j})$ have mean bounded in absolute value by a multiple of a_j^{-1} and Orlicz ψ_2-norm bounded by $a_j^{-1/2}$. The second implies that $\sum_{j>k}\big[\log(2V_{x_1\cdots x_j}) - \mathrm{E}(\log(2V_{x_1\cdots x_j}))\big]$ possesses Orlicz ψ_2-norm bounded by a multiple of the square root of $\sum_{j>k} a_j^{-1}$, which is bounded by $C2^{-k}$ by assumption. (See Lemma 2.2.1 and Proposition A.1.6 in van der Vaart and Wellner 1996) for the relevant results on Orlicz norms.) It follows that for $2^{-k} \lesssim \epsilon^2/16$ there exists a constant D such that

$$\Pi(\mathcal{P}_{n,2}) = \mathrm{P}\Big(\Big|\sum_{j>k}\log(2V_{x_1\cdots x_j})\Big| > \frac{\epsilon^2}{8}\Big) \le e^{-D\epsilon^4 2^k}.$$

Therefore condition (ii) of Theorem 6.23 is satisfied for $2^k \gtrsim n$. In particular this is true for the maximal value $2^k \sim D_\epsilon n$ found in the preceding paragraph. □

The exponential growth condition for the parameters a_m of the preceding theorem seems a bit too strong for practical use. In Section 6.8.5 the posterior mean was seen to be consistent under only the condition $\sum_m a_m^{-1} < \infty$ which is needed even for absolute continuity.

7.2.4 Exponential Densities

In Theorem 7.10 a prior on a density $p = p_f$ of the form (7.3) induced by a prior Π on f is seen to possess the Kullback-Leibler property as soon as $\log p_0$ belongs to the uniform support of Π. Because the Hellinger distance $d_H(p_f, p_g)$ is bounded by a continuous function of $\|f - g\|_\infty$ (see Lemma 2.5), the entropy condition of Theorem 6.23 similarly is implied by an entropy condition on the functions f under the uniform norm.

Theorem 7.17 *If Π is the distribution of a stochastic process $(f(x): x \in [0,1]^d)$ with sample paths belonging to $\mathfrak{C}^\alpha([0,1]^d)$ and with $\mathrm{P}(\|f\|_{\mathfrak{C}^\alpha} > M) \le e^{-CM^r}$, for some constants C, r and $\alpha \ge d/r$, then the posterior distribution $\Pi_n(\cdot| X_1, \ldots, X_n)$ for p in the model $X_1, \ldots, X_n| p \stackrel{iid}{\sim} p$, $p = p_f$, $f \sim \Pi$ is strongly consistent at every p_0 such that* $\log p_0$ *belongs to the uniform support of* Π.

Proof The uniform entropy $\log N(\epsilon, \mathcal{F}_n, \|\cdot\|_\infty)$ of the set of functions $\mathcal{F}_n = \{f: \|f\|_{\mathfrak{C}^\alpha} \le cn^{1/r}\}$ is bounded above by a multiple of $(cn^{1/r}/\epsilon)^{d/\alpha}$. Hence for given ϵ and c' it is bounded above by $c'n\epsilon^2$, if c is chosen sufficiently small. In view of Lemma 2.5 the ϵ-entropy of the induced set of densities $\{\bar p_f: f \in \mathcal{F}_n\}$ relative to the Hellinger distance is bounded above the 2ϵ-entropy of $\mathcal{F}_n$ relative to the uniform norm. By assumption $\Pi(\mathcal{F}_n^c) \le e^{-Cc^r n}$. The theorem follows from Theorem 6.23. □

Example 7.18 (Gaussian process) According to *Borell's inequality* a Gaussian random element f in a separable Banach space satisfies $\mathrm{P}(\|f\| > M) \le 2e^{-M^2/2}$. Thus if the prior process f is a Gaussian, Borel measurable map in a separable subspace of $\mathfrak{C}^\alpha([0,1]^d)$, then the preceding theorem applies with $r = 2$. For a d-dimensional domain its condition is then that the sample paths of f are $d/2$-smooth ($\alpha \ge d/2$).[1]

Example 7.19 (Exponential family) Let Π be the distribution of $f := \sum_j \theta_j \psi_j$ for given basis functions $\psi_j : [0,1]^d \to \mathbb{R}$ and independent random variables θ_j. If the basis functions are contained in $\mathfrak{C}^\alpha([0,1]^d)$ with norms b_j, then $\|\sum_j \theta_j \psi_j\|_{\mathfrak{C}^\alpha} \le \sum_j |\theta_j| b_j$. Thus Π concentrates on $\mathfrak{C}^\alpha([0,1]^d)$ as soon as $\sum_j \mathrm{E}|\theta_j| b_j < \infty$. The tail condition is certainly verified if $\mathrm{P}(\sum_j |\theta_j| b_j > M) \le e^{-CM^r}$, but due to cancellation of positive and negative terms in the series $\sum_j \theta_j \psi_j$, this condition may be pessimistic.

For instance, if $\theta_j \overset{\text{ind}}{\sim} \mathrm{Nor}(0, \tau_j^2)$, then the process is Gaussian, and satisfies a tail bound with $r = 2$ as soon as $\sum_j \tau_j b_j < \infty$.

For $d = 1$ and the trigonometric basis $\psi_0(x) = 1$, and $\psi_j(x) = \sqrt{2}\cos(j\pi x)$ for $j \in \mathbb{N}$, we can choose $b_0 = 1$ and $b_j \sim j^\alpha$, for $\alpha \in (0,1]$. For Gaussian variables we choose $\alpha = 1/2$ and obtain the condition that $\sum_j \tau_j \sqrt{j} < \infty$. E.g. $\tau_j \sim j^{-3/2-\delta}$ works for any $\delta > 0$. For further discussion of Gaussian priors, see Chapter 11, or Section 6.8.4.

7.3 Other Nonparametric Models

In this section we consider a few other nonparametric function estimation problems and investigate conditions for consistency in appropriate distance measures.

7.3.1 Nonparametric Binary Regression

In nonparametric binary regression the relationship between a binary response $Y \in \{0,1\}$ and a covariate X is modeled as

$$\mathrm{P}_f(Y = 1 \mid X = x) = H(f(x)),$$

for a known "link function" H and an unknown function f that ranges over a large model $\mathcal{F}$. We assume that H is strictly monotone and continuously differentiable, with bounded derivative h. The logistic function is the most important example of a link function and is also the most convenient to deal with.

We consider two versions of the problem, both with as complete set of observations a sequence of pairs $(X_1, Y_1), \ldots, (X_n, Y_n)$. In *random design* these are a random sample from a fixed distribution P_f, determined by the marginal distribution G of the covariates and the conditional Bernoulli distribution $Y_i \mid X_i \sim \mathrm{Bin}(1, H(f(X_i)))$ of the response. In *fixed design* the covariates are given and only the responses are random; they are assumed independent, with Bernoulli distributions as before.

In both cases we construct a prior on the distribution of the observations by endowing f with a prior Π.

The random design model is a special case of the density model of Section 6.4. Theorem 6.23 translates into the following result.

[1] Sufficient is that its sample paths are in $\mathfrak{C}^\beta([0,1]^d)$, for some $\beta > \alpha$; see Lemma I.7.

Theorem 7.20 (Random design) *If for every $\epsilon > 0$ there exists a partition $\mathcal{F} = \mathcal{F}_{n,1} \cup \mathcal{F}_{n,2}$ and a constant $C > 0$, such that* $\log N(\epsilon, \mathcal{F}_{n,1}, \mathbb{L}_2(G)) \le \|h\|_\infty^2 n\epsilon^2$, *and* $\Pi(\mathcal{F}_{n,2}) \le e^{-Cn}$, *then the posterior distribution is strongly consistent relative to the distance $d(p_f, p_g) = \|H(f) - H(g)\|_{G,2}$ at every p_{f_0} in the Kullback-Leibler support of the prior. For the logistic link function the latter is true for every f_0 in the $\mathbb{L}_2(G)$-support of Π. For a general link function it suffices that f_0 is uniformly bounded and is in the uniform support of Π.*

Proof We apply Theorem 6.23 with d equal to $(1/2)$ times the $\mathbb{L}_2$-distance on the densities p_f, which is bounded by the Hellinger distance, as the densities p_f are bounded by 1 (see Lemma B.1). By Lemma 2.8(i) this distance is equivalent to the distance $\|H(f) - H(g)\|_{2,G}$ of the theorem, and bounded above by $\|h\|_\infty$ times the $\mathbb{L}_2(G)$-distance on the regression functions. Thus (i)–(ii) of Theorem 6.23 are verified by the present conditions.

By Lemma 2.8 the Kullback-Leibler divergence between the densities p_f is bounded above the $\mathbb{L}_2(G)$-distance between the functions f, always in case of the logistic link function, and if the functions f are bounded for general link functions. This yields the Kullback-Leibler property. □

The fixed design model can be handled by Theorem 6.41. Let $G_n = n^{-1}\sum_{i=1}^n \delta_{X_i}$ be the empirical distribution of the design points.

Theorem 7.21 (Fixed design) *If for every $\epsilon > 0$ there exist partitions $\mathcal{F} = \mathcal{F}_{n,1} \cup \mathcal{F}_{n,2}$ and a constant $C > 0$, such that* $\log N(\epsilon, \mathcal{F}_{n,1}, \mathbb{L}_2(G_n)) \le 3\|h\|_\infty^2 n\epsilon^2$ *and $\Pi(\mathcal{F}_{n,2}) \le e^{-Cn}$, then $\Pi_n(\|H(f) - H(g)\|_{2,G_n} > \epsilon \mid Y_1, \ldots, Y_n) \to 0$ in probability for every $\epsilon > 0$ in the random design model with regression function f_0, for any f_0 such that* $\liminf \Pi_n(\|f - f_0\|_{2,G_n} < \epsilon) > 0$ *in case of the logistic link function, and such that* $\liminf \Pi_n(\|f - f_0\|_\infty < \epsilon) > 0$ *in general.*

Proof By Lemma 2.8(ii), the root average square Hellinger distance $d_{n,H}$ on the (multivariate Bernoulli) density of $(Y_1, \ldots, Y_n)$ is bounded by the $\mathbb{L}_2(G_n)$-distance on the functions f. Thus (i)–(ii) of Theorem 6.41 are verified by the present conditions. Similarly by Lemma 2.8(iii)–(iv) the Kullback-Leibler divergence and Kullback-Leibler variation are bounded above by a multiple of the square $\mathbb{L}_2(G_n)$-distance on the functions f, which allows to verify the first condition of Theorem 6.41. The theorem gives consistency for $d_{n,H}$, but then also for the root average square $\mathbb{L}_2$-distance, as this is bounded by a multiple of the root average square Hellinger distance. This distance is equivalent to the distance $\|H(f) - H(g)\|_{2,G_n}$. □

The marginal distribution G or G_n in the consistency assertion can be removed in favor of a neutral, not problem-dependent measure, such as the Lebesgue measure, if the former is sufficiently regular (see e.g. Problem 7.11). This may not be true of link function in the assertion: estimating f in its tails may be harder than estimating it in the middle as the distributions of the observations depends on f through $H(f)$ and thus is insensitive to perturbations of f in its tails.

Gaussian processes provide concrete examples of priors. They are discussed in detail in Chapter 11.

7.3.2 Nonparametric Regression with Normal Errors

In the nonparametric regression problem the observations are pairs $(X_1, Y_1), \dots, (X_n, Y_n)$ that follow the model, for an unknown function f and mean zero errors $\varepsilon_1, \dots, \varepsilon_n$,

$$Y_i = f(X_i) + \varepsilon_i.$$

We consider that the errors are a random sample from a normal distribution Nor$(0, \sigma^2)$, and only equip the regression function f with a prior. In particular we treat the standard deviation σ of the errors as known; modification to unknown σ is possible for light-tailed priors.

We restrict to the *random design* model, where the observations (X_i, Y_i) are i.i.d.; the fixed design case is similar. The marginal distribution G of $X_1, \dots, X_n$ cancels from the posterior distribution for f, even if it were equipped with an (independent) prior, and hence can be assumed known without loss of generality.

Let $[f]_M := (f \vee -M) \wedge M$ be f truncated to the interval $[-M, M]$.

Theorem 7.22 *If for every $\epsilon > 0$ there exist partitions $\mathcal{F} = \mathcal{F}_{n,1} \cup \mathcal{F}_{n,2}$ and a constant $C > 0$ such that $\log N(2\sigma\epsilon, \mathcal{F}_{n,1}, \|\cdot\|_{2,G}) \le n\epsilon^2$ and $\Pi(\mathcal{F}_{n,2}) \le e^{-cn}$, then $\Pi_n(f{:}\, \|[f]_M - [f_0]_M\|_{2,G}) > \epsilon | (X_1, Y_1), \dots, (X_n, Y_n)) \to 0$ a.s. $[P_{f_0}^{\infty}]$, for every M, for any f_0 that belongs to the $\mathbb{L}_2(G)$-support of Π.*

Proof The inequality $K(\phi_{\theta_1}; \phi_{\theta_2}) \lesssim |\theta_1 - \theta_2|^2$ for the Kullback-Leibler divergence between two Nor(θ_i, σ^2)-densities ϕ_{θ_1} and ϕ_{θ_2}, implies $K(p_{f_0}; p_f) \le \|f_0 - f\|_{2,G}^2$ for the density p_f of (X, Y). Therefore, the Kullback-Leibler property of p_{f_0} follows from the assumption that f_0 is in the $\mathbb{L}_2(G)$-support of Π.

Similarly the inequality $d_H(\phi_{\theta_1}; \phi_{\theta_2}) \le \frac{1}{2}|\theta_1 - \theta_2|/\sigma$ implies that $d_H(p_f, p_g) \le \frac{1}{2}\|f - g\|_{2,G}/\sigma$, whence $N(\epsilon, \{p_f{:}\, f \in \mathcal{F}_{n,1}\}, d_H) \le N(2\sigma\epsilon, \mathcal{F}_{n1}, \|\cdot\|_{2,G})$. Consequently, the entropy condition of Theorem 6.23 is implied by the present condition.

We conclude that the posterior distribution is consistent relative to the Hellinger distance. Finally $\|[f]_M - [f_0]_M\|_{G,2} \lesssim d_H(f, f_0)$. ☐

7.3.3 Spectral Density Estimation

The second order moments of a stationary time series $(X_t{:}\, t \in \mathbb{Z})$ are completely described by its *spectral density* defined by, for $\gamma(h) = \mathrm{cov}(X_{t+h}, X_t)$ the autocovariance function,

$$f(\omega) = \frac{1}{2\pi} \sum_{h=-\infty}^{\infty} \gamma(h) e^{-ih\omega}, \qquad 0 \le \omega \le \pi.$$

This may be estimated from data $X_1, \dots, X_n$ by (weighted) averages of the *periodogram* I_n, defined as $I_n(\omega) = (2\pi n)^{-1} \big| \sum_{t=1}^n X_t e^{-it\omega} \big|^2$. Under general conditions the periodogram $I_n(\omega)$ at a frequency $\omega \in (0, \pi)$ tends in distribution to an exponential distribution with mean $f(\omega)$, and periodogram values at sufficiently separated frequencies are asymptotically independent. Furthermore, the periodogram values $U_l = I_n(\omega_l)$ at the *natural frequencies* (or *Fourier frequencies*) $\omega_l = 2\pi l/n$, for $l = 1, \dots, \nu = \lfloor (n-1)/2 \rfloor$, are asymptotically distributed as independent exponential variables with means $f(\omega_l)$. This motivates the

Whittle likelihood, which is an approximate likelihood for f based on the assumption that the variables $U_1, \ldots, U_\nu$ are exactly exponentially distributed:

$$L_n(f \mid X_1, \ldots, X_n) = \prod_{l=1}^{\nu} \frac{1}{f(\omega_l)} e^{-U_l/f(\omega_l)}, \tag{7.4}$$

If the time series is Gaussian with $\sum_{r=0}^{\infty} r^\alpha |\gamma(r)| < \infty$ for some $\alpha > 0$, and the spectral density is bounded away from 0, then by Theorem L.8 the sequence of true distributions of $(U_1, \ldots, U_\nu)$ and the sequence of distributions under the assumption that the U_i are independent exponential variables with means $f(\omega_l)$ are mutually *contiguous*. This means that any convergence in probability, such as consistency, automatically holds under both true and "Whittle" distributions, if it holds under one of them.

Here we consider Bayesian estimation using a prior on f and forming the posterior distribution $\Pi_n(\cdot \mid U_1, \ldots, U_\nu)$ with the Whittle likelihood in place of the true likelihood. As a metric on the spectral densities consider

$$d_n(f, g) = \sqrt{\frac{1}{\nu} \sum_{l=1}^{\nu} \frac{(\sqrt{f(\omega_l)} - \sqrt{g(\omega_l)})^2}{f(\omega_l) + g(\omega_l)}}.$$

Theorem 7.23 *If for every $\epsilon > 0$ there exist partitions $\mathcal{F} = \mathcal{F}_{n,1} \cup \mathcal{F}_{n,2}$ and a constant $C > 0$ such that $\log N(\epsilon, \mathcal{F}_{n,1}, d_n) \le n\epsilon^2$ and $\Pi(\mathcal{F}_{n,2}) \le e^{-cn}$, then $\Pi_n(f: d_n(f, f_0) > \epsilon \mid U_1, \ldots, U_\nu) \to 0$ a.s. under every f_0 that is bounded away from 0 and ∞ and belongs to the $\|\cdot\|_\infty$-support of the prior Π. This is true both if $U_1, \ldots, U_\nu$ are independent exponential variables with means $f(\omega_l)$, and if these variables are the periodogram at the natural frequencies of a stationary Gaussian times series with $\sum_{h=0}^{\infty} h^\alpha |\gamma(h)| < \infty$, for some $\alpha > 0$.*

Proof The final assertion follows from Theorem L.8; it suffices to prove the theorem for $(U_1, \ldots, U_\nu)$ exponential variables. For p_θ the density of the exponential distribution with mean θ,

$$\begin{aligned}
d_H(p_{\theta_1}, p_{\theta_2}) &= \frac{\sqrt{2}|\sqrt{\theta_1} - \sqrt{\theta_2}|}{\sqrt{\theta_1 + \theta_2}}, \\
K(p_{\theta_0}; p_\theta) &= -\log \frac{\theta_0}{\theta} - 1 + \frac{\theta_0}{\theta} \le \frac{|\theta - \theta_0|^2}{(\theta \wedge \theta_0)^2}, \\
V_{2,0}(p_{\theta_0}; p_\theta) &= \frac{|\theta - \theta_0|^2}{\theta^2}.
\end{aligned}$$

From the second and third, it follows that if $\|f - f_0\|_\infty < \epsilon$ and f_0 is bounded away from zero, then the corresponding distributions $P_{f,n}$ of $(U_1, \ldots, U_\nu)$ possess Kullback-Leibler divergence and variation $K(P_{f_0,n}; P_{f,n})$ and $V_{2,0}(P_{f_0,n}; P_{f,n})$ bounded by a multiple of ϵ^2. The first inequality shows that the root average square Hellinger distance on the distributions $P_{f,n}$ is equal to the distance d_n on the spectral densities f. Thus the theorem follows from Theorem 6.41. $\square$

The integral of the spectral density is the variance $\operatorname{var} X_0 = \int f(\omega)\,d\omega$ of the time series. Norming the spectral density to a probability density entails dividing it by this variance, and gives the Fourier transform of the auto*correlation* function. Consider estimating this function in the case that the variance is known. (As the variance may be estimated by $n^{-1}\sum_{t=1}^n X_t^2$, this also gives a data-dependent prior for estimating the full spectral density.)

Theorem 7.24 *If Π is a prior on $q = f/\|f\|_1$ such that $\Pi(\|q\|_{\mathrm{Lip}} > L_n) \lesssim e^{-Cn}$, for some $C > 0$ and $L_n = o(n)$, then the posterior distribution as in Theorem 7.23 is consistent relative to the $\mathbb{L}_1$-norm on q, at any Lipschitz continuous q_0 that is bounded away from zero and is in the support of Π for the uniform norm.*

Proof That the conditions of Theorem 6.41 concerning $K(P_{f_0,n};\, P_{f,n})$ and $V_{2,0}(P_{f_0,n};\, P_{f,n})$ hold follow as in the preceding proof. Rather than invoking the general construction of tests through the metrics d_n we use a direct argument.

We may assume that $\|q\|_{\mathrm{Lip}} \le L_n$. If Q_n is the probability measure putting equal mass at each of the Fourier frequencies, then $\int q\, dQ_n = 1 + O(L_n/n) = 1 + o(1)$, and $\|q - q_0\|_1 > 2\epsilon$ implies that $\|q - q_0\|_{1,Q_n} > \epsilon$ for sufficiently large n. Thus it follows from part (ii) of Lemma K.9 that $Q_n(q < q_0 - \epsilon) > \epsilon'$ for some $\epsilon' > 0$: the fraction of Fourier frequencies where $f(\omega_j) < f_0(\omega_j) - \epsilon$ is of the order of n. Since this gives a separation of the means under q and q_0 of at least cn of the observations U_j, a test with exponentially small error probabilities can be constructed by Lemma D.10. For all f inside a ball $\{f\colon \|f - f_1\|_\infty < \epsilon\}$ around some f_1, the same set of Fourier frequencies pertains and hence the same test can be used. We can cover $\mathcal{F}_{n,1} = \{f\colon \|q\|_{\mathrm{Lip}} \le L_n\}$ with a multiple of $L_n/\epsilon \ll n$ balls of radius ϵ. Thus the overall error probabilities are of the order $e^{o(n)}e^{-Cn}$. Furthermore, by assumption $\mathcal{F}_{n,2} = \{f\colon \|q\|_{\mathrm{Lip}} > L_n\}$ has exponentially small prior mass. □

Example 7.25 The conditions of Theorem 7.24 hold for a Bernstein polynomial prior on q (with range rescaled to $[0, \pi]$) provided the prior ρ on the index k satisfies $0 < \rho(k) \lesssim e^{-Ck^2 \log k}$. In this case we may choose $\mathcal{F}_{n,1} = \{q\colon k \le \delta\sqrt{n/\log n}\}$, which leads to $L_n = \delta' n/\log n$. The condition is also satisfied for the prior $q \propto e^W$, where W is a Gaussian process with sufficiently regular paths.

7.4 Semiparametric Models

In semiparametric problems, such as location or linear regression problems with unknown error distribution, the Euclidean part of the parameter is usually of most interest. In a proof of consistency of the posterior of this parameter the topology on the functional part of the parameter may then be chosen for convenience. For instance, in the location or regression problems one may work with the weak topology on the error density.

7.4.1 Location Problem

Suppose we observe a random sample from the model $X = \theta + \varepsilon$, where $\varepsilon \sim p$ and p is a probability density with a fixed known quantile. Specifically, we shall consider the more involved case that p is symmetric about zero, but similar results are true if the median or mean of the error is fixed.

The key to consistency is a prior on the error density that has the Kullback-Leibler property, so as to avoid misbehavior as exhibited in Example 6.14. Let $\bar{\Pi}$ be the prior induced by $p \sim \Pi$ on symmetric densities $\bar{p}$ under the map $p \mapsto \bar{p}$, given by $\bar{p}(x) = \frac{1}{2}(p(x)+p(-x))$.

Theorem 7.26 *If $(\theta, p) \sim \mu \times \bar{\Pi}$ for $\theta_0 \in \text{supp}(\mu)$, then the posterior distribution for (θ, p) is consistent at (θ_0, p_0) with respect to the product of the Euclidean and the weak topology if for every $C > 1$ there exist a density r with $p_0 \leq Cr$, $r_\theta \in \text{KL}(\Pi)$ for every sufficiently small $|\theta|$ and $K(r; r_\theta) \to 0$ as $\theta \to 0$. In particular, if the Kullback-Leibler support of Π contains all densities p with $K(p; \lambda) < \infty$ for some density λ, then this is true if one of the following conditions holds:*

(a) $K(p_{0,\theta}; \lambda) < \infty$ *for sufficiently small* $|\theta|$, *and* $K(p_0; p_{0,\theta}) \to 0$ *as* $\theta \to 0$.
(b) p_0 *is continuous with compact support and* $-\int \phi_{\mu,\sigma^2} \log \lambda < \infty$, *for all* (μ, σ^2).

Proof Let Π^* be the prior on the shifted symmetric densities p_θ induced by the prior $(p, \theta) \sim \bar{\Pi} \times \mu$ under the map $(p, \theta) \mapsto p_\theta$. By Proposition A.9 the latter map is continuously invertible relative to the weak topology on p and p_θ and the Euclidean topology on θ. Hence it suffices that the posterior distribution for p_θ is weakly consistent at p_{0,θ_0}. By Example 6.20 this is the case as soon as $p_{0,\theta_0} \in \text{KL}(\Pi^*)$.

If $p_0 \leq Cr$, then $p_{0,\theta_0} \leq Cr_{\theta_0}$, whence it suffices to show that $r_{\theta_0} \in \text{KL}(\Pi^*)$, in view of Proposition 6.34. Now $r_\theta \in \text{KL}(\Pi)$ implies $\overline{r_\theta} \in \text{KL}(\bar{\Pi})$, by Proposition 6.32. This being true for every $\theta \approx 0$ and also $K(r; r_\theta) \to 0$ as $\theta \to 0$ by assumption, the desired result follows from Proposition 6.36.

(a) It suffices to apply the above argument with $r = p_0$.

(b) First assume that the support of p_0 is an interval $[-a, a]$. For $\eta \in (0, 1)$ define $r = \phi_{-a,\sigma^2}\mathbb{1}\{(-\infty, -a)\} + (1 - 2\eta)p_0 + \eta\phi_{-a,\sigma^2}\mathbb{1}\{(a, \infty)\}$, where $\sigma = \sigma(\eta)$ is adjusted so that r is continuous, at the points $\pm a$ and hence everywhere. For $C(1 - 2\eta) > 1$ we have $p_0 \leq Cr$, and $K(r_\theta; \lambda) = R \log r - R_\theta \log \lambda < \infty$. Furthermore $\log(r/r_\theta)(x) \to 0$ as $\theta \to 0$ for every x, while the function is dominated by a multiple of $1 + |x|$.

Next assume that the support of p_0 is contained in $[-a, a]$, but not necessarily the whole interval. Define r to be proportional to $(p_0 \vee \eta)\mathbb{1}\{[-a, a]\}$. Then $p_0 \leq Cr$ for sufficiently small η, and r has support $[-a, a]$. Next, proceed as for the special case. □

7.4.2 Linear Regression with Unknown Error Density

Consider the univariate regression problem with observations $Y_1, \ldots, Y_n$ following the model $Y_i = \alpha + \beta x_i + \varepsilon_i$, for fixed covariates $x_1, \ldots, x_n$ not all equal and errors $\varepsilon_i \stackrel{\text{iid}}{\sim} f$ with symmetric, but unknown density. The main interest is in estimating the regression parameters $(\alpha, \beta) \in \mathbb{R}^2$. Let $P_{\alpha,\beta,f,i}$ be the distribution of Y_i.

Given a prior Π on (α, β, f) assume that, for every $\epsilon > 0$,

$$\liminf_{n\to\infty} \Pi_n\Big((\alpha, \beta, f)\colon \frac{1}{n}\sum_{i=1}^n K(P_{\alpha_0,\beta_0,f_0,i};\, P_{\alpha,\beta,f,i}) < \epsilon, \tag{7.5}$$

$$\frac{1}{n^2}\sum_{i=1}^n V_{2,0}(P_{\alpha_0,\beta_0,f_0,i};\, P_{\alpha,\beta,f,i}) < \epsilon\Big) > 0.$$

Theorem 7.27 *If* $\liminf n^{-1}\sum_{i=1}^n (x_i - \bar{x}_n)^2 > 0$, *then the posterior distribution for* (α, β) *is consistent at every* (α_0, β_0, f_0) *at which* (7.5) *holds. If* (α, β) *and* f *are a priori independent and the prior of* (α, β) *possesses full support, then sufficient conditions for the latter to be true are that the covariates are uniformly bounded and either (i) or (ii) holds:*

(i) *For every* $C > 1$ *there exists* $\eta > 0$ *and a symmetric density* g *with* $\sup_{|\theta|<\eta} f_0(\cdot - \theta) \le Cg(\cdot)$ *and* $\Pi(f\colon K(g; f) < \epsilon, V_2(g; f) < \infty) > 0$ *for every* $\epsilon > 0$.
(ii) *The maps* $\theta \mapsto K(f_0; f(\cdot - \theta))$ *and* $\theta \mapsto V^+(f_0; f(\cdot - \theta))$ *are continuous at* $\theta = 0$ *and* $\Pi(f\colon K(f_0; f) < \epsilon, V_2(f_0; f) < \infty) > 0$ *for every* $\epsilon > 0$.

If the prior on f *has the form* $f = \int \phi_h(\cdot - z)\, dQ(z)$ *with* $Q \sim \tilde{\Pi}$, *then (ii) is verified if* $\int\int t^2 dQ(t)\, d\tilde{\Pi}(Q) < \infty$ *and* $\int (y^4 + \log_+^2 f_0(y)) f_0(y)\, dy < \infty$.

Proof As (7.5) repeats the condition on existence of sets B_n of Theorem 6.41, for the first assertion it suffices to show the existence of exponentially powerful tests of (α_0, β_0, f_0) versus $\tilde{\Theta}_n := \{(\alpha, \beta, f)\colon |\alpha - \alpha_0| > \epsilon, \text{ or } |\beta - \beta_0| > \epsilon\}$, so that the posterior distribution of these sets tends to zero by (a) of that theorem.

For $\psi = -g'/g = (2G - 1)$ the score function of the logistic distribution, consider the test statistics $T = n^{-1}\sum_{i=1}^n \psi(Y_i - \alpha_0 - \beta_0 x_i)(1, x_i)^\top$. By the symmetry of the error density $\mathrm{E}_0 T = 0$, while

$$\mathrm{E}_{\alpha,\beta,f} T = \frac{1}{n}\sum_{i=1}^n \int \psi(y + \alpha - \alpha_0 + (\beta - \beta_0)x_i) f(y)\, dy \begin{pmatrix} 1 \\ x_i \end{pmatrix} =: H_f(\alpha - \alpha_0, \beta - \beta_0).$$

Because $\psi' = 2g$, the Hessian matrix of the function H_f can be computed as

$$H_f'(\alpha, \beta) = \frac{1}{n}\sum_{i=1}^n \int 2g(y + \alpha + \beta x_i) f(y)\, dy \begin{pmatrix} 1 & x_i \\ x_i & x_i^2 \end{pmatrix}.$$

We show that the expression above is positive definite. To see this, fix y and abbreviate $2g(y + \alpha + \beta x_i) f(y)$ by c_i. Then for any $(a, b) \neq (0, 0)$, we observe that $a^2\sum_{i=1}^n c_i + 2ab\sum_{i=1}^n c_i x_i + b^2 \sum_{i=1}^n c_i x_i^2 > 0$ because by the Cauchy-Schwarz inequality,

$$2ab\sum_{i=1}^n c_i x_i \le 2ab\sqrt{\sum_{i=1}^n c_i}\sqrt{\sum_{i=1}^n c_i x_i^2} \le a^2\sum_{i=1}^n c_i^2 + b^2\sum_{i=1}^n c_i x_i^2$$

and equality is ruled out as all x_i values are not equal. Clearly integration over y preserves the positivity, and hence H_f' is positive definite. Hence for any w in the unit sphere, the function $t \mapsto w^\top H_f(tw)$, which has derivative $w^\top H_f'(tw)w$, is nondecreasing on $[0, \infty)$. This shows that the infimum of $H_f(\alpha, \beta)$ over (α, β) outside a sphere around the origin is assumed on the sphere. In view of the form of $H_f'(0)$ and continuity, it follows that on a sufficiently small sphere the infimum is bounded away from 0, uniformly in f such that $\int gf\,d\lambda$ is bounded away from 0. By Hoeffding's inequality the test that rejects when $\|T\|$ is bigger than a small constant has exponential power.

We are left with testing (α_0, β_0, f_0) versus alternatives (α, β, f) with $\int gf\,d\lambda < \eta$, for arbitrary constant $\eta > 0$. The statistic $T = n^{-1}\sum \mathbb{1}\{|Y_i - \alpha_0 - \beta_0 x_i| > M\}$ possesses means $\mathrm{E}_{\alpha_0,\beta_0,f_0} T = \mathrm{P}_{f_0}(|\varepsilon| > M)$ and

$$\mathrm{E}_{\alpha,\beta,f} T = \frac{1}{n}\sum_{i=1}^{n} \mathrm{P}_f(|\varepsilon_i + \alpha - \alpha_0 + (\beta - \beta_0)x_i| > M) \geq \mathrm{P}_f(\varepsilon > M),$$

as follows by splitting the sum over the terms with $\alpha - \alpha_0 + (\beta - \beta_0)x_i \geq 0$ (when the event contains $\varepsilon_i < -M$) and $\alpha - \alpha_0 + (\beta - \beta_0)x_i < 0$ (when the event contains $\varepsilon_i > M$), combined with the symmetry of f. Because $g(M)\mathbb{1}\{[-M, M]\} \leq g$, we have that $\mathrm{P}_f(\varepsilon > M) \geq \frac{1}{2}(1 - \int gf\,d\lambda/g(M))$, for any $M > 0$. Thus the expectations under null and alternative hypotheses are separated for M such that $\frac{1}{2}(1 - \eta/g(M)) > \mathrm{P}_{f_0}(|\varepsilon| > M)$. For sufficiently small η such M exists (e.g. $\mathrm{P}_{f_0}(|\varepsilon| > M) = 1/8$ and $\eta = \frac{1}{2}g(M)$).

To prove (7.5) under (i) we note that for bounded x_i all shifts $\alpha - \alpha_0 + (\beta - \beta_0)x_i$ are arbitrarily small if (α, β) ranges over a sufficiently small neighborhood of (α_0, β_0). The densities $f(\cdot - \alpha - \beta x_i)$ are then bounded above by $Cg(\cdot - \alpha_0 - \beta_0 x_i)$, whence $K(P_{\alpha_0,\beta_0,f_0,i}; P_{\alpha,\beta,f,i}) = K(f(\cdot + \alpha - \alpha_0 + (\beta - \beta_0)x_i); f) \leq \log C + C(\epsilon + \sqrt{\epsilon/2})$ if $K(g; f) < \epsilon$, by Lemma B.14. The Kullback-Leibler variation $V_{2,0} \leq V = V^+ + V^-$ is similarly bounded by Lemma B.14 (for V^+) and Lemma B.13 (for $V^- \leq 4K$).

For a proof of (7.5) under (ii) let $f_\theta = f(\cdot - \theta)$ and $A_\eta := \bigcap_{|\theta|<\eta} \{f : K(f_0; f_\theta) < \epsilon, V_2(f_0; f_\theta) < \infty\}$, for given $\epsilon > 0$. Then $\bigcup_{\eta>0} A_\eta = \{f : K(f_0; f) < \epsilon, V_2(f_0; f) < \infty\}$ by the assumed continuity, and hence $\Pi(f \in A_\eta) > 0$ for some $\eta > 0$, by the assumption on Π. For all shifts $\alpha - \alpha_0 + (\beta - \beta_0)x_i$ bounded by η, we have $K(P_{\alpha_0,\beta_0,f_0,i}; P_{\alpha,\beta,f,i}) < \epsilon$ and $V_2(P_{\alpha_0,\beta_0,f_0,i}; P_{\alpha,\beta,f,i}) < \infty$, and then (ii) holds.

By Jensen's inequality a density of the form $f(y) = \int \phi_t(y - \theta)\,dQ(t)$ satisfies the inequality $\log(f_0/f) \leq \log f_0 - \int \log \phi_t\,dQ(t)$. For ϕ_t a shifted scaled normal kernel the absolute value of the second term is bounded above by a multiple of $y^2 + \int t^2\,dQ(t)$, uniformly in small shifts. The required continuity therefore follows by continuity of these functions relative to shifts and the dominated convergence theorem. ☐

Part (i) of the theorem applies, for instance, to symmetrized canonical Pólya tree priors with $\sum_{m=1}^\infty a_m^{-1/2} < \infty$, and part (ii) to Dirichlet mixtures with mixing measure $Q \sim \mathrm{DP}(\alpha)$ satisfying $\int t^2\,d\alpha(t) < \infty$.

Generalizations of Theorem 7.27 to multidimensional covariates are considered in Problems 7.18 and 7.19.

7.4.3 Binary Nonparametric Monotone Regression

Suppose we observe $Y_1, \ldots, Y_n$ for independent Bernoulli variables $Y_i \overset{\text{ind}}{\sim} \mathrm{Bin}(1, H(x_i))$, with success probabilities $H(x_i)$ depending on real deterministic covariates $x_1, \ldots, x_n$, where $H: \mathbb{R} \to [0, 1]$ is a monotone increasing unknown link function. We are interested in finding the covariate value x for which the probability of success $H(x)$ is equal to a given value ξ, i.e. the ξth quantile $x = H^{-1}(\xi)$ of H.

As a prior consider modeling the function H as $H(x) = F(\alpha + \beta x)$, and equipping the parameters F, α and β with priors. The latter parameters are not identifiable, but this is of no concern when estimating $H^{-1}(\xi)$, which is identifiable.

Let $\mathbb{P}_n = n^{-1}\sum_{i=1}^n \delta_{x_i}$ be the empirical measure of the covariates.

Theorem 7.28 *Assume that H_0 is strictly increasing on a neighborhood of $H_0^{-1}(\xi)$. Let the covariates be contained within the support of H_0 and assume that* $\liminf \mathbb{P}_n[a, b] > 0$ *for every $a < b$ in a neighborhood of $H_0^{-1}(\xi)$. If $F \sim \Pi$ and $(\alpha, \beta) \sim \mu$ are a priori independent with priors with (full) supports $\mathfrak{M}$ and $\mathbb{R} \times [0, \infty)$, respectively, then $\Pi_n(|H^{-1}(\xi) - H_0^{-1}(\xi)| > \epsilon \mid Y_1, \ldots, Y_n) \to 0$ a.s. under H_0, for every $\epsilon > 0$.*

Proof If $H_m \rightsquigarrow H_0$, then by Pólya's theorem $H_m \to H_0$ uniformly. Therefore for every $\epsilon > 0$ there exists a weak neighborhood $\mathcal{U}$ of H_0 such that, for all $H \in \mathcal{U}$,

$$\frac{1}{n}\sum_{i=1}^n K(\mathrm{Bin}(1, H_0(x_i)), \mathrm{Bin}(1, H(x_i))) \le \frac{1}{n}\sum_{i=1}^n |H_0(x_i) - H(x_i)|^2 < \epsilon^2,$$

$$\sum_{i=1}^\infty \frac{1}{i^2} V_2(\mathrm{Bin}(1, H_0(x_i)), \mathrm{Bin}(1, H(x_i))) \le \sum_{i=1}^\infty \frac{1}{i^2}|H_0(x_i) - H(x_i)|^2 < \infty.$$

(See Problem B.2.) By the assumptions on the supports of the priors and Kolmogorov's strong law of large numbers the condition of Theorem 6.40 is verified.

It now suffices to show that there exist exponentially powerful tests of the null hypothesis H_0 versus the alternative $\{H: |H_0^{-1}(\xi) - H^{-1}(\xi)| > \epsilon\}$. We can split the alternative into two one-sided alternatives; consider the right alternative $\{H: H^{-1}(\xi) > H_0^{-1}(\xi) + \epsilon\}$. Because H_0 is strictly increasing, it is strictly greater than ξ on the interval $\mathfrak{X}_0 := [H_0^{-1}(\xi) + \epsilon/2, H_0^{-1}(\xi) + \epsilon]$; let $\eta = H_0(H_0^{-1}(\xi) + \epsilon/2) - \xi$ be the minimum gain. Because a function H attains the level ξ not before the right end point of $\mathfrak{X}_0$, we have $\mathrm{E}_H(Y_i) \le \xi$ for every i with $x_i \in \mathfrak{X}$. Since $\mathrm{E}_{H_0}(Y_i) = H_0(x_i) \ge \xi + \eta$, the test that rejects the null hypothesis if $n^{-1}\sum Y_i \mathbb{1}\{x_i \in \mathfrak{X}_0\}\mathbb{P}_n(\mathfrak{X}_0) < \xi + \eta/2$ is exponentially powerful, by Hoeffding's inequality and the assumption that $\liminf \mathbb{P}_n(\mathfrak{X}_0) > 0$. □

An extension to generalized linear models is considered in Problem 7.21.

7.5 Historical Notes

The Kullback-Leibler property of Dirichlet process mixtures was proved by Ghosal et al. (1999b) and subsequently improved by Tokdar (2006) for normal mixtures. Consistency for the Bernstein polynomial prior was shown by Petrone and Wasserman (2002); and of Bayesian histograms by Gasparini (1996), by direct methods. The Kullback-Leibler property of general kernel mixtures was studied by Wu and Ghosal (2008a). Simplified versions of their results are presented in Section 7.1.2 as well as in Problems 7.2–7.7. Bhattacharya and Dunson (2010) studied the Kullback-Leibler property of kernel mixture priors on manifolds, considering the complex Bingham and Watson kernels. The Kullback-Leibler property and consistency for the multivariate normal kernel were studied by Wu and Ghosal (2010). Consistency for the infinite-dimensional exponential family prior was studied by Barron et al. (1999) in the case of a polynomial basis and a priori normally distributed coefficients, although their proof appears to be based on an incorrect estimate. The inconsistency example in Example 7.12 is due to Barron et al. (1999). Walker et al. (2005) called this phenomenon data tracking, and provided more insight on it. Theorem 7.13 is due to Ghosal et al. (1999b). We learned the construction of a Dirichlet mixture sieve based on the stick-breaking property of a Dirichlet process used in the proof of Theorem 7.15 from Tokdar (personal communication). The results in Sections 7.2.3 and 7.2.4 are based on Barron et al. (1999). Consistency for nonparametric binary regression, or estimation of response probability, was addressed by Ghosal and Roy (2006), who specifically used Gaussian process priors. Coram and Lalley (2006) established consistency for priors supported on step functions using a delicate large deviation technique. Similar results for categorical, but infinitely many covarites were obtained earlier by Diaconis and Freedman (1993). Normal regression was considered by Choi and Schervish (2007), who also used Gaussian process priors, but used somewhat different methods. Spectral density estimation using Whittle's likelihood is common in the frequentist literature. In the Bayesian context, Carter and Kohn (1997), Gangopadhyay et al. (1999) and Liseo et al. (2001) used the Whittle likelihood with the integrated Brownian motion or free-knot spline based priors. Consistency under this setting using the contiguity technique is due to Choudhuri et al. (2004a), who specifically used the Bernstein polynomial prior. Theorem 7.26 is due to Ghosal et al. (1999a), who pointed out that the Diaconis-Freedman-type inconsistency can be avoided in the location problem by using a prior with the Kullback-Leibler property. Theorems 7.27 and 7.28 are due to Amewou-Atisso et al. (2003). Consistency proofs for many other problems, including the Cox model and Problems 7.18–7.22 can be found in Wu and Ghosal (2008b). The estimate in Problem 7.23 is adapted from Walker (2004) who considered density estimation in the i.i.d. case and used an infinite-dimensional exponential family.

Problems

7.1 (Barron et al. 1999) For infinite-dimensional exponential families in Section 2.3, positivity of the prior probability of Kullback-Leibler neighborhoods of p_0 can be verified under weaker conditions on p_0. Show that the uniform approximation requirement and continuity can be relaxed to $K(p_0; \lambda) < \infty$ and $\mathbb{L}_1$-approximation of functions bounded by B by finite basis expansion satisfying a bound rB, where $r > 1$ can be chosen to work for all possible B.

7.2 (Wu and Ghosal 2008a) Show that for the double exponential and the logistic kernel, the conditions of Theorem 7.3 hold under finiteness of $(1+\eta)$-moment of p_0 for some $\eta > 0$.

7.3 (Wu and Ghosal 2008a) Show that for the t-kernel, conditions of Theorem 7.3 hold under $\int \log_+ |x| p_0(x)\, dx < \infty$.

7.4 (Wu and Ghosal 2008a) Consider the triangular density kernel given by

$$\psi(x;m,n)=\begin{cases}\begin{cases}2n-2n^2x, & x\in(0,\frac{1}{n}),\\ 0, & \text{otherwise},\end{cases} & m=0,\\[2ex] \begin{cases}n^2(x-\frac{m}{n})+n, & x\in(\frac{m-1}{n},\frac{m}{n}),\\ -n^2(x-\frac{m}{n})+n, & x\in(\frac{m}{n},\frac{m+1}{n}),\\ 0, & \text{otherwise},\end{cases} & m=1,2,\ldots,n-1,\\[2ex] \begin{cases}2n+2n^2(x-1), & x\in(0,\frac{1}{n}),\\ 0, & \text{otherwise},\end{cases} & m=n.\end{cases}$$

Construct a kernel mixture prior by mixing both m and n according to a prior with weak support $\mathfrak{M}(\{(m,n): m\le n\})$. If p_0 is a continuous density on $[0,1]$, then $p_0\in \mathrm{KL}(\Pi^*)$.

7.5 (Wu and Ghosal 2008a) For the Weibull kernel, using log transformation to reduce the problem to a location-scale mixture, show that a nonzero, bounded, continuous density on $(0,\infty)$, f_0, is in the Kullback-Leibler support of the mixture prior if $\log f_0(x)$, $e^{2|\log x|^{1+\eta}}$ and $\log(f_0(x)/\phi_\delta(x))$ are f_0-integrable for some $\eta,\delta>0$, where $\phi_\delta(x)=\inf_{|t-x|<\delta} f_0(t)$.

7.6 (Wu and Ghosal 2008a) Consider (a reparameterized) inverse-gamma density kernel defined by $\psi(x;\beta,z)=(\beta z)^\beta x^{-\beta-1}e^{-\beta z/x}/\Gamma(\beta)$. Consider a kernel mixture prior by mixing (β,z) according to a prior Π with weak support $\mathfrak{M}((2,\infty)\times\mathbb{R}^+)$. Find conditions for a continuous density p_0 on $[0,\infty)$ to satisfy $p_0\in\mathrm{KL}(\Pi^*)$, where Π^* is the induced prior distribution of $p_F=\int\psi(\cdot;\beta,z)\,dF(\beta,z)$ for $F\sim\Pi$.

7.7 (Wu and Ghosal 2008a) Consider a kernel mixture prior based on the scaled uniform kernel $\psi(x;\theta)=\theta^{-1}\mathbb{1}\{0\le x\le\theta\}$, where $\theta>0$ is mixed according to F having a prior Π with weak support $\mathfrak{M}(\mathbb{R}^+)$. If p_0 is a continuous and decreasing density function on $\mathbb{R}^+$ such that $\int p_0(x)|\log p_0(x)|\,dx<\infty$, then show that $p_0\in\mathrm{KL}(\Pi^*)$, where Π^* is the induced prior distribution of $p_F=\int\psi(\cdot;\theta)\,dF(\theta)$ for $F\sim\Pi$.

7.8 (Tokdar 2006) If the kernel is univariate normal location-scale mixture and the mixing distribution follows the Dirichlet process, show that the moment condition $\int|x|^{2(1+\eta)}p_0(x)\,dx<\infty$ for some $\eta>0$, in Example 7.4 can be weakened to $\int|x|^\eta p_0(x)\,dx<\infty$ for some $\eta>0$.

7.9 (Wu and Ghosal 2010) Consider a multivariate normal mixture prior $p\sim\Pi^*$ defined by $p(x)=\int\phi_d(x;\theta,\Sigma)\,dF(\theta)$, $F\sim\Pi$ and $\Sigma\sim\mu$, where Π has support $\mathfrak{M}(\mathbb{R}^d)$ and the prior μ for the scale matrix Σ has σI_d in its support for all sufficiently small $\sigma>0$. Assume that the true density p_0 is everywhere positive, bounded above, that the integrals $\int p_0(x)|\log p_0(x)|\,dx$ and $\int p_0(x)\log(p_0(x)/\phi_\delta(x))\,dx$ are finite, for some $\delta>0$, where $\phi_\delta(x)=\inf\{p_0(t): \|t-x\|<\delta\}$ and $\int\|x\|^{2(1+\eta)}p_0(x)\,dx<\infty$ for some $\eta>0$. Show that $p_0\in\mathrm{KL}(\Pi^*)$.

7.10 (Wu and Ghosal 2010) Consider the setting of Problem 7.9 with Π being α and Σ^m, $m\ge 2d$, has a Wishart distribution with nonsingular scale matrix A and degrees of freedom $q>d$, truncated to satisfy $\mathrm{tr}(\Sigma)\le M$ for some fixed $M>0$. Show that the posterior is consistent at p_0. Further, if p_0 has sub-Gaussian tails, show that

$m > d$ suffices. [Hint: show that $\Pi^*(F(\|\theta\|_\infty > 2a_n) \geq \epsilon_n | X_1, \ldots, X_n) \to 0$ in P_0^n-probability whenever $a_n \to \infty$, $n^{-1}a_n^{2d}\epsilon_n \to \infty$ and $n^{-1}\epsilon_n \inf\{g(t): \|t\| \leq a_n\}e^{a_n^2/(2M)} \to \infty$, where g is the density of $\bar{\alpha}$. Now use Problem C.1.]

7.11 (Ghosal and Roy 2006) In Theorem 7.20, assume all conditions hold and the covariate is deterministic and one-dimensional. Further assume that given $\delta > 0$, there exist a constant K_1 and an integer N such that for $n > N$, we have that $\sum_{i: S_{i,n} > K_1 n^{-1}} S_{i,n} \leq \delta$, where $S_{i,n} = x_{i+1,n} - x_{i,n}$ are the spacings between consecutive covariate values. Show that the posterior for f is consistent at f_0 in the usual $\mathbb{L}_1$-distance $\int |f_1(x) - f_2(x)|\, dx$.

7.12 Derive a consistency result for nonparametric binary regression from that on the densities using Theorems 6.23 and 6.41 for stochastic and deterministic regressors, respectively.

7.13 Formulate and prove a consistency result for nonparametric Poisson regression given by $Y | X = x \sim \text{Poi}(\lambda(x))$.

7.14 (Coram and Lalley 2006) This example shows sometimes it is beneficial to directly bound the posterior probability of the complement of the sieve rather than using Theorem 6.17. In this case, delicate large deviation estimates are used.

Consider the binary regression problem $Y | X = x \sim \text{Bin}(1, f(x))$, where X has a nonsingular distribution on $[0, 1]$. Let (X_i, Y_i), $i = 1, 2, \ldots$ be i.i.d. observations. Consider a series prior for f defined as follows: A positive integer-valued random variable $N \sim \lambda$, where $\lambda_m > 0$ for infinitely many m, $U_1, \ldots, U_m | N = m \stackrel{\text{iid}}{\sim} \text{Unif}(0, 1)$, $W_1, \ldots, W_m | N = m \stackrel{\text{iid}}{\sim} \text{Unif}(0, 1)$ and $f(x) = \sum_{j=1}^{m+1} W_j \mathbb{1}\{U_{j-1:m} < x \leq U_{j:m}\}$, where $U_{j:m}$ is the jth order statistic among m, $j = 1, \ldots, m$, and $U_{0:m} = 0$, $U_{m+1:m} = 1$. Let the true response probability f_0 be a measurable function from $[0, 1]$ to $[0, 1]$ not identically $1/2$.

The posterior distribution for f given $\mathscr{F}_n := \sigma\langle (X_1, Y_1), \ldots, (X_n, Y_n) \rangle$ can be written as

$$\Pi(\cdot | \mathscr{F}_n) = \frac{\sum \lambda_m Z_{m,n} \Pi_m(\cdot | \mathscr{F}_n)}{\sum \lambda_m Z_{m,n}},$$

where $\Pi_m(\cdot | \mathscr{F}_n)$ is the "posterior based on model $N = m$" and "$Z_{m,n}$ is the marginal likelihood of the model $N = m$."

Show that if $m/n \to \alpha$, then $n^{-1} \log Z_{m,n} \to \psi(\alpha) < 0$ for all $\alpha > 0$, where the function $\psi(\alpha)$ has a unique maximum 0 at $\alpha = 0$. In other words, models with more than ϵn discontinuities for any $\epsilon > 0$ will have marginal likelihood decaying exponentially.

Show that if $K \leq m < \epsilon n$, then the likelihood of $N = m$ given values of U's decays exponentially like $\exp[-n\{f_u \log f_u + (1 - f_u) \log(1 - f_u)\}]$, where f_u is the step function obtained by averaging f_0 over the intervals defined by $U_1, \ldots, U_m$.

Show that the entropy of all models with $N \leq K$ grows logarithmically, and hence Schwartz's theory applies to give consistency at f_0 using only these models.

Combining the above three conclusions, obtain consistency of the posterior at any measurable f_0 not identically $1/2$.

7.15 (Diaconis and Freedman 1993) Consider a binary response variable Y and infinitely many binary predictors $X = (X_1, X_2, \ldots)$ with response function at $X = x$ given by $f(x)$, where $f: \{0, 1\}^\infty \to [0, 1]$ is measurable. Consider a series prior for f defined as follows: A positive integer-valued random variable $N \sim \lambda$, where $\lambda_m > 0$ for infinitely many m. Given $N = m$, $f(x)$ depends only on $(x_1, \dots, x_m)$ and let Π_m be the joint prior distribution of $(\theta_\varepsilon : \varepsilon \in \{0,1\}^m)$, where $\theta_\varepsilon = f(\varepsilon)$. Let the true response probability be f_0. Suppose that at stage n we observe Y for every of the 2^n combinations of covariate values. Let the metric on f be given by the $\mathbb{L}_1$-distance $\int |f(x) - g(x)|\, d\lambda(x)$, where λ is the infinite product of i.i.d. Bin(1, 1/2).

Show that if f_0 is not identically $\frac{1}{2}$, then the posterior is consistent.

Show that if λ_ms are given by geometric with success probability $1 - r$, then the posterior is consistent at $f_0 \equiv 1/2$ whenever $r < 2^{-1/2}$ and inconsistent whenever $r > 2^{-1/2}$.

7.16 In the semiparametric linear regression problem, if random fs are not symmetrized around zero, then α is not identifiable. Nevertheless, show that uniformly consistent tests for β can be obtained by considering the difference $Y_i - Y_j$, which has a density that is symmetric around $\beta(x_i - x_j)$, and hence consistency for β may be obtained.

7.17 (Ghosal et al. 1999a) In the location problem, consider a symmetrized Dirichlet mixture of normal prior Π on the error density f. Assume that the true error density f_0 is in the Kullback-Leibler support of Π, $\int z^2 f_0(z)\, dz < \infty$, $\int f_0 \log_+^2 f_0 < \infty$ and the base measure of the Dirichlet process has finite second moment. Show that the posterior for θ is consistent at any θ_0 belonging to the support of the prior for θ.

7.18 (Wu and Ghosal 2008b) Consider a multiple regression model $Y_i = \alpha + X_i^\top \beta + \varepsilon_i$, $X_i \stackrel{\text{iid}}{\sim} Q \in \mathfrak{M}(\mathbb{R}^d)$, independently $\varepsilon_i \stackrel{\text{iid}}{\sim} f$, $i = 1, 2, \dots$, $\beta \in \mathbb{R}^d$, and $f \in \mathcal{F} = \{f : \int_{-\infty}^0 f(x)\, dx = a\}$ for some $0 < a < 1$ (i.e. has the same value of some fixed quantile). Let $f \sim \tilde{\Pi}$, independently $(\alpha, \beta) \sim \mu$. Let Π stand for $\tilde{\Pi} \times \mu$. Let $f_{\alpha,\beta} = f(y - \alpha - x^\top \beta)$, $p_0 = f_{0,\alpha_0,\beta_0}$, and $g_\theta(y) = g(y - \theta)$. Assume that the covariate X is compactly supported and f_0 is continuous with $f_0(0) > 0$.

Assume that

(a) For some $\zeta > 0$, $Q(\gamma_j X_j > \zeta, j = 1, \dots, d) > 0$ for all $\gamma_j = \pm 1$;

Show that there exist exponentially consistent tests for testing $H_0 : (f, \alpha, \beta) = (f_0, \alpha_0, \beta_0)$ against $H_1 : \{(f, \alpha, \beta) : f \in \mathcal{U}, |\alpha - \alpha_0| < \epsilon, \|\beta - \beta_0\| < \epsilon\}$, where $\mathcal{U}$ is a weak neighborhood of f_0.

Further, assume that

(b) for $\eta > 0$ sufficiently small, there exists $g_\eta \in \mathcal{F}$ and constant $C_\eta > 0$ such that for $|\eta'| < \eta$, $f_0(y - \eta') < C_\eta g_\eta(y)$ for all y and $C_\eta \to 1$ as $\eta \to 0$;

(c) for all sufficiently small η and all $\xi > 0$, $\tilde{\Pi}\{K(g_\eta; f) < \xi\} > 0$ and $(\alpha_0, \beta_0) \in \text{supp}(\mu)$.

Show that for any weak neighborhood $\mathcal{U}$ of f_0, we have $\Pi\{(f, \alpha, \beta) : f \in \mathcal{U}, |\alpha - \alpha_0| < \epsilon, \|\beta - \beta_0\| < \epsilon | (X_1, Y_1), \dots, (X_n, Y_n)\} \to 1$ a.s. P_0^∞. Let the covariate X be deterministic and assume values $x_1, \dots, x_n$, replace Condition (a) by the following deterministic version: for some $\zeta > 0$

$$\liminf n^{-1} \#\{i : \gamma_j x_{ij} > \zeta, j = 1, \dots, d\} > 0 \text{ for all } \gamma_j = \pm 1. \tag{7.6}$$

Show that exponentially consistent tests for testing the pair H_0 and H_1 defined above, exist. If, moreover, $V_2(g_\eta; f) < \infty$ a.s. $[\Pi]$, then the posterior for (f, α, β) is consistent at (f_0, α_0, β_0).

7.19 (Wu and Ghosal 2008b) In the multiple regression model of Problem 7.18 with stochastic covariates, suppose that we use a symmetrized Dirichlet mixture of normal prior given by $f(y) = f_{h,G}(y) = (\int \phi_h(y-t)\,dG(t) + \int \phi_h(y+t)\,dG(t))/2$, $G \sim \mathrm{DP}(\pi)$ and $h \sim \nu$. Show that Condition (ii) can be replaced by simpler moment conditions, namely, $\int y^2 p_0(y)\,dy < \infty$, $\int p_0|\log p_0| < \infty$ and $\int t^2 d\pi(t) < \infty$. For deterministic covariates, show that Conditions (b) and (c) hold if $\int y^4 p_0(y)\,dy < \infty$ and $\int p_0(y)\log^2(y)p_0(y)\,dy < \infty$.

7.20 (Wu and Ghosal 2008b) Consider the exponential frailty model described by

$$(X_i, Y_i)|\, W_i \overset{\text{ind}}{\sim} (\mathrm{Exp}(W_i), \mathrm{Exp}(\lambda W_i)) \text{ independently, } W_i \overset{\text{iid}}{\sim} F,$$

and priors are given by $F \sim \tilde{\Pi}$ and $\lambda \sim \mu$ independently. Let the p.d.f. of (X, Y) be $p_{\lambda,F}(x, y) = \int \lambda_0 w^2 e^{-w(x+\lambda_0 y)} dF(w)$ and $p_0 = p_{\lambda_0,F_0}$, where F_0 is the true value of the frailty distribution F and λ_0 the true value of the frailty parameter λ.

Show that there exist exponentially consistent tests for testing $H_0\colon \lambda = \lambda_0$ against $H_1\colon |\lambda - \lambda_0| > \epsilon$ for any $\epsilon > 0$, and also for $H_0\colon \lambda = \lambda_0$, $F = F_0$ against $H_1\colon |\lambda - \lambda_0| \le \epsilon$, $F \in \mathcal{U}^c$, $F(r_n) - F(r_n^{-1}) \ge 1 - \delta$ for some $\delta > 0$, weak neighborhood $\mathcal{U}$ of F_0 and sequence $r_n < \sqrt{n\beta}$ for some sufficiently small $\beta > 0$.

Further assume that $\log(f_0(x, y))$, x and y are f_0-integrable, $F_0 \in \mathrm{supp}(\tilde{\Pi})$, $\lambda_0 \in \mathrm{supp}(\mu)$ and $w_E := \int w^2 dF_0(w) < \infty$. If for any $\delta > 0$ and $\beta > 0$ there exists a sequence $r_n < \sqrt{n\beta}$ and $\beta_0 > 0$ such that $\tilde{\Pi}\{F\colon F(r_n) - F(r_n^{-1}) < 1 - \delta\} < e^{-n\beta_0}$, then show that the posterior for (λ, F) is consistent at (λ_0, F_0).

7.21 (Wu and Ghosal 2008b) Consider a generalized linear model $X_i \overset{\text{iid}}{\sim} Q$, $Y_i|\, X_i \sim \exp[\theta_i y - b(\theta_i)]a(y)$, where $\mathrm{E}(Y_i|\, X_i) = b'(\theta_i) = g(X_i^\top \beta)$, for $i = 1, \ldots, n$, and $\|\beta\| = 1$ and g is differentiable and strictly increasing. Assume that $\|X\| \le L$. Let $\mathcal{G}$ stand for the space of all possible link functions, and $M := \{\mu\colon \mu = g(x^\top \beta), \|x\| \le L, \|\beta\| = 1\}$ be a compact interval. Let $T\colon M \to [0, 1]$ be a fixed strictly increasing and differentiable function and put a prior on g by $g = F \circ T^{-1}$, where F is a c.d.f. on $[-L, L]$ following a prior $\tilde{\Pi}$ distributed independently of β. Let $\Pi = \tilde{\Pi} \times \pi$. Assume that the covariates satisfy Condition (a) in Problem 7.18. Furthermore, assume that the following conditions hold:

- $\int \tilde{f}_0(t)|\log \tilde{f}_0(t)|dt < \infty$ and $\int \tilde{f}_c(t+\xi)|\log \bar{\tilde{F}}_0(t)|dt < \infty$ for all $\xi \in \mathbb{R}$;
- for some $\delta > 0$, $\int \tilde{f}_0(t)\log(\tilde{f}_0/\phi_\delta)(t)\,dt < \infty$, where $\phi_\delta(t) := \inf\{\tilde{f}_0(s)\colon |s-t| < \delta\}$;
- there exists $\eta > 0$ such that $\int (e^{2|\log t|^\eta} + e^{(\log t - a)/b}) f_0(t)\,dt < \infty$, for any $a \in \mathbb{R}$ and $b \in (0, \infty)$;
- for Π-almost all P and any given $h > 0$, $\int (e^{-t/h} + t/h)\,dP(t) < \infty$;
- the weak support of $\tilde{\Pi}$ is the space of all probability measures on $\mathbb{R}$;
- for some $\delta > 0$ and any $\xi \in \mathbb{R}$, $\int \tilde{f}_c(t+\xi)\log(\bar{\tilde{F}}_{0,\beta_0}/\phi_\delta)(t)\,dt < \infty$;
- there exists $\eta > 0$ such that $\int f_c(t)(e^{2(\log t)^{1+\eta}} + e^{(\log t - a)/b})\,dt < \infty$, for $a \in \mathbb{R}$ and $b > 0$.

(a) Show that g is identifiable.

(b) Show that for any $\epsilon > 0$, there exists an exponentially consistent sequence of tests for testing $H_0\colon (\beta, F) = (\beta_0, F_0)$ against $H_1\colon \|\beta - \beta_0\| > \epsilon$, $F \in \mathcal{U}$, for some weak neighborhood $\mathcal{U}$ of F_0.

(c) Show that for any weak neighborhood $\mathcal{U}$ of F_0, almost surely $[P^{\infty}_{\beta_0, f_0}]$,

$$\Pi\Big((\beta, F): \|\beta - \beta_0\| < \epsilon, \tilde{F} \in \mathcal{U}, | (Z_1, X_1, \Delta_1), \dots, (Z_n, X_n, \Delta_n)\Big) \to 1.$$

(d) Show that the Kullback-Leibler property holds at the true density of Z.
(e) Conclude that the posterior is consistent at (β_0, f_0).

[Hint: Show that there is an exponentially consistent sequence of tests for testing

- $H_0: (\beta, F) = (\beta_0, F_0)$ against $H_1: F \notin \mathcal{U}$, for any weak neighborhood $\mathcal{U}$ of F_0;
- $H_0: (\beta, F) = (\beta_0, F_0)$ against $H_1: d_{KS}(F, F_0) < \Delta, \|\beta - \beta_0\| > \delta$.]

7.22 (Wu and Ghosal 2008b) Consider a partial linear model $Y_i = X_i^\top \beta + f(Z_i) + \varepsilon_i$, where the covariates $X_i \in \mathbb{R}^d$ and $Z_i \in \mathbb{R}$, errors $\varepsilon_i \stackrel{\text{iid}}{\sim} \text{Nor}(0, \sigma^2)$ for $i = 1, \dots, n$, and f is an odd function. Assume that X has distribution $Q_X(x)$ and Z has symmetric p.d.f. $q_Z(z)$. Let $q_{f,Z}$ denote the p.d.f. of $f(Z)$. Assume that X and Z are compactly supported and Condition (a) in Exercise 7.18 holds. Consider fully supported priors μ for β, Π^* for $f(Z)$ and ρ for σ. Let $\tilde{\Pi}$ stand for the prior induced on $q_{f,Z} * \phi_\sigma$ and $\Pi = \tilde{\Pi} \times \mu$. Let f_0, β_0 and σ_0 denote the true values of f, β and σ respectively. Assume that

(a) there exists $\eta > 0$, C_η and a density g_η, such that $q_{f_0,Z} * \phi_{\sigma_0}(y - \eta') < C_\eta g_\eta(y)$, for $|\eta'| < \eta$;
(b) for all sufficiently small η and for all $\delta > 0$, $\tilde{\Pi}\{f: K(g_\eta; f) < \delta\} > 0$.

Show that the posterior is consistent at (β_0, f_0, σ_0). [Hint: Consider $\xi = f(Z) + \varepsilon$ as the error variable in a multiple regression problem $Y = X^\top \beta + \xi$ and apply Exercise 7.18. The oddness of f and symmetry of $q_Z(z)$ ensures that ξ is symmetrically distributed about 0.]

7.23 Consider a Markov process with transition density $f(y|\, x) = g(y - \rho x)$, where g is a density function on $[0, 1]$ and $-1 < \rho < 1$ is known. Assume that the true density g_0 is continuous and bounded away from 0, so that the corresponding Markov process is ergodic. Consider a finite random series prior based on trigonometric polynomials on g as described in Example 7.19: $g(u; \theta) = \exp[\sum_{j=0}^N \theta_j \psi_j(u) - c(\theta)]$, for $\psi_0(u) = 1$ and $\psi_j(u) = \sqrt{2}\cos(j\pi x)$ if $j \in \mathbb{N}$; $\text{P}(N = k) = \rho_k$ for $\sum_{k=1}^\infty \sqrt{\rho_k} < \infty$; and $\theta_j \stackrel{\text{ind}}{\sim} \text{Nor}(0, \tau_j^2)$, where $\tau_j \lesssim j^{-q}$ for some $q > 1$.

For given $N = k$ consider a countable partition $\{\mathcal{A}_{n_1,\dots,n_k}, n_j \in \mathbb{Z}, k \in \mathbb{N}\}$ of the form $\{g(y - \rho x): g \in \mathcal{B}_{n_1,\dots,n_k}, n_j \in \mathbb{Z}, k \in \mathbb{N}\}$. Let $\gamma_j = cj^{-\beta}$ be such that $\sum_{j=1}^\infty \gamma_j \leq 1$. Let $B_{n_1,\dots,n_k} = \prod_{j=1}^k (\delta n_j \gamma_j, \delta(n_j + 1)\gamma_j]$ and $\mathcal{B}_{n_1,\dots,n_k} = \{g(\cdot; (\theta_1, \dots, \theta_k)): (\theta_1, \dots, \theta_k) \in B_{n_1,\dots,n_k}\}$.

Show that $d_H(g, g^*) \leq 2\sqrt{\delta}$ if $g, g^* \in \mathcal{B}_{n_1,\dots,n_k}$.

[Hint: Integrate $|\exp[\sum \theta_j \psi_j(x)] - \exp[\sum \theta_j^* \psi_j(x)]|$ to obtain $|e^{c(\theta)} - e^{c(\theta^*)}| < 2\delta e^{c(\theta^*)}$ for $\delta < \log 2$. Hence obtain $|c(\theta) - c(\theta^*)| < 2\delta$. Now bound the Kullback-Leibler divergence and hence the Hellinger distance.]

Show consistency of the posterior by verifying that

$$\sum_{k=1}^\infty \sum_{n_1 \in \mathbb{Z}} \cdots \sum_{n_k \in \mathbb{Z}} \rho_k^{1/2} \prod_{j=1}^k \sqrt{\Pi(\delta n_j \gamma_j < \theta_j \leq \delta(n_j + 1)\gamma_j]} < \infty.$$

8

Contraction Rates: General Theory

A posterior contraction rate quantifies the speed at which a posterior distribution approaches the true parameter of the distribution of the data. Characterising a contraction rate can be seen as a significant refinement of establishing posterior consistency. It links nonparametric Bayesian analysis to the minimax theory of general statistical estimation. In this chapter we present rates of posterior contraction for dominated experiments. We begin with the general notion of a contraction rate, and its implications for Bayesian point estimation. We then present a basic rate theorem for i.i.d. observations, several refinements and a number of examples of priors, which illustrate optimal rates. Next we deal with general observations, with emphasis on independent, non-identically observations and Markov processes, and present a theory for misspecified models, illustrated by nonparametric regression under a misspecified error distribution. Finally we show that the α-posterior possesses a contraction rate that is solely determined by the prior concentration rate (for $\alpha < 1$), and conclude with an information-theoretic analysis.

8.1 Introduction

The contraction rate of a posterior distribution may be viewed as elaborating on its consistency. Both are asymptotic properties, and therefore we consider the same setup as in Chapter 6, of a sequence of statistical experiments $(\mathfrak{X}^{(n)}, \mathscr{X}^{(n)}, P_\theta^{(n)}: \theta \in \Theta_n)$ with parameter spaces Θ_n and observations $X^{(n)}$ indexed by a parameter $n \to \infty$. To measure a speed of contraction, the parameter spaces Θ_n must carry semimetrics d_n[1], unlike in Chapter 6, where a neighborhood system was enough for characterizing consistency. The parameter space and semimetric may depend on n, but we shall make this explicit only when needed, as in Section 8.3.

As in Chapter 6 we fix particular versions of the posterior distributions $\Pi_n(\cdot|\, X^{(n)})$ relative to a prior Π_n, which we refer to as "the" posterior distribution.

Definition 8.1 (Contraction rate) A sequence ϵ_n is a *posterior contraction rate* at the parameter θ_0 with respect to the semimetric d if $\Pi_n(\theta: d(\theta, \theta_0) \geq M_n\epsilon_n|\, X^{(n)}) \to 0$ in

[1] A semimetric satisfies the axiom of symmetry and the triangle inequality like a metric, but two distinct elements need not be at a positive distance. A semimetric space can always be converted to a metric space by identifying any pair of elements with zero distance.

$P_{\theta_0}^{(n)}$-probability, for every $M_n \to \infty$. If all experiments share the same probability space and the convergence to zero takes place almost surely $[P_{\theta_0}^{(\infty)}]$, then ϵ_n is said to be a *posterior contraction rate in the strong sense*.

We defined "*a*" rather than *the* rate of contraction, and hence logically any rate slower than a contraction rate is also a contraction rate. Naturally we are interested in a fastest decreasing sequence ϵ_n, but in general this may not exist or may be hard to establish. Thus our rate is an upper bound for a targeted rate, and generally we are happy if our rate is equal to or close to an "optimal" rate. With an abuse of terminology we often make statements like "ϵ_n is *the* rate of contraction."

In most infinite-dimensional models, the constants M_n in the definition of a contraction rate can be fixed to a single large constant M without changing ϵ_n, which gives a contraction rate in a slightly stronger sense. However, for typical finite-dimensional models, the sequence M_n must be allowed to grow indefinitely to obtain the "usual" contraction rate (such as $n^{-1/2}$ in a "smooth" model). The requirement for "every $M_n \to \infty$" should be understood as "for any arbitrarily slow $M_n \to \infty$."

We begin with some examples where the rate of contraction can be directly calculated, often using the following simple observation.

Lemma 8.2 *If* $\Theta \subset \mathbb{R}$ *and* $\mathrm{E}(\theta|\, X^{(n)}) = \theta_0 + O_p(\epsilon_n)$ *and* $\mathrm{var}(\theta|\, X^{(n)}) = O_p(\epsilon_n^2)$, *with respect to the distribution generated by the true parameter* θ_0, *then* ϵ_n *is a rate of contraction at* θ_0 *with respect to the Euclidean distance.*

Proof By Chebyshev's inequality, the probability $\Pi_n(\theta\colon |\theta - \mathrm{E}(\theta|\, X^{(n)})| > M_n\epsilon_n|\, X^{(n)})$ is bounded above by $\mathrm{var}(\theta|\, X^{(n)})/(M_n\epsilon_n)^2$ and hence tends to zero for any $M_n \to \infty$. Furthermore, the variable $\Pi_n(\theta\colon |\mathrm{E}(\theta|\, X^{(n)}) - \theta_0| > M_n\epsilon_n|\, X^{(n)})$ is the indicator of the event where $|\mathrm{E}(\theta|\, X^{(n)}) - \theta_0| > M_n\epsilon_n$, which has probability tending to zero. ∎

Example 8.3 (Bernoulli) If $X_1, \ldots, X_n|\,\theta \overset{\text{iid}}{\sim} \mathrm{Bin}(1, \theta)$ and $\theta \sim \mathrm{Be}(a, b)$, then by elementary calculation the posterior is $\theta|\, X_1, \ldots, X_n \sim \mathrm{Be}(a + n\bar{X}_n, b + n - n\bar{X}_n)$. Thus

$$\mathrm{E}(\theta|\, X_1, \ldots, X_n) = \frac{a + n\bar{X}_n}{a + b + n} = \theta_0 + O_p\Big(\frac{1}{\sqrt{n}}\Big),$$

$$\mathrm{var}\,(\theta|\, X_1, \ldots, X_n) = \frac{(a + n\bar{X}_n)(b + n - n\bar{X}_n)}{(a + b + n)^2(a + b + n + 1)} \le \frac{1}{n}.$$

Hence the rate of contraction is $\epsilon_n = n^{-1/2}$.

Example 8.4 (Uniform) If $X_1, \ldots, X_n|\,\theta \overset{\text{iid}}{\sim} \mathrm{Unif}(0, \theta)$ and θ has the improper density proportional to θ^{-a}, then the posterior density is proportional to $\theta^{-(n+a)}\mathbb{1}\{\theta > X_{(n)}\}$, where $X_{(n)}$ is the maximum of the observations. Thus

$$\mathrm{E}(\theta|\, X_1, \ldots, X_n) = \frac{n + a - 1}{n + a - 2} X_{(n)} = \theta_0 + O_p\Big(\frac{1}{n}\Big),$$

$$\operatorname{var}(\theta|\,X_1,\dots,X_n)=\Big(\frac{n+a-1}{n+a-3}-\Big(\frac{n+a-1}{n+a-2}\Big)^2\Big)X_{(n)}^2=O_p\Big(\frac{1}{n^2}\Big).$$

Hence the rate of contraction is $\epsilon_n=n^{-1}$.

In the preceding examples the prior is conjugate to the model, enabling explicit expressions for the posterior mean and variance. However, the conclusions go through for all priors that possess a continuous and positive density at θ_0. In fact, for most parametric models the posterior distribution is asymptotically free of the prior, provided that it satisfies this weak restriction. For regular models, such as Example 8.3, this follows from the Bernstein–von Mises theorem (see e.g. Theorem 10.1 of van der Vaart 1998), which shows that the posterior distribution of θ is asymptotically a normal distribution with mean centered at an efficient estimator (e.g. the maximum likelihood estimator) and with variance proportional to n^{-1}, giving an $n^{-1/2}$ contraction rate. More generally, the rate of contraction can often be identified with a *local scaling rate* r_n that ensures that the local likelihood ratio process $Z_n(u)=\prod_{i=1}^n(p_{\theta_0+r_nu}/p_{\theta_0})(X_i)$ converges in (marginal) distribution to a limit (see Chapter I of Ibragimov and Has'minskiĭ 1981, and the proof of Proposition 1 of Ghosal et al. 1995).

Example 8.5 (Dirichlet process) The posterior distribution in the model $X_1,\dots,X_n|\,P\overset{\text{iid}}{\sim}P$ and $P\sim\mathrm{DP}(\alpha)$ is the $\mathrm{DP}(\alpha+n\mathbb{P}_n)$-distribution, for $\mathbb{P}_n$ the empirical distribution of the observations. The posterior mean and variance of $P(A)$ are given by (4.11) and (4.12), and are $P_0(A)+O_P(n^{-1/2})$ and $O_P(n^{-1})$, respectively. Thus the posterior rate of contraction relative to the semimetric $d(P_1,P_2)=|P_1(A)-P_2(A)|$ is $n^{-1/2}$. (This result can also be obtained from Example 8.3.)

As an example of a global metric d, consider sample space $\mathbb{R}$ and the $\mathbb{L}_2(\nu)$-distance between the distribution functions of the measures, for a σ-finite positive measure ν on $\mathbb{R}$: $d^2(F_1,F_2)=\int(F_1(t)-F_2(t))^2\,d\nu(t)$. By Markov's inequality followed by Fubini's theorem

$$\Pi_n(F\colon d(F,F_0)\geq\epsilon|\,X_1,\dots,X_n)\leq\frac{1}{\epsilon^2}\int \mathrm{E}\Big((F(t)-F_0(t))^2|\,X_1,\dots,X_n\Big)\,d\nu(t).$$

The integrand can be further expanded as the sum of $\{\mathrm{E}(F(t)|\,X_1,\dots,X_n)-F_0(t)\}^2$ and $\operatorname{var}(F(t)|\,X_1,\dots,X_n)$, which are given by (4.11) and (4.12). Thus the preceding display is bounded by, for $\alpha=MG$,

$$\frac{2M^2\int(G-F_0)^2\,d\nu}{(M+n)^2}+\frac{2n^2\int(\mathbb{F}_n-F_0)^2\,d\nu}{(M+n)^2}+\frac{Mn\int F_0(1-F_0)\,d\nu}{(M+n)^2(M+n+1)}.$$

For a finite measure ν all terms are of the order $O_P(n^{-1})$, whence the rate of contraction is $n^{-1/2}$. Even without finiteness of ν the rate is $n^{-1/2}$ provided the integrals $\int G^2(1-G)^2\,d\nu$ and $\int F_0(1-F_0)\,d\nu$ are finite. This follows since

$$\begin{aligned}\int(G-F_0)^2d\nu&\leq 2\int_{-\infty}^0 G^2d\nu+2\int_{\infty}^0 F_0^2+\int_0^\infty(1-G)^2d\nu+2\int_0^\infty(1-F_0)^2d\nu\\&\lesssim\int G^2(1-G)^2d\nu+\int F_0^2(1-F_0)^2d\nu\end{aligned}$$

and $F_0^2(1-F_0)^2 \leq F_0(1-F_0)$. In particular, this applies to the Lebesgue measure, if both G and F_0 have finite integrals.

Example 8.6 (White noise model) Suppose we observe an infinite-dimensional random vector $X^{(n)} = (X_{n,1}, X_{n,2}, \ldots)$, where $X_{n,i}\mid\theta \overset{\text{ind}}{\sim} \text{Nor}(\theta_i, n^{-1})$ and the parameter $\theta = (\theta_1, \theta_2, \ldots)$ belongs to ℓ_2, with square norm $\|\theta\|^2 = \sum_{i=1}^\infty \theta_i^2$. This model is equivalent to the white noise model considered in Section 8.3.4, and hence can be interpreted as concerned with estimating a function with Fourier coefficients (θ_i).

The priors $\theta_i \overset{\text{ind}}{\sim} \text{Nor}(0, i^{-2\alpha-1})$, for some fixed $\alpha > 0$, are conjugate in this model. The posterior distribution can be computed by considering the countably many coordinates separately, and is explicitly given by

$$\theta_i\mid X^{(n)} \overset{\text{ind}}{\sim} \text{Nor}\Big(\frac{nX_{n,i}}{n+i^{2\alpha+1}}, \frac{1}{n+i^{2\alpha+1}}\Big). \tag{8.1}$$

We claim that the rate of posterior contraction is $\epsilon_n = n^{-\min(\alpha,\beta)/(1+2\alpha)}$ at true parameters θ_0 with finite square Sobolev norm $\|\theta_0\|_{2,2,\beta}^2 = \sum_{i=1}^\infty i^{2\beta}\theta_{i,0}^2$.

By Chebyshev's inequality, $\Pi_n(\|\theta-\theta_0\| \geq \epsilon\mid X^{(n)})$ is bounded above by ϵ^{-2} times

$$\int \|\theta-\theta_0\|^2\, d\Pi_n(\theta\mid X^{(n)}) = \sum_{i=1}^\infty (\text{E}(\theta_i\mid X^{(n)}) - \theta_{0,i})^2 + \sum_{i=1}^\infty \text{var}(\theta_i\mid X^{(n)}).$$

The expectations of the terms of the first series on the right are the mean square errors of the univariate estimators $\text{E}(\theta_i\mid X^{(n)}) = nX_{n,i}/(n+i^{2\alpha+1})$ of the coordinates $\theta_{i,0}$, and can be easily evaluated as the sum of a square bias and variance, while the second series on the right is deterministic and can be directly read off from (8.1). We conclude that the expectation of the right side of the preceding display is given by

$$\sum_{i=1}^\infty \frac{\theta_{i,0}^2 i^{4\alpha+2}}{(n+i^{2\alpha+1})^2} + \sum_{i=1}^\infty \frac{n}{(n+i^{2\alpha+1})^2} + \sum_{i=1}^\infty \frac{1}{n+i^{2\alpha+1}}.$$

By Lemma K.7, the second and third series are of the order $n^{-2\alpha/(2\alpha+1)}$; and the first is of the order $n^{-2\min(\alpha,\beta)/(2\alpha+1)}$ if the true parameter θ_0 satisfies $\sum_{i=1}^\infty i^{2\beta}\theta_{i,0}^2 < \infty$. Lemma K.7 also shows that the latter bound is best possible if it is to be uniform in θ_0 such that $\sum_{i=1}^\infty i^{2\beta}\theta_{i,0}^2 \leq 1$. We may interpret the preceding as saying that $n^{-\min(\alpha,\beta)/(2\alpha+1)}$ is the (best possible) contraction rate if the only information on θ_0 is that it has a finite Sobolev norm of order β.

For given β, the prior corresponding to the choice $\alpha = \beta$ gives the best contraction rate for these parameters θ_0. Interestingly, under this prior $\sum_{i=1}^\infty i^{2\beta}\theta_i^2 = \infty$ a.s., which might be interpreted as indicating that the prior gives probability zero to the parameters it seems to target. This seems embarrassing to the Bayesian method. However we may argue that the support of any of the priors under consideration is the whole of ℓ_2, and approximation of a true parameter by elements of the support is of greater importance for a posterior contraction rate than belonging to a set of prior probability 1. We revisit this situation in Section 12.4.

Contraction of the posterior distribution at a rate implies the existence of point estimators that converge at the same rate. The same construction as in Proposition 6.7 applies.

Theorem 8.7 *If the posterior contraction rate at θ_0 is ϵ_n, then the center $\hat\theta_n$ of the smallest ball that contains posterior mass at least $1/2$ satisfies $d(\hat\theta_n, \theta_0) = O_P(\epsilon_n)$ in $P_{\theta_0}^{(n)}$-probability (or almost surely if the posterior contraction rate is in the strong sense).*

Proof The proof of Proposition 6.7 with ϵ replaced by $M_n\epsilon_n$ shows that $d(\hat\theta_n, \theta_0) \leq 2M_n\epsilon_n + 1/n$, where the $1/n$ term can be replaced by an arbitrarily small, positive quantity. Since this is true for every $M_n \to \infty$, the sequence $d(\hat\theta_n, \theta_0)/\epsilon_n$ is tight for every $M_n \to \infty$. This implies the claim. □

For the posterior mean, if this is defined, the argument of Theorem 6.8 gives the following theorem.

Theorem 8.8 *If d is bounded and $\theta \mapsto d^s(\theta, \theta_0)$ is convex for some $s \geq 1$, then the posterior mean $\hat\theta_n = \int \theta \, d\Pi_n(\theta| X^{(n)})$ satisfies $d(\hat\theta_n, \theta_0) \leq M_n\epsilon_n + \|d\|_\infty^{1/s} \Pi_n\,(\theta\colon d(\theta, \theta_0) > M_n\epsilon_n| X^{(n)})^{1/s}$.*

Thus the rate of posterior contraction is transferred to the posterior mean provided the posterior probability of the complement of (a multiple) of an ϵ_n-ball around θ_0 converges to zero sufficiently fast. For instance, for the Hellinger distance we can choose $s = 2$, and $\Pi_n(\theta\colon d(\theta, \theta_0) > M_n\epsilon_n| X^{(n)})$ should be bounded by a multiple of ϵ_n^2. This is usually the case; in fact, the contraction of the posterior is typically exponentially fast.

The applicability of Lemma 8.2 is typically limited to the situation where the posterior distribution can be computed explicitly. In general, closed-form expressions are not available, and the posterior needs to be analyzed through more abstract arguments. In the subsequent sections we develop a theory of posterior contraction rates for dominated models. We describe the rates in terms of a prior concentration rate and the size of the model, which may be viewed as quantitative analogs of Schwartz's conditions for posterior consistency.

8.2 Independent Identically Distributed Observations

In this section we take the parameter equal to a probability density p, belonging to a class $\mathcal{P}$ of probability densities relative to a given dominating measure ν on a sample space $(\mathfrak{X}, \mathscr{X})$. We consider the posterior distribution $\Pi_n(\cdot| X_1, \dots X_n)$ based on a sequence of priors Π_n on $\mathcal{P}$ and observations satisfying $X_1, \dots, X_n| p \overset{\text{iid}}{\sim} p$, where $p \sim \Pi_n$ in the Bayesian setup and the posterior distribution is studied under the assumption that $p = p_0$. As usual we denote a density and the corresponding probability measure by lower- and uppercase letters (e.g. p and P).

We derive posterior contraction rates relative to a metric d on the parameter set $\mathcal{P}$ that satisfies the requirement: for every $n \in \mathbb{N}$ and $\epsilon > 0$ and p_1 with $d(p_1, p_0) > \epsilon$ there exists a test ϕ_n with, for some universal constants $\xi, K > 0$,

$$P_0^n\phi_n \leq e^{-Kn\epsilon^2}, \qquad \sup_{d(p,p_1)<\xi\epsilon} P^n(1-\phi_n) \leq e^{-Kn\epsilon^2}. \tag{8.2}$$

Thus the "null hypothesis" P_0 can be separated from every d-ball at some distance from it, by a test with exponentially small error probabilities, smaller if the alternative is farther from the null hypothesis. It is shown in Proposition D.8 that this requirement is fulfilled (with $\xi = 1/2$ and $K = 1/8$) by any semimetric d that generates convex balls and satisfies $d(p_0, p) \leq d_H(p_0, p)$ for every density p, where d_H is the Hellinger metric. Examples are the Hellinger metric itself, but also the total variation metric, and for a bounded set of densities a multiple of the $\mathbb{L}_2(\nu)$-metric (namely this metric divided by twice the upper bound $\sup_{p\in\mathcal{P}} \|\sqrt{p}\|_\infty$ on the densities; see Lemma B.1).

The following theorem describes the posterior contraction rate in terms of a prior concentration rate and the entropy of the model. The concentration rate is measured in terms of a combination of the Kullback-Leibler divergence $K(p_0; p) = P_0 \log(p_0/p)$ and the kth Kullback-Leibler variation $V_{k,0}(p_0; p) = P_0|\log(p_0/p) - K(p_0; p)|^k$. For every $\epsilon > 0$ define neighborhoods of p_0 by

$$B_0(p_0, \epsilon) = \{p: K(p_0; p) < \epsilon^2\},$$

$$B_k(p_0, \epsilon) = \{p: K(p_0; p) < \epsilon^2, V_{k,0}(p_0; p) < \epsilon^k\}, \qquad (k > 0). \tag{8.3}$$

Theorem 8.9 (Basic contraction rate) *Given a distance d for which* (8.2) *is satisfied, suppose that there exist partitions $\mathcal{P} = \mathcal{P}_{n,1} \cup \mathcal{P}_{n,2}$ and a constant $C > 0$, such that, for constants $\bar\epsilon_n \leq \epsilon_n$ with $n\bar\epsilon_n^2 \to \infty$,*

(i) $\Pi_n(B_2(p_0, \bar\epsilon_n)) \geq e^{-Cn\bar\epsilon_n^2}$. (8.4)

(ii) $\log N(\xi\epsilon_n, \mathcal{P}_{n,1}, d) \leq n\epsilon_n^2$. (8.5)

(iii) $\Pi_n(\mathcal{P}_{n,2}) \leq e^{-(C+4)n\bar\epsilon_n^2}$. (8.6)

Then the posterior rate of contraction at p_0 is ϵ_n. If $\epsilon_n \gtrsim n^{-\alpha}$ for some $\alpha \in (0, 1/2)$ and (i) holds with $B_2(p_0, \bar\epsilon_n)$ replaced by $B_k(p_0, \bar\epsilon_n)$ for some k such that $k(1 - 2\alpha) > 2$, then the contraction rate is also in the almost sure sense $[P_0^\infty]$.

Proof Assumption (8.2) implies assumption (D.6) with K in the latter condition equal to the present Kn and $c = 1$. Hence by assumption (ii) and Theorem D.5 applied with $\epsilon = M\epsilon_n$ and $j = 1$ in its assertion, there exist tests ϕ_n with error probabilities

$$P_0^n\phi_n \leq e^{n\epsilon_n^2}\frac{e^{-KnM^2\epsilon_n^2}}{1 - e^{-KnM^2\epsilon_n^2}}, \qquad \sup_{p\in\mathcal{P}_{n,1}: d(p,p_0)>M\epsilon_n} P^n(1 - \phi_n) \leq e^{-KnM^2\epsilon_n^2}.$$

For $KM^2 > 1$ both error probabilities tend to zero. In view of Bayes's formula, for A_n the event that $\int \prod_{i=1}^n (p/p_0)(X_i)\, d\Pi_n(p) \geq e^{-(2+C)n\bar\epsilon_n^2}$, we can bound the posterior mass of the set $\{p: d(p, p_0) > M\epsilon_n\}$ by

$$\phi_n + \mathbb{1}\{A_n^c\} + e^{(2+C)n\bar\epsilon_n^2}\int_{d(p,p_0)>M\epsilon_n} \prod_{i=1}^n \frac{p}{p_0}(X_i)\, d\Pi_n(p)(1 - \phi_n). \tag{8.7}$$

The expected value under P_0^n of the first term tends to zero by construction of the tests ϕ_n. The same is true for the second term, by assumption (i) and Lemma 8.10 (below, with the constant D of the lemma taken equal to 1). The expected value of the third term is bounded above by

$$e^{(2+C)n\bar\epsilon_n^2}\int_{d(p,p_0)>M\epsilon_n} P^n(1-\phi_n)\,d\Pi_n(p) \le e^{(2+C)n\bar\epsilon_n^2}\Big(e^{-KnM^2\epsilon_n^2}+\Pi_n(\mathcal{P}_{n,2})\Big).$$

The last inequality follows by splitting the integral in parts over $\mathcal{P}_{n,1}$ and $\mathcal{P}_{n,2}$, and next using the bounds on the tests in the first part and the trivial bound $P^n(1-\phi_n)\le 1$ in the second part. For $KM^2 > 2+C$ the right side tends to zero by assumption (iii).

For the final assertion on almost sure contraction we note that for a general $k\ge 2$ Lemma 8.10 gives the bound $\mathrm{P}(A_n^c)\lesssim(\sqrt{n}\epsilon_n)^{-k}$, so that $\sum_n \mathrm{P}(A_n^c)<\infty$ under the stated conditions on ϵ_n and k, whence $P_0^\infty(\limsup A_n)=0$, by the Borel-Cantelli lemma, which means that $\mathbb{1}\{A_n^c\}\to 0$ almost surely. The other terms in (8.7) converge to zero exponentially fast, and can be treated by the same reasoning. ☐

From inspection of the proof, it is clear that if the constant C can be chosen the same for a collection of values of the true density p_0, then the contraction rate sequence ϵ_n can be chosen to be the same for all p_0 in the collection. Hence it makes sense to compare the posterior contraction rate with the corresponding minimax rate for the collection of true p_0.

The theorem copies the form of Theorem 6.23 on posterior consistency. The Kullback-Leibler property of p_0 has been quantified in condition (i), and the fixed ϵ is replaced by the sequences $\bar\epsilon_n$ and ϵ_n. Conditions (i) and (ii) (for given $\mathcal{P}_{n,1}$) are monotone in $\bar\epsilon_n$ or ϵ_n in the sense that they are satisfied for every larger value of these rates as soon at they are true for given values. They can thus be seen as defining minimal possible values of $\bar\epsilon_n$ and ϵ_n. The rate of posterior contraction is the slowest of the minimal values satisfying (ii) and the pair (i), (iii), respectively.

The *prior mass condition* (i) requires that the priors put a sufficient amount of mass near the true density p_0. In the basic theorem "near" is measured through the Kullback-Leibler divergence and variation $K(p_0; p)$ and $V_{2,0}(p_0; p)$, which define the neighborhoods $B_2(p_0; \epsilon)$. Later results will show that the variation may be superfluous and the neighborhoods enlarged to the Kullback-Leibler balls $B_0(p_0; \epsilon)$.

The *entropy condition* (ii) is the other main determinant of the contraction rate. It bounds the complexity of the model $\mathcal{P}_{n,1}$, which is the bulk of the "support" of the prior, in view of condition (iii). It is known that a rate ϵ_n satisfying (ii) for d the Hellinger metric gives the *minimax optimal rate* of contraction for estimators of p relative to the Hellinger metric, given the model $\mathcal{P}_{n,1}$ (see Birgé 1983a, Yang and Barron 1999). In the present situation the contraction rate can be seen to be uniform in p_0 that satisfy the conditions for given constants, so that the contraction is uniform and meets the minimax criterion. Technically the entropy condition ensures the existence of good tests of p_0 versus alternatives at some distance. For more flexibility it can be replaced by a testing condition; see Theorem 8.12.

The conditions (i) and (ii) are related. For the sake of the argument, assume that the square distance d^2 and the discrepancies K and $V_{2,0}$ are equivalent (which is the case if $d=d_H$ and the likelihood ratios p_0/p are uniformly bounded; see (8.8) below). Let ϵ_n be the minimal value that satisfies (ii), i.e. the optimal rate of contraction for the model $\mathcal{P}_{n,1}$. Thus, a minimal ϵ_n-cover of $\mathcal{P}_{n,1}$ consists of $e^{n\epsilon_n^2}$ balls. If the prior Π_n would spread its mass uniformly over $\mathcal{P}_{n,1}$, then every ball would receive prior mass approximately 1 over the number of balls in the cover, say $e^{-Cn\epsilon_n^2}$; the constant C may express the use of multiple distances and possible overlap between the balls in the cover. On the other hand, if Π_n is not evenly spread, then we should expect (i) to fail for some $p_0 \in \mathcal{P}_{n,1}$.

"Uniform priors" do not exist in infinite-dimensional models, and actually condition (i) is stronger than needed and will be improved ahead in Theorem 8.11. Nevertheless, a rough implication of this argument is that Π_n should be "not very unevenly spread" in order for the posterior distribution to attain the optimal rate of contraction at all elements p_0 in the parameter space.

Condition (iii), relative to (ii), can be interpreted as saying that a part of the model that barely receives prior mass need not have bounded complexity. It is trivially satisfied for the partition with $\mathcal{P}_{n,1} = \mathcal{P}$; we can make this choice if the entropy of the full model is bounded as in (ii).

The prior mass and entropy conditions of Theorem 8.9 use the Kullback-Leibler divergence and the metric d, respectively. Mixing two types of neighborhoods is not pretty, but seems to be unavoidable in general. An advantage is that the metric d reappears in the assertion of the theorem, giving a stronger sense of contraction when using a stronger metric (at the cost of a more restrictive complexity condition). It is of interest to note that for sufficiently regular models, the square Hellinger distance and Kullback-Leibler discrepancies are equivalent. In particular, by Lemmas B.2 and B.3, for any $k \geq 2$,

$$\left\{p: d_H(p, p_0)\left\|\frac{p_0}{p}\right\|_\infty < \sqrt{2k!}\,\epsilon\right\} \subset B_k(p_0, \epsilon) \subset \{p: d_H(p, p_0) < \epsilon\}. \tag{8.8}$$

Thus, given bounded likelihood ratios, a theorem can be stated solely in terms of the Hellinger distance (or, equivalently the Kullback-Leibler divergence). The same lemmas also apply under just moment conditions on the ratios p_0/p, but only up to factors $\log_- \epsilon$, giving equivalent rates of contraction up to logarithmic factors.

The testing condition (8.2) links the metric d to the likelihood. In the case of i.i.d. observations the Hellinger distance d_H seems to be a canonical choice. For a metric d that is stronger than the Hellinger distance the condition may not hold. Posterior contraction may still take place, but may involve properties of prior and model additional to the prior mass and entropy considered in the basic theorem. One possible approach to handling a different metric is to show, by additional arguments, that the posterior distribution gives mass tending to one to sets $\mathcal{S}_n \subset \mathcal{P}$ on which the new metric is comparable to the Hellinger distance. Specifically, if $d(p_0, p) \leq m_n d_H(p_0, p)$ for all $p \in \mathcal{S}_n$ and some sequence of numbers m_n, and $\Pi_n(\mathcal{S}_n^c | X_1, \ldots, X_n) \to 0$ in probability, then a posterior contraction rate ϵ_n relative to the Hellinger distance implies a contraction rate $m_n \epsilon_n$ with respect to d. One method to show that the posterior distributions concentrate on $\mathcal{S}_n$ is to prove that $\Pi_n(\mathcal{S}_n^c)$ is exponentially small, and apply Theorem 8.20 (below, which gives a generalization of (iii) in the basic theorem above). This simple device has been used to deal with the supremum distance and certain inverse problems.

The following lemma was used in the proof of Theorem 8.9. It gives a lower bound on the denominator of the posterior measure (the "evidence"), similar to Lemma 6.26. Employment of the stronger neighborhoods $B_k(p_0; \epsilon)$, rather than Kullback-Leibler balls $B_0(p_0; \epsilon)$ as in the latter lemma, gives better control of the probability of failure of the lower bound. Further variations, which give exponential bounds on these probabilities, are given in Problems 8.6 and 8.8.

Lemma 8.10 (Evidence lower bound) *For every $k \geq 2$ there exists a constant $d_k > 0$ (with $d_2 = 1$) such that for any probability measure Π on $\mathcal{P}$, and any positive constants ϵ, D, with P_0^n-probability at least $1 - d_k(D\sqrt{n}\epsilon)^{-k}$,*

$$\int \prod_{i=1}^{n} \frac{p}{p_0}(X_i)\, d\Pi(p) \geq \Pi(B_k(p_0, \epsilon))e^{-(1+D)n\epsilon^2}.$$

Proof The integral becomes smaller by restricting it to the set $B := B_k(p_0, \epsilon)$. By next dividing the two sides of the inequality by $\Pi(B)$, we can rewrite the inequality in terms of the prior Π restricted and renormalized to a probability measure on B. Thus we may without loss of generality assume that $\Pi(B) = 1$. By Jensen's inequality applied to the logarithm,

$$\log \int \prod_{i=1}^{n} \frac{p}{p_0}(X_i)\, d\Pi(p) \geq \sum_{i=1}^{n} \int \log \frac{p}{p_0}(X_i)\, d\Pi(p) =: Z.$$

The right side has mean $\mathrm{E}Z = -n \int K(p_0; p)\, d\Pi(p) > -n\epsilon^2$, by the definition of B. Furthermore, by the Marcinkiewicz-Zygmund inequality (Lemma K.4), followed by Jensen's inequality, Fubini's theorem, and again the definition of B,

$$\mathrm{E}\Big|\frac{Z - \mathrm{E}Z}{\sqrt{n}}\Big|^k \lesssim P_0 \Big|\int \Big(\log \frac{p}{p_0} - P_0 \log \frac{p}{p_0}\Big)\, d\Pi(p)\Big|^k \leq \int V_{k,0}(p_0; p)\, d\Pi(p) \leq \epsilon^k.$$

The constant d_k in the lemma is the proportionality constant in the first inequality (e.g. $d_2 = 1$). Finally $P_0^n(Z < -(1+D)n\epsilon^2) \leq P_0^n(Z - \mathrm{E}Z < -Dn\epsilon^2)$ is bounded by $\mathrm{E}|Z - \mathrm{E}Z|^k/(Dn\epsilon^2)^k$, by Markov's inequality, which can be further bounded as in the lemma. ☐

One deficit of Theorem 8.9 is that it does not satisfactorily cover finite-dimensional models. For models of fixed dimension it yields the rate $n^{-1/2}$ times a logarithmic factor rather than the correct $n^{-1/2}$, and when applied to sieves of finite, increasing dimensions it incurs a similar unnecessary logarithmic term. To improve this situation, both the prior mass (i) and entropy (ii) conditions must be refined.

The following theorem is also more precise in employing the neighborhoods $B_0(p_0; \epsilon)$, which are solely in terms of the Kullback-Leibler divergence and not the variation.

Theorem 8.11 (Refined contraction rate) *Given a distance d for which (8.2) is satisfied, suppose that there exist partitions $\mathcal{P} = \mathcal{P}_{n,1} \cup \mathcal{P}_{n,2}$, such that, for constants $\epsilon_n, \bar{\epsilon}_n \geq n^{-1/2}$, and every sufficiently large j,*

$$\text{(i)}\quad \frac{\Pi_n(P\colon j\epsilon_n < d(p, p_0) \leq 2j\epsilon_n)}{\Pi_n(B_0(p_0, \epsilon_n))} \leq e^{Kn\epsilon_n^2 j^2/2}, \tag{8.9}$$

$$\text{(ii)}\quad \sup_{\epsilon \geq \epsilon_n} \log N\Big(\xi\epsilon, \{p \in \mathcal{P}_{n,1}\colon d(p, p_0) \leq 2\epsilon\}, d\Big) \leq n\epsilon_n^2, \tag{8.10}$$

$$\text{(iii)}\quad \frac{\Pi_n(\mathcal{P}_{n,2})}{\Pi_n(B_0(p_0, \bar{\epsilon}_n))} = o(e^{-D_n n\bar{\epsilon}_n^2}), \qquad \textit{for some } D_n \to \infty. \tag{8.11}$$

Then the posterior rate of contraction at p_0 is ϵ_n. This remains true if (iii) is replaced by

$$\text{(iii}'\text{)}\quad \frac{\Pi_n(\mathcal{P}_{n,2})}{\Pi_n(B_2(p_0, \bar{\epsilon}_n))} = o(e^{-2n\bar{\epsilon}_n^2}). \tag{8.12}$$

Proof The proof is similar to the proof of Theorem 8.9, but we use Lemma 6.26 next to Lemma 8.10, more refined tests, and split the sieves in "shells." We prove first that the posterior probability of the set $\{p \in \mathcal{P}_{n,1}: d(p, p_0) > M\epsilon_n\}$ tends to zero for large M or $M \to \infty$ (under (i) and (ii)), and next separately the same for the sets $\mathcal{P}_{n,2}$ (under either (iii) or (iii$'$)).

By (8.10) and Theorem D.5, applied with $\epsilon = M\epsilon_n$, a given $M > 1$, and N the constant function $N(\epsilon) = n\epsilon_n^2$, there exist tests ϕ_n such that, for every $j \in \mathbb{N}$,

$$P_0^n \phi_n \le e^{n\epsilon_n^2} \frac{e^{-KnM^2\epsilon_n^2}}{1 - e^{-KnM^2\epsilon_n^2}}, \qquad \sup_{p \in \mathcal{P}_{n,1}: d(p,p_0) > Mj\epsilon_n} P^n(1 - \phi_n) \le e^{-KnM^2\epsilon_n^2 j^2}.$$

The set $\{p \in \mathcal{P}_{n,1}: d(p, p_0) > M\epsilon_n\}$ can be partitioned into the countably many "shells" $\mathcal{S}_{n,j} = \{p \in \mathcal{P}_{n,1}: M\epsilon_n j < d(p, p_0) \le M\epsilon_n(j + 1)\}$, for $j \in \mathbb{Z}$. For every $j \in \mathbb{N}$, by Fubini's theorem and the preceding display,

$$P_0^n\Big[\int_{\mathcal{S}_{n,j}} \prod_{i=1}^n \frac{p}{p_0}(X_i)\, d\Pi_n(p)(1 - \phi_n)\Big] \le e^{-KnM^2\epsilon_n^2 j^2}\Pi_n(\mathcal{S}_{n,j}).$$

For A_n the event that $\int \prod_{i=1}^n (p/p_0)(X_i)\, d\Pi_n(p) \ge e^{-2Dn\epsilon_n^2}\Pi_n(B_0(p_0, \epsilon_n))$, we now replace the upper bound (8.7) on the posterior probability of the set $\{p \in \mathcal{P}_{n,1}: d(p, p_0) > M\epsilon_n\}$ by

$$\phi_n + \mathbb{1}\{A_n^c\} + \frac{1}{e^{-2Dn\epsilon_n^2}\Pi_n(B_0(p_0, \epsilon_n))} \int_{p \in \mathcal{P}_{n,1}: d(p,p_0) > M\epsilon_n} \prod_{i=1}^n \frac{p}{p_0}(X_i)\, d\Pi_n(p)(1 - \phi_n).$$

By construction of the tests, the expected value of the first term tends to zero if $n\epsilon_n^2 \to \infty$ and $KM^2 > 1$, and can be made arbitrarily small by choosing M big if $n\epsilon_n^2$ is only known to be bounded away from zero. The expected value of the second term is bounded above by $1/D$ by Lemma 6.26, and can be made arbitrarily small by choice of D. The domain of the integral in the third term is contained in $\cup_{j \ge 1}\mathcal{S}_{n,j}$, and hence the expected value of this term is bounded above by

$$\sum_{j \ge 1} \frac{e^{-KnM^2\epsilon_n^2 j^2}\Pi_n(\mathcal{S}_{n,j})}{e^{-2Dn\epsilon_n^2}\Pi_n(B_0(p_0, \epsilon_n))} \le \sum_{j \ge 1} e^{-n\epsilon_n^2(KM^2j^2 - 2D - \frac{1}{2}KM^2j^2)},$$

by (8.9). This converges to zero as $n\epsilon_n^2 \to \infty$ and $KM^2 > 4D$, and can be made arbitrarily small by choosing large M if $n\epsilon_n^2$ is only known to be bounded away from zero.

Next we consider the posterior probability of the sets $\mathcal{P}_{n,2}$. Given validity of (iii), we define the events A_n as previously, but with ϵ_n replaced by $\bar\epsilon_n$. Reasoning as before we obtain that the posterior probability of $\mathcal{P}_{n,2}$ is bounded above by

$$\mathbb{1}\{A_n^c\} + \frac{\Pi_n(\mathcal{P}_{n,2})}{e^{-2Dn\bar\epsilon_n^2}\Pi_n(B_0(p_0, \bar\epsilon_n))}.$$

The expected value of the first term is bounded by $1/D$ and hence can be made arbitrarily small by choosing a large D. The second term tends to zero under assumption (iii), for any D. If not (iii) but (iii$'$) is assumed, then we redefine the A_n as the events that $\int \prod_{i=1}^n (p/p_0)(X_i)\, d\Pi_n(p) \ge e^{-2Dn\bar\epsilon_n^2}\Pi_n(B_2(p_0, \bar\epsilon_n))$. The posterior probability of $\mathcal{P}_{n,2}$

is then bounded by the preceding display, but with B_0 replaced by B_2. By Lemma 8.10 the expectation of the first term $\mathbb{1}\{A_n^c\}$ is bounded above by $(2D-1)^{-2}(n\bar\epsilon_n^2)^{-1}$. To finish the proof, we may assume that either $n\bar\epsilon_n^2 \to \infty$ or $n\bar\epsilon_n^2$ is bounded; otherwise we argue along subsequences. If $n\bar\epsilon_n^2 \to \infty$, then we choose $D = 1$ and the expectation of $\mathbb{1}\{A_n^c\}$ tends to zero, as does the second term in the preceding display, by assumption (iii′). If $n\bar\epsilon_n^2$ remains bounded then the factor $e^{-2Dn\bar\epsilon_n^2}$ is bounded away from zero and infinity for every fixed D, since $n\bar\epsilon_n^2 \geq 1$, by assumption. By (iii′) the second term of the preceding display then tends to zero for every fixed $D > 0$, and the expectation of the first can be made arbitrarily small by choosing a large D. □

The left side of (8.10) is called the *local entropy* or *Le Cam dimension* of the set $\mathcal{P}_{n,1}$ at p_0. It measures the number of balls of radius $\xi\epsilon$ needed to cover a ball (or a shell) of radius 2ϵ. That the quotient of the latter two radii is independent of ϵ makes the Le Cam dimension smaller for "finite-dimensional" models, where it indeed behaves as a dimension (see Appendix D.1). For genuinely infinite-dimensional models the local entropy in the left side of (8.10) is typically comparable in magnitude to $\log N(\xi\epsilon_n, \mathcal{P}_{n,1}, d)$. As the latter quantity is always an upper bound on the left side of (8.10), this condition is implied by (8.5).

The numerator of (8.9) is trivially bounded above by 1, and hence this condition is implied by (8.4) (where B_2 may even be replaced by the larger set B_0). Similarly, given (8.4), condition (8.12) relaxes the corresponding condition (8.6) of Theorem 8.9.

8.2.1 Further Refinements

The entropy conditions appearing in the preceding theorems are used to construct tests against (nonconvex) alternatives of the form $\{p \in \mathcal{P}_{n,1}: \epsilon < d(p, p_0) \leq 2\epsilon\}$. Tests are more fundamental than entropies for contraction rates, and in some applications can be constructed by direct methods. Thus the testing conditions in the following theorem give more flexibility and maneuverability. The proof is essentially contained in that of Theorem 8.11 with the choice $M = 1$.

Theorem 8.12 (Rates by testing) *If there exist partitions $\mathcal{P} = \mathcal{P}_{n,1} \cup \mathcal{P}_{n,2}$ and a constant $C > 0$, such that* (8.9) *and* (8.12) *hold for constants $\bar\epsilon_n \leq \epsilon_n$ with $n\bar\epsilon_n^2 \geq 1$ and every sufficiently large j, and in addition there exists a sequence of tests ϕ_n such that for some constant $K > 0$ and for every sufficiently large j*

$$P_0^n\phi_n \to 0, \qquad \sup_{p\in\mathcal{P}_{n,1}: j\epsilon_n < d(p,p_0) \leq 2j\epsilon_n} P^n(1-\phi_n) \leq e^{-Kn\epsilon_n^2 j^2}, \qquad (8.13)$$

then the posterior rate of contraction at p_0 is ϵ_n.

Like for consistency, testing conditions for a posterior contraction rate result are almost necessary. Given prior concentration, posterior contraction at an exponential speed implies the existence of tests and a sieve, as hypothesized in Theorem 8.12.

Theorem 8.13 (Necessity of testing) *If $P_0^n \Pi_n(p: d(p_0, p) \geq M\epsilon_n | X_1, \ldots, X_n) \leq e^{-Cn\epsilon_n^2}$, for given positive constants C and M, then there exists a partition $\mathcal{P} = \mathcal{P}_{1,n} \cup \mathcal{P}_{2,n}$ and a sequence of tests ϕ_n such that $P_0^n \phi_n \to 0$, $\sup\{P^n(1-\phi_n): p \in \mathcal{P}_{1,n}\} \leq e^{-Cn\epsilon_n^2/4}$ and $\Pi_n(\mathcal{P}_{2,n}) \leq e^{-Cn\epsilon_n^2/4}$.*

Proof Define an event $S_n = \{\Pi_n(p: d(p_0, p) \geq M\epsilon_n | X_1, \ldots, X_n) > e^{-Cn\epsilon_n^2/2}\}$ and proceed as in the proof of Theorem 6.22. The test $\phi_n = \mathbb{1}_{S_n}$ has type I error probability $P_0^n \phi_n = P_0^n(S_n) \leq e^{-Cn\epsilon_n^2/2} \to 0$ and $P^n(1-\phi_n) = P^n(S_n^c) \leq e^{-Cn\epsilon_n^2/2}$ for $p \in \mathcal{P}_{n,1} = \mathcal{P} \setminus \mathcal{P}_{n,2}$, where $\mathcal{P}_{n,2} = \{p: P^n(S_n^c) \geq e^{-Cn\epsilon_n^2/2}\}$. Finally $\Pi(\mathcal{P}_{n,2}) \leq e^{-Cn\epsilon_n^2/4}$ by arguments as before. □

In Theorems 8.9 and 8.11, the sieves $\mathcal{P}_{n,1}$ are controlled by their entropy, without taking account of the prior mass they carry, whereas their complements $\mathcal{P}_{n,2}$ are measured solely by their prior probability, regardless of their complexity. These are two extreme sides of what matters for posterior concentration. In the following theorem, which parallels Theorems 8.9, the parameter space is divided into more than two pieces to make a smoother transition from one method to another. Theorem 8.11 can be similarly refined.

Theorem 8.14 (Rate by partition entropy) *Given a distance d for which* (8.2) *is satisfied suppose that there exists disjoint subsets $\mathcal{P}_{n,j}$ of $\mathcal{P}$ such that, for constants ϵ_n with $n\epsilon_n^2 \to \infty$,*

$$\sum_{j=1}^{\infty} \sqrt{N(\xi\epsilon_n, \mathcal{P}_{n,j}, d)} \sqrt{\Pi_n(\mathcal{P}_{n,j})} \leq e^{n\epsilon_n^2}. \tag{8.14}$$

If (8.4) *holds for some constant $C > 0$ and $\bar{\epsilon}_n = \epsilon_n$, then $\Pi_n(p \in \cup_{j=1}^{\infty} \mathcal{P}_{n,j}: d(p, p_0) > M\epsilon_n | X_1, \ldots, X_n) \to 0$ in P_0^n-probability, for every sufficiently large M.*

Proof Without loss of generality, we may assume that the priors charge only $\cup_{j\geq 1} \mathcal{P}_{n,j}$. On the event $A_n = \{\int \prod_{i=1}^n (p/p_0)(X_i)\, d\Pi_n(p) \geq e^{-(C+2)n\epsilon_n^2}\}$ the posterior mass of the set $\{p: d(p, p_0) > M\epsilon_n\}$ can be bounded above, for arbitrary tests $\phi_{n,j}$, by

$$\sum_{j=1}^{\infty} \phi_{n,j} + \mathbb{1}\{A_n^c\} + e^{(C+2)n\epsilon_n^2} \sum_{j=1}^{\infty} \int_{p \in \mathcal{P}_{n,j}: d(p, p_0) > M\epsilon_n} \prod_{i=1}^n \frac{p}{p_0}(X_i)\, d\Pi_n(p)^n (1 - \phi_{n,j}).$$

The event A_n has probability tending to one, by Lemma 8.10 and assumption (8.4). By Theorem D.5 applied for every $j \in \mathbb{Z}$ separately with $c \leftarrow c_j$ and $\mathcal{Q} = \{p \in \mathcal{P}_{n,j}: d(p_0, p) \geq M\epsilon_n\}$, there exist tests $\phi_{n,j}$ so that the expected value under P_0^n of the preceding display is bounded above by

$$\sum_{j=1}^{\infty} c_j\, N(\xi\epsilon_n, \mathcal{P}_{n,j}, d) \frac{e^{-KnM^2\epsilon_n^2}}{1 - e^{-KnM^2\epsilon_n^2}} + P_0^n(A_n^c) + e^{(C+2)n\epsilon_n^2} \sum_{j=1}^{\infty} \frac{1}{c_j} e^{-KnM^2\epsilon_n^2} \Pi_n(\mathcal{P}_{n,j}).$$

Now choose $c_j^2 = \Pi_n(\mathcal{P}_{n,j})/N(\xi\epsilon_n, \mathcal{P}_{n,j}, d)$ and $KM^2 > C + 2$, and sum over j to obtain that this tends to zero, in view of (8.14). □

Theorem 8.14 contains Theorem 8.9 as the special case of a partition in a single set $\mathcal{P}_{n,1}$: if $\mathcal{P}_{n,j} = \varnothing$ for $j \geq 2$, then (8.14) reduces to the bound $\log N(\xi\epsilon_n, \mathcal{P}_{n,1}, d) \leq n\epsilon_n^2$. (The present set $\mathcal{P}_{n,0}$ plays the role of $\mathcal{P}_{n,2}$ in Theorem 8.9, and is presently not addressed.) At the other extreme, if the model is decomposed fine enough so that each $\mathcal{P}_{n,j}$ has diameter smaller than $\xi\epsilon_n$, then the covering numbers appearing in (8.14) are all 1, and hence (8.14) reduces to

$$\sum_{j=1}^{\infty} \sqrt{\Pi_n(\mathcal{P}_{n,j})} \leq e^{n\epsilon_n^2}. \tag{8.15}$$

This condition is a quantitative analog of the summability condition for posterior consistency in Example 6.24 (with $\alpha = 1/2$). (A contraction rate theorem based on (8.15) may also be derived using the martingale approach discussed in Section 6.8.4; see Problem 8.11.)

Theorem 8.14 can be written more neatly in terms of a new measure of size, called *Hausdorff entropy*, which takes into account both number and prior probabilities of ϵ-coverings; see Problem 8.13.

Surprisingly, Theorem 8.14 can also be proved as a corollary to Theorem 8.9 by a clever definition of the sieves; see Problem 8.13.

8.2.2 Priors Based on Finite Approximating Sets

The rate of posterior contraction is determined by the complexity of a model and the prior concentration. Although in the Bayesian setup the "model" can be viewed as being implicitly determined by the prior, complexity is predominantly a non-Bayesian quality. In fact, under general conditions a rate ϵ_n satisfying the entropy inequality $\log N(\epsilon_n, \mathcal{P}, d_H) \leq n\epsilon_n^2$ is *minimax optimal* for estimating a density in the model $\mathcal{P}$, relative to the Hellinger distance, for any method of point estimation. In this section we show that this rate can always be obtained (perhaps up to a logarithmic factor) also as a rate of posterior contraction, for some prior. This prior is constructed as a convex combination of discrete uniform measures on a sequence of ϵ_n-nets over the model.

We start by assuming the slightly stronger entropy inequality $\log N_](\epsilon_n, \mathcal{P}, d_H) \leq n\epsilon_n^2$, defined in terms of upper brackets, rather than covers by balls. The *ϵ-upper bracketing number* $N_](\epsilon, \mathcal{P}, d_H)$ is defined as the minimal number of functions $u_{1,n}, \ldots, u_{N_n,n}$ such that for every $p \in \mathcal{P}$ there exists j with $p \leq u_{j,n}$ and $d_H(p, u_{j,n}) \leq \epsilon$. Given a minimal set of $N_n = N_](\epsilon_n, \mathcal{P}, d_H)$ upper brackets at discretization level ϵ_n, let Π_n be the discrete uniform measure on the set $\mathcal{P}_n = \{u_j / \int u_j \, d\nu \colon j = 1, \ldots, N_n\}$. Next, as overall prior form the mixture $\Pi = \sum_{n \in \mathbb{N}} \lambda_n \Pi_n$, for (λ_n) a strictly positive probability vector on $\mathbb{N}$.

Normalizing the upper brackets $u_{j,n}$ is necessary to ensure that the prior is supported on genuine probability densities, as upper brackets are functions with (typically) integrals larger than unity. The final prior Π is not supported on the target model $\mathcal{P}$, but this does not pose a challenge for applying the rate theorems.

Theorem 8.15 (Prior on nets) *If $\epsilon_n \downarrow 0$ is a sequence such that* $\log N_](\epsilon_n, \mathcal{P}, d_H) \leq n\epsilon_n^2$ *for every n and $n\epsilon_n^2/\log n \to \infty$ and the weights λ_n are strictly positive with* $\log \lambda_n^{-1} = O(\log n)$, *then the posterior converges relative to d_H at the rate ϵ_n almost surely, at every $p_0 \in \mathcal{P}$.*

Proof Let $\mathcal{P}_n$ be the support of Π_n and let $\mathcal{Q} = \cup_{n=1}^{\infty}\mathcal{P}_n$, whence $\Pi(\mathcal{Q}) = 1$ by construction. It suffices to control the Hellinger entropy of $\mathcal{Q}$ and verify the prior mass condition.

If $p \le u$ and $\|\sqrt{u} - \sqrt{p}\|_2 = d_H(p, u) < \epsilon$, then by the triangle inequality $1 \le \|\sqrt{u}\|_2 \le \|\sqrt{u} - \sqrt{p}\|_2 + \|\sqrt{p}\|_2 \le \epsilon + 1$, and hence

$$d_H\Big(p, \frac{u}{\int u\, dv}\Big) \le d_H(p, u) + d_H\Big(u, \frac{u}{\int u\, dv}\Big) = \|\sqrt{u} - \sqrt{p}\|_2 + \|\sqrt{u}\|_2 - 1 \le 2\epsilon.$$

Therefore, by construction every point of $\mathcal{P}$ is within $2\epsilon_n$ distance of some point in $\mathcal{P}_n$. Since for $j > n$, every $p \in \mathcal{P}_j$ is within distance $2\epsilon_j \le 2\epsilon_n$ of $\mathcal{P}$, it follows that $\mathcal{P}_n$ is a $4\epsilon_n$-net over $\mathcal{P}_j$. Consequently, $\mathcal{P}_n$ is a $4\epsilon_n$-net over $\cup_{j \ge n}\mathcal{P}_j$ and hence $\cup_{j \le n}\mathcal{P}_j$ is a $4\epsilon_n$-net over $\mathcal{Q}$. The cardinality of $\cup_{j \le n}\mathcal{P}_j$ is bounded above by $nN_n \le \exp(\log n + n\epsilon_n^2) \le e^{2n\epsilon_n^2}$, for sufficiently large n. This verifies the entropy condition (ii) of Theorem 8.9 with the sieve taken equal to $\mathcal{Q}$ and ϵ_n taken equal to four times the present ϵ_n. For this sieve the remaining mass condition (iii) is trivially satisfied.

If u is the upper limit of the ϵ_n-bracket containing p_0, then

$$\frac{p_0}{u/\int u\, dv} \le \int u\, dv \le (1 + \epsilon_n)^2 \le 2.$$

It follows that for large n, the set of points p such that $d_H^2(p, p_0)\|p_0/p\|_\infty \le 8\epsilon_n^2$ contains at least the function $u/\int u\, dv$ and hence has prior mass at least

$$\lambda_n \frac{1}{N_n} \ge \exp[-n\epsilon_n^2 - O(\log n)] \ge e^{-2n\epsilon_n^2},$$

for large n. In virtue of (8.8), this verifies the prior mass condition (i) of Theorem 8.9 for ϵ_n, a multiple of the present ϵ_n and with the neighborhoods $B_k(p_0; \epsilon_n)$ instead of $B_2(p_0; \epsilon_n)$, for any $k \ge 2$. Thus Theorem 8.9 gives the contraction rate ϵ_n in the strong sense. ☐

Example 8.16 (Smooth densities) Suppose that $\mathcal{P}$ consists of all densities p such that $\sqrt{p}$ belongs to a fixed multiple of the unit ball of the Hölder class $\mathfrak{C}^\alpha[0, 1]$, for some fixed $\alpha > 0$ (see Definition C.4). By Proposition C.5, the entropy of this unit ball relative to the uniform norm, and hence also relative to the weaker $\mathbb{L}_2(\nu)$-norm, is bounded by a multiple of $\epsilon^{-1/\alpha}$. Thus the Hellinger bracketing entropy (see Appendix C) of $\mathcal{P}$ possesses the same upper bound, and $\epsilon_n \asymp n^{-\alpha/(2\alpha+1)}$ satisfies the relation $\log N_{[\,]}(\epsilon_n, \mathcal{P}, d_H) \le n\epsilon_n^2$ and so does the upper bracketing entropy. Thus the prior based on upper brackets achieves the posterior contraction rate $n^{-\alpha/(2\alpha+1)}$. This rate is known to be the frequentist optimal rate for point estimators of a density in a Hölder ball.

Example 8.17 (Monotone densities) Suppose that $\mathcal{P}$ consists of all monotone decreasing densities on a compact interval in $\mathbb{R}$, bounded above by a fixed constant. Since the square root of a monotone density is again monotone, the bracketing entropy of $\mathcal{P}$ for the Hellinger distance is bounded by the bracketing $\mathbb{L}_2$-entropy of the set of monotone functions, which grows like ϵ^{-1}, in view of Proposition C.8. This leads to an $n^{-1/3}$-rate of contraction of the posterior. Again this agrees with the optimal frequentist rate for the problem, and hence it cannot be improved.

Theorem 8.15 implicitly requires that the model $\mathcal{P}$ is totally bounded for d_H. When $\mathcal{P}$ is a countable union of totally bounded models, a simple modification works, provided that we use a sequence of priors. To this end, suppose that there exist subsets $\mathcal{P}_n \uparrow \mathcal{P}$ with finite upper bracketing numbers, and let ϵ_n be numbers such that $\log N_](\epsilon_n, \mathcal{P}_n, d_H) \le n\epsilon_n^2$, for every n. Now construct Π_n as before with $\mathcal{P}$ replaced by $\mathcal{P}_n$. Next, do not form a mixture prior, but use Π_n itself as the prior distribution (different for different sample sizes). Then, as before, the corresponding posteriors achieve the contraction rate ϵ_n (see Theorem 8.24).

The entropy condition (ii) of Theorem 8.9 is in terms of metric entropy only, whereas presently we use brackets. The brackets give control over likelihood ratios, and are used for verifying the prior mass condition only. If likelihood ratios can be controlled by other means, then ordinary entropy will do also in Theorem 8.15. For instance, it suffices that the densities are uniformly bounded away from zero and infinity; the quotients p_0/p are then uniformly bounded. In the absence of such uniform bounds, ordinary metric entropy can still be used if the maximum of the set of densities is integrable, but a small (logarithmic) price has to be paid in the contraction rate.

To show this, let m be a ν-integrable *envelope function* for the set of densities $\mathcal{P}$; thus $p(x) \le m(x)$ for every $p \in \mathcal{P}$ and every x. Let $\{p_{1,n}, \ldots, p_{N_n,n}\}$ be a minimal ϵ_n-net over $\mathcal{P}$, and put $g_{j,n} = (p_{j,n}^{1/2} + \epsilon_n m^{1/2})^2 / c_{j,n}$, where $c_{j,n}$ is the constant that normalizes $g_{j,n}$ to a probability density. Let Π_n be the uniform discrete measure on $g_{1,n}, \ldots, g_{N_n,n}$, and let $\Pi = \sum_{n=1}^{\infty} \lambda_n \Pi_n$ be a convex combination of the Π_n.

Theorem 8.18 *Assume that $\mathcal{P}$ has a ν-integrable envelope m, and construct Π as indicated for $\epsilon_n \downarrow 0$ with $n\epsilon_n^2/\log n \to \infty$ satisfying $\log N(\epsilon_n, \mathcal{P}, d_H) \le n\epsilon_n^2$, and strictly positive weights λ_n with $\log \lambda_n^{-1} = O(\log n)$. Then the posterior contracts at the rate $\epsilon_n(\log n)^{1/2}$ in probability, relative to the Hellinger distance d_H.*

Proof It can be verified that for $p \in \mathcal{P}$ with $d_H(p, p_{j,n}) \le \epsilon_n$,

$$d_H(p, g_{j,n}) = O(\epsilon_n), \qquad \frac{p}{g_{j,n}} = O(\epsilon_n^{-2}). \tag{8.16}$$

Then Lemma B.2 implies that $K(p; g_{j,n}) = O(\epsilon_n^2 \log_- \epsilon_n)$. This verifies the prior mass condition (i) of Theorem 8.11 with ϵ_n replaced by a multiple of $\epsilon_n(\log n)^{1/2}$, since $\log_- \epsilon_n = O(\log n)$ by the condition $n\epsilon_n^2/\log n \to \infty$. The rest of the proof is the same as the proof of Theorem 8.15. $\square$

8.3 General Observations

In this section we return to the general setup of the introduction of the present chapter, involving experiments $(\mathfrak{X}^{(n)}, \mathscr{X}^{(n)}, P_\theta^{(n)}: \theta \in \Theta_n)$, with parameter spaces Θ_n, observations $X^{(n)}$ and true parameters $\theta_{n,0} \in \Theta_n$. The following theorem applies whenever exponentially powerful tests exist, and it is not otherwise restricted to a particular structure of the experiments.

For each n, let d_n and e_n be two semimetrics on Θ_n with the property: there exist universal constants $\xi, K > 0$ such that for every $\epsilon > 0$ and every $\theta_{n,1} \in \Theta_n$ with $d_n(\theta_{n,1}, \theta_{n,0}) > \epsilon$, there exists a test ϕ_n such that

$$P_{\theta_{n,0}}^{(n)}\phi_n \le e^{-Kn\epsilon^2}, \qquad \sup_{\theta\in\Theta_n: e_n(\theta,\theta_{n,1})<\xi\epsilon} P_\theta^{(n)}(1-\phi_n) \le e^{-Kn\epsilon^2}. \tag{8.17}$$

Typically, we have $d_n \le e_n$, and in many cases we choose $d_n = e_n$, but using two semimetrics adds flexibility. Apart from this, the condition is a direct generalization of (8.2) to general experiments.

As in the i.i.d. case, the behavior of posterior distributions depends on the concentration rate of the prior at the true parameter and the *Le Cam dimension* of the parameter set. For K and $V_{k,0}$ the Kullback-Leibler divergence and variation, let

$$B_{n,0}(\theta_{n,0},\epsilon) = \Big\{\theta\in\Theta_n: K(p_{\theta_{n,0}}^{(n)}; p_\theta^{(n)}) \le n\epsilon^2\Big\}, \tag{8.18}$$

$$B_{n,k}(\theta_{n,0},\epsilon) = \Big\{\theta\in\Theta_n: K(p_{\theta_{n,0}}^{(n)}; p_\theta^{(n)}) \le n\epsilon^2, V_{k,0}(p_{\theta_{n,0}}^{(n)}; p_\theta^{(n)}) \le n^{k/2}\epsilon^k\Big\}, \quad (k>0). \tag{8.19}$$

Theorem 8.19 (Contraction rate) *Let d_n and e_n be semimetrics on Θ_n for which there exist tests satisfying (8.17), and let $\Theta_{n,1}\subset\Theta_n$ be arbitrary. If for constants ϵ_n with $n\epsilon_n^2\ge 1$, for every sufficiently large $j\in\mathbb{N}$,*

$$\text{(i)}\quad \frac{\Pi_n(\theta\in\Theta_{n,1}: j\epsilon_n < d_n(\theta,\theta_{n,0}) \le 2j\epsilon_n)}{\Pi_n(B_{n,0}(\theta_{n,0},\epsilon_n))} \le e^{Kn\epsilon_n^2 j^2/2}, \tag{8.20}$$

$$\text{(ii)}\quad \sup_{\epsilon>\epsilon_n}\log N\Big(\xi\epsilon, \{\theta\in\Theta_{n,1}: d_n(\theta,\theta_{n,0}) < 2\epsilon\}, e_n\Big) \le n\epsilon_n^2, \tag{8.21}$$

then $\Pi_n(\theta\in\Theta_{n,1}: d_n(\theta,\theta_{n,0}) \ge M_n\epsilon_n| X^{(n)}) \to 0$, in $P_{\theta_{n,0}}^{(n)}$-probability, for every $M_n\to\infty$. Furthermore, if (8.20) holds with $B_{n,0}(\theta_{n,0},\epsilon_n)$ replaced by $B_{n,k}(\theta_{n,0},\epsilon_n)$ for some $k>1$, then

(a) *If all $X^{(n)}$ are defined on the same probability space, $\epsilon_n \gtrsim n^{-\alpha}$ for some $\alpha\in(0,1/2)$ and $k(1-2\alpha)>2$, then the contraction holds also in the almost sure sense.*
(b) *If $n\epsilon_n^2\to\infty$, then $\Pi_n(\theta\in\Theta_{n,1}: d_n(\theta,\theta_{n,0})\ge M\epsilon_n| X^{(n)}) = O_P(e^{-n\epsilon_n^2})$, for sufficiently large M.*

Theorem 8.19 employs subsets $\Theta_{n,1}\subset\Theta_n$ to alleviate the entropy condition (8.21), but returns an assertion about the posterior distribution on $\Theta_{n,1}$ only. A complementary assertion about the sets $\Theta_{n,2} = \Theta_n \setminus \Theta_{n,1}$ may be obtained either by a direct argument or by the following result.

Theorem 8.20 (Remaining mass) *For any sets $\Theta_{n,2}\subset\Theta_n$ we have $\Pi_n(\Theta_{n,2}| X^{(n)})\to 0$, in $P_{\theta_{n,0}}^{(n)}$-probability if, for arbitrary ϵ_n, either (iii) or (iii′) holds:*

$$\text{(iii)}\quad \frac{\Pi_n(\Theta_{n,2})}{\Pi_n(B_{n,0}(\theta_{n,0},\epsilon_n))} = o(e^{-D_n n\epsilon_n^2}), \qquad \textit{for some } D_n\to\infty,$$

$$\text{(iii}'\text{)}\quad \frac{\Pi_n(\Theta_{n,2})}{\Pi_n(B_{n,k}(\theta_{n,0},\epsilon_n))} = o(e^{-2n\epsilon_n^2}), \qquad \textit{for some } k>1.$$

Under (iii′) the convergence to zero can be strengthened to almost sure convergence and to be of order $O_P(e^{-n\epsilon_n^2})$ under conditions (a)–(b) of Theorem 8.19.

The proofs of Theorems 8.19 and 8.20 are very similar to the proof of Theorem 8.11, apart from changes in the notation and the fact that we cover d_n-shells by e_n-balls to exploit the two metrics. The lower bound on the norming constant of the posterior distribution is based on the following lemma, which is proved in the same way as Lemmas 6.26 (when $k = 0$) and 8.10 (when $k > 1$).

Lemma 8.21 (Evidence lower bound) *For any $D > 0$, and $\epsilon \geq n^{-1/2}$,*

$$P_{\theta_{n,0}}^{(n)}\left(\int \frac{p_\theta^{(n)}}{p_{\theta_{n,0}}^{(n)}}\,d\Pi_n(\theta) \leq \Pi_n(B_{n,k}(\theta_{n,0},\epsilon))e^{-(1+D)n\epsilon^2}\right) \leq \begin{cases} 2/(1+D), & \text{if } k=0,\\ (D\sqrt{n}\epsilon)^{-k}, & \text{if } k>1. \end{cases}$$

As in the i.i.d. case weaker, but simpler theorems are obtained by using an absolute lower bound on the prior mass and by replacing the local by the global entropy. In particular, conditions (8.20) and (8.21) are implied by, for some $C > 0$,

$$\Pi_n(B_{n,0}(\theta_{n,0},\epsilon_n)) \geq e^{-Cn\epsilon_n^2}, \tag{8.22}$$

$$\log N(\xi\epsilon_n, \Theta_{n,1}, e_n) \leq n\epsilon_n^2. \tag{8.23}$$

Condition (8.22) leads to (8.20) (for $Kj^2/2 \geq C$) by bounding the numerator in the latter condition by the trivial bound one. Simple sufficient conditions in terms of only $\Pi_n(\Theta_{n,2})$ for (iii) and (iii′) in Theorem 8.20 can be similarly derived.

Bounding entropy to construct tests is not always the most fruitful strategy. As sometimes ad hoc tests can be easily constructed, a theorem using testing conditions is useful.

Theorem 8.22 *Given a semimetric d_n on Θ_n and subsets $\Theta_{n,1} \subset \Theta_n$, suppose that for $k > 1$ and ϵ_n with $n\epsilon_n^2 \geq 1$ condition* (8.20) *holds and there exist tests ϕ_n such that, for some $K > 0$ and sufficiently large j,*

$$P_{\theta_{n,0}}^{(n)}\phi_n \to 0, \qquad \sup_{\theta\in\Theta_{n,1}: j\epsilon_n < d_n(\theta,\theta_{n,0}) \leq 2j\epsilon_n} P_\theta^{(n)}(1-\phi_n) \lesssim e^{-Kj^2n\epsilon_n^2}, \tag{8.24}$$

Then $\Pi_n(\theta \in \Theta_{n,1}: d_n(\theta,\theta_{n,0}) \geq M_n\epsilon_n \mid X^{(n)}) \to 0$, in $P_{\theta_{n,0}}^{(n)}$-probability, for every $M_n \to \infty$.

8.3.1 Independent Observations

For $X^{(n)}$ consisting of n independent observations, with laws $P_{n,\theta,1},\ldots,P_{n,\theta,n}$, tests satisfying (8.17) always exist relative to $d_n = e_n$ equal to the *root average square Hellinger metric*

$$d_{n,H}(\theta_1,\theta_2) = \sqrt{\frac{1}{n}\sum_{i=1}^n d_H^2(P_{n,\theta_1,i}, P_{n,\theta_2,i})}. \tag{8.25}$$

(See Proposition D.9.) Furthermore, by Lemma B.8, the Kullback-Leibler divergence is additive and the variation $V_{k,0}$ subadditive up to an $O(n^{k/2-1})$ factor (for $k \geq 2$). It follows that the neighborhoods $B_{n,0}(\theta_{n,0}, \epsilon)$ and $B_{n,k}(\theta_{n,0}, \epsilon)$ contain the sets

$$B_{n,0}^*(\theta_{n,0}, \epsilon) = \Big\{\theta \in \Theta_n: \frac{1}{n}\sum_{i=1}^n K(P_{i,\theta_{n,0},n}; P_{i,\theta,n}) \leq \epsilon^2\Big\},$$

$$B_{n,k}^*(\theta_{n,0}, \epsilon) = \Big\{\theta \in \Theta_n: \frac{1}{n}\sum_{i=1}^n K(P_{i,\theta_{n,0},n}; P_{i,\theta,n}) \leq \epsilon^2, \frac{1}{n}\sum_{i=1}^n V_{k,0}(P_{i,\theta_{n,0},n}; P_{i,\theta,n}) \leq d_k\epsilon^k\Big\}.$$

The choices $k = 0$ or $k = 2$ (with $d_2 = 1$) are convenient and good enough for many applications. This leads to the following special case of Theorem 8.19.

Theorem 8.23 *Let $P_\theta^{(n)}$ be the product of measures $P_{n,\theta,1}, \ldots, P_{n,\theta,n}$, and suppose that there exists a partition $\Theta_n = \Theta_{n,1} \cup \Theta_{n,2}$ such that for sequences $\bar\epsilon_n, \epsilon_n \geq n^{-1/2}$ and all sufficiently large j,*

$$\text{(i)}\quad \frac{\Pi_n(\theta \in \Theta_{n,1}: j\epsilon_n < d_{n,H}(\theta, \theta_{n,0}) \leq 2j\epsilon_n)}{\Pi_n(B_{n,0}^*(\theta_{n,0}, \epsilon_n))} \leq e^{n\epsilon_n^2 j^2/4}, \tag{8.26}$$

$$\text{(ii)}\quad \sup_{\epsilon > \epsilon_n} \log N\Big(\epsilon/36, \{\theta \in \Theta_{n,1}: d_{n,H}(\theta, \theta_{n,0}) < \epsilon\}, d_{n,H}\Big) \leq n\epsilon_n^2, \tag{8.27}$$

$$\text{(iii)}\quad \frac{\Pi_n(\Theta_{n,2})}{\Pi_n(B_{n,k}^*(\theta_{n,0}, \bar\epsilon_n))} = o(e^{-D_{n,k}n\bar\epsilon_n^2}), \tag{8.28}$$

for $k = 0$ and some $D_{n,0} \to \infty$, or for some $k \geq 2$ and $D_{n,k} = 2$. Then the rate of posterior contraction at $\theta_{n,0}$ relative to $d_{n,H}$ is ϵ_n.

The root average square Hellinger distance $d_{n,H}$ always works with independent observations, but it is not always the best choice of metric to derive contraction rates. Rates relative to a stronger metric d_n may be obtainable by applying Theorem 8.19 directly. This is possible as soon as tests as in (8.17) exist for these metrics (and metrics e_n). An important example where this is relevant is nonparametric regression with normal errors, as considered in Section 8.3.2.

Discrete Uniform Priors

As in the i.i.d. case, priors attaining the minimax rate of contraction may be constructed using bracketing methods. Consider a sequence of models $\{P_\theta^{(n)}: \theta \in \Theta\}$ with fixed parameter set Θ, where $P_\theta^{(n)}$ has a density $(x_1, \ldots, x_n) \mapsto \prod_{i=1}^n p_{\theta,i}(x_i)$ with respect to a product measure $\otimes_{i=1}^n \nu_i$. For a given n and $\epsilon > 0$ define the *componentwise Hellinger upper bracketing number* $N_]^{n\otimes}(\epsilon, \Theta_n, d_{n,H})$ for a set $\Theta_n \subset \Theta$ as the smallest number N of vectors $(u_{j,1}, \ldots, u_{j,n})$ of ν_j-integrable nonnegative functions $u_{j,i}$ (for $j = 1, 2, \ldots, N$), with the property that for every $\theta \in \Theta_n$ there exists some j such that $p_{\theta,i} \leq u_{j,i}$ for all $i = 1, 2, \ldots, n$ and $\sum_{i=1}^n d_H^2(p_{\theta,i}, u_{j,i}) \leq n\epsilon^2$.

Given a sequence of sets $\Theta_n \uparrow \Theta$ and $\epsilon_n \to 0$ such that $\log N_{]}^{n\otimes}(\epsilon_n, \Theta_n, d_{n,H}) \le n\epsilon_n^2$, let $\{(u_{j,1}, \ldots, u_{j,n}): j = 1, 2, \ldots, N\}$ be a minimal componentwise Hellinger upper bracketing for Θ_n. Let Π_n be the uniform discrete measure on the joint densities

$$p_j^{(n)}(x_1, \ldots, x_n) = \prod_{i=1}^n \frac{u_{j,i}(x_i)}{\int u_{j,i}\, d\nu_i}.$$

Theorem 8.24 *If $\Theta_n \uparrow \Theta$ are sets satisfying $\log N_{]}^{n\otimes}(\epsilon_n, \Theta_n, d_{n,H}) \le n\epsilon_n^2$ for some ϵ_n with $n\epsilon_n^2 \to \infty$ and the prior is constructed as indicated, then the posterior contraction rate relative to $d_{n,H}$ is ϵ_n at any $\theta_0 \in \Theta$.*

Proof The collection $\{p_j^{(n)}: j = 1, 2, \ldots, N\}$ forms a sieve for the models $\{P_\theta^{(n)}: \theta \in \Theta\}$. Formally, we can add the measures $p_j^{(n)}$ to the model, identified by new parameters $j = 1, \ldots, N_n$, and apply Theorem 8.23 with the partition of the enlarged parameter space in $\Theta_{n,1} = \{1, \ldots, N_n\}$ the set of new parameters and $\Theta_{n,2} = \Theta_n \setminus \Theta_{n,1}$.

The covering number of $\Theta_{n,1} = \{1, \ldots, N_n\}$ certainly does not exceed its cardinality, which is $N_{]}^{n\otimes}(\epsilon_n, \Theta_n, d_{n,H})$ by construction. Therefore (8.27), with ϵ_n a multiple of the present ϵ_n, holds by assumption. Condition (8.28) holds trivially as $\Pi_n(\Theta_{n,2}) = 0$.

Because $\Theta_n \uparrow \Theta$, any $\theta_0 \in \Theta$ is in Θ_n for all sufficiently large n. For such large n, let j_0 be the index for which $p_{\theta_0,i} \le u_{j_0,i}$ for every i, and $\sum_{i=1}^n d_H^2(p_{\theta_0,i}, u_{j_0,i}) \le n\epsilon_n^2$. If p is a probability density, u is an integrable function such that $u \ge p$, and $v = u/\int u\, d\nu$, then

$$d_H^2(p, v) = 2 - \frac{2\int \sqrt{pu}\, d\nu}{\sqrt{\int u\, d\nu}} \le \frac{1 + \int u\, d\nu - 2\int \sqrt{pu}\, d\nu}{\sqrt{\int u\, d\nu}} \le \frac{d_H^2(p, u)}{\sqrt{\int u\, d\nu}}.$$

Because the likelihood ratios satisfy $p_{\theta_0,i}/v_{j_0,i} \le \int u_{j_0,i}\, d\nu \le 2$, it follows with the help of Lemma B.2 that $n^{-1}\sum_{i=1}^n K(p_{\theta_0,i}; v_{j_0,i}) \lesssim \epsilon_n^2$ and $n^{-1}\sum_{i=1}^n V_2(p_{\theta_0,i}; v_{j_0,i}) \lesssim \epsilon_n^2$. As $\otimes_{i=1}^n v_{j_0,i}$ receives prior probability equal to $N^{-1} \ge e^{-n\epsilon_n^2}$, the prior mass condition (8.26) holds, for a multiple of the present ϵ_n. ☐

Finite-dimensional Models

Application of Theorem 8.23 to finite-dimensional models yields the following result.

Theorem 8.25 *Let $X^{(n)}$ consist of independent observations $X_1, \ldots, X_n$ following densities $p_{\theta,i}$, indexed by parameters in $\Theta \subset \mathbb{R}^k$ such that for constants $\alpha, c, C > 0$, and $c_i \le C_i$ with $c \le \bar{c}_n \le \bar{C}_n \le C$, for every $\theta, \theta_1, \theta_2 \in \Theta$,*

$$K(p_{\theta_0,i}; p_{\theta,i}) \le C_i \|\theta - \theta_0\|^{2\alpha}, \qquad V_{2,0}(p_{\theta_0,i}; p_{\theta,i}) \le C_i \|\theta - \theta_0\|^{2\alpha}, \tag{8.29}$$

$$c_i \|\theta_1 - \theta_2\|^{2\alpha} \le d_H^2(p_{\theta_1,i}, p_{\theta_2,i}) \le C_i \|\theta_1 - \theta_2\|^{2\alpha}. \tag{8.30}$$

If the prior Π possesses a bounded density that is bounded away from zero on a neighborhood of θ_0, then the posterior converges at the rate $n^{-1/(2\alpha)}$ with respect to the Euclidean metric.

Proof By assumption (8.30), it suffices to show that the posterior contraction rate with respect to $d_{n,H}$, as defined in (8.25) is $n^{-1/2}$. Now by Proposition C.2,

$$N(\epsilon/18, \{\theta \in \Theta: d_{n,H}(\theta,\theta_0) < \epsilon\}, d_{n,H})$$
$$\leq N\Big(\frac{\epsilon^{1/\alpha}}{(36C)^{1/(2\alpha)}}, \{\theta \in \Theta: \|\theta - \theta_0\| < \frac{(\sqrt{2}\epsilon)^{1/\alpha}}{c^{1/(2\alpha)}}\}, \|\cdot\|\Big) \leq 3^k\Big(\frac{72C}{c}\Big)^{k/(2\alpha)}.$$

This verifies (8.27). For (8.26), note that

$$\frac{\Pi(\theta: d_{n,H}(\theta,\theta_0) \leq j\epsilon)}{\Pi(B^*_{n,2}(\theta_0,\epsilon))} \leq \frac{\Pi(\theta: \|\theta-\theta_0\| \leq (2j^2\epsilon^2/c)^{1/(2\alpha)})}{\Pi(\theta: \|\theta-\theta_0\| \leq (\epsilon^2/(2C))^{1/(2\alpha)})} \leq Aj^{k/\alpha},$$

for sufficiently small $\epsilon > 0$, where A is a constant depending on d, c, C and the upper and lower bound on the prior density. It follows that (8.26) is satisfied for ϵ_n equal to a large multiple of $n^{-1/2}$. □

For regular parametric families, the conditions of the theorem are satisfied for $\alpha = 1$, and the usual $n^{-1/2}$ rate is obtained. When the densities have discontinuities or other singularities depending on the parameter (for instance the uniform distribution on $(0,\theta)$, or the examples discussed in Chapters V and VI of Ibragimov and Has'minskiĭ, 1981), then the conditions of Theorem 8.25 hold for $\alpha < 1$, and contraction rates faster than $n^{-1/2}$ pertain.

In an unbounded parameter space the Hellinger distance cannot be bounded below by a power of the Euclidean distance and hence the lower inequality of assumption (8.30) fails. In this case Theorem 8.22 may be applied after first proving that the posterior distribution concentrates on bounded sets. By the consistency theorems of Chapter 6 the latter can be achieved by showing the existence of uniformly exponentially consistent tests for $\theta = \theta_0$ against the complement of a bounded set. Often such tests exist by bounds on the log affinity, in view of Lemma D.11.

8.3.2 Gaussian Regression with Fixed Design

The observation $X^{(n)}$ in the Gaussian, fixed-design regression model consists of independent random variables $X_1, \ldots, X_n$ distributed as $X_i = \theta(z_i) + \varepsilon_i$, for an unknown regression function θ, deterministic covariates $z_1, \ldots, z_n$, and $\varepsilon_i \overset{\text{iid}}{\sim} \text{Nor}(0,\sigma^2)$, for $i = 1, \ldots, n$. Write $\|\theta\|_{2,n}$ for the $\mathbb{L}_2$-norm of a function $\theta: \mathfrak{Z} \to \mathbb{R}$ relative to the empirical measure of the covariates:

$$\|\theta\|^2_{2,n} = \frac{1}{n}\sum_{i=1}^n \theta^2(z_i).$$

By Lemma 2.7, the average square Hellinger distance and the Kullback-Leiber divergence and variation on the distributions of the observations are all bounded above by the square of the $\|\cdot\|_{2,n}$-norm. However, for the Hellinger distance this can be reversed (in general) only if the regression functions are bounded. This makes it better to deduce a rate of contraction directly from the general rate theorem, Theorem 8.19, with an ad hoc construction of tests, rather than to refer to Theorem 8.23.

Theorem 8.26 (Regression) *Let $P_\theta^{(n)}$ be the product of the measures* $\mathrm{Nor}(\theta(z_i), \sigma^2)$, *for* $i = 1, \ldots, n$. *If there exists a partition $\Theta_n = \Theta_{n,1} \cup \Theta_{n,2}$ such that for a sequence $\epsilon_n \to 0$ with $n\epsilon_n^2 \geq 1$ and all sufficiently large j,*

$$\text{(i)}\ \frac{\Pi_n(\theta \in \Theta_{n,1}: j\epsilon_n < \|\theta - \theta_{n,0}\|_{2,n} \leq 2j\epsilon_n)}{\Pi_n(\theta \in \Theta_n: \|\theta - \theta_{n,0}\|_{2,n} < \epsilon_n)} \leq e^{n\epsilon_n^2 j^2/16}, \tag{8.31}$$

$$\text{(ii)}\ \sup_{\epsilon > \epsilon_n} \log N\Big(\epsilon/2, \{\theta \in \Theta_{n,1}: \|\theta - \theta_{n,0}\|_{2,n} < 2\epsilon\}, \|\cdot\|_n\Big) \leq n\epsilon_n^2, \tag{8.32}$$

$$\text{(iii)}\ \frac{\Pi_n(\Theta_{n,2})}{\Pi_n(\theta \in \Theta_n: \|\theta - \theta_{n,0}\|_{2,n} < \epsilon_n)} = o(e^{-2n\epsilon_n^2}), \tag{8.33}$$

then the rate of posterior contraction at $\theta_{n,0}$ relative to $\|\cdot\|_{2,n}$ is ϵ_n.

Proof We combine Theorems 8.19 and 8.20, with $k = 2$. Because by Lemma 2.7 the neighborhoods $B_{n,2}(\theta_{n,0}, \epsilon)$ in these theorems contain balls $\{\theta: \|\theta - \theta_{n,0}\|_{2,n} < \epsilon\}$, it suffices to exhibit tests as in (8.17), with d_n and e_n taken equal to the metric induced by $\|\cdot\|_{2,n}$. A likelihood ratio test for testing $\theta_{n,0}$ versus $\theta_{n,1}$ with appropriate cutoff is shown to do the job in Lemma 8.27(i) below, with $\xi = 1/2$ and $K = 1/32$. ☐

The following lemma verifies the existence of tests as in (8.17). Part (i) is used in the proof of the preceding theorem. Part (ii) may be used to extend the theorem to the case of unknown error variance. If the prior of σ has bounded support $[0, \bar\sigma]$, for a constant $\bar\sigma$, then Theorem 8.19 applies with $d_n = e_n$ and $d_n((\theta_1, \sigma_1), (\theta_2, \sigma_2))$ taken equal to a multiple of $\|\theta_1 - \theta_2\|_{2,n} + |\sigma_1 - \sigma_2|$.

Lemma 8.27 *For $\theta \in \mathbb{R}^n$ and $\sigma > 0$, let $P_{\theta,\sigma}^{(n)} = \mathrm{Nor}_n(\theta, \sigma^2 I)$, and let $\|\theta\|$ be the Euclidean norm.*

(i) *For any $\theta_0, \theta_1 \in \mathbb{R}^n$ and $\sigma > 0$, there exists a test ϕ such that $P_{\theta_0,\sigma}^{(n)}\phi \vee P_{\theta,\sigma}^{(n)}(1 - \phi) \leq e^{-\|\theta_0 - \theta_1\|^2/(32\sigma^2)}$, for every $\theta \in \mathbb{R}^n$ with $\|\theta - \theta_1\| \leq \|\theta_0 - \theta_1\|/2$.*
(ii) *For any $\theta_0, \theta_1 \in \mathbb{R}^n$ and $\sigma_0, \sigma_1 > 0$, there exists a test ϕ such that $P_{\theta_0,\sigma_0}^{(n)}\phi \vee P_{\theta,\sigma}^{(n)}(1 - \phi) \leq e^{-K[\|\theta_0 - \theta_1\|^2 + n|\sigma_0 - \sigma_1|^2]/(\sigma_0^2 \vee \sigma_1^2)}$, for every $\theta \in \mathbb{R}^n$ and $\sigma > 0$ with $\|\theta - \theta_1\| \leq \|\theta_0 - \theta_1\|/2$ and $|\sigma - \sigma_1| \leq |\sigma_0 - \sigma_1|/2$, and a universal constant K.*

Proof (i). By shifting the observation $X \sim \mathrm{Nor}(\theta, \sigma^2 I)$ by $-\theta_0$ and dividing by σ, we can reduce to the case that $\theta_0 = 0$ and $\sigma = 1$. If $\|\theta - \theta_1\| \leq \|\theta_1\|/2$, then $\|\theta\| \geq \|\theta_1\|/2$ and hence $\theta_1^\top\theta = (\|\theta\|^2 + \|\theta_1\|^2 - \|\theta - \theta_1\|^2)/2 \geq \|\theta_1\|^2/2$. Therefore, the test $\phi = \mathbb{1}\{\theta_1^\top X > D\|\theta_1\|\}$ satisfies, with Φ the standard normal distribution function,

$$P_{\theta_0}^{(n)}\phi = 1 - \Phi(D), \quad P_\theta^{(n)}(1 - \phi) = \Phi((D\|\theta_1\| - \theta_1^\top\theta)/\|\theta_1\|) \leq \Phi(D - \rho),$$

for $\rho = \|\theta_1\|/2$. The infimum over D of $1 - \Phi(D) + \Phi(D - \rho)$ is attained at $D = \rho/2$, for which $D - \rho = -\rho/2$. We substitute this in the preceding display and use the bound $1 - \Phi(x) \leq e^{-x^2/2}$, valid for $x \geq 0$ (see Lemma K.6(i)).

(ii). By shifting by $-\theta_0$ we again reduce to the case that $\theta_0 = 0$.

By the proof as under (i), the test $\phi = \mathbb{1}\{\theta_1^\top X > D\|\theta_1\|\}$ can be seen to have error probabilities satisfying $P^{(n)}_{\theta_0,\sigma_0}\phi \le 1 - \Phi(D/\sigma_0)$ and $P^{(n)}_{\theta,\sigma}(1-\phi) \le \Phi((D-\rho)/\sigma)$, for $\rho = \|\theta_1\|/2$. For $D = \rho/2$, both error probabilities are bounded by $e^{-\|\theta_0-\theta_1\|^2/(32\sigma_0^2\vee\sigma_1^2)}$.

Assume first that $\sigma_1 > \sigma_0$. The test $\psi = \mathbb{1}\{n^{-1}\|X\|_1 - c\sigma_0 > D\}$, where $c = \sqrt{2/\pi}$ is the absolute mean of a standard normal variable, has error probability of the first kind satisfying $P^{(n)}_{\theta_0,\sigma_0}\psi = \mathrm{P}(\|Z\|_1 - \mathrm{E}\|Z\|_1 > Dn/\sigma_0)$ for $Z \sim \mathrm{Nor}_n(0, I)$. Since $|\,\|X-\theta\|_1 - \|X\|_1| \le \|\theta\|_1 \le \sqrt{n}\|\theta\|$, by the Cauchy-Schwarz inequality, the error probability of the second kind satisfies

$$P^{(n)}_{\theta,\sigma}(1-\psi) \le P^{(n)}_{\theta,\sigma}(n^{-1}\|X-\theta\|_1 < D + c\sigma_0 + n^{-1/2}\|\theta\|)$$
$$\le \mathrm{P}(\|Z\|_1 - \mathrm{E}\|Z\|_1 \le Dn/\sigma + nc(\sigma_0 - \sigma)/\sigma + \sqrt{n}\|\theta\|/\sigma).$$

For $D = c(\sigma_1 - \sigma_0)/4$, the argument in the last probability is bounded above by $-cn(\sigma_1 - \sigma_0)/(4\sigma) + \sqrt{n}(3/2)\|\theta_1\|/\sigma$, whenever $|\sigma - \sigma_1| \le |\sigma_0 - \sigma_1|/2$ and $\|\theta - \theta_1\| \le \|\theta_1\|/2$, which is further bounded above by $-cn(\sigma_1 - \sigma_0)/(8\sigma) \le -cn(\sigma_1 - \sigma_0)/(12\sigma_1)$, if also $\|\theta_1\| \le \sqrt{n}(\sigma_1 - \sigma_0)/12$. Because $\|Z\|_1 - \mathrm{E}\|Z\|_1$ is the sum of independent, mean-zero sub-Gaussian variables, it satisfies a sub-Gaussian tail bound of the type $\mathrm{P}(|\,\|Z\|_1 - \mathrm{E}\|Z\|_1| > x) \le e^{-K'x^2/n}$, for some constant $K' > 0$. Conclude that the error probabilities of ψ are bounded above by $e^{-Kn(\sigma_1-\sigma_0)^2/(\sigma_0\vee\sigma_1)^2}$ for some constant $K > 0$.

For $\sigma_1 < \sigma_0$, a test of the form $\psi = \mathbb{1}\{n^{-1}\|X\|_1 - c\sigma_0 < D\}$ can be seen to satisfy the same error bounds.

Finally, consider two cases. If $\|\theta_0 - \theta_1\| > \sqrt{n}|\sigma_0 - \sigma_1|/12$, then we use only the test ϕ, and it has the desired error bounds, for sufficiently small K. If $\|\theta_0 - \theta_1\| \le \sqrt{n}|\sigma_0 - \sigma_1|/12$, then we use only the test ψ, and it has again the desired error bounds. □

8.3.3 Markov Chains

Let $P^{(n)}_\theta$ stand for the law of $(X_0, \ldots, X_n)$, for a discrete time Markov process $(X_n{:}\, n \in \mathbb{Z})$ with stationary transition probabilities, in a given state space $(\mathfrak{X}, \mathscr{X})$. Assume that the transition kernel is given by a transition density p_θ relative to a given dominating measure ν, and let $Q_{\theta,i}$ be the distribution of X_i. Thus $(x, y) \mapsto p_\theta(y|\,x)$ are measurable maps, and $p_\theta(\cdot|\,x)$ is a probability density relative to ν, for every x.

Assume that there exist finite measures $\underline{\mu}$ and $\overline{\mu}$ on $(\mathfrak{X}, \mathscr{X})$ such that, for some $k, l \in \mathbb{N}$, every $\theta \in \Theta$, and every $x \in \mathfrak{X}$ and $A \in \mathscr{X}$,

$$\underline{\mu}(A) \le \frac{1}{k}\sum_{j=1}^k \mathrm{P}_\theta(X_j \in A|\,X_0 = x), \qquad \mathrm{P}_\theta(X_l \in A|\,X_0 = x) \le \overline{\mu}(A). \tag{8.34}$$

Here P_θ is a generic notation for a probability law governed by θ. This holds in particular if there exists $r \in \mathbb{L}_1(\nu)$ such that $r \lesssim p_\theta(\cdot|\,x) \lesssim r$ for every x, with the measures $\underline{\mu}$ and $\overline{\mu}$ equal to multiples of $A \mapsto \int_A r\,d\nu$. For $\mu \in \{\underline{\mu}, \overline{\mu}\}$ define semimetrics $d_{H,\mu}$ by

$$d^2_{H,\mu}(\theta_1, \theta_2) = \int\!\!\int \Big[\sqrt{p_{\theta_1}(y|\,x)} - \sqrt{p_{\theta_2}(y|\,x)}\Big]^2\, d\nu(y)\, d\mu(x). \tag{8.35}$$

It is shown in Proposition D.14 that tests as in (8.17) exist for $d_n = d_{H,\underline{\mu}}$ and $e_n = d_{H,\overline{\mu}}$.

The following lemma relates the Kullback-Leibler divergence and variation on the laws $P_\theta^{(n)}$ to the corresponding characteristics of the transition densities. For the variation we assume that the Markov chain is α-mixing, for the divergence itself this is not necessary. The kth *α-mixing coefficient* α_k under θ_0 is defined as

$$\alpha_k = \sup_{m\in\mathbb{N}} \sup_{A,B\in\mathscr{X}} |\mathrm{P}_{\theta_0}(X_m \in A, X_{m+k} \in B) - \mathrm{P}_{\theta_0}(X_m \in A)\mathrm{P}_{\theta_0}(X_{m+k} \in B)|.$$

Lemma 8.28 *For every $s > 2$ and $D_s = 8s \sum_{h=0}^\infty \alpha_h^{1-2/s}$,*

(i) $K(p_{\theta_0}^{(n)}; p_\theta^{(n)}) \le n \sup_{i\ge 0} \int K(p_{\theta_0}(\cdot|\, x); p_\theta(\cdot|\, x))\, dQ_{\theta_0,i}(x) + K(q_{\theta_0,0}; q_{\theta,0})$.

(ii) $V_2(p_{\theta_0}^{(n)}; p_\theta^{(n)}) \le nD_s \sup_{i\ge 0} \big[\int V_s(p_{\theta_0}(\cdot|\, x); p_\theta(\cdot|\, x))\, dQ_{\theta_0,i}(x)\big]^{2/s} + 2V_2(q_{\theta_0,0}; q_{\theta,0})$.

Proof Assertion (i) is immediate from the expression for the likelihood:

$$\log \frac{p_{\theta_0}^{(n)}}{p_\theta^{(n)}}(X^{(n)}) = \sum_{i=1}^n \log \frac{p_{\theta_0}}{p_\theta}(X_i|\, X_{i-1}) + \log \frac{q_{\theta_0,0}}{q_{\theta,0}}(X_0) =: \sum_{i=1}^n Y_i + Z_0.$$

The time series $Y_i = \log(p_{\theta_0}/p_\theta)(X_i|\, X_{i-1})$ is α-mixing with mixing coefficients α_{k-1}. Consequently, the variance $\mathrm{var}(\sum_{i=1}^n Y_i)$ of their sum can be further bounded by the desired expression, in view of Lemma K.5. □

Let $\Theta_{0,1} \subset \Theta$ be the set of parameter values such that $K(q_{\theta_0,0}; q_{\theta,0})$ and $V_2(q_{\theta_0,0}; q_{\theta,0})$ are bounded by 1. Then by the preceding lemma, for large n and $n\epsilon^2 \ge 4$, the neighborhoods $B_{n,0}(\theta_0, \epsilon)$ and $B_{n,2}(\theta_0, \epsilon)$ contain the sets, respectively,

$$B_0^*(\theta_0, \epsilon) = \Big\{\theta \in \Theta_{0,1}: \sup_i \int K(p_{\theta_0}(\cdot|\, x); p_\theta(\cdot|\, x))\, dQ_{\theta_0,i}(x) \le \tfrac{1}{2}\epsilon^2\Big\},$$

$$B_s^*(\theta_0, \epsilon) = \Big\{\theta \in B_0^*(\theta_0, \epsilon): \sup_i \int V_s(p_{\theta_0}(\cdot|\, x); p_\theta(\cdot|\, x))\, dQ_{\theta_0,i}(x) \le (2D_s)^{s/2}\epsilon^s\Big\}.$$

Therefore, Theorem 8.19 implies the following result.

Theorem 8.29 *Let $P_\theta^{(n)}$ be the distribution of a Markov chain $(X_0, X_1, \ldots, X_n)$ with transition densities p_θ satisfying (8.34) and marginal densities $q_{\theta,i}$. If there exists a partition $\Theta = \Theta_{n,1} \cup \Theta_{n,2}$ such that for sequences $\epsilon_n, \bar\epsilon_n \ge 2n^{-1/2}$ and every sufficiently large j,*

$$\text{(i)}\quad \frac{\Pi_n(\theta \in \Theta_{n,1}: d_{H,\underline{\mu}}(\theta, \theta_0) \le j\epsilon_n)}{\Pi_n(B_0^*(\theta_0, \epsilon_n))} \le e^{Kn\epsilon_n^2 j^2/8}, \tag{8.36}$$

$$\text{(ii)}\quad \sup_{\epsilon>\epsilon_n} \log N\Big(\epsilon/16, \{\theta \in \Theta_{n,1}: d_{H,\underline{\mu}}(\theta, \theta_0) < \epsilon\}, d_{H,\bar\mu}\Big) \le n\epsilon_n^2, \tag{8.37}$$

$$\text{(iii)}\quad \frac{\Pi_n(\Theta_{n,2})}{\Pi_n(B_0^*(\theta_0, \bar\epsilon_n))} = o(e^{-M_n n\bar\epsilon_n^2}), \qquad \textit{for some } M_n \to \infty, \tag{8.38}$$

then the posterior rate of contraction at θ_0 *relative to* $d_{H,\underline{\mu}}$ *is* ϵ_n. *If under* θ_0 *the chain is* α*-mixing with coefficients* α_h *satisfying* $\sum_{k=0}^{\infty}\alpha_k^{1-2/s} < \infty$ *for some* $s > 2$, *then this conclusion remains true if (iii) is replaced by*

$$\text{(iii}'\text{)}\quad \frac{\Pi_n(\Theta_{n,2})}{\Pi_n(B_s^*(\theta_0,\bar\epsilon_n))} = o(e^{-2n\bar\epsilon_n^2}). \tag{8.39}$$

Condition (iii′) may yield a slightly faster rate, but at the cost of introducing mixing numbers and higher moments of the log likelihood. Condition (iii) is easier to apply, but we finish the section with some remarks on the verification of (iii′).

As usual, there is a trade-off between mixing and moment conditions: for quickly mixing sequences, with convergence of the series $\sum_k \alpha_k^{1-2/s}$ for small s, existence of lower-order moments suffices.

From the Markov property it can be seen that, with $Q_{\theta_0,\infty}$ any measure,

$$\alpha_k \le 2\sup_x d_{TV}(\mathrm{P}_{\theta_0}(X_k \in \cdot \mid X_0 = x), Q_{\theta_0,\infty}).$$

A Markov chain is said to be *uniformly ergodic* if the right side of this display, with $Q_{\theta_0,\infty}$ an invariant probability measure of the chain, tends to zero as $k \to \infty$. The convergence is then automatically exponentially fast: the right side is bounded by a multiple of c^k for some constant $c < 1$ (see Meyn and Tweedie 1993, Theorem 16.0.2). Then $\sum_{k=0}^{\infty}\alpha_k^{1-2/s} < \infty$ for every $s > 2$, and (iii′) of the preceding theorem is applicable with an arbitrary fixed $s > 2$.

A simple (but strong) condition for uniform ergodicitiy is:

$$\sup_{x_1,x_2\in\mathfrak{X}} d_{TV}(p_{\theta_0}(\cdot\mid x_1), p_{\theta_0}(\cdot\mid x_2)) < 2.$$

This follows by integrating the display with respect to x_2 relative to the stationary measure Q_{θ_0}, and next applying condition (16.8) of Theorem 16.0.2 of Meyn and Tweedie (1993).

8.3.4 White Noise Model

For $\theta \in \Theta \subset \mathbb{L}_2[0,1]$, let $P_\theta^{(n)}$ be the distribution on $\mathfrak{C}[0,1]$ of the stochastic process $X^{(n)}$ defined structurally relative to a standard Brownian motion W as

$$X_t^{(n)} = \int_0^t \theta(s)\,ds + \frac{1}{\sqrt{n}}W_t, \qquad 0 \le t \le 1.$$

This model arises as an approximation of many different types of experiments, including density estimation and nonparametric regression (see Nussbaum 1996 and Brown and Low 1996).

An equivalent experiment is obtained by expanding $dX^{(n)}$ on an arbitrary orthonormal basis $e_1, e_2, \ldots$ of $\mathbb{L}_2[0,1]$, giving the variables $X_{n,i} = \int e_i(t)\,dX_t^{(n)}$. (The "differentials" $dX^{(n)}$ cannot be understood as elements of $\mathbb{L}_2[0,1]$ and hence "expanding" must be understood in the generalized sense of Itô integrals.) These variables are independent and normally distributed with means the coefficients $\theta_i := \int_0^1 e_i(t)\theta(t)\,dt$ in the expansion of $\theta = \sum_i \theta_i e_i$ relative to the basis (e_i), and variance $1/n$. Observing $X^{(n)}$ or the infinite vector $(X_{n,1}, X_{n,2}, \ldots)$ are equivalent by sufficiency of the latter vector in the experiment consisting of observing $X^{(n)}$. The parameter $(\theta_1, \theta_2, \ldots)$ belongs to ℓ_2 and the norms $\|\theta\|_2$ of the function $\theta \in \mathbb{L}_2[0,1]$ and the vector $(\theta_1, \theta_2, \ldots) \in \ell_2$ are equal.

By Lemma D.16, likelihood ratio tests satisfy the requirements of (8.17), for the metrics $d_n = e_n$ taken equal to the $\mathbb{L}_2$-norm. The Kullback-Leibler divergence and variation $V_{2,0}$ turn out to be multiples of the $\mathbb{L}_2$-norm as well.

Lemma 8.30 *For every* $\theta, \theta_0 \in \Theta \subset \mathbb{L}_2[0,1]$,

(i) $K(P_{\theta_0}^{(n)}; P_\theta^{(n)}) = \frac{1}{2} n \|\theta - \theta_0\|_2^2$.
(ii) $V_{2,0}(P_{\theta_0}^{(n)}; P_\theta^{(n)}) = n \|\theta - \theta_0\|_2^2$.

Proof The log likelihood ratio for the sequence formulation of the model takes the form

$$\log \frac{p_{\theta_0}^{(n)}}{p_\theta^{(n)}}(X^{(n)}) = n \sum_{i=1}^\infty (\theta_0 - \theta)_i X_{n,i} - \frac{n}{2}\|\theta_0\|_2^2 + \frac{n}{2}\|\theta\|_2^2.$$

The mean and variance under θ_0 of this expression are the Kullback-Leibler divergence and variation. □

Lemma 8.30 implies that $B_{n,2}(\theta_0, \epsilon) = \{\theta \in \Theta: \|\theta - \theta_0\|_2 \le \epsilon\}$, and we obtain the following corollary to Theorem 8.19.

Theorem 8.31 *Let $P_\theta^{(n)}$ be the distribution on $\mathfrak{C}[0,1]$ of the solution to the diffusion equation $dX_t = \theta(t)\,dt + n^{-1/2}\,dW_t$ with $X_0 = 0$. If there exists a partition $\Theta = \Theta_{n,1} \cup \Theta_{n,2}$ such that for ϵ_n with $n\epsilon_n^2 \ge 1$, and for every $j \in \mathbb{N}$,*

$$\text{(i)}\quad \frac{\Pi_n(\theta \in \Theta_{n,1}: \|\theta - \theta_0\|_2 \le j\epsilon_n)}{\Pi_n(\theta \in \Theta: \|\theta - \theta_0\|_2 \le \epsilon_n)} \le e^{n\epsilon_n^2 j^2/64}, \tag{8.40}$$

$$\text{(ii)}\quad \sup_{\epsilon > \epsilon_n} \log N(\epsilon/8, \{\theta \in \Theta_{n,1}: \|\theta - \theta_0\|_2 < \epsilon\}, \|\cdot\|_2) \le n\epsilon_n^2, \tag{8.41}$$

$$\text{(iii)}\quad \frac{\Pi_n(\Theta_{n,2})}{\Pi_n(\theta \in \Theta: \|\theta - \theta_0\|_2 \le \epsilon_n)} = o(e^{-n\epsilon_n^2}), \tag{8.42}$$

then the posterior rate of contraction relative to $\|\cdot\|_2$ at θ_0 is given by ϵ_n.

The prior that models the coordinates θ_i as independent Gaussian variables is conjugate in this model, and allows explicit calculation of the posterior distribution. In Example 8.6 and Section 9.5.4 we calculate the rate of contraction by a direct method.

8.3.5 Gaussian Time Series

For $f \in \mathcal{F} \subset \mathbb{L}_2(-\pi, \pi)$, let $P_f^{(n)}$ be the distribution of $(X_1, \ldots, X_n)$ for a stationary, mean-zero Gaussian time series $(X_t: t \in \mathbb{Z})$ with spectral density f. Denote the corresponding autocovariance function by

$$\gamma_f(h) = \int_{-\pi}^{\pi} e^{ih\lambda} f(\lambda)\,d\lambda.$$

Assume that $\|\log f\|_\infty \le M$ and $\sum_{h=0}^{\infty} h\gamma_f^2(h) \le N$, for every $f \in \mathcal{F}$ and given constants M and N. The first condition is reasonable as it is known that the structure of the time series changes dramatically if the spectral density approaches zero or infinity. The second condition imposes smoothness on the spectral density.

Under these restrictions Proposition D.15 shows that condition (8.17) is satisfied, for d_n the metric of $\mathbb{L}_2(-\pi,\pi)$, and e_n the uniform metric on the spectral densities, and constants ξ and K depending on M and N only.

The following lemma relates the Kullback-Leibler divergence and variation to the $\mathbb{L}_2$-norm on the spectral densities, and implies that the neighborhoods $B_{n,2}(f_0,\epsilon)$ contain $\mathbb{L}_2$-neighborhoods.

Lemma 8.32 *There exists a constant C depending only on M, such that for every spectral densities f, g with $\|\log f\|_\infty \le M$ and $\|\log g\|_\infty \le M$,*

(i) $K(p_f^{(n)}; p_g^{(n)}) \le Cn\|f-g\|_2^2$.
(ii) $V_{2,0}(p_f^{(n)}; p_g^{(n)}) \le Cn\|f-g\|_2^2$.

Proof The covariance matrix $T_n(f)$ of $X^{(n)} = (X_1, \ldots, X_n)$ given the spectral density f is linear in f, and $T_n(1) = 2\pi I$. Using the matrix identities $\det(AB^{-1}) = \det(I + B^{-1/2}(A - B)B^{-1/2})$ and $A^{-1} - B^{-1} = A^{-1}(A-B)B^{-1}$, we can write the log likelihood ratio in the form

$$\log\frac{p_f^{(n)}}{p_g^{(n)}}(X^{(n)}) = -\tfrac{1}{2}\log\det\left(I + T_n(g)^{-1/2}T_n(f-g)T_n(g)^{-1/2}\right)$$
$$-\tfrac{1}{2}(X^{(n)})^\top T_n(f)^{-1}T_n(g-f)T_n(g)^{-1}X^{(n)}.$$

As the mean and variance of a quadratic form in a random vector X with mean zero and covariance matrix Σ take the forms $\mathrm{E}(X^\top AX) = \mathrm{tr}(\Sigma A)$ and $\mathrm{var}(X^\top AX) = \mathrm{tr}(\Sigma A\Sigma A) + \mathrm{tr}(\Sigma A\Sigma A^\top)$, it follows that

$$\begin{aligned}2K(p_f^{(n)}; p_g^{(n)}) &= -\log\det\left(I + T_n(g)^{-1/2}T_n(f-g)T_n(g)^{-1/2}\right)\\ &\quad - \mathrm{tr}(T_n(g-f)T_n(g)^{-1}),\\ 4V_{2,0}(p_f^{(n)}; p_g^{(n)}) &= \mathrm{tr}(T_n(g-f)T_n(g)^{-1}T_n(g-f)T_n(g)^{-1})\\ &\quad + \mathrm{tr}(T_n(g-f)T_n(g)^{-1}T_n(f)T_n(g)^{-1}T_n(g-f)T_n(f)^{-1}).\end{aligned}$$

For $\|A\|$ the *Frobenius norm* (given by $\|A\|^2 = \sum_k\sum_l a_{k,l}^2 = \mathrm{tr}(AA^\top)$) and $|A|$ the operator norm (given by $|A| = \sup\{\|Ax\|: \|x\| = 1\}$) of a matrix A, we have $\mathrm{tr}(A^2) \le \|A\|^2$ and $\|AB\| \le |A|\|B\|$. Furthermore $-\frac{1}{2}\mathrm{tr}(A^2) \le \log\det(I+A) - \mathrm{tr}(A) \le 0$ for any nonnegative-definite matrix A, as a result of the inequality $-\frac{1}{2}\mu^2 \le \log(1+\mu) - \mu \le 0$ for $\mu \ge 0$. Because $x^\top T_n(f)x = \int_{-\pi}^{\pi}|\sum_{k=1}^n x_k e^{ik\lambda}|^2 f(\lambda)\,d\lambda$, we have that the map $f \mapsto T_n(f)$ is monotone in the sense of positive-definiteness. In particular, $T_n(1)\inf_{-\pi<x<\pi} f(x) \le T_n(f) \le T_n(1)\|f\|_\infty$. Because $T_n(1)$ is equal to 2π times the identity matrix, we can derive that

$$|T_n(f)| \le 2\pi\|f\|_\infty \qquad |T_n(f)^{-1}| \le (2\pi)^{-1}\|1/f\|_\infty.$$

(For the second inequality we use that $|A^{-1}| \leq c^{-1}$ if $\|Ax\| \geq c\|x\|$ for all x.) Furthermore,

$$\|T_n(f)\|^2 = \sum_{|h|<n} (n - |h|)\gamma_f^2(h) \leq 2\pi n \int_{-\pi}^{\pi} f^2(\lambda)\,d\lambda. \tag{8.43}$$

Given the preceding bounds and the identity $\mathrm{tr}(AB) = \mathrm{tr}(BA)$, derivation of the desired bounds on the mean and variance of $\log(p_f^{(n)}/p_g^{(n)})(X^{(n)})$ is now straightforward. □

The lemma allows us to reduce the prior mass condition to $\mathbb{L}_2$-neighborhoods. The following theorem gives the rate of contraction of the posterior distribution for the spectral density, relative to the $\mathbb{L}_2$-distance. The theorem is an immediate corollary of Theorem 8.19 and the preceding lemma.

Theorem 8.33 (Gaussian time series) *Let $P_f^{(n)}$ be the distribution of $(X_1, \ldots, X_n)$ for a stationary Gaussian time series $(X_t: t \in \mathbb{Z})$ with spectral density $f \in \mathcal{F}$. If there exist constants M and N such that $\|\log f\|_\infty \leq M$ and $\sum_h |h|\gamma_f^2(h) \leq N$ for every $f \in \mathcal{F}$, and there exist partitions $\mathcal{F} = \mathcal{F}_{n,1} \cup \mathcal{F}_{n,2}$ such that for numbers $\epsilon_n \geq n^{-1/2}$ and every $j \in \mathbb{N}$,*

(i) $\dfrac{\Pi(f \in \mathcal{F}_{n,1}: \|f - f_0\|_2 \leq j\epsilon_n)}{\Pi(f \in \mathcal{F}: \|f - f_0\|_2 \leq \epsilon_n)} \lesssim e^{Kn\epsilon_n^2 j^2/8}$,

(ii) $\sup_{\epsilon \geq \epsilon_n} \log N(\xi\epsilon/2, \{f \in \mathcal{F}_{n,1}: \|f - f_0\|_2 \leq \epsilon\}, \|\cdot\|_\infty) \leq n\epsilon_n^2$,

(iii) $\dfrac{\Pi(\mathcal{F}_{n,2})}{\Pi(f \in \mathcal{F}: \|f - f_0\|_2 \leq \epsilon_n)} = o(e^{-n\epsilon_n^2})$,

then the posterior rate of contraction relative to the $\mathbb{L}_2$-norm is ϵ_n at every $f_0 \in \mathcal{F}$.

8.4 Lower Bounds

The theory so far gives upper bounds for posterior contraction rates, without a guarantee that the actual posterior contraction rate at a given true parameter is not faster. In applications, the theory is typically applied to obtain a rate of contraction for every parameter (uniformly) in a given model. If this compares to the minimax rate, then clearly the rate is sharp in the minimax sense. However, such a minimax rate refers to a worst parameter in a model, and the rate at a specific parameter may well be faster. To study this we define *lower bounds* on contraction rates as follows.

We adopt the general setting of a given sequence $(\mathfrak{X}^{(n)}, \mathscr{X}^{(n)}, P_\theta^{(n)}: \theta \in \Theta_n)$ of statistical experiments, with observations $X^{(n)}$, a prior distribution Π_n on the parameter set Θ_n, and a semimetric d on Θ_n.

Definition 8.34 (Contraction rate, lower bound) A sequence δ_n is a lower bound for the contraction rate at the parameter θ_0 with respect to the semimetric d if $\Pi_n(\theta: d(\theta, \theta_0) \leq m_n\delta_n |\, X^{(n)}) \to 0$ in $P_{\theta_0}^{(n)}$-probability as $n \to \infty$, for any $m_n \to 0$.

If δ_n is a lower bound for the posterior contraction rate at θ_0, then a ball around θ_0 of radius slightly smaller than δ_n is too small to hold any appreciable posterior mass. Naturally, it is desirable to obtain a lower bound that is as large as possible, ideally one that matches an upper bound ϵ_n up to a constant, or a slowly decaying factor such a (negative) power of logarithm. A contraction rate ϵ_n and lower bound δ_n together imply that the posterior distribution puts asymptotically all its mass inside an annular region of the type $\{\theta \in \Theta: m_n\delta_n \leq d(\theta,\theta_0) \leq M_n\epsilon_n\}$, for any sequences $m_n \to 0$ and $M_n \to \infty$. (In nonparametric situations the two sequences m_n and M_n can often be chosen fixed, but the current definition including the two sequences yields the correct rate $n^{-1/2}$ in parametric situations, and also the correct rate in other situations where the rescaled posterior distribution concentrates asymptotically on an unbounded set.)

A simple, but potentially effective, tool to establish a lower convergence rate is given by Theorem 8.20. This shows that the posterior mass of sets of asymptotically vanishing prior probability tends to zero. Applied in the present context, this gives the following sufficient criterion.

Theorem 8.35 (Contraction rate, lower bound) *A sequence δ_n is (certainly) a lower bound for the rate of contraction if, for some ϵ_n and $k > 1$,*

$$\frac{\Pi_n(\theta: d(\theta,\theta_0) \leq \delta_n)}{\Pi_n(B_{n,k}(\theta_0,\epsilon_n))} = o(e^{-2n\epsilon_n^2}), \tag{8.44}$$

for $B_{n,k}(\theta_0,\epsilon)$ the neighborhoods of the true parameter defined in (8.19).

The sequence ϵ_n may be arbitrary, but will typically be (an upper bound on) the rate of contraction, for which the prior mass $\Pi_n(B_{n,k}(\theta_{n,0},\epsilon_n))$ is at least $e^{-Cn\epsilon_n^2}$, under the prior mass condition (8.22). The condition on δ_n is then satisfied if $\Pi_n(\theta: d(\theta,\theta_0) \leq \delta_n) \ll e^{-(C+2)n\epsilon_n^2}$.

This criterion is particularly attractive if the neighborhoods $B_{n,k}(\theta_{n,0},\epsilon_n)$ can be replaced by balls for the metric d. If these neighborhoods contain the balls $\{\theta: d(\theta,\theta_0) < c\epsilon_n\}$, then δ_n is a lower bound on the contraction rate if, for some ϵ_n and constants with $C_2 > C_1 + 2c^{-2}$,

$$\Pi_n(\theta: d(\theta,\theta_0) \leq \delta_n) \leq e^{-C_2 n\epsilon_n^2} \leq e^{-C_1 n\epsilon_n^2} \leq \Pi_n(\theta: d(\theta,\theta_0) \leq \epsilon_n). \tag{8.45}$$

This requires a lower and an upper bound on the prior mass near the true parameter and may be difficult to obtain. However, it is not unreasonable to expect these bounds to hold for δ_n of (nearly) the same order as ϵ_n, as infinite-dimensional priors tend to concentrate near boundaries of balls rather than spread continuously. Intuitively this can be understood from the fact that high-dimensional balls have a large surface.

8.5 Misspecification

Consider placing a prior on a (dominated) model $\mathcal{P}$ that does not contain or approximate the true density p_0 of the observations. As the posterior distributions also concentrate on $\mathcal{P}$, they cannot be consistent. From a frequentist point of view studying such *model misspecification* is relevant, because a model is typically only perceived as an approximation to reality. From an objective Bayesian point of view model misspecification is of interest, because in principle the Bayesian paradigm does not restrict the prior, and hence this may rule out a given density p_0. For both it is relevant, because a model may be adopted for convenience only, such as Gaussian errors in nonparametric regression.

The main result of this section shows that the posterior distributions, even though inconsistent, will still settle down eventually, near a Kullback-Leibler projection p^* of p_0 into the model. Furthermore, the rate at which the posterior concentrates near p^* is characterized by similar quantities as before.

We restrict to i.i.d. observations following the Bayesian model $X_1, \ldots, X_n | p \overset{\text{iid}}{\sim} p$ and $p \sim \Pi_n$, where Π_n is a prior distribution on a set $\mathcal{P}$ of probability densities relative to a dominating measure ν on a sample space $(\mathfrak{X}, \mathscr{X})$. We adapt the entropy and the neighborhoods for the prior mass condition to misspecified models, as follows.

First we define the *covering number for testing under misspecification* of $\mathcal{P}$ relative to a semimetric d, denoted by $N_{t,\text{mis}}(\epsilon, \mathcal{P}, d; p_0, p^*)$, as the minimal number N of sets in a partition $\mathcal{P}_1, \ldots, \mathcal{P}_N$ of $\{p \in \mathcal{P}: \epsilon < d(p, p^*) < 2\epsilon\}$ such that, for every partitioning set

$$\inf_{p \in \text{conv}(\mathcal{P}_l)} \sup_{0<\alpha<1} -\log P_0\Big(\frac{p}{p^*}\Big)^\alpha \geq \tfrac{1}{2}\epsilon^2. \tag{8.46}$$

The logarithm of this number is referred to as “entropy.”

Second, we consider modified Kullback-Leibler neighborhoods of p, given by

$$B(p^*, \epsilon; p_0) = \Big\{p \in \mathcal{P}: P_0 \log \frac{p^*}{p} \leq \epsilon^2,\ P_0(\log \frac{p^*}{p})^2 \leq \epsilon^2\Big\}. \tag{8.47}$$

Theorem 8.36 *If there exist partitions $\mathcal{P} = \mathcal{P}_{n,1} \cup \mathcal{P}_{n,2}$ and a constant $C > 0$ such that for some $p_n^* \in \mathcal{P}$ with $K(p_0; p_n^*) < \infty$ and $P_0(p/p_n^*) < \infty$ for all $p \in \mathcal{P}$ and constants $\bar\epsilon_n \leq \epsilon_n$ with $n\bar\epsilon_n^2 \geq 1$,*

$$\text{(i)}\ \frac{\Pi_n(p \in \mathcal{P}_{n,1}: j\epsilon_n < d(p, p_n^*) \leq 2j\epsilon_n)}{\Pi_n(B(p_n^*, \bar\epsilon_n; p_0))} \leq e^{n\bar\epsilon_n^2 j^2/8}, \tag{8.48}$$

$$\text{(ii)}\ \sup_{\epsilon \geq \epsilon_n} \log N_{t,\text{mis}}(\epsilon, \mathcal{P}_{n,1}, d; p_0, p_n^*) \leq n\epsilon_n^2, \tag{8.49}$$

$$\text{(iii)}\ \frac{\int_{\mathcal{P}_{n,2}} (P(p_0/p_n^*))^n \, d\Pi_n(p)}{\Pi_n(B(p_n^*, \bar\epsilon_n; p_0))} = o(e^{-2n\bar\epsilon_n^2}), \tag{8.50}$$

then $\Pi_n(p \in \mathcal{P}: d(p, p_n^*) \geq M_n\epsilon_n | X_1, \ldots, X_n) \to 0$ in P_0^n-probability, for every $M_n \to \infty$, and for every sufficiently large $M_n = M$ if $n\epsilon_n^2 \to \infty$.

Proof For $p \in \mathcal{P}$, the function $q(p) = (p_0/p_n^*)p$ is integrable by assumption, and hence defines a finite measure. Its Hellinger transform with p_0 satisfies $\rho_\alpha(p_0; q(p)) = P_0(p/p_n^*)^{1-\alpha}$. By Lemma D.6, for any set $\mathcal{P}_l$ of densities,

$$\rho_\alpha\Big(p_0^n; \text{conv}(q(p)^n: p \in \mathcal{P}_l)\Big) \leq \sup_{p \in \text{conv}(\mathcal{P}_l)} \rho_\alpha(p_0; q(p))^n = \sup_{p \in \text{conv}(\mathcal{P}_l)} \Big[P_0\Big(\frac{p}{p_n^*}\Big)^{1-\alpha}\Big]^n.$$

If we take the infimum over $\alpha \in (0, 1)$ across this inequality, then on the right side the order of the infimum can be exchanged with the supremum over p, by the minimax theorem Theorem L.5, applicable, because $\alpha \mapsto P_0(p/p_n^*)^{1-\alpha}$ is convex and can be continuously extended to the compact set $[0, 1]$, and $p \mapsto P_0(p/p_n^*)^{1-\alpha}$ is concave. Thus when applied to a set $\mathcal{P}_l$ as in the definition of the covering number for testing under misspecification, the right and hence left side is bounded above by $e^{-n\epsilon^2/2}$. It follows that the sets

$\mathcal{Q}_l := \{q(p)^n : p \in \mathcal{P}_l\}$ in the induced partition of $\mathcal{Q} := \{q(p)^n : \epsilon < d(p, p_n^*) \le 2\epsilon\}$ satisfy (D.5) with $K = n/2$. Therefore by Theorem D.4, applied with $\log N(\epsilon)$ equal to the right side of (8.49) and $\epsilon = M\epsilon_n$, there exist tests ϕ_n such that

$$P_0^n \phi_n \le e^{n\epsilon_n^2} \frac{e^{-nM^2\epsilon_n^2/2}}{1 - e^{-nM^2\epsilon_n^2/2}}, \qquad Q(p)^n(1 - \phi_n) \le e^{-nM^2\epsilon_n^2/2}.$$

Next we partition $\{p \in \mathcal{P}_{n,1} : d(p, p_n^*) > M\epsilon_n\}$ into the "shells" $\mathcal{S}_{n,j} = \{p \in \mathcal{P}_{n,1} : M\epsilon_n j < d(p, p_n^*) \le M\epsilon_n(j+1)\}$ for $j \in \mathbb{N}$. For every j we obtain, by Fubini's theorem,

$$\begin{aligned} P_0^n\Big[\int_{\mathcal{S}_{n,j}} \prod_{i=1}^n \frac{p}{p_n^*}(X_i)\, d\Pi_n(p)(1 - \phi_n)\Big] &\le \int_{\mathcal{S}_{n,j}} Q(p)^n(1 - \phi_n)\, d\Pi_n(p) \\ &\le e^{-KnM^2\epsilon_n^2 j^2/2}\Pi_n(\mathcal{S}_{n,j}). \end{aligned}$$

For A_n the event $\{\int \prod_{i=1}^n (p/p_n^*)(X_i)\, d\Pi_n(p) \ge e^{-2Cn\bar\epsilon_n^2}\Pi_n(B(p_n^*, \bar\epsilon_n; p_0))\}$ the posterior probability of the set $\{p : d(p, p_n^*) > M\epsilon_n\}$ is bounded above by

$$\phi_n + \mathbb{1}\{A_n^c\} + \frac{1}{e^{-Cn\bar\epsilon_n^2}\Pi_n(B(p_n^*, \bar\epsilon_n; p_0))} \int_{d(p, p_n^*) > M\epsilon_n} \prod_{i=1}^n \frac{p}{p_n^*}(X_i)\, d\Pi_n(p)(1 - \phi_n).$$

The expected value under P_0^n of the third term is bounded by

$$\sum_{j \ge M} \frac{e^{-KnM^2\epsilon_n^2 j^2/2}\Pi_n(\mathcal{S}_{n,j})}{e^{-2Cn\bar\epsilon_n^2}\Pi_n(B_2(p_n^*, \bar\epsilon_n; p_0))} + \frac{\int_{\mathcal{P}_{n,2}} (P(p_0/p_n^*))^n\, d\Pi_n(p)}{e^{-2Cn\bar\epsilon_n^2}\Pi_n(B_2(p_n^*, \bar\epsilon_n; p_0))}.$$

We finish as in the proof of Theorem 8.11, where we substitute Lemma 8.37 below for Lemma 8.10. □

Lemma 8.37 *For any $p^* \in \mathcal{P}$ with $P_0 \log(p_0/p^*) < \infty$, every probability measure Π on $\mathcal{P}$, and every constants $\epsilon, C > 0$, with P_0^n-probability at least $1 - (C\sqrt{n}\epsilon)^{-2}$,*

$$\int \prod_{i=1}^n \frac{p}{p^*}(X_i)\, d\Pi(p) \ge \Pi(B(p^*, \epsilon; p_0))e^{-(1+C)n\epsilon^2}.$$

Since $P_0 \log(p_n^*/p) = K(p_0; p) - K(p_0; p^*)$, the first inequality in the definition of the neighborhoods $B(p^*, \epsilon; p_0)$ can be written in the form

$$K(p_0; p) \le K(p_0; p^*) + \epsilon^2.$$

Therefore the neighborhood contains only p that are within ϵ^2 of the Kullback-Leibler divergence $K(p_0; p^*)$, which may be regarded the minimal divergence of p_0 to the model. For $p^* = p_0$ the prior mass condition (8.48) reduces to the analogous condition employed previously for correctly specified models.

On the other hand, the entropy condition (8.49) is more abstract than the corresponding condition in the correctly specified case. The technical motivation is the same, existence of good tests, but presently the tests concern the null hypothesis p_0 versus the alternatives $q(p)$ defined in the proof of Theorem 8.36, which may have total mass bigger than one. The form of the condition is explained in Appendix D.1, and also that (8.49) can be valid only if p_n^* is at minimal Kullback-Leibler divergence of p_0 to $\mathcal{P}_{n,1}$, which is not otherwise included in the conditions of the theorem.

The testing entropy can be bounded in terms of ordinary entropy relative to appropriate semimetrics. For convex models $\mathcal{P}$, this is possible for convex semimetrics d satisfying

$$d^2(p, p^*) \leq \sup_{0<\alpha<1} -\log P_0\Big(\frac{p}{p^*}\Big)^\alpha. \tag{8.51}$$

More generally, this is possible for semimetrics that satisfy the following strengthening of (8.51): for some fixed constants $c, C > 0$ and for every $m \in \mathbb{N}$, $\lambda_1, \ldots, \lambda_m \geq 0$ with $\sum_i \lambda_i = 1$ and every $p, p_1, \ldots, p_m \in \mathcal{P}$ with $d(p, p_i) \leq c\, d(p, p^*)$ for all i,

$$\sum_i \lambda_i \big[d^2(p_i, p^*) - C d^2(p_i, p)\big] \leq \sup_{0<\alpha<1} -\log P_0\Big(\frac{\sum_i \lambda_i p_i}{p^*}\Big)^\alpha. \tag{8.52}$$

Evaluating this inequality with $m = 1$ and $p = p_1$ leads back to (8.51).

Lemma 8.38 *We have $N_{t,\mathrm{mis}}(\epsilon, \mathcal{P}, d; p_0, p^*) \leq N(A\epsilon, \{p \in \mathcal{P}: \epsilon < d(p, p^*) < 2\epsilon\}, d)$, for all $\epsilon > 0$, and some constant $A > 0$, if one of the following conditions holds:*

(i) (8.51) *is valid for every $p \in \mathrm{conv}(\mathcal{P})$ and the map $p \mapsto d^2(p, p')$ is defined and convex on $\mathrm{conv}(\mathcal{P})$, for every $p' \in \mathcal{P}$.*
(ii) (8.52) *holds.*

Proof (ii). For a given constant $A > 0$ we can cover the set $\mathcal{P}_\epsilon := \{p \in \mathcal{P}: \epsilon < d(p, p^*) < 2\epsilon\}$ with $N = N(A\epsilon, \mathcal{P}_\epsilon, d)$ balls of radius $A\epsilon$. If the centers of these balls are not contained in $\mathcal{P}_\epsilon$, then we can replace them by N balls of radius $2A\epsilon$ with centers in $\mathcal{P}_\epsilon$ whose union also covers $\mathcal{P}_\epsilon$. It suffices to show that (8.46) is valid for $\mathcal{P}_l$ equal to a typical ball B in this cover. Choose $2A < 1$. If $p \in \mathcal{P}_\epsilon$ is the center of B and $p_i \in B$ for every i, then $d(p_i, p^*) \geq d(p, p^*) - 2A\epsilon$ by the triangle inequality, and hence by assumption (8.52) the left side of (8.46) is bounded below by $\sum_i \lambda_i((\epsilon - 2A\epsilon)^2 - C(2A\epsilon)^2)$. This is bounded below by $\epsilon^2/2$ for sufficiently small A.

(i). It suffices to show that condition (i) implies condition (ii) for $d/2$ instead of d. By repeatedly applying the triangle inequality we find

$$\begin{aligned}\sum_i \lambda_i d^2(p_i, p^*) &\leq 3 \sum_i \lambda_i \Big[d^2(p_i, p) + d^2\Big(p, \sum_j \lambda_j p_j\Big) + d^2\Big(\sum_j \lambda_j p_j, p^*\Big)\Big] \\ &\leq 6 \sum \lambda_i d^2(p_i, p) + 3d^2\Big(\sum_j \lambda_j p_j, p^*\Big),\end{aligned}$$

by the convexity of d^2. It follows that

$$d^2\Big(\sum_i \lambda_i p_i, p^*\Big) \geq \frac{1}{3} \sum_i \lambda_i d^2(p_i, p^*) - 2 \sum_i \lambda_i d^2(p_i, p).$$

If (8.51) holds for $p = \sum_i \lambda_i p_i$, then we obtain (8.52) with d^2 replaced by $d^2/4$ (or even $d^2/3$) and $C = 6$. □

8.5.1 Convex Models

The case of convex models $\mathcal{P}$ is of interest, in particular for non- or semiparametrics, and permits some simplification. For a convex model the point of minimal Kullback-Leibler divergence (if it exists) is automatically unique (up to a P_0-null set). Moreover, the expectations $P_0(p/p^*)$ are automatically finite, and condition (8.52) is automatically satisfied for the *weighted Hellinger distance*, with square,

$$d_{H,w}^2(p_1, p_2) = \int (\sqrt{p_1} - \sqrt{p_2})^2 \frac{p_0}{p^*}\, d\nu.$$

Theorem 8.39 (Convex model) *If $\mathcal{P}$ is convex and $K(p_0; p^*) \le K(p_0, p)$, for $p^* \in \mathcal{P}$ and every $p \in \mathcal{P}$, then $P_0(p/p^*) \le 1$ for every $p \in \mathcal{P}$, and there exists a constant $A > 0$ such that $N_{t,\mathrm{mis}}(\epsilon, \mathcal{P}, d_{H,w}; p_0, p^*) \le N(A\epsilon, \{p \in \mathcal{P}: \epsilon < d_{H,w}(p, p^*) < 2\epsilon\}, d_{H,w})$, for all $\epsilon > 0$.*

Proof For $p \in \mathcal{P}$, define a family of convex combinations $\{p_\lambda: \lambda \in [0,1]\} \subset \mathcal{P}$ by $p_\lambda = \lambda p + (1-\lambda)p^*$. Since $p^* \in \mathcal{P}$ minimizes the Kullback-Leibler divergence with respect to p_0 over $\mathcal{P}$, for every $\lambda \in [0,1]$,

$$0 \le f(\lambda) := P_0 \log \frac{p^*}{p_\lambda} = -P_0 \log\Bigl(1 + \lambda\Bigl(\frac{p}{p^*} - 1\Bigr)\Bigr).$$

For every fixed $y \ge 0$ the function $\lambda \mapsto \log(1+\lambda y)/\lambda$ is nonnegative and increases monotonically to y as $\lambda \downarrow 0$. The function is bounded in absolute value by 2 for $y \in [-1, 0]$ and $\lambda \le \frac{1}{2}$. Therefore, by the monotone and dominated convergence theorems applied to the positive and negative parts of the integrand on the right side the right derivative of f at 0 is given by $f'(0+) = 1 - P_0(p/p^*)$. Now $f'(0+) \ge 0$, since $0 = f(0) \le f(\lambda)$ for every λ, whence $P_0(p/p^*) \le 1$.

Because $-\log x \ge 1 - x$, we have $-\log P_0(p/p^*)^{1/2} \ge 1 - P_0(p/p^*)^{1/2}$. Now

$$\int (\sqrt{p^*} - \sqrt{p})^2 \frac{p_0}{p^*}\, d\mu = 1 + P_0 \frac{p}{p^*} - 2P_0\sqrt{\frac{p}{p^*}} \le 2 - 2P_0\sqrt{\frac{p}{p^*}},$$

by the first part of the proof. It follows that the semimetric $d_{H,w}/2$ satisfies (8.51), whence the theorem follows by Lemma 8.38(i). ∎

8.5.2 Nonparametric Regression

Let P_f be the distribution of a random vector (X, Y) following the regression model $Y = f(X) + e$, for independent random variables X and e taking values in a measurable space $(\mathfrak{X}, \mathscr{X})$ and in $\mathbb{R}$, respectively, and an unknown measurable function $f: \mathfrak{X} \to \mathbb{R}$ belonging to some model $\mathcal{F}$. We form the posterior distribution for f in the Bayesian model $(X_1, Y_1), \ldots, (X_n, Y_n) \mid f \overset{\text{iid}}{\sim} P_f$ and $f \sim \Pi$, where the corresponding errors are assumed to have either a standard normal or Laplace distribution and the covariates a fixed distribution. The latter covariate distribution cancels from the posterior distribution, and hence can be assumed known. On the other hand, the normal or Laplace distribution for the errors is considered a working model only, but is essential to the definition of the posterior distribution. We investigate the posterior distribution under the assumption that the observations in reality follow the model $Y = f_0(X) + e_0$, for a given regression function f_0, and an error e_0 that has mean or median (in the normal and Laplace cases) zero, but may not be Gaussian or Laplace. Thus the error distribution is typically misspecified. We also allow that the true regression function f_0 is misspecified in that it does not belong to the model $\mathcal{F}$, but this is of secondary interest, as in our nonparametric situation $\mathcal{F}$ will typically be large.

The main finding is that misspecification of the error distribution does not have serious consequences for estimation of the regression function. Thus the nonparametric Bayesian approach possesses the same robustness to misspecification as minimum contrast estimation using least squares or minimum absolute deviation. The results below require that the error distribution be light-tailed in the normal case, but not so for the Laplace posterior. Thus the tail robustness of minimum absolute deviation versus the nonrobustness of least squares appears to extend also to Bayesian regression.

As in the general theory the true distribution of an observation is denoted P_0. This should not be confused with P_f for $f = 0$, which does not appear in the following.

Normal Errors

If the working (misspecified) error distribution is standard normal, and a typical observation follows the model $Y = f_0(X) + e_0$ for a a true error e_0 with mean zero, then

$$\log \frac{p_{f_0}}{p_f}(X, Y) = \tfrac{1}{2}(f - f_0)^2(X) - e_0(f - f_0)(X),$$
$$K(p_0; p_f) = \tfrac{1}{2} P_0(f - f_0)^2.$$

It follows that the Kullback-Leibler divergence of p_0 to the model is minimized at p_{f^*} for $f^* \in \mathcal{F}$ minimizing the map $f \mapsto P_0(f - f_0)^2$. In particular $f^* = f_0$, in the case that the model for the regression functions is correctly specified ($f_0 \in \mathcal{F}$). If $\mathcal{F}$ is convex and closed in $\mathbb{L}_2$, then a unique minimizer f^* exists by the Hilbert space projection theorem, and is characterized by the inequalities $P_0(f_0 - f^*)(f^* - f) \geq 0$, for every $f \in \mathcal{F}$. If the minimizer f^* is also the minimizer over the linear span of $\mathcal{F}$, then the latter inequalities are equalities.

Replace the "neighborhoods" $B(p_{f^*}, \epsilon; p_0)$ in the prior mass condition (8.48) by

$$\tilde{B}(p_{f^*}, \epsilon; f_0) = \Big\{ f \in \mathcal{F}: \|f - f_0\|_2^2 \leq \|f^* - f_0\|_2^2 + \epsilon^2, \|f - f^*\|_2^2 \leq \epsilon^2 \Big\}.$$

If $P_0(f - f^*)(f^* - f_0) = 0$ for every $f \in \mathcal{F}$, then this reduces to an $\mathbb{L}_2$-ball around f^*, since $\|f - f_0\|_2^2 = \|f - f^*\|_2^2 + \|f^* - f_0\|_2^2$, by Pythagoras's theorem.

Theorem 8.40 *Let $\mathcal{F}$ be a class of uniformly bounded functions $f: \mathfrak{X} \to \mathbb{R}$ such that either $f_0 \in \mathcal{F}$ or $\mathcal{F}$ is convex and closed in $\mathbb{L}_2$. If f_0 is uniformly bounded,* $\mathrm{E}_0 e_0 = 0$ *and* $\mathrm{E}_0 e^{M|e_0|} < \infty$ *for every $M > 0$, and ϵ_n are positive numbers with $n\epsilon_n^2 \to \infty$ such that for a constant $C > 0$,*

$$\Pi(\tilde{B}(f^*, \epsilon_n; f_0)) \geq e^{-Cn\epsilon_n^2},$$
$$\log N(\epsilon_n, \mathcal{F}, \|\cdot\|_{P_0,2}) \leq n\epsilon_n^2,$$

then $\Pi_n(f \in \mathcal{F}: P_0(f - f^*)^2 \geq M\epsilon_n^2 | X_1, \ldots, X_n) \to 0$ *in* P_0^n*-probability, for every sufficiently large constant* M.

The theorem is a corollary of Theorem 8.36 and the following lemma, which shows that (8.52) is satisfied for a multiple of the $\mathbb{L}_2$-distance on $\mathcal{F}$, and that the neighborhoods (8.52) contain the neighborhoods induced by the map $f \mapsto p_f$ and the present neighborhoods (with ϵ replaced by a multiple).

Lemma 8.41 *Let* $\mathcal{F}$ *be a class of uniformly bounded functions* $f: \mathfrak{X} \to \mathbb{R}$ *such that either* $f_0 \in \mathcal{F}$ *or* $\mathcal{F}$ *is convex and closed in* $\mathbb{L}_2(P_0)$. *If* f_0 *is uniformly bounded,* $\mathrm{E}_0 e_0 = 0$ *and* $\mathrm{E}_0 e^{M|e_0|} < \infty$ *for every* $M > 0$, *then there exist positive constants* C_1, C_2, C_3 *such that, for all* $m \in \mathbb{N}$, $f, f_1, \ldots, f_m \in \mathcal{F}$ *and* $\lambda_1, \ldots, \lambda_m \geq 0$ *with* $\sum_i \lambda_i = 1$,

$$P_0\Big(\log \frac{p_{f^*}}{p_f}\Big)^2 \leq C_1 P_0(f - f^*)^2,$$

$$\sup_{0<\alpha<1} -\log P_0\Big(\frac{\sum_i \lambda_i p_{f_i}}{p_{f^*}}\Big)^\alpha \geq C_2 \sum_i \lambda_i \Big(P_0(f_i - f^*)^2 - C_3 P_0(f - f_i)^2\Big).$$

Proof The second term on the right in

$$\log \frac{p_{f^*}}{p_f}(X, Y) = \tfrac{1}{2}[(f_0 - f)^2 - (f_0 - f^*)^2](X) + e_0(f^* - f)(X) \tag{8.53}$$

has mean zero by assumption. The first term has expectation $\frac{1}{2}P_0(f^* - f)^2$ if $f_0 = f^*$, as is the case if $f_0 \in \mathcal{F}$. Furthermore, if $\mathcal{F}$ is convex the minimizing property of f^* implies that $P_0(f_0 - f^*)(f^* - f) \geq 0$ for every $f \in \mathcal{F}$ and then the expectation of the first term on the right is bounded above by $\frac{1}{2}P_0(f^* - f)^2$. Therefore, in both cases $P_0 \log(p_{f^*}/p_f) \geq \frac{1}{2}P_0(f - f^*)^2$.

From (8.53) we also have, with M a uniform upper bound on $\mathcal{F}$ and f_0,

$$P_0(\log \frac{p_f}{p_{f^*}})^2 \leq 2P_0[(f^* - f)^2(2f_0 - f - f^*)^2] + 2P_0 e_0^2 P_0(f^* - f)^2,$$

$$P_0(\log \frac{p_f}{p_{f^*}})^2(\frac{p_f}{p_{f^*}})^\alpha \leq 2P_0[(f^* - f)^2(2f_0 - f - f^*)^2 + 2e_0^2(f^* - f)^2]e^{2\alpha(M^2 + M|e_0|)}.$$

Both right sides can be further bounded by a constant times $P_0(f - f^*)^2$, where the constant depends on M and the distribution of e_0 only.

In view of Lemma B.6 with $p = p_{f^*}$ and $q_i = p_{f_i}$, we see that there exists a constant $C > 0$ depending on M only such that for all $\lambda_i \geq 0$ with $\sum_i \lambda_i = 1$,

$$\Big|1 - P_0\Big(\frac{\sum_i \lambda_i p_{f_i}}{p_{f^*}}\Big)^\alpha - \alpha P_0 \log \frac{p_{f^*}}{\sum_i \lambda_i p_{f_i}}\Big| \leq 2\alpha^2 C \sum_i \lambda_i P_0(f_i - f^*)^2. \tag{8.54}$$

By Lemma B.6 with $\alpha = 1$ and $p = p_f$ and similar arguments also, for any $f \in \mathcal{F}$,

$$\Big|1 - P_0\Big(\frac{\sum_i \lambda_i p_{f_i}}{p_f}\Big) - P_0 \log \frac{p_f}{\sum_i \lambda_i p_{f_i}}\Big| \leq 2C \sum_i \lambda_i P_0(f_i - f)^2.$$

For $\lambda_i = 1$ this becomes $|1 - P_0(p_{f_i}/p_f) - P_0 \log(p_f/p_{f_i})| \le 2C P_0(f_i - f)^2$. Subtracting the convex combination of these inequalities from the preceding display gives

$$\Big| P_0 \log \frac{p_f}{\sum_i \lambda_i p_{f_i}} - \sum_i \lambda_i P_0 \log \frac{p_f}{p_{f_i}} \Big| \le 4C \sum_i \lambda_i P_0(f_i - f)^2.$$

By the fact that $\log(ab) = \log a + \log b$ for every $a, b > 0$, this inequality remains true if f on the left is replaced by f^*. We combine the resulting inequality with (8.54) to find that

$$\begin{aligned} 1 - P_0\Big(\frac{\sum_i \lambda_i p_{f_i}}{p_{f^*}}\Big)^\alpha &\ge \alpha \sum_i \lambda_i P_0 \log \frac{p_{f^*}}{p_{f_i}} - 2\alpha^2 C \sum_i \lambda_i P_0(f^* - f_i)^2 \\ &\quad - 4C \sum_i \lambda_i P_0(f_i - f)^2 \\ &\ge \Big(\frac{\alpha}{2} - 2\alpha^2 C\Big) \sum_i \lambda_i P_0(f^* - f_i)^2 - 4C \sum_i \lambda_i P_0(f_i - f)^2. \end{aligned}$$

In the last step we used that $P_0 \log(p_{f^*}/p_f) \ge \frac{1}{2} P_0(f - f^*)^2$. For sufficiently small $\alpha > 0$ and suitable constants C_2, C_3 the right side is bounded below by the right side of the lemma. Finally the left side of the lemma can be bounded by the supremum over $\alpha \in (0, 1)$ of the left side of the last display, since $-\log x \ge 1 - x$ for every $x > 0$. □

Laplace Errors

If the working (misspecified) error distribution is Laplace, and a typical observation follows the model $Y = f_0(X) + e_0$ for a true error e_0 with median zero, then

$$\begin{aligned} \log \frac{p_{f_0}}{p_f}(X, Y) &= (|e_0 + f_0(X) - f(X)| - |e_0|), \\ K(p_0; p_f) &= P_0 \Phi(f - f_0), \qquad \Phi(v) := P_0(|e_0 - v| - |e_0|). \end{aligned}$$

The function Φ is minimized over $v \in \mathbb{R}$ at the median of e_0, which is zero by assumption. Thus the Kullback-Leibler divergence $f \mapsto K(p_0; p_f)$ is minimized over $f \in \mathcal{F}$ at f_0, if $f_0 \in \mathcal{F}$. If $\mathcal{F}$ is a compact, convex subset of $\mathbb{L}_1(P_0)$, then in any case there exists $f^* \in \mathcal{F}$ that minimizes the Kullback-Leibler divergence.

If the distribution of e_0 is smooth, then so is the function Φ. Because it is minimal at $v = 0$ it is reasonable to expect that, for v in a neighborhood of 0 and some positive constant C_0,

$$\Phi(v) = P_0(|e_0 - v| - |e_0|) \ge C_0 |v|^2. \tag{8.55}$$

Because Φ is convex, it is also reasonable to expect that its second derivative, if it exists, is strictly positive.

Define neighborhoods $\tilde{B}(f^*, \epsilon; f_0)$ as in Theorem 8.40. The following theorem is a corollary of Theorem 8.36, in the same way as Theorem 8.40. (See Kleijn and van der Vaart 2006 for a detailed proof.)

Theorem 8.42 *Let $\mathcal{F}$ be a class of uniformly bounded functions $f: \mathfrak{X} \to \mathbb{R}$ and suppose that either $f_0 \in \mathcal{F}$ and (8.55) hold, or $\mathcal{F}$ is convex and compact in $\mathbb{L}_1(P_0)$ and Φ is twice continuously differentiable with strictly positive second derivative. If f_0 is uniformly* *bounded, e_0 has median 0, and ϵ_n are positive numbers with $n\epsilon_n^2 \to \infty$ such that for a constant $C > 0$,*

$$\Pi(\tilde{B}(f^*, \epsilon_n; f_0)) \geq e^{-Cn\epsilon_n^2},$$
$$\log N(\epsilon_n, \mathcal{F}, \|\cdot\|_{P_0,2}) \leq n\epsilon_n^2,$$

then $\Pi_n(f \in \mathcal{F}: P_0(f - f^*)^2 \geq M\epsilon_n^2 | X_1, \ldots, X_n) \to 0$ in P_0^n-probability, for every sufficiently large constant M.

8.6 α-Posterior

The α-posterior distribution is defined for i.i.d. observations in (6.19). In the context of general statistical experiments $(\mathfrak{X}, \mathscr{X}, P_\theta: \theta \in \Theta)$ with observations X, the *α-posterior distribution* is defined as

$$\Pi^{(\alpha)}(\theta \in B | X) = \frac{\int_B p_\theta^\alpha(X)\, d\Pi_n(\theta)}{\int p_\theta^\alpha(X)\, d\Pi_n(\theta)}. \tag{8.56}$$

Let $R_\beta(p; q) = -\log \rho_\beta(p; q)$ denote the *Renyi divergence* between densities p and q, as defined in (B.5). By Lemma B.5 this is an upper bound on the square Hellinger distance. The following theorem gives a bound on the expected Renyi divergence from a true density under the posterior distribution.

Theorem 8.43 *For any numbers $\epsilon > 0$, $\beta \in (0, 1)$, $\gamma \geq 0$, for $\alpha = (\gamma + \beta)/(\gamma + 1)$,*

$$P_{\theta_0} \int R_\beta(p_{\theta_0}; p_\theta)\, d\Pi^{(\alpha)}(\theta | X) \leq (\gamma + 1)\epsilon^2\alpha - \log \Pi(\theta: K(p_{\theta_0}; p_\theta) < \epsilon^2).$$

Proof The nonnegativity of the Kullback-Leibler divergence implies that for any given nonnegative measurable function $v: \Theta \to \mathbb{R}$ and every probability density w relative to Π,

$$-\int (\log w) w\, d\Pi + \int (\log v) w\, d\Pi \leq \log \int v\, d\Pi. \tag{8.57}$$

Furthermore, equality is attained for any function w with $w \propto v$. Applying this twice, first for a fixed observation X with $v(\theta) \propto (p_\theta/p_{\theta_0})^\delta(X)$ and $\delta = 1$, and second with $v(\theta) \propto (p_\theta/p_{\theta_0})^\beta(X)/\rho_\beta(p_{\theta_0}; p_\theta)$, we find for any probability density w relative to Π,

$$-\int (\log w) w\, d\Pi + \delta \int \Big(\log \frac{p_\theta}{p_{\theta_0}}(X)\Big) w(\theta)\, d\Pi(\theta) \leq \log \int \Big(\frac{p_\theta}{p_{\theta_0}}\Big)^\delta (X)\, d\Pi(p_\theta),$$
$$-\int (\log w) w\, d\Pi + \beta \int \Big(\log \frac{p_\theta}{p_{\theta_0}}(X)\Big) w(\theta)\, d\Pi(\theta) - \int \log \rho_\beta(p_{\theta_0}; p_\theta)\, w(\theta)\, d\Pi(\theta)$$
$$\leq \log c_\beta(X),$$

where $c_\beta(X) = \int (p_\theta/p_{\theta_0})^\beta(X)/\rho_\beta(p_{\theta_0}; p_\theta)\, d\Pi(\theta)$ is the normalizing constant. By the convexity of the negative of the logarithm, Jensen's inequality and Fubini's theorem, the expectation of the right side of the first inequality is at most $\log \int \rho_\delta(p_{\theta_0}; p_\theta)\, d\Pi(\theta) \leq 0$. Similarly the expectation of the right side of the second inequality is nonnegative. We replace the right sides by 0 and add the second inequality to γ times the first inequality. The resulting inequality can be reorganized into

$$\int R_\beta(p_\theta, p_{\theta_0})\, w(\theta)\, d\Pi(\theta) \le (\gamma+1)\int (\log w) w\, d\Pi - (\gamma\delta+\beta)\int \Bigl(\log\frac{p_\theta}{p_{\theta_0}}(X)\Bigr) w(\theta)\, d\Pi(\theta).$$

By (8.57) the right side is minimized with respect to probability densities w, for fixed X, by $w(\theta) \propto p_\theta^\alpha(X)$. For this minimizing function $w(\theta)\, d\Pi(\theta)$ in the left side becomes $d\Pi_\alpha(\theta|\, X)$. It follows that

$$\begin{aligned}
&\frac{1}{\gamma+1} P_{\theta_0}\int R_\beta(p_{\theta_0}; p_\theta)\, d\Pi_\alpha(\theta|\, X)\\
&\quad \le P_{\theta_0}\inf_w\Bigl[\int (\log w) w\, d\Pi - \alpha\int \Bigl(\log\frac{p_\theta}{p_{\theta_0}}(X)\Bigr) w(\theta)\, d\Pi(\theta)\Bigr]\\
&\quad \le \inf_w\Bigl[\int (\log w) w\, d\Pi + \alpha\int P_{\theta_0}\Bigl(\log\frac{p_{\theta_0}}{p_\theta}(X)\Bigr) w(\theta)\, d\Pi(\theta)\Bigr]\\
&\quad = \inf_w\Bigl[\int (\log w) w\, d\Pi - \int \Bigl(\log e^{-\alpha K(p_{\theta_0}; p_\theta)}\Bigr) w(\theta)\, d\Pi(\theta)\Bigr]\\
&\quad = -\log\int e^{-\alpha K(p_{\theta_0}; p_\theta)}\, d\Pi(\theta).
\end{aligned}$$

Here the last step follows again by (8.57). Finally we use Markov's inequality to bound this by $(\gamma+1)^{-1}$ times the right side of the theorem. □

An attractive feature of the preceding theorem is that it makes no assumption on the sampling model: the family p_θ is completely general. The theorem will typically be applied with t such that the two terms on its right side have the same order of magnitude. If $\Pi(\theta\colon K(p_{\theta_0}; p_\theta) < t^2) \ge \exp(-t^2)$, then the order of magnitude is t^2. This is a version of the prior mass condition.

The parameter α is necessarily strictly smaller than 1, and hence the theorem applies to the pseudo-posterior, but not to the true posterior distribution. The theorem shows that for the pseudo-posterior distribution, the prior mass condition alone gives posterior concentration.

Example 8.44 (Pseudo-posterior, i.i.d. observations) The Rényi and Kullback-Leibler divergences are additive on product measures. Therefore, if p_θ in the theorem are replaced by product densities p_θ^n and ϵ by $\sqrt{n}\epsilon_n$, then we obtain

$$P_{\theta_0}^n\int R_\beta(p_{\theta_0}; p_\theta)\, d\Pi_n^{(\alpha)}(\theta|\, X_1,\ldots,X_n) \le (\gamma+1)\epsilon_n^2\alpha - \frac{1}{n}\log\Pi(\theta\colon K(p_{\theta_0}; p_\theta) < \epsilon_n^2).$$

Under the prior mass condition $\Pi(\theta\colon K(p_{\theta_0}; p_\theta) < \epsilon_n^2) \ge \exp(-n\epsilon_n^2)$, the right side is of the order ϵ_n^2. On the left side we can lower bound R_β by $\min(\beta, 1-\beta)d_H^2$, in view of Lemma B.5. Thus, by Markov's inequality the "usual" measure of posterior contraction $\Pi_n(d_H(p_{\theta_0}, p_\theta) > M_n\epsilon_n|\, X_1,\ldots,X_n)$ is bounded by $(M_n\epsilon_n)^2$ times the left side, and hence tends to zero for any $M_n\to\infty$.

It follows that the usual prior mass condition (8.4) alone guarantees a contraction rate of the α-posterior.

8.7 Historical Notes

Posterior contraction rates for i.i.d. observations were studied almost simultaneously by Ghosal et al. (2000) and Shen and Wasserman (2001). While the first used metric entropy to quantify the size of a parameter space and showed the fundamental role of tests, as in Schwartz's theory of consistency and inspired by Birgé (1983a), the second employed stronger conditions, involving bracketing numbers and entropy integrals to ensure that the likelihood ratio is uniformly exponentially small in view of results of Wong and Shen (1995); see Problem 8.5 for a formulation of posterior contraction rate using bracketing entropy integral. That the prior mass condition need not involve the Kullback-Leibler variation, but only the divergence, is a small innovation in the results presented here relative to the results in Ghosal et al. (2000). The idea of constructing priors based on finite approximation of a compact parameter space was first used by Ghosal et al. (1997) in the context of constructing default priors for nonparametric problems; see Theorem 6.48. Ghosal et al. (2000) refined their approach through the use of brackets and controlled ratios to obtain bounds for prior concentration. The resulting conditions for the posterior contraction rate are essentially the same as the conditions for convergence of maximum likelihood estimators or minimum contrast estimators, derived by Wong and Shen (1995), Theorem 1. Contraction rates in finite-dimensional models were studied by various authors, including Le Cam (1986) and Ibragimov and Has'minskiĭ (1981) and Ghosal et al. (1995), who also treated non-regular models. Posterior contraction rates in non-i.i.d. settings, including the special cases of independent, non-identically distributed observations, Markov chains, white noise model and Gaussian time series, were studied by Ghosal and van der Vaart (2007a). Zhao (2000) studied conjugate priors in the white noise model, of the type considered in Example 8.6, but also block-based priors, which have certain advantages. Posterior contraction rates under misspecification were studied by Kleijn and van der Vaart (2006). The information theoretic approach to posterior contraction rate was developed in Zhang (2006), Catoni (2004) and Kruijer and van der Vaart (2013). A martingale-based approach in the spirit of Theorem 6.24 was considered by Walker et al. (2007); see Problem 8.11. Although the approach does not require the explicit construction of a sieve, existence of a sieve is implied. Xing (2010) introduced an elegant approach, where the size of a model is measured by the so called "Hausdorff entropy" introduced by Xing and Ranneby (2009), which also takes into account the role of the prior in size calculations; see Problem 8.13.

Problems

8.1 When θ is multidimensional, show that Lemma 8.2 holds if $\|\mathrm{E}(\theta|\, X^{(n)}) - \theta_0\| = O_p(n^{-1/2})$ and for each component θ_j of θ, $\mathrm{var}(\theta_j|\, X^{(n)}) = O_p(n^{-1})$.

8.2 Verify the calculations in Example 8.4.

8.3 If $\Pi_n(\theta'\colon d(\theta_0,\theta') < \epsilon_n|\, X^{(n)}) \to 1$ in $P_{\theta_0}^{(n)}$-probability (respectively, a.s.) and an estimator $\hat{\theta}_n$ is defined as the near maximizer (up to a tolerance that tends to zero) of $\theta \mapsto \Pi_n(\theta'\colon d(\theta,\theta') < \epsilon_n|\, X^{(n)})$, show that $d(\hat{\theta}_n,\theta_0) \le 2\epsilon_n$ in $P_{\theta_0}^{(n)}$-probability (respectively, a.s.).

8.4 Show the following analog of Theorem 8.8 for a general family indexed by an abstract parameter $\theta \in \Theta$: If Θ is a convex set with a semi-distance d_n on it (possibly depending on n), d_n^s is a convex function in one argument keeping the other fixed for some $s > 0$, d_n is bounded above by some universal constant K and $\Pi_n\left(d_n(\theta,\theta_0) \ge \epsilon_n|\, X^{(n)}\right) = O_p(\epsilon_n^2)$, then $d_n(\hat{\theta}_n,\theta_0) = O_p(\epsilon_n)$, where $\hat{\theta}_n = \int \theta\, d\Pi_n(\theta|\, X^{(n)})$.

8.5 (Shen and Wasserman 2001) Let $X_1, X_2, \ldots \overset{\text{iid}}{\sim} p$, $p \in \mathcal{P}$, and $p \sim \Pi$. Let d be a semi-distance on $\mathcal{P}$ and let p_0 stand for the true density. Let $t_n > 0$, $s_n \ge r_n > 0$ be sequences such that $\Pi(p\colon K(p_0,p) \le t_n, V_{2,0}(p_0,p) \le t_n) \gtrsim e^{-2nt_n}$ and $P_0^*(\sup\{\mathbb{P}_n \log(p/p_0)\colon d(p,p_0) \ge s_n\} > -cs_n^2) \to 0$ for some $c > 0$. (An empirical process inequality [cf. Wong and Shen 1995] implies that for the Hellinger distance, the condition holds if the bracketing entropy integral satisfies the inequality $\int_{s_n^2}^{s_n} \sqrt{\log N_{[\,]}(u,\mathcal{P},d_H)}\,du \lesssim s_n^2$.) Show that the posterior contracts at the rate $\max(r_n,\sqrt{t_n})$ at p_0 in P_0^n-probability.

8.6 (Ghosal et al. 2000) Show that for $B = \{p\colon d_H^2(p,p_0)\|p_0/p\|_\infty \le \epsilon^2\}$ there exists a universal constant $C > 0$, such that

$$P_0^n\Big(\int \prod_{i=1}^n \frac{p}{p_0}(X_i)\, d\Pi(p) \le \Pi(B)e^{-3n\epsilon^2}\Big) \le e^{-Cn\epsilon^2}. \tag{8.58}$$

8.7 (Exponential rate) Verify that the proof of Theorem 8.9 yields the stronger assertions that $\Pi_n(p \in \mathcal{P}_{n,1}\colon d(p,p_0) > M\epsilon_n|\, X_1,\ldots,X_n) = O_P(e^{-n\epsilon_n^2})$, for every sufficiently large constant M, under its assumptions (i) and (ii), and $\Pi_n(\mathcal{P}_{n,2}|\, X_1,\ldots,X_n) = O_P(e^{-n\epsilon_n^2})$, under its assumptions (i) an (iii), both in P_0^n-probability. Combine this with the preceding problem to see that given uniformly bounded likelihood ratios these assertions are also true in mean, whence also $P_0^n\Pi_n(p\colon d(p,p_0) > M\epsilon_n|\, X_1,\ldots,X_n) = O(e^{-n\epsilon_n^2})$.

8.8 (Ghosal et al. 2000) For a given function m let $\Phi^{-1}(\epsilon) = \sup\{M\colon \Phi(M) \ge \epsilon\}$ be the inverse function of $\Phi(M) = P_0 m \mathbb{1}\{m \ge M\}/M$. Then for every $\epsilon \in (0, 0.44)$ and probability measure Π on the set

$$\Big\{P\colon p_0/p \le m,\, 18h^2(P,P_0)\Big(1 + \log_+ \frac{\sqrt{P_0 m}}{h(P,P_0)} + \Phi^{-1}(h^2(P,P_0))\Big) \le \epsilon^2\Big\}$$

show that for a universal constant $B > 0$,

$$P_0^n\Big(\int \prod_{i=1}^n \frac{p}{p_0}(X_i)\, d\Pi(P) \le e^{-2n\epsilon^2}\Big) \le e^{-Bn\epsilon^2}. \tag{8.59}$$

8.9 (Ghosal et al. 2000) Suppose that for a given function m with $P_0 m < \infty$ and $\Phi^{-1}(\epsilon) = \sup\{M\colon \Phi(M) \ge \epsilon\}$ the inverse of the function $\Phi(M) = P_0 m\mathbb{1}\{m \ge M\}/M$,

$$\Pi_n\big(P\colon 18d_H^2(P,P_0)\big(1 + \log_+ \big(\sqrt{P_0 m}/d_H(P,P_0)\big) + \Phi^{-1}(d_H^2(P,P_0))\big) \le \epsilon_n^2,\, p_0/p \le m\big)$$

is at least $e^{-n\epsilon_n^2 C}$ for some $C > 0$. If conditions (i) and (ii) of Theorem 8.9 hold and in addition $\sum_n e^{-Bn\epsilon_n^2} < \infty$ for every $B > 0$, then $\Pi_n(P: d(P, P_0) \geq M\epsilon_n | X_1, \ldots, X_n) \to 0$ almost surely $[P_0^n]$, for sufficiently large M.

8.10 Prove (8.16).

8.11 (Walker et al. 2007) Let $X_1, X_2, \ldots \stackrel{\text{iid}}{\sim} p$, $p \in \mathcal{P}$, $p \sim \Pi$, and let p_0 stand for the true density. Let $\epsilon_n \to 0$ be a positive sequence such that $\Pi(p: K(p_0, p) \leq \epsilon_n^2, V_2(p_0, p) \leq \epsilon_n^2) \gtrsim e^{-cn\epsilon_n^2}$ for some $c > 0$. Assume that there exists a sequence $\delta_n \leq \rho\epsilon_n$, $\rho < 1$, such that $n\delta_n^2 \to \infty$ and that there exists a countable partition $\{\mathcal{P}_{j,n}: j \in \mathbb{N}\}$ of $\mathcal{P}$, each with d_H-diameter at most δ_n such that $e^{-n\delta_n^2/16} \sum_{j=1}^{\infty} \sqrt{\Pi(\mathcal{P}_{j,n})} \to 0$. Then the posterior contracts at the rate $\max(r_n, \sqrt{t_n})$ at p_0 in P_0^n-probability with respect to d_H.

8.12 If the conditions of Problem 8.11 hold, show that there exists a sequence $\mathcal{P}_n \subset \mathcal{P}$ such that the conditions of Theorem 8.9 also hold.

8.13 (Xing 2010) Let $\mathcal{P}$ be a set of densities, Π a prior on $\mathcal{P}$, $\alpha \geq 0$, $\epsilon > 0$ and $\mathcal{P}' \subset \mathcal{P}$. Define the *Hausdorff entropy* by

$$J(\epsilon, \mathcal{P}', \Pi, \alpha) = \log\inf\Big\{\sum \Pi(\mathcal{P}_j)^{\alpha}: \cup_{j\in\mathbb{N}}\mathcal{P}_j \supset \mathcal{P}', \operatorname{diam}(\mathcal{P}_j) \leq \epsilon\Big\}.$$

Note that for $\alpha = 0$, this exactly reduces to the ordinary metric entropy. Show that in condition (ii) of Theorem 8.9, metric entropy can be replaced by the Hausdorff entropy for any $0 < \alpha < 1$ to reach the same conclusion. (For $\alpha = 1/2$, the result implies the conclusion of Problem 8.11.)

8.14 (Kleijn and van der Vaart 2006) Show that for the well-specified case, the condition of prior concentration (8.48) in Theorem 8.36 reduces to the prior mass condition (i) in Theorem 8.11.

8.15 (Kleijn and van der Vaart 2006) If all points in a finite subset $\mathcal{P}^* \subset \mathcal{P}$ are at minimal Kullback-Leibler divergence, prove the following version of Theorem 8.36. Redefine covering numbers for testing under misspecification $N_{t,\text{mis}}(\epsilon, \mathcal{P}, d; P_0, \mathcal{P}^*)$ as the minimal number N of convex sets $B_1, \ldots, B_N$ of probability measures on $(\mathfrak{X}, \mathscr{X})$ needed to cover the set $\{P \in \mathcal{P}: \epsilon < d(P, \mathcal{P}^*) < 2\epsilon\}$ such that

$$\sup_{P^* \in \mathcal{P}^*} \inf_{P \in B_i} \sup_{0<\alpha<1} -\log P_0\Big(\frac{p}{p^*}\Big)^{\alpha} \geq \frac{\epsilon^2}{4}.$$

Suppose that there exists a sequence of strictly positive numbers ϵ_n with $\epsilon_n \to 0$ and $n\epsilon_n^2 \to \infty$ and a constant $L > 0$, such that for all n and all $\epsilon > \epsilon_n$:

$$\inf_{P^* \in \mathcal{P}^*} \Pi(B(\epsilon_n, P^*; P_0)) \geq e^{-Ln\epsilon_n^2}, \tag{8.60}$$

$$N_{t,\text{mis}}(\epsilon, \mathcal{P}, d; P_0, \mathcal{P}^*) \leq e^{n\epsilon_n^2}. \tag{8.61}$$

Then for every sufficiently large constant $M > 0$, $\Pi_n(P \in \mathcal{P}: d(P, \mathcal{P}^*) \geq M\epsilon_n | X_1, \ldots, X_n \to 0$ as $n \to \infty$ in P_0^n-probability.

8.16 (Kleijn and van der Vaart 2006) Consider a mixture model $p_F(x) = \int \psi(x; z)\, dF(z)$, $F \in \mathfrak{F}$, where ψ is continuous in z for all x and $\mathfrak{F}$ is a compact subset of $\mathfrak{M}$. Let p_0 be a density such that $K(p_0; p_F) < \infty$ for some $F \in \mathfrak{F}$. Then there exists $F^* \in \mathfrak{F}$ which minimizes $K(p_0; p_F)$ over $F \in \mathfrak{F}$, and F^* is unique if the family $\{p_F: F \in \mathfrak{F}\}$ is identifiable. (For example, the normal location model is identifiable [Teicher 1961].)

8.17 (Kleijn and van der Vaart 2006) Consider the setup of the last problem. Assume that p_0/p_{F^*} is bounded and $\Pi(B(\epsilon, P_{F^*}; P_0)) > 0$ for every $\epsilon > 0$. Then $P_0^n \Pi_n(F: d(P_F, P_{F^*}) \geq \epsilon | X_1, \ldots, X_n) \to 0$, where d is the weighted Hellinger distance defined by $d^2(P_{F_1}, P_{F_2}) = \frac{1}{2} \int (\sqrt{p_{F_1}} - \sqrt{p_{F_2}})^2 (p_0/p_{F^*})\, d\mu$.

8.18 Suppose that for every n and $D > 0$ there exists a partition $\mathcal{P}_n = \mathcal{P}_{n,1} \cup \mathcal{P}_{n,2}$ such that (ii) of Theorem 8.11 holds and $\Pi_n(\mathcal{P}_{n,2}) = o(e^{-Dn\bar{\epsilon}_n^2})$. Then there also exist partitions such that (ii) and (iii) hold. [Hint: if $a_n(D) \to 0$ for every $D > 0$, then there exists $D_n \to \infty$ with $a_n(D_n) \to 0$.]

9

Contraction Rates: Examples

In this chapter we obtain explicit posterior contraction rates by applying the general theorems from the preceding chapter to common priors and models: log-spline models, Dirichlet process mixtures of normal or Bernstein polynomial kernels for density estimation, nonparametric and semiparametric regression models with priors obtained from bracketing or the Dirichlet process on the link function or error density, spectral density estimation and current status and interval censoring with the Dirichlet process prior.

9.1 Log-Spline Priors

Spline functions $f\colon [0,1] \to \mathbb{R}$ of order q are piecewise polynomials of degree $q-1$ whose pieces connect smoothly of order $q-2$ at the boundaries of their domains. They possess excellent approximation properties for smooth functions, and permit a numerically stable representation as linear combinations of the *B-spline* basis (see Section E.2 for an introduction). Splines can be used for density estimation by exponentiaton and normalizing, as in Section 2.3.1, thus leading to an exponential family with the B-spline basis functions as the sufficient statistics. A prior density is then induced through a prior on the parameter vector of the family.

In the nonparametric setup it is necessary to let the dimension of the family increase, so that it can approximate an arbitrary continuous true density. In this section we show that a prior with of the order $n^{1/(2\alpha+1)}$ basis functions yields a posterior distribution that contracts at the optimal rate for a true density in the Hölder space $\mathfrak{C}^\alpha[0,1]$. (In Chapter 10 the construction will be extended with a prior on α or the dimension, in order to achieve this rate simultaneously for every $\alpha > 0$.)

Fix some *order* $q \in \mathbb{N}$, and for given $K \in \mathbb{N}$ partition $[0,1)$ into the K equal-length subintervals $\big[(k-1)/K, k/K\big)$, for $k = 1, \ldots, K$. The B-spline functions $B_{J,1}, \ldots, B_{J,J}$, described in Section E.2, generate the $J = q + K - 1$-dimensional linear space of all *splines of order* q relative to this partition. Consider probability densities relative to Lebesgue measure on $[0,1]$ given by the exponential family

$$p_{J,\theta}(x) = e^{\theta^\top B_J(x) - c(\theta)}, \quad \theta^\top B_J := \sum_{j=1}^{J} \theta_j B_{J,j}, \quad c(\theta) = \log \int_0^1 e^{\theta^\top B_J(x)}\, dx, \quad \theta \in \mathbb{R}^J.$$

Because the B-splines add up to unity, the family is actually of dimension $J-1$ and we can restrict the parameter to $\{\theta \in \mathbb{R}^J : \theta^\top 1 = 0\}$. We place a prior on the latter set, and this next induces a prior on probability densities through the map $\theta \mapsto p_{J,\theta}$. In the special case

$q = 1$, the splines are piecewise constant functions, and the prior sits on histograms with cell boundaries k/K, for $k = 0, 1, \ldots, K$.

The true density p_0 need not be of the form $p_{J,\theta}$ for any θ and/or J, and therefore needs to be approximated by some $p_{J,\theta}$ for the posterior distribution to be consistent. To ensure this for a large class of p_0, the dimension $J - 1$ of the log-spline model must tend to infinity with n. The minimal rate at which $J = J_n$ must grow is determined by the approximation properties of the spline functions. For $\log p_0$ belonging to the Hölder space $\mathfrak{C}^\alpha[0,1]$ and $q \geq \alpha$, Lemma E.5 asserts, for a constant C depending only on q and α,

$$\inf_{\theta \in \mathbb{R}^J} \|\theta^\top B_J - \log p_0\|_\infty \leq C J^{-\alpha} \|\log p_0\|_{\mathfrak{C}^\alpha}.$$

This bound cannot be improved for a general $p_0 \in \mathfrak{C}^\alpha[0,1]$: it is sharp for p_0 that are "exactly of smoothness α." Consequently, as it certainly cannot be smaller than the distance of the model to the true density, the posterior contraction rate ϵ_n always satisfies $\epsilon_n \gtrsim J^{-\alpha}$ for some p_0. This favors high-dimensional models (large J), and for consistency it is needed that $J = J_n \to \infty$. On the other hand, the local entropy of the spline model turns out to be a multiple of its dimension J, whence the entropy condition (ii) of Theorem 8.11 takes the form $J \lesssim n\epsilon_n^2$, or $\epsilon_n \gtrsim \sqrt{J/n}$. The maximum of the two lower bounds $J^{-\alpha}$ and $\sqrt{J/n}$ is minimized by $J \sim n^{1/(2\alpha+1)}$, which leads to $\epsilon_n \sim n^{-\alpha/(2\alpha+1)}$, the minimax rate of estimation for $\mathfrak{C}^\alpha[0,1]$. The following theorem shows that the posterior distribution corresponding to flat priors on the coefficients attains this rate.

We restrict to true densities that are bounded away from zero, and use a compactly supported prior for the coefficients θ.

Theorem 9.1 (Contraction rate, log-splines) *Let $\theta \sim \pi_n$ for π_n a Lebesgue density on $\{\theta \in [-M, M]^{J_n} : \theta^\top 1 = 0\}$ such that $\underline{c}^{J_n} \leq \pi_n(\theta) \leq \overline{c}^{J_n}$, for some $0 < \underline{c} < \overline{c} < \infty$ and all θ, where $M \geq 1$ is fixed and $J_n \sim n^{1/(2\alpha+1)}$. If $p_0 \in \mathfrak{C}^\alpha[0,1]$, with $q \geq \alpha \geq 1/2$ and $\|\log p_0\|_\infty \leq d_0 M/2$, where d_0 is the universal constant in Lemma 9.2, then the posterior distribution relative to the prior $p_{J_n,\theta}$ contracts at p_0 at the rate $n^{-\alpha/(2\alpha+1)}$ with respect to the Hellinger distance.*

Proof The proof is based on Theorem 8.11 in combination with a sequence of lemmas listed below.

In Lemma 9.5 the local entropy relative to the Hellinger distance of the full support of the prior is estimated to be bounded above by a multiple of J_n, and hence the entropy equation in (ii) of Theorem 8.11 is satisfied if $J_n \lesssim n\epsilon_n^2$, which is the case for the proposed dimension and ϵ_n a multiple of the claimed contraction rate.

By Lemma 9.4(iii), a Kullback-Leibler ball around p_0 contains a Hellinger ball around p_0 of a multiple of the radius. By Lemma 9.6(ii) and (i) there exists θ_J with $\|\theta_J\|_\infty \leq M$ such that $d_H(p_0, p_{J,\theta_J}) \lesssim J^{-\alpha}$, which is $n^{-\alpha/(2\alpha+1)}$ for $J = J_n$. Next by Lemma 9.4(ii) a Hellinger ball of radius a multiple of ϵ_n contains a set of the form $\Theta(J_n, \epsilon_n)$, for $\Theta(J, \epsilon) = \{\theta \in [-M, M]^J : \|\theta - \theta_J\|_2 \leq \sqrt{J}\epsilon\}$.

By Lemma 9.4(i) Hellinger balls around p_0 of radius a multiple of $2j\epsilon_n$ are contained in a set $\Theta(J_n, Cj\epsilon_n)$, for some constant C depending on M only. We conclude that, for ϵ_n a multiple of $n^{-a(2\alpha+1)}$ and c and C constants that depend on M only,

$$\frac{\Pi_n(p_{J_n,\theta}: d_H(p_{J_n,\theta}, p_0) \le 2j\epsilon_n)}{\Pi_n(B_2(p_0, \epsilon_n))} \le \frac{\Pi_n(\Theta(J_n, Cj\epsilon_n))}{\Pi_n(\Theta(J_n, c\epsilon_n))}$$
$$\le \frac{\sup_\theta \pi_n(\theta)\,(\sqrt{J_n}Cj\epsilon_n)^{J_n-1}\mathrm{vol}\{x \in \mathbb{R}^{J_n}: \|x\| \le 1\}}{\inf_\theta \pi_n(\theta)\,(\sqrt{J_n}c\epsilon_n)^{J_n-1}\mathrm{vol}\{x \in \mathbb{R}^{J_n}: \|x\| \le 1\}} \lesssim \Big(\frac{\bar{c}Cj}{\underline{c}}\Big)^{J_n-1}.$$

This easily satisfies the bound in (i) of Theorem 8.11. ☐

The following lemmas compare the statistical distances on the log-spline densities to distances on their parameters. They were used in the preceding proof, and will be useful later on as well. Define

$$\begin{aligned}
\Theta_{J,M} &= \{\theta \in [-M, M]^J: \theta^\top 1 = 0\},\\
B_{J,M}(p_0, \epsilon) &= \{\theta \in \Theta_{J,M}: K(p_0; p_{J,\theta}) < \epsilon^2, V_2(p_0; p_{J,\theta}) < \epsilon^2\}, \qquad (9.1)\\
C_{J,M}(p_0, \epsilon) &= \{\theta \in \Theta_{J,M}: d_H(p_0, p_{J,\theta}) < \epsilon\}.
\end{aligned}$$

Abusing notation, we shall use the notations $B_{J,M}(p_0, \epsilon)$ and $C_{J,M}(p_0, \epsilon)$ also for the set of densities $p_{J,\theta}$ when θ ranges over the sets as in the display.

Lemma 9.2 $d_0\|\theta\|_\infty \le \|\log p_{J,\theta}\|_\infty \le 2\|\theta\|_\infty$, *for any* $\theta \in \mathbb{R}^J$ *with* $\theta^\top 1 = 0$, *and a universal constant* d_0.

Lemma 9.3 $c_0 e^{-M}(\|\theta_1 - \theta_2\| \wedge \sqrt{J}) \le \sqrt{J}\, d_H(p_{J,\theta_1}, p_{J,\theta_2}) \le C_0 e^M \|\theta_1 - \theta_2\|$, *for every* $\theta_1, \theta_2 \in \Theta_{J,M}$, *and universal constants* $0 < c_0 < C_0 < \infty$.

Proofs Since $\|\theta^\top B_J\|_\infty \le \|\theta\|_\infty =: M$, by the second inequality in Lemma E.6, the number $e^{c(\theta)} = \int_0^1 e^{\theta^\top B_J(x)}dx$ is contained in the interval $[e^{-M}, e^M]$. Hence $|c(\theta)| \le M$, and next $\|\log p_{J,\theta}\|_\infty = \|\theta^\top B_J - c(\theta)\|_\infty \le 2M$, by the triangle inequality, which is the upper bound in the first lemma. For the lower bound, we use Lemma E.6 to see that $\|\theta\|_\infty \lesssim \|\theta^\top B_J\|_\infty$, which is bounded above by $\|\log p_{J,\theta}\|_\infty + |c(\theta)|$, by the triangle inequality. Since $\theta^\top 1 = 0$, the second term can be rewritten as $|(\theta - c(\theta)1)^\top 1|/J \le \|\theta - c(\theta)1\|_\infty$. Finally we use Lemma E.6 again to bound this by $\|(\theta - c(\theta)1)^\top B_J\|_\infty = \|\log p_{J,\theta}\|_\infty$.

For the proof of the second lemma we derive, by direct calculation and Taylor's theorem,

$$\begin{aligned}
d_H^2(p_{J,\theta_1}, p_{J,\theta_2}) &= 2\Big(1 - \exp\big[c(\tfrac{1}{2}\theta_1 + \tfrac{1}{2}\theta_2) - \tfrac{1}{2}c(\theta_1) - \tfrac{1}{2}c(\theta_2)\big]\Big)\\
&= 2\Big(1 - \exp\big[-\tfrac{1}{16}(\theta_1 - \theta_2)^\top(\ddot{c}(\tilde{\theta}) + \ddot{c}(\bar{\theta}))(\theta_1 - \theta_2)\big]\Big), \qquad (9.2)
\end{aligned}$$

for $\tilde{\theta}, \bar{\theta}$ on the line segment connecting θ_1 and θ_2, and $\ddot{c}(\theta)$ the Hessian of c. By well-known properties of exponential families $t^\top \ddot{c}(\theta)t = \mathrm{var}_\theta(t^\top B_J)$. Because $1^\top B_J$ is degenerate, this is equal to the minimum over μ of $\int_0^1 \big[(t - \mu 1)^\top B(x)\big]^2 p_{J,\theta}(x)\,dx$, which is bounded below and above by $\|(t - \mu 1)^\top B_J\|_2^2$ times the infimum and supremum of $p_{J,\theta}(x)$ over $x \in [0, 1]$ and θ. By Lemma 9.2 the latter infimum and supremum can be bounded below and above by multiples of e^{-2M} and e^{2M}, respectively, while the square norm is comparable to $\|t - \mu 1\|_2^2/J$, by Lemma E.6. We minimize the latter with respect to μ to reduce it to $\|t\|^2/J$ when $1^\top t = 0$.

To finish the proof, we insert the last bound in (9.2), and apply the elementary inequalities $(x \wedge 1)/2 \le 1 - e^{-x} \le x$ for $x \ge 0$, and $(cx) \wedge 1 \ge c(x \wedge 1)$ for $x \ge 0$ and $c \le 1$. ☐

Lemma 9.4 *If $\theta_{J,M}$ minimizes $\theta \mapsto d_H(p_{J,\theta}, p_0)$ over $\Theta_{J,M}$, then, for universal constants c_0, C_0, d_0, D_0,*

(i) $C_{J,M}(p_0, \epsilon) \subset \{\theta \in \Theta_{J,M}: \|\theta - \theta_{J,M}\|_2 \le 2e^M c_0^{-1}\sqrt{J}\epsilon\}$, *for* $2\epsilon < c_0 e^{-M}$.
(ii) $C_{J,M}(p_0, 2\epsilon) \supset \{\theta \in \Theta_{J,M}: \|\theta - \theta_{J,M}\|_2 \le e^{-M} C_0^{-1}\sqrt{J}\epsilon\}$, *for* $\epsilon \ge d_H(p_0, p_{J,\theta_{J,M}})$.
(iii) $C_{J,M}(p_0, \epsilon) \subset B_{J,M}(p_0, D_0 M\epsilon)$, *if* $\|\log p_0\|_\infty \le M$.

Proof The set $C_{J,M}(p_0, \epsilon)$ is void if $\epsilon < d_H(p_0, p_{J,\theta_{J,M}})$, so for the proof of (i) we may assume the opposite. Then the triangle inequality gives that $d_H(p_{J,\theta}, p_{J,\theta_{J,M}}) \le 2\epsilon$, for any $p_{J,\theta} \in C_{J,M}(p_0, \epsilon)$. If also $2\epsilon < c_0 e^{-M}$, then the lower bound of Lemma 9.3 shows that $c_0 e^{-M}\|\theta - \theta_{J,M}\|_2 \le 2\sqrt{J}\epsilon$. Statement (ii) follows similarly, and more easily, from the upper bound in Lemma 9.3 combined with the triangle inequality. Inclusion (iii) is a consequence of the bound $2M$ on $\|\log p_{J,\theta}\|_\infty$ for $\theta \in \Theta_{J,M}$, by Lemma 9.2, and the equivalence of Hellinger distance and Kullback-Leibler divergence and variation, for densities with bounded likelihood ratios, as quantified in Lemma B.2. □

Lemma 9.5 $\log N(\epsilon/5, C_{J,M}(p_0, \epsilon), d_H) \le (2M + \log(30C_0/c_0))J$, *for every* $\epsilon > 0$.

Proof If $2\epsilon < c_0 e^{-M}$, then by Lemma 9.4(i) it suffices to bound the entropy of the set of functions $p_{J,\theta}$ with $\theta \in \{\theta \in \Theta_{J,M}: \|\theta - \theta_{J,M}\|_2 \le 2\sqrt{J}\epsilon\, e^M/c_0\}$. By Lemma 9.3 on this set the Hellinger distance is bounded by $C_0 e^M/\sqrt{J}$ times the Euclidean norm. Therefore, the $\epsilon/5$-entropy in the lemma is bounded by the $\sqrt{J}e^{-M}\epsilon/(5C_0)$-entropy of a Euclidean ball of radius $2\sqrt{J}\epsilon\, e^M/c_0$, for the Euclidean metric, which follows from Proposition C.2.

The entropy at a value of $\epsilon/5$ with $2\epsilon \ge c_0 e^{-M}$ is bounded by the entropy at the values below this threshold, as just obtained. □

Lemma 9.6 *If* $\log p_0 \in \mathfrak{C}^\alpha[0, 1]$, *then for every J there exists $\theta_J \in \mathbb{R}^J$ with, for universal constants d_0, d_1,*

(i) $d_0\|\theta_J\|_\infty \le \|\log p_0\|_\infty + d_1\|\log p_0\|_{\mathfrak{C}^\alpha} J^{-\alpha}$.
(ii) $d_H(p_{J,\theta_J}, p_0) \le d_1\|\log p_0\|_{\mathfrak{C}^\alpha} J^{-\alpha} e^{d_1\|\log p_0\|_{\mathfrak{C}^\alpha} J^{-\alpha}}$.
(iii) $K(p_0; p_{J,\theta_J}) \le d_1\|\log p_0\|_{\mathfrak{C}^\alpha} J^{-2\alpha} e^{d_1\|\log p_0\|_{\mathfrak{C}^\alpha} J^{-\alpha}}$.
(iv) $V_2(p_0; p_{J,\theta_J}) \le d_1\|\log p_0\|_{\mathfrak{C}^\alpha} J^{-2\alpha} e^{d_1\|\log p_0\|_{\mathfrak{C}^\alpha} J^{-\alpha}}$.

Proof By Lemma E.5 there exists θ_J such that $\|\theta_J^\top B_J - \log p_0\|_\infty \le DJ^{-\alpha}$, for D a universal multiple of $\|\log p_0\|_{\mathfrak{C}^\alpha}$. Inequalities (ii)–(iv) are next immediate from Lemma 2.5, with $f = \theta_J^\top B_J$ and $g = \log p_0$, where we increase the universal constant in D, if necessary. For (i) we apply the triangle inequality to obtain that $\|\theta_J^\top B_J\|_\infty \le \|\log p_0\|_\infty + DJ^{-\alpha}$, and next combine this with Lemma E.6. □

9.2 Priors Based on Dirichlet Processes

In this section, we obtain posterior contraction rates for some priors based on Dirichlet processes.

Example 9.7 (Current status censoring) Consider the problem of estimating a cumulative distribution function F on $[0, \infty)$ when only indirect data is available: for a random sample $Y_1, \ldots, Y_n$ from F and an independent sample of "observation times" $C_1, \ldots, C_n$ we only observe the observation times C_i and the information $\Delta_i = \mathbb{1}\{Y_i \le C_i\}$ whether Y_i had realized or not. Thus our observations are a random sample $X_1, \ldots, X_n$ of pairs $X_i = (\Delta_i, C_i)$. Censored data models will be treated in more details in Chapter 13.

If the times C_i have distribution G with density g relative to a measure μ, then the density p_F of the X_i with respect to the product ν of counting measure on $\{0, 1\}$ and μ is given by

$$p_F(\delta, c) = F(c)^\delta (1 - F(c))^{1-\delta} g(c), \qquad \delta \in \{0, 1\}, c > 0. \tag{9.3}$$

Since this factorizes into expressions depending on F and g only, for a product prior on the pair (F, g) the factors involving g will cancel from the expression for the posterior distribution of F. Consequently, as we are interested in F, we may treat g as known, and need not specify a prior for it.

We assume that g is supported on a compact interval $[a, b]$, and that the true distribution F_0 is continuous and contains $[a, b]$ in the interior of its support (equivalently, $F_0(a-) > 0$ and $F_0(b) < 1$). We assign a Dirichlet prior $F \sim \mathrm{DP}(\alpha)$ with base measure α that has a positive, continuous density on a compact interval containing $[a, b]$. We shall show that the conditions of Theorem 8.9 are satisfied for ϵ_n a large multiple of $n^{-1/3}(\log n)^{1/3}$, and d the $\mathbb{L}_2$-metric on F. Let $\mathcal{F}$ stand for the set of all distribution functions on $(0, \infty)$.

Since $\|p_F - p_{F_0}\|_{2,\nu} = 2\|F - F_0\|_{2,G}$, we have,

$$N(\epsilon, \{p_F: F \in \mathcal{F}\}, \|\cdot\|_{2,\mu}) \le N(\epsilon/2, \mathcal{F}, \|\cdot\|_{2,G}).$$

The corresponding entropy is bounded above by a multiple of ϵ^{-1}, by Proposition C.8. Therefore the entropy condition (ii) of Theorem 8.9 with the $\mathbb{L}_2$-metric is verified for $\epsilon_n \asymp n^{-1/3}$.

Under our conditions F_0 is bounded away from zero and one on the interval $[a, b]$ that contains all observation times $C_1, \ldots, C_n$. Consequently, the quotients p_{F_0}/p_F are uniformly bounded away from zero and infinity, uniformly in F that are uniformly close to F_0 on the interval $[a, b]$. For such F,

$$\begin{aligned} d_H^2(p_F, p_{F_0}) &\le \int |F^{1/2} - F_0^{1/2}|^2 \, dG + \int |(1-F)^{1/2} - (1-F_0)^{1/2}|^2 \, dG \\ &\le C \sup_{c \in [a,b]} |F(c) - F_0(c)|^2, \end{aligned}$$

for a constant C that depends on F_0 only. Thus by (8.8) a neighborhood $B_2(p_0, \epsilon')$ as in the prior mass condition (i) of Theorem 8.9 contains a Kolmogorov-Smirnov neighborhood $\{F: d_{KS}(F, F_0) \le \epsilon\}$, for ϵ a multiple of ϵ'.

Given $\epsilon > 0$, partition the positive half line in intervals $E_1, \ldots, E_N$ such that $F_0(E_i) < \epsilon$ and $A\epsilon \le \alpha(E_i) < 1$ for every i and such that $N \le B\epsilon^{-1}$, for some fixed $A, B > 0$. If $\sum_{i=1}^N |F(E_i) - F_0(E_i)| < \epsilon$, then $|F(c) - F_0(c)| < \epsilon$ at every end point c of an interval E_i, and then $d_{KS}(F, F_0) < 2\epsilon$ by monotonicity and the fact that by construction F_0 varies by at most ϵ on each interval. Because $A\epsilon \le \alpha(E_i) < 1$ for every i, Lemma G.13 gives that $\Pi(F: \sum_{i=1}^N |F(E_i) - F_0(E_i)| < \epsilon) \ge \exp(-c\epsilon^{-1} \log \epsilon^{-1})$, for some $c > 0$. We conclude that the prior mass condition (i) is satisfied for $\epsilon_n \asymp n^{-1/3}(\log n)^{1/3}$.

The rate $n^{-1/3}(\log n)^{1/3}$ is close to the optimal rate of contraction $n^{-1/3}$ in this model. The small discrepancy may be an artifact of the prior mass estimation rather than a deficit of the Dirichlet prior. In Section 8.2.2 the optimal rate $n^{-1/3}$ was obtained for a prior based on bracketing approximation.

9.3 Bernstein Polynomials

Mixtures of beta densities are natural priors for the density p of observations $X_1, X_2, \ldots$ in the unit interval $[0, 1]$. In this section we consider the two types of mixtures:

type I $$b(x; k, w) = \sum_{j=1}^{k} w_{j,k}\mathrm{be}(x; j, k-j+1),$$

type II $$\tilde{b}(x; k^2, w) = \sum_{i=1}^{k} w_{i,k}\frac{1}{k}\sum_{j=(i-1)k+1}^{ik} \mathrm{be}(x; j, k^2-j+1).$$

In both types k is equal to the number of parameters $w_{j,k}$; the degrees of the beta polynomials are k and k^2, respectively. We can form priors for p by equipping k and the coefficients $w_{j,k}$ with priors. For instance,

$$w_k := (w_{1,k}, \ldots, w_{k,k}) |\, k \sim \mathrm{Dir}(k; \alpha_{1,k}, \ldots, \alpha_{k,k}), \qquad k \sim \rho.$$

In particular, in Example 5.10 it was suggested to link the Dirichlet priors across k, by letting $F \sim \mathrm{DP}(\alpha)$, and setting $w_k = (F(0, 1/k], F(1/k, 2/k], \ldots, F((k-1)/k, 1])$, giving *Bernstein-Dirichlet processes*. For a given F the first type of mixture is then the derivative of the Bernstein polynomial corresponding to F. The second type of mixture is of interest, as it can repair the suboptimal approximation properties of these polynomials. The k^2 beta densities in these mixtures are grouped in k groups and the random coefficients $w_{j,k}$ are assigned to the averages of these groups rather than to the individual beta densities. See Lemma E.4 for further discussion.

The Bernstein polynomial prior is a role model for mixture priors for densities on bounded intervals with a discrete smoothing parameter (see Problem 9.20 for the example of a triangular density kernel). The prior on the smoothing parameter k is crucial for the concentration rate. Its natural discreteness makes that a point mass may be assigned to each value of the smoothing parameter.

Theorem 9.8 *Let the true density p_0 be strictly positive and be contained in $\mathfrak{C}^\alpha[0, 1]$, for $\alpha \in (0, 2]$. Assume that $A_1 k^{-b} \le \alpha_{j,k} \le A_2$ and that $B_1 e^{-\beta_1 k} \le \rho(k) \le B_2 e^{-\beta_2 k}$, for all j and k, where $A_1, A_2, b, B_1, B_2, \beta_1, \beta_2$ are fixed positive constants.*

(i) *If $\alpha \in (0, 2]$, then the posterior distribution relative to the type I mixtures contracts at the rate $n^{-\alpha/(2+2\alpha)}(\log n)^{(1+2\alpha)/(2+2\alpha)}$ with respect to the Hellinger distance.*
(ii) *If $\alpha \in (0, 1]$, then the posterior distribution relative to the type II mixtures contracts at the rate $n^{-\alpha/(1+2\alpha)}(\log n)^{(1+4\alpha)/(2+4\alpha)}$ with respect to the Hellinger distance.*
(iii) *If $p_0 = b(\cdot; k, w^0)$, for some $w^0 \in \mathbb{S}_k$, then the posterior distribution relative to the type I mixtures contracts at the rate $n^{-1/2}\log n$ with respect to the Hellinger distance.*

Proof We only give the details of the proof of (ii). The proof of the other parts is similar, and slightly simpler. The slower rate in (i) is due to the weaker approximation rate $k^{-\alpha/2}$ of order k type I Bernstein polynomials, as given in Lemma E.3, as opposed to the rate for type II mixtures, given in Lemma E.4.

For given $w_k \in \mathbb{S}_k$ we have $\tilde{b}(\cdot; k^2, w_k) = b(\cdot; k^2, \tilde{w}_k)$, for $\tilde{w}_k$ the vector in $\mathbb{S}_{k^2}$ given by $(w_{k,1}, \ldots, w_{k,1}, w_{k,2}, \ldots, w_{k,2}, \ldots, w_{k,k})/k$, obtained by repeating each coordinate $w_{k,j}$ of w_k exactly k times and renormalizing. Therefore, Lemma E.2 implies that, for any $w_k, w_k' \in \mathbb{S}_k$,

$$\|\tilde{b}(\cdot; k^2, w_k) - \tilde{b}(\cdot; k^2, w_k')\|_\infty \le k^2 \|w_k - w_k'\|_1. \tag{9.4}$$

We now verify the conditions of Theorem 8.9.

For $w_k^0 \in \mathbb{S}_k$ the vector with coordinates $w_{k,j}^0 = P_0((j-1)/k, j/k]$, for $j = 1, \ldots, k$, Lemma E.4 gives that $\|p_0 - \tilde{b}(\cdot; k^2, w_k^0)\|_\infty \lesssim k^{-\alpha}$. Thus, by the triangle inequality, for any $w_k \in \mathbb{S}_k$,

$$\|p_0 - \tilde{b}(\cdot; k^2, w_k)\|_\infty \lesssim k^{-\alpha} + k^2 \|w_k - w_k^0\|_1.$$

If we choose $k = k_n$ the integer part of a big multiple of $\epsilon_n^{-1/\alpha}$, then the right side will be smaller than a multiple of ϵ_n, for any w_k with $\|w_k - w_k^0\|_1 \le \epsilon_n^{1+2/\alpha}$. In particular, the function $\tilde{b}(\cdot; k^2, w_k)$ will be bounded away from zero, for sufficiently large n, and we can apply Lemma B.1 (v) and (vii) and the last two assertions of Lemma B.2 to obtain that $\tilde{b}(\cdot; k^2, w_k) \in B_2(p_0, C_1\epsilon_n)$, for B_2 the Kullback-Leibler neighborhood defined in (8.3). It follows that the left side of the prior mass condition (8.4) is bounded below by

$$\rho(k_n)\, \Pi(w_{k_n}: \|w_{k_n} - w_{k_n}^0\|_1 \le \epsilon_n^{1+2/\alpha}) \gtrsim e^{-\beta_1 k_n} e^{-c_1 k_n \log_- \epsilon_n},$$

for some $c_1 > 0$, in view of Lemma G.13. (The condition $\epsilon \le 1/(Mk)$ in the lemma is easily satisfied, by virtue of the relation $\epsilon_n^{1+2/\alpha} \lesssim k_n^{-1}$.) For $\epsilon_n = n^{-\alpha/(1+2\alpha)}(\log n)^{\alpha/(1+2\alpha)}$ the right side is bounded below by a constant multiple of $e^{-c_2 n\epsilon_n^2}$, for some constant $c_2 > 0$, whence (8.4) is satisfied.

To verify (8.5) we use the sieve $\mathcal{P}_{n,1} = \cup_{j=1}^{s_n} \mathcal{C}_j$, where $\mathcal{C}_k = \{\tilde{b}(\cdot; k^2, w_k): w_k \in \mathbb{S}_k\}$, and s_n is the integer part of a multiple of $n^{1/(1+2\alpha)}(\log n)^{2\alpha/(1+2\alpha)}$. In view of (9.4) we have

$$D(\epsilon, \mathcal{C}_k, \|\cdot\|_1) \le D(\epsilon/k^2, \mathbb{S}_k, \|\cdot\|_1) \le \Big(\frac{5k^2}{\epsilon}\Big)^k,$$

for $\epsilon/k^2 < 1$, by Proposition C.1. Because the Hellinger distance is bounded above by the root of the $\mathbb{L}_1$-distance, it follows that $D(\epsilon, \mathcal{P}_{n,1}, d_H) \le \sum_{k=1}^{s_n} (5k^2/\epsilon^2)^k \le s_n(5s_n^2/\epsilon^2)^{s_n}$, for $\epsilon < 1$. Thus condition (8.5) is satisfied for $\bar{\epsilon}_n = n^{-\alpha/(1+2\alpha)}(\log n)^{(1+4\alpha)/(2+4\alpha)}$.

Finally, since $\Pi(\mathcal{P}_{n,1}^c) = \sum_{k>s_n} \rho(k) \lesssim e^{-\beta_2 s_n}$, condition (8.6) is verified by the choice of s_n. □

Part (ii) of the theorem shows that the type II mixtures give the minimax rate up to a logarithmic factor, while part (i) suggests a suboptimal rate. Part (ii) of the theorem is restricted to $\alpha \le 1$, but for $\alpha = 1$ it gives a better contraction rate than part (i) for $1 \le \alpha < 2$. Thus part (ii) gives a better result than part (i) even for $\alpha \in [1, 2)$.

9.4 Dirichlet Process Mixtures of Normal Kernel

Mixtures of normal densities are popular for estimating densities on a full Euclidean space. Finite mixtures are appropriate for a population that consists of finitely many subclasses. Kernel density estimators are often normal mixtures centered at the observations. In the Bayesian setup nonparametric normal mixtures with a Dirichlet process prior on the mixing distribution lead to elegant computational algorithms, as seen in Chapter 5, and turn out to be remarkably flexible.

In this section we study the contraction rate of the posterior distribution of the density p of an i.i.d. sample $X_1, \ldots, X_n$ in $\mathbb{R}^d$ for a Dirichlet mixture of normal prior, in two scenarios for the true density: either the true density itself is a mixture of normal densities, or it belongs to a Hölder class of functions. In the first situation, the "supersmooth case," a nearly parametric rate of contraction is obtained, while in the second situation, the "ordinary smooth case," a nonparametric rate prevails. Remarkly the same Dirichlet mixture of normal prior yields an almost minimax contraction rate in either situation, and in the ordinary smooth case also for any smoothness level $\beta > 0$. This is in striking contrast with classical kernel density estimation, where the normal kernel, which is a second order kernel, gives the minimax rate only for smoothness up to order 2, and better rates require higher-order kernels that are not probability densities. The Bayesian procedure achieves this by spreading the mass of the posterior mixing distribution in a clever way, rather than putting equal masses at the observations. A key lemma in the derivations below is that a β-smooth density can be approximated closely by a normal mixture of the form $f_\sigma * \phi_{0,\sigma^2 I}$, where f_σ belongs to the support of prior and is different from the usual choice p_0, for which the approximation rate is only σ^2 (see Problem 9.10).

Let $\phi(\cdot; \Sigma)$ be the density of the $\mathrm{Nor}_d(0, \Sigma)$-distribution, where Σ is a positive-definite $(d \times d)$-matrix, and denote the normal mixture with mixing distribution F on $\mathbb{R}^d$ by

$$p_{F,\Sigma}(x) = \int \phi(x - z; \Sigma)\, dF(z).$$

We induce a prior on densities on $\mathbb{R}^d$ by equipping F with a $\mathrm{DP}(\alpha)$ prior and independently Σ with a prior G. We assume that the base measure α and the measure G have continuous and positive densities in the interior of their supports, and satisfy, for positive constants $a_1, a_2, a_3, a_4, a_5, b_1, b_2, b_3, b_4$, and C_1, C_2, C_3, κ,

$$1 - \bar{\alpha}([-z, z]^d) \leq b_1 e^{-C_1 z^{a_1}}, \quad z > 0, \tag{9.5}$$

$$G(\Sigma: \mathrm{eig}_d(\Sigma^{-1}) \geq s) \leq b_2 e^{-C_2 s^{a_2}}, \quad s > 0, \tag{9.6}$$

$$G(\Sigma: \mathrm{eig}_1(\Sigma^{-1}) \leq s) \leq b_3 s^{a_3}, \quad s > 0, \tag{9.7}$$

$$G\Big(\Sigma: \bigcap_{1\leq j\leq d} \{s_j < \mathrm{eig}_j(\Sigma^{-1}) < s_j(1+t)\}\Big) \geq b_4 s_1^{a_4} t^{a_5} e^{-C_3 s_d^{\kappa/2}}, 0 < s_1 \leq \cdots \leq s_d, \quad 0 < t < 1. \tag{9.8}$$

Here $\mathrm{eig}_1(\Sigma) \leq \cdots \leq \mathrm{eig}_d(\Sigma)$ denote the ordered eigenvalues of a matrix Σ.

If Σ is a diagonal matrix $\mathrm{diag}(\sigma_1^2, \ldots, \sigma_d^2)$ with $\sigma_j \stackrel{\text{iid}}{\sim} G_0$, then (9.6) and (9.7) hold if G_0 has a polynomial tail at 0 and an exponential tail at infinity. Furthermore, (9.8) is satisfied with $\kappa = 2$ if G_0 is an inverse-gamma distribution, and with $\kappa = 1$ if G_0 is the distribution of the square of an inverse-gamma random variable. In Lemma 9.16 inequalities (9.6)–(9.8) are shown to hold with $\kappa = 2$ if Σ^{-1} is distributed according to a Wishart distribution with positive-definite scale matrix.

For a multi-index $k = (k_1, \ldots, k_d)$ of nonnegative integers k_i, define $k. = \sum_{j=1}^d k_j$, and let $D^k = \partial^{k.}/\partial x_1^{k_1} \cdots \partial x_d^{k_d}$ denote the mixed partial derivative operator. We assume that the true density p_0 of the observations satisfies one of the two conditions:

- *Supersmooth case*: $p_0(x) = \int \phi(x - z; \Sigma_0)\, dF_0(z)$, for some positive-definite matrix Σ_0, and a probability measure F_0 on $\mathbb{R}^d$ satisfying $1 - F_0([z, z]^d) \lesssim e^{-c_0 z^{r_0}}$, for all $z > 0$ and some fixed $c_0 > 0$ and $r_0 \geq 2$. (The case that F_0 is compactly supported can be recovered by formally substituting $r_0 = \infty$.)
- *β-smooth case*: p_0 has mixed partial derivatives $D^k p_0$ of order up to $k. \leq \underline{\beta} := \lceil \beta - 1 \rceil$, satisfying for a function $L\colon \mathbb{R}^d \to [0, \infty)$,

$$|(D^k p_0)(x + y) - (D^k p_0)(x)| \leq L(x)\, e^{c_0 \|y\|^2} \|y\|^{\beta - \underline{\beta}}, \quad k. = \underline{\beta},\ x, y \in \mathbb{R}^d, \tag{9.9}$$

$$P_0\Big[\Big(\frac{L}{p_0}\Big)^2 + \Big(\frac{|D^k p_0|}{p_0}\Big)^{2\beta/k.}\Big] < \infty, \qquad k. \leq \underline{\beta}. \tag{9.10}$$

Furthermore, $p_0(x) \leq c e^{-b\|x\|^\tau}$, for every $\|x\| > a$, for positive constants a, b, c, τ.

Theorem 9.9 *If the Dirichlet mixture of normal prior satisfies conditions* (9.5)–(9.8) *and the true density p_0 is either supersmooth with $r_0 \geq 2$ such that $a_1 > r_0(d + 1) + 1$, or β-smooth, then the posterior distribution contracts at the rate $\epsilon_n = n^{-1/2}(\log n)^{(d+1+1/r_0)/2}$ in the supersmooth case and at the rate $\epsilon_n = n^{-\beta/(2\beta+d^*)}(\log n)^t$ in the β-smooth case, where $d^* = d \vee \kappa$ and $t > (\beta d^* + \beta d^*/\tau + d^* + \beta)/(2\beta + d^*)$, relative to the Hellinger distance.*

In the ordinary smooth case the minimax rate $n^{-\beta/(2\beta+d)}$ is attained up to a logarithmic factor provided that $\kappa \leq d$. This is true for the inverse-Wishart prior on Σ if $d \geq 2$, but not for $d = 1$. The prior based on diagonal Σ with each entry distributed as the square of an inverse-gamma variable (interestingly, not the inverse-gamma, which is conjugate) gives the optimal rate for all dimensions, including $d = 1$.

An essentially identical posterior contraction rate is obtained with a prior on the mixing distribution supported on a random number of random points with random weights, instead of the Dirichlet process prior (see Problem 9.16). Such a prior is intuitively appealing, but posterior computation may need to involve a reversible jump MCMC procedure. The advantage of the generalized Pólya urn scheme is available only for the Dirichlet process and similarly structured priors.

The theorem can be extended to anisotropic smoothness classes, where the smoothness level is different for different coordinates. A naive application of Theorem 9.9 would only give a rate based on the minimal smoothness level of the coordinates, and hence would not take advantage of higher smoothness in other components. A generalization of the theorem, presented in Problem 9.17, gives a rate corresponding to the harmonic mean of the smoothness levels for the different coordinates. This rate coincides with the minimax rate for an anisotropic class (up to a logarithmic factor).

The remainder of this section is devoted to the proof of the theorem, which encompasses approximation results, entropy and prior mass bounds, and an application of the basic contraction theorem, Theorem 8.9 with suitable sieves $\mathcal{P}_{n,1}$.

9.4.1 Approximation

In this section we prove first that any smooth density can be closely approximated by a normal mixture with appropriate bandwidth, and second that general normal mixtures can be closely approximated by finite normal mixtures with few components.

For a nonnegative multi-integer $k = (k_1, \ldots, k_d)$, let $m_k = \int \prod_{j=1}^d x_j^{k_j} \phi(x; I)\, dx$ denote the kth mixed moment of the standard normal distribution on $\mathbb{R}^d$; in particular $m_k = 0$ if one or more of the coordinates of k are odd. Recursively define sequences c_n and d_n by setting $c_n = d_n = 0$ for $n. = 1$ and, for $n. \geq 2$ and $k! = \prod_{j=1}^d k_j!$,

$$c_n = -\sum_{\substack{l+k=n\\ l.\geq 1, k.\geq 1}} \frac{(-1)^{k.}}{k!} m_k d_l, \qquad d_n = \frac{(-1)^{n.} m_n}{n!} + c_n. \tag{9.11}$$

For positive constants β, c and a function $L\colon \mathbb{R}^d \to [0, \infty)$, let $\mathfrak{C}^{\beta,L,c}(\mathbb{R}^d)$ be the set of functions $f\colon \mathbb{R}^d \to \mathbb{R}$ whose mixed partial derivatives of order $k. = \underline{\beta}$ exist and satisfy

$$|(D^k f)(x+y) - (D^k f)(x)| \leq L(x)\, e^{c\|y\|^2} \|y\|^{\beta - \underline{\beta}}, \qquad x, y \in \mathbb{R}^d.$$

For f belonging to this set and $\sigma > 0$, define, with the sum ranging over nonnegative multi-integers $k = (k_1, \ldots, k_d)$,

$$T_{\beta,\sigma} f = f - \sum_{2 \leq k. \leq \underline{\beta}} d_k \sigma^{k.} D^k f.$$

Lemma 9.10 *For any $\beta, c > 0$ there exists a positive constant M such that $|\phi(\cdot; \sigma^2 I) * T_{\beta,\sigma} f(x) - f(x)| \leq M L(x) \sigma^\beta$, for all $x \in \mathbb{R}^d$ and all $\sigma \in (0, (3c)^{-1/2})$, and any $f \in \mathfrak{C}^{\beta,L,c}(\mathbb{R}^d)$.*

Proof Abbreviate $\phi(\cdot; \sigma^2 I)$ to ϕ_σ. A multivariate Taylor expansion allows to decompose $f(x-y) - f(x) = \sum_{1 \leq k. \leq \underline{\beta}} (-y)^{k.}/k!(D^k f)(x) + R(x, y)$, where the remainder satisfies $|R(x, y)| \lesssim L(x) \exp(c\|y\|^2)\|y\|^\beta$, for $x, y \in \mathbb{R}^d$. Consequently, the difference $\phi_\sigma * T_{\beta,\sigma} f(x) - f(x)$ can be written in the form

$$\int \phi_\sigma(y)(f(x-y) - f(x))\, dy - \sum_{2 \leq k. \leq \underline{\beta}} d_k \sigma^{k.} (\phi_\sigma * D^k f)(x)$$

$$= \int \phi_\sigma(y) R(x, y)\, dy + \sum_{2 \leq k. \leq \underline{\beta}} \Big[\frac{(-\sigma)^{k.} m_k}{k!} (D^k f)(x) - d_k \sigma^{k.} (\phi_\sigma * D^k f)(x) \Big].$$

For $\sigma \in (0, (3c)^{-1/2})$ the first term is bounded by a multiple of $L(x)\sigma^\beta$. For $\underline{\beta} < 2$, the second sum is empty, and the result follows. For $\underline{\beta} \geq 2$, we decompose the second term as

$$\sum_{2 \leq k. \leq \underline{\beta}} \frac{(-\sigma)^{k.} m_k}{k!} \big[D^k f - \phi_\sigma * T_{\beta-k.,\sigma} D^k f\big]$$

$$+ \sum_{2 \leq k. \leq \underline{\beta}} \phi_\sigma * \Big[\frac{(-\sigma)^{k.} m_k}{k!} T_{\beta-k.,\sigma} D^k f \quad - d_k \sigma^{k.} D^k f\Big].$$

The function $D^k f - \phi_\sigma * T_{\beta-k.,\sigma} D^k f$ in the first sum is of the same form as in the lemma, but with the function $D^k f$ instead of f, which is approximated with the help of $T_{\beta-k.,\sigma}$. Since $D^k f \in \mathfrak{C}^{\beta-k.,L,c}(\mathbb{R}^d)$, an induction argument on $\underline{\beta}$ can show that it is bounded above by a multiple of $L(x)\sigma^{\beta-k.}$, for every $x \in \mathbb{R}^d$. The second sum is actually identically zero. To see this we substitute the definitions of $T_{\beta-k.,\sigma}$ and d_k, and see that the sum of the functions within square brackets, without the convolution, is identical to

$$\sum_{2 \leq k. \leq \underline{\beta}} \Big[\frac{-(-\sigma)^{k.} m_k}{k!} \sum_{1 \leq j. \leq \underline{\beta}-k.} d_j \sigma^{j.} D^{j+k} f - c_k \sigma^{k.} D^k f\Big]$$

$$= \sum_{3 \leq n. \leq \underline{\beta}} \Big[\sum_{\substack{j+k=n \\ j. \geq 1, k. \geq 2}} \frac{(-1)^{k.+1}}{k!} m_k d_j - c_n \Big] \sigma^{n.} D^n f.$$

The right side vanishes by the definition (9.11) of c_n. □

When Lemma 9.10 is applied to a probability density p_0, then the resulting approximation $T_{\beta,\sigma} p_0$ is not necessarily a probability density. This is remedied in the following result, which also gives approximation in the Hellinger distance.

Lemma 9.11 (Smooth approximation) *For any probability density p_0 satisfying (9.9)–(9.10) there exist constants a_0, τ, K, s_0 and for any $0 < \sigma < s_0$ a probability density f_σ supported within the interval $[-a_0(\log_- \sigma)^{1/\tau}, a_0(\log_- \sigma)^{1/\tau}]$ such that $d_H(p_0, \phi(\cdot; \sigma^2 I) * f_\sigma) \leq K\sigma^\beta$.*

Proof Let λ stand for the Lebesgue measure. Write T_σ for $T_{\beta,\sigma}$ and set

$$f_\sigma = \frac{|T_\sigma p_0| \mathbb{1}\{E_\sigma\}}{\int_{E_\sigma} |T_\sigma p_0| \, d\lambda}, \qquad E_\sigma = \big[- a_0(\log_- \sigma)^{1/\tau}, a_0(\log_- \sigma)^{1/\tau}\big].$$

This first makes the approximation given in Lemma 9.10 nonnegative, and next truncates it to the interval E_σ. We shall show that the effects of these operations are small.

Because $\int D^k p_0 \, d\lambda = 0$ for $k \neq 0$, the definition of T_σ gives that $\int T_\sigma p_0 \, d\lambda = 1$. We shall first prove that $\int |T_\sigma p_0| \, d\lambda = 1 + O(\sigma^{2\beta})$. By the definition of T_σ,

$$|T_\sigma p_0 - p_0| \leq \sum_{k. \leq \underline{\beta}} |d_k| \sigma^{k.} |D^k p_0|.$$

Thus $|T_\sigma p_0 - p_0| \lesssim \eta p_0$ on the set $A_\sigma = \{\sigma^{k.}|D^k p_0| \leq \eta p_0, \forall k. \leq \underline{\beta}\}$, whence $T_\sigma p_0 \geq 0$, for sufficiently small $\eta > 0$. Since also $|T_\sigma p_0| - T_\sigma p_0 \leq 2|T_\sigma p_0 - p_0|$ by the nonnegativity of p_0,

$$\int (|T_\sigma p_0| - T_\sigma p_0)\, d\lambda = \int_{A_\sigma^c} (|T_\sigma p_0| - T_\sigma p_0)\, d\lambda \leq 2 \int_{A_\sigma^c} |T_\sigma p_0 - p_0|\, d\lambda,$$
$$\leq 2 \sum_{k. \leq \underline{\beta}} |d_k| \sigma^{k.} \|D^k p_0 / p_0\|_{2\beta/k., p_0} P_0(A_\sigma^c)^{1-k./(2\beta)},$$

by the preceding display and Hölder's inequality. The definition of A_σ^c, the moment condition (9.10), and Markov's inequality give that $P_0(A_\sigma^c) \lesssim \sigma^{2\beta}$. Thus the preceding display is of this same order, completing the proof that $\int |T_\sigma p_0|\, d\lambda = 1 + O(\sigma^{2\beta})$.

Next

$$d_H^2(p_0, \phi_\sigma * |T_\sigma p_0|) \leq \int \frac{(p_0 - \phi_\sigma * |T_\sigma p_0|)^2}{p_0 + \phi_\sigma * |T_\sigma p_0|}\, d\lambda$$
$$\leq 2 \int \frac{(p_0 - \phi_\sigma * T_\sigma p_0)^2}{p_0}\, d\lambda + 2 \int \frac{(\phi_\sigma * T_\sigma p_0 - \phi_\sigma * |T_\sigma p_0|)^2}{\phi_\sigma * |T_\sigma p_0|}\, d\lambda$$
$$\lesssim \int \frac{L^2}{p_0} \sigma^{2\beta}\, d\lambda + \int \phi_\sigma * (|T_\sigma p_0| - T_\sigma p_0)\, d\lambda,$$

where we use Lemma 9.10 for the first term, and the fact that $|\phi_\sigma * (g - |g|)| \leq \phi_\sigma * (|g| - g) \leq 2\phi_\sigma * |g|$, for any measurable function g, in the second term. Since convolution retains total integral, the second term further evaluates to $\int (|T_\sigma p_0| - T_\sigma p_0)\, d\lambda$, which was seen to be also of the order $O(\sigma^{2\beta})$.

By convexity of the function $(u, v) \mapsto (\sqrt{u} - \sqrt{v})^2$, we have that $|(\phi_\sigma * f)^{1/2} - (\phi_\sigma * g)^{1/2}|^2 \leq \phi_\sigma * (\sqrt{f} - \sqrt{g})^2$, for any nonnegative measurable functions f and g. Therefore

$$d_H^2(\phi_\sigma * |T_\sigma p_0|, \phi_\sigma * f_\sigma) \leq d_H^2(|T_\sigma p_0|, f_\sigma) = \int |T_\sigma p_0|\, d\lambda + 1 - 2\Big(\int_{E_\sigma} |T_\sigma p_0|\, d\lambda\Big)^{1/2}.$$

The first term on the right side was seen to be $1 + O(\sigma^{2\beta})$. By the tail condition on p_0, we have $P_0(E_\sigma^c) = O(\sigma^{2\beta})$ if a_0 is chosen sufficiently large (easily, with a bigger power of σ if a_0 is chosen larger). By the same argument as used for A_σ^c we obtain that $\int_{E_\sigma^c} |T_\sigma p_0 - p_0|\, d\lambda = O(\sigma^{2\beta})$ and hence $\int_{E_\sigma} |T_\sigma p_0|\, d\lambda = 1 + O(\sigma^{2\beta})$. It follows that the right side of the display is $O(\sigma^{2\beta})$. □

General normal mixtures can be approximated by discrete normal mixtures with a small number of support points.

Lemma 9.12 (Finite approximation) *Given a probability measure F_0 on $[-a, a]^d$, a positive-definite matrix Σ, and $\epsilon > 0$, there exists a discrete probability measure F^* on $[-a, a]^d$ with no more than $D(a/\underline{\sigma} \vee 1)^d (\log_- \epsilon)^d$ support points such that $\|p_{F_0,\Sigma} - p_{F^*,\Sigma}\|_\infty \lesssim \epsilon/\underline{\sigma}^d$ and $\|p_{F_0,\Sigma} - p_{F^*,\Sigma}\|_1 \lesssim \epsilon(\log_- \epsilon)^{d/2}$, where D is a constant that depends on d and $\underline{\sigma}$ is the smallest eigenvalue of $\Sigma^{1/2}$. Without loss of generality the support points can be chosen in the set $\underline{\sigma}\epsilon\, \mathbb{Z}^d \cap [-a, a]^d$.*

Proof The identity $p_{F,\Sigma}(x) = \det \Sigma^{-1/2} p_{F\circ\Sigma^{-1},I}(\Sigma^{-1/2}x)$ gives that the distances in the lemma are bounded above by $\|p_{F_0\circ\Sigma^{-1/2},I} - p_{F^*\circ\Sigma^{-1/2},I}\|_1$ and $\underline{\sigma}^{-d}\|p_{F_0\circ\Sigma^{-1/2},I} - p_{F^*\circ\Sigma^{-1/2},I}\|_\infty$, respectively. The measures $F \circ \Sigma^{-1/2}$ concentrate on the set $\Sigma^{-1/2}[-a,a]^d \subset [-\sqrt{d}\,a/\underline{\sigma}, \sqrt{d}\,a/\underline{\sigma}]^d$, since $\|\Sigma^{-1/2}x\|_\infty \le \|\Sigma^{-1/2}x\|_2 \le \|x\|_2/\underline{\sigma} \le \sqrt{d}\,\|x\|_\infty/\underline{\sigma}$. Thus the problem can be reduced to mixtures of the standard normal kernel relative to mixing distributions on $[-\sqrt{d}\,a/\underline{\sigma}, \sqrt{d}\,a/\underline{\sigma}]^d$.

We can partition the latter cube into fewer than $D_1(a/\underline{\sigma} \vee 1)^d$ rectangles $I_1, \ldots, I_k$ with sides of length at most 1. Decomposing a probability measure on the cube as $F = \sum_{i=1}^k F(I_i)F_i$, where each F_i is a probability measure on I_i, we have $p_{F,I} = \sum_{i=1}^k F(I_i)p_{F_i,I}$. We shall show that for each $i = 1, \ldots, k$ there exists a discrete distribution F_i^* on I_i with at most $D_2(\log_- \epsilon)^d$ many support points such that the $\mathbb{L}_\infty$- and $\mathbb{L}_1$-norms between $p_{F_i^*,I}$ and $p_{F_{0,i},I}$ are bounded above by ϵ and $\epsilon(\log_- \epsilon)^{d/2}$. Then $F^* = \sum_{i=1}^k F_0(I_i)F_i^*$ will be the appropriate approximation of F_0. Because we can shift the rectangles I_i to the origin and the two distances are invariant under shifting, it is no loss of generality to construct the approximation only for I_i equal to the unit cube. For simplicity of notation, consider a probability measure F_0 on $[0,1]^d$ and the mixture $p_{F_0,I}$.

For $\|x\|_\infty \ge \sqrt{8\log_- \epsilon}$ and sufficiently small ϵ, we have that $\phi(x-z) \le \epsilon$ for all $z \in [0,1]^d$, so that $p_{F,I}(x) \le \epsilon$ for every F concentrated on $[0,1]^d$. For $\|x\|_\infty \le \sqrt{8\log_- \epsilon}$ Taylor's expansion of the exponential function, gives

$$\phi(x-z) = \frac{1}{(2\pi)^{d/2}} \prod_{j=1}^d \left[\sum_{r=0}^{k-1} \frac{\left[-(x_j-z_j)^2/2\right]^r}{r!} + R(x_j - z_j)\right],$$

where the remainder satisfies $|R(x)| \le (x^2/2)^k/k!$. If $|x_j| \le \sqrt{8\log_- \epsilon}$ and $|z_j| \le 1$, then $|R(x_j - z_j)| \le (8e\log_- \epsilon/k)^k$, in view of the inequality $k! \ge k^k e^{-k}$, which will be bounded by 1 for sufficiently large k. Since the univariate standard normal density is uniformly bounded, it follows that the sums, which are the differences of the density and the remainder terms, are also uniformly bounded. Hence the remainder can be pulled out of the product, and for $\|x\|_\infty \le \sqrt{8\log_- \epsilon}$ we can write $\phi(x-z)$ as

$$\frac{1}{(2\pi)^{d/2}} \prod_{j=1}^d \left[\sum_{r=0}^{k-1} \frac{(-1)^r}{2^r r!} \sum_{l=0}^{2r} \binom{2r}{l} x_j^{2r-l} z_j^l\right] + \bar{R}(x-z), \qquad |\bar{R}| \lesssim \left|\frac{(8e\log_- \epsilon)}{k}\right|^k.$$

Let F^* be a probability measure on $[0,1]^d$ such that, for integers l_j,

$$\int z_1^{l_1} \cdots z_d^{l_d}\, dF^*(z) = \int z_1^{l_1} \cdots z_d^{l_d}\, dF_0(z), \qquad 0 \le l_1, \ldots, l_d \le 2k-2,$$

Then integrating the second last display with respect to $F^* - F_0$, giving $p_{F^*,I} - p_{F_0,I}$, leaves only the integral over $\bar{R}$. For k the smallest integer exceeding $(1+c^2)\log_- \epsilon$, the latter integral is bounded by a multiple of ϵ. The preceding display requires matching $(2k-1)^d$ expectations, and hence F^* can be chosen a discrete distribution with at most $(2k-1)^d + 1$ support points, by Lemma L.1.

This concludes the proof for the $\mathbb{L}_\infty$-norm. For the $\mathbb{L}_1$-norm we note that $p_{F,I}(x) \le e^{-x_j^{2/8}}$ if $|x_j| \ge 2$, for any probability measure F on $[0,1]^d$, whence $\int_{\|x\|_\infty \ge T} p_{F,I}(x)\,dx \le$

$2^d e^{-T^2/8}$, for sufficiently large T. Letting F^* as for the $\mathbb{L}_\infty$-bound, we have that $|p_{F^*,I} - p_{F_0,I}| \lesssim \epsilon$ on $[-T,T]^d$. Integrating this and combining with the first bound we see that $\|p_{F^*,I} - p_{F_0,I}\|_1 \lesssim e^{-T^2/8} + T^d\epsilon$. We choose T a suitable multiple of $(\log_- \epsilon)^{1/2}$ to establish the second assertion of the lemma.

The final assertion of the lemma follows from the fact that moving the support points of F^* to a closest point in the given lattice increases the $\mathbb{L}_1$-error by at most a multiple of ϵ, as $z \mapsto \phi(x-z;\Sigma)$ is Lipschitz continuous with constant $1/\underline{\sigma}$ relative to the $\mathbb{L}_1$-norm; and increases the $\mathbb{L}_\infty$-error by at most $\epsilon/\underline{\sigma}^d$, as $\|\phi'(\cdot;\Sigma)\|_\infty \lesssim 1/\underline{\sigma}^{d+1}$. □

9.4.2 Prior Concentration

In this section we estimate the prior mass of a ball $B_2(p_0,\epsilon)$, as in (8.3), around a true density from below.

Lemma 9.13 *For a measurable partition $\mathbb{R}^d = \cup_{j=0}^N U_j$ and points $z_j \in U_j$, for $j=1,\ldots,N$, let $F^* = \sum_{j=1}^N w_j\delta_{z_j}$ be a probability measure. Then, for any probability measure F on $\mathbb{R}^d$ and any positive-definite matrix Σ with eigenvalues bounded below by $\underline{\sigma}^2$,*

$$\|p_{F,\Sigma} - p_{F^*,\Sigma}\|_\infty \lesssim \frac{1}{\underline{\sigma}^{d+1}} \max_{1\le j\le N} \mathrm{diam}(U_j) + \frac{1}{\underline{\sigma}^d}\sum_{j=1}^N |F(U_j) - w_j|,$$

$$\|p_{F,\Sigma} - p_{F^*,\Sigma}\|_1 \lesssim \frac{1}{\underline{\sigma}} \max_{1\le j\le N} \mathrm{diam}(U_j) + \sum_{j=1}^N |F(U_j) - w_j|.$$

Proof We can decompose $p_{F,\Sigma}(x) - p_{F^*,\Sigma}(x)$ as

$$\int_{U_0} \phi(x-z;\Sigma)\,dF(z) + \sum_{j=1}^N \int_{U_j} \big[\phi(x-z;\Sigma) - \phi(x-z_j;\Sigma)\big]\,dF(z)$$
$$+ \sum_{j=1}^N \phi(x-z_j;\Sigma)\big[F(U_j) - w_j\big].$$

Here the mass of the set $U_0 = \mathbb{R}^d \setminus \cup_{j\ge1} U_j$ is bounded as $F(U_0) \le \sum_{j=1}^N |F(U_j) - w_j|$, since $(w_1,\ldots,w_N)$ is assumed to be a probability vector. The inequality for the $\mathbb{L}_\infty$-norm next follows from the estimates $\|\phi(\cdot;\Sigma)\|_\infty \lesssim \underline{\sigma}^{-d}$ and $\|\phi'(\cdot;\Sigma)\|_\infty \lesssim \underline{\sigma}^{-(d+1)}$, while for the inequality for the $\mathbb{L}_1$-norm we employ the estimates $\|\phi(\cdot - z;\Sigma) - \phi(\cdot - z';\Sigma)\|_1 \lesssim \underline{\sigma}^{-1}\|z - z'\|_\infty$ and $\|\phi(\cdot;\Sigma)\|_1 = 1$. □

Proposition 9.14 (Prior mass) *If α has a continuous, positive density and satisfies (9.5), then there exist constants $A, C > 0$ such that $(\mathrm{DP}_\alpha \times G)((F,\Sigma): p_{F,\Sigma} \in B_2(p_0, A\epsilon_n)) \ge e^{-Cn\epsilon_n^2}$, where ϵ_n is given by*

(i) $n^{-1/2}(\log n)^{(d+1+1/r_0)/2}$ *if $p_0 = p_{F_0,\Sigma_0}$ is supersmooth with Σ_0 an interior point of the support of G,*

(ii) $n^{-\beta/(2\beta+d^*)}(\log n)^{t_0}$ *if* p_0 *is* β*-smooth and* G *satisfies* (9.6)–(9.8), *where* $d^* = d \vee \kappa$ *and* $t_0 = (\beta d^* + \beta d^*/\tau + d^* + \beta)/(2\beta + d^*)$.

Proof (i). The assumption on the tail of F_0 gives that $1 - F_0([-a,a]^d) \le \epsilon^2$, for $a = a_0(\log_- \epsilon)^{1/r_0}$, sufficiently small $\epsilon > 0$ and large a_0, whence the renormalized restriction $\tilde F_0$ of F_0 to $[-a,a]^d$ satisfies $\|p_{F_0,\Sigma_0} - p_{\tilde F_0,\Sigma_0}\|_1 \le \epsilon^2$, by Lemma K.10. Next Lemma 9.12 gives a discrete distribution $F^* = \sum_{j=1}^N w_j\delta_{z_j}$ on $[-a,a]^d$ with at most $N \lesssim a(\log_- \epsilon)^d \lesssim (\log_- \epsilon)^{d+1/r_0}$ support points such that $\|p_{\tilde F_0,\Sigma_0} - p_{F^*,\Sigma_0}\|_1 \lesssim \epsilon^2$. The points z_j can be chosen ϵ^2-separated without loss of generality, whence they possess disjoint neighborhoods $U_j \subset [-a,a]^d$ of diameter of the order ϵ^2. If F is a probability measure with $\sum_{j=1}^N |F(U_j) - w_j| \le \epsilon^2$, then $\|p_{F^*,\Sigma_0} - p_{F,\Sigma_0}\|_1 \lesssim \epsilon^2$, by Lemma 9.13.

Combining the preceding with the triangle inequality gives $\|p_{F_0,\Sigma_0} - p_{F,\Sigma_0}\|_1 \lesssim \epsilon^2$ and hence $d_H(p_{F_0,\Sigma_0}, p_{F,\Sigma_0}) \lesssim \epsilon$.

For any probability measure F and positive-definite matrix Σ,

$$d_H^2(p_{F,\Sigma}, p_{F,\Sigma_0}) \le d_H^2(\phi(\cdot;\Sigma), \phi(\cdot;\Sigma_0)) = 2\big[1 - \det\big[2(\Sigma+\Sigma_0)^{-1}\Sigma^{1/2}\Sigma_0^{1/2}\big]^{1/2}\big]. \tag{9.12}$$

If $\|\Sigma - \Sigma_0\| \le \epsilon$, then the right side is bounded above by a multiple of ϵ^2.

When combined the preceding paragraphs give that

$$\Big\{(F,\Sigma): d_H(p_{F,\Sigma}, p_{F_0,\Sigma_0}) \lesssim \epsilon\Big\} \supset B := \Big\{(F,\Sigma): \sum_{j=1}^N |F(U_j) - w_j| \le \epsilon^2,\ \|\Sigma - \Sigma_0\| \le \epsilon\Big\}.$$

By Lemma G.13, applied to the Dirichlet vector $(F(U_0), F(U_1), \ldots, F(U_N))$ where $U_0 = \mathbb{R}^d \setminus \cup_j U_j$, and the assumption that the prior density for Σ is bounded away from zero, the prior mass of the set on the right is bounded below by $e^{-cN\log(1/\epsilon)}\epsilon^q$, for $q > 0$ depending on the dimension of the support of Σ.

For any F and Σ and $\underline{\sigma}^2$ the minimal eigenvalue of Σ,

$$p_{F,\Sigma}(x) \ge \frac{1}{\underline{\sigma}^d}\int_{\|z\|\le a} e^{-\|x-z\|^2/2\underline{\sigma}^2}\,dF(z) \gtrsim \begin{cases} \frac{1}{\underline{\sigma}^d} e^{-2da^2/\underline{\sigma}^2} F[-a,a]^d, & \|x\|_\infty \le a,\\ \frac{1}{\underline{\sigma}^d} e^{-2d\|x\|^2/\underline{\sigma}^2} F[-a,a]^d, & \|x\|_\infty > a. \end{cases} \tag{9.13}$$

If (F,Σ) is in the set on the right side of the second last display, then the probability $F[-a,a]^d$ and the smallest eigenvalue $\underline{\sigma}^2$ are bounded away from zero. Since p_{F_0,Σ_0} is uniformly bounded with sub-Gaussian tails, it then follows that $P_0(p_{F_0,\Sigma_0}/p_{F,\Sigma})^\delta$ is uniformly bounded, for sufficiently small $\delta > 0$. Consequently $K(p_{F_0,\Sigma_0}; p_{F,\Sigma})$ and $V_2(p_{F_0,\Sigma_0}; p_{F,\Sigma})$ are bounded above by a multiple of $d_H^2(p_{F,\Sigma}, p_{F_0,\Sigma_0})\log_- d_H(p_{F,\Sigma}, p_{F_0,\Sigma_0})$, by Lemma B.2. The lower bound on the prior mass found in the preceding paragraph, $e^{-cN\log(1/\epsilon)}\epsilon q$, translates into a lower bound on the prior mass of the $B_2(p_{F_0,\Sigma_0}, \eta)$-neighborhood, with $\eta \sim \epsilon(\log_- \epsilon)$. Replacing ϵ by $\epsilon/(\log_- \epsilon)$ in the exponent of the lower bound, does not affect its form, and assertion (i) follows upon equating this exponent to $n\epsilon_n^2$.

(ii). Given small $\sigma > 0$ and large a_0, set $a_\sigma = a_0(\log_- \sigma)^{1/\tau}$. By Lemma 9.11 there exists a probability distribution F_σ on $[-a_\sigma, a_\sigma]^d$ such that $d_H(p_0, p_{F_\sigma,\sigma^2 I}) \lesssim \sigma^\beta$. By Lemma 9.12, applied with $\epsilon^2/(\log_- \epsilon)^d$ instead of ϵ, there exists a discrete probability measure $F_\sigma^* = \sum_{j=1}^N w_j \delta_{z_j}$ with at most $N \lesssim (a_\sigma/\sigma)^d (\log_- \epsilon)^d$ support points inside $[-a_\sigma, a_\sigma]^d$ such that $\| p_{F_\sigma,\sigma^2 I} - p_{F_\sigma^*,\sigma^2 I}\|_1 \lesssim \epsilon^2$ and hence $d_H(p_{F_\sigma,\sigma^2 I}, p_{F_\sigma^*,\sigma^2 I}) \lesssim \epsilon$.

The support points of F_σ^* can be chosen on a lattice of mesh width $\sigma\epsilon^2$ without loss of generality. Then there exist disjoint neighborhoods $U_1, \ldots, U_N$ of $z_1, \ldots, z_N$ of diameters of the order $\sigma\epsilon^2$, which can be extended into a partition $\{U_1, \ldots, U_{N'}\}$ of $[-a_\sigma, a_\sigma]^d$ in at most $N' \lesssim N$ sets of diameter at most σ, and next into a partition $\{U_1, \ldots, U_M\}$ of $\mathbb{R}^d$ in $M \lesssim N$ sets, all with the property that $(\sigma\epsilon^2)^d \lesssim \alpha(U_j) \lesssim 1$, for all $j = 1, \ldots, M$. By Lemma 9.13 applied with $U_0 = \cup_{j>N} U_j$ and $w_{N+1} = \cdots = w_M = 0$, if F is a probability measure with $\sum_{j=1}^M |F(U_j) - w_j| \le \epsilon^2$ then $\|p_{F_\sigma^*,\sigma^2 I} - p_{F,\sigma^2 I}\|_1 \lesssim \epsilon^2$, and hence $d_H(p_{F_\sigma^*,\sigma^2 I}, p_{F,\sigma^2 I}) \lesssim \epsilon$.

By (9.12), for any probability measure F,

$$d_H^2(p_{F,\Sigma}, p_{F,\sigma^2 I}) \le 2 - 2\prod_{j=1}^d \left(1 - \frac{(\mathrm{eig}_j(\Sigma)^{1/2} - \sigma)^2}{\mathrm{eig}_j(\Sigma) + \sigma^2}\right)^{1/2}.$$

If all eigenvalues of Σ are contained in the interval $[\sigma^2/(1+\sigma^\beta), \sigma^2]$, then the right side is bounded by a multiple of $\sigma^{2\beta}$.

When combined the preceding paragraphs give that

$$\Big\{(F, \Sigma): d_H(p_0, p_{F,\Sigma}) \lesssim \sigma^\beta + \epsilon\Big\} \tag{9.14}$$

$$\supset B := \Big\{(F, \Sigma): \sum_{j=1}^M |F(U_j) - w_j| \le \epsilon^2, \min_{1\le j\le M} F(U_j) \ge \epsilon^4, 1 \le \sigma^2 \mathrm{eig}(\Sigma^{-1}) \le 1 + \sigma^\beta\Big\}.$$

By Lemma G.13 and (9.8), the prior mass of the set B on the right side is bounded below by a multiple of $e^{-c_1 M \log_- \epsilon}\, \sigma^{-2a_4} \sigma^{\beta a_5} e^{-C_3 \sigma^{-\kappa}}$. For M as chosen previously and small σ and ϵ this is bounded below by $e^{-c' E(\sigma,\epsilon)}$, for $E(\sigma, \epsilon) = (\log_- \sigma)^{d/\tau} \sigma^{-d} (\log_- \epsilon)^{d+1} + \sigma^{-\kappa}$.

Because $\{U_1, \ldots, U_{N'}\}$ forms a partition of $[-a_\sigma, a_\sigma]^d$ in sets of diameter smaller than σ, for every $x \in [-a_\sigma, a_\sigma]^d$ there exists a set $U_{j(x)}$ that is contained in the ball of radius σ around x. Hence, for any (F, Σ) in the set B of (9.14) and $\|x\|_\infty \le a_\sigma$,

$$p_{F,\Sigma}(x) \ge \frac{1}{\sigma^d} \int_{\|x-z\|\le\sigma} e^{-\|x-z\|^2/\sigma^2}\, dF(z) \ge \frac{e^{-1}}{\sigma^d} F(U_{j(x)}) \ge \frac{e^{-1}\epsilon^4}{\sigma^d}.$$

On the other hand, for $\|x\|_\infty > a_\sigma$, and $(F, \Sigma) \in B$, the second estimate in (9.13), with $a = a_\sigma$, gives that $p_{F,\Sigma}(x) \gtrsim \sigma^{-d} e^{-2d\|x\|^2/\sigma^2} F[-a_\sigma, a_\sigma]^d$, where $F[-a_\sigma, a_\sigma]^d$ is bounded away from zero. This shows that $\log p_{F,\Sigma}(x) \lesssim \log_- \sigma + \|x\|^2/\sigma^2$, whence

$$P_0\Big[\Big(\log \frac{p_0}{p_{F,\Sigma}}\Big)^2 \mathbb{1}\Big\{\frac{p_{F,\Sigma}}{p_0} < \frac{e^{-1}\epsilon^d}{\|p_0\|_\infty \sigma^d}\Big\}\Big] \lesssim \int_{\|x\|>a_\sigma} \Big[(\log_- \sigma)^2 + \frac{\|x\|^4}{\sigma^4}\Big] p_0(x)\, dx.$$

By the definition of a_σ and the tail condition on p_0 the right side is smaller than any given power of σ, if a_0 is sufficiently large. By Lemma B.2 we see that both $K(p_0; p_{F,\Sigma})$ and $V_2(p_0; p_{F,\Sigma})$ are bounded above by a multiple of $d_H^2(p_0; p_{F,\Sigma})(\log_-(\epsilon^4/\sigma^d))^2 + o(\sigma^{2\beta})$, provided that $e^{-1}\epsilon^4/\sigma^d \leq 0.44\|p_0\|_\infty$. This implies that the prior mass of the neighborhood $B_2(p_0, \epsilon_n)$ is bounded below by the prior mass in the set B of (9.14) if ϵ and σ are chosen so that $(\sigma^\beta + \epsilon)\log_-(\epsilon^4/\sigma^d) \lesssim \epsilon_n$ and ϵ^4/σ^d is sufficiently small. The prior mass is bounded below by $e^{-Cn\epsilon_n^2}$ for some C if

$$(\sigma^{2\beta}+\epsilon^2)\big[\log_-(\epsilon^4/\sigma^d)\big]^2 \lesssim \epsilon_n^2, \quad \epsilon^4/\sigma^d \ll 1, \quad (\log_-\sigma)^{d/\tau}\sigma^{-d}(\log_-\epsilon)^{d+1}+\sigma^{-\kappa} \leq n\epsilon_n^2.$$

We choose $\epsilon^4 \asymp \sigma^d \wedge \sigma^{2\beta}$ in order to satisfy the second requirement and reduce the first requirement to $\sigma^{2\beta}(\log_-\sigma)^2 \lesssim \epsilon_n^2$. If $\kappa \leq d$, then we further choose $\sigma^{2\beta+d} \sim n^{-1}(\log n)^{d/\tau+d-1}$ to satisfy the remaining requirements with the rate ϵ_n as given in the theorem. If $\kappa > d$, then we choose $\sigma^{2\beta+\kappa} \sim n^{-1}(\log n)^{-2}$, which leads to a slightly smaller ϵ_n than given in the theorem. □

9.4.3 Entropy Estimate and Controlling Complexity

The following lemma bounds the metric entropy of a suitable sieve of discrete normal mixtures and, the prior probability of its complement. For positive constants ϵ, a, σ and integers M, N, define

$$\begin{aligned}\mathcal{F} &= \Big\{\sum_{j=1}^{\infty} w_j\delta_{z_j} \colon \sum_{j=N+1}^{\infty} w_j < \epsilon^2, \quad z_1, \ldots, z_N \in [-a, a]^d\Big\},\\ \mathcal{S} &= \Big\{\Sigma \colon \sigma^2 \leq \mathrm{eig}_1(\Sigma) \leq \mathrm{eig}_d(\Sigma) < \sigma^2(1+\epsilon^2)^M\Big\}.\end{aligned} \tag{9.15}$$

Lemma 9.15 (Entropy) *For $a \geq \sigma\epsilon$ and $M \geq d$, the sets $\mathcal{F}$ and $\mathcal{S}$ satisfy, for some $A > 0$,*

(i) $N(A\epsilon, \{p_{F,\Sigma} \colon F \in \mathcal{F}, \Sigma \in \mathcal{S}\}, d_H) \lesssim (5/\epsilon^2)^N (3a/(\sigma\epsilon^2))^{dN} (5/\epsilon^2)^{d^2}(1+\epsilon^2)^{Md^2} M^d$.
(ii) $\mathrm{DP}_\alpha(\mathcal{F}^c) \leq (2e|\alpha|\log_-\epsilon/N)^N + N(1-\bar{\alpha}([-a,a]^d))$.

Proof Let $\mathcal{F}_\epsilon$ be the set of all mixtures $F = \sum_{j=1}^N w_j\delta_{z_j}$, with weight vector $(w_1, \ldots, w_N)$ belonging to a fixed maximal ϵ^2-net in the N-dimensional unit simplex $\mathbb{S}_N$, and support points z_j belonging to a fixed maximal $\sigma\epsilon^2$-net in $[-a, a]^d$ (relative to the $\mathbb{L}_1$- and maximum norms). For $\eta = \epsilon^2/(1+\epsilon^2)^M$, let $\mathcal{S}_\epsilon$ be the set of all matrices $O\Lambda O^\top$ for O belonging to a fixed maximal η-net over the set of all orthogonal matrices (relative to the Frobenius norm), and Λ a diagonal matrix with diagonal entries belonging to the set of M points $\sigma^2(1+\epsilon^2)^m$, for $m = 0, 1, \ldots, M-1$.

The cardinality of the set $\mathcal{F}_\epsilon \times \mathcal{S}_\epsilon$ is bounded above by the upper bound in (i) (see Propositions C.1 and C.2). We shall show that the corresponding set of $p_{F,\Sigma}$ forms an $A\epsilon$-net for d_H over the set of all such densities with $(F, \Sigma) \in \mathcal{F} \times \mathcal{S}$, for a sufficiently large constant A.

Given probability measures $F = \sum_{j=1}^\infty w_j\delta_{z_j}$ and $F' = \sum_{j=1}^N w'_j\delta_{z'_j}$ arguments similar as in the proof of Lemma 9.13 show that $\|p_{F,\Sigma} - p_{F',\Sigma}\|_1 \leq \sigma^{-1}\|z - z'\|_\infty + \sum_{j=1}^\infty |w_j - w'_j|$, for $\Sigma \in \mathcal{S}$ (so that its eigenvalues are bounded below by σ^2). For z and w corresponding to some $F \in \mathcal{F}$, the right side can be reduced to $(\sigma^{-1})\sigma\epsilon^2 + 3\epsilon^2$ by choice of z' and w' from the nets. Thus for every $F \in \mathcal{F}$ and $\Sigma \in \mathcal{S}$, there exists $F' \in \mathcal{F}_\epsilon$ such that $d_H(p_{F,\Sigma}, p_{F',\Sigma}) \leq 2\epsilon$.

For an orthogonal matrix O and diagonal matrices $\Lambda = \mathrm{diag}(\lambda_j)$ and $\Lambda' = \mathrm{diag}(\lambda'_j)$, the matrices $O\Lambda O^\top$ and $O\Lambda' O^\top$ possess the same eigenvectors. Therefore, by (9.12), for any probability measure F,

$$d_H^2(p_{F,O\Lambda O^\top}, p_{F,O\Lambda' O^\top}) \le 2 - 2\prod_{i=1}^d \Bigl(1 - \frac{(\sqrt{\lambda_i} - \sqrt{\lambda'_i})^2}{\lambda_i + \lambda'_i}\Bigr)^{1/2}.$$

For $1 \le \lambda'_i/\lambda_i \le 1 + \epsilon^2$ this is of the order ϵ^4. For any diagonal matrix Λ and orthogonal matrices O and O', by the first inequality in (9.12) followed by Lemma B.2(iv) and a calculation of the Kullback-Leibler divergence, since $\det(O\Lambda O^\top) = \det(O'\Lambda(O')^\top)$,

$$\begin{aligned} d_H^2(p_{F,O\Lambda O^\top}, p_{F,O'\Lambda(O')^\top}) &\le K(\phi(\cdot; O\Lambda O^\top), \phi(\cdot; O'\Lambda(O')^\top)) \\ &= \mathrm{tr}([O'\Lambda(O')^\top]^{-1} O\Lambda O^\top - I). \end{aligned}$$

The right side can be written $\mathrm{tr}(Q\Lambda^{-1}Q^\top\Lambda - I)$, for $Q = O^\top O'$, which can be further rewritten as $2\mathrm{tr}(Q - I) + \mathrm{tr}((Q - I)\Lambda^{-1}(Q^\top - I)\Lambda)$. Since $|\mathrm{tr}(A)| \lesssim \|A\|$, this is bounded above by $2\|Q - I\| + \|Q - I\|^2\| \|\Lambda\| \|\Lambda^{-1}\|$. Since $\|Q - I\| = \|Q - O^\top O\| \le \|O' - O\|$, it follows that there exists O' in the net such that the right side is bounded above by $2\eta + \eta^2\|\Lambda\| \|\Lambda^{-1}\| \le 2\eta + \eta^2(1 + \epsilon^2)^M$, if $\Lambda \in \mathcal{S}$, which is bounded by a multiple of ϵ^2 by the definition of η.

Combined, the preceding inequalities show that for any $(F, \Sigma) \in \mathcal{F} \times \mathcal{S}$ there exists $(F', \Sigma') \in \mathcal{F}_\epsilon \times \mathcal{S}_\epsilon$ such that $d_H(p_{F,\sigma}, p_{F',\sigma'}) \lesssim \epsilon$.

For the proof of (ii) we use the stick-breaking representation $F = \sum_{j=1}^\infty W_j \delta_{Z_j}$ of a Dirichlet process, given in Theorem 4.12. Then $W_j = V_j \prod_{l=1}^{j-1}(1 - V_l)$ for $V_j \overset{\text{iid}}{\sim} \mathrm{Be}(1, |\alpha|)$, and $R := -\log\sum_{j>N} W_j = -\sum_{j=1}^N \log(1 - V_j)$ is distributed as a $\mathrm{Ga}(N, |\alpha|)$-variable. Hence

$$\begin{aligned} \mathrm{DP}_\alpha(\mathcal{F}^c) &\le \mathrm{P}(R < -\log\epsilon^2) + \sum_{j=1}^N \mathrm{P}(Z_j \notin [-a, a]^d) \\ &\le \frac{(2|\alpha| \log_- \epsilon)^N}{N!} + N(1 - \bar\alpha([-a, a]^d)). \end{aligned}$$

The bound in (ii) follows upon using that $N^N/N! \le e^N$. ☐

9.4.4 Proof of Theorem 9.9

We apply Theorem 8.9 with $\bar\epsilon_n$ the rate found in Proposition 9.14 on the prior mass, different in the supersmooth and β-smooth cases, and the sieve $\mathcal{P}_{n,1}$ consisting of all densities $p_{F,\Sigma}$ with F in $\mathcal{F}$ and S in $\mathcal{S}$ as given in (9.15), with the choices, for a large constant C,

$$N = \frac{Cn\bar\epsilon_n^2}{\log(n\bar\epsilon_n^2)}, \quad n\epsilon^2 = C\,N\log n, \quad M = n, \quad a^{a_1} = n\epsilon^2, \quad \sigma^{-2a_2} = n\epsilon^2.$$

This gives that $\epsilon_n^2 := \epsilon^2 = C^2\bar\epsilon_n^2 \log n/\log(n\bar\epsilon_n^2)$, which implies that $\epsilon_n \gg \bar\epsilon_n$ in the supersmooth case and $\epsilon_n \ge \sqrt{C}\bar\epsilon_n$ in the β-smooth case.

The prior mass condition (8.4) of Theorem 8.9 is satisfied by choice of $\bar\epsilon_n$ and Proposition 9.14. By Lemma 9.15,

$$\log N(A\epsilon_n, \mathcal{P}_{n,1}, d_H) \lesssim N\Big[\log \frac{3a}{\sigma\epsilon_n^2} + \log \frac{5}{\epsilon_n}\Big] + M\log(1+\epsilon_n^2) + \log M.$$

From the second to fifth definitions the right side can be seen to be bounded by a multiple of $n\epsilon_n^2$, as required by (8.5) of Theorem 8.9. By (ii) of Lemma 9.15 and the conditions (9.5)–(9.7), the prior probability of $\mathcal{P}_{n,1}^c$ is bounded by a multiple of

$$\Big(\frac{2e|\alpha|\log_- \epsilon_n}{N}\Big)^N + Ne^{-C_1 a^{a_1}} + e^{-C_2\sigma^{-2a_2}} + \sigma^{-2a_3}(1+\epsilon_n^2)^{-a_3 M}.$$

It is immediate from the definitions that the second and third terms are bounded above by $e^{-C_i n\epsilon_n^2} \le e^{-C_i C^2 n\bar\epsilon_n^2}$. In the fourth term $(1+\epsilon_n^2)^{-M} \le e^{-n\epsilon_n^2/2}$ is similarly small, and dominates σ^{-2a_3}. Finally in the first term the quotient $U_n = \log_- \epsilon_n/N = C\log_- \epsilon_n \log n/(n\epsilon_n^2)$ is bounded above by $C(\log n)^2/n\epsilon_n^2$. In the supersmooth case $n\epsilon_n^2 \ge n\bar\epsilon_n^2 = (\log n)^{d+1+1/r_0}$ and hence $U_n \le C(\log n)^{-\gamma}$, for some $\gamma > 0$, whence $U_n^N \le e^{-N[\gamma \log\log n - \log C]} \le e^{-Cn\bar\epsilon_n^2/2}$. In the ordinary smooth case $(\log n)^2/n\epsilon_n^2$ is bounded above by a power of $1/n$ and hence $U_n^N \le e^{-N\delta \log n} \le e^{-C\delta_1 n\bar\epsilon_n^2}$. In both cases it follows that (8.6) of Theorem 8.9 is satisfied as well.

9.4.5 Wishart Prior

Conditions (9.6)–(9.8) hold for the example of a conjugate inverse-Wishart prior.

Lemma 9.16 *If* $\Sigma^{-1} \sim \mathrm{Wis}(\nu, \Psi)$, *with positive-definite scale matrix* Ψ, *then* (9.6)–(9.8) *hold with* $\kappa = 2$.

Proof First consider the case Ψ is the identity matrix. Then $\mathrm{tr}(\Sigma^{-1}) \sim \chi^2_{\nu d}$, and hence $\mathrm{P}(\mathrm{tr}(\Sigma^{-1}) > s) \le e^{-s/4}$, for all sufficiently large s, whence (9.6) holds, as $\mathrm{eig}_d(\Sigma^{-1}) \le \mathrm{tr}(\Sigma^{-1})$. The joint probability density of $\mathrm{eig}_1(\Sigma^{-1}), \ldots, \mathrm{eig}_d(\Sigma^{-1})$ satisfies (cf. Muirhead 1982, page 106)

$$f(s_1, \ldots, s_d) \propto e^{-\sum_j s_j/2} \prod_{j=1}^d s_j^{(\nu+1-d)/2} \prod_{1\le j<k\le d} (s_k - s_j), \quad 0 < s_1 < \cdots < s_d. \tag{9.16}$$

Since $\prod_{j<k}(s_k - s_j) \le \prod_{j<k} s_k = \prod_{k=2}^d s_k^{k-1}$, the marginal density of $\mathrm{eig}_1(\Sigma^{-1})$ is bounded by a multiple of

$$s_1^{(\nu+1-d)/2} e^{-s_1/2} \prod_{k=2}^d \int_0^\infty s_k^{(\nu+1-d)/2+k-1} e^{-s_k/2}\, ds_k \propto s_1^{(\nu+1-d)/2} e^{-s_1/2}, \qquad s_1 > 0.$$

This leads to relation (9.7). For (9.8) it suffices to lower bound the probability of the event $\cap_{j=1}^d \{\mathrm{eig}_j(\Sigma^{-1}) \in I_j\}$, where $I_j = (s_j(1+(j-1/2)t/d), s_j(1+jt/d))$, for $j = 1, \ldots, d$. If

$s_j \in I_j$ and $s_k \in I_k$, for $j < k$, then $s_k - s_j > s_k(1+(k-1/2)t/d) - s_j(1+jt/d) \geq s_1 t/(2d)$, and hence, for

$$\begin{aligned}
\mathrm{P}(s_j < \mathrm{eig}_j(\Sigma^{-1}) < s_j(1+t), \quad j = 1,\ldots,d) &\geq \int_{I_1} \cdots \int_{I_1} f(s_1,\ldots,s_d)\, ds_1 \cdots ds_d \\
&\geq c_{d,\nu} e^{-ds_d} s_1^{d\nu/2} (t/(2d))^{d(d-1)/2} \int_{I_d} \cdots \int_{I_1} dx_1 \cdots dx_d \\
&= c_{d,\nu} (2d)^{-d(d+1)/2} e^{-ds_d} s_1^{d(\nu+2)/2} t^{d(d+1)/2},
\end{aligned}$$

where $c_{d,\nu}$ is the norming constant in (9.16). This gives (9.8) for positive constants a_4, a_5, b_4, C_3.

If $\Psi \neq I$, then the random matrix $\Omega = \Psi^{1/2}\Sigma\Psi^{1/2}$ follows the inverse-Wishart distribution with identity scale matrix, and hence satisfies (9.6) and (9.7) by the special case. Since $\mathrm{eig}_d(\Sigma^{-1}) = \|\Sigma^{-1}\| = \|\Psi^{1/2}\Omega^{-1}\Psi^{1/2}\| \leq \|\Psi\|\,\|\Omega^{-1}\|$, it follows that (9.6) holds for Σ as well, with a different constant C_2. The smallest eigenvalue satisfies $\mathrm{eig}_1(\Sigma^{-1}) = \|\Sigma\|^{-1} \geq \|\Psi^{-1}\|^{-1}\|\Omega\|^{-1} = \|\Psi^{-1}\|^{-1}\mathrm{eig}_1(\Omega^{-1})$. Hence condition (9.7) follows again from the special case, with a different choice of b_3. Finally, to check (9.8), it suffices to show that the joint density of the eigenvalues admits a lower bound of the form (9.16). The joint density is proportional to (cf. Muirhead (1982), page 106) by

$$\prod_{j=1}^{d} s_j^{(\nu+1-d)/2} \prod_{j<k} (s_k - s_j) \int_{\mathcal{O}(d)} e^{-\nu \mathrm{tr}(\Psi^{-1} H \Delta(s) H^\top/2)}\, dH,$$

where $\Delta(s) = \mathrm{diag}(s_1,\ldots,s_d)$ and $\mathcal{O}(d)$ is the group of $d \times d$ orthogonal matrices. Here $\mathrm{tr}(\Psi^{-1/2} H\Delta(s)H^\top \Psi^{-1/2})$ increases if we replace $\Delta(s)$ by $s_d I$, whence this expression is bounded above by $s_d \mathrm{tr}(\Psi^{-1/2} H H^\top \Psi^{-1/2}) = s_d \mathrm{tr}(\Psi^{-1})$, in view of the orthogonality of H. □

9.5 Non-i.i.d. Models

In this section we present a number of applications of Theorem 8.19 in the non-i.i.d. setup. We consider various combinations of models and prior distributions.

9.5.1 Finite Sieves

In Theorem 8.24 we constructed a sequence of prior distributions using bounds for bracketing numbers such that the posterior converges at the optimal rate. We illustrate the construction for two concrete models.

Example 9.17 (Nonparametric Poisson regression) Suppose we observe $X_1,\ldots,X_n$, where $X_i \stackrel{\mathrm{ind}}{\sim} \mathrm{Poi}(\psi(z_i))$ for an an unknown increasing function $\psi: \mathbb{R} \to (0,\infty)$ and real, deterministic covariates $z_1, z_2, \ldots$. We assume that $L \leq \psi \leq U$ for some constants $0 < L < U < \infty$.

If $l \leq \psi \leq u$, then $e^{-\psi(z)}\psi(z)^x/x! \leq e^{-l(z)}u(z)^x/x!$, for all z, x. Thus for $q_{l,u,z}(x)$ defined as the right side of this inequality, the vector $(q_{l,u,z_1},\ldots,q_{l,u,z_n})$ is a componentwise upper bracket for the density of an observation with parameter ψ. For any constants $L < \lambda_1, \lambda_2, \mu_1, \mu_2 < U$,

$$\sum_{x=0}^{\infty}\Big[\Big(e^{-\lambda_1}\frac{\mu_1^x}{x!}\Big)^{1/2} - \Big(e^{-\lambda_2}\frac{\mu_2^x}{x!}\Big)^{1/2}\Big]^2 = \Big[e^{-(\lambda_1+\mu_1)/2} - e^{-(\lambda_2+\mu_2)/2}\Big]^2$$
$$+ 2e^{-(\lambda_1+\lambda_2)/2}\Big[e^{(\mu_1+\mu_2)/2} - e^{\sqrt{\mu_1\mu_2}}\Big]$$
$$\leq \Big(\frac{1}{2}+\frac{1}{4L}\Big)e^{U-L}\Big(|\lambda_1-\lambda_2|^2 + |\mu_1-\mu_2|^2\Big).$$

Therefore, for given pairs of functions $l_1 \leq u_1$ and $l_2 \leq u_2$ taking their values in the interval $[L, U]$,

$$\frac{1}{n}\sum_{i=1}^{n} d_H^2(q_{l_1,u_1,z_i}, q_{l_2,u_2,z_i}) \lesssim \int (|l_1 - l_2|^2 + |u_1 - u_2|^2)\, d\mathbb{P}_n^z,$$

for $\mathbb{P}_n^z = n^{-1}\sum_{i=1}^n \delta_{z_i}$ the empirical distribution of the covariates. Hence an ϵ-bracketing for the class of functions ψ in the $\mathbb{L}_2(\mathbb{P}_n^z)$-metric yields a componentwise Hellinger upper bracketing for the model of size a multiple of ϵ. The ϵ-bracketing entropy number in $\mathbb{L}_2(\mathbb{P}_n^z)$ of the class of all monotone, uniformly bounded functions is bounded by a multiple of ϵ^{-1}, in view of Proposition C.8. Equating this to $n\epsilon^2$, we obtain the rate $\epsilon_n = n^{-1/3}$ relative to the root mean square Hellinger distance, for the posterior distribution based on the discrete prior on a minimal set of renormalized upper brackets.

The normalization of the upper bracket $q_{l,u,z}$ is the density of the Poisson distribution with mean $u(z)$. Hence in this example the bracketing prior charges the original model, which makes its interpretation, and that of the posterior contraction rate, more transparent. As the parameter space is fixed, proceeding as in Theorem 8.15, a prior not depending on n can be constructed such that the posterior converges at the same $n^{-1/3}$ rate.

Example 9.18 (Semiparametric Poisson regression) Suppose we observe $X_1, \ldots, X_n$ for $X_i \overset{\text{ind}}{\sim} \text{Poi}(\psi(\beta^\top z_i))$, where $z_1, z_2, \ldots$ are deterministic covariates in $\mathbb{R}^d$, β is an unknown vector of the same dimension and $\psi\colon \mathbb{R} \to (0, \infty)$ is an unknown function. We assume that the function ψ is smooth, but it need not be monotone. The parameters ψ and β are not jointly identifiable, but this will not concern us as we measure distances based on the distributions of the observations rather than the parameters. Assume that the possible values of the regressors $z_1, z_2, \ldots$ and the regression coefficient β lie in a known compact set. Let ψ lie in the Hölder class $\mathfrak{C}^\alpha(\mathbb{K})$, defined in Definition C.4, with norm $\|\psi\|_{\mathfrak{C}^\alpha}$ bounded by M, for given $\alpha \geq 1$ and $M > 0$, and $\mathbb{K}$ a compact interval containing the range of all possible linear combinations $\beta^\top z_i$.

We consider the discrete prior on a minimal set of renormalized componentwise upper brackets. To construct these we bracket the set of functions $z \mapsto \psi(\beta^\top z)$ and can next use the Poisson brackets as in the preceding example. For B a bound for the norm of the regressors find an $\epsilon/(3MB)$ net $\beta_1, \beta_2, \ldots, \beta_{N_1}$ for β and an $\epsilon/3$-bracketing $(l_1, u_1), \ldots, (l_{N_2}, u_{N_2})$ for ψ and consider the collection $(l_k(\beta_j^\top z) - \epsilon/3, u_k(\beta_j^\top z) + \epsilon/3)$, $j = 1, 2, \ldots, N_1, k = 1, 2, \ldots, N_2$. It is straightforward to check that this is an ϵ-bracketing for the functions $z \mapsto \psi(\beta^\top z)$. As we can choose $\log N_1 \lesssim \log_- \epsilon$ and $\log N_2 \lesssim \epsilon^{-1/\alpha}$, the ϵ-bracketing entropy numbers for the desired class is bounded by a multiple of $\epsilon^{-1/\alpha}$. Therefore the posterior contracts at the rate $n^{-\alpha/(1+2\alpha)}$ with respect to the root mean square Hellinger distance.

9.5.2 Whittle Estimation of a Spectral Density

Consider estimating the spectral density f of a stationary Gaussian time series $(X_t: t \in \mathbb{Z})$ using the Whittle likelihood, as considered in Section 7.3.3. Thus we act as if the periodogram values $U_l = I_n(\omega_l)$, for $l = 1, \ldots, \nu$, are independent exponential variables with means $f(\omega_l)$, for $\omega_l = 2\pi l/n$ the natural frequencies, and form a posterior distribution based on the corresponding (pseudo) likelihood. By Theorem L.8, the sequence of actual distributions of the periodogram vectors $(U_1, \ldots, U_\nu)$ and their exponential approximation are contiguous. Thus a rate of contraction for this (pseudo) posterior distribution is also valid relative to the original distribution of the time series.

Because the observations $U_1, \ldots, U_n$ are independent exponential variables, rates of contraction can be obtained with the help of Theorem 8.23. The Hellinger distance and Kullback-Leibler discrepancies were computed in Section 7.3.3. If the spectral densities are bounded away from zero and infinity, then the root average square Hellinger distance d_n of the joint densities can be bounded by the distance $\bar{d}_n$ on spectral densities defined by $\bar{d}_n^2(f_1, f_2) = \nu^{-1} \sum_{l=1}^{\nu} (f_1(\omega_l) - f_2(\omega_l))^2$. Indeed if $m \le f_1, f_2 \le M$, then for $U_l \overset{\text{ind}}{\sim} P_{f,l}$, $l = 1, \ldots, \nu$,

$$\frac{1}{4M^2}\bar{d}_n^2(f_1, f_2) \le d_{n,H}^2(p_{f_1}, p_{f_2}) \le \frac{1}{4m^2}\bar{d}_n^2(f_1, f_2) \le \frac{1}{4m^2}\|f_1 - f_2\|_\infty^2,$$

$$\frac{1}{\nu}\sum_{l=1}^{\nu} K(P_{f_0,l}; P_{f,l}) \vee V_{2,0}(P_{f_0,l}; P_{f,l}) \lesssim \bar{d}_n^2(f_0, f) \lesssim \|f - f_0\|_\infty^2.$$

This shows that the entropy and prior mass conditions may be verified for the metric $\bar{d}_n$, or even the uniform distance. If, furthermore, the spectral densities are Lipschitz, then we may also use the $\mathbb{L}_2$-distance $\|\cdot\|_2$ relative to the Lebesgue measure on $[0, \pi]$. This is because $\bar{d}_n(f_1, f_2)$ is equal to $\sqrt{2\pi\nu/n}$ times the $\mathbb{L}_2$-distance $\|f_{1,n} - f_{2,n}\|_2$ between the discretizations $f_{i,n} = \sum_{l=1}^{\nu} \mathbb{1}_{(\omega_{l-1},\omega_l]} f_i(\omega_l)$ of the functions f_i, while $\|f - f_n\|_2 \lesssim \|f\|_{\text{Lip}}/n$. If all spectral densities in the sieve are Lipschitz with constant L_n and $\epsilon_n \gg L_n/n$, then we may replace d_n also by the $\mathbb{L}_2$-norm $\|\cdot\|_2$ when verifying (8.27). Similarly we may estimate the prior mass of an $\mathbb{L}_2$-ball of Lipschitz functions.

Example 9.19 (Bernstein polynomial prior for spectral density) Consider the prior on $f = \tau\bar{f}$ induced by equipping the scalar τ with a nonsingular prior density on $(0, \infty)$ and the probability density $\bar{f}$ with the Dirichlet-Bernstein prior of Example 5.10 (rescaled to the domain $[0, \pi]$), and then restricting the prior of $\tau\bar{f}$ to the set $\mathcal{F} = \{f: m < f < M\}$, for given $0 < m < M < \infty$. Let α be the base measure of the Dirichlet process, and let the order of the Bernstein polynomial possess prior $k \sim \rho$ such that $e^{-\beta_1 k \log k} \lesssim \rho(k) \lesssim e^{-\beta_2 k}$. Let Π stand for the resulting prior. Assume that $f_0 \in \mathfrak{C}^\alpha[0, \pi]$, for some $\alpha \in (0, 2]$.

If $f_0 \in \mathcal{F}$, then restricting the prior to $\mathcal{F}$ can only increase the prior probability of the set $\{f: \|f - f_0\|_\infty < \epsilon\}$, whence the restriction can be disregarded for lower bounding the prior concentration. For any k such that $\|f_0 - b(\cdot;\, f_0, k)\|_\infty < \epsilon/2$, a lower bound for the prior mass of the set $\{f: \|f - f_0\|_\infty < \epsilon\}$ is given by $\rho(k)\mathrm{DP}_\alpha(\|b(\cdot;\, f_0, k) - b(\cdot;\, F, k)\|_\infty < \epsilon/2)$. As in Section 9.3, this can be bounded below by $\rho(k)e^{-Ck\log_-\epsilon}$ for some constant $C > 0$. If $f_0 \in \mathfrak{C}^\alpha[0, \pi]$, then the uniform distance $\|f_0 - b(\cdot;\, f_0, k)\|_\infty$ is bounded by a multiple of $k^{-\alpha/2}$, by Lemma E.3. Optimizing over all k such that $\|f_0 - b(\cdot;\, f_0, k)\|_\infty < \epsilon/2$, which forces $k \gtrsim \epsilon^{-2/\alpha}$, gives a lower bound of the form $e^{-c\epsilon^{-2/\alpha}\log_-\epsilon}$, for some $c > 0$. Therefore (8.26) is satisfied for ϵ_n of the order $(n/\log n)^{-\alpha/(2+2\alpha)}$.

Let $\mathcal{F}_n \subset \mathcal{F}$ be the set of consisting of only Bernstein polynomials of order k_n or less. By similar calculations as in Section 9.3, based on the entropy of the unit simplex of dimension k_n, it can be shown that $\log D(\epsilon, \mathcal{F}_n, \|\cdot\|_\infty) \lesssim k_n \log k_n + k_n \log_- \epsilon$. Thus the entropy condition (8.27) is satisfied for $\epsilon_n \gtrsim \sqrt{k_n \log k_n}/\sqrt{n}$.

Finally the prior of the complement of the sieve satisfies $\Pi(\mathcal{F}_n^c) = \rho(k > k_n) \lesssim e^{-\beta_2 k_n}$. This is $o(e^{-Cn\epsilon_n^2})$ for an (arbitrarily) large constant C if k_n is a sufficiently large multiple of $n^{1/(1+\alpha)}(\log n)^{\alpha/(1+\alpha)}$. The posterior probability of $\mathcal{F}_n^c$ then goes to 0 by Lemma 8.20. For this choice of k_n the contraction rate on $\mathcal{F}_n$, and hence on $\mathcal{F}$, is $n^{-\alpha/(2+2\alpha)}(\log n)^{(1+2\alpha)/(2+2\alpha)}$.

This is the contraction rate relative to root average square Hellinger distance, or equivalently the discrete $\mathbb{L}_2$-distance d_n. Because the functions in $\mathcal{F}_n$ are Lipschitz with Lipschitz constant at most k_n^2, this distance differs at most of the order k_n^2/n from the ordinary $\mathbb{L}_2$-distance, uniformly in $\mathcal{F}_n$. It follows that the discrete $\mathbb{L}_2$-contraction rate carries over into a rate $k_n^2/n \vee n^{-\alpha/(2+2\alpha)}(\log n)^{(1+2\alpha)/(2+2\alpha)}$ for the ordinary $\mathbb{L}_2$-distance. For k_n as chosen previously, the term k_n^2/n dominates for all $\alpha \leq 2$, and tends to zero only for $\alpha > 1$, thus giving a slower rate for the ordinary $\mathbb{L}_2$-distance in these cases. (For $\alpha = 2$ the difference is only in the logarithmic term.)

For $f_0 \in \mathfrak{C}^\alpha[0, \pi]$, for $\alpha \in (0, 1]$, a coarsened Bernstein polynomial prior as used in Section 9.3 will lead to the posterior contraction rate $n^{-\alpha/(1+2\alpha)}(\log n)^{(1+4\alpha)/(2+4\alpha)}$, which is nearly minimax. Other priors such as those based on finite-dimensional approximation lead to a nearly optimal rate for any α; see Section 10.4.6.

9.5.3 Nonlinear Autoregression

Consider a nonlinear autoregressive stationary time series $\{X_t: t \in \mathbb{Z}\}$ given by

$$X_i = f(X_{i-1}) + \varepsilon_i, \quad i = 1, 2, \ldots, n, \tag{9.17}$$

where f is an unknown function and $\varepsilon_1, \ldots, \varepsilon_n \stackrel{\text{iid}}{\sim} \mathrm{Nor}(0, 1)$. Then X_n is a Markov chain with transition density $p_f(y|x) = \phi(y - f(x))$, where ϕ is the standard normal density. Assume that $f \in \mathcal{F}$, a class of functions such that $|f(x)| \leq M$ and $|f(x) - f(y)| \leq L|x - y|$ for all x, y and $f \in \mathcal{F}$.

Set $r(y) = (\phi(y - M) + \phi(y + M))/2$. Then $r(y) \lesssim p_f(y|x) \lesssim r(y)$ for all $x, y \in \mathbb{R}$ and $f \in \mathcal{F}$. Further, $\sup\{\int |p(y|x_1) - p(y|x_2)|\, dy: x_1, x_2 \in \mathbb{R}\} < 2$. Hence the chain is α-mixing with exponentially decaying mixing coefficients by the discussion following Theorem 8.29, has a unique stationary distribution Q_f whose density q_f satisfies $r \lesssim q_f \lesssim r$. Let $\|f\|_s = (\int |f|^s dr)^{1/s}$.

Because $d_H^2(\mathrm{Nor}(\mu_1, 1), \mathrm{Nor}(\mu_2, 1)) = 2[1 - e^{-|\mu_1-\mu_2|^2/8}]$ (see Problem B.1), it easily follows that for $f_1, f_2 \in \mathcal{F}$, d defined in (8.35) and $d\nu = r\,d\lambda$ that $\|f_1 - f_2\|_2 \lesssim d(f_1, f_2) \lesssim \|f_1 - f_2\|_2$. Thus we may verify (8.37) relative to the $\mathbb{L}_2(r)$-metric. It can also be computed that

$$P_{f_0} \log \frac{p_{f_0}(X_2|X_1)}{p_f(X_2|X_1)} = \frac{1}{2}\int (f_0 - f)^2 q_{f_0}\, d\lambda \lesssim \|f - f_0\|_2^2.$$

$$P_{f_0} \left|\log \frac{p_{f_0}(X_2|X_1)}{p_f(X_2|X_1)}\right|^s \lesssim \int |f_0 - f|^s q_{f_0}\, d\lambda \lesssim \|f - f_0\|_s^s.$$

Thus $B^*(f_0, \epsilon; s) \supset \{f : \|f - f_0\|_s \le c\epsilon\}$ for some constant $c > 0$, where $B^*(f_0, \epsilon; s)$ is as in Theorem 8.29. Thus it suffices to verify (8.36) with $s > 2$.

Example 9.20 (Random histograms) As a prior on the regression functions f, consider a random histogram as follows. For a given number $K \in \mathbb{N}$, partition a given compact interval in $\mathbb{R}$ into K subintervals $I_1, \ldots, I_K$ and let $I_0 = (\cup_k I_k)^c$. Let the prior Π_n on f be induced by the map $\alpha \mapsto f_\alpha$ given by $f_\alpha = \sum_{k=1}^K \alpha_k \mathbb{1}_{I_k}$, where $\alpha = (\alpha_1, \ldots, \alpha_K) \in \mathbb{R}^K$, and a priori $\alpha_j \overset{\text{iid}}{\sim} \mathrm{Unif}[-M, M]$, $j = 1, \ldots, K$, and $K = K_n$ is to be chosen later. Let $r(I_k) = \int_{I_k} r\, d\lambda$.

The support of Π_n consists of all functions with values in $[-M, M]$ that are piecewise constant on each interval I_k for $k = 1, \ldots, K$, and vanish on I_0. For any pair f_α and f_β of such functions we have, for any $s \in [2, \infty]$, $\|f_\alpha - f_\beta\|_s = \|\alpha - \beta\|_s$, where $\|\alpha\|_s$ is the r-weighted ℓ_s-norm of α given by $\|\alpha\|_s^s = \sum_k |\alpha_k|^s r(I_k)$. The dual use of $\|\cdot\|_s$ should not lead to any confusion, as it will be clear from the context whether $\|\cdot\|_s$ is a norm on functions or that on vectors. The $\mathbb{L}_2(r)$-projection of f_0 onto this support is the function f_{α_0} for $\alpha_{0,k} = \int_{I_k} f_0 r\, d\lambda / r(I_k)$, whence, by Pythagoras' theorem, $\|f_\alpha - f_0\|_2^2 = \|f_\alpha - f_{\alpha_0}\|_2^2 + \|f_{\alpha_0} - f_0\|_2^2$, for any $\alpha \in [-M, M]^K$. In particular, $\|f_\alpha - f_0\|_2 \ge c\|\alpha - \alpha_0\|_2$ for some constant c, and hence, with $\mathcal{F}_n$ denoting the support of Π_n,

$$\begin{aligned} &N(\epsilon, \{f \in \mathcal{F}_n : \|f - f_0\|_2 \le 16\epsilon\}, \|\cdot\|_2) \\ &\quad \le N(\epsilon, \{\alpha \in \mathbb{R}^K : \|\alpha - \alpha_0\|_2 \le 16c\epsilon\}, \|\cdot\|_2) \le (80c)^K \end{aligned}$$

by Proposition C.1. Thus (8.37) holds if $n\epsilon_n^2 \gtrsim K$.

To verify (8.36), note that, for $\lambda = (\lambda(I_1), \ldots, \lambda(I_K))$

$$\|f_{\alpha_0} - f_0\|_s^s = \int_{I_0} |f_0|^s r\, d\lambda + \sum_{k=1}^K \int_{I_k} |\alpha_{0,k} - f_0|^s r\, d\lambda \le M^s r(I_0) + L^s \|\lambda\|_s^s.$$

Hence as $f_0 \in \mathcal{F}$, for every $\alpha \in [-M, M]^K$,

$$\|f_\alpha - f_0\|_s \lesssim \|\alpha - \alpha_0\|_s + r(I_0)^{1/s} + \|\lambda\|_s \le \|\alpha - \alpha_0\|_\infty + r(I_0)^{1/s} + \|\lambda\|_s,$$

where $\|\cdot\|_\infty$ is the ordinary maximum norm on $\mathbb{R}^K$. For $r(I_0)^{1/s} + \|\lambda\|_s \le \epsilon/2$, we have that $\{f : \|f - f_0\|_s \le \epsilon\} \supset \{f_\alpha : \|\alpha - \alpha_0\|_\infty \le \epsilon/2\}$. Using $\|\alpha - \alpha_0\|_2 \le c\|f_\alpha - f_0\|_2$, for any $\epsilon > 0$ such that $r(I_0)^{1/s} + \|\lambda\|_s \le \epsilon/2$,

$$\frac{\Pi_n(f : \|f - f_0\|_2 \le j\epsilon)}{\Pi_n(f : \|f - f_0\|_s \le \epsilon)} \le \frac{\Pi_n(\alpha : \|\alpha - \alpha_0\|_2 \le j\epsilon)}{\Pi_n(\alpha : \|\alpha - \alpha_0\|_\infty \le \epsilon c/2)}.$$

We show that the right-hand side is bounded by $e^{Cn\epsilon^2/8}$ for some C.

For $\{I_1, \ldots, I_K\}$ a regular partition of an interval $[-A, A]$, we have that $\|\lambda\|_s = 2A/K$ and, since $r(I_k) \geq \lambda(I_k) \inf\{r(x): x \in I_k\}$, $k \geq 1$, the norm $\|\cdot\|_2$ is bounded below by the Euclidean norm multiplied by $\sqrt{2A\phi(A)/K} \gtrsim \sqrt{\phi(A)/K}$. In this case, the preceding display is bounded above by

$$\frac{(Cj\epsilon\sqrt{K/\phi(A)}/(2M))^K \operatorname{vol}\{x \in \mathbb{R}^K: \|x\| \leq 1\}}{(\epsilon c/(4M))^K} \sim \left(\frac{2Cj\sqrt{2\pi e}}{c\sqrt{\phi(A)}}\right)^K \frac{1}{\sqrt{\pi K}},$$

by Lemma K.13. The probability $r(I_0)$ is bounded above by $1 - 2\Phi(A) \lesssim \phi(A)$. Hence (8.36) will hold if $K \log_- \phi(A) \lesssim n\epsilon_n^2$, $\phi(A) \lesssim \epsilon_n^s$, and $A/K \lesssim \epsilon_n$. With $K \asymp \epsilon_n^{-1}(\log_- \epsilon_n)^{1/2}$ and $A \asymp (\log_- \epsilon_n)^{1/2}$, all conditions are met for ϵ_n a sufficiently large multiple of $n^{-1/3}(\log n)^{1/2}$. This is only marginally weaker than the minimax rate which is $n^{-1/3}$ for this problem provided that the autoregressive functions are assumed to be only Lipschitz continuous.

The logarithmic factor in the contraction rate appears to be a consequence of the fact that the regression functions are defined on the full real line. The present prior is a special case of a spline-based prior to be discussed in Subsection 9.5.5. If f has smoothness beyond Lipschitz continuity, then the use of higher order splines should yield a faster contraction rate.

9.5.4 White Noise with Conjugate Priors

The observation in the white noise model is an infinite-dimensional vector $X^{(n)} = (X_{n,1}, X_{n,2}, \ldots)$, where $X_{n,i} \stackrel{\text{ind}}{\sim} \text{Nor}(\theta_i, n^{-1})$, and the prior Π_n is given by $\theta_i \stackrel{\text{ind}}{\sim} \text{Nor}(0, \sigma_{i,k}^2)$, $i = 1, \ldots, k$, and $\theta_{k+1} = \theta_{k+2} = \cdots = 0$, where $k = k_n = \lfloor n^{1/(2\alpha+1)} \rfloor$ for some $\alpha > 0$. The posterior distribution and rates of contraction can be computed simply and explicitly, as shown Example 8.6, but as illustration we derive the rate in this section from the general theorem in Section 8.3.4.

Assume that

$$\min\{\sigma_{i,k}^2 i^{2\alpha}: 1 \leq i \leq k\} \sim k^{-1}. \tag{9.18}$$

This is valid if $\sigma_{i,k}^2 = k^{-1}$, for $i = 1, \ldots, k$, but also if $\sigma_{i,k}^2 = i^{-(2\alpha+1)}$, for $i = 1, \ldots, k$. Even though quite different both priors give the optimal contraction rate if the true parameter θ_0 is "α-regular."

Theorem 9.21 *If* (9.18) *holds, then the posterior contracts at the rate* $\epsilon_n = n^{-\alpha/(2\alpha+1)}$ *for any* θ_0 *such that* $\sum_{i=1}^{\infty} \theta_{0,i}^2 i^{2\alpha} < \infty$.

Proof The support Θ_n of the prior is the set of all $\theta \in \ell_2$ with $\theta_i = 0$ for $i > k$, and can be identified with $\mathbb{R}^k$. Moreover, the ℓ_2-norm $\|\cdot\|$ on the support can be identified with the Euclidean norm on $\mathbb{R}^k$, temporarily to be denoted by $\|\cdot\|_k$ to make the dependence on k explicit. Let $B_k(x, \epsilon)$ denote the k-dimensional Euclidean balls of radius ϵ and $x \in \mathbb{R}^k$. For any true parameter $\theta_0 \in \ell_2$ we have $\|\theta - \theta_0\| \geq \|\operatorname{Proj}\theta - \operatorname{Proj}\theta_0\|_k$, where Proj is the projection operator on Θ_n and hence

$$N(\epsilon/8, \{\theta \in \Theta_n: \|\theta - \theta_0\| \le \epsilon\}, \|\cdot\|) \le N(\epsilon/8, B_k(\operatorname{Proj}\theta_0, \epsilon), \|\cdot\|_k) \le (40)^k$$

in view of Proposition C.1. It follows that (8.41) is satisfied if $n\epsilon_n^2 \gtrsim k$, equivalently, if $\epsilon_n \gtrsim n^{-\alpha/(2\alpha+1)}$.

By Pythagoras's theorem we have that $\|\theta - \theta_0\|^2 = \|\operatorname{Proj}\theta - \operatorname{Proj}\theta_0\|^2 + \sum_{i>k}\theta_{0,i}^2$ for any $\theta \in \operatorname{supp}(\Pi_n)$. Hence for $\sum_{i>k}\theta_{0,i}^2 \le \epsilon_n^2/2$ we have that

$$\Pi_n(\theta \in \Theta_n: \|\theta - \theta_0\| \le \epsilon_n) \ge \Pi_n(\theta \in \mathbb{R}^k: \|\theta - \operatorname{Proj}\theta_0\|_k \le \epsilon_n/2).$$

By the definition of the prior, the right-hand side involves a quadratic form in Gaussian variables. Set $\Sigma = \operatorname{diag}(\sigma_{i,k}^2: i = 1, \ldots, k)$ and let Φ_k refer to the probability content of a k-dimensional normal distribution. The quotient on the left-hand side of (8.40) can be bounded as

$$\frac{\Pi_n(\theta \in \Theta_n: \|\theta - \theta_0\| \le j\epsilon_n)}{\Pi_n(\theta \in \Theta_n: \|\theta - \theta_0\| \le \epsilon_n)} \le \frac{\Phi_k(-\operatorname{Proj}\theta_0, \Sigma)(B(0, j\epsilon_n))}{\Phi_k(-\operatorname{Proj}\theta_0, \Sigma)(B(0, \epsilon_n/2))}.$$

The probability in the numerator increases if we center the normal distribution at 0 rather than at $-\operatorname{Proj}\theta_0$, by Anderson's lemma (Lemma K.12). Furthermore, for any $\mu \in \mathbb{R}^k$, the normal densities satisfy

$$\frac{\phi_{\mu,\Sigma}(\theta)}{\phi_{0,\Sigma/2}(\theta)} = \frac{e^{-\sum_{i=1}^k(\theta_i-\mu_i)^2/(2\sigma_{i,k}^2)}}{\sqrt{2}^k e^{-\sum_{i=1}^k \theta_i^2/\sigma_{i,k}^2}} \ge 2^{-k/2}\exp\Big(-\sum_{i=1}^k \frac{\mu_i^2}{\sigma_{i,k}^2}\Big).$$

Therefore, we may recenter the denominator at 0 at the cost of adding the factor on the right, with $\mu = \theta_0$, and dividing the covariance matrix by 2. We obtain that the left-hand side of (8.40) is bounded above by

$$\begin{aligned} &2^{k/2} e^{\sum_{i=1}^k \theta_{0,i}^2/\sigma_{i,k}^2} \frac{\Phi_k(0, \Sigma)(B(0, j\epsilon_n))}{\Phi_k(0, \Sigma/2)(B(0, \epsilon_n/2))} \\ &\le 2^{k/2}\exp\Big(\sum_{i=1}^k \frac{\theta_{0,i}^2}{\sigma_{i,k}^2}\Big)\Big(\frac{\bar\sigma_k}{\underline{\sigma}_k}\Big)^k \frac{\Phi_k(0, \bar\sigma_k^2 I)(B(0, j\epsilon_n))}{\Phi_k(0, \underline{\sigma}_k^2 I/2)(B(0, \epsilon_n/2))}, \end{aligned}$$

where $\bar\sigma_k$ and $\underline{\sigma}_k$ denote the maximum and the minimum of $\sigma_{i,k}$ for $i = 1, 2, \ldots, k$. The probabilities on the right are left-tail probabilities of chi-square distributions with k degrees of freedom, and can be expressed as integrals. The preceding display is bounded above by

$$2^{k/2}\exp\Big\{\sum_{i=1}^k \theta_{0,i}^2/\sigma_{i,k}^2\Big\}\Big(\frac{\bar\sigma_k}{\underline{\sigma}_k}\Big)^k \frac{\int_0^{j^2\epsilon_n^2/\bar\sigma_k^2} x^{k/2-1}e^{-x/2}\,dx}{\int_0^{\epsilon_n^2/(2\underline{\sigma}_k^2)} x^{k/2-1}e^{-x/2}\,dx}. \tag{9.19}$$

The exponential in the integral in the numerator is bounded by 1 and hence this integral is bounded above by $j^k\epsilon_n^k/(k\bar\sigma_k^k)$.

Now consider two separate cases.

(i) If $\epsilon_n^2/\underline{\sigma}_k^2$ remains bounded, then the exponential in the integral in the denominator of the expression in (9.19) is bounded below by a constant, and we have that the expression in (9.19) is bounded above by a multiple of $4^k j^k \exp(\sum_{i=1}^k \theta_{0,i}^2/\sigma_{i,k}^2)$.

(ii) If $\epsilon_n^2/\underline{\sigma}_k^2 \to \infty$, then we bound the integral in the denominator of the expression in (9.19) by $(\eta/2)^{k/2-1}\int_{\eta/2}^{\eta} e^{-x/2}\,dx$ for $\eta = \epsilon_n^2/(2\underline{\sigma}_k^2)$. This leads to the upper bound a multiple of $8^k j^k \exp\big(\sum_{i=1}^k \theta_{0,i}^2\sigma_{i,k}^{-2}\big)\,\epsilon_n^2\underline{\sigma}_k^{-2}\exp\big(\epsilon_n^2\underline{\sigma}_k^{-2}/8\big)$.

By assumption (9.18) we have that $\underline{\sigma}_k^2 \gtrsim k^{-(2\alpha+1)} \asymp n^{-1}$. We also have that $k \asymp n\epsilon_n^2$. It follows that $\epsilon_n^2/\underline{\sigma}_k^2 \lesssim n\epsilon_n^2$, and $\underline{\sigma}_k^{-2}$ is bounded by a polynomial in k. Thus (8.40) holds if ϵ_n satisfies $\sum_{i=1}^k \theta_{0,i}^2/\sigma_{i,k}^2 \lesssim n\epsilon_n^2$, and $\sum_{i>k}\theta_{0,i}^2 \leq \epsilon_n^2/2$. Since $\sum_{i=1}^{\infty}\theta_{0,i}^2 i^{2\alpha} < \infty$, then for ϵ_n a sufficiently large multiple of $n^{-\alpha/(2\alpha+1)}$, all required conditions hold, proving the theorem. □

9.5.5 Nonparametric Regression Using Splines

Consider the nonparametric regression model, where we observe independent random variables $X_1, \ldots, X_n$ distributed as $X_i = f(z_i) + \varepsilon_i$ for an unknown regression function $f\colon [0,1] \to \mathbb{R}$, deterministic covariates $z_1, \ldots, z_n$ in $[0,1]$, and $\varepsilon_i \overset{\text{iid}}{\sim} \text{Nor}(0, \sigma^2)$, for $i = 1, \ldots, n$. For simplicity, we assume that the error variance σ^2 is known. Let $\mathbb{P}_n^z = n^{-1}\sum_{i=1}^n \delta_{z_i}$ be the empirical measure of the covariates, and let $\|\cdot\|_{2,n}$ denote the norm of $\mathbb{L}_2(\mathbb{P}_n^z)$.

Assume that the true regression function f_0 belongs to the unit ball of a Hölder class $\mathfrak{C}^\alpha[0,1]$ defined in Definition C.4 for some $\alpha \geq \frac{1}{2}$, and without loss of generality by rescaling. We shall construct priors based on a spline series representation like in Section E.2, namely, $f_\beta(z) = \beta^\top B(z)$ and induce a prior on f from a prior on $\beta = (\beta_1, \ldots, \beta_J)$, for instance by $\beta_j \overset{\text{iid}}{\sim} \text{Nor}(0,1)$, $j = 1, \ldots, J$.[1]

We need the regressors $z_1, z_2, \ldots, z_n$ to be sufficiently regularly distributed in the interval $[0,1]$. In view of the spatial separation property of the B-spline functions, the precise condition can be expressed in the covariance matrix $\Sigma_n = ((\int B_i B_j\, d\mathbb{P}_n^z))$, namely

$$J^{-1}\|\beta\|^2 \lesssim \beta^\top \Sigma_n \beta \lesssim J^{-1}\|\beta\|^2, \tag{9.20}$$

where $\|\cdot\|$ is the Euclidean norm on $\mathbb{R}^J$.

Under condition (9.20) we have that, for all $\beta_1, \beta_2 \in \mathbb{R}^J$,

$$C\|\beta_1 - \beta_2\| \leq \sqrt{J}\|f_{\beta_1} - f_{\beta_2}\|_{2,n} \leq C'\|\beta_1 - \beta_2\|, \tag{9.21}$$

for some constants C and C'. This enables us to perform all calculations in terms of the Euclidean norms on the spline coefficients.

Theorem 9.22 *If $J = J_n \asymp n^{1/(1+2\alpha)}$, then the posterior contracts at the minimax rate $n^{-\alpha/(1+2\alpha)}$ relative to $\|\cdot\|_{2,n}$.*

Proof We verify the conditions of Theorem 8.23. Let f_{β_n} be the $\mathbb{L}_2(\mathbb{P}_n^z)$-projection of f_0 onto the J-dimensional space of splines $f_\beta = \beta^\top B$. Then $\|f_{\beta_n} - f_\beta\|_{2,n} \leq \|f_0 - f_\beta\|_{2,n}$ for every $\beta \in \mathbb{R}^J$ and hence, by (9.21), for every $\epsilon > 0$, we have $\{\beta\colon \|f_\beta - f_0\|_{2,n} \leq \epsilon\} \subset \{\beta\colon \|\beta - \beta_n\| \leq C'\sqrt{J}\epsilon\}$. It follows that the ϵ-covering numbers of the set $\{f_\beta\colon \|f_\beta -$

[1] Unlike the case for densities, in the present situation a restriction to a compact interval is unnecessary.

$f_0\|_{2,n} \le \epsilon\}$ for $\|\cdot\|_{2,n}$ are bounded by the $C\sqrt{J}\epsilon$-covering numbers of a Euclidean ball of radius $C'\sqrt{J}\epsilon$, which are of the order D^J for some constant D. Thus the entropy condition (8.27) is satisfied provided that $J \lesssim n\epsilon_n^2$.

By the projection property, with β_∞ as in Lemma E.5,

$$\|f_{\beta_n} - f_0\|_{2,n} \le \|f_{\beta_\infty} - f_0\|_{2,n} \le \|f_{\beta_\infty} - f_0\|_\infty \lesssim J^{-\alpha}. \tag{9.22}$$

Combination with (9.21) shows that there exists a constant C'' such that for every $\epsilon \gtrsim 2J^{-\alpha}$, $\{\beta: \|f_\beta - f_0\|_n \le \epsilon\} \supset \{\beta: \|\beta - \beta_n\| \le C''\sqrt{J}\epsilon\}$. Thus

$$\frac{\Pi_n(f: \|f - f_0\|_n \le j\epsilon)}{\Pi_n(f: \|f - f_0\|_n \le \epsilon)} \le \frac{\Phi_J(0,I)(\beta: \|\beta - \beta_n\| \le C'j\sqrt{J}\epsilon)}{\Phi_J(0,I)(\beta: \|\beta - \beta_n\| \le C''\sqrt{J}\epsilon)}$$
$$\le \frac{\Phi_J(0,I)(\beta: \|\beta\| \le C'j\sqrt{J}\epsilon)}{2^{-J/2}e^{-\|\beta_n\|^2}\Phi_J(0,I)(\beta: \|\beta\| \le C''\sqrt{J}\epsilon/\sqrt{2})},$$

where Φ_J refers to the normal probability content in $\mathbb{R}^J$. This follows since in the last step, the numerator increases if we replace the centering β_n by the origin in view of Lemma K.12, whereas the normal densities satisfy

$$\frac{\phi_{\beta_n,I}(\beta)}{\phi_{0,I/2}(\beta)} = \frac{e^{-\|\beta-\beta_n\|^2/2}}{2^{J/2}e^{-\|\beta\|^2}} \ge 2^{-J/2}e^{-\|\beta_n\|^2}.$$

Here, by the triangle inequality, (9.21) and (9.22), we have that $\|\beta_n\| \lesssim \sqrt{J}\|f_{\beta_n}\|_{2,n} \lesssim \sqrt{J}(J^{-\alpha} + \|f_0\|_\infty) \lesssim \sqrt{J}$. Furthermore, the two Gaussian probabilities are left tail probabilities of the chi-square distribution with J degrees of freedom. The quotient can be evaluated as

$$2^{J/2}e^{\|\beta_n\|^2}\frac{\int_0^{(C')^2j^2J\epsilon^2} x^{J/2-1}e^{-x/2}\,dx}{\int_0^{(C'')^2J\epsilon^2/2} x^{J/2-1}e^{-x/2}\,dx}.$$

This is bounded above by $(Cj)^J$ for some constant C, by rescaling the domain of integration in the numerator to match with the denominator and using the fact that $\|\beta_n\| \le \sqrt{J}\|\beta\|_\infty = O(\sqrt{J})$. Hence to satisfy (8.26) it suffices again that $n\epsilon_n^2 \gtrsim J$.

We conclude the proof by choosing $J = J_n \asymp n^{1/(1+2\alpha)}$. □

Nonparametric Regression Using Orthonormal Series

The arguments in the preceding subsection use the special nature of the B-spline basis only through the approximation result Lemma E.5 and the comparison of norms (9.21). Theorem 9.22 thus may be extended to many other possible bases. For instance, we can use a sequence of orthonormal bases with good approximation properties for a given class of regression functions f_0. Then (9.20) should be replaced by

$$\|\beta_1 - \beta_2\| \lesssim \|f_{\beta_1} - f_{\beta_2}\|_{2,n} \lesssim \|\beta_1 - \beta_2\|. \tag{9.23}$$

This is trivially true if the bases are orthonormal in $\mathbb{L}_2(\mathbb{P}_n^z)$, such as the discrete wavelet bases relative to the design points. (Then the basis functions must necessarily change with the design points $z_1, \dots, z_n$.)

9.5.6 Binary Nonparametric Regression with a Dirichlet Process Prior

We revisit the example in Section 7.4.3 and obtain the contraction rate. The observations are independent Bernoulli variables $Y_1, \dots, Y_n$ with $Y_i \overset{\text{ind}}{\sim} \text{Bin}(1, H(x_i))$ for deterministic, known real-valued covariates $x_1, \dots, x_n$ and an unknown, monotone function $H: \mathbb{R} \to (0, 1)$. The prior is constructed as $H(x) = F(\alpha + \beta x)$, where $F \sim \text{DP}(\gamma)$ is independent of (α, β), which receives a prior on $\mathbb{R} \times (0, \infty)$. Thus H possesses a mixture of Dirichlet process prior, and given (α, β), follows the Dirichlet process prior with base measure $\gamma((\cdot - \alpha)/\beta)$.

Assume that the true function H_0 is continuous and that $x_1, \dots, x_n$ lie in an interval $[a, b]$ strictly within its support. Furthermore assume that γ is absolutely continuous with a positive, continuous density on its support.

Theorem 9.23 *If there exists a compact set $\mathbb{K}$ inside the support of the prior for (α, β) such that the support of the base measure $\gamma((\cdot - \alpha)/\beta)$ strictly contains the interval $[a, b]$ for every $(\alpha, \beta) \in \mathbb{K}$, then the posterior distribution of H contracts at the rate $n^{-1/3}(\log n)^{1/3}$ with respect to the root mean square Hellinger distance d_n.*

Proof Let $\mathbb{P}_n$ be the empirical distribution of $x_1, \dots, x_n$. In view of Problem B.2,

$$d_{n,H}^2(H_1, H_2) \le \int |H_1^{1/2} - H_2^{1/2}|^2 \, d\mathbb{P}_n + \int |(1 - H_1)^{1/2} - (1 - H_2)^{1/2}|^2 \, d\mathbb{P}_n.$$

Both the class of all functions $H^{1/2}$ and the class of all functions $(1 - H)^{1/2}$ when H ranges over all cumulative distribution functions, possess ϵ-entropy bounded by a multiple of $1/\epsilon$ by Proposition C.1. Thus any $\epsilon_n \gtrsim n^{-1/3}$ satisfies (8.27).

By Problem B.2 again, for K_i and $V_{2;i}$ the Kullback-Leibler divergence and variation for the distributions of Y_i,

$$K_i(H_0; H) = H_0(x_i) \log \frac{H_0(x_i)}{H(x_i)} + (1 - H_0(x_i)) \log \frac{1 - H_0(x_i)}{1 - H(x_i)},$$
$$V_i(H_0; H) = H_0(x_i) \log^2 \frac{H_0(x_i)}{H(x_i)} + (1 - H_0(x_i)) \log^2 \frac{1 - H_0(x_i)}{1 - H(x_i)}.$$

By assumption, the numbers $H_0(x_i)$ are bounded away from 0 and 1. Therefore, by a Taylor expansion it can be seen that, with $\|H - H_0\|_\infty = \sup\{|H(z) - H_0(z)|: z \in [a, b]\}$,

$$\frac{1}{n} \sum_{i=1}^{n} K_i(H_0; H) \vee V_{2;i}(H_0; H) \lesssim \|H - H_0\|_\infty^2.$$

Hence, in order to verify (8.26), it suffices to lower bound the prior probability of the set $\{H: \|H - H_0\|_\infty \le \epsilon\}$.

For a given $\epsilon > 0$, partition the line into $N \lesssim \epsilon^{-1}$ intervals $E_1, \dots, E_N$ such that $H_0(E_j) \le \epsilon$ and such that $A\epsilon \le \gamma((E_j - \alpha)/\beta) \le 1$ for all $j = 1, \dots, N$ and $(\alpha, \beta) \in \mathbb{K}$, for some $A > 0$. Existence of such a partition follows from the continuity of H_0 and absolute continuity of γ. It can be seen that $\|H - H_0\|_\infty \lesssim \epsilon$ for every H such that $\sum_{j=1}^N |H(E_j) - H_0(E_j)| \le \epsilon$. To compute the prior probability of $\sum_{j=1}^N |H(E_j) - H_0(E_j)| \le \epsilon$, first condition on a fixed value of (α, β) in $\mathbb{K}$. By

Lemma G.13, the prior probability is at least $\exp(-c\epsilon^{-1}\log_- \epsilon)$ for some constant c. A uniform estimate works for all $(\alpha, \beta) \in \mathbb{K}$. Hence (8.26) holds for ϵ_n the solution of $n\epsilon^2 = \epsilon^{-1}\log_- \epsilon$, or $\epsilon_n = n^{-1/3}(\log n)^{1/3}$. ☐

9.5.7 Interval Censoring Using a Dirichlet Process Prior

Let $T_1, T_2, \ldots, T_n$ be an i.i.d. sample from a life distribution F on $(0, \infty)$. However, the observations are subject to interval censoring by deterministic intervals $(l_1, u_1), \ldots, (l_n, u_n)$. Along with the values of these interval end points, we observe $(\delta_1, \eta_1), \ldots, (\delta_n, \eta_n)$, where $\delta_i = \mathbb{1}\{T_i \le l_i\}$ and $\eta_i = \mathbb{1}\{l_i < T_i < u_i\}$, $i = 1, 2, \ldots, n$. The observations (δ_i, η_i), $i = 1, 2, \ldots, n$ are therefore i.n.i.d. multinomial with probability mass function

$$(F(l_i))^{\delta_i}(F(u_i) - F(l_i))^{\eta_i}(1 - F(u_i))^{1-\delta_i-\eta_i}. \tag{9.24}$$

Theorem 9.24 *Consider a prior $F \sim \mathrm{DP}(\alpha)$. Let the true distribution function F_0 be continuous. Assume that $(l_1, u_1), \ldots, (l_n, u_n)$ lie in an interval $[a, b]$ where $F_0(a-) > 0$ and $F_0(b) < 1$ and the density of α is positive and continuous on $[a, b]$. Then the posterior contracts at the rate $n^{-1/3}(\log n)^{1/3}$ with respect to the root average square Hellinger distance d_n.*[2]

Proof If F_1 and F_2 are two life distributions, then the squared Hellinger distance between the distributions of (δ_i, η_i) under F_1 and F_2 is given by

$$\begin{aligned}(\sqrt{F_1(l_i)} - \sqrt{F_2(l_i)})^2 + (\sqrt{F_1(u_i) - F_1(l_i)} - \sqrt{F_2(u_i) - F_2(l_i)})^2 \\ + (\sqrt{1 - F_1(u_i)} - \sqrt{1 - F_2(u_i)})^2,\end{aligned}$$

and hence $d_n^2(F_1, F_2)$ is given by

$$\begin{aligned}\int (\sqrt{F_1(l)} - \sqrt{F_2(l)})^2\, d\mathbb{P}_n(u, l) + \int (\sqrt{F_1(u) - F_1(l)} - \sqrt{F_2(u) - F_2(l)})^2\, d\mathbb{P}_n(u, l) \\ + \int (\sqrt{1 - F_1(u)} - \sqrt{1 - F_2(u)})^2\, d\mathbb{P}_n(u, l),\end{aligned} \tag{9.25}$$

where $\mathbb{P}_n$ stands for the empirical distribution of $(l_1, u_1), \ldots, (l_n, u_n)$. Therefore, to bound the ϵ-entropy with respect to d_n, it suffices to bound the $\mathbb{L}_2(\mathbb{P}_n)$ entropies of the classes $(l, u) \mapsto (F(l))^{1/2}$, $(l, u) \mapsto (F(u) - F(l))^{1/2}$ and $(l, u) \mapsto (1 - F(u))^{1/2}$. As in Section 9.5.6, it follows from Proposition C.1 that the first and the third are of the order ϵ^{-1}. It therefore suffices to bound the second.

Let $\mathbb{H}_n = (2n)^{-1}\sum_{i=1}^n (\delta_{l_i} + \delta_{u_i})$. Let $(\underline{H}_1, \overline{H}_1), \ldots, (\underline{H}_N, \overline{H}_N)$ be an ϵ-bracketing for the class of life distributions with respect to $\mathbb{L}_1(\mathbb{H}_n)$. By Proposition C.1 again, we may choose $\log N \lesssim \epsilon^{-1}$. We shall show that for any bracket $(\underline{H}, \overline{H})$ any distributions F_1, F_2 enclosed by the bracket, the second term of (9.25) is at most 2ϵ. Then it will follow that the

[2] If the censoring intervals are stochastic and are enclosed inside $[a, b]$ with $0 < F_0(a-) < F_0(b) < 1$, then by similar arguments it also follows that the posterior contracts at the rate $n^{-1/3}(\log n)^{1/3}$ for the Hellinger distance.

ϵ-entropy of the class of all life distributions for the metric d_n is bounded by a multiple of ϵ^{-1}. Hence (8.27) will hold for $\epsilon_n = n^{-1/3}$.

Without loss of generality, we may assume that $\underline{H}$ and $\overline{H}$ are monotone increasing and $0 \leq \underline{H} \leq \overline{H} \leq 1$. Take any two distributions F_1, F_2 enclosed by $(\underline{H}, \overline{H})$. Then for $j = 1, 2$, $l \leq u$, we have $(\underline{H}(u) - \overline{H}(l))_+ \leq F_j(u) - F_j(l) \leq \overline{H}(u) - \underline{H}(l)$, and hence using $(a-b)^2 \leq a^2 - b^2$ for $a \geq b \geq 0$, we obtain

$$\Big(\sqrt{F_1(u) - F_1(l)} - \sqrt{F_2(u) - F_2(l)}\Big)^2 \leq (\overline{H}(u) - \underline{H}(l)) - (\underline{H}(u) - \overline{H}(l))_+. \quad (9.26)$$

If $\underline{H}(u) \geq \overline{H}(l)$, then the right-hand side of (9.26) is equal to $(\overline{H}(u) - \underline{H}(u)) + (\overline{H}(l) - \underline{H}(l))$. If $\underline{H}(u) < \overline{H}(l)$, then the right-hand side of (9.26) is equal to

$$\begin{aligned}\overline{H}(u) - \underline{H}(l) &= (\overline{H}(u) - \underline{H}(u)) + (\underline{H}(u) - \overline{H}(l)) + (\overline{H}(l) - \underline{H}(l)) \\ &\leq (\overline{H}(u) - \underline{H}(u)) + (\overline{H}(l) - \underline{H}(l)).\end{aligned}$$

Therefore, the second terms in (9.25) can be bounded by $2\int(\overline{H}(x) - \underline{H}(x))\, d\mathbb{H}_n(x) < 2\epsilon$.

To estimate the prior probability in (8.26), we proceed as in Subsection 9.5.6. By similar arguments, the required probability is bounded below by the probability of $\{F\colon \|F - F_0\|_\infty \leq c\epsilon\}$ under the Dirichlet process, which is at least a multiple of $\exp[-\beta\epsilon^{-1}\log_- \epsilon]$. □

9.6 Historical Notes

Log-spline models were introduced for maximum likelihood density estimation by Stone (1990), and generalized to the multivariate case by Stone (1994). Ghosal et al. (2000) used the log-spline model to construct a prior distribution on smooth densities, as in Theorem 9.1. Example 9.7 is also from this paper. The rate of contraction for Bernstein polynomial mixtures was obtained by Ghosal (2001) for supersmooth Bernstein polynomials and for twice continuously differentiable densities. The improved contraction rate using coarsened Bernstein polynomials was derived by Kruijer and van der Vaart (2008) for smoothness level $0 < \alpha < 2$; they established adaptation to smoothness and a nearly minimax rate in the range $0 < \alpha \leq 1$. The minimax rate of estimation for analytic functions was obtained by Ibragimov and Has'minskiĭ (1982) with respect to $\mathbb{L}_p$-distances. Ghosal and van der Vaart (2001) obtained the posterior contraction rate with respect to the Hellinger distance of Dirichlet process mixtures of the location and location-scale family of univariate normals when the true density is a normal mixture. The rate matches the minimax rate up to a logarithmic factor. They introduced the moment-matching technique for entropy calculation. However the prior on σ was supported in a bounded interval, ruling out the common conjugate inverse-gamma prior. Posterior contraction rates for ordinary smooth normal mixtures were first obtained Ghosal and van der Vaart (2007b) for univariate twice continuously differentiable densities and a sequence of priors on the scale. Their result was subsequently improved in steps. Kruijer et al. (2010), obtained the (nearly) optimal rate for any smoothness level of the underlying density (i.e., adaptive posterior contraction rate) using discrete mixtures of normal densities with a random number of points, but their results did not cover the case of Dirichlet process mixtures. The primary technique used in their result was an adaptive construction of a KL-approximation in the model for the true density that improves with the level of smoothness of the true density. A construction of this type was first used by Rousseau (2010) in the context of beta mixtures. De Jonge and van Zanten (2010) constructed a similar approximation in the multivariate case for global Hölder class of functions. Finally the (adaptive) posterior contraction rate for ordinary smooth normal mixtures using Dirichlet process mixtures of normals with a general prior on scale was obtained by Shen et al. (2013). They also treated anisotropic smooth classes of multivariate densities (see Problem 9.17 below). Their construction of the adaptive KL-approximation uses only local Hölder classes, and can be applied to the log-density. They also introduced the entropy calculation based on the stick-breaking representation of the Dirichlet process. The unification of the the smooth and supersmooth cases (which in particular contains multivariate supersmooth mixtures) presented in the present chapter appears to be new. Posterior rates of contraction for the examples in Section 9.5 were obtained by Ghosal and van der Vaart (2007a).

Problems

9.1 (Ghosal and van der Vaart 2001) Let F^* be any probability measure with sub-Gaussian tails and σ^* be fixed. Show that the density $p_{F^*,\sigma^*}(x) = \int \phi_{\sigma^*}(x-z)\,dF^*(z)$ also has sub-Gaussian tails.

9.2 (Ghosal and van der Vaart 2001) Let $X_i \stackrel{\text{iid}}{\sim} p$, $i = 1, 2, \ldots$, where p is a density in $\mathbb{R}$ that can be written as (i) $p(x) = \int \phi_{\sigma_0}(x-z)\,dF(z)$, (ii) $p(x) = \int\int \phi_\sigma(x-z)\,dF(z)\,dG(\sigma)$, or (iii) $p(x) = \int \phi_\sigma(x-z)\,dH(z,\sigma)$. In model (i), let F be any probability measure supported on $[-a, a]$ where $a \lesssim (\log n)^\gamma$ and σ take any arbitrary value on the interval $[\underline{\sigma}, \overline{\sigma}]$. For model (ii), we assume that F is as above and the distribution G of σ is supported on $[\underline{\sigma}, \overline{\sigma}]$. In model (iii), the distribution H for (z, σ) is supported on $[-a, a] \times [\underline{\sigma}, \overline{\sigma}]$, where a is as above. Assume that the true F_0 has compact support in case of model (i) and (ii); for model (iii), H_0 is assumed to have compact support. Show that for a sufficiently large constant M, the MLE $\hat{p}_n$ satisfies $P_0(d(\hat{p}_n, p_0) > M\epsilon_n) \lesssim e^{-c\log^2 n}$ where $\epsilon_n = (\log n)^{\max(\gamma,\frac{1}{2})+\frac{1}{2}}/\sqrt{n}$ for models (i) and (ii) while $\epsilon_n = (\log n)^{2\max(\gamma,\frac{1}{2})+\frac{1}{2}}/\sqrt{n}$ for model (iii), and c is a constant. In particular, $\hat{p}_n$ converges to p_0 in Hellinger distance at the rate ϵ_n in P_0-probability, and a.s. $[P_0]$. (It is known that the MLE exists; see Theorem 18 of Lindsay 1995.)

9.3 (Ghosal and van der Vaart 2001) Consider the setting of Problem 9.2. For model (i), let $\hat{p}_{n,k}$ be the maximizer of the likelihood on $\{p_{F,\sigma} : F = \sum_{j=1}^k w_j \delta_{z_j}, w_j \geq 0, \sum_{j=1}^k w_j = 1, z_j \in [-a, a]\}$. Define $\hat{p}_{n,k}$ in model (ii) similarly by restricting the mixing distributions F and G to have at most k support points, while in model (iii), let H to be supported on k^2 points. Show that if $k \geq C \log n$ for some sufficiently large C, then $\hat{p}_{n,k}$ converges at the rate ϵ_n given in Problem 9.2.

9.4 (Ghosal and van der Vaart 2001) Let $X_i \stackrel{\text{iid}}{\sim} p$, $i = 1, 2, \ldots$, where p belongs to one of the models (i), (ii) or (iii). Consider the sieve $\mathcal{P}_n = \{g_1, \ldots, g_N\}$, where $N = N_{[\,]}(\epsilon_n, \mathcal{P}, d)$, ϵ_n is the solution of the entropy equation $\log N_{[\,]}(\epsilon, \mathcal{P}, d) \leq n\epsilon^2$, $g_j = u_j / \int u_j$, $j = 1, \ldots, N$, and $[l_1, u_1], \ldots, [l_N, u_N]$ is a Hellinger bracketing for $\mathcal{P}$ of size ϵ_n. If we choose $a \lesssim \sqrt{\log n}$, then $\epsilon_n \lesssim n^{-1/2}\log n$ for models (i) and (ii), $\epsilon_n \lesssim n^{-1/2}(\log n)^{3/2}$ for model (iii) and the sieve MLE $\hat{p}_n$ satisfies $P_0(d(\hat{p}_n, p_0) > M\epsilon_n) \lesssim e^{-c\log^2 n}$.

9.5 (Ghosal and van der Vaart 2001) The sieve in Problem 9.4 can also be used to construct a prior for which the posterior converges at rate ϵ_n. Put the uniform distribution Π_j on $\mathcal{P}_j$ and consider the prior $\Pi = \sum_{j=1}^{\infty} \lambda_j \Pi_j$, where $\lambda_j > 0$, $\sum_{j=1}^{\infty} \lambda_j = 1$ and $\log \lambda_j^{-1} = O(\log j)$ as $j \to \infty$. Alternatively, for a sample of size n, simply consider the prior Π_n. Using Theorem 8.15, show that the posterior contracts at the intended rate ϵ_n.

9.6 (Ghosal and van der Vaart 2001) Assume the setup and conditions of Theorem 9.9, except that $m = \alpha(\mathbb{R})$ varies with the sample size. If $1 \lesssim m \lesssim \log n$ and p_0 is supersmooth, then for a sufficiently large constant M, show that the posterior contracts at the same rate as given in the theorem. Formulate a similar result for the β-smooth case.

9.7 (Ghosal and van der Vaart 2001) Assume the setup of Theorem 9.9, but the base measure α of the Dirichlet process prior for F depends on a parameter θ, where θ is given an arbitrary prior and α_θ are base measures satisfying the following conditions:

(i) $\alpha_\theta(\mathbb{R})$ is bounded above and below in θ,
(ii) There exist constants B, b and $\delta > 0$ such that $\alpha_\theta(z\colon |z| > t) \le B e^{-bt^\delta}$ for all $t > 0$ and θ,
(iii) Every α_θ has a density α'_θ such that for some $\epsilon > 0$, $\alpha'_\theta(x) \ge \epsilon$ for all θ and $x \in [-k_0, k_0]$.

(These conditions usually hold if the parametric family α_θ is "well behaved" and θ has a compact range. However, conditions (ii) and (iii) are not expected to hold if the range of θ is unbounded.) If p_0 is supersmooth, show that the posterior contracts at the same rate given in the theorem. Formulate a similar result for the β-smooth case.

9.8 (Ghosal and van der Vaart 2001) In certain situations, the conclusion of Problem 9.7 may still hold even if the hyperparameters are not compactly supported. Let $d = 1$ and the base measure α be $\mathrm{Nor}(\mu, \tau)$, $\mu \sim \mathrm{Nor}(\mu_0, A)$ and τ is either given, or has compactly supported prior distribution. Then the posterior contracts at the same rate.

9.9 (Walker et al. 2007, Xing 2010) Consider a model consisting of univariate supersmooth normal location mixtures with a known σ and the let true density be $\phi_\sigma * F_0$ for some compactly supportd F_0. For all $n \in \mathbb{N}$, define $a_{nj} \uparrow \infty$ and consider a covering for the space of densities given by $\mathcal{G}_{\underline{\sigma}, a_{n1}, \delta_n^2} = \{\phi_\sigma * F\colon F([-a_{n1}, a_{n1}]) \ge 1 - \delta_n^2\}$, $\mathcal{G}_{\underline{\sigma}, a_{nj}, \delta_n^2} = \{\phi_\sigma * P\colon F([-a_{nj}, a_{nj}]) \ge 1 - \delta_n^2, F([-a_{n,j-1}, a_{n,j-1}]) \ge 1 - \delta_n^2\}$, $j \ge 2$. Show that for $\delta_n = n^{-1/2} \log n$, we have $e^{-n\delta_n^2} \sum_{j=1}^{\infty} \sqrt{\Pi(\mathcal{G}_{\underline{\sigma}, a_{nj}, \delta_n^2})} \to 0$. Hence using Problem 8.11, conclude that the posterior contracts at the rate $n^{-1/2} \log n$ a.s. Reach the same conclusion using Problem 8.13.

9.10 (Ghosal and van der Vaart 2007b) Let p_0 be a twice continuously differentiable probability density.

(i) If $\int (p_0''/p_0)^2\, p_0\, d\lambda < \infty$ and $\int (p_0'/p_0)^4\, p_0\, d\lambda < \infty$, then $d_H(p_0, p_0 * \phi_\sigma) \lesssim \sigma^2$.
(ii) If p_0 is a bounded with $\int |p_0''|\, d\lambda < \infty$, then $\|p_0 - p_0 * \phi_\sigma\|_1 \lesssim \sigma^2$.

In both cases the constants in "$\lesssim$" depend on the given integrals only.

9.11 (Ghosal and van der Vaart 2007b, Wu and Ghosal 2010) Let $X_1, \ldots, X_n \stackrel{\text{iid}}{\sim} P$, where P has density p, and let the true value of p be p_0, which is bounded. Let p be given a Dirichlet mixture prior Π: $p = F * \phi_\sigma$, $F \sim \mathrm{DP}(\alpha)$, $\sigma/\sigma_n \sim G$. Let α have positive and continuous density on $[-a, a]$, where a is fixed. Then for any $\epsilon > 0$ and $0 < b < a\sigma_n^{-1}$, there exists K not depending on n such that

$$P_0[\Pi_n(F[-2a, 2a]^c > \epsilon | X_1, \ldots, X_n)\mathbb{1}\{\max_{1\le i\le n} |X_i| \le a\}]$$
$$\lesssim P_0\Pi_n(\sigma > b\sigma_n | X_1, \ldots, X_n) + \frac{\alpha[-2a, 2a]^c}{\epsilon(\alpha(\mathbb{R}) + n)} + Kn\epsilon^{-1}e^{-a^2/4b^2\sigma_n^2}.$$

Moreover, if P_0 is compactly supported, α has positive and continuous density on an interval containing $\mathrm{supp}(P_0)$, $b_n \to \infty$ such that $b_n\sigma_n \to 0$, $n\epsilon_n^{-2}e^{-a^2/4b_n^2\sigma_n^2} \to 0$ and $\mathrm{P}(\sigma > b_n\sigma_n) = o(e^{-n\epsilon_n^2})$ for some sequence ϵ_n satisfying Condition (8.4) on prior concentration rate, then

$$P_0^n\Pi_n(F\colon F[-2a, 2a]^c > \epsilon_n^2 | X_1, \ldots, X_n) \to 0. \tag{9.27}$$

Generalize the result when $a = a_n \to \infty$.

9.12 (Ghosal and van der Vaart 2007b) Let $X_i \stackrel{\text{iid}}{\sim} p$, $i = 1, 2, \ldots$, $p \in \mathcal{P}$. Consider the sieve $\mathcal{P}_n = \{p_{F,\sigma}\colon F[-a_n, a_n] = 1, b_1\sigma_n \le \sigma \le b_2\sigma_n\}$, where $a_n \ge e$ and $\sigma_n \to 0$ are positive sequences such that $\log n \lesssim \log(a_n/\sigma_n) \lesssim \log n$. Let $\hat{p}_n = \arg\max\{\prod_{i=1}^n p(X_i)\colon p \in \mathcal{P}_n\}$. Assume that P_0 has compact support and $[-a_n, a_n] \supset \mathrm{supp}(P_0)$ for all sufficiently large n. Using Theorem F.4, show that the sieve-MLE $\hat{p}_n$ converges at the rate $\epsilon_n = \max\{(n\sigma_n)^{-1/2}a_n \log n, \sigma_n^2\}$.

9.13 (Ghosal and van der Vaart 2007b) Consider the setting of Problem 9.12 except that P_0 is not compactly supported. Consider the sieve $\mathcal{P}_n = \{p_{F,\sigma}\colon F[-a, a]^c \le A(a)$ for all $a > 0, b_1\sigma_n \le \sigma \le b_2\sigma_n\}$, where $A(a) = e^{-da^{1/\delta}}$, $d, \delta > 0$ constants and $\log n \lesssim \log\sigma_n^{-1} \lesssim \log n$. Assume that $P_0[-a, a]^c \le A(a)$ for every $a > 0$ and $p_0/(p_0 * \phi_{\sigma_n})$ are uniformly bounded. Show that the sieve-MLE $\hat{p}_n$ converges at the rate $\epsilon_n = \max\{(n\sigma_n)^{-1/2}(\log n)^{1+(1\vee 2\delta)/4}, \sigma_n^2\}$.

9.14 (Ghosal and van der Vaart 2007b) If p_0 is increasing on $(-\infty, a]$, bounded below on $[a, b]$ and decreasing on $[b, \infty)$ for some $a < b$, show that $p_0/(p_0 * \phi_{\sigma_n})$ are uniformly bounded in Problem 9.13.

9.15 (Ghosal and van der Vaart 2007b) The sieves $\mathcal{P}_n$ in Problem 9.12 and Problem 9.13 can also be used to construct a prior for which the posterior contracts at the same rate ϵ_n as the corresponding sieve-MLEs. Consider a minimal collection of Hellinger ϵ_n-brackets that cover $\mathcal{P}_n$ and impose the uniform prior Π_n on the renormalized upper brackets. Using Theorem 8.15, show that under the same assumptions on p_0, the resulting posterior converges at the rate ϵ_n.

9.16 (Shen et al. 2013) For a finite mixture prior specification Π, where the density function f is represented by $f(x) = \sum_{j=1}^N \omega_j\phi_\Sigma(x - \mu_j)$ and priors are assigned on N, Σ, $\omega = (\omega_1, \ldots, \omega_N)$ and $\mu_1, \ldots, \mu_N$. We assume $\Sigma \sim G$, which satisfies (9.6), (9.7) and (9.8), and that there exist positive constants $a_4, b_4, b_5, b_6, b_7, C_4, C_5, C_6, C_7$ such that for sufficiently large $x > 0$,

$$b_4 \exp\{-C_4 t(\log t)^{\tau_1}\} \le \Pi(N \ge t) \le b_5 \exp\{-C_5 t(\log t)^{\tau_1}\}$$

while for every fixed value of N,

$$\Pi(\mu_i \notin [-x, x]^d) \le b_6 \exp(-C_6 x^{a_4}), \quad \text{for sufficiently large } x > 0, \quad i = 1, \dots, N,$$
$$\Pi(\|\omega - \omega_0\| \le \epsilon) \ge b_7 \exp\{-C_7 h \log_- \epsilon\}, \quad \text{for all } 0 < \epsilon < 1/N \text{ and all } \omega_0 \in \mathbb{S}_N.$$

Show that the conclusion of Theorem 9.9 also holds for this prior.

9.17 (Shen et al. 2013) For any $a = (a_1, \dots, a_d)$ and $b = (b_1, \dots, b_d)$, let $\langle a, b\rangle$ denote $\sum_{j=1}^d a_j b_j$ and for $y = (y_1, \dots, y_d)$, let $\|y\|_1$ denote the ℓ_1-norm $\sum_{j=1}^d |y_j|$. For a $\beta > 0$, an $\alpha = (\alpha_1, \dots, \alpha_d) \in (0, \infty)^d$ with sum $\alpha = d$ and an $L\colon \mathbb{R}^d \to (0, \infty)$ satisfying $L(x + y) \le L(x) \exp(c_0 \|y\|_1^2)$ for all $x, y \in \mathbb{R}^d$ and some $c_0 > 0$, the α-anisotropic β-Hölder class with envelope L is defined as the set of all functions $f\colon \mathbb{R}^d \to \mathbb{R}$ that have continuous mixed partial derivatives $D^k f$ of all orders $k \in \mathbb{N}_0^d$, $\beta - \alpha_{\max} \le \langle k, \alpha\rangle < \beta$, with

$$|D^k f(x + y) - D^k f(x)| \le L(x) e^{c_0 \|y\|_1^2} \sum_{j=1}^d |y_j|^{\min(\beta/\alpha_j - k_j, 1)}, \quad x, y \in \mathbb{R}^d,$$

where $\alpha_{\max} = \max(\alpha_1, \dots, \alpha_d)$. We denote this set of functions by $\mathfrak{C}^{\alpha,\beta,L,c_0}(\mathbb{R}^d)$; here β refers to the mean smoothness and α refers to the anisotropy index. An $f \in \mathfrak{C}^{\alpha,\beta,L,c_0}$ has partial derivatives of all orders up to $\underline{\beta}_j$ along axis j where $\beta_j = \beta/\alpha_j$, and β is the harmonic mean $d/(\sum_{j=1}^d \beta_j^{-1})$ of these axial smoothness coefficients. (In the special case of $\alpha = (1, \dots, 1)$, the anisotropic set $\mathfrak{C}^{\alpha,\beta,L,c_0}(\mathbb{R}^d)$ equals the isotropic set $\mathfrak{C}^{\beta,L,c_0}(\mathbb{R}^d)$.)

Suppose that $p_0 \in \mathfrak{C}^{\alpha,\beta,L,c_0}(\mathbb{R}^d)$ is a probability density function satisfying

$$P_0\Big(|D^k p_0|/p_0\Big)^{(2\beta+\epsilon)/\langle k,\alpha\rangle} < \infty, \quad k \in \mathbb{N}_0^d, \langle k, \alpha\rangle < \beta, \quad P_0\,(L/p_0)^{(2\beta+\epsilon)/\beta} < \infty \tag{9.28}$$

for some $\epsilon > 0$ and that $p_0(x) \le c e^{-b\|x\|^\tau}$, $\|x\| > b$, holds for some constants $a, b, c, \tau > 0$. If Π is as in Section 9.4, then the posterior contraction rate at p_0 in the Hellinger or the $\mathbb{L}_1$-metric is $\epsilon_n = n^{-\beta/(2\beta+d^*)}(\log n)^t$, where $t > \{d^*(1 + \tau^{-1} + \beta^{-1}) + 1\}/(2 + d^*/\beta)$, and $d^* = \max(d, \kappa\alpha_{\max})$.

9.18 (Kleijn and van der Vaart 2006) In a supersmooth normal location mixture model (assume standard deviation σ is known to be 1), consider a Dirichlet mixture prior with compact base measure. If the true density p_0 lies outside the model, then using the results of Section 8.5, show that the posterior contracts at $p_{F^*} = \phi * F^*$ at the rate $n^{-1/2}\log n$ provided that p_0/p_{F^*} is bounded, where $F^* = \operatorname{argmin} K(p_0; \phi * F)$.

9.19 (Petrone and Wasserman 2002, Ghosal 2001) Let $X_i \overset{\text{iid}}{\sim} p$, $i = 1, 2, \dots$, $p \in \mathcal{P}$, where p is a density on the unit interval. An extended Bernstein polynomial prior, which is supported on extended Bernstein polynomial densities $\sum_{r=1}^k \sum_{j=1}^r w_{j,r}\beta(x; j, r - j + 1)$. Show that if the true density is of the extended Bernstein type, then posterior contracts at the rate $n^{-1/2}\log n$.

9.20 (McVinish et al. 2009) Let $X_i \overset{\text{iid}}{\sim} p$, $i = 1, 2, \dots$, $p \in \mathcal{P}$, where p is a density on the unit interval. Consider a prior described by the triangular kernel given below:
Let $0 = t_0 < t_1 < \cdots < t_{k-1} < t_k = 1$ and let $\Delta_j(x)$ be the triangular density on $[t_{j-1}, t_{j+1}]$ with mode of height $2/(t_{j+1} - t_{j-1})$ at $t = t_j$, $j = 1, \dots, k - 1$. Also

define $\Delta_0(x) = 2(t_1 - x)/(t_1 - t_0)$, $t_0 \le x \le t_1$, and $\Delta_k(x) = 2(x - t_{k-1})/(t_k - t_{k-1})$, $t_{k-1} \le x \le t_k$. Let $p(x) = \sum_{j=0}^{k} w_j \Delta_j(x)$. Induce a prior on p in one of the following two ways:

Type I mixture: $t_j = j/k$, $(w_0, \ldots, w_k)|k \sim \mathrm{Dir}(k+1: \alpha_{0,k}, \ldots, \alpha_{k+1,k})$, $k \sim \rho(\cdot)$;
Type II mixture: $w_j = 1/(k+1)$, $(t_1, \ldots, t_{k-1})|k \sim h_k(\cdot)$, a positive density h_k on the set $\{(t_1, \ldots, t_{k-1}): t_1 < t_2 < \cdots < t_{k-1}\}$, $k \sim \rho(\cdot)$.

For the type I mixture prior, assume that $\{\alpha_{j,k}\}$ is uniformly bounded, $e^{-c_1 k \log k} \lesssim \rho(k) \lesssim e^{-c_2 k}$, and the true density $p_0 \in \mathfrak{C}^\beta[0,1]$, $\beta \le 2$, $p_0(x) \ge x(1-x)$. Then the posterior contracts at the rate $n^{-\beta/(2\beta+1)}(\log n)^{(4\beta+1)/4\beta}$ a.s.

For the type II mixture prior, assume that $e^{-c_1 k \log k} \lesssim \sum_{j=k+1}^{\infty} \rho(j) \lesssim e^{-c_2 k \log k}$, $\mathrm{P}(\max_j |t_j - t_{j-1}| < e^{-c_3 n^\gamma} |k) \le \exp(-c_4 n^{1/(2\beta+1)} \log n)$, $k \le k_0 n^{1/(2\beta+1)}$, and $h_k \gtrsim (\Gamma(k))^{-r}$ for some $r > 0$, and suppose that the true density $p_0 \in \mathfrak{C}^\beta(0,1)$, $\beta \le 2$, p_0 bounded away from 0. Then the posterior contracts at the rate $n^{-\beta/(2\beta+1)} \log n$ a.s.

9.21 (Scricciolo 2006) Consider the estimation of a periodic density p on $[0,1]$, $p \in \mathfrak{W}^\beta(L)$, the Sobolev space of smoothness index β with Sobolev norm $\|p\|_{2,2,\beta}$ bounded by L (see Section E.3) and a prior on p induced by the infinite-dimensional exponential family of trigonometric series

$$p(x) = \frac{\exp[\sqrt{2}\sum_{k=1}^{\infty}\{a_k \sin(2\pi k x) + b_k \cos(2\pi k x)\}]}{\int_0^1 \exp[\sqrt{2}\sum_{k=1}^{\infty}\{a_k \sin(2\pi k y) + b_k \cos(2\pi k y)\}]dy}, \qquad a_k, b_k \overset{\text{ind}}{\sim} \mathrm{Nor}(0, k^{-2q})$$

conditioned on $\sum_{k=1}^{\infty} k^{2p}(a_k^2 + b_k^2) < \pi^{-2p} L^2$. Then the prior is supported within $\mathfrak{W}^\beta(L)$ and the posterior for p contracts at the rate $n^{-p/(2p+1)}$ at p_0 with respect to d_H.

9.22 (Ghosal and van der Vaart 2007a) For the Gaussian white noise model with conjugate prior on it, show that Theorem 9.21 can be modified to give the same rate of contraction for a fixed prior of the mixture type $\sum_n \lambda_n \Pi_n$ under suitable conditions on the weights λ_n.

9.23 (Jiang 2007) The posterior contraction rate theorems apply also to finite-dimensional models where the dimension grows with the sample size. Consider a generalized linear model with K_n-dimensional predictor $X \in [-1,1]^{K_n}$, where $K_n \to \infty$, and $Y|X = x$ has density $p(y|x) = \exp\{a(x^\top \beta)y + b(x^\top \beta) + c(y)\}$ with respect to a dominating σ-finite measure ν. Suppose that $X_i \overset{\text{iid}}{\sim} G$, the true value of β is β_0, the true value of p is p_0, and assume that the ℓ_1-sparsity condition $\limsup_{n\to\infty} \sum_{j=1}^{K_n} |\beta_{0j}| < \infty$ holds. Define a metric for p by $d^2(p_1, p_2) = \int\int (\sqrt{p_1}(y|x) - \sqrt{p_2}(y|x))^2 \, d\nu(y)\, dG(x)$.

Let a prior be specified through the following variable selection scheme. Fix sequences $r_n, \bar{r}_n \to \infty$ with $r_n/K_n \to 0$ and $1 \le r_n \le \bar{r}_n \le K_n$. Let $\gamma_1, \ldots \gamma_{K_n}$ be indicators of inclusion of the corresponding predictors in the model, and set $|\gamma| = \sum_{j=1}^{K_n} \gamma_j$. Let $\gamma_1, \ldots \gamma_{K_n}$ be a priori i.i.d. $\mathrm{Bin}(1, r_n/K_n)$ conditioned on $|\gamma| \le \bar{r}_n$. Given $\gamma_1, \ldots \gamma_{K_n}$ let $\beta_j = 0$ if $\gamma_j = 0$ and let $(\beta_j: \gamma_j = 1) \sim \mathrm{Nor}(0, V_\gamma)$.

Define $\Delta(r_n) = \inf\{\sum_{j \notin \gamma} |\beta_{0j}|: |\gamma| = r_n\}$, $\underline{B}_n = \sup\{\mathrm{eig}(V_\gamma^{-1}): |\gamma| \le \bar{r}_n\}$, $\bar{B}_n = \sup\{\mathrm{eig}(V_\gamma): |\gamma| \le \bar{r}_n\}$, and $D(R) = 1 + R\sup\{|a'(h)|: |h| \le R\}\sup\{|b'(h)/a'(h)|: |h| \le R\}$. Suppose that ϵ_n satisfy $0 < \epsilon_n < 1$, $n\epsilon_n^2 \gg 0$ and

(i) $\bar{r}_n \log_- \epsilon_n = o(n\epsilon_n^2)$;
(ii) $\bar{r}_n \log K_n = o(n\epsilon_n^2)$;
(iii) $\bar{r}_n \log D(\bar{r}_n \sqrt{n}\bar{B}_n \epsilon_n)$;
(iv) $\Delta(r_n) = o(\epsilon_n^2)$;
(v) $\underline{B}_n = o(n\epsilon_n^2)$;
(vi) $r_n \log \bar{B}_n = o(n\epsilon_n^2)$.

Then the posterior for p contracts at p_0 with respect to d_H at the rate ϵ_n at p_0 a.s. In particular, if $\max\{\underline{B}_n, \bar{B}_n\} \lesssim \bar{r}_n^v$ for some $v \geq 0$, $\bar{r}_n = o(n^{1/(4+v)})$, $\bar{r}_n / K_n \to 0$ and $\bar{r}_n \log K_n = o(n)$, then the posterior for p is consistent at p_0 with respect to d.

Under appropriate conditions $\log K_n$ can be as large as n^ξ, for $\xi < 1$, with ϵ_n of the order $n^{-(1-\xi)/2}(\log n)^{k/2}$ if $\bar{r}_n = o(\log^k n)$. The rate is close to the parametric rate $n^{-1/2}$ if ξ can be taken arbitrarily close to 0, for instance when K_n has polynomial growth.

9.24 (Jiang 2007) Under the setting of Problem 9.23, consider specific regression models to obtain explicit conditions. Let $\max\{\underline{B}_n, \bar{B}_n\} \lesssim r_n^v$ for some $v \geq 0$.

(a) *Poisson regression:* $Y | X \sim \mathrm{Poi}(e^{X^\top \beta})$. Assume (ii), (iv) and $\bar{r}_n^{4+v} = o(n\epsilon_n^2)$. Then the posterior contracts at the rate ϵ_n.
(b) *Normal regression:* $Y | X \sim \mathrm{Nor}(X^\top \beta, 1)$. Assume (ii), (iv) and $\bar{r}_n^v = o(n\epsilon_n^2)$. Then the posterior contracts at the rate ϵ_n.
(c) *Exponential regression:* $Y | X \sim \mathrm{Exp}(e^{X^\top \beta})$. Assume (ii), (iv) and $\bar{r}_n^{4+v} = o(n\epsilon_n^2)$. Then the posterior contracts at the rate ϵ_n.
(d) *Logistic binary regression:* $Y | X \sim \mathrm{Bin}(1, (1 + e^{-X^\top \beta})^{-1})$. Assume (ii), (iv) and $\bar{r}_n^v = o(n\epsilon_n^2)$. Then the posterior contracts at the rate ϵ_n.
(e) *Probit binary regression:* $Y | X \sim \mathrm{Bin}(1, \Phi(X^\top \beta))$. Assume (ii), (iv) and $\bar{r}_n^v = o(n\epsilon_n^2)$. Then the posterior contracts at the rate ϵ_n.

9.25 (Jiang 2007) This example shows that in high dimensional problems, even consistency for the posterior distribution of p may not hold without the variable selection step in the prior. Adopt the notations in Problem 9.23. Let Z_i be i.i.d. with $\mathrm{P}(Z = j/K) = K^{-1}$, $j = 1, \ldots, K$. Let $X_{ij} = \mathbb{1}\{Z_i = j/K\}$, $K > n$. Let $Y_i | X_i \sim \mathrm{Nor}(X_i^\top \beta, 1)$. Consider a prior $\beta \sim \mathrm{Nor}_K(0, I_K)$. Let $\beta_0 = 0$. Show that $\Pi(d_H(p, p_0) \geq \sqrt{\eta} | X_1, Y_1, \ldots, X_n, Y_n) \geq 1 - \eta^{-2} n^{-1}$ a.s. for $\eta = \frac{1}{2} - \frac{1}{\sqrt{5}}$.

10

Adaptation and Model Selection

Many nonparametric priors possess a given regularity or complexity, and are appropriate when the true parameter of the data is of similar regularity. In this chapter we investigate general constructions that combine priors of varying regularity into a mixture with the aim of achieving optimal contraction rates for a variety of regularity levels simultaneously. Mixture priors arise naturally from placing a prior on a hyperparameter that expresses regularity or complexity. In this chapter we obtain parallels of the main results of Chapter 8 for mixture priors, for fairly general priors on the hyperparameters. Examples include finite-dimensional models, sieves, the white noise model, log-spline densities and random series with a random number of terms. Adaptation is connected, but certainly does not require consistent selection of the complexity parameter of a model, as is explored in a final section of the chapter. The general message of the chapter is that adaptation is easily achieved, in particular if one is satisfied with optimal rates up to logarithmic factors. To some extent this conclusion also emanates from the adaptivity of special priors, such as Dirichlet process mixtures or Gaussian processes, which is explored by direct methods in the respective chapters.

10.1 Introduction

Many nonparametric procedures involve a *bandwidth* or *regularization* parameter, which adapts the procedure to the "regularity level" or "smoothness" of the function being estimated. This parameter is typically not known, and within the Bayesian framework it is natural to put a prior on it and let the data decide on a correct value through the corresponding posterior distribution. The resulting procedure then combines suitable priors Π_α on statistical models of regularities α, with a prior λ on the regularity α itself, resulting in the "overall" mixture prior $\int \Pi_\alpha \, d\lambda(\alpha)$. The parameter α may concretely refer to a smoothness level (e.g. the number of derivatives of a Gaussian process), but may also be more abstract (e.g. the number of terms in a series expansion, the bandwidth of a kernel, or the scale of a stochastic process prior), as long as the collection of priors Π_α for varying α covers (or can approach) all targeted true densities.

Such a hierarchical Bayesian procedure fits naturally within the framework of *adaptive estimation*, which focuses on constructing estimators that are rate-optimal simultaneously across a scale of models. Informally, given a collection of models, an estimator is said to be *rate-adaptive* if it attains at any given true parameter in the union of all models the rate of contraction that would have been attained had only the best model been used. For instance, the minimax rate for estimating a probability density or regression function on $[0, 1]^d$ that is known to have α derivatives (is "α-smooth"), is $n^{-\alpha/(2\alpha+d)}$, where n is the number of observations. For smoother functions, faster rates of estimation are possible, with the rate $n^{-\alpha/(2\alpha+d)}$ approaching the "parametric rate" $n^{-1/2}$ as $\alpha \to \infty$. An estimator would be rate-adaptive in this context if its construction is free of α and it attains the rate $n^{-\alpha/(2\alpha+d)}$ whenever the true density is α-smooth, for any $\alpha > 0$. An adaptive estimator possesses an *oracle property* in that it performs as if it were an "oracle" in possession of a priori knowledge on the smoothness of the true function.

In the Bayesian context we may investigate whether the contraction rate of a posterior distribution adapts to the regularity of a true parameter. In an abstract setting we have parameter sets $\Theta_{n,\alpha}$ indexed by a parameter α ranging over a set A, and assume that the true parameter belongs to the union $\Theta_n = \cup_\alpha \Theta_{n,\alpha}$. We are given an observation $X^{(n)}$ in statistical experiments $(\mathfrak{X}^{(n)}, \mathscr{X}^{(n)}, P_\theta^{(n)}: \theta \in \Theta_n)$, and form a posterior distribution $\Pi_n(\cdot|\, X^{(n)})$ based on a prior Π_n on Θ_n. Furthermore, for each model $\Theta_{n,\alpha}$ we have a targeted contraction rate $\epsilon_{n,\alpha}$ relative to some semimetric d_n on Θ_n.

Definition 10.1 (Adaptation) The prior Π_n is *rate-adaptive* (or adaptive, in short) if for every $\alpha \in A$ and every $M_n \to \infty$,

$$\sup_{\theta_0 \in \Theta_{n,\alpha}} P_{\theta_0}^{(n)} \Pi_n(\theta: d_n(\theta, \theta_0) \geq M_n \epsilon_{n,\alpha} |\, X^{(n)}) \to 0.$$

Without specifying the rates $\epsilon_{n,\alpha}$, the definition can merely serve as a template for the desired results. One typical specialization would be to a mixture prior $\Pi_n = \int \Pi_{n,\alpha}\, d\lambda(\alpha)$, with $\Pi_{n,\alpha}$ prior distributions on the models $\Theta_{n,\alpha}$, and $\epsilon_{n,\alpha}$ the contraction rates of the posterior distributions $\Pi_{n,\alpha}(\cdot|\, X^{(n)})$ induced by $\Pi_{n,\alpha}$, at parameters in the model $\Theta_{n,\alpha}$. It is natural to require the rate of contraction to be uniform over true parameters θ_0 in a model, as in the definition, but for simplicity we often formulate pointwise results.

The examples in Chapters 9 and 11 may guide the choice of the priors $\Pi_{n,\alpha}$, given a collection of models $\Theta_{n,\alpha}$. The mixing weights λ are a new element. As the models $\Theta_{n,\alpha}$ may have very different complexities, weighting them equally may not be a good choice. In this chapter we shall see that there may be a wide range of choices for λ. In particular, if we are willing to add a logarithmic factor to the rates $\epsilon_{n,\alpha}$ for a given model, then simple, universal choices may do.

The posterior distribution induced by the mixture $\Pi_n = \int \Pi_{n,\alpha}\, d\lambda(\alpha)$ is itself a mixture, given by

$$\Pi_n(\theta \in B|\, X^{(n)}) = \int \Pi_{n,\alpha}(\theta \in B|\, X^{(n)})\, d\lambda(\alpha|\, X^{(n)}), \tag{10.1}$$

with $\Pi_{n,\alpha}(\cdot|\, X^{(n)})$ the posterior distribution when using the prior $\Pi_{n,a}$ and $\lambda(\cdot|\, X^{(n)})$ the posterior distribution of α. Adaptation in this setting would easily be seen to occur if the latter weights concentrate on a "true" index α. However, the notion of a true smoothness or regularity level is generally misguided. For a given true parameter θ_0, multiple models $\Theta_{n,\alpha}$ may contain parameters that approximate θ_0 closely, and adaptation will occur when the posterior distribution concentrates on these parameters, without the posterior distribution of α settling down on specific values. In Section 10.5 we further investigate the posterior distribution of the model index.

The main interest in the present chapter is to exhibit general choices of the weights λ. Specific priors may come with natural choices of hyperpriors on the bandwidth parameter. In particular, with Dirichlet process mixtures of normal kernel, an inverse-gamma prior on the scale of the normal kernel is natural, and for Gaussian process priors changing the length scale is customary. These adaptation schemes are described elsewhere together with their general constructions (see e.g. Sections 9.4 and 11.6).

The following result shows that a rate-adaptive prior induces rate-adaptive Bayesian point estimators. The theorem is an immediate corollary of Theorem 8.7.

Theorem 10.2 (Point estimator) *If the prior Π_n is rate-adaptive in the sense of Definition 10.1, then the center $\hat\theta_n$ of the (nearly) smallest d_n-ball that contains posterior mass at least 1/2 satisfies $d_n(\hat\theta_n, \theta_{n,0}) = O_P(\epsilon_{n,\alpha})$ in $P^{(n)}_{\theta_{n,0}}$-probability, for any $\theta_{n,0} \in \Theta_{n,\alpha}$, for every α.*

10.2 Independent Identically Distributed Observations

In this section we obtain adaptive versions of the posterior contraction results for i.i.d. observations in Section 8.2. We take the parameter equal to the probability density p of the observations $X_1, \ldots, X_n$, and consider a countable collection of models $\mathcal{P}_{n,\alpha}$ for this density, indexed by a parameter α ranging through a set A_n, which may depend on n. The index α may refer to the regularity of the elements of $\mathcal{P}_{n,\alpha}$, but in general may be arbitrary. Thus $\mathcal{P}_{n,\alpha}$ is an arbitrary set of probability densities with respect to a σ-finite measure ν on the sample space $(\mathfrak{X}, \mathscr{X})$ of the observations, for every α in the arbitrary countable set A_n.

Consider a metric d on the set of all ν-probability densities for which tests as in (8.2) exist. For instance, d may be the Hellinger or $\mathbb{L}_1$-distance, or the $\mathbb{L}_2$-distance if the densities are uniformly bounded.

Let $\Pi_{n,\alpha}$ be a probability measure on $\mathcal{P}_{n,\alpha}$, and $\lambda_n = (\lambda_{n,\alpha}\colon \alpha \in A_n)$ a probability measure on A_n, viewed as prior distributions for p within the model $\mathcal{P}_{n,\alpha}$ and for the index α, respectively, at stage n. Thus the overall prior is a probability measure on the set of probability densities, given by

$$\Pi_n = \sum_{\alpha \in A_n} \lambda_{n,\alpha} \Pi_{n,\alpha}.$$

The corresponding posterior distribution is given by Bayes's rule as

$$\begin{aligned} \Pi_n(B \mid X_1, \ldots, X_n) &= \frac{\int_B \prod_{i=1}^n p(X_i)\, d\Pi_n(p)}{\int \prod_{i=1}^n p(X_i)\, d\Pi_n(p)} \\ &= \frac{\sum_{\alpha \in A_n} \lambda_{n,\alpha} \int_{p \in \mathcal{P}_{n,\alpha}: p \in B} \prod_{i=1}^n p(X_i)\, d\Pi_{n,\alpha}(p)}{\sum_{\alpha \in A_n} \lambda_{n,\alpha} \int_{p \in \mathcal{P}_{n,\alpha}} \prod_{i=1}^n p(X_i)\, d\Pi_{n,\alpha}(p)}. \end{aligned} \tag{10.2}$$

Here we have implicitly assumed the existence of a suitable measurability structure on $\mathcal{P}_{n,\alpha}$, for which the priors $\Pi_{n,\alpha}$ are defined and the maps $(x, p) \mapsto p(x)$ are jointly measurable.

The true density p_0 of the observations need not belong to any of the models $\mathcal{P}_{n,\alpha}$. However, we denote by β_n parameters in A_n that should be thought of as giving "best" models $\mathcal{P}_{n,\beta_n}$. The purpose is to state conditions under which the mixture posterior distribution

(10.2) attains the same contraction rate as obtained with the single prior Π_{n,β_n}, whenever the true density p_0 belongs to $\mathcal{P}_{n,\beta_n}$. Then the hierarchical Bayesian procedure would automatically adapt to the set of models $\mathcal{P}_{n,\alpha}$ in that it performs at par with an oracle that uses the knowledge of the "correct model" $\mathcal{P}_{n,\beta_n}$.

Similarly as in (8.3), define

$$B_{n,\alpha}(p_0,\epsilon) = \{p \in \mathcal{P}_{n,\alpha}: K(p_0;p) \leq \epsilon^2, V_2(p_0;p) \leq \epsilon^2\}, \tag{10.3}$$
$$C_{n,\alpha}(p_0,\epsilon) = \{p \in \mathcal{P}_{n,\alpha}: d(p_0,p) \leq \epsilon\}. \tag{10.4}$$

Let $\epsilon_{n,\alpha}$ be given positive sequences of numbers tending to zero, to be thought of as the rate attached to the model $\mathcal{P}_{n,\alpha}$ if this is (approximately) correct.

For $\beta_n \in A_n$, thought of as the index of a best model for a given p_0, and a fixed constant $H \geq 1$, decompose A_n into

$$A_{n,\succeq\beta_n} = \{\alpha \in A_n: \epsilon^2_{n,\alpha} \leq H\epsilon^2_{n,\beta_n}\},$$
$$A_{n,\prec\beta_n} = \{\alpha \in A_n: \epsilon^2_{n,\alpha} > H\epsilon^2_{n,\beta_n}\}.$$

These two sets are thought of as indices of the collections of models consisting of densities that are "more regular" or "less regular" than the given true density p_0. Thus, as models they are less complex (or lower dimensional) or more complex, respectively. Even though we do not assume that A_n is ordered, we shall write $\alpha \succeq \beta_n$ and $\alpha \prec \beta_n$ if α belongs to the sets $A_{n,\succeq\beta_n}$ or $A_{n,\prec\beta_n}$, respectively. The set $A_{n,\succeq\beta_n}$ contains β_n and hence is never empty, but the set $A_{n,\prec\beta_n}$ can be empty (if β_n is the "smallest" possible index). In the latter case, conditions involving $\alpha \prec \beta_n$ are understood to be automatically satisfied.

The assumptions of the following theorem are analogous to those in Theorem 8.9 on posterior contraction rate in a single model. They entail a bound on the complexity of the models and a condition on the concentration of the priors. The complexity bound is exactly as in Section 8.2, namely, for some constants E_α,

$$\sup_{\epsilon \geq \epsilon_{n,\alpha}} \log N\Big(\epsilon/3, C_{n,\alpha}(2\epsilon), d\Big) \leq E_\alpha n\epsilon^2_{n,\alpha}, \qquad \alpha \in A_n. \tag{10.5}$$

The conditions on the priors involve comparisons of the prior masses of balls of various sizes in various models. These conditions are split in conditions on the models that are smaller or bigger than the best model: for given constants $\mu_{n,\alpha}$, L, H, I,

$$\frac{\lambda_{n,\alpha}}{\lambda_{n,\beta_n}} \frac{\Pi_{n,\alpha}(C_{n,\alpha}(p_0, i\epsilon_{n,\alpha}))}{\Pi_{n,\beta_n}(B_{n,\beta_n}(p_0,\epsilon_{n,\beta_n}))} \leq \mu_{n,\alpha} e^{Li^2 n\epsilon^2_{n,\alpha}}, \qquad \alpha \prec \beta_n, \quad i \geq I, \tag{10.6}$$
$$\frac{\lambda_{n,\alpha}}{\lambda_{n,\beta_n}} \frac{\Pi_{n,\alpha}(C_{n,\alpha}(p_0, i\epsilon_{n,\beta_n}))}{\Pi_{n,\beta_n}(B_{n,\beta_n}(p_0,\epsilon_{n,\beta_n}))} \leq \mu_{n,\alpha} e^{Li^2 n\epsilon^2_{n,\beta_n}}, \qquad \alpha \succeq \beta_n, \quad i \geq I. \tag{10.7}$$

A final condition requires that the prior mass in a ball of radius of the order $\epsilon_{n,\alpha}$ in a bigger model (i.e., smaller α) is significantly smaller than in a small model: for a constant $M > I$,

$$\sum_{\alpha \prec \beta_n} \frac{\lambda_{n,\alpha}}{\lambda_{n,\beta_n}} \frac{\Pi_{n,\alpha}(C_{n,\alpha}(p_0, M\epsilon_{n,\alpha}))}{\Pi_{n,\beta_n}(B_{n,\beta_n}(p_0,\epsilon_{n,\beta_n}))} = o(e^{-2n\epsilon^2_{n,\beta_n}}). \tag{10.8}$$

Let K be the universal testing constant appearing in (8.2), and assume that there exists a finite constant (with an empty supremum defined as 0) such that

$$E \geq \sup_{\alpha \prec \beta_n} E_\alpha \vee \sup_{\alpha \succeq \beta_n} E_\alpha \frac{\epsilon_{n,\alpha}^2}{\epsilon_{n,\beta_n}^2}.$$

Theorem 10.3 (Adaptive contraction rates) *Assume that* (10.5)–(10.8) *hold for constants* $\mu_{n,\alpha} > 0$, $E_\alpha > 0$, $L > 0$, $H \geq 1$, *and* $M > I > 2$ *such that* $KM^2/I^2 > E + 1$, *and* $M^2H(K - 2L) > 3$, *and* $\sum_{\alpha \in A_n} \sqrt{\mu_{n,\alpha}} \leq e^{n\epsilon_{n,\beta_n}^2}$. *If* $\beta_n \in A_n$ *for every* n *and satisfies* $n\epsilon_{n,\beta_n}^2 \to \infty$, *then* $P_0^n \Pi_n(p: d(p, p_0) \geq M\sqrt{H}\, \epsilon_{n,\beta_n} | X_1, \cdots, X_n) \to 0$.

Remark 10.4 In view of Theorem 8.20, the entropy condition (10.5) can, as usual, be relaxed to the same condition on submodels $\mathcal{P}'_{n,\alpha} \subset \mathcal{P}_{n,\alpha}$ that carry most of the prior mass, in the sense that

$$\frac{\sum_\alpha \lambda_{n,\alpha} \Pi_{n,\alpha}(\mathcal{P}_{n,\alpha} \setminus \mathcal{P}'_{n,\alpha})}{\lambda_{n,\beta_n} \Pi_{n,\beta_n}(B_{n,\beta_n}(p_0, \epsilon_{n,\beta_n}))} = o(e^{-2n\epsilon_{n,\beta_n}^2}).$$

Proof of Theorem 10.3 Abbreviate $J_{n,\alpha} = n\epsilon_{n,\alpha}^2$, so that $E \geq \sup\{E_\alpha J_{n,\alpha}/J_{n,\beta_n} : \alpha \succeq \beta_n\}$. For $\alpha \succeq \beta_n$ we have $B\epsilon_{n,\beta_n} \geq B/\sqrt{H}\epsilon_{n,\alpha} \geq \epsilon_{n,\alpha}$, whence, in view of the entropy bound (10.5) and Lemma D.3 with $\epsilon = B\epsilon_{n,\beta_n}$ and $\log N(\epsilon) = E_\alpha J_{n,\alpha}$ (constant in ϵ), there exists for every $\alpha \succeq \beta_n$ a test $\phi_{n,\alpha}$ with,

$$P_0^n \phi_{n,\alpha} \leq \sqrt{\mu_{n,\alpha}} \frac{e^{E_\alpha J_{n,\alpha} - KB^2 J_{n,\beta_n}}}{1 - e^{-KB^2 J_{n,\beta_n}}} \lesssim \sqrt{\mu_{n,\alpha}} e^{(E - KB^2) J_{n,\beta_n}},$$
$$\sup_{p \in \mathcal{P}_{n,\alpha}: d(p,p_0) \geq iB\epsilon_{n,\beta_n}} P^n(1 - \phi_{n,\alpha}) \leq \frac{1}{\sqrt{\mu_{n,\alpha}}} e^{-KB^2 i^2 J_{n,\beta_n}}. \tag{10.9}$$

For $\alpha \prec \beta_n$ we have $\epsilon_{n,\alpha} > \sqrt{H}\epsilon_{n,\beta_n}$ and we cannot similarly test balls of radius proportional to ϵ_{n,β_n} in $\mathcal{P}_{n,\alpha}$. However, Lemma D.3 with $\epsilon = B'\epsilon_{n,\alpha}$ and $B' := B/\sqrt{H} > 1$, still gives tests $\phi_{n,\alpha}$ such that for every $i \in \mathbb{N}$,

$$P_0^n \phi_{n,\alpha} \leq \sqrt{\mu_{n,\alpha}} \frac{e^{(E_\alpha - KB'^2) J_{n,\alpha}}}{1 - e^{-KB'^2 J_{n,\alpha}}} \lesssim \sqrt{\mu_{n,\alpha}} e^{(E - KB'^2) J_{n,\alpha}},$$
$$\sup_{p \in \mathcal{P}_{n,\alpha}: d(p,p_0) > iB'\epsilon_{n,\alpha}} P^n(1 - \phi_{n,\alpha}) \leq \frac{1}{\sqrt{\mu_{n,\alpha}}} e^{-KB'^2 i^2 J_{n,\alpha}}. \tag{10.10}$$

Let $\phi_n = \sup\{\phi_{n,\alpha} : \alpha \in A_n\}$ be the supremum of all tests so constructed.

The test ϕ_n is more powerful than all the tests $\phi_{n,\alpha}$, and has error of the first kind $P_0^n \phi_n$ bounded by $P_0^n \sum_{\alpha \in A_n} \phi_{n,\alpha} \lesssim \sum_{\alpha \in A_n} \sqrt{\mu_{n,\alpha}} e^{-cJ_{n,\beta_n}}$, for c equal to the minimum of $KB^2 - E$ and $KB^2 - EH$. This tends to zero, because the sum is bounded by $e^{(1-c)J_{n,\beta_n}}$, by assumption, and $c > 1$ and $J_{n,\beta_n} \to \infty$, by assumption. Consequently, for any M we have that $P_0^n\big[\Pi_n(p: d(p, p_0) > M\sqrt{H}\epsilon_{n,\beta_n} | X_1, \ldots, X_n)\phi_n\big] \leq P_0^n \phi_n \to 0$. We shall complement this with an analysis of the posterior multiplied by $1 - \phi_n$.

For $i \in \mathbb{N}$, define

$$\mathcal{S}_{n,\alpha,i} = \{p \in \mathcal{P}_{n,\alpha}: iB'\epsilon_{n,\alpha} < d(p, p_0) \le (i+1)B'\epsilon_{n,\alpha}\}, \qquad \alpha \prec \beta_n,$$
$$\mathcal{S}_{n,\alpha,i} = \{p \in \mathcal{P}_{n,\alpha}: iB\epsilon_{n,\beta_n} < d(p, p_0) \le (i+1)B\epsilon_{n,\beta_n}\}, \qquad \alpha \succeq \beta_n.$$

Then the set $\{p: d(p, p_0) > IB\epsilon_{n,\beta_n}\}$ is contained in the union of the set $\cup_\alpha \cup_{i \ge I} \mathcal{S}_{n,\alpha,i}$ and the set $\cup_{\alpha \prec \beta_n} C_{n,\alpha}(p_0, IB'\epsilon_{n,\alpha})$. We shall estimate the posterior mass in the sets of the first union with the help of the tests, and handle the sets in the second union using their prior mass only.

We estimate numerator and denominator of the posterior distribution separately with the help of the inequalities, for any set C and suitable events A_n,

$$P_0^n \int_C \prod_{i=1}^n \frac{p(X_i)}{p_0(X_i)} (1 - \phi_n)\, d\Pi_{n,\alpha}(p) \le \sup_{p \in C} P^n(1-\phi_n)\, \Pi_{n,\alpha}(C),$$
$$\int \prod_{i=1}^n \frac{p(X_i)}{p_0(X_i)} d\Pi_n(p) \mathbb{1}_{A_n} \ge e^{-2J_{n,\beta_n}} \lambda_{n,\beta_n} \Pi_{n,\beta_n}(B_{n,\beta_n}(p_0, \epsilon_{n,\beta_n})).$$

The first inequality follows by Fubini's theorem. Because $\Pi_n \ge \lambda_{n,\beta_n} \Pi_{n,\beta_n}$, the second is true on events A_n such that $P_0^n(A_n) \ge 1 - (n\epsilon_{n,\beta_n}^2)^{-1} \to 1$, by Lemma 8.10. Combining these two inequalities, and also the first inequality but with $\phi_n \equiv 0$, with (10.9) and (10.10), we see that

$$\begin{aligned} P_0^n\Big[\Pi_n\Big(d(p, p_0) > IB\epsilon_{n,\beta_n} | X_1, \ldots, X_n\Big)(1-\phi_n)\mathbb{1}_{A_n}\Big] \\ \le \sum_{\alpha \in A_n: \alpha \succeq \beta_n} \sum_{i \ge I} \frac{\lambda_{n,\alpha}}{\lambda_{n,\beta_n}} \frac{e^{-KB^2 i^2 J_{n,\beta_n}} \Pi_{n,\alpha}(\mathcal{S}_{n,\alpha,i})}{e^{-2J_{n,\beta_n}} \Pi_{n,\beta_n}(B_{n,\beta_n}(p_0, \epsilon_{n,\beta_n}))} \frac{1}{\sqrt{\mu_{n,\alpha}}} \\ + \sum_{\alpha \in A_n: \alpha \prec \beta_n} \sum_{i \ge I} \frac{\lambda_{n,\alpha}}{\lambda_{n,\beta_n}} \frac{e^{-KB'^2 i^2 J_{n,\alpha}} \Pi_{n,\alpha}(\mathcal{S}_{n,\alpha,i})}{e^{-2J_{n,\beta_n}} \Pi_{n,\beta_n}(B_{n,\beta_n}(p_0, \epsilon_{n,\beta_n}))} \frac{1}{\sqrt{\mu_{n,\alpha}}} \\ + \sum_{\alpha \in A_n: \alpha \prec \beta_n} \frac{\lambda_{n,\alpha}}{\lambda_{n,\beta_n}} \frac{\Pi_{n,\alpha}(C_{n,\alpha}(p_0, IB'\epsilon_{n,\alpha}))}{e^{-2J_{n,\beta_n}} \Pi_{n,\beta_n}(B_{n,\beta_n}(p_0, \epsilon_{n,\beta_n}))}. \end{aligned} \tag{10.11}$$

The third term on the right-hand side tends to zero by assumption (10.8) if M is defined as $M = IB' = IB/\sqrt{H}$. Because for $\alpha \succeq \beta_n$ and for $i \ge I \ge 3$, we have $\mathcal{S}_{n,\alpha,i} \subset C_{n,\alpha}(p_0, \sqrt{2} i B\epsilon_{n,\beta_n})$, the assumption (10.7) shows that the first term is bounded by

$$\sum_{\alpha \in A_n: \alpha \succeq \beta_n} \sum_{i \ge I} \frac{\lambda_{n,\alpha}}{\lambda_{n,\beta_n}} \frac{e^{-KB^2 i^2 J_{n,\beta_n}} \Pi_{n,\alpha}(C_{n,\alpha}(p_0, \sqrt{2} i B\epsilon_{n,\beta_n}))}{e^{-2J_{n,\beta_n}} \Pi_{n,\beta_n}(B_{n,\beta_n}(p_0, \epsilon_{n,\beta_n}))} \frac{1}{\sqrt{\mu_{n,\alpha}}}$$
$$\le \sum_{\alpha \in A_n: \alpha \succeq \beta_n} \sqrt{\mu_{n,\alpha}} e^{2J_{n,\beta_n}} \sum_{i \ge I} e^{(2L-K)B^2 J_{n,\beta_n} i^2} \le \sum_{\alpha \in A_n: \alpha \succeq \beta_n} \frac{e^{(2L-K)B^2 J_{n,\beta_n} I^2}}{1 - e^{(2L-K)B^2 J_{n,\beta_n}}}.$$

This tends to zero if $(K-2L)I^2B^2 > 3$, which is ensured by the assumption that $M^2H(K-2L) > 3$. Similarly, for $\alpha \prec \beta_n$ the second term is bounded by, in view of (10.6),

$$\sum_{\alpha\in A_n:\alpha\prec\beta_n}\sum_{i\geq I}\frac{\lambda_{n,\alpha}}{\lambda_{n,\beta_n}}\frac{e^{-KB'^2i^2J_{n,\alpha}}\Pi_{n,\alpha}(C_{n,\alpha}(p_0,\sqrt{2}iB'\epsilon_{n,\alpha}))}{e^{-2J_{n,\beta_n}}\Pi_{n,\beta_n}(B_{n,\beta_n}(p_0,\epsilon_{n,\beta_n}))}\frac{1}{\sqrt{\mu_{n,\alpha}}}$$

$$\leq \sum_{\alpha\in A_n:\alpha\prec\beta_n}\sqrt{\mu_{n,\alpha}}e^{2J_{n,\beta_n}}\sum_{i\geq I}e^{(2L-K)B'^2i^2J_{n,\alpha}} \leq \sum_{\alpha\in A_n:\alpha\prec\beta_n}\frac{e^{(2L-K)B'^2J_{n,\alpha}I^2}}{1-e^{(2L-K)B'^2J_{n,\alpha}}}.$$

Here $J_{n,\alpha} > HJ_{n,\beta_n}$ for every $\alpha \prec \beta_n$, and hence this tends to zero, again because $(K-2L)B^2I^2 > 3$. □

The conditions of Theorem 10.3 look complicated, but the theorem should be interpreted as saying that adaptation is easy. It appears that even the model weights $\lambda_{n,\alpha}$ need not be very specific. To see this, we simplify the theorem along the same lines as the posterior contraction theorems in Section 8.2.

In analogy to the "crude" prior mass condition (8.4), consider the lower bound

$$\Pi_{n,\beta_n}(B_{n,\beta_n}(p_0,\epsilon_{n,\beta_n})) \geq e^{-Fn\epsilon^2_{n,\beta_n}}. \tag{10.12}$$

Combined with the trivial bound $\Pi_{n,\alpha}(C) \leq 1$, for any set C, we see that (10.6) and (10.7) hold (for sufficiently large I) if, for all $\alpha \in A_n$,

$$\frac{\lambda_{n,\alpha}}{\lambda_{n,\beta_n}} \leq \mu_{n,\alpha}e^{n(H^{-1}\epsilon^2_{n,\alpha}\vee\epsilon^2_{n,\beta_n})}. \tag{10.13}$$

As the right side tends to infinity fast, this does not seem restrictive. Similarly a sufficient condition for (10.8) is that

$$\sum_{\alpha\in A_n:\alpha\prec\beta_n}\frac{\lambda_{n,\alpha}}{\lambda_{n,\beta_n}}\Pi_{n,\alpha}(C_{n,\alpha}(p_0,M\epsilon_{n,\alpha})) = o(e^{-(F+2)n\epsilon^2_{n,\beta_n}}). \tag{10.14}$$

This condition is more involved, but ought also to be true for fairly general model weights, as the prior probabilities $\Pi_{n,\alpha}(C_{n,\alpha}(p_0,M\epsilon_{n,\alpha}))$ ought to be small. In particular, if a reverse of the bound (10.12) holds, for α instead of β_n, then these prior probabilities would be of the order $e^{-F_0n\epsilon^2_{n,\alpha}}$. Since $\epsilon^2_{n,\alpha} \geq H\epsilon^2_{n,\beta_n}$ for $\alpha \prec \beta_n$, this would easily give the condition, for sufficiently large H. These observations are summarized in the following corollary.

Corollary 10.5 (Adaptive contraction rates) *Assume that* (10.5) *and* (10.12)–(10.14) *hold for constants* $\mu_{n,\alpha} > 0$, $E_\alpha > 0$, $F > 0$, *and* $H \geq 1$ *such that* $K^2M^2 > 3(1+F)(K+E+1)$, *and* $KM^2H > 9$, *and* $\sum_{\alpha\in A_n}\sqrt{\mu_{n,\alpha}} \leq e^{n\epsilon^2_{n,\beta_n}}$. *If* $\beta_n \in A_n$ *for every* n *and satisfies* $n\epsilon^2_{n,\beta_n} \to \infty$, *then* $P_0^n\Pi_n(p: d(p,p_0) \geq M\sqrt{H}\,\epsilon_{n,\beta_n}\,|\,X_1,\cdots,X_n) \to 0$.

Proof Choose $L = K/3$, so that $K - 2L = K/3$. Then conditions (10.6)–(10.7) are satisfied for $i^2 \geq I^2 := 3(1+F)$, in view of (10.12) and (10.13). The conditions on the constants in Theorem 10.3 now translate in the conditions as given. □

10.2.1 Universal Weights

Even though the prior probabilities in (10.14) will typically be small, the condition may also be forced by the model weights $\lambda_{n,\alpha}$, for *any* priors Π_α. For given nonnegative numbers (λ_α), consider the "universal weights"

$$\lambda_{n,\alpha} = \frac{\lambda_\alpha e^{-Cn\epsilon_{n,\alpha}^2}}{\sum_{\gamma\in A_n}\lambda_\gamma e^{-Cn\epsilon_{n,\gamma}^2}}. \tag{10.15}$$

These weights put more weight on less complex models (with small $\epsilon_{n,\alpha}$), and thus down-weight the bigger models (with bigger $\epsilon_{n,\alpha}$).

Corollary 10.6 (Adaptive contraction rates, universal weights) *Assume that* (10.5) *and* (10.12) *hold for positive constants E_α and F such that and $\sum_\alpha(\lambda_\alpha/\lambda_\beta)e^{-Cn\epsilon_{n,\alpha}^2/4} = O(1)$ and $\sup_\alpha E_\alpha < \infty$. If $\beta_n \in A_n$ eventually and satisfies $n\epsilon_{n,\beta_n}^2 \to \infty$, then the posterior distribution based on the weights* (10.15) *attains contraction rate ϵ_{n,β_n} at p_0.*

Proof We have $\lambda_{n,\alpha}/\lambda_{n,\beta} = (\lambda_\alpha/\lambda_\beta)e^{-C(n\epsilon_{n,\alpha}^2 - n\epsilon_{n,\beta}^2)}$ by the definition (10.15) of the weights. Combining this with (10.12) we see that

$$\begin{aligned}\frac{\lambda_{n,\alpha}}{\lambda_{n,\beta_n}}\frac{1}{\Pi_{n,\beta_n}(B_{n,\beta_n}(p_0,\epsilon_{n,\beta_n}))} &\le \frac{\lambda_\alpha}{\lambda_{\beta_n}}e^{-Cn\epsilon_{n,\alpha}^2+(F+C)n\epsilon_{n,\beta_n}^2}\\ &\le \frac{\lambda_\alpha}{\lambda_\beta}e^{-Cn\epsilon_{n,\alpha}^2/2}\begin{cases} e^{(-C/2+(F+C)/H)n\epsilon_{n,\alpha}^2}, & \text{if } \alpha \prec \beta_n,\\ e^{(-CH/2+F+C)n\epsilon_{n,\beta_n}^2}, & \text{if } \alpha \succeq \beta_n.\end{cases}\end{aligned}$$

This is a bound on the left sides of both (10.6) and (10.7), and on the terms of the sum in (10.8). The first two inequalities are valid, with $\mu_{n,\alpha} = (\lambda_\alpha/\lambda_\beta)e^{-Cn\epsilon_{n,\alpha}^2}$, for $i^2 \ge F + C$. Condition (10.8) is seen to be satisfied if H is chosen sufficiently large that $-C/2 + (F + C + 2)/H < 0$. □

10.2.2 Parametric Rate

Theorem 10.3 excludes the case that ϵ_{n,β_n} is equal to the "parametric rate" $n^{-1/2}$. To cover this case the statement of the theorem must be slightly modified.

Theorem 10.7 (Adaptive contraction, parametric rate) *Assume there exist constants $B > 0$, $E_\alpha > 0$, $0 < L < K/2$, $H \ge 1$ and $I > 0$ such that* (10.5)–(10.8) *hold for every sufficiently large I. Furthermore, assume that $\sum_{\alpha\in A_n}\sqrt{\mu_{n,\alpha}} = O(1)$. If $\beta_n \in A_n$ for every n and $\epsilon_{n,\beta_n} = n^{-1/2}$, then for every $I_n \to \infty$, $P_0^n\Pi_n(p\colon d(p,p_0) \ge I_n\epsilon_{n,\beta_n}|\, X_1,\cdots,X_n) \to 0$.*

Proof We follow the line of argument of the proof of Theorem 10.3, the main difference being that presently $J_{n,\beta_n} = 1$ and hence does not tend to infinity. To make sure that $P_0^n\phi_n$ is small, we choose the constant B sufficiently large, and to make $P_0^n(A_n)$ sufficiently large we apply Lemma 8.10 with C a large constant instead of $C = 1$. This gives a factor $e^{-(1+C)J_{n,\beta_n}}$

instead of $e^{-2J_{n,\beta_n}}$ in the denominators of (10.11), but this is fixed for fixed C. The arguments then show that for an event A_n with probability arbitrarily close to 1 the expectation $P_0^n[\Pi_n(d(p,p_0) > IB\epsilon_{n,\beta_n} | X_1,\ldots,X_n)\mathbb{1}_{A_n}]$ can be made arbitrarily small by choosing sufficiently large I and B. ☐

10.2.3 Two Models

It is instructive to apply the theorems to the situation of two models, say $\mathcal{P}_{n,1}$ and $\mathcal{P}_{n,2}$ with rates $\epsilon_{n,1} > \epsilon_{n,2}$. For simplicity we shall also assume (10.12) and use universal constants.

Corollary 10.8 *Assume that* (10.5) *holds for* $\alpha \in A_n = \{1,2\}$ *and sequences* $\epsilon_{n,1} > \epsilon_{n,2}$ *such that* $n\epsilon_{n,2}^2 \to \infty$.

(i) *If* $\Pi_{n,1}(B_{n,1}(p_0,\epsilon_{n,1})) \geq e^{-n\epsilon_{n,1}^2}$ *and* $\lambda_{n,2}/\lambda_{n,1} \leq e^{n\epsilon_{n,1}^2}$, *then the posterior contracts at the rate* $\epsilon_{n,1}$.

(ii) *If* $\Pi_{n,1}(B_{n,2}(p_0,\epsilon_{n,2})) \geq e^{-n\epsilon_{n,2}^2}$ *and* $\lambda_{n,2}/\lambda_{n,1} \geq e^{-n\epsilon_{n,1}^2}$, *and for some* M *with* $M^2 \geq 3 \vee (9/K) \vee ((6/K^2 \vee (3/K)(E+1))$ *we have that* $\Pi_{n,1}(C_{n,1}(p_0,M\epsilon_{n,1})) \leq (\lambda_{n,2}/\lambda_{n,1})o(e^{-3n\epsilon_{n,2}^2})$, *then the posterior contracts at the rate* $\epsilon_{n,2}$.

Proof We apply the preceding theorems with $\beta_n = 1$, $A_{n,\prec\beta_n} = \emptyset$ and $A_{n,\succeq\beta_n} = \{1,2\}$ in case (i) and $\beta_n = 2$, $A_{n,\prec\beta_n} = \{1\}$ and $A_{n,\succeq\beta_n} = \{2\}$ in case (ii), both times with $H = 1$ and $\mu_{n,1} = \mu_{n,2} = 1$ and $L = K/3$. ☐

Statement (i) of the corollary gives the slower $\epsilon_{n,1}$ of the two rates under the assumption that the bigger model satisfies the prior mass condition (10.12) and a condition on the weights $\lambda_{n,i}$ that ensures that the smaller model is not overly down-weighted. The latter condition is very mild, as it allows the weights of the two models to be very different. Apart from this, Statement (i) is not surprising, and could also be obtained from Theorem 8.9.

Statement (ii) gives the faster rate $\epsilon_{n,2}$ under the same two conditions, but with the two models swapped, and an additional condition on the prior weight $\Pi_{n,1}(C_{n,1}(p_0,M\epsilon_{n,1}))$ that the bigger model attaches to a neighborhood of the true distribution. If this were of the expected order $e^{-2n\epsilon_{n,1}^2}$ and $\epsilon_{n,1} \gg \epsilon_{n,2}$, then the union of the conditions on the weights $\lambda_{n,1}$ and $\lambda_{n,2}$ in (i) and (ii) could be summarized as

$$e^{-n\epsilon_{n,1}^2} \leq \frac{\lambda_{n,2}}{\lambda_{n,1}} \leq e^{n\epsilon_{n,1}^2}.$$

This is a remarkably big range of weights, showing that Bayesian methods are very robust to the prior specification of model weights.

10.3 Examples

10.3.1 Priors Based on Finite Approximating Sets

Priors based on finite approximating sets were shown in Section 8.2.2 to yield posterior distributions that contract at the rate determined by the (bracketing) entropy approximations.

In this section we extend the construction to multiple models, and combine the resulting priors using the universal weights described in Section 10.2.1.

For each $\alpha \in A_n$ let $\mathcal{Q}_{n,\alpha}$ be a set of nonnegative, integrable functions on the sample space with finite upper bracketing numbers relative to the Hellinger distance d_H (not necessarily probability densities). Let target rates $\epsilon_{n,\alpha}$ satisfy, for every $\alpha \in A_n$,

$$\log N_](\epsilon_{n,\alpha}, \mathcal{Q}_{n,\alpha}, d_H) \lesssim n\epsilon_{n,\alpha}^2.$$

For each α choose a set $\mathcal{U}_{n,\alpha} = \{u_{1,n,\alpha}, \dots, u_{N_n,n,\alpha}\}$ of $\epsilon_{n,\alpha}$-upper brackets over $\mathcal{Q}_{n,\alpha}$, and let $\Pi_{n,\alpha}$ be the uniform discrete probability measure on the set of re-normalized functions $u/\int u\,d\nu$, for $u \in \mathcal{U}_{n,\alpha}$. The collection $\mathcal{U}_{n,\alpha}$ may be a minimal set of $\epsilon_{n,\alpha}$-upper brackets over $\mathcal{Q}_{n,\alpha}$, but the following theorem only assumes a bound on its cardinality, in agreement with its entropy, and that every bracket intersects the set $\mathcal{Q}_{n,\alpha}$.

We combine the priors $\Pi_{n,\alpha}$ with the universal model weights (10.15). The following theorem shows that the resulting mixture prior is appropriate whenever the true density is contained in the union $\cup_{M>0}(M\mathcal{Q}_{n,\alpha})$, for some α.

Theorem 10.9 (Priors on nets) *Construct the priors $\Pi_{n,\alpha}$ as described, using at most $\exp(En\epsilon_{n,\alpha}^2)$ upper brackets, for some constant E. If there exist $\beta_n \in A_n$ and M_0 such that $p_0 \in M_0\mathcal{Q}_{n,\beta_n}$ eventually and such that $\sum_\alpha(\lambda_\alpha/\lambda_{\beta_n})e^{-Cn\epsilon_{n,\alpha}^2/4} = O(1)$ and $n\epsilon_{n,\beta_n}^2 \to \infty$, then the posterior distributions relative to the model weights* (10.15) *contract to p_0 at the rate ϵ_{n,β_n} relative to the Hellinger distance.*

Proof Because the support $\mathcal{P}_{n,\alpha}$ of $\Pi_{n,\alpha}$ has cardinality at most $\exp(En\epsilon_{n,\alpha}^2)$, by construction, the entropy condition (10.5) with d equal to the Hellinger distance is trivially satisfied. If we can also show that (10.12) holds, then the theorem follows from Corollary 10.6.

Because $p_0/M_0 \in \mathcal{Q}_{n,\beta_n}$, there exists $u_n \in \mathcal{U}_{n,\beta_n}$ such that $p_0/M_0 \le u_n$ and $\|\sqrt{p_0} - \sqrt{M_0 u_n}\|_2 \le \sqrt{M_0}\epsilon_{n,\beta_n}$. It follows that $\|\sqrt{M_0 u_n}\|_2 \ge \|\sqrt{p_0}\|_2 = 1$, and by the triangle inequality also that $\|\sqrt{M_0 u_n}\|_2 \le \sqrt{M_0}\epsilon_{n,\beta_n} + 1$. By construction the function $p_n = u_n/\int u_n d\nu$ belongs to $\mathcal{P}_{n,\beta_n}$. Furthermore, by the triangle inequality,

$$\begin{aligned} d_H(p_0, p_n) &\le d_H(p_0, M_0u_n) + d_H(M_0u_n, p_n) \\ &= d_H(p_0, M_0u_n) + |\|\sqrt{M_0u_n}\|_2 - 1| \le 2\sqrt{M_0}\epsilon_{n,\beta_n}. \end{aligned}$$

The inequality in the second line follows from the fact that $\|r - r/\|r\|\| = |1 - \|r\||$ for every norm and function r, applied with $r = \sqrt{M_0u_n}$. We also have $p_0/p_n \le M_0\int u_n\,d\nu = \|\sqrt{M_0u_n}\|_2^2$, which is bounded by the square of $\sqrt{M_0}\epsilon_{n,\beta_n}$. Therefore, in view of the comparison of Hellinger and Kuillback-Leibler discrepancies given in Lemma B.2, it follows that $p_n \in B_{n,\beta_n}(p_0, D\sqrt{M_0}\epsilon_{n,\beta_n})$ for a sufficiently large constant D, whence

$$\Pi_{n,\beta_n}(B_{n,\beta_n}(p_0, D\sqrt{M_0}\epsilon_{n,\beta_n})) \ge \Pi_{n,\beta_n}(\{p_n\}) \ge \frac{1}{\#\mathcal{P}_{n,\beta_n}} \ge e^{-En\epsilon_{n,\beta_n}^2}.$$

It follows that the prior mass condition (10.12) holds, but for a multiple of ϵ_{n,β_n} instead of ϵ_{n,β_n}. By redefining the rates $\epsilon_{n,\alpha}$, we can still draw the desired conclusion from Corollary 10.6. □

If the base collection $\mathcal{Q}_{n,\alpha}$ is the unit ball in a Banach space, then $\cup_{M>0}(M\mathcal{Q}_{n,\alpha})$ is the full space, and the preceding theorem simply assumes that p_0 is contained in the Banach space. For instance, if the Banach space is defined by a regularity norm, then the assumption is that the true density is "α-regular," for some α, without having to satisfy quantitative regularity bounds.

Example 10.10 (Hölder spaces) Consider for each $\alpha > 0$ a Banach space $\mathbb{B}^\alpha(\mathfrak{X})$ of measurable functions $f: \mathfrak{X} \to \mathbb{R}$ whose unit ball $\mathbb{B}_1^\alpha(\mathfrak{X})$ processes finite upper bracketing numbers relative to the $\mathbb{L}_2(\nu)$-norm. Let the constants E_α and functions $H_\alpha: (0, \infty) \to (0, \infty)$ satisfy

$$\log N_{]}(\epsilon, \mathbb{B}_1^\alpha(\mathfrak{X}), \|\cdot\|_2) \leq E_\alpha H_\alpha(\epsilon). \tag{10.16}$$

For $\alpha \in (0, \infty)$, define $\mathcal{Q}_\alpha$ as the set of all nonnegative functions $p: \mathfrak{X} \to \mathbb{R}$ such that the root $\sqrt{p}$ is contained in the unit ball $\mathbb{B}_1^\alpha(\mathfrak{X})$. Because the Hellinger distance on $\mathcal{Q}_\alpha$ corresponds to the $\mathbb{L}_2(\nu)$-distance on the roots of the elements of $\mathcal{Q}_\alpha$, the inequality (10.16) implies that there exists a set of $\epsilon_{n,\alpha}$-brackets of cardinality as in the preceding theorem, for the rates $\epsilon_{n,\alpha}$ satisfying

$$H_\alpha(\epsilon_{n,\alpha}) = n\epsilon_{n,\alpha}^2.$$

The root $\sqrt{p_0}$ of the true density belongs to the Banach space $\mathbb{B}^\beta(\mathfrak{X})$, if and only if $p_0 \in \cup_{M>0}(M\mathcal{Q}_\beta)$. Therefore, the priors chosen as in Theorem 10.9 yield the rate of contraction ϵ_{n,β_n} for any β_n such that $\sqrt{p_0} \in \mathbb{B}^{\beta_n}(\mathfrak{X})$ and $\beta_n \in A_n$ eventually. The prior construction does not use information about the norm of $\sqrt{p_0}$ in $\mathbb{B}^{\beta_n}(\mathfrak{X})$; it suffices that the square root of p_0 be contained in $\mathbb{B}^{\beta_n}(\mathfrak{X})$.

A typical concrete example is a Hölder space $\mathfrak{C}^\alpha([0,1]^d)$ of α-smooth functions $f: [0,1]^d \to \mathbb{R}$. Let By Proposition C.5 the ϵ-entropy of the unit ball of these spaces relative to the uniform norm are of the order $\epsilon^{-d/\alpha}$, whence (10.16) is satisfied with $H_\alpha(\epsilon) = \epsilon^{-d/\alpha}$. This yields posterior contraction at the rate $n^{-\alpha/(2\alpha+d)}$ whenever $\sqrt{p_0} \in \mathfrak{C}^\alpha([0,1]^d)$.

There are many ways of constructing an ϵ-net for the uniform norm over $C_1^\alpha[0,1]^d$, some of which are only of theoretical interest, but others of a constructive nature. Splines of an appropriate degree and dimension are one example.

10.3.2 White Noise Model

In the white noise model, considered previously in Example 8.6 and Section 9.5.4, the observation is (equivalent with) an infinite-dimensional random vector $X^{(n)} = (X_{n,1}, X_{n,2}, \ldots)$, whose coordinates are independent variables $X_{n,i} \overset{\text{ind}}{\sim} \text{Nor}(\theta_i, n^{-1})$, for a parameter vector $\theta = (\theta_1, \theta_2, \ldots)$ belonging to ℓ_2. This fits into the i.i.d. setup by writing the observation as the average of n independent variables distributed as $X^{(1)}$. In the latter model, the average is sufficient and hence the resulting posterior distribution is the same as the posterior distribution given $X^{(n)}$.

Consider the prior Π_α that models the coordinates $\theta_1, \theta_2, \ldots$ as independent variables with $\theta_i \sim \text{Nor}(0, i^{-2\alpha-1})$, for $i = 1, 2, \ldots$ and $\alpha > 0$. In Example 8.6 and Section 9.5.4 the posterior contraction rate for a given α was seen to be the minimax rate $n^{-\alpha/(2\alpha+1)}$ if θ_0 belongs to the Sobolev space $\mathfrak{W}^\alpha$ of sequences θ with $\sum_i i^{2\alpha}\theta_i^2 < \infty$, but to be a possibly slower rate if θ is contained in a Sobolev space of a different order β. By combining the priors into an average $\sum_\alpha \lambda_\alpha \Pi_\alpha$ we may achieve the optimal rate for a Sobolev space of any order.

Theorem 10.3 is limited to a countable set A_n of indices α, but this is not a real restriction, as two rate sequences $n^{-\alpha/(2\alpha+1)}$ and $n^{-\alpha_n/(2\alpha_n+1)}$ are equivalent up to constants whenever $|\alpha - \alpha_n| \lesssim 1/\log n$. Thus a collection of α in a grid of mesh width of the order $1/\log n$ suffices to construct a prior that is rate-adaptive for any $\alpha > 0$.

Theorem 10.11 (White noise model) *If there exists $\beta_n \in A_n$ with $|\beta_n - \beta| \lesssim 1/\log n$ and $\lambda_{\beta_n} e^{n^{1/(2\beta+1)}} \gtrsim 1$, then $\Pi_n(\|\theta - \theta_0\|_2 \leq M n^{-\beta/(2\beta+1)} | X^{(n)}) \to 1$, for any $\theta_0 \in \mathfrak{W}^\beta$, for a constant M that depends on β and $\sum_i i^{2\beta}\theta_{i,0}^2$ only.*

Proof We use estimates on prior masses that may be obtained by direct methods, or derived from general results on Gaussian measures, as discussed in Section 11.4.5. First, by Lemmas 11.47 and 11.48 there exist positive constants d_1 and F_1 depending only on β and $\sum_i i^{2\beta}\theta_{i,0}^2$, if this is finite, so that, for $0 < \epsilon < d_1$,

$$\Pi_{\beta_n}(\theta: \|\theta - \theta_0\|_2 < \epsilon) \geq e^{-F_1 \epsilon^{-1/\beta_n}}. \tag{10.17}$$

Second, there exist universal constants $d > 0$ and $0 < c < C < \infty$ such that, for $0 < \epsilon < d^\alpha$ and every $\alpha > 0$, and B_1 the unit ball of ℓ_2,

$$e^{-C(\epsilon/\sqrt{2})^{-1/\alpha}} \leq \Pi_\alpha(\theta: \|\theta\|_2 < \epsilon) \leq e^{-c(\epsilon\sqrt{2})^{-1/\alpha}}, \tag{10.18}$$

$$\Pi_\alpha(M\mathfrak{W}_1^{\alpha+1/2} + \epsilon B_1) \geq \Phi\Big(-\sqrt{2C(\epsilon/2^{1/2})^{-1/\alpha}} + M\Big). \tag{10.19}$$

The two "small ball probability" inequalities are given in Lemma 11.47. Inequality (10.19) is a consequence of Borell's inequality, the lower bound on the small ball probability given by (10.18), and Lemma K.6(ii).

We derive the theorem as a corollary of Theorem 10.3, with d the ℓ_2-norm, making the choices, for a constant $D > 0$ to be determined:

$$\epsilon_{n,\alpha} = \begin{cases} 6\sqrt{2}\, n^{-\alpha/(2\alpha+1)}, & \text{if } \alpha < \beta, \\ 6\sqrt{2}\, n^{-\beta/(2\beta+1)}, & \text{if } \alpha \geq \beta, \end{cases}$$

$$M_\alpha^2 = D n^{1/(2\alpha+1)} = D n \epsilon_{n,\alpha}^2/72,$$

$$\mathcal{P}_{n,\alpha} = \begin{cases} M_\alpha \mathfrak{W}_1^{\alpha+1/2} + \frac{1}{6}\epsilon_{n,\alpha} B_1, & \text{if } \alpha < \beta, \\ M_\beta \mathfrak{W}_1^{\beta+1/2} + \frac{1}{6}\epsilon_{n,\beta} B_1, & \text{if } \alpha \geq \beta. \end{cases}$$

Condition (10.12) is satisfied, by (10.17) and the assumption on λ_{β_n}, with $F = F_1 + 1$. We verify the remaining conditions of Corollary 10.5.

By Proposition C.10, for $\alpha < \beta$,

$$\log N\Big(\frac{\epsilon_{n,\alpha}}{3}, \mathcal{P}_{n,\alpha}, \|\cdot\|_2\Big) \leq \log N\Big(\frac{\epsilon_{n,\alpha}}{6}, M_\alpha \mathfrak{W}_1^{\alpha+1/2}, \|\cdot\|_2\Big) \lesssim \beta\Big(\frac{18 M_\alpha}{\epsilon_{n,\alpha}}\Big)^{1/(\alpha+1/2)}.$$

The right side evaluates as $E_\alpha n \epsilon_{n,\alpha}^2$, for the constants $E_\alpha = \beta(3\sqrt{D/2})^{1/\alpha+1/2}/72$, which are bounded in $\alpha \leq \beta$. For $\alpha \geq \beta$, the statement reduces to the single statement for $\alpha =$

β, by the definitions, and hence is also true. This verifies (10.5) for the models $\mathcal{P}_{n,\alpha}$. As these models do not fully support the priors Π_α, we supplement this with showing that the prior mass in their complements is negligible, following Remark 10.4. For $\alpha < \beta$ we have $(\epsilon_{n,\alpha}/6\sqrt{2})^{-1/\alpha} = n^{1/(2\alpha+1)} = M_\alpha^2/D$ and hence, by (10.19), for $D > 8C$,

$$\Pi_\alpha(\mathcal{P}_{n,\alpha}^c) \leq 1 - \Phi\Big(-\sqrt{2CM_\alpha^2/D} + M_\alpha\Big) \leq e^{-M_\alpha^2/8} \leq e^{-Dn\epsilon_{n,\beta}^2/576}.$$

For $\alpha > \beta$ we have $(\epsilon_{n,\beta}/6\sqrt{2})^{-1/\alpha} \leq (\epsilon_{n,\beta}/6\sqrt{2})^{-1/\beta} = M_\beta^2/D$. Since the Sobolev unit balls $\mathfrak{W}_1^\alpha$ are decreasing in α, it follows that $\mathcal{P}_{n,\alpha} \supset M_\beta \mathfrak{W}_1^{\alpha+1/2} + \epsilon_{n,\beta}B_1/6$ and hence, by (10.19), for $D > 8C$,

$$\Pi_\alpha(\mathcal{P}_{n,\alpha}^c) \leq 1 - \Phi\Big(-\sqrt{2CM_\beta^2/D} + M_\beta\Big) \leq e^{-Dn\epsilon_{n,\beta}^2/576}.$$

Combining the preceding two displays we see that $\sum_\alpha \lambda_\alpha \Pi_\alpha(\mathcal{P}_{n,\alpha}^c) = o(e^{-(F+2)n\epsilon_{n,\beta}^2})$, if D is chosen to satisfy $D > 576(F+2)$.

We are left to verify conditions (10.13) and (10.14). The first is easily satisfied, with $\lambda_{n,\alpha} = \lambda_\alpha = \mu_{n,\alpha}$, by the assumption on λ_{β_n}. By the upper inequality in (10.18) the left side of the second is bounded above by

$$\sum_{\alpha \prec \beta_n} \frac{\lambda_\alpha}{\lambda_{\beta_n}} e^{-c(M\epsilon_{n,\alpha}\sqrt{2})^{-1/\alpha}} = \sum_{\alpha \prec \beta_n} \frac{\lambda_\alpha}{\lambda_{\beta_n}} e^{-c(12M)^{-1/\alpha} n\epsilon_{n,\alpha}^2/72}.$$

For $\alpha \prec \beta_n$ minus the exponent is bigger than $c(12M)^{-1/\alpha}Hn\epsilon_{n,\beta}^2/72$, which can be made bigger than $(3+F)n\epsilon_{n,\beta}^2$, for every α and fixed M, by choosing a large H. Condition (10.14) follows. □

10.3.3 Finite-Dimensional Approximations

Consider models $\mathcal{P}_{n,J,M}$ indexed by a parameter $\alpha = (J, M)$ consisting of a "dimension parameter" $J \in \mathbb{N}$ and a second parameter $M \in \mathfrak{M}$, such that, for every (J, M) and constants A_M, for every $\epsilon > 0$,

$$\log N\Big(\frac{\epsilon}{5}, C_{n,J,M}(p_0, 2\epsilon), d\Big) \leq A_M J. \tag{10.20}$$

Thus the models $\mathcal{P}_{n,J,M}$ are J-dimensional in the sense of the Le Cam entropy. Such finite-dimensional models may arise from approximation of a collection of target densities through a set of basis functions (e.g. trigonometric functions, splines or wavelets), where a model of dimension J is generated by a selection of J basis functions. The index M may refer to a restriction on the coefficients used with this selection, but also to multiple selections of dimension J.

In this context an abstract definition of "regularity" of order β of a true density p_0, given the list of models $\mathcal{P}_{n,J,M}$, could be that, for some $M_0 \in \mathfrak{M}$,

$$d(p_0, \mathcal{P}_{n,J,M_0}) \lesssim J^{-\beta}.$$

If p_0 is β-regular in this sense, then one might hope that a suitable estimation scheme using the model $\mathcal{P}_{n,J,M_0}$ would lead to a bias of order $J^{-\beta}$, and to a variance term of order J/n.

The best dimension J would balance the square bias and the variance, leading to an optimal dimension J satisfying $J^{-2\beta} \asymp J/n$. This is solved by $J \asymp n^{1/(2\beta+1)}$ and would lead to an "optimal" rate of contraction $n^{-\beta/(2\beta+1)}$.

For super-regular densities satisfying $d(p_0, \mathcal{P}_{n,J,M_0}) \lesssim \exp(-J^\beta)$, or even $p_0 \in \mathcal{P}_{n,J_0,M_0}$ for some J_0 and M_0, a similar argument would lead to rates closer to $1/\sqrt{n}$.

The theorem in this section shows that an adaptive Bayesian scheme, using fairly simple priors, can yield these optimal rates up to a logarithmic factor. As finite-dimensional models are widely applicable, this illustrates the ease by which adaptation is achievable if one is willing to accept a logarithmic factor in the rate. This factor can be avoided by using other schemes (e.g. based on a discretization of the coefficient space as in Section 10.3.1, a smooth prior on restricted coefficient space, or more complicated model weights as in Section 10.3.4).

Le Cam's definition of dimension is combinatorial rather than geometric. A "geometrically J-dimensional" model can be described smoothly by a J-dimensional parameter $\theta \in \mathbb{R}^J$. In that case it is natural to construct a prior on $\mathcal{P}_{n,J,M}$ by putting a prior on the parameter θ. If this prior is chosen to be smooth on $\mathbb{R}^J$, and a ball of d-radius ϵ in $\mathcal{P}_{n,J,M}$ corresponds to a ball of radius $\bar{B}_J\bar{C}_M\epsilon$ on the coefficients $\theta \in \mathbb{R}^J$ (for some constants $\bar{B}_J\bar{C}_M$), then we may expect that, for some constant D_M,

$$\Pi_{n,J,M}(B_{n,J,M}(p_0, \epsilon)) \geq (B_J C_M \epsilon)^J, \qquad \text{if } \epsilon > D_M\, d(p_0, \mathcal{P}_{n,J,M}). \tag{10.21}$$

Here the constants B_J and C_M incorporate the constants $\bar{B}_J$ and $\bar{C}_M$, the prior density on $\mathbb{R}^J$, and the volume of a J-dimensional ball. A restriction of the type $\epsilon \gtrsim d(p_0, \mathcal{P}_{n,J,M})$ is necessary, because by their definition the sets $B_{n,J,M}(p_0, \epsilon)$ are centered around p_0, and this may be at a positive distance to $\mathcal{P}_{n,J,M}$. If $\epsilon > 2d(p_0, \mathcal{P}_{n,J,M})$, then a ball of radius $\epsilon/2$ around a projection of p_0 into $\mathcal{P}_{n,J,M}$ is contained in $C_{n,J,M}(p_0, \epsilon)$. The general constant D_M in (10.21), instead of the universal constant 2, is meant to make up for the difference between the neighborhoods $B_{n,J,M}(p_0, \epsilon)$ and $C_{n,J,M}(p_0, \epsilon)$.

For a large constant A, an arbitrary positive constant C and finite sets $\mathcal{J}_n \subset \mathbb{N}$ and $\mathcal{M}_n \subset \mathcal{M}$, define

$$\epsilon_{n,J,M} = \sqrt{\frac{J \log n}{n} A_M A},$$

$$\lambda_{n,J,M} = \frac{\exp[-Cn\epsilon_{n,J,M}^2]}{\sum_{(J,M)\in\mathcal{J}_n\times\mathcal{M}_n} \exp[-Cn\epsilon_{n,J,M}^2]} \mathbb{1}_{\mathcal{J}_n\times\mathcal{M}_n}(J, M).$$

The $\lambda_{,n,J,M}$ are the "universal weights" (10.15), with $\alpha = (J, M)$ and $\lambda_{J,M} = 1$, restricted to the set $A_n = \mathcal{J}_n \times \mathcal{M}_n$.

Theorem 10.12 (Finite-dimensional approximation) *Suppose that* (10.20)–(10.21) *hold for every J and M, where $A_M A \geq 1$, $C_M^2 A_M A \geq e$ and $B_J\sqrt{J} \geq e$. Let $\mathcal{J}_n$ and $\mathcal{M}_n$ be such that $\sum_{M\in\mathcal{M}_n} e^{-LA_M} = O(1)$ for some $L > 0$. Then for every sequences $J_n \in \mathcal{M}_n$ and $M_n \in \mathcal{M}_n$ with $D_{M_n} d(p_0, \mathcal{P}_{n,J_n,M_n}) \leq \epsilon_{n,J_n,M_n}$, the posterior distribution relative to the model weights $\lambda_{n,J,M}$ attains rate of contraction ϵ_{n,J_n,M_n} at p_0 relative to d.*

Proof We apply Corollary 10.6 with α equal to the pair (J, M) and $\beta_n = (J_n, M_n)$. Condition (10.5) is satisfied (easily with an extra logarithmic factor) by (10.20) in virtue of the definition of the numbers $\epsilon_{n,J,M}$, with $E_\alpha = 1$, for which $n\epsilon_{n,J,M}^2 = J(\log n)A_M A$.

Because $D_{M_n} d(p_0, \mathcal{P}_{n,J_n,M_n}) \leq \epsilon_{n,J_n,M_n}$ by assumption, condition (10.21) implies that the prior mass in the left side of (10.12) can be bounded below by

$$\left(B_{J_n} C_{M_n} \epsilon_{n,J_n,M_n}\right)^{J_n} = \exp\{J_n \log(B_{J_n}\sqrt{J_n}) + \frac{1}{2} J_n \log(C_{M_n}^2 A_{M_n} A)\} \exp\{-\frac{1}{2} J_n \log(n/\log n)\}.$$

The first factor on the right is bounded below by 1 in view of the assumptions on the constants. Because $n\epsilon_{n,J,M}^2 = J(\log n)A_M A$ and $A_M A \geq 1$, it follows that (10.12) is satisfied with $F = 1$.

Finally, we verify that $\sum_\alpha (\lambda_\alpha/\lambda_\beta) e^{-Cn\epsilon_{n,\alpha}^2/4} = O(1)$. Because presently $\lambda_\alpha = 1$, the left side of this equation takes the form

$$\sum_{J \in \mathcal{J}_n} \sum_{M \in \mathcal{M}_n} e^{-CJ(\log n)A_M A/4} \leq \sum_{M \in \mathcal{M}_n} e^{-LA_M},$$

for any constant L and n sufficiently large. The right side is bounded for some L, by assumption. □

Example 10.13 (Supersmooth true density) If $p_0 \in \mathcal{P}_{n,J_0,M_0}$ for some pair of constants (J_0, M_0), then we can apply the preceding theorem with $(J_n, M_n) = (J_0, M_0)$, yielding a rate $(\log n)^{1/2}/\sqrt{n}$.

If $d(p_0, \mathcal{P}_{n,J,M_0}) \lesssim e^{-J^\beta}$ for every J, then we can apply the preceding theorem with J_n a multiple of $(\log n)^{1/\beta}$, yielding a rate $(\log n)^{1/(2\beta)+1/2}/\sqrt{n}$.

Example 10.14 (Regular true density) If $d(p_0, \mathcal{P}_{n,J,M_0}) \lesssim J^{-\beta}$ for every J and some M_0, then we can apply the preceding theorem with J_n a multiple of $(n/\log n)^{1/(2\beta+1)}$, yielding a rate $(n/\log n)^{-\beta/(2\beta+1)}$.

As the two examples illustrate, the logarithmic factor present in the definition of $\epsilon_{n,J,M}$ typically works through in the contraction rate, which tends to be suboptimal by a logarithmic factor. This appears to be inherent in the construction, using smooth priors on the coefficients of the model. The discrete priors of Section 10.3.1 use similar model weights, and do not have this deficit. This point is explored further in the concrete situation of spline approximations in Section 10.3.4.

10.3.4 Log-Spline Models

In Section 9.1 log-spline models of dimension $J_{n,\alpha} \sim n^{1/(2\alpha+1)}$ were seen to yield the minimax posterior contraction rate for a true density whose logarithm belongs to $\mathfrak{C}^\alpha[0,1]$. The prior, which depends on the target smoothness α through the dimension of the model, can be made independent of α with a hyperprior on the dimension. As long as this gives the right amount of weight to (neighborhoods) of the dimension $J_{n,\alpha}$, the posterior distribution should adapt to α.

In the following four subsections we discuss several strategies, which differ in their choices of prior for the coefficients θ and prior for the dimension. The first strategy simply follows the general finite-dimensional construction of Section 10.3.3. It combines non-informative priors on the coefficients, with the universal model weights (10.15), which down-weight higher-dimensional models. The second construction uses similar flat priors on the coefficients, but fixed model weights. Notwithstanding the big difference, both constructions yield adaptation up to a logarithmic factor in the contraction rate. The third and fourth constructions are focused on removing this factor. The third is limited to finitely many smoothness levels, and somewhat surprisingly, manages to remove the logarithmic factor by down-weighting lower-dimensional models. The fourth construction uses again the universal model weights on the dimension, but changes the prior on the coefficients to a discrete prior, in the spirit of Section 10.3.1. One may conclude from these examples that a variety of adaptation schemes may work, in particular if an unnecessary logarithmic factor is taken for granted.

Spline functions only have good approximation properties for α-smooth functions if their order is at least α. For adaptation to a finite range of α, the order can, of course, be chosen an upper bound to this range. Furthermore, if the hyperprior on dimension is placed indirectly and induced by a prior on α and the map $\alpha \mapsto J_{n,\alpha}$, then the splines for that particular model can be chosen of order at least α. For a construction with a hyperparameter directly on the dimension, independent of the sample size, it is in general not possible to link the order of the splines to a target smoothness. Adaptation may be limited to smoothness smaller than the order of the splines, unless an additional layer of models consisting of spline bases of different orders is introduced.

Throughout this section we assume that the true density p_0 is contained in the Hölder space $\mathfrak{C}^\beta[0,1]$, for some $\beta > 0$, and strictly positive. In the first construction the posterior adapts to the minimal and maximal values of the true density, but in the second to fourth constructions the prior depends on known lower and upper bounds on the true density. An additional layer of adaptation might remove this dependency, but is not explored here.

We adopt the notation of Section 9.1. In particular, a log-spline density is denoted by $p_{J,\theta}$, where J is the dimension, and $\theta \in \mathbb{R}^J$ the parameter of the exponential family. We also adopt the notations $B_{J,M}(p_0,\epsilon)$ and $C_{J,M}(p_0,\epsilon)$ for a Kullback-Leibler and Hellinger ball of radius ϵ around p_0 within the restricted log-spline model with parameter set $\Theta_{J,M} = \{\theta \in [-M,M]^J : \theta^\top 1 = 0\}$, as given in (9.1).

Throughout we denote by $\Pi_{n,J,M}$ the prior distribution both for the parameter $\theta \in \Theta_{J,M}$ and for the induced density $p_{J,\theta}$.

Smooth Coefficients, Universal Model Weights

We combine the uniform probability measures $\Pi_{n,J,M}$ on the sets $\Theta_{J,M}$, for $(J,M) \in \mathbb{N}^2$, with the model weights $\lambda_{n,J,M}$ proportional to $e^{-CJ\log(nM)}$. This corresponds to the general construction of Section 10.3.3, with the models $\mathcal{P}_{J,M}$ consisting of the spline densities $p_{J,\theta}$ with $\theta \in \Theta_{J,M}$. We thus obtain the following result.

Corollary 10.15 *If* $\log p_0 \in \mathfrak{C}^\beta[0,1]$, *for some* $\beta > 0$, *then the rate of posterior contraction for the Hellinger distance is* $(n/\log n)^{-\beta/(2\beta+1)}$.

Proof From Lemmas 9.4–9.6 it can be seen that conditions (10.20)–(10.21) of Theorem 10.12 are satisfied with

$$A_M = 2M + \log(30C_0/c_0), \qquad B_J = \sqrt{J}(\mathrm{vol}\{x \in \mathbb{R}^J : \|x\| \le 1\})^{1/J},$$
$$C_M = M^{-2}e^M/(8D_0C_0), \qquad D_M = 2D_0M.$$

The constants B_J tend to a fixed value as $J \to \infty$, by Lemma K.13. If we choose A_M to satisfy $A_M = C_M^{-2}$, then the conditions of Theorem 10.12 are satisfied. Since $d_H(p_0, \mathcal{P}_{J,M}) \lesssim J^{-\beta}$, for every $M \ge 2\|\log p_0\|_\infty/d_0$, by Lemma 9.6(ii) and (i), the theorem applies with $J_n \asymp n^{1/(2\beta+1)}$ and fixed large $M = M_n$. □

Flat Priors, Fixed Model Weights

We combine the uniform probability measures $\Pi_{n,\alpha}$ on the sets $\Theta_{J_{n,\alpha},M}$, for $\alpha \in A := \{\alpha \in \mathbb{Q}^+ : \alpha \ge \underline{\alpha}\}$ for some constant $\underline{\alpha} > 0$ and a fixed $M > 0$, with fixed model weights $\lambda_{n,\alpha} = \lambda_\alpha > 0$ concentrated on A.

Corollary 10.16 *If $p_0 \in \mathfrak{C}^\beta[0,1]$ for some $\beta \in \mathbb{Q}^+ \cap [\underline{\alpha}, \infty)$ and $\|\log p_0\|_\infty < d_0M/2$ and $\sum_\alpha \sqrt{\lambda_\alpha} < \infty$, then then the rate of posterior contraction for the Hellinger distance is $n^{-\beta/(2\beta+1)}\sqrt{\log n}$.*

Proof We shall show that the conditions of Theorem 10.3 hold for $\epsilon_{n,\alpha} := n^{-\alpha/(2\alpha+1)}\sqrt{\log n}$. These numbers satisfy $n\epsilon_{n,\alpha}^2 \asymp J_{n,\alpha}\log n$ and $\epsilon_{n,\alpha} \gtrsim J_{n,\alpha}^{-\alpha}$, for $J_{n,\alpha} \asymp n^{1/(2\alpha+1)}$, and all $\alpha > 0$.

By Lemma 9.5 relation (10.5) holds whenever $n\epsilon_{n,\alpha}^2 \gtrsim J_{n,\alpha}$, with constants E_α that are independent of α.

Because $\|\log p_0\|_\infty < d_0M/2$ by assumption, $d_H(p_0, \mathcal{P}_{J,M}) \lesssim J^{-\beta}$, by Lemma 9.6. Since $\epsilon_{n,\beta} \gtrsim J_{n,\beta}^{-\beta}$, Lemma 9.4 implies that for some positive constants c and C, and sufficiently large n,

$$\Pi_{n,\alpha}(C_{n,\alpha}(p_0,\epsilon)) \le \Pi_{n,\alpha}(\theta \in \Theta_{J_{n,\alpha},M} : \|\theta - \theta_{J_{n,\alpha},M}\| \le c\sqrt{J_{n,\alpha}}\epsilon),$$
$$\Pi_{n,\beta}(B_{n,\beta}(p_0,\epsilon_{n,\beta})) \ge \Pi_{n,\beta}(\theta \in \Theta_{J_{n,\beta},M} : \|\theta - \theta_{J_{n,\beta}}\| \le 2CA\sqrt{J_{n,\alpha}}\epsilon_{n,\beta}).$$

For $\theta_{J,M} \in \Theta_{J,M}$ defined in Lemma 9.6, $\mathrm{vol}(\theta \in \Theta_{J,M} : \|\theta - \theta_{J,M}\| \le \epsilon) \ge 2^{-J}\mathrm{vol}(\theta : \|\theta - \theta_J\|_2 \le \epsilon)$. Thus by Lemma K.13, for any $\alpha, \beta \in A$ and every small ϵ,

$$\frac{\Pi_{n,\alpha}(C_{n,\alpha}(p_0, i\epsilon))}{\Pi_{n,\beta}(B_{n,\beta}(p_0,\epsilon_{n,\beta}))} \le \frac{(cDi\epsilon\sqrt{J_{n,\alpha}})^{J_{n,\alpha}}\mathrm{vol}\{x \in \mathbb{R}^{J_{n,\alpha}} : \|x\| \le 1\}}{(Cd\epsilon_{n,\beta}\sqrt{J_{n,\beta}})^{J_{n,\beta}}\mathrm{vol}\{x \in \mathbb{R}^{J_{n,\beta}} : \|x\| \le 1\}} \lesssim \frac{(Ai\epsilon)^{J_{n,\alpha}}}{(a\epsilon_{n,\beta})^{J_{n,\beta}}}, \tag{10.22}$$

for suitable constants a and A.

If $\alpha \prec \beta$, then with $\epsilon = \epsilon_{n,\alpha}$, inequality (10.22) yields

$$\frac{\Pi_{n,\alpha}(C_{n,\alpha}(p_0, i\epsilon_{n,\alpha}))}{\Pi_{n,\beta}(B_{n,\beta}(p_0,\epsilon_{n,\beta}))} \lesssim \exp\Big[J_{n,\alpha}\Big(\log(Ai\epsilon_{n,\alpha}) - \frac{J_{n,\beta}}{J_{n,\alpha}}\log(a\epsilon_{n,\beta})\Big)\Big]$$
$$\le \exp\Big[J_{n,\alpha}\Big(\log(Ai) + H^{-1}|\log a| + \log\epsilon_{n,\alpha} - H^{-1}\log\epsilon_{n,\beta}\Big)\Big],$$

because $J_{n,\alpha} > H J_{n,\beta}$ for $\alpha < \beta$ for sufficiently large n, no matter which fixed constant H is chosen. Also for some sufficiently large H, simultaneously for all $\alpha \geq \underline{\alpha}$,

$$\log \epsilon_{n,\alpha} - \frac{1}{H} \log \epsilon_{n,\beta} = \Big(\frac{1}{H}\frac{\beta}{2\beta+1} - \frac{\alpha}{2\alpha+1}\Big) \log n + \tfrac{1}{2}\Big(1 - \frac{1}{H}\Big) \log\log n$$

is bounded by a negative multiple of $\log n$, so (10.6) holds with $\mu_{n,\alpha} = \mu_\alpha/\mu_\beta$ and arbitrarily small $L > 0$.

If $\alpha \prec \beta$, then with $\epsilon = IB\epsilon_{n,\alpha}$, inequality (10.22) and similar calculations yield

$$\begin{aligned} &e^{2n\epsilon_{n,\beta}^2} \frac{\Pi_{n,\alpha}(C_{n,\alpha}(p_0, IB\epsilon_{n,\alpha}))}{\Pi_{n,\beta}(B_{n,\beta}(p_0, \epsilon_{n,\beta}))} \\ &\quad\lesssim \exp\Big[J_{n,\alpha}\Big(\log(AIB) + \log \epsilon_{n,\alpha} - \frac{J_{n,\beta}}{J_{n,\alpha}} \log(a\epsilon_{n,\beta}) + 2\frac{J_{n,\beta}}{J_{n,\alpha}} \log n\Big)\Big] \\ &\quad\leq \exp\Big[J_{n,\alpha}\Big(\log(AIB) + \frac{1}{H}|\log a| + \log \epsilon_{n,\alpha} - \frac{1}{H}\log \epsilon_{n,\beta} + \frac{2}{H}\log n\Big)\Big]. \end{aligned}$$

As before, for sufficiently large H, the exponent is smaller than $-J_{n,\alpha}c \log n$ for a positive constant c, simultaneously for all $\alpha \geq \underline{\alpha}$, eventually. This implies that Condition (10.8) is satisfied.

With $\epsilon = \epsilon_{n,\beta}$, inequality (10.22) yields

$$\frac{\Pi_{n,\alpha}(C_{n,\alpha}(p_0, i\epsilon_{n,\beta}))}{\Pi_{n,\beta}(B_{n,\beta}(p_0, \epsilon_{n,\beta}))} \lesssim \exp\Big[J_{n,\beta}\Big(\frac{J_{n,\alpha}}{J_{n,\beta}}(\log(Ai) + \log \epsilon_{n,\beta}) - \log(a\epsilon_{n,\beta})\Big)\Big].$$

If $\alpha \succeq \beta$ the right-hand side is bounded above by

$$\exp\big[J_{n,\beta}(H|\log(Ai)| - \log(a\epsilon_{n,\beta}))\big] \leq e^{J_{n,\beta}Li^2 \log n},$$

for sufficiently large i, for any arbitrarily small constant L. Thus condition (10.7) holds. ☐

Flat Priors, Decreasing Model Weights

We combine the uniform probability measures $\Pi_{n,\alpha}$ on the sets $\Theta_{J_{n,\alpha},M}$, for α in a finite set $\{\alpha_1, \alpha_2, \ldots, \alpha_N\} \subset (0, \infty)$ with the weights

$$\lambda_{n,\alpha} \propto \prod_{\gamma \in A: \gamma < \alpha} (C\epsilon_{n,\gamma})^{J_{n,\gamma}}.$$

These weights vary with n and are decreasing in α, unlike the universal weights (10.15).

Corollary 10.17 *If $p_0 \in \mathfrak{C}^\beta[0,1]$ for some $\beta \in \{\alpha_1, \alpha_2, \ldots, \alpha_N\}$ and $\|\log p_0\|_\infty < d_0 M/2$, then the rate of posterior contraction for the Hellinger distance is $n^{-\beta/(2\beta+1)}$.*

Proof Let $\epsilon_{n,\alpha} = n^{-\alpha/(2\alpha+1)}$, so that $J_{n,\alpha} \asymp n\epsilon_{n,\alpha}^2$, and $J_{n,\alpha'}/J_{n,\alpha} \ll n^{-c}$ for some $c > 0$ whenever $\alpha' > \alpha$. Assume without loss of generality that $\alpha_1 < \cdots < \alpha_N$.

If $r < s$, and hence $\alpha_r < \alpha_s$, then, by inequality (10.22),

$$\frac{\lambda_{n,\alpha_r}\Pi_{n,\alpha_r}(C_{n,\alpha_r}(p_0, i\epsilon_{n,\alpha_r}))}{\lambda_{n,\alpha_s}\Pi_{n,\alpha_s}(B_{n,\alpha_s}(p_0, \epsilon_{n,\alpha_s}))}$$

$$\lesssim \exp\Big[J_{n,\alpha_r}\Big(\log(Ai\epsilon_{n,\alpha_r}) - \frac{J_{n,\alpha_s}}{J_{n,\alpha_r}}\log(a\epsilon_{n,\alpha_s}) - \sum_{k=r}^{s-1}\frac{J_{n,\alpha_k}}{J_{n,\alpha_r}}\log(C\epsilon_{n,\alpha_k})\Big)\Big]$$

$$= \exp\Big[J_{n,\alpha_r}\Big(\log\Big(\frac{Ai}{C}\Big) - \frac{J_{n,\alpha_s}}{J_{n,\alpha_r}}\log(a\epsilon_{n,\alpha_s}) - \sum_{k=r+1}^{s-1}\frac{J_{n,\alpha_k}}{J_{n,\alpha_r}}\log(C\epsilon_{n,\alpha_k})\Big)\Big].$$

The exponent takes the form $J_{n,\alpha_r}(\log(Ai/C) + o(1))$. Applying this with $\alpha_r = \alpha < \beta = \alpha_s$, it follows that, for any chosen C, (10.6) holds with $\mu_{n,\alpha} = 1$ and any arbitrarily small $L > 0$, for sufficiently large i and n.

Similarly, again with $\alpha_r = \alpha < \beta = \alpha_s$,

$$e^{2n\epsilon_{n,\alpha_s}^2}\frac{\lambda_{n,\alpha_r}\Pi_{n,\alpha_r}(C_{n,\alpha_r}(p_0, IB\epsilon_{n,\alpha_r}))}{\lambda_{n,\alpha_s}\Pi_{n,\alpha_s}(B_{n,\alpha_s}(p_0, \epsilon_{n,\alpha_s}))}$$

$$= \exp\Big[J_{n,\alpha_r}\Big(\log\Big(\frac{AIB}{C}\Big) - \frac{J_{n,\alpha_s}}{J_{n,\alpha_r}}\log(a\epsilon_{n,\alpha_s}) - \sum_{k=r+1}^{s-1}\frac{J_{n,\alpha_k}}{J_{n,\alpha_r}}\log(C\epsilon_{n,\alpha_k}) + \frac{2J_{n,\alpha_s}}{J_{n,\alpha_r}}\Big)\Big].$$

This tends to 0 if $C > AIB$. Hence, for C big enough, condition (10.8) is fulfilled as well.

Finally, choose $\alpha_r = \beta < \alpha = \alpha_s$ and note that

$$\frac{\lambda_{n,\alpha_s}\Pi_{n,\alpha_s}(C_{n,\alpha_s}(p_0, i\epsilon_{n,\alpha_r}))}{\lambda_{n,\alpha_r}\Pi_{n,\alpha_r}(B_{n,\alpha_r}(p_0, \epsilon_{n,\alpha_r}))}$$

$$\lesssim \exp\Big[J_{n,\alpha_r}\Big(\frac{J_{n,\alpha_s}}{J_{n,\alpha_r}}(\log(Ai\epsilon_{n,\alpha_r}) - \log(a\epsilon_{n,\alpha_r}) + \sum_{k=r}^{s-1}\frac{J_{n,\alpha_k}}{J_{n,\alpha_r}}\log(C\epsilon_{n,\alpha_k})\Big)\Big]$$

$$= \exp\Big[J_{n,\alpha_r}\Big(\frac{J_{n,\alpha_s}}{J_{n,\alpha_r}}\log(Ai\epsilon_{n,\alpha_r}) + \log\Big(\frac{C}{a}\Big) + \sum_{k=r+1}^{s-1}\frac{J_{n,\alpha_k}}{J_{n,\alpha_r}}\log(C\epsilon_{n,\alpha_k})\Big)\Big].$$

Here the exponent is of the order $J_{n,\alpha_r}(\log(C/a) + o(1)\log i + o(1))$. We conclude that the condition (10.7) holds.

Now the assertion follows from Theorem 10.3, with $A_n = A$. ☐

Discrete Priors, Universal Model Weights

We combine discrete priors $\Pi_{n,\alpha}$ on the sets $\Theta_{J_{n,\alpha},M}$, for $\alpha \in \mathbb{Q}^+$, with model weights, for fixed λ_α and $C > 0$,

$$\lambda_{n,\alpha} \propto \lambda_\alpha e^{-Cn^{1/(2\alpha+1)}}.$$

The support of the prior $\Pi_{n,\alpha}$ are finitely many spline functions, constructed as follows. By Proposition C.5 there exists an $\epsilon_{n,\alpha}$-net $f_1, \ldots, f_{N_{n,\alpha}}$ in the collection of $f \in \mathfrak{C}^\alpha[0,1]$ with $\|f\|_{\mathfrak{C}^\alpha} \le M$ of cardinality $N_{n,\alpha} \lesssim (M/\epsilon_{n,\alpha})^{1/\alpha}$. By Lemma E.5 there exist $\theta_i \in \mathbb{R}^{J_{n,\alpha}}$ such that $\|\theta_i^\top B_{J_{n,\alpha}} - f_i\|_\infty \lesssim MJ_{n,\alpha}^{-\alpha}$, for $i = 1, 2, \ldots, N_{n,\alpha}$. The points $p_{J_{n,\alpha},\theta_i}$ form the support of $\Pi_{n,\alpha}$.

This construction follows the general construction using discrete priors, described in Section 10.3.1, with model weights of the form (10.15). The following is a corollary of Theorem 10.9.

Corollary 10.18 *If $p_0 \in \mathfrak{C}^\beta[0,1]$ for some $\beta \in \mathbb{Q}^+$, $\|\log p_0\|_{\mathfrak{C}^\beta} < M$ and $\|\log p_0\|_\infty < d_0 M/2$, then the rate of posterior contraction for the Hellinger distance is $n^{-\beta/(2\beta+1)}$.*

10.4 Finite Random Series

In this section we consider combining priors defined by finite random, linear combinations of given basis functions into an overall prior by equipping the number J of terms with a hyperprior. Adaptation then occurs provided some suitable linear combination of the basis functions approximates the "true" function well enough and the prior on J is chosen correctly. Given the great variety of possible bases, this provides a flexible method of prior construction, which we illustrate with several examples. The priors are a concrete implementation of the finite-dimensional approximations of Section 10.3.3. The resulting contraction rates likewise suffer from logarithmic factors that can only be avoided with more carefully constructed priors.

For given $J \in \mathbb{N}$, fix basis functions $\psi_{J,1}, \ldots, \psi_{J,J}: \mathfrak{X} \to \mathbb{R}$ on a domain $\mathfrak{X}$, which is typically a bounded convex set in $\mathbb{R}^d$. Construct a prior on functions $w: \mathfrak{X} \to \mathbb{R}$ by choosing a random dimension J and random coefficients $\theta_J = (\theta_{J,1}, \ldots, \theta_{J,J}) \in \mathbb{R}^J$, and next forming the function

$$\theta_J^\top \psi_J := \sum_{j=1}^J \theta_{J,j}\psi_{J,j}.$$

As indicated in the notation, the basis functions, and the priors on coefficients and dimension, may depend on J, and in the asymptotic results they may also further change from one stage to the next. However, the constants in the following conditions should be universal. With some abuse of notation we denote by $\theta^\top\psi$ a generic function, for an unspecified dimension J, and denote by Π both the prior on the pair (θ, J) and on the functions $\theta^\top\psi$.

Throughout the section we impose the following conditions on these priors:

(A1) $A(j) \le \Pi(J = j) \le \Pi(J \ge j) \le B(j)$, for sufficiently large j, for nonnegative, strictly decreasing functions $A, B: \mathbb{N} \to \mathbb{R}$ with limit $A(\infty) = B(\infty) = 0$ at ∞.

(A2) $\Pi(\|\theta_j - \theta_{0,j}\| \le \epsilon) \ge e^{-c_2 j \log_- \epsilon}$, for every $\theta_{0,j} \in \mathbb{R}^j$ with $\|\theta_{0,j}\|_\infty \le H$, for given positive constants c_2 and H, and every sufficiently small $\epsilon > 0$.

(A3) $\Pi(\theta_j \notin [-M, M]^j) \le j e^{-R(M)}$, for every j and M, for an increasing function $R: [0,\infty) \to [0,\infty)$ with limit $R(\infty) = \infty$ at ∞.

Geometric and negative binomial distributions on J satisfy (A1) with $A(j) \asymp B(j) \asymp e^{-cj}$, while Poisson distributions satisfy (A1) with $A(j) \asymp B(j) \asymp e^{-cj\log j}$, for some $c > 0$. Priors that satisfy (A2) include the gamma and exponential distributions assigned independently to every coordinate of θ_j, and multivariate normal and Dirichlet distributions if the parameters lie in a fixed compact set (see Lemma G.13).

Let d be a metric such that, for a positive increasing function $\rho: \mathbb{N} \to \mathbb{R}$, and every $j \in \mathbb{N}$,

$$d(\theta_1^\top\psi_j, \theta_2^\top\psi_j) \le \rho(j)\,\|\theta_1 - \theta_2\|, \qquad \text{every } \theta_1, \theta_2 \in \mathbb{R}^j. \tag{10.23}$$

Example 10.19 (Metrics) For any uniformly bounded basis functions on a domain of finite dominating measure μ the conformity (10.23) of the metric d on the functions $\theta^\top\psi$ with the Euclidean metric on the coefficients is valid for d equal to any $\mathbb{L}_r(\mu)$-metric (where $1 \le r \le \infty$) and $\rho(j) \asymp \sqrt{j}$. This follows from the estimate $\|(\theta_1 - \theta_2)^\top \psi\|_r \le \|\theta_1 - \theta_2\|_1 \max_{1\le j \le J} \|\psi_j\|_r$, and the relation $\|\cdot\|_1 \le \sqrt{j}\|\cdot\|_2$ between the ℓ_1- and ℓ_2-norms on $\mathbb{R}^j$. This observation applies for instance to the classical Fourier base, B-splines and many wavelet bases. For B-splines and d equal to the uniform norm, this estimate is sharp, while for d equal to the $\mathbb{L}_2$-norm, inequality (10.23) is even valid with $\rho(j) \asymp j^{-1/2}$ (see Lemma E.6).

For orthonormal basis functions and d equal to the $\mathbb{L}_2$-norm, the correspondence is of course valid with $\rho(j) = 1$.

We shall see below that in the case of a polynomial function ρ its degree is essentially irrelevant for the contraction rate. In particular $\rho(j) = \sqrt{j}$ is not worse than a constant function ρ.

The following lemma reveals the roles of the preceding assumptions in a proof of a contraction rate, with sieves of the form $\mathcal{W}_{J,M} = \{\theta^\top \psi_j : \theta \in \mathbb{R}^j, j \le J, \|\theta\|_\infty \le M\}$.

Lemma 10.20 *If (A1), (A2) and (A3) and* (10.23) *are satisfied, and for a given function* $w_0 : \mathfrak{X} \to \mathbb{R}$ *there exist integers* $\bar{J}_n \in \mathbb{N}$ *and vectors* $\theta_0 \in [-H, H]^{\bar{J}_n}$ *with* $d(w_0, \theta_0^\top \psi_{\bar{J}_n}) \le \bar{\epsilon}_n$*, then, for* $\bar{\epsilon}_n/\rho(\bar{J}_n)$ *sufficiently small and* $\epsilon_n \le J_n \rho(J_n) M_n$,

$$\Pi(\theta^\top \psi : d(w_0, \theta^\top \psi) \le 2\bar{\epsilon}_n) \ge A(\bar{J}_n)(\bar{\epsilon}_n/\rho(\bar{J}_n))^{c_2 \bar{J}_n},$$

$$\log D(\epsilon_n, \mathcal{W}_{J_n, M_n}, d) \lesssim J_n \log(3 J_n \, \rho(J_n) \, M_n/\epsilon_n),$$

$$\Pi(\mathcal{W}_{J_n, M_n}^c) \le B(J_n) + J_n e^{-R(M_n)}.$$

Proof To prove the first inequality we use the triangle inequality and the assumption on $\bar{J}_n$ to see that the set in its left side contains the set of all functions $\theta^\top \psi_{\bar{J}_n}$ with $d(\theta^\top \psi_{\bar{J}_n}, \theta_0^\top \psi_{\bar{J}_n}) \le \bar{\epsilon}_n$. By (10.23) the latter inequality can be translated into the coefficients θ, giving that the prior probability of this set of functions is bounded below by $\Pi(J = \bar{J}_n)\Pi(\|\theta_{\bar{J}_n} - \theta_0\| \le \bar{\epsilon}_n/\rho(\bar{J}_n))$. The inequality then follows by (A2) and (A3).

Using that the packing number of a union of sets is smaller than the sum of the packing numbers of the individual sets and again assumption (10.23), we obtain that $D(\epsilon, \mathcal{W}_{J_n, M_n}, d) \le \sum_{j=1}^{J_n} D(\epsilon/\rho(j), [-M_n, M_n]^j, \|\cdot\|)$, for any $\epsilon > 0$. Since ρ is increasing by assumption, the right side is bounded above by $J_n D(\epsilon/\rho(J_n), [-M_n, M_n]^{J_n}, \|\cdot\|)$, which is further bounded by $J_n D(\epsilon/(\sqrt{J_n}\rho(J_n)), [-M_n, M_n]^{J_n}, \|\cdot\|_\infty)$. The second assertion next follows from Proposition C.2.

The third Inequality is immediate from (A1) and (A3), upon noting that $\Pi(\mathcal{W}_{J_n, M_n}^c) \le \Pi(J > J_n) + \sum_{j=1}^{J_n} \Pi(\theta \notin [-M_n, M_n]^j)$. □

In the intended application, $\bar{\epsilon}_n$ will be (close to) the targeted rate of contraction as in Theorem 8.9 or Theorem 8.19. The dimension $\bar{J}_n$ will be required to be sufficiently large in order to have sufficiently good approximation to w_0 (i.e. control of the bias of the model), but a too large value will verify the prior mass condition only for a slow rate. The dimension J_n will typically have to be larger than $\bar{J}_n$ to control the remaining mass $\Pi(\mathcal{W}_{J_n, M_n}^c)$, but small enough to make the sieve not too complex. In the upper bounds given by the lemma, the proportionality function ρ from (10.23) influences the log-prior probability and entropy only through a logarithmic factor. This means that as long as ρ is polynomial, its exact degree hardly matters for the final result, the contraction rate.

The following theorem combines the lemma with Theorem 8.19 to obtain contraction rates. Following the setup of Section 8.3, we consider a general observation $X^{(n)}$. We assume that its distribution $P_{w,\eta}^{(n)}$ is parameterized by a pair of a function $w: \mathfrak{X} \to \mathbb{R}$ and possibly an additional nuisance parameter $\eta \in \mathbb{R}^{d_0}$ (which may be vacuous). For simplicity we assume that the two metrics d_n and e_n on the parameters (w, η), which should satisfy the testing assumption (8.17), are identical, and write it as d_n. This notation should be discriminated from the metric d on the set of functions w introduced in (10.23). For instance, in the case that $X^{(n)}$ is a vector of n independent observations, the metric d_n can be taken equal to the root average squared Hellinger distance $d_{n,H}$, defined in (8.25), as shown in Theorem 8.23.

We equip the function w with the finite random series prior as considered previously, and independently equip η with an additional prior.

Theorem 10.21 (Adaptation, finite random series) *Let (A1)–(A3) hold with* $\log A(j) \asymp -j(\log j)^{t_1}$ *and* $\log B(j) \asymp -j(\log j)^{t_2}$, *for* $t_1 \geq t_2 \geq 0$, *and let* (10.23) *hold with* $\rho(j) = j^k$, *for some* $k \geq 0$. *Let* $\epsilon_n \geq \bar{\epsilon}_n$ *be sequences of positive numbers with* $\epsilon_n \to 0$ *and* $n\bar{\epsilon}_n^2 \to \infty$, *and let* $J_n, \bar{J}_n \geq 3$, $M_n > 0$ *be such that, for some sequence* $b_n \to \infty$,

$$\inf_{\theta \in [-H,H]^{\bar{J}_n}} d(w_0, \theta^\top \psi_{\bar{J}_n}) \leq \bar{\epsilon}_n, \tag{10.24}$$

$$\bar{J}_n (\log \bar{J}_n)^{t_1 \vee 1} + \bar{J}_n \log_- \bar{\epsilon}_n \lesssim n\bar{\epsilon}_n^2, \tag{10.25}$$

$$J_n \log \frac{J_n M_n n}{\epsilon_n} \lesssim n\epsilon_n^2, \tag{10.26}$$

$$J_n (\log J_n)^{t_2} \geq b_n n\bar{\epsilon}_n^2, \tag{10.27}$$

$$\log J_n + b_n n\bar{\epsilon}_n^2 \leq R(M_n). \tag{10.28}$$

Suppose that the prior on η *satisfies, for some* $\mathcal{H}_n \subset \mathbb{R}^{d_0}$,

$$\Pi(\eta: \|\eta - \eta_0\| < \bar{\epsilon}_n) \geq e^{-n\bar{\epsilon}_n^2}, \tag{10.29}$$

$$\log_+ \operatorname{diam}(\mathcal{H}_n) \leq n\epsilon_n^2, \tag{10.30}$$

$$\Pi(\eta \notin \mathcal{H}_n) \leq e^{-b_n n\bar{\epsilon}_n^2}. \tag{10.31}$$

Furthermore, assume that $d_n = e_n$ *is a metric for which tests as in* (8.17) *exist, and such that, for some* $a, c > 0$ *and every* $w_1, w_2 \in \mathcal{W}_{J_n, M_n}$ *and* $\eta_1, \eta_2 \in \mathcal{H}_n$,

$$n^{-c}\, d_n((w_1, \eta_1), (w_2, \eta_2)) \lesssim d(w_1, w_2)^a + \|\eta_1 - \eta_2\|^a, \tag{10.32}$$

and for all w, η *such that* $d(w_0, w)$ *and* $\|\eta - \eta_0\|$ *are sufficiently small,*

$$n^{-1} K(P_{w_0,\eta_0}^{(n)}; P_{w,\eta}^{(n)}) \lesssim d^2(w_0, w) + \|\eta - \eta_0\|^2. \tag{10.33}$$

Then $P_{w_0,\eta_0}^{(n)} \Pi_n(d_n((w, \eta), (w_0, \eta_0)) > K_n \epsilon_n) \to 0$, *for every* $K_n \to \infty$.

Proof We verify the conditions of Theorems 8.19 and 8.20, with $\Theta_{n,1} = \mathcal{W}_{J_n, M_n} \times \mathcal{H}_n$ and $\Theta_{n,2}$ its complement, where $\mathcal{W}_{J_n, M_n}$ is defined in Lemma 10.20.

The latter lemma combined with (10.24) gives that $\Pi(\theta^\top\psi: d(w_0, \theta^\top\psi) \leq 2\bar{\epsilon}_n) \geq e^{-c_3 n\bar{\epsilon}_n^2}$, in view of (10.25), for some constant $c_3 > 0$. The prior of η satisfies a similar lower bound by assumption (10.29). The prior independence of w and η and (10.33) show that the simplified prior mass condition (8.22) is satisfied with ϵ_n in the latter condition taken equal to a multiple of the sequence $\bar{\epsilon}_n$. This implies condition (i) of Theorem 8.19.

Lemma 10.20 and condition (10.26) give the bound $\log D(n^{-c/a}\epsilon_n^{1/a}, \mathcal{W}_{J_n,M_n}, d) \lesssim n\epsilon_n^2$. Condition (10.30) and again (10.26) give a similar bound on $\log D(n^{-c/a}\epsilon_n^{1/a}, \mathcal{H}_n, \|\cdot\|)$. Combined with (10.32) this verifies the global entropy condition (8.23), with $e_n = d_n$ and ϵ_n replaced by a multiple, and hence the local entropy condition (ii) of Theorem 8.19.

Lemma 10.20 and conditions (10.27) and (10.28) show that $\Pi(\mathcal{W}_{J_n,M_n}^c) \leq 2e^{-b_n n\bar{\epsilon}_n^2}$. A similar bound on the remaining mass for the prior on η holds by (10.31). Together with the simplified prior mass condition (8.22) found previously, this verifies condition (iii) as given in Theorem 8.20. ◻

Conditions (10.32) and (10.33) relate the statistical problem to the metric structure of the parameter space. They are verified for several examples below. It is notable that condition (10.33) requires a correspondence between the Kullback-Leibler divergence (a square statistical discrepancy) and the square distance d^2, whereas the condition (10.32) allows any polynomial relation between the statistical distance d_n and the distance d, and even an extra factor n^{-c}. A technical explanation for this difference is that the entropy of the present finite-dimensional models is mostly driven by their dimensions, in the form of estimates $J_n \log(1/\epsilon)$, with J_n the dimension. Relation (10.32) is employed to bound the d_n-entropy of the statistical models in terms of the d-entropy of the parameter space, but the nature of the relation between the metrics (if at least polynomial) will affect the end result only through the logarithmic factor $\log(1/\epsilon)$. (Sharp contraction results, without logarithmic factors, would require a more precise analysis, and possibly different priors, as discussed in Sections 10.3.4 and Section 9.1.)

It is helpful that the more stringent of the two conditions is needed only at the true parameter, where the Kullback-Leibler divergence is often better behaved.

Conditions (10.29)–(10.31) concern the prior on the parameter η, and do not usually play a determining role for the contraction rate.

In many situations a best approximation of a true function w_0 by a linear combination of J basis functions satisfies, for some $\alpha > 0$ and every $J \in \mathbb{N}$,

$$\inf_{\theta\in\mathbb{R}^J} d(w_0, \theta^\top\psi_J) \lesssim J^{-\alpha}. \tag{10.34}$$

The parameter α depends on w_0 and can be considered the regularity level of this function, relative to the basis functions. If the coordinates of the minimizing vectors θ can be limited to the fixed interval $[-H, H]$ of condition (A2), then (10.24) will be satisfied for $\bar{\epsilon}_n \geq \overline{J}_n{}^{-\alpha}$. In many examples M_n and $R(M_n)$ can be taken (large) powers of n. It may be verified that inequalities (10.24)–(10.28) are then satisfied for the choices, for $\bar{t}_1 = t_1 \vee 1$,

$$\overline{J}_n \asymp (n/\log^{\bar{t}_1} n)^{1/(2\alpha+1)}, \qquad J_n \gg n^{1/(2\alpha+1)}(\log n)^{2\alpha/(2\alpha+1)\bar{t}_1 - t_2}$$

$$\bar{\epsilon}_n \asymp (n/\log^{\bar{t}_1} n)^{-\alpha/(2\alpha+1)}, \qquad \epsilon_n \gg n^{-\alpha/(2\alpha+1)}(\log n)^{\alpha/(2\alpha+1)\bar{t}_1 - t_2/2+1/1}.$$

The maximum of $\bar{\epsilon}_n$ and ϵ_n is the posterior contraction rate under the preceding theorem; it misses the usual rate $n^{-\alpha/(2\alpha+1)}$ by only a logarithmic factor.

The preceding applies to functions on univariate and multivariate domains alike, albeit that when α is taken to be the usual smoothness of a function of d variables (e.g. the number of bounded derivatives), then the index α in (10.34) should be replaced by α/d. In terms of the usual smoothness index, the resulting rate is then $n^{-\alpha/(2\alpha+d)}$ times a logarithmic factor. A function $w_0\colon [0,1]^d \to \mathbb{R}$ may possess different regularity levels in its d coordinates. For such an *anisotropic* function, an approximation exploiting different dimensions for the different coordinates is appropriate. If $\alpha_1, \ldots, \alpha_d$ are the regularity levels, with $\alpha^* = d/(\sum_{k=1}^d \alpha_k^{-1})$ their *harmonic mean*, and $J_n(k)$ is the dimension used for the kth coordinate, a typical approximation rate takes the form

$$\inf_{\theta\in\mathbb{R}^{J_n(1)}\times\cdots\times\mathbb{R}^{J_n(d)}} d(w_0, \theta^\top \psi_J) \lesssim \sum_{k=1}^d J_n(k)^{-\alpha_k}.$$

For $J_n = \prod_{k=1}^d J_n(k)$ the overall dimension, optimal choices satisfying inequalities (10.26)–(10.25) are then

$$J_n(k) \asymp (n/\log n)^{(\alpha^*/\alpha_k)/(2\alpha^*+d)}, \quad \overline{J}_n = \prod_{k=1}^d J_n(k) \asymp \bar{\epsilon}_n^{-d/\alpha^*},$$
$$\epsilon_n \asymp (n/\log n)^{-\alpha^*/(2\alpha^*+d)}(\log n)^{(1-t_1)/2}.$$

Example 10.22 (Classical regularity bases) Relation (10.34) and the conclusion of the preceding paragraph applies to most of the classical regularity spaces: Hölder functions $w_0 \in \mathfrak{C}^\alpha[0,1]$ in combination with classical Fourier series (with d the $\mathbb{L}_2$- or the uniform distance; see Jackson 1912), polynomials (with d the uniform norm; see Hesthaven et al. 2007)), B-splines (with d the $\mathbb{L}_2$ or supremum metric; see Section E.2), and wavelets (see Cohen et al. 1993 and Section E.3).

In all examples tensor products of the basis functions can be used to approximate functions on a multi-dimensional domain. If $w_0 \in \mathfrak{C}^\alpha([0,1]^d)$, then relation (10.34) is again valid, but with α replaced by α/d (e.g. see Lemma E.7 for multivariate B-splines).

The approximation rate of such tensor products for a function w_0 in an anisotropic Hölder class of functions on $[0,1]^d$, with regularity levels $\alpha_1, \ldots, \alpha_d$ (see (E.15)), is $\sum_{k=1}^d J(k)^{-\alpha_k}$, when the tensors are formed by using the first $J(k)$ basis functions for the kth coordinate, for $k = 1, \ldots, d$ (see Section E.2 for a precise statement for tensor B-splines). This leads to the rates described previously, with the harmonic mean of the regularity levels replacing α.

Example 10.23 (Bernstein polynomials) Bernstein polynomials, as in Example 5.10, possess the slower approximation rate of $J^{-\alpha/2}$, for $\alpha \leq 2$ (see Lemma E.3). This leads to the same rates as in the preceding discussion, but with α replaced by $\alpha/2$ throughout. In particular, the contraction rate, even though dependent on α, will be suboptimal for all α. For $\alpha \leq 1$, this can be repaired using the coarsened Bernstein polynomials, as in Lemma E.4. This is similar to adaptation by the Dirichlet priors considered in Section 9.3.

10.4.1 Density Estimation

A linear combination of basis functions is not naturally nonnegative or integrates to one. However, a finite random series prior $\theta^\top\psi$ can be transformed into a prior for a probability density by applying a link function $\Psi: \mathbb{R} \to (0, \infty)$ followed by renormalization, giving the prior

$$\frac{\Psi(\theta^\top\psi)}{\int \Psi(\theta^\top\psi)\, d\mu}.$$

The exponential link function $\Psi(x) = e^x$ is particularly attractive, as the statistical distances and discrepancies of the transformed densities relate to the uniform norm on the functions $\theta^\top\psi$, as shown in Lemma 2.5, but other link functions may do as well. (Link functions such that $\log\Psi$ is Lipschitz behave exactly as the exponential link, as shown in Problem 2.4; for properties of other link functions see Problem 2.5.) The special case of the B-spline basis with the exponential link function is discussed in detail in Section 10.3.4.

For nonnegative basis functions employing a link function is unnecessary if the linear combinations are restricted to vectors θ with positive coordinates, and the prior can be constructed through renormalization only, giving $\theta^\top\psi/\theta^\top c$, for $c = \int \psi\, d\mu$. It seems that in general the approximation (10.34) might suffer from a restriction to nonnegative coefficients, as basis functions may need to partially cancel each other for good approximation, but for localized bases, such as wavelets or B-splines, such a problem should not occur. For the B-spline basis, this is verified in Lemma E.5, part (d), under the assumption that the true function is bounded away from zero. The B-spline basis, normalized to integrate to one, has the further elegant property that $\int \theta^\top B^*_{J,j}(x)\, dx = \sum_j \theta_j$, so that renormalization becomes unnecessary if the parameter vector θ is further restricted to the unit simplex $\mathbb{S}_J$. A Dirichlet prior on θ has the further attraction of allowing an analytic expression for posterior moments (see Problem 10.11).

The preceding observations are valid for functions on a univariate and on multivariate domains alike. The domains would typically have to be compact to ensure the approximation property (10.34).

10.4.2 Nonparametric Normal Regression

Consider independent observations $(X_1, Y_1), \ldots, (X_n, Y_n)$ following the regression model $Y_i = w(X_i) + \varepsilon_i$, where $\varepsilon_i \stackrel{\text{iid}}{\sim} \text{Nor}(0, \sigma^2)$, and w and σ are unknown parameters. The covariates $X_1, \ldots, X_n$ can be either fixed or random; in the latter case they are assumed i.i.d. and independent of the errors.

We model the regression function w by a finite random series prior $\theta^\top\psi$, and put an independent prior distribution on σ. We assume that for any $\sigma_0 > 0$, there exist positive numbers t_4, t_5 and t_6 such that for sufficiently small $\sigma_1, \sigma_2, \sigma_3 > 0$,

$$\Pi(|\sigma - \sigma_0| < \sigma_3) \gtrsim \sigma_3^{t_4},$$
$$\Pi(\sigma < \sigma_1) + \Pi(\sigma > \sigma_2^{-1}) \lesssim \exp(-\sigma_1^{-t_5}) + \sigma_2^{t_6}.$$

For instance, these conditions hold if some positive power of σ is inverse-gamma distributed. The first condition (easily) ensures (10.29) as soon as $n\bar\epsilon_n^2$ is some power of n, while the second makes (10.31) valid when the sieve for σ is taken equal to $\mathcal{H}_n = (n^{-1/t_5}, e^{n\bar\epsilon_n^2/t_6})$.

The random design model is an i.i.d. model, and hence the Hellinger distance on the law of the observations is a possible choice for the metric d_n. For known σ this distance is bounded above by the $\mathbb{L}_2(G)$-distance on the regression functions, for G the marginal distribution of the design points, while the Kullback-Leibler discrepancies are equal to the square of this distance, by Lemma 2.7. Relations (10.32) and (10.33) are then verified for d the $\mathbb{L}_2(G)$-distance, or a stronger distance, such as the uniform distance. For unknown σ the situation is more complicated, but Lemma 2.7(i) shows that for σ restricted to the sieve $\mathcal{H}_n$, so that $\sigma \geq n^{-1/t_5}$, the Hellinger distance is bounded above by a multiple of n^{1/t_5} times the sum of the $\mathbb{L}_2(G)$ distance and the Euclidean distance on σ. Thus relation (10.32) is still valid in the weaker form with $c = 1/t_5$. In the neighborhood of the true parameter (f_0, σ_0) the Kullback-Leibler divergence still translates by Lemma 2.7(ii) into the sum of the $\mathbb{L}_2(G)$- and the Euclidean metric, whence (10.33) remains valid. Thus under the conditions of Theorem 10.21 for the finite series prior we obtain a rate of contraction relative to the Hellinger distance on the densities of the observations. If the prior would be restricted to uniformly bounded regression functions, or the posterior could be shown to concentrate on such functions, then this translates into a contraction rate relative to the $\mathbb{L}_2(G)$-distance on the regression functions, but in general this requires additional arguments and possibly conditions.

For the fixed design model we can reach the same conclusion, but with metric d_n taken equal to the root average square Hellinger distance $d_{n,H}$ and with the empirical distribution $\mathbb{G}_n$ of the design points replacing the marginal distribution G. (It may then be attractive to bound this by the uniform distance, which does not change with n.) The rate of contraction will then also be relative to $d_{n,H}$, which entails convergence to zero of

$$\frac{1}{n}\sum_{i=1}^{n} d_H^2(\mathrm{Nor}(f_0(x_i), \sigma_0^2), \mathrm{Nor}(f(x_i), \sigma^2))$$

$$= 2\Big[1 - \sqrt{1 - \frac{(\sigma_0 - \sigma)^2}{\sigma_0^2 + \sigma^2}} \int e^{-(f_0 - f)^2/(4\sigma_0^2 + 4\sigma^2)}\, d\mathbb{G}_n\Big].$$

This readily gives consistency and a contraction rate for σ, but as in the random design case translates into a rate relative to the more natural $\mathbb{L}_2(\mathbb{G}_n)$-metric only under additional conditions. For the fixed design case a better alternative it to not employ $d_{n,H}$. For known σ, it is shown in Section 8.3.2 that the general contraction theorem is valid when $d_n = e_n$ is taken equal to the $\mathbb{L}_2(\mathbb{G}_n)$-norm of the regression functions. Lemma 8.27 in the same section shows that this extends to the case of unknown σ if this is restricted to a bounded interval. Since we already concluded that the posterior distribution of σ is consistent for σ_0, the latter restriction can be made without loss of generality. Thus in the fixed design case we obtain under the conditions of Theorem 10.21 for the finite series prior (relative to the uniform norm on the series) a rate of contraction relative to the $\mathbb{L}_2(\mathbb{G}_n)$-distance on the regression functions.

10.4.3 Nonparametric Binary Regression

Consider independent observations $(X_1, Y_1), \ldots, (X_n, Y_n)$ in the binary regression model discussed in Section 2.5. Thus Y_i takes is values in $\{0, 1\}$, and given X_i is Bernoulli distributed with success probability $p(X_i)$, for a given function p. The covariates $X_1, \ldots, X_n$ can be either i.i.d. or fixed values in a set $\mathfrak{X}$.

For a given link function $H\colon \mathbb{R} \to (0, 1)$ and vector of basis functions ψ, we model the success probability p through a prior of the form $H(\theta^\top \psi)$.

The logistic link function is convenient as it allows a direct relationship between the Hellinger distance, Kullback-Leibler divergence, and the $\mathbb{L}_2$-norm of the functions $\theta^\top \psi$: by Lemma 2.8(ii)–(iv), relations (10.32) and (10.33) hold (in its most natural form with $c = 0$ and $a = 2$) for d the $\mathbb{L}_2(G)$-distance, where G is the marginal distribution of a covariate X_i in the case of a random design, and the empirical distribution of the design points given fixed design. Theorem 10.21 therefore gives adaptive contraction rates relative to the Hellinger distance on the densities of the observations as defined in (2.6). In view of Problem B.2(i)+(iii), this also implies the same contraction rates relative to the $\mathbb{L}_1$ and Hellinger distances on the response probability function p relative to G.

Other link functions are possible as well. For instance, by Lemma 2.8(ii) the Hellinger distance is bounded above by the $\mathbb{L}_2(G)$-distance for any link function with finite Fisher information for location, while by part (i) of the lemma and Lemma B.1(ii) the Hellinger distance is bounded above by the root of the $\mathbb{L}_1(G)$-distance for the slightly larger class of all Lipschitz link functions. This readily gives relation (10.32) for d one of these distances, with $a = 1/2$ in the second case. A fortiori the relation is true for d the uniform distance. For the latter choice we can also retain (10.33), at least when the true regression function is bounded away from zero and one. Indeed a function in a uniform neighborhood of this true function is then also bounded and for bounded functions the Kullback-Leibler divergence of the observations are upper bounded by the uniform distance, by the last assertion of Lemma 2.8.

For $\mathfrak{X} = (0, 1)$ the B-spline basis may be used with the identity link function, and the coordinates of the parameter vector θ restricted to the interval $(0, 1)$. In view of part (c) of Lemma E.5 this restriction does not affect the approximation of the splines. Problem 10.11 suggests that independent beta priors on the coefficients allow an explicit expression of posterior moments.

10.4.4 Nonparametric Poisson Regression

Consider independent observations $(X_1, Y_1), \ldots, (X_n, Y_n)$ in the Poisson regression model discussed in Section 2.6. Thus $Y_i \mid X_i \overset{\text{ind}}{\sim} \text{Poi}(H(w(X_i)))$, for a fixed link function $H\colon \mathbb{R} \to (0, \infty)$ and an unknown function $w\colon \mathfrak{X} \to \mathbb{R}$. The covariates $X_1, \ldots, X_n$ can be either i.i.d. or fixed values in the set $\mathfrak{X}$; denote by G and G_n the marginal density or the empirical distribution of $X_1, \ldots, X_n$ in these two cases.

We model the conditional mean value function $H(w)$ through a prior of the form $H(\theta^\top \psi)$.

For a link function such that the function $H'/\sqrt{H}$ is bounded, the Hellinger distance on the densities p_w of an observation (X_i, Y_i) in the random design model is bounded above by the $\mathbb{L}_2(G)$-distance on the functions w, by Lemma 2.9(i). Furthermore, if the true mean function $H(w_0)$ is bounded, then part (ii) of the lemma shows that the Kullback-Leibler divergence is bounded above by a multiple of the square of the $\mathbb{L}_2(G)$-distance. This verifies relations (10.32) (in its most natural form with $c = 0$ and $a = 2$ and vacuous η) and (10.33) for d the $\mathbb{L}_2(G)$-distance. Theorem 10.21 therefore gives adaptive contraction rates relative to the Hellinger distance on the densities of the observations. If the design is fixed rather than random, the same reasoning, but using the root average square Hellinger distance, leads to the same conclusion, with G_n taking the place of G.

Link functions such that the function $H'/\sqrt{H}$ is only locally bounded, such as the exponential link function, are not covered by the preceding paragraph. We may still reach the same conclusion if $|\sqrt{H}(\theta_1^\top\psi) - \sqrt{H}(\theta_2^\top\psi)| \le n^c\|(\theta_1 - \theta_2)^\top\psi\|_\infty$, for some constant $c \ge 0$ for every $\|\theta_1\|, \|\theta_2\| \le M_n$, for M_n a positive sequence satisfying $R(M_n) \le bn$ for some $b > 0$ (and R the function in (A2)). This inequality will then replace Lemma 2.9(i) and ensures (10.32) with d the uniform distance on the functions $\theta^\top\psi$ (and $a = 1$ and vacuous η), while part (ii) of the lemma still verifies (10.33) with this distance d as long as the true mean function $H(w_0)$ is bounded.

Contraction relative to the Hellinger distance on the densities p_w does not generally imply contraction relative to (other) natural distances on $H(w)$. However, the same contraction rate will be implied for the Hellinger distance $\|\sqrt{H(w)} - \sqrt{H(w_0)}\|_{2,G}$ (or G_n) on the mean functions if it can be shown that the posterior distribution concentrates asymptotically all its mass on a uniformly bounded set of functions (i.e. $\Pi_n(\|H(w) - H(w_0)\|_\infty > B|\, X_1, Y_1, \ldots, X_n, Y_n)$ tends to zero in probability for some B) . This follows because the Hellinger distance between two Poisson measures is bounded below by a constant times the distance between the roots of their means, where the constant is uniform in bounded means. There are at least two ways to meet the extra condition — showing that the posterior is consistent for the uniform distance on $H(w)$, or ensuring that the prior on $H(w)$ charges only functions that are bounded away from zero and infinity.

If $\mathfrak{X}$ is the unit interval and the B-spline basis is used for the prior $\theta^\top\psi$, then we can use the identity function as the link function and restrict the coefficients θ to the positive half line, in view of in view of part (b) of Lemma E.5. Problem 10.11 suggests an explicit expression of the Bayes estimator if these coefficients are given independent gamma priors. This remark extends to multivariate covariates and tensor products of B-splines.

10.4.5 Functional Regression

In the (linear) *functional regression* model the covariates X and the parameter w are functions $X: \mathfrak{T} \to \mathbb{R}$ and $w: \mathfrak{T} \to \mathbb{R}$ on a set $\mathfrak{T} \subset \mathbb{R}^d$, and the usual linear regression is replaced by the integral $\int_{\mathfrak{T}} X(t)w(t)\,dt$, which we observe with a random error. Given Gaussian errors $\varepsilon_i \overset{\text{iid}}{\sim} \text{Nor}(0, \sigma^2)$, for $i = 1, \ldots, n$, the observations are an i.i.d. sample of pairs $(X_1, Y_1), \ldots, (X_n, Y_n)$ following the model

$$Y_i = \int_{\mathfrak{T}} X_i w\, d\lambda + \varepsilon_i. \tag{10.35}$$

Induce a prior on w through an expansion $w = \theta^\top\psi$. For simplicity we consider σ as fixed; alternatively it can be treated as in Section 10.4.2. Assume that $\mathrm{E}\int_{\mathfrak{T}} X^2\, d\lambda < \infty$ and write $\|\cdot\|_2$ for the $\mathbb{L}_2$-norm relative to the Lebesgue measure λ on $\mathfrak{T}$.

The square Hellinger distance and Kullback-Leibler divergence on the laws P_w of a single observation (X_i, Y_i) satisfy, by the Cauchy-Schwartz inequality,

$$d_H^2(P_v, P_w) \le \frac{1}{2\sigma^2}\mathrm{E}_X\Big(\int_{\mathfrak{T}} X\,(v-w)\,d\lambda\Big)^2 \lesssim \|v-w\|_2^2,$$

$$K(P_v; P_w) = \frac{1}{2\sigma^2}\mathrm{E}_X\Big(\int_{\mathfrak{T}} X\,(v-w)\,d\lambda\Big)^2 \lesssim \|v-w\|_2^2.$$

Therefore (10.32) and (10.33) are verified for the $\mathbb{L}_2$-distance, and under the conditions of Theorem 10.21 we obtain an adaptive rate of contraction relative to the Hellinger distance on P_w.

Instead of the preceding model we can also consider a longitudinal version, in which instead of Y_i we observe $Y_i(T_i) = X_i(T_i)w(T_i) + \varepsilon_i$, for $T_1, \ldots, T_n$ i.i.d. random points in $\mathfrak{T}$, independent of $X_1, \ldots, X_n, \varepsilon_1, \ldots, \varepsilon_n$. The statistical distances for this model are identical to the preceding model and hence we can draw the same conclusions.

10.4.6 Whittle Estimation of a Spectral Density

Consider estimating the spectral density f of a stationary Gaussian time series $(X_t : t \in \mathbb{Z})$ using the Whittle likelihood, as considered in Section 7.3.3. Thus we act as if the periodogram values $U_l = I_n(\omega_l)$, for $l = 1, \ldots, \nu$, are independent exponential variables with means $f(\omega_l)$, for $\omega_l = 2\pi l/n$ the natural frequencies, and form a posterior distribution based on the corresponding (pseudo) likelihood. By Theorem L.8, the sequence of actual distributions of the periodogram vectors $(U_1, \ldots, U_\nu)$ and their exponential approximation are contiguous. Thus a rate of contraction for this (pseudo) posterior distribution is also valid relative to the original distribution of the time series.

As a prior model for the spectral density we set $f = H(\theta^\top \psi)$ for a finite series prior $\theta^\top \psi$ and a given monotone link function $H: \mathbb{R} \to (0, \infty)$, which should have the property that $\log H$ is Lipschitz on bounded intervals.

As the observations are independent, the root average square Hellinger distance $d_{n,H}$ is the natural candidate for the metric d_n in the application of Theorem 10.21. In Section 7.3.3 this is seen to be equivalent to the root of

$$e_n(f_1, f_2)^2 = \frac{1}{\nu}\sum_{i=1}^{\nu} \frac{(f_1(\omega_l) - f_2(\omega_l))^2}{(f_1(\omega_l) + f_2(\omega_l))^2}.$$

Because $\log u - \log v = \int_u^v s^{-1}\,ds \ge \frac{1}{2}(v-u)/(v+u)$, for any $0 < u < v$, we see that

$$d_n(\theta_1^\top \psi, \theta_2^\top \psi) := e_n(H(\theta_1^\top \psi), H(\theta_2^\top \psi)) \le \|\log H(\theta_1^\top \psi) - \log H(\theta_2^\top \psi)\|_\infty.$$

Assume that the right side is bounded above by $n^c \|(\theta_1 - \theta_2)^\top \psi\|_\infty$, for some constant $c \ge 0$ and every $\|\theta\|_\infty \le M_n$, for some sequence $M_n > 0$ satisfying $R(M_n) \le bn$ for some $b > 0$ and R as in (A2). Then (10.32) is verified with d taken equal to the uniform distance on the functions $\theta^\top \psi$.

In Section 7.3.3 the Kullback-Leibler divergence between exponential densities is bounded above by a multiple of the square of the difference of their means, if the means are bounded away from zero. If we assume that the true spectral density $f_0 = H(w_0)$ is bounded away from zero, then this gives, for every w such that $\|w_0 - w\|_\infty$ is small enough,

$$\frac{1}{\nu}\sum_{l=1}^{\nu} K(P_{w_0,l}; P_{w,l}) \lesssim \|\log H(w_0) - \log H(w)\|_\infty^2 \lesssim \|w - w_0\|_\infty^2,$$

in view of the assumed Lipschitz continuity of $\log H$. Hence (10.33) is verified for the uniform distance on w.

The root average square Hellinger distance is perhaps not the most natural distance on the spectral densities. On the set of spectral densities that are bounded away from zero, we have that $e_n(f_1, f_2) \gtrsim \nu^{-1}\sum_{l=1}^{\nu}\bigl|\log f_1(\omega_l) - \log f_2(\omega_l)\bigr|$. (This follows, because the inequality involving the Hellinger distance between exponential distributions used in the preceding can be reversed if the means are bounded away from zero.) Thus if the posterior probability of the event $\{\|\log f_0 - \log f\|_\infty > B\}$ tends to zero for some B, then the contraction rate with respect to e_n implies the same contraction rate for the average $\mathbb{L}_1$-distance on the log spectral densities. The additional condition certainly holds if the posterior is consistent for the uniform distance on $\log f$, or if the prior on f charges only functions that are bounded away from zero and infinity.

The B-spline basis may be used with the reciprocal function as the link and the coefficients restricted to the positive half line. Problem 10.11 suggests explicit expressions of posterior moments when using independent gamma priors on the coefficients.

10.5 Model Selection Consistency

When multiple models are under consideration one may ask whether the posterior distribution can identify the "correct model" from the given class of models. Within our asymptotic framework this is equivalent to asymptotic contraction of the posterior distribution of the model index to the "true index." However, a useful precision of this idea is not so straightforward to formulate. This is true in particular for nonparametric model, where a "true model" is often defined only by approximation. It may not be unique, and the true density may not belong to any of the models under consideration.

The issue of consistency of model selection is connected to adaptation, in that consistency for model selection will force rate adaptation. This is clear from representation (10.1) of the posterior distribution. Model selection consistency would entail that the posterior $\lambda(\cdot| X^{(n)})$ on the model index contract to the "true index" β, whence the full posterior distribution will be asymptotic to the posterior $\Pi_{n,\beta}(\cdot| X^{(n)})$ if only the true model had been used. However, this argument cannot be reversed: adaptation may well occur without consistency of model selection. The main reason is that a "true model" may not exist, as noted previously. Under the (reasonable) conditions of the main theorem of this chapter, the model index prior will choose models of low complexity given equal approximation properties, but may choose a model of too low complexity if this approximates the true parameter well enough.

In the situation that there are precisely two models $\mathcal{P}_0$ and $\mathcal{P}_1$ with equal prior weights, model selection consistency can be conveniently described by the behavior of the *Bayes factor* of the two models, the quotient of the marginal likelihoods of the two models. The posterior distribution is consistent for model selection if and only if the Bayes factor of $\mathcal{P}_0$

versus $\mathcal{P}_1$ has a dichotomous behavior: it tends to infinity if the true density p_0 belongs to $\mathcal{P}_0$, and it converges to zero otherwise, so-called *Bayes factor consistency*. This consistency property is especially relevant for Bayesian goodness of fit testing against a nonparametric alternative. A computationally convenient method of such goodness of fit test using a mixture of Pólya tree prior on the nonparametric alternative is described in Problem 10.16.

10.5.1 Testing a Point Null

Model selection with a singleton model is essentially a problem of testing goodness of fit. It is natural to formulate it as testing a null hypothesis H_0: $p = p_*$ against the alternative H_1: $p \neq p_*$. If $\lambda \in (0, 1)$ is the prior weight of the null model and Π_1 is the prior distribution on p under the alternative model, then the overall prior is the mixture $\Pi = (1-\lambda)\delta_{p_*} + \lambda\Pi_1$. The posterior probability of the null model and the Bayes factor are given by

$$\Pi_n(p = p_*|\, X_1, \ldots, X_n) = \frac{(1-\lambda)\prod_{i=1}^n p_*(X_i)}{(1-\lambda)\prod_{i=1}^n p_*(X_i) + \lambda \int \prod_{i=1}^n p(X_i)\, d\Pi_1(p)}$$

and

$$\mathbb{B}_n = \frac{\prod_{i=1}^n p_*(X_i)}{\int \prod_{i=1}^n p(X_i)\, d\Pi_1(p)}.$$

Theorem 10.24 *As as $n \to \infty$,*

(i) $\mathbb{B}_n \to \infty$ *a.s.* $[P_*^\infty]$.
(ii) $\mathbb{B}_n \to 0$ *a.s.* $[P^\infty]$ *for any* $p \neq p_*$ *with* $p \in \mathrm{KL}(\Pi_1)$.

Proof The Bayes factor tends to 0 or ∞ if and only if the posterior probability $\Pi_n(p = p_*|\, X_1, \ldots, X_n)$ of the null hypothesis tends to 0 or 1.

Since the prior mass of $\{p_*\}$ is positive, Doob's theorem, Theorem 6.9, applied with the function $f(p) := \mathbb{1}_{\{p_0\}}(p)$, gives that $\Pi_n(p = p_*|\, X_1, \ldots, X_n) \to 1$ a.s. under the null hypothesis. This implies (i).

If $p \in \mathrm{KL}(\Pi_1)$, then also $p \in \mathrm{KL}(\Pi)$ and hence $\Pi_n(\mathcal{W}|\, X_1, \ldots, X_n) \to 1$ a.s. $[P^\infty]$ for any weak neighborhood $\mathcal{W}$ of p, by Schwartz's theorem and Example 6.20. If $p \neq p_*$, then the neighborhood can be chosen to exclude p_* and hence the posterior mass of the null hypothesis tends to zero. This proves (ii). □

The role of Doob's theorem in giving a painless proof for the relatively complicated issue of Bayes factor consistency is intriguing. It is not difficult to see that the result goes through when the null hypothesis is a countable closed set in which every point receives positive prior mass. (Also Schwartz's theorem is not needed if the alternative hypothesis has a similar form.) Unfortunately, the method of proof does not seem to generalize any further.

10.5.2 General Case

Consider the setting of Section 10.2, where we consider a collection of models $\mathcal{P}_{n,\alpha}$, for $\alpha \in A_n$, for the density of i.i.d. observations $X_1, X_2, \ldots$ We retain the notations $A_{n,\prec\beta_n}$ and

$A_{n,\succeq\beta_n}$, and, as before, consider a distance d on the set of densities for which tests as in (8.2) exist.

Somewhat abusing notation, we denote the posterior distribution of the model index by the same symbol as the posterior distribution: for any set $B \subset A_n$ of indices,

$$\Pi_n(\alpha \in B | X_1, \ldots, X_n) = \frac{\sum_{\alpha \in B} \lambda_{n,\alpha} \int \prod_{i=1}^n p(X_i)\, d\Pi_{n,\alpha}(p)}{\sum_{\alpha \in A_n} \lambda_{n,\alpha} \int \prod_{i=1}^n p(X_i)\, d\Pi_{n,\alpha}(p)}.$$

Theorem 10.25 *Under the conditions of Theorem 10.3,*

$$P_0^n \Pi_n(\alpha \in A_{n,\prec\beta_n} | X_1, \cdots, X_n) \to 0, \tag{10.36}$$

$$P_0^n \Pi_n(\alpha \in A_{n,\succeq\beta_n} : d(p_0, \mathcal{P}_{n,\alpha}) > M\sqrt{H}\, \epsilon_{n,\beta_n} | X_1, \cdots, X_n) \to 0. \tag{10.37}$$

Under the conditions of Theorem 10.7 assertion (10.36) *is satisfied and* (10.37) *holds with* $M\sqrt{H}$ *replaced by* I_n, *for any* $I_n \to \infty$.

Proof Theorems 10.3 and 10.7 show that the posterior concentrates all its mass on balls of radius $M\sqrt{H}\epsilon_{n,\beta_n}$ or $I_n\epsilon_{n,\beta_n}$ around p_0, respectively. Consequently the posterior cannot charge any model that does not intersect these balls and (10.37), or its adapted version under the conditions of Theorem 10.7, follows.

The first assertion can be proved using exactly the proof of Theorems 10.3 and 10.7, except that the references to $\alpha \succeq \beta_n$ can be omitted. In the notation of the proof of Theorem 10.3, we have that

$$\bigcup_{\alpha \prec \beta_n} \mathcal{P}_{n,\alpha} \subset \Big(\bigcup_{\alpha \prec \beta_n} \bigcup_{i \geq I} S_{n,\alpha,i}\Big) \bigcup \Big(\bigcup_{\alpha \prec \beta_n} C_{n,\alpha}(p_0, IB'\epsilon_{n,\alpha})\Big).$$

It follows that $P_0[\Pi_n(A_{n,\prec\beta_n} | X_1, \ldots, X_n)(1 - \phi_n)\mathbb{1}_{A_n}]$ can be bounded by the sum of the second and third terms on the right-hand side of (10.11), which tend to zero under the conditions of Theorem 10.3, or can be made arbitrarily small under the conditions of Theorem 10.7 by choosing B and I sufficiently large. ☐

The first assertion of the theorem is pleasing. It can be interpreted as saying that the models that are bigger than the model $\mathcal{P}_{n,\beta_n}$ that contains the true distribution eventually receive negligible posterior weight. The second assertion makes a similar claim about the smaller models, but it is restricted to the smaller models that keep a certain distance to the true distribution. Such a restriction appears not unnatural, as a small model that can represent the true distribution well ought to be favored by the posterior, which looks at the data through the likelihood and hence will judge a model by its approximation properties rather than its parameterization. That big models with similarly good approximation properties are not favored is caused by the fact that (under our conditions) the prior mass on the big models is more spread out, yielding relatively less prior mass near good approximants within the big models.

It is insightful to specialize the theorem to the case of two models, and simplify the prior mass conditions to (10.12). The behavior of the posterior of the model index can then be described through the posterior odds ratio $\mathbb{D}_n = (\lambda_{n,2}/\lambda_{n,1})\mathbb{B}_n$, which is the prior odds times the Bayes factor

$$\mathbb{B}_n = \frac{\int \prod_{i=1}^n p(X_i)\, \Pi_{n,2}(p)}{\int \prod_{i=1}^n p(X_i)\, \Pi_{n,1}(p)}.$$

Corollary 10.26 *Adopt the notation* (10.3)–(10.4). *Assume that* (10.5) *holds for* $\alpha \in A_n = \{1, 2\}$ *and sequences* $\epsilon_{n,1} > \epsilon_{n,2}$ *such that* $n\epsilon_{n,2}^2 \to \infty$.

(i) *If* $\Pi_{n,1}(B_{n,1}(p_0, \epsilon_{n,1})) \geq e^{-n\epsilon_{n,1}^2}$ *and* $\lambda_{n,2}/\lambda_{n,1} \leq e^{n\epsilon_{n,1}^2}$ *and* $d(p_0, \mathcal{P}_{n,2}) \geq I_n \epsilon_{n,1}$ *for every* n *and some* $I_n \to \infty$, *then* $\mathbb{D}_n \to 0$ *in* P_0^n*-probability.*

(ii) *If* $\Pi_{n,2}(B_{n,2}(p_0, \epsilon_{n,2})) \geq e^{-n\epsilon_{n,2}^2}$ *and* $\lambda_{n,2}/\lambda_{n,1} \geq e^{-n\epsilon_{n,1}^2}$, *and also* $\Pi_{n,1}(C_{n,1}(p_0, M\epsilon_{n,1})) \leq (\lambda_{n,2}/\lambda_{n,1}) o(e^{-3n\epsilon_{n,2}^2})$ *for a sufficiently large constant* M, *then* $\mathbb{D}_n \to \infty$ *in* P_0^n*-probability.*

Proof The posterior odds ratio $\mathbb{D}_n$ converges respectively to 0 or ∞ if the posterior probability of model $\mathcal{P}_{n,2}$ or $\mathcal{P}_{n,1}$ tends to zero, respectively. Therefore, we can apply Theorem 10.25 with the same choices as in the proof of Corollary 10.8. ☐

10.5.3 Testing Parametric versus Nonparametric Models

Suppose that there are two models, with the bigger model $\mathcal{P}_{n,1}$ infinite dimensional, and the alternative model a fixed parametric model $\mathcal{P}_{n,2} = \mathcal{P}_2 = \{p_\theta : \theta \in \Theta\}$, for $\Theta \subset \mathbb{R}^d$, equipped with a fixed prior $\Pi_{n,2} = \Pi_2$. We shall show that the posterior odds ratio $\mathbb{D}_n$ is typically consistent in this situation: $\mathbb{D}_n \to \infty$ if $p_0 \in \mathcal{P}_2$, and $\mathbb{D}_n \to 0$ if $p_0 \notin \mathcal{P}_2$.

If the prior Π_2 is smooth in the parameter and the parameterization $\theta \mapsto p_\theta$ is regular, then, for any $\theta_0 \in \Theta$ where the prior density is positive and continuous, as $\epsilon \to 0$,

$$\Pi_2\Big(\theta : K(p_{\theta_0}; p_\theta) \leq \epsilon^2, V_2(p_{\theta_0}; p_\theta) \leq \epsilon^2\Big) \sim C(\theta_0)\epsilon^d$$

for some positive constant $C(\theta_0)$ depending on θ_0. Therefore, if the true density p_0 is contained in $\mathcal{P}_2$, then $\Pi_{n,2}(B_{n,2}(p_0, \epsilon_{n,2})) \gtrsim \epsilon_{n,2}^d > e^{-n\epsilon_{n,2}^2}$ for $\epsilon_{n,2}^2 = Dn^{-1}\log n$, for $D \geq d/2$ and sufficiently large n.[1]

For this choice of $\epsilon_{n,2}$ we have $e^{n\epsilon_{n,2}^2} = n^D$. Therefore, it follows from (ii) of Corollary 10.26, that if $p_0 \in \mathcal{P}_2$, then the posterior odds ratio $\mathbb{D}_n$ tends to ∞, as $n \to \infty$, as soon as there exists $\epsilon_{n,1} > \epsilon_{n,2}$ such that, for every I,

$$\Pi_{n,1}(p : d(p, p_0) \leq I\epsilon_{n1}) = o(n^{-3D}). \tag{10.38}$$

For an infinite-dimensional model $\mathcal{P}_{n,1}$, relation (10.38) is typically true. In fact, for $p_0 \in \mathcal{P}_{n,1}$ the left-hand side is typically of the order $e^{-Fn\epsilon_{n,1}^2}$, for $\epsilon_{n,1}$ the rate attached to the model $\mathcal{P}_{n,1}$. For a true infinite-dimensional model this rate is not faster than n^{-a} for some $a < 1/2$, leading to an upper bound of the type e^{-n^b} for some $b > 0$, which is certainly $o(n^{-3D})$. If $p_0 \notin \mathcal{P}_{n,1}$, then the prior mass in (10.38) ought to be even smaller.

[1] The logarithmic factor enters because we use the crude prior mass condition (10.12) instead of the comparisons of prior mass in the main theorems, but it does not matter for the purpose of this subsection.

If p_0 is not contained in the parametric model, then typically $d(p_0, \mathcal{P}_2) > 0$, and hence $d(p_0, \mathcal{P}_2) > I_n \epsilon_{n,1}$ for any $\epsilon_{n,1} \to 0$ and sufficiently slowly increasing I_n, as required in (i) of Corollary 10.26. To ensure that $\mathbb{D}_n \to 0$, it suffices that for some $\epsilon_{n,1} > \epsilon_{n,2}$,

$$\Pi_{n,1}\Big(p\colon K(p_0; p) \le \epsilon_{n,1}^2, V_2(p_0; p) \le \epsilon_{n,1}^2\Big) \ge e^{-n\epsilon_{n,1}^2}. \tag{10.39}$$

This is the usual prior mass condition (8.4) for obtaining the rate of contraction $\epsilon_{n,1}$ using the prior $\Pi_{n,1}$ on the model $\mathcal{P}_{n,1}$.

We illustrate the preceding in three examples. In each, we assume that the prior model weights are fixed, positive numbers.

Example 10.27 (Bernstein polynomial prior for density estimation) Let $\mathcal{P}_{n,1}$ be the set of densities realized by the Bernstein polynomial prior in Section 9.3, with a geometric or Poisson prior on the index parameter k. The rate of contraction for this model is given by $\epsilon_{n,1} = n^{-1/3}(\log n)^{1/3}$. As the prior spreads its mass over an infinite-dimensional set that can approximate any smooth function, condition (10.38) will be satisfied for most true densities p_0. In particular, if k_n is the minimal degree of a polynomial that is within Hellinger distance $n^{-1/3}\log n$ of p_0, then the left-hand side of (10.38) is bounded by the prior mass of all Bernstein-Dirichlet polynomials of degree at least k_n, which is e^{-ck_n} for some constant c, by construction. Thus the condition is certainly satisfied if $k_n \gg \log n$. Consequently, the Bayes factor is consistent for true densities that are not well approximable by polynomials of low degree.

Example 10.28 (Log-spline prior for density estimation) Let $\mathcal{P}_{n,1}$ be the set of log-spline densities described of dimension $J \asymp n^{1/(2\alpha+1)}$, in Section 10.3.4, equipped with the prior obtained by equipping the coefficients with the uniform distribution on $[-M, M]^J$. The corresponding rate can then be taken to be $\epsilon_{n,1} = n^{-\alpha/(2\alpha+1)}\sqrt{\log n}$, by the calculations in Section 10.3.4. Condition (10.38) can be verified easily by computations on the uniform prior, after translating the distances on the spline densities into the Euclidean distance on the coefficients as in Lemma 9.4.

Example 10.29 (White noise model) Let $\mathcal{P}_{n,1}$ be the set of $\mathrm{Nor}_\infty(\theta, I)$-distributions with $\theta \in \mathfrak{W}^\alpha$, as in Section 10.3.2, with prior given by $\theta_i \overset{\text{ind}}{\sim} \mathrm{Nor}(0, i^{-(2\alpha+1)})$. Relations (10.38)–(10.39) follow from the estimates (10.17) and (10.18). Thus the Bayes factor is consistent for testing $\mathcal{P}_{n,1}$ against finite-dimensional alternatives.

10.6 Historical Notes

Adaptive estimation in a minimax sense was first studied by Efroĭmovich and Pinsker (1984) in the context of the white noise model. Earlier Pinsker (1980) had derived the the minimax risk for squared error loss over ellipsoids in this model. Efroĭmovich and Pinsker (1984) constructed an estimator by adaptively determining optimal damping coefficients in an orthogonal series. There has been considerable work on adaptive estimation in this and many other models since then. The white noise model with the conjugate prior $\theta_i \overset{\text{ind}}{\sim} \mathrm{Nor}(0, \tau_i^2)$, for $i = 1, 2, \ldots$, was studied by Cox (1993) and Freedman (1999), with particular interest for the posterior distribution of the nonlinear functional $\|\theta - \hat{\theta}\|^2$, where $\hat{\theta}$ is the Bayes estimator. Zhao (2000) noted the minimaxity for ellipsoids and the prior with $\tau_i^2 = i^{-(2\alpha+1)}$, for $i = 1, 2, \ldots$, and also pointed out that this prior gives probability zero to any ellipsoid of order α. She then also studied mixtures of finite-dimensional normal priors to rectify the problem and retain minimaxity. Such priors are also considered by Shen and Wasserman (2001) and Ghosal and van der Vaart (2007a). Bayesian adaptation in this framework was studied by Belitser and Ghosal (2003), with direct methods, for a discrete set of α. Bayesian adaptation for i.i.d. models was studied by Ghosal et al. (2003) in the setting of a finite collection of smoothness indices and log-spline models. They put fixed but arbitrary weights on each model and obtained minimax rate up to a logarithmic factor. Huang (2004) used both log-spline and wavelet series priors and avoided logarithmic factors by carefully constructing the priors on the coefficient spaces, combined with appropriate model weights. Scricciolo (2006) proved adaptation over Sobolev balls using exponential family models, without a logarithmic factor, but with an additional condition on the true parameters. Theorem 10.3 on adaptation in i.i.d. models was obtained by Ghosal et al. (2008), together with the case study on log-spline models presented in Section 10.3.4. The universal weights discussed in Section 10.2.1 were described by Lember and van der Vaart (2007). The applications to priors based on nets and finite-dimensional models given in Sections 10.3.1 and 10.3.1 are taken from their paper. The example of priors on nets, which do not lead to additional logarithmic factors in the rates, suggests that these factors are due to an interaction between the way mass is spread over a model or divided between the models. Section 10.5 is adapted from Shen and Ghosal (2015). Other references on adaptation include Scricciolo (2007, 2014) and van der Vaart (2010). An alternative approach developed in Xing (2008), using the concept of Hausdorff entropy (see Problem 8.13), gives contraction results in the a.s. sense, based on a stronger condition on concentration in terms of a modification of the Hellinger distance. Some of these results are presented in Problems 10.8–10.10 and 10.13. Certain kernel mixture priors are able to adapt automatically to the smoothness level of the true density if augmented by an appropriate prior on a bandwidth parameter. In order for this to happen the mixture class must approximate the true density with increasingly higher accuracy for smoother true densities. The mixture with the true distribution as mixing distribution may not be a close projection into the class of mixtures. Rousseau (2010) devised a technique of constructing better approximations to smoother true densities in the context of a certain type of nonparametric mixtures of beta kernel (see Problem 10.15). In Section 9.4 a similar idea is applied to Dirichlet mixtures of normal distributions. Bayesian model selection in the parametric context is the subject of a rich literature, of which Schwarz (1978) is notable for introducing the Bayesian Information Criterion (BIC) to approximate Bayes factors. The first paper dealing with model selection consistency in the nonparametric context is Dass and Lee (2004), who looked at point null hypotheses (Subsection 10.5.1). Verdinelli and Wasserman (1998) considered testing a uniform density against an infinite-dimensional exponential family, where they showed that the prior satisfies the KL property, and showed by direct calculations that the Bayes factor is consistent. The general case, which requires a different proof, was studied by Ghosal et al. (2008). They also noted the (dis)connection between Bayesian adaptation and model selection consistency. Section 10.5.2 and the examples of Section 10.5.3 are due to Ghosal et al. (2008). Berger and Guglielmi (2001) developed a method of testing a parametric model using a mixture of Pólya tree prior on the nonparametric alternative (see Problem 10.16).

Problems

10.1 (Belitser and Ghosal 2003) For the white noise model with a prior λ_α on a discrete set of α without limit points in $(0, \infty)$, show that, for any l,

$$\mathrm{E}_{\theta_0}\Pi(\alpha = q_m | X^{(n)}) \le \frac{\lambda_m}{\lambda_l} \exp\left[\frac{1}{2}\sum_{i=1}^{\infty} \frac{(i^{-(2q_l+1)} - i^{-(2q_m+1)})(i^{-(2q_l+1)} - \theta_{i0}^2)}{i^{-2(q_l+q_m+1)} + 2n^{-1}i^{-(2q_l+1)} + n^{-2}}\right].$$

10.2 (Belitser and Ghosal 2003) Consider the infinite-dimensional normal model of Section 10.3.2. Suppose that $q_0 \in \mathcal{Q}$, where $\mathcal{Q}$ is an arbitrary subset of $(0, \infty)$. Choose a countable dense subset $\mathcal{Q}^*$ of $\mathcal{Q}$, and put a prior λ on it such that $\lambda(q = s) > 0$ for each $s \in \mathcal{Q}^*$. For any sequence $M_n \to \infty$ and $\delta > 0$, show that $\Pi\{\theta: n^{q_0/(2q_0+1)-\delta}\|\theta - \theta_0\| > M_n | X\} \to 0$ in P_{θ_0}-probability as $n \to \infty$.

10.3 (Belitser and Ghosal 2003) Consider the infinite-dimensional normal model of Section 10.3.2. For any $\theta^* \in \Theta_{q_0}$, show that there exists $\epsilon > 0$ such that

$$\sup_{\theta_0 \in \mathcal{E}_{q_0}(\theta^*, \epsilon)} \mathrm{E}_{\theta_0}\Pi\{\theta: n^{q_0/(2q_0+1)}\|\theta - \theta_0\| > M_n | X\} \to 0,$$

where $\mathcal{E}_{q_0}(\theta^*, \epsilon) = \{\theta_0: \sum_{i=1}^{\infty} i^{2q_0}(\theta_{i0} - \theta_i^*)^2 < \epsilon\}$.

10.4 (Belitser and Ghosal 2003) In the infinite-dimensional normal model of Section 10.3.2, let $\mathcal{Q} = \{q_0, q_1\}$ and $\theta_{i0} = i^{-p}$. Show that if $p \ge q_1 + 1 - \frac{2q_1+1}{2(2q_0+1)}$, then $\Pi(q = q_1 | X) \to 1$ in probability even if $\theta_0 \in \Theta_{q_0} \setminus \Theta_{q_1}$.

Show that the prior Π_{q_1} leads to a slower contraction rate than Π_{q_0} only when $p < q_1 + 1 - (2q_1 + 1)/(2(2q_0 + 1))$.

10.5 (Huang 2004) Consider $X_1, \ldots, X_n \stackrel{\text{iid}}{\sim} p$, p an unknown density on $[0, 1]$, and let the true density $p_0 \in \mathfrak{W}^\alpha[0, 1]$ be such that $\|\log p_0\|_\infty \le M_0$. Let $J = \{(l, L): l \in \mathbb{N} \cup \{0\}, L \in \mathbb{N}\}$. For $j \in J$, let $m_j = 2^{l+1}$ and $B_{j,i}$, $i = 1, \ldots, m_j - 1$ stand for the Haar wavelets ψ_{j_1,k_1}, $0 \le j_1 \le l$, $0 \le k_1 \le 2^{j_1} - 1$, respectively. Define a prior on p by the relation $p(x) \propto \exp[\sum_{i=1}^{m_j-1} \theta_i B_{j,i}(x)]$, $\theta := (\theta_1, \ldots, \theta_{m_j-1}) \in \Theta_j$, where $\Theta_j = \{\theta \in \mathbb{R}^{m_j-1}: \|\sum_{i=1}^{m_j-1} \theta_j B_{j,i}\|_\infty \le L\}$. Given j, let θ be uniformly distributed over Θ_j. Choose $A_j = 19.28 \cdot 2^{(l+1)/2}(2L + 1)e^L + 0.06$, $C_j = m_j + L$ and assign a prior on various models by $a_j = \alpha \exp\{-(1 + 0.5\sigma^{-2} + 0.0056\sigma^{-1})\eta_j\}$, where $\eta_j = 4m_j \log(1072.5A_j)c^{-1}(1 - 4\gamma)^{-1} + C_j \max(1, 8c^{-1}(1 - 4\gamma)^{-1})$, $c = 0.5\min\{M^{-2}(1 - e^{-M^2/(2\sigma^2)}, \sigma^{-2}), 0.13\gamma(1 - 4\gamma)^{-1/2} = 0.0056$. Show that the posterior distribution for p contracts at p_0 at the rate $n^{-\alpha/(1+2\alpha)}\sqrt{\log n}$.

10.6 (Huang 2004) Consider the regression problem $Y_i = f(X_i) + \varepsilon_i$, $X_i \stackrel{\text{iid}}{\sim} G$, $\varepsilon_i \stackrel{\text{iid}}{\sim} \mathrm{Nor}(0, \sigma^2)$, σ known, p_f the density of (X, Y). Consider a sieve-prior on f given by $f = f_{\theta,j}$, $\theta \in \Theta_j$, a subset of a Euclidean space, $\theta | j \sim \pi_j$, $j \in J$, a countable set, $j \sim a(\cdot)$. Let d_j the metric on Θ_j, $d_{j,\infty}(\theta, \eta) = \|f_{\theta,j} - f_{\eta,j}\|_\infty$ for $\theta, \eta \in \Theta_j$, $B_{2,j}(\eta, r) = \{\theta \in \Theta_j: \|f_{\theta,j} - f_{\eta,j}\|_{2,G} \le r\}$ and $\|f_{\theta,j}\|_\infty \le M$ for all $j, \theta \in \Theta_j$. Let f_0 be the true regression function. Assume that the following conditions hold:

(a) There are $0 < A_j < 0.0056$, $m_j \geq 1$ such that $N(B_{2,j}(\theta, r), \delta, d_{j,\infty}) \leq (A_j r/\delta)^{m_j}$ for all $r > 0$ and $\delta < 0.00056r$.
(b) Let $a_j = \alpha \exp\{-(1 + 0.5\sigma^{-2} + 0.0056\sigma^{-1})\eta_j\}$, $c = 0.5 \min\{M^{-2}(1 - e^{-M^2/(2\sigma^2)}, \sigma^{-2}), 0.13\gamma(1-4\gamma)^{-1/2} = 0.0056$, $\eta_j = 4m_j \log(1072.5A_j)c^{-1}(1-4\gamma)^{-1} + C_j \max(1, 8c^{-1}(1-4\gamma)^{-1})$, $\sum_{j=1}^{\infty} a_j = 1$, $\sum_{j=1}^{\infty} e^{-C_j} \leq 1$. There exist j_n and $\epsilon_n > 0$, $\epsilon_n \to 0$, $n\epsilon_n^2 \to \infty$, $\beta_n \in \Theta_{j_n}$ such that $\max\{K(p_{f_0}, p_{\beta_n, j_n}), V_2(p_{f_0}, p_{\beta_n, j_n})\} + n^{-1}\eta_{j_n}^2 \leq \epsilon_n^2$.
(c) $\|f_{\theta, j_n} - f_{\eta, j_n}\|_{2,G} \lesssim d_{j_n}(\theta, \eta)$ for all $\theta, \eta \in \Theta_{j_n}$.
(d) $\log N(\Theta_{j_n}, \epsilon_n, d_n) \leq m_{j_n}(K + b \log A_{j_n})$ for some constants K and b.

Show that the posterior for f contracts at the rate ϵ_n with respect to $\|f_{\theta,j} - f_{\eta,j}\|_{2,G}$.

10.7 (Huang 2004) In the setting of the nonparametric regression in Problem 10.6, let the true regression function $f_0 \in \mathfrak{W}^s[0,1]$, $\|f_0\|_\infty \leq M$ with known M, G be Unif[0, 1] and $J = \{(k, q, L): k \in \mathbb{N} \cup \{0\}, q \in \mathbb{N}, L \in \mathbb{N}\}$. For $j = (k, q, L)$, let $m_j = k + q$, $B_{j,i}$, $i = 1, \ldots, k$, be the normalized B-splines of order q and k regularly spaced knots, $f_{\theta,j} = \sum_{i=1}^{m_j} \theta_i B_{j,i}$, $(\theta_1, \ldots, \theta_{m_j}) \in \mathbb{R}^{m_j}$. Choose $A_j = 9.64\sqrt{q}(2q+1)9^{q-1} + 0.06$, $C_j = m_j + L$ and define η_j and a_j as in Problem 10.6. Show that the posterior distribution for f contracts at f_0 at the rate $n^{-s/(1+2s)}$ with respect to $\mathbb{L}_2$-distance.

Since the prior does not depend on the true smoothness level s, the posterior is rate adaptive.

10.8 (Xing 2008) Consider the setting and notations of Section 10.2. Suppose that there exist positive constants $H \geq 1$, E_α, $\mu_{n,\alpha}$, G, J, L, C and $0 < \gamma < 1$ such that $1 - \gamma > 18\gamma L$, $n\epsilon_{n,\beta_n}^2 \geq (1 + C^{-1})\log n$, $E_\alpha \epsilon_{n,\alpha}^2 \leq G\epsilon_{n,\beta_n}^2$ for all $\alpha \in A_{n,\succeq\beta_n}$, $E_\alpha \leq G$ for all $\alpha \in A_{n,\succ\beta_n}$ and $\sum_{\alpha \in A_n} \mu_{n,\alpha}^\gamma = O(e^{Jn\epsilon_{n,\beta_n}^2})$. Let $M \geq \sqrt{H} + 1 + 18(C + J + G + 3\gamma + 2\gamma C)/(1 - \alpha - 18\alpha L)$ such that

(a) $\log N(\epsilon/3, C_{n,\alpha}(p_0, 2\epsilon), d) \leq E_\alpha n\epsilon_{n,\alpha}^2$ for all $\alpha \in A_n$, $\epsilon \geq \epsilon_n$;
(b) $\dfrac{\lambda_{n,\alpha}\Pi_{n,\alpha}(C_{n,\alpha}(p_0, j\epsilon_{n,\alpha}))}{\lambda_{n,\beta_n}\Pi_{n,\beta_n}(\mathcal{H}_{n,\beta_n}(\epsilon_{n,\beta_n}))} \leq \mu_{n,\alpha} e^{Lj^2 n\epsilon_{n,\alpha}^2}$ for all $\alpha \in A_{n,\succ\beta_n}$ and $j \geq M$, where $\mathcal{H}_{n,\beta_n}(\epsilon) = \{p \in \mathcal{P}_{n,\beta_n}: \|(\sqrt{p} - \sqrt{p_0})(2\sqrt{p_0/p} + 1)/\sqrt{3}\|_2 < \epsilon\}$;
(c) $\dfrac{\lambda_{n,\alpha}\Pi_{n,\alpha}(C_{n,\alpha}(p_0, j\epsilon_{n,\alpha}))}{\lambda_{n,\beta_n}\Pi_{n,\beta_n}(\mathcal{H}_{n,\beta_n}(\epsilon_{n,\beta_n}))} \leq \mu_{n,\alpha} e^{Lj^2 n\epsilon_{n,\beta_n}^2}$ for all $\alpha \in A_{n,\succeq\beta_n}$ and $j \geq M$;
(d) $\displaystyle\sum_{n=1}^{\infty} \sum_{\alpha \in A_{n,\succ\beta_n}} \frac{\lambda_{n,\alpha}\Pi_{n,\alpha}(C_{n,\alpha}(p_0, j\epsilon_{n,\alpha}))}{\lambda_{n,\beta_n}\Pi_{n,\beta_n}(\mathcal{H}_{n,\beta_n}(\epsilon_{n,\beta_n}))} e^{(1+2C)n\epsilon_{n,\beta_n}^2} < \infty$.

Show that $\Pi_n(p: d(p_0, p) \geq M\epsilon_{n,\beta_n} | X_1, \ldots, X_n) \to 0$ a.s. $[P_0^\infty]$ as $n \to \infty$.

10.9 (Xing 2008) In Problem 10.8, replace Condition (iii) by the following condition: There exists $K \geq 1$ independent of n, α and j such that $\dfrac{\lambda_{n,\alpha}\Pi_{n,\alpha}(C_{n,\alpha}(p_0, j\epsilon_{n,\alpha}))}{\lambda_{n,\beta_n}\Pi_{n,\beta_n}(\mathcal{H}_{n,\beta_n}(K\epsilon_{n,\beta_n}))} \leq \mu_{n,\alpha} e^{Lj^2 n\epsilon_{n,\beta_n}^2}$ for all $\alpha \in A_{n,\succeq\beta_n}$ and $j \geq M$. Assume that $M \geq \sqrt{H} + 1 + 18\frac{C+J+G+3\gamma K+2\gamma CK}{1-\alpha-18\alpha L}$. Show that $\Pi_n(p: d(p_0, p) \geq M\epsilon_{n,\beta_n} | X_1, \ldots, X_n) \to 0$ a.s. $[P_0^\infty]$ as $n \to \infty$.

10.10 (Xing 2008) Applying Problem 10.9 to the log-spline prior with a finite index set A_n, show that the posterior adapts to the correct rate $n^{-\beta/(1+2\beta)}$ a.s., whenever $f \in \mathfrak{C}^\beta[0,1]$ for the $\mathbb{L}_2$-distance.

10.11 (Shen and Ghosal 2015) Show that in each of the following occasions, the posterior mean (and other moments) are analytically expressible (but may have a large number of terms):

- White noise model, identity link function, random series prior using multivariate normal distribution on the coefficients of B-spline functions — $f(x) = \sum_{j=1}^J \theta_j B_j(x)$, $(\theta_1, \ldots, \theta_J) \sim \mathrm{Nor}_J(\mu_J, \Sigma_{J\times J})$, $J \sim \pi$;
- Density estimation, identity link function, random series prior using Dirichlet distribution on the coefficients of normalized B-spline functions — $p(x) = \sum_{j=1}^J \theta_j B_j^*(x)$, $(\theta_1, \ldots, \theta_J) \sim \mathrm{Dir}(J; \alpha_1, \ldots, \alpha_J)$, $J \sim \pi$;
- Nonparametric normal regression, identity link function, random series prior using multivariate normal distribution on the coefficients of B-spline functions — $f(x) = \sum_{j=1}^J \theta_j B_j(x)$, $(\theta_1, \ldots, \theta_J) \sim \mathrm{Nor}_J(\mu_J, \Sigma_{J\times J})$, $J \sim \pi$;
- Nonparametric Poisson regression, identity link function, random series prior using independent gamma distribution on the coefficients of B-spline functions — $f(x) = \sum_{j=1}^J \theta_j B_j(x)$, $\theta_j \overset{\text{ind}}{\sim} \mathrm{Ga}(\alpha_j, \beta_j)$, $J \sim \pi$;
- Functional regression, identity link function, random series prior using multivariate normal distribution on the coefficients of B-spline functions — $\beta(x) = \sum_{j=1}^J \theta_j B_j(x)$, $(\theta_1, \ldots, \theta_J) \sim \mathrm{Nor}_J(\mu_J, \Sigma_{J\times J})$, $J \sim \pi$;
- Spectral density estimation, reciprocal link function, random series prior using independent gamma distribution on the coefficients of B-spline functions — $1/f(x) = \sum_{j=1}^J \theta_j B_j(x)$, $\theta_j \overset{\text{ind}}{\sim} \mathrm{Ga}(\alpha_j, \beta_j)$, $J \sim \pi$.

10.12 (Walker et al. 2005) Consider two models $\mathcal{P}_1$ and $\mathcal{P}_2$ for density estimation. Suppose that for a prior distribution Π, $\mathrm{KL}(\Pi) \cap \mathcal{P}_2 = \varnothing$. Show that the Bayes factor of model $\mathcal{P}_1$ to model $\mathcal{P}_2$ goes to infinity a.s. $[P_0^\infty]$ for every $p_0 \in \mathcal{P}_1 \cap \mathrm{KL}(\Pi)$.

10.13 (Xing 2008) Under the setting of Problem 10.8, show that $\Pi_n(A_{n,\prec\beta_n} | X_1, \ldots, X_n) \to 0$ a.s. $[P_0^\infty]$. If further

$$\sum_{n=1}^\infty \sum_{\alpha \in A_{n,\succeq\beta_n}} \frac{\lambda_{n,\alpha}\Pi_{n,\alpha}(C_{n,\alpha}(p_0, M\epsilon_{n,\beta_n}))e^{(3+2C)n\epsilon_{n,\beta_n}^2}}{\lambda_{n,\beta_n}\Pi_{n,\beta_n}(\mathcal{H}_{n,\beta_n}(\epsilon_{n,\beta_n}))} < \infty,$$

then show that $\Pi_n(\alpha \in A_n: H^{-1/2}\epsilon_{n,\beta_n}\epsilon_{n,\alpha} \le \sqrt{H}\epsilon_{n,\beta_n} | X_1, \ldots, X_n) \to 1$ a.s. $[P_0^\infty]$.

10.14 (Xing 2008) Assume the setting of Corollary 10.26 of two nested models. Let the assumption of Problem 10.8 hold and further assume that $\epsilon_{n,1} > \epsilon_{n,2} \ge n^{-1/2}\sqrt{(1+C^{-1})\log n}$ for some $C > 0$ and $M > 700(2C+G+2)$. Show that

(a) if $\Pi_{n,2}(\mathcal{H}_{n,2}(\epsilon_{n,2}) \ge e^{-n\epsilon_{n,2}^2}$, $\frac{\lambda_{n,1}}{\lambda_{n,2}}\Pi_{n,1}(C_{n,1}(p_0, M\epsilon_{n,1})) = O(e^{-(4+3C)n\epsilon_{n,2}^2})$ and $\frac{\lambda_{n,1}}{\lambda_{n,2}} \le e^{n\epsilon_{n,1}^2}$, then $\mathbb{B}_n \to \infty$ a.s. $[P_0^\infty]$;

(b) if $\Pi_{n,1}(\mathcal{H}_{n,1}(\epsilon_{n,1}) \geq e^{-n\epsilon_{n,1}^2}$, $\frac{\lambda_{n,1}}{\lambda_{n,2}}\Pi_{n,2}(C_{n,2}(p_0, M\epsilon_{n,2})) = O(e^{-(4+3C)n\epsilon_{n,2}^2})$ and $\frac{\lambda_{n,2}}{\lambda_{n,1}} \leq e^{n\epsilon_{n,1}^2}$, then $\mathbb{B}_n \to 0$ a.s. $[P_0^\infty]$.

10.15 (Rousseau 2010) A key step in the derivation of a posterior contraction rate at a smooth density p_0 using a kernel mixture prior $\int \psi(x;\theta,\sigma)\,dF(\theta)$ is to construct a density p_1 of mixture type such that $\max\{K(p_0;p_1), V_2(p_0;p_1)\} \leq \epsilon_n^2$. Typically p_1 is chosen as $\int \psi(x;\theta,\sigma)p_0(\theta)\,d\theta$, or $\int \psi(x;\theta,\sigma)p_0^*(\theta)\,d\theta$, where p_0^* is some suitably truncated version of p_0, and the order of approximation is then $\max(\sigma^\beta, \sigma^2)$, where β is the smoothness level of p_0. In particular, the order does not improve if $\beta > 2$, leading to no improvement in contraction rate. In some situations it is possible to consider a different KL approximation $\int \psi(x;\theta,\sigma)p_1(\theta)\,d\theta$ which will lead to smaller approximation error $O(\sigma^\beta)$ for $\beta \geq 2$. The density p_1 used in the approximation is related to p_0, but the relation depends on the smoothness level of p_0. The following describes a situation where the higher order approximation error is achievable.

Consider estimating a density on $[0,1]$. Let $\psi(x;\theta,\sigma) = x^{a-1}(1-x)^{b-1}/B(a,b)$, where $0 < x < 1$, $a^{-1} = \sigma(1-\theta)$, $b^{-1} = \sigma\theta$, $0 < \theta < 1$, $\sigma > 0$. Assume that $p_0 \in \mathfrak{C}^\beta[0,1]$ with $p^{(m_1}(0) > 0$, $p^{(m_2)}(1) > 0$ for some $m_1, m_2 \in \mathbb{N}$. Show that there exists functions $r_1(x), r_2(x), \ldots$ such that $p_0(x)[1+\sum_{j=2}^{\lfloor\beta\rfloor-1}\sigma^{j/2}r_j(x)]$ is proportional to a probability density $f_1(x)$ on $(0,1)$ and $\int p_0(x)|\log(p_0/p_1)|^m \lesssim \sigma^\beta$, where $p_1(x) = \int \psi(x;\theta,\sigma)f_1(x)\,dx$.

10.16 (Berger and Guglielmi 2001) Let $X_1, X_2, \ldots \stackrel{\text{iid}}{\sim} p$. Consider the problem of testing a parametric model $p \in \{p_\theta: \theta \in \Theta\}$ versus the nonparametric alternative that p is an arbitrary density. Let π be a prior density on θ in the parametric model. For a prior on the nonparametric alternative, consider a mixture of Pólya trees of the second kind introduced in Subsection 3.7.2 as follows: Let F_θ be the c.d.f. of p_θ, P_θ the corresponding probability measure,

$$\begin{aligned}\mathcal{T}_m &= \{B_{\varepsilon_1\cdots\varepsilon_m}: (\varepsilon_1\cdots\varepsilon_m) \in \{0,1\}^m\}\\ &= \{(F^{-1}((k-1)/2^m), F^{-1}(k/2^m)]: k = 1,\ldots,2^m\},\end{aligned}$$

for some fixed distribution with a density (for instance, p_{θ^*}, where θ^* is a prior guess for θ in the parametric family). Let

$$\begin{aligned}c_{\varepsilon_1\cdots\varepsilon_{m-1}}(\theta) &= P_\theta(B_{\varepsilon_1\cdots\varepsilon_{m-1}1})/P_\theta(B_{\varepsilon_1\cdots\varepsilon_{m-1}0}),\\ \alpha_{\varepsilon_1\cdots\varepsilon_{m-1}0}(\theta) &= h^{-1}a_m/\sqrt{c_{\varepsilon_1\cdots\varepsilon_{m-1}}(\theta)},\\ \alpha_{\varepsilon_1\cdots\varepsilon_{m-1}1}(\theta) &= h^{-1}a_m\sqrt{c_{\varepsilon_1\cdots\varepsilon_{m-1}}(\theta)},\end{aligned}$$

a_m a sequence with $\sum_{m=1}^\infty a_m^{-1} < \infty$ (like $a_m = m^2$) and h a hyperparameter for scaling. Let θ have density π in the alternative as well.

Let $m^*(X_i) = \min\{m\colon X_{i'} \notin B_{\varepsilon_1\cdots\varepsilon_m(X_i)}, i' < i\}$, where $\varepsilon_1 \cdots \varepsilon_m(x)$ denotes the first m digits in the binary expansion for $F(x)$. Let

$$\psi(\theta|\, X_1, \ldots, X_n) = \prod_{i=2}^{n} \prod_{m=1}^{m^*(X_i)} \frac{\alpha'_{\varepsilon_1\cdots\varepsilon_m(X_i)}(\theta)[\alpha_{\varepsilon_1\cdots\varepsilon_{m-1}(X_i)0}(\theta) + \alpha_{\varepsilon_1\cdots\varepsilon_{m-1}(X_i)1}(\theta)]}{\alpha_{\varepsilon_1\cdots\varepsilon_m(X_i)}(\theta)[\alpha'_{\varepsilon_1\cdots\varepsilon_{m-1}(X_i)0}(\theta) + \alpha'_{\varepsilon_1\cdots\varepsilon_{m-1}(X_i)1}(\theta)]}.$$

Show that the Bayes factor for testing the parametric model versus the nonparametric alternative is given by

$$\mathbb{B}_n(X_1, \ldots, X_n) = \frac{\int \prod_{i=1}^n p_\theta(X_i)\pi(\theta)\, d\theta}{\int m_1(X_1, \ldots, X_n|\, \theta)\pi(\theta)\, d\theta},$$

where $m_1(X_1, \ldots, X_n|\, \theta) = \prod_{i=1}^n p_\theta(X_i)\psi(\theta|\, X_1, \ldots, X_n)$.

This allows the Bayes factor to be computed numerically by simulating $\theta_1, \ldots, \theta_N$ from a proposal density $q(\cdot)$ and computing the importance sampling Monte Carlo approximation

$$\frac{\sum_{j=1}^N \prod_{i=1}^n p_{\theta_j}(X_i)\pi(\theta_j)/q(\theta_j)}{\sum_{j=1}^N \psi(\theta_j|\, X_1, \ldots, X_n)\pi(\theta_j)/q(\theta_j)}.$$

11

Gaussian Process Priors

Gaussian processes are random functions and can serve as priors on function spaces. The variety of Gaussian processes, given by their covariance kernels, provides many modelling choices, which can be further modified by including hyperparameters in their covariance kernels. The associated geometrical structure of Gaussian processes, encoded in their reproducing kernel Hilbert spaces, makes for an elegant theory of posterior contraction. We begin the chapter with a review and examples of Gaussian processes. Next we present general results on posterior contraction rates, and apply them in several inference problems: density estimation, binary and normal regression, white noise. We proceed with methods of rescaling and adaptation using mixtures of Gaussian processes. The chapter closes with a review of computational techniques.

11.1 Definition and Examples

Definition 11.1 (Gaussian process) A *Gaussian process* is a stochastic process $W = (W_t: t \in T)$ indexed by an arbitrary set T such that the vector $(W_{t_1}, \ldots, W_{t_k})$ possesses a multivariate normal distribution, for every $t_1, \ldots, t_k \in T$ and $k \in \mathbb{N}$. A Gaussian process W indexed by $\mathbb{R}^d$ is called *self-similar* of index α if $(W_{\sigma t}: t \in \mathbb{R}^d)$ is distributed like $(\sigma^\alpha W_t: t \in \mathbb{R}^d)$, for every $\sigma > 0$, and *stationary* if $(W_{t+h}: t \in \mathbb{R}^d)$ has the same distribution as $(W_t: t \in \mathbb{R}^d)$, for every $h \in \mathbb{R}^d$.

The vectors $(W_{t_1}, \ldots, W_{t_k})$ in the definition are called marginals, and their distributions *marginal distributions* or *finite-dimensional distributions*. Since a multivariate normal distribution is determined by its mean vector and covariance matrix, the finite-dimensional distributions of a Gaussian process are determined by the *mean function* and *covariance kernel*, defined by

$$\mu(t) = \mathrm{E}(W_t), \qquad K(s,t) = \mathrm{cov}(W_s, W_t), \qquad s, t \in T.$$

The mean function is an arbitrary function $\mu: T \to \mathbb{R}$, which for prior modeling is often taken equal to zero: a shift to a nonzero mean can be incorporated in the statistical model. The covariance kernel is a bilinear, symmetric nonnegative-definite function $K: T \times T \to \mathbb{R}$, and determines the properties of the process.[1] By Kolmogorov's extension theorem there exists a Gaussian process for any mean function and covariance kernel.

[1] Symmetric nonnegative-definite means that every matrix $((K(t_i, t_j)))_{i,j=1,\ldots,k}$ possesses the property with the same name, for every $t_1, \ldots, t_k$.

Definition 11.1 defines a Gaussian process as a collection of random variables W_t restricted only to be marginally normally distributed. It does not refer to the *sample paths* $t \mapsto W_t$, which define the process as a *random function*. The properties of the sample paths are often of importance for interpreting a prior, and are not fully determined by the mean and covariance functions, as the marginal distributions do not change by changing the variables W_t on null sets, whereas the sample paths do. Often this discrepancy is resolved by considering the version of the process with continuous sample paths, if that exists.[2] Here a process $\tilde{W}$ is a *version* of W if $\tilde{W}_t = W_t$, almost surely for every $t \in T$. Consideration of the sample paths makes it also possible to think of the process as a map $W: \Omega \to \mathbb{B}$ from the underlying probability space to a function space $\mathbb{B}$, such as the space of continuous functions on T or a space of differentiable functions. This connects to the following abstract definition.

Let $\mathbb{B}^*$ denote the *dual space* of a Banach space $\mathbb{B}$: the collection of continuous, linear maps $b^*: \mathbb{B} \to \mathbb{R}$.

Definition 11.2 (Gaussian random element) A *Gaussian random element* is a Borel measurable map W into a Banach space $(\mathbb{B}, \|\cdot\|)$ such that the random variable $b^*(W)$ is normally distributed, for every $b^* \in \mathbb{B}^*$.

The two definitions can be connected in two ways. First a Gaussian random element W always induces the Gaussian stochastic process $(b^*(W): b^* \in T)$, for any subset T of the dual space. Second if the sample paths $t \mapsto W_t$ of the stochastic process $W = (W_t: t \in T)$ belong to a Banach space $\mathbb{B}$ of functions $z: T \to \mathbb{R}$, then under reasonable conditions the process will be a Gaussian random element. The following lemma is sufficient for most purposes. For instance, it applies to the space $\mathfrak{C}(T)$ of continuous functions on a compact metric space T whenever the Gaussian process has continuous sample paths. For a proof and additional remarks, see Lemmas I.5 and I.6 in the appendix.

Lemma 11.3 *If the sample paths $t \mapsto W_t$ of the stochastic process $W = (W_t: t \in T)$ belong to a separable subset of a Banach space and the norm $\|W - w\|$ is a random variable for every w in the subset, then W is a Borel measurable map in this space. Furthermore, if W is a Gaussian process and the Banach space is $\mathfrak{L}_\infty(T)$ equipped with the supremum norm, then W is a Gaussian random element in this space.*

Example 11.4 (Random series) If $Z_1, \ldots, Z_m \overset{\text{iid}}{\sim} \text{Nor}(0, 1)$ variables and $a_1, \ldots, a_m$ are arbitrary functions, then $W_t = \sum_{i=1}^m a_i(t) Z_i$ defines a mean zero Gaussian process with covariance kernel $K(s, t) = \sum_{i=1}^m a_i(s) a_i(t)$.

Special cases are the *polynomial process* given by the functions $a_i(t) = t^i$, and the *trigonometric process* given by the trigonometric functions.

Another case of interest is given by functions $a_i(t) = \sigma^{-d} \psi((t - t_i)/\sigma)$, constructed by shifting and dilating a kernel function ψ, for instance a probability density ψ on $T = [0, 1]^d$ and a regular grid $t_1, \ldots, t_m$ over T.

[2] A weaker restriction is that the process is *separable*, which means that there is a countable subset $T_0 \subset T$ such that the suprema over open sets are equal to suprema over the intersections of these sets with T_0.

If $a_1, a_2, \ldots$ are functions such that $\sum_{i=1}^{\infty} a_i^2(t) < \infty$ for all t, then the construction extends to the case $m = \infty$: the series $W_t = \sum_{i=1}^{\infty} a_i(t) Z_i$ converges a.s. Theorem I.25 shows that all Gaussian processes can be expressed as an infinite series.

Example 11.5 (Brownian motion) The standard *Brownian motion* or *Wiener process* on $[0, 1]$ (or $[0, \infty)$) is the mean zero Gaussian process with continuous sample paths and covariance function $K(s, t) = \min(s, t)$. The process has stationary, independent increments: $B_t - B_s \sim \text{Nor}(0, t - s)$ and is independent of $(B_u : u \leq s)$, for any $s < t$. Hence it is a Lévy process, i.e. a process with stationary and independent increments. Brownian motion is self-similar of index $1/2$.

The sample paths of Brownian motion are Lipschitz continuous of any order $\alpha < 1/2$, and hence W can be viewed as a map in the Hölder space $\mathfrak{C}^{\alpha}[0, 1]$, for any $0 \leq \alpha < 1/2$. That it is also a *Gaussian random element* in the sense of Definition 11.2 is not immediate, but can be shown along the lines of Lemma 11.3 (see Lemma I.7).

The sample paths also belong to the Besov space $\mathfrak{B}_{1,\infty}^{1/2}[0, 1]$, and hence the process can also be viewed to be of exact regularity $1/2$, relative to a weaker norm.

The process $B_t - t B_1$ is called the *Brownian bridge*. It has covariance function $K(s, t) = \min(s, t) - st$, and can also be derived through conditioning Brownian motion B on the event $B_1 = 0$.

Example 11.6 (Integrated Brownian motion, Riemann-Liouville) In view of its rough appearance Brownian motion is often considered a poor prior model for a function. A simple method to smooth it is to take consecutive primitives. If $I_{0+} f$ denotes the primitive function $I_{0+} f(t) = \int_0^t f(s)\, ds$ of a function f, and $I_{0+}^k f = I_{0+}^{k-1} I_{0+} f$ its k-fold primitive, and B is Brownian motion, then $W = I_{0+}^k B$ is a Gaussian process with sample paths that are k times differentiable with a kth derivative that is Lipschitz of order almost $1/2$. Thus W is nearly $k + 1/2$-smooth.

This operation creates processes of smoothness levels $1/2, 3/2, \ldots$. It is possible to interpolate between these values by *fractional integration*. By partial integration it can be seen that, for $k \in \mathbb{N}$ and a continuous function f,

$$I_{0+}^k f(t) = \frac{1}{\Gamma(k)} \int_0^t (t - s)^{k-1} f(s)\, ds. \tag{11.1}$$

The formula on the right makes good sense also for noninteger values of k, and then is called the fractional integral of f. The process $I_{0+}^{\alpha} B$ resulting from integrating Brownian motion B is called the Riemann-Liouville process with Hurst parameter $\alpha + 1/2$, for $\alpha \geq 0$. The sample paths of this process are nearly $\alpha + 1/2$-smooth. To cover also the smoothness levels in $[0, 1/2]$, the *Riemann-Liouville process* with *Hurst parameter* $\alpha > 0$ is defined more generally as the stochastic integral

$$R_t^{\alpha} = \frac{1}{\Gamma(\alpha + 1/2)} \int_0^t (t - s)^{\alpha - 1/2}\, dB_s, \qquad t \geq 0. \tag{11.2}$$

The process R^{α} can be viewed as the the $(\alpha + 1/2)$-fractional integral of the "derivative dB of Brownian motion." It is Gaussian with zero mean and has nearly α-smooth sample paths.

These primitive processes are "tied at zero," in that their function value as well as their derivatives at zero vanish. This is undesirable for prior modeling. The easiest method to "release" the processes at zero is to add a polynomial $t \mapsto \sum_{j=0}^{k} Z_j t^j$, for independent centered Gaussian random variables $Z_0, \ldots, Z_k$, independent of B. The resulting process is then still Gaussian with mean zero.

The self-similarity property can be seen to "integrate": k times integrated Brownian motion $I_{0+}^k B$ is self-similar of index $k + 1/2$, and the Riemann-Liouville process is self-similar of index α.

Example 11.7 (Ornstein-Uhlenbeck) The standard *Ornstein-Uhlenbeck process* with parameter θ is the (only) mean zero, stationary, Markovian Gaussian process with time set $T = [0, \infty)$ and continuous sample paths. It can be constructed from a Brownian motion B through the relation $W_t = (2\theta)^{-1/2} e^{-\theta t} B_{e^{2\theta t}}$, for $t \geq 0$, and its covariance kernel is given by $K(s, t) = (2\theta)^{-1} e^{-\theta |s-t|}$. It can also be characterized as the solution to the stochastic differential equation $dW_t = -\theta\, W_t\, dt + dB_t$, which exhibits the process as a continuous time autogressive process.

Example 11.8 (Stationary process, square exponential, Matérn) Stationary Gaussian processes with index set $T \subset \mathbb{R}^d$ are characterized by covariance kernels of the form $K(s, t) = K(s - t)$, where $K: \mathbb{R}^d \to \mathbb{R}$ is a positive-definite function, and we abuse notation by giving this the same name as the kernel. Provided the index set is rich enough *Bochner's theorem* gives an alternative characterization in terms of the *spectral measure* of the process. This is a symmetric, finite measure μ on $\mathbb{R}$ such that

$$K(s - t) = \int e^{-i\langle s-t, \lambda\rangle}\, d\mu(\lambda). \tag{11.3}$$

Popular examples are the *square exponential process* and the family of *Matérn processes*, given by the spectral measures

$$d\mu(\lambda) = 2^{-d} \pi^{-d/2} e^{-\|\lambda\|^2/4}\, d\lambda, \tag{11.4}$$

$$d\mu(\lambda) = (1 + \|\lambda\|^2)^{-\alpha-d/2}\, d\lambda, \qquad \alpha > 0. \tag{11.5}$$

The covariance function of the square exponential process takes the simple explicit form $\mathrm{E}[W_s W_t] = e^{-\|s-t\|^2}$. For the Matérn process it can be represented in terms of special functions (see e.g. Rasmussen and Williams 2006, page 84).

As the notation suggests, the sample paths of the Matérn process are α-smooth. The sample paths of the square exponential process are infinitely often differentiable, and even analytic. These claims may be inferred from Proposition I.4.

Example 11.9 (Fractional Brownian motion) The *fractional Brownian motion* (fBm) with *Hurst parameter* $\alpha \in (0, 1)$ is the mean zero Gaussian process $W = (W_t: t \in [0, 1])$ with continuous sample paths and covariance function

$$\mathrm{E}(W_s W_t) = \tfrac{1}{2}(s^{2\alpha} + t^{2\alpha} - |t - s|^{2\alpha}). \tag{11.6}$$

The choice $\alpha = 1/2$ yields the ordinary Brownian motion. To obtain a process of a given smoothness $\alpha > 1$, we can take an ordinary integral of fractional Brownian motion of order $\alpha - \langle\alpha\rangle$.

Example 11.10 (Kriging) For a given Gaussian process $(W_t : t \in T)$ and fixed, distinct points $t_1, \ldots, t_m \in T$ the conditional expectations $W_t^* = \mathrm{E}(W_t |\, W_{t_1}, \ldots, W_{t_m})$ define another Gaussian process. It can be written as $W_t^* = \sum_{i=1}^m a_i(t) W_{t_i}$, where, for K the covariance function of W,

$$\begin{pmatrix} a_1(t) \\ \vdots \\ a_m(t) \end{pmatrix} = \Sigma^{-1}\sigma(t), \qquad \Sigma = ((K(t_i, t_j)))_{i,j=1,\ldots,m}, \qquad \sigma(t) = \begin{pmatrix} K(t, t_1) \\ \vdots \\ K(t, t_m) \end{pmatrix}.$$

The process W^* takes its randomness from the finitely many random variables $W_{t_1}, \ldots, W_{t_m}$, and coincides with these variables at the points $t_1, \ldots, t_m$. In spatial statistics this interpolation operation is known as *kriging*. The process W^* is referred to as the *interpolating Gaussian process*, the *Gaussian sieve process*, or the *predictive process*.

The covariance kernel of the process W^* is given by

$$K^*(s,t) = \sum_{i=1}^m \sum_{j=1}^m a_i(s) a_j(t) K(t_i, t_j) = \sigma(s)^\top \Sigma \sigma(t). \tag{11.7}$$

If W has continuous sample paths, then so does W^*. In that case the process W^* converges to W when $m \to \infty$ and the interpolating points $t_1, \ldots, t_m$ grow dense in T.

Example 11.11 (Scaling) If $W = (W_t : t \in \mathbb{R}^d)$ is a Gaussian process with covariance kernel K, then the process $(W_{at} : t \in \mathbb{R}^d)$ is another Gaussian process, with covariance kernel $K(as, at)$, for any $a > 0$. A scaling factor $a < 1$ stretches the sample paths, whereas a factor $a > 1$ shrinks them. Intuitively, if restricted to a compact domain stretching decreases the variability of the sample paths, whereas shrinking increases the variability. This is illustrated in Figure 11.1. Even though the smoothness of the sample paths does not change in a qualitative analytic sense, these operations may completely change the properties of the process as a prior, as we shall see in Section 11.5.

11.2 Reproducing Kernel Hilbert Space

Every Gaussian process comes with an intrinsic Hilbert space, determined by its covariance kernel. This space determines the support and shape of the process, and therefore is crucial for the properties of the Gaussian process as a prior. The definition is different for Gaussian processes (as in Definition 11.1) and Gaussian random elements (as in Definition 11.2).

For a Gaussian process $W = (W_t : t \in T)$, let $\overline{\mathrm{lin}}(W)$ be the closure of the set of all linear combinations $\sum_i \alpha_i W_{t_i}$ in the $\mathbb{L}_2$-space of square-integrable variables. The space $\overline{\mathrm{lin}}(W)$ is a Hilbert space, called the *first order chaos* of W.

Definition 11.12 (Stochastic process RKHS) The *reproducing kernel Hilbert space (RKHS)* of the mean zero, Gaussian process $W = (W_t : t \in T)$ is the set of all functions

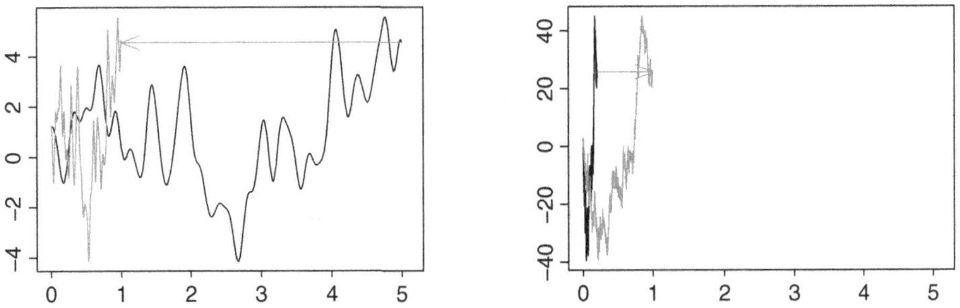

Figure 11.1 Shrinking the time scale of the square exponential process (left) yields a process of more variability on the desired domain [0, 1], whereas stretching a Brownian motion (right) creates a smoother prior.

$z_H: T \to \mathbb{R}$ defined by $z_H(t) = \mathrm{E}(W_t H)$, for H ranging over $\overline{\mathrm{lin}}(W)$. The corresponding inner product is

$$\langle z_{H_1}, z_{H_2} \rangle_{\mathbb{H}} = \mathrm{E}(H_1 H_2).$$

By the definition of the inner product, the correspondence $z_H \leftrightarrow H$ is an isometry between the RKHS $\mathbb{H}$ and the first-order chaos $\overline{\mathrm{lin}}(W)$. This shows that the definition is well posed (the correspondence is one-to-one), and also that $\mathbb{H}$ is indeed a Hilbert space.

The function corresponding to $H = \sum_i \alpha_i W_{s_i}$ is

$$z_H(t) = \mathrm{E}(W_t H) = \sum_{i=1}^k \alpha_i K(s_i, t).$$

In particular, the function $K(s, \cdot)$ is contained in the RKHS, for every fixed $s \in T$, and corresponds to the variable W_s. For a general function $z_H \in \mathbb{H}$ we can compute $\langle z_H, K(s, \cdot)\rangle_{\mathbb{H}} = \mathrm{E}(H W_s) = z_H(s)$. In other words, for every $h \in \mathbb{H}$,

$$h(t) = \langle h, K(t, \cdot)\rangle_{\mathbb{H}}. \tag{11.8}$$

This is the *reproducing formula* that gives the RKHS its name. In general, a reproducing kernel Hilbert space is defined to be a Hilbert space of functions for which there exists a "kernel" K that makes the reproducing formula valid.

A Gaussian random element in a separable Banach space (in the sense of Definition 11.2) also comes with a reproducing kernel Hilbert space. This has a more abstract definition, but will be fundamental in the sequel. First we define the map $S: \mathbb{B}^* \to \mathbb{B}$ by

$$Sb^* = \mathrm{E}\big[b^*(W)W\big].$$

The right side of this equation is the *Pettis integral* of the $\mathbb{B}$-valued random variable $b^*(W)W$. By definition this is the unique element $Sb^* \in \mathbb{B}$ such that $b_2^*(Sb^*) = \mathrm{E}\big[b_2^*(W)b^*(W)\big]$, for every $b_2^* \in \mathbb{B}^*$.[3]

Definition 11.13 (RKHS) The *reproducing kernel Hilbert space* (RKHS) of the Gaussian random element W is the completion of the range $S\mathbb{B}^*$ of the map $S: \mathbb{B}^* \to \mathbb{B}$ for the inner product

[3] This exists by Lemma I.9, as $\|Wb^*(W)\| \le \|b^*\|\,\|W\|^2$, and $\|W\|$ has finite moments.

$$\langle Sb_1^*, Sb_2^* \rangle_{\mathbb{H}} = \mathrm{E}\big[b_1^*(W) b_2^*(W)\big].$$

Even though there is nothing "Gaussian" about the RKHS of a Gaussian process, as this depends on the covariance function only, it may be shown that this completely characterizes the Gaussian measure: every RKHS corresponds to exactly one centered Gaussian measure on a Banach space. Furthermore, to a certain extent the RKHS does not depend on the Banach space in which the process is embedded, but is intrinsic to the process (see Lemma I.17).

In Lemma 11.3 Gaussian processes are related to Gaussian random elements. There is a similar correspondence between the two types of RKHSs. If the sample paths $t \mapsto W_t$ of a mean zero stochastic process $W = (W_t: t \in T)$ belong to a Banach space $\mathbb{B}$ of functions $b: T \to \mathbb{R}$, and the coordinate projections $\pi_t: b \mapsto b(t)$ are elements of $\mathbb{B}^*$, then

$$K(s,t) = \mathrm{E}[W_s W_t] = \mathrm{E}\,[\pi_s(W)\pi_t(W)] = \langle S\pi_s, S\pi_t \rangle_{\mathbb{H}}. \tag{11.9}$$

The three equalities are immediate from the definitions of K, π_t and S, respectively. By the reproducing formula (11.8), the left side is equal to $\langle K(s,\cdot), K(t,\cdot)\rangle_{\mathbb{H}}$, and hence we infer the correspondence $K(t,\cdot) \leftrightarrow S\pi_t$ between the stochastic process and Banach space RKHSs. The following lemma makes this precise for the case of most interest.

Lemma 11.14 *If W is a mean zero Gaussian random element in a separable subspace of $\mathfrak{L}_\infty(T)$ equipped with the supremum norm, then the stochastic process RKHS of Definition 11.12 and the Banach space RKHS of Definition 11.13 coincide, with the correspondence given by $K(t,\cdot) = S\pi_t$.*

Proof As argued in the proof of Lemma I.6, the assumptions imply that W can be viewed as a Gaussian random element in the subspace $\mathfrak{UC}(T,\rho)$ of $\mathfrak{L}_\infty(T)$ for some semimetric ρ on T under which T is totally bounded. Furthermore, it is shown in this proof that every element of the dual of the space is the pointwise limit of a sequence of linear combinations of coordinate projections. In other words, the linear span $\mathbb{B}_0^*$ of the coordinate projections is weak-* dense in $\mathfrak{UC}(T,\rho)^*$, and hence the RKHS is the completion of $S\mathbb{B}_0^*$, by Lemma I.11 and the identification given in equation (11.9). □

Example 11.15 (Euclidean space) To gain insight in the RKHS it helps to consider a Gaussian random vector $W \sim \mathrm{Nor}_k(0,\Sigma)$ in $\mathbb{R}^k$. This can be identified with the stochastic process $W = (W_i: i = 1,\ldots,k)$ on the time set $T = \{1,2,\ldots,k\}$ and is of course also a random element in the Banach space $\mathbb{R}^k$. The covariance kernel is $K(i,j) = \Sigma_{i,j}$ and the RKHS is the space of functions $z_\alpha: \{1,\ldots,k\} \to \mathbb{R}$ given by $z_\alpha(i) = \mathrm{E}\big[W_i(\alpha^T W)\big] = (\Sigma\alpha)_i$ indexed by the (coefficients of the) linear combinations $\alpha^\top W \in \mathrm{lin}(W_1,\ldots,W_k)$, with inner product $\langle z_\alpha, z_\beta\rangle_{\mathbb{H}} = \mathrm{E}\big[(\alpha^\top W)(\beta^\top W)\big] = \alpha^\top \Sigma \beta$. We can identify z_α with the vector $\Sigma\alpha$, and the inner product then satisfies $\langle \Sigma\alpha, \Sigma\beta\rangle_{\mathbb{H}} = \alpha^\top\Sigma\beta$. The map S in the definition of the Banach space RKHS is equal to Σ.

In other words, the RKHS is the range of the covariance matrix, with inner product given through the (generalized) inverse of the covariance matrix. If the covariance matrix is nonsingular, then the RKHS is $\mathbb{R}^k$, but equipped with the inner product generated by the inverse covariance matrix Σ^{-1}.

The RKHS reflects the familiar ellipsoid contours of the density of the multivariate normal distribution.

Example 11.16 (Series) Any mean zero Gaussian random element in a separable Banach space can be represented as an infinite series

$$W = \sum_{i=1}^{\infty} Z_i h_i,$$

for i.i.d. standard normal variables $Z_1, Z_2, \ldots$ and $h_1, h_2, \ldots$ (deterministic) elements of the Banach space $\mathbb{B}$. The series converges almost surely in the norm of $\mathbb{B}$, and the elements h_i have norm one in the RKHS (if they are chosen linearly independent). For a Banach function space and $h_1, h_2, \ldots$ equal to multiples of the eigen functions of the covariance kernel, this is known as the *Karhunen-Loève expansion*, but in fact any orthonormal basis $h_1, h_2, \ldots$ of the RKHS will do. (See Section I.6 for discussion of infinite Gaussian series, and Theorem I.25 for the existence of series representations.)

The series representation invites to identify W with the infinite vector $(Z_1, Z_2, \ldots)$, whose random coordinates Z_i give the spread of W in the deterministic, pre-chosen directions h_i. One should note here that the RKHS norm is stronger than the original Banach space norm, and necessarily $\|h_i\| \to 0$ as $i \to \infty$, even if $\|h_i\|_{\mathbb{H}} = 1$, for every i. If the directions h_i are placed in the order of decreasing norms, then the spread decreases as $i \to \infty$, in some sense to zero, even though Z_i is unbounded. This reveals the Gaussian distribution as an ellipsoid with smaller and smaller axes.

Assume that the functions $\{h_i : i \in \mathbb{N}\}$ are chosen linearly independent in $\mathbb{B}$ in the sense that $w = 0$ is the only element $w \in \ell_2$ such that $\sum_{i=1}^{\infty} w_i h_i = 0$. Then the RKHS of W can also be described in terms of infinite sums: it consists of all linear combinations $\sum_{i=1}^{\infty} w_i h_i$ for $w = (w_1, w_2, \ldots)$ ranging over ℓ_2, and the correspondence $\sum_{i=1}^{\infty} w_i h_i \leftrightarrow w$ is an isometry between $\mathbb{H}$ and ℓ_2, i.e.

$$\Big\langle \sum_{i=1}^{\infty} v_i h_i, \sum_{i=1}^{\infty} w_i h_i \Big\rangle_{\mathbb{H}} = \sum_{i=1}^{\infty} v_i w_i.$$

Thus the RKHS gives the ellipsoid shape of the distribution, as for the multivariate-normal distribution.

Three properties of Gaussian variables relating to their RKHS are important for the analysis of Gaussian priors. (For proofs and references for the following, see Appendices I.5 and I.7.) The first relates to the shape of the Gaussian distribution, and is described in *Borell's inequality*. Let $\mathbb{B}_1$ and $\mathbb{H}_1$ stand for the unit balls of the spaces $\mathbb{B}$ and $\mathbb{H}$, respectively, under their norms, and let Φ be the cumulative distribution function of the standard normal distribution. Furthermore, let $\varphi_0(\epsilon)$ be the *small ball exponent*, defined through, for $\epsilon > 0$,

$$\mathrm{P}(\|W\| < \epsilon) = e^{-\varphi_0(\epsilon)}. \tag{11.10}$$

Proposition 11.17 (Borell's inequality) *For any mean zero Gaussian random element W in a separable Banach space and every $\epsilon, M > 0$,*

$$\mathrm{P}(W \in \epsilon\mathbb{B}_1 + M\mathbb{H}_1) \geq \Phi\left(\Phi^{-1}(e^{-\varphi_0(\epsilon)}) + M\right).$$

For $M = 0$ the inequality in the proposition is an equality, and just reduces to the definition of the small ball exponent. For $M \to \infty$ and fixed $\epsilon > 0$, the right side tends to 1 like the tail of the normal distribution. This shows that the bulk of the Gaussian distribution is contained in an ϵ-shell of a big multiple of the unit ball of the RKHS. We should keep in mind here the ellipsoid shapes found in Examples 11.15 and 11.16, which is coded in general in the shape of $\mathbb{H}_1$ within $\mathbb{B}$. Not only does the Gaussian distribution concentrate most mass close to zero (it has small tails), but also it distributes the mass unevenly in the infinitely many possible directions, as determined by the shape of the RKHS.

The addition of the small ball $\epsilon\mathbb{B}_1$ creates an ϵ-cushion around the multiple $M\mathbb{H}_1$. This is necessary to capture the mass of W, because the RKHS itself may have probability zero. Since $M\mathbb{H}_1 \uparrow \mathbb{H}$ as $M \uparrow \infty$, we have the equality $\mathrm{P}(W \in \epsilon\mathbb{B}_1 + \mathbb{H}) = 1$, for any $\epsilon > 0$, and hence W is supported on the closure $\bar{\mathbb{H}}$ of the RKHS in $\mathbb{B}$.

The second property of a Gaussian variable relates to the change in measure under changes of location. Zero-mean Gaussian variables put most mass near 0; in the Euclidean case this results from the mode of the density being at the mean, while *Anderson's lemma* (see Lemma K.12) expresses this in general. The decrease of mass in a ball of a given radius if the center of this ball is moved away from 0 may be studied quantitatively using Radon-Nikodym derivatives.

The distributions of the two Gaussian variables W and $W + h$ can be shown to be either mutually absolutely continuous or orthogonal, depending on whether the shift h is contained in the RKHS or not. In the first case, when $h \in \mathbb{H}$, a density of the law of $W + h$ relative to the law of W can be shown to take the form

$$\frac{dP_{W+h}}{dP_W}(W) = e^{Uh - \frac{1}{2}\|h\|_{\mathbb{H}}^2},$$

where $U: \mathbb{H} \to \overline{\mathrm{lin}}(W)$ is a linear isometry that extends the map defined by $U(Sb^*) = b^*(W)$, for $b^* \in \mathbb{B}^*$. The variable Uh is normally distributed, and the formula should recall the formula for the quotient of two normal densities with different means and equal variances. The formula allows us to compute the small ball probability $\mathrm{P}(\|W + h\| < \epsilon)$ of the shifted variable (a *decentered small ball probability*) in terms of the distribution of W, and leads to the following lemma (see Lemma I.27 for a proof of a stronger result).

Lemma 11.18 (Decentered small ball) *For any $h \in \mathbb{H}$ and every $\epsilon > 0$ we have*

$$\mathrm{P}(\|W + h\| < \epsilon) \geq e^{-\frac{1}{2}\|h\|_{\mathbb{H}}^2}\mathrm{P}(\|W\| < \epsilon).$$

The lemma only applies to shifts within the reproducing kernel Hilbert space, but it can be extended to general shifts by approximation. Here we restrict to shifts w_0 inside the closure of the RKHS, as otherwise a small enough ball around w_0 will have probability zero. Define the *concentration function* of W at w by

$$\varphi_w(\epsilon) = \inf_{h\in\mathbb{H}:\|h-w\|\le\epsilon} \tfrac{1}{2}\|h\|_{\mathbb{H}}^2 - \log \mathrm{P}\left(\|W\| < \epsilon\right). \tag{11.11}$$

For $w = 0$ this reduces to the small ball exponent $\varphi_0(\epsilon)$ defined in (11.10) (the infimum is achieved for $h = 0$), and it measures concentration of W at 0. The extra term if $w \neq 0$, which will be referred to as the *decentering function*, measures the decrease in mass when shifting from the origin to w. Indeed, up to constants, the concentration function is the exponent of the small ball around w, for every w in the support of W. See Lemma I.28 for a proof of the following result.

Proposition 11.19 (Small ball exponent) *For any mean zero Gaussian random element W in a separable Banach space, and any w in the closure of its RKHS, and any $\epsilon > 0$,*

$$\varphi_w(\epsilon) \le -\log \mathrm{P}\left(\|W - w\| < \epsilon\right) \le \varphi_w(\epsilon/2).$$

11.3 Posterior Contraction Rates

In this section we first give a generic result on Gaussian priors, which characterizes rates such that the prior has sufficient mass near a given "true function" w_0 and almost all its mass in a set of bounded complexity. The formulation of the result reminds of Theorems 8.9 or 8.19 (and Lemma 8.20) on posterior contraction rates. However, the result is purely in terms of the norm of the Banach space in which the prior process lives. In subsequent sections we apply the generic result to obtain rates of posterior contraction for Gaussian process priors in standard statistical settings by relating the statistically relevant norms and discrepancies to the Banach space norm.

The theorem is based on Borell's inequality and the concentration function, whence we assume that the prior is the law of a random element in an appropriate separable Banach space $(\mathbb{B}, \|\cdot\|)$. We write its RKHS as $(\mathbb{H}, \|\cdot\|_{\mathbb{H}})$, and denote by w_0 a fixed element of the Banach space, considered to be the "true" parameter, where posterior concentration is expected to occur. As the Gaussian prior is supported on the closure of the RKHS in $\mathbb{B}$, it is necessary that this parameter belongs to this closure.

Theorem 11.20 (Gaussian contraction rate) *Let W be a mean zero Gaussian random element in a separable Banach space $\mathbb{B}$ with RKHS $\mathbb{H}$ and let $w_0 \in \bar{\mathbb{H}}$. If $\epsilon_n > 0$ is such that*

$$\varphi_{w_0}(\epsilon_n) \le n\epsilon_n^2, \tag{11.12}$$

then for any $C > 1$ such that $Cn\epsilon_n^2 > \log 2$, there exists a measurable set $B_n \subset \mathbb{B}$ such that

$$\log N(3\epsilon_n, B_n, \|\cdot\|) \le 6Cn\epsilon_n^2, \tag{11.13}$$

$$\mathrm{P}(W \notin B_n) \le e^{-Cn\epsilon_n^2}, \tag{11.14}$$

$$\mathrm{P}\left(\|W - w_0\| < 2\epsilon_n\right) \ge e^{-n\epsilon_n^2}. \tag{11.15}$$

Proof Inequality (11.15) is an immediate consequence of (11.12) and Proposition 11.19. We need to prove existence of the set B_n such that the first and second inequalities in the theorem hold.

If $B_n = \epsilon_n \mathbb{B}_1 + M_n \mathbb{H}_1$, where $\mathbb{B}_1$ and $\mathbb{H}_1$ are the unit balls of $\mathbb{B}$ and $\mathbb{H}$, respectively, and M_n is a positive constant, then $\mathrm{P}(W \notin B_n) \le 1 - \Phi(\alpha_n + M_n)$ for α_n given by $\Phi(\alpha_n) = \mathrm{P}(W \in \epsilon_n \mathbb{B}_1) = e^{-\varphi_0(\epsilon_n)}$, by Borell's inequality. Since $\varphi_0(\epsilon_n) \le \varphi_{w_0}(\epsilon_n) \le n\epsilon_n^2$ and $C > 1$, we have that $\alpha_n \ge -M_n/2$ if $M_n = -2\Phi^{-1}(e^{-Cn\epsilon_n^2})$. It follows that for this choice $\mathrm{P}(W \notin B_n) \le 1 - \Phi(M_n/2) = e^{-Cn\epsilon_n^2}$.

It remains to verify the complexity estimate (11.13). If $h_1, \ldots, h_N \in M_n \mathbb{H}_1$ are $2\epsilon_n$-separated in terms of the Banach space norm $\|\cdot\|$, then the ϵ_n-balls $h_1 + \epsilon_n \mathbb{B}_1, \ldots, h_1 + \epsilon_n \mathbb{B}_1$ are disjoint and hence, by Lemma 11.18,

$$1 \ge \sum_{j=1}^{N} \mathrm{P}(W \in h_j + \epsilon_n \mathbb{B}_1) \ge \sum_{j=1}^{N} e^{-\|h_j\|_{\mathbb{H}}^2/2} \mathrm{P}(W \in \epsilon_n \mathbb{B}_1) \ge N e^{-M_n^2/2} e^{-\varphi_0(\epsilon_n)}.$$

For a maximal $2\epsilon_n$-separated set $h_1, \ldots, h_N$, the balls around $h_1, \ldots, h_N$ of radius $2\epsilon_n$ cover the set $M_n \mathbb{H}_1$ and hence we obtain the estimates $N(2\epsilon_n, M_n \mathbb{H}_1, \|\cdot\|) \le N \le e^{M_n^2/2} e^{\varphi_0(\epsilon_n)}$. Since any point of B_n is within ϵ_n of an element of $M_n \mathbb{H}_1$, this is also a bound on $N(3\epsilon_n, B_n, \|\cdot\|)$. To complete the proof, observe that $M_n^2/2 \le 5Cn\epsilon_n^2$, in view of Lemma K.6. □

Ignoring multiplicative constants in the rate, we can rewrite relation (11.12) as the pair of inequalities

$$-\log \mathrm{P}(\|W\| < \epsilon_n) \le n\epsilon_n^2, \qquad \inf_{h \in \mathbb{H}: \|h - w_0\| \le \epsilon_n} \|h\|_{\mathbb{H}}^2 \le n\epsilon_n^2.$$

Both inequalities have a minimal solution ϵ_n, and the final rate ϵ_n satisfying (11.12) is the maximum of the two minimal solutions, up to a constant. The first inequality in the preceding display concerns the small ball probability at 0. It depends on the prior, but not on the true parameter w_0: priors that put little mass near 0 will give slow rates ϵ_n, whatever the true parameter w_0. The second inequality measures the decrease of prior mass around the true parameter w_0 relative to the zero parameter (on a logarithmic scale). A prior that puts much mass around 0 may still give bad performance for a nonzero w_0, depending on its position relative to the RKHS. The most favorable situation is that the true parameter is contained in the RKHS: then the choice $h = w_0$ is eligible in the infimum, whence the infimum is bounded by $\|w_0\|_{\mathbb{H}}^2$, and the condition merely says that ϵ_n must not be smaller than a multiple of the "parametric rate" $n^{-1/2}$. However, the RKHS can be a very small space, and hence this favorable situation is rare.

By Proposition 11.19 the concentration function $\varphi_{w_0}(\epsilon)$ measures the prior mass around w_0. Thus the theorem shows that for Gaussian priors the rate of contraction is driven by the prior mass condition only. The existence of sieves of a prescribed complexity, required in (8.5) and (8.4) of the general rate theorems, is implied by the prior mass condition. The preceding theorem establishes this only for entropy, neighborhoods and metric of convergence all described in terms of the Banach space norm, but in the sequel we show that for the standard inference problems this can be translated into the statistically relevant quantities used in Theorems 8.9 and 8.19.

Lower Bounds

The concentration function φ_{w_0} delivers both lower and upper bounds on the concentration of a Gaussian prior. The lower bound is instrumental to verify the prior mass condition and derive an upper bound on the rate of contraction. The upper bound may be used to show that this rate of contraction is sharp, through an application of Theorem 8.35. By Lemma 11.19 the probability $\mathrm{P}(\|W - w_0\| < \epsilon)$ is sandwiched between $e^{-\varphi_{w_0}(\epsilon/2)}$ and $e^{-\varphi_{w_0}(\epsilon)}$. Therefore, inequalities (8.45) with d the metric induced by the norm are satisfied if

$$\phi_{w_0}(\epsilon_n/2) \le C_1 n\epsilon_n^2 \le C_2 n\epsilon_n^2 \le \phi_{w_0}(\delta_n).$$

Then Theorem 8.35 says that δ_n is a lower bound for the contraction rate. If the concentration function is "regularly varying", then these inequalities ought to be true for δ_n, a multiple of ϵ_n. If then the Banach space norm also relates properly to the Kullback-Leibler discrepancies, the bounds on the contraction rates obtained in the following will be sharp.

11.3.1 Density Estimation

Consider estimating a probability density p relative to a σ-finite measure ν on a measurable space $(\mathfrak{X}, \mathscr{X})$, based on a sample of observations $X_1, \ldots, X_n |\, p \overset{\text{iid}}{\sim} p$. Construct a prior Π for p as the exponential transform of a Gaussian process $W = (W_x : x \in \mathfrak{X})$

$$p(x) = \frac{e^{W_x}}{\int_{\mathfrak{X}} e^{W_y}\, d\nu(y)}.$$

This transform was considered in Section 2.3.1, and statistical distances were seen to be controlled by the uniform norm on the exponent.

Theorem 11.21 *Let W be a mean-zero Gaussian random element in a separable subspace of $\mathfrak{L}_\infty(\mathfrak{X})$ with measurable sample paths. If $w_0 = \log p_0$ belongs to the support of W, and ϵ_n satisfies the rate equation $\varphi_{w_0}(\epsilon_n) \le n\epsilon_n^2$, then $\Pi_n(p : d_H(p, p_0) > M\epsilon_n |\, X_1, \ldots, X_n) \to 0$ in P_0^n-probability, for some sufficiently large constant M. Furthermore, if $\tilde\varphi_{\tilde w_0}(\delta_n) \ge C_2 n\epsilon_n^2$, for a sufficiently large constant C_2, where $\tilde\varphi_{\tilde w_0}$ is the concentration function of the process $W - W(x_0)$ at $\tilde w_0 = w_0 - w_0(x_0)$ for some $x_0 \in \mathfrak{X}$, then $\Pi_n(p : \|p - p_0\|_\infty \le m\delta_n |\, X_1, \ldots, X_n) \to 0$ in P_0^n-probability, for a sufficiently small constant m.*

Proof The Kullback-Leibler divergence and variation, and the square Hellinger distance between densities $p_{w_1} \propto e^{w_1}$ and $p_{w_2} \propto e^{w_2}$ are bounded by the square of the uniform norm of the difference between the exponents w, by Lemma 2.5. Therefore the prior mass, remaining mass and entropy conditions of Theorem 8.9 are verified at a multiple of ϵ_n in view of Theorem 11.20.

By the same lemma, a ball $\{w : \|p_w - p_{w_0}\|_\infty < \delta\}$ is contained in a ball $\{w : \|w - w_0\|_\infty < 2D(w, w_0)\delta\}$, if all functions w and w_0 are restricted to take the value zero at a given point $x_0 \in \mathfrak{X}$. Here $2D(w, w_0)$ is bounded above by a constant D_0 that depends only on w_0, for every w with $\|p_w - p_{w_0}\|_\infty < 1/2$. Because a constant shift in w cancels under forming the density p_w, the (prior) distributions of the densities $p_{\tilde{W}}$ and p_W are identical. Thus $\mathrm{P}\big(\|p_W - p_{w_0}\|_\infty < \delta_n/D_0\big)$ is bounded above by $\mathrm{P}(\|W - W(x_0) - w_0 - w_0(x_0)\|_\infty < \delta_n) \le e^{-\tilde{\varphi}_{\tilde{w}_0}(\delta_n)}$, by Proposition 11.19. For δ_n chosen such that the right side is bounded above by $e^{-C_2 n\epsilon_n^2}$, the posterior mass of this set tends to zero, by Theorem 8.35. □

11.3.2 Nonparametric Binary Regression

Consider estimating a binary regression function $p(x) = \mathrm{P}(Y = 1|\, X = x)$ based on an i.i.d. sample $(X_1, Y_1), \ldots, (X_n, Y_n)$ from the distribution of (X, Y), where $Y \in \{0, 1\}$, and X takes its values in some measurable space $(\mathfrak{X}, \mathscr{X})$ following a distribution G. Given a fixed link function $\Psi: \mathbb{R} \to (0, 1)$ construct a prior for p through a Gaussian process $W = (W_x: x \in \mathfrak{X})$ by

$$p(x) = \Psi(W_x).$$

We shall assume that Ψ is strictly increasing from $\Psi(-\infty) = 0$ to $\Psi(\infty) = 1$, and differentiable with bounded derivative.

The likelihood for (X, Y) factorizes as $p(x)^y\,(1 - p(x))^{1-y}\, dG(x)$. Because the contribution of G cancels out from the posterior distribution for p, we can consider G known and need not specify a prior for it. Assume that p_0 is never zero and let $w_0 = \Psi^{-1}(p_0)$, for p_0 the true value of p.

The following theorem considers two settings: the first assumes that w_0 is bounded and the second not. In the second case the theorem is true under an additional condition on Ψ, which is satisfied by the logistic link function, but rules out the normal link. Of course, on compact domains $\mathfrak{X}$, boundedness is implied by continuity of w_0, and the first case suffices.

Theorem 11.22 *Let W be a mean-zero Gaussian random element W in a separable Banach space $\mathbb{B}$. If w_0 belongs to the support of W and ϵ_n satisfies the rate equation $\varphi_{w_0}(\epsilon_n) \le n\epsilon_n^2$, then $\Pi_n(\|p - p_0\|_{2,G} > M\epsilon_n|\, X_1, Y_1, \ldots, X_n, Y_n) \to 0$ in P_0^n-probability, for some $M > 0$, in the following cases:*

(i) *$\mathbb{B} \subset \mathfrak{L}_\infty(\mathfrak{X})$ and w_0 is bounded.*
(ii) *$\mathbb{B} = \mathbb{L}_2(G)$ and the function $\psi/\,(\Psi(1 - \Psi))$ is bounded.*

Proof This follows from combining Lemma 2.8, Theorem 8.9, and Theorem 11.20. □

11.3.3 Nonparametric Normal Regression

Consider estimating a regression function f based on observations $Y_1, \ldots, Y_n$ in the normal regression model with fixed covariates $Y_i = f(x_i) + \varepsilon_i$, where $\varepsilon_i \overset{\text{iid}}{\sim} \mathrm{Nor}(0, \sigma_0^2)$ and the covariates $x_1, \ldots, x_n$ are fixed elements from a set $\mathfrak{X}$.

A prior on f is induced by setting $f(x) = W_x$, for a Gaussian process $(W_x: x \in \mathfrak{X})$. If σ is unknown, then we also put a prior on σ, which we assume to be supported on an interval $[a, b] \subset (0, \infty)$ with a density π that is bounded away from zero.

Let $\|\cdot\|_n$ be the $\mathbb{L}_2(\mathbb{P}_n^x)$-norm for the empirical measure $\mathbb{P}_n^x$ of the design points $x_1, \ldots, x_n$, and let $\varphi_{w_0,n}$ be the concentration function of W viewed as a map in $\mathbb{L}_2(\mathbb{P}_n^x)$. The inconvenience that this depends on n can be removed by bounding the empirical norm by the uniform norm, which gives a corresponding bound on the concentration function.

Theorem 11.23 *Let W be a mean-zero Gaussian random element in $\mathbb{L}_2(\mathbb{P}_n^x)$, for every n, and suppose that the true values f_0 of f and σ_0 of σ belong to the supports of W and π. If ϵ_n satisfies the rate equation $\varphi_{w_0,n}(\epsilon_n) \leq n\epsilon_n^2$, then $\Pi_n\left((w, \sigma): \|w - w_0\|_n + |\sigma - \sigma_0| > M\epsilon_n \mid Y_1, \ldots, Y_n\right) \to 0$ in P_0^n-probability, for some $M > 0$. Furthermore, if $\varphi_{w_0,n}(\delta_n) \geq C_2 n\epsilon_n^2$ for a sufficiently large constant C_2, then $\Pi_n(w: \|w - w_0\|_n \leq m\delta_n \mid Y_1, \ldots, Y_n) \to 0$ in P_0^n-probability, for sufficiently small $m > 0$.*

Proof It is noted in Section 8.3.2 that the Kullback-Leibler divergence and variation are equivalent to the empirical squared distance $\|\cdot\|_{n,2}^2$. Furthermore, likelihood ratio tests have the appropriate properties relative to this distance. Since σ is restricted to a compact interval containing the truth and its prior is bounded away from zero, it plays a minor role in the rate of contraction. Thus the upper bound of the theorem follows from combining an extension of Theorem 8.26 to include σ and Theorem 11.20.

The lower bound follows from combining the lower bounds from Theorem 8.35 and Proposition 11.19. □

11.3.4 White Noise Model

Consider estimation of the signal θ based on the observation $X^{(n)} = (X_t^{(n)}: 0 \leq t \leq 1)$ in the white noise model described in Section 8.3.4. As a prior on θ we take a Gaussian random element W in $\mathbb{L}_2[0, 1]$, which is conjugate with the Gaussian likelihood.

Theorem 11.24 *Let W be a mean zero Gaussian random element in $\mathbb{L}_2[0, 1]$. If the true value θ_0 of θ is contained in the support of W, and ϵ_n satisfies the rate equation $\varphi_{\theta_0}(\epsilon_n) \leq n\epsilon_n^2$ with respect to the $\mathbb{L}_2[0, 1]$-norm, then $\Pi_n\left(\|\theta - \theta_0\|_2 > M\epsilon_n \mid X^{(n)}\right) \to 0$ in P_0^n-probability, for some $M > 0$. Furthermore, if $\varphi_{w_0}(\delta_n) \geq C_2 n\epsilon_n^2$ for a sufficiently large constant C_2, then $\Pi_n\left(\|\theta - \theta_0\|_2 \leq m\delta_n \mid X^{(n)}\right) \to 0$ in P_0^n-probability, for sufficiently small $m > 0$.*

Proof The upper bound follows from combining Theorems 8.31 and 11.20; the lower bound from combining Theorem 8.35 and Proposition 11.19. □

11.4 Specific Gaussian Processes as Priors

In this section we calculate posterior contraction rates associated with specific Gaussian process priors. In each example this involves an estimate of the small ball exponent $\varphi_0(\epsilon)$, a characterization of the RKHS of the process, and the approximation of a true function w_0 by elements of the RKHS. The small ball exponent can be calculated by a probabilistic method, but can also be obtained from the metric entropy of the unit ball $\mathbb{H}_1$ of the RKHS. Under regularity conditions it is true that, as $\epsilon \downarrow 0$,[4]

$$\varphi_0(\epsilon) \asymp \log N\Big(\frac{\epsilon}{\sqrt{2\varphi_0(\epsilon)}}, \mathbb{H}_1, \|\cdot\|_{\mathbb{B}}\Big).$$

Given an upper bound for the covering number $N(\epsilon, \mathbb{H}_1, \|\cdot\|_{\mathbb{B}})$, this may be solved for the small ball probability. The following lemma gives the solution in a common case.

Lemma 11.25 (Small ball and entropy) *For $\alpha > 0$ and $\beta \in \mathbb{R}$, as $\epsilon \downarrow 0$, $\varphi_0(\epsilon) \asymp \epsilon^{-\alpha}(\log_- \epsilon)^{\beta}$ if and only if $\log N(\epsilon, \mathbb{H}_1, \|\cdot\|) \asymp \epsilon^{-2\alpha/(2+\alpha)}(\log_- \epsilon)^{2\beta/(2+\alpha)}$.*

11.4.1 Brownian Motion and Its Primitives

Brownian motion B is a good starting point for modeling functions on [0, 1]. The RKHS and small ball probabilities of Brownian motion are classical. Let $\mathfrak{W}^k[0,1]$ be the *Sobolev space* of all functions $f \in \mathbb{L}_2[0,1]$ that are $k-1$ times differentiable with $f^{(k-1)}$ absolutely continuous with derivative $f^{(k)}$ belonging to $\mathbb{L}_2[0,1]$ (see Definition C.6).

Lemma 11.26 (RKHS) *The RKHS of Brownian motion is equal to $\{f \in \mathfrak{W}^1[0,1]: f(0) = 0\}$ with the inner product $\langle f, g\rangle_{\mathbb{H}} = \int_0^1 f'(t)g'(t)\,dt$.*

Lemma 11.27 (Small ball) *The small ball exponent of Brownian motion viewed as a map into $\mathfrak{C}[0,1]$ or $\mathbb{L}_r[0,1]$, for some $r \geq 1$, satisfies, as $\epsilon \downarrow 0$,*

$$\varphi_0(\epsilon) \asymp \epsilon^{-2}.$$

Proof For a proof of the second lemma apply Lemma 11.25 together with the entropy estimate of the unit ball of the RKHS given by Proposition C.7, or see Li and Shao (2001).

Since $K(s,t) = s \wedge t$, the RKHS is equal to the completion of the linear span of the functions $\{t \mapsto s \wedge t : s \in [0,1]\}$ under the inner product determined by

$$\langle s_1 \wedge \cdot, s_2 \wedge \cdot\rangle_{\mathbb{H}} = s_1 \wedge s_2 = \int (s_1 \wedge t)'(s_2 \wedge t)'\,dt.$$

Here $t \mapsto (s \wedge t)' = \mathbb{1}\{[0,s]\}(t)$ is the Radon-Nikodym derivative of the function $s \wedge \cdot$. The equation shows that the RKHS inner product is indeed the inner product of $\mathfrak{W}^1[0,1]$. It suffices to show that the linear span of all functions of this type is dense in $\{f \in \mathfrak{W}^1[0,1]: f(0) = 0\}$. Now the linear span contains every function that is 0 at 0, continuous, and piecewise linear on a partition $0 = s_0 < s_1 \cdots < s_N = 1$, since such a function with slopes α_j on the intervals (s_{j-1}, s_j), for $j = 1, \ldots, N$, can be constructed as a linear combination by first determining the coefficient of $(s_N \wedge \cdot)$ to have the correct slope on (s_{N-1}, s_N), next determining the coefficient of $(s_{N-1} \wedge \cdot)$ to have the correct slope on (s_{N-2}, s_{N-1}), etc. The derivatives of these piecewise linear functions are piecewise constant, and the set of piecewise constant functions is dense in $\mathbb{L}_2[0,1]$. Thus the completion of the linear span is as claimed. ☐

[4] See Lemma I.29 for a precise statement.

Standard Brownian motion is zero at zero, and hence for use as a prior it is preferable to "release it at zero," by adding a variable that gives a prior for the unknown function at zero. Adding an independent Gaussian variable $Z \sim \text{Nor}(0, 1)$ yields another Gaussian process of the form $t \mapsto Z + B_t$. The RKHS of the constant process $t \mapsto Z$ consists of the constant functions, and general rules for the formation of reproducing Hilbert spaces (see Lemma I.18) shows that the RKHS of $Z + B$ is the direct sum of the constant functions and the RKHS of B. In other words, the point zero is also "released" in the RKHS: the RKHS is equal to the Sobolev space $\mathfrak{W}^1[0, 1]$ with inner product

$$\langle f, g\rangle_{\mathbb{H}} = f(0)g(0) + \int_0^1 f'(s)g'(s)\, ds.$$

The addition of the single variable Z makes the small ball probability at 0 smaller, but it does not change the rate ϵ^{-2} obtained in the preceding lemma, as $-\log \mathrm{P}(|Z| < \epsilon) \asymp \log_- \epsilon \ll \epsilon^{-2}$.

The concentration function $\varphi_{w_0}(\epsilon)$ of $Z+B$ depends on the position of the true parameter w_0 relative to the RKHS. It can be computed using a kernel smoother $w_0 * \psi_\sigma$, which is contained in the RKHS if ψ is a smooth kernel.

Lemma 11.28 (Decentering) *If $w_0 \in \mathfrak{C}^\beta[0, 1]$ for some $\beta \in (0, 1]$, then the decentering of the concentration function of Brownian motion released at zero viewed as map in $\mathfrak{C}[0, 1]$ satisfies, for $\epsilon \downarrow 0$,*

$$\inf_{h:\|h-w_0\|_\infty<\epsilon} \|h'\|_2^2 \lesssim \epsilon^{-(2-2\beta)/\beta}.$$

Proof If ψ_σ is the density of the $\text{Nor}(0, \sigma^2)$-distribution, then $\|w_0 * \psi_\sigma - w_0\| \lesssim \sigma^\beta$, as $\sigma \to 0$, and the squared RKHS-norm of $w_0 * \psi_\sigma$ is given by $(w_0 * \psi_\sigma)(0)^2 + \|(w_0 * \psi_\sigma)'\|_2^2 \asymp \sigma^{-(2-2\beta)}$. Choosing $\sigma \asymp \epsilon^{1/\beta}$, we obtain the assertion. ☐

Combining the preceding we see that the concentration function of released Brownian motion satisfies, if $w_0 \in \mathfrak{C}^\beta[0, 1]$,

$$\varphi_{w_0}(\epsilon) \lesssim \epsilon^{-(2-2\beta)/\beta} + \epsilon^{-2}.$$

For $\beta \geq 1/2$, the second term ϵ^{-2} dominates, and hence the minimal solution to the rate equation $\varphi_{w_0}(\epsilon_n) \leq n\epsilon_n^2$ satisfies $\epsilon_n^{-2} \asymp n\epsilon_n^2$, or $\epsilon_n \asymp n^{-1/4}$. For $\beta \in (0, 1/2)$, the first term dominates, leading to $\epsilon_n^{-(2-2\beta)/\beta} \lesssim n\epsilon_n^2$, with minimal solution $\epsilon_n \asymp n^{-\beta/2}$.

The resulting contraction rate can be summarized as $n^{-(\beta\wedge 1/2)/2}$. It is equal to the minimax rate of estimation $n^{-\beta/(2\beta+1)}$ for functions in a Hölder space of order β if and only if $\beta = 1/2$. This is intuitively understandable, as the sample paths of a Brownian motion are regular of that order: only matching of prior and true smoothness yields optimal results. For any positive $\beta \neq 1/2$ the posterior distribution is consistent, but the performance of the Brownian motion prior is suboptimal. The discrepancy is most felt for smooth w_0: the contraction rate is $n^{-1/4}$, no matter how smooth w_0 is. Technically this stems from the small ball probability of Brownian motion. This prior places a tiny fraction $\exp(-C\epsilon^{-2})$ of its mass in a ball of radius ϵ around the zero function, which may be viewed as the smoothest function of all. No amount of data will fully wash out this prior preference for non-smooth functions.

This shortcoming of Brownian motion as a prior for smooth functions can be remedied by integrating its sample paths. The k-fold integrated Brownian motion $I_{0+}^k B$ is smooth of order (nearly) $k+1/2$. Its k vanishing derivatives at zero can be released by adding a random polynomial, yielding the process

$$W_t = \sum_{i=0}^k Z_i \frac{t^i}{i!} + (I_{0+}^k B)_t, \qquad Z_0, \ldots, Z_k \overset{\text{iid}}{\sim} \text{Nor}(0,1) \perp\!\!\!\perp B. \tag{11.16}$$

Lemma 11.29 (RKHS) *The RKHS of the process W given in* (11.16)*, for B a Brownian motion independent of $Z_1, \ldots, Z_k \overset{iid}{\sim} \text{Nor}(0,1)$, is the Sobolev space $\mathfrak{W}^{k+1}[0,1]$, with inner product*

$$\langle f, g \rangle_{\mathbb{H}} = \sum_{i=0}^k f^{(i)}(0) g^{(i)}(0) + \int_0^1 f^{(k+1)}(s) g^{(k+1)}(s)\, ds.$$

Lemma 11.30 (Small ball) *The small ball exponents of k-fold integrated Brownian motion $I_{0+}^k B$ and the process W given in* (11.16)*, for B a Brownian motion independent of $Z_1, \ldots, Z_k \overset{iid}{\sim} \text{Nor}(0,1)$, viewed as maps into $\mathfrak{C}[0,1]$ satisfy, as $\epsilon \downarrow 0$,*

$$\varphi_0(\epsilon) \asymp \epsilon^{-2/(2k+1)}.$$

Proof The RKHS of $I_{0+}^k B$ can be deduced from the RKHS of Brownian motion and the general principle for the transformation of a RKHS under a continuous, linear transformation given in Lemma I.16. Because I_{0+}^k is a continuous, linear, one-to-one map from $\mathfrak{C}[0,1] \to \mathfrak{C}[0,1]$, the RKHS of $I_{0+}^k B$ is given by $\mathbb{H} = \{I_{0+}^k f : f \in \mathbb{H}_B\}$, where $\mathbb{H}_B$ is the RKHS of the Brownian motion given in Lemma 11.26, and the inner product satisfies $\langle I_{0+}^k f, I_{0+}^k g \rangle_{\mathbb{H}} = \int_0^1 f'(s) g'(s)\, ds$. In other words, the RKHS is the subset of the Sobolev space $\mathfrak{W}^{k+1}[0,1]$ of functions f with $f(0) = \cdots = f^{(k)}(0) = 0$, under its natural inner product.

The RKHS of the process W can likewise be obtained by applying a general result, which gives the RKHS of a sum of independent Gaussian random elements (see Lemma I.18). Since the supports of the polynomial process $t \mapsto \sum_{i=0}^k Z_i t^i / i!$ and $I_{0+}^k B$ in $\mathfrak{C}[0,1]$ intersect nontrivially, the lemma does not apply directly. However, we may also consider these processes as Borel measurable random elements in the space $\mathfrak{C}^k[0,1]$. The RKHSs remain the same in view of Lemma I.17. Since the only kth degree polynomial with k vanishing derivatives at zero is the zero function, the supports of the two processes in $\mathfrak{C}^k[0,1]$ have trivial intersection, and hence the RKHS of W is the direct sum of the RKHSs of the polynomial process and $I_{0+}^k B$, by Lemma I.18. The first process is a finite series and hence its RKHS are the polynomials with norm the Euclidean norm of the coefficients. The norm can alternatively be written in terms of the derivates at zero; see Example I.24.

The small ball probability is obtained in Li and Linde (1998), or follows from Lemma I.29 together with the entropy estimate of the unit ball of the RKHS given in Proposition C.7. □

Lemma 11.31 (Decentering) *If $w_0 \in \mathfrak{C}^\beta[0,1]$ for $\beta \le k+1$, then the decentering function of the process in* (11.16)*, for B a Brownian motion independent of $Z_1,\ldots,Z_k \overset{iid}{\sim} \mathrm{Nor}(0,1)$, viewed as map in $\mathfrak{C}[0,1]$ satisfies, as $\epsilon \downarrow 0$,*

$$\inf_{h::\|h-w_0\|_\infty<\epsilon} \|h\|_{\mathbb{H}}^2 \lesssim \epsilon^{-(2k-2\beta+2)/\beta}.$$

Proof The convolution $w_0 * \psi_\sigma$, where ψ_σ is the scaled version of a smooth kth order kernel (an integrable function ψ satisfying $\int \psi(t)\,dt = 1$ and $\int t^r\psi(t)\,dt = 0$ for $r = 1,\ldots,k$, and $\int |t|^{k+1}\psi(t)\,dt < \infty$), satisfies $\|w_0 * \psi_\sigma - w_0\|_\infty \lesssim \sigma^\beta$, as follows by the well known estimates from the literature on kernel density estimation. The function $w_0 * \psi_\sigma$ belongs to the RKHS and satisfies $(w_0 * \psi_\sigma)^{(l)} = w_0^{(\underline{\beta})} * \psi_\sigma^{(l-\underline{\beta})}$, for $\underline{\beta}$ the largest integer strictly smaller than β. Hence $\|(w_0 * \psi_\sigma)^{(l)}\|_\infty \lesssim \sigma^{-(l-\beta)}$ if $w_0 \in \mathfrak{C}^\beta[0,1]$ and $l \ge \beta$. Consequently $\int (w_0 * \psi_\sigma)^{(k+1)}(t)^2\,dt \lesssim \sigma^{-(2k-2\beta+2)}$ and the square derivatives at 0 up to order k are bounded or of smaller order in $1/\sigma$. It follows that $\|w_0 * \psi_\sigma\|_{\mathbb{H}} \lesssim \sigma^{-(k-\beta+1)}$, if $w_0 \in \mathfrak{C}^\beta[0,1]$. The choice $\sigma \asymp \epsilon^{1/\beta}$ leads to $\|w_0 * \psi_\sigma - w_0\|_\infty \lesssim \epsilon$ and $\|w_0 * \psi_\sigma\|_{\mathbb{H}}^2 \lesssim \epsilon^{-(2k-2\beta+2)/\beta}$. □

It follows that the concentration function of "released integrated Brownian motion" (11.16) takes the form

$$\varphi_{w_0}(\epsilon) \lesssim \epsilon^{-(2k-2\beta+2)/\beta} + \epsilon^{-2/(2k+1)}.$$

For $\beta \ge k + 1/2$ the second term dominates, and the rate inequality becomes $\epsilon_n^{-2/(2k+1)} \le n\epsilon_n^2$, with as minimal solution $\epsilon_n \asymp n^{-(2k+1)/(4k+4)}$. For $\beta \le k + 1/2$ the first term in the concentration function dominates, and the rate inequality $\epsilon_n^{-(2k-2\beta+2)/\beta} \lesssim n\epsilon_n^2$ has the minimal solution $\epsilon_n \asymp n^{-\beta/(2k+2)}$. The posterior contraction rate can be summarized as the maximum $n^{-(\beta\wedge k+1/2)/(2k+2)}$ of the two rates. This is the minimax rate for β-regular functions if and only if $\beta = k + 1/2$.

11.4.2 Riemann-Liouville Process

The Riemann-Liouville process with Hurst parameter $\alpha > 0$, defined in Example 11.6, is a random element in $\mathfrak{C}[0,1]$. We add an independent Gaussian polynomial of degree the smallest integer $\lfloor\alpha\rfloor + 1$ strictly greater than α to "release" it at zero, and consider, with $Z_1,\ldots,Z_n \overset{\text{iid}}{\sim} \mathrm{Nor}(0,1)$ independent of R^α,

$$W_t = \sum_{k=0}^{\lfloor\alpha\rfloor+1} Z_k t^k + R_t^\alpha. \tag{11.17}$$

Let I_{0+}^α be the fractional integral, as defined in (11.1).

Lemma 11.32 (RKHS) *The RKHS of the Riemann-Liouville process of order α is the space $I_{0+}^{\alpha+1/2}(\mathbb{L}_2[0,1])$ with the inner product*

$$\langle I_{0+}^{\alpha+1/2} f, I_{0+}^{\alpha+1/2} g\rangle_{\mathbb{H}} = \langle f, g\rangle_2.$$

Lemma 11.33 (Small ball) *The small ball exponent of the Riemann-Liouville process released at zero viewed as a map into* $\mathfrak{C}[0,1]$, *satisfies, as* $\epsilon \downarrow 0$,

$$\varphi_0(\epsilon) \asymp \epsilon^{-1/\alpha}.$$

Proof The Riemann-Liouville process is $R_t^\alpha = \int_0^t f_t(u)\,dB_u$, for the function $f_t(u) = (t-u)_+^{\alpha-1/2}/\Gamma(\alpha+1/2)$. The isometry property of stochastic integrals gives that

$$\mathrm{E}(R_s^\alpha R_t^\alpha) = \langle f_s, f_t\rangle_2 = I_{0+}^{\alpha+1/2} f_s(t).$$

It follows that the function $I_{0+}^{\alpha+1/2} f_s$ is contained in the RKHS, for every $s \in [0,1]$, and has RKHS norm $\|f_s\|_2$. By definition the RKHS is the completion of the linear span of these functions under this norm. It suffices to show that the linear span of the functions f_s is dense in $\mathbb{L}_2[0,1]$. Now $f \perp f_s$, for every s, implies that $I_{0+}^{\alpha-1/2} f(t) = \int f(u)(t-u)^{\alpha-1/2}\,du = 0$, and this implies that $f = 0$, by the injectivity of the operator $I_{0+}^{\alpha-1/2}\colon \mathbb{L}_2[0,1] \to \mathbb{L}_2[0,1]$.

The small ball estimate follows from Theorem 2.1 of Li and Linde (1998), or a calculation on the entropy of the unit ball of the RKHS. □

Lemma 11.34 (Decentering) *If* $w_0 \in \mathfrak{C}^\beta[0,1]$, *then the concentration function of the Riemann-Liouville process released at zero viewed as a map into* $\mathfrak{C}[0,1]$, *satisfies, as* $\epsilon \downarrow 0$,

$$\varphi_{w_0}(\epsilon) \lesssim \begin{cases} \epsilon^{-1/\alpha}, & \text{if } 0 < \alpha \le \beta, \\ \epsilon^{-(2\alpha-2\beta+1)/\beta}, & \text{if } \alpha > \beta \text{ and } (\langle\alpha\rangle = \frac{1}{2} \text{ or } \alpha \notin \beta + \frac{1}{2} + \mathbb{N}), \\ \epsilon^{-(2\alpha-2\beta+1)/\beta}\log_- \epsilon, & \text{otherwise.} \end{cases}$$

Proof For $\alpha < \beta$ the small ball probability dominates the concentration function. For $\alpha \ge \beta$, the proof must use properties of fractional integrals, which makes it technical. We refer to van der Vaart and van Zanten (2008a) or Castillo (2008) for details. □

In view of the preceding lemma the rate equation $\varphi_{w_0}(\epsilon_n) \le n\epsilon_n^2$ is satisfied by $\epsilon_n \ge n^{-(\alpha\wedge\beta)/(2\alpha+1)}$ if either $\langle\alpha\rangle = \frac{1}{2}$ or $\alpha - \beta - \frac{1}{2} \notin \mathbb{N}$, and by $\epsilon_n \gtrsim n^{-\beta/(2\alpha+1)}\log n$ otherwise. Up to the logarithmic factor in the last case, this is the minimax rate for β-regular functions if and only if $\alpha = \beta$. (The extra $\log n$-factor in the case that $\langle\alpha\rangle \ne \frac{1}{2}$ or $\alpha - \beta - \frac{1}{2} \in \mathbb{N}$ appears, because the fractional integral I_{0+}^α maps $\mathfrak{C}^\lambda[0,1] \to \mathfrak{C}^{\lambda+\alpha}[0,1]$ only if $\alpha + \lambda \ne 1$, and appears not to be an artifact of the method of proof.)

11.4.3 Fractional Brownian Motion

The fractional Brownian motion fBm^α of Hurst index α is related to the Riemann-Liouville process R^α as $\mathrm{fBm}_t^\alpha = c_\alpha R_t^\alpha + c_\alpha Z_t$, where $c_\alpha > 0$ is a constant, R^α and Z are independent Gaussian processes in $\mathfrak{C}[0,1]$, and Z possesses small ball probability, as $\epsilon \downarrow 0$,

$$\varphi_0(\epsilon; Z) = -\log \mathrm{P}(\|Z\| < \epsilon) = o(\epsilon^{-1/\alpha}). \qquad (11.18)$$

(See Mandelbrot and Van Ness 1968 and Lemma 3.2 of Li and Linde 1998.) Because this is of smaller order than the small ball probability of the Riemann-Liouville process, the concentration functions φ_{w_0} of fBm^α and R^α behave similarly (for any w_0), and hence the results on the Riemann-Liouville process extend to the fractional Brownian motion.

11.4.4 Stationary Processes

The RKHS of a stationary Gaussian process $(W_t : t \in T)$ indexed by a subset $T \subset \mathbb{R}^d$, as in Example 11.8, can be characterized in terms of its spectral measure μ.

For $t \in T$ let $e_t : \mathbb{R}^d \to \mathbb{C}$ denote the function $e_t(\lambda) = e^{i\lambda^T t}$, and for $\psi \in \mathbb{L}_2(\mu)$, let $H\psi : T \to \mathbb{R}$ be the function defined by $(H\psi)(t) = \int e_t \psi \, d\mu$. If μ has Lebesgue density m, then this function is the Fourier transform of the (integrable) function ψm.

Lemma 11.35 (RKHS) *The RKHS of the stationary Gaussian process with spectral measure μ is the set of functions $H\psi$ for ψ ranging over $\mathbb{L}_2(\mu)$, with RKHS-norm equal to $\|H\psi\|_{\mathbb{H}} = \|P\psi\|_{2,\mu}$, where P is the projection onto the closed linear span of the set of functions $(e_t : t \in T)$ in $\mathbb{L}_2(\mu)$. If $T \subset \mathbb{R}^d$ has an interior point and $\int e^{\delta\|\lambda\|} \, d\mu(\lambda) < \infty$ for some $\delta > 0$, then this closed linear span is $\mathbb{L}_2(\mu)$ and the RKHS norm is $\|H\psi\|_{\mathbb{H}} = \|\psi\|_{2,\mu}$.*

Proof The spectral representation (11.3) can be written $\mathrm{E} W_s W_t = \langle e_t, e_s \rangle_{\mu,2}$. By definition the RKHS is therefore the set of functions $H\psi$ with ψ running through the closure $\mathbb{L}_T$ in $\mathbb{L}_2(\mu)$ of the linear span of the set of functions $(e_s : s \in T)$, and the norm equal to the norm of ψ in $\mathbb{L}_T \subset \mathbb{L}_2(\mu)$. Here the "linear span" is taken over the reals. If instead we take the linear span over the complex numbers, we obtain complex functions whose real parts give the RKHS.

The set of functions obtained by letting ψ range over the full space $\mathbb{L}_2(\mu)$ rather than $\mathbb{L}_T$ is precisely the same, as a general element $\psi \in \mathbb{L}_2(\mu)$ gives exactly the same function as its projection $P\psi$ onto $\mathbb{L}_T$. However, the associated norm is the $\mathbb{L}_2(\mu)$ norm of $P\psi$. This proves the first assertion of the lemma. For the second we must show that $\mathbb{L}_T = \mathbb{L}_2(\mu)$ under the additional conditions.

The partial derivative of order $(k_1, \ldots, k_d)$ with respect to $(t_1, \ldots, t_d)$ of the map $t \mapsto e_t$ at t_0 is the function $\lambda \mapsto (i\lambda_1)^{k_1} \cdots (i\lambda_d)^{k_d} e_{t_0}(\lambda)$. Appealing to the dominated convergence theorem, we see that this derivative exists as a derivative in $\mathbb{L}_2(\mu)$. Because t_0 is an interior point of T by assumption, we conclude that the function $\lambda \mapsto (i\lambda)^k e_{t_0}(\lambda)$ belongs to $\mathbb{L}_T$ for any multi-index k of nonnegative integers. Consequently, the function $p e_{t_0}$ belongs to $\mathbb{L}_T$ for any polynomial $p : \mathbb{R}^d \to \mathbb{C}$ in d arguments. It suffices to show that these functions are dense in $\mathbb{L}_2(\mu)$.

Equivalently, it suffices to prove that the polynomials themselves are dense in $\mathbb{L}_2(\mu)$. Indeed, if $\psi \in \mathbb{L}_2(\mu)$ is orthogonal to all functions of the form $p e_{t_0}$, then $\psi \overline{e_{t_0}}$ is orthogonal to all polynomials. Denseness of the set of polynomials then gives that $\psi \overline{e_{t_0}}$ vanishes μ-almost everywhere, whence ψ vanishes μ-almost everywhere.

Finally we prove that the polynomials are dense in $\mathbb{L}_2(\mu)$ if μ has an exponential moment. Suppose that $\psi \in \mathbb{L}_2(\mu)$ is orthogonal to all polynomials. Since μ is a finite measure, the complex conjugate $\overline{\psi}$ is μ-integrable, and hence we can define a complex measure ν by

$$\nu(B) = \int_B \overline{\psi(\lambda)} \, \mu(d\lambda).$$

It suffices to show that ν is the zero measure, so that $\psi = 0$ almost everywhere relative to μ.

By the Cauchy-Schwarz inequality and the assumed exponential integrability of the spectral measure, with $|\nu|$ the (total) variation measure of ν, $\int e^{\delta\|\lambda\|/2} \, |\nu|(d\lambda) < \infty$. By a standard argument, based on the dominated convergence theorem (see e.g. Bauer 2001, Theorem 8.3.5), this implies that the function $z \mapsto \int e^{\langle\lambda,z\rangle}\,\nu(d\lambda)$ is analytic on the strip $\Omega = \{z \in \mathbb{C}^d\colon |\operatorname{Re} z_1| < \delta/(2\sqrt{d}), \ldots, |\operatorname{Re} z_d| < \delta/(2\sqrt{d})\}$. Also for z real and in this strip, by the dominated convergence theorem,

$$\int e^{\langle\lambda,z\rangle}\,\nu(d\lambda) = \int \sum_{n=0}^{\infty} \frac{\langle\lambda,z\rangle^n}{n!}\,\nu(d\lambda) = \sum_{n=0}^{\infty} \int \frac{\langle\lambda,z\rangle^n}{n!}\,\overline{\psi}(\lambda)\,\mu(d\lambda).$$

The right side vanishes, because ψ is orthogonal to all polynomials by assumption.

We conclude that the function $z \mapsto \int e^{\langle\lambda,z\rangle}\,\nu(d\lambda)$ vanishes on the set $\{z \in \Omega\colon \operatorname{Im} z = 0\}$. Because this set contains a nontrivial interval in $\mathbb{R}$ for every coordinate, we can apply (repeated) analytic continuation to see that this function vanishes on the complete strip Ω. In particular the Fourier transform $t \mapsto \int e^{i\lambda^\top t}\,\nu(d\lambda)$ of ν vanishes on all of $\mathbb{R}^d$, whence ν is the zero-measure. □

The functions $H\psi$ in the RKHS are always uniformly continuous, as ψ is μ-integrable. Their exact regularity is determined by the tails of the spectral measure. If these are light, then $H\psi$ is a mixture of mainly low frequency trigonometric functions, and hence smooth.

To characterize the small ball probability and concentration function we must specialize to particular spectral measures. We consider the Matérn and square exponential measures in the following subsections.

Matérn Process

The spectral measure of the Matérn process is given in (11.5). Its polynomial tails make the sample paths of the process regular of finite order. The small ball exponent of the process is similar to that of the Riemann-Liouville process of the same regularity.

Lemma 11.36 (Small ball) *The small ball exponent of the Matérn process W viewed as a map in $\mathfrak{C}[0,1]$ satisfies, as $\epsilon \downarrow 0$,*

$$\varphi_0(\epsilon) \lesssim \epsilon^{-d/\alpha}.$$

Proof The Fourier transform of $H\psi$ is, up to a constant, the function $\phi = \psi m$, if $d\mu(\lambda) = m\,d\lambda$. For ψ the choice of minimal norm in the definition of $H\psi$, this function satisfies

$$\int |\phi(\lambda)|^2 (1 + \|\lambda\|^2)^{\alpha+d/2}\,d\lambda = \|H\psi\|_{\mathbb{H}}^2.$$

In other words, the unit ball $\mathbb{H}_1$ of the RKHS is contained in a Sobolev ball of order $\alpha + d/2$. The metric entropy relative to the uniform norm of such a Sobolev ball is bounded by a constant times $\epsilon^{-d/(\alpha+d/2)}$, by Proposition C.7. The lemma next follows from Lemma 11.25, which characterizes the small ball probability in terms of the entropy of the RKHS-unit ball. □

To estimate the infimum in the definition of the concentration function φ_{w_0} for a nonzero response function w_0, we approximate w_0 by elements of the RKHS. The idea is to write w_0 in terms of its Fourier inverse $\hat{w}_0$ as, with $d\mu(\lambda) = m(\lambda)\,d\lambda$,

$$w_0(x) = \int e^{i\lambda^\top x}\hat{w}_0(\lambda)\,d\lambda = \int e^{i\lambda^\top x}\frac{\hat{w}_0}{m}(\lambda)\,d\mu(\lambda). \tag{11.19}$$

If $\hat{w}_0/m$ were contained in $\mathbb{L}_2(\mu)$, then w_0 would be contained in the RKHS, with RKHS-norm bounded by the $\mathbb{L}_2(\mu)$-norm of $\hat{w}_0/m$, i.e. the square root of $\int(|\hat{w}_0|^2/m)(\lambda)\,d\lambda$. In general this integral may be infinite, but we can remedy this by truncating the tails of $\hat{w}_0/m$.

A natural a priori condition on the true response function $w_0\colon [0,1]^d \to \mathbb{R}$ is that this function is contained in a Sobolev space of order β. The *Sobolev space* $\mathfrak{W}^\alpha[0,1]^d$ is here defined as the set of functions $w\colon [0,1]^d \to \mathbb{R}$ that are restrictions of a function $w\colon \mathbb{R}^d \to \mathbb{R}$ with Fourier transform $\hat{w}(\lambda) = (2\pi)^{-d}\int e^{i\lambda^\top t}w(t)\,dt$ such that

$$\|w\|^2_{2,2,\alpha} := \int (1+\|\lambda\|^2)^\alpha |\hat{w}(\lambda)|^2\,d\lambda < \infty.$$

Roughly speaking, for integer α, a function belongs to $\mathfrak{W}^\alpha([0,1]^d)$ if it has partial derivatives up to order α that are all square integrable. This follows, because the αth derivative of a function w has Fourier transform $\lambda \mapsto (i\lambda)^\alpha\hat{w}(\lambda)$,

Lemma 11.37 (Decentering) *If $w_0 \in \mathfrak{C}^\beta([0,1]^d) \cap \mathfrak{W}^\beta([0,1]^d)$ for $\beta \leq \alpha$, then the decentering function of the Matérn process satisfies, for $\epsilon < 1$,*

$$\inf_{h:\|h-w_0\|_\infty<\epsilon} \|h\|^2_{\mathbb{H}} \lesssim \epsilon^{-(2\alpha+d-2\beta)/\beta}.$$

Proof Let $\kappa\colon \mathbb{R} \to \mathbb{R}$ be a function with a real, symmetric Fourier transform $\hat{\kappa}$, which equals $1/(2\pi)$ in a neighborhood of 0 and which has compact support. From $\hat{\kappa}(\lambda) = (2\pi)^{-1}\int e^{i\lambda t}\kappa(t)\,dt$ it then follows that $\int\kappa(t)\,dt = 1$ and $\int(it)^k\kappa(t)\,dt = 0$ for $k \geq 1$. For $t = (t_1,\ldots,t_d)$, define $\phi(t) = \kappa(t_1)\cdots\kappa(t_d)$. Then ϕ integrates to 1, has finite absolute moments of all orders, and vanishing moments of all orders bigger than 0.

For $\sigma > 0$ set $\phi_\sigma(x) = \sigma^{-d}\phi(x/\sigma)$ and $h = \phi_\sigma * w_0$. Because ϕ is a higher order kernel, standard arguments from the theory of kernel estimation show that $\|w_0 - \phi_\sigma * w_0\|_\infty \lesssim \sigma^\beta$.

The Fourier transform of h is the function $\lambda \mapsto \hat{h}(\lambda) = \hat{\phi}(\sigma\lambda)\hat{w}_0(\lambda)$, and therefore, by (11.19),

$$\begin{aligned}\|h\|^2_{\mathbb{H}} &\lesssim \int |\hat{\phi}(\sigma\lambda)\hat{w}_0(\lambda)|^2\frac{1}{m(\lambda)}\,d\lambda \lesssim \sup_\lambda\Big[(1+\|\lambda\|^2)^{\alpha+d/2-\beta}|\hat{\phi}(\sigma\lambda)|^2\Big]\|w_0\|^2_{2,2,\beta}\\ &\lesssim C(\sigma)\sup_\lambda\Big[(1+\|\lambda\|^2)^{\alpha+d/2-\beta}|\hat{\phi}(\lambda)|^2\Big]\|w_0\|^2_{2,2,\beta},\end{aligned}$$

for

$$C(\sigma) = \sup_\lambda\Big(\frac{1+\|\lambda\|^2}{1+\|\sigma\lambda\|^2}\Big)^{\alpha+d/2-\beta} \lesssim \Big(\frac{1}{\sigma}\Big)^{2\alpha+d-2\beta},$$

if $\sigma \leq 1$. The assertion of the lemma follows upon choosing $\sigma \sim \epsilon^{1/\beta}$. ☐

It follows that the rate equation $\varphi_{w_0}(\epsilon_n) \leq n\epsilon_n^2$ has the minimal solution $\epsilon_n \asymp n^{-\beta/(2\alpha+d)}$, for $\beta \leq \alpha$. Thus again the rate is minimax if and only if prior and smoothness match.

Square Exponential Process

The spectral measure of the *square exponential process* is the Gaussian measure given in (11.4). The analytic sample paths of the process make the small ball probability of this process much larger than that of the processes considered so far: it is nearly "parametric."

Lemma 11.38 (Small ball) *The small ball exponent of the square exponential process viewed as a map in $\mathfrak{C}([0,1]^d)$ satisfies, for a constant C depending only on d, as $\epsilon \downarrow 0$,*

$$\varphi_0(\epsilon) \leq C(\log_- \epsilon)^{1+d/2}.$$

Proof Every function $H\psi$ in the (complex) RKHS as described in Lemma 11.35 can be extended to an entire function $H\psi: \mathbb{C} \to \mathbb{C}$ defined by $H\psi(z) = \int e^{\langle \lambda, z\rangle} \psi(\lambda)\, d\mu(\lambda)$. If the function $H\psi$ is contained in the unit ball of the RKHS, then $\|\psi\|_{2,\mu} \leq 1$, and an application of the Cauchy-Schwarz inequality gives that $|H\psi(z)|^2 \leq \int e^{2\|\lambda\|\,|z|}\, d\mu(\lambda) \leq e^{2C|z|^2}$, for some universal constant C. In particular, the functions $H\psi$ can be extended to analytic functions on the strip $\{z \in \mathbb{C}: \|z\|_\infty \leq A\}$ that are uniformly bounded by e^{CA^2}, for any A. It follows by Proposition C.9 that $N(\epsilon, \mathbb{H}_1, \|\cdot\|_\infty) \leq A^{-d} \log(e^{CA^2}/\epsilon)^{1+d}$. Choosing A of the order $\log_- \epsilon$ leads to the bound $N(\epsilon, \mathbb{H}_1, \|\cdot\|_\infty) \leq (\log_- \epsilon)^{1+d/2}$.

The result now follows from the characterization of the small ball exponent by the entropy of the RKHS unit ball, Lemma I.29. ☐

Lemma 11.39 (Decentering) *If $w_0 \in \mathfrak{W}^\beta([0,1]^d)$ for $\beta > d/2$, then the concentration function of the square exponential process satisfies, for $\epsilon < 1$, a constant C that depends only on w_0,*

$$\inf_{h: \|h - w_0\|_\infty < \epsilon} \|h\|_{\mathbb{H}}^2 \lesssim \exp\left(C\epsilon^{-2/(\beta - d/2)}\right).$$

Proof For given $K > 0$ let $\psi(\lambda) = (\hat{w}_0/m)(\lambda)\mathbb{1}_{\|\lambda\| \leq K}$, for m the density in (11.4). The function $H\psi$ satisfies

$$\|H\psi - w_0\|_\infty \leq \int_{\|\lambda\| > K} |\hat{w}_0(\lambda)|\, d\lambda \leq \|w_0\| \left[\int_{\|\lambda\| > K} (1 + \|\lambda\|^2)^{-\beta}\, d\lambda\right]^{1/2}$$
$$\lesssim \|w_0\|_{2,2,\beta} K^{-(\beta - d/2)}.$$

Furthermore, the squared RKHS-norm of $H\psi$ is given by

$$\|H\psi\|_{\mathbb{H}}^2 = \int_{\|\lambda\| \leq K} \frac{|\hat{w}_0|^2}{m}(\lambda)\, d\lambda \leq \sup_{\|\lambda\| \leq K} \left[m(\lambda)^{-1}(1 + \|\lambda\|^2)^{-\beta}\right]$$
$$\times \|w_0\|_{2,2,\beta}^2 \lesssim e^{K^2/4} \|w_0\|_{2,2,\beta}^2.$$

We conclude the proof by choosing $K \asymp \epsilon^{-1/(\beta - d/2)}$. ☐

Combining the preceding we see that the concentration function of the square exponential process satisfies

$$\varphi_{w_0}(\epsilon) \lesssim \exp\left(C\epsilon^{-2/(\beta - d/2)}\right) + (\log_- \epsilon)^{1+d/2}.$$

The first term (decentering function) dominates the second (centered small ball exponent) for any $\beta > 0$, and the contraction rate for a β-smooth function satisfies $\epsilon_n \asymp (\log n)^{-(\beta/2-d/4)}$. This extremely slow rate is the result of the discrepancy between the infinite smoothness of the prior and the finite smoothness of the true parameter. A remedy for this mismatch is to rescale the sample paths and is discussed in Sections 11.5 and 11.6.

Actually, the preceding establishes only an upper bound on the contraction rate. The next lemma can be used to show that the logarithmic rate is real: balls of logarithmic radius around a parameter w_0 that is exactly of finite smoothness asymptotically receive zero posterior mass. The lemma shows that the upper bound on the decentered concentration function obtained in Lemma 11.39 can be reversed for w_0 that are exactly finitely smooth, at least for the $\mathbb{L}_2$-norm. The implication for the posterior contraction rate can then be obtained from Theorem 8.35. The "exact" finite smoothness of w_0 is operationalized by assuming that its Fourier transform has polynomial tails. (See van der Vaart and van Zanten 2011, Theorem 8, for a proof of the lemma, and the details of an application to posterior rates in the regression model.)

Lemma 11.40 (Lower bound) *If w_0 is contained in $\mathfrak{W}^\beta([0,1]^d)$ for some $\beta > d/2$, has support within $(0,1)^d$ and possesses a Fourier transform satisfying $|\hat{w}_0(\lambda)| \gtrsim \|\lambda\|^{-k}$ for some $k > 0$ and every $\|\lambda\| \geq 1$, then there exists a constants $b, v > 0$ such that*

$$\inf_{h::\|h-w_0\|_2<\epsilon} \|h\|_{\mathbb{H}}^2 \geq e^{b\epsilon^{-v}}.$$

As the square exponential process prior puts all of its mass on analytic functions, perhaps it is not fair to study its performance only for β-regular functions. We shall now show that for "supersmooth," analytic true parameters the prior works very well.

For $r \geq 1$ and $\lambda > 0$, we define $\mathcal{A}^{\gamma,r}(\mathbb{R}^d)$ as the space of functions $w: \mathbb{R}^d \to \mathbb{R}$ with Fourier transform $\hat{w}$ satisfying

$$\|w\|_{\mathcal{A}}^2 := \int e^{\gamma\|\lambda\|^r} |\hat{w}|^2(\lambda)\, d\lambda < \infty.$$

Finiteness of this norm requires exponential decrease of the Fourier transform, in contrast to polynomial decrease for Sobolev smoothness. The functions in $\mathcal{A}^{\gamma,r}(\mathbb{R}^d)$ are infinitely often differentiable and "increasingly smooth" as γ or r increase. They extend to functions that are analytic on a strip in $\mathbb{C}^d$ containing $\mathbb{R}^d$ if $r = 1$, and to entire functions if $r > 1$ (see e.g. Bauer 2001, 8.3.5).

Lemma 11.41 (Decentering) *Suppose that w_0 is the restriction to $[0,1]^d$ of an element of $\mathcal{A}^{\gamma,r}(\mathbb{R}^d)$.*

(i) *If $r > 2$ or ($r \geq 2$ and $\gamma \geq 1/4$), then $w_0 \in \mathbb{H}$.*
(ii) *If $r < 2$, then there exists a constant C depending on w_0 such that, for $\epsilon < 1$,*

$$\inf_{h:\|h-w\|_\infty\leq\epsilon} \|h\|_{\mathbb{H}}^2 \leq C \exp\left(\frac{(\log_- \epsilon)^{2/r}}{4\gamma^{2/r}}\right).$$

Proof The first assertion follows, because $\psi = \hat{w}_0/m$ is contained in $\mathbb{L}_2(\mu)$, if w_0 satisfies the conditions, and $H\psi = w_0$.

The second assertion is proved in the same way as Lemma 11.39, where this time, with $\|w_0\|_{\mathcal{A}}$ the norm of w_0 in $\mathcal{A}^{\gamma,r}(\mathbb{R}^d)$,

$$\|H\psi - w_0\|_\infty^2 \leq \int_{\|\lambda\|>K} e^{-\gamma\|\lambda\|^r}\, d\lambda\, \|w_0\|_{\mathcal{A}}^2 \leq e^{-\gamma K^r} K^{-r+1} \|w_0\|_{\mathcal{A}}^2,$$

$$\|H\psi\|_{\mathbb{H}}^2 \leq \sup_{\|\lambda\|\leq K} e^{\|\lambda\|^2/4-\gamma\|\lambda\|^r} \|w_0\|_{\mathcal{A}}^2 \leq e^{K^2/4} \|w_0\|_{\mathcal{A}}^2.$$

We finish by choosing $K \asymp (\gamma^{-1}\log_- \epsilon)^{1/r}$. ☐

Combination of Lemmas 11.38 and 11.41 leads to the rate equation $\varphi_{w_0}(\epsilon_n) \leq n\epsilon_n^2$ of the form

$$\left(\log_- \epsilon_n\right)^{1+d/2} + \exp\left(C(\log_- \epsilon_n)^{2/r}\right) \leq n\epsilon_n^2.$$

Solving this leads to a posterior contraction rate $n^{-1/2}(\log n)^{1/r\vee(1/2+d/4)}$ for any supersmooth true parameter $w_0 \in \mathcal{A}^{\gamma,r}(\mathbb{R}^d)$. This is up to a logarithmic factor equal to the posterior rate of contraction $n^{-1/2}$ for finite-dimensional models. This "almost parametric rate" is explainable from the fact that spaces of analytic functions are only slightly bigger than finite-dimensional spaces in terms of their metric entropy.

11.4.5 Series Priors

Any Gaussian process can be represented (in many ways) as an infinite series with independent standard normal coefficients and deterministic "coordinate functions." Since the RKHS can be characterized using the same coordinate functions (see Example 11.16) this may be a useful device to derive properties of the prior. Conversely, we may fix our favorite coordinate functions from the beginning, and form the process as a random series. The properties of the resulting prior will depend on the coordinate functions. In this section we consider two examples, one with a truncated wavelet basis and one with an infinite, single-indexed sequence of basis functions. Finite series are attractive for computation, but for good approximation their truncation point must increase to infinity.

We start with a general observation that relates a truncated and an infinite series prior. A truncated Gaussian series is another Gaussian process, to which the general theory on Gaussian processes applies. Naturally, if the truncation point is sufficiently high, the concentration functions of the truncated and full series are equivalent, and the resulting theory for the two prior processes is identical. The following lemma quantifies the truncation point.

The lemma applies more generally to approximation of a Gaussian variable W by a sequence W_n, defined on the same probability space.

Lemma 11.42 *Let W be a mean-zero Gaussian random element in a separable Banach space $\mathbb{B}$ with RKHS $\mathbb{H}$ and concentration function φ_{w_0} at $w_0 \in \bar{\mathbb{H}}$. If W_n are mean-zero Gaussian random elements in $\mathbb{B}$ defined on the same probability space such that* $10\mathrm{E}\|W_n - W\|^2 \leq n^{-1}$, *then for $\epsilon_n > 0$ satisfying $n\epsilon_n^2 \geq 4\log 4$ and $\varphi_{w_0}(\epsilon_n) \leq n\epsilon_n^2$ and any $C > 4$,* *then there exist measurable sets* $B_n \subset \mathbb{B}$ *such that* (11.13)–(11.15) *hold with* W *replaced by* W_n *and* ϵ_n *replaced by* $2\epsilon_n$.

Proof By Borell's inequality for the norm, Proposition I.8, applied to $W_n - W$,

$$\mathrm{P}(\|W_n - W\| \geq \epsilon_n) \leq 2e^{-\epsilon_n^2/8\mathrm{E}\|W_n - W\|^2} \leq 2e^{-5n\epsilon_n^2/4}.$$

In view of the inequalities $\mathrm{P}(\|W_n - w_0\| < 3\epsilon_n) \geq \mathrm{P}(\|W - w_0\| < 2\epsilon_n) - \mathrm{P}(\|W_n - W\| \geq \epsilon_n)$, the small ball probability of W_n can be bounded in terms of that of W. Next we choose the sets $B_n = 2\epsilon_n \mathbb{B}_1 + M_n \mathbb{H}_1^n$, where $\mathbb{H}_1^n$ is the unit ball of the RKHS of W_n, and follow steps as in the proof of Theorem 11.20. □

Truncated Wavelet Expansions

Let $\{\psi_{j,k}: j \in \mathbb{N}, k = 1, \ldots, 2^{jd}\}$ be an orthonormal basis of $\mathbb{L}_2([0,1]^d)$ of compactly supported wavelets, with j referring to the resolution level and k to the dilation. Let $w = \sum_{j=1}^{\infty} \sum_{k=1}^{2^{jd}} w_{j,k} \psi_{j,k}$ be the expansion of a given function $w \in \mathbb{L}_2([0,1]^d)$, and consider the norms

$$\|w\|_{1,2} = \sum_{j=1}^{\infty} \Big(\sum_{1 \leq k \leq 2^{jd}} |w_{j,k}|^2 \Big)^{1/2},$$

$$\|w\|_{1,\infty} = \sum_{j=1}^{\infty} 2^{jd/2} \max_{1 \leq k \leq 2^{jd}} |w_{j,k}|,$$

$$\|w\|_{\infty,\infty,\beta} = \sup_{1 \leq j < \infty} 2^{j\beta} 2^{jd/2} \max_{1 \leq k \leq 2^{jd}} |w_{j,k}|.$$

For a suitable smooth basis these norms are upper bounds on the $\mathbb{L}_2$-norm and the supremum norm, and equivalent to the Besov (∞, ∞, β)-norm of w, respectively. Actually, that the functions $\psi_{j,k}$ are a wavelet basis is only important to ensure this interpretation. All the following is valid for arbitrary bases, provided the norms of the functions are defined by the preceding formulas.

Consider a Gaussian prior of the type, for given positive constants μ_j, and i.i.d. random variables $Z_{j,k} \sim \mathrm{Nor}(0,1)$,

$$W = \sum_{j=1}^{J_\alpha} \sum_{k=1}^{2^{jd}} \mu_j Z_{j,k} \psi_{j,k}. \tag{11.20}$$

The number of terms in the sum is $(2^{J_\alpha d} - 1)/(1 - 2^{-d})$. For interpretability of α we set J_α so that this is the usual dimension for estimating an α-regular function: determine J_α to be the integer that is closest to the solution of the equation $2^{Jd} = n^{d/(2\alpha+d)}$. We next study the posterior rate of contraction when the true parameter has regularity β, which may be different from the "nominal smoothness level" α.

Since $W_j := \sum_{k=1}^{2^{jd}} \mu_j Z_{j,k} \psi_{j,k}$ contributes the variance $\mathrm{E}\|W_j\|_2^2 = \mu_j^2 2^{jd}$ to the prior, the choice $\mu_j = 2^{-jd/2}$ gives all resolution levels the same prior variation. If $\mu_j 2^{jd/2} \downarrow 0$ as $j \to \infty$, then higher resolution levels receive less weight and hence the prior gives increasingly more weight to dimensions lower than the nominal dimension $2^{J_\alpha d}$. This is advantageous if the true regularity satisfies $\beta > \alpha$, but in the opposite case $\beta < \alpha$ the nominal dimension $2^{J_\alpha d}$ is already too small, and this would be exacerbated by putting lower weight on the higher levels. The following results show that the choice $\mu_j 2^{jd/2} = 2^{-j\beta}$ is a good compromise: it (nearly) obtains the optimal rate $n^{-\beta/(2\beta+d)}$ for $\beta \geq \alpha$, and the "optimal rate $n^{-\beta/(2\alpha+d)}$ when using a $2^{J_\alpha d}$-dimensional model" in the other case.

Lemma 11.43 (RKHS) *The RKHS of W given by* (11.20) *is the set of functions* $w = \sum_{j=1}^{J_\alpha} \sum_k w_{j,k} \psi_{j,k}$ *for which the norm* $\|w\|_{\mathbb{H}}^2 = \sum_{j=1}^{J_\alpha} \sum_k \mu_j^{-2} w_{j,k}^2$ *is finite.*

Lemma 11.44 (Small ball) *The centered small ball probability of W given by* (11.20) *with* $\mu_j 2^{jd/2} = 2^{-ja}$ *for some* $a \geq 0$, *viewed as a map in* $\mathfrak{L}_\infty([0,1]^d)$, *satisfies, as* $\epsilon_n \to 0$

$$\varphi_0(\epsilon_n) \lesssim \begin{cases} 2^{J_\alpha d}(\log_- \epsilon_n + J_\alpha), & \text{if } \epsilon_n 2^{J_\alpha a} \lesssim J_\alpha^2, \\ \epsilon_n^{-d/a}(\log_- \epsilon_n)^{2d/a}, & \text{if } \epsilon_n 2^{J_\alpha a} \gtrsim J_\alpha^2. \end{cases}$$

Proof The first lemma is immediate from Example 11.16. For the proof of the second, take any numbers $\alpha_j \geq 0$ with $\sum_{j=1}^{J_\alpha} \alpha_j \leq 1$, and note that

$$\mathrm{P}(\|W\|_\infty < \epsilon) = \mathrm{P}\Big(\sum_{j=1}^{J_\alpha} 2^{jd/2} \max_{1\leq k\leq 2^{jd}} |\mu_j Z_{j,k}| < \epsilon\Big) \geq \prod_{j=1}^{J_\alpha} \prod_{k=1}^{2^{jd}} \mathrm{P}(|\mu_j 2^{jd/2} Z_{j,k}| < \alpha_j \epsilon).$$

Therefore for $\mu_j 2^{jd/2} = 2^{-ja}$, and $\alpha_j = (K + d^2 j^2)^{-1}$, it follows that

$$\varphi_0(\epsilon_n) \leq -\sum_{j=1}^{J_\alpha} 2^{jd} \log\Big(2\Phi(\alpha_j \epsilon_n 2^{ja}) - 1\Big) \lesssim \int_1^{2^{J_\alpha d}} -\log\Big(2\Phi\Big(\frac{\epsilon_n x^{a/d}}{K + \log_2^2 x}\Big) - 1\Big)\, dx,$$

provided that K is large enough to make the map $x \mapsto x^{a/d}/(K + \log_2^2 x)$ nondecreasing on $[1, \infty)$. It may be verified that the function f given by $f(y) = -\log(2\Phi(y) - 1)$ appearing in the right side is decreasing, with $f(0) = \infty$, $f(\infty) = 0$, $f(y) \leq 1 + |\log y|$ on $[0, c]$, and $f(y) \leq e^{-y^2/2}$ for $y \geq c$.

We bound the right side of the preceding display separately for the two cases considered in the lemma. If $\epsilon_n 2^{J_\alpha a} \leq K + J_\alpha^2 d^2$, then $\epsilon_n x^{a/d}/(K + \log_2^2 x) = O(1)$ on $[1, 2^{J_\alpha d}]$ and hence the right side is bounded by a multiple of

$$\int_1^{2^{J_\alpha d}} \Big(1 + \Big|\log\Big(\frac{\epsilon_n x^{a/d}}{K + \log_2^2 x}\Big)\Big|\Big)\, dx \lesssim 2^{J_\alpha d}(\log_- \epsilon_n + J_\alpha).$$

Alternatively, if $\epsilon_n 2^{J_\alpha a} > (K + J_\alpha^2 d^2)$, then the right side is bounded by

$$\epsilon_n^{-d/a} \int_{\epsilon_n}^{\epsilon_n 2^{J_\alpha a}} f\Big(\frac{y}{K + (d/a)^2(\log_2 y + \log_2 \epsilon_n^{-1})^2}\Big) \frac{d}{a} y^{d/a-1}\, dy$$

$$\leq \Big[\int_0^{\epsilon_n^{-1}} f\Big(\frac{y}{K + (2d/a)^2 \log_2^2 \epsilon_n^{-1}}\Big) + \int_{\epsilon_n^{-1}}^\infty f\Big(\frac{y}{K + (2d/a)^2 \log_2^2 y}\Big)\Big] \frac{d}{a} y^{d/a-1}\, dy$$

$$\leq \mu_n^{d/a} \int_0^{\epsilon_n^{-1}\mu_n^{-1}} f(x) \frac{d}{a} x^{d/a-1}\, dx + \int_0^\infty f\Big(\frac{y}{K + (2d/a)^2 \log_2^2 y}\Big) \frac{d}{a} y^{d/a-1}\, dy,$$

for $\mu_n = K + (2d/a)^2(\log_2 \epsilon_n^{-1})^2$. The first term on the right side is bounded by a constant, whence the whole expression is bounded by a multiple of $(\log_- \epsilon_n)^{2d/a}$. ☐

Lemma 11.45 (Decentering) *If $\|w\|_{\infty,\infty,\beta} < \infty$, then the decentering function of W given by* (11.20) *with $\mu_j 2^{jd/2} = 2^{-ja}$ for some $a \geq 0$ satisfies, for any $J \leq J_\alpha$ and $\epsilon \geq 2^{-J\beta}\|w\|_{\infty,\infty,\beta}/(2\beta - 1)$,*

$$\inf_{h:\|h-w\|_\infty<\epsilon} \|h\|_{\mathbb{H}}^2 \lesssim 2^{J(2a-2\beta+d)}\|w\|_{\infty,\infty,\beta}^2.$$

Proof If w^J stands for the projection of w on the space spanned by the wavelet basis up to resolution level J, then

$$\|w - w^J\|_\infty \leq \sum_{j=J+1}^{\infty} 2^{jd/2} \max_k |w_{j,k}| \leq \sum_{j=J+1}^{\infty} 2^{-j\beta}\|w\|_{\infty,\infty,\beta} \lesssim \frac{2^{-J\beta}}{2^\beta - 1}\|w\|_{\infty,\infty,\beta}.$$

For $J \leq J_\alpha$ the function w^J belongs to the RKHS, and hence the infimum in the lemma is smaller than the square RKHS norm of w^J if $2^{-J\beta}\|w\|_{\infty,\infty,\beta} \leq \epsilon(2^\beta - 1)$. This norm is equal to $\sum_{j=1}^{J}\sum_{k=1}^{2^{jd}} w_{j,k}^2/\mu_j^2 \leq \sum_{j=1}^{J} 2^{j(2a-2\beta+d)}\|w\|_{\infty,\infty,\beta}^2$. ☐

To solve the rate equation $\varphi_{w_0}(\epsilon_n) \leq n\epsilon_n^2$ for w_0 with finite norm $\|w_0\|_{\beta,\infty,\infty}$, we determine ϵ_n and J such that $2^{-J\beta} \lesssim \epsilon_n$ and $2^{J(2a-2\beta+d)} \leq n\epsilon_n^2$ and $\varphi_0(\epsilon_n) \leq n\epsilon_n^2$. It can be verified that the minimal value of ϵ_n is obtained when choosing $J = J_a \wedge J_\alpha$, and results in the following lower bounds on ϵ_n:

- $n^{-\beta/(2\alpha+d)} \log n$ if $a \leq \beta \leq \alpha$;
- $n^{-\alpha/(2\alpha+d)} \log n$ if $a \leq \alpha \leq \beta$;
- $n^{-a/(2a+d)}(\log n)^{d/(2a+d)}$ if $\alpha \leq a \leq \beta$;
- $n^{-\beta/(2a+d)}(\log n)^{d/(2a+d)}$ if $\alpha \leq \beta \leq a$.

Infinite Series

Let $\phi_1, \phi_2, \ldots$ be a sequence of basis functions, such that the series $w = \sum_{j=1}^\infty w_j\phi_j$ is a well-defined function for every $w \in \ell_2$, where the correspondence between the function and the coefficients is one-to-one. We define the square norms

$$\|w\|_2^2 = \sum_{j=1}^{\infty} w_j^2, \qquad \|w\|_{2,2,\beta}^2 = \sum_{j=1}^{\infty} j^{2\beta} w_j^2.$$

For an orthonormal basis of $\mathbb{L}_2[0,1]$, the first norm is simply the $\mathbb{L}_2$-norm. An example is given by the trigonometric functions $\phi_1(t) = 1$, $\phi_{2j}(t) = \cos(2\pi jt)$, and $\phi_{2j+1}(t) = \sin(2\pi jt)$, for $j \in \mathbb{N}$. For this basis the norm $\|\cdot\|_{2,2,\beta}$ is a *Sobolev norm*, which measures smoothness of the function w. For a general basis (ϕ_j), we may think of this norm as measuring regularity of order β in a general sense.

We consider the Gaussian random element $W = \sum_{j=1}^\infty j^{-\alpha-1/2} Z_j\phi_j$, and $Z_j \overset{\text{iid}}{\sim}$ Nor(0, 1). Then $\mathrm{E}\|W\|_{2,2,\beta}^2 = \sum_j j^{2\beta-2\alpha-1} < \infty$ for every $\beta < \alpha$, and hence the prior can be viewed as (almost) regular of order β. In particular, the prior is well defined as an almost surely converging sequence whenever $\alpha > 0$.

Lemma 11.46 (RKHS) *The RKHS of the variable* $W = \sum_{j=1}^{\infty} j^{-\alpha-1/2} Z_j \phi_j$ *is the set of functions* $\sum_j w_j \phi_j$ *with* $w \in \mathfrak{W}^{\alpha+1/2}[0,1]$ *and the RKHS norm is* $\|w\|_{2,2,\alpha+1/2}$.

Lemma 11.47 (Small ball) *The small ball probability of* $W = \sum_{j=1}^{\infty} j^{-\alpha-1/2} Z_j \phi_j$ *relative to the norm* $\|\cdot\|_2$ *satisfies, for universal constants* $0 < c < C < \infty$ *and* $d > 0$, *and every* $\epsilon^{-1/\alpha} \geq d$,

$$(c2^{-1/(2\alpha)})\,\epsilon^{-1/\alpha} \leq \varphi_0(\epsilon) \leq (C2^{1/(2\alpha)})\,\epsilon^{-1/\alpha}.$$

Lemma 11.48 (Decentering) *If* $\|w\|_{2,2,\beta} < \infty$ *for* $\beta \leq \alpha + 1/2$, *then, for* $\epsilon \leq \|w\|_{2,2,\beta}$,

$$\inf_{h:\|h-w\|_2<\epsilon} \|h\|_{\mathbb{H}}^2 \leq \|w\|_{2,2,\beta}^{(2\alpha+1)/\beta} \epsilon^{-(2\alpha-2\beta+1)/\beta}.$$

Proof The first lemma is immediate from Example 11.16.

For the proof of the upper bound in the second lemma we first note that, because normal densities with standard deviations $\sigma \geq \tau$ satisfy $\phi_\sigma(x)/\phi_\tau(x) \geq \tau/\sigma$, for every x,

$$\mathrm{P}\Big(\sum_{j\leq J} Z_j^2 j^{-2\alpha-1} < \epsilon^2\Big) \geq \prod_{j=1}^{J}\Big(\frac{J^{-\alpha-1/2}}{j^{-\alpha-1/2}}\Big)\mathrm{P}\Big(\sum_{j=1}^{J} Z_j^2 J^{-2\alpha-1} < \epsilon^2\Big).$$

The leading factor is $(J!/J^J)^{\alpha+1/2} \geq e^{-J(\alpha+1/2)}$ and the probability on the far right tends to a number not smaller than $1/2$ if $\epsilon^2 J^{2\alpha} \geq 1$, by the central limit theorem, as $J \to \infty$. Second, by Markov's inequality,

$$\mathrm{P}\Big(\sum_{j>J} Z_j^2 j^{-2\alpha-1} < \epsilon^2\Big) \geq 1 - \frac{1}{\epsilon^2}\sum_{j>J} \mathrm{E}(Z_j^2 j^{-2\alpha-1}) \geq 1 - \frac{1}{2\alpha\epsilon^2 J^{2\alpha}}.$$

The right side is at least $1/2$ if $J \geq \epsilon^{-1/\alpha}\alpha^{-2\alpha}$. Since $\alpha^{-2\alpha} \geq e^{2e^{-1}}$, for any $\alpha > 0$, the inequalities in both preceding displays are satisfied for $J \geq \epsilon^{-1/\alpha} e^{2e^{-1}}$, and then $\mathrm{P}(\sum_j Z_j^2 j^{-2\alpha-1} < 2\epsilon^2)$ is bounded below by a multiple of $e^{-J(\alpha+1/2)} \geq e^{-J/2}$.

The small ball probability is upper bounded by $\mathrm{P}(\sum_{j=1}^{J} Z_j^2 J^{-2\alpha-1} < \epsilon^2)$, for every J. If $\epsilon^2 J^{2\alpha} \leq 1/2$, then this is bounded above by the probability that a chi-squared variable with J degrees of freedom is bounded above by $J/2$, which can be written as, for $I = J/2$,

$$\frac{1}{\Gamma(I)}\int_0^{I/2} x^{I-1} e^{-x}\,dx = \frac{e^{-I/2} I^{I-1}}{\Gamma(I)2^I}\int_0^I \Big(1 - \frac{y}{I}\Big)^{I-1} e^{y/2}\,dy,$$

by the substitution $2x = I - y$. By Stirling's formula the leading factor is bounded above by $(\sqrt{e}/2)^I/\sqrt{\pi I}$, while for $I \geq 1$ the integrand is bounded above by $e^{-y/2+y/I}$ whence the integral is bounded by $1/(1/2 - 1/I)$ if $I > 2$. We conclude by choosing $2I = J$ equal to the biggest integer smaller than $(\sqrt{2}\epsilon)^{-1/\alpha}$.

For every $J \in \mathbb{N}$ the truncated function $w^J := \sum_{j=1}^{J} w_j \phi_j$ is contained in the RKHS. Its square distance to w and square RKHS-norm satisfy

$$\|w^J - w\|_2^2 = \sum_{j>J} w_j^2 \le J^{-2\beta} \|w\|_{2,2,\beta}^2,$$

$$\|w^J\|_{\mathbb{H}}^2 = \sum_{j=1}^{J} j^{2\alpha+1} w_j^2 \le \|w\|_{2,2,\beta}^2 \max_{1\le j\le J} j^{2\alpha-2\beta+1}.$$

Choosing a minimal integer J such that $J^\beta \ge \|w\|_{2,2,\beta}/\epsilon$ readily gives the third lemma. □

If the true parameter w_0 has finite norm $\|w_0\|_{2,2,\beta}$, then the minimal solution to the rate equation $\varphi_{w_0}(\epsilon_n) \le n\epsilon_n^2$ can be seen to be $\epsilon_n \asymp n^{-(\alpha\wedge\beta)/(2\alpha+1)}$. This is the minimax rate $n^{-\beta/(2\beta+1)}$ if and only if prior and true function match (i.e. $\alpha = \beta$), as usual. We shall now show that the suboptimal rates in the other cases are sharp, separately in the case that $\beta > \alpha$ (rate $n^{-\alpha/(2\alpha+1)}$) and the case that $\beta < \alpha$ (rate $n^{-\beta/(2\alpha+1)}$).

Because the centered small ball exponent is of the exact order $\epsilon^{-1/\alpha}$, we have $\varphi_{w_0}(\delta_n) \gtrsim \delta_n^{-1/\alpha}$, for any w_0. For $\delta_n = C_2^{-\alpha} n^{-\alpha/(2\alpha+1)}$ this is equal to $C_2 n^{1/(2\alpha+1)} = C_2 n\epsilon_n^2$ if $\alpha \le \beta$. By Theorem 8.35 we conclude that the posterior rate of contraction is not faster than δ_n for any w_0 that is β-regular of order $\beta \ge \alpha$. Thus for any w_0 that is regular of order $\beta \ge \alpha$ bigger than the prior regularity α, the posterior rate of contraction is of the exact order $n^{-\alpha/(2\alpha+1)}$. The suboptimality of this rate (relative to the minimax rate $n^{-\beta/(2\beta+1)}$) when β is strictly bigger than α is due to the roughness of the prior, which puts less mass near 0 than is desirable for smooth true functions.

For w_0 of regularity $\beta < \alpha$ the rate $\epsilon_n \asymp n^{-\beta/(2\alpha+1)}$ may be thought of as sharp as well, but only for w_0 that are *exactly* of regularity β. Finiteness of the Sobolev norm $\|w_0\|_{2,2,\beta}$ allows w_0 also to be *more* regular than β and then the posterior will contract at a faster rate. Unfortunately, in the Sobolev case the concept of "exact regularity" appears difficult to define. In the following examples we obtain nearly sharp rates for certain fixed w_0 "near" the boundary of the Sobolev ball, and exhibit a sequence $w_{0,n}$ that "moves to the boundary" for which the rate $n^{-\beta/(2\alpha+1)}$ is sharp.

Example 11.49 (Lower bound) Let $\beta < \alpha$ and $a > 0$, and consider the function w with Fourier coefficients $w_1 = 0$ and $w_j = j^{-\beta-1/2}(\log j)^{-a}$, for $j \in \mathbb{N}$ and $j \ge 2$. For $a > 1/2$ the function satisfies $\|w\|_{2,2,\beta} < \infty$, whereas $\|w\|_{2,2,b} = \infty$, for every $b > \beta$. Hence w can be interpreted as being of "nearly exact" regularity β (up to a logarithmic factor).

The decentering part of the concentration function, as in Lemma 11.48, takes the form

$$d(\epsilon) := \inf_{\sum_{j=1}^\infty (h_j - w_j)^2 \le \epsilon^2} \sum_{j=1}^{\infty} j^{2\alpha+1} h_j^2.$$

By introducing a Lagrange multiplier λ, we can see that the solution of this problem takes the form $h_j = \lambda w_j/(j^{2\alpha+1} + \lambda)$, where $\lambda = \lambda(\epsilon)$ solves $\sum_j (h_j - w_j)^2 = \epsilon^2$, for small $\epsilon > 0$. For $\{w_j\}$ as given, some calculus (see Lemma K.8) gives the solution

$$d(\epsilon) \asymp \lambda(\epsilon)\epsilon^2, \quad \lambda(\epsilon) \asymp \epsilon^{-(2\alpha+1)/\beta}(\log_- \epsilon)^{-a(2\alpha+1)/\beta}, \quad \epsilon \asymp \lambda(\epsilon)^{-\beta/(2\alpha+1)}(\log \lambda(\epsilon))^{-a}.$$

The equation $d(\epsilon_n) \asymp n\epsilon_n^2$ is solved for $\lambda(\epsilon_n) \asymp n$ and $\epsilon_n \asymp n^{-\beta/(2\alpha+1)}(\log n)^{-a}$. Since $\epsilon_n^{-1/\alpha} \ll d(\epsilon_n)$, the rate equation $\varphi_0(\epsilon_n) \asymp n\epsilon_n^2$ gives the same solution, whence we obtain the rate of contraction $n^{-\beta/(2\alpha+1)}$ found in the preceding up to a logarithmic factor. By the same calculation it also follows that for every constant C_2 there exists a constant m such that $d(m\epsilon_n) \geq C_2 n\epsilon_n^2$. This shows that $m\epsilon_n$ is a lower bound for the rate of contraction, and hence the rate $n^{-\beta/(2\alpha+1)}(\log n)^{-a}$ is sharp.

Example 11.50 (Lower bound, sequence) Let $\beta < \alpha$ and consider the sequence w^n with a single nonzero coordinate $w_j = j^{-\beta}$ for $j = J_n \sim C n^{1/(2\alpha+1)}$ (and $w_j = 0$ for $j \neq J_n$). The sequence w^n satisfies $\|w^n\|_{2,2,\beta} = 1$, for every n, and hence is uniformly of regularity β. The minimizing h in the left side of Lemma 11.48 has $h_j = (w_j - \epsilon)_+$ for $j = J_n$ and $h_j = 0$ otherwise and hence the decentering function is given by $d(\epsilon) = J_n^{2\alpha+1}(J_n^{-\beta} - \epsilon)_+^2$.

The corresponding concentration function is $\varphi_{w^n}(\epsilon) \asymp \epsilon^{-1/\alpha} + d(\epsilon)$. For $\epsilon_n = J_n^{-\beta}$ we have $\varphi_{w^n}(\epsilon_n) \lesssim \epsilon_n^{-1/\alpha} = J_n^{\beta/\alpha} \lesssim n^{(\beta/\alpha)/(2\alpha+1)}$, and for $\delta_n = J_n^{-\beta}/2$ we have $\varphi_{w^n}(\delta_n) \geq d(\delta_n) \gtrsim J_n^{2\alpha+1} J_n^{-2\beta} \sim C^{2\alpha+1-2\beta} n^{(2\alpha-2\beta+1)/(2\alpha+1)}$. Since $n\epsilon_n^2 \sim C^{-2\beta} n^{(2\alpha-2\beta+1)/(2\alpha+1)}$ we have $\varphi_{w^n}(\epsilon_n) \ll n\epsilon_n^2$ and $C_2 n\epsilon_n^2 \leq \varphi_{w^n}(\delta_n)$, if C is chosen sufficiently large.

It follows that the rate $\delta_n \asymp n^{-\beta/(2\alpha+1)}$ is sharp along the sequence w^n.

11.5 Rescaled Gaussian Processes

The general finding in Section 11.3 is that priors based on Gaussian processes lead to optimal or near-optimal posterior contraction rates provided the smoothness of the Gaussian process matches that of the target function. Both oversmoothing and undersmoothing lead to suboptimal contraction rates. Nevertheless, it is common practice to use a Gaussian process of a specific form, for instance a supersmooth stationary process such as the square exponential process. The properties of this process are then adapted to the target function by hyperparameters in the covariance kernel. Inserting a scale parameter a in the covariance function is equivalent to scaling the time parameter of the process, thus considering a prior process $t \mapsto W_t^a := W_{at}$ instead of the original process, and is known as changing the *length scale* of the process. If the scale parameter is limited to a compact subset of $(0, \infty)$, then the contraction rate does not change (see e.g. Theorem 2.4 in van der Vaart and van Zanten 2008a). However, while the qualitative smoothness of the sample paths will not change for any a, a dramatic impact on the posterior contraction rate can be observed when $a = a_n$ decreases to 0 or increases to infinity with the sample size n. This is illustrated in this section for the classes of stationary and self-similar prior processes, for deterministic rescaling rates. In Section 11.6 we follow this up by considering a hyperprior on the rescaling rates, thus allowing for random, data-dependent rescaling.

We take the index set for the Gaussian process prior W^a equal to $[0,1]^d$, and construct this process as $W_t^a = W_{at}$, given a fixed process $W = (W_t : t \in [0, 1/a]^d)$. For $a < 1$ this entails shrinking a process on a bigger time set to the time set $[0,1]^d$, whereas $a > 1$ corresponds to stretching. Intuitively shrinking makes the sample paths more variable, as the randomness on a bigger time set is packed inside $[0,1]^d$, whereas stretching creates a smoother process. One finding in the following is that shrinking can make a given process arbitrarily rough, but the smoothing effect of stretching is limited to the smoothness of functions in the RKHS of the initial process. This may motivate to start with a very smooth process.

The scaling map $w \mapsto (t \mapsto w(at))$ is typically a continuous, linear map between the encompassing Banach spaces (e.g. $\mathfrak{C}(T)$, $\mathbb{L}_2(T)$, $\mathfrak{L}_\infty(T)$). Then by general arguments (see

Lemma I.16) the same map is an isometry between the RKHSs of the rescaled and original processes W^a and W.

Lemma 11.51 (RKHS) *If the map $w \mapsto (t \mapsto w(at))$ is a continuous, linear map from the Banach spaces $\mathbb{B}_a$ into $\mathbb{B}$, and W is a random element in $\mathbb{B}_a$, then the RKHS of the process W^a given by $W_t^a = W_{at}$ consists of the functions $t \mapsto h(at)$ for h in the RKHS of W, with identical norms.*

11.5.1 Self-Similar Processes

A stochastic process W is self-similar of order α if the processes $(W_{at}: 0 \le t \le 1)$ and $(a^\alpha W_t: 0 \le t \le 1)$ are equal in distribution. We shall understand the latter as referring to the Borel laws of the two processes in the encompassing Banach spaces, and assume that the stochastic process and Banach space RKHSs are equivalent. Thus the rescaling of the time-axis is equivalent to a rescaling of the vertical axis. This observation makes the following two lemmas evident.

Lemma 11.52 (RKHS) *The RKHS $\mathbb{H}^a$ of the rescaled process W^a corresponding to a self-similar process W of order α is the RKHS $\mathbb{H}$ of W, but equipped with the norm $\|h\|_{\mathbb{H}^a} = a^{-\alpha}\|h\|_{\mathbb{H}}$.*

Lemma 11.53 (Small ball) *The small ball exponent φ_0^a of the rescaled process W^a corresponding to a self-similar process W of order α satisfies $\varphi_0^a(\epsilon) = \varphi_0(a^{-\alpha}\epsilon)$, for φ_0 the small ball exponent of W.*

Proofs The function $t \mapsto \mathrm{E}[W_{at}Z] = a^\alpha \mathrm{E}[W_t Z]$ is contained in both $\mathbb{H}^a$ and $\mathbb{H}$ and has square norm $\mathrm{E}[Z^2]$ in $\mathbb{H}^a$ and $a^{2\alpha}\mathrm{E}[Z^2]$ in $\mathbb{H}$. The second lemma is immediate from the fact that the events $\{\|W^a\| < \epsilon\}$ and $\{\|a^\alpha W\| < \epsilon\}$ are equal in probability. ☐

It follows that the concentration function $\varphi_{w_0}^a$ of the rescaled process W^a can be written as

$$\varphi_{w_0}^a(\epsilon) = \varphi_0(a^{-\alpha}\epsilon) + a^{-2\alpha} \inf_{\|h-w_0\|\le\epsilon} \|h\|_{\mathbb{H}}^2.$$

Increasing the scaling factor a makes the first term on the right side bigger, but decreases the second term. We may now choose $a = a_n$ such that the solution ϵ_n of the rate equation $\varphi_{w_0}^{a_n}(\epsilon_n) \asymp n\epsilon_n^2$ is smallest. In the case that $\varphi_0(\epsilon) \asymp \epsilon^{-r}$ and $\inf\{\|h\|_{\mathbb{H}}^2: \|h-w_0\| \le \epsilon\} \asymp \epsilon^{-s}$, for some $r, s > 0$, the optimal scaling value and resulting contraction rate can be seen to be

$$a_n = n^{(s-r)/(4\alpha+4r\alpha+rs\alpha)}, \qquad \epsilon_n = n^{-(2+r)/(4+4r+rs)}.$$

It may be noted that the exponent of self-similarity appears in the scaling factor, but not in the contraction rate.

Example 11.54 (Rescaled integrated Brownian motion) The k-fold integrated Brownian motion $I_{0+}^k B$ of Example 11.6 is self-similar of order $k + 1/2$. Its small ball probability is of the order $\epsilon^{-1/(k+1/2)}$, and its decentering function for a function w_0 belonging to the Hölder space of order $\beta \le k+1$ and having vanishing derivatives at 0 is of the order $\epsilon^{-(2k-2\beta+2)/\beta}$.

Substitution of $\alpha = k + 1/2$, $r = 1/(k + 1/2)$ and $s = (2k - 2\beta + 2)/\beta$ in the preceding display yields the rescaling rate $a_n = n^{(k+1/2-\beta)/((k+1/2)(2\beta+1))}$ and the contraction rate $n^{-\beta/(2\beta+1)}$, for $\beta \leq k + 1$. Thus the minimax rate is obtained for all $\beta \leq k + 1$. For $\beta < k + 1/2$ the integrated Brownian motion is shrunk, and every possible smoothness level is attained. For $\beta > k + 1/2$ the process is stretched, but this is successful only up to smoothness level $k + 1$.

In order to drop the restriction that w_0 has vanishing derivatives at 0 we added a random polynomial of order k to the process in Example 11.6. Because this polynomial process is not self-similar, the preceding argument does not apply to this extension. Actually rescaling the polynomial part in the same way as the integrated Brownian motion may also not be natural. Now it may be checked that the decentering function of the process $a_n^{k+1/2} I_{0+}^k B + b_n \sum_{i=1}^k Z_i t^i$ satisfies, for $w_0 \in \mathfrak{C}^\beta[0, 1]$ and $\beta \leq k + 1$,

$$\inf_{\|h-w\|_\infty \leq \epsilon} \|h\|_{\mathbb{H}^{a,b}}^2 \lesssim a_n^{-(2k+1)} \epsilon^{-(2k-2\beta+2)/\beta} + b_n^{-2} [\epsilon^{-(2k-2\beta)/\beta} \vee 1].$$

This is dominated by the first term (arising from the integrated Brownian motion) if $b_n \geq a_n^{k+1/2} \epsilon_n^{((k+1-\beta)\wedge 1)/\beta}$. Since the small ball exponent of the process is hardly determined by the polynomial part, under the latter condition the preceding derivation goes through without essential changes for any w_0 in the Hölder space of order $\beta \leq k + 1$.

11.5.2 Stationary Gaussian Processes

In the preceding section it is seen that certain finitely smooth processes may be roughened to an arbitrary degree to make it a suitable prior for a function of lesser smoothness than the process, but cannot be smoothened a lot. This may motivate to take a process with infinitely smooth sample paths as the base prior. In this section we discuss the example of a stationary process with a spectral measure that possesses exponential moments.

If W is a stationary Gaussian process with spectral measure μ, then the rescaled process W^a is a stationary Gaussian process with spectral measure μ^a given by $\mu^a(B) = \mu(a^{-1}B)$. Furthermore, if μ is absolutely continuous with spectral density m, then μ^a has density $t \mapsto m(t/a)/a^d$.

The RKHS of W^a is already described in Lemma 11.35, where μ must be replaced by μ^a. The small ball probability and decentering function require new calculations that take the scaling into account.

Lemma 11.55 (Small ball) *Let W be a stationary process with spectral measure μ such that $\int e^{\delta\|\lambda\|}\, \mu(\lambda) < \infty$, for some $\delta > 0$. For any $a_0 > 0$ there exist constants C and ϵ_0 depending only on a_0, μ and d only such that the small ball exponent of the process W^a satisfies, for $a \geq a_0$ and $\epsilon < \epsilon_0$,*

$$\varphi_0^a(\epsilon) \leq C a^d (\log(a/\epsilon))^{1+d}.$$

Proof This is similar to the proof of Lemma 11.38. The entropy of the unit ball of the RKHS $\mathbb{H}^a$ of W^a, can be seen to satisfy $\log N(\epsilon, \mathbb{H}_1^a, \|\cdot\|_\infty) \lesssim a^d (\log_- \epsilon)^{1+d}$. This upper bound has exponent $1+d$ instead of $1+d/2$ in Lemma 11.38, because presently only exponential and not square exponential moments on μ are assumed finite. The proof must take proper care of the dependence of all entities on a. In the application of the characterization of the small ball exponent through entropy this requires following the steps in the proof of Lemma I.29 in Kuelbs and Li (1993) and Li and Linde (1998). See van der Vaart and van Zanten (2007) for details. □

Lemma 11.56 (Decentering) *If the spectral density m is bounded away from 0 in a neighborhood of 0 and $\int e^{\delta\|\lambda\|}\, m(\lambda)\, d\lambda < \infty$ for some $\delta > 0$, then for any $\beta > 0$ and $w \in \mathfrak{C}^\beta([0,1]^d)$ there exist constants C and D depending only on m and w such that, for $a > 0$,*

$$\inf_{h:\|h-w\|_\infty \le Ca^{-\beta}} \|h\|^2_{\mathbb{H}^a} \le Da^d.$$

Proof Let $\underline{\beta}$ be the biggest integer strictly smaller than β, and let G be a bounded neighborhood of the origin on which m is bounded away from 0. Take a function $\psi: \mathbb{R} \to \mathbb{C}$ with a symmetric, real-valued, infinitely smooth Fourier transform $\hat\psi$ that is supported on an interval I such that $I^d \subset G$ and which equals $1/(2\pi)$ in a neighborhood of zero. Then ψ has moments of all orders and

$$\int (it)^k \psi(t)\, dt = 2\pi \hat\psi^{(k)}(0) = \begin{cases} 0, & k \ge 1, \\ 1, & k = 0. \end{cases}$$

Define $\phi: \mathbb{R}^d \to \mathbb{C}$ by $\phi(t) = \psi(t_1) \times \cdots \times \psi(t_d)$. Then we have that $\int \phi(t)\, dt = 1$, and $\int t^k \phi(t)\, dt = 0$, for any nonzero multi-index $k = (k_1, \ldots, k_d)$ of nonnegative integers. Moreover, we have that $\int \|t\|^\beta |\phi(t)|\, dt < \infty$, and the functions $|\hat\phi|/m$ and $|\hat\phi|^2/m$ are uniformly bounded.

By Whitney's theorem we can extend $w: [0,1]^d \to \mathbb{R}$ to a function $w: \mathbb{R}^d \to \mathbb{R}$ with compact support and $\|w\|_\beta < \infty$. (See Whitney 1934 or Stein 1970, Chapter VI; we can multiply an arbitrary smooth extension by an infinitely smooth function that vanishes outside a neighborhood of $[0,1]^d$ to ensure compact support.)

By Taylor's theorem we can write, for $s, t \in \mathbb{R}^d$,

$$w(t+s) = \sum_{j: j. \le \underline{\beta}} D^j w(t) \frac{s^j}{j!} + S(t,s),$$

where $|S(t,s)| \le C\|s\|^\beta$, for a positive constant C that depends on w but not on s and t. If we set $\phi_a(t) = \phi(at)$ we get, in view of the fact that ϕ is a higher order kernel, for any $t \in \mathbb{R}^d$,

$$a^d(\phi_a * w)(t) - w(t) = \int \phi(s)(w(t - s/a) - w(t))\, ds = \int \phi(s) S(t, -s/a)\, ds.$$

Combining the preceding displays shows that $\|a^d \phi_a * w - w\|_\infty \le KCa^{-\beta}$, for $K = \int \|s\|^\beta |\phi|(s)\, ds$.

For $\hat w$ the Fourier transform of w, we can write

$$\frac{1}{(2\pi)^d}(\phi_a * w)(t) = \int e^{-i(t,\lambda)} \hat w(\lambda) \hat\phi_a(\lambda)\, d\lambda = \int e^{-i(t,\lambda)} \frac{\hat w(-\lambda)\hat\phi_a(\lambda)}{m_a(\lambda)}\, d\mu_a(\lambda).$$

Therefore, by Lemma 11.35 the function $a^d\phi_a * w$ is contained in the RKHS $\mathbb{H}^a$, with square norm a multiple of

$$a^{2d}\int\Big|\frac{\hat{w}\hat{\phi}_a}{m_a}\Big|^2\,d\mu_a \le a^d\int|\hat{w}(\lambda)|^2\,d\lambda\,\Big\|\frac{|\hat{\phi}|^2}{m}\Big\|_\infty.$$

Here $(2\pi)^d\int|\hat{w}(\lambda)|^2\,d\lambda = \int|w(t)|^2\,dt$ is finite, and $|\hat{\phi}|^2/m$ is bounded by the construction of $\hat{\phi}$. □

Combining the preceding we see that for $\epsilon \gtrsim a^{-\beta}$ the concentration function of W^a at $w \in \mathfrak{C}^\beta([0,1]^d)$ satisfies

$$\varphi_w^a(\epsilon) \lesssim Da^d + a^d(\log(a/\epsilon))^{1+d}.$$

Thus for $a = a_n$ the rate equation $\varphi_w^a(\epsilon_n) \le n\epsilon_n^2$ is satisfied by ϵ_n such that

$$\epsilon_n \gtrsim a_n^{-\beta}, \qquad a_n^d(\log(a_n/\epsilon_n))^{1+d} \lesssim n\epsilon_n^2, \qquad a_n^d \lesssim n\epsilon_n^2.$$

For a_n bounded away from zero the third inequality is redundant and the first two inequalities can be seen to be satisfied by

$$a_n = n^{1/(2\beta+d)}(\log n)^{-(1+d)/(2\beta+d)}, \qquad \epsilon_n = n^{-\beta/(2\beta+d)}(\log n)^{\beta(1+d)/(2\beta+d)}.$$

The contraction rate ϵ_n is the minimax rate for estimating a β-regular function up to a logarithmic factor, while the inverse scaling rate a_n^{-1} agrees with the usual bandwidth for kernel smoothing up to a logarithmic factor.

11.6 Adaptation

In the preceding section it is seen that an appropriate length scale can turn a Gaussian process with smooth sample paths into an appropriate prior for true functions of arbitrary smoothness. However, inspection of the formulas reveals that the optimal length scale depends on the smoothness of the function of interest, which will typically not be known a priori. It is natural to try and choose the length scale using the data. A popular *empirical Bayes* method is to maximize the marginal (Bayesian) likelihood for the rescaling constant a, but the theoretical properties of these procedures have become known only recently and for the Gaussian white noise model only; see Szabó et al. (2013). In this section we consider the full Bayes alternative to put a (hyper) prior on a, and show that the resulting mixture prior gives adaptation in the sense of Chapter 10.

We adopt the setting of Section 11.5.2, with a stationary Gaussian process $W = (W_t: t \in \mathbb{R}^d)$ with spectral density with exponentially small tails, so that the process has analytic sample paths. Rather than scaling the process deterministically, we now consider the process $W^A = (W_{At}: t \in [0,1]^d)$, for A a random variable independent of W. The resulting prior is a mixture of Gaussian processes. Specifically we study the case that the variable A^d follows a gamma distribution. The parameters of this gamma distribution are inessential, and the gamma distribution can be replaced by another distribution with tails of the same weights, but the power d in A^d appears important.

The following theorem gives the contraction rate for the prior process W^A in the abstract setting of a mixed Gaussian prior, and may be compared with Theorem 11.20. The theorem can be translated to contraction rates in concrete settings, such as density estimation, classification, regression and white noise, in the same way as Theorems 11.21–11.24 are derived from Theorem 11.20. The theorem gives the existence of sets B_n and rates ϵ_n and $\bar\epsilon_n$ such that

$$\log N(\epsilon_n, B_n, \|\cdot\|_\infty) \le n\epsilon_n^2, \tag{11.21}$$

$$\mathrm{P}(W^A \notin B_n) \le e^{-4n\bar\epsilon_n^2}, \tag{11.22}$$

$$\mathrm{P}(\|W^A - w_0\|_\infty \le \bar\epsilon_n) \ge e^{-n\bar\epsilon_n^2}. \tag{11.23}$$

The rates are specified for two situations: the ordinary smooth case, where the true function w_0 belongs to a Hölder class, and the supersmooth case, where w_0 is analytic. For the first situation it was seen in Section 11.5.2 that a suitable deterministic shrinking rate turns the process W in an appropriate prior. For the second situation the unscaled process W was seen to be an appropriate prior at the end of Section 11.4.4. The following theorem shows that choosing the length scale according to dth root of a gamma variable, keeps the best of both worlds.

Theorem 11.57 (Mixed Gaussian contraction rate) *If $(W_t: t \in \mathbb{R}^d)$ is a stationary Gaussian process with spectral density m such that $a \mapsto m(a\lambda)$ is decreasing on $(0, \infty)$, for every $\lambda \in \mathbb{R}^d$, and $\int e^{\delta\|\lambda\|}\, m(\lambda)\, d\lambda < \infty$, for some $\delta > 0$, and A^d is an independent gamma variable, then there exist measurable sets $B_n \subset \mathfrak{C}([0,1]^d)$ such that* (11.21)–(11.23) *hold for the process $W_t^A = W_{At}$ and ϵ_n and $\bar\epsilon_n$ defined as follows.*

(i) *If $w_0 \in \mathfrak{C}^\beta([0,1]^d)$, then $\bar\epsilon_n \asymp n^{-\beta/(2\beta+d)}(\log n)^{(1+d)\beta/(2\beta+d)}$ and $\epsilon_n \asymp \bar\epsilon_n(\log n)^{(1+d)/2}$.*
(ii) *If w_0 is the restriction of a function in $\mathcal{A}^{\gamma,r}(\mathbb{R}^d)$ to $[0,1]^d$ and $m(\lambda) \ge C_3\exp(-D_3\|\lambda\|^\nu)$ for constants $C_3, D_3, \nu > 0$, then $\bar\epsilon_n \asymp n^{-1/2}(\log n)^{(d+1)/2}$ and $\epsilon_n \asymp \bar\epsilon_n(\log n)^\kappa$, where $\kappa = 0$ for $r \ge \nu$, and $\kappa = d/(2r)$ for $r < \nu$.*

Proof If $\varphi^a_{w_0}$ is the concentration function of W^a and g the density of A, then by applying Lemma 11.19 we see

$$\mathrm{P}(\|W^A - w_0\| \le 2\epsilon) \ge \int_0^\infty e^{-\varphi^a_{w_0}(\epsilon)}\, g(a)\, da. \tag{11.24}$$

By Lemma 11.55 we have that $\varphi_0^a(\epsilon) \le C_4 a^d\big(\log(a/\epsilon)\big)^{1+d}$, for $a > a_0$ and $\epsilon < \epsilon_0$, where the constants a_0, ϵ_0, C_4 depend only on μ and w.

For $\mathbb{B}_1$ the unit ball of $\mathfrak{C}([0,1]^d)$ and given constants $M, r, \delta, \epsilon > 0$, set

$$B = \Big(M\sqrt{\frac{r}{\delta}}\mathbb{H}_1^r + \epsilon\mathbb{B}_1\Big)\bigcup\Big(\bigcup_{a<\delta}(M\mathbb{H}_1^a) + \epsilon\mathbb{B}_1\Big). \tag{11.25}$$

By Lemma 11.59, $B \supset M\mathbb{H}_1^a + \epsilon\mathbb{B}_1$ for any $a \in [\delta, r]$, and also for $a < \delta$, by definition. Thus by Borell's inequality, for any $a \le r$,

$$\begin{aligned}\mathrm{P}(W^a \notin B) &\le \mathrm{P}(W^a \notin M\mathbb{H}_1^a + \epsilon\mathbb{B}_1) \le 1 - \Phi(\Phi^{-1}(e^{-\varphi_0^a(\epsilon)}) + M)\\ &\le 1 - \Phi(\Phi^{-1}(e^{-\varphi_0^r(\epsilon)}) + M),\end{aligned}$$

because $e^{-\varphi_0^a(\epsilon)} = \mathrm{P}(\sup_{t\in[0,a]^d}|W_t| \le \epsilon)$ is decreasing in a. Let ϵ_1 be such that $e^{-\varphi_0^1(\epsilon_1)} < 1/4$, which ensures $e^{-\varphi_0^r(\epsilon)} \le e^{-\varphi_0^1(\epsilon)} < 1/4$ for all $r > 1$ and $\epsilon < \epsilon_1$. Now using Lemma K.6 (ii), for $M \ge 4\sqrt{\varphi_0^r(\epsilon)}$, $r > 1$ and $\epsilon < \epsilon_1$, we have that $M \ge -2\Phi^{-1}(e^{-\varphi_0^r(\epsilon)})$. Consequently, the right-hand side of the preceding display is bounded by $1 - \Phi(M/2) \le e^{-M^2/8}$, in view of Lemma K.6 (i). By Lemma 11.55, we conclude that $\mathrm{P}(W^a \notin B) \le e^{-M^2/8}$ for any $a \le r$, provided that

$$M^2 \ge 16C_4 r^d(\log(r/\epsilon))^{1+d}, \qquad r > 1, \qquad \epsilon < \epsilon_1 \wedge \epsilon_0, \tag{11.26}$$

where ϵ_0 is obtained from Lemma 11.55. Since $A^d \sim \mathrm{Ga}(p,q)$, for $r > r_0$ and a constant r_0 that depends on d, p, q only,

$$\mathrm{P}(W^A \notin B) \le \mathrm{P}(A > r) + \int_0^r \mathrm{P}(W^a \notin B)\, g(a)\, da \lesssim r^{d(p-1)} e^{-qr^d} + e^{-M^2/8}. \tag{11.27}$$

This holds for any $B = B_{M,r,\delta,\epsilon}$ with M, r, δ, ϵ satisfying (11.26).

By the proof of Lemma 11.38, for $M\sqrt{r/\delta} > 2\epsilon$ and $r > a_0$,

$$\begin{aligned}\log N\Big(2\epsilon, M\sqrt{\frac{r}{\delta}}\mathbb{H}_1^r + \epsilon\mathbb{B}_1, \|\cdot\|_\infty\Big) &\le \log N\Big(\epsilon, M\sqrt{\frac{r}{\delta}}\mathbb{H}_1^r, \|\cdot\|_\infty\Big)\\ &\le Kr^d\bigg(\log\Big(\frac{M\sqrt{r/\delta}}{\epsilon}\Big)\bigg)^{1+d}.\end{aligned}$$

By Lemma 11.58, any function $h \in M\mathbb{H}_1^a$, for $a < \delta$ is within uniform distance $\delta\sqrt{d}\tau M$ of a constant function, where the constant value belongs to the interval $[-M\sqrt{\|\mu\|}, M\sqrt{\|\mu\|}]$, and $\tau^2 = \int \|\lambda\|^2\, d\mu(\lambda)$. It follows that, for $\epsilon > \delta\sqrt{d}\tau M$,

$$N\Big(3\epsilon, \bigcup_{a<\delta}(M\mathbb{H}_1^a) + \epsilon\mathbb{B}_1, \|\cdot\|_\infty\Big) \le \frac{2M\sqrt{\|\mu\|}}{\epsilon}.$$

Choose $\delta = \epsilon/(2\sqrt{d}\tau M)$. Combining the last two displays, and using the elementary inequality $\log(x+y) \le \log x + 2\log y$ for $x \ge 1$, $y \ge 2$, we obtain, for $\epsilon \le M\sqrt{\|\mu\|}$,

$$\log N(3\epsilon, B, \|\cdot\|_\infty) \le Kr^d\bigg(\log\Big(\frac{M^{3/2}\sqrt{2\tau r}d^{1/4}}{\epsilon^{3/2}}\Big)\bigg)^{1+d} + 2\log\frac{2M\sqrt{\|\mu\|}}{\epsilon}. \tag{11.28}$$

This inequality is valid for any $B = B_{M,r,\delta,\epsilon}$ with $\delta = \epsilon/(2\sqrt{d}\tau M)$, and any M, r, ϵ with

$$M^{3/2}\sqrt{2\tau r}d^{1/4} > 2\epsilon^{3/2}, \qquad r > a_0, \qquad M\sqrt{\|\mu\|} > \epsilon. \tag{11.29}$$

In the remainder of the proof we make special choices for these parameters, depending on the assumption on w_0, to complete the choice of B_n and obtain $\bar\epsilon_n$ and ϵ_n.

(i). *Hölder smoothness.* Let $w_0 \in \mathfrak{C}^\beta([0,1]^d)$ for some $\beta > 0$. In view of Lemmas 11.56 and 11.55, for every a_0, there exist $\epsilon_0 < \frac{1}{2}$, $C, D, K > 0$ depending on w and μ only such that, for $a > a_0$, $\epsilon < \epsilon_0$, and $\epsilon > Ca^{-\beta}$,

$$\varphi_{w_0}^a(\epsilon) \le Da^d + C_4a^d\Big(\log\frac{a}{\epsilon}\Big)^{1+d} \le K_1a^d\Big(\log\frac{a}{\epsilon}\Big)^{1+d},$$

for K_1 depending on a_0, μ and d only. Therefore, for $\epsilon < \epsilon_0 \wedge Ca_0^{-\beta}$ (so that $(C/\epsilon)^{1/\beta} > a_0$), by (11.24),

$$\mathrm{P}(\|W^A - w_0\|_\infty \le 2\epsilon) \ge \int_{(C/\epsilon)^{1/\beta}}^{2(C/\epsilon)^{1/\beta}} e^{-K_1 a^d \log^{1+d}(a/\epsilon)}\, g(a)\, da$$
$$\ge C_1 e^{-K_2 \epsilon^{-d/\beta}(\log_- \epsilon)^{(1+d)}} \Big(\frac{C}{\epsilon}\Big)^{(dp-1)/\beta} \Big(\frac{C}{\epsilon}\Big)^{1/\beta},$$

for a constant K_2 that depends only on K_1, C, D_1, d, β. Therefore, for all sufficiently large n, we have that $\mathrm{P}(\|W^A - w_0\|_\infty \le \bar\epsilon_n) \ge e^{-n\bar\epsilon_n^2}$ for $\bar\epsilon_n$ a large multiple of $n^{-1/(2+d/\beta)}(\log n)^\gamma$, where $\gamma = (1+d)/(2+d/\beta)$.

Inequalities (11.26)–(11.27) give that $\mathrm{P}(W^A \notin B) \lesssim e^{-C_0 n\bar\epsilon_n^2}$ for an arbitrarily large constant C_0 if (11.26) holds and

$$D_2 r^d (\log r)^q \ge 2C_0 n\bar\epsilon_n^2, \qquad r^{p-d+1} \le e^{C_0 n\bar\epsilon_n^2}, \qquad M^2 \ge 8C_0 n\bar\epsilon_n^2. \tag{11.30}$$

Given C_0, choose $r = r_n$ equal to the minimal solution of the first equation in (11.30), and then choose $M = M_n$ to satisfy the third. The second equation is then automatically satisfied, for large n.

Let B_n be the set B from (11.25) with the preceding choices of M_r and r_n, and ϵ_n bounded below by a power of n. Then the right side of (11.28) is bounded above by a multiple of $r_n^d(\log n)^{1+d} + \log n \le n\epsilon_n^2$ for ϵ_n^2 a large multiple of $(r_n^d/n)(\log n)^{1+d}$. Inequalities (11.29) are clearly satisfied.

(ii)-1. *Infinite smoothness: $r \ge \nu$.*

By combining the first part of Lemma 11.41 and Lemma 11.55, we see that there exist positive constants $a_0 < a_1$, ϵ_0, K_1 and C_4 that depend on w and μ only such that $\varphi_{w_0}^a(\epsilon) \le K_1 + C_4 a^d(\log(a/\epsilon))^{1+d}$, for $a \in [a_0, a_1]$ and $\epsilon < \epsilon_0$. Consequently, by (11.24),

$$\mathrm{P}(\|W^A - w_0\|_\infty \le 2\epsilon) \ge e^{-K_1 - C_4 a_1^d(\log(a_1/\epsilon))^{1+d}}\, \mathrm{P}(a_0 < A < a_1).$$

This gives $\mathrm{P}(\|W^A - w_0\|_\infty \le \bar\epsilon_n) \ge e^{-n\bar\epsilon_n^2}$ for $\bar\epsilon_n$ a large multiple of $n^{-1/2}(\log n)^{(d+1)/2}$, provided that n is sufficiently large.

Next we choose B_n as before, with r and M solving (11.30) and satisfying (11.26), i.e. r_n^d and M_n^2 large multiples of $(\log n)^{d+1}$. Then (11.26)–(11.27) imply $\mathrm{P}(W^A \notin B_n) \gtrsim e^{-C_0 n\bar\epsilon_n^2}$, and the right side of (11.28) is bounded by a multiple of $r_n^d(\log_- \epsilon + \log\log n)^{1+d} + \log_- \epsilon + \log\log n$. For $\epsilon = \epsilon_n$ a large multiple of $n^{-1/2}(\log n)^{d+1}$, this is bounded above by $n\epsilon_n^2$.

(ii)-2. *Infinite smoothness: $r < \nu$.*

Combining the second part of Lemma 11.41 and Lemma 11.55, we see that there exist $a_0, \epsilon_0, C, D, K_1, C_4 > 0$ depending only on w and μ, and $\gamma' > \gamma$ such that, for $a > a_0$, $\epsilon < \epsilon_0$ and $Ce^{-\gamma' a^r} < \epsilon$, we have $\varphi_{w_0}^a(\epsilon) \le Da^d + C_4 a^d(\log(a/\epsilon))^{1+d}$. Consequently, by (11.24), with D_1, D_2 depending on w and μ only,

$$\mathrm{P}(\|W^A - w_0\|_\infty \le 2\epsilon) \ge \int_{(\log(C/\epsilon)/\gamma')^{1/r}}^{\infty} e{-}\varphi_{w_0}^a(\epsilon) g(a)\, da \ge D_2 e^{-D_1(\log_- \epsilon)^{d/r+d+1}}.$$

For all sufficiently large n this gives that $\mathrm{P}(\|W^A - w_0\|_\infty \le \bar\epsilon_n) \ge e^{-n\bar\epsilon_n^2}$, for $\bar\epsilon_n$ a large multiple of $n^{-1/2}(\log n)^{d/(2r)+(d+1)/2}$.

Next we choose B_n of the form as before, with $r = r_n$ and $M = M_n$ solving (11.30), i.e., r_n^d and M_n^2 chosen equal to large multiples of $(\log n)^{d/r+d+1}$. Then (11.26)–(11.27) imply that $\mathrm{P}(W^A \notin B) \leq e^{-C_0 n\epsilon_n^2}$, and the right side of (11.28) is bounded above by a multiple of $r_n^d(\log_- \epsilon + \log\log n)^{1+d} + \log_- \epsilon + \log\log n$. For $\epsilon = \epsilon_n$ a large multiple of $n^{-1/2}(\log n)^{d+1+d/(2r)}$, this is bounded above by $n\epsilon_n^2$. □

The following three lemmas are used in the preceding proof. The third is similar to Lemma 11.41, and extends this to the case of analytic true functions.

Lemma 11.58 *For any* $h \in \mathbb{H}_1^a$ *and* $t \in \mathbb{R}^d$, *we have* $|h(0)|^2 \leq \|\mu\|$ *and* $|h(t) - h(0)|^2 \leq a^2\|t\|^2 \int \|\lambda\|^2\, d\mu(\lambda)$.

Proof An element $H\psi$ of $\mathbb{H}^a$ takes the form $H\psi(t) = \int e^{i\langle t,\lambda\rangle}\psi(\lambda)\, d\mu_a(\lambda)$ and has square norm $\|H\psi\|_{\mathbb{H}^a}^2 = \int |\psi|^2\, d\mu_a$. It follows that $|H\psi(0)| \leq \int |\psi|\, d\mu_a$ and

$$|H\psi(t) - H\psi(0)| \leq \int |e^{i\langle t,\lambda\rangle} - 1|\, |\psi(\lambda)|\, d\mu_a(\lambda) \leq \int |\langle t,\lambda\rangle|\, |\psi(\lambda)|\, d\mu_a(\lambda).$$

By two successive applications of the Cauchy-Schwarz inequality, the square of the right side is bounded above by $\|t\|^2 \int \|\lambda\|^2 d\mu_a(\lambda) \int |\psi|^2\, d\mu_a$. This gives the result. □

Lemma 11.59 *If the spectral density of W is radially decreasing and* $\mathbb{H}^a$ *is the RKHS of the rescaled stationary Gaussian process* W^a, *then* $\sqrt{a}\,\mathbb{H}_1^a \subset \sqrt{b}\,\mathbb{H}_1^b$, *for any* $a \leq b$.

Proof The assumption that $a \mapsto m(a\lambda)$ is decreasing in $a \in (0, \infty)$ implies that $m_a(\lambda) \leq (b/a)m_b(\lambda)$, for every λ. By Lemma 11.35 an arbitrary element of $\mathbb{H}^a$ takes the form

$$(H^a\psi)(t) = \int e^{i\langle t,\lambda\rangle}\psi(\lambda)\, d\mu_a(\lambda) = \int e^{i\langle t,\lambda\rangle}\Big(\psi\frac{m_a}{m_b}\Big)(\lambda)\, d\mu_b(\lambda).$$

Because $(\psi m_a/m_b) \in \mathbb{L}_2(\mu_b)$ if $\psi \in \mathbb{L}_2(\mu_a)$, it follows that $H^a\psi = H^b(\psi m_a/m_b)$ is contained in $\mathbb{H}^b$. The square norms of this function in the two spaces satisfy $\|H^a\psi\|_{\mathbb{H}^b}^2 = \|\psi m_a/m_b\|_{\mu_b,2}^2 \leq (b/a)\|\psi\|_{\mu_a,2}^2 = \|H\psi\|_{\mathbb{H}^a}^2$. This gives the nesting of the two scaled unit balls. □

Lemma 11.60 (Decentering) *Suppose that* $m(\lambda) \geq C_3 \exp(-D_3\|\lambda\|^\nu)$ *for some constants* $C_3, D_3, \nu > 0$, *and let w be the restriction to* $[0, 1]^d$ *of a function in* $\mathcal{A}^{\gamma,r}(\mathbb{R}^d)$.

(a) *If* $r \geq \nu$, *then* $w \in \mathbb{H}^a$ *for all sufficiently large a with uniformly bounded norm* $\|w\|_{\mathbb{H}^a}$.
(b) *If* $r < \nu$, *then there exist* $a_0, C, D > 0$ *depending only on* μ *and w such that, for* $a > a_0$,

$$\inf_{\|h-w\|_\infty \leq Ce^{-\gamma a^r}/a^{-r+1}} \|h\|_{\mathbb{H}^a}^2 \leq Da^d.$$

11.7 Computation

This section provides a brief review of algorithms to compute or approximate aspects of the posterior distribution corresponding to a Gaussian prior.

11.7.1 Kernel Methods and the Posterior Mode

In regression problems the posterior mode corresponding to a Gaussian process prior can be computed as the solution to a minimization problem involving the corresponding RKHS-norm. In particular, for the integrated Brownian motion prior the posterior mode is a penalized spline estimator.

Consider a regression model with independent observations $(X_1, Y_1), \ldots, (X_n, Y_n)$ following a density $(x, y) \mapsto p_w(x, y)$ that depends on the unknown parameter w only through its value $w(x)$. Define a loss $L(y, w(x)) = -2 \log p_w(x, y)$ and consider the criterion

$$x \mapsto \sum_{i=1}^n L(Y_i, w(X_i)) + n\lambda \|w\|_{\mathbb{H}}^2. \tag{11.31}$$

Here the *smoothing parameter* λ is fixed, and $\|\cdot\|_{\mathbb{H}}$ is the RKHS-norm corresponding to a given Gaussian process W.

Theorem 11.61 *The minimizer $\hat{w}$ of the criterion* (11.31) *over $w \in \mathbb{H}$ is the posterior mode under the prior $w \sim (\sqrt{n\lambda})^{-1} W$, for W the Gaussian process with RKHS $\mathbb{H}$. This minimizer can be written as a linear combination of the functions $x \mapsto K(X_i, x)$, for $i = 1, \ldots, n$ and K the covariance kernel of W.*

Proof We shall show that $\hat{w}$ and the posterior mode agree at the points $X_1, \ldots, X_n$ and also at an arbitrary additional point X_0, whence at all points.

The reproducing formula gives that $w(X_i) = \langle K(X_i, \cdot), w\rangle_{\mathbb{H}}$, for every $i = 0, 1, \ldots, n$. If Pw is the orthogonal projection of w in $\mathbb{H}$ onto the linear span of the functions $K(X_i, \cdot)$, for $i = 0, \ldots, n$, then $w(X_i) = \langle K(X_i, \cdot), Pw\rangle_{\mathbb{H}}$, and hence the function Pw attains the same value on the loss function $\sum_{i=1}^n L(Y_i, w(X_i))$ as the function w. Since $\|Pw\|_{\mathbb{H}} \leq \|w\|_{\mathbb{H}}$ it follows that Pw attains a smaller value in the criterion (11.31) than w. Hence the minimizer of (11.31) can be written $\hat{w} = \sum_{j=0}^n \hat{v}_j K(X_j, \cdot)$, for numbers $\hat{v}_0, \ldots, \hat{v}_n$ that minimize

$$\begin{aligned}
&\sum_{i=1}^n L\Big(Y_i, \sum_{j=0}^n v_j K(X_j, X_i)\Big) + n\lambda \Big\|\sum_{j=0}^n v_j K(X_j, \cdot)\Big\|_{\mathbb{H}}^2 \\
&\quad = \sum_{i=1}^n L\big(Y_i, (K_{n+1} v)_i\big) + n\lambda v^\top K_{n+1} v,
\end{aligned}$$

where K_{n+1} is the matrix with (i, j)th element $K(X_i, X_j)$ and $v = (v_0, \ldots, v_n)^\top$.

By assumption, the likelihood depends on w only through (the last n coordinates) of the vector $w_{n+1} = (w(X_0), w(X_1), \ldots, w(X_n))^\top$. Under the prior the vector w_{n+1} possesses a $\mathrm{Nor}_{n+1}(0, K_{n+1}/(n\lambda))$-distribution, and hence is distributed as $K_{n+1} v$ for $v \sim \mathrm{Nor}_{n+1}(0, K_{n+1}^{-1}/(n\lambda))$. The corresponding posterior for v is proportional to, with w the vector $K_{n+1} v$ minus its first coordinate,

$$\prod_{i=1}^n p_w(X_i, Y_i) e^{-\frac{1}{2} n\lambda v^\top K_{n+1} v} = \exp\Big(-\tfrac{1}{2}\Big[\sum_{i=1}^n L(Y_i, (K_{n+1} v)_i) + n\lambda v^\top K_{n+1} v\Big]\Big).$$

Clearly, the posterior mode for ν is the same as the minimizer $\hat\nu$ found previously. The resulting mode for w_{n+1} is $K_{n+1}\hat\nu$.

Thus the penalized estimator $\hat w$ and the posterior mode agree at the points $X_0, X_1, \ldots, X_n$. ☐

Example 11.62 The preceding theorem applies to the Gaussian regression problem, with loss function $L(y, w(x)) = (y - w(x))^2$. The posterior distribution is then Gaussian, and hence the posterior mode is identical to the posterior mean.

The theorem also applies to the binary regression model. For instance, for logistic regression with responses coded as $y \in \{-1, 1\}$, we can set $L(y, w(x)) = \log(1 + e^{-w(x)y})$.

In both cases the theorem shows that the computation of the posterior mode can be reduced to solving a finite-dimensional optimization problem.

The RKHS norm of $(k-1)$ times integrated Brownian motion $I_{0+}^{k-1}B$ is the Sobolev norm of order k (see Lemma 11.29). For this prior the criterion in (11.31) takes the form

$$\sum_{i=1}^n L(Y_i, w(X_i)) + n\lambda \int_0^1 |w^{(k)}(t)|^2\, dt. \tag{11.32}$$

By the preceding theorem, the posterior mode for this prior is the minimizer of this expression over the RKHS of integrated Brownian motion. This consists of all functions in the Sobolev space $\mathfrak{W}^k[0, 1]$ with $w(0) = w'(0) = \cdots = w^{(k-1)}(0) = 0$. We may also minimize the preceding display over all functions in the Sobolev space, not necessarily with derivatives tied to zero at zero. This minimizer can be related to the integrated Brownian motion prior released at zero.

Theorem 11.63 *The minimizer of* (11.32) *is the limit as $b \to \infty$ of the posterior mode under the prior $w \sim (\sqrt{n}\lambda)^{-1}W$, for $W_t = I_{0+}^{k-1}B_t + b\sum_{j=0}^{k-1} Z_j t^j$, where B is a Brownian motion independent of $Z_j \overset{iid}{\sim} \mathrm{Nor}(0, 1)$.*

Proof As in the preceding proof let $w_{n+1} = (w(X_0), w(X_1), \ldots, w(X_n))^\top$. The first part of the preceding theorem still applies, but in the second part we introduce the extra parameter vector $\mu \sim \mathrm{Nor}_k(0, b^2 I)$ to form the prior for w_{n+1} through $w_{n+1} = P_{n+1}\mu + K_{n+1}\nu$, for P_{n+1} the matrix with ith row $(1, X_i, \ldots, X_i^{k-1})$, for $i = 0, 1, \ldots, n$. The posterior distribution for (μ, ν) is proportional to

$$\exp\Bigl(-\tfrac{1}{2}\Bigl[\sum_{i=1}^n L(Y_i, (K_{n+1}\nu)_i) + \mu^\top\mu/b^2 + n\lambda\nu^\top K_{n+1}\nu\Bigr]\Bigr).$$

For $b \to \infty$ the second term on the right disappears. The maximizer of the expression then converges to the minimizer of (11.32). ☐

Example 11.64 (Integrated Brownian motion and splines) The kernel of Brownian motion can be written $\mathrm{E}[B_s B_t] = s \wedge t = \int_0^1 \mathbb{1}_{[0,s]}(u)\mathbb{1}_{[0,t]}(u)\, du$. By Fubini's theorem the kernel of integrated Brownian motion is then

$$\mathrm{E}(I_{0+}B_s)(I_{0+}B_t) = \int_0^1 \int_0^s \mathbb{1}_{[0,x]}(u)\,dx \int_0^t \mathbb{1}_{[0,y]}(u)\,dy\,du = \int_0^1 (s-u)_+(t-u)_+\,du.$$

Repeating this argument we find that the kernel of the RKHS of $I_{0+}^{k-1}B$ is equal to

$$K(s,t) = \int_0^1 \frac{(s-u)_+^{k-1}}{(k-1)!}\frac{(t-u)_+^{k-1}}{(k-1)!}\,du.$$

This shows that the function $t \mapsto K(s,t)$ for fixed s is a spline of order $2k$ (i.e. pieces of polynomial of degree $2k-1$) with knot s. By Theorem 11.61 the posterior mode is a linear combination of these functions with $s \in \{X_1, \ldots, X_n\}$, and hence itself a spline function with knots $X_1, \ldots, X_n$.

11.7.2 Density Estimation

Suppose the data consists of a random sample $X_1, \ldots, X_n$ from the density p_w considered in Section 11.3.1. The likelihood function is given by

$$L(w|\,X_1, \ldots, X_n) = \prod_{i=1}^n \frac{e^{w(X_i)}}{\int e^w\,d\nu}.$$

Importance Sampling

The posterior expectation of a function $f(W)$, such as the density $p_W(x)$ at a point, is given by

$$\frac{\mathrm{E}_W\big[f(W)L(W|\,X_1, \ldots, X_n)\big]}{\mathrm{E}_W\big[L(W|\,X_1, \ldots, X_n)\big]} = \frac{\mathrm{E}_{W^*}\big[f(W^*)(\int e^{W^*(x)}\,d\nu(x))^{-n}\big]}{\mathrm{E}_{W^*}\big[(\int e^{W^*(x)}\,d\nu(x))^{-n}\big]},$$

where E_W denotes the expectation relative to the prior, and E_{W^*} the expectation with respect to the process W^* whose law has density proportional to $\prod_{i=1}^n e^{W(X_i)}$ relative to the law of W.

An elementary calculation shows that W^* is also a Gaussian process, with mean function $\mathrm{E}[W^*(x)] = \mathrm{E}[W(x)] + \sum_{i=1}^n K(x, X_i)$ and the same covariance kernel K as W. The right side of the preceding display can be approximated by generating a large number of copies of W^*s and replacing the two expectations by averages. In practice we generate the Gaussian processes on a grid, and approximate the integrals by sums.

Finite Interpolation

Let W^* be the kriging of W on a fine grid, as discussed in Example 11.10. If the original process W has continuous sample paths, and the mesh size of the grid is fine enough, then W^* will be close to W, and hence p_{W*} close to p_W, by Lemma 2.5.

The kriging W^* is a finite linear combination $W_t^* = \sum_{j=1}^m a_j(t) W_{t_j}$ of the values of W at the grid points, and hence to compute the posterior distribution with respect to the prior W^* it suffices to update the prior distribution of the vector $W^m = (W_{t_1}, \ldots, W_{t_m})$ with the data. For K_m the covariance matrix of W^m, its posterior distribution satisfies

$$p(W^m|\, X_1,\dots,X_n) \propto \prod_{i=1}^n p_{\sum_j a_j W_{t_j}}(X_i)\, \pi(W^m)$$
$$= \frac{e^{\sum_{i=1}^n \sum_j a_j(X_i) W_{t_j}}}{\prod_{i=1}^n \int e^{\sum_j a_j(x) W_{t_j}}\, d\nu(x)} e^{-(W^m)^\top K_m W^m/2}.$$

This can be the basis for execution of a Metropolis-Hastings algorithm, where at each step the right side is evaluated by (numerically) computing the integrals. A random walk Metropolis-Hastings step is particularly easy to implement. The formula can be adapted to include a tuning parameter in the kernel function, thus allowing a conditional Gaussian prior of W^m given this tuning parameter.

For increased accuracy and efficiency an adaptive choice of the grid can be incorporated into the algorithm by treating also the locations $t_1,\dots,t_m$ as variables, which then are updated in addition to W^m and tuning parameters. It is natural to move through the grid points by birth, death or swap moves, with corresponding proposal probabilities. Acceptance probabilities for movements across spaces of different dimensions can be calculated using the general recipe of reversible jump MCMC.

11.7.3 Nonparametric Binary Regression

The posterior distribution in the binary regression model of Section 2.5 is not analytically tractable, due to the nonlinearity of the link function. This can be overcome by an MCMC procedure based on data augmentation.

Because the likelihood depends on the process W only through its values at the observed covariates $X^n = (X_1,\dots,X_n)$, the posterior distribution of W given $W^n = (W_{X_1},\dots,W_{X_n})$ and the data $Y^n = (Y_1,\dots,Y_n)$ is the same as the prior of W given W^n. Since this is Gaussian and can be computed from the joint multivariate normal distribution, we may focus attention on the posterior distribution of W^n. Let μ_n and K_n be the prior mean vector and covariance matrix of this vector.

The case of the *probit link* function is the most straighforward. In this case the data Y^n can be viewed as arising from the following hierarchical scheme:

$$Y_i = \mathbb{1}\{Z_i > 0\}, \qquad Z_i|\, X^n, W^n \overset{\text{ind}}{\sim} \mathrm{Nor}(W_{X_i}, 1), \qquad W^n|\, X^n \sim \mathrm{Nor}_n(\mu_n, K_n).$$

Given (X^n, W^n) the resulting variables $Y_1,\dots,Y_n$ are independent with $\mathrm{P}(Y_i = 1|\, X^n, W^n) = \Phi(W_{X_i})$, as in the original model. The advantage of the latent variables $Z^n = (Z_1,\dots,Z_n)$ (and the probit link) is that their joint distribution with W^n is Gaussian, so that the conditional distributions needed for a Gibbs sampler are Gaussian as well, and given by standard formulas. In its simplest form, with distinct covariates X_i, the steps of this sampler are:

(i) $W^n|\, X^n, Y^n, Z^n \sim \mathrm{Nor}_n(\mu^*, K^*)$, for $K^* = (I + K_n^{-1})^{-1}$, $\mu^* = K^*(Z^n - \mu_n) + \mu_n$.
(ii) $Z_1,\dots,Z_n|\, W^n, X^n, Y^n$ are independent and distributed according to the $\mathrm{Nor}(W_{X_i}, 1)$-distribution truncated to $[0,\infty)$ if $Y_i = 1$, and according the same distribution truncated to $(-\infty, 0]$ if $Y_i = 0$.

If a covariate X_i appears multiple times, then the corresponding observations and values W_{X_i} must be grouped. In the computation of K^* for large n, direct computation of K^{-1} must be avoided but instead the spectral decomposition should be used.

If the mean function and the covariance kernel have additional hyperparameters, these should be updated also in additional Gibbs steps, possibly using nested Metropolis-Hastings samplers.

The case of the *logit link* function is often resolved by approximating the logistic distribution by a scale mixture of Gaussian distributions. The distribution function of a mixture with, for instance, five (well-chosen and fixed) normal components is practically indistinguishable from the logistic distribution function. The advantage of data augmentation with a Gaussian variable can then be retained.

11.7.4 Expectation Propagation

In many problems the parameter is a (high-dimensional) vector $\theta = (\theta_1, \ldots, \theta_n)$, and the likelihood depends on the parameter as a product of terms depending on the individual coordinates θ_i of the vector. The *expectation propagation* (EP) algorithm tries to approximate the marginal posterior distributions of the θ_i.

If $\theta \sim \mathrm{Nor}_n(\mu, \Sigma)$ under the prior, with prior density denoted by ϕ, and the likelihood takes the form $\prod_{j=1}^n t_j(\theta_j)$, where the observations are suppressed from the notation, then the posterior density p satisfies

$$p(\theta) \propto \prod_{j=1}^{n} t_j(\theta_j)\, \phi(\theta). \tag{11.33}$$

We assume that the numerical value of the right side of (11.33) is computable, for every θ. The problem we wish to solve is that its normalizing constant is an n-dimensional integral and hence is expensive to compute, unless it is accessible to analytical computation.

The main interest is in computing the marginal densities

$$p_i(\theta_i) \propto \int \prod_{j=1}^{n} t_j(\theta_j)\, \phi(\theta)\, d\theta_{-i} = t_i(\theta_i)\phi_i(\theta_i) \int \prod_{j\neq i} t_j(\theta_j)\, \phi_{-i|i}(\theta_{-i}|\, \theta_i)\, d\theta_{-i}. \tag{11.34}$$

Here θ_{-i} denotes the vector $\theta \in \mathbb{R}^n$ without its ith coordinate, and ϕ_i and $\phi_{-i|i}$ are the marginal and conditional (prior) density of θ_i, and of θ_{-i} given θ_i. These are a univariate and multivariate normal density, respectively. The integral is $(n-1)$-dimensional and hence numerically expensive to compute. On the other hand, the normalizing constant for the right side as a function of θ_i is a low-dimensional integral, whose computation should be feasible.

The EP-algorithm can be viewed as a method to compute an approximation to the normalizing constant in (11.33). If applied to the "conditional" density $\theta_{-i} \mapsto \prod_{j\neq i} t_j(\theta_j)\, \phi_{-i|i}(\theta_{-i}|\, \theta_i)$ it gives an approximation to the integral in (11.34).

The approximation is based on a *Gaussian approximation* to the posterior (11.33), i.e. a function q that is proportional to a Gaussian density that hopefully is close to (11.33). Because the integral of such a "Gaussian function" is easy to compute analytically, it is irrelevant whether q is normalized or not.

The approximation is computed recursively, where we cycle repeatedly through the index $i = 1, \ldots, n$. Every approximation q is of the form, for functions $\tilde{t}_i$ (called *term approximations*) that are proportional to univariate Gaussian densities,

$$q(\theta) \propto \prod_{i=1}^{n} \tilde{t}_i(\theta_i)\,\phi(\theta). \tag{11.35}$$

Given a Gaussian approximation q of the form (11.35) and some i we form a new density $\tilde{q}$ by replacing the current $\tilde{t}_i$ by the true t_i:

$$\tilde{q}(\theta) \propto q(\theta)\frac{t_i(\theta_i)}{\tilde{t}_i(\theta_i)}.$$

Unless the factors t_i are "Gaussian", the resulting density $\tilde{q}$ will not be Gaussian. We define q^{new} as the multivariate Gaussian density with the same moments as the distribution with density proportional to $\theta \mapsto \tilde{q}(\theta)$. Depending on the functions t_i, the appropriate moments may have to be computed numerically. To keep the validity of (11.35) we update $\tilde{t}_i$ to $\tilde{t}_i^{\mathrm{new}}$ so that $q^{\mathrm{new}}(\theta) \propto \prod_{j\neq i} \tilde{t}_j(\theta_j)\tilde{t}_i^{\mathrm{new}}(\theta_i)\phi(\theta)$, i.e., in view of (11.35),

$$\tilde{t}_i^{\mathrm{new}}(\theta_i) \propto \frac{q^{\mathrm{new}}}{q}(\theta)\tilde{t}_i(\theta_i). \tag{11.36}$$

Here there is a function of θ_i on the left and seemingly a function of the full vector θ on the right. However, for the (Gaussian) $q/\tilde{t}_i$ satisfying (11.35) it turns out that the function on the right depends on θ_i only. (See the lemma below, with the $\mathrm{Nor}(\mu, \Sigma)$-density taken equal to the density propertional to $q/\tilde{t}_i$ and $U\theta = \theta_i$; also see Problem 11.8 for explicit formulas.) This algorithm is iterated until "convergence," although apparently it is not known whether convergence will always take place.

Lemma 11.65 *Let $Z \sim \mathrm{Nor}_n(\mu, \Sigma)$, $U\colon \mathbb{R}^n \to \mathbb{R}^m$ be a linear map of full rank and $t\colon \mathbb{R}^m \to [0, \infty)$ be a measurable map such that $t(UZ)$ has finite second moment. If μ_1 and Σ_1 are the mean and covariance matrix of the distribution with density proportional to $z \mapsto t(Uz)\phi_{\mu,\Sigma}(z)$, then $(\phi_{\mu_1,\Sigma_1}/\phi_{\mu,\Sigma})(z)$ is a function of Uz.*

Proof Minimization of the Kullback-Leibler divergence $K(Q; N)$ between a given measure Q on $\mathbb{R}^n$ over the set of all n-variate normal distributions can be shown to yield the normal distribution with mean and covariance equal to the mean and covariance of Q. Therefore the normal distribution $\mathrm{Nor}(\mu_1, \Sigma_1)$ is closest in Kullback-Leibler divergence to the distribution with density proportional to $z \mapsto t(Uz)\phi_{\mu,\Sigma}(z)$, i.e. (μ_1, Σ_1) minimizes

$$\int \Big[\log \frac{t(Uz)\phi_{\mu,\Sigma}(z)}{\phi_{\mu_1,\Sigma_1}(s)}\Big]\, t(Uz)\,\phi_{\mu,\Sigma}(z)\,dz.$$

Let ψ_0 and ψ_1 be the densities of $Y = UZ$ under (μ, Σ) and (μ_1, Σ_1), respectively. We can factorize $\phi_{\mu,\Sigma}(z) = \psi_0(y)\phi_0(z|\,y)$ and $\phi_{\mu_1,\Sigma_1}(z) = \psi_1(y)\phi_1(z|\,y)$, for $y = Uz$ and ϕ_0 and ϕ_1 conditional densities given a suitable dominating measure ν. The preceding display can be written as $\int \big[\log t(y)\big]\, t(y)\,\psi_0(y)\,dy$ plus

$$\int \Big[\log \frac{\psi_0(y)}{\psi_1(y)}\Big]\, t(y)\,\psi_0(y)\,dy + \int \Big(\int \Big[\log \frac{\phi_0(z|\,y)}{\phi_1(z|\,y)}\Big]\,\phi_0(z|\,y)\,d\nu(z)\Big)\, t(y)\psi_0(y)\,dy.$$

The marginal and conditional densities ψ_1 and ϕ_1 range over all Gaussian (conditional) densities if (μ_1, Σ_1) ranges over all possible combinations of a mean vector and covariance matrix. Thus we can minimize the sum in the display over (μ_1, Σ_1) by minimizing both terms separately. The inner integral in the second term is nonnegative for every fixed y and hence it can be minimized (to 0) by choosing $\phi_1(\cdot|\, y) = \phi_0(\cdot|\, y)$. Thus minimization is equivalent to minimizing the first term. The minimizing values satisfy $(\phi_{\mu_1,\Sigma_1}/\phi_{\mu,\Sigma})(z) = (\psi_1/\psi_0)(y)$. ☐

Since q approximates the posterior, its marginal density is an approximation to the marginal density $\theta_i \mapsto p_i(\theta_i)$ in (11.34). This gives a Gaussian approximation, but it appears to be not accurate.

An alternative is to apply EP as described in the preceding to compute the norming constant not of the full posterior, but of the conditional density $\theta_{-i} \mapsto \prod_{j\neq i} t_j(\theta_j)\, \phi_{-i|i}(\theta_{-i}|\, \theta_i)$. This may next be substituted in the far right side of (11.34). (The "conditional prior" $\psi_{-i|i}$ is Gaussian, so that the preceding applies.)

Other approximations are suggested by the identities (for the first one use (11.35) to eliminate ϕ)

$$\begin{aligned} p_i(\theta_i) &\propto \int \prod_{j=1}^n t_j(\theta_j)\, \phi(\theta)\, d\theta_{-i} \propto \int \prod_{j=1}^n \frac{t_j(\theta_j)}{\tilde{t}_j(\theta_j)}\, q(\theta)\, d\theta_{-i} \\ &= \frac{t_i(\theta_i)}{\tilde{t}_i(\theta_i)}\, q_i(\theta_i) \int \prod_{j\neq i} \frac{t_j(\theta_j)}{\tilde{t}_j(\theta_j)}\, q_{-i|i}(\theta_{-i}|\, \theta_i)\, d\theta_{-i}. \end{aligned}$$

Here q_i and q_{-i} are the marginal and conditional density resulting from q. Both are Gaussian and can be computed analytically from q. The function $\theta_i \mapsto q_{-i|i}(\theta_{-i}|\, \theta_i)$ is proportional to a univariate Gaussian density, and hence the integral is a mixture of Gaussian densities. A crude approximation is to approximate every $t_j/\tilde{t}_j$ by 1 and evaluate the integral also to 1. This gives the approximation

$$\frac{t_i(\theta_i)}{\tilde{t}_i(\theta_i)}\, q_i(\theta_i).$$

This is the marginal density of the Gaussian approximation q, but with the "correct" t_i put back.

Better would be to approximate the integral. This is possible using EP applied to the density that is proportional to

$$\theta_{-i} \mapsto \prod_{j\neq i} \frac{t_j(\theta_j)}{\tilde{t}_j(\theta_j)}\, q_{-i|i}(\theta_{-i}|\, \theta_i).$$

In other words, the original ϕ is replaced by $q_{-i|i}$ and the original t_j are replaced by the quotients $t_j/\tilde{t}_j$. Because this procedure requires nested approximations, it is expensive. It is therefore recommended not to run the iterations of EP to convergence, but perform only one round of updates of the coordinates $t_j/\tilde{t}_j$.

Several other modifications have been suggested in the literature.

Many Gaussian priors come with a low-dimensional hyperparameter τ. If this is determined by an empirical Bayes method, then this causes no additional difficulty. In a full Bayes approach the marginal densities take the form

$$p_i(\theta_i) = \int p_i(\theta_i \mid \tau)\, p_\tau(\tau)\, d\tau. \tag{11.37}$$

Here $\theta_i \mapsto p_i(\theta_i \mid \tau)$ is the marginal density given τ, and p_τ is the *posterior* density of τ (dependence on the data is not shown!). The first was approximated in the preceding, with τ fixed to a particular value. The second is a conditional density given the data, and has the form, for π the prior on τ and the likelihood proportional to $\prod_{j=1}^n t_j(\theta_j)$,

$$p_\tau(\tau) \propto \int \prod_{j=1}^n t_j(\theta_j)\, \phi(\theta \mid \tau)\, d\theta\, \pi(\tau).$$

The integral is the normalizing constant of the right-hand side in the equation (11.33) for the posterior density. Hence up to the normalizing constant the expression can be approximated using the EP approximation.

Thus we can approximate the two terms in the integrand of (11.37), up to normalizing constants, for every τ. If τ is low-dimensional, then the integral can be approximated by a discretization. The normalizing constant to (11.37) can next also be determined by numerical integration.

11.7.5 Laplace Approximation

The *Laplace approximation* to the posterior distribution is the Gaussian approximation that also appears in the Bernstein–von Mises theorem. The approximation to the posterior density can be obtained by exponentiating and renormalizing a second order Taylor expansion of the sum of the log likelihood and log prior density around its mode.

Consider this in detail for the situation of a likelihood that factorizes in the coordinates of a parameter vector, as in (11.33), combined with a Gaussian prior with density ϕ. Let $\underline{p}$ be the right side of (11.33), so that $p \propto \underline{p}$, and compute a second order Taylor expansion in θ to

$$\log \underline{p}(\theta) = \sum_{j=1}^n \log t_j(\theta_j) + \log \phi(\theta),$$

around its point of maximum (the *mode* of p). Putting the quadratic in the exponent gives a Gaussian approximation to the full posterior, up to the normalizing constant.

The Gaussian term $\log \phi$ is quadratic already, so it suffices to expand the univariate functions $\log t_j$. The expansion of $\log \underline{p}$ around its mode takes the form

$$\log \underline{p}(\theta) = \log \underline{p}(\hat\theta) - \tfrac{1}{2}(\theta - \hat\theta)^\top \hat A(\theta - \hat\theta) + o(\|\theta - \hat\theta\|^2),$$

where the matrix $\hat A$ is minus the second derivative matrix

$$\hat A = -\frac{\partial^2}{\partial\theta^2} \log \underline{p}(\theta)|_{\theta=\hat\theta} = \operatorname{diag}(\hat c) + \Sigma^{-1}, \tag{11.38}$$

for Σ the covariance matrix of the prior, and $\operatorname{diag}(\hat c)$ the diagonal matrix with minus the second derivatives of the functions $\log t_j$ at the mode $\hat\theta$:

$$\hat{c}_j = -\frac{\partial^2}{\partial\theta_j^2}\log t_j(\theta_j)|_{\theta_j=\hat{\theta}_j}.$$

The mode $\hat{\theta}$ can be computed by Newton-Raphson: maximize a quadratic expansion (with linear term) of $\log \underline{p}$ around an initial guess $\tilde{\theta}$ to obtain a new guess, and iterate. The iterations are $\tilde{\theta}^{\text{new}} = \tilde{\theta} + \tilde{A}^{-1}\tilde{b}$, for $\tilde{b}$ the vector of first derivatives of the functions $\log t_j$ at $\tilde{\theta}$, and $\tilde{A} = \tilde{c} + \Sigma^{-1}$ the second derivative matrix at this point.

The normalizing constant of $\underline{p}$ can be approximated by, with d the dimension of θ,

$$\int \underline{p}(\theta)\,d\theta \approx \underline{p}(\hat{\theta})\int e^{-\frac{1}{2}(\theta-\hat{\theta})^\top \hat{A}(\theta-\hat{\theta})}\,d\theta = \underline{p}(\hat{\theta})\,(\det \hat{A})^{-1/2}\,(2\pi)^{d/2}. \tag{11.39}$$

This requires only the computation of the determinant.

The Gaussian approximation to the full posterior may be marginalized to give Gaussian approximations of the marginals, but these appear to be inaccurate in many situations. A better approximation is obtained by noting that the marginal density $\theta_i \mapsto p_i(\theta_i)$ is the denominator in the definition of the conditional density $\theta_{-i} \mapsto p_{-i|i}(\theta_{-i}|\,\theta_i)$, i.e. the norming constant of the function $\theta_{-i} \mapsto \underline{p}(\theta)$, with θ_i fixed. We can apply the Laplace approximation to the norming constant (11.39) with $\underline{p}$ in that equation replaced by $\underline{p}$ viewed as a function of θ_{-i} (and integrated with respect to this variable), for fixed θ_i. This leads to

$$\underline{p}(\hat{\theta}_{-i}(\theta_i),\theta_i)\Big(\det\Big(-\frac{\partial^2}{\partial\theta_{-i}^2}\log \underline{p}(\theta_{-i},\theta_i)\Big)|_{\theta_{-i}=\hat{\theta}_{-i}(\theta_i)}\Big)^{-1/2}(2\pi)^{d_1/2}. \tag{11.40}$$

Here $\hat{\theta}_{-i}(\theta_i)$ is the mode of $\theta_{-i} \mapsto \underline{p}(\theta)$, and we write $\underline{p}(\theta_{-i},\theta_i)$ for $\underline{p}(\theta)$.

A different way to derive this approximation, due to Tierney and Kadane (1986), starts from the identity

$$p_i(\theta_i) = \frac{p(\theta_{-i},\theta_i)}{p_{-i|i}(\theta_{-i}|\,\theta_i)}. \tag{11.41}$$

This is the definition of the conditional density $p_{-i|i}$, written "upside down," and it is an identity for *all* θ_{-i} (θ_{-i} appears on the right, but not on the left side!). We obtain an approximation of p_i by replacing the denominator $p_{-i|i}(\theta_{-i}|\,\theta_i)$ by an approximation, and next evaluating the resulting quotient at a suitable value θ_{-i}, for instance the mode $\hat{\theta}_{-i}(\theta_i)$ of $\theta_{-i} \mapsto p_{-i|i}(\theta_{-i}|\,\theta_i) \propto \underline{p}(\theta_{-i},\theta_i)$. Because the constant cancels in the quotients (11.41) the approximation may be up to the normalizing constant. The value of the density of the multivariate distribution $\mathrm{Nor}_d(\mu,\Lambda)$ at its mode is equal to $(\det\Lambda)^{-1/2}(2\pi)^{-d/2}$, and the Laplace approximation to $\theta_{-i} \mapsto \log p(\theta_{-i},\theta_i)$ is a Gaussian with the second derivative matrix in the right side of (11.40) as its inverse covariance matrix. Therefore, if we use the Laplace approximation for $\theta_{-i} \mapsto p_{-i|i}(\theta_{-i}|\,\theta_i)$, we are lead back to (11.40).

A hyperparameter in the prior can be handled in the same way as described in Section 11.7.4.

11.8 Historical Notes

Gaussian processes appear to have first been considered as priors by Kimeldorf and Wahba (1970) and Wahba (1978), in connection to spline smoothing, as explained in Section 11.7.1.

Wood and Kohn (1998) and Shively et al. (1999) extended the idea to binary regression. Gaussian processes as priors for density estimation were first used by Leonard (1978) and later by Lenk (1988), who introduced the computational method based on importance sampling in Section 11.7.2. The computational method based on kriging presented here and its convergence properties were obtained by Tokdar (2007). The MCMC technique for (parametric) probit regression using latent Gaussian variables was introduced by Albert and Chib (1993), and modified to the nonparametric context in Choudhuri et al. (2007). Expectation propagation has its roots in mathematical phyics; we learned it from Cseke and Heskes (2011) and the thesis by Cseke. The Laplace approximation is the basis for Integrated Nested Laplace Approximation or *INLA* (see Rue et al. 2009), which is available as an R-package. An extensive review of Gaussian process priors from the point of view of machine learning is given in Rasmussen and Williams (2006). Appendix I gives an overview of Gaussian process theory, including many references. The key results mentioned in the present chapter are Borell's inequality, discovered by Borell (1975), and the bounds on small ball probabilities and shifted normal distributions due to Kuelbs and Li (1993) and Li and Linde (1998). Posterior consistency for Gaussian process priors was studied by Tokdar and Ghosh (2007), Ghosal and Roy (2006) and Choi and Schervish (2007), respectively, for density estimation, nonparametric binary regression and nonparametric normal regression models. The theory of contraction rates for Gaussian process priors was developed in the paper van der Vaart and van Zanten (2008a), and elaborated for various particular examples and rescaled processes in van der Vaart and van Zanten (2007, 2009, 2011). Lemma 11.34 is taken from Castillo (2008), who generalized a similar result by van der Vaart and van Zanten (2008a) for the special case $\alpha = \beta$. The lower bound statements in the rate theorems are also due to Castillo (2008). From many recent papers, many of which concerned with adaptation using Gaussian priors, we mention De Jonge and van Zanten (2010) on Gaussian mixtures, Panzar and van Zanten (2009) on diffusion models, van Waaij and van Zanten (2016) on adaptation, Yang and Tokdar (2015) on choice of variables in Gaussian regression models, Sniekers and van der Vaart (2015b,a) on adaptive regression and Knapik et al. (2016) on inverse problems.

Problems

11.1 (van der Vaart and van Zanten 2008a) A version of Theorem 11.22 is possible even if w_0 and $\psi/(\Psi(1-\Psi))$ are both unbounded, by using appropriate norms on the Gaussian process. Using Problem 2.6, show that for the probit link function, contraction rate holds for the sum of $\mathbb{L}_2((w_0^2 \vee 1)\cdot G)$ and $\mathbb{L}_4(G)$-norms, provided that $w_0 \in \mathbb{L}_4(G)$.

11.2 We know that the support of Brownian motion in $\mathfrak{C}[0,1]$ is the set of all functions with $w(0) = 0$. Show that the support of the standard Brownian motion as a random element in $\mathbb{L}_r[0,1]$ is the full space $\mathbb{L}_r[0,1]$ for any $r < \infty$.

11.3 (van der Vaart and van Zanten 2008a) If $w_0 \in \mathfrak{C}^\beta[0,1]$, $0 < \beta \le 1$, and ϕ is a differentiable kernel function, then show that $(w_0 * \phi_\sigma)(0)^2 + \|(w_0 * \phi_\sigma)'\|_2^2 \lesssim \sigma^{-(2-2\beta)}$.

11.4 (van der Meulen et al. 2006) Consider a (sequence of) diffusion process defined by the stochastic differential equation

$$dX_t^{(n)} = \beta_\theta^{(n)}(t, X^{(n)})\,dt + \sigma^{(n)}(t, X^{(n)})\,dB_t^{(n)}, \quad t \in [0, T_n], \quad X_0^{(n)} = X_0, \quad (11.42)$$

where $B^{(n)}$ is a (sequence of) Brownian motion, and the functional forms of the drift coefficient $\beta_\theta^{(n)}(t, X^{(n)})$ and diffusion coefficient $\sigma^{(n)}(t, X^{(n)})$ are given continuous functions and $\theta \in \Theta$ is the unknown parameter. Let $\theta \sim \Pi$ and $\theta_0 \in \Theta$ stand for the true value of the parameter. Let $h_n^2(\theta, \theta_0) = \int_0^{T_n} (\beta_\theta^{(n)}(t, X^{(n)}) - \beta_{\theta_0}^{(n)}(t, X^{(n)})^2 (\sigma^{(n)}(t, X^{(n)}))^{-2} dt$. (This is equal to the squared Hellinger distance between the Gaussian process distributions of $(X_t^{(n)}: t \in T_n)$ under θ and θ_0, respectively.) Let $B_n(\theta_0, \epsilon) = \{\theta: h_n(\theta, \theta_0) < \epsilon\}$. Suppose that for some sequence ϵ_n, $\log N(a\epsilon, B_n(\theta_0, \epsilon), h_n) \lesssim \epsilon_n^2$ for all $\epsilon > \epsilon_n$ and that for any $\xi > 0$, there is $J \in \mathbb{N}$ such that $\Pi(B_n(\theta_0, j\epsilon_n))/\Pi(B_n(\theta_0, \epsilon_n)) \leq e^{\xi j^2 \epsilon_n^2}$ for all $j \geq J$. Show that the posterior contracts at θ_0 at the rate ϵ_n with respect to h_n.

11.5 (van der Meulen et al. 2006) Consider a special case of Problem 11.4 given by signal plus white noise model $dX_t^{(n)} = \theta(t)\, dt + \sigma_n dB_t$, $t \in [0, T]$, $X_0^{(n)} = x_0$, $\sigma_n \to 0$. Assume that for some sequence ϵ_n, $\log N(\epsilon/8, \{\theta: \|\theta - \theta_0\|_2 < \epsilon\}, \|\cdot\|_2) \lesssim \sigma_n^{-2}\epsilon_n^2$ for all $\epsilon \geq \epsilon_n$ and for some $J \in \mathbb{N}$, $\Pi(\|\theta - \theta_0\|_2 < j\epsilon_n))/\Pi(\|\theta - \theta_0|_2 < \epsilon_n) \leq e^{j^2\epsilon_n^2\sigma_n^{-2}/9216}$ for all $j \geq J$. Show that the posterior contracts at θ_0 at the rate ϵ_n with respect to $\|\cdot\|_2$.

11.6 (van der Meulen et al. 2006) Consider a special case of Problem 11.4 given by the perturbed dynamical system $dX_t^{(n)} = \theta(X_t^{(n)})\, dt + \sigma_n dB_t^{(n)}$, $t \in [0, T]$, $X_0^{(n)} = X_0$, $\sigma_n \to 0$. Let $d^2(\theta, \theta_0) = \int |\theta(x_t) - \theta_0(x_t)|^2 dt$, where x_t is the unique solution of the ordinary differential equation $dx_t = \theta_0(x_t)$, $x_t = x_0$ for $t = 0$. Suppose that all θ are uniformly bounded and uniformly Lipschitz continuous. Assume that for some sequence ϵ_n, $\log N(\epsilon/24, \{\theta: d(\theta, \theta_0) < \epsilon\}, d) \lesssim \sigma_n^{-2}\epsilon_n^2$ for all $\epsilon \geq \epsilon_n$, and for some $J \in \mathbb{N}$, $\Pi(d(\theta, \theta_0) < j\epsilon_n))/\Pi(d(\theta, \theta_0) < \epsilon_n) \leq e^{j^2\epsilon_n^2\sigma_n^{-2}/20736}$ for all $j \geq J$. Show that the posterior contracts at θ_0 at the rate ϵ_n with respect to d.

If θ lies in a bounded subset of the Besov space $\mathfrak{B}_{p,\infty}^\alpha$ of α-smooth functions, $p\alpha > 1$, construct a prior based on an $\sigma_n^{2\alpha/(2\alpha+1)}$-net in terms of the uniform distance as in Subsection 8.2.2. Show that the posterior contracts at the rate $\sigma_n^{2\alpha/(2\alpha+1)}$ with respect to d.

If θ lies in a bounded subset of the Besov space $\mathfrak{B}_{\infty,\infty}^\alpha$ of α-smooth functions, construct a prior based on a wavelet expansion $\theta = \sum_{j=1}^J \sum_{k=1}^{2^j} 2^{-j/2} Z_{j,k}\psi_{j,k}$, where $\psi_{j,k}$ are wavelet functions and $Z_{j,k} \stackrel{\text{iid}}{\sim} \mathrm{Nor}(0, 1)$. Show that the posterior contracts at the rate $\sigma_n^{2\alpha/(2\alpha+1)}$ with respect to d.

11.7 (van der Meulen et al. 2006) Consider a special case of Problem 11.4 given by the ergodic diffusion model $dX_t = \theta(X_t)\, dt + \sigma(X_t) dB_t$, $t \in [0, T_n]$, $T_n \to \infty$. Let m_{θ_0} be the speed measure defined by having a density $\sigma^{-2}(x) \exp\{2 \int_{x_0}^x \theta_0(z)\sigma^{-2}(z)\, dz\}$, where x_0 is fixed but arbitrary. Let I be a compact subinterval of $\mathbb{R}$ and $\|f\|_{2,\mu_0,I}^2 = \int_I |f(x)|^2 d\mu_0(x)$ and $\|\cdot\|_{2,\mu_0} = \|\cdot\|_{2,\mu_0,\mathbb{R}}$. Assume that $m_{\theta_0}(I) < \infty$ and $\mu_0 = m_0/m_{\theta_0}(I)$. Suppose that for some sequence $\epsilon_n > 0$, $T_n\epsilon_n^2 \gg 0$, we have $\{\log N(a\epsilon, \{\theta: \|(\theta - \theta_0)/\sigma\|_{2,I} < \epsilon\}, \|\cdot\|_{2,\mu_0}): \epsilon \geq \epsilon_n\} \lesssim T_n\epsilon_n^2$ for every $a > 0$ and for all $\xi > 0$, there exists some $J \in \mathbb{N}$, $\Pi(\|\theta - \theta_0\|_{2,\mu_0,I} < j\epsilon_n)/\Pi(\|\theta - \theta_0\|_{2,\mu_0} < \epsilon_n) \leq e^{\xi j^2 T_n \epsilon_n^2}$ for all $j \geq J$. Show that the posterior contracts at θ_0 at the rate ϵ_n with respect to $\|\cdot\|_{2,\mu_0}$.

In the above setting, let there be a k-dimensional parameter θ defining the process by $dX_t = \beta_\theta(X_t)\,dt + \sigma(X_t)dB_t$, where $\underline{\beta}(x)\|\theta - \theta^*\| \le |\beta_\theta(x) - \beta_{\theta^*}(x)| \le \bar{\beta}(x)$ for some functions $\underline{\beta}$ and $\bar{\beta}$ satisfying $0 < \int(\underline{\beta}/\sigma)^2 d\mu_0(x) \le \int(\bar{\beta}/\sigma)^2 d\mu_0(x) < \infty$. Let θ have a prior density bounded and bounded away from zero, and let θ_0 be the true value of θ. Show that the posterior for θ contracts at θ_0 at the rate $T_n^{-1/2}$ with respect to the Euclidean distance.

11.8 The parameters μ_1 and Σ_1 in Lemma 11.65 can be computed from the factorization $\phi_{\mu_1,\Sigma_1}(z) = \psi_1(Uz)\phi_1(z|\,Uz)$, where ψ_1 is the normal density with mean vector $\bar{\mu}_1 = \mathrm{E}Yt(Y)/\mathrm{E}t(Y)$ and covariance matrix $\mathrm{E}(Y - \bar{\mu}_1)(Y - \bar{\mu}_1)^\top t(Y)/\mathrm{E}t(Y)$, for $Y \sim \mathrm{Nor}(U\mu, U\Sigma U^\top)$, and $\phi_1 = \phi_0$ is the conditional density of Z given $Y = UZ$. The latter is a normal distribution with mean $\mathrm{E}(Z|\,Y) = AU(Z - \mu) + \mu$, for $A = \Sigma U^\top (U\Sigma U^\top)^{-1}$ and covariance matrix $\mathrm{Cov}\,(Z - \mathrm{E}(Z|\,Y)) = \Sigma - AU\Sigma U^\top A^\top$. Alternatively these parameters can be computed directly using

$$\mu_1 = \frac{\mathrm{E}Zt(UZ)}{\mathrm{E}t(UZ)} = \frac{\mathrm{E}\mathrm{E}(Z|\,UZ)t(UZ)}{\mathrm{E}t(UZ)},$$

$$\Sigma_1 = \frac{\mathrm{E}(Z - \mu_1)(Z - \mu_1)^\top t(UZ)}{\mathrm{E}t(UZ)} = \frac{\mathrm{E}\mathrm{E}((Z - \mu_1)(Z - \mu_1)^\top |\,UZ)t(UZ)}{\mathrm{E}t(UZ)}$$

$$= \mathrm{Cov}(Z|\,UZ) + \frac{\mathrm{E}(\mathrm{E}(Z|\,UZ) - \mu_1)(\mathrm{E}(Z|\,UZ) - \mu_1)^\top t(UZ)}{\mathrm{E}t(UZ)}.$$

The last term on the right is the covariance matrix of the vector $AU(Z^* - \mu) + \mu$, with Z^* given the tilted normal distribution with density proportional to $z \mapsto t(Uz)\phi_{\mu,\Sigma}(z)$.

12

Infinite-Dimensional Bernstein–von Mises Theorem

Although nonparametric theory is mostly concerned with rates of estimation, distributional approximations are possible for statistical procedures that focus on special aspects of a parameter. Within the Bayesian framework these may take the form of a normal approximation to the marginal posterior distribution of a functional of the parameter. We begin this chapter with a review of the classical result in this direction: the Bernstein–von Mises theorem for parametric models. Next we present a result in the same spirit for the posterior distribution based on the Dirichlet process prior, and corresponding strong approximations. After a brief introduction to semiparametric models, we proceed with Bernstein–von Mises theorems for functionals on such models, with strict semiparametric models as a special case. These are illustrated with applications to Gaussian process priors and the Cox proportional hazard model. A discussion of the Bernstein–von Mises theorem in the context of the white noise model illustrates the possibilities and difficulties of extending the theorem to the fully infinite-dimensional setting.

12.1 Introduction

In many statistical experiments with a Euclidean parameter space, good estimators $\hat{\theta}_n$ are asymptotically normally distributed with mean the parameter θ and covariance matrix proportional to the inverse Fisher information. More formally, for r_n^{-1} a rate of convergence, the sequence $r_n(\hat{\theta}_n - \theta)$ tends in distribution to a Nor$(0, I_\theta^{-1})$-distribution. Under regularity conditions this is true for the maximum likelihood estimator and most Bayes estimators; it is usually proved by approximating the estimator $\hat{\theta}_n$ by an average and next applying the central limit theorem. The Fisher-Cramér-Rao–Le Cam theory designates these estimators as asymptotically efficient, the inverse Fisher information being the minimal attainable asymptotic variance in the local minimax sense.

For a frequentist the randomness in $r_n(\hat{\theta}_n - \theta)$ comes through the observation $X^{(n)}$, hidden in $\hat{\theta}_n = \hat{\theta}_n(X^{(n)})$, which is considered drawn from a distribution $P_\theta^{(n)}$ indexed by a fixed parameter θ. To a Bayesian the quantity $r_n(\hat{\theta}_n - \theta)$ has a quite different interpretation. In the Bayesian setup θ is the variable and the randomness is evaluated according to the posterior distribution $\Pi_n(\cdot| X^{(n)})$, given a fixed observation $X^{(n)}$. It is remarkable that these two assessments of randomness, with diametrically opposite interpretations, are approximately equal in large samples: by the *Bernstein–von Mises theorem* the posterior distribution of $r_n(\hat{\theta}_n - \theta)$ also tends to a Nor$(0, I_\theta^{-1})$-distribution, for most observations $X^{(n)}$.

The symmetry between the two statements may be nicely exposed by writing them both in the Bayesian framework, where both θ and $X^{(n)}$ are random:

$$r_n(\hat{\theta}_n - \theta)|\,\theta = \theta_0 \rightsquigarrow \mathrm{Nor}_d(0, I_{\theta_0}^{-1}),$$
$$r_n(\theta - \hat{\theta}_n)|\,X^{(n)} \rightsquigarrow \mathrm{Nor}_d(0, I_{\theta_0}^{-1}).$$

The first line gives the frequentist statement, which conditions on a given value of the parameter, whereas the left side of the second line refers to the posterior distribution of θ, centered at $\hat{\theta}_n$ and scaled by r_n. In the first case the distribution of $\hat{\theta}_n = \hat{\theta}_n(X^{(n)})$ is evaluated for $X^{(n)}$ following the parameter value θ_0, whereas in the second case $\hat{\theta}_n$ is fixed by conditioning on $X^{(n)}$, but the distribution of the random (posterior) measure on the left side is again evaluated for $X^{(n)}$ following the parameter θ_0.

Besides having conceptual appeal, the Bernstein–von Mises theorem is of great importance for the interpretation of Bayesian credible regions: sets in the parameter space of given posterior probability. From the normal approximation it can be seen that a region of highest posterior probability density will be asymptotically centered at $\hat{\theta}_n$ and spread in such a way that it is equivalent to a frequentist confidence region based on $\hat{\theta}_n$. Consequently, the frequentist confidence level of a Bayesian credible set will be approximately equal to its credibility. This provides a frequentist justification of Bayesian credible regions.

The Bernstein–von Mises theorem extends beyond the setting of i.i.d. observations, to models that are not regular in the parameter (such as uniform distributions) and lead to non-normal limit experiments, and also to parameters of increasing dimension. The mode of approximation by a normal distribution can also be made precise in various ways (e.g. using Kullback-Leibler divergence or higher order approximations). For reference, we state only a version for i.i.d. observations that requires minimal regularity of the model and employs a testing condition similar to the ones used in posterior consistency and convergence rate theorems.

A set $\{p_\theta: \theta \in \Theta\}$ of probability densities with respect to a σ-finite dominating measure ν on a measurable space $(\mathfrak{X}, \mathscr{A})$ indexed by an open subset $\Theta \subset \mathbb{R}^d$ is said to be *differentiable in quadratic mean* at θ if there exists a measurable map $\dot{\ell}_\theta: \mathfrak{X} \to \mathbb{R}$ such that, as $\|h\| \to 0$,

$$\int \Big[\sqrt{p_{\theta+h}} - \sqrt{p_\theta} - \frac{1}{2} h^\top \dot{\ell}_\theta \sqrt{p_\theta}\Big]^2 \, d\nu = o(\|h\|^2). \tag{12.1}$$

The function $\dot{\ell}_\theta$ is a version of the *score function* of the model, and its covariance matrix $I_\theta = P_\theta(\dot{\ell}_\theta \dot{\ell}_\theta^\top)$ is the *Fisher information matrix*. For simplicity we assume that this is nonsingular. For $X_1, X_2, \ldots \overset{\text{iid}}{\sim} p_\theta$, set

$$\Delta_{n,\theta} = \frac{1}{\sqrt{n}} \sum_{i=1}^n I_\theta^{-1} \dot{\ell}_\theta(X_i).$$

This sequence is well defined and asymptotically normally distributed as soon as the model is differentiable in quadratic mean.

Theorem 12.1 (Bernstein–von Mises–Le Cam) *If the model $(p_\theta: \theta \in \Theta)$ is differentiable in quadratic mean at θ_0 with nonsingular Fisher information matrix I_{θ_0}, and for any $\epsilon > 0$ there exists a sequence of tests ϕ_n based on $(X_1, \ldots, X_n)$ such that*

$$P^n_{\theta_0}\phi_n \to 0, \qquad \sup_{\theta:\|\theta-\theta_0\|\geq\epsilon} P^n_\theta(1-\phi_n)\to 0,$$

then for any prior distribution on θ that possesses a density with respect to the Lebesgue measure in a neighborhood of θ_0 that is bounded away from zero,

$$\mathrm{E}_{\theta_0}\Big\|\Pi_n(\theta{:}\,\sqrt{n}(\theta-\hat\theta_n)\in\cdot\,|\,X_1,\ldots,X_n)-\mathrm{Nor}_d(\Delta_{n,\theta_0},I_{\theta_0}^{-1})\Big\|_{TV}\to 0.$$

For a proof, see van der Vaart (1998), pages 140–143. To translate back to the preceding we note that under regularity conditions the scaled and centered maximum likelihood estimators $\sqrt{n}(\hat\theta_n-\theta)$ will be asymptotically equivalent to the sequence Δ_{n,θ_0} when the observations are independently sampled from P_{θ_0}. Because the total variation norm is invariant under a shift of location and a change of scale, the sequence Δ_{n,θ_0} may be replaced by $\sqrt{n}(\hat\theta_n-\theta_0)$ in that case, and the centering shifted by $\sqrt{n}\theta_0$ and the scale by $\sqrt{n}$, leading to

$$\mathrm{E}_{\theta_0}\Big\|\Pi_n(\theta{:}\,\theta\in\cdot\,|\,X_1,\ldots,X_n)-\mathrm{Nor}_d(\hat\theta_n,(nI_{\theta_0})^{-1})\Big\|_{TV}\to 0.$$

This is equivalent to the assertion in the theorem, except for the fact that the good behavior of the maximum likelihood estimator requires additional regularity conditions.

In most of the nonparametric examples in this book, a Bernstein–von Mises theorem is not valid in the same way. This appears to be due at least partly to the fact that a bias-variance trade-off is at the core of these examples, whereas in the (parametric) Bernstein–von Mises theorem the bias is negligible. It has thus been claimed in the literature that the "infinite-dimensional Bernstein–von Mises theorem does not hold." However, for smoother aspects of the parameter the theorem does hold, and one may say that the theorem holds provided the topology on the parameter space is chosen appropriately (and the prior does not unnecessarily introduce a bias). In the next sections we illustrate this by a Bernstein–von Mises theorem for estimating a measure using the Dirichlet prior, and Bernstein–von Mises theorems for smooth functionals on semiparametric models.

12.2 Dirichlet Process

By Theorem 4.6 the posterior distribution of P in the model

$$P\sim \mathrm{DP}(\alpha), \qquad X_1,\ldots,X_n|\,P\overset{\text{iid}}{\sim}P,$$

is the Dirichlet process with base measure $\alpha+n\mathbb{P}_n$, for $\mathbb{P}_n$ the empirical measure of $X_1,\ldots,X_n$. We shall show that this satisfies a Bernstein–von Mises theorem.

The empirical measure $\mathbb{P}_n$ is the (nonparametric) maximum likelihood estimator in this problem, which suggests to use it as the centering measure, which will then concern the conditional distribution of the process $\sqrt{n}(P-\mathbb{P}_n)$, for $P|\,X_1,X_2,\ldots\sim\mathrm{DP}(\alpha+\sum_{i=1}^n\delta_{X_i})$. The posterior mean in this problem is $(\alpha+n\mathbb{P}_n)/(|\alpha|+n)$, and differs from the empirical measure $\mathbb{P}_n$ only by $(|\alpha|/(|\alpha|+n))(\bar\alpha-\mathbb{P}_n)$, which is of order $1/n$, if $n\to\infty$ and α remains fixed. This difference is negligible even after scaling by $\sqrt{n}$, whence in the Bernstein–von Mises theorem the posterior distribution can equivalently be centered at its mean.

For a fixed measurable set A the posterior distribution of $P(A)$ is the beta distribution $\text{Be}(\alpha(A) + n\mathbb{P}_n(A), \alpha(A^c) + n\mathbb{P}_n(A^c))$. This is the same posterior distribution as in the reduced model where we observe a sample of indicators $\mathbb{1}\{X_1 \in A\}, \ldots, \mathbb{1}\{X_n \in A\}$ from the Bernoulli distribution $\text{Bin}(1, P(A))$, with a $\text{Be}(\alpha(A), \alpha(A^c))$ prior distribution on the success probability. The inverse Fisher information for a binomial proportion p is equal to $p(1-p)$, while the maximum likelihood estimator is the sample proportion $\mathbb{P}_n(A)$. Hence the posterior distribution of $\sqrt{n}(P(A) - \mathbb{P}_n(A))$ is asymptotically equivalent to a $\text{Nor}(0, P(A)(1 - P(A)))$-distribution, by the parametric Bernstein–von Mises theorem. Alternatively, this may be verified directly from the beta-distribution: a weak approximation to the normal distribution is immediate from the central limit theorem and the delta-method (see Problem 12.1), whereas an approximation in the total variation norm needs some work.

The argument extends to the case of multiple measurable sets $A_1, \ldots, A_k$: the posterior distribution of the vectors $\sqrt{n}(P(A_1) - \mathbb{P}_n(A_1), \ldots, P(A_k) - \mathbb{P}_n(A_k))$ tends to a multivariate normal distribution $\text{Nor}_k(0, \Sigma)$, where Σ is the matrix with elements $P(A_i \cap A_j) - P(A_i)P(A_j)$.

This normal limit distribution is identical (of course) to the limit distribution of the scaled maximum likelihood estimator $\sqrt{n}(\mathbb{P}_n(A_1) - P(A_1), \ldots, \mathbb{P}_n(A_k) - P(A_k))$, the *empirical process* at the sets $A_1, \ldots, A_k$. The empirical process $\sqrt{n}(\mathbb{P}_n - P)$ has been studied in detail, and is known to converge also in stronger ways than in terms of its marginal distributions. In particular, it can be viewed as a map $f \mapsto \sqrt{n}(\mathbb{P}_n f - Pf)$ that attaches to every integrable function f the expected value $n^{-1/2}\sum_{i=1}^n (f(X_i) - Pf)$. A collection of functions $\mathcal{F} \subset \mathbb{L}_2(P)$ is said to be *Donsker* if this map is bounded in $f \in \mathcal{F}$ and converges in distribution to a tight limit in the space $\mathfrak{L}_\infty(\mathcal{F})$ of bounded functions $z: \mathcal{F} \to \mathbb{R}$, equipped with the supremum norm $\|z\| = \sup\{|z(f)|: f \in \mathcal{F}\}$. (See Appendix F.) The limit process $\mathbb{G}$ is the Gaussian process with mean zero and covariance function $\text{cov}(\mathbb{G}(f), \mathbb{G}(g)) = P[fg] - P[f]\,P[g]$, for $f, g \in \mathcal{F}$, known as the *P-Brownian bridge*.

The following theorem shows that the Bernstein–von Mises theorem for the Dirichlet posterior is also valid in this stronger, uniform sense. (For the most general version it is necessary to understand the convergence in distribution in terms of outer expectations and to be precise about the definition of the posterior process. The following theorem refers to the version outlined in the proof, and the conditional convergence is understood in terms of the bounded Lipschitz metric, as in the reference given. If there exists a Borel measurable version of the process, then the theorem applies in particular to this version.)

Theorem 12.2 (Bernstein–von Mises theorem for Dirichlet process) *For any P_0-Donsker class $\mathcal{F}$ of functions with envelope function F such that $(P_0 + \alpha)^*[F^2] < \infty$, the process $\sqrt{n}(P - \mathbb{P}_n)$ with $P \sim \text{DP}(\alpha + n\mathbb{P}_n)$ converges conditionally in distribution given $X_1, X_2, \ldots$ in $\mathfrak{L}_\infty(\mathcal{F})$ to a Brownian bridge process a.s. $[P_0^\infty]$, as $n \to \infty$. The same conclusion is valid if the centering $\mathbb{P}_n$ is replaced by the posterior mean $\tilde{\mathbb{P}}_n = (\alpha + n\mathbb{P}_n)/(|\alpha| + n)$.*

Proof By Proposition G.10, the $\text{DP}(\alpha + n\mathbb{P}_n)$-distribution can be represented as $V_n Q + (1 - V_n)\mathbb{B}_n$, where the variables $Q \sim \text{DP}(\alpha)$, $\mathbb{B}_n \sim \text{DP}(n\mathbb{P}_n)$ and $V_n \sim \text{Be}(|\alpha|, n)$ are independent. Assume further that these three variables are defined on a product probability space, with their independence expressed by being functions of separate coordinates in the product, and that $\mathbb{B}_n$ is defined as $\mathbb{B}_n f = \sum_{i=1}^n W_{n,i} f(X_i)$, where $X_1, X_2, \ldots$ are coordinate projections on further factors of the product probability space, and $(W_{n,1}, \ldots, W_{n,n})$ is a Dir$(n; 1, \ldots, 1)$-vector independent of the other variables and defined on yet one more factor of the underlying product probability space. By Theorem 3.6.13 in van der Vaart and Wellner (1996), it then follows that $\sqrt{n}(\mathbb{B}_n - \mathbb{P}_n)$ tends conditionally given $X_1, X_2, \ldots$ in distribution to a Brownian bridge process, a.s. $[P_0^\infty]$.

Since $\sqrt{n}V_n \to 0$ in probability, and $f \mapsto Qf$ is a well defined element of $\mathfrak{L}_\infty(\mathcal{F})$, the process $\sqrt{n}V_n Q$ tends to zero in the latter space. By Slutsky's lemma the sum $\sqrt{n}V_n Q + \sqrt{n}(\mathbb{B}_n - \mathbb{P}_n)$ has the same limit as the second term. □

Example 12.3 (Cumulative distribution function) The class $\mathcal{F}$ of functions consisting of the set of indicator functions of cells $(-\infty, t]$ in Euclidean space is Donsker for any underlying measure. The measure generated by it can be identified with the corresponding cumulative distribution function $F(t) = P(-\infty, t]$. Thus the process $t \mapsto \sqrt{n}(F - \mathbb{F}_n)(t)$ converges conditionally in distribution given $X_1, X_2, \ldots$ to a Brownian bridge process, where $\mathbb{F}_n$ is the *empirical distribution function* of $X_1, \ldots, X_n$.

The preceding Bernstein–von Mises theorem may be combined with the continuous mapping theorem or the delta-method to obtain further consequences. The delta-method for random processes is treated in Section 3.9.3 of van der Vaart and Wellner (1996). Here we only note two consequences based on the continuous mapping theorem. If $\psi: \mathfrak{L}_\infty(\mathcal{F}) \to \mathbb{R}$ is a continuous map, then this theorem shows that $\psi(\sqrt{n}(P - \mathbb{P}_n)) \rightsquigarrow \psi(W)$, for W the limiting Brownian bridge. If the distribution of $\psi(W)$ can be evaluated analytically, then the asymptotic weak limit of $\psi(\sqrt{n}(P - \mathbb{P}_n))$ is obtained, which may be useful in constructing approximate credible sets. The following corollary describes two such occasions, involving one-sided and two-sided Kolmogorov-Smirnov distances, whose limiting distributions, as distributions of the maximum of Brownian bridges and its absolute value process, respectively, are well known in the literature.

Corollary 12.4 *If $P \sim \mathrm{DP}(\alpha)$ on the sample space $\mathbb{R}$, and $X_1, \ldots, X_n | P \overset{iid}{\sim} P$, then for any $\lambda > 0$, a.s. $[P_0^\infty]$ as $n \to \infty$,*

(i) $\mathrm{P}(\sqrt{n} \sup_{x \in \mathbb{R}} |F(x) - \tilde{\mathbb{F}}_n(x)| > \lambda | X_1, \ldots, X_n) \to 2 \sum_{j=1}^\infty (-1)^{j+1} e^{-2j^2\lambda^2}$;

(ii) $\mathrm{P}(\sqrt{n} \sup_{x \in \mathbb{R}} (F(x) - \tilde{\mathbb{F}}_n(x)) > \lambda | X_1, \ldots, X_n) \to e^{-2\lambda^2}$.

12.2.1 Strong Approximation

The weak convergence $X_n \rightsquigarrow X$ of a sequence of random variables involves the distributions of these variables only, and not the underlying probability space(s) on which they are defined. By the *almost sure representation theorem* there always exist variables X_n^* and X^* with identical distributions that are defined on a single probability space and are such that

$X_n^* \to X^*$, almost surely.[1] As almost sure convergence is a stronger and simpler property than convergence in distribution, this result is sometimes helpful. Strong approximations can be viewed as strengthening this result, by giving a rate to the almost sure convergence.

A *strong approximation* at rate ϵ_n of a given sequence of weakly converging variables $X_n \rightsquigarrow X$ is a pair of two sequences X_n^* and $\tilde{X}_n$ of random elements, all defined on a single underlying probability space, such that $X_n^* =_d X_n$, $\tilde{X}_n =_d X$, for all $n \in \mathbb{N}$, and $d(X_n^*, \tilde{X}_n) = O(\epsilon_n)$ a.s.

Strong approximations are useful for deriving approximations to transformed variables $\psi_n(X_n)$, where the function ψ_n depends on n. In particular, if the functions ψ_n satisfy a Lipschitz condition with suitable Lipschitz constants L_n which can be related to the rate of approximation ϵ_n, then it may be possible to analyze the distribution of $\psi_n(X_n)$ through its strong approximation $\psi_n(X_n^*)$, which is equal in distribution. For instance, this technique is useful for approximating the distribution of a kernel smoother.

A classical strong approximation is the *KMT construction* (after Komlós, Major and Tusnády) to the empirical process. The empirical process $\sqrt{n}(\mathbb{F}_n - F)$ of a random sample of n variables from the cumulative distribution function F on the real line tends in distribution to a Brownian bridge process $B \circ F$, for B a standard Brownian bridge on $[0, 1]$ (a mean zero Gaussian process with covariance kernel $\mathrm{E}[B_s B_t] = s \wedge t - st$). The KMT construction provides a random sample of variables and a sequence of Brownian bridges $B_n =_d B$ on a suitable probability space such that

$$\|\sqrt{n}(\mathbb{F}_n - F) - B_n \circ F\|_\infty = O(n^{-1/2}(\log n)^2), \qquad \text{a.s.}$$

Thus the distance between the empirical process and its "limit" is nearly $n^{-1/2}$. It is known that this rate cannot be improved.

The Brownian bridges B_n can also be related amongst themselves, by tying them to a *Kiefer process*. This is a mean-zero Gaussian process $\mathbb{K} = \{\mathbb{K}(s, t)\colon (s, t) \in [0, 1] \times [0, \infty)\}$ with continuous sample paths and covariance kernel $\mathrm{E}[\mathbb{K}(s_1, t_1)\mathbb{K}(s_2, t_2)] = [s_1 \wedge s_2 - s_1 s_2](t_1 \wedge t_2)$. The process $s \mapsto t^{-1/2}\mathbb{K}(s, t)$ is a Brownian bridge, for every $t > 0$, and the Brownian bridges in the KMT theorem can be taken as $B_n = n^{-1/2}\mathbb{K}(\cdot, n)$, for every n, and a suitable Kiefer proces.

The following theorem gives an analogous result for the posterior process corresponding to a sample from the Dirichlet process.

Theorem 12.5 *On a suitable probability space there exist random elements F and $\mathbb{K}$ and $X_1, X_2, \ldots \overset{iid}{\sim} F_0$ such that $F | X_1, \ldots, X_n \sim \mathrm{DP}(\alpha + n\sum_{i=1}^n \delta_{X_i})$ for every n, and $\mathbb{K}$ is a Kiefer process independent of $X_1, X_2, \ldots$, and such that*

$$\sup_{x \in \mathbb{R}} \Big| \sqrt{n}(F - \mathbb{F}_n)(x) - \frac{\mathbb{K}(F_0(x), n)}{n^{1/2}} \Big| = O\Big(\frac{(\log n)^{1/2}(\log\log n)^{1/4}}{n^{1/4}} \Big), \qquad \text{a.s.}$$

The theorem remains true if the empirical distribution function $\mathbb{F}_n$ is replaced by the posterior mean $(\alpha + n\mathbb{F}_n)/(|\alpha| + n)$. Furthermore, the choice $\alpha = 0$, for which F follows the Bayesian bootstrap, is allowed. The strong approximation rate is close to $n^{-1/4}$, much

[1] See e.g. van der Vaart and Wellner (1996), Theorem 1.10.3, or 1.10.4 for a stronger version assuming less measurability.

weaker than the rate in the KMT construction for the empirical distribution function. For a proof, which is long as for all strong approximation results, see Lo (1987).

The theorem can be applied to obtain the limiting distribution of a smoothed Dirichlet posterior process. For a given kernel w and $F \sim \mathrm{DP}(\alpha + n\mathbb{F}_n)$, set

$$f_n(x) = \int \frac{1}{h_n} w\Bigl(\frac{x-\theta}{h_n}\Bigr)\, dF(\theta), \qquad \hat{f}_n(x) = \frac{1}{nh_n}\sum_{i=1}^n w\Bigl(\frac{x-X_i}{h_n}\Bigr).$$

The function $\hat{f}_n$ is the usual kernel density estimator, and serves as the centering in the following theorem. Assume that the kernel w integrates to 1, is symmetric about 0, and absolutely continuous on the convex hull of its support, with $\|w'\|_2 < \infty$ and $\int z^2 w(z)\, dz < \infty$ and $\int_3^\infty z^{3/2}(\log\log z)^{1/2}(|w'(z)| + |w(z)|)\, dz < \infty$.

Theorem 12.6 *If $X_1, X_2, \ldots \overset{iid}{\sim} F_0$ for a distribution F_0 with strictly positive, twice differentiable density f_0 such that f_0, $f_0'/f_0^{1/2}$ and f_0'' are bounded, and $F \mid X_1, \ldots, X_n \sim \mathrm{DP}(\alpha + n\sum_{i=1}^n \delta_{X_i})$ for every n, then, for $h_n = n^{-\delta}$ for some $0 < \delta < \frac{1}{2}$ and $n \to \infty$,*

$$\mathrm{P}\Bigl(\sqrt{2\log h_n^{-1}}\Bigl[\sup_{x\in\mathbb{R}} \frac{|f_n(x) - \hat{f}_n(x)|}{h_n f_0(x)^{1/2}\|w\|_2} - a_n\Bigr] \le t \mid X_1, X_2, \ldots\Bigr) \to e^{-2e^{-t}}, \qquad \text{a.s.,}$$

where $a_n = \sqrt{2\log h_n^{-1}} + 2\log(\|w'\|_2/\|w\|_2)) - \log\pi$.

The choice $\alpha = 0$, leading to the *smoothed Bayesian bootstrap process*, is allowed. Furthermore, the true density f_0 in the denominator may be replaced by the kernel estimator $\hat{f}_n$. The proof can be based on Theorem 12.5 and convergence to the extreme value distribution of the maximum of stationary Gaussian processes.

The quantile process of an absolutely continuous distribution also admits strong approximation by Kiefer processes. Under the conditions that the true density f_0 is differentiable and positive on its domain (a, b), and that the function $F_0(1 - F_0)|f_0'|/f_0^2$ is bounded, there exists a Kiefer process $\mathbb{K}$ and observations $X_1, X_2, \ldots$ on a suitable probability space such that, with $\delta_n = 25n^{-1}\log\log n$,

$$\sup_{\delta_n \le u \le 1-\delta_n} \Bigl|\sqrt{n} f_0 \circ F_0^{-1}(\mathbb{F}_n^{-1} - F_0^{-1})(u) - \frac{\mathbb{K}(u, n)}{\sqrt{n}}\Bigr| = O\Bigl(\frac{(\log n)^{1/2}(\log\log n)^{1/4}}{n^{1/4}}\Bigr), \text{ a.s.}$$

The restriction to the interval $[\delta_n, 1 - \delta_n]$ removes the left and right tails of the quantile process, where its population counterpart can be large. See Csörgő and Révész (1981), Theorem 6 for a proof.

There is a similar strong approximation to the Dirichlet posterior and the Bayesian bootstrap quantile functions.

Theorem 12.7 *Assume that the true density f_0 is differentiable and nonzero in its domain (a, b), and that the function $F_0(1 - F_0)|f_0'|/f_0^2$ is bounded. Then on a suitable probability space there exist random elements F and $\mathbb{K}$ and $X_1, X_2, \ldots \overset{iid}{\sim} F_0$ such that*

$F \mid X_1, \ldots, X_n \sim \mathrm{DP}(\alpha + n \sum_{i=1}^n \delta_{X_i})$ for every n, and $\mathbb{K}$ is a Kiefer process independent of $X_1, X_2, \ldots$, and such that

$$\sup_{\delta_n \le u \le 1-\delta_n} \Big| f_0 \circ F_0^{-1} \sqrt{n}(F^{-1} - \mathbb{F}_n^{-1})(u) - \frac{\mathbb{K}(u,n)}{\sqrt{n}} \Big| = O\Big(\frac{(\log n)^{1/2} (\log \log n)^{1/4}}{n^{1/4}} \Big), \text{ a.s.}$$

The proof of the result is based on the same techniques used to prove the corresponding result for the empirical quantile process and using Theorem 12.5 instead of the KMT theorem; see Gu and Ghosal (2008). For an application to approximating the Bayesian bootstrap distribution of the Receiver Operating Characteristic function, see Problem 12.4.

12.3 Semiparametric Models

A *semiparametric model* in the narrow sense is a set of densities $p_{\theta,\eta}$ parameterized by a pair of a Euclidean parameter θ and an infinite-dimensional *nuisance parameter* η, and the problem of most interest is to estimate θ. The parameterization is typically smooth in θ, and in case the nuisance parameter were known, the difficulty of estimating θ given a random sample of observations would be measured by the *score function* for θ:

$$\dot{\ell}_{\theta,\eta}(x) = \frac{\partial}{\partial \theta} \log p_{\theta,\eta}(x).$$

In particular, the covariance matrix of this function is the *Fisher information* matrix, whose inverse is a bound on the smallest asymptotic variance attainable by a (regular) estimator. When the nuisance parameter is unknown, estimating θ is a harder problem. The increased difficulty can be measured through the part of the score function for θ that can also be explained through *nuisance scores*. These are defined as the score functions

$$B_{\theta,\eta} b(x) := \frac{\partial}{\partial t} \log p_{\theta,\eta_t}(x)|_{t=0} \tag{12.2}$$

of suitable one-dimensional submodels $t \mapsto p_{\theta,\eta_t}$ with $\eta_0 = \eta$. If η ranges over an infinite-dimensional set, then there typically exist infinitely many submodels along which the derivative as in the display exists, and they can often be naturally identified by "directions" b in which η_t approaches η. The expression $B_{\theta,\eta} b$ on the left side of the display may be considered a notation only, but it was chosen to suggest an operator $B_{\theta,\eta}$ working on a direction b in which η_t approaches η. In many examples this *score operator* can be identified with a derivative of the map $\eta \mapsto \log p_{\theta,\eta}$. The linear span of $\dot{\ell}_{\theta,\eta}$ and all nuisance scores $B_{\theta,\eta} b$ is known as the *tangent space* of the model at the parameter (θ, η) (or at the density $p_{\theta,\eta}$).

For $\mathrm{Proj}_{\theta,\eta}$ the orthogonal projection in $\mathbb{L}_2(p_{\theta,\eta})$ onto the closed, linear span of all score functions $B_{\theta,\eta} b$ for the nuisance parameters, the *efficient score function* for θ is defined as

$$\tilde{\ell}_{\theta,\eta}(x) = \dot{\ell}_{\theta,\eta}(x) - \mathrm{Proj}_{\theta,\eta} \dot{\ell}_{\theta,\eta}(x).$$

The covariance matrix $\tilde{I}_{\theta,\eta} = P_{\theta,\eta} \tilde{\ell}_{\theta,\eta} \tilde{\ell}_{\theta,\eta}^\top$ of the efficient score function is known as the *efficient information* matrix. Being the covariance matrix of a projection, it is smaller than the ordinary information matrix, and hence possesses a larger inverse. This inverse can be shown to give a lower bound for estimators of θ in the situation that η is unknown. In fact, an estimator sequence $\hat{\theta}_n$ is considered to be asymptotically efficient at (θ, η) in the situation that η is unknown if

$$\sqrt{n}(\hat{\theta}_n - \theta) = \frac{1}{\sqrt{n}} \sum_{i=1}^{n} \tilde{I}_{\theta,\eta}^{-1} \tilde{\ell}_{\theta,\eta}(X_i) + o_{P_{\theta,\eta}}(1). \tag{12.3}$$

In many situations such efficient estimators have been constructed using estimating equations or variants of maximum likelihood estimators (see e.g. van der Vaart 1998, Chapter 25, for an overview). Posterior means for suitable priors could also play this role, as will be seen in the following.

As the efficient score function plays the same role in semiparametric models as the score function in ordinary parametric models, it is reasonable to expect that the Bernstein–von Mises theorem extends to semiparametric models, with ordinary score and information replaced by efficient score and information. In the semiparametric setting we equip both parameters θ and η with priors and given observations $X_1, \ldots, X_n$ form a posterior distribution for the joint parameter (θ, η) as usual. Since interest is in θ, we study the marginal posterior distribution induced on this parameter, written as $\Pi_n(\theta \in \cdot \mid X_1, \ldots, X_n)$. A semiparametric Bernstein–von Mises theorem should now assert that

$$\mathrm{E}_{\theta_0,\eta_0} \left\| \Pi_n(\theta \in \cdot \mid X_1, \ldots, X_n) - \mathrm{Nor}(\hat{\theta}_n, n^{-1} \tilde{I}_{\theta_0,\eta_0}^{-1}) \right\|_{TV} \to 0,$$

where $\hat{\theta}_n$ satisfies (12.3) at $(\theta, \eta) = (\theta_0, \eta)$. We derive sufficient conditions for this assertion below. A main interest is to identify the properties of priors on the nuisance parameter that lead to this property. In analogy with the parametric situation the prior on θ should wash out as $n \to \infty$, as long as it has a positive and continuous density in a neighborhood of the true value θ_0.

If the scores for the nuisance parameter are given as the range of an operator $B_{\theta,\eta}$, and this range is closed, then the projection operator can be found as $\mathrm{Proj}_{\theta,\eta} = B_{\theta,\eta}(B_{\theta,\eta}^{\top} B_{\theta,\eta})^{-1} B_{\theta,\eta}^{\top}$, for $B_{\theta,\eta}^{\tau}$ the adjoint of $B_{\theta,\eta}$. The projection of the score function $\dot{\ell}_{\theta,\eta}$ onto the linear span of the nuisance scores then takes the form $B_{\theta,\eta}\tilde{b}_{\theta,\eta}$, for

$$\tilde{b}_{\theta,\eta} = (B_{\theta,\eta}^{\top} B_{\theta,\eta})^{-1} B_{\theta,\eta}^{\top} \dot{\ell}_{\theta,\eta}.$$

This is known as the *least favorable direction*, and a corresponding submodel $t \mapsto \eta_t$ as a least-favorable submodel, because the submodel $t \mapsto p_{\theta+t,\eta_t}$ has the smallest information about t (at $t = 0$). In many situations the *information operator* $B_{\theta,\eta}^{\top} B_{\theta,\eta}$ is not invertible, and the preceding formulas are invalid. However, the projection of the θ-score onto the closed, linear span of the nuisance space always exists, and can be approximated by (linear combinations of) scores for the nuisance parameters.

The phrase "semiparametric estimation" is also attached more generally to estimating a Euclidean-valued functional on an infinite-dimensional model. This invites to consider the more general setup of densities p_η indexed by a single parameter η and a Euclidean parameter of interest $\chi(\eta)$. The models give rise to score functions $B_\eta b$ defined as in (12.2), but with θ removed throughout. The parameter χ is said to be *differentiable* at η if there exists a function $x \mapsto \tilde{\kappa}_\eta(x)$ such that, for every submodel $t \mapsto \eta_t$ as in (12.2) (which approaches $\eta = \eta_0$ from the "direction" b),

$$\chi(\eta_t) = \chi(\eta) + t P_\eta[\tilde\kappa_\eta\,(B_\eta b)] + o(t). \tag{12.4}$$

If this is true, then the quantity $P_\eta[\tilde\kappa_\eta\,(B_\eta b)]$ on the right side is the ordinary derivative of the map $t \mapsto \chi(\eta_t)$ at $t = 0$, but it is assumed that this derivative can be written as an inner product between the function $\tilde\kappa_\eta$ and the score function $B_\eta b$. (This representation must hold for efficient estimators for $\chi(\eta)$ to exist; see van der Vaart 1991.) The function $\tilde\kappa_\eta$ is unique only up to projection onto the closure of the tangent space; the unique projection is known as the *efficient influence function*. Efficient estimators $\hat\chi_n$ for $\chi(\eta)$ should satisfy

$$\sqrt{n}(\hat\chi_n - \chi(\eta)) = \frac{1}{\sqrt{n}}\sum_{i=1}^n \tilde\kappa_\eta(X_i) + o_{P_\eta}(1).$$

A comparison with (12.3) shows that $\tilde\kappa_\eta$ plays the same role as $\tilde I_{\theta,\eta}^{-1}\tilde\ell_{\theta,\eta}$ in the narrow semiparametric case, and hence the covariance matrix of $\tilde\kappa_\eta$ is comparable to the inverse of the efficient information matrix. With these substitutions one expects a Bernstein–von Mises theorem for a posterior distribution of $\chi(\eta)$ of the same type as before, with centering at an efficient estimator $\hat\chi_n$ and with the covariance matrix of $\tilde\kappa_\eta$ in place of the inverse information matrix:

$$\mathrm{E}_{\eta_0}\Big\|\Pi_n(\chi(\eta)\in\cdot\,|\,X_1,\ldots,X_n) - \mathrm{Nor}\Big(\hat\chi_n, n^{-1}P_{\eta_0}\big(\tilde\kappa_{\eta_0}\tilde\kappa_{\eta_0}^\top\big)\Big)\Big\|_{TV} \to 0. \tag{12.5}$$

A direction $\tilde b_\eta$ such that $B_\eta\tilde b_\eta = \tilde\kappa_\eta$ is the *least favorable direction*, in that if the parameter were known to belong to a one-dimensional model in this direction, then the influence function would still be $\tilde\kappa_\eta$. Hence statistical inference for χ in the submodel is "as hard" as it is in the full model. (A least-favorable direction may not exist, because the range of B_η need not be closed; the "supremum of the difficulty over the submodels" may not be assumed.)

The strict semiparametric setup can be incorporated in the general setup by replacing the parameter (θ,η) of the former setup by η and defining $\chi(\theta,\eta) = \theta$. It can be shown that the latter functional is differentiable as soon as the efficient influence function $\tilde\ell_{\theta,\eta}$ for θ exists, with efficient influence function $\tilde\kappa_{\theta,\eta} = \tilde I_{\theta,\eta}^{-1}\tilde\ell_{\theta,\eta}$ (see van der Vaart 1998, Lemma 25.25). Thus it suffices to work in the general setup.

12.3.1 Functionals

We shall give sufficient conditions for the Bernstein–von Mises theorem (12.5) in the general infinite-dimensional setup. As the conditions are in terms of the likelihood of the full observation, there is no advantage to restricting to the i.i.d. setup. Consider a general sequence of statistical models with observations $X^{(n)}$ following a jointly measurable density $p_\eta^{(n)}$ with respect to a σ-finite measure on some sample space, and a parameter η belonging to a Polish parameter set $\mathcal{H}$. For simplicity we take the functional of interest $\chi:\mathcal{H}\to\mathbb{R}$ real-valued.

The Bernstein–von Mises theorem is based on two approximations, which both involve approximately "least-favorable transformations." These are defined as arbitrary measurable maps $\eta\mapsto\tilde\eta_n(\eta)$ that are one-to-one on each set of the form $\{\eta\in\mathcal{H}_n:\chi(\eta)=\theta\}$, and satisfy the conditions (12.7) and (12.8) below. They should be thought of as approaching η approximately in the direction of a scaled version of the least-favorable direction $\tilde b_{\eta_0}$, as explained in Section 12.3.1 below. For instance, if $\mathcal{H}$ is embedded in a linear space, then they often take the form $\tilde\eta_n(\eta) = \eta - (\chi(\eta) - \theta_0)\tilde I_n \tilde b_n$, for $\tilde b_n$ approximations to the least-favorable direction $\tilde b_{\eta_0}$ at the true parameter value and $\tilde I_n$ the efficient information (the inverse of the efficient variance). The transformation (and particular its scaling) will typically be chosen such that the value $\chi(\tilde\eta_n(\eta))$ of the functional is close to the constant value $\chi(\eta_0)$, although this is not an explicit part of the requirements; only (12.7) and (12.8) need to be satisfied.

Assume that the statistical models are locally asymptotically normal along the approximately least favorable directions in the following sense: for $\theta_0 = \chi(\eta_0)$,

$$\log \frac{p_\eta^{(n)}}{p_{\tilde\eta_n(\eta)}^{(n)}}(X^{(n)}) = \sqrt{n}(\chi(\eta) - \theta_0)\,\tilde G_n - \tfrac{1}{2} n \tilde I_n |\chi(\eta) - \theta_0|^2 + R_n(\eta), \tag{12.6}$$

where $\tilde G_n$ is a tight sequence of random variables, $\tilde I_n$ are positive numbers that are bounded away from zero, and $(R_n(\eta){:}\ \eta \in \mathcal{H})$ are stochastic processes such that $\rho_{n,1} \to 0$, for

$$\rho_{n,1} = \sup_{\eta \in \mathcal{H}_n} \frac{|R_n(\eta)|}{1 + n|\chi(\eta) - \theta_0|^2}. \tag{12.7}$$

Here $\mathcal{H}_n$ are sets in the parameter space on which the posterior distribution concentrates with probability tending to one. The variables $\tilde G_n$ and numbers $\tilde I_n$ may depend on η_0, but not on η, suggesting that the sets $\mathcal{H}_n$ must shrink to η_0.

A second main condition concerns an invariance of the prior of the nuisance parameter under a shift in the direction of the least-favorable direction. Assume that the conditional distribution $\Pi_{n,\theta}$ of $\tilde\eta_n(\eta)$ given $\chi(\eta) = \theta$ under the prior is absolutely continuous relative to this distribution Π_{n,θ_0} at $\theta = \theta_0$, with density $d\Pi_{n,\theta}/d\Pi_{n,\theta_0}$ satisfying $\rho_{n,2} \to 0$, for

$$\rho_{n,2} = \sup_{\eta \in \mathcal{H}_n} \frac{|\log(d\Pi_{n,\chi(\eta)}/d\Pi_{n,\theta_0}(\eta))|}{1 + n|\chi(\eta) - \theta_0|^2}. \tag{12.8}$$

Theorem 12.8 (Semiparametric Bernstein–von Mises, functionals) *Suppose that there exist measurable sets $\mathcal{H}_n \subset \mathcal{H}$ and maps $\tilde\eta_n{:}\ \mathcal{H} \to \mathcal{H}$ that are one-to-one and bimeasurable on each set $\mathcal{H}_{n,\theta}{:}= \{\eta \in \mathcal{H}_n{:}\ \chi(\eta) = \theta\}$, equal to the identity on $\mathcal{H}_{n,\theta_0}$, for which* (12.7) *and* (12.8) *hold, and such that the sets $\Theta_n{:}= \{\chi(\eta){:}\ \eta \in \mathcal{H}_n\}$ shrink to $\theta_0 = \chi(\eta_0)$ with $\sqrt{n}(\Theta_n - \theta_0) \to \mathbb{R}$ and $\Pi_n(\eta \in \mathcal{H}_n|\, X^{(n)}) \to 1$ and $\inf_{\theta \in \Theta_n} \Pi_n(\eta \in \tilde\eta_n(\mathcal{H}_{n,\theta})|\, X^{(n)}, \chi(\eta) = \theta_0) \to 1$. If the induced prior for $\chi(\eta)$ possesses a positive, continuous Lebesgue density in a neighborhood of $\chi(\eta_0)$, then*

$$\mathrm{E}_{\eta_0} \Big\| \Pi_n(\chi(\eta) \in \cdot\,|\, X^{(n)}) - \mathrm{Nor}\Big(\chi(\eta_0) + \tilde I_n^{-1} \tilde G_n, n^{-1} \tilde I_n^{-1}\Big) \Big\|_{TV} \to 0.$$

Proof Because the posterior probability that $\eta \in \mathcal{H}_n$ tends to one by assumption, it suffices to establish the normal approximation to $\Pi_n(\chi(\eta) \in \cdot\,|\, X^{(n)}, \eta \in \mathcal{H}_n)$. In particular, we may assume that $\chi(\eta)$ belongs (with posterior probability tending to one) to the sets Θ_n. Let $\Pi(d\theta)$ denote the marginal law of $\chi(\eta)$ and let $\Pi_\theta(d\eta)$ denote the conditional law of η given $\chi(\eta) = \theta$, under the prior law of η. By (12.6), for any Borel set B,

$$\Pi_n(\chi(\eta) \in B|\, X^{(n)}, \eta \in \mathcal{H}_n) = \frac{\int_{B \cap \Theta_n} e^{\sqrt{n}(\theta - \theta_0)\tilde G_n - \frac{1}{2} n \tilde I_n |\theta - \theta_0|^2}\, Q_n(\theta)\, \Pi(d\theta)}{\int_{\Theta_n} e^{\sqrt{n}(\theta - \theta_0)\tilde G_n - \frac{1}{2} n \tilde I_n |\theta - \theta_0|^2}\, Q_n(\theta)\, \Pi(d\theta)},$$

where Q_n is defined as

$$Q_n(\theta) = \int_{\mathcal{H}_n} e^{R_n(\eta)} p^{(n)}_{\tilde\eta_n(\eta)}(X^{(n)})\, \Pi_\theta(d\eta).$$

The essential part of the proof is to show that Q_n is asymptotically independent of θ. If it were free of θ for any n, then it could be canceled out from the expression for the marginal posterior distribution, and the remaining expression would be the posterior distribution for a one-dimensional Gaussian location model, and hence satisfy the Bernstein–von Mises theorem. We show below that this is approximately true in that there exist constants $\rho_n \to 0$ such that, for $\theta \in \Theta_n$, with probability tending to one,

$$e^{-\rho_n(1+n|\theta-\theta_0|^2)} \le \frac{Q_n(\theta)}{Q_n(\theta_0)} \le e^{\rho_n(1+n|\theta-\theta_0|^2)}. \tag{12.9}$$

Substituting this in the preceding display, and using that the restriction of $\Pi(d\theta)$ to a neighborhood of Θ_n possesses a Lebesgue density that is bounded below and above by positive constants, which can be taken arbitrarily close if Θ_n is sufficiently small, we see that $\Pi_n(\chi(\eta) \in B|\, X^{(n)}, \eta \in \mathcal{H}_n)$ is lower and upper bounded by

$$(1+o(1))e^{\pm\rho_n} \frac{\int_{B\cap\Theta_n} e^{\sqrt{n}(\theta-\theta_0)\tilde G_n - \frac{1}{2}n(\tilde I_n \mp \rho_n)|\theta-\theta_0|^2}\, d\theta}{\int_{\Theta_n} e^{\sqrt{n}(\theta-\theta_0)\tilde G_n - \frac{1}{2}n(\tilde I_n \pm \rho_n)|\theta-\theta_0|^2}\, d\theta}.$$

From the assumption that the neighborhoods Θ_n shrink to θ_0 at slower rate than $1/\sqrt{n}$, it can be seen that replacing them by $\mathbb{R}$ multiplies the numerator and denominator by $1+o(1)$ terms. The resulting expression can be seen to be asymptotically equivalent to $\mathrm{Nor}(I_n^{-1}\tilde G_n, n^{-1}\tilde I_n^{-1})(B-\theta_0)$, uniformly in B, by explicit calculation of the integrals.

It remains to establish (12.9). By (12.7) the term $e^{R_n(\eta)}$ in the definition of $Q_n(\theta)$ can be lower and upper bounded by $e^{\pm\rho_{n,1}(1+n|\theta-\theta_0|^2)}$. Next, by the assumption that $\tilde\eta_n$ is one-to-one on the set $\mathcal{H}_{n,\theta}$, we have that $\eta \in \mathcal{H}_n$ and $\chi(\eta) = \theta$ if and only if $\tilde\eta_n(\eta) \in \tilde\eta_n(\mathcal{H}_{n,\theta})$ and $\chi(\eta) = \theta$. Therefore by making the substitution $\eta \mapsto \tilde\eta_n(\eta)$ in the integral, we see that $Q_n(\theta)$ is lower and upper bounded by $e^{\pm\rho_{n,1}(1+n|\theta-\theta_0|^2)}\tilde Q_n(\theta)$, for $\tilde Q_n(\theta)$ defined by

$$\tilde Q_n(\theta) = \int_{\tilde\eta_n(\mathcal{H}_{n,\theta})} p^{(n)}_\eta(X^{(n)})\, \Pi_{n,\theta}(d\eta),$$

where $\Pi_{n,\theta}$ is the law of $\tilde\eta_n(\eta)$ if $\eta \sim \Pi_\theta$. This is the same law as in the statement of the theorem, so that we can use assumption (12.8) to change $\Pi_{n,\theta}$ to $\Pi_{n,\theta_0} = \Pi_{\theta_0}$, at the cost of inserting a further multiplicative factor $e^{\pm\rho_{n,2}(1+n|\theta-\theta_0|^2)}$ in the lower and upper bounds. The resulting expression is the numerator of the posterior probability

$$\Pi_n(\eta \in \tilde\eta_n(\mathcal{H}_{n,\theta})|\, X^{(n)}, \chi(\eta) = \theta_0) = \frac{\int_{\tilde\eta_n(\mathcal{H}_{n,\theta})} p^{(n)}_\eta(X^{(n)})\, \Pi_{\theta_0}(d\eta)}{\int p^{(n)}_\eta(X^{(n)})\, \Pi_{\theta_0}(d\eta)}.$$

Since this posterior probability tends to 1 by assumption, it is between $e^{-\rho_{n,3}}$ and 1 for some $\rho_{n,3} \to 0$, uniformly in $\theta \in \Theta_n$. In other words, with probability tending to one, $\tilde Q_n(\theta)/\tilde Q_n(\theta_0)$ is bounded below and above by $e^{\pm\rho_{n,2}(1+n|\theta-\theta_0|^2)}e^{\pm\rho_{n,3}}$.

This concludes the proof of claim (12.9), with $\rho_n = \sum_i \rho_{n,i}$. ∎

Although this is not included in the condition, for a true Bernstein–von Mises theorem the variables $\tilde{G}_n$ ought to be asymptotically normally distributed with mean zero and variance (the limit of) $\tilde{I}_n$. This is typically the case (and almost implied by the LAN expansion (12.6)).

The sieves $\mathcal{H}_n$ in the theorem are meant to make the expansions and approximations (12.7) and (12.8) possible, which favors small sieves, but must asymptotically contain posterior mass one. Typically they would be neighborhoods that shrink to the true parameter η_0 at the rate of contraction of the posterior distribution. Then an application of the theorem would be preceded by a derivation of a rate of contraction of the posterior distribution, possibly by the methods of Chapter 8.

In the case of i.i.d. observations, the LAN expansion (12.7) is implied by the pair of approximations, with $\mathbb{G}_n = \sqrt{n}(\mathbb{P}_n - P_{\eta_0})$ the empirical process of the observations and $\tilde{G}_n = \mathbb{G}_n(\tilde{\ell}_n)$: for any $\hat{\eta} \in \mathcal{H}_n$ and $\hat{\theta} = \chi(\hat{\eta})$,

$$\mathbb{G}_n\big[\log p_{\hat{\eta}} - \log p_{\tilde{\eta}_n(\hat{\eta})} - (\hat{\theta} - \theta_0)\tilde{\ell}_{\eta_0}\big] = o_P(\hat{\theta} - \theta_0),$$
$$P_{\eta_0}\big[\log p_{\hat{\eta}} - \log p_{\tilde{\eta}_n(\hat{\eta})}\big] = -\tfrac{1}{2}\tilde{I}_{\eta_0}(\hat{\theta} - \theta_0)^2(1 + o_P(1)) + o_P(|\hat{\theta} - \theta_0|n^{-1/2}).$$

The first can be verified by showing that the functions within square brackets divided by $\hat{\theta} - \theta_0$ are with probability tending to one contained in a Donsker class, and have second moments tending to zero. Typically the function $\tilde{\ell}_\eta$ will be the efficient score function (the efficient influence function $\tilde{\kappa}_\eta$ divided by its variance) at $\eta = \eta_0$. The second condition is ostensibly a Taylor expansion to the second order. However, because the expectation is relative to the true parameter η_0 and the difference on the left is a difference between perturbations of $\hat{\eta}$, the linear term in the expansion does not necessarily vanish, but takes the form $(\hat{\theta} - \theta_0)P_{\eta_0}\tilde{\ell}_{\hat{\eta}}$. Then the fulfilment of the condition seems to demand that $P_{\eta_0}\tilde{\ell}_{\hat{\eta}} = o_P(n^{-1/2})$. In general this requires a rate for the convergence of $\tilde{\ell}_{\hat{\eta}}$ to the efficient influence function. This is similar to the "no-bias" condition found in the analysis of semiparametric maximum likelihood estimators (see Murphy and van der Vaart 2000 or equation (25.75) and the ensuing discussion in van der Vaart 1998). Analogous reasoning suggests that the bias is quadratic in the discrepancy and is negligible if the rate of contraction is at least $o_P(n^{-1/4})$. If not, then the prior may create a bias and the Bernstein–von Mises theorem may not hold.

Full LAN Expansion

The conditions of the preceding theorem can alternatively be phrased in terms of a LAN expansion of the full likelihood. This is more involved, but makes it evident that for the condition to be satisfied the transformations $\eta \mapsto \tilde{\eta}_n(\eta)$ in (12.6) must be approximately in the least favorable direction and $\tilde{I}_n$ the efficient information.

Suppose that the linear space $\mathbb{B}_0$ that encompasses the parameter set $\mathcal{H}_n$ is a Hilbert space, with inner product $\langle\cdot,\cdot\rangle_0$ and norm $\|\cdot\|_0$, and make two assumptions:

(i) There exist elements $\tilde{b}_n \in \mathbb{B}_0$ such that

$$\sup_{\substack{\eta\in\mathcal{H}_n\\ \theta=\chi(\eta),\theta_0=\chi(\eta_0)}} \frac{n\big|\langle \eta - \eta_0 - (\theta - \theta_0)\tilde{b}_n, (\theta - \theta_0)\tilde{b}_n\rangle_0\big|}{1 + n|\theta - \theta_0|^2} \to 0. \tag{12.10}$$

(ii) The full statistical model is LAN in the sense that

$$\log \frac{p_\eta^{(n)}}{p_{\eta_0}^{(n)}}(X^{(n)}) = \sqrt{n}\,\mathbb{G}_n(\eta-\eta_0) - \tfrac{1}{2}n\|\eta-\eta_0\|_0^2 + R_n(\eta),$$

for linear stochastic processes $(\mathbb{G}_n(b)\colon b\in\mathbb{B}_0)$, and stochastic processes $(R_n(\eta)\colon \eta\in\mathcal{H})$ such that

$$\sup_{\substack{\eta\in\mathcal{H}_n\\ \theta=\chi(\eta),\theta_0=\chi(\eta_0)}} \frac{\bigl|R_n(\eta)-R_n(\eta-(\theta-\theta_0)\tilde{b}_n)\bigr|}{1+n|\theta-\theta_0|^2}\to 0. \tag{12.11}$$

Assumptions (i)–(ii) can be seen to imply (12.6), with the transformations $\tilde{\eta}_n(\theta)=\eta-(\chi(\eta)-\theta_0)\tilde{b}_n$, the variables $\tilde{G}_n=\mathbb{G}_n(\tilde{b}_n)$ and $\tilde{I}_n=\|\tilde{b}_n\|_0^2$, and the remainder $R_n(\eta)$ of (12.6) equal to the sum of twice the numerator of (12.10) and the present $R_n(\eta)-R_n(\tilde{\eta}_n(\eta))$. That condition (12.11) is placed on the latter difference of two remainders makes the two conditions in fact almost equivalent, apart from the introduction of the Hilbert space structure and the linearity of the transformation and the process $\mathbb{G}_n$ in (12.10)–(12.11).

Relation (12.10) is certainly satisfied if the inner product in its numerator vanishes, for all η. This is the case if $(\theta-\theta_0)\tilde{b}_n$ is the orthogonal projection of $\eta-\eta_0$ onto the one-dimensional linear space spanned by $\tilde{b}_n$, i.e. $\theta-\theta_0=\langle\eta-\eta_0,\tilde{b}_n\rangle_0/\tilde{I}_n$. We shall give a heuristic argument that this implies that $\tilde{b}_n$ must be close to a least favorable direction.

The Hilbert space $(\mathbb{B}_0,\langle\cdot,\cdot\rangle_0)$ would typically derive from the information structure of the statistical model at the true parameter value η_0. In the general notation of the introduction to this section,

$$\langle a,b\rangle_0 = P_{\eta_0}\bigl[(B_{\eta_0}a)(B_{\eta_0}b)\bigr].$$

If the efficient influence function can be written in the form $\tilde{\kappa}_{\eta_0}=B_{\eta_0}\tilde{b}_{\eta_0}$, where by definition $\tilde{b}_{\eta_0}$ is a least favorable direction, then the differentiability (12.4) of the functional χ at η_0 gives, for a path η_t that approaches η_0 in the "direction" b,

$$\chi(\eta_t)=\chi(\eta_0)+t\,P_{\eta_0}[\tilde{\kappa}_{\eta_0}(B_{\eta_0}b)]+o(t)=\chi(\eta_0)+t\langle\tilde{b}_{\eta_0},b\rangle_0+o(t).$$

Thus $\tilde{b}_{\eta_0}$ is a "derivative" of χ at η_0 relative to the information inner product $\langle\cdot,\cdot\rangle_0$, and informally we have that $\chi(\eta)-\chi(\eta_0)\doteq\langle\tilde{b}_{\eta_0},\eta-\eta_0\rangle_0$. If this were true exactly, and $\eta-\eta_0-(\chi(\eta)-\theta_0)\tilde{b}_n$ would be orthogonal to $\tilde{b}_n$, for every η, then it can be seen that $\tilde{b}_n=\tilde{b}_{\eta_0}/\|\tilde{b}_{\eta_0}\|_0^2$ (see Problem 12.8).

If an exact least favorable direction exists, then assumption (12.10) can also be formulated in terms of the distance between $\tilde{b}_n$ and this direction. This gives the alternative condition:

(i′) there exists $\tilde{b}_0\in\mathbb{B}_0$ such that $\langle\eta-\eta_0-(\theta-\theta_0)\tilde{b}_0,\tilde{b}_0\rangle_0=0$ and

$$\sqrt{n}\|\tilde{b}_n-\tilde{b}_0\|_0\sup_{\eta\in\mathcal{H}_n}\|\eta-\eta_0\|_0\to 0,\qquad\text{and}\qquad \|\tilde{b}_n-\tilde{b}_0\|_0\to 0. \tag{12.12}$$

That this condition is stronger than (12.10) can be seen by comparing the numerator of (12.10) with its zero value when $\tilde{b}_n$ is replaced by $\tilde{b}_0$ and using the inequality $\bigl|\langle a,b\rangle_0-\langle a',b'\rangle_0\bigr|\le\|a-a'\|_0\|b\|_0+\|b-b'\|_0\|a'\|_0$, for any directions a,b. The function $\tilde{b}_0$ will be a scaled version $\tilde{b}_0=\tilde{b}_{\eta_0}/\|\tilde{b}_{\eta_0}\|_0^2$ of the least favorable direction $\tilde{b}_{\eta_0}$ encountered previously.

12.3.2 Strict Semiparametric Model

A semiparametric model in the narrow sense is indexed by a partitioned parameter (θ, η), and interest is in the functional $\chi(\theta, \eta) = \theta$. Score functions take the form $a\dot\ell_{\theta,\eta} + B_{\theta,\eta}b$, and are also indexed by pairs (a, b) of "directions." A least favorable direction, if it exists, is one that gives the efficient score function $\tilde\ell_{\theta,\eta}$ as score function, and is given by a direction of the form $(1, -\tilde b_{\theta,\eta})$, yielding the score $\dot\ell_{\theta,\eta} - B_{\theta,\eta}\tilde b_{\theta,\eta}$. For a linear parameter space this motivates to choose the transformation of η of the form $\eta + (\theta - \theta_0)\tilde b_{\theta_0,\eta_0}$, but it may be profitable to replace $\tilde b_{\theta_0,\eta_0}$ by an approximation. (The plus sign arises because in the likelihood ratio (12.13) the transformation on η is inserted in the denominator, whereas θ is perturbed in the numerator.)

We choose the least favorable transformation as in (12.7) of a parameter (θ, η) to take the functional value θ_0, i.e. we use a transformation of the form $(\theta, \eta) \mapsto (\theta_0, \tilde\eta_n(\theta, \eta))$. Condition (12.7) then requires that there exists a tight sequence of variables $\tilde G_n$ and positive numbers $\tilde I_n$ so that the remainder

$$\log \frac{p^{(n)}_{\theta,\eta}}{p^{(n)}_{\theta_0,\tilde\eta_n(\theta,\eta)}}(X^{(n)}) = \sqrt{n}(\theta - \theta_0)\tilde G_n - \tfrac{1}{2}n\tilde I_n|\theta - \theta_0|^2 + R_n(\theta, \eta)$$

to the quadratic expansion of the log likelihood satisfies

$$\sup_{\substack{\theta\in\Theta_n\\ \eta\in\mathcal{H}_n}} \frac{R_n(\theta, \eta)}{1 + n|\theta - \theta_0|^2} \to 0. \tag{12.13}$$

The variables $\tilde G_n$ and numbers $\tilde I_n$ may depend on (θ_0, η_0), but not on (θ, η). The numbers $\tilde I_n$ must be bounded away from zero.

It is natural to choose the two parameters θ and η independent under the prior. Then the measure $\Pi_{n,\theta}$ in (12.8), the distribution of $(\theta_0, \tilde\eta_n(\theta, \eta))$ given that $\chi(\theta, \eta) = \theta$,[2] is a product of the Dirac measure at θ_0 times the law of $\tilde\eta_n(\theta, \eta)$ under the prior on η, for fixed θ. The Dirac factors cancel and the condition reduces to exactly the same condition, but with $\Pi_{n,\theta}$ the distribution of $\tilde\eta_n(\theta, \eta)$. Assume that $\rho_{n,2} \to 0$, for

$$\rho_{n,2} = \sup_{\substack{\theta\in\Theta_n\\ \eta\in\mathcal{H}_n}} \frac{|\log(d\Pi_{n,\theta}/d\Pi_{n,\theta_0}(\eta))|}{1 + n|\theta - \theta_0|^2}. \tag{12.14}$$

In these conditions Θ_n and $\mathcal{H}_n$ may be arbitrary measurable sets of parameters, such that the posterior distribution for (θ, η) concentrates on $\Theta_n \times \mathcal{H}_n$ with probability tending to one. The sets Θ_n are also assumed to shrink at θ_0 at a rate such that $\sqrt{n}(\Theta_n - \theta_0)$ increases to $\mathbb{R}$.

Theorem 12.8 specializes to this setup in the following form. Let $\tilde\eta_n(\theta, \mathcal{H}_n)$ be the set of all parameters $\tilde\eta_n(\theta, \eta)$ as η ranges over $\mathcal{H}_n$.

Theorem 12.9 (Semiparametric Bernstein–von Mises) *Suppose that there exist measurable sets Θ_n and $\mathcal{H}_n$ for every θ a one-to-one bimeasurable map $\eta \mapsto \tilde\eta_n(\theta, \eta)$ for which* (12.13) *and* (12.14) *hold, for $\Pi_{n,\theta}$ the prior distribution of $\tilde\eta_n(\theta, \eta)$ given fixed θ, and such*

[2] The notation is awkward; the last appearance of θ is a fixed value, the first two appearances in this assertion refer to the prior random variable θ.

that $\Pi_n(\theta \in \Theta_n, \eta \in \mathcal{H}_n |\, X^{(n)}) \to 1$ *and* $\inf_{\theta\in\Theta_n} \Pi_n(\eta \in \tilde{\eta}_n(\theta, \mathcal{H}_n) |\, X^{(n)}, \theta = \theta_0) \to 1$. *If the prior for* θ *possesses a continuous, positive Lebesgue density in a neighborhood of* θ_0, *then*

$$\mathrm{E}_{\theta_0,\eta_0} \Big\| \Pi_n(\theta \in \cdot \,|\, X^{(n)}) - \mathrm{Nor}(\theta_0 + \tilde{I}_n^{-1}\tilde{G}_n, n^{-1}\tilde{I}_n^{-1}) \Big\|_{TV} \to 0.$$

Example 12.10 (Adaptive model) A strict semiparametric model is called *adaptive* if there is no loss of information in not knowing the nuisance parameter. In such a model the efficient score function $\tilde{\ell}_{\theta,\eta}$ coincides with the ordinary score function $\dot{\ell}_{\theta,\eta}$ and the least-favorable direction $\tilde{b}_{\theta,\eta}$ in the nuisance parameter space is equal to zero.

For $\tilde{\eta}_n(\theta, \eta) = \eta$ condition (12.14) is trivially satisfied, and (12.13) simplifies into an ordinary LAN condition on the parametric models $\theta \mapsto p_{\theta,\eta}$, uniformly in $\eta \in \mathcal{H}_n$, where the centering variables $\tilde{G}_n$ may *not* depend on η. The latter suggests that the sets $\mathcal{H}_n$ must shrink to a point as $n \to \infty$.

Example 12.11 (Gaussian prior) A shifted Gaussian prior is absolutely continuous relative to the original Gaussian process if and only if the shift belongs to its reproducing kernel Hilbert space. In that case the Cameron-Martin theorem (see Lemma I.20) gives an explicit description of the density. This shows where to choose the approximate least favorable directions $\tilde{b}_n$ and enables to verify (12.14) for a Gaussian prior on η.

If η is a centered Gaussian random element in a separable Banach space and $\tilde{b}_n$ is in its RKHS $\mathbb{H}$, then the law $\Pi_{n,\theta}$ of $\eta + (\theta - \theta_0)\tilde{b}_n$ has log-density

$$\log \frac{d\Pi_{n,\theta}}{d\Pi_{n,\theta_0}}(\eta) = (\theta - \theta_0) U(\tilde{b}_n, \eta) - \tfrac{1}{2}|\theta - \theta_0|^2 \|\tilde{b}_n\|_{\mathbb{H}}^2,$$

where $U(\tilde{b}_n, \eta)$ is a centered Gaussian random variable with variance $\|\tilde{b}_n\|_{\mathbb{H}}^2$, under $\Pi_{n,\theta_0} = \Pi$. If we set $\mathcal{H}_n = \{\eta \in \mathcal{H}: |U(\tilde{b}_n, \eta)| \leq 2\sqrt{n}\epsilon_n \|\tilde{b}_n\|_{\mathbb{H}}\}$, then

$$\sup_{\substack{\eta\in\mathcal{H}_n\\ \theta=\chi(\eta)}} \frac{|\log(d\Pi_{n,\theta}/d\Pi_{n,\theta_0}(\eta))|}{1 + n|\theta - \theta_0|^2} \leq 2\epsilon_n \|\tilde{b}_n\|_{\mathbb{H}} + \frac{1}{n}\|\tilde{b}_n\|_{\mathbb{H}}^2.$$

Thus (12.14) is satisfied provided $\epsilon_n \|\tilde{b}_n\|_{\mathbb{H}} \to 0$, for some $\epsilon_n \gg n^{-1/2}$. To ensure that the sieve $\mathcal{H}_n$ has posterior probability tending to one, we choose ϵ_n large enough, so that the prior mass of $\mathcal{H}_n^c$ is small enough. Since $U(\tilde{b}_n, \eta)$ is Gaussian, the tail bound for the normal distribution gives $\Pi(\mathcal{H}_n^c) \leq e^{-2n\epsilon_n^2}/(\sqrt{n}\epsilon_n)$. Then Theorem 8.20 shows that the posterior mass of $\mathcal{H}_n$ tends to one if

$$\Pi\Big((\theta, \eta)\colon K(p_{\theta_0,\eta_0}^{(n)}; p_{\theta,\eta}^{(n)}) \leq n\epsilon_n^2, V_{2,0}(p_{\theta_0,\eta_0}^{(n)}; p_{\theta,\eta}^{(n)}) \leq n\epsilon_n^2\Big) \geq e^{-n\epsilon_n^2}.$$

Often this will be satisfied for ϵ_n the rate of contraction for the full densities, which in turn will typically be determined by the concentration function of the Gaussian prior at η_0.

In the situation that there exists an exact least-favorable direction $\tilde{b}_0$, the resulting rate $\epsilon_n \|\tilde{b}_n\|_{\mathbb{H}} \to 0$ must be traded to the rate of approximation of $\tilde{b}_n$ to $\tilde{b}_0$, as measured in (12.12). For a Gaussian prior this rate can be bounded by the concentration function $\varphi_{\tilde{b}_0}(\cdot)$ given in (11.11) relative to the information norm: for any δ_n there exists $\tilde{b}_n$ with $\|\tilde{b}_n - \tilde{b}_0\|_0 \leq \delta_n$ and $\|\tilde{b}_n\|_{\mathbb{H}}^2 \leq \varphi_{\tilde{b}_0}(\delta_n)$. Then the pair of conditions (12.12) and (12.14) are satisfied if, for

ϵ_n as in the preceding display and $\epsilon_{n,0}$ the rate of posterior concentration relative to the information metric $\|\cdot\|_0$, and some δ_n,

$$\epsilon_n^2 \varphi_{\tilde{b}_0}(\delta_n) \to 0 \qquad \text{and} \qquad \sqrt{n}\delta_n \epsilon_{n,0} \to 0. \tag{12.15}$$

In nice cases the rates ϵ_n and $\epsilon_{n,0}$ will agree, and be given by the rate equation $\varphi_{\eta_0}(\epsilon_n) \asymp n\epsilon_n^2$. This leads to the condition that there exists a solution $\delta_n \downarrow 0$ to

$$\varphi_{\eta_0}(\epsilon_n) \asymp n\epsilon_n^2 \quad \text{and} \quad \epsilon_n^2 \varphi_{\tilde{b}_0}(\delta_n) \to 0 \quad \text{and} \quad \delta_n^2 \varphi_{\eta_0}(\epsilon_n) \to 0.$$

The two moduli in this display will typically depend on the "regularity" of the functions η_0 and b_0. Here the true parameter η_0 is a given function of a given "regularity," but the least-favorable direction $\tilde{b}_0$ depends on the surrounding parameters in the full "model," as defined implicitly by the prior. This makes the final message opaque in general. Fine properties of the prior and model may matter.

12.3.3 Cox Proportional Hazard Model

In the Cox model under random right censoring (also see Section 13.6) we observe a random sample of observations distributed as (T, Δ, Z), for $T = X \wedge C$ the minimum of a survival time X and a censoring time C, an indicator $\Delta = \mathbb{1}\{X \leq C\}$ that records if the observation is censored ($\Delta = 0$) or not, and a covariate variable Z. For simplicity we assume that the latter variable is univariate and takes its values in a compact interval in the real line. The survival and censoring times X and C are assumed to be conditionally independent given Z, and the Cox model postulates the conditional hazard function of X given Z to take the form

$$h(x|\,Z) = e^{\theta Z} h(x),$$

for an unknown *baseline hazard* function h, and an unknown real-valued parameter θ. (Section 13.1 gives a brief introduction to survival analysis, including the definition (13.2) of a hazard function.)

For H the cumulative hazard function corresponding to h, a density of (T, Δ, Z) relative to a suitable dominating measure is given by

$$\Big(e^{\theta z} h(t) e^{-e^{\theta z} H(t)} \bar{F}_{C|Z}(t-|\,z)\Big)^{\delta} \Big(e^{-e^{\theta z} H(t)} f_{C|Z}(t|\,z)\Big)^{1-\delta} p_Z(z). \tag{12.16}$$

The terms involving the conditional distribution of C given Z and the marginal distribution of Z factor out of this likelihood seen as a function of (θ, h), and can be ignored for inference on (θ, h) if these parameters are chosen a priori independent. Thus we equip only θ and h with priors, which we shall choose independent also. We shall further assume that the censoring distribution is supported on a compact interval $[0, \tau]$, with a positive atom at its right end point τ. (The atom simplifies the technical arguments, but is natural too: it corresponds to ending the study at the finite time τ.) Then the hazard functions h can be restricted to $[0, \tau]$, and the prior on h will be a prior on integrable functions $h\colon [0, \tau] \to [0, \infty)$.

The score function for θ takes the form

$$\dot{\ell}_{\theta,h}(t, \delta, z) = \delta z - z e^{\theta z} H(t).$$

For any measurable function $b\colon [0,\tau] \to \mathbb{R}$, the path defined by $h_u = he^{ub}$ defines a submodel passing through h at $u = 0$. Its score function at $u = 0$ takes the form

$$B_{\theta,h}b(t,\delta,z) = \delta b(t) - e^{\theta z}\int_{[0,t]} b\,dH.$$

This can be seen to be a mapping $B_{\theta,h}\colon \mathbb{L}_2(h) \to \mathbb{L}_2(p_{\theta,h})$, and hence it has an adjoint mapping $B_{\theta,h}^\top\colon \mathbb{L}_2(p_{\theta,h}) \to \mathbb{L}_2(h)$. For continuous H it can be shown that (for details on this and subsequent formulas, see e.g. van der Vaart 1998, Section 25.12.1)

$$B_{\theta,h}^\top B_{\theta,h}h(t) = h(t)\,M_{0,\theta,h}(t), \qquad B_{\theta,h}^\top \dot\ell_{\theta,h}(t) = M_{1,\theta,h}(t),$$

where

$$M_{k,\theta,h}(t) = \mathrm{E}_{\theta,h}\mathbb{1}_{T\geq t}Z^k e^{\theta Z}, \qquad k = 0, 1, \ldots$$

Thus the information operator $B_{\theta,h}^\top B_{\theta,h}$ is simply a multiplication by the function $M_{0,\theta,h}$. If this function is bounded away from zero on its domain, then the operator is invertible, with the inverse being division by the same function. By the general semiparametric theory in the introduction of the section a least-favorable direction is given by

$$\tilde b_{\theta,h}(t) = (B_{\theta,h}^\top B_{\theta,h})^{-1}B_{\theta,h}^\top \dot\ell_{\theta,h}(t) = \frac{M_{1,\theta,h}}{M_{0,\theta,h}}(t).$$

The efficient score function takes the form $\tilde\ell_{\theta,h} = \dot\ell_{\theta,h} - B_{\theta,h}\tilde b_{\theta,h}$, and is explicitly given as

$$\tilde\ell_{\theta,h}(t,\delta,z) = \delta\Big(z - \frac{M_{1,\theta,h}}{M_{0,\theta,h}}(t)\Big) - e^{\theta z}\int_{[0,t]}\Big(z - \frac{M_{1,\theta,h}}{M_{0,\theta,h}}(s)\Big)\,dH(s).$$

The square information norm at parameter (θ, h) is

$$\begin{aligned}\|(a,b)\|_{\theta,h}^2 &= a^2 I_{\theta,h} + 2a\langle\dot\ell_{\theta,h}, B_{\theta,h}b\rangle_{\theta,h} + \langle B_{\theta,h}b, B_{\theta,h}b\rangle_{\theta,h} \qquad (12.17)\\ &= \int_{[0,\tau]}(a^2 M_{2,\theta,h}(t) + 2a\,b(t)M_{1,\theta,h}(t) + b^2(t)M_{0,\theta,h}(t))\,dH(t).\end{aligned}$$

Finally, the efficient information for θ can be computed as

$$\tilde I_{\theta,h} = \mathrm{E}_Z\int\Big(Z - \frac{M_{1,\theta,h}}{M_{0,\theta,h}}(t)\Big)^2\,\bar F_{C|Z}(t-|\,Z)\,e^{\theta Z}e^{-e^{\theta Z}H(t)}\,dH(t).$$

This is positive as soon as Z is not almost surely equal to a function of T.

We consider a Gaussian prior on the function $\eta = \log h\colon [0,\tau] \to \mathbb{R}$. Let $\varphi_w(\epsilon)$ be its concentration function (11.11) at w relative to the uniform norm on functions on $[0,\tau]$.

Theorem 12.12 *Suppose that* $\mathrm{P}(C \geq \tau) = \mathrm{P}(C = \tau) > 0$, *that the conditional distribution of* C *given* Z *admits a continuous strictly positive Lebesgue density on* $[0,\tau)$, *and that* $\mathrm{P}_{\theta_0,h_0}(X \geq \tau) > 0$. *If there exist* ϵ_n *and* δ_n *with* $\varphi_{\eta_0}(\epsilon_n) \asymp n\epsilon_n^2$ *and* $\epsilon_n^2\varphi_{\tilde b_{\theta_0,\eta_0}}(\delta_n) \to 0$ *and* $n^{3/2}\epsilon_n^4 \to 0$ *and* $\delta_n\epsilon_n\sqrt{n} \to 0$, *then the Bernstein–von Mises theorem for* θ *holds at* (θ_0, h_0).

Proof The theorem is a corollary of Theorem 12.9, applied to the transformations $\tilde{\eta}_n(\theta, \eta) = \eta + (\theta - \theta_0)\tilde{b}_n$, where $\tilde{b}_n$ will be chosen to approximate the least-favorable direction $\tilde{b}_{\theta_0,\eta_0}$. The proof uses various comparisons between metrics provided in Lemma 12.14 below. We reparameterize the model from (θ, h) to (θ, η), but keep the notation H for the cumulative hazard function corresponding to $h = e^\eta$. We choose Θ_n equal to open intervals that shrink to θ_0 slowly and $\mathcal{H}_n = \{\eta: d_H(p_{\theta,\eta}, p_{\theta_0,\eta_0}) < \epsilon_n, \|\eta\|_\infty^2 \lesssim n\epsilon_n^2\}$, where ϵ_n is a big multiple of the rate of contraction of the posterior distribution relative to the Hellinger distance. In view of Lemma 12.14(i)–(iii)+(v) the Hellinger distance is bounded above by the sum of the Euclidean distance on θ and the uniform distance on η, and the Kullback-Leibler neighborhoods contain neighborhoods for this sum-distance. It follows from Theorems 11.20 and 8.9 that an upper bound on ϵ_n is given by the solution to the inequality $\varphi_{\eta_0}(\epsilon_n) \lesssim n\epsilon_n^2$. Furthermore, the prior mass of a Kullback-Leibler type neighborhood of p_{θ_0,η_0} of radius a multiple of ϵ_n is at least $e^{-n\epsilon_n^2}$. By Borell's inequality the prior probability that $\|\eta\|_\infty$ exceeds $C\sqrt{n}\epsilon_n$ is bounded above by $e^{-C_1 n\epsilon_n^2}$, for a constant C_1 that can be made arbitrarily large by choosing C large. Therefore, by Theorem 8.20 the posterior probability that $\|\eta\|_\infty^2 \leq Cn\epsilon_n^2$ tends to one. Combined the preceding shows that the posterior mass of $\mathcal{H}_n$ tends to one as well.

By Lemma 12.14(xi) the posterior consistency for the Hellinger distance implies consistency for the supremum norm: $\|H - H_0\|_\infty \to 0$, for any $\eta \in \mathcal{H}_n$.

By the definition of the concentration functions there exists, for every $\delta_n > 0$, elements $\tilde{b}_n$ such that

$$\|\tilde{b}_n - \tilde{b}_{\theta_0,h_0}\|_\infty < \delta_n, \qquad \|\tilde{b}_n\|_{\mathbb{H}}^2 \leq \varphi_{\tilde{b}_{\theta_0,h_0}}(\delta_n).$$

We use these approximative least-favorable directions. Then condition (12.14) is valid provided $\epsilon_n \|\tilde{b}_n\|_{\mathbb{H}} \to 0$, as explained in Example 12.11.

The verification of (12.13) is based on establishing the pair of assertions, where $\mathbb{G}_n = \sqrt{n}(\mathbb{P}_n - P_{\theta_0,\eta_0})$ is the empirical process of the observations (T_i, Δ_i, Z_i): for any random sequences $\hat{\theta} \in \Theta_n$ and $\hat{\eta} \in \mathcal{H}_n$,

$$\mathbb{G}_n\big[\log p_{\hat{\theta},\hat{\eta}} - \log p_{\theta_0,\hat{\eta}+(\hat{\theta}-\theta_0)\tilde{b}_n} - (\hat{\theta} - \theta_0)(\tilde{\ell}_{\theta_0,\eta_0} - B_{\theta_0,\eta_0}\tilde{b}_n)\big] = o_P(|\hat{\theta} - \theta_0|), \quad (12.18)$$

$$P_{\theta_0,\eta_0}\big[\log p_{\hat{\theta},\hat{\eta}} - \log p_{\theta_0,\hat{\eta}+(\hat{\theta}-\theta_0)\tilde{b}_n}\big] = -\tfrac{1}{2}\tilde{I}_{\theta_0,\eta_0}(\hat{\theta} - \theta_0)^2(1 + o_P(1)) + o_P(|\hat{\theta} - \theta_0|n^{-1/2}). \quad (12.19)$$

For $\theta_u = \theta_0 + u(\theta - \theta_0)$ we can write the function in square brackets in (12.18), divided by $\theta - \theta_0$ and with $\hat{\theta}$ and $\hat{\eta}$ replaced by θ and η, as

$$\int_0^1 \big[(\dot{\ell}_{\theta_u,\eta} - \dot{\ell}_{\theta_0,\eta} - (B_{\theta_0,\eta+u(\theta-\theta_0)\tilde{b}_n} - B_{\theta_0,\eta_0})(\tilde{b}_n)\big]\,du$$
$$= \int_0^1 \Big[-ze^{\theta_u z}H(t) + z^{\theta_0 z}H_0(t) + e^{\theta_0 z}\int_0^t \tilde{b}_n\,(e^{u(\theta-\theta_0)\tilde{b}_n}\,dH - dH_0)\Big]\,du.$$

The functions $z \mapsto ze^{\theta z}$, $t \mapsto H(t)$ and $t \mapsto \int_0^t \tilde{b}_n e^{u(\theta-\theta_0)\tilde{b}_n}\,dH$ are all contained in bounded, universal Donsker classes, if $\|H\|_\infty$ remain bounded. Hence the functions appearing in the integral on the right are contained in a Donsker class by preservation of the Donsker property under Lipschitz transformations (see van der Vaart and Wellner 1996,

Chapter 2.10). If $\theta \to \theta_0$, $H \rightsquigarrow H_0$ and $\|\tilde b_n - b_{\theta_0,h_0}\|_\infty$, then the second moments of these functions tend to zero. Therefore (12.18) follows by Lemma 3.3.5 in van der Vaart and Wellner (1996).

By Taylor's theorem with integral remainder, $g(1) = g(0) + g'(0) + \int_0^1 g''(u)(1-u)\,du$, applied to $g(u) = \log(p_{\theta_u,\eta}/p_{\theta_0,\eta+(\theta_u-\theta_0)\tilde b_n})$, we can write the left side of (12.19) with $\hat\theta$ and $\hat\eta$ replaced by θ and η, as

$$(\theta-\theta_0)P_{\theta_0,\eta_0}\big[\dot\ell_{\theta_0,\eta} - B_{\theta_0,\eta}\tilde b_n\big] \\ + |\theta-\theta_0|^2\int_0^1 P_{\theta_0,\eta_0}\Big[-z^2e^{\theta_u z}H(t) + e^{\theta_0 z}\int_0^t \tilde b_n^2 e^{u(\theta-\theta_0)\tilde b_n}\,dH\Big](1-u)\,du.$$

The integral can be seen to converge to $-\int_0^1 \tilde I_{\theta_0,\eta_0}(1-u)\,du = -\frac{1}{2}\tilde I_{\theta_0,\eta_0}$. The proof of (12.19) is complete if we show that the expectation in the linear term is $o(n^{-1/2} + |\theta-\theta_0|)$, uniformly in $\eta \in \mathcal{H}_n$. Because $\tilde\ell_{\theta_0,\eta} = \dot\ell_{\theta_0,\eta} - B_{\theta_0,\eta}b_{\theta_0,h}$ is by its definition orthogonal in $\mathbb{L}_2(p_{\theta_0,\eta})$ to the range of $B_{\theta_0,\eta}$, we can write this expectation as

$$P_{\theta_0,\eta_0}\big[B_{\theta_0,\eta}(\tilde b_{\theta_0,h} - \tilde b_n)\big] + P_{\theta_0,\eta}\tilde\ell_{\theta_0,\eta}\Big[\frac{p_{\theta_0,\eta_0}}{p_{\theta_0,\eta}} - 1 - B_{\theta_0,\eta}(\eta_0-\eta)\Big].$$

Because $\|B_{\theta_0,\eta}b\|_\infty \lesssim \|b\|_\infty$, uniformly in η with uniformly bounded $\|H\|_\infty$, the first term can be seen to be bounded by $\|\tilde b_{\theta_0,h} - \tilde b_n\|_\infty d_H(p_{\theta_0,\eta}, p_{\theta_0,\eta_0})$, where the first factor is bounded by a multiple of $\delta_n + \|\tilde b_{\theta_0,h_0} - \tilde b_{\theta_0,h}\|_\infty$, by Lemma 12.14(x) and the second factor is bounded by a multiple of $|\theta-\theta_0| + d_H(p_{\theta,\eta}, p_{\theta_0,\eta_0})$. Hence this part of the linear term is $o(n^{-1/2} + |\theta-\theta_0|)$ if $\sqrt{n}(\delta_n+\epsilon_n)\epsilon_n \to 0$. The function in square brackets in the second term can be written

$$e^{\delta(\eta_0-\eta)(t) - e^{\theta_0 z}(H_0 - H(t))} - 1 - \delta(\eta_0-\eta)(t) + e^{\theta_0 z}\int_0^t (\eta_0-\eta)\,dH.$$

Applying the inequality $|e^x - 1 - x| \le (e^x \vee 1)x^2$, valid for any $x \in \mathbb{R}$, to the leading exponential and to the exponential in $H_0(t) - H(t) = \int_0^t (e^{\eta_0-\eta} - 1)\,dH$, the expectation of this function can be seen to be bounded by a multiple of $P_{\theta_0,\eta}\big[(e^{\eta_0-\eta}\vee 1)(\eta_0-\eta)^2\big] \lesssim \int(\eta-\eta_0)^2\,d(H+H_0)$. By Lemma 12.14(viii) and (vii)+(ix)+(iii), this is bounded above by $\|\eta-\eta_0\|_\infty^2 d_H^2(p_{\theta_0,\eta}, p_{\theta_0,\eta_0})$. Hence it is $o(n^{-1/2}+|\theta-\theta_0|)$ if $\sqrt{n}\,n\epsilon_n^2\epsilon_n^2 \to 0$. ☐

Example 12.13 (Riemann-Liouville) The concentration function of the released Riemann-Liouville process with parameter $\alpha > 0$ given in (11.17) at a function $w \in \mathfrak{C}^\beta[0,\tau]$ is proportional to $\epsilon^{-1/\alpha}$ if $\beta \ge \alpha$ and $\epsilon^{-(2\alpha-2\beta+1)/\beta}$ if $\alpha > \beta$ and $\alpha - \beta \notin \mathbb{N} + 1/2$, and otherwise equal to this function times a logarithmic factor. If $h_0 \in \mathfrak{C}^\beta[0,\tau]$, for some $\beta > 0$, and is bounded away from zero, then the requirement $\varphi_{h_0}(\epsilon_n) \lesssim n\epsilon_n^2$ results in the "usual" rate $\epsilon_n = n^{-\alpha\wedge\beta/(2\alpha+1)}$. In order that also $n^{3/2}\epsilon_n^4 \to 0$, it is necessary that $3/2 < \alpha < 4\beta/3 - 1/2$. The remaining two requirements $\epsilon_n^2\varphi_{\tilde b_{\theta_0,h_0}}(\delta_n) \to 0$ and $\delta_n\epsilon_n\sqrt{n}$ can be understood as giving a lower and upper bound for δ_n in terms of ϵ_n and n. If $\tilde b_{\theta_0,h_0} \in \mathfrak{C}^B[0,\tau]$ for some $B > 0$, then the lower bound is smaller than the upper bound if $B \ge \alpha$, or if $1/2 < B \le \alpha$ and $B + \beta > \alpha + 1/2$.

The restriction $\alpha < 4\beta/3 - 1/2$ arises through the "no-bias" condition. A prior on the nuisance parameter h of too large smoothness α relative to the true regularity β of h, may cause a bias in the posterior distribution for the parameter of interest θ.

The additional condition on B can be interpreted as requiring that the least-favorable direction be sufficiently smooth. If the censoring variable C and covariate Z are independent, then the least-favorable direction can be seen to have the same Hölder regularity $B = \beta$ as h_0 and the extra condition on B is automatic.

Lemma 12.14 *For every $\theta, \theta^* \in [-M, M]$, every pair of functions $h = e^\eta$, $h^* = e^{\eta^*}$, and $\lesssim$ and $\gtrsim$ denoting inequalities up to multiplicative constants that depend on $\|Z\|_\infty$ and M only, and $\|\eta\|_\infty = \sup_{0\le t\le\tau} |\eta(t)|$, the following assertions hold:*

(i) $d_H(p_{\theta,h}, p_{\theta^*,h^*}) \le 2(\mathrm{E}Z^2)^{1/2}|\theta - \theta^*| + \|\eta - \eta^*\|_\infty$.
(ii) $K(p_{\theta,h}; p_{\theta^*,h^*}) \lesssim (1 + \|\eta - \eta^*\|_\infty + \|H\|_\infty + \|H^*\|_\infty) d_H^2(p_{\theta,h}, p_{\theta^*,h^*})$.
(iii) $V_2(p_{\theta,h}; p_{\theta^*,h^*}) \lesssim (1 + \|\eta - \eta^*\|_\infty + \|H\|_\infty + \|H^*\|_\infty)^2 d_H^2(p_{\theta,h}, p_{\theta^*,h^*})$.
(iv) $\|\log(p_{\theta^*,h^*}/p_{\theta,h})\|_\infty \lesssim 1 + \|\eta^* - \eta\|_\infty + \|H\|_\infty + \|H^*\|_\infty$.
(v) $\|H - H^*\|_\infty \le \|H\|_\infty \|\eta - \eta^*\|_\infty e^{\|\eta-\eta^*\|_\infty}$.
(vi) $\|(a,b)\|_{\theta,h}^2 \lesssim a^2 + \int b^2\, dH$.
(vii) $\|(a,b)\|_{\theta,h}^2 \gtrsim C(a^2 + \int b^2\, dH)$, *for every h such that $\|H\|_\infty \le K$ and a constant C that depends on K only.*
(viii) $\|b\|_{H,2} \lesssim C\|b\|_\infty d_H(p_{\theta,h}, p_{\theta^*,h^*}) \vee C\|b\|_{H^*,2}$, *for every h and h^* with $\|H\|_\infty \vee \|H^*\|_\infty \le K$ and a constant C that depends on K only.*
(ix) $\|\theta - \theta^*, \eta - \eta^*\|_{\theta,h}^2 \lesssim C\, V_2(p_{\theta,h}; p_{\theta^*,h^*})$, *for every h such that $\|H\|_\infty \le K$, for a constant C that depends on K only.*
(x) $\|\tilde b_{\theta,h} - \tilde b_{\theta^*,h}\|_\infty \lesssim C d_H(p_{\theta,h}, p_{\theta^*,h})$, *for every h such that $\|H\|_\infty \le K$, for a constant C that depends on K only.*
(xi) *If $d_H(p_{\theta,h}, p_{\theta_0,h_0}) \to 0$, for fixed θ_0 and h_0, then $\theta \to \theta_0$ and $\|H - H_0\|_\infty \to 0$.*

Proof (i). For $a = \theta^* - \theta$ and $b = \eta^* - \eta$ the path $(\theta_u, h_u) := (\theta + ua, e^{\eta+ub})$ is equal to (θ, h) at $u = 0$ and equal to (θ^*, h^*) at $u = 1$. The derivative of the path $u \mapsto p_{\theta_u,h_u}^{1/2}$ is equal to $\frac{1}{2}(a\dot\ell_{\theta_u,h_u} + B_{\theta_u,h_u} b)\, p_{\theta_u,h_u}^{1/2}$. Therefore, by the mean value theorem in Hilbert space,

$$\|p_{\theta,h}^{1/2} - p_{\theta^*,h^*}^{1/2}\|_2 \le \sup_{0\le u\le 1} \|\tfrac{1}{2}(a\dot\ell_{\theta_u,h_u} + B_{\theta_u,h_u} b)\, p_{\theta_u,h_u}^{1/2}\|_2.$$

The right side is the information norm $\|(a,b)\|_{\theta_u,h_u}$, whose square is given as a sum of three terms in (12.17). Inserting the definitions of the functions $M_{k,\theta,h}$, we can write the square of $\|(a,b)\|_{\theta,h}$ as

$$\mathrm{E}_Z \int (a^2 Z^2 + 2ab(t)Z + b^2(t))\bar F_{C|Z}(t|\,Z)\, e^{\theta Z} e^{-e^{\theta Z} H(t)}\, dH(t). \tag{12.20}$$

Inequality (i) follows by bounding b and $\bar F_{C|Z}(t|\,Z)$ by $\|b\|_\infty$ and $\mathbb{1}\{t \le \tau\}$, and next evaluating the integral as $\int_0^\sigma e^{-u}\, du \le 1$, for $\sigma = e^{\theta Z} H(\tau)$.

(ii), (iii) and (iv) The log-likelihood ratio satisfies, for any $\theta, \theta^*, h = e^\eta, h^* = e^{\eta^*}$,

$$\log \frac{p_{\theta^*,h^*}}{p_{\theta,h}}(t, \delta, z) = \delta((\theta^* - \theta)z + (\eta^* - \eta)(t)) + H(t)e^{\theta z} - H^*(t)e^{\theta^* z}.$$

This is bounded as in (iv). Next (ii) and (iii) follow by the general Lemma B.2.

(v). Since $|e^x - e^y| \leq e^{x\vee y}|x-y|$, and $x \vee y \leq x + |x-y|$, for every $x, y \in \mathbb{R}$, $|H(t) - H^*(t)| \leq \int_0^t e^{\eta(s)} e^{|\eta(s)-\eta^*(s)|}|\eta(s) - \eta^*(s)|\,ds$. The result follows.

(vi) and (vii). These are clear from (12.20).

(viii). By the inequality $g - g^* \leq 2\sqrt{g}(\sqrt{g} - \sqrt{g^*})$, for any $g, g^* \geq 0$,

$$\begin{aligned}
&\mathrm{E}_Z \int b^2\,(e^{-e^{\theta Z}H} e^{\theta Z}\,dH - e^{-e^{\theta^* Z}H^*} e^{\theta^* Z}\,dH^*)\\
&\leq \|b\|_\infty \mathrm{E} \int \Big[|b|\, 2e^{-e^{\theta Z}H/2} e^{\theta Z/2}\sqrt{h}(e^{-e^{\theta Z}H/2} e^{\theta Z/2}\sqrt{h} - e^{-e^{\theta^* Z}H^*/2} e^{\theta^* Z/2}\sqrt{h^*})\Big]\,d\nu\\
&\lesssim \|b\|_\infty \|b\|_{H,2}\, d_H(p_{\theta,h}, p_{\theta^*,h^*}).
\end{aligned}$$

The factors $e^{-e^{\theta Z}H}$ and $e^{-e^{\theta^* Z}H^*}$ are bounded away from zero if $\|H\|_\infty$ and $\|H^*\|_\infty$ are bounded above. We conclude that $\int b^2\,dH$ is bounded by a multiple of the maximum of $\int b^2\,dH^*$ and the right side of the display. This is equivalent to assertion (viii).

(ix). We have

$$\begin{aligned}
V_2(p_{\theta,h}, p_{\theta^*,h^*}) &= \mathrm{E} \int (H(t)e^{\theta Z} - H^*(t)e^{\theta^* Z})^2 e^{-e^{\theta Z}H(t)}\,dF_{C|Z}(t|\,Z)\\
&+ \mathrm{E} \int \big[(\theta^*-\theta)Z + (\eta^*-\eta)(t) + H(t)e^{\theta Z} - H^*(t)e^{\theta^* Z}\big]^2 \bar{F}_{C|Z}(t|\,Z) e^{-e^{\theta Z}H(t)} e^{\theta Z}\,dH(t).
\end{aligned}$$

The two integrals are bounded below by constants, depending on h and the distribution of (C, Z) only, times the integrals obtained by replacing the measures given by $e^{-e^{\theta Z}H(t)}\,dF_{C|Z}(t|\,Z)$ and $\bar{F}_{C|Z}(t|\,Z)e^{-e^{\theta Z}H(t)}e^{\theta Z}\,dH(t)$ by the Lebesgue measure on $[0, \tau]$. Since $\|g + g^*\|^2 + \|g^*\|^2 \geq \frac{1}{2}\|g^*\|^2$, for any g and g^*, it follows that $V_2(p_{\theta,h}, p_{\theta^*,h^*})$ is bounded below by a multiple of $\mathrm{E}\int_0^\tau ((\theta^*-\theta)Z + (\eta^*-\eta)(t))^2\,dt$. We combine this with (vi).

(x). The functions $M_{k,\theta,h}$ can be seen to be bounded away from zero and infinity uniformly in bounded $\|H\|_\infty$, and $\|M_{k,\theta,h} - M_{k,\theta^*,h}\|_\infty \lesssim d_{TV}(P_{\theta,h}, P_{\theta^*,h})$.

(xi). Because the $\mathbb{L}_1$-distance is bounded by twice the Hellinger distance,

$$\mathrm{E}_Z \int |e^{\theta Z} h(t) e^{-e^{\theta Z}H(t)} - e^{\theta_0 Z} h_0(t) e^{-e^{\theta_0 Z}H_0(t)}|\, \bar{F}_{C|Z}(t|\,Z)\,dt \to 0.$$

The integral $\int_0^t e^{\theta z}h(s)e^{-e^{\theta z}H(s)}\,ds$ can be explicitly evaluated as $\exp(-e^{\theta z}H(t))$, and the survival function $\bar{F}_{C|Z}(t|\,z)$ is bounded away from zero on $[0, \tau]$, for all z in a set $\mathfrak{Z}$ of positive measure under the law of Z. We conclude, that for all $A \subset \mathfrak{Z}$ and all $t \in [0, \tau]$,

$$\mathrm{E}\big[\mathbb{1}_A(Z)e^{-e^{\theta Z}H(t)}\big] \to \mathrm{E}\big[\mathbb{1}_A(Z)e^{-e^{\theta_0 Z}H_0(t)}\big].$$

Because $e^{\theta_0 Z}H_0(t)$ is uniformly bounded, the right side is contained in a compact interval in $(0, \infty)$. Since $e^{-e^{\theta Z}H(t)} \leq e^{-MH(t)}$, for some $M > 0$, it follows that $H(t)$ must be contained in a bounded interval, uniformly in $t \in [0, \tau]$. Thus we can apply Helly's theorem to select from any sequence of (θ, H) with $\|p_{\theta,h} - p_{\theta_0,h_0}\|_1 \to 0$ a subsequence (θ_m, H_m) with $\theta_m \to \tilde{\theta}$ and $H_m \rightsquigarrow \tilde{H}$, for some $\tilde{\theta}$ and $\tilde{H}$. Then the left side of the preceding display at (θ_m, H_m) converges to its value at $(\tilde{\theta}, \tilde{H})$, for all continuity points $t \in [0, \tau]$ of $\tilde{H}$, and hence

$\mathrm{E}\big[\mathbb{1}_A(Z)e^{-e^{\tilde\theta Z}\tilde H(t)}\big] = \mathrm{E}\big[\mathbb{1}_A(Z)e^{-e^{\theta_0 Z}H_0(t)}\big]$, for all such t and all A as before. This implies that $e^{\tilde\theta z}\tilde H(t) = e^{\theta_0 z}H_0(t)$, for all such t and almost all $z \in \mathfrak{Z}$. Since Z is nondegenerate on $\mathfrak{Z}$, this is possible only if $\tilde\theta = \theta_0$. Then also $\tilde H = H_0$, and the convergence $H_m \rightsquigarrow H_0$ is uniform on every interval $[0, t]$ with $t < \tau$, by the continuity of H_0. We can extend to the full interval $[0, \tau]$ by also considering the second, "censored" part of the sum that defines $\|p_{\theta,h} - p_{\theta_0,h_0}\|_1$, where we use that $F_{C|Z}(\{\tau\}|\, z) > 0$, for z in a set of positive probability. □

12.4 White Noise Model

In the white noise model, discussed previously in Example 8.6 and Section 8.3.4, we observe a sequence $X^{(n)} = (X_{n,1}, X_{n,2}, \ldots)$ of independent variables $X_{n,i} \overset{\text{ind}}{\sim} \mathrm{Nor}(\theta_i, n^{-1})$. A conjugate analysis in this simple problem gives insight into the infinite-dimensional Bernstein–von Mises theorem.

The parameter $\theta = (\theta_1, \theta_2, \ldots)$ belongs to $\mathbb{R}^\infty$, but it is often restricted to a small subset of sequences with $\theta_i \to 0$ as $i \to \infty$. This reflects the fact that the coordinates θ_i typically arise as the Fourier coefficients of a function, in which case they minimally satisfy that $\theta \in \ell_2$, and often are further restricted to belong to a smoothness class, such as the Sobolev space $\mathfrak{W}^\beta$ of all sequences with $\|\theta\|^2_{2,2,\beta} := \sum_{i=1}^\infty i^{2\beta}\theta_i^2 < \infty$, for some $\beta > 0$.

A Gaussian prior of the type $\theta_i \overset{\text{ind}}{\sim} \mathrm{Nor}(0, \tau_i^2)$ is conjugate, and leads to the posterior distribution

$$\theta_i|\, X^{(n)} \overset{\text{ind}}{\sim} \mathrm{Nor}\Big(\frac{\tau_i^2 X_{n,i}}{n^{-1}+\tau_i^2}, \frac{n^{-1}\tau_i^2}{n^{-1}+\tau_i^2}\Big). \tag{12.21}$$

Any prior variances $\tau_i^2 > 0$ lead to a proper prior with full support on $\mathbb{R}^\infty$. Prior variances τ_i^2 that decrease to 0 as $i \to \infty$ are natural to model parameters θ with decreasing coordinates. For instance, the choice $\tau_i^2 = i^{-1-2\alpha}$, for some fixed $\alpha > 0$, leads to a prior Π with $\Pi(\mathfrak{W}^\beta) = 1$ whenever $\beta < \alpha$, and $\Pi(\mathfrak{W}^\beta) = 0$ if $\beta \geq \alpha$. In Example 8.6 these priors were seen to lead to the posterior contraction rate $n^{-(\alpha\wedge\beta)/(2\alpha+1)}$ if the true parameter θ belongs to $\mathfrak{W}^\beta$, with the (embarrassing) finding that the fastest rate is obtained by the prior with $\alpha = \beta$, which gives prior mass zero to $\mathfrak{W}^\beta$.

In this section we study the Bernstein–von Mises phenomenon, first for the full posterior distribution, and next for the induced marginal posterior distributions of linear functionals of the parameter. We assume throughout that $\tau_i > 0$, for every i.

12.4.1 Full Parameter

In the white noise model with an infinite-dimensional parameter set, such as the Sobolev space $\mathfrak{W}^\beta$, there is no notion of an efficient asymptotically normal estimator, and a maximum likelihood estimator fails to exist. This makes the formulation of a Bernstein–von Mises theorem not obvious, in particular regarding the centering variables. We might take a general Bernstein–von Mises theorem to make the assertion that, for given centering variables $\hat\theta = \hat\theta(X^{(n)})$, the frequentist distribution of $\hat\theta - \theta|\,\theta$ and the Bayesian distribution of $\theta - \hat\theta|\, X^{(n)}$ approach each other as $n \to \infty$, for a given true θ that governs the distribution of $\hat\theta$ in the first case and the distribution of the conditioning variable $X^{(n)}$ in the Bayesian case. In the finite-dimensional case choosing $\hat{\theta}$ the maximum likelihood estimator is typical, but the posterior mean is also possible. If presently we choose $\hat{\theta}$ equal to the posterior mean, then the two relevant distributions, scaled by $\sqrt{n}$ for convenience, are given by

$$\sqrt{n}(\mathrm{E}(\theta|\, X^{(n)}) - \theta)|\,\theta \quad \sim \bigotimes_{i=1}^{\infty} \mathrm{Nor}\Big(-\frac{n^{-1/2}\theta_i}{n^{-1}+\tau_i^2}, \frac{\tau_i^4}{(n^{-1}+\tau_i^2)^2}\Big), \tag{12.22}$$

$$\sqrt{n}(\theta - \mathrm{E}(\theta|\, X^{(n)}))|\, X^{(n)} \sim \bigotimes_{i=1}^{\infty} \mathrm{Nor}\Big(0, \frac{\tau_i^2}{n^{-1}+\tau_i^2}\Big). \tag{12.23}$$

These two distributions differ both in mean (unless $\theta = 0$) and variance, where the mean of the Bayesian distribution (12.23) vanishes by our choice of centering statistic. In general these differences do not disappear as $n \to \infty$. In fact, these infinite product measures (on $\mathbb{R}^\infty$) are orthogonal (and hence at maximum distance) unless their variances are sufficiently close, which is possible only if τ_i^2 tends to infinity fast enough.

Theorem 12.15 (Bernstein–von Mises) *For any $\theta \in \mathbb{R}^\infty$ the following assertions are equivalent:*

(i) *The total variation distance between the measures on the right sides of* (12.22) *and* (12.23) *tends to* 0 *as $n \to \infty$.*
(ii) $\sum_{i=1}^\infty (1/\tau_i^4) < \infty$ *and* $\sum_{i=1}^\infty (\theta_i^2/\tau_i^4) < \infty$.

If these conditions hold, then both measures as in (i) tend to $\otimes_{i=1}^\infty \mathrm{Nor}(0, 1)$.

Proof Convergence in total variation distance to zero is equivalent to convergence in the Hellinger distance to zero, and hence equivalent to the Hellinger affinities between the measures converging to 1. As the measures are product measures, this translates into the convergence

$$\prod_{i=1}^{\infty} \rho_{1/2}\Big(\mathrm{Nor}\Big(0, \frac{\tau_i^2}{n^{-1}+\tau_i^2}\Big), \mathrm{Nor}\Big(-\frac{n^{-1/2}\theta_i}{n^{-1}+\tau_i^2}, \frac{\tau_i^4}{(n^{-1}+\tau_i^2)^2}\Big)\Big) \to 1.$$

This can be shown to be equivalent to the conditions under (ii). The final statement follows similarly by computing affinities. ☐

Condition (ii) of the theorem requires that the prior variances τ_i^2 tend to infinity (at a rate), which goes in the opposite direction of the smoothing priors with $\tau_i^2 \downarrow 0$ that were argued to be natural. Thus the theorem may be safely remembered as saying that a Bernstein–von Mises theorem for the full parameter holds only for unrealistic priors, at least in the sense of (i) of the theorem. It may also be noted that the centering $\hat{\theta}$ hardly influences this conclusion, because even if $\theta = 0$, when both distributions in (12.22) and (12.23) are centered at zero, it is still required that $\tau_i^2 \to \infty$.

This negative conclusion can be attributed to the infinite-dimensional setting, where "optimal" estimation of a parameter typically involves a bias-variance trade-off, with the bias controlled through a priori assumptions. More precisely, the conclusion can be linked to the use of the total variation distance, which measures the infinitely many dimensions equally. In the next subsection we shall see that the difficulties may disappear if the full posterior is projected onto finite-dimensional marginals. Alternatively, it is also possible to keep the infinite-dimensional formulation, but down-weight the increasing dimensions. To set this up, consider a positive weight sequence $(w_i) \in \ell_1$, and the Hilbert space $H_w = \{\theta \in \mathbb{R}^\infty: \sum_i w_i\theta_i^2 < \infty\}$, with inner product

$$\langle \theta, \theta' \rangle_w = \sum_{i=1}^{\infty} w_i\theta_i\theta_i'.$$

As necessarily $w_i \to 0$, the space H_w is bigger than ℓ_2 and it has a weaker norm. The space is big enough to carry the law $\otimes_{i=1}^\infty \mathrm{Nor}(0, 1)$ of an infinite sequence $(Z_1, Z_2, \ldots)$ of independent standard normal variables (as $\sum_i w_i Z_i^2 < \infty$ a.s.), which arises as the limit in the preceding theorem. The measures on the right sides of (12.22) and (12.23) also concentrate on H_w, and they both converge to the latter law with respect to the weak topology, under mild conditions.

Theorem 12.16 (Bernstein–von Mises, weak) *Fix $w \in \ell_1$. For every θ with $\sum_i w_i\theta_i^2/\tau_i^2 < \infty$, the measures on the right sides of* (12.22) *and* (12.23) *tend to* $\otimes_{i=1}^\infty \mathrm{Nor}(0, 1)$ *relative to the weak topology on the space of Borel probability measures on H_w. In particular, this is true for almost every θ from* $\otimes_{i=1}^\infty \mathrm{Nor}(0, \tau_i^2)$.

Proof The distribution on the right of (12.22) can be represented as the distribution of the vector Z_n with coordinates $Z_{n,i} = -n^{-1/2}\theta_i/(n^{-1} + \tau_i^2) + Z_i\tau_i^2/(n^{-1} + \tau_i^2)$, for $Z = (Z_1, Z_2, \ldots)$ a sequence of independent standard normal variables. Under the stated conditions $\|Z_n - Z\|_w \to 0$ almost surely, by the dominated convergence theorem, whence $Z_n \rightsquigarrow Z$. Convergence of the distributions on the right side of (12.23) follows similarly. ☐

Thus an infinite-dimensional Bernstein–von Mises theorem does hold provided the mode of approximation is chosen weak enough.

12.4.2 Linear Functionals

The posterior distribution of a real-valued functional $L\theta$ of the parameter $\theta = (\theta_1, \theta_2, \ldots)$ is the marginal distribution $\Pi_n(\theta: L\theta \in \cdot \mid X^{(n)})$ of the full posterior distribution of θ, described in (12.21). A continuous, linear functional $L: \ell_2 \to \mathbb{R}$ can be represented by an element $l \in \ell_2$ in the form $L\theta = \sum_i l_i\theta_i$, for $\theta \in \ell_2$. If the prior is concentrated on ℓ_2 (i.e. $\sum_i \tau_i^2 < \infty$), then it is immediate from (12.21) that

$$L\theta \mid X^{(n)} \sim \mathrm{Nor}\Big(\sum_i \frac{l_i\tau_i^2 X_{n,i}}{n^{-1} + \tau_i^2}, \sum_i \frac{l_i^2 n^{-1}\tau_i^2}{n^{-1} + \tau_i^2}\Big). \tag{12.24}$$

This formula is true more generally for *measurable linear functionals relative to the prior* $\otimes_i \mathrm{Nor}(0, \tau_i^2)$. These are defined as Borel measurable maps $L: \mathbb{R}^\infty \to \mathbb{R}$ that are linear on a measurable linear subspace of probability one under the prior. For T the coordinatewise multiplication $\theta \mapsto (\tau_1\theta_1, \tau_2\theta_2, \ldots)$ by the prior standard deviations, the map $LT: \ell_2 \to \mathbb{R}$ is then automatically continuous and linear, whence representable by some element $m \in \ell_2$, and formula (12.24) is true with $l_i = m_i/\tau_i$. (This satisfies $\sum_i \tau_i^2 l_i^2 < \infty$, but need not be contained in ℓ_2; see Knapik et al. 2011, Proposition 3.2, and Skorohod 1974, pages 25–27, for details.)

The marginal normality (12.24) of the posterior distribution provides one aspect of a Bernstein–von Mises theorem for $L\theta$. However, the theorem ought to include a relationship between the frequentist and Bayesian distributions. For $\widehat{L\theta} := \mathrm{E}(L\theta |\, X^{(n)})$ the posterior mean we say that in the current setting the Bernstein–von Mises theorem *holds* at θ if the total variation distance between the laws of the following variables tends to zero

$$\sqrt{n}(\widehat{L\theta} - L\theta)|\,\theta \sim \mathrm{Nor}\Big(-\sum_i \frac{l_i n^{-1/2}\theta_i}{n^{-1}+\tau_i^2}, \sum_i \frac{l_i^2\tau_i^4}{(n^{-1}+\tau_i^2)^2}\Big),$$

$$\sqrt{n}(L\theta - \widehat{L\theta})|\, X^{(n)} \sim \mathrm{Nor}\Big(0, \sum_i \frac{l_i^2\tau_i^2}{n^{-1}+\tau_i^2}\Big).$$

Here the first law is the frequentist law of the centered posterior mean, considered under θ-probability, while the second is the posterior distribution of $L\theta$, centered at (posterior) mean zero. The scaling by $\sqrt{n}$ is for convenience. Both distributions are Gaussian, and hence their total variation distance tends to zero if the quotient of their two variances tends to 1 and the square difference between their means is negligible relative to this variance. This depends on the combination of l, the prior variances, and θ. The following theorem gives some cases of interest.

Theorem 12.17 (Bernstein–von Mises, functionals) *Let l represent L as indicated.*

(i) *If $l \in \mathfrak{W}^q$ for some $q \geq 0$, and $\tau_i^2 = i^{-2r}$ for some $r > 0$, and $\beta + q > r$, then the Bernstein–von Mises theorem holds at any $\theta \in \mathfrak{W}^\beta$.*
(ii) *If $l \in \ell_1$, then the Bernstein–von Mises theorem holds at almost every θ sampled from $\otimes_{i=1}^\infty \mathrm{Nor}(0, \tau_i^2)$, for any τ_i.*
(ii) *If $l \in \ell_2$, then the Bernstein–von Mises theorem holds in probability if θ is sampled from $\otimes_{i=1}^\infty \mathrm{Nor}(0, \tau_i^2)$, for any τ_i.*
(iv) *If $l_i \asymp i^{-q-1/2}S(i)$ for a slowly varying function S, and $\tau_i^2 = i^{-2r}$ for some $r > 0$, then the Bernstein–von Mises theorem holds at some θ only if $q \geq 0$, in which case $n \mapsto n\,\mathrm{var}(\theta |\, X^{(n)})$ is slowly varying.*

Proof As indicated before the statement of the theorem the Bernstein–von Mises theorem holds if the frequentist variance of the posterior mean is asymptotically equivalent to the variance of the posterior distribution and the square bias of the posterior mean as an estimator of $L\theta$ vanishes relative to this common variance. These quantities are given in the display and hence we are lead to the pair of conditions

$$\Big|\sum_{i=1}^\infty \frac{l_i n^{-1/2}\theta_i}{n^{-1}+\tau_i^2}\Big|^2 \ll \sum_{i=1}^\infty \frac{l_i^2\tau_i^2}{n^{-1}+\tau_i^2}, \quad \text{and} \quad \sum_{i=1}^\infty \frac{l_i^2\tau_i^4}{(n^{-1}+\tau_i^2)^2} \sim \sum_{i=1}^\infty \frac{l_i^2\tau_i^2}{n^{-1}+\tau_i^2}.$$

If $l \in \ell_2$, then the last two series, which provide the variances, tend to $\|l\|_2^2$, by the dominated convergence theorem, whence the variances behave as they should, under every of the assumptions (i)–(iii).

If $\theta \in \mathfrak{W}^\beta$, then the leftmost expression in the display, the square bias, can be bounded above by $\{\sum_i l_i^2 n^{-1} i^{-2\beta}/(n^{-1}+\tau_i^2)^2\}\|\theta\|_{2,2,\beta}^2$, by the Cauchy-Schwarz inequality. For $\tau_i^2 = i^{-2r}$, this is seen to be of order $nn^{-(\beta+q)/r\wedge 2}$ by Lemma K.7, and hence $o(1)$ if $\beta + q > r$.

If θ is sampled from the prior, then the mean of the bias series $\sum_i l_i n^{-1/2}\theta_i/(n^{-1} + \tau_i^2)$ vanishes and its variance is equal to $\sum_i l_i^2 n^{-1}\tau_i^2/(n^{-1} + \tau_i^2)^2$. If $l \in \ell_2$, then the latter is $o(1)$, by the dominated convergence theorem, as $n^{-1}\tau_i^2/(n^{-1} + \tau_i^2)^2$ is bounded by 1 and tends to zero as $n \to \infty$, for every i. This gives the Bernstein–von Mises theorem in probability (iii). If $l \in \ell_1$, then we bound the absolute value of the bias series above by $\sum_i |l_i n^{-1/2}\tau_i Z_i|/(n^{-1} + \tau_i^2)$, for $Z_i = \theta_i/\tau_i$. The terms of this series are bounded above by the terms $|l_i Z_i|$ of the converging series $\sum_i |l_i Z_i|$ and tend to zero for every i, as $n \to \infty$, surely. Thus $\sum_i |l_i n^{-1/2}\tau_i Z_i|/(n^{-1} + \tau_i^2) \to 0$, almost surely, by the dominated convergence theorem.

For the proof of (iv) we first note that the frequentist variance (given by the third series in the display) is always smaller than the posterior variance, as the terms of the two series differ by factors $\tau_i^2/(n^{-1} + \tau_i^2) \leq 1$. For i such that $n\tau_i^2 \leq c$ for some $c > 0$, the extra factor is strictly bounded away from 1. This implies that the quotient of the two series can converge to 1 only if

$$\sum_{i:n\tau_i^2\leq c} \frac{l_i^2\tau_i^2}{n^{-1} + \tau_i^2} \ll \sum_{i=1}^{\infty} \frac{l_i^2\tau_i^2}{n^{-1} + \tau_i^2}.$$

For l satisfying the assumption in (iv) and $\tau_i^2 = i^{-2r}$, the last assertion of Lemma K.8 gives that $(2r + 2q)/(2r) \geq 1$, i.e. $q \geq 0$. The same lemma then gives that the series on the right is a slowly varying function of n (namely $\sum_{i\leq n^{1/(2r)}} S^2(i)/i$ if $q = 0$ and converging to a constant if $q > 0$). ☐

Parts (ii) and (iii) are comforting to the Bayesian who really believes his prior: the approximation holds for a large class of functionals at almost every true parameter deemed possible by the prior. Part (iv) suggests that the condition that $l \in \ell_2$ cannot be much relaxed (although the theorem may hold for l with $q = 0$ that are "almost but not really in ℓ_2"). Perhaps the most interesting part is (i), which shows the interplay between the regularity β of the true parameter, the smoothness q of the functional, and the regularity of the prior. If $r = \alpha + 1/2$, then the condition becomes $\beta + q > \alpha + 1/2$ and can be interpreted as saying that the prior smoothness α should not exceed the smoothness of the true parameter plus the smoothness of the functional minus 1/2.

It is shown in the preceding proof that under conditions (i)–(iii) the posterior variance and the variance of the posterior mean are of order n^{-1}. The rate of posterior contraction of the marginal distribution of $L\theta$ is then the "parametric rate" $n^{-1/2}$. Under condition (iv) this is true, possibly up to a slowly varying factor. The fast rate of estimation in these cases is possible by the assumed smoothness of the functional L and parameter θ. It is known that for a functional that is "regular" of (exact) order $q < 0$, the minimax rate over balls in $\mathfrak{W}^\beta$ is equal to $n^{-(\beta+q)/(2\beta)} \gg n^{-1/2}$. The following proposition shows that this is the posterior contraction rate for a prior with variances $\tau_i^2 = i^{-1-2\alpha}$ if $\alpha = \beta - 1/2$. (See Knapik et al. 2011 for further discussion and a proof.) This is in contrast with the "parametric case": there any prior that is not overly smooth (precisely: $\alpha < \beta + q - 1/2$) performs well, whereas in the "nonparametric case" only a single prior smoothness is optimal.

Proposition 12.18 (Contraction rate) *If $l \in \mathfrak{W}^q$ and $\theta_0 \in \mathfrak{W}^\beta$ for some $q \geq -\beta$ and $\beta > 0$, then the marginal posterior contraction rate relative to the prior with $\tau_i^2 = i^{-1-2\alpha}$ is given by $\epsilon_n = n^{-(\beta\wedge(\alpha+1/2)+q)/(2\alpha+1)} \vee n^{-1/2}$: for any $M_n \to \infty$,*

$$\mathrm{E}_{\theta_0} \Pi_n(\theta\colon |L\theta - L\theta_0| \geq M_n\epsilon_n |\, X^{(n)}) \to 0.$$

12.5 Historical Notes

Versions of Theorem 12.2 were proved by Lo (1983) and Lo (1986). He treated the classical Donsker classes consisting of indicator functions, and proved tightness by verifying Markov properties and increment bounds; see Problems L.3 and L.4. For the classical Donsker classes Theorem 12.2 is also a special case of Corollary 13.23, which allows more general priors and censored data. A thorough treatment of the theory of strong approximation is given in Csörgő and Révész (1981). The original references Komlós et al. (1975); Csörgő and Révész (1975) were improvements of work by Strassen and Kiefer. Theorems 12.5 and 12.6 are Theorems 2.1 and Proposition 5.7 of Lo (1987). The proof of the latter result is similar to the results in the non-Bayesian setting due to Bickel and Rosenblatt (1973). The main results of Section 12.3 on the semiparametric Bernstein–von Mises theorem are based on Castillo (2012b). We have extended his main result from the strict semiparametric case to functionals, allow general priors, and have formulated the theorem in terms of an LAN expansion of a least favorable submodel, somewhat parallel to the treatment of maximum (or profile) likelihood in Murphy and van der Vaart (2000). The original formulation in Castillo (2012b) used a full LAN expansion as in Section 12.3.1, and was restricted to Gaussian priors. The notation and general discussion of semiparametric models (including the Cox model) are taken from Chapter 25 of van der Vaart (1998). The formulas for the information in the Cox model go back to Begun et al. (1983). Castillo discovered the role of the shift in the prior on the nuisance parameter in the least-favorable direction, a property that is underemphasized in Bickel and Kleijn (2012). For further results see Rivoirard and Rousseau (2012) and Castillo and Rousseau (2015). Castillo (2014) used nonparametric Bernstein–von Mises theorem results to derive posterior contraction rates with respect to the stronger uniform norm for the Gaussian white noise model and density estimation using wavelet series. Theorem 12.15 is Lemma 2 of Leahu (2011) and Theorem 12.16 is a slight extension of his Theorem 1. Leahu (2011) clarified and generalized previous work by Cox (1993) and Freedman (1999). Other versions of the weak infinite-dimensional Bernstein–von Mises Theorem 12.16 were recently developed by Castillo and Nickl (2013), Castillo and Nickl (2014) and Ray (2014). The main results in Section 12.4.2 are based on Knapik et al. (2011), who also consider inverse problems and scaling of the priors, and applied the results to study the coverage of credible sets. The almost sure part (ii) of Theorem 12.17 is due to Leahu (2011).

Problems

12.1 Show that $\sqrt{a_n+b_n}(V_n - \mathrm{E}(V_n)) \rightsquigarrow \mathrm{Nor}(0, \lambda(1-\lambda))$, whenever $V_n \sim \mathrm{Be}(a_n, b_n)$, and $a_n, b_n \to \infty$ in such a way that $a_n/(a_n+b_n) \to \lambda$. Use this to give a direct proof of the Bernstein–von Mises theorem for $P(A)$ based on observing a random sample from the Dirichlet process. [Hint: use a representation by gamma variables, and the delta-method.]

12.2 (James 2008) Let $P \sim \mathrm{DP}(\alpha)$ and let $\mathcal{F}$ be a Donsker class of functions with square integrable envelope with respect to α. Show that $\sqrt{|\alpha|}(P - \bar\alpha) \rightsquigarrow \mathbb{G}$ in $\mathfrak{L}_\infty(\mathcal{F})$, as $|\alpha| \to \infty$, where $\mathbb{G}$ is an $\bar\alpha$-Brownian bridge indexed by $\mathcal{F}$.

12.3 (Kiefer process) Given a Kiefer process $\mathbb{K}$, show that:

(a) $s \mapsto \mathbb{K}(s, n+1) - \mathbb{K}(s, n)$ are independent Brownian bridges, for $n \in \mathbb{N}$.
(b) $t \mapsto \mathbb{K}(s,t)/\sqrt{s(1-s)}$ is a standard Brownian motion, for any $0 < s < 1$.
(c) $\mathbb{K}$ is equal in distribution to the process $W(s,t) - sW(1,t)$, for W a two-parameter standard Wiener process (which has covariance kernel $\min(s_1, s_2)\min(t_1, t_2)$).

12.4 (Gu and Ghosal 2008) The receiver operating characteristic (ROC) function of two cumulative distribution functions F and G is defined by $R(t) = \bar G(\bar F^{-1}(t))$, for $t \in [0,1]$, and the area under the curve (AUC) is $A = \int_0^1 R(t)\,dt$. The ROC is invariant under continuous, strictly increasing transformations.The empirical ROC $\mathbb{R}_{m,n}$ and AUC $\mathbb{A}_{m,n}$ of two independent samples $X_1, \ldots, X_m \overset{\text{iid}}{\sim} F$ and $Y_1, \ldots, Y_n \overset{\text{iid}}{\sim} G$ are obtained by replacing F and G by their empirical versions. Induce the Bayesian bootstrap distribution of R from those of F and G. Assume that the true distributions F_0 and G_0 possess continuous densities f_0 and g_0, and that the functions $F_0\bar F_0|f_0'|/f_0^2$ and $G_0\bar G_0|g_0'|/g_0^2$ are bounded. Show that there exist independent Kiefer processes K_1 and K_2 such that a.s. as $m/(m+n) \to \lambda \in (0,1)$ and $\alpha_m \gg m^{-1/4}$, the Bayesian bootstrap distribution of R can be represented as

$$R(t) = \mathbb{R}_{m,n}(t) + \frac{1}{m}R_0'(t)K_1(t,m) + \frac{1}{n}K_2(R_0(t), n) + O\Big(\frac{(\log m)^{1/2}(\log\log m)^{1/4}}{m^{3/4}\alpha_m}\Big),$$

uniformly on $t \in (\alpha_m^*, 1-\alpha_m^*)$, where $\alpha_m^* = \alpha_m + m^{-1/2}\sqrt{\log\log m}$. Furthermore, show that the AUC functional satisfies the Bernstein–von Mises theorem:

$$\sqrt{m+n}(A - \mathbb{A}_{m,n})|\, X_1, \ldots, X_m, Y_1, \ldots, Y_n \rightsquigarrow \mathrm{Nor}(0, \sigma^2), \qquad \text{a.s.},$$

as $m, n \to \infty$, where

$$\sigma^2 = \int_0^1\int_0^1 \Big[\frac{1}{\lambda}R_0'(s)R_0'(t)[s \wedge t - st] + \frac{1}{(1-\lambda)}[R_0(s)\wedge R_0(t) - R_0(s)R_0(t)]\Big]\,dt\,ds,$$

and the ranges of the integrals in the definitions of A and $\mathbb{A}_{m,n}$ are taken $(\alpha_m^*, 1-\alpha_m^*)$ rather than $(0,1)$.

12.5 (Castillo 2012b) Consider the Gaussian white noise model $dX_t^{(n)}(t) = f(t-\theta) + n^{-1/2}dB_t$, for $t \in [-\frac12, \frac12]$, where the signal $f \in \mathbb{L}_2[-\frac12, \frac12]$, symmetric about zero, 1-periodic and satisfies $\int_0^1 f(t)\,dt = 0$. Let $f_k = \sqrt{2}\int f(t)\cos(2\pi kt)\,dt$, $k \in \mathbb{N}$, be the

Fourier coefficients. Assume that the true signal f_0 satisfies $\sum_{k=1}^{\infty} k^{2\beta} f_k^2 < \infty$. Let the prior density on θ be positive and continuous, and let f be given a prior through $f_k \stackrel{\text{ind}}{\sim} \text{Nor}(0, k^{-2\alpha-1})$. Show that the Bernstein–von Mises theorem for θ holds if $\beta > 3/2$ and $1 + \sqrt{3}/2 < \alpha < (3\beta - 2)/(4 - 2\beta)$.

If the prior for f is modified to $f_k \stackrel{\text{ind}}{\sim} \text{Nor}(0, k^{-2\alpha-1})$ for $k \le \lfloor n^{1/(1+2\alpha)} \rfloor$ and $f_k = 0$ otherwise, show that the Bernstein–von Mises theorem for θ holds if $\beta > \frac{3}{2}$ and $3/2 < \alpha < (3\beta - 2)/(4 - 2\beta)$.

12.6 (Castillo 2012b) Consider a pair of functional regression models $Y_i \stackrel{\text{ind}}{\sim} \text{Nor}(f(i/n), 1)$ and $Z_i \stackrel{\text{ind}}{\sim} \text{Nor}(f(i/n - \theta), 1)$, $i = 1, \ldots, n$, θ lies in a compact interval $[-c_0, c_0] \subset (-1/2, 1/2)$, and f is square integrable and 1-periodic. Assume that the true f_0 has complex Fourier coefficients $\hat{f}_0(k) = \int e^{-2\pi i k t} f(t)\,dt$ satisfying $\hat{f}_0(0) = 0$, $\hat{f}_0(1) \neq 0$, $\sum_{k\in\mathbb{Z}} k|\hat{f}_0(k)| < \infty$ and $\sum_{k\in\mathbb{Z}} k^{2\beta}|\hat{f}_0(k)|^2 < \infty$. Put a prior density on θ which is positive and continuous, and on f, define a prior by the relations $f = Z_0 + \sum_{j=1}^{k_n} \big[Z_j \cos(2\pi j t) + W_j \sin(2\pi j t)\big]$, $Z_0 \sim \text{Nor}(0, 1)$, $Z_j, W_j \stackrel{\text{ind}}{\sim} \text{Nor}(0, j^{-(1+2\alpha)})$, $j = 1, \ldots, k_n := \lfloor n^{1/(1+2\alpha)} \rfloor$. Show that the Bernstein–von Mises theorem holds for θ if $\beta > 3/2$ and $3/2 < \alpha < 2\beta - 3/2$.

12.7 (Castillo 2012a) Consider two independent Gaussian white noise models $dX(t) = f(t)\,dt + n^{-1/2}\,dB_1(t)$ and $dY(t) = f(t - \theta)\,dt + n^{-1/2}\,dB_2(t)$, $t \in [0, 1]$, where f is an unknown periodic function and θ an unknown location parameter taking values in some interval $[-\tau, \tau]$, $0 < \tau < 1/2$. Let θ be given the uniform prior and for f consider the following two priors Π_1 and Π_2 under which the distributions of f can be represented respectively as

$$f(x) = \sqrt{2} \sum_{k=1}^{\infty} \big[(2k)^{-\alpha-1/2} \xi_{2k} \cos(2\pi k x) + (2k)^{-\alpha-1/2} \xi_{2k} \sin(2\pi k x)\big],$$

$$f(x) = \sqrt{2} \sum_{k=1}^{\infty} \big[(2k)^{-\alpha-1/2} \xi_{2k} \cos(2\pi k x) + (2k+1)^{-\alpha-1/2} \xi_{2k} \sin(2\pi k x)\big],$$

where $\xi_1, \xi_2, \ldots \stackrel{\text{iid}}{\sim} \text{Nor}(0, 1)$. Assume that the true value of θ is 0 and the true value f_0 of f is given by

$$f_0(x) = \sqrt{2} \sum_{k=1}^{\infty} \big[(2k)^{-\beta-1/2} \cos(2\pi k x) + (2k+1)^{-\beta-1/2} \sin(2\pi k x)\big],$$

where $\beta > 3/2$. Show that both priors lead to the same posterior convergence rate $n^{-\alpha\wedge\beta/(2\alpha+1)}$, the Bernstein–von Mises theorem holds for the prior Π_1, but the Bernstein–von Mises theorem fails for the prior Π_2.

12.8 Suppose that a and b are elements of a Hilbert space such that $h - \langle h, a\rangle b \perp b$, for every h. Show that $a = b/\|b\|^2$ and $b = a/\|a\|^2$.

13

Survival Analysis

In survival analysis the data are times of occurrences of events, with the special feature that these times may be only partially observable. In this chapter we consider Bayesian nonparametric methods for survival data, allowing for right censoring of the survival times. After a general introduction, we consider the Dirichlet process prior for the survival distribution and the beta process prior for the cumulative hazard function, and study their properties. Next we introduce general "neutral to the right" priors on the survival distribution, or equivalently independent increment process priors for the cumulative hazard function, and derive their conjugacy, their asymptotic consistency, and Bernstein–von Mises theorems. We obtain similar results for the Cox proportional hazard model. We also discuss priors for the hazard function, and the Bayesian bootstrap for censored data.

13.1 Introduction

Survival analysis is concerned with inference on the distribution of times of occurrence of events of interest. These are referred to as *survival times*, as in most applications the event is associated with failure of an object, such as the death of a subject, the onset of a disease, or the end of functioning of equipment. Survival times are generally nonnegative, whence the analysis concerns inference on distributions on the positive half line $(0, \infty)$.

The distribution of a random variable X on $(0, \infty)$ is given by its cumulative distribution function F, but in survival analysis it is customary to make inference on the *survival function* $\bar{F}$, defined as $\bar{F}(x) = 1 - F(x)$, for $x \geq 0$. The main complicating factor is that the survival time X may not be directly observable, but be subject to censoring. Censoring blocks the actual value of the observation, and only reveals partial information in the form of a comparison with a *censoring variable* C. Various types of censoring are possible.

(i) *Right censoring*: X is observed if $X \leq C$; otherwise C is observed and $X \geq C$ is noted. In other words, we observe the pair $T = \min(X, C)$ and $\Delta = \mathbb{1}\{X \leq C\}$.
(ii) *Left censoring*: the pair $T = \max(X, C)$ and $\Delta = \mathbb{1}\{X \geq C\}$ is observed.
(iii) *Interval censoring, type I*: given two censoring variables $C_1 \leq C_2$, there are three possibilities: X is fully observed if $C_1 \leq X \leq C_2$; C_1 is observed and the fact that $X < C_1$; or C_2 is observed and the fact that $X \geq C_2$. In other words, we observe $C_1, C_2, \mathbb{1}\{X < C_1\}, X\mathbb{1}\{C_1 \leq X \leq C_2\}, \mathbb{1}\{X > C_2\}$.
(iv) *Interval censoring, type II*: two censoring variables $C_1 \leq C_2$, and the indicators $\mathbb{1}\{X < C_1\}, \mathbb{1}\{C_1 \leq X \leq C_2\}, \mathbb{1}\{X > C_2\}$ are observed.

(v) *Current status censoring*: a censoring variable C, and only the "current status" of the observation relative to C, i.e. whether $X \leq C$ or not are observed.

In addition, the observation could be subject to *truncation*: not at all observed if the observation fails to cross a certain threshold. Whereas a censored observation gives partial information, a truncated observation is completely lost. In the latter case the distribution of the observed data may not be the quantity of interest: the sample is biased towards larger values.

Instead of modeling survival times through their survival function $\bar{F}$ it is often insightful to use the *cumulative hazard function*

$$H(t) = \int_{(0,t]} \frac{dF}{1 - F-}. \tag{13.1}$$

This is a nondecreasing right-continuous function, and hence can be thought of as a Lebesgue-Stieltjes measure. It is finite on finite intervals strictly within the support of F, but its total mass may exceed 1. In fact its total mass necessarily increases to infinity unless F has an atom at the right end of its support.[1] We shall use F and H interchangeably as functions and measures. If F possesses a density f relative to the Lebesgue measure, then H is absolutely continuous also, with density $h(t) = f(t)/(1 - F(t))$. This is referred to as the *hazard function* of X, and can be interpreted as the "instantaneous rate of failure at time t given survival up to time t," since

$$h(t) = \lim_{\delta \downarrow 0} \frac{1}{\delta} \mathrm{P}(t \leq X \leq t + \delta \,|\, X \geq t) = \frac{f(t)}{1 - F(t)}. \tag{13.2}$$

Survival and cumulative hazard functions are in one-to-one correspondence. The survival function can be recovered uniquely from the cumulative hazard function through the relation

$$1 - F(t) = \prod_{(0,t]}(1 - dH) = e^{-H^c(t)} \prod_{u \in (0,t]} (1 - \Delta H(u)). \tag{13.3}$$

Here the expression in the middle is formal notation for the *product integral*, for which the expression on the far right is one definition, where H^c is the continuous part of H and $\Delta H(u) = H(u) - H(u-)$ the jump of H at u. An alternative definition of the product integral is

$$\prod_{(s,t]}(1 - dH) = \lim_{n \to \infty} \prod_{i=1}^{n} (1 - (H(u_i) - H(u_{i-1}))), \tag{13.4}$$

where the limit is taken over any sequence of meshes $s = u_0 < u_1 < \cdots < u_n = t$ with maximum mesh width tending to zero. If F (or equivalently H) is continuous, then $H = H^c$ and (13.3) reduces to the simpler relation $\bar{F} = e^{-H}$. It will be useful to define for general F the function $A = -\log \bar{F}$, so that $\bar{F} = e^{-A}$ always; and $A = H$ if F has no atoms.

In this chapter we mostly discuss *random right censoring*: right censoring by observation times C that are independent of the event times. In the i.i.d. model one has an independent random sample of survival times $X_i \stackrel{\text{iid}}{\sim} F$, independent censoring times $C_i \stackrel{\text{iid}}{\sim} G$, for $i = 1, \ldots, n$, and observes the n pairs of observations $(T_1, \Delta_1), \ldots, (T_n, \Delta_n)$, where

[1] Proofs and further background to the results in this section can be found for instance in Andersen et al. (1993).

$T_i = \min(X_i, C_i)$ and $\Delta_i = \mathbb{1}\{X_i \leq C_i\}$. We denote the data by D_n. The parameters in this model are identifiable, in that the relationship between the pair (F, G) and the distribution of the pair (T, Δ) is one-to-one if restricted to the interval where $(1-F)(1-G) > 0$.[2] The observations can be summarized through the two *counting processes* $N = (N(t)\colon t \geq 0)$ and $(Y(t)\colon t \geq 0)$ defined by

$$N(t) = \sum_{i=1}^{n} \mathbb{1}\{T_i \leq t\}\Delta_i = \sum_{i=1}^{n} \mathbb{1}\{X_i \leq t, X_i \leq C_i\}, \tag{13.5}$$

$$Y(t) = \sum_{i=1}^{n} \mathbb{1}\{T_i \geq t\} = \sum_{i=1}^{n} \mathbb{1}\{X_i \geq t, C_i \geq t\}. \tag{13.6}$$

The processes N and Y give the numbers of observed failures and of subjects still alive (or *at risk*) at time t, respectively. It can be shown that the process $N - \int_0^{\cdot} Y\, dH$ is a martingale (the "predictable" process $\int_0^{\cdot} Y\, dH$ is the "compensator" of the counting process N), from which it can be concluded that the process $\int_0^{\cdot} \mathbb{1}\{Y > 0\}Y^{-1}\, dN - H$ is a (local) martingale as well. Since it is zero at zero, it has zero expectation, provided this exists. This may motivate the following estimator for the cumulative hazard function, known as the *Nelson-Aalen estimator*,

$$\hat{H}(t) = \int_{(0,t]} \mathbb{1}\{Y > 0\}\, \frac{dN}{Y}. \tag{13.7}$$

The classical estimator of the survival function $\bar{F}$, known as the *Kaplan-Meier estimator*, is the survival function corresponding to $\hat{H}$:

$$1 - \hat{F}(t) = \prod_{(0,t]} \Big(1 - \mathbb{1}\{Y > 0\}\frac{dN}{Y}\Big). \tag{13.8}$$

This estimator can also be derived as the *nonparametric maximum likelihood estimator*, defined as the maximizer of the *empirical likelihood*

$$F \mapsto \prod_{i=1}^{n} F\{X_i\}^{\Delta_i}(1 - F(C_i))^{1-\Delta_i}.$$

Alternatively, it can be motivated by a factorization of the survival probability $\bar{F}(t)$ as the product $\prod_i \mathrm{P}(X > t_i \mid X > t_{i-1})$ of conditional survival probabilities over a grid of time points $0 = t_0 < t_1 < \cdots < t_k = t$. A natural estimator of $\mathrm{P}(X > t_i \mid X > t_{i-1})$ is one minus the ratio of the number of deaths $N(t_{i-1}, t_i]$ in the interval $(t_{i-1}, t_i]$ and the number at risk $Y(t_{i-1})$ at the beginning of the interval. For a fine enough partition each interval will contain at most one distinct time of death (which may shared by several individuals), and each interval without a death contributes a factor 1 to the product. In the limit of finer and finer partitions the estimator thus takes the form

$$1 - \hat{F}(t) = \prod_{j:T_j^* \leq t} \Big(1 - \frac{\sum_{i:X_i = T_j^*} \Delta_i}{\sum_{i:X_i \geq T_j^*} 1}\Big), \tag{13.9}$$

[2] See e.g. Lemma 25.74 in van der Vaart (1998).

where $T_1^* < T_2^* < \cdots < T_k^*$ are the distinct values of the uncensored observations. This is identical to (13.8), and for this reason the Kaplan-Meier estimator is also known as the *product limit estimator*. If the largest observation is censored, then the nonparametric maximum likelihood estimator is non-unique, as any placement of the mass beyond the last observation will give the same likelihood. The Kaplan-Meier estimator (13.8) will then be an improper distribution, in that $\hat{F}(t) = \hat{F}(t_{(n)}) < 1$ for $t > T_{(n)}$.

13.2 Dirichlet Process Prior

The random right censoring model has two parameters: the survival distribution F and the censoring distribution G. By the independence of the survival and censoring times, a prior on G has no role in the posterior computation for F if G is chosen a priori independent of F. The most obvious prior on the survival distribution F is the Dirichlet process $\mathrm{DP}(\alpha)$.

By Theorem 4.6 the posterior distribution of F given the complete set of survival times $X_1, \ldots, X_n$ is $\mathrm{DP}(\alpha + \sum_{i=1}^n \delta_{X_i})$. Adding the censoring times to the data does not change the posterior. Hence by the towering rule of conditioning, the posterior distribution of F given the observed data D_n is the $\mathrm{DP}(\alpha + \sum_{i=1}^n \delta_{X_i})$ distribution "conditioned on D_n." This is a mixture of Dirichlet processes with the mixing distribution equal to the conditional distribution of $X_1, \ldots, X_n$ given D_n, in the model $F \sim \mathrm{DP}(\alpha)$ and $X_1 \ldots, X_n | F \overset{\text{iid}}{\sim} F$.

In the latter setup the variables $X_1, \ldots, X_n$ are a "sample from the Dirichlet process" and their distribution is described through their predictive distributions in Section 4.1.4. The dependence between $X_1, \ldots, X_n$ makes the conditioning nontrivial. Denoting the distribution of $(X_1, \ldots, X_n)$ by m, and the conditional distributions of groups of variables within m by $m(\cdot | \cdot)$, we can write the posterior distribution of F given D_n as

$$\int \cdots \int \prod_{i:\Delta_i=0} \mathbb{1}_{x_i \geq C_i} \mathrm{DP}\Big(\alpha + \sum_{i=1}^n \delta_{X_i} \Delta_i + \sum_{i=1}^n \delta_{x_i} (1 - \Delta_i)\Big)$$
$$dm((x_i : \Delta_i = 0) | (X_i : \Delta_i = 1)).$$

However, this representation is cumbersome, as m itself is a mixture of many components. It turns out that there is a tractable representation in terms of a Pólya tree processes.

We can view the censoring times $C_1, \ldots, C_n$ as constants. Fix any sequence $\mathcal{T}_m = \{A_\varepsilon : \varepsilon \in \mathcal{E}^m\}$ of successive, binary, measurable partitions of the sample space $\mathbb{R}^+$, as in Section 3.5, that includes the sets (C_i, ∞) for all observed censoring times C_i and generates the Borel sets. (Such a tree can be constructed by including the ordered times C_i as splitting points; see the proof below for an example.) The Dirichlet process prior is a Pólya tree process for this sequence of partitions, as it is for *any* sequence of partitions (see Section 4.1.2). The following theorem shows that it remains a Pólya tree process with respect to this special type of tree after updating it with censored data.

Theorem 13.1 *If D_n are random right censored data resulting from a sample of survival times $X_1, \ldots, X_n | F \overset{iid}{\sim} F$ with $F \sim \mathrm{DP}(\alpha)$, then the posterior distribution $F | D_n$ follows a Pólya tree prior $\mathrm{PT}(\mathcal{T}_m, \alpha(A_\varepsilon) + N(A_\varepsilon) + M(A_\varepsilon))$ on any sequence of nested partitions that includes the sets (C_i, ∞) for the observed censoring times C_i, where*

$$N(A)=\sum_{i=1}^n \Delta_i \mathbb{1}\{T_i\in A\},\qquad M(A)=\sum_{i=1}^n (1-\Delta_i)\mathbb{1}\{(T_i,\infty)\subset A\}.$$

In particular, for any partition $0=t_0<t_1<\cdot<t_k=t$ *that contains the observed censoring times* $\{C_i\colon \Delta_i=0\}$ *smaller than t, with* $\bar M(t)=\sum_{i=1}^n(1-\Delta_i)\mathbb{1}\{C_i\ge t\}$,

$$\mathrm{E}\big[1-F(t)|\,D_n\big]=\prod_{i=1}^k\Big(1-\frac{\alpha(t_{i-1},t_i]+N(t_{i-1},t_i]}{\alpha(t_{i-1},\infty)+N(t_{i-1},\infty)+\bar M(t_i)}\Big).$$

Proof The data consists of full observations X_i, for i such that $\Delta_i=1$, and censored observations, for which it is only known that $X_i\in(C_i,\infty)$. Updating the prior with only the full observations gives a Dirichlet posterior process $\mathrm{DP}(\alpha+\sum_{i=1}^n\Delta_i\delta_{X_i})$, by Theorem 4.6. On the given partition sequence $\{\mathcal{T}_m\}$ this can be represented as a Pólya tree process $\mathrm{PT}(\mathcal{T}_m,\alpha'_\varepsilon)$ with parameters $\alpha'_\varepsilon=\alpha_\varepsilon+N(A_\varepsilon)$. Each censored observation contributes information that is equivalent to a Bernoulli experiment with possible outcomes that the survival time is in (C_i,∞) or not, with probabilities $\bar F(C_i)$ and $F(C_i)$, and that is realized in giving the first of these two outcomes. Thus this observation contributes the term $\bar F(C_i)$ to the likelihood. By construction the set (C_i,∞) is contained in the partitioning tree, whence $\bar F(C_i)$ can be written as the product of the splitting variables along the path in the partitioning tree corresponding to the set. The prior distribution of the splitting variables is given by a product of beta densities. Multiplying this product by the likelihood increases the powers in the beta densities by 1 if they refer to a set A_ε that contains (C_i,∞) (so that A_ε is in its path) and leaves the powers unchanged otherwise. In other words, the posterior is again a Pólya tree process with parameters α'_ε or $\alpha'_\varepsilon+1$ depending on whether A_ε contains the set (C_i,∞) or not. Repeating this procedure for all censored observations, we obtain the first assertion of the theorem.

To derive the formula for the posterior mean, we apply the preceding with the partition that uses the numbers $t_1,t_2,\ldots,t_k$ as the first k splitting points for the right most branch of the splitting tree: we first split at t_1 giving $A_0=(-\infty,t_1]$ and $A_1=(t_1,\infty)$, next split A_0 arbitrarily and A_1 into $A_{10}=(t_1,t_2]$ and $A_{11}=(t_2,\infty)$, next split A_{00}, A_{01}, A_{10} arbitrarily and A_{11} at t_3, etc., thus ensuring for every i that $A_{11\cdots1}=(t_i,\infty)$ for the string $11\cdots1\in\{0,1\}^i$. We continue in this manner until $t_k=t$, and let all subsequent splits be arbitrary, except that they must include the remaining censoring times in order to meet the condition in the first part of the theorem. By their definition the splitting variables V_0, V_{10}, $V_{110},\ldots$ give the conditional probabilities $F(A_0)$, $F(A_{10}|\,A_1)$, $F(A_{110}|\,A_{11}),\ldots$, whence for this special partition

$$1-F(t)=F(A_{11\cdots1})=(1-V_0)(1-V_{10})\times\cdots\times(1-V_{11\cdots10}),$$

where V_0 is attached to A_0, V_{10} to $(t_1,t_2]$, etc., and the last variable $V_{11\cdots10}$ in the product is attached to $(t_{k-1},t_k]$. Under the (Pólya tree) posterior distribution these variables are independent and possess beta distributions. By the first part of the proof the beta distribution of $V_{11\cdots10}$ with $i-1$ symbols 1 has parameters $\alpha(t_{i-1},t_i]+N(t_{i-1},t_i]$ and $\alpha(t_i,\infty)+N(t_i,\infty)+M(t_i,\infty)$. Here $M(t_i,\infty)=\bar M(t_i)$, since (t_i,∞) contains (C_j,∞) if and only if $C_j\ge t_i$. Finally we replace every term in the product representing $\bar F(t)$ by its expectation. □

The variable $M(A)$ is zero on every set A that is bounded to the right, and counts the number of observed censoring times C_i with $C_i > c$ for a partitioning set of the form $A_{11\cdots 1} = (c, \infty)$. In the latter case the survival time X_i is known to be to the right of C_i. Hence $M(A)$ counts the number of censored survival times that are certain to belong to A, and the sum $N(A) + M(A)$ counts all survival times that are certain to belong to A. This is intuitively plausible: survival times are counted in a partitioning set A_ε if and only if they are certain to have fallen in this set.

The formula for the posterior mean may be applied with different partitions, to produce seemingly different representations. If the grid contains all observed values $T_j < t$ and t, then $N(t_{i-1}, t_i] = \Delta N(t_i)$, for N the counting process defined in (13.5). Since in this case also $N(t_{i-1}, \infty) + \bar{M}(t_i) = Y(t_i)$ for the at risk process Y defined in (13.6), the formula reduces to

$$\mathrm{E}[1 - F(t)|\, D_n] = \prod_{i=1}^{k}\Big(1 - \frac{\alpha(t_{i-1}, t_i] + \Delta N(t_i)}{\alpha(t_{i-1}, \infty) + Y(t_i)}\Big).$$

For $|\alpha| \to 0$ this reduces to the Kaplan-Meier estimator. The minimal permissible grid for the formula for the posterior mean in Theorem 13.1 consists of all observed censoring times (C_i with $\Delta_i = 0$) smaller than t augmented with t. The formula then reduces to a product "over the censored observations." This appears not to be especially useful, but is somewhat unexpected as the Kaplan-Meier estimator is a product over the *uncensored* observations.

The processes N and Y both tend to infinity at the order n, on any interval within the supports of the survival and censoring times. Therefore the difference between the posterior mean and the Kaplan-Meier estimator should be asymptotically negligible, resulting in the consistency and asymptotic normality of the posterior mean. The following result follows from a general theorem in Section 13.4.1, but is included here with a sketch of a direct proof.

Theorem 13.2 *As $n \to \infty$ almost surely $[P^\infty_{F_0,G}]$,*

(i) $\mathrm{E}[F(t)|\, D_n] \to F_0(t)$,
(ii) $var[F(t)|\, D_n] \to 0$.

In particular, the posterior distribution of $\bar{F}(t)$ is consistent.

Proof For A_n the discrete measure with $A_n\{t_i\} = \alpha(t_{i-1}, t_i]$, we can write the posterior mean of the survival distribution as

$$\mathrm{E}[1 - F(t)|\, D_n] = \prod_{i=1}^{k}\Big(1 - \frac{\Delta(A_n + N)(t_i)}{A_n(t_i-) + Y(t_i)}\Big).$$

By the uniform law of large numbers we have that $n^{-1}(A_n + N)$ tends almost surely to $n^{-1}\mathrm{E}[N] = \int_{(0,\cdot]} \bar{G}_-\, dF$, and that $n^{-1}(A_{n-} + Y)$ tends almost surely to $\mathrm{E}[n^{-1}Y] = \bar{F}_-\bar{G}_-$, where for a monotone function H, $H_-(t) = H(t-)$. The continuity of the maps $(U, V) \mapsto \Lambda = \int_{(0,\cdot]} V^{-1}\, dU \mapsto \prod_{(0,\cdot]}(1 - d\Lambda)$ on a domain where the function V is bounded away from zero now gives result (i).

To prove (ii) it now suffices to show that $\mathrm{E}[(1-F(t))^2|\, D_n] \to \bar{F}_0(t)^2$, almost surely. This may be proved by calculation of second moments of product of independent beta variables. We omit the details. □

13.3 Beta Process Prior

The beta process appears to be the canonical prior for the cumulative hazard function, just as the Dirichlet process is for the distribution function. The beta process is an example of an independent increment process, which are discussed in the next section in general. In this section we derive the beta process as a limit of beta priors on the hazard function in a discrete time setup, where definition and computation are elementary.

13.3.1 Discrete Time

For a given grid width $b > 0$ consider survival and censoring variables taking values in the set of points $0, b, 2b, \ldots$ Let $f(jb) = \mathrm{P}(X = jb)$ be the probability of failure at time jb, and let $F(jb) = \mathrm{P}(X \leq jb) = \sum_{l=0}^{j} f(lb)$. Define a corresponding *discrete hazard function* and cumulative hazard function by, for $j = 0, 1, \ldots,$

$$h(jb) = \mathrm{P}(X = jb|\, X \geq jb) = \frac{f(jb)}{\bar{F}(jb-)}, \qquad H(jb) = \sum_{l=0}^{j} h(lb).$$

The discrete hazard function takes its values in $[0, 1]$, and $h(jb) < 1$ except at the final atom of f, if there is one. The functions f and F can be recovered from h (or H) by

$$1 - F(jb) = \prod_{l=0}^{j} (1 - h(lb)), \qquad f(jb) = \Big[\prod_{l=0}^{j-1} (1 - h(lb))\Big] h(jb). \tag{13.10}$$

Suppose that we observe survival times $X_1, \ldots, X_n$ subject to random right censoring at the points $C_1, \ldots, C_n$, with all possible values belonging to the lattice with span b. For inference on f the censoring times may be considered fixed. A typical censored observation $(X_i \wedge C_i, \mathbb{1}\{X_i \leq C_i\})$ then falls in the sample space consisting of the point $(C_i, 0)$ and the points $(jb, 1)$ for $j = 0, 1, \ldots, C_i$, and the likelihoods of these points are $1 - F(C_i)$ and $f(jb)$, respectively. The total likelihood is the product over the likelihoods for the individual observations. In terms of the counting process N and the at-risk process Y, defined in (13.5) and (13.6), this likelihood can be written as[3]

$$\prod_{j=0}^{\infty} (1 - h(jb))^{Y(jb) - \Delta N(jb)} h(jb)^{\Delta N(jb)}. \tag{13.11}$$

[3] It suffices to show this for $n = 1$. For a censored observation $Y(jb) = 1$ or 0 as $jb \leq C$ or $jb > C$, respectively, while $\Delta N \equiv 0$, reducing the display to $\prod_{j:jb \leq C}(1 - h(jb)) = \bar{F}(X)$, by (13.10). For an uncensored observation $Y(jb) = 1$ or 0 as $jb \leq X$ or $jb > X$, while $\Delta N(X) = 1$ and $\Delta N(jb) = 0$ otherwise, reducing the display to $\prod_{j:jb<X}(1 - h(jb))h(X) = f(X)$, by (13.10).

This also has an intuitive interpretation in terms of binomial experiments: at each time jb we record the number $\Delta N(jb)$ of successes in $Y(jb)$ independent experiments with success probability $h(jb)$. Conjugacy of the binomial and beta distributions suggests to put independent beta priors on the parameters $h(jb)$: for given constants $c_{j,b} > 0$ and $0 < \lambda_{j,b} < 1$,

$$h(jb) \overset{\text{ind}}{\sim} \text{Be}\Big(c_{j,b}\lambda_{j,b}, c_{j,b}(1 - \lambda_{j,b})\Big). \tag{13.12}$$

The parameters $\lambda_{j,b}$ can be interpreted as prior guesses, since $\text{E}[h(jb)] = \lambda_{j,b}$, while the parameters $c_{j,b}$ control the prior variability of the discrete hazard, since $\text{var}[h(jb)] = \lambda_{j,b}(1-\lambda_{j,b})/(1+c_{j,b})$. It is attractive to generate the hyper parameters by a single function $c\colon [0, \infty) \to \mathbb{R}^+$ and cumulative hazard function Λ on $[0, \infty)$ through

$$c_{j,b} = c(jb), \qquad \lambda_{j,b} = \Lambda(jb) - \Lambda((j-1)b). \tag{13.13}$$

The induced prior on the cumulative hazard function H is then called a *discrete beta process* with parameters c and Λ.

By the binomial-beta conjugacy (see Proposition G.8) the joint posterior distribution of the parameters $h(jb)$ is

$$h(jb)|\, D_n \overset{\text{ind}}{\sim} \text{Be}\Big(c_{j,b}\lambda_{j,b} + \Delta N(jb), c_{j,b}(1 - \lambda_{j,b}) + Y(jb) - \Delta N(jb)\Big).$$

It follows that the posterior distribution of the cumulative hazard function H also follows a discrete beta process. If the parameters satisfy (13.13) and either c is constant on every interval $((j-1)b, jb\big]$ or Λ is discrete and supported on the grid points, then the new parameters can again be written in the form (13.13) but with c and Λ updated to $c + Y$ and $t \mapsto \int_{[0,t]}(c+Y)^{-1}\,(c d\Lambda + dN)$. The posterior mean and variance are given by

$$\hat{H}(jb) := \text{E}[H(jb)|\, D_n] = \sum_{l=0}^{j} \frac{c_{l,b}\lambda_{l,b} + \Delta N(lb)}{c_{l,b} + Y(lb)},$$

$$\text{var}[H(jb)|\, D_n] = \sum_{l=0}^{j} \frac{\Delta\hat{H}(lb)(1 - \Delta\hat{H}(lb))}{c_{l,b} + Y(lb) + 1}.$$

By the prior independence of the $h(jb)$ and the factorization of the likelihood over these parameters, it follows that the variables $h(jb)$ are a posteriori independent, which leads to

$$1 - \hat{F}(jb) := \text{E}[1 - F(jb)|\, D_n] = \prod_{l=0}^{j}\Big(1 - \frac{c_{l,b}\lambda_{l,b} + \Delta N(lb)}{c_{l,b} + Y(lb)}\Big). \tag{13.14}$$

The "noninformative" limits as $c_{j,b} \to 0$ of the posterior means $\hat{H}(jb)$ and $\hat{F}(jb)$ are the discrete-time Nelson-Aalen estimator $\sum_{l=0}^{j} \Delta N(lb)/Y(lb)$ and the discrete time Kaplan-Meier estimator, respectively. Like in the case of a Dirichlet process, the parameters $c_{j,b}$, or function c, measure the "strength of prior belief."

13.3.2 Continuous Time

Consider now a passage to the limit as the mesh width b tends to zero. We can view the discrete beta process as a process in continuous time, that jumps at the grid points jb only. By construction it has independent, nonnegative increments; it is cadlag, provided that we define it as such on the intermediate time values. In the terminology introduced in Appendix J all jumps occur at "fixed times," and hence its continuous intensity measure ν^c vanishes. Its discrete intensity measure has the beta distributions (13.12) as its jump heights distributions $\nu^d(\{bj\}, ds)$. The following proposition shows that as $b \to 0$ the discrete beta process with parameters specified as in (13.13) tends to a *beta process* with parameters c and Λ, which is defined as follows.

Definition 13.3 (Beta process) A *beta process* with parameters (c, Λ) is an independent increment process with intensity measure $\nu = \nu^c + \nu^d$ on $(0, \infty) \times (0, 1)$ of the form

$$\nu^c(dx, ds) = c(x)s^{-1}(1-s)^{c(x)-1}\, d\Lambda^c(x)\, ds,$$
$$\nu^d(\{x\}, \cdot) = \text{Be}(c(x)\Delta\Lambda(x), c(x)(1 - \Delta\Lambda(x))).$$

Here $c: [0, \infty) \to [0, \infty)$ is a measurable function, and Λ is a cumulative hazard function, with $\Lambda^c = \Lambda - \Lambda^d$ its continuous part. The beta distribution in the second line is understood to be degenerate at zero if $\Delta\Lambda(x) = 0$.

The interpretation of the definition is as follows. A beta process is a sum

$$H(t) = \sum_{x:x\le t} \Delta H(x) + \sum_{j:x_j\le t} \Delta H(x_j),$$

of two independent processes that both increase by jumps ΔH only. The second component jumps only at "fixed times" $x_1, x_2, \ldots$, given by the atoms of Λ, and with "jump heights" $\Delta H(x_j) \overset{\text{ind}}{\sim} \text{Be}(c(x_j)\Lambda(\{x_j\}), c(x_j)(1 - \Lambda(\{x_j\}))$. The first component can be obtained by simulating a Poisson process with intensity measure ν^c on the set $(0, \infty) \times (0, 1)$ (i.e. the number of points in a set A is $\text{Poi}(\nu^c(A))$) and creating a jump at x of height $s = \Delta H(x)$ for each point (x, s) in the Poisson process with $x \le t$. In the next section this type of process is introduced in general.

The beta process may be viewed as a canonical prior for a cumulative hazard function. In Example 13.11 it will be seen that a Dirichlet process prior on the survival function $\bar{F}$ yields a beta process of a special parameterization on the cumulative hazard function.

From the general theory (see (13.18) and (13.19)), it follows that

$$\text{E}[H(t)] = \Lambda(t), \qquad \text{var}[H(t)] = \int_{(0,t]} \frac{1-\Delta\Lambda}{c+1}\, d\Lambda. \tag{13.15}$$

Proposition 13.4 *For c a cadlag, nonnegative function and Λ a continuous cumulative hazard function, the discrete time beta process with parameters specified as in* (13.13) *tends as $b \to 0$ in distribution in the Skorohod space $\mathfrak{D}[0, \infty)$ equipped with the Skorohod topology on compacta to a beta process with parameters c and Λ.*

Proof The discrete time beta process has intensity measure $\nu_b = \sum_j \delta_{jb} \times B_{j,b}$, for $B_{j,b}$ the beta distribution specified in (13.12) and (13.13). We first show that the measures $s\,\nu_b(dx, ds)$ converge weakly on compact time sets to the measure $s\,\nu(dx, ds)$, for ν given in Definition 13.3. For any continuous function with compact support f and $k \geq 0$, by the formula for the $(k+1)$st moment of a beta distribution, and with $a_{j,b} = c_{j,b}\lambda_{j,b}$,

$$\int\int f(x)s^k\, s\,\nu_b(dx, ds) = \sum_j f(jb)\frac{(a_{j,b}+k)\cdots(a_{j,b}+1)a_{j,b}}{(c_{j,b}+k)\cdots(c_{j,b}+1)c_{j,b}}.$$

Since $a_{j,b}/c_{j,b} = \lambda_{j,b} = \Lambda((j-1)b, jb]$ and $c_{j,b} = c(jb)$, the right side can be written as $\int f_b g_b\, d\Lambda$, for $f_b = \sum_j f(jb)\mathbb{1}\{((j-1)b, jb]\}$ and

$$g_b = \sum_j \frac{(c(jb)\Lambda((j-1)b, jb]+k)\cdots(c(jb)\Lambda((j-1)b, jb]+1)}{(c(jb)+k)\cdots(c(jb)+1)}(jb)\mathbb{1}\{((j-1)b, jb]\}.$$

By continuity $f_b \to f$ pointwise, and similarly the functions g_b converge pointwise to a limit g. By the dominated convergence theorem $\int f_b g_b\, d\Lambda$ tends to

$$\int f(x)\frac{(c\Delta\Lambda+k)\cdots(c\Delta\Lambda+1)}{(c(x)+k)\cdots(c(x)+1)}\, d\Lambda(x) = \int\int f(x)s^k\, s\,\nu(dx, ds).$$

For continuous Λ the jumps $\Delta\Lambda$ vanish, and the numerator in the left side reduces to $k!$. Then the last equality follows by evaluating the integral over s in the right side as the beta integral $B(k+1, c(x))$. By convergence of moments we conclude that the measures $A \mapsto \int\int \mathbb{1}\{s \in A\} f(x) s\,\nu_b(dx, ds)$ converge weakly to the same measures with ν replacing ν_b, for every f. Then the weak convergence $s\,\nu_b(dx, ds) \rightsquigarrow s\,\nu(dx, ds)$ on compact time sets follows.

Under the assumption that Λ is continuous the proposition follows by Proposition J.18. □

As its discrete counterpart, the beta process possesses the attractive property of conjugacy with respect to randomly right censored data. In the case that the prior parameter Λ is continuous the following theorem is a consequence of the general conjugacy of independent increment processes given in Theorem 13.15.

Theorem 13.5 (Conjugacy) *If the cumulative hazard function H follows a beta process with parameters (c, Λ), where c is continuous and bounded away from zero and Λ has at most finitely many discontinuities in any bounded interval, then the posterior distribution of H given random right censored data D_n is again a beta process, with parameters $(c+Y, \Lambda^*)$, where $\Lambda^*(t) = \int_{(0,t]}(c+Y)^{-1}(c\,d\Lambda + dN)$.*

Proof We update the parameters of the intensity measure $\nu(dx, ds) = \rho(ds|\, x)\,\Lambda(dx)$ as in Theorem 13.15 with $\rho(ds|\, x) = c(x)s^{-1}(1-s)^{c(x)-1}\, ds$. The continuous part of the intensity measure is updated to $c(x)s^{-1}(1-s)^{c(x)+Y(x)-1}\, ds\, \Lambda^c(dx)$, in which we can rewrite $c(x)\,\Lambda^c(dx)$ as $(c(x)+Y(x))\,(\Lambda^*)^c(dx)$. Thus we obtain the continuous part of the intensity measure of a beta process with parameters $(c+Y, \Lambda^*)$. In the fixed jump part, the measure $\rho(\cdot|\, x)$ must be updated to $s^{\Delta N(x)}(1-s)^{Y(x)-\Delta(x)}\rho(ds|\, x)$. If Λ had no atom at x, then this means creating a new fixed jump with a beta distribution with parameters equal to $(\Delta N(x), Y(x) - \Delta N(x) + c(x))$; in the other case the parameters $(c(x)\Delta\Lambda(x), c(x)(1 - \Delta\Lambda(x)))$ must be updated by adding $(\Delta N(x), Y(x) - \Delta N(x))$, giving the beta distribution with parameters $(\Delta N(x) + c(x)\Delta\Lambda(x), Y(x) - \Delta N(x) + c(x)(1 - \Delta\Lambda(x)))$. In both cases this corresponds to a fixed jump distribution of a beta process with parameters $(c + Y, \Lambda^*)$.

If Λ is continuous, then the assumptions of Theorem 13.15 are fulfilled. As is noted in its proof, Theorem 13.15 remains valid if Λ has finitely many jumps. □

13.3.3 Sample Path Generation

Because the (marginal) distributions of the increments of a beta process do not possess a simple, closed form, it is not straightforward to simulate sample paths from the prior or posterior distributions. In this section we list some algorithms that give approximations. Most of the algorithms extend to other independent increment processes by substituting the correct intensity measure.

As explained following Definition 13.3 the beta process can be split in a part with a continuous parameter Λ and a fixed jump part. As the parts are independent and the fixed jump part is easy to generate, we concentrate on the case that $\Lambda = \Lambda^c$ is continuous.

The first algorithm is to simulate the discrete time beta process; this tends to the beta process as the time step tends to zero by Proposition 13.4. The other algorithms are based on the representation of a beta process through a counting measure: $H(t) = \sum_{x:x\le t} \Delta H(x)$, where for $t \le \tau$ the sum has countably many terms given by the points $(x, \Delta H(x))$ of a Poisson process on $(0, \tau] \times (0, 1)$ with intensity measure given by

$$\nu(dx, ds) = c(x)s^{-1}(1 - s)^{c(x)}\, ds\, d\Lambda(x).$$

We obtain an approximation by generating a large (but regrettably necessarily finite) set of points (X_i, S_i) according to this Poisson process, or some approximation, and forming the process $H(t) = \sum_i S_i \mathbb{1}\{X_i \le t\}$. This can be achieved in various ways, leading to several algorithms, which differ in complexity, dependent on the availability of numerical routines to compute certain special functions. Algorithm (d) appears to give a particularly good trade-off between ease, efficiency and accuracy.

Algorithm a (Time discretization) For a sufficiently small b generate H as a discrete time beta process on the points $0, b, 2b, \ldots$, with parameters given by (13.13). This construction is justified by Proposition 13.4.

Algorithm b (Inverse Lévy measure I) For $L_x(s) = \Lambda(\tau)\int_s^1 c(x)u^{-1}(1 - u)^{c(x)-1}\, du$, and a sufficiently large number m, generate, for $i = 1, \ldots, m$,

$$X_i \overset{\text{iid}}{\sim} \frac{\Lambda(\cdot \wedge \tau)}{\Lambda(\tau)}, \qquad E_i \overset{\text{iid}}{\sim} \text{Exp}(1), \qquad V_j = \sum_{i=1}^{j} E_i, \qquad S_i = L_{X_i}^{-1}(V_i).$$

Then $H(t) = \sum_{i=1}^{m} S_i \mathbb{1}\{X_i \le t\}$ is approximately a beta process.

This algorithm is based on the fact that the points (X_i, S_i), for $i = 1, 2, \ldots$, are a realization from the Poisson process with intensity measure ν (see Example J.9). For $m = \infty$ the algorithm would be exact.

Algorithm c (Inverse Lévy measure II) For Q_s the probability distribution on $(0, \tau]$ satisfying $dQ_s(x) \propto c(x)(1-s)^{c(x)-1}\, d\Lambda(x)$, the nondecreasing function L given by $L(s) = \int_0^\tau \int_{(s,1]} u^{-1} c(x)(1-u)^{c(x)-1} du\, d\Lambda(x)$, and a sufficiently large number m, generate, for $i = 1, \ldots, m$,

$$E_i \overset{\text{iid}}{\sim} \text{Exp}(1), \qquad V_j = \sum_{i=1}^{j} E_i, \qquad S_i = L^{-1}(V_i), \qquad X_i \mid S_i \overset{\text{ind}}{\sim} Q_{S_i}.$$

Then $H(t) = \sum_{i=1}^m S_i \mathbb{1}\{X_i \le t\}$ is approximately a beta process.

As Algorithm (b), this algorithm generates the points (X_i, S_i) of the Poisson process with intensity measure ν, but this intensity measure is disintegrated in the other direction. Marginally the jump heights of the beta process follow a Poisson process on $[0, 1]$ with intensity measure $\nu((0, \tau] \times \cdot)$. As the latter process explodes at 0, it is convenient to order its points in reverse order of magnitude, downwards from 1 to 0. The function L is the cumulative intensity function of this reversed process and hence its points can be obtained by transforming the points $V_1, V_2, \ldots$ of a standard Poisson process by the inverse L^{-1}. Given the jump heights, the locations X_i in $[0, \tau]$ are independent, and can be generated from the intensity measure conditioned to the horizontal line at the specified height w, which is given by Q_w.

Algorithm d (Poisson weighting) For positive conditional probability densities $s \mapsto g(s \mid x)$ on $[0, 1]$, and $f(x, s) = c(x)s^{-1}(1-s)^{c(x)-1}$, and a sufficiently large number m, generate, for $i = 1, \ldots, m$,

$$X_i \overset{\text{iid}}{\sim} \frac{\Lambda(\cdot \wedge \tau)}{\Lambda(\tau)}, \qquad S_i \mid X_i \overset{\text{ind}}{\sim} g(\cdot \mid X_i), \qquad K_i \mid X_i, S_i \overset{\text{ind}}{\sim} \text{Poi}\Big(\frac{\Lambda(\tau) f(X_i, S_i)}{m\, g(S_i \mid X_i)}\Big).$$

Then $H(t) = \sum_{i=1}^m S_i K_i \mathbb{1}\{X_i \le t\}$ is approximately a beta process. A convenient particular special choice for g is the density of the Be$(1, c(x))$-distribution, for which the quotient $f(x, s)/g(s \mid x)$ reduces to $1/s$. However, the Be$(\epsilon, c(x))$-distribution for $\epsilon < 1$ appears to work more efficiently.

The algorithm can be understood as placing Poisson numbers K_i of points at the locations (X_i, S_i) in the jump space $(0, \tau] \times [0, 1]$. Allowing more than one point at a location differentiates it from the other algorithms and is unlike the Poisson process corresponding to the beta process. However, for large m this may be viewed as a (random) discretization, and the mean number of points is chosen so that the expected number of points is exactly right, as, for any Borel set A,

$$m\text{E}[K_i \mathbb{1}\{(X_i, S_i) \in A\}] = m \int_0^\tau \int_0^1 \mathbb{1}_A(x, s) \frac{\Lambda(\tau) f(x, s)}{m\, g(s \mid x)}\, g(s \mid x)\, ds\, \frac{d\Lambda(x)}{\Lambda(\tau)} = \nu(A).$$

A rigorous justification for the algorithm is that the process H converges in the Skorohod space to a beta process, as $m \to \infty$ (see Lee 2007 or Damien et al. 1995).

Algorithm e (ϵ-approximation) For a sufficiently small $\epsilon > 0$, generate

$$K \sim \mathrm{Poi}\Big(\frac{1}{\epsilon}\int_0^\tau c\,d\Lambda\Big), \qquad X_1,\ldots,X_K|\,K \overset{\text{iid}}{\sim} \frac{c\,d\Lambda}{\int_{[0,\tau]} c\,d\Lambda}, \qquad S_i|\,X_i \overset{\text{ind}}{\sim} \mathrm{Be}(\epsilon, c(X_i)).$$

Then $H(t) = \sum_{i=1}^m S_i \mathbb{1}\{X_i \leq t\}$ is approximately a beta process.

The scheme actually generates a realization of the independent increment process with intensity measure $\nu_\epsilon(dx, ds) = \epsilon^{-1} c(x)\,\mathrm{be}(\epsilon, c(x))(s)\,ds\,d\Lambda(x)$, where $\mathrm{be}(\alpha, \beta)$ is the density of the beta distribution. Indeed, for reasonable functions c the measure ν_ϵ is finite with total mass $\nu_\epsilon([0,\tau]\times[0,1]) = \epsilon^{-1}\int_0^\tau c\,d\Lambda$ and hence the corresponding Poisson process can be simulated by first generating its total number of points and next their locations, where we may first generate the x-coordinates from the second marginal of the renormalized ν_ϵ and next the corresponding s-coordinates. The justification of the algorithm is that the intensity measures ν_ϵ converge in the appropriate sense to ν, as $\epsilon \to 0$, so that the corresponding independent increment processes converge in the Skorohod topology to the beta process with intensity measure ν, by Proposition J.18. Note that $\epsilon^{-1}\mathrm{be}(\epsilon, c)(s) \sim s^{-1}(1-s)^{c-1}$, as $\epsilon \downarrow 0$, since $\epsilon B(\epsilon, c) \to 1$.

13.3.4 Mixtures of Beta Processes

As in the case of a Dirichlet or Pólya tree process, the parameters c and Λ of a beta process may depend on a hyperparameter θ, which may be given a prior ν. This results in a *mixture of beta processes* prior for H.

The posterior distribution given right censored data D_n is again a mixture.

Theorem 13.6 *If $H|\,\theta$ follows a mixture of beta process prior with continuous parameters c_θ and Λ_θ and $\theta \sim \nu$, then the posterior distribution of H given right censored data is again a mixture of beta process: $H|\,(D_n, \theta)$ follows a beta process with parameters $(c_\theta^*, \Lambda_\theta^*)$, and $\theta|\,D_n \sim \nu^*$, with parameter updates given by $c_\theta^* = c_\theta + Y$, $d\Lambda_\theta^* = (c_\theta + Y)^{-1}(c_\theta\,d\Lambda_\theta + dN)$ and*

$$\nu^*(d\theta) \propto \exp\Big[-\sum_{i=1}^n \int_0^{T_{(i)}} \frac{c_\theta}{c_\theta + n - i}\,d\Lambda_\theta\Big] \prod_{i=1}^{K_{n,j}} \frac{c_\theta(U_j) h_\theta(U_j)}{\prod_{i=1}^j (c_\theta(U_i) + Y(U_i) - i)}\,\nu(d\theta),$$

where K_n is the number of distinct uncensored observations, $U_1, \ldots, U_{K_n}$ are the distinct uncensored observations and $K_{n,j}$ is the number of uncensored observations greater than U_j, for $j = 1, \ldots, K_n$.

The proof of the theorem is based on a calculation of the marginal distribution of the sample using the posterior conjugacy of the beta process (see Theorem 13.5), and then applying Bayes's theorem, as in case of a mixture of Dirichlet or Pólya tree process posterior; see Kim (2001) for details.

13.4 Neutral to the Right and Independent Increment Processes

The lack of conjugacy of the Dirichlet process prior for right censored data diminishes its role in survival analysis. Since the family of survival distributions is not (assumed) dominated, Bayes's theorem is not applicable, which leaves conjugacy as the most important path to computation of the posterior distribution. A large class of priors defined below will be seen to have the desirable conjugacy property.

Definition 13.7 (Neutral to the right process) A random distribution function F (or the corresponding survival function) is said to follow a *neutral to the right* (NTR) process if for every finite partition $0 = t_0 \leq t_1 < \cdots < t_k < \infty$ and $k \in \mathbb{N}$, the random variables

$$F(t_1), \frac{F(t_2) - F(t_1)}{1 - F(t_1)}, \ldots, \frac{F(t_k) - F(t_{k-1})}{1 - F(t_{k-1})}$$

(or equivalently the variables $\bar{F}(t_j)/\bar{F}(t_{j-1})$) are mutually independent.

This definition can be alternatively and perhaps more easily posed in terms of the cumulative hazard function: a random survival function is neutral to the right if the corresponding cumulative hazard function has independent increments. We say that a stochastic process $H = (H(t): t \geq 0)$ is an *independent increment process* (or *PII*) if for every finite partition $0 = t_0 \leq t_1 < \cdots < t_k < \infty$ and $k \in \mathbb{N}$, the random variables $H(t_1) - H(t_0), H(t_2) - H(t_1), \ldots, H(t_k) - H(t_{k-1})$ are jointly independent.

Theorem 13.8 *For a random distribution F, the corresponding cumulative hazard function H, and the function $A = -\log \bar{F}$, the following assertions are equivalent:*

(i) *F follows a neutral to the right process.*
(ii) *H is an independent increment process.*
(iii) *A is an independent increment process.*

In this case, the means $F_0(t) = \mathrm{E}[F(t)]$ and $H_0(t) = \mathrm{E}[H(t)]$ are corresponding survival and cumulative hazard functions: $H_0(t) = \int_{(0,t]} dF_0/\bar{F}_{0-}$ and $\bar{F}_0(t) = \prod_{(0,t]}(1 - dH_0)$.

Proof The equivalence of (i) and (iii) is immediate from the definition of A and the fact that the logarithm turns quotients into differences. We prove the equivalence of (i) and (ii).

Fix a partition $0 = t_0 < t_1 < \cdots < t_k < \infty$, and a countable dense subset $\{s_1, s_2, \ldots\}$ of $\mathbb{R}^+$. For a given m, let $s_{1:m} < \cdots < s_{m:m}$ be the ordering of $\{s_1, \ldots, s_m\}$. Then

$$H(t_j) - H(t_{j-1}) = \lim_{m \to \infty} \sum_{t_{j-1} < s_{i:m} \leq t_j} \frac{F(s_{i:m}) - F(s_{i-1:m})}{1 - F(s_{i-1:m})}. \tag{13.16}$$

If F follows an NTR process, then the summands on the right are mutually independent. Since for disjoint intervals $(t_{j-1}, t_j]$ the sums have no common terms, it follows that the variables $H(t_j) - H(t_{j-1})$, for $j = 1, \ldots k$, are mutually independent. Conversely, by the product-integral representation (13.3) of F in terms of H,

$$\frac{\bar{F}(t_j)}{\bar{F}(t_{j-1})} = \lim_{m\to\infty} \prod_{t_{j-1}<s_{i:m}\le t_j} (1 - H(s_{i-1:m}, s_{i:m}]).$$

If H has independent increments, then the factors of the product are mutually independent. Since there is no common factor for disjoint intervals $(t_{j-1}, t_j]$, the variables $\bar{F}(t_j)/\bar{F}(t_{j-1})$, for $j = 1, \ldots, k$, are mutually independent.

To prove the final statement, observe that $\mathrm{E}(U/V) = \mathrm{E}(U)/\mathrm{E}(V)$ whenever the random variables U/V and V are independent. Applying this to the variables $U = F(s_{i:m}) - F(s_{i-1:m})$ and $V = 1 - F(s_{i-1:m})$ in (13.16), we see that this display remains valid if H and F are replaced by their means H_0 and F_0. This leads to the representation of H_0 in terms of F_0. The converse follows as the relation between cumulative hazard and survival functions is one-to-one. □

An independent increment process with cadlag, nondecreasing sample paths is the distribution function of a random measure on $[0, \infty)$, which is a *completely random measure* (CRM) as defined in Appendix J (see Problem J.5). By Proposition J.6 it can be represented as

$$H(t) = \int_{(0,t]} \int s\, N^c(dx, ds) + \sum_{j:x_j\le t} \Delta H(x_j),$$

for a Poisson random measure N^c on $[0, \infty) \times (0, \infty)$ and arbitrary points $x_1, x_2, \ldots$ in $[0, \infty)$, called the *fixed jump times* of H. The Poisson process N^c is fully characterized by its *intensity measure* $\nu_H^c(A) = \mathrm{E}[N^c(A)]$, and is independent of the "fixed jump" heights $\Delta H(x_1), \Delta H(x_2), \ldots$, which are arbitrary independent nonnegative variables. The representation shows that the sample paths of a PII with nondecreasing sample paths necessarily increase by jumps only, and hence $H(t) = \sum_{x\le t} \Delta H(x)$, where for every sample path the series has at most countable many jumps. The fixed jump times are deterministic locations at which *every* sample path of H increases (by the random heights $\Delta H(x_j)$), but typically most of the jumps occur at random locations, different for different sample paths. These locations are given by the x-coordinates of the points (x, s) in the Poisson process N^c, with s the jump height $\Delta H(x)$ at location x.

For unity of notation it is attractive to encode the fixed jumps also in a counting measure, and write the representation in the form

$$H(t) = \int_{(0,t]} \int s\, N(dx, ds), \qquad N = N^c + N^d, \qquad N^d = \sum_j \delta_{x_j, \Delta H(x_j)}.$$

The sum N has mean measure the sum $\nu_H = \nu_H^c + \nu_H^d$ of the mean measures of N^c and N^d. The measure ν_H^d concentrates on the set $\cup_j\{x_j\} \times (0, \infty)$, while ν_H^c gives zero probability to this set; in particular, the measures ν_H^c and ν_H^d are identifiable from their sum. The latter measure gives the law of the fixed jump heights: for $j = 1, 2, \ldots$ and $D \subset (0, \infty)$,[4]

$$\mathrm{P}(\Delta H(x_j) \in D) = \nu_H(\{x_j\} \times D) = \nu_H^d(\{x_j\} \times D).$$

The measures N^d and N (unless $N^d = 0$) are *not* Poisson random measures, unlike N^c.

[4] The variable $\Delta H(x_j)$ in a general PII may have an atom at zero of size $1 - \nu_H^d(\{x_j\} \times [0, \infty))$, but in the present chapter this atom is always zero.

For theoretical manipulation the Laplace transform of the variable $H(t)$ is handy, as this allows explicit expression in terms of the intensity measure. For $\theta > 0$, if the fixed jump heights are strictly positive a.s.,

$$\mathrm{E}[e^{-\theta H(t)}] = e^{-\int_{(0,t]}\int (1-e^{-\theta s})\,\nu_H^c(dx,ds)} \prod_{x:x\leq t} \int e^{-\theta s}\,\nu_H^d(\{x\}, ds). \tag{13.17}$$

Here $\int f(s)\,\nu_H^d(\{x\}, ds)$ denotes the integral of the function f with respect to the measure $A \mapsto \nu_H^d(\{x\}, A)$, and the product is understood to be over the fixed jump times x, and to be 1 if there are no fixed jumps. Differentiating at $\theta = 0$ (or see (J.11) and (J.12)), we find

$$\mathrm{E}[H(t)] = \int_{(0,t]}\int s\,\nu_H(dx, ds), \tag{13.18}$$

$$\mathrm{var}[H(t)] = \int_{(0,t]}\int s^2\,\nu_H(dx, ds) - \sum_{x:x\leq t}\Big(\int s\,\nu_H^d(\{x\}, ds)\Big)^2. \tag{13.19}$$

This aids in the interpretation of the intensity measure. For prior modeling one will typically not include fixed atoms, but we shall see that these arise naturally in the posterior distribution.

To construct a stochastic process H with independent, nondecreasing sample paths one may start from any measure $\nu_H = \nu_H^c + \nu_H^d$ on $[0,\infty) \times (0,\infty)$, where the x-marginals of ν_H^c and ν_H^d are atomless and discrete, respectively, satisfying, for $t, x \geq 0$,

$$\int_{(0,t]}\int (s \wedge 1)\,\nu_H^c(dx, ds) < \infty, \qquad \nu_H^d(\{x\} \times [0,\infty)) \leq 1. \tag{13.20}$$

A cumulative hazard function H is restricted to have jump heights ΔH smaller than 1, and corresponds to a proper probability distribution only if either $H(t) \to \infty$ as $t \to \infty$ and all jumps are strictly smaller than 1, or H has a single jump of size 1 at the end of its support. The jump heights can be controlled by restricting ν_H to $[0,\infty) \times [0,1)$; the following lemma gives a sufficient condition for the distribution function to be proper.

Lemma 13.9 *Any measure ν_H on $[0,\infty) \times [0,\infty)$ satisfying* (13.20), *for every $t > 0$, is the intensity measure of a stochastic process with independent, nonnegative increments. If ν_H concentrates on $[0,\infty) \times [0,1)$ and satisfies $\int_0^\infty \int_0^1 s\,\nu_H(dx, ds) = \infty$, then this process is a cumulative hazard function corresponding to a proper probability distribution F on $(0,\infty)$, almost surely. In particular, this is true for an intensity measure of the form $\nu_H(dx, ds) = \rho(ds|\,x)\,d\alpha(x)$, with $\inf_x \int_0^1 s\,\rho(ds|\,x) > 0$ and $\alpha(\mathbb{R}^+) = \infty$.*

Proof That (13.20) characterizes intensity measures follows from the general theory explained in Appendix J. In view of (13.18) the condition $\int_0^\infty \int_0^1 s\,\nu_H(dx, ds) = \infty$ implies that $\mathrm{E}[H(t)] \uparrow \infty$ as $t \to \infty$, whence $\mathrm{E}[\bar F(t)] \to 0$ by Theorem 13.8. By monotonicity $\bar F(t) \to 0$ almost surely.

For the special choice of intensity, the double integral is for every $t > 0$ bounded below by $\inf_{x:x\leq t} \int s\,\rho(ds|\,x)\,\alpha(0,t]$, and hence is infinite. □

If F follows a neutral to the right process, then both H and A corresponding to F are processes with independent increments, and hence allow representations through jump measures. The following proposition connects their intensity measures.

Proposition 13.10 *The intensity measures of the independent increment processes H and A corresponding to a neutral to the right survival distribution F satisfy*

$$\nu_A(C \times D) = \nu_H(C \times \{s: -\log(1-s) \in D\}),$$
$$\nu_H(C \times D) = \nu_A(C \times \{s: 1 - e^{-s} \in D\}).$$

Proof The jumps of A and H are related by $\Delta A = -\log(\bar{F}/\bar{F}_-) = -\log(1 - \Delta H)$. It follows that the Poisson process N_A^c corresponding to A has a jump at $(x, -\log(1-s))$ if and only if the Poisson process N_H^c corresponding to H has a jump at (x, s). This implies the relationship between the mean measures ν_A^c and ν_H^c of the Poisson processes, as stated. The mean measures ν_A^d and ν_H^d of the fixed jumps transform similarly.[5] ☐

Example 13.11 (Dirichlet process) It follows from Theorem 4.28 that a Dirichlet process prior on a distribution F on $(0, \infty)$ is also a neutral to the right process. In fact, the theorem shows that the Dirichlet process is "completely neutral," which may be described as "neutral to the right" also if the intervals $(t_{j-1}, t_j]$ are placed in arbitrary and not their natural order.

It turns out that a survival distribution F follows a $\mathrm{DP}(MF_0)$-process with prior strength M and center measure F_0 if and only if the corresponding cumulative hazard function follows a beta process with parameters $(M\bar{F}_{0-}, H_0)$, for H_0 the cumulative hazard function that goes with F_0. Thus the Dirichlet processes form a subclass of the beta processes, with the two parameters linked.

An insightful method to verify the claim is to approximate the prior that the Dirichlet process induces on the cumulative hazard function by a discrete beta process. As $b \downarrow 0$ the discrete cumulative hazard functions H_b with jumps $\Delta H_b(jb) = F((j-1)b, jb]/F((j-1)b)$ tend pointwise to H, as seen in (13.16). If $F \sim \mathrm{DP}(MF_0)$, then the jumps $\Delta H_b(jb)$ are independent $\mathrm{Be}(M\bar{F}_0((j-1)b)\lambda_{b,j}, M\bar{F}_0((j-1)b)\,(1-\lambda_{b,j}))$ variables, for $\lambda_{j,b} := F_0((j-1)b, jb]\,/F_0((j-1)b)$ by the self-similarity of the Dirichlet process (see (4.9)). In other words, H_b is a discrete beta process with parameters $c(jb) = M\bar{F}_0((j-1)b)$ and $\lambda_{j,b}$, as given (note that $c(jb)$ is identified as the sum of the two parameters, and next $\lambda_{j,b}$ as the first parameter divided by $c(jb)$). An argument analogous to that of Proposition 13.4 will show that the processes H_b converge in distribution to a beta process with parameters $c = M\bar{F}_{0-}$ and $\Lambda = H_0$.

Example 13.12 (Beta-Stacy process) A survival distribution F is said to follow a *beta-Stacy process* with parameters (c_0, F_0) if the corresponding cumulative hazard function H follows a beta process with parameters $(c_0\bar{F}_{0-}, H_0)$ (as in Definition 13.3). In other words, the intensity measure is given by

[5] Since a Poisson process remains a Poisson process under a change of variables, the argument gives an elegant proof of Theorem 13.8 as well.

$$\nu_H^c(dx, ds) = c_0(x)s^{-1}(1-s)^{c_0(x)\bar{F}_0(x-)-1}\, dF_0^c(x)\, ds,$$
$$\nu_H^d(\{x\}, ds) = \mathrm{be}(c_0(x)F_0\{x\}, c_0(x)\bar{F}_0(x-))\, ds.$$

The parameters c_0 and F_0 are a measurable function $c_0\colon \mathbb{R}^+ \to \mathbb{R}^+$ and a cumulative distribution function F_0, and H_0 is the cumulative hazard function that goes with F_0. In view of Example 13.11 the beta-Stacy process generalizes the Dirichlet process by replacing the prior strength M by a function c_0, thus providing a "location-dependent" prior strength. On the other hand, it offers not more than a reparameterization of the beta process.

Mean and variance of the beta-Stacy process can be computed using the general equations (13.18) and (13.19), leading to (13.15) with $\Lambda = H_0$ and $c = c_0 F_0 -$. In particular, the parameter H_0 is the prior mean of H, whence F_0 is the prior mean of F, in view of Theorem 13.8.

By Proposition 13.10 the process $A = -\log \bar{F}$ follows a neutral to the right process with intensity measure

$$\nu_A^c(dx, ds) = (1-e^{-s})^{-1} e^{-sc_0(x)\bar{F}_0(x-)} ds\, c_0(x)\, dF_0^c(x),$$
$$\nu_A^d(\{x\}, ds) \propto (1-e^{-s})^{c_0(x)F_0\{x\}-1} e^{-sc_0(x)\bar{F}_0(x-))}\, ds.$$

Here the discrete components $\nu_A^d(\{x\}, \cdot)$ are proper probability measures on $(0, \infty)$, and restricted to the jump points x of F_0.

Example 13.13 (Extended gamma process) A prior process is said to follow an *extended gamma process* with parameters (c_0, A_0) if its associated process $A = -\log \bar{F}$ is an independent increment process with intensity measure

$$\nu_A^c(dx, ds) = s^{-1} c_0(x) e^{-c_0(x)s}\, ds\, dA_0(x).$$

By Proposition 13.10 the corresponding cumulative hazard function H has intensity measure

$$\nu_H^c(dx, ds) = \frac{c_0(x)(1-s)^{c_0(x)-1}}{\log_-(1-s)}\, ds\, dA_0(x).$$

For c_0 a constant function we obtain the *standard gamma process*.

The increments of the standard gamma process possess gamma distributions. The extended process can be constructed from a standard process ξ as the integral $A(x) = \int_0^x c_0^{-1}\, d\xi$, where ξ must have intensity $c_0\, dA_0$, i.e. the process ξ must have independent increments with $\xi(x) \sim \mathrm{Ga}(\int_0^x c\, dA_0, 1)$, for every $x > 0$ (see Example J.15).

The support of a neutral to the right process is determined by the support of the intensity measure ν_H of the independent increment process H, and is typically very large in terms of the weak topology.

Theorem 13.14 (Support) *If H does not have fixed jumps and the support of its intensity measure ν_H is equal to $\mathbb{R}^+ \times [0, 1]$, then the weak support of the corresponding neutral to the right process F is the full space of probability measures $\mathfrak{M}(\mathbb{R}^+)$.*

Proof Since the continuous distributions are dense in $\mathfrak{M}(\mathbb{R}^+)$, it suffices to show that any F_0 with continuous cumulative distribution is in the support. We shall show that for every continuous F_0 the prior gives positive probability to every Kolmogorov-Smirnov ball $\{F: d_{KS}(F, F_0) < \epsilon\}$, for $\epsilon > 0$. We can restrict to compact support, because if $1 - F_0(\tau) < \epsilon/2$ and F is within uniform distance $\epsilon/2$ to F_0 on the interval $[0, \tau]$, then F is within uniform distance ϵ to F_0 on $\mathbb{R}^+$. Because the product integral $H \mapsto F = 1 - \prod(1 - dH)$ is continuous relative to the uniform norms on compact intervals,[6] it further suffices to show that for every $\tau, \delta > 0$ the set $\{H: \sup_{x \le \tau} |H(x) - H_0(x)| < \delta\}$ receives positive prior mass, for H_0 the cumulative hazard function of F_0.

By the uniform continuity of H_0 on $[0, \tau]$, there exists a partition $0 = t_0 < t_1 < \cdots < t_k = \tau$ such that $H_0(t_{i-1}, t_i] < \epsilon/2$, for every i. If $|H(t_{i-1}, t_i] - H_0(t_{i-1}, t_i]| < \epsilon/(2k)$, for every i, then $|H(x) - H_0(x)| < \epsilon$, for every $x \le \tau$. Now $H(t_{i-1}, t_i]$ is the sum of the heights of the points of the associated Poisson process N inside the strip $(t_{i-1}, t_i] \times [0, 1)$. For D_i equal to the interval $(H_0(t_{i-1}, t_i] - \epsilon/(2k), H_0(t_{i-1}, t_i] + \epsilon/(2k)]$, we split this process as

$$N = \sum_{(x,s): s<\delta} \delta_{(x,s)} + \sum_{i=1}^{k} \sum_{(x,s)\in(t_{i-1},t_i]\times D_i, s\ge\delta} \delta_{(x,s)} + \sum_{i=1}^{k} \sum_{(x,s)\in(t_{i-1},t_i]\times D_i^c, s\ge\delta} \delta_{(x,s)}.$$

Because $\int_{0,\tau]} \int_0^\delta s \nu_H(dx, ds) \downarrow 0$, as $\delta \downarrow 0$, the contribution $\int_{(0,\tau]} \int_0^\delta s\, N(dx, ds)$ of the first term on the right to H can be made arbitrarily small by choosing δ small. Let G be the event that this contribution is smaller than ϵ. Let E_i and F_i be the events that the i terms in the two sums on the right have exactly one term and are empty, respectively, i.e. the process N has exactly one point inside every set $(t_{i-1}, t_i] \times (D_i \cap [\delta, 1))$ and zero points inside $(0, \tau] \times (D_i^c \cap [\delta, 1))$, respectively.

The events $E_1, \ldots, E_k, F_1, \ldots, F_k, G$ are independent and have positive probabilities, as they refer to disjoint parts of the Poisson process, and $N(E_i)$ and $N(F_i)$ are Poisson distributed with means $\nu_H(E_i) > 0$ and $\nu_H(F_i) > 0$, by the assumption on ν_H. On the intersection $\cap_i (E_i \cap F_i) \cap G$ the process H is within distance 2ϵ of H_0. □

The main result of this section is the conjugacy of neutral to the right process priors for random right censored data: the posterior distribution of F corresponding to an neutral to the right prior is again neutral to the right. Because the neutral to the right property is equivalent to the independent increment property of the cumulative hazard function, this fact can also be formulated in terms of PIIs and their intensity measures. The following theorem explicitly derives the intensity measure of the posterior process.

A typical prior intensity measure will be absolutely continuous. However, the posterior intensity measure will have fixed jumps at the uncensored observations. This is somewhat analogous to the Kaplan-Meier estimator, which also jumps (only) at uncensored observations.

Recall the notations N and Y for the observed death process and at risk process relating to randomly right censored data $D_n = \{(T_1, \Delta_1), \ldots, (T_n, \Delta_n)\}$, given in (13.5) and (13.6).

Theorem 13.15 (Conjugacy) *If F follows a neutral to the right process, then the posterior distribution of F given randomly right censored data D_n also follows a neutral to the right*

[6] See Theorem 7 of Gill and Johansen (1990).

process. If the corresponding cumulative hazard process H possesses intensity measure ν_H with disintegration $\nu_H(dx, ds) = \rho(ds|\, x)\, \Lambda(dx)$ such that $x \mapsto s\rho(ds|\, x)$ is weakly continuous and Λ is without atoms, then its posterior distribution possesses intensity measure $\nu_{H|D_n}$ given by

$$\nu^c_{H|D_n}(dx, ds) = (1-s)^{Y(x)} \rho(ds|\, x)\, d\Lambda(x),$$
$$\nu^d_{H|D_n}(\{x\}, ds) \propto s^{\Delta N(x)}(1-s)^{Y(x)-\Delta N(x)} \rho(ds|\, x),$$

where the set of fixed jump times is equal to the set $\{T_i\colon \Delta_i = 1\}$ of uncensored observations, and the fixed jump distributions $\nu^d_{H|D_n}(\{x\}, \cdot)$ are probability distributions on $(0, 1)$.

Proof To highlight the structural property of a neutral to the right process, we start with a simple proof of the preservation of the neutral to the right property in the posterior. (The property also follows from the explicit calculations in the second part of the proof.) It suffices to prove the result for $n = 1$, since then the conclusion can be repeated n times.

First consider the case that the observation is not censored. Denote it by T and fix a partition $0 < u_1 < \cdots < u_k < \infty$ of $\mathbb{R}^+$. Given a finer partition $0 = t_0 < t_1 < \cdots < t_{m+1} = \infty$, define $M_i = \mathbb{1}\{t_{i-1} < X \le t_i\}$ and $\theta_i := \bar{F}(t_i)/\bar{F}(t_{i-1})$. The likelihood function for observing $M = (M_1, \ldots, M_m)$ can be written as

$$\prod_{i=1}^m \mathrm{P}_\theta(M_i = m_i|\, M_{i-1} = m_{i-1}, \ldots, M_1 = m_1) = \prod_{i=1}^m (1-\theta_i)^{m_i} \theta_i^{1-\sum_{l=1}^{i-1} m_l},$$

which factorizes in separate functions of $\theta_1, \ldots, \theta_m$. Since F follows a neutral to the right process prior, the random variables $\theta_1, \ldots, \theta_m$ are independent in the prior. Since the likelihood factorizes, these variables are then also independent in the posterior given M. The same is true for the variables $\bar{F}(u_i)/\bar{F}(u_{i-1})$, which are products of disjoint sets of θ_i. By making the partitions $0 < t_1 < \cdots < t_{m+1} = \infty$ finer and finer, the latter posterior distribution tends to the posterior distribution of the variables $\bar{F}(u_i)/\bar{F}(u_{i-1})$ given T, by the martingale convergence theorem, where the independence is preserved.

If the observation is censored, then the same proof works, but we choose the partitions with t_m equal to the censoring variable C (which is independent of everything and may be considered a fixed number), so that by successively refining the partition $0 = t_0 < \cdots < t_m$ the vector M eventually contains the same information as the censored observation. The grid points u_i bigger than t_m cannot appear in the latter partitions, but the corresponding variables $\bar{F}(u_i)/\bar{F}(u_{i-1})$ also do not enter the likelihood for M. Hence the likelihood trivially factorizes and the argument can be finished as before.

For the proof of the updating formula for the intensities, we may equivalently use the intensities of the process $A = -\log \bar{F}$ or of the cumulative hazard function H, in view of Proposition 13.10. We shall use the former, as this allows a more accessible expression for the likelihood function.

Again it suffices to consider the case of a single observation, provided we allow the prior intensity measure to have the form of the posterior. In particular, we allow it to have finitely many fixed atoms, with strictly positive jump sizes. For a given partition $0 = t_0 < t_1 < \cdots < t_m \le t_{m+1} = \infty$, chosen to have $t_m = C$ if the observation is censored, let $M = (M_1, \ldots, M_{m+1})$ have coordinates $M_i = \mathbb{1}\{(t_{i-1}, t_i]\}(T)$. Then M is less informative than the observation (T, Δ), but in the limit as the meshwidth of the partition tends to zero (and $m \to \infty$) it generates the same σ-field. By the martingale convergence theorem the posterior distribution given M tends to the posterior distribution given (T, Δ).

The vector M possesses a multinomial distribution with parameters 1 and $F(t_{i-1}, t_i]$, for $i = 1, \ldots, m + 1$. Its likelihood can be written in terms of A as $L_M(A) = \prod_{i=1}^{m+1}(e^{-A(t_{i-1})} - e^{-A(t_i)})^{M_i}$. We can identify the posterior distribution of A given M by evaluating expectations of the form, for bounded, continuous functions $f: \mathbb{R}^+ \to \mathbb{R}^+$,

$$\mathrm{E}(e^{-\int f\,dA} \mid M) = \frac{\mathrm{E}_A\big[e^{-\int f\,dA} L_M(A)\big]}{\mathrm{E}_A\big[L_M(A)\big]}.$$

Here the expectations E_A are relative to the prior distribution of A, for fixed M.

Exactly one coordinate of M is nonzero, and this is fixed in the conditioning event. On the event that this is the jth coordinate, we can write $e^{-\int f\,dA} L_M(A) = e^{-\int (f+e_{j-1})\,dA} - e^{-\int (f+e_j)\,dA}$, for the function $e_i = \mathbb{1}\{(0, t_i]\}$. Two applications of formula (13.17) give, with the products taken over the set of fixed atoms of A,

$$\begin{aligned}
&\mathrm{E}_A(e^{-\int f\,dA} L_M(A)) \\
&= e^{-\int\int (1-e^{-(f+e_{j-1})(x)s})\,\nu_A^c(dx,ds)} \prod_x \int e^{-(f+e_{j-1})(x)s}\,\nu_A^d(\{x\}, ds) \\
&\quad - e^{-\int\int (1-e^{-(f+e_j)(x)s})\,\nu_A^c(dx,ds)} \prod_x \int e^{-(f+e_j)(x)s}\,\nu_A^d(\{x\}, ds) \\
&= e^{-\int\int (1-e^{-(f+e_{j-1})(x)s})\,\nu_A^c(dx,ds)} \prod_x \int e^{-(f+e_{j-1})(x)s}\,\nu_A^d(\{x\}, ds) \\
&\quad \times \Big[1 - e^{-\int_{(t_{j-1},t_j]}\int e^{-f(x)s}(1-e^{-s})\,\nu_A^c(dx,ds)} \prod_{t_{j-1}<x\le t_j} \frac{\int e^{-f(x)s-s}\,\nu_A^d(\{x\},ds)}{\int e^{-f(x)s}\,\nu_A^d(\{x\},ds)}\Big].
\end{aligned}$$

The denominator $\mathrm{E}_A[L_M(A)]$ can be expressed in exactly the same way, but with f taken equal to 0. Suppose that the meshwidth of the partition tends to zero. If the observation is uncensored, so that $t_j - t_{j-1} \to 0$, then $e_{j-1} \to \mathbb{1}\{(0, T)\}$; if the observation is censored, then $t_{j-1} = C = T$ and hence $e_{j-1} = \mathbb{1}\{(0, T]\}$. If in both cases the limit function is written as e_T, then the quotient (arising from numerator and denominator of the posterior) of the leading terms, outside the square brackets, tends to

$$e^{-\int\int (1-e^{-f(x)s})e^{-e_T(x)s}\,\nu_A^c(dx,ds)} \prod_x \frac{\int e^{-f(x)s}e^{-e_T(x)s}\,\nu_A^d(\{x\},ds)}{\int e^{-e_T(x)s}\,\nu_A^d(\{x\},ds)}. \tag{13.21}$$

This shows that the intensity measure is updated by multiplying it with the density $e^{-e_T(x)s}$. It remains to analyze the expression within square brackets.

If the observation is censored, then $t_j = \infty$ and the first exponential inside the square brackets is equal to $e^{-\infty} = 0$ (this corresponds to $e^{-A(\infty)} = 0$). The proof is then complete. If the observation is uncensored, we split in two cases: T is a fixed jump time of A, or it is not.

If T is a fixed jump time of A, then for a sufficiently fine partition, it will be the only one in the interval $(t_{j-1}, t_j]$, and the product in square brackets will contain exactly one term. Since the x-marginal of ν_A^c has no atoms (by definition), the first exponential within square brackets tends to $e^{-0} = 1$. The quotient (arising from the numerator and denominator of the posterior) of the two terms in square brackets tends to

$$\frac{1 - \int e^{-f(T)s-s}\, \nu_A^d(\{T\}, ds) / \int e^{-f(T)s}\, \nu_A^d(\{T\}, ds)}{1 - \int e^{-s}\, \nu_A^d(\{T\}, ds) / \int \nu_A^d(\{T\}, ds)}.$$

This combines with the contribution of the fixed jump T to the product in the leading term (see (13.21), where $e_T(T) = 0$) to give the contribution of this atom as

$$\frac{\int e^{-f(T)s}\, \nu_A^d(\{T\},ds) - \int e^{-f(T)s} e^{-s}\, \nu_A^d(\{T\},ds)}{\int \nu_A^d(\{T\},ds) - \int e^{-s}\, \nu_A^d(\{T\},ds)} = \frac{\int e^{-f(T)s}(1-e^{-s})\, \nu_A^d(\{T\},ds)}{\int (1-e^{-s})\, \nu_A^d(\{T\},ds)}.$$

This exhibits the updated intensity measure to be proportional to $(1 - e^{-s})\nu_A^d(\{T\}, ds)$.

If T is not a fixed jump time of A, then for a sufficiently fine partition, the interval $(t_{j-1}, t_j]$ will be free of fixed atoms and the product is empty and should be interpreted as 1. The term within square brackets tends to zero, but so does the corresponding term from the denominator of the posterior distribution. By Taylor's expansion, as the meshwidth of the partition (left of the censoring time) tends to zero, the quotient of the terms satisfies

$$\frac{1 - e^{-\int_{(t_{j-1},t_j]} \int e^{-f(x)s}(1-e^{-s})\, \nu_A^c(dx,ds)}}{1 - e^{-\int_{(t_{j-1},t_j]} \int (1-e^{-s})\, \nu_A^c(dx,ds)}} \to \frac{\int e^{-f(T)s}(1 - e^{-s})\, \nu_A^c(ds|\, T)}{\int (1 - e^{-s})\, \nu_A^c(ds|\, T)},$$

where $\nu_A^c(ds|\, x) = \rho(ds|\, x)$ is the conditional distribution of the second coordinate in ν_A^c given the first coordinate. Comparison to (13.17) shows that a (strictly positive) fixed jump is added at T, with intensity measure proportional to $(1 - e^{-s})\, \nu_A^c(ds|\, T)$. □

Because the jump process ΔN is nonzero only at finitely many time points, it vanishes almost everywhere under Λ, and the updating formula for the continuous part of the intensity measure in Theorem 13.15 can also be unified to

$$\nu_{H|D_n}(dx, ds) = s^{\Delta N(x)}(1 - s)^{Y(x)-\Delta N(x)}\, [\nu_H(dx, ds) + Z(x)\, \nu_H(ds|\, x)\, dN(x)],$$

where ν_H is the (continuous part of the) intensity of the prior process, and Z gives the norming constants for the fixed jump components: $Z^{-1} = (\Delta N)^{-1} \int s^{\Delta N}(1-s)^{Y-\Delta N}\, \nu_H(ds|\, \cdot)$. Because of the different roles of continuous and fixed jump parts of the process, this rewrite may be of moderate help.

When combined with the general formulas (13.18) and (13.19), the theorem also gives formulas for the posterior mean $\mathrm{E}[H(t)|\, D_n]$ and posterior variance $\mathrm{var}[H(t)|\, D_n]$, in terms of the posterior intensity measure $\nu_{H|D_n}$.

Example 13.16 (Beta-Stacy process) Within the class of neutral to the right processes, the subclass of beta-Stacy processes also forms a conjugate class for randomly right censored data. Because a beta-Stacy process is a reparameterized beta-process, this follows from Theorem 13.5.

Example 13.17 (Extended beta process) The *extended beta process* with parameters (a, b, Λ) is the independent increment process without fixed jumps with intensity measure given by

$$\nu_H^c(dx, ds) = s^{-1}\mathrm{be}(s; a(x), b(x))\, ds\, d\Lambda(x).$$

The parameters are positive functions a and b and a continuous cumulative hazard function Λ. The posterior process given randomly right censored data D_n is again an independent increment process, with intensity measure given by

$$\nu_{H|D_n}^c(dx, ds) = s^{-1}(1-s)^{Y(x)}\mathrm{be}(s; a(x), b(x))\, d\Lambda(x)\, ds,$$
$$\nu_{H|D_n}^d(\{x\}, ds) = \mathrm{be}(s; a(x) + \Delta N(x) - 1, b(x) + Y(x) - \Delta N(x))\, ds.$$

Here the fixed atoms x appear only at the uncensored survival times. In view of (13.18) and (13.19) the posterior mean and variance of $H(t)$ are given by

$$\mathrm{E}[H(t)|\, D_n] = \int_{(0,t]} \frac{\Gamma(a+b)\Gamma(b+Y)}{\Gamma(b)\Gamma(a+b+Y)}\, d\Lambda + \int_{(0,t]} \frac{a+\Delta N - 1}{a+b+Y-1}\, dN,$$
$$\mathrm{var}[H(t)|\, D_n] = \int_{(0,t]} \frac{a\Gamma(a+b)\Gamma(b+Y)\, d\Lambda}{\Gamma(b)\Gamma(a+b+Y+1)} + \int_{(0,t]} \frac{(a+\Delta N-1)(b+Y-\Delta N)\, dN}{(a+b+Y-1)^2(a+b+Y)}.$$

13.4.1 Consistency

From the variety of independent increment processes, only a small subset leads to an asymptotically consistent posterior distribution. In this section we give simple sufficient conditions in terms of the intensity measure. As is customary in survival analysis we study consistency of the cumulative hazard function H on a finite interval $[0, \tau]$ within the supports of the survival and censoring distributions. We say that the posterior distribution $\Pi_n(\cdot|\, D_n)$ is *consistent* at (F_0, G_0) in this setup if, for every $\epsilon > 0$,

$$\Pi_n\Big(H\colon \sup_{0\le t\le\tau} |H(t) - H_0(t)| > \epsilon|\, D_n\Big) \to 0, \qquad \text{a.s. } [P_{F_0,G_0}^\infty]. \tag{13.22}$$

Here (F_0, G_0) are the true survival and censoring distribution, and P_{F_0,G_0} refers to the distribution of a sample of random right censored data.

Instead of the cumulative hazard function, we may also consider the survival function, but this leads to exactly the same definition of consistency. Indeed, by the continuity of the product integral $H \mapsto \bar{F} = \prod_{(0,\cdot)}(1-dH)$ the consistency (13.22), for every $\epsilon > 0$, implies that, for every $\epsilon > 0$,

$$\Pi_n\Big(F\colon \sup_{0\le t\le\tau} |F(t) - F_0(t)| > \epsilon|\, D_n\Big) \to 0, \qquad \text{a.s. } [P_{F_0,G_0}^\infty].$$

Because the inverse map $F \mapsto H$ is also continuous, the two types of consistency are equivalent.[7]

Since the realizations of independent increment processes are discrete distributions supported on random countable sets and are not dominated (by a σ-finite measure), Schwartz's consistency theory is inapplicable. However, for the neutral to the right priors we can utilize the characterization of the posterior distribution established in Theorem 13.15. This allows computation of posterior mean and variance, and next an appeal to Lemma 6.4. Below we

[7] See Theorem 7 of Gill and Johansen (1990).

add the sample size n as an index to the notations N_n and Y_n for the counting process N and the at-risk process Y.

To illustrate that consistency is far from automatic, we first study the special case of extended beta processes, introduced in Example 13.17.

Proposition 13.18 (Consistency extended beta process) *If the prior for H follows an extended beta process prior with parameters (a, b, Λ), for continuous a, b and Λ, then the posterior distribution for H is consistent at a continuous H_0 (i.e. (13.22) holds for any τ such that $(\bar{F}_0\bar{G}_0)(\tau-) > 0$), if and only if a is identically 1, equivalently, if and only if the prior is a beta process.*

Proof The posterior mean and variance of $H(t)$ are given in Example 13.17. Since H_0 is continuous without probability one there are no ties in the survival times, and hence the jumps ΔN_n in these formulas can be replaced by 1. By the Glivenko-Cantelli theorem the processes $n^{-1}Y_n$ tend almost surely uniformly to $(F_0G_0)_-$, which is bounded away from zero on $[0, \tau]$ by assumption, with a reciprocal of bounded variation. The processes $n^{-1}N_n$ are of variation bounded by 1 and tend uniformly to the function $\int_{(0,\cdot]} G_{0-}\, dF_0$. Application of Lemma 13.19 shows that the first term in the posterior mean converges to 0 almost surely, while the second term tends almost surely $\int_{(0,\cdot]} a/(F_0G_0)_-\, G_{0-}\, dF_0 = \int_{(0,\cdot]} a\, dH_0$. By the same arguments both terms in the expression for the posterior variance tend to zero almost surely.

It follows that the posterior distribution of $H(t)$ tends to $\int_0^t a\, dH_0$, for every $t \in [0, \tau]$. Because pointwise convergence of functions of bounded variation to a continuous function of bounded variation implies uniform convergence, we can apply Lemma 6.4, to conclude that the posterior distribution also tends to $\int_0^t a\, dH_0$ as a process, in the sense of (13.22).

The limit is equal to H_0 if and only if a is identically equal to 1. □

Lemma 13.19 *If A_n are cadlag functions and B_n are cadlag functions of uniformly bounded variation and $A_n \to A$ and $B_n \to B$ uniformly on $[0, \tau]$ for a function of bounded variation A, then $\int_{(0,\cdot]} A_n\, dB_n \to \int_{(0,\cdot]} A\, dB$ uniformly on $[0, \tau]$.*

Proof The difference $\int A_n\, dB_n - \int A\, dB$ is the sum of $\int (A_n - A)\, dB_n$ and $\int A\, d(B_n - B)$. The first is bounded above in absolute value by $\|A_n - A\|_\infty \|B_n\|_{TV}$, and after partial integration, the second can be seen to be bounded by $2\|B_n - B\|_\infty \|A\|_{TV}$. □

Example 13.20 (Curious case of inconsistency) The extended beta process priors Π_1 and Π_2 on the cumulative hazard function H with parameters $(\frac{1}{2}, 1, H_0)$ and $(1, 1, H_0)$ both have prior mean H_0, and hence are centered perfectly. Their prior variances are given by $\mathrm{var}_1[H(t)] = \frac{1}{3}H_0(t)$, for Π_1 and $\mathrm{var}_2[H(t)] = \frac{1}{2}H_0(t)$, for Π_2. The prior Π_1 might seem preferable over Π_2 if H_0 is indeed the true cumulative hazard function, since they have the same, correct mean, but Π_1 is more concentrated. However, by Proposition 13.18 the prior Π_1 leads to an inconsistent posterior distribution, whereas Π_2 is consistent. The proof of the proposition shows that under Π_1 the posterior settles down to $\frac{1}{2}H_0$.

This example demonstrates that the prior mean and (pointwise) variance have minimal roles in determining consistency, which is rather dependent on the structure of the full prior process.

Proposition 13.18 illustrates that consistency depends on the fine properties of the intensity measure as $s \to 0$: within the class of extended beta processes the density of ν_H^c must blow up exactly at the rate s^{-1}, as is true (only) for the ordinary beta processes. In the general case of independent increment processes we obtain consistency under the same condition on the intensity.

Theorem 13.21 (Consistency) *Let the prior for H follows an independent increment process with intensity measure of the form $\nu_H(dx, ds) = s^{-1}q(x, s)\, d\Lambda(x)\, ds$ for a continuous cumulative hazard function Λ and a function q. Assume that q is continuous in x and such that $\sup_{x\in[0,\tau],s\in(0,1)}(1-s)^{\kappa}q(x, s) < \infty$ for some $\kappa > 0$ and $\sup_{x\in[0,\tau]} |q(x, s)-q_0(x)| \to 0$ as $s \to 0$ for a function q_0 that is bounded away from zero and infinity. Then the posterior distribution for H is consistent at every continuous H_0:* (13.22) *holds for every τ such that* $(\bar F_0\bar G_0)(\tau-) > 0$.

Proof We show that $\mathrm{E}[H(t)|\, D_n] \to H_0(t)$ and $\mathrm{var}[H(t)|\, D_n] \to 0$, for every $t \in [0, \tau]$. Since pointwise convergence to a continuous function implies uniform convergence, by Pólya's theorem, we can then appeal to Lemma 6.4 to obtain the result.

Because F_0 is assumed continuous, all jumps ΔN_n in the process N_n have size 1, and a sum over the jumps is the same as an integral with respect to N_n.

By Theorem 13.15 and (13.18), the posterior mean $\mathrm{E}[H(t)|\, D_n]$ is given by

$$\int_0^t\int_0^1 (1-s)^{Y_n(x)}q(x, s)\, ds\, \Lambda(dx) + \int_0^t \frac{\int_0^1 s(1-s)^{Y_n(x)-1}q(x, s)\, ds}{\int_0^1 (1-s)^{Y_n(t)-1}q(x, s)\, ds}\, dN_n(x).$$

The sequence of processes $n^{-1}Y_n$ tends uniformly almost surely to the function $G_{0-}F_{0-}$, which is bounded away from zero on $[0, \tau]$. This implies that the first term is bounded above by $\int_0^t\int_0^1 (1-s)^{cn}q(x, s)\, ds\, d\Lambda(x)$ for some $c > 0$. By the assumption on q this integral is uniformly bounded in $t \le \tau$ by a multiple of $\int_0^1 (1-s)^{cb-\kappa}\, ds = O(1/n)$ and hence negligible.

The processes $n^{-1}N_n$ in the second term are of variation bounded by 1 and tend almost surely to $\int_0^{\cdot} G_{0-}\, dF_0$. To analyze the quotient of integrals inside the outer integral, we first note that $\int_\epsilon^1 (1-s)^{Y_n(x)-1}q(x, s)\, ds \lesssim (1-\epsilon)^{cn}$ almost surely, for some $c > 0$ and uniformly in x, for any $\epsilon > 0$. For $\epsilon_n \to 0$ slowly, this is of smaller order than any power of n^{-1}. We thus see that the quotient of integrals can be written, uniformly in x,

$$\frac{\int_0^{\epsilon_n} s(1-s)^{Y_n(x)-1}q(x, s)\, ds + o(n^{-2})}{\int_0^{\epsilon_n} (1-s)^{Y_n(x)-1}q(x, s)\, ds + o(n^{-2})} \tag{13.23}$$
$$= \frac{Y_n(x)^{-2}\int_0^{Y_n(x)\epsilon_n} u(1-u/Y_n(x))^{Y_n(x)-1}\, du\, (q_0(x)+o(1)) + o(n^{-2})}{Y_n(x)^{-1}\int_0^{Y_n(x)\epsilon_n} (1-u/Y_n(x))^{Y_n(x)-1}\, du (q_0(x)+o(1)) + o(n^{-2})}.$$

Because $n^{-1}Y_n$ converges to a positive limit, for $n\epsilon_n \to \infty$ the integrals in numerator and denominator tend to $\int_0^\infty ue^{-u}\,du$ and $\int_0^\infty e^{-u}\,du$, respectively, which are both equal to 1. We conclude that the left side is asymptotically equivalent to $Y_n(x)^{-1}$, uniformly in x. Substituting this in the preceding display and appealing to Lemma 13.19, we conclude that the posterior mean is equivalent to $\int_{(0,t]} Y_n^{-1}\,dN_n$, which converges to $H_0(t)$.

By (13.19) the posterior variance $\text{var}[H(t)|\,D_n]$ can be written

$$\int_0^t \int_0^1 s(1-s)^{Y_n(x)} q(x,s)\,ds\,d\Lambda(x) \tag{13.24}$$

$$+\int_0^t \Big[\frac{\int_0^1 s^2(1-s)^{Y_n(x)-1}q(x,s)\,ds}{\int_0^1 (1-s)^{Y_n(x)-1}q(x,s)\,ds} - \Big[\frac{\int_0^1 s(1-s)^{Y_n(x)-1}q(x,s)\,ds}{\int_0^1 (1-s)^{Y_n(x)-1}q(x,s)\,ds}\Big]^2\Big]\,dN(x).$$

The first term tends to zero by the same argument as for the posterior mean. Similarly the quotient of integrals in the second term can be seen to be equivalent to $2Y_n(x)^{-2}$, uniformly in x, by arguments as before, while the quotient inside square brackets in the term was already seen to be equivalent to $Y_n(x)^{-1}$. It follows that both terms tend to zero. Hence the posterior variance tends to zero. □

Example 13.22 (Extended gamma process) The extended gamma process prior with parameters (c, A_0) is described in Example 13.13, and has q-function

$$q(x,s) = c(x)\frac{s}{\log_-(1-s)}(1-s)^{c(x)-1}. \tag{13.25}$$

If the function c is continuous and bounded away from zero and infinity on $[0,\tau]$, then this satisfies the conditions of Theorem 13.21.

To see this we note that $\log_-(1-s) \sim s$ and $(1-s)^c - 1 \sim 1$ as $s \to 0$, uniformly in c in bounded intervals $[c_0, c_1] \subset (0,\infty)$, so that $q(x,s) \sim c(x)$ uniformly in x, as $s \to 0$. Furthermore $(1-s)^\epsilon/\log_-(1-s) \to 0$ as $s \to 1$ for $\epsilon > 0$ and hence $(1-s)q(x,s)$ is uniformly bounded.

13.4.2 Bernstein–von Mises Theorem

Under slightly stronger conditions on the intensity measure the posterior distribution converges at the rate $n^{-1/2}$, and satisfies a Bernstein–von Mises theorem. Given the data D_n, the posterior distribution of $\sqrt{n}(H(t) - \mathrm{E}[H(t)|\,D_n])$ converges in distribution to the same (centered) Gaussian distribution as the sequence of centered and scaled posterior means $\sqrt{n}(\mathrm{E}[H(t)|\,D_n] - H_0(t))$, almost surely. This Gaussian distribution is the same as the limit distribution of the Nelson-Aalen estimator $\hat{H}_n$, the nonparametric maximum likelihood estimator of H, given in (13.1). In fact the difference $\sqrt{n}(\mathrm{E}[H(t)|\,D_n] - \hat{H}_n(t))$ tends to zero.

These results are true for a fixed t within the supports of F_0 and G_0, but also in a uniform sense, for the estimators as processes, viewed as elements of the Skorokhod space $\mathfrak{D}[0,\tau]$ of cadlag functions equipped with the uniform distance.

Theorem 13.23 (Bernstein–von Mises) *If the prior for H follows an independent increment process with intensity measure of the form $\nu_H(dx, ds) = s^{-1}q(x, s)\, dx\, ds$ for a function q that is continuous in x and such that $\sup_{x\in[0,\tau], s\in(0,1)}(1-s)^{\kappa}q(x, s) < \infty$ for some $\kappa > 0$ and and $\sup_{x\in[0,\tau]}|q(x, s) - q_0(x)| = O(s^{\alpha})$ as $s \to 0$ for a function q_0 that is bounded away from zero and infinity and some $\alpha \in (1/2, 1]$, then for τ such that $(\bar F_0 \bar G_0)(\tau-) > 0$ and continuous F_0:*

(i) $\sqrt{n}(H - \mathrm{E}[H|\, D_n])|\, D_n \rightsquigarrow B \circ U_0$ *a.s.* $[P^{\infty}_{F_0,G_0}]$ *where B is a standard Brownian motion and* $U_0 = \int_{[0,\cdot)}(\bar F_0 \bar G_0)_-^{-1}\, dH_0$.
(ii) $\sqrt{n}(\mathrm{E}[H(t)|\, D_n] - \hat H_n(t)) = O_P(n^{-(\alpha-1/2)})$ *a.s.* $[P^{\infty}_{F_0,G_0}]$.

Consequently, if $\alpha > 1/2$, then $\sqrt{n}(H - \hat H_n)|\, D_n \rightsquigarrow B \circ U_0$ a.s.$[P^{\infty}_{F_0,G_0}]$.

Proof The posterior distribution is the law of an independent increment process with intensity measure consisting of a Poisson and a fixed jump part, given in Theorem 13.15. The independent increment process is the sum $H = H^p + H^f$ of two independent increment processes corresponding to these two components. We show that the continuous part and its (posterior) mean tend to zero at high speed in n, whereas the discrete part gives rise to the limit law.

By (13.18) and (13.19) (and Theorem 13.15), the posterior mean $\mathrm{E}[H^p(t)|\, D_n]$ and variance $\mathrm{var}[H^p(t)|\, D_n]$ are given by $\int_0^t \int_0^1 s^k(1-s)^{Y_n(x)}q(x, s)\, ds\, dx$, for $k = 0$ and $k = 1$, respectively. From the assumption on the function and the fact that $n^{-1}Y_n$ tends to infinity almost surely, it follows that these integrals are of the order $\int_0^1 s^k(1-s)^{cn}\, ds$, for some $c > 0$, which is $O(n^{-1-k})$. It follows that $\sqrt{n}H^p(t)$ and $\sqrt{n}\mathrm{E}[H^p(t)|\, D_n]$ tend to zero almost surely. Since both processes are nondecreasing, the pointwise convergence at $t = \tau$ actually gives uniform convergence in $\mathfrak{D}[0, \tau]$.

The fixed jump part H^f is a sum of finitely many independent random variables, one at each uncensored survival time. We prove its asymptotic normality (conditional given D_n) using Lyapunov's central limit theorem, applied with the fourth moment (see e.g. Billingsley 1979, Theorem 27.3). The conditional distribution of $\Delta H(x)$ is given by $\nu^d_{H|D_n}$ in Theorem 13.15, where we can take $\Delta N = 1$ by the assumed continuity of F_0. The fourth moments given D_n are given by

$$\mathrm{E}\big[|\Delta H(x)|^4|\, D_n\big] = \frac{\int_0^1 s^4(1-s)^{Y_n(x)-1}q(x, s)\, ds}{\int_0^1 (1-s)^{Y_n(x)-1}q(x, s)\, ds}.$$

By the same arguments as in the proof of Theorem 13.21 this expression can be seen to be equal to $Y_n(x)^{-4}\Gamma(5)$ up to lower order terms. Since $n^{-1}Y_n$ tends almost surely to a positive limit, the sum of the fourth moments over the uncensored survival times x is of order $O(n^{-3})$. Thus $\sum_x \sqrt{n}(\Delta H(x) - \mathrm{E}[\Delta H(x)|\, D_n])$ easily satisfies Lyapunov's condition.

The variance $\mathrm{var}[\Delta H(x)|\, D_n]$ is given by the second integral in (13.24). Extending the argument we see that the two quotients inside the integral are asymptotic to $2Y_n(x)^{-2}$ and

$Y_n(x)^{-1}$, respectively, so that the difference of the first minus the square of the second is equivalent to $Y_n(x)^{-2}$. This shows that

$$\sum_{x:x\le t}\int_{(0,t]}\operatorname{var}[\sqrt{n}\Delta H(x)|\,D_n] = \int_{(0,t]}\frac{n}{Y_n^2}\,dN_n + o(1) \to \int_{(0,t]}\frac{G_{0-}\,dF_0}{(F_{0-}G_{0-})^2} = U_0(t).$$

This convergence also applies to increments of the process H^f. By the Lyapounov central limit theorem the posterior process $\sqrt{n}(H^f(t) - \mathrm{E}[H^f(t)|\,D_n])$ converges marginally given D_n to a Gaussian independent increment process with variance process U_0, i.e. the process $B \circ U_0$, almost surely.

The convergence of the variances of the increments is uniform in t, and the limiting variance process U_0 is continuous. Together with the independence of the increments this gives tightness in view of Theorem V.19 of Pollard (1984). Thus assertion (i) is established.

To prove assertion (ii) we refine the approximation for the posterior mean of $H^f(t)$ in (13.23). Letting $r(x,s) = q(x,s)/q_0(x) - 1$, so that $|r(x,s)| \le Cs^\alpha$ for $s \to 0$, we can approximate $\mathrm{E}[H_n|\,D_n] - \hat{H}_n$ as, for $\epsilon_n \to 0$ sufficiently slowly,

$$\int \frac{1}{Y_n}\Big[\frac{\int_0^{Y_n\epsilon_n} u(1-u/Y_n)^{Y_n-1}(1+r(\cdot,u/Y_n))\,du + O(n^{-3})}{\int_0^{Y_n\epsilon_n}(1-u/Y_n)^{Y_n-1}(1+r(\cdot,u/Y_n))\,du + O(n^{-3})} - 1\Big]\,dN_n.$$

Because the variation of $\int Y_n^{-1}\,dN_n$ is uniformly bounded, it suffices to show that $\sqrt{n}$ times the supremum over the omitted argument x of the expression within square brackets tends to zero in probability. Since $\int_0^{y\epsilon} u^k(1-u/y)^{y-1}\,du = \Gamma(k+1) + O(1/y)$ if $y \to \infty$ such that $e^{-y\epsilon} \gtrsim y^{-1}$, after replacing r by its upper bound the numerator and denominator of the quotient expand as $\Gamma(2) + \Gamma(2+\alpha)/Y_n^\alpha + O(1/Y_n)$ and $\Gamma(1) + \Gamma(1+\alpha)/Y_n^\alpha + O(1/Y_n)$, respectively. Their quotient differs from 1 by a terms of the order $O(1/Y_n^\alpha) = O(n^{-\alpha})$, almost surely. □

Corollary 13.24 *Under the conditions of Theorem 13.23 with $\alpha > 1/2$, we have $\sqrt{n}(\bar{F} - \hat{\bar{F}}_n)|\,D_n) \rightsquigarrow -\bar{F}_0\, B \circ U_0$ in $[P^\infty_{F_0,G_0}]$-probability, where $\hat{\bar{F}}_n$ is the Kaplan-Meier estimator for the survival function.*

Proof This is an immediate consequence of the Hadamard differentiability of the product integral $H \mapsto \bar{F} = \prod(1-dH)$ and the delta-method for conditional distributions (Theorem 3.9.11 in van der Vaart and Wellner 1996). □

Remark 13.25 If $\frac{\partial}{\partial s}g(x,s)$ exists and is bounded in a (right) neighborhood of 0, then the function $q_0(x) = q(x,0)$ satisfies the condition in Theorem 13.23, with $\alpha = 1$.

Example 13.26 (Beta process) The beta process prior on the cumulative hazard function with parameter (c, Λ) possesses q-function given by $q(x,s) = c(x)(1-s)^{c(x)-1}$. If c is continuous and bounded away from 0 and ∞, then q satisfies the conditions of Theorem 13.23 with $\alpha = 1$.

It suffices to note that $(1-s)^{c-1} = 1-(c-1)s+o(s)$, as $s \to 0$, uniformly in c belonging to a bounded interval in $(0,\infty)$, whence $q(x,s) = c(x) + O(s)$, as $s \to 0$.

Example 13.27 (Dirichlet process) By Example 13.11 the cumulative hazard function H corresponding to $F \sim \mathrm{DP}(MF_0)$ is a beta process with parameters parameters $c = M\bar{F}_{0-}$ and H_0. By the preceding example the conditions of Theorem 13.23 are satisfied if H_0 is continuous.

For this special example the Bernstein–von Mises theorem for the survival function was earlier obtained in a more abstract setting in Theorem 12.2 (but without censoring).

Example 13.28 (Extended gamma process) The extended gamma process prior with parameters (c, A_0) is described in Examples 13.13 and 13.22. If the function c in its intensity function (13.25) is continuous and bounded away from zero and infinity on $[0, \tau]$, then this satisfies the conditions of Theorem 13.23. See Example 13.22.

13.5 Smooth Hazard Processes

In view of the discreteness of its sample paths, an independent increment process is not an adequate model for a smooth cumulative hazard function. Smoothing through a kernel may be natural, yielding analogs of Dirichlet mixture processes. The resulting priors for smooth hazard functions may be useful even outside the setting of survival analysis.

For a given measurable *kernel function* $k\colon [0,\infty) \times (0,\infty) \to \mathbb{R}^+$ and a finite Borel measure Φ on $(0,\infty)$, consider a hazard function h_Φ of the form

$$h_\Phi(t) = \int k(t, v)\, d\Phi(v).$$

We do not require that $k(\cdot, v)$ is a probability density or that Φ is a probability measure, but do assume that the mixture is finite for all t. We obtain a prior on hazard functions by placing a prior on the mixing measure Φ. Specifically, we consider processes with independent nonnegative increments Φ, identified with measures through their cumulative distribution function, as discussed in Section 13.4. Although we take the mixing variable v in this section to be a positive number, the setup can be extended to general Polish spaces with Φ a general completely random measure.

The cumulative hazard function for h_Φ is a mixture over the integrated kernel:

$$H_\Phi(t) = \int K(t, v)\, d\Phi(v), \qquad K(t, v) := \int_0^t k(u, v)\, du.$$

The corresponding survival function and probability densities are $1 - F_\Phi(t) = e^{-H_\Phi(t)}$ and $f_\Phi(t) = h_\Phi(t)e^{-H_\Phi(t)}$. If $K(t, v) \to \infty$ as $t \to \infty$ for every v, then $H_\Phi(t) \to \infty$ as $t \to \infty$, and F_Φ is a proper probability distribution on $(0,\infty)$.

Common choices of kernels are

(i) Dykstra-Laud (DL) kernel: $k(t, v) = \mathbb{1}\{t \geq v\}$;
(ii) Rectangular kernel: $k(t, v) = \mathbb{1}\{|v - t| \leq \tau\}$, for "bandwidth" $\tau > 0$;
(iii) Ornstein-Uhlenbeck (OU) kernel: $k(t, v) = 2\kappa e^{-\kappa(t-v)}\mathbb{1}\{t \geq v\}$;
(iv) Exponential kernel: $k(t, v) = v^{-1}e^{-t/v}$.

The Dykstra-Laud kernel generates increasing hazard functions, whereas the exponential kernel makes the hazard decreasing. The rectangular kernel is a "local smoother," whereas the Dykstra-Laud and Ornstein-Uhlenbeck kernels smooth "over the entire past," and the exponential kernel smooths "over the entire time axis." Among these kernels, only the exponential kernel gives infinitely smooth sample paths.

Example 13.29 (Dykstra-Laud and extended gamma) Within the present class of priors on hazard functions, the Dykstra-Laud kernel combined with the extended gamma process (see Example J.15) as the mixing measure is particularly tractable. The extended gamma process can be obtained as $\Phi(v) = \int_{(0,v]} b\,d\xi$ from a gamma process ξ and a given positive function b. Combination with the Dykstra-Laud kernel yields the prior hazard function $h_\Phi(t) = \int_{(0,t]} b\,d\xi$ and hence $\mathrm{E}[h_\Phi(t)] = \int_{(0,t]} b\,d\alpha$ and $\mathrm{var}[h_\Phi(t)] = \int_{(0,t]} b^2\,d\alpha$, for $\alpha(t) = \mathrm{E}[\xi(t)]$ the mean function of ξ. The explicit expressions for prior expectation and variance are helpful for eliciting b and α. The cumulative hazard function can be similarly expressed in the integrated kernel as $H_\Phi(t) = \int K(t,v)\,b(v)\,d\xi(v)$. For the Dykstra-Laud kernel we have $K(t,v) = (t-v)_+$, whence by Example J.7, for $\theta > 0$,

$$\log \mathrm{E}[e^{-\theta H_\Phi(t)}] = -\int_0^t \log\left(1 + \theta(t-v)_+ b(v)\right) d\alpha(v). \tag{13.26}$$

The prior mean $\mathrm{E}[\bar{F}_\Phi(t)]$ of the survival function is obtained by setting $\theta = 1$ in this formula and exponentiating.

The tractability of the Laplace transform of H_Φ allows us to derive a formula for the posterior distribution given (censored) data from F_Φ, which expresses the cumulative hazard function as a mixture of extended gamma processes, with varying α measures. This representation next leads to an expression for the posterior expectation of $h_\Phi(t)$, similar to the one in Proposition 4.7 for the Dirichlet process. The large number of terms makes the result difficult to apply. Simulation methods, which are also available for other kernels and priors, often are preferable.[8]

Consider the posterior distribution based on random right censored data D_n when the cumulative hazard function of the survival times is a priori modelled as H_Φ. Hence, D_n consists of a sample $(T_1, \Delta_1), \ldots, (T_n, \Delta_n)$, where T_i is the minimum of a survival X_i time and a censoring time, Δ_i indicates censoring, and given Φ the survival times $X_1, \ldots, X_n$ are i.i.d. with cumulative hazard function H_Φ. It is convenient to treat the mixing random measure Φ rather than the hazard function as the primary parameter. The likelihood of Φ given the censored data D_n can be written as

$$\exp\Big[-\sum_{i=1}^n \int K(T_i,v)\,d\Phi(v)\Big] \prod_{i=1:\Delta_i=1}^n \int k(T_i,v)\,d\Phi(v).$$

The integrals with respect to v can be eliminated if we introduce latent variables $V_1, \ldots, V_n$ "realized from the CRM Φ" as additional observations. Since Φ is a CRM, and hence is a.s. discrete, there will typically be repetitions in $V_1, \ldots, V_n$. Let $\tilde{V}_1, \ldots, \tilde{V}_k$ stand for the distinct values, appearing with multiplicities $N_1, \ldots, N_k$. They define a partition of

[8] See Theorem 3.3 of Dykstra and Laud (1981) for details.

$\{1, 2, \ldots, n\}$, given by the sets $S_j = \{i: V_i = \tilde{V}_j\}$, for $j = 1, \ldots, k$. The following theorem describes the posterior distribution of Φ in a form that is also suitable for MCMC computations.

Theorem 13.30 (Posterior distribution) *If the prior for Φ follows a CRM with intensity measure $\nu(dv, ds) = \rho(ds|\, v)\,\alpha(dv)$, where α is a σ-finite measure on $\mathbb{R}^+$, then the posterior distribution of Φ given D_n is described as follows, where $K_n(v) = \sum_{i=1}^n K(T_i, v)$ and $\tau_m(v) = \int_0^\infty s^m e^{-sK_n(v)}\,\rho(ds|\, v)$:*

(i) *The conditional distribution of Φ given D_n and the auxiliary variables $V_1, \ldots, V_n$ is equal to the distribution of the CRM with intensity measure*

$$\nu^c_{\Phi|D_n,V}(dv, ds) = e^{-sK_n(v)}\rho(ds|\, v)\,\alpha(dv),$$

$$\nu^d_{\Phi|D_n,V}(\{\tilde{V}_j\}, ds) \propto s^{N_j} e^{-sK_n(\tilde{V}_j)}\rho(ds|\, \tilde{V}_j).$$

(ii-1) *The conditional distribution of $(\tilde{V}_1, \ldots, \tilde{V}_k)$ given D_n and configuration $(k, S_1, \ldots, S_k)$ has density proportional to*

$$(v_1, \ldots, v_k) \propto \prod_{j=1}^k \tau_{N_j}(v_j) \prod_{i\in S_j, \Delta_i=1} k(T_i, v_j).$$

(ii-2) *The probability of the configuration $(k, S_1, \ldots, S_k)$ given D_n is proportional to*

$$\prod_{j=1}^k \int \tau_{N_j}(v) \prod_{i\in S_j, \Delta_i=1} k(T_i, v)\,\alpha(dv).$$

Proof See Theorem 4.1 of James (2005). The proof uses arguments similar to those used in the proofs of Theorem 14.56 and Lemma 14.62 and Theorem 5.3 (for part (ii)). □

Based on corresponding results on Dirichlet mixtures one may expect the posterior distribution of a smooth hazard prior to be consistent if the true hazard is also smooth, even at a rate, that will depend on a bandwidth in the kernel. The following theorem is only a first step in this direction: it gives weak consistency for the cumulative hazard function (or, equivalently, the survival function). Consistency is of course limited to an interval within the range of the censoring variables. We assume that the censoring distribution G is supported on a compact interval within the support of the survival function and then obtain consistency for the survival function on the same interval.

Theorem 13.31 (Consistency) *Suppose that the censoring distribution G is supported on $[0, \tau]$ with density bounded away from zero. Consider a smooth hazard process h_Φ for a CRM Φ such that* $\liminf_{t\downarrow 0} t^{-r} h_\Phi(t) = \infty$ *almost surely* $[\Pi]$ *for some $r > 0$ and such that* $\Pi(\Phi\colon \sup_{0<t\le\tau} |h_\Phi(t) - h_0(t)| < \delta) > 0$, *for all $\delta > 0$. Then the posterior distribution of the cumulative distribution function F is consistent with respect to the Kolmogorov-Smirnov distance on* $[0, \tau]$ *at any F_0 with $F_0(\tau) < 1$ and with a hazard function h_0 that is bounded away from zero on every interval* $[\delta, \tau]$, *for every $\delta > 0$, and is such that* $\int_0^1 (\log h_0(t) + \log_- t)\, dF_0(t) < \infty$.

Proof The distribution P_F of an observation (T, Δ) on $(0, \infty) \times \{0, 1\}$ has a density $p_F(t, \delta) = f(t)^\delta \bar{F}(t)^{1-\delta} g(t)^{1-\delta} \bar{G}(t)^\delta$ with respect to the product of Lebesgue measure and counting measure. Here we consider the censoring distribution G and density g as fixed and given. Weak convergence of a sequence P_{F_m} to P_{F_0} can be seen to imply the weak convergence of F_m to F_0 in the space of subprobability measures on $[0, \tau]$. Since F_0 is continuous by assumption, this next implies the uniform convergence of F_m to F_0 on $[0, \tau]$, by Pólya's theorem. Therefore it is sufficient to show the weak consistency of the posterior of P_F. By Example 6.20 this is ensured by the Kullback-Leibler property of the prior of P_F at P_{F_0}.

The censoring model is an information loss model, of the type considered in Lemma B.11, where the distribution F of X is transformed to the distribution P_F of the observation (X, Δ). Since G is supported on $[0, \tau]$ every observation is censored at τ and hence the full data model can also be taken to have survival time $X \wedge \tau$ instead of X; in other words rather than $X \sim F$ we can take $X \sim \tilde{F}$, for $\tilde{F}$ the mixture distribution of the density f on $[0, \tau)$ and a point mass of size $1 - F(\tau)$ at τ. Then it follows, for h_0, h, H_0, H the hazard functions and cumulative hazard functions corresponding to F_0, F,

$$K(P_{F_0}; P_F) \le K(\tilde{F}_0; \tilde{F}) = \int_{[0,\tau)} \log \frac{f_0}{f}\, dF_0 + \log \frac{1 - F_0(\tau)}{1 - F(\tau)}\, \bar{F}_0(\tau)$$
$$= \int_0^\tau \log \frac{h_0}{h}\, dF_0 - \int_0^\infty (H_0 - H)(t \wedge \tau)\, dF_0(t).$$

It suffices to show that the right side evaluated at $h = h_\Phi$ is bounded above by ϵ with positive probability under the prior of Φ, for every $\epsilon > 0$.

For given positive numbers δ, η let $\mathfrak{H}(\delta, \eta)$ be the set of all hazard functions h such that $\inf_{t\in[0,\delta]} (h(t)t^{-r}) \ge 1$ and $\sup_{t\in[0,\tau]} |h(t) - h_0(t)| < \eta$. For $h \in \mathfrak{H}(\delta, \eta)$ we have $|\log(h_0/h)| \le |h - h_0|/(h \wedge h_0) \le \eta/(c_0(\delta, \tau) - \eta)$, on the interval $[\delta, \tau]$, for $c_0(\delta, \tau)$ the minimum value of h_0 on the interval. Furthermore, on the same interval $|H - H_0| \le \eta\tau$. By splitting the integral in the expression for $K(\tilde{F}_0; \tilde{F})$ over the ranges $(0, \delta)$ and $[\delta, \tau]$, we see that for $h \in \mathfrak{H}(\delta, \eta)$ (where $\delta < 1$):

$$K(\tilde{F}_0; \tilde{F}) \le \int_0^\delta (\log h_0)\, dF_0 + r \int_0^\delta \log_- t\, dF_0(t) + \int_\delta^\tau \frac{\eta}{c_0(\delta, T) - \eta}\, dF_0 + \eta\tau.$$

The first and second integrals on the right become smaller than an arbitrary positive constant, as $\delta \downarrow 0$. For given δ the third integral and the last term become arbitrarily small if η is small. We conclude that the Kulback-Leibler property holds at P_{F_0} if h_Φ belongs with positive prior probability to $\mathfrak{H}(\delta, \eta)$, for sufficiently small $\delta, \eta > 0$.

By assumption $\liminf_{t\downarrow 0} h_\Phi(t)t^{-r} = \infty$ almost surely under the prior. Therefore the event $\cup_{\delta>0} A_\delta$, for A_δ the event that $h_\Phi(t) \ge t^r$ for every $t \in (0, \delta]$, has prior probability equal to 1. This implies that the prior probabilities of the events $h_\Phi \in \mathfrak{H}(\delta, \eta)$ tend to the prior probability that $\sup_{t\in[0,\tau]} |h_\Phi(t) - h_0(t)| < \eta$. Since the latter is positive by assumption, the proof is complete. □

Both the Dykstra-Laud and Ornstein-Uhlenbeck kernels satisfy $h_\Phi(t) \gtrsim \Phi((0, t])$. Then the condition that $\liminf_{t\downarrow 0} t^{-r} h_\Phi(t) = \infty$ almost surely is implied by the similar condition on the CRM that $\liminf_{t\downarrow 0} t^{-r} \Phi((0, t]) = \infty$ almost surely. This holds for the generalized extended gamma process with intensity measure $\nu(dv, ds) = (\Gamma(1-\sigma))^{-1}s^{-(1+\sigma)}e^{-c(v)s}\,dv\,ds$. For the other two kernels $h_\Phi(0) > 0$ almost surely and hence the required condition holds automatically.

13.6 Proportional Hazard Model

The *proportional hazard model* or *Cox model* is frequently used to investigate the dependence of survival on a covariate. It was considered previously in Section 12.3.3. The Cox model postulates that the hazard function of an individual characterized by the covariate vector $z \in \mathbb{R}^d$ is equal to $e^{\beta^\top z}$ times a *baseline hazard function*. Presently we study priors on the cumulative hazard function and do not assume existence of a hazard function. In this situation one possible definition of proportional hazards is that the cumulative hazard function of the survival time X given the covariate Z takes the form $x \mapsto e^{\beta^\top Z}H(x)$, for an unknown "baseline cumulative hazard function" H. For a function H with jumps this would be awkward, as the jumps of a cumulative hazard function are bounded by 1, which would only be achieved by limiting the jumps of H by the minimum of the multiplicative factors $e^{-\beta^\top Z}$. A definition of proportional hazards that avoids this difficulty is that the negative log-conditional survival function (in the preceding denoted by the symbol A instead of H) satisfies the proportionality requirement: for a baseline negative log-survival function A we postulate that the survival distribution of X given Z satisfies

$$-\log(1 - F(x|\, Z)) = e^{\beta^\top Z}A(x). \tag{13.27}$$

For large samples from a continuous survival distribution, the difference between these possible definitions should be minor as the posterior would place most of its weight on cumulative hazard functions with small jumps. Because it is conceptually cleaner, in the following we adopt the second definition, as given in the preceding display.

We allow for right censoring, and take the data as $D_n = \{(T_i, \Delta_i, Z_i): i = 1, \ldots, n\}$, for $T_i = \min(X_i, C_i)$ and $\Delta_i = \mathbb{1}\{X_i \leq C_i\}$ and a random sample of triplets (X_i, C_i, Z_i) of a survival time, censoring time and covariate.

We assume that the survival time X and censoring time C are conditionally independent given the covariate Z. We choose a prior such that the pair of parameters (θ, H), the conditional distribution of C given Z, and the marginal distribution of Z are a priori independent. Then the priors of the latter two components do not enter the posterior distribution of (θ, H) and will be left unspecified. We consider an independent increment process as a prior on the baseline cumulative hazard function H, given by an intensity measure of the form $\nu_H(dt, ds) = g(x, s)\,dx\,ds$, and an independent prior with density π on β. By Theorem 13.8 the first is equivalent to choosing an independent increment process prior on the function A.

13.6.1 Posterior Distribution

The expression for the posterior distribution in the Cox model extends Theorem 13.15, to which it reduces if no covariate is present.

Theorem 13.32 (Cox posterior) *If the prior on H is an independent increment process with intensity measure $\nu_H(dx, ds) = g(x, s)\,dx\,ds$ independent of θ for a function g that is continuous in x, then the conditional posterior distribution of H given β and the data D_n is the law of an independent increment process with intensity measure*

$$\nu^c_{H|D_n,\beta}(dx, ds) = \prod_{i:T_i\geq x} (1-s)^{e^{\beta^\top Z_i}} g(x, s)\,dx\,ds,$$

$$\nu^d_{H|D_n,\beta}(\{x\}, ds) \propto \prod_{i:T_i=x,\Delta_i=1} [1-(1-s)^{e^{\beta^\top Z_i}}] \prod_{i\in R_n^+(x)} (1-s)^{e^{\beta^\top Z_i}} g(x, s)\,ds,$$

where the fixed jump measures are probability measures on $(0, 1)$, *the fixed jump times range over the set* $\{T_i\colon \Delta_i = 1\}$ *of uncensored observations, and* $R_n^+(t) = \{i\colon T_i \geq t\} \setminus \{i\colon T_i = t, \Delta_i = 1\}$. *Furthermore, the marginal posterior density of β given D_n satisfies*

$$\pi(\beta|\, D_n) \propto \pi(\beta) e^{-\rho_n(\beta)} \prod_{x:\Delta N(x)>0} k(x, \beta),$$

where $k(x, \beta)$ is the norming constant for the the jump measure $\nu^d_{H|D_n,\beta}(\{x\}, ds)$ as given (the integral over $(0, 1)$ *of the right side of the proportionality equation), and*

$$\rho_n(\beta) = \sum_{i=1}^n \int_0^{T_i}\int_0^1 [1-(1-s)^{e^{\beta^\top Z_i}}] \prod_{j<i:T_j\geq x} (1-s)^{e^{\beta^\top Z_j}} g(x, s)\,dx\,ds.$$

Proof By adding the observations one-by-one we can reduce to the case of a single observation (T, Δ, Z), provided we allow a prior for H of the postulated form of the posterior distribution for H. Thus given β the cumulative hazard function is an independent increment process with intensity measure $\nu_{H|\beta}$ possibly depending on β, and it may contain a fixed jump part.

In the first step we condition on β and Z and hence the observation is equivalent to just a single observation (T, Δ) in the random censoring model. In view of our definition of proportional hazards, it is convenient to parameterize the model by $A = -\log \bar F$, for $\bar F$ the baseline survival function. By (13.27) the negative log survival function given β and Z of the survival time X is equal to $e(\beta)A$, for $e(\beta) = e^{\beta^\top Z}$ and hence given β and Z is a priori an independent increment process with intensity measure given by $\nu_{e(\beta)A|\beta}(dx, ds) = \nu_{A|\beta}(dx, ds/e(\beta))$. By Theorem 13.15, the posterior distribution of $e(\beta)A$ (given β and Z and (T, Δ)) is also an independent increment process, and its intensity process can be expressed in $\nu_{e(\beta)A|\beta}$. This can next be translated back to see that the posterior distributions of A and H are independent increment processes, where the intensity measure for H is given in the statement of the theorem.

For the derivation of the marginal posterior distribution of β we retake the proof of Theorem 13.15, which consists of discretizing the observation to a multinomial vector M and taking limits as the discretization becomes fully informative. The likelihood for M takes the form $L_M(A, \beta) = \prod_{i=1}^{m+1} (e^{-A(t_{i-1})e(\beta)} - e^{-A(t_i)e(\beta)})^{M_i}$. By Bayes's rule a version of the posterior density of β is given by

$$\pi(\beta|\, M) = \frac{\mathrm{E}_{A|\beta} L_M(A,\beta)\pi(\beta)}{\mathrm{E}_\beta \mathrm{E}_{A|\beta} L_M(A,\beta)} \propto \mathrm{E}_{A|\beta} L_M(A,\beta)\,\pi(\beta).$$

Here the expectations $\mathrm{E}_{A|\beta}$ are relative to the prior distribution of A given β, for fixed M. On the event that the jth coordinate of M is nonzero, the right side is equal to, by two applications of formula (13.17) and arguing as in the proof of Theorem 13.15,

$$e^{-\int_{(0,t_{j-1}]}\int (1-e^{-e(\beta)s})\,\nu^c_{A|\beta}(dx,ds)} \prod_{x\le t_{j-1}} \int e^{-e(\beta)s}\,\nu^d_{A|\beta}(\{x\},ds)$$
$$\times \Big[1 - e^{-\int_{(t_{j-1},t_j]}\int (1-e^{-e(\beta)s})\,\nu^c_{A|\beta}(dx,ds)} \prod_{t_{j-1}<x\le t_j} \int e^{-e(\beta)s}\,\nu^d_A(\{x\},ds)\Big].$$

If the observation is censored, then the partitions are constructed so that $t_{j-1} = C$ and $t_j = \infty$. The expression is fixed and no refinement limit need be taken. If the observation is uncensored, then the leading term, outside brackets, tends to the same expression but with t_{j-1} replaced by its limit T. For the limit of the term within square brackets, we split in two cases. If the observation is uncensored and equal to a fixed jump time of $\nu_{A|\beta}$, then the exponential within square brackets tends to $e^{-0} = 1$ and the product is eventually equal to the product over the single fixed jump time T. If the observation is uncensored and not equal to a fixed jump time of $\nu_{A|\beta}$, then the interval $(t_{j-1}, t_j]$ will eventually be free of fixed jump times and the product inside square brackets should be read as 1. Because the integral in the exponent tends to zero, the term in square brackets can be expanded as

$$\int_{(t_{j-1},t_j]}\int (1-e^{-e(\beta)s})\,\nu^c_{A|\beta}(dx,ds)(1+o(1)).$$

In the sequential updating scheme the intensity measure $\nu^c_{A|\beta}$ will correspond to the conditional posterior of A based on the preceding observations and hence possess density $(x,s) \mapsto \prod_{i:T_i\ge x} e^{-se_i(\beta)}$ relative to the original prior intensity ν_A, which is free of β. As a function of x it changes values only at the T_i. If T is not an existing fixed jump, then there is no other uncensored observation in $(t_{j-1}, t_j]$, eventually. If we construct the partitions to include the preceding censored times, then a censored T_i in $(t_{j-1}, t_j]$ will be necessarily equal to t_j. In that case the density is constant in its first argument and equal to $q(s,\beta) = \prod_{i:T_i\ge T} e^{-se_i(\beta)}$ throughout the interval. The preceding display becomes

$$\int_{(t_{j-1},t_j]}\int (1-e^{-e(\beta)s}) g(s,\beta)\,\nu^c_A(dx,ds)(1+o(1)).$$

The quotient of this expression with its value at $\beta = 0$ is asymptotically proportional (as a function of β) to $\int (1-e^{-e(\beta)s}) g(s,\beta)\,\nu^c_A(ds|\, T)$.

This finishes the derivation of the updating formula by Scheffe's theorem, since the expression at $\beta = 0$ does not change with refining discretization. By some bookkeeping, n rounds of updating can be seen to lead to the formulas as claimed. □

Example 13.33 (Extended gamma process) Let the baseline function $A = -\log \bar F$ follow an extended gamma process with parameter A_0 such that $dA_0(x) = \lambda(x)\,dx$, as described by Example 13.13, and let B be the corresponding cumulative hazard function. By Theorem 13.32 the intensity measure of H in the posterior is given by

$$\nu^c_{H|D_n,\beta}(dx, ds) = \frac{c(x)}{\log_-(1-s)} \prod_{i\in R_n(x)} (1-s)^{c(x)-1+e^{\beta^\top Z_i}} \lambda(x)\, dx\, ds,$$

$$\nu^d_{H|D_n,\beta}(\{x\}, ds) \propto \frac{1}{\log_-(1-s)} \prod_{i\in D_n(x)} \left[1-(1-s)^{e^{\beta^\top Z_i}}\right] \prod_{i\in R_n^+(x)} (1-s)^{c(x)-1+e^{\beta^\top Z_i}}.$$

The posterior process for H is not extended gamma, since the jump sizes are not gamma distributed.

Example 13.34 (Beta process) For H a beta process prior with parameters (c, Λ), with Λ having density λ, Theorem 13.32 yields the posterior intensity measure

$$\nu^c_{H|D_n,\beta}(dx, ds) = c(x)s^{-1} \prod_{i\in R_n(x)} (1-s)^{c(x)-1+e^{\beta^\top Z_i}} \lambda(x)\, dx\, ds,$$

$$\nu^d_{H|D_n,\beta}(\{x\}, ds) \propto s^{-1} \prod_{i\in D_n(x)} [1-(1-s)^{e^{\beta^\top Z_i}}] \prod_{i\in R_n^+(x)} (1-s)^{c(x)-1+e^{\beta^\top Z_i}}.$$

The marginal posterior density of β is obtained from the expression for $\rho_n(\beta)$ given by

$$\sum_{i=1}^n \int_0^{T_i} \int_0^1 c(x)s^{-1}[1-(1-s)^{e^{\beta^\top Z_i}}] \prod_{j<i:T_j\geq x} (1-s)^{c(x)-1+e^{\beta^\top Z_j}} \lambda(x)\, dx\, ds.$$

Prior information about the regression coefficient β is often not available, and hence a uniform improper prior for β is natural. Under mild conditions on the covariates and the density of the intensity measure the posterior distribution is proper. The conditions on the prior intensity measure in the following proposition are satisfied by the gamma and beta processes. The remaining condition is a form of "linear independence of the covariates." This is also used to show the uniqueness of the maximum likelihood estimator and the log-concavity of the partial likelihood function (see Andersen et al. 1993).

Proposition 13.35 *If the function $(x, s) \mapsto s(1-s)^{1-c_1} g(x, s)$ is bounded over $[0, T_{(n)}] \times [0, 1]$ and $s\, g(x, s) \geq M(1-s)^{c_2-1} a_0(x)$ for some $M, c_1, c_2 > 0$ and a continuous function a_0, and every $z \in \mathbb{R}^d$ is expressible as $\sum_{k:\Delta_k=1} \sum_{i:T_i=T_k,\Delta_i=1} \sum_{j\in R_n^+(T_k)} \lambda_{ijk}(Z_i - Z_j)$ for some choice of $\lambda_{ijk} \geq 0$, then the posterior for β under the improper uniform prior is proper.*

Proof We only describe the main idea behind the proof, and refer to Kim and Lee (2003a) for the details. Under the given conditions the posterior density of β can be bounded by a constant multiple of the minimum over $\exp\left[\beta^\top(Z_i - Z_j)\right] \wedge 1$ for all pairs (i, j) such that either $\Delta_i = 1$ and $T_j > T_i$ or $T_j = T_i$ and $\Delta_j = 0$. By the third condition of the proposition, it follows that the posterior has exponentially decaying tail in every direction. ☐

13.6.2 Bernstein–von Mises Theorem

For large samples the joint posterior distribution of β and H obtained in Theorem 13.32 possesses a Gaussian approximation, under a similar condition on the small jump sizes in the intensity measure of the cumulative hazard function as in Theorem 13.23. When restricted to the marginal posterior distribution of β this gives a Bernstein–von Mises theorem in the spirit of Section 12.3.3, but using an independent increment process prior on the baseline hazard. As this prior is not supported on a dominated model, the techniques developed in Chapter 12 are not applicable, but the theorem is derived using the explicit, conjugate form of the posterior distribution established in Theorem 13.32.

The semiparametric maximum likelihood estimator for (β, H) are the maximizer $\hat\beta_n$ of *Cox partial likelihood*

$$L_n(\beta) = \prod_{\substack{i=1\\ \Delta_i=1}}^{n} \frac{\exp(\beta^\top Z_i)}{\sum_{j:T_j\ge T_i}\exp(\beta^\top Z_j)}. \tag{13.28}$$

and *Breslow's estimator*

$$\hat H_n(t) = \int_{(0,t]} \Big(\sum_{i:T_i\ge x} e^{\hat\beta_n^\top Z_i}\Big)^{-1}\, dN(x).$$

These are known to be asymptotically normally distributed in the sense that $\sqrt{n}(\hat\beta_n - \beta_0, \hat H_n - H_0) \rightsquigarrow (V, W\circ U_0 - V^\top e_0)$, for $V\sim \mathrm{Nor}_d(0, \tilde I_\beta^{-1})$ and W a standard Brownian motion independent of V, where

$$\begin{aligned}
U_0(t) &= \int_{(0,t]} \frac{1}{\mathrm{E}_0[e^{\beta^\top Z}\mathbb{1}\{T\ge x\}]}\, dH_0(x),\\
e_0(t) &= \int_{(0,t]} \frac{\mathrm{E}_0[Ze^{\beta^\top Z}\mathbb{1}\{T\ge x\}]}{\mathrm{E}_0[e^{\beta^\top Z}\mathbb{1}\{T\ge x\}]}\, dH_0(x),\\
\tilde I_\beta &= \int_{(0,\tau]} \mathrm{E}_0\big[(ZZ^\top - e_0(x)e_0(x)^\top)e^{\beta^\top Z}\mathbb{1}\{T\ge x\}\big]\, dH_0(x).
\end{aligned} \tag{13.29}$$

The last quantity is the *efficient Fisher information* for estimating β.

Theorem 13.36 (Bernstein–von Mises) *Assume that the covariates lie in a bounded subset of $\mathbb{R}^d$ and are not concentrated in any lower-dimensional subset, and assume that the prior density on β is positive and continuous at β_0. If the prior for H follows an independent increment process with intensity measure of the form $\nu_H(dx, ds) = s^{-1}q(x,s)\,dx\,ds$ for a function q that is continuous in x and such that $\sup_{x\in[0,\tau], s\in(0,1)}(1-s)q(x,s) < \infty$ and $\sup_{x\in[0,\tau]}|q(x,s) - q_0(x)| = O(s^\alpha)$ as $s\to 0$ for a function q_0 that is bounded away from zero and infinity and some $\alpha > 1/2$, then for τ such that $(\bar F_0\bar G_0)(\tau-) > 0$ and $G_0(\tau) = 1$, and continuous F_0:*

$$\sqrt{n}(\beta - \hat\beta_n, H - \hat H_n)|\, D_n \rightsquigarrow (V, W\circ U_0 - V^\top e_0), \qquad a.s.\ [P^\infty_{\beta_0,H_0}],$$

where $V\sim \mathrm{Nor}_d(0, \tilde I_\beta^{-1})$ and W is a standard Brownian motion independent of V, and the convergence is in the product of the $\mathbb{R}^d$ and the Skorohod space $\mathfrak{D}[0,\tau]$ equipped with the uniform norm.

Proof The proof of the theorem is long; we refer to Kim (2006) for details. The proof can be based on the characterization of the posterior distribution given in Theorem 13.32. In particular, the theorem gives an explicit expression for the marginal posterior density of β. The asymptotic analysis of the posterior density of $\sqrt{n}(\beta - \hat{\beta}_n)$ proceeds as in the proof of the parametric Bernstein–von Mises theorem for smoothly parameterized models, with two main differences. First the log-likelihood is not a sum of independent terms, and more importantly, the difference between the derivatives of the marginal log-likelihood and the partial log-likelihood need to be shown to be uniformly $o(n)$ in order to expand the partial log-likelihood in a Taylor expansion and develop a normal approximation.

The joint asymptotics of $\sqrt{n}(\beta - \hat{\beta}_n)$ and $\sqrt{n}(H - \hat{H}_n)$ can be derived from the characterization of the conditional posterior distribution of $\sqrt{n}(H - \hat{H}_n)$ given $\sqrt{n}(\beta - \hat{\beta}_n)$, using the structure of the independent increment process prior. □

The conditions of the theorem hold in particular for the extended beta and gamma process priors on H; see Examples 13.26 and 13.28.

13.7 The Bayesian Bootstrap for Censored Data

In this section we study analogs of the Bayesian bootstrap appropriate for censored data.

13.7.1 Survival Data without Covariates

The Bayesian bootstrap (BB) was introduced in Section 4.7 for i.i.d. complete data. It can be viewed as a smooth alternative to Efron's bootstrap, as the noninformative limit of the posterior distribution based on a conjugate prior, or as the posterior distribution based on the empirical likelihood with a Dirichlet prior. In this section we extend these three approaches to the censored data setting. Since censored data contain complete data as a special case, it is imperative that the definition reduces to the ordinary Bayesian bootstrap if no observations are censored, but otherwise we allow some freedom of definition.

The Bayesian bootstrap for uncensored data replaces the weights $(n^{-1}, \ldots, n^{-1})$ in the empirical distribution $\sum_{i=1}^{n} n^{-1}\delta_{X_i}$ by a random vector $(W_1, \ldots, W_n)$ from the $\mathrm{Dir}(n; 1, 1, \ldots, 1)$-distribution, the uniform distribution on the n-simplex. In both cases (and also in the case of Efron's bootstrap, which has $(nW_1, \ldots, nW_n) \sim \mathrm{MN}_n(n; n^{-1}, \ldots, n^{-1})$) the expected weight of every observation is $1/n$. One extension to censored data is to replace the weights in the Kaplan-Meier estimator (13.9), the analog of the empirical distribution, by random variables with the same expectations. This leads to the definition of the Bayesian bootstrap for censored data:

$$\bar{F}^*(t) = \prod_{j:T_j^* \le t} \left(1 - \frac{\sum_{i:X_i = T_j^*} \Delta_i \Gamma_i}{\sum_{i:X_i \ge T_j^*} \Gamma_i}\right), \tag{13.30}$$

where $\Gamma_i \overset{\text{iid}}{\sim} \mathrm{Ex}(1)$ and $T_1^* < T_2^* < \cdots < T_k^*$ are the distinct values of the uncensored observations. This definition reduces to the ordinary Bayesian bootstrap when all observations are uncensored; see Problem 13.15. Furthermore, this definition has a stick-breaking representation; see Problem 13.16.

The second approach is to define the Bayesian bootstrap as a noninformative limit of the posterior distribution corresponding to conjugate priors. In the context of survival analysis, beta processes form a natural conjugate family of priors for the cumulative hazard function. In Definition 13.3, this family is defined to have two parameters c and Λ; the noninformative limit is to let $c \to 0$. By Theorem 13.5 in this case the posterior distribution tends to a beta process with parameters Y and $\hat{H}$, for $\hat{H}$ the Nelson-Aalen estimator, given by (13.7). A beta process with a discrete Λ is the cumulative sum of its fixed jump heights, which are beta distributed with parameter $(c\Delta\Lambda, c(1-\Delta\Lambda)$. Since $\hat{H}$ is supported (only) on the uncensored observations and $Y\Delta\hat{H} = \Delta N$, the Bayesian bootstrap should be defined by

$$H^*(t) = \sum_{j=1}^{k} W_j \mathbb{1}\{T_j^* \leq t\}, \tag{13.31}$$

for $W_j \overset{\text{ind}}{\sim} \text{Be}(\Delta N(T_j^*), Y(T_j^*) - \Delta N(T_j^*))$. As in the case of uncensored data, the limit does, not depend on the second parameter Λ of the beta process prior. The corresponding distribution on the survival function $\bar{F}$ coincides with that given by (13.30); see Problem 13.17.

The third approach to the Bayesian bootstrap is to apply Bayes's theorem to the empirical likelihood with a noninformative prior on the empirical likelihood weights. In the uncensored case, letting $F = \sum_{i=1}^n W_i\delta_{X_i}$ and updating the posterior distribution of $(W_1, \dots, W_n)$ from the improper prior density at $(w_1, \dots, w_n)$ proportional to $w_1^{-1} \cdots w_n^{-1}$ on $\mathbb{S}_n$ to the posterior proportional to $(\prod_{i=1}^n w_i) \times (\prod_{i=1}^n w_i^{-1}) \equiv 1$ on $\mathbb{S}_n$, we obtain the Bayesian bootstrap distribution. This can be modifed to censored data, as we illustrate for the proportional hazard model in the next section.

The second interpretation of the Bayesian bootstrap shows that the Bayesian bootstrap is a special case of the beta process posterior. Hence, in view of Theorem 13.23, it gives rise to a Bernstein–von Mises theorem; see Problem 13.18 for a precise statement.

13.7.2 Cox Proportional Hazard Model

A probability density f and its survival function $\bar{F}$ can be expressed in the corresponding hazard and cumulative hazard functions as $f = he^{-H}$ and $\bar{F} = e^{-H}$. As in the Cox model these hazard functions are multiplied by $e^{\beta^\top Z}$, the likelihood of a single right censored observation (T, Δ, Z) in this model is $e^{\beta^\top Z}h(T)e^{-e^{\beta^\top Z}H(T)}$ if the observation is complete $(\Delta = 1)$ and $e^{-e^{\beta^\top Z}H(T)}$ otherwise. This likelihood makes sense in the absolutely continuous case, but to define a Bayesian bootstrap we require an *empirical likelihood*, which is valid also for general distributions. There are two possibilities in the literature, the Poisson and the binomial likelihood.

For the Poisson empirical likelihood, the baseline hazard function h is replaced by the jump ΔH of the cumulative hazard function, leading to

$$\left(e^{\beta^\top Z_i}\Delta H(T)\right)^{\Delta} e^{-e^{\beta^\top Z}H(T)}.$$

The product of this expression evaluated at n observations (T_i, Δ_i, Z_i) gives an overall *empirical likelihood*, which we shall refer to as the "Poisson form." Maximization of this empirical likelihood over (β, H) leads to the Cox partial likelihood estimator and the Breslow estimator, and in fact the resulting profile likelihood for β is exactly the Cox partial likelihood function given in (13.28) (see e.g. Murphy and van der Vaart 2000). In particular, the maximizer for H is supported only on the uncensored observations $\{T_i : \Delta_i = 1\}$. For discrete H with jumps only in the latter set, the "Poisson empirical likelihood" can be written as

$$L_n^P(\beta, H) \quad \prod_{t:\Delta N(t)>0} \Big[e^{\sum_{i:T_i=t,\Delta_i=1} e^{\beta^\top Z_i}} \Delta H(t)^{\Delta N(t)} e^{-\Delta H(t) \sum_{i:T_i \geq t} e^{\beta^\top Z_i}} \Big]. \tag{13.32}$$

The parameters in this likelihood are β and the jump sizes $\Delta H(t)$. The form of the likelihood suggests a prior for the jump sizes proportional to $\prod_t \Delta H(t)^{-\alpha}$. Theorem 13.21 suggests to take $\alpha = 1$ to ensure consistency. Under this prior, by Bayes's rule, given β the jumps $\Delta H(t)$ are also independent under the posterior distribution, with

$$\Delta H(t) | \, (\beta, D_n) \overset{\text{ind}}{\sim} \text{Ga}\Big(\Delta N(t), \sum_{i \in R_n(t)} e^{\beta^\top Z_i}\Big). \tag{13.33}$$

By integrating out $\Delta H(t)$ from the likelihood under the prior, the posterior density of β is found as

$$\pi_n(\beta | \, D_n) \propto \pi(\beta) \prod_{t \in \mathcal{T}_n} \frac{\exp(\sum_{i \in D_n(t)} \beta^\top Z_i)}{(\sum_{i \in R_n(t)} \exp(\beta^\top Z_i))^{\Delta N_n(t)}}. \tag{13.34}$$

This is precisely the prior density $\pi(\beta)$ times Cox's partial likelihood (13.28).

An alternative to the Poisson empirical likelihood derives from a different discretization of the continuous likelihood. The starting point is the "counting process form" of the continuous likelihood for one observation (T, Δ, Z):[9]

$$\prod_t d\Lambda(t)^{dN(t)} (1 - d\Lambda(t))^{Y(t) - dN(t)}.$$

Here Λ is the cumulative intensity of the counting process $N(t) = \mathbb{1}\{T \leq t, \Delta = 1\}$, which in the Cox model is given by $t \mapsto \int_{(0,t]} \mathbb{1}\{T \geq x\} e^{\beta^\top Z} \, dH(x)$. The "counting process likelihood" is exact for absolutely continuous H, but suggests a "discretized" extension in which the infinitesimal $d\Lambda$ are replaced by the jump heights $\Delta\Lambda$. The resulting formula resembles (13.11). Deviating from the preceding paragraph we take "proportional hazards" to mean that the survival function of the life time X is of the form $\bar{F}_X = \bar{F}^{\exp(\beta^\top Z)}$, for $\bar{F}$ the baseline survival function, which implies $1 - \Delta\Lambda = \bar{F}_X / \bar{F}_{X-} = (1 - \Delta H)^{\exp(\beta^\top Z)}$ (rather than $\Delta\Lambda = e^{\beta^\top Z} \Delta H$). As before, given n observations we consider cumulative baseline hazard distributions H supported on only the points $\{T_i : \Delta_i = 1\}$. Substituting $(1 - \Delta H)^{\exp(\beta^\top Z)}$ for $1 - d\Lambda$ in the counting process likelihood, we find the "binomial empirical likelihood"

$$\prod_{i=1}^n \Big(1 - (1 - \Delta H(T_i))^{e^{\beta^\top Z_i}}\Big)^{\Delta N_i(T_i)} (1 - \Delta H(T_i))^{e^{\beta^\top Z_i}(Y_i(T_i) - \Delta N_i(T_i))}.$$

9 The formula must be interpreted as formal notation for product integration, see Andersen et al. (1993).

The binomial form of this likelihood suggests the prior distribution on the jump heights $\Delta H(t)$ given by

$$\prod_t \Delta H(t)^{-1}(1 - \Delta H(t))^{-1}. \tag{13.35}$$

This results in a posterior that can also be found as the limit as $c \to 0$ of the posterior distribution based on a beta process prior on H (see Problem 13.19). The marginal posterior distribution of β does not reduce to a closed form expression, but it can be calculated numerically using a Metropolis-Hastings algorithm.

Since the prior on H is improper, the marginal posterior distribution for β may be improper even when the prior distribution on β is proper. It is known that Cox's partial likelihood is log-concave under the general assumption that the positive linear span of the differences $\{Z_i - Z_j : T_i = t, \Delta_i = 1, j \in R_n^+(t)\}$ is the full space $\mathbb{R}^d$ (Jacobsen 1989), where $R_n^+(t) = \{i : T_i \geq t\} \setminus \{i : T_i = t, \Delta_i = 1\}$. Under this condition the Poisson likelihood leads to a proper posterior if π is improper uniform. Propriety of the posterior distribution corresponding to the binomial likelihood is more complicated; see Problem 13.21 for a counterexample.

Given the Bayesian bootstrap distribution for H the corresponding distribution on the survival function can be obtained by the transformations $\bar{F}(t) = \exp[-\sum_{T_i \leq t, \Delta_i = 1} \Delta H(T_i)]$ and $\bar{F}(t) = \prod_{T_i \leq t, \Delta_i = 1}(1 - \Delta H(T_i))$, in the Poisson and binomial cases, respectively. In the Poisson case the increments $\Delta H(t)$ may be larger than 1 (and hence H is not a genuine cumulative hazard function), which renders the second correspondence between H and $\bar{F}$ unfeasible.

As in Section 13.6.2 the posterior distribution described by the Bayesian bootstrap process admits a Bernstein–von Mises theorem. We concentrate on the parametric part β, even though a joint Bernstein–von Mises theorem for (β, H) in the spirit of Theorem 13.36 is possible also.

Theorem 13.37 *If the true cumulative hazard function H_0 is absolutely continuous with $H_0(\tau) < \infty$ for some $\tau > 0$ with $G(\tau-) < 1 = G(\tau)$, and the support of the covariate Z is a compact subset of $\mathbb{R}^d$ with nonempty interior, then the Bayesian bootstrap posterior distribution given by the density $\pi_n(\cdot| D_n)$ in* (13.34) *for a proper prior density π that is continuous and positive at β_0, satisfies $\|\Pi_n(\cdot| D_n) - \mathrm{Nor}(\hat{\beta}_n, n^{-1}\tilde{I}_{\beta_0}^{-1})\|_{TV} \to 0$ a.s. $[P_{H_0,\beta_0}^{\infty}]$ as $n \to \infty$, for $\hat{\beta}_n$ the partial likelihood estimator (which maximizes* (13.28)*) and $\tilde{I}_{\beta_0}$ is given by* (13.29).

Proof The proof proceeds as for the classical Bernstein–von Mises theorem for the parametric case, by expanding the partial likelihood in a Taylor series and using Cramér-type regularity conditions to bound the terms. A noticeable difference is that the partial likelihood (13.28) is not of a product of i.i.d. terms, which necessitates a nonstandard law of large numbers.

The posterior density (13.34) is proportional to $\pi(\beta)L_n(\beta)$, for L_n the partial likelihood function (13.28). It suffices to show that $\int |g_n(u) - \pi(\beta_0)\exp\{-u^\top \tilde{I}_{\beta_0} u/2\}|\, du \to 0$ a.s., where $g_n(u) = \pi(\hat{\beta}_n + n^{-1/2}u)L_n(\hat{\beta}_n + n^{-1/2}u)/L_n(\hat{\beta}_n)$. For given $B, \delta > 0$ we can bound this $\mathbb{L}_1$-distance by the sum $I_1 + I_2 + I_3 + I_4$ given by

$$\int_{\|u\|\le B} |g_n(u) - \pi(\beta_0)e^{-u^\top \tilde I_{\beta_0} u/2}|\,du + \int_{B<\|u\|\le\delta\sqrt{n}} g_n(u)\,du$$
$$+ \int_{\|u\|>\delta\sqrt{n}} g_n(u)\,du + \int_{\|u\|>B} \pi(\beta_0)e^{-u^\top \tilde I_{\beta_0} u/2}\,du.$$

The term I_4 can clearly be made arbitrarily small by choice of B. We shall argue that I_1 tends to zero for every B, the term I_2 can be arbitrarily small by choice of δ, and I_3 tends to zero for any δ and B.

To bound I_1 we expand $u \mapsto \log L_n(\hat\beta_n + n^{-1/2}u)$ in a second order Taylor expansion around 0 with third order remainder. The linear term of this expansion vanishes as $\hat\beta_n$ maximizes L_n. Because n^{-1} times the third order partial derivatives of $\beta \mapsto \log L_n(\beta)$ are uniformly bounded in a neighborhood of 0, and $\hat\beta_n$ is consistent, the remainder term is bounded by a multiple of $n^{-1/2}$. Since also π is continuous at β_0, the term I_1 can be shown to tend to zero by standard arguments used in proving parametric Bernstein–von Mises theorems.

The matrix of second derivatives of $\beta \mapsto \log L_n(\beta)$ at β_0 approaches the positive definite matrix $\tilde I_{\beta_0}$, by a law of large numbers (see Tsiatis 1981 or Andersen et al. 1993 for details). Together with the boundedness of n^{-1} the third-order partial derivatives, this shows that there exists $\lambda > 0$ such that $\log L_n(\hat\beta_n + n^{-1/2}u) - \lambda\|u\|^2/2$, for $B < \|u\| \le \delta\sqrt{n}$ and sufficiently small $\delta > 0$. In view of the boundedness of the ratio $\pi(\beta)/\pi(\beta_0)$ over a small neighborhood around β_0, this implies that I_2 can be made small.

To bound I_3, we establish an exponential bound on $L_n(\beta)/L_n(\hat\beta_n)$, for $\|\beta\| > \delta$. The function $l_n(\beta)$ is concave, attains its maximum at $\hat\beta_n$, and for every fixed β the sequence $n^{-1}l_n(\beta)$ tends almost surely to

$$l(\beta) = \beta^\top \mathrm{E}(Z\mathbb{1}\{\Delta = 1\}) - \int_0^\tau \mathrm{E}[e^{\beta_0^\top Z}\mathbb{1}\{T \ge t\}]\log(\mathrm{E}[e^{\beta^\top Z}\mathbb{1}\{T \ge t\}])\,dH_0(t).$$

By concavity, the convergence is automatically uniform over compact sets. Let $\rho = \inf\{l(\beta_0) - l(\beta)\colon \|\beta - \beta_0\| = \delta\}$, and determine a sufficiently large n such that

$$\sup_{\|\beta-\beta_0\|\le\delta} |n^{-1}l_n(\beta) - l(\beta)| + |n^{-1}l_n(\hat\beta_n) - l(\beta_0)| < \frac{\rho}{2}.$$

By concavity, the supremum of $l_n(\beta) - l_n(\hat\beta_n)$ over $\|\beta - \beta_0\| \ge \delta$ is attained on the circle $\|\beta - \beta_0\| = \delta$. By the triangle inequality, it follows $l_n(\beta) - l_n(\hat\beta_n) < -n\rho/2$, for $\|\beta - \beta_0\| \ge \delta$, eventually. Hence $I_3 \le n^{d/2}e^{-n\rho/2} \to 0$, as π is proper. ☐

13.8 Historical Notes

Bayesian estimation under censoring was addressed by Susarla and Van Ryzin (1976). They obtained the mixture of Dirichlet process representation for the posterior distribution given a Dirichlet prior, and computed the corresponding posterior mean. The Pólya tree representation of the Dirichlet process posterior was obtained by Ghosh and Ramamoorthi (1995). The gamma process was first used in Bayesian survival analysis by Kalbfleisch (1978), the generalized gamma process by Ferguson and Phadia (1979), and the extended gamma process by Dykstra and Laud (1981). The beta process was constructed as a conjugate prior distribution for the cumulative hazard function for randomly right censored data by Hjort (1990). His motivation of a beta process through a discrete time setting leads also to the time discretization algorithm (a) in Section 13.3.3. Wolpert and Ickstadt (1998) proposed the inverse Lévy measure algorithm I. The Poisson weighting algorithm is based on the general method for sampling from an infinitely divisible distribution by Damien et al. (1995, 1996); Section 13.3.3 provides the refined version by Lee (2007). The ϵ-approximation algorithm is due to Lee and Kim (2004). Neutral to the right processes were introduced by Doksum (1974). Ferguson and Phadia (1979) obtained conjugacy for the posterior distribution under random right censoring. Several aspects of neutral to the right processes including Proposition 13.10 were obtained by Dey et al. (2003); see also the problems section. The beta-Stacy process was introduced by Walker and Muliere (1997). Conjugacy of independent increment processes in the form of Theorem 13.15 is due to Hjort (1990). It was generalized by Kim (1999) to the more general framework of multiplicative counting processes, which allows for observing a general counting process N with compensator $\int_0^t Y\,dH$ for some predictable increasing process Y, a setup that can treat left-truncated data as well. The proof presented here is indicated as an alternative, more-intuitive approach in Hjort (1990) and inspired by Prünster (2012). Theorem 13.18 characterizing consistency for the extended beta process prior and the example of inconsistency are due to Kim and Lee (2001). They also obtained Theorem 13.21 for a general independent increment process prior on the cumulative hazard function. Dey et al. (2003) obtained a somewhat similar result for neutral to the right process prior on distributions. The Bernstein–von Mises theorem for the cumulative hazard function was derived by Kim and Lee (2004). Kernel smoothing on a hazard function to construct a prior was first used by Dykstra and Laud (1981). They obtained the expression for the posterior moment generating function of the cumulative hazard function using the extended gamma process and the Dykstra-Laud kernel. Consistency for such a prior distribution was obtained by Drăghici and Ramamoorthi (2003). The representation of the posterior distribution given by Theorem 13.30 was obtained by James (2005). Consistency for a general smooth hazard process prior was obtained by De Blasi et al. (2009), together with a pathwise central limit theorem for the tail of linear and quadratic functionals of the posterior smooth hazard function. An analogous result on the prior process was obtained earlier by Peccati and Prünster (2008). Consistency of the posterior distribution in the Cox model for independent increment process priors was obtained by Kim and Lee (2003a), and the Bernstein–von Mises theorem in Kim (2006), using an elegant approach based on counting process likelihoods. The Bayesian bootstrap for censored data with covariates was first defined by Lo (1993). The Bayesian bootstrap was extended to the Cox model by Kim and Lee (2003b). They also obtained the Bernstein–von Mises theorem for the Bayesian bootstrap in the Cox model.

Problems

13.1 (Ghosh and Ramamoorthi 2003) Let $\mathfrak{M}_0$ be the set of all pairs of distribution functions (F, G) such that $F(t) < 1, G(t) < 1$ for all t and F and G have no common points of discontinuity. Let $\varphi(x, y) = (x \wedge y, \mathbb{1}\{x \leq y\})$ and $\mathfrak{M}_0^* =$

$\{(F, G)\varphi^{-1} : (F, G) \in \mathfrak{M}_0\}$. Let Π^* be the prior on $(F, G)\varphi^{-1}$ induced from a prior Π on (F, G), and if $\Pi^*(\cdot|D_n)$ is (strongly) consistent at $(F_0, G_0)\varphi^{-1}$ with respect to the weak topology, then show that $\Pi(\cdot| D_n)$ is also (strongly) consistent at (F_0, G_0) with respect to the weak topology.

Let $S_u(t) = \mathrm{P}(X > t, X \le C| X \sim F, C \sim G)$ and $S_c(t) = \mathrm{P}(C > t, X > C| X \sim F, C \sim G)$ be the sub-survival functions corresponding respectively to uncensored and censored observations. If the induced posterior for S_u and S_c are (strongly) consistent with respect to the Kolmogorov-Smirnov norm, then show that the posterior for $(F, G)\varphi^{-1}$ is (strongly) consistent with respect to the weak topology.

13.2 (Ghosh and Ramamoorthi 2003) Consider an interval censoring mechanism where one observes only $X_i \in (L_i, R_i]$, $i = 1, \ldots, n$, and $(L_1, R_1), \ldots, (L_n, R_n)$ are independent. Let the distribution F for X follow a Dirichlet process. Obtain the expressions for the posterior distribution and Bayes estimate of F given $(L_1, R_1), \ldots, (L_n, R_n)$.

13.3 (Walker and Muliere 1997, Ramamoorthi et al. 2002) Let X be an observation following cumulative distribution function F and F be given a prior Π. Show that Π is an NTR process prior if and only if for any $t > 0$, $\Pi(F(t)| X) = \Pi(F(t)| \mathbb{1}\{X \le s\}, 0 < s < t)$.

13.4 (Dey et al. 2003) The connection between the means of the survival function and the cumulative hazard function given by Theorem 13.8 characterizes NTR processes. Let Π be a prior for a survival distribution F such that $\Pi(0 < \bar{F}(i)/\bar{F}(i-1) < 1) = 1$ for all i. Let $H(F)$ be the cumulative hazard function associated with F. Suppose that for all $n \in \mathbb{N}$ and $X_1, \ldots, X_n$, $\mathrm{E}\big[H(F)| X_1, \ldots, X_n\big] = H(\mathrm{E}[F| X_1, \ldots, X_n])$ and $\mathrm{E}\big[H(F)| X > x\big] = H(\mathrm{E}[F| X > x])$ for all x. Show that Π is NTR.

13.5 (Dey et al. 2003) Let Π_1 and Π_2 be two NTR processes, and let ν_1 and ν_2 be the intensity measures for the corresponding independent increment processes on the c.h.f. Assume that $\nu_1 \ll \nu_2$ and $f = d\nu_1/d\nu_2$. Then Π_1 and Π_2 are mutually singular if either one of the following conditions hold:

(a) for some $c > 0$, $\int_{|f-1|>c} |f - 1|\, d\nu_2 = \infty$;
(b) for all $c > 0$, $\int_{|f-1|<c} |f - 1|^2\, d\nu_2 = \infty$.

13.6 (Dey et al. 2003) Let H_1 and H_2 be continuous c.h.f.s and let c_1 and c_2 be positive measurable functions such that the measure $c_1 H_1$ and $c_2 H_2$ are not identical. Then the beta processes with parameters (c_1, H_1) and (c_2, H_2) are mutually singular.

13.7 Show that if the distribution of the discrete hazard rates $h(jb)$ are given by (13.12), with $c(j)(1 - h(j)) = c(j + 1)$ for all j, then $(f(jb): j = 0, 1, \ldots)$ follows $\mathrm{DP}_{(c(0), c(1), \ldots)}$.

13.8 (Dey et al. 2003) Let $X_1, \ldots, X_n \overset{\text{iid}}{\sim} F$, a cumulative distribution function on $[0, \infty)$. Let F be given an NTR prior distribution Π. If the posterior $\Pi(\cdot| X_1, \ldots, X_n)$ is consistent at every continuous true distribution F_0, then show that $\Pi(\cdot| D_n)$ is consistent at every continuous true distribution F_0.

13.9 (Lo 1993) Let α, β be finite measures on $[0, \infty)$ and let Γ_α and Γ_β be standard gamma processes with parameters α and β respectively. Show that the process $H(t) = \int_0^t (\Gamma_\alpha(x, \infty) + \Gamma_\beta(x, \infty))^{-1} \Gamma_\alpha(dx)$ follows a beta process with parameters (c, H_0) given by $c(t) = \alpha(t, \infty) + \beta(t, \infty)$ and $H_0(t) = \int_0^t (\alpha(x, \infty) + \beta(x, \infty))^{-1} \alpha(dx)$.
Note that $c(t)$ is monotone decreasing, so not all beta processes can be represented in this way.

13.10 If a beta process with parameter (c, H_0) is homogeneous, i.e. $c(x) = c > 0$, a constant, then show that $L(J_j) = \sum_{i=1}^{j} V_i / H_0(\tau)$, where $[0, \tau]$ is a given interval where jumps are being studied, $V_1, V_2, \ldots$ are i.i.d. standard exponential and $L(s) = \nu_H([0, \tau \times (s, 1))$, for ν_H the intensity measure of the beta process. In particular, if $c = 1$, show that J_j is a product of j i.i.d. $\mathrm{Be}(H_0(\tau), 1)$.

13.11 (Dey et al. 2003) For independent increment process priors, if observations are all uncensored, then the following shows that consistency of the posterior mean of the distribution actually implies posterior consistency. (Problem 13.8 shows that the assumptions that all observations are uncensored can be dropped if F_0 is continuous.)

Let $X_i \stackrel{\text{iid}}{\sim} F$, a distribution on $[0, \infty)$. Let the c.h.f. H follow an independent increment process prior Π with intensity measure given by $\nu(dt, ds) = a(x, s)\, ds\, \Lambda(dt)$, where Λ itself is a c.h.f. If $\mathrm{E}[F(t) |\, X_1, \ldots, X_n] \to F_0(t)$ a.s., then $\mathrm{var}[F(t) |\, X_1, \ldots, X_n] \to 0$ a.s.

If the condition holds for all t, then in view of Lemma 6.4, $\Pi(\cdot |\, X_1, \ldots, X_n) \rightsquigarrow \delta_{F_0}$ a.s.

13.12 (Kim and Lee 2004) Adopt the notations of Section 13.4.2. Consider an independent increment process prior on a c.h.f. H with intensity measure given by $\nu(dt, ds) = s^{-1}(1 + s^\alpha)\, ds\, dt$. Let $J_\alpha(t) = \alpha \Gamma(\alpha + 1) \int_0^t (Q(u))^{-\alpha} d H_0(u)$. Show that a.s. as $n \to \infty$:

(a) If $0 < \alpha \leq \frac{1}{2}$, then $\sup\{|n^\alpha [\mathrm{E}(H_d(t) |\, D_n) - \hat{H}_n(t)] - J_\alpha(t)| : t \in [0, \tau]\} \to 0$;
(b) If $0 < \alpha < \frac{1}{2}$, then $n^\alpha (H - \hat{H}_n) |\, D_n \rightsquigarrow \delta_{J_\alpha(\cdot)}$;
(c) If $\alpha = \frac{1}{2}$, then $n^{1/2}(H - \hat{H}_n) |\, D_n \rightsquigarrow B(U_0(\cdot) + J_\alpha(\cdot))$;
(d) If $\alpha > \frac{1}{2}$, then $n^{1/2}(H - \hat{H}_n) |\, D_n \rightsquigarrow B(U_0(\cdot))$.

Thus the value of the index α determining the rate at which $g(t, s)$ approaches $q(t)$ as $s \to 0$ is critical for the Bernstein–von Mises theorem.

13.13 (De Blasi et al. 2009) Consider a CRM Φ on $[0, \infty)$ with intensity $\nu(dv, ds) = \rho(ds |\, v)\, \alpha(dv)$ satisfying $\rho(\mathbb{R}^+ |\, v) = \infty$ a.s. $[\alpha]$, and $\mathrm{supp}(\alpha) = \mathbb{R}^+$. Show that for any Lebesgue-Stieltjes distribution function G_0 $\mathrm{P}(\sup_{x \leq M} |\Phi([0, x]) - G_0(x)| < \epsilon) > 0$ and $M, \epsilon > 0$.

13.14 (Kim and Lee 2003a) Let A be an independent increment process with intensity measure of the form $g(t, s)\, ds\, dt + \sum_{i=1}^{k} \delta_{v_j}(dt)\, G_j(ds)$ and let $h: [0, 1] \to [0, 1]$ be a strictly increasing continuously differentiable bijection. Show that $B(t) = \sum_{x \leq t} h(\Delta A(x)) \mathbb{1}\{\Delta A(x) > 0\}$ is an independent increment process with intensity measure $(h^{-1})'(s) g(t, h^{-1}(s))\, ds\, dt + \sum_{j=1}^{k} \delta_{v_j}(dt)\, (G_j \circ h^{-1})(ds)$.

13.15 Show that (13.30) reduces to Rubin's Bayesian bootstrap when all the observations are uncensored.

13.16 Show that the probability measure corresponding to (13.30) can be written as $\sum_{i=1}^{n} W_j \delta_{T_j^*}$, where the variables W_j are the stick-breaking weights based on the relative stick lengths given by $V_j \overset{\text{ind}}{\sim} \text{Be}(\sum_{i:X_i=T_j^*} \Delta_i, \#\{i: X_i \geq T_j^*\})$, for $j = 1, \ldots, k$, and $V_k = 1$.

13.17 Show that the distribution on $\bar{F}$ induced from (13.31) on H is identical with (13.30).

13.18 Taking the limit $c(\cdot) \to 0$ in Theorem 13.23 for a beta process, obtain a Bernstein–von Mises theorem for the Bayesian bootstrap for censored data.

13.19 (Kim and Lee 2003b) Show that the limit of the beta process posterior as $c(\cdot) \to 0$ in the Cox model with binomial likelihood coincides with the posterior based on the binomial likelihood and the improper prior (13.35).

13.20 (Kim and Lee 2003b) Show that the Bayesian bootstrap distribution based on the binomial form of the likelihood reduces to (13.31) when no covariate is present.

13.21 (Kim and Lee 2003b) Consider observations of the form (T, Δ, Z) and form the binomial likelihood. Suppose that $n = 3$ and the observations are $(1, 1, -1)$, $(2, 1, -1.9)$ and $(3, 1, -1.5)$. Show that the likelihood is bounded below, and hence the posterior is improper for the improper uniform prior. Further, show that the posterior is improper for some proper priors with sufficiently thick tails.

14

Discrete Random Structures

Random, exchangeable partitions of a finite or countable set arise naturally through the pattern of tied observations when sampling repeatedly from a random discrete measure. They may be used as prior models for clustering, or more generally for sharing of features. The Chinese restaurant process is a distinguished example of a random partition, and arises from the Dirichlet process. After presenting general theory on random partitions, we discuss many other examples of species sampling processes, such as Gibbs processes, and the Poisson-Kingman and Pitman-Yor processes. These random discrete distributions can be viewed as generalizations of the Dirichlet process. Normalized completely random measures are another generalization, with the normalized inverse-Gaussian process as a particularly tractable example. Random discrete distributions with atoms or locations that depend on a covariate can be used as priors for conditional distributions. We discuss some constructions that lead to tractable processes. Finally we discuss a prior distribution on infinite binary matrices, known as the Indian buffet process, as a model for overlapping clusters and for sharing a potentially unlimited number of features.

14.1 Exchangeable Partitions

A *partition* $\{A_1, \ldots, A_k\}$ of the finite set $\mathbb{N}_n = \{1, \ldots, n\}$ is a decomposition of this set in disjoint (nonempty) subsets: $\mathbb{N}_n = \cup_i A_i$ and $A_i \cap A_j = \varnothing$ for $i \neq j$. In this section we discuss probability measures on the collection of all partitions of $\mathbb{N}_n$, where $n \in \mathbb{N}$ may vary. Exchangeable probability measures that can be defined "consistently" across n are of special interest. These probability measures may serve as prior distributions in methods for clustering data.

The cardinalities $n_i = |A_i|$ of the sets in a partition of $\mathbb{N}_n$ are said to form a *partition of* n: an unordered set $\{n_1, \ldots, n_k\}$ of natural numbers such that $n = \sum_{i=1}^k n_i$. The sets A_i in a partition are considered to be unordered, unless specified otherwise, but if the sets are listed in a particular order, then so are their cardinalities. An "ordered partition" $(n_1, \ldots, n_k)$ of n is called a *composition* of n, and the set of all compositions of n is denoted by $\mathcal{C}_n$. The particular order by the sizes of the minimal elements of the sets (so $\min A_1 < \min A_2 < \cdots < \min A_k$) is called the *order of appearance*; thus in this case A_1 contains the element 1, A_2 contains the smallest number from $\mathbb{N}_n$ not in A_1, etc.

A *random partition* of $\mathbb{N}_n$ is a random element defined on some probability space taking values in the set of all partitions of $\mathbb{N}_n$. (Measurability is understood relative to the discrete σ-field on the (finite) set of all partitions.) Its induced distribution is a probability measure on the set of all partitions of $\mathbb{N}_n$. We shall be interested in "exchangeable partitions."

Write $\sigma(A) = \{\sigma(i): i \in A\}$ for the image of a subset $A \subset \mathbb{N}_n$ under a permutation σ of $\mathbb{N}_n$.

Definition 14.1 (Exchangeable partition) A random partition $\mathcal{P}_n$ of $\mathbb{N}_n$ is called *exchangeable* if its distribution is invariant under the action of any permutation $\sigma: \mathbb{N}_n \mapsto \mathbb{N}_n$, i.e. for every partition $\{A_1, \ldots, A_k\}$ of $\mathbb{N}_n$ the probability $\mathrm{P}(\mathcal{P}_n = \{\sigma(A_1), \ldots, \sigma(A_k)\})$ is the same for every permutation σ of $\mathbb{N}_n$. Equivalently, a random partition $\mathcal{P}_n$ of $\mathbb{N}_n$ is *exchangeable* if there exists a symmetric function $p: \mathcal{C}_n \to [0, 1]$ such that, for every partition $\{A_1, \ldots, A_k\}$ of $\mathbb{N}_n$,

$$\mathrm{P}(\mathcal{P}_n = \{A_1, \ldots, A_k\}) = p(|A_1|, \ldots, |A_k|). \tag{14.1}$$

The function p is called the *exchangeable partition probability function* (EPPF) of $\mathcal{P}_n$.

The EPPF is defined as a function on compositions $(n_1, \ldots, n_k)$, but, as it is necessarily symmetric, could have been defined on the corresponding *partitions* $\{n_1, \ldots, n_k\}$. Its value $p(n_1, \ldots, n_k)$ is the probability of a *particular* partition $\{A_1, \ldots, A_k\}$ with $|A_i| = n_i$ for every i, and its number of arguments varies from 1 to n.[1]

Example 14.2 (Product partition model) An EPPF is said to lead to a *product partition model* if the function p in (14.1) factorizes as $p(|A_1|, \ldots, |A_k|) = V_{n,k} \prod_{i=1}^{k} \rho(|A_i|)$. The function $\rho: \mathbb{N} \to [0, 1]$ is then called the *cohesion function* of the product partition model.

Example 14.3 (Ties in exchangeable vector) Given an ordered list $(x_1, \ldots, x_n)$ of n arbitrary elements, the equivalence classes for the equivalence relation $i \sim j$ if and only if $x_i = x_j$ form a partition of $\mathbb{N}_n$. The partition defined in this way by an exchangeable random vector $(X_1, \ldots, X_n)$ is exchangeable.

Example 14.4 (General finite partition) Any exchangeable random partition of $\mathbb{N}_n$ can be obtained by the following scheme.[2] First generate a composition $(n_1, \ldots, n_k)$ from a given but arbitrary distribution on the set of all compositions of n. Next, given $(n_1, \ldots, n_k)$, form a random vector $(X_1, \ldots, X_n)$ by randomly permuting[3] the set of n symbols

$$(\underbrace{1, \ldots, 1}_{n_1}, \underbrace{2, \ldots, 2}_{n_2}, \ldots, \underbrace{k, \ldots, k}_{n_k}).$$

Finally define a partition of $\mathbb{N}_n$ by the equivalence classes under the relationship $i \sim j$ if and only if $X_i = X_j$.

An alternative way of coding the information in a partition $\{n_1, \ldots, n_k\}$ of n generated by a partition $\{A_1, \ldots, A_k\}$ of $\mathbb{N}_n$ is to list the corresponding *multiplicity class* $(m_1, \ldots, m_n)$,

[1] For instance $p(1, 2)$ is the probability of the partition of $\mathbb{N}_3$ into $\{1\}$ and $\{2, 3\}$, and also of the partition in $\{2\}$ and $\{1, 3\}$; it is equal to $p(2, 1)$, which is the probability of the partition into e.g. $\{1, 3\}$ and $\{2\}$; $p(1, 1, 1)$ is the probability of the partition $\{\{1\}, \{2\}, \{3\}\}$; $p(3)$ is the probability of $\{\{1, 2, 3\}\}$.

[2] See Section 11 of Aldous (1985) for this and further discussion. We shall not use this characterization in the following.

[3] This is to say by a permutation drawn from the uniform distribution on the set of all $n!$ permutations of $\mathbb{N}_n$.

where m_i is defined as the number of sets A_j of cardinality i in the partition. The entries m_i of a multiplicity vector are nonnegative integers satisfying $\sum_{i=1}^n i m_i = n$ (with necessarily many zero m_i).

This coding was used already in the statement of Ewens's sampling formula given in Proposition 4.11. The following proposition generalizes this formula to general partitions.

Proposition 14.5 (Sampling formula) *The probability that a random exchangeable partition of $\mathbb{N}_n$ consists of m_i sets of cardinality i, for $i = 1, 2, \ldots, n$, is equal to, for any composition $(n_1, \ldots, n_k)$ compatible with the multiplicity class $(m_1, \ldots, m_n)$,*

$$\frac{n!}{\prod_{i=1}^n m_i!(i!)^{m_i}} p(n_1, \ldots, n_k).$$

Furthermore, the probability that a random exchangeable partition of $\mathbb{N}_n$ consists of k sets, for $k = 1, \ldots n$, is equal to, with the sum over all compositions $(n_1, \ldots, n_k)$ of n in k elements,

$$\sum_{(n_1,\ldots,n_k)} \frac{1}{k!}\binom{n}{n_1 \cdots n_k} p(n_1, \ldots, n_k).$$

Proof The EPPF gives the probability of a particular partition. The factor preceding it is the cardinality of the number of partitions giving the multiplicity class. See the proof of Proposition 4.10 for details.

For the proof of the second formula it is helpful to create an *ordered* random partition of $\mathbb{N}_n$ by randomly ordering the sets of an ordinary (i.e. unordered) random partition. The probability that the partition contains k sets is the same, whether the partition is ordered or not. For a given composition $(n_1, \ldots, n_k)$ of n, the probability of a particular ordered partition with set sizes $n_1, \ldots, n_k$ is $p(n_1, \ldots, n_k)/k!$. We can construct all ordered partitions of $\mathbb{N}_n$ with composition $(n_1, \ldots, n_k)$ by first ordering the elements $1, 2, \ldots, n$ in every possible way and next defining the first set to consist of the first n_1 elements, the second set of the next n_2 elements, etc. There are $n!$ possible orderings of $1, 2, \ldots, n$, but permuting the first n_1 elements, the next n_2 elements, etc. gives the same ordered partition. Thus there are $n!/\prod_i n_i!$ ordered partitions with composition $(n_1, \ldots, n_k)$. By exchangeability they all have the same probability, $p(n_1, \ldots, n_k)/k!$. □

Rather than elaborating on exchangeable partitions of a finite set, we turn to *partition structures*, which link partitions across n. We want these structures to be *consistent* in the following sense. From a given partition $\{A_1, \ldots A_k\}$ of $\mathbb{N}_n$, we can obtain a partition of $\mathbb{N}_m$ for a given $m < n$ by removing the elements $\{m+1, \ldots, n\}$ from the sets A_i, and removing possible empty sets thus created. If this is applied to an exchangeable partition $\mathcal{P}_n$, then the resulting partition will be an exchangeable partition of $\mathbb{N}_m$, since a permutation of just the elements of $\mathbb{N}_m$ can be identified with the permutation of $\mathbb{N}_n$ that leaves $m+1, \ldots, n$ invariant.

Definition 14.6 (Exchangeable partition) An *infinite exchangeable random partition* (or *exchangeable random partition of* $\mathbb{N}$) is a sequence $(\mathcal{P}_n : n \in \mathbb{N})$ of exchangeable random partitions of $\mathbb{N}_n$ that are consistent in the sense that $\mathcal{P}_{n-1}$ is equal to the partition obtained from $\mathcal{P}_n$ by leaving out the element n, almost surely, for every n. The function $p\colon \cup_{n=1}^{\infty} \mathcal{C}_n \to [0, 1]$ whose restriction to $\mathcal{C}_n$ is equal to the EPPF of $\mathcal{P}_n$ is called the *exchangeable partition probability function* (EPPF) of $(\mathcal{P}_n\colon n \in \mathbb{N})$.

That the definition does not refer directly to infinite partitions, but focuses on consistent finite partitions instead, avoids measurability issues. However, a partition of $\mathbb{N}$ is implied as it can be ascertained from $\mathcal{P}_{i\vee j}$ whether i and j belong to the same set.

An infinite exchangeable random partition may be obtained from an (infinite) exchangeable sequence of random variables $X_1, X_2, \ldots$, by defining $\mathcal{P}_n$ by the pattern of ties in the collection $(X_1, \ldots, X_n)$, for every n, as in Example 14.3. The sequence $X_1, X_2, \ldots$ is said to *generate* the infinite random partition in this case. The following theorem shows that *all* infinite exchangeable random partitions are generated by an exchangeable sequence. It is similar to de Finetti's representation of exchangeable sequences, to which it can also be reduced.

Theorem 14.7 (Kingman's representation) *For any infinite exchangeable random partition $(\mathcal{P}_n\colon n \in \mathbb{N})$ defined on a probability space that is rich enough to support an independent i.i.d. sequence of uniform variables, there exists a random probability measure P on* [0, 1] *and a sequence of random variables $X_1, X_2, \ldots$ defined on the same probability space with $X_1, X_2, \ldots \mid P \overset{iid}{\sim} P$ that generates $(\mathcal{P}_n\colon n \in \mathbb{N})$. Furthermore, the size $N_{(j),n}$ of the jth largest set in $\mathcal{P}_n$ satisfies $n^{-1}N_{(j),n} \to W_{(j)}$* a.s. *as $n \to \infty$, for $W_{(1)} \geq W_{(2)} \geq \cdots$ the sizes of the atoms of P ordered in decreasing size.*

Proof Let $\xi_1, \xi_2, \ldots$ be an i.i.d. sequence of uniform random variables defined on the same probability space as $(\mathcal{P}_n)$ and independent of it. By deleting a null set it can be ensured that these variables "surely" assume different values. For any $i \in \mathbb{N}$ let $j(i)$ be the smallest number in the partitioning set to which i belongs and define $X_i = \xi_{j(i)}$. (As $i \in \mathbb{N}_i$ and $(\mathcal{P}_n)$ is consistent, it suffices to consider the partition $\mathcal{P}_i$, and $j(i) \leq i$.) Then $X_1, X_2, \ldots$ generates the partition $(\mathcal{P}_n)$. We can write $X_i = g_i((\xi_j), (\mathcal{P}_n))$, for some measurable map g_i, and then have $X_{\sigma(i)} = g_i((\xi_{\sigma(j)}), (\sigma\langle\mathcal{P}_n\rangle))$, for any permutation σ of $\mathbb{N}$ that permutes only a finite set of coordinates, where $(\sigma\langle\mathcal{P}_n\rangle)$ is the sequence of partitions of $\mathbb{N}_n$ obtained from $\sigma\langle\mathcal{P}_n\rangle$ by swapping the elements of $\mathbb{N}$ according to σ. Because the distribution of $((\xi_{\sigma(j)}), (\sigma\langle\mathcal{P}_n\rangle)$ is invariant under the permutation σ, the sequence $X_1, X_2, \ldots$ is exchangeable. By de Finetti's theorem there exists a random measure P such that given P it is an i.i.d. sequence.

To prove the last assertion we note that the classes of the partition $\mathcal{P}_n$ generated by $X_1, \ldots, X_n$ are among the sets $\{i \in \mathbb{N}_n\colon X_i = x\}$, when x varies (most of which are empty), and hence the numbers $N_{(j),n}$ are the first n elements of the numbers $N_n(x) := \#\{i \in \mathbb{N}_n\colon X_i = x\}$ ordered in decreasing size (hence with the uncountably many zeros at the end). By the (ergodic) law of large numbers $n^{-1}N_n(x) \to \mathrm{P}(X_1 = x \mid P)$, a.s., which implies the result. ◻

The unit interval and the random probability measure P in the theorem are rather arbitrary. Only the sizes (W_i) of the atoms of P play a role for the pattern of equal X_i; their locations merely serve as labels. The atomless part of P is even more arbitrary: it produces labels that are used only once and hence give rise to singletons in the induced partition. This will motivate to make in the next section a convenient choice of P, in the form of a "species sampling model," which gives a random measure on an arbitrary Polish space.

By marginalization, the EPPFs of the random exchangeable partitions of $\mathcal{P}_{n-1}$ and $\mathcal{P}_n$ obtained from an infinite exchangeable random partition satisfy the relationships

$$p(n_1,\ldots,n_k)=\sum_{j=1}^{k}p(n_1,\ldots,n_j+1,\ldots,n_k)+p(n_1,\ldots,n_k,1). \tag{14.2}$$

Here we abuse notation by using the same letter p for both EPPFs, with p on the left side referring to $\mathcal{P}_{n-1}$ and every p on the right side to $\mathcal{P}_n$. This is customary and should not cause confusion as the sum of the arguments reveals the set $\mathbb{N}_n$ to which it refers.[4] If for $\boldsymbol{n}=(n_1,\ldots,n_k)$ we introduce the notations $\boldsymbol{n}^{j+}=(n_1,\ldots,n_{j-1},n_j+1,n_{j+1},\ldots,n_k)$, for $1\le j\le k$, and $\boldsymbol{n}^{(k+1)+}=(n_1,\ldots,n_k,1)$, then the preceding formula can also be written as $p(\boldsymbol{n})=\sum_{j=1}^{k+1}p(\boldsymbol{n}^{j+})$. The $j+$ refers to the set A_j from which n was deleted when reducing $\mathcal{P}_n$ to $\mathcal{P}_{n-1}$, or perhaps better added when extending the latter to the former.

Equation (14.2) can be viewed as expressing consistency of the partitions in a distributional sense, which is a consequence of the almost sure consistency that is baked into Definition 14.6. As we shall be mainly interested in distributions and not exact sample space constructions, this distributional consistency will be good enough for our purposes. In fact, given a sequence of symmetric functions $p\colon\cup_{n=1}^{\infty}\mathcal{C}_n\to[0,1]$ that satisfy the marginalization equation (14.2), one can always construct an infinite exchangeable partition that has p as its EPPF. To see this we may first construct a joint distribution of a sequence $(\mathcal{P}_n\colon n\in\mathbb{N})$ of partitions of $\mathbb{N}_n$ by recursively extending $\mathcal{P}_{n-1}$ to $\mathcal{P}_n$, for $n=2,3,\ldots$, as indicated in Figure 14.1. Equation (14.2) ensures that this construction leads to a proper probability distribution on the set of all partitions of $\mathbb{N}_n$. Symmetry of p in its arguments implies the exchangeability of the partitions. Finally we may appeal to an abstract theorem by Ionescu-Tulcea on the existence of a joint distribution given consistent marginal distributions to establish the distribution of the whole infinite sequence $\{\mathcal{P}_n\}$.

Instead of as a sequence of partitions of $\mathbb{N}_n$ it is convenient to think of an infinite exchangeable partition as a sequential process that adds the points $2,3,\ldots$ in turn to the existing partition formed by the preceding points. The (conditional) probability that the point $n+1$ is added to the jth set in the partition of $\mathcal{P}_n$ of $\mathbb{N}_n$ with composition $\boldsymbol{n}=(n_1,\ldots,n_k)$ is equal to

$$p_j(\boldsymbol{n})=\frac{p(\boldsymbol{n}^{j+})}{p(\boldsymbol{n})},\qquad j=1,\ldots,k+1. \tag{14.3}$$

The collection of functions $p_j\colon\cup_n\mathcal{C}_n\to[0,1]$ is called the *prediction probability function* (PPF). Clearly $(p_1(\boldsymbol{n}),\ldots,p_{k+1}(\boldsymbol{n}))$ is a probability vector for every composition $\boldsymbol{n}=(n_1,\ldots,n_k)$. In Figure 14.1 these probability vectors are placed as weights on the branches emanating from the node $\boldsymbol{n}$, so that the probabilities of the terminal nodes are

[4] For instance $p(1,2)$ refers to a (particular) partition of $\mathbb{N}_3$ in two sets of sizes 1 and 2; by the preceding formula it is equal to $p(2,2)+p(1,3)+p(1,2,1)$, whose three terms refer to (particular) partitions of $\mathbb{N}_4$.

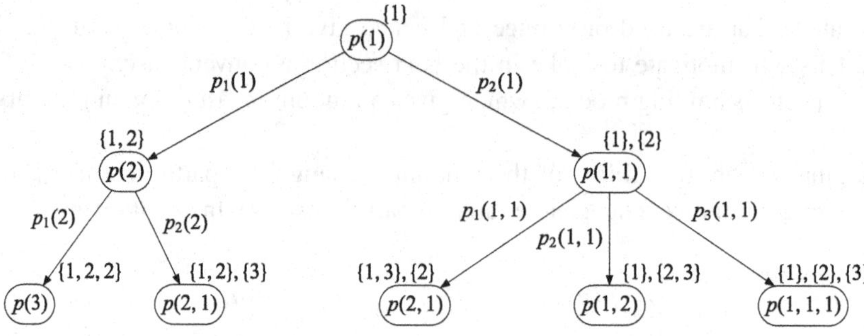

Figure 14.1 Recursive partitioning. The three levels show all partitions of $\mathbb{N}_1$, $\mathbb{N}_2$ and $\mathbb{N}_3$, respectively. The partitions at each consecutive level are constructed by adding the next number (2 and 3 for the levels shown, 4 for the next level not shown) to every of the sets in a partition of the preceding level, thus splitting a node into as many branches as sets present in the partition at the node plus 1. Equation (14.2) ensures that the probability of a node splits along the branches. In the tree as shown exchangeability of the partition means (only) that the middle three leaves (the partitions $\{1, 2\}, \{3\}$ and $\{1, 3\}, \{2\}$ and $\{1\}, \{2\}, \{3\}$ of $\mathbb{N}_3$) have equal probability.

the products of the weights on the branches along the path connecting them to the root of the tree.

Rather than deriving a PPF from an EPPF as in (14.3), we might start from a collection of functions p_j, and sequentially generate a sequence of partitions. To obtain an infinite exchangeable partition, the functions p_j must correspond to an EPPF p. The following lemma gives necessary and sufficient conditions that the implied p is an EPPF.

Lemma 14.8 *A collection of functions p_j is a PPF of an infinite exchangeable partition if and only if $(p_1(\boldsymbol{n}), \ldots, p_{k+1}(\boldsymbol{n}))$ is a probability vector, for every composition $\boldsymbol{n} = (n_1, \ldots, n_k)$, and, for all $i, j = 1, 2, \ldots,$*

$$p_i(\boldsymbol{n})p_j(\boldsymbol{n}^{i+}) = p_j(\boldsymbol{n})p_i(\boldsymbol{n}^{j+}), \tag{14.4}$$

and for every permutation σ of $\mathbb{N}_k$, the following permutation equivariance condition holds:

$$p_i(n_1, \ldots, n_k) = p_{\sigma^{-1}(i)}(n_{\sigma(1)}, \ldots, n_{\sigma(k)}). \tag{14.5}$$

Proof The necessity of (14.4) follows upon substituting (14.3) for all four occurrences of a p_j, which reduces both sides of the equation to $p((\boldsymbol{n}^{i+})^{j+})/p(\boldsymbol{n})$.

Conversely, given the p_j we may use (14.3) to define p recursively by $p(1) = 1$, and $p(\boldsymbol{n}^{j+}) = p_j(\boldsymbol{n})p(\boldsymbol{n})$. Because a given composition $\boldsymbol{n}$ can be reached by many different paths starting at 1 and augmenting the composition by steps of 1, it must be checked that this definition is well posed, and the function so obtained satisfies the consistency condition (14.2) of an EPPF. Both are guaranteed by (14.4).

Equation (14.5) can be seen to be necessary (e.g. by (14.3)) and sufficient for the function p so constructed to be permutation symmetric. □

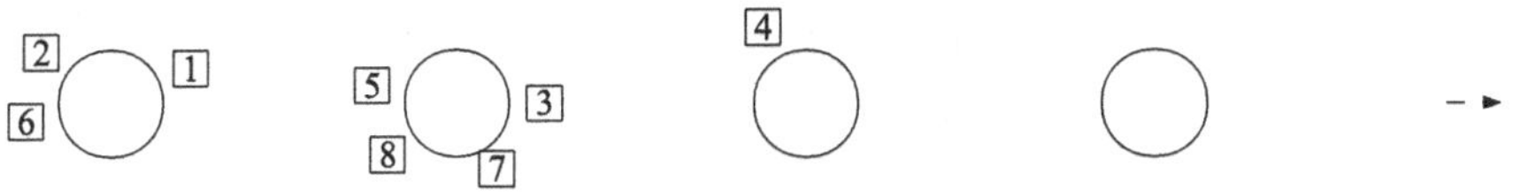

Figure 14.2 Chinese restaurant process. The ninth customer will sit at the tables from left to right with probabilities $3/(M+8)$, $4/(M+8)$, $1/(M+8)$ and $M/(M+8)$. There is an infinite list of empty tables.

14.1.1 The Chinese Restaurant Process

The infinite exchangeable random partition generated by a sample from the Dirichlet process $\mathrm{DP}(MG)$ with atomless center measure G, as in Section 4.1.4, is called the *Chinese restaurant process* (CRP) with parameter M. By Proposition 4.11 its EPPF is given by

$$p(n_1, \ldots, n_k) = \frac{M^k \prod_{i=1}^k (n_i - 1)!}{M^{[n]}}, \tag{14.6}$$

where $a^{[k]} = a(a+1)\cdots(a+k-1)$ stands for the ascending factorial. This is a product partition model with cohesion function given by $\rho(n_i) = M(n_i - 1)!$. The PPF of the process is given by the generalized Pólya urn (or Blackwell-MacQueen scheme):

$$p_j(n_1, \ldots, n_k) = \frac{n_j}{M+n}, \qquad j = 1, \ldots, k, \qquad p_{k+1}(n_1, \ldots, n_k) = \frac{M}{M+n}.$$

Thus the probability of adding a next element to an existing set in the partition is proportional to the number of elements already in the partition, and a new set is "opened" with probability proportional to M.

The name derives from the following metaphor. Suppose that customers arrive sequentially in a Chinese restaurant with an infinite number of tables, each with infinite seating capacity. The first customer chooses an arbitrary table. The second customer has two options, sit at the table opened by customer 1 or open a new table, between which he decides with probabilities $1/(M+1)$ and $M/(M+1)$. More generally, the $(n+1)st$ customer finds n customers seated at k tables in groups of $n_1, \ldots, n_k$, where $\sum_{j=1}^k n_j = n$, and chooses to sit at the jth open table with probability $n_j/(M+n)$, or open a new table with probability $M/(M+n)$. The gravitational effect of this scheme – more massive tables, apparently with more known faces, attract a newcomer with a higher probability – is valuable for clustering variables together in groups.

The following result shows that the simple form of the PPF of the Chinese restaurant process as proportional to a function of the respective cardinalities, characterizes this process.

Proposition 14.9 *The Chinese restaurant process is the only exchangeable random partition with PPF $(p_j\colon j \in \mathbb{N})$ of the form, for some function $f\colon \mathbb{N} \to (0, \infty)$ not depending on j, some $M > 0$, and every composition $(n_1, \ldots, n_k)$,*

$$p_j(n_1, \ldots, n_k) \propto \begin{cases} f(n_j), & j = 1, \ldots, k, \\ M, & j = k+1. \end{cases}$$

Proof By equation (14.4) applied with $\boldsymbol{n} = (n_1, \ldots, n_k)$ and $i \neq j \in \{1, \ldots, k\}$, the number

$$\frac{f(n_i)}{\sum_{l=1}^k f(n_l) + M} \times \frac{f(n_j)}{\sum_{l=1:l\neq i}^k f(n_l) + f(n_i + 1) + M}$$

is invariant under swapping i and j. This is possible only if $f(n_j) + f(n_i + 1)$ is invariant, so that $f(n_i + 1) - f(n_i)$ is constant in i. Hence f must be of the form $f(n) = an + b$ for some real constants a, b.

The same reasoning with $i = 1, \ldots, k$ and $j = k + 1$ leads to the relation $f(n_i + 1) = f(n_i) + f(1)$, whence we must have $b = 0$, that is $f(n) = an$. Thus (p_j) is the PPF of a Dirichlet process. □

14.1.2 The Chinese Restaurant Franchise Process

The *Chinese restaurant franchise* process is a hierarchical partition structure based on the hierarchical Dirichlet process (HDP) introduced in Example 5.12. The hierarchical Dirichlet process consists of a "global" random probability measure G generated from a $\mathrm{DP}(M_0 G_0)$-distribution, "local" random probability measures $G_1, G_2, \ldots$ generated i.i.d. from $\mathrm{DP}(MG)$ for given G, and a random sample $X_{i,1}, X_{i,2}, \ldots$ from every G_i independent across i.

The measure G can be written in a stick-breaking representation $G = \sum_{j=1}^\infty W_j \delta_{\theta_j}$, for an i.i.d. sequence $\theta_1, \theta_2, \ldots$ from G_0. Similarly every of the measures G_i has a stick-breaking representation $G_i = \sum_{j=1}^\infty W_{i,j} \delta_{\theta_{j,i}}$, but with the support points $\theta_{1,i}, \theta_{2,i}, \ldots$ now chosen from G, which means that they are subsets of the variables $\theta_1, \theta_2, \ldots$ Finally the observations $X_{i,1}, X_{i,2}, \ldots$ are subsets of the sequences $\theta_{1,i}, \theta_{2,i}, \ldots$ As a result, observations from different samples are linked through the hierarchical structure, and when the samples $X_{i,1}, X_{i,2}, \ldots$ are used as before to induce a partitioning by ties, then this partition structure will be dependent across the samples.

This structure becomes better visible in the predictive distributions of the samples. Given G the variables $X_{i,1}, X_{i,2}, \ldots$ are a sample from the Dirichlet process $\mathrm{DP}(MG)$ in the sense of Section 4.1.4, and can be generated by the Pólya urn scheme

$$X_{i,j} | X_{i,1}, \cdots, X_{i,j-1}, G \sim \sum_{t=1}^{K_{i,j}} \frac{N_{i,t}}{M + j - 1} \delta_{\tilde{X}_{i,t}} + \frac{M}{M + j - 1} G, \tag{14.7}$$

where $\tilde{X}_{i,1}, \ldots, \tilde{X}_{i,K_{i,j}}$ are the distinct values in $X_{i,1}, \ldots, X_{i,j-1}$ and $N_{i,1}, \ldots, N_{i,K_{i,j}}$ their multiplicities. The G_i do not appear in this scheme, but "have been integrated out." The variable $X_{i,j}$ generated in the display is either a previous value $\tilde{X}_{i,t}$ from the same sample, or a "new" value from G. The latter measure is itself a realization from the $\mathrm{DP}(M_0 G_0)$-process. The same realization is used in the sequential scheme for every sample $X_{i,1}, X_{i,2}, \ldots$, for $i = 1, 2, \ldots$, but the sampling in the display is independent across i. Every time the variable $X_{i,j}$ must be "new," then this new value must be independently drawn from G, both within a sample of observations $X_{i,1}, X_{i,2}, \ldots$ as across i. Because only a sample from G is needed, rather than realizing the complete measure we can appeal a second time to the Pólya urn scheme of Section 4.1.4, and generate a sample from G, in the form

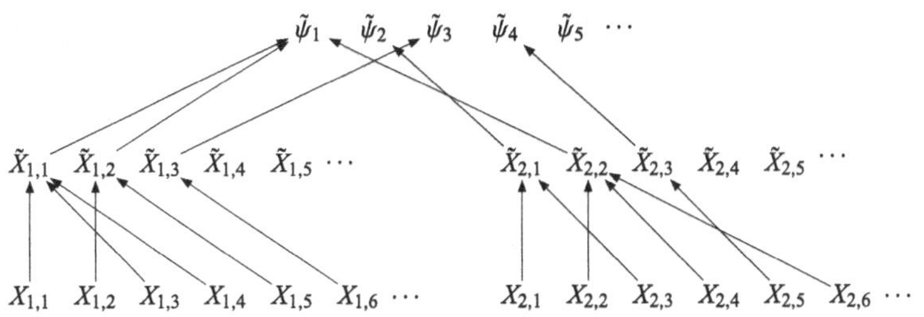

Figure 14.3 Chinese restaurant franchise process. At the top level are the distinct values in a sample $\psi_1, \psi_2, \ldots$ from the DP(MG_0) process. The middle level shows the distinct values in two samples of observations shown at the bottom level. The arrows show a possible dependency structure in a single realization: the variable at the bottom is a copy from the variable at the top.

$$\psi_k | \psi_1, \ldots, \psi_{k-1} \sim \sum_{s=1}^{L_k} \frac{P_{k,s}}{M_0 + k - 1} \delta_{\tilde\psi_s} + \frac{M_0}{M_0 + k - 1} G_0,$$

where $\tilde\psi_1, \ldots, \tilde\psi_{L_k}$ are the distinct values in $\psi_1, \ldots, \psi_{k-1}$ and $P_{k,s}$ are their multiplicities. Only a single such sequential scheme is needed to generate all samples $X_{i,1}, X_{i,2}, \ldots$: every time it is required to sample a new value (from G) in (14.7), we substitute the first value ψ_k that has not been used before. In practice we shall generate the sample values $X_{i,j}$ in a given order and generate the next observation ψ_k whenever it is needed to extend one of the samples $X_{i,1}, X_{i,2}, \ldots$ This next observation will then be equal to an existing value $\tilde\psi_s$ or a new value generated from G_0, as in a Chinese restaurant process. In the end every $X_{i,j}$ will be one of the distinct values $\tilde\psi_s$ in the sequence $\psi_1, \psi_2, \ldots$, which may well occur multiple times in the sample $X_{i,1}, X_{1,2}, \ldots$, as well as in the other samples. This practical implementation presupposes an ordering in which the observations $X_{i,j}$ are sampled. This is a bit arbitrary, but unimportant for the resulting clustering. Figure 14.3 illustrates the scheme.

The following culinary metaphor explains the name "Chinese restaurant franchise process." Consider a number of Chinese restaurants that serve a common "franchise-wide" menu. Customers arrive and are seated in each restaurant according to independent Chinese restaurant processes with parameter M. Their overall order of arrivals is arbitrary, but fixed. All customers at a given table eat the same dish, which is chosen by the first customer at the table from the dishes of the previously opened tables in the franchise or from the franchise-wide menu with probabilities proportional to the total number of tables eating this dish in the franchise and M_0 for the new dish, respectively. In the preceding notation $X_{i,1}, X_{i,2}, \ldots$ are the dishes eaten by the customers in restaurant i, $\psi_1, \psi_2, \ldots$ are dishes that are new to a restaurant, and $\tilde\psi_1, \tilde\psi_2, \ldots$ are dishes that are new to the franchise. The resulting process induces a random partition by table in each restaurant, which is nested within a partition by dish across the franchise.

14.2 Species Sampling Processes

As noted following its statement only the sizes (W_i) of the atoms of the directing measure P in Kingman's representation theorem, Theorem 14.7, play a role for the pattern of equal X_i. It is convenient to represent P in the special form of a *species sampling model*.

Definition 14.10 (Species sampling model) A *species sampling model* (SSM) is a pair of a sequence (X_i) of random variables and a random measure P such that $X_1, X_2, \ldots \mid P \overset{\text{iid}}{\sim} P$ and P takes the form

$$P = \sum_{j=1}^{\infty} W_j \delta_{\theta_j} + (1 - \sum_{j=1}^{\infty} W_j) G, \tag{14.8}$$

for $\theta_1, \theta_2, \ldots \overset{\text{iid}}{\sim} G$ for an atomless probability distribution G on a Polish space $\mathfrak{X}$ and an independent random subprobability vector (W_j). The random distribution P in a SSM is called a *species sampling process* (SSP). If $\sum_{j=1}^{\infty} W_j = 1$, the SSP is called *proper*.

Because G is atomless, the "labels" θ_j are all different (almost surely), and hence the species sampling process P in (14.8) has atoms (W_j), as the random measure P in Kingman's theorem. Thus the partitions generated by the output $X_1, X_2, \ldots$ of a species sampling model can be viewed as a general model for an infinite exchangeable partition.

One advantage of this special construction is that the predictive distributions of $X_1, X_2, \ldots$ take a simple form. Write $\tilde{X}_1, \tilde{X}_2, \ldots$ for the distinct values in the sequence of variables $X_1, X_2, \ldots$ in *order of appearance*, and for each n let $N_{j,n}$ be the multiplicity of $\tilde{X}_j$ in $X_1, \ldots, X_n$, and K_n the total number of distinct values in the latter set.[5] Write $N_n = (N_{1,n}, \ldots, N_{K_n,n})$ for the vector of counts of distinct values.

Lemma 14.11 *The predictive distributions of variables $X_1, X_2, \ldots$ in the species sampling model* (14.8) *take the form $X_1 \sim G$ and, for $n \geq 1$,*

$$X_{n+1} \mid X_1, \ldots, X_n \sim \sum_{j=1}^{K_n} p_j(N_n) \delta_{\tilde{X}_j} + p_{K_n+1}(N_n) G, \tag{14.9}$$

where the functions $p_j \colon \cup_n \mathcal{C}_n \to [0, 1]$ are the PPF of the infinite exchangeable random partition generated by $X_1, X_2, \ldots$ (cf. (14.3)*). A sequence $X_1, X_2, \ldots$ with the same distribution as the original sequence can be generated by first generating the infinite random exchangeable partition defined by this PPF and next attaching to the partitioning sets in order of appearance the values of an independent i.i.d. sequence $\tilde{X}_1, \tilde{X}_2, \ldots$ from G.*

Proof The variable X_1 is either a θ_j or generated from G. As $\theta_1, \theta_2, \ldots \overset{\text{iid}}{\sim} G$ it follows that marginally $X_1 \sim G$.

For the proof of the general formula we introduce latent variables (I_i) that indicate the partitioning classes of the (X_i). We generate random variables hierarchically in the order: $(W_j)_{j\geq 0}$, where $W_0 = 1 - \sum_j W_j$, next $(I_i)_{i\geq 1} \mid (W_j) \overset{\text{iid}}{\sim} (W_j)$ on $\mathbb{N} \cup \{0\}$, and $(\theta_j)_{j\geq 1} \overset{\text{iid}}{\sim} G$ and $(\theta_i')_{i\geq 1} \overset{\text{iid}}{\sim} G$ independently from each other and from (W_j) and (I_i). Finally we set $X_i = \theta_{I_i}$ if $I_i \geq 1$ and $X_i = \theta_i'$ if $I_i = 0$. Then, for $X_{1:n}$ abbreviating $X_1, \ldots, X_n$,

[5] So $\tilde{X}_1 = X_1$, $\tilde{X}_2 = X_{\hat{\imath}(2)}$, for $\hat{\imath}(2) = \min\{i \geq 2 \colon X_i \neq X_1\}$, etc. In the statement of the following lemma the ordering of the distinct values is actually irrelevant, but it is made explicit to unify notation.

$$P(X_{n+1} \in B \mid X_{1:n}) = \sum_{l=1}^{\infty} P(\theta_l \in B \mid X_{1:n}, I_{n+1} = l) P(I_{n+1} = l \mid X_{1:n})$$
$$+ P(\theta'_{n+1} \in B \mid X_{1:n}, I_{n+1} = 0) P(I_{n+1} = 0 \mid X_{1:n}).$$

Since θ'_{n+1} is independent of everything the leading probability in the second term on the right can be reduced to $G(B)$. The corresponding probabilities in the first term can be decomposed as the expectation

$$E(P(\theta_l \in B \mid X_{1:n}, I_1, \ldots, I_n, I_{n+1} = l) \mid X_{1:n}, I_{n+1} = l).$$

If $I_i = 0$, then $X_i = \theta'_i$ and is independent of everything and can be removed from the inner conditioning. If $I_i = k \geq 1$ and $k \neq l$, then $X_i = \theta_k$ and can be removed from the inner conditioning as well. Then either $l \notin \{I_1, \ldots, I_n\}$ and all $X_1, \ldots, X_n$ can be removed from the inner conditioning and the display reduces to $G(B)$, or $l = I_i$ for some i and then the display reduces to $\mathbb{1}\{X_i \in B\}$. For $\tilde{I}_1, \tilde{I}_2, \ldots$ the sequence $I_1, I_2, \ldots$ with nonzero duplicates removed the latter indicator can also be written $\sum_{j=1}^{K_n} \mathbb{1}\{\tilde{I}_j = l, \tilde{X}_j \in B\}$. Multiplying this and the indicator of $l \notin \{I_1, \ldots, I_n\}$ by $P(I_{n+1} = l \mid X_{1:n})$ and summing over l, we obtain that $P(X_{n+1} \in B \mid X_{1:n})$ is equal to

$$\sum_{l=1}^{\infty} \Big[\sum_{j=1}^{K_n} \mathbb{1}\{\tilde{I}_j = l, \tilde{X}_j \in B\} + G(B)\mathbb{1}\{l \notin \{I_1, \ldots, I_n\}\}\Big] + G(B) P(I_{n+1} = 0 \mid X_{1:n})$$
$$= \sum_{j=1}^{K_n} \mathbb{1}\{\tilde{X}_j \in B\} P(I_{n+1} = \tilde{I}_j \mid X_{1:n}) + G(B) P(I_{n+1} \notin \{\tilde{I}_1, \ldots, \tilde{I}_{K_n}\} \mid X_{1:n}).$$

The events $\{I_{n+1} = \tilde{I}_j\}$ describe that the point $n+1$ is added to the jth set when enlarging the partition generated by $X_1, X_2, \ldots$ from $\mathbb{N}_n$ to $\mathbb{N}_{n+1}$ if $j \leq K_n$, or opens a new set if $j = K_n + 1$. The probabilities $P(I_{n+1} = \tilde{I}_j \mid X_{1:n})$ can therefore be interpreted as the transition probabilities p_j of the transition $\mathcal{P}_n$ to $\mathcal{P}_{n+1}$. We must still show that they depend on $X_{1:n}$ through N_n only.

The conditioning variable $X_{1:n}$ is equal in information to the partition $\mathcal{P}_n$ of $\mathbb{N}_n$, the common values $\tilde{X}_j$ shared by the variables in the partitioning sets, and the ordering of these values repeated by their multiplicities. By the exchangeability of $X_1, X_2, \ldots$ the ordering is random and not informative on the events $\{I_{n+1} = \tilde{I}_j\}$. The common values are not informative either, as they are i.i.d. variables with law G and independent of the partition. In the case of a proper species sampling model this follows, as for any partition π of $\mathbb{N}_n$ with composition $(n_1, \ldots, n_k)$ and any measurable sets $B_1, \ldots, B_k$,

$$P(\mathcal{P}_n = \pi, \tilde{X}_1 \in B_1, \ldots, \tilde{X}_k \in B_k) = E \sum_{1 \leq i_1 \neq \cdots \neq i_k < \infty} \prod_{j=1}^{k} W_{i_j}^{n_j} \prod_{j=1}^{k} \mathbb{1}\{\theta_{i_j} \in B_j\}.$$

(The indices $i_1, \ldots, i_k$ represent the indices of the k different θ_i chosen.) As the weights (W_j) and locations (θ_j) are independent, the expectation factorizes, and the right side can be seen to be equal to $P(\mathcal{P}_n = \pi) \prod_{j=1}^{k} G(B_j)$. This argument can be extended to a possibly improper species sampling process. We conclude that the conditioning information $X_{1:n}$ can be reduced to the partition $\mathcal{P}_n$; equivalently by exchangeability, to the multiplicities of the partitioning sets.[6]

That $X_1, X_2, \ldots$ can be generated as in the final assertion of the lemma follows from the preceding. □

Species sampling processes take their name from a probabilistic model for discovery of new types of objects, or *species*. Interpret W_j as the relative frequency of the jth species, which carries label θ_j, in a population, and suppose we explore the population by sequentially sampling individuals. After sampling n individuals we record the number K_n of distinct species discovered, along with their multiplicities $N_n = (N_{1,n}, \ldots, N_{K_n,n})$ in the sample, and may wish to predict the type of the next individual, or estimate the frequency distribution of the species in the population. The predictive probabilities p_j (see (14.9)) are the solution to the first question, with $p_1(N_n), \ldots, p_{K_n}(N_n)$ describing the probabilities of sampling one of the K_n previously discovered species, and $p_{K_n+1}(N_n)$ the probability of finding a thus far undiscovered species.

The preceding lemma expresses the close link between a species sampling model and the (exchangeable) partition generated by the observed species. In accordance, a species sampling model can be characterized through any of the following three pairs of objects:

(i) G and the distribution of $(W_1, W_2, \ldots)$;
(ii) G and the EPPF;
(iii) G and the PPF.

The measure G is the mean measure $\mathrm{E}(P) = G$ of the random measure (see Section 3.4.2), and is also called the *center measure* of the species sampling process.

A random (sub)probability measure $(W_1, W_2, \ldots)$ on $\mathbb{N}$ can easily be obtained by the methods of Section 3.3. For the definition of the species sampling model, the ordering of the weights W_j is irrelevant, but for calculations one ordering may be more convenient than another. An easy way to identify the weights is to put them in decreasing order, as in Kingman's representation, Theorem 14.7. However, this may lead to a complicated description of their joint distribution. An alternative is a description in "order of appearance," which is a limit of the predictive distributions (14.9), and gives the weights in *size-biased order*.

Definition 14.12 (Size-biased permutation) The *size-biased permutation* of a probability distribution (w_j) on $\mathbb{N}$ is the random vector $(\tilde{w}_1, \tilde{w}_2, \ldots)$ for $\tilde{w}_j = w_{\tilde{I}_j}$ and $\tilde{I}_1, \tilde{I}_2, \ldots$ the distinct values in an i.i.d. sequence $I_1, I_2, \ldots$ from (w_j), in order of appearance. A random probability distribution $(\tilde{W}_j)$ on $\mathbb{N}$ is said to be *invariant under size-biased permutation* (ISBP), or to be in *size-biased order*, if its distribution is invariant under size-biased permutation by a sequence $I_1, I_2, \ldots$ that given $(\tilde{W}_j)$ is i.i.d. from $(\tilde{W}_j)$.

Because size-biased permutation takes a full probability vector (w_j) on $\mathbb{N}$ as input, listed in arbitrary order, and the size-biased version $(\tilde{w}_j)$ is a (random) permutation of the original sequence, a second independent size-biased permutation of a deterministic sequence does not change its distribution. It follows that any random sequence (W_j) can be changed in an

[6] Hansen and Pitman (2000) prove that for *any* exchangeable $X_1, X_2, \ldots$ with predictive distributions of the form (14.9) with p_j possibly dependent on $X_{1:n}$, these p_j necessarily depend on the partition only.

ISBP sequence $(\tilde{W}_j)$ by a single (independent) size-biased permutation, and its distribution is the same irrespective of the ordering of the original sequence (W_j).

If one thinks of the (w_j) as the relative frequencies of species in a population (labeled $1, 2, \ldots$), and $I_1, I_2, \ldots$ as a random sample of individuals, then $\tilde{I}_1, \tilde{I}_2, \ldots$ are the distinct species in order of appearance. The name "size-biased" expresses that more frequent species (with bigger w_i) will tend to show up earlier.

Rather than as the distinct values in an i.i.d. sequence, a sequence $\tilde{I}_1, \tilde{I}_2, \ldots$ with the same distribution can also be generated sequentially by the algorithm:

- Choose $\tilde{I}_1$ from $\mathbb{N}$ according to the probability vector (w_j).
- Given $\tilde{I}_1, \ldots, \tilde{I}_{j-1}$, choose $\tilde{I}_j$ from $\mathbb{N} \setminus \{\tilde{I}_1, \ldots, \tilde{I}_{j-1}\}$ according to the probability vector (w_j) conditioned to the latter set.

The following theorem reveals the role of size-biasing. Recall the notation $\tilde{X}_1, \ldots, \tilde{X}_{K_n}$ for the distinct species in $X_1, \ldots, X_n$, and $N_{1,n}, \ldots N_{K_n,n}$ for their multiplicities.

Theorem 14.13 (Pitman's representation) *If $X_1, X_2, \ldots$ follow the species sampling model* (14.8), *then for every j the sequence $N_{j,n}/n$ tends a.s. surely to a limit $\tilde{W}_j$, and the predictive distributions* (14.9) *tend a.s. in total variation norm to $\tilde{P} = \sum_{j=1}^{\infty} \tilde{W}_j \delta_{\tilde{X}_j} + (1 - \sum_{i=1}^{\infty} \tilde{W}_j)G$. Here $\tilde{X}_1, \tilde{X}_2, \ldots \stackrel{iid}{\sim} G$ are independent of $(\tilde{W}_j)$, and for every $n \geq 1$,*

$$X_{n+1} | X_1, \ldots, X_n, \tilde{W}_1, \tilde{W}_2, \ldots \sim \sum_{j=1}^{K_n} \tilde{W}_j \delta_{\tilde{X}_j} + (1 - \sum_{j=1}^{K_n} \tilde{W}_j)G. \tag{14.10}$$

If the species sampling model is proper, then $(\tilde{W}_j)$ is a size-biased permutation of (W_j).

Proof Let $I_1, I_2, \ldots$ denote the indices of the components δ_{θ_j} from which $X_1, X_2, \ldots$ is chosen, with $I_i = 0$ if X_i is chosen from the continuous component G (and $X_i = \theta_{I_i}$ otherwise). Furthermore, let $\tilde{I}_1, \tilde{I}_2, \ldots$ be the distinct values in this sequence, in order of appearance.

Let $\hat{j}$ be the time of appearance of the jth species, so that $X_{\hat{j}} = \tilde{X}_j$. From the fact that the variables $X_1, X_2, \ldots$ are conditionally i.i.d. given P, it can be seen that the variables $Y_1, Y_2, \ldots$, for $Y_i = X_{\hat{j}+i}$, are conditionally i.i.d. given P and $X_1, \ldots, X_{\hat{j}}$. Since $N_{j,n} = 1 + \sum_{i=1}^{n-\hat{j}} \mathbb{1}\{Y_i = \tilde{X}_j\}$, the law of large numbers gives that $n^{-1} N_{j,n} \to W_{\tilde{I}_j} =: \tilde{W}_j$, almost surely. If the species sampling model is proper, then $I_1, I_2, \ldots$ is an i.i.d. sequence from (W_j) and hence the sequence $(\tilde{W}_j)$ is a size-biased permutation of the sequence (W_j).

In the proof of Lemma 14.11 the distinct values $\tilde{X}_1, \tilde{X}_2, \ldots$ were seen to be independent and identically distributed according to G and independent of the partition generated by $X_1, X_2, \ldots$. Thus they are also independent of the variables N_n and hence of the limits $\tilde{W}_j$.

In the proof of Lemma 14.11 the predictive probabilities were expressed as $p_j(N_n) = \mathrm{P}(I_{n+1} = \tilde{I}_j | I_{1:n})$, for $j \leq K_n$, where $I_{1:n} = (I_1, \ldots I_n)$ can also be replaced by N_n in the conditioning. Consider first the conditional probabilities $\mathrm{P}(I_{n+1} = \tilde{I}_j | I_{1:n}, N_m)$, for $m > n$. The extra conditioning information indicates that $N_{j,m} - N_{j,n}$ variables out of $X_{n+1}, \ldots, X_m$ must be realized as equal to $\tilde{X}_j$, to make the cardinality of the jth species grow from $N_{j,n}$ to $N_{j,m}$. By exchangeability the indices of these variables are equally distributed over $n+1,\ldots,m$, so that $\mathrm{P}(I_{n+1}=\tilde I_j|\, I_{1:n}, N_m) = (N_{j,m}-N_{j,n})/(m-n)$. The latter variable has the same limit as $N_{j,m}/m$ if $m\to\infty$ and n is fixed, which was identified as $\tilde W_j$ in the preceding paragraph. Applying the dominated convergence theorem to the conditional expectations $\mathrm{E}(\mathrm{P}(I_{n+1}=\tilde I_j|\, I_{1:n}, N_m)|\, I_{1:n})$, we conclude that $p_j(N_n)=\mathrm{E}(\tilde W_j|\, I_{1:n})$. Finally, martingale convergence gives that $p_j(N_n)\to\tilde W_j$, almost surely as $n\to\infty$.

If we condense the continuous part G of the predictive distributions to an additional atom x_0, then the predictive distributions and $\tilde P$ concentrate on the countable set $\{x_0, \tilde X_1, \tilde X_2, \ldots\}$ and the sizes of the atoms of the predictive distributions on all but the first element of this set have been shown to tend to those of $\tilde P$. For any subsequence along which the sizes of the atoms at x_0 also converge, Scheffe's theorem gives convergence in total variation. Because the predictive distributions and the limit $\tilde P$ are all probability measures, the whole sequence must converge in total variation to $\tilde P$.

As limits of the sequence $N_{j,n}/n$, the variables $\tilde W_j$ are clearly measurable relative to $\mathcal{F}:=\cap_m \mathcal{F}_m$, for $\mathcal{F}_m=\sigma\langle N_m, N_{m+1},\ldots\rangle$. By the Markov property the conditional probability $\mathrm{P}(I_{n+1}=\tilde I_j|\, I_{1:n}, N_m)$, for $m>n$, does not change if we add $\mathcal{F}_m$ to the conditioning, and the reverse submartingale thus obtained tends almost surely to $\mathrm{P}(I_{n+1}=\tilde I_j|\, I_{1:n},\mathcal{F})$, as $m\to\infty$, for fixed j and n. As the limit was already seen to be equal to $\tilde W_j$, it follows that taking the conditional expectation given the smaller σ-field generated by $I_{1:n}$ and the full sequence $(\tilde W_i)$ does not change the limit, whence $\mathrm{P}(I_{n+1}=\tilde I_j|\, I_{1:n},(\tilde W_i))=\tilde W_j$. This shows that the partition structure is generated according to equation (14.10) with the X_i replaced by the I_i. Together with the independence of the partition structure and the values $\tilde X_j$ attached to the partitioning sets, this implies the latter equation. □

The constructive part of the theorem is the displayed equation (14.10), which shows how to generate the observations $X_1, X_2, \ldots$ sequentially, given the sequence $(\tilde W_j)$. The $(n+1)$st observation X_{n+1} is with probability $\tilde W_j$ equal to the existing value $\tilde X_j$ (for $j=1,\ldots,K_n$), and newly generated from G otherwise.

The observations $X_1, X_2, \ldots$ generate a sequence of partitions, which grows likewise. The first partition is $\mathcal{P}_1=\{1\}$. Given the partition $\mathcal{P}_n=\{\mathcal{P}_{1,n},\ldots,\mathcal{P}_{K_n,n}\}$ of $\mathbb{N}_n$ generated by $(X_1,\ldots,X_n)$, listed in *order of appearance*, the partition $\mathcal{P}_{n+1}$ is created by

(i) attaching the element $n+1$ to $\mathcal{P}_{j,n}$, with probability $\tilde W_j$, for $j=1,\ldots,K_n$;
(ii) opening the new set $\mathcal{P}_{K_{n+1},n+1}=\{n+1\}$, with probability $1-\sum_{j=1}^{K_n}\tilde W_j$.

Once the sequence of partitions $\mathcal{P}_n$ is created, the values $X_1, X_2, \ldots$ can be created by attaching an i.i.d. sequence $\tilde X_1, \tilde X_2, \ldots$ from G to the partitioning sets, one $\tilde X_i$ to each newly opened set, and equating every X_i to the $\tilde X_j$ of its partitioning set. This was already seen in Lemma 14.11.

The weights $\tilde W_j$ in the theorem are the almost sure limits of the multiplicities of the ties in $X_1, X_2, \ldots$, but if we are interested only in distributions (and not "strong" definitions as maps on a probability space), then we may turn the construction around. Given a proper sequence (W_j), we determine (the distribution of) its size-biased permutation $(\tilde W_j)$, and generate the partitions and variables by the preceding scheme. This gives a realization from the species sampling model, as the distribution of the right side of (14.10) is determined by the *distribution* of $(\tilde{W}_j)$ only.

The size-bias of the sequence $(\tilde{W}_j)$ is crucial for this representation. In the predictive formula (14.9) the weights $p_j(N_n)$ change with the generation of every new variable, whereas the size-biased weights $\tilde{W}_j$ in (14.10) are fixed at the beginning. Intuitively, they can be fixed precisely because they are size-biased. Importantly, to generate the beginning $X_1, \ldots, X_n$ of the sequence of observations, only the initial weights $\tilde{W}_1, \ldots \tilde{W}_{K_n}$ are needed.

As a first illustration of the usefulness of the size-biased representation, the following corollary expresses the EPPF in the weights $(\tilde{W}_j)$. The formula may be contrasted with the formula using arbitrary weights, given in Problem 14.1, which involves infinite sums.

Corollary 14.14 *The EPPF of an infinite exchangeable partition given by a weight sequence $(\tilde{W}_j)$ listed in size-biased order satisfies, for every composition $(n_1, \ldots, n_k)$,*

$$p(n_1, \ldots, n_k) = \mathrm{E}\Big[\prod_{j=1}^{k} \tilde{W}_j^{n_j-1} \prod_{j=2}^{k}\Big(1 - \sum_{i<j} \tilde{W}_i\Big)\Big]. \tag{14.11}$$

Proof In view of Theorem 14.13, a random partition $\mathcal{P}_n$ of $\mathbb{N}_n$ in the order of appearance is obtained by following the scheme (i)–(ii). The specification of an event $\{\mathcal{P}_n = \{A_1, \ldots, A_k\}\}$ for a partition $\{A_1, \ldots, A_k\}$ in order of appearance completely determines the steps (i)–(ii) by which the elements $2, 3, \ldots, n$ are added to the existing partition: the partition of $\mathbb{N}_1$ is $\{1\}$; next if $2 \in A_1$, the element 2 was attached to $\{1\}$, which has probability $\tilde{W}_1$, or otherwise $2 \in A_2$, which has probability $1 - \tilde{W}_1$; next there are three possibilities for 3: $3 \in A_1$, which has probability $\tilde{W}_1$, $3 \in A_2$, which has probability $\tilde{W}_2$, or otherwise $3 \in A_3$, which has probability $1 - \tilde{W}_1 - \tilde{W}_2$; etc. In total, we attach $n_j - 1$ times a new element to the jth opened set, and open $k - 1$ times a new set. Multiplying the probabilities gives the probability that $\mathcal{P}_n$ is this particular partition. Because $(\mathcal{P}_n)$ is assumed exchangeable, this probability is equal to the probability $p(n_1, \ldots, n_k)$ of any partition with the same composition. □

The construction (i)–(ii) can also be applied with an *arbitrary* sub-probability sequence (W_j), but the resulting partition is then not necessarily exchangeable: it leads to the wider class of *partially exchangeable random partitions*, defined as infinite random partitions $(\mathcal{P}_n)$ such that for any partition $\{A_1, \ldots, A_k\}$ of $\mathbb{N}_n$, in the *order of appearance*,

$$\mathrm{P}(\mathcal{P}_n = \{A_1, \ldots, A_k\}) = p(|A_1|, \ldots, |A_k|),$$

for some function $p: \cup_{n=1}^{\infty} \mathcal{C}_n \to [0, 1]$. By the proof of the preceding lemma this function p will then be given by (14.11); the partition is exchangeable if and only if this function is symmetric in its arguments. This leads to the following characterization of size-biasedness.

Lemma 14.15 (Size-bias) *A random probability vector $(\tilde{W}_j)$ is in size-biased order, if $\tilde{W}_1 > 0$ a.s. and the measure $A \mapsto \mathrm{E}[\mathbb{1}_A(\tilde{W}_1, \ldots, \tilde{W}_k) \prod_{j=2}^{k}(1 - \sum_{i<j} \tilde{W}_i)]$ on $\mathbb{R}^k$ is symmetric, for every $k \in \mathbb{N}$.*

Proof If the given measure is symmetric, then the function p defined by (14.11) in Corollary 14.14 is symmetric. The preceding scheme (i)–(ii) then generates an infinite exchangeable partition. Comparison of (i)–(ii) with Theorem 14.13 shows that $(\tilde{W}_j)$ from (i)–(ii) must be equal in distribution to the sequence $(\tilde{W}_j)$ in this theorem. The theorem asserts that this sequence is size-biased if the corresponding species sampling model is proper. If this would not be the case, then the first observation would with positive probability be generated from the atomless component G. On this event, the first set in the partition would remain a singleton, whence $n^{-1}N_{1,n} = n^{-1}$ for every n. By Theorem 14.13 this sequence tends almost surely to $\tilde{W}_1$, and hence this event is excluded by the assumption that $\tilde{W}_1 > 0$, almost surely. □

Example 14.16 (Dirichlet process) A random sample from a Dirichlet process DP(MG) is a proper species sampling model with center measure G. This follows from the description of a Dirichlet process as a random discrete measure in Theorem 4.12.

The distribution of the weights (W_j) in decreasing magnitude $W_1 \geq W_2 \geq \cdots$ is known as the *Poisson-Dirichlet distribution*. The stick-breaking weights in Theorem 4.12 are the size-biased permutation of this sequence, as will be seen in greater generality in Theorem 14.33. Its distribution is also known as the *GEM distribution*, named after Griffiths, Engen and McClosky.

Example 14.17 (Fisher process) The *Fisher process* is the species sampling process with weight sequence given by $(W_1, \ldots, W_m) \sim \mathrm{Dir}(m; -\sigma, \ldots, -\sigma)$, for some $\sigma < 0$, and $W_j = 0$, for $j > m$. The corresponding model gives partitions in at most m sets. Using Kingman's formula (see Problem 14.1) and Corollary G.4 its EPPF can be computed as, for $k \leq m$,

$$p(n_1, \ldots, n_k) = \frac{m!}{(m-k)!} \frac{\Gamma(-m\sigma)}{\Gamma(-m\sigma + n)} \prod_{j=1}^{k} \frac{\Gamma(-\sigma + n_j)}{\Gamma(-\sigma)}$$

$$= \frac{\prod_{i=1}^{k-1}(-m\sigma + i\sigma)}{(-m\sigma + 1)^{[n-1]}} \prod_{j=1}^{k} (1 - \sigma)^{[n_j - 1]}.$$

The second rendering exhibits the model as a Pitman-Yor process, discussed in Section 14.4. This connection is also the motivation to write the (positive!) parameter as $-\sigma$.

The weight vector $(W_1, \ldots, W_m)$ is exchangeable, and not size-biased. In Theorem 14.33 it will be seen, by an application of Corollary 14.14 and Lemma 14.15, that the size-biased weights are given by the stick-breaking scheme $\tilde{W}_j = V_j \prod_{i<j}(1 - V_i)$, for $V_1, \ldots, V_{m-1}$ independent with $V_j \sim \mathrm{Be}(1 - \sigma, -(m - j)\sigma)$, for $j = 1, \ldots, m - 1$, and $\tilde{W}_m = 1 - \sum_{j<m} \tilde{W}_j$. (This is *not* the stick-breaking representation of the $\mathrm{Dir}(m; -\sigma, \ldots, -\sigma)$-distribution.)

14.2.1 Posterior Distribution

A species sampling process P may be used as a prior for the distribution of a random sample $X_1, \ldots, X_n$ of observations. As always the posterior mean is given by the predictive distribution, which for a species sampling process has an attractive expression in terms of the PPF: $\mathrm{E}(P|\, X_1, \ldots, X_n)$ is equal to the distribution on the right side of (14.9). The full posterior distribution is a modified species sampling process. The following theorem shows that it has an easy description in terms of the weight sequence in size-biased order.

Theorem 14.18 (Posterior distribution) *The posterior distribution of P in the model with observations $X_1, \ldots, X_n|\, P \stackrel{iid}{\sim} P$ with P following a proper species sampling prior* (14.8) *with the weight sequence $(W_j) = (\tilde{W}_j)$ in size-biased order is the distribution of*

$$\sum_{j=1}^{K_n} \widehat{W}_j \delta_{\tilde{X}_j} + \sum_{j=K_n+1}^{\infty} \widehat{W}_j \delta_{\widehat{X}_j},$$

where $\tilde{X}_1, \ldots, \tilde{X}_{K_n}$ are the distinct values in $X_1, \ldots, X_n$ in order of appearance, the variables $\widehat{X}_{K_n+1}, \widehat{X}_{K_n+2}, \ldots \stackrel{iid}{\sim} G$, and $\widehat{W} = (\widehat{W}_j)$ is an independent vector with distribution given by, for every bounded measurable function f,

$$\mathrm{E}(f(\widehat{W})|\, X_1, \ldots, X_n) = \frac{1}{p(N_n)} \mathrm{E}\Big[f(\tilde{W}) \prod_{j=1}^{K_n} \tilde{W}_j^{N_{j,n}-1} \prod_{j=2}^{K_n} \Big(1 - \sum_{i<j} \tilde{W}_i\Big)\Big],$$

where p is the EPPF of the species sampling model, and the expectation on the right side is under the prior distribution of $(\tilde{W}_i)$.

Proof Generate $X_1, \ldots, X_n$ by the scheme of Theorem 14.13: first sequentially generate a random partition of $\mathbb{N}_n$ by the scheme (i)–(ii) described following the theorem (also see the proof of Corollary 14.14) and next attach to each partitioning set a different value from an i.i.d. sequence $\tilde{X}_1, \tilde{X}_2, \ldots$ from G. Now suppose that we have observed $X_1, \ldots, X_n$. Then the value of K_n and the values $\tilde{X}_1, \ldots, \tilde{X}_{K_n}$ can be recovered exactly from the observations $X_1, \ldots, X_n$, whereas the values $\tilde{X}_{K_n+1}, \tilde{X}_{K_n+2}, \ldots$ remain i.i.d. from G, also conditionally given these observations. By (the proof of) Corollary 14.14 the likelihood of the observed partition given the weight sequence $(\tilde{W}_j)$ is given by $\prod_{j=1}^{K_n} \tilde{W}_j^{n_j-1} \prod_{j=2}^{K_n} (1 - \sum_{i<j} \tilde{W}_i)$. By Bayes's rule the posterior distribution of the weight sequence is obtained by reweighting its prior with the likelihood, and renormalizing. The normalizing constant is $p(N_n)$, by Corollary 14.14. □

Under the mild conditions of Lemma 3.5 on the weight sequence (W_j) a proper species sampling process has full weak support in the set of all probability measures on $\mathfrak{X}$. This might suggest that it leads to posterior consistency under the weak topology when used as a prior on the distribution of i.i.d. data. This is *not* true in general. The following theorem gives (somewhat abstract) sufficient conditions for posterior consistency. We shall encounter several examples where these conditions are not met and the posterior distribution is inconsistent. As the prior is neither tail-free, nor possesses the Kullback-Leibler property (as the family of distributions under consideration is not even dominated), it should not be a complete surprise that examples of inconsistency are abundant.

Theorem 14.19 (Consistency) *Let S be the support of the discrete part P_0^d of the probability measure $P_0 = P_0^c + P_0^d$. If P follows a species sampling process prior with PPF (p_j) satisfying, for nonnegative numbers $\alpha_n = O(1)$ and numbers $\delta_n = O(1)$,*

$$\sum_{j=1:\tilde{X}_j \in S}^{K_n} \Big| p_j(N_n) - \frac{\alpha_n N_{j,n} + \delta_n}{n} \Big| \to 0, \qquad \text{a.s. } [P_0^\infty], \tag{14.12}$$

$$\sum_{i=1}^{K_n} \sum_{j=1}^{K_n} \Big| p_i(N_n) p_j(N_n^{i+}) - p_i(N_n) p_j(N_n) \Big| \to 0, \qquad \text{a.s. } [P_0^\infty], \tag{14.13}$$

then the posterior distribution of P in the model $X_1, \ldots, X_n |\, P \stackrel{iid}{\sim} P$ is strongly consistent at P_0 relative to the topology of pointwise convergence on bounded measurable functions if both $\alpha_n \to 1$ and $p_{K_n+1}(N_n) \to 0$ a.s. The latter two conditions are necessary if $P_0^d \neq 0$ or $G \neq P_0^c / \| P_0^c \|$, respectively; and equivalent if P_0 is discrete. Furthermore, if $\alpha_n \to \alpha$ and $p_{K_n+1}(N_n) \to \gamma$ and either $p_{K_n+2}(N_n^{(K_n+1)+}) \to \gamma$ or $\gamma = 0$, then the posterior distribution tends to $\alpha P_0^d + \beta P_0^c + \gamma G$, for $\beta = (1 - \alpha \| P_0^d \| - \gamma)/\| P_0^c \|$.

Proof For posterior consistency it is necessary and sufficient that the posterior mean of Pf tends to $P_0 f$ and the posterior variance to zero, for every bounded measurable function f (see Lemma 6.4). By conditioning on P and using the towering property of conditional expectation, the posterior mean and second moment can be expressed in the predictive distributions as $\mathrm{E}(Pf |\, X_{1:n}) = \mathrm{E}(f(X_{n+1}) |\, X_{1:n})$ and $\mathrm{E}((Pf)^2 |\, X_{1:n}) = \mathrm{E}(f(X_{n+2}) f(X_{n+1}) |\, X_{1:n})$. In conjunction with the predictive formula (14.9) this permits to express posterior mean and variance in the PPF of the species sampling model.

Observations X_i not in the support S of the discrete part P_0^d of P_0 appear only once. Thus they contribute distinct values $\tilde{X}_j$ of multiplicity $N_{j,n} = 1$ in $X_1, \ldots, X_n$, and can be viewed as sampled i.i.d. from the continuous part P_0^c of P_0 conditioned to S^c. The predictive probabilities $p_j(N_n)$ corresponding to these distinct values are all the same. Indeed by symmetry of the EPPF, the value of $p_j(\boldsymbol{n}) = p(\boldsymbol{n}^{j+})/p(\boldsymbol{n})$ (see (14.3)) depends on $\boldsymbol{n}$, but is the same for every j with $n_j = 1$: it does not matter which 1 is raised to 2 to obtain $\boldsymbol{n}^{j+}$.

For every given m, the fraction of values $X_1, \ldots, X_n$ landing on atoms θ_j with $j > m$ tends to $\sum_{j>m} W_j$ almost surely, by the law of large numbers. Therefore the number K_n^d of *distinct values* in $X_1, \ldots, X_n$ that belong to S is bounded above by $m + n \sum_{j>m} W_i (1 + o(1))$. As this is true for every m, it follows that $K_n^d = o(n)$ almost surely.

We have that $f(\tilde{X}_j) N_{j,n}$ is equal to $\sum_{i=1}^n f(X_i) \mathbb{1}\{X_i = \tilde{X}_j\}$, which sums to $n\mathbb{P}_n(f \mathbb{1}_S)$ when summed over j with $\tilde{X}_j \in S$, for $\mathbb{P}_n$ the empirical distribution of $X_1, \ldots, X_n$. Using this after applying (14.12), we see, with $\beta_n \geq 0$ the common value of $np_j(N_n)$ for j such that $\tilde{X}_j \notin S$, and $\gamma_n = p_{K_n+1}(N_n)$,

$$\begin{aligned} \mathrm{E}[Pf |\, X_{1:n}] &= \sum_{j=1:\tilde{X}_j \in S}^{K_n} p_j(N_n) f(\tilde{X}_j) + \sum_{j=1:\tilde{X}_j \notin S}^{K_n} p_j(N_n) f(\tilde{X}_j) + p_{K_n+1}(N_n) Gf \\ &= \alpha_n \mathbb{P}_n(f \mathbb{1}_S) + \delta_n o(1) + o(1) + \beta_n \mathbb{P}_n(f \mathbb{1}_{S^c}) + \gamma_n Gf. \end{aligned}$$

Up to the $o(1)$ terms the right side is a mixture of three (sub)probability measures (where $\beta_n > 1$ is not a priori excluded). By the law of large numbers $\mathbb{P}_n(f\mathbb{1}_S) \to P_0^d f$ and $\mathbb{P}_n(f\mathbb{1}_{S^c}) \to P_0^c f$, almost surely, and hence the right side of the display reduces to $\alpha_n P_0^d f + \beta_n P_0^c f + \gamma_n Gf + o(1) + o(\beta_n)$, almost surely. If $\alpha_n \to 1$ and $\gamma_n \to 0$, then the first term tends to $P_0^d f$ and the third term to zero. Choosing $f \equiv 1$ in the display shows that the total mass of the measure remains one, whence $\beta_n \to 1$. We conclude that the limit is $P_0^d f + P_0^c f = P_0 f$, and hence consistency pertains.

Since only the first of the three measures can give rise to P_0^d, for consistency in the case that the latter measure is nonzero, it is necessary that $\alpha_n \to 1$. Similarly if G is not proportional to P_0^c, then consistency requires the third measure $\gamma_n G$ to disappear and hence $\gamma_n \to 0$.

If $\alpha_n \to \alpha$ and $\gamma_n \to \gamma$, then the total masses of the first and third measures tend to limits. Consequently, if $P_0(S^c) > 0$, then β_n must also converge to a limit; in the other case β_n is irrelevant for the limit of the mixture, as $\mathbb{P}_n(f\mathbb{1}_{S^c})$ vanishes for every n. The limit of the mixture takes the form $\alpha P_0^d + \beta P_0^c + \gamma G$, where the weight β can be identified from the fact that the mixture is a probability measure.

It remains to consider the posterior variance of Pf. The posterior second moment satisfies

$$\mathrm{E}[(Pf)^2|\, X_{1:n}] = \int \mathrm{E}(f(X_{n+2})f(x)|\, X_{n+1} = x, X_{1:n})\, dP^{X_{n+1}|X_{1:n}}(x).$$

We use (14.9) twice, the second time with $n+1$ instead of n, to replace first the integrating measure and next the integrand by a mixture. We use that K_{n+1} is equal to K_n if $X_{n+1} = \tilde{X}_j$ for some j and equal to $K_n + 1$ otherwise, and apply (14.4) to combine the terms with coefficients $p_i(N_n)p_j(N_n^{i+})$ and $p_j(N_n)p_i(N_n^{j+})$ into a single one. Subtracting the square of the posterior mean we finally find that

$$\begin{aligned}
\mathrm{var}[Pf|\, X_{1:n}] &= \sum_{i=1}^{K_n}\sum_{j=1}^{K_n}\Big(p_i(N_n)p_j(N_n^{i+}) - p_i(N_n)p_j(N_n)\Big) f(\tilde{X}_i)f(\tilde{X}_j)\\
&\quad + 2\sum_{i=1}^{K_n}\Big(p_i(N_n)p_{K_n+1}(N_n^{i+}) - p_i(N_n)p_{K_n+1}(N_n)\Big) f(\tilde{X}_i)Gf\\
&\quad + \Big(p_{K_n+1}(N_n)p_{K_n+2}(N_n^{(K_n+1)+}) - p_{K_n+1}(N_n)^2\Big)(Gf)^2\\
&\quad + p_{K_n+1}(N_n)p_{K_n+1}(N_n^{(K_n+1)+})G(f^2).
\end{aligned}$$

The first term on the right tends to zero by assumption (14.13). The second one also, as is seen by rewriting the probabilities $p_{k+1}(\boldsymbol{n})$ as $1 - \sum_{j=1}^k p_j(\boldsymbol{n})$. The third and fourth terms tend to zero if $p_{K_n+1}(N_n) \to 0$.

The third term also tends to zero if $p_{K_n+1}(N_n)$ and $p_{K_n+2}(N_n^{(K_n+1)+})$ have the same limit. By evaluating the display with $f \equiv 1$, we see that the fourth term can be written as minus the sum of the first three terms for $f = 1$. This shows that it also tends to zero. □

The product $p_i(N_n)p_j(N_n^{i+})$ in (14.13) is the probability that X_{n+1} is added to the ith set and next X_{n+2} to the jth set in the partitions generated by $X_1, X_2, \ldots$ Thus condition (14.13) requires that for large n two consecutive steps of the Markov chain of partitions are almost independent. This seems reasonable as the change in the partition from time n to $n+1$ is minor.

Condition (14.12) with $\alpha_n \approx 1$ requires that the predictive probabilities settle down on the empirical frequencies. This seems natural, also in view of Theorem 14.13, which shows that this is true under the prior law. However, presently the convergence must take place under the "true" distribution of $X_1, X_2, \ldots$, whence the condition is independent of the latter theorem, and certainly not automatic. The condition is empty if P_0 is continuous.

The theorem shows that a species sampling process can be an appropriate prior (in that leads to posterior consistency) only if $p_{K_n+1}(N_n) \to 0$, except in the accidentally lucky case that the continuous component P_0^c is proportional to the prior center measure G. The convergence $p_{K_n+1}(N_n) \to 0$ requires that eventually new observations should not open new sets in the partition. In view of the form of the predictive distribution (or posterior mean) this is intuitively natural, as otherwise the prior center measure G does not wash out. In many examples $p_{K_n+1}(N_n)$ converges to a positive limit γ (or even $\gamma = 1$) if P_0 has a nonzero absolutely continuous component and hence leads to inconsistency (see the next sections and Problems 14.2 and 14.3). This includes the Pitman-Yor process, discussed in Section 14.4, whenever this does not coincide with the Dirichlet process.

14.2.2 Species Sampling Process Mixtures

Even though inconsistency can easily occur when using a species sampling process prior directly on the distribution of the observations, putting the prior on the mixing distribution of a kernel mixture will typically work. For a species sampling process with large weak support the Kullback-Leibler support of the corresponding kernel mixture prior contains many true densities, by Theorem 7.2. Hence a kernel mixture of a species sampling process prior will lead to posterior consistency for the weak topology, and under entropy conditions also to consistent density estimation, just as for the Dirichlet mixture process prior, considered in Chapter 5. Consistency under $\mathbb{L}_1$-norm can also be obtained if more structure is imposed; see Problem 14.16.

A generalization of the Gibbs sampling scheme for Dirichlet process mixtures, considered in Chapter 5, may be used for computing the posterior distribution. For families of probability densities $x \mapsto \psi_i(x; \theta_i)$ indexed by parameters (or latent variables) θ_i, we assume that P follows a species sampling process and

$$X_i \,|\, \theta_i \overset{\text{ind}}{\sim} \psi_i(\cdot; \theta_i), \qquad \theta_i \,|\, P \overset{\text{iid}}{\sim} P, \qquad i = 1, 2, \ldots$$

Assume that the species sampling process has an atomless center measure G (as usual) and PPF $\{p_j\}$. Then as in Theorem 5.3, which specializes to the Dirichlet process, the posterior distribution of the latent variables θ_i satisfies

$$\theta_i \,|\, \theta_{-i}, X_1, \ldots, X_n \sim \sum_{j=1}^{K_{-i}} q_{i,j} \delta_{\theta^*_{-i,j}} + q_{i,0} G_{b,i}, \tag{14.14}$$

where $\theta^*_{-i,j}$ is the jth distinct value in $\{\theta_1, \ldots, \theta_{i-1}, \theta_{i+1}, \ldots, \theta_n\}$, K_{-i} is the total number of distinct values and

$$q_{i,j} = cp_j(N_{-i,1}, \dots, N_{-i,K_{-i}})\psi_i(X_i; \theta^*_{-i,j}), \qquad j \neq 0,$$
$$q_{i,0} = cp_{K_{-i}+1}(N_{-i,1}, \dots, N_{-i,K_{-i}}) \int \psi_i(X_i; \theta)\, dG(\theta), \tag{14.15}$$

where $N_{-i,1}, \dots, N_{-i,K_{-i}}$ are the multiplicities of the distinct values $\theta^*_{-i,j}$, c is chosen to satisfy $q_{i,0} + \sum_{j\neq i} q_{i,j} = 1$, and $G_{b,i}$ is the "baseline posterior measure" proportional to $\psi_i(X_i; \theta)\, dG(\theta)$.

14.3 Gibbs Processes

Partition models in which the EPPF depends multiplicatively on the sizes of the partitioning sets are attractive for their simplicity. Lemma 14.21 below shows that these are all of Gibbs type, as defined in the following definition.

Definition 14.20 (Gibbs process) A *Gibbs partition* of *type* $\sigma \in (-\infty, 1)$ is an infinite exchangeable random partition with EPPF of the form[7]

$$p(n_1, \dots, n_k) = V_{n,k} \prod_{j=1}^{k} (1-\sigma)^{[n_j-1]}, \qquad n = \sum_{i=1}^{k} n_i, \tag{14.16}$$

where $V_{n,k}$, for $n = 1, 2, \dots, k = 1, \dots, n$, are nonnegative numbers satisfying the backward recurrence relation

$$V_{n,k} = (n - \sigma k) V_{n+1,k} + V_{n+1,k+1}, \qquad V_{1,1} = 1. \tag{14.17}$$

The species sampling process corresponding to a Gibbs partition and an atomless measure G is called a *Gibbs process*.

For a given type σ, multiple arrays $(V_{n,k})$ will satisfy relation (14.17), and define different Gibbs processes. The Dirichlet processes form a family of Gibbs processes of type $\sigma = 0$, indexed by the precision parameter M (note that $1^{[n]} = n!$ and see (14.6) and/or apply the following lemma). Other prominent examples of Gibbs processes are the Pitman-Yor processes, discussed in Section 14.4. Mixtures of Gibbs processes of the same type are Gibbs processes of again the same type.

The recurrence relation (14.17) arises naturally, as it ensures that p given by (14.16) satisfies the consistency equation (14.2), and thus that p is indeed an EPPF. At first sight the defining relation (14.16) is odd, but it turns out to be the only possible form of a product partition model, apart from trivial partition models.

Lemma 14.21 (Product partition) *Any infinite exchangeable partition with EPPF of the form $p(n_1, \dots, n_k) = V_{n,k} \prod_{i=1}^{k} \rho(n_i)$ for every composition $(n_1, \dots, n_k)$ of n, for a given array of constants $V_{n,k}$ and a strictly positive, nonconstant function ρ, is a Gibbs partition.*

Proof Equation (14.2) applied to the product representation $p(n_1, \dots, n_k) = V_{n,k} \prod_{i=1}^{k} \rho(n_i)$ can be rearranged to give $V_{n,k} = V_{n+1,k} \sum_{j=1}^{k} \rho(n_j + 1)/\rho(n_j) +$

[7] Recall that $a^{[n]} = a(a+1)\cdots(a+n-1)$ denotes the ascending factorial, and $a^{[0]} = 1$.

$\rho(1)V_{n+1,k+1}$. Set $f(n) = \rho(n+1)/\rho(n)$ and apply this equation with $k = 2$ to find that $V_{n,2} = V_{n+1,2}[f(n_1)+f(n_2)]+\rho(1)V_{n+1,3}$, for every partition $n = n_1+n_2$. If $V_{n+1,2} > 0$ for every n, then we can conclude that the sum $f(n_1) + f(n_2)$ depends on $n_1 + n_2$ only, whence $f(n+1)-f(n) = f(m+1)-f(m)$, for every m, n. This implies that $f(n) = bn-a$, for some $b > 0$ and $a < b$ (as ρ is strictly positive and nonconstant), and hence

$$\rho(n) = \rho(1)\prod_{i=1}^{n-1} f(i) = \rho(1)(b-a)(2b-a)\cdots((n-1)b-a) = \rho(1)b^{n-1}(1-a/b)^{[n-1]}.$$

Substituting this in the product formula for $p(n_1, \ldots, n_k)$ we see that this is of Gibbs form, with $V_{n,k}$ taken equal to $V_{n,k}b^{n-k}\rho(1)^k$ and $\sigma = a/b$.

If $V_{n,2} = 0$ for some $n \geq 2$, then $p(n_1, n_2) = 0$ for every partition $n = n_1 + n_2$ of n. This implies that a partition of $\mathbb{N}_n$ in more than one block is impossible. The consistency of the exchangeable partition shows then this is true for every $n \geq 1$, which indicates that the species sampling model can have one species only. This is the Gibbs partition with $V_{n,k} = 0$ for $k \geq 2$, and $V_{n,1} = (1-\sigma)^{[n-1]}$, for any $\sigma < 1$. □

From (14.3) the PPF of a Gibbs process is obtained as

$$p_j(n_1, \ldots, n_k) = \begin{cases} \dfrac{V_{n+1,k}}{V_{n,k}}(n_j - \sigma), & j = 1, \ldots, k, \\[2ex] \dfrac{V_{n+1,k+1}}{V_{n,k}}, & j = k+1. \end{cases} \tag{14.18}$$

Thus the probability of discovering a new species after observing K_n different species in n individuals is $V_{n+1,K_n+1}/V_{n,K_n}$. This depends on n and the total number K_n of observed species, but not on their multiplicities $N_{1,n}, \ldots, N_{K_n,n}$. Ignoring the multiplicities and bringing to bear only K_n may be considered a deficiency of the Gibbs model, but also a reasonable compromise between flexibility and tractability. In terms of flexibility it is still a step up from the Dirichlet process $\mathrm{DP}(MG)$ (which is a Gibbs process of type $\sigma = 0$) for which the probability of observing a new species is $M/(M+n)$, and hence depends only on n. The following proposition shows that the latter property characterizes the Dirichlet process within the Gibbs processes.

Proposition 14.22 *In a nontrivial Gibbs process, the probability $p_{k+1}(n_1, \ldots, n_k)$ of discovering a new species depends only on the total number of individuals $n = \sum_i n_i$, for all $n = 1, 2, \ldots$ and $k = 1, 2, \ldots n$, if and only if the partition corresponds to the Dirichlet process.*

Proof For a Dirichlet process, the probability of observing a new species is $M/(M+n)$ and hence depends on n only. Conversely, by (14.18) (or (14.17)) the probability of a new species in a Gibbs partition satisfies $V_{n+1,k+1}/V_{n,k} = 1-(n-\sigma k)V_{n+1,k}/V_{n,k}$ and hence does not depend on k if and only if $V_{n+1,k}/V_{n,k} = c_n/(n-\sigma k)$ for some sequence c_n depending only on n, in which case $V_{n+1,k+1}/V_{n,k} = 1-c_n$. Applying the latter two equations, also with n replaced by $n+1$ and k by $k+1$, we can rewrite the obvious identity

$$\frac{V_{n+2,k+1}}{V_{n+1,k+1}}\frac{V_{n+1,k+1}}{V_{n,k}} = \frac{V_{n+2,k+1}}{V_{n+1,k}}\frac{V_{n+1,k}}{V_{n,k}}$$

in the form $\sigma\big[k(d_{n+1}-d_n)+d_{n+1}\big]=(n+1)d_{n+1}-nd_n$, for $d_n=(1-c_n)/c_n$. This can be true for all $k=1,\ldots n$ and $n=1,2,\ldots$ only if either $d_{n+1}=d_n$ or $\sigma=0$. In the first case the equation implies $\sigma=1$, which is excluded by definition, or $d_n=0$ for every n, which implies $V_{n,k}=0$ for all $n,k\geq 2$, so that the partition is the trivial one in one block. In the second case we find $d_{n+1}=d_n n/(n+1)$. Iterating the latter equation shows that $d_n=d_1/n$ or $c_n=n/(d_1+n)$. Thus the PPF of the partition is that of a Dirichlet process with $M=d_1$, and the proposition follows by application of Proposition 14.9. □

The following proposition shows that, given the number of sets K_n in a Gibbs partition, the distribution of the composition of the partition is free of the $V_{n,k}$ parameters, but it does depend on the type parameter σ. In other words, in a Gibbs process of known type the number of sets K_n is a sufficient statistic.

The first formula in the following proposition generalizes (4.16) for a Dirichlet process.

Proposition 14.23 (Number of species) *The distribution of the number K_n of sets in a Gibbs partition $\mathcal{P}_n$ and the partition probabilities conditioned on this number are given by*

$$\mathrm{P}(K_n=k)=B_{n,k}V_{n,k},$$

$$\mathrm{P}(\mathcal{P}_n=\{A_1,\ldots,A_k\}\,|\,K_n=k)=\frac{1}{B_{n,k}}\prod_{j=1}^k(1-\sigma)^{[|A_j|-1]},$$

where $\sigma^k B_{n,k}$ is a generalized factorial coefficient, *given by, with the sum over all compositions $(n_1,\ldots,n_k)$ of n of size k,*

$$B_{n,k}=\sum_{(n_1,\ldots,n_k)}\frac{n!}{k!\prod_{j=1}^k n_j!}\prod_{j=1}^k(1-\sigma)^{[n_j-1]}=\frac{1}{\sigma^k k!}\sum_{j=1}^k(-1)^j\binom{k}{j}(-j\sigma)^{[n]}.$$

Proof The second formula follows from the formula for conditional probability, the first formula, and expression (14.16) for the EPPF. Proving the first formula is equivalent to verifying the formula for $B_{n,k}$. The first expression for $B_{n,k}$ follows from substituting (14.16) in the general formula for $\mathrm{P}(K_n=k)$ given in the second part of Proposition 14.5. To see that this is identical to the second expression we consider the generating function of the sequence $\sigma^k B_{n,k}/n!$, for $n=1,2,\ldots$, for a fixed value of k. Since $k\leq n$, the first $k-1$ values of this sequence vanish, and hence the generating function at argument s is given by

$$\frac{(-1)^k}{k!}\sum_{n=k}^\infty s^n\sum_{n_1+\cdots+n_k=n}\prod_{j=1}^k\frac{(-\sigma)^{[n_j]}}{n_j!}=\frac{(-1)^k}{k!}\prod_{j=1}^k\sum_{n_j=1}^\infty\frac{(-\sigma)^{[n_j]}s^{n_j}}{n_j!}.$$

In view of the identity $1-(1-s)^t=-\sum_{m=1}^\infty(-t)^{[m]}s^m/m!$, the last display is equal to

$$\frac{1}{k!}\big[1-(1-s)^\sigma\big]^k=\frac{1}{k!}\sum_{j=0}^k(-1)^j\binom{k}{j}(1-s)^{j\sigma}=\frac{1}{k!}\sum_{j=0}^k(-1)^j\binom{k}{j}\sum_{n=0}^\infty\frac{(-j\sigma)^{[n]}s^n}{n!},$$

by the binomial formula and a second application of the identity, in the other direction. Exchanging the order of summation and comparing the coefficient of s^n to $\sigma^k B_{n,k}/n!$ gives the desired identity. □

The type parameter σ has a key role in determining the structure of Gibbs processes. One may discern three different regimes.

Proposition 14.24 *The following characterizations of Gibbs processes hold.*

(i) *Gibbs partitions of type* $\sigma < 0$ *are mixtures over* m *of finite-dimensional Dirichlet* $\mathrm{Dir}(m; -\sigma, \ldots, -\sigma)$ *partitions (also called "Fisher" or "two-parameter partitions"* $(\sigma, -\sigma m)$ *and discussed in Example 14.17 and Section 14.4).*
(ii) *Gibbs partitions of type* $\sigma = 0$ *are mixtures over* M *of Dirichlet* $\mathrm{DP}(MG)$ *partitions.*
(iii) *Gibbs partitions of type* $0 < \sigma < 1$ *are Poisson-Kingman* $\mathrm{PK}(\rho_\sigma, \eta)$ *partitions for* ρ_σ *a* σ*-stable Lévy measure (equivalently mixtures over* y *of* $\mathrm{PK}(\rho_\sigma \mid y)$ *partitions), as discussed in Section 14.5 (and Example 14.43).*

Proof See Gnedin and Pitman (2006), Theorem 12. □

Here, "mixture" refers to a decomposition $p(\boldsymbol{n}) = \int p(\boldsymbol{n}|\, m)\, \pi(dm)$ of the EPPF p of the partition into the EPPFs $p(\cdot|\, m)$ of the basic partitions, relative to some mixing measure π. Equivalently, the representing species sampling process (in particular its weight sequence) is a mixture of the species sampling processes of the basic partitions. In (i) and (ii) "Dirichlet partition" is short for the partition generated by a random sample $X_1, X_2, \ldots$ from the random Dirichlet distribution or process. In (ii) the partition is the Chinese restaurant process, whereas in (i) the Dirichlet distribution $\mathrm{Dir}(m; -\sigma, \ldots, -\sigma)$ is understood as a model for a random probability distribution on the finite set $\{1, 2, \ldots, m\}$ and $X_1, X_2, \ldots$ are variables with values in this finite set. This model can also be framed as the species sampling model with the infinite weight sequence whose first m elements are distributed according to $\mathrm{Dir}(m; -\sigma, \ldots, -\sigma)$ and whose remaining elements are zero. More formally, a Gibbs partition of type (i) can be described as the random partition generated by a species sampling model with weight sequence $(W_1, W_2, \ldots)$ satisfying

$$(W_1, \ldots, W_m, W_{m+1}, \ldots)|\, m \sim \mathrm{Dir}(m; -\sigma, \ldots, -\sigma) \times \delta_0^\infty, \qquad m \sim \pi. \tag{14.19}$$

Here π can be any probability measure on $\mathbb{N}$.

The three types can be differentiated by the distribution of the number of distinct species K_n among the first n individuals. In the basic type (i) the total number of species is m and hence K_n is bounded; mixing over m relieves this restriction. The number of species in a type (ii) Gibbs processes (i.e. the number of tables in the Chinese restaurant process) is of the order $\log n$ by Proposition 4.8. Finally, the number of distinct species in a Gibbs partition of type (iii) is of the order n^σ (see Theorem 14.50). This has obvious consequences for modeling purposes. If the number of clusters is a priori thought to be large, then type (iii) partitions may be more useful than types (i)–(ii). In contrast, type (i) is natural as a model for discovery of a finite number of species, where moderate mixing over m allows an a priori unknown total number.[8]

[8] For instance, for a type (i) Gibbs process with mixing measure $\pi(j) = \gamma(1-\gamma)^{[j-1]}/j!$, for $j = 1, 2, \ldots$ and some $\gamma \in (0, 1)$, the total number of species K_n remains finite a.s., but its limit has infinite expectation; see Gnedin (2010).

Gibbs processes of positive type admit an explicit stick-breaking representation with *dependent* stick-breaking variables.

Theorem 14.25 (Stick-breaking) *A Gibbs process of type $0 < \sigma < 1$ and atomless center measure G can be represented as $P = \sum_{i=1}^{\infty} \tilde{W}_i \delta_{\theta_i}$, for $\theta_1, \theta_2, \ldots \stackrel{iid}{\sim} G$, and $\tilde{W}_i = V_i \prod_{j=1}^{i-1}(1 - V_j)$, where $V_i |\, V_1 = v_1, \ldots, V_{i-1} = v_{i-1}$ has density function on $(0, 1)$ given by*

$$v_i \mapsto \frac{\sigma/\Gamma(1-\sigma)}{v_i^{\sigma}\prod_{j=1}^{i-1}(1-v_j)^{\sigma}} \frac{\int_0^{\infty} y^{-i\sigma} f_{\sigma}(y \prod_{j=1}^{i}(1-v_j))/f_{\sigma}(y)\, d\eta(y)}{\int_0^{\infty} y^{-(i-1)\sigma} f_{\sigma}(y \prod_{j=1}^{i-1}(1-v_j))/f_{\sigma}(y)\, d\eta(y)}.$$

Here f_σ is the density of a positive σ-stable random variable (with Laplace transform equal to $e^{-\lambda^{\sigma}}$), and η is the mixing measure appearing in Proposition 14.24 (iii).

Proof In view of Proposition 14.24, a Gibbs process with $0 < \sigma < 1$ is a Poisson-Kingman $\mathrm{PK}(\rho_\sigma, \eta)$ process, for ρ_σ a σ-stable Lévy measure (see Example 14.43). The result is a special case of Theorem 14.49. ☐

Gibbs process priors, and species sampling models in general, are infinite series priors, which were seen to have full weak support under mild conditions in Lemma 3.5. These conditions are satisfied by most Gibbs processes. In the following theorem we assume that the sample space is a Polish space.

Theorem 14.26 (Support) *The weak support of a Gibbs process prior with type-parameter $\sigma \geq 0$ and center measure G is equal to $\{P \in \mathfrak{M}: \mathrm{supp}(P) \subset \mathrm{supp}(G)\}$. The same is true for type $\sigma < 0$, provided the mixing distribution π in (14.19) has unbounded support in $\mathbb{N}$.*

Proof Without loss of generality we may assume that G has full support; otherwise we redefine the sample space. By Proposition 14.24 the Gibbs prior is a mixture. In the cases $\sigma = 0$ and $0 < \sigma < 1$ the mixture components are the Dirichlet process and the Poisson-Kingman processes $\mathrm{PK}(\rho_\sigma |\, y)$, for ρ_σ a normalized σ-stable measure. These can each be seen to have full support by Lemma 3.5, and this is then inherited by the mixture. If $\sigma < 0$ the Gibbs process is a mixture of finite-dimensional Dirichlet distributions of varying dimension. The full support follows again from Lemma 3.5, provided this dimension is unbounded under the mixing distribution (the prior π (14.19)). ☐

The following theorem specializes Theorem 14.19 for posterior consistency to Gibbs processes. The main requirement that the probability of a new species tends to zero translates into convergence of the sequence $V_{n+1,K_n+1}/V_{n,K_n}$ to zero. The posterior limit is further dependent on the limit of K_n/n, which is the total mass of the continuous part of P_0 and can be anywhere in the interval $[0, 1]$.

Theorem 14.27 (Consistency) *If P follows a Gibbs process prior with coefficients such that both $V_{n+1,K_n+1}/V_{n,K_n} \to \gamma$ and $V_{n+2,K_n+2}/V_{n+1,K_n+1} \to \gamma$ almost surely, for some $0 < \gamma \leq 1$, or $V_{n+1,K_n+1}/V_{n,K_n} \to 0$, then the posterior distribution of P in the* *model $X_1, \ldots, X_n | P \stackrel{iid}{\sim} P$ converges almost surely under P_0 relative to the weak topology to $\alpha P_0^d + \beta P_0^c + \gamma G$, for $\alpha = (1-\gamma)/(1-\sigma\|P_0^c\|)$ and $\beta = (1-\gamma)(1-\sigma)/(1-\sigma\|P_0^c\|)$. In particular, unless P_0^c is proportional to G, the posterior distribution is consistent if and only if $\gamma = 0$ and one of the three possibilities is true: $\sigma = 0$ or P_0 is discrete or P_0 is atomless.*

Proof Observations $X_1, \ldots, X_n$ not in the support of the discrete part of P_0 occur only once, and hence contribute a count of 1 to the total number of distinct values. As shown in the proof of Theorem 14.19 observations from the discrete part contribute $o(n)$ distinct values. Therefore $K_n/n \to \|P_0^c\| =: \xi$, almost surely.

By (14.18) $p_{K_n+1}(N_n) = V_{n+1,K_n+1}/V_{n,K_n}$, which tends to γ by assumption. Furthermore $p_j(N_n) = (\alpha_n N_{j,n} + \delta_n)/n$, for $\alpha_n = nV_{n+1,K_n}/V_{n,K_n}$ and $\delta_n = -\sigma\alpha_n$, where by (14.17) (or the fact that the p_j add up to 1) $(n - \sigma K_n)V_{n+1,K_n}/V_{n,K_n} = 1 - V_{n+1,K_n+1}/V_{n,K_n} \to 1 - \gamma$, so that $\alpha_n \to (1-\sigma\xi)^{-1}(1-\gamma)$. We conclude that condition (14.12) of Theorem 14.19 is satisfied, trivially with the right side equal to 0.

In view of (14.18), the left side of (14.13) is equal to

$$\sum_{i=1}^{K_n} p_i(N_n) \sum_{j=1}^{K_n} \left| \frac{V_{n+2,K_n}}{V_{n+1,K_n}}(N_{j,n}+\mathbb{1}\{i=j\}-\sigma) - \frac{V_{n+1,K_n}}{V_{n,K_n}}(N_{j,n}-\sigma)\right|$$

$$\leq \left|\frac{V_{n+2,K_n}}{V_{n+1,K_n}} - \frac{V_{n+1,K_n}}{V_{n,K_n}}\right|(n + K_n\sigma) + \frac{V_{n+2,K_n}}{V_{n+1,K_n}}.$$

As noted the assumptions imply that the three quotients in the right side are asymptotically equivalent to $(1-\gamma)/(n-\sigma K_n)$, where $n - \sigma K_n \sim n(1-\sigma\xi)$, almost surely. It follows that the expression tends to zero.

For the final assertion we solve the equation $\alpha P_0^d + \beta P_0^c + \gamma G = P_0$. If the measure G, which is continuous by assumption, is not proportional to P_0^c, then this implies $\gamma = 0$. Next if both the discrete and continuous parts of P_0 are nonzero, the equation is valid if and only if $\alpha = \beta = 1$, which is true if and only if $\sigma = 0$. If P_0^d vanishes, then $\xi = 1$ and $\beta = 1$, and the equation is valid. Finally, if P_0^c vanishes, then $\xi = 0$ and $\alpha = 1$ and the equation is valid. ☐

It is remarkable that if P_0 possesses both a discrete and a continuous component, then consistency under the conditions of the preceding theorem pertains only if $\sigma = 0$. Otherwise (if $\gamma = 0$) the limit measure is a mixture of the discrete and continuous components of P_0, but with incorrect weights.

The following result specializes to the case of Gibbs process of type $\sigma < 0$. Consistency then depends on the mixing distribution π in (14.19). For most distributions the probabilities $\pi(j)$ are decreasing for sufficiently large j; this is sufficient for consistency at discrete distributions P_0. Faster decrease guarantees consistency also at continuous distributions.

Lemma 14.28 (Consistency, negative type) *If P follows a Gibbs process prior of type $\sigma < 0$ with fully supported mixing measure π as in* (14.19), *then the posterior distribution based on $X_1 \ldots, X_n | P \stackrel{iid}{\sim} P$ is consistent relative to the weak topology at any discrete P_0 if*

$\pi(j+1) \leq \pi(j)$, for all sufficiently large j. Furthermore, it is consistent at any atomless P_0 if $\pi(j+1) \lesssim j^{-1}\pi(j)$, for all sufficiently large j.

Proof It suffices to show that the limit γ in Theorem 14.27 is equal to 0. The coefficients of the Gibbs processes corresponding to sampling from a Dir$(m; -\sigma, \ldots, -\sigma)$-distribution can be written in the form,

$$V_{n,k}^m = |\sigma|^{k-n} \frac{(m-1)(m-2)\ldots(m-k+1)}{(m+s)(m+2s)\cdots(m+(n-1)s)},$$

where $s = -1/\sigma$. The coefficients with $k > m$ are understood to vanish (as also follows from the formula). The parameter m in the expression for the corresponding EPPF is attached only to $V_{n,k}^m$, and hence the mixture over m relative to the prior π is Gibbs with coefficients $V_{n,k} = \sum_{m\geq k} \pi(m) V_{n,k}^m$. We need to consider the quotient

$$\frac{V_{n+1,k+1}}{V_{n,k}} = \frac{\sum_{m\geq k+1} \pi(m) V_{n+1,k+1}^m}{\sum_{m\geq k} \pi(m) V_{n,k}^m} = \frac{\sum_{m\geq k} \pi(m+1) V_{n+1,k+1}^{m+1}}{\sum_{m\geq k} \pi(m) V_{n,k}^m}.$$

By a little algebra

$$\frac{V_{n+1,k+1}^{m+1}}{V_{n,k}^m} = \frac{m}{m+1+ns} \prod_{i=1}^{n-1} \frac{m+is}{m+1+is}.$$

This quotient is smaller than 1 for every m, as all terms of the product are, and bounded by $c/(c+s)$ if $m \leq cn$.

If $\pi(m+1) \lesssim m^{-1}\pi(m)$, then we bound $\pi(m+1)V_{n+1,k+1}^{m+1}$ by $m^{-1}\pi(m)V_{n,k}^m$, and it immediately follows that $V_{n+1,k+1}/V_{n,k} \lesssim 1/k$. In the case of an atomless P_0, the number of distinct observations K_n is equal to n and tends to infinity, whence the proof of the second assertion of the lemma is complete.

If only $\pi(m+1) \lesssim \pi(m)$, then we can still replace the terms $\pi(m+1)V_{n+1,k+1}^{m+1}$ for $m \leq cn$ by $c/(c+s)$ times $\pi(m)V_{n,k}^m$, where $c/(c+s)$ can be made arbitrarily small by choice of a small c. Arguing similarly, we see that the first part of the series can be made arbitrarily small and it only remains to prove that $\sum_{m\geq cn} \pi(m) V_{n,K_n}^m / V_{n,K_n} \to 0$, for every $c > 0$. Now $V_{n,k} \geq \pi(k) V_{n,k}^k$ and, again by some algebra

$$\frac{V_{n,k}^m}{V_{n,k}^k} = \binom{m}{k} \prod_{i=0}^{n-1} \frac{k+is}{m+is}.$$

The binomial coefficient is bounded by $(me/k)^k$. The terms of the product are increasing in i, and bounded by 1 for $m \geq k$. Hence the product of the first $2k$ terms is bounded above by $(k+2ks)^{2k}/(m+2ks)^{2k}$, which we combine with the bound on the binomial coefficient, and the product of the remaining $n-2k$ terms is bounded above by the product of any number of these terms. Under the assumption that $n \geq 4k-1$, we can use the two remaining terms given by $i = 3k$ and $i = 4k$, and arrive at the bound

$$\sum_{m\geq cn} \pi(m) \frac{V_{n,k}^m}{V_{n,k}} \lesssim \sum_{m\geq cn} \frac{V_{n,k}^m}{V_{n,k}^k} \leq \sum_{m\geq cn} \left[\frac{me(k+2ks)^2}{k(m+2ks)^2}\right]^k \frac{k+3ks}{m+3ks} \frac{k+4ks}{m+4ks}.$$

If $k = o(n)$, then the expression in square brackets tends to zero, uniformly in $m \geq cn$, and the right side can be bounded by a multiple of $r^k k^2 \sum_{m \geq cn} m^{-2}$, for some $r < 1$, which tends to zero as $n \to \infty$. If P_0 is discrete, then this suffices, as $K_n/n \to 0$ almost surely, in this case. □

Example 14.29 (Geometric) For the geometric distribution π conditioned to be positive the sequence $\pi(j+1)/\pi(j)$ tends to a positive limit smaller than 1. This ensures consistency at every discrete P_0, but not at continuous P_0. By directly using Theorem 14.27 the posterior can be shown to be inconsistent at atomless true distributions (see Problem 14.3).

Example 14.30 (Poisson) The Poisson distribution π conditioned to be positive (i.e. $\pi(j) = (1-e^{-\lambda})^{-1} e^{-\lambda} \lambda^j / j!$, for $j = 1, 2, \ldots$) satisfies $\pi(j+1)/\pi(j) = \lambda/(j+1) \lesssim j^{-1}$. This ensures consistency at every discrete or continuous P_0.

14.4 Pitman-Yor Process

The Pitman-Yor processes form an important subclass of Gibbs processes, which share and generalize essential properties of the Dirichlet process.

Definition 14.31 (Pitman-Yor process, two-parameter family) The *Pitman-Yor partition* or *two-parameter family* is the Gibbs process of type σ with

$$V_{n,k} = \frac{\prod_{i=1}^{k-1}(M + i\sigma)}{(M+1)^{[n-1]}}. \tag{14.20}$$

The parameters σ and M are restricted to either ($\sigma < 0$ and $M \in \{-2\sigma, -3\sigma, \ldots\}$) or ($\sigma \in [0, 1)$ and $M > -\sigma$). The distribution of the weight sequence $(W_1, W_2, \ldots)$ of the corresponding species sampling model, listed in size-biased order, is known as the *Pitman-Yor distribution* and the corresponding random measure as the *Pitman-Yor process* $\mathrm{PY}(\sigma, M, G)$, where G is the center measure.

The relation (14.20) may seem odd. It turns out to be the only possibility so that the numbers $V_{n,k}$ factorize over n and k.

Lemma 14.32 *Any array of nonnegative numbers $V_{n,k}$ that satisfies the backward recurrence relation* (14.17) *with $V_{1,1} = 1$ and factorize as $V_{n,k} = V_k/c_n$, for a nonnegative sequence $V_1, V_2, \ldots$ with $V_2 > 0$ and a sequence of positive numbers $c_1, c_2, \ldots$, is of the form* (14.20) *for numbers $\sigma < 1$ and M such that either ($\sigma < 0$ and $M \in \{-2\sigma, -3\sigma, \ldots\}$) or ($\sigma \in [0, 1)$ and $M > -\sigma$).*

Proof For $V_{n,k}$ that factorize as given, relation (14.17) can be written in the form $V_{k+1} = V_k(c_{n+1}/c_n - n + \sigma k)$, for $k \leq n$ and $n = 1, 2, \ldots$ This shows that either $V_k > 0$ for all k or there exists k_0 with $V_k > 0$ for $k \leq k_0$ and $V_k = 0$ for $k > k_0$. The assumption $V_{1,1} = 1$ implies $V_1 > 0$, which shows that k_0 solves $c_{n+1}/c_n - n + \sigma k_0 = 0$. In both cases the equation can be rearranged to $V_{k+1}/V_k - \sigma k = c_{n+1}/c_n - n$, for $k \leq k_0 \leq n$, where the left side is constant in n and the right side constant in k. Thus the two sides have a common value

M, and hence $V_{k+1} = V_k(M+\sigma)$ and $c_{n+1} = c_n(M+n)$, for $k \le k_0 \le n$ and $n = 1, 2, \ldots$ Iterating these equations gives $V_{k+1} = V_1 \prod_{i=1}^{k}(M + i\sigma)$ and $c_{n+1} = c_1(M+1)^{[n]}$, and hence relation (14.17), since $V_1/c_1 = V_{1,1} = 1$, by assumption. If $V_k > 0$ for all $k \ge 1$, then $M + i\sigma > 0$ for every $i \in \mathbb{N}$, which forces $\sigma \ge 0$ and $M > -\sigma$. In the other case, that there exists k_0 with $V_k > 0$ for $k \le k_0$ and $V_k = 0$ for $k > k_0$, we have already seen that $M = -\sigma k_0$. The assumption $V_2 > 0$ forces $k_0 \ge 2$ and $M + \sigma > 0$ and hence $\sigma < 0$, as $M + \sigma = \sigma(1 - k_0)$. □

Besides "two-parameter" and "Pitman-Yor," various other names are attached to the process. First the process is also referred to as the *two-parameter Poisson-Dirichlet model* and as the *Ewens-Pitman model*, with as special cases:

(i) For $\sigma < 0$ the model is also known as the *Fisher model.* For $M = -m\sigma$ the exchangeable random partition is then generated by a random sample $X_1, X_2, \ldots$ from the $\mathrm{Dir}(m; -\sigma, \ldots, -\sigma)$ distribution, and hence eventually will consist of m sets (or "species").
(ii) For $\sigma = 0$ the model is generated by a sample from the Dirichlet process $\mathrm{DP}(MG)$, whence the underlying partition is the Chinese restaurant process, which is also known as the *Ewens model.*

Proofs of these statements can be based on Theorem 14.33 below, or deduced from the resulting expression for the EPPF. Second the Pitman-Yor model is also a Poisson-Kingman model obtained from a σ-stable subordinator or generalized gamma process; see Section 14.5.

The EPPF and PPF of this model are given by, with $n = \sum_{j=1}^{k} n_j$,

$$p(n_1, \ldots, n_k) = \frac{\prod_{i=1}^{k-1}(M + i\sigma)}{(M+1)^{[n-1]}} \prod_{j=1}^{k} (1-\sigma)^{[n_j - 1]},$$

$$p_j(n_1, \ldots, n_k) = \begin{cases} (n_j - \sigma)/(M+n), & j = 1, \ldots, k, \\ (M + k\sigma)/(M+n), & j = k+1. \end{cases}$$

This follows from the definition and the expression (14.18) for the PPF of a Gibbs process. For $\sigma = 0$ we recognize the PPF of the Dirichlet process.

The following theorem describes the (size-biased) weight sequence $(\tilde{W}_1, \tilde{W}_2, \ldots)$ of a Pitman-Yor process explicitly through stick breaking (3.2). It generalizes Theorem 4.12 for the Dirichlet process (which is the special case $\sigma = 0$). The stick-breaking algorithm is also called a *residual allocation model* (RAM) in this context.[9]

Theorem 14.33 (Stick-breaking) *Let* $V_j \stackrel{ind}{\sim} \mathrm{Be}(1-\sigma, M + j\sigma)$ *and set* $\tilde{W}_j = V_j \prod_{l=1}^{j-1}(1 - V_l)$, *for* $j = 1, 2, \ldots$

(i) *If* $\sigma \ge 0$ *and* $M > -\sigma$, *then the sequence* $(\tilde{W}_1, \tilde{W}_2, \ldots)$ *is in size-biased order and possesses the Pitman-Yor distribution.*

9 The independence of the allocations distinguishes the Pitman-Yor process from other species sampling processes: the Pitman-Yor process is the only species sampling process for which the size-biased random permutation of the atoms admit independent relative stick lengths; see Pitman (1996a).

(ii) *If* $\sigma < 0$ *and* $M = -m\sigma$, *then the vector* $(\tilde W_1, \ldots, \tilde W_{m-1}, 1 - \sum_{j<m} \tilde W_j)$ *is in size-biased order and this vector augmented with zeros possesses the Pitman-Yor distribution.*

Proof (i). Since $\log(1-x) \le -x$, for $x < 1$, we have $\sum_j \log \mathrm{E}(1 - V_j) \le -\sum_j (1-\sigma)/(M+1+(j-1)\sigma) = -\infty$. Therefore, the sequence $(\tilde W_j)$ is a proper probability vector, by Lemma 3.4.

To show that the sequence $(\tilde W_j)$ is in size-biased order it suffices to verify that the measure λ_k given by $\lambda_k(A) = \mathrm{E}\mathbb{1}_A(\tilde W_1, \ldots, \tilde W_k)\prod_{j=2}^k (1-\sum_{i<j} \tilde W_i)$, given in Lemma 14.15, is permutation symmetric, for every k. Since $(\tilde W_1, \ldots, \tilde W_k)$ is a transformation of $(V_1, \ldots, V_k)$, we can express the expectation in the definition of λ_k as an integral relative to the density g_k of $(V_1, \ldots, V_k)$ and next transform into an integral over $(w_1, \ldots, w_k) \in A$ by substituting the inverse stick-breaking map $v_1 = w_1$, $v_2 = w_2/(1 - v_1)$, $v_3 = w_3/(1 - w_1 - w_2)$, etc. (Note that the remaining stick lengths satisfy $1 - \sum_{i<j} w_i = \prod_{i<j}(1 - v_i)$.) The Jacobian $|\partial v/\partial w|$ of this transformation cancels the factor $\prod_{j=2}^k (1 - \sum_{i<j} w_j)$ and hence

$$\lambda_k(A) = \int_A g_k\Big(w_1, \frac{w_2}{1-w_1}, \ldots, \frac{w_k}{1-\sum_{i<k} w_i}\Big)\, dw_1 \cdots dw_k.$$

Evaluating g_k as a product of the given beta densities reduces the integrand to a multiple of $(w_1 w_2 \cdots w_k)^{-\sigma}(1 - w_1 - w_2 - \cdots - w_k)^{M+k\sigma-1}$, which indeed is symmetric in $(w_1, \ldots, w_k)$. Thus λ_k is symmetric and hence the distribution of $(\tilde W_j)$ is in size-biased order, by Lemma 14.15.

Using Corollary 14.14 we can now compute the EPPF as

$$p(n_1, \ldots, n_k) = \mathrm{E}\prod_{j=1}^k \tilde W_j^{n_j - 1} \prod_{j=2}^k \Big(1 - \sum_{i<j} \tilde W_i\Big) = \prod_{l=1}^k \mathrm{E}[(1 - V_l)^{\sum_{j=l+1}^k n_l} V_l^{n_l - 1}].$$

The right side can be evaluated with the help of Corollary G.4 to give (after a little algebra on the factorials), the EPPF in the Gibbs form of Definition 14.31.

(ii). Again it suffices to show that the measure λ_k is symmetric for all k. For $k > m$ the measure vanishes, as $1 - \sum_{i=1}^m \tilde W_i = 0$, by construction. For $k < m$ the proof given under (i) is valid. Only the case $k = m$ needs further consideration. For any measurable sets $A_1, \ldots, A_m$,

$$\lambda_k(A_1 \times \cdots \times A_m) = \int_{\substack{w_i \in A_i, i<m \\ 1-\sum_{i<m} w_i \in A_m}} g_{m-1}\Big(w_1, \frac{w_2}{1-w_1}, \ldots, \frac{w_{m-1}}{1-\sum_{i<m-1} w_i}\Big) \times (1 - \sum_{i<m} w_i)\, dw_1 \cdots dw_{m-1}.$$

Unlike in the preceding proof, the factor $1 - \sum_{i<m} w_i$ from the definition of λ_m has not cancelled, because the integral only involves $m-1$ variables and the inverse Jacobian is $\prod_{j=2}^{m-1}(1 - \sum_{i<j} w_j)$. However, since $M = -m\sigma$, the density g_{m-1} evaluated at the relative stick lengths now takes the form $(w_1 w_2 \cdots w_{m-1})^{-\sigma}(1 - w_1 - w_2 - \cdots - w_{m-1})^{-\sigma-1}$, and

hence the integrand reduces to $(w_1 w_2 \cdots w_{m-1} w_m)^{-\sigma}$, for $w_m = 1 - \sum_{i<m} w_i$. Thus λ_m is the $\mathrm{Dir}(m; 1-\sigma, \ldots, 1-\sigma)$-distribution and hence is symmetric.

The EPPF can be derived as under (i). □

The following propositions describe the mean and covariances and an integral transform of integrals $\int \psi \, dP$ of functions with respect to a random probability measure P distributed according to a Pitman-Yor process. In particular, they give expressions for means and covariances of probabilities of sets. The assertions of the first proposition reduce to those given for the Dirichlet process in Propositions 4.2 and 4.3 as $\sigma \to 0$.

Proposition 14.34 (Moments) *If $P \sim \mathrm{PY}(\sigma, M, G)$, then for any real-valued, bounded, measurable functions ϕ, ψ,*

$$\begin{aligned} \mathrm{E}[P(\psi)] &= G(\psi), \\ \mathrm{E}[P(\psi)P(\phi)] &= \frac{1-\sigma}{M+1} G(\psi\phi) + \frac{M+\sigma}{M+1} G(\psi)G(\phi), \\ cov(P(\psi), P(\phi)) &= \frac{1-\sigma}{M+1}[G(\psi\phi) - G(\psi)G(\phi)]. \end{aligned}$$

Proof The first assertion holds as $\mathrm{PY}(\sigma, M, G)$ is a species sampling process with center measure G, and the third assertion follows from combining the first and the second.

To prove the second assertion write $\mathrm{E}[P(\psi)P(\phi)] = \mathrm{E}\big[\psi(X_1)\phi(X_2)\big]$ for $X_1, X_2 |\, P \stackrel{\text{iid}}{\sim} P$, and use (14.9) to obtain

$$\mathrm{E}[\phi(X_2)|\, X_1] = \frac{1-\sigma}{M+1}\phi(X_1) + \frac{M+\sigma}{M+1} G(\phi).$$

Next multiply both sides by $\psi(X_1)$ and integrate with respect to the marginal distribution $X_1 \sim G$, to find the formula. □

Proposition 14.35 (Cauchy-Stieltjes transform) *If $P \sim \mathrm{PY}(\sigma, M, G)$ for $\sigma > 0$ and $M > 0$, then for any nonnegative, measurable function ψ with $\int \psi^\sigma \, dG < \infty$,*

$$\Big(\mathrm{E}\Big[(1 + \int \psi \, dP)^{-M}\Big]\Big)^{-1/M} = \Big(\int (1+\psi)^\sigma \, dG\Big)^{1/\sigma}.$$

Consequently, if $P_i \stackrel{\text{ind}}{\sim} \mathrm{PY}(\sigma, M_i, G)$ are independent of $(W_1, \ldots, W_k) \sim \mathrm{Dir}(k; M_1, \ldots, M_k)$, then $\sum_{i=1}^k W_i P_i \sim \mathrm{PY}(\sigma, \sum_i M_i, G)$.

Proof For the derivation of the formula for $M \neq 0$ we use the characterization of the Pitman-Yor process as a Poisson-Kingman process, as derived in Example 14.47: for (Y_j) a σ-stable Poisson point process, with total mass $Y = \sum_j Y_j$, and $\Phi = \sum_j Y_j \delta_{\theta_j}$ for $\theta_j \stackrel{\text{iid}}{\sim} G$, the $\mathrm{PY}(\sigma, M, G)$-process is equal in distribution to the mixture of the conditional distribution of Φ/Y given $Y = y$ over the distribution with density $y \mapsto c_{\sigma,M} y^{-M} f_\sigma(y)$, where $c_{\sigma,M} = \sigma\Gamma(M)/\Gamma(M/\sigma)$ is the norming constant and f_σ is the density of Y. Given $M > 0$ the $(-M)$th power of the left side of the formula is

$$\int \mathrm{E}\Big[\Big(1+\frac{\Phi[\psi]}{Y}\Big)^{-M} \,|\, Y=y\Big]\frac{c_{\sigma,M}}{y^M} f_\sigma(y)\,dy = \mathrm{E}\Big[(Y+\Phi[\psi])^{-M}\Big] c_{\sigma,M}$$
$$= \frac{c_{\sigma,M}}{\Gamma(M)}\mathrm{E}\Big[\int_0^\infty \lambda^{M-1} e^{-\lambda(Y+\Phi[\psi])}\,d\lambda\Big],$$

since $y^{-M} = \int_0^\infty \lambda^{M-1}e^{-y\lambda}\,d\lambda/\Gamma(M)$. Exchanging the expectation and integral and applying formula (J.6) for the Laplace transform of a completely random measure to $Y+\Phi[\psi] = \int(1+\psi)\,d\Phi$, we see that the preceding display is equal to

$$\frac{c_{\sigma,M}}{\Gamma(M)}\int_0^\infty \lambda^{M-1} e^{-\int\int(1-e^{s\lambda(1+\psi(x))}) s^{-\sigma-1}\,ds\,dG(x)\,\sigma/\Gamma(\sigma)}\,d\lambda$$
$$= \frac{c_{\sigma,M}}{\Gamma(M)}\int_0^\infty \lambda^{M-1} e^{-\lambda^\sigma \int(1+\psi)^\sigma\,dG}\,d\lambda,$$

by evaluation of the integral over s with the help of partial integration on $\sigma s^{-\sigma-1}\,ds = ds^{-\sigma}$. The integral on the right side can be reduced to gamma form by changing the variable λ^σ, and then transforms into to the right side of the proposition to the power $-M$.

For the proof of the last assertion, set $V_i \overset{\text{ind}}{\sim} \mathrm{Ga}(M_i,1)$, so that $V = \sum_i V_i \sim \mathrm{Ga}(\sum_i M_i, 1)$ and independent of $(W_1,\ldots,W_k) \sim (V_1,\ldots,V_k)/V$, by Proposition G.2. By the formula for the Laplace transform of a gamma variable and the formula of the present proposition, for $\lambda \geq 0$,

$$\mathrm{E}\big[e^{-\lambda\sum_i V_i P_i[\psi_i]}\big] = \prod_i \mathrm{E}\Big[\Big(1+\lambda P_i[\psi_i]\Big)^{-M_i}\Big] = \Big(\int(1+\lambda\psi)^\sigma\,dG\Big)^{-\sum_i M_i/\sigma}.$$

By the same calculation (but without the product) it follows that the right side is the Laplace transform of the variable $VP[\psi]$, for $P \sim \mathrm{PY}(\sigma, \sum_i M_i, G)$. By uniqueness of Laplace transforms (Feller 1971, Theorem XIII.1.1), it follows that $VP[\psi] \sim \sum_i V_i P_i[\psi] = V\sum_i W_i P_i[\psi]$. Here the variable V at the right is independent of $\sum_i W_i P_i$, Since the Fourier transform of $\log V$ is never zero, the logarithm of this distributional equation can be deconvolved, and the last assertion of the proposition follows. □

Example 14.36 (Mean functional) In terms of the probability density function $h_{M,\sigma,G,\psi}$ of $P(\psi)$ Proposition 14.35 gives the identity

$$\int(1+\lambda x)^{-M} h_{M,\sigma,G,\psi}(x)\,dx = \Big[\int(1+\lambda\psi)^\sigma\,dG\Big]^{-M/\sigma}.$$

The left-hand side is the generalized *Cauchy-Stieltjes transform* of order M of the density function $h_{M,\sigma,G,\psi}$, which may be inverted to get $h_{M,\sigma,G,\psi}$. In the limit $\sigma \to 0$ this becomes the *Cifarelli-Regazzini identity* or *Markov-Krein identity*

$$\int(1+\lambda x)^{-M} h_{M,0,G,\psi}(x)\,dx = \exp\Big[-M\int\log(1+\lambda\psi)\,dG\Big].$$

Here $h_{M,0,G,\psi}$ is the density of a mean functional of the Dirichlet process, and the formula is a close analog of (4.26). Another special case is obtained by letting $M \to 0$ when $\sigma \in (0,1)$ is fixed, leading to

$$\exp\Big[-\int \log(1+\lambda x) h_{0,\sigma,G,\psi}(x)\,dx\Big] = \Big[\int (1+\lambda\psi)^{\sigma}\,dG\Big]^{-1/\sigma}.$$

The left side of this equation coincides with the right side of the preceding equation when $M = 1$, $\psi = \iota$ is the identity, and G is replaced by the probability measure $H_{\sigma,0,G,\psi}$ with density $h_{\sigma,0,G,\psi}$. In particular, the generalized Cauchy-Stieltjes transform of order 1 of $h_{0,\sigma,G,\psi}$ coincides with that of $h_{1,0,\sigma,H_{\sigma,0,G,\psi},\iota}$. In view of the uniqueness of the inversion of generalized Cauchy-Stieltjes transform of order 1, this implies the curious distributional equality: $P(\psi) =_d Q(\iota)$ if $P \sim \mathrm{PY}(1,\sigma,G)$ and $Q \sim \mathrm{DP}(H_{\sigma,0,G,\psi})$.

Unlike the Dirichlet process, the Pitman-Yor process is not conjugate under observing independent samples. However, the posterior distribution still has a neat characterization, in terms of Pitman-Yor process with updated parameters.

Theorem 14.37 (Posterior distribution) *If $P \sim \mathrm{PY}(\sigma, M, G)$ for $\sigma \geq 0$, then the posterior distribution of P based on observations $X_1,\ldots,X_n|\, P \sim P$ is the distribution of the random measure*

$$R_n \sum_{j=1}^{K_n} \hat{W}_j \delta_{\tilde{X}_j} + (1-R_n) Q_n, \tag{14.21}$$

where $R_n \sim \mathrm{Be}(n-K_n\sigma, M+K_n\sigma)$, $(\hat{W}_1,\ldots,\hat{W}_{K_n}) \sim \mathrm{Dir}(K_n; N_{1,n}-\sigma,\ldots,N_{K_n,n}-\sigma)$, and $Q_n \sim \mathrm{PY}(\sigma, M+\sigma K_n, G)$, all independently distributed. Here $\tilde{X}_1,\ldots,\tilde{X}_{K_n}$ are the distinct values in $X_1,\ldots,X_n$ and $N_{1,n},\ldots,N_{K_n,n}$ their multiplicities.

Proof Theorem 14.33 gives the size-biased weight sequence of the Pitman-Yor process in the stick-breaking form $\tilde{W}_j = V_j \prod_{l=1}^{j-1}(1-V_l)$, for $V_l \overset{\text{ind}}{\sim} \mathrm{Be}(1-\sigma, M+l\sigma)$, $l = 1,2,\ldots$ Since the remaining stick lengths satisfy $1-\sum_{i=1}^{l}\tilde{W}_i = \prod_{l=1}^{j}(1-V_l)$, on the event $K_n = k$ and $N_{1,n} = n_1,\ldots,N_{k,n} = n_k$

$$\prod_{j=1}^{k} \tilde{W}_j^{n_j-1} \prod_{j=2}^{k}\Big(1-\sum_{i<j}\tilde{W}_i\Big) = \prod_{l=1}^{k}(1-V_l)^{\sum_{j=l+1}^{k} n_l} V_l^{n_l-1}.$$

By Theorem 14.18 the posterior distribution is a species sampling model whose weight sequence $(\hat{W}_j)$ has density relative to the prior of $(\tilde{W}_j)$ proportional to this expression. The stick-breaking variables $V_{k+1}, V_{k+2},\ldots$ do not enter in this expression and are independent of the earlier variables. Thus their posterior distribution is the same as their prior distribution and they remain independent of $(V_1,\ldots,V_k)$. To obtain the posterior distribution of the latter vector we multiply the display by its prior density, which is proportional to $\prod_{l=1}^{k} V_l^{-\sigma}(1-V_l)^{M+\sigma l-1}$. The resulting product factorizes as a product of beta densities, up to the normalizing constant, whence we conclude that the weights $V_1,\ldots,V_k$ are again independent under the posterior, with $V_j \sim \mathrm{Be}(n_j-\sigma, M+j\sigma+\sum_{i=j+1}^{k} n_i)$, for $j \leq k$.

The posterior weights $\hat{W}_j$ for $j > k$ can be decomposed as $\hat{W}_j = (1-R_n)\prod_{l=k+1}^{j-1}(1-V_l)V_j$, for $R_n = 1-\prod_{l=1}^{k}(1-V_l) = \sum_{i=1}^{k}\hat{W}_i$. Thus $\sum_{j>k}\hat{W}_j\delta_{\hat{X}_j}$ is equal to $1-R_n$ times a species sampling process with weights $\prod_{l=k+1}^{j-1}(1-V_l)V_j$ for $j > k$. This entails a shift

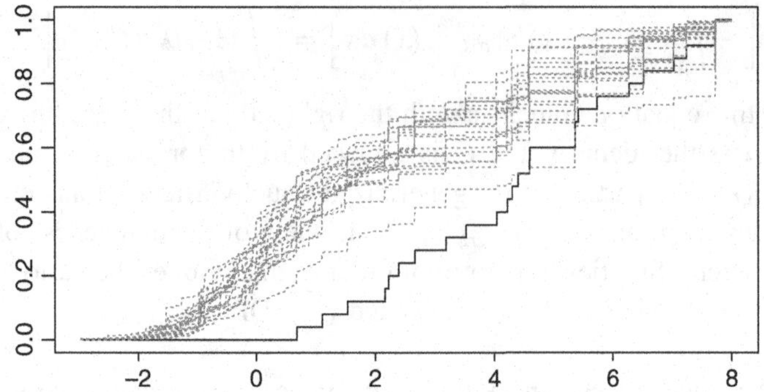

Figure 14.4 Pitman-Yor process. Realization (solid curve) of the empirical distribution of a sample of size 25 from the normal distribution with mean 4 and variance 4, and 20 realizations (dashed curves) from the corresponding posterior distribution relative to the Pitman-Yor prior process with $\sigma = 1/2$, $M = 1$ and standard normal base measure. The posterior inconsistency is clearly visible.

by k in the parameter j of the beta distributions of these variables, which can be viewed as shift by $k\sigma$ in the parameter M. In view of Theorem 14.33 the weights $(\hat{W}_j)_{j>k}$ are those of a $\mathrm{PY}(\sigma, M + k\sigma)$ process.

The vector $(\hat{W}_1, \ldots, \hat{W}_k, 1 - R_n)$ possesses a $\mathrm{Dir}(k + 1; n_1 - \sigma, \ldots, n_k - \sigma, M + k\sigma)$-distribution in view of of the stick-breaking representation of a discrete Dirichlet process, as given in Corollary G.5 with $k + 1$ instead of k, $\alpha_j = n_j - \sigma$, for $j = 1, \ldots, k$, and $\alpha_{k+1} = M + k\sigma$. Equivalently, by the aggregation property of Dirichlet distributions, given in Proposition G.3, the vector $(\hat{W}_1, \ldots, \hat{W}_k)/R_n$ possesses a $\mathrm{Dir}(k; n_1 - \sigma, \ldots, n_k - \sigma)$-distribution, and is independent of R_n, which has a $\mathrm{Be}(n - k\sigma, M + k\sigma)$-distribution. □

The Pitman-Yor process prior inherits the (in)consistency of the general Gibbs processes as a prior for estimating a distribution P of a sample of observations. It is consistent at discrete distributions, but inconsistent for distributions with a continuous component, except when the center measure is (accidentally) proportional to the continuous part of the true distribution, or in the case $\sigma = 0$ that the Pitman-Yor process coincides with the Dirichlet process.

Theorem 14.38 (Consistency) *If P follows a Pitman-Yor* $\mathrm{PY}(\sigma, M, G)$ *process, then the posterior distribution of P in the model $X_1, \ldots, X_n | P \overset{iid}{\sim} P$ converges under P_0 relative to the weak topology to $P_0^d + (1 - \sigma)P_0^c + \sigma\xi G$, for $\xi = \|P_0^c\|$. In particular, the posterior distribution is consistent if and only if P_0 is discrete or G is proportional to P_0^c or $\sigma = 0$.*

Proof A $\mathrm{PY}(\sigma, M, G)$ process is Gibbs, with probability of a new observation equal to $V_{n+1,K_n+1}/V_{n,K_n} = (M + K_n\sigma)/(M + n)$ and $V_{n+2,K_n+2}/V_{n+1,K_n+1} = (M + K_n\sigma + \sigma)/(M + n + 1)$. As seen in the proof of Theorem 14.27 $K_n/n \to \xi$, almost surely, whence both quotients converge to $\gamma = \sigma\xi$. The theorem is a consequence of Theorem 14.27. □

In particular, in the case of an atomless P_0 the Pitman-Yor posterior contracts to the (typically wrong) distribution $(1 - \sigma)P_0 + \sigma G$. The following lemma of Bernstein–von Mises type gives a preciser description of the asymptotics in this case.

Lemma 14.39 (Bernstein–von Mises) *If P follows a $\mathrm{PY}(\sigma, M, G)$ process with $\sigma \geq 0$, then the posterior distribution of P in the model $X_1, \ldots, X_n| P \stackrel{iid}{\sim} P$ satisfies*

$$\sqrt{n}(P - (1-\sigma)\mathbb{P}_n - \sigma G)| X_1, \ldots, X_n \rightsquigarrow \sqrt{1-\sigma}\, \mathbb{G}_{P_0} + \sqrt{\sigma(1-\sigma)}\, \mathbb{G}_G + \sqrt{\sigma(1-\sigma)}\, Z(P_0 - G)$$

in $\mathfrak{L}_\infty(\mathcal{F})$ a.s. $[P_0^\infty]$, for every atomless measure P_0 and for every P_0-Donsker class of functions $\mathcal{F}$ for which the $\mathrm{PY}(\sigma, \sigma, G)$-process satisfies the central limit theorem in $\mathfrak{L}_\infty(\mathcal{F})$. Here $\mathbb{G}_{P_0}$ and $\mathbb{G}_G$ are independent P_0- and G-Brownian bridges independent of the variable $Z \sim \mathrm{Nor}(0, 1)$. In particular this is is true for every sufficiently measurable class of functions $\mathcal{F}$ with finite P_0- and G-bracketing integrals or a finite uniform entropy integral and envelope function that is square integrable under both P_0 and G.

Proof The Pitman-Yor process for $\sigma = 0$ is the Dirichlet process and hence the theorem follows from Theorem 12.2. We may assume that $\sigma > 0$.

Because P_0 is atomless, all observations $X_1, \ldots, X_n$ are distinct, whence $K_n = n$, the multiplicities $N_{j,n}$ are all 1, and $\tilde{X}_1, \ldots, \tilde{X}_n$ are the original observations. Therefore, by Theorem 14.37, the posterior distribution can be represented as $R_n\tilde{\mathbb{P}}_n + (1-R_n)Q_n$, for $\tilde{\mathbb{P}}_n = \sum_{i=1}^n W_{n,i}\delta_{X_i}$, where $R_n \sim \mathrm{Be}(n - n\sigma, M + n\sigma)$, $W_n = (W_{n,1}, \ldots, W_{n,n}) \sim \mathrm{Dir}(n; 1 - \sigma, \ldots, 1-\sigma)$ and $Q_n \sim \mathrm{PY}(\sigma, M + \sigma n, G)$ are independent variables.

By representing R_n as a quotient of gamma variables, as in Proposition G.2, and applying the delta-method and the central limit theorem, we see that $\sqrt{n}(R_n - 1 + \sigma) \rightsquigarrow \sqrt{\sigma(1-\sigma)}Z$.

The vector nW_n can be represented as $(Y_1, \ldots, Y_n)/\bar{Y}_n$ for $Y_i \stackrel{iid}{\sim} \mathrm{Ga}(1-\sigma, 1)$, and hence is exchangeable as in Example 3.7.9 of van der Vaart and Wellner (1996) with $c = (1-\sigma)^{-1/2}$. Consequently $\sqrt{n}(\tilde{\mathbb{P}}_n - \mathbb{P}_n)| X_1, X_2, \ldots \rightsquigarrow c\mathbb{G}_{P_0}$, almost surely, by the exchangeable bootstrap theorem, Theorem 3.7.13 of the same reference.

Applying the delta-method we find that

$$\sqrt{n}(R_n\tilde{\mathbb{P}}_n - (1-\sigma)\mathbb{P}_n) = \sqrt{n}(R_n - 1 + \sigma)P_0 + (1-\sigma)\sqrt{n}(\tilde{\mathbb{P}}_n - \mathbb{P}_n) + o_P(1),$$

where the remainder term tends to zero given $X_1, X_2, \ldots$ almost surely. If we can show that $\sqrt{n}(Q_n - G) \rightsquigarrow \sqrt{(1-\sigma)/\sigma}\, \mathbb{G}_G$, then the same reasoning gives

$$\sqrt{n}((1 - R_n)Q_n - \sigma G) = -\sqrt{n}(R_n - 1 + \sigma)G + \sigma\sqrt{n}(Q_n - G) + o_P(1).$$

The lemma follows by adding the last two displays.

By the last assertion of Proposition 14.35 the Pitman-Yor process Q_n can be represented as $\sum_{i=0}^n W_{n,i}P_i$, for $(W_{n,0}, \ldots, W_{n,n}) \sim \mathrm{Dir}(n+1; M, \sigma, \ldots, \sigma)$ independent of the Pitman-Yor processes $P_0, \ldots, P_n$ with parameters (σ, M, G) for $i = 0$ and (σ, σ, G) for $i > 0$. Since $Q_n - G = W_{n,0}(P_0 - G) + (1 - W_{n,0})\sum_{i=1}^n \tilde{W}_{n,i}(P_i - G)$, where $\sqrt{n}W_{n,0} \to 0$ in probability and the variables $\tilde{W}_{n,i} = W_{n,i}/(1 - W_{n,0})$ form a Dirichlet vector of dimension n with parameters $(\sigma, \ldots, \sigma)$, the term for $i = 0$ is negligible, and the limit distribution of $\sqrt{n}(Q_n - G)$ is as if the term for $i = 0$ did not exist. For simplicity of notation, assume the latter. Represent the remaining Dirichlet coefficients as V_i/V, for $V_i \stackrel{iid}{\sim} \mathrm{Ga}(\sigma, 1)$, for $i = 1, \ldots, n$, and $V = \sum_{i=1}^n V_i$.

For a given $f \in \mathbb{L}_2(G)$, the sequence of variables $n^{-1/2}\sum_{i=1}^n V_i(P_i f - Gf)$ converges by the univariate central limit theorem in distribution to a normal distribution with mean zero and variance $\text{var}\,(V_1(P_1f - Gf)) = \text{E}[V_1^2](1-\sigma)/(\sigma+1)\,\text{var}\left[\mathbb{G}_G(f)\right]$, by Proposition 14.34. By Slutsky's lemma the sequence $\sqrt{n}\sum_{i=1}^n (V_i/V)(P_if - Gf)$ converges in distribution to $1/\text{E}[V_1]$ times the same limiting process. Here $\text{E}[V_1^2]/(\text{E}[V_1])^2 = (1+\sigma)/\sigma$, giving the final variance equal to $(1-\sigma)/\sigma\ \text{var}\left[\mathbb{G}_G(f)\right]$.

This shows that the processes $\sqrt{n}\sum_{i=1}^n (V_i/V)(P_i - G)$ converge marginally in distribution to the process $((1-\sigma)/\sigma)^{1/2}\mathbb{G}_G$. We finish by showing that this sequence of processes, or equivalently the processes $n^{-1/2}\sum_{i=1}^n V_i(P_i - G)$, are asymptotically tight (i.e. equicontinuous) in $\mathfrak{L}_\infty(\mathcal{F})$. The assumption that the $\text{PY}(\sigma, \sigma, G)$-process satisfies the central limit theorem in $\mathfrak{L}_\infty(\mathcal{F})$ entails that the sequence $n^{-1/2}\sum_{i=1}^n (P_i - G)$ converges to a tight Gaussian limit and hence it is asymptotically tight. By Lemma 2.9.1 (also see the proof of Theorem 2.9.2) of van der Vaart and Wellner (1996), this asymptotic tightness is inherited by the sequence of processes $n^{-1/2}\sum_{i=1}^n V_i(P_i - G)$.

For the final assertion of the theorem, it suffices to show that the sequence $n^{-1/2}\sum_{i=1}^n (P_i - G)$ converges in distribution to a tight limit in $\mathfrak{L}_\infty(\mathcal{F})$. This follows because the processes $Z_i = P_i$ satisfy the conditions of Theorems 2.11.9 or 2.11.1 of van der Vaart and Wellner (1996), where Lemma 2.11.6 in the same reference can be used to verify the main condition of the second theorem. ☐

Samples from Pitman-Yor processes of different types $\sigma \geq 0$ cluster in different manners. The numbers K_n of species forms a Markov chain, which increases by jumps of size one if a new species is added. Since a new species is added with probability $(M + K_n\sigma)/(M+n)$,

$$\text{E}(K_{n+1}|\,X_1,\ldots,X_n) = K_n + \frac{M + K_n\sigma}{M+n} = a_nK_n + b_n,$$

for $a_n = 1 + \sigma/(M+n)$ and $b_n = M/(M+n)$. It follows that the process

$$\frac{K_n}{\prod_{i=1}^{n-1}[1+\sigma/(M+i)]} - \sum_{i=1}^{n-1}\frac{M/(M+i)}{\prod_{j=1}^{i}[1+\sigma/(M+j)]}$$

is a martingale. If $\sigma > 0$, then the numbers $\prod_{i=1}^{n-1}[1+\sigma/(M+i)]$ are of the order n^σ, while the sum in the display is bounded in n. It follows that the mean number of species $\mu_n = \text{E}(K_n)$ is of the order n^σ. By using the martingale convergence theorem it can further be derived that K_n/n^σ converges almost surely to a finite random variable, as $n \to \infty$. The limit random variable, known as the *σ-diversity*, can be shown to have a positive Lebesgue density. Thus the number of clusters formed by a Pitman-Yor process is of the order n^σ, which is much higher than the logarithmic order of the number of clusters formed by the Dirichlet process, established in Proposition 4.8. For this reason, in some applications, the Pitman-Yor process model is favored over the Dirichlet process model. For further discussion, within the more general context of Poisson-Kingman partitions, see Theorem 14.50.

14.5 Poisson-Kingman Processes

A *Poisson point process* on $(0,\infty)$ with intensity (or Lévy) measure ρ can be defined as a collection $\{Y_j: j \in \mathbb{N}\}$ of random variables with values in $(0,\infty)$ such that the numbers of variables (or "points") $\#\{j: Y_j \in A_i\}$ in disjoint Borel sets A_i are independent Poisson variables with means $\rho(A_i)$. Consider intensity measures ρ such that

$$\int (s \wedge 1)\, \rho(ds) < \infty, \qquad \rho(0, \infty) = \infty. \tag{14.22}$$

Then the number of points is infinite (only clustering at zero) and their sum $Y := \sum_{i=1}^{\infty} Y_i$ is finite almost surely. The points Y_i can be viewed as the "jumps" in the series Y. By placing these jumps at (possibly random) locations in $(0, \infty)$ one obtains a process with independent increments or "completely random measure." See Appendix J for background on these processes.

By Lemma J.2 the Laplace transform of $Y = \sum_{j=1}^{\infty} Y_j$ is given by, for f the probability density function of Y,

$$\int e^{-\lambda y} f(y)\, dy = e^{-\psi(\lambda)}, \qquad \psi(\lambda) = \int (1 - e^{-\lambda s})\, \rho(ds).$$

The function ψ is called the *Laplace exponent* of Y or the point process (or also *Lévy exponent*).

Definition 14.40 (Poisson-Kingman process) The (special) *Poisson-Kingman distribution* $\mathrm{PK}(\rho)$ with intensity measure (or Lévy measure) ρ is the distribution of $(Y_{(1)}, Y_{(2)}, \ldots)/Y$ for $Y_{(1)} \geq Y_{(2)} \geq \cdots$ the ranked values of a Poisson point process on $(0, \infty)$ with intensity measure ρ and $Y = \sum_{i=1}^{\infty} Y_i$ (which is assumed finite). The *Poisson-Kingman distribution* $\mathrm{PK}(\rho|\, y)$ is the conditional distribution of the same vector given $Y = y$. Given a probability measure η on $(0, \infty)$, the *Poisson-Kingman distribution* $\mathrm{PK}(\rho, \eta)$ is the mixture $\int \mathrm{PK}(\rho|\, y)\, d\eta(y)$. The corresponding species sampling models $P = \sum_{i=1}^{\infty} (Y_i/Y)\delta_{\theta_i}$, for an independent random sample $\theta_1, \theta_2, \ldots$ from an atomless probability distribution G, are said to follow *Poisson-Kingman processes.*

Example 14.41 (Scaling) By the renormalization of the vector (Y_j) by its sum, the scale of the variables Y_j is irrelevant in the definition of the Poisson-Kingman distribution. Thus any measure of the scale family $\rho_\tau(A) = \rho(A/\tau)$ of a given intensity measure gives the same Poisson-Kingman distribution. (On the other hand, multiplying the intensity measure with a constant, to obtain $\tau\rho$, gives a different process.)

Example 14.42 (Gamma process) The sum Y of all points of the point process with intensity measure given by $\rho(ds) = M s^{-1} e^{-s}\, ds$ possesses a gamma distribution with shape parameter M and scale 1. For this process, the ranked normalized jump sizes $(Y_{(1)}, Y_{(2)}, \ldots)/Y$ are independent of the sum Y (see below). The Poisson-Kingman distributions $\mathrm{PK}(\rho|\, y)$ and $\mathrm{PK}(\rho, \eta)$ are therefore free of y and η, and hence coincide with $\mathrm{PK}(\rho)$. This turns out to be the distribution of the ranked jump sizes of a Dirichlet process, so that the corresponding species sampling process can be identified with the Dirichlet prior. In the present context, the distribution of the ranked normalized jump sizes is often called the *one-parameter Poisson-Dirichlet distribution* with parameter M (which should not be confused with $\mathrm{PK}(\rho)$ with a single but general parameter ρ).

The present point process (Y_j) and its intensity measure ρ can be obtained by marginalization to the second coordinate of the Poisson point process $\{(X_j, Y_j): j \in \mathbb{N}\}$ with intensity $s^{-1}e^{-s}\,dx\,ds$ on the space $(0, M] \times (0, \infty)$. The process $t \mapsto \sum_j Y_j \mathbb{1}\{X_j \le t\}$ is the gamma process, which has marginal distributions $\mathrm{Ga}(t, 1)$ and independent increments (see Example J.14). The ranked jump sizes $(Y_{(1)}, Y_{(2)}, \ldots)$ can be seen to be the coordinatewise limits of the ranked increments over increasingly finer uniform partitions of $[0, M]$. This may be combined with the independence of the sum and ratios of independent gamma variables (Proposition G.2), to show that the ranked normalized jump sizes $(Y_{(1)}, Y_{(2)}, \ldots)/Y$ are independent of the sum Y. The connection to the gamma process also shows that the normalized jumps are those of the Dirichlet process (see Section 4.2.3).

Example 14.43 (Stable process) The Laplace exponent $\psi_\sigma(\lambda) = \int (1 - e^{-\lambda s})\, \rho_\sigma(ds)$ of the process with intensity measure $\rho_\sigma(ds) = (\sigma/\Gamma(1-\sigma))\, s^{-1-\sigma}\, ds$, for $\sigma \in (0, 1)$, takes the simple form $\psi_\sigma(\lambda) = \lambda^\sigma$. (One way to see this is to compute the derivative as $\psi'_\sigma(\lambda) = \sigma\lambda^{\sigma-1}$, after differentiating under the integral.) It follows that the density f_σ of Y has Laplace transform $e^{-\lambda^\sigma}$, and hence Y possesses the σ-stable distribution. The corresponding process with independent increments generated by the intensity function $\rho_\sigma(ds)\, dx$ is known as the *σ-stable process* (see Example J.13).

The EPPFs of Poisson-Kingman processes can be expressed in the intensity measure ρ. The formulas are somewhat involved, but lead to simple representations for several examples. The density f in the following theorem is the density of the sum Y of all weights.

Theorem 14.44 (EPPF) *Let $\psi(\lambda) = \int_0^\infty (1 - e^{-\lambda s})\, \rho(ds)$ be the Laplace exponent of an absolutely continuous measure ρ on $(0, \infty)$ satisfying* (14.22), *let $\psi^{(n)}$ be its nth derivative, and let f be the probability density with Laplace transform $e^{-\psi}$. Then the EPPF of the Poisson-Kingman processes* $\mathrm{PK}(\rho)$ *and* $\mathrm{PK}(\rho|\, y)$ *are given by, respectively, for $n = \sum_{j=1}^k n_j$,*

$$p(n_1, \ldots, n_k) = \frac{(-1)^{n-k}}{\Gamma(n)} \int_0^\infty \lambda^{n-1} e^{-\psi(\lambda)} \prod_{j=1}^k \psi^{(n_j)}(\lambda)\, d\lambda,$$

$$p(n_1, \ldots, n_k|\, y) = \int \cdots \int_{\sum_{j=1}^k y_j < y} \frac{f(y - \sum_{j=1}^k y_j)}{y^n f(y)} \prod_{j=1}^k y_j^{n_j} \prod_{j=1}^k \rho(dy_j).$$

Proof By Lemma J.2, f is the density of the sum $Y = \sum_i Y_i$ of the jumps of the Poisson process. If $\mathcal{P}_n$ is the partition generated by the species sampling model defined by the renormalized sequence $W_i = Y_i/Y$, and $\tilde{W}_i = \tilde{Y}_i/Y$ is this sequence in size-biased order, then for any partition $A_1, \ldots, A_k$ of $\mathbb{N}_n$ in order of appearance,

$$\mathrm{P}(\mathcal{P}_n = \{A_1, \ldots, A_k\}, \tilde{Y}_1 \in B_1, \ldots, \tilde{Y}_k \in B_k, Y \in C)$$
$$= \int_{B_1} \cdots \int_{B_k} \int_C y^{-n} f\Big(y - \sum_{j=1}^k y_j\Big)\, dy \prod_{j=1}^k y_j^{|A_j|} \prod_{j=1}^k \rho(dy_j).$$

An intuitive argument for this formula is as follows. (For a longer but precise proof, see the proof of Lemma 14.62.) Form the partition by covering $(0, Y)$ with contiguous intervals of lengths Y_i, generating an independent sample $U_1, \ldots, U_n \stackrel{\text{iid}}{\sim} \text{Unif}(0, Y)$, and letting i and j belong to the same partitioning set if U_i and U_j belong to the same interval. Then $\prod_{j=1}^k \rho(dy_j)$ is the probability that $\tilde{Y}_j = y_j$ are the jumps in the Poisson process corresponding to the k sets in the partition, $f(y - \sum_{j=1}^k y_j)\,dt$ is the conditional probability that the sum of all jumps is $Y = y$ given these points, and $\prod_{j=1}^k (y_j/y)^{|A_j|}$ is the probability that the U_i fall in the intervals in the manner that produces the partition.

The EPPFs follow by marginalizing out the jumps $\tilde{Y}_i$, and marginalizing out (for p) or conditioning on $Y = y$ (for $p(\cdot|\, y)$), respectively. To obtain the formula for p we further write $y^{-n} = \int_0^\infty \lambda^{n-1} e^{-\lambda y}\, d\lambda / \Gamma(n)$, apply Fubini's theorem, make the change of variables $t = y - \sum_{j=1}^k y_j$, and substitute the identities $\int e^{-\lambda t} f(t)\, dt = e^{-\psi(\lambda)}$ and $\psi^{(n)}(\lambda) = (-1)^{n-1} \int_0^\infty s^n e^{-\lambda s}\, \rho(ds)$. ∎

Because the Poisson-Kingman distribution $\text{PK}(\rho, \eta)$ is a mixture of $\text{PK}(\rho|\, y)$ distributions, so is its EPPF. Thus the formulas in the preceding theorem extend to general Poisson-Kingman processes. In the following two examples, which involve polynomial or exponential *tilting* of the intensity measure, the resulting integrations can be reduced to a simple form.

Example 14.45 (Polynomial tilting) For a given absolutely continuous infinite Lévy measure ρ, and given $M \geq 0$, let f_M be the probability density of the form $f_M(y) = c_M y^{-M} f(y)$, for f the density of $Y = \sum_{j=1}^\infty Y_j$ associated with ρ. Then the EPPF of the $\text{PK}(\rho, f_M)$ process is given by

$$p(n_1, \ldots, n_k) = \frac{c_M(-1)^{n-k}}{\Gamma(M+n)} \int_0^\infty \lambda^{M+n-1} e^{-\psi(\lambda)} \prod_{j=1}^k \psi^{(n_j)}(\lambda)\, d\lambda.$$

For $M = 0$ this reduces to the formula of the EPPF of the $\text{PK}(\rho)$ process, given in Theorem 14.44. The formula can be derived in exactly the same manner as the latter formula, by computing $\int p(n_1, \ldots, n_k|\, y) f_M(y)\, dy$.

Example 14.46 (Exponential tilting) For a given absolutely continuous Lévy measure ρ, and given $\tau \geq 0$, the measure $\rho_\tau(ds) = e^{-\tau s}\rho(ds)$ is another Lévy measure. It turns out that the associated Poisson-Kingman distributions $\text{PK}(\rho_\tau|\, t)$ are identical to the $\text{PK}(\rho|\, t)$-distribution, for every t. As a consequence, the Poisson-Kingman distributions $\text{PK}(\rho_\tau, \eta)$ are also independent of this *exponential tilting* of the intensity measure.

On the other hand, the special Poisson-Kingman distributions $\text{PK}(\rho_\tau)$ do depend on τ, in general, as the associated mixing distribution (the distribution of Y) does.

To verify the claim, first note that the Laplace exponent of the tilted intensity measure satisfies $\psi_\tau(\lambda) = \psi_0(\lambda + \tau) - \psi_0(\tau)$, and hence the density of Y satisfies $f_\tau(y) = f_0(y) e^{\psi_0(\tau) - \tau y}$. When substituting these expressions in the formula for $p(\cdot|\, t)$ in Theorem 14.44, the tilting terms can be seen to cancel out.

Another, perhaps more insightful argument, is to note that the laws P_τ of the Poisson processes N_τ with intensities ρ_τ are mutually absolutely continuous, with the density of P_τ with respect to P_0 equal to $dP_\tau/dP_0 = e^{\psi_0(\tau)-\tau Y}$, for $Y = \sum_{i=1}^\infty Y_i$, as before. This can be seen from the identity, for any bounded measurable function f,

$$\mathrm{E}e^{-\int f\,dN_0}e^{-\tau Y} = \mathrm{E}e^{-\int (f(s)+\tau s)\,N(ds)} = e^{-\int (1-e^{f(s)+\tau s})\,\rho(ds)} = e^{-\psi_0(\tau)}\mathrm{E}e^{-\int f\,dN_\tau}.$$

This identity follows by two applications of the formula for the Laplace transform of a Poisson process in Lemma J.2. The fact that the density of P_τ depends on the process only through the sum Y of its points means that the latter variable is statistically sufficient for the parameter τ, and hence the conditional law of the process given Y is free of τ.

Example 14.47 (Pitman-Yor) The Pitman-Yor process with parameter $(\sigma, M) \in (0,1) \times (0,\infty)$ is identical to the Poisson-Kingman distribution $\mathrm{PK}(\rho_\sigma, \eta_{\sigma,M})$ with intensity and mixing measures given by

$$\rho_\sigma(ds) = \frac{\sigma s^{-1-\sigma}}{\Gamma(1-\sigma)}\,ds, \qquad \eta_{\sigma,M}(dy) = \frac{\sigma\Gamma(M)}{\Gamma(M/\sigma)}\,y^{-M} f_\sigma(y)\,dy,$$

where f_σ is the probability density (associated with ρ_σ) with Laplace transform $e^{-\lambda^\sigma}$.

To see this, note that the intensity measure is the σ-stable measure and has Laplace exponent $\psi_\sigma(\lambda) = \lambda^\sigma$ (see Example 14.43) and the measure $\eta_{\sigma,M}$ in the display is a polynomial tilting of the associated density f_σ. Hence the EPPF can be computed as in Example 14.45, with $\psi_\sigma^{(n)}(\lambda) = \sigma(-1)^{n-1}(1-\sigma)^{[n-1]}\lambda^{\sigma-n}$.

As noted in Example 14.46 the Poisson-Kingman distributions $\mathrm{PK}(\rho|\,t)$ are invariant under exponential tilting of the intensity measure, and hence the Pitman-Yor process can also be obtained as a mixture of a tilted stable process, i.e. a generalized gamma process. See (iv) of Example 14.48 below.

Example 14.48 (Generalized gamma) The *generalized gamma process* has intensity measure $\rho(ds) = (\sigma/\Gamma(1-\sigma))\,s^{-1-\sigma}e^{-\tau s}\,ds$, where $\sigma \in (0,1)$ and $\tau > 0$. The process is an exponentially tilted version of the σ-stable process. By the invariance of the Poisson-Kingman distribution under the scaling $s \mapsto s/\tau$, the exponential term can without loss of generality be simplified to the unit form e^{-s} if at the same time the full intensity measure is multiplied by τ^σ, giving $(\sigma/\Gamma(1-\sigma))\tau^\sigma\,s^{-1-\sigma}e^{-s}\,ds$.

The EPPF of the Poisson-Kingman process based on a generalized gamma process can be found with the help of Theorem 14.44. First we compute the Laplace exponent as $\psi(\lambda) = \int(1-e^{-\lambda s})\,\rho(ds) = (\lambda+\tau)^\sigma - \tau^\sigma$. (Write $1-e^{-\lambda s}$ as the integral of its derivative and apply Fubini's theorem.) This leads to the derivatives $\psi^{(n)}(\lambda) = \sigma(\sigma-1)\cdots(\sigma-n+1)(\lambda+\tau)^{\sigma-n} = \sigma(-1)^{n-1}(1-\sigma)^{[n]}(\lambda+\tau)^{\sigma-n}$, and hence to the EPPF

$$p(n_1,\ldots,n_k) = \int_0^\infty \frac{\lambda^{n-1}}{\Gamma(n)}(\lambda+\tau)^{k\sigma-n}e^{-\psi(\lambda)}\,d\lambda\,\sigma^k(-1)^{n-k}\prod_{j=1}^k(1-\sigma)^{[n_j]}.$$

It follows that the special Poisson-Kingman process based on a generalized gamma process (i.e. a normalized generalized gamma process) is a Gibbs process of type σ with coefficients

$$V_{n,k} = \sigma^k(-1)^{n-k}\int_0^\infty \frac{\lambda^{n-1}}{\Gamma(n)}(\lambda+\tau)^{k\sigma-n}e^{-\psi(\lambda)}\,d\lambda$$
$$= \frac{e^{\tau^\sigma}\sigma^{k-1}(-1)^{n-k}}{\Gamma(n)}\sum_{j=0}^{n-1}\binom{n-1}{j}(-\tau)^j\Gamma(k-j/\sigma;\tau^\sigma),$$

where $\Gamma(n;x) = \int_x^\infty s^{n-1}e^{-s}ds$ is the incomplete gamma function, and the second expression follows after some algebra.

The class of generalized gamma processes contains several interesting special cases.

(i) For $\tau \to 0$ the process reduces to the σ-stable process.
(ii) For $\sigma \to 0$ the process with intensity measure $\sigma^{-1}\rho$ tends to the intensity measure of the gamma process, and hence the resulting Poisson-Kingman process reduces to the Dirichlet process.
(iii) The choice $\sigma = 1/2$ leads to tractable marginal distributions of the normalized process; it is called the *normalized inverse-Gaussian process* and is discussed in Section 14.6.
(iv) The Pitman-Yor process $\mathrm{PY}(\sigma, M, G)$ is a mixture over ξ of the generalized gamma processes with parameters σ and $\tau = \xi^{1/\sigma}$, with a $\mathrm{Ga}(M/\sigma, 1)$-mixing distribution on ξ. (See Problem 14.9.) By scaling and normalization these generalized gamma processes are equivalent to the processes with intensity measure $\xi(\sigma/\Gamma(1-\sigma))\,s^{-1-\sigma}e^{-s}\,ds$.

One reason to be interested in Poisson-Kingman distributions is that the weight sequence in size-biased order is relatively easy to characterize. In the following theorem let $(\tilde Y_j)$ be the random permutation of the point process (Y_j) such that $(\tilde Y_1, \tilde Y_2, \ldots)$ is in size-biased order. Then $\tilde W_j = \tilde Y_j/Y$ are the normalized weights of the Poisson-Kingman species sampling process in size-biased order. The theorem gives the joint density of the stick-breaking weights (V_j) corresponding to this sequence: the variables such that $\tilde W_i = V_i\prod_{l=1}^{i-1}(1-V_l)$, for $i = 1, 2, \ldots$.

Theorem 14.49 (Residual allocation, stick-breaking) *If ρ is absolutely continuous, and $\tilde H(ds|\,t) = (s/t)f(t-s)/f(t)\,\rho(ds)$, for f the density of $Y = \sum_{j=1}^\infty Y_j$, then the points of the Poisson-Kingman process in size-biased order satisfy*

$$\tilde Y_i\,|\,Y, \tilde Y_1, \ldots, \tilde Y_{i-1} \sim \tilde H\Big(\cdot\,|\sum_{j\ge i}\tilde Y_j\Big), \qquad i = 1, 2, \ldots \tag{14.23}$$

Furthermore, the joint density of $(Y, V_1, \ldots, V_i)$ of Y and the stick-breaking variables (V_j) is given by, for r the density of ρ,

$$(y, v_1, \ldots, v_i) \mapsto \prod_{j=1}^i\Big[v_j\prod_{l=1}^{j-1}(1-v_l)\,r\Big(yv_j\prod_{l=1}^{j-1}(1-v_l)\Big)\Big]\,y^i\,f\Big(y\prod_{j=1}^i(1-v_j)\Big).$$

Consequently, in the PK(ρ, η) *process the conditional density of* V_i *given* $(V_1, \ldots, V_{i-1}) = (v_1, \ldots, v_{i-1})$ *is*

$$v_i \mapsto \frac{v_i \prod_{j=1}^{i-1}(1 - v_j) \int \prod_{j=1}^{i} r(yv_j \prod_{l=1}^{j-1}(1 - v_l))\, y^i\, f(y \prod_{j=1}^{i}(1 - v_j))/f(y)\, d\eta(y)}{\int \prod_{j=1}^{i-1} r(yv_j \prod_{l=1}^{j-1}(1 - v_l))\, y^{i-1}\, f(y \prod_{j=1}^{i-1}(1 - v_j))/f(y)\, d\eta(y)}.$$

Proof As noted in the proof of Theorem 14.44, for any measurable sets B and C,

$$\mathrm{P}(\tilde{Y}_1 \in B, Y \in C) = \int_B \int_C f(y - y_1) \frac{y_1}{y}\, \rho(dy_1)\, dy.$$

Since f is the density of Y, the identity can also be obtained from Lemma J.4(i), which implies that (Y_1, Y) has density $(y_1, y) \mapsto f(y - y_1)y_1/y$ relative to the product of ρ and Lebesgue measure. The equation readily gives the conditional density of $\tilde{Y}_1$ given Y and proves (14.23) for $i = 1$.

Define N_1 as the point process obtained from the point process $N = (Y_j)$ by removing the point $\tilde{Y}_1$, and let $T_1 = Y - \tilde{Y}_1 = \sum_{j \geq 2} \tilde{Y}_j$ be its total mass. By general properties of Poisson processes (see Lemma J.4) the conditional distribution of N_1 given $(T_1 = t_1, \tilde{Y}_1)$ is identical to the conditional distribution of N given $Y = t_1$. Since the point $\tilde{Y}_2$ can be viewed as the first element of N_1 in order of appearance, we can repeat the conclusion of the preceding paragraph to obtain (14.23) for $i = 2$. By further induction we obtain the identity for every i.

Equation (14.23) shows that the vector $(\tilde{Y}_1, \ldots, \tilde{Y}_i)$ has conditional density given $Y = y$ of the form, for $\tilde{h}$ the density of $\tilde{H}$,

$$\prod_{j=1}^{i} \tilde{h}\Big(y_j |\, y - \sum_{l<j} y_l\Big) = \prod_{j=1}^{i} \Big[\frac{y_j}{y - \sum_{l<j} y_l} r(y_j) \frac{f(y - \sum_{l\leq j} y_l)}{f(y - \sum_{l<j} y_l)}\Big].$$

Given $Y = y$ the weights $(\tilde{W}_1, \ldots, \tilde{W}_i)$ are a simple scaling of $(\tilde{Y}_1, \ldots, \tilde{Y}_i)$ by y, and the stick-breaking variables satisfy $V_j = \tilde{W}_j/(1 - \sum_{l<j} \tilde{W}_j) = \tilde{Y}_j/(y - \sum_{l<j} \tilde{Y}_l)$, since $1 - \sum_{l<j} \tilde{W}_l = \prod_{l<j}(1 - V_l)$ is the remaining stick length. Thus the joint density of $(V_1, \ldots, V_i)$ given $Y = y$ can be obtained from the joint density of $(\tilde{Y}_1, \ldots, \tilde{Y}_i)$ given $Y = y$ by the change of variables $v_j = y_j/(y - \sum_{l<j} y_l)$, which has inverse $y_j = yv_j \prod_{l<j}(1 - v_l)$, since $y - \sum_{l<j} y_j = y(1 - \sum_{l<j} w_l)$. The Jacobian of the transformation is $|\partial y/\partial v| = y^i \prod_{j=1}^{i} \prod_{l<j}(1 - v_l)$ and hence the joint density of $(V_1, \ldots, V_i)$ given $Y = y$ is given by

$$\prod_{j=1}^{i} \Big[v_j r\Big(yv_j \prod_{l<j}(1 - v_l)\Big) \frac{f(y \prod_{l\leq j}(1 - v_l))}{f(y \prod_{l<j}(1 - v_l))}\Big] y^i \prod_{j=1}^{i} \prod_{l<j}(1 - v_l).$$

The product of quotients involving f telescopes out to a simple quotient, with $f(y)$ in the denominator. The latter factor cancels upon multiplying with the density of Y to obtain the joint density of $(V_1, \ldots, V_i, Y)$, as given in the theorem.

The PK(ρ, η) distribution is a mixture over $y \sim \eta$ of the PK$(\rho |\, y)$-distributions. The density of its stick-breaking weights can be obtained by mixing the corresponding densities of the PK($\rho|\, y$)-distribution. We obtain the marginal density of $(V_1, \ldots, V_i)$ by returning to the conditional density of this vector given $Y = y$ (hence divide by $f(y)$), and next integrating the variable Y out relative to its marginal distribution η. Finally the conditional density of V_i is the quotient of the joint densities of $(V_1, \ldots, V_i)$ and $(V_1, \ldots, V_{i-1})$. □

Although Theorem 14.49 gives an explicit expression for the joint density of the stick-breaking proportions (V_j), it is clear from (14.23) that a description of the Poisson-Kingman process, in size-biased order, in terms of the unnormalized weights is simpler. The conditioning variable $\sum_{i\geq j} \tilde{Y}_i = Y - \sum_{i<j} \tilde{Y}_i$ in the right side of (14.23) is the "remaining (unnormalized) mass" at stage j. The conditioning variables $Y, \tilde{Y}_1, \ldots, \tilde{Y}_{i-1}$ on the left side of the equation jointly contain the same information as the collection of remaining lengths $Y = \sum_{j\geq 1} \tilde{Y}_j, \sum_{j\geq 2} \tilde{Y}_j, \ldots, \sum_{j\geq i} \tilde{Y}_j$, and the equation says that only the last variable is relevant for the distribution of $\tilde{Y}_i$. Equation (14.23) expresses that the masses $\tilde{Y}_1, \tilde{Y}_2, \ldots$ of the Poisson-Kingman species sampling process, in size-biased order, are allocated consecutively by chopping them off the remaining mass, where the remaining mass decreases with every new allocation, but the distribution for allocating the next mass keeps the same general form. (An equivalent way of formulating the equation is that the remaining masses form a stationary Markov chain. See Problem 14.8.)

The parameter σ in a stable or generalized gamma process controls the distribution of the number K_n of distinct species when sampling n individuals from the process. A partition model is said to possess *σ-diversity S* if S is a finite strictly positive random variable such that

$$\frac{K_n}{n^\sigma} \to S, \qquad \text{a.s.} \tag{14.24}$$

The Dirichlet process is outside the range of this definition, since it has $K_n/\log n$ converging to a constant. In contrast, by the following theorem stable or generalized gamma partitions with parameter $\sigma > 0$ have diversity equal to $U^{-\sigma}$, for U the mixing variable. If this mixing variable is chosen heavier-tailed (such as is true for the default choice of the special Poisson-Kingman distribution), then the distribution of K_n will be more spread out. This is attractive for modeling clustering in some applications, where the Dirichlet process gives a sufficient number of clusters only if the precision parameter M is sufficiently large, which makes the distribution relatively concentrated and/or makes the analysis sensitive to the prior on this parameter.

Theorem 14.50 (Diversity) *The* PK(ρ_σ, η) *partition for ρ_σ the σ-stable intensity for* $0 < \sigma < 1$, *or equivalently the generalized gamma intensity, has σ-diversity equal to $U^{-\sigma}$, for $U \sim \eta$. In particular, the* PK($\rho_\sigma|\, y$) *partition has σ-diversity equal to the constant $y^{-\sigma}$.*

A proof of the theorem can be based on a result of Karlin (1967), which connects the σ-diversity for sampling from a (random) distribution on $\mathbb{N}$ with the limiting behavior of the ranked probabilities $p_{(1)} \geq p_{(2)} \geq \cdots$. It can be shown that $K_n \sim Sn^\sigma$ if $p_{(i)} \sim (S/\Gamma(1-\sigma)i)^{1/\sigma}$. For the process PK($\rho_\sigma|\, y$) this can be shown for $S = y^{-\sigma}$ (see Kingman 1975).

14.6 Normalized Inverse-Gaussian Process

The normalized inverse-Gaussian process is both a special Poisson-Kingman process, and a specific Gibbs process of type $1/2$. It is of special interest, because the marginal distributions $(P(A_1), \dots, P(A_k))$ of its species sampling process P admit explicit expressions. This is similar to the Dirichlet process, and allows to define the process in the same way as a Dirichlet process.

Appendix H gives an introduction to finite-dimensional inverse-Gaussian distributions.

Definition 14.51 (Normalized inverse-Gaussian process) A random probability measure P on a Polish space $(\mathfrak{X}, \mathscr{X})$ is said to follow a *normalized inverse-Gaussian process* with base measure α if for every finite measurable partition $A_1, \dots, A_k$ of $\mathfrak{X}$,

$$(P(A_1), \dots, P(A_k)) \sim \mathrm{NIGau}(k; \alpha(A_1), \dots, \alpha(A_k)).$$

In the definition α is a finite Borel measure on $(\mathfrak{X}, \mathscr{X})$. By Proposition H.5 its normalization $\bar{\alpha} = \alpha/|\alpha|$, where $|\alpha| = \alpha(\mathfrak{X})$ is the total mass, is the mean measure of the normalized inverse-Gaussian process; it is called the *center measure.*

To see that the definition is well posed and the random measure exists we may employ Theorem 3.1 and consistency properties of the normalized inverse-Gaussian distribution. Alternatively, in the following theorem the process is constructed as a species sampling process.

Theorem 14.52 *The Poisson-Kingman species sampling process with intensity measure $\rho(ds) = \frac{1}{2}M\pi^{-1/2}s^{-3/2}e^{-s}\,ds$ and mean measure G is the normalized inverse-Gaussian process with $\alpha = MG$.*

Proof Both the normalized inverse-Gaussian distribution and the Poisson-Kingman distribution arise by normalization. Therefore it suffices to show that a multiple of the unnormalized Poisson-Kingman process $\Phi := \sum_{i=1}^{\infty} Y_i \delta_{\theta_i}$ is an inverse-Gaussian process. We shall show that this is true for 2Φ; it suffices to verify that $2\Phi(A_i) \overset{\text{ind}}{\sim} \mathrm{IGau}(\alpha(A_i), 1)$, for every partition $A_1, \dots, A_k$ of $\mathfrak{X}$.

The marginal distribution of $\Phi(A)$ is the sum of all points in the process obtained by thinning the process $(Y_1, Y_2, \dots)$ by the indicator variables $(\mathbb{1}\{\theta_1 \in A\}, \mathbb{1}\{\theta_2 \in A\}, \dots)$. Since these indicators are independent Bernoulli variables with parameter $G(A)$, the thinned process is a Poisson process with intensity measure $G(A)\rho$. Hence by Lemma J.2 the log-Laplace transform of the sum $\Phi(A)$ of its points is given by

$$\log \mathrm{E}e^{-\lambda\Phi(A)} = -G(A)\int (1 - e^{-\lambda s})\,\rho(ds) = -\frac{\alpha(A)}{2\sqrt{\pi}}\int_0^{\infty} (1 - e^{-\lambda s}) s^{-3/2} e^{-s}\,ds.$$

The remaining integral can be evaluated by first taking its derivative with respect to λ, which is $\int_0^{\infty} e^{-(\lambda+1)s} s^{-1/2}\,ds = \Gamma(1/2)(\lambda + 1)^{-1/2}$, and next integrating this with respect to λ, resulting in $2\Gamma(1/2)((\lambda+1)^{1/2} - 1)$. Thus the preceding display with λ replaced by 2λ evaluates to $-\alpha(A)((2\lambda + 1)^{1/2} - 1)$, which is the exponent of the $\mathrm{IGau}(\alpha(A), 1)$-distribution, by Proposition H.4.

The variables $\Phi(A_1), \ldots, \Phi(A_k)$ are the sums of all points in the k Poisson processes obtained by separating (or thinning) the process (Y_j) in k processes using the independent multinomial vectors $(\mathbb{1}\{\theta_j \in A_1\}, \ldots, \mathbb{1}\{\theta_j \in A_k\})$. By general properties of Poisson processes these k processes are independent, and hence so are the sums of their points. □

The intensity measure ρ in Theorem 14.52 corresponds to the generalized gamma process with $\sigma = 1/2$ and $\tau = M^2$, considered in Example 14.48. In particular, the normalized inverse-Gaussian process is Gibbs of type 1/2. Its EPPF and PPF admit explicit, albeit a little complicated, expressions. They follow the general forms for Gibbs processes, given in (14.16) and (14.18), with the coefficients $V_{n,k}$ given in Example 14.48.

The number of distinct species in a sample of size n from a normalized inverse-Gaussian process is of the order $\sqrt{n}$, as for every generalized gamma process of type 1/2. The following proposition specializes the formula for the exact distribution of the number of species, given in Proposition 14.23 for general Gibbs processes, to the normalized inverse-Gaussian process.

Proposition 14.53 *If P follows a normalized inverse-Gaussian process with atomless base measure α, then the number K_n of distinct values in a sample $X_1, \ldots, X_n \mid P \stackrel{iid}{\sim} P$ satisfies, for $k = 1, \ldots, n$,*

$$\mathrm{P}(K_n = k) = \binom{2n-k-1}{n-1} \frac{e^{|\alpha|}(-|\alpha|)^{2n-2}}{2^{2n-k-1}\Gamma(k)} \sum_{r=0}^{n-1} \binom{n-1}{r} (-1)^r |\alpha|^{-2r} \Gamma(k+2+2r-2n; |\alpha|).$$

Being a Poisson-Kingman process, the normalized inverse-Gaussian process allows an easy characterization through residual allocation. The ensuing stick-breaking algorithm given in Theorem 14.49 can be described in terms of auxiliary variables, with standard (albeit somewhat exotic) distributions, as in the following proposition.[10]

Proposition 14.54 (Stick-breaking) *Let $\xi_1 \sim \mathrm{GIGau}(M^2, 1, -1/2)$ be independent of the sequence $\zeta_1^{-1}, \zeta_2^{-1}, \ldots \stackrel{iid}{\sim} \mathrm{Ga}(1/2, 1/2)$, and for $i = 2, 3, \ldots$,*

$$\xi_i \mid \xi_1, \ldots, \xi_{i-1}, \zeta_1, \ldots, \zeta_{i-1} \sim \mathrm{GIGau}\Big(M^2 \prod_{j=1}^{i-1}\Big(1 + \frac{\xi_j}{\zeta_j}\Big), 1, -\frac{i}{2}\Big).$$

Then $\tilde{W}_j = V_j \prod_{l=1}^{j-1}(1 - V_l)$ for $V_j = \xi_j/(\xi_j + \zeta_j)$ give the weight sequence of the Poisson-Kingman process with generalized gamma intensity measure with parameters $\sigma = 1/2$ and $\tau = M^2$. Consequently, the process $\sum_{j=1}^{\infty} \tilde{W}_j \delta_{\theta_j}$, with $\theta_j \stackrel{iid}{\sim} G$, for $j = 1, 2, \ldots$, follows a normalized inverse-Gaussian process with base measure $\alpha = MG$.

The normalized inverse-Gaussian process provides an alternative to the Dirichlet process as a prior on mixing distributions. Empirical studies show that compared with the Dirichlet process the resulting procedure is less sensitive to the choice of the total mass parameter $|\alpha|$.

[10] See Definition H.6 for the "generalized inverse-Gaussian distribution." The positive 1/2-stable distribution is the distribution of of $1/Z$ for $Z \sim \mathrm{Ga}(1/2, b/2)$. It has density $z \mapsto (2\pi)^{-1/2} b^{1/2} z^{-3/2} e^{-b/(2z)}$ on $(0, \infty)$. For details of the derivation, see Favaro et al. (2012b).

As for the Dirichlet process, the prior and posterior distributions of the mean of a normalized inverse-Gaussian process can be obtained explicitly; see Problems 14.13 and 14.14.

14.7 Normalized Completely Random Measures

A completely random measure (CRM) on a Polish space $\mathfrak{X}$ is a random element Φ in the space $(\mathfrak{M}_\infty, \mathscr{M}_\infty)$ of Borel measures on $\mathfrak{X}$ such that the variables $\Phi(A_1), \ldots, \Phi(A_k)$ are independent, for every measurable partition $A_1, \ldots, A_k$ of $\mathfrak{X}$. Appendix J gives a review of CRMs, and shows that, apart from a deterministic component, a CRM can be represented as

$$\Phi(A) = \int_A \int_0^\infty s\, N^c(dx, ds) + \sum_j \Phi_j \delta_{a_j}(A),$$

for a Poisson process N^c on $\mathfrak{X} \times (0, \infty]$, arbitrary independent, nonnegative random variables $\Phi_1, \Phi_2, \ldots$, and given elements $a_1, a_2, \ldots$ of $\mathfrak{X}$, called "fixed atoms" or "fixed jumps." For N^d the point process consisting of the points (a_j, Φ_j) and $N = N^c + N^d$, the equation can be written succinctly as $\Phi(A) = \int_A \int_0^\infty s\, N(dx, ds)$, but N is not a Poisson process, but called an "extended Poisson process." The distribution of N, and hence of Φ, is determined by its intensity measure ν on $\mathfrak{X} \times [0, \infty)$, which splits as $\nu = \nu^c + \nu^d$, where ν^c is the mean measure of N^c and gives mass 0 to every strip $\{x\} \times [0, \infty]$, and ν^d is concentrated on $\cup_j \{a_j\} \times [0, \infty]$, with $\nu^d(\{a_j\} \times B) = \mathrm{P}(\Phi_j \in B)$, for every j. The representation shows in particular that the realizations of a CRM are discrete measures.

If $0 < \Phi(\mathfrak{X}) < \infty$ a.s., then the CRM can be renormalized to the random probability measure $\Phi/\Phi(\mathfrak{X})$, which is also discrete.

Definition 14.55 (Normalized completely random measure) The random probability measure $\Phi/\Phi(\mathfrak{X})$, for a given completely random measure Φ with $0 < \Phi(\mathfrak{X}) < \infty$ a.s., is called a *normalized completely random measure* (NCRM), and for $\mathfrak{X}$ an interval in $\mathbb{R}$, also a *normalized random measure with independent increments* (NRMI).

The "independent increments" in the second name refers to the situation that a CRM can be identified with a process with independent increments through its cumulative distribution function.

Since there is a one-to-one correspondence between CRMs and their intensity measures, every intensity measure defines an NCRM. However, the normalization to a probability measure cancels the scale of the jump sizes so that the measure $\nu(dx, c\, ds)$ gives the same NCRM as $\nu(dx, ds)$, for every $c > 0$.

For a prior specification we shall usually choose an intensity measure without discrete component (i.e. $\nu^d = 0$), but "fixed jumps" will appear in the posterior distribution at the observations, as shown in the theorem below.

A completely random measure without fixed atoms is said to be *homogeneous* if its intensity measure is equal to a product measure $\nu = G \times \rho$, for an atomless probability measure G on $\mathfrak{X}$. Such a measure can be generated by independently laying down point masses $(Y_1, Y_2, \ldots)$ according to a Poisson point process on $(0, \infty)$ with intensity measure ρ and locations $\theta_1, \theta_2, \ldots \stackrel{\text{iid}}{\sim} G$ in $\mathfrak{X}$, and next forming $\Phi = \sum_i Y_i \delta_{\theta_i}$ (the corresponding Poisson process sits at the points (θ_i, Y_i); see Example J.8). The normalized completely random measure is then $\sum_i (Y_i/Y)\delta_{\theta_i}$, for $Y = \sum_i Y_i$, which is exactly the species sampling model corresponding to the PK(ρ)-distribution (see Definition 14.40). Therefore, homogeneous NCRM do not add beyond the special Poisson-Kingman distribution, and are proper species sampling models. On the other hand, non-homogeneous normalized completely random measures are not species sampling models.

Although for constructing a prior distribution we shall usually choose a homogeneous intensity measure, the perspective of NCRMs leads to a different representation of the posterior distribution. Even though the NCRM does not have a structural conjugacy property, except in the special case of the Dirichlet process.[11] the posterior can be described as a *mixture* of NCRMs. The representation in the following theorem allows simulating the posterior distribution by the general *Ferguson-Klass algorithm* for simulating a completely random measure.[12]

As before let $\tilde{X}_1, \tilde{X}_2, \ldots$ be the distinct values in $X_1, X_2, \ldots$, and let $(N_{1,n}, \ldots, N_{K_n,n})$ give the number of times they appear in $D_n := (X_1, \ldots, X_n)$.

Theorem 14.56 (Posterior distribution) *If P follows a completely random measure with intensity measure $\rho(ds|\, x)\,\alpha(dx)$ for $x \mapsto \rho(ds|\, x)$ weakly continuous and α atomless, then the posterior distribution of P in the model $X_1, \ldots, X_n|\, P \stackrel{iid}{\sim} P$ is a mixture over λ of the distributions of the NCRM $\Phi_\lambda/\Phi_\lambda(\mathfrak{X})$ with intensity measures*

$$\nu^c_{\Phi|\lambda, D_n}(dx, ds|\, \lambda) = e^{-\lambda s}\, \rho(ds|\, x)\, \alpha(dx),$$
$$\nu^d_{\Phi|\lambda, D_n}(\{\tilde{X}_j\}, ds|\, \lambda) \propto s^{N_{j,n}} e^{-\lambda s}\, \rho(ds|\, \tilde{X}_j),$$

where $\nu^d_{\Phi|\lambda, D_n}(\{x\}, \cdot|\, \lambda)$ are probability measures on $(0, \infty)$, with mixing density proportional to, with $\psi(\lambda) = \int\int (1 - e^{-\lambda s})\, \rho(ds|\, x)\, \alpha(dx)$,

$$\lambda \mapsto \lambda^{n-1} e^{-\psi(\lambda)} \prod_{j=1}^{K_n} \int_0^\infty s^{N_{j,n}} e^{-\lambda s} \rho(ds|\, \tilde{X}_j).$$

Proof For a given partition $A_1, \ldots, A_k$ of the sample space the vector $M = (M_1, \ldots, M_k)$ of counts $M_j = \#\{i\colon X_i \in A_j\}$ possesses a multinomial distribution with cell probabilities $\Phi(A_j)/\Phi(\mathfrak{X})$ and hence has likelihood $L_M(\Phi) \propto \Phi(\mathfrak{X})^{-n} \prod_{j=1}^k \Phi(A_j)^{M_j}$. By Bayes's rule, for any bounded function f,

$$\mathrm{E}(e^{-\int f\, d\Phi}|\, M) = \frac{\mathrm{E}_\Phi(e^{-\int f\, d\Phi} L_M(\Phi))}{\mathrm{E}_\Phi(L_M(\Phi))}.$$

Here the expectations on the right side are taken relative to the prior of Φ. We shall explicitly evaluate this expression, and then take the limit along partitions that become finer and finer and whose union generate the Borel σ-field of $\mathfrak{X}$. The martingale convergence theorem then

[11] The Dirichlet process prior is the only NCRM whose posterior distribution is again an NCRM; see Theorem 1 of James et al. (2006).

[12] See Barrios et al. (2013) and Favaro and Teh (2013) for descriptions.

implies that the left side tends to $\mathrm{E}(e^{-\int f\,d\Phi}|\, X_1,\ldots,X_n)$, and the posterior distribution of Φ can be identified from the limit of the right side.

We start by rewriting the term $\Phi(\mathfrak{X})^{-n}$ in the likelihood as $\int \lambda^{n-1}\prod_j e^{-\lambda\Phi(A_j)}\,d\lambda/\Gamma(n)$. Using Fubini's theorem and the independence of Φ on the disjoint sets A_j, we see that on the event $M=m$,

$$\mathrm{E}_\Phi(e^{-\int f\,d\Phi}L_M(\Phi)) = \int_0^\infty \frac{\lambda^{n-1}}{\Gamma(n)}\prod_j \mathrm{E}\Big[e^{-\int_{A_j}(f+\lambda)\,d\Phi}\,\Phi(A_j)^{m_j}\Big]\,d\lambda.$$

The expected values in the product take the general form

$$\frac{d^m}{d\lambda^m}\mathrm{E}\Big[e^{-\int_A(f+\lambda)\,d\Phi}\Big](-1)^m = \frac{d^m}{d\lambda^m}e^{-\int_A\int(1-e^{-s(f(x)+\lambda)})\,\nu(dx,ds)}(-1)^m,$$

by the expression for the Laplace transform of a CRM, given in Proposition J.6. The derivative on the right side can be computed, and be written in the form

$$e^{-\int_A\int(1-e^{-s(f(x)+\lambda)})\,\nu(dx,ds)}\sum_{(i_1,\ldots,i_r)} a_{i_1,\ldots,i_r}\prod_{l=1}^r\int_A\int e^{-s(f(x)+\lambda)}s^{i_l}\,\nu(dx,ds),$$

where the sum is over all partitions $(i_1,\ldots,i_r)$ of m and is to be read as 1 if $m=0$. As $\alpha(A)\to 0$, all the integrals $\int_A\int e^{-s(f(x)+\lambda)}s^i\,\nu(dx,ds)$ tend to zero, and hence the sum is dominated by the term for which the product contains only one term ($r=1$), which corresponds to the partition of m in a single set, and has coefficient $a_m=(-1)^m$.

Only finitely many of the partitioning sets contain data points and have $m_j>0$. If the partition is fine enough so that the distinct values $\tilde X_i$ are in different sets, then the nonzero m_i are equal to the multiplicities $N_{i,n}$, and if f is uniformly continuous, then $f(x)\to f(\tilde X_i)$ along the sequence of partitions uniformly in x in the partitioning set $\tilde A_i$ that contains $\tilde X_i$. We conclude that, as the partitions become finer, the numerator $\mathrm{E}_\Phi(e^{-\int f\,d\Phi}L_M(\Phi))$ is asymptotic to

$$\int_0^\infty \frac{\lambda^{n-1}}{\Gamma(n)}e^{-\int\int(1-e^{-s(f(x)+\lambda)})\,\nu(dx,ds)}\prod_i\Big[\int e^{-s(f(\tilde X_i)+\lambda)}s^{N_{i,n}}\,\rho(ds|\,\tilde X_i)\alpha(\tilde A_i)\Big]\,d\lambda.$$

When divided by $\mathrm{E}_\Phi L_M(\Phi)$, the factors $\alpha(\tilde A_i)$ cancel and both numerator and denominator tend to a limit. It remains to write this in the form corresponding to the claim of the theorem.

In view of (J.10), the product over i corresponds to fixed jumps, where the norming factors $\int e^{-s\lambda}s^{N_{i,n}}\,\rho(ds|\,\tilde X_i)$ of the integrals can be compensated in the integral over λ. The leading exponential can be written as $e^{-\psi(\lambda)}\exp\big[-\int\int(1-e^{-sf(x)})\,e^{-\lambda s}\,\nu(dx,ds)\big]$, and contributes a factor $e^{-\psi(\lambda)}$ to the integral over λ and the continuous part of the CRM. The denominator provides the norming factor to the integral over λ. ☐

Example 14.57 (Normalized generalized gamma CRM) If P follows the normalized process of the generalized gamma CRM described in Example 14.48, the continuous part $\nu^c(\cdot|\,\lambda)$ of the CRM Φ_λ in the theorem has intensity measure

$$\tau^\sigma(\sigma/\Gamma(1-\sigma))\,s^{-(1+\sigma)}\,e^{-(1+\lambda)s}\alpha(dx)\,ds,$$

again of generalized gamma form; and the distributions in the discrete part are $\mathrm{Ga}(N_{i,n} - \sigma, \lambda + 1)$. Furthermore, the mixing density is proportional to $\lambda \mapsto \lambda^{n-1}(1 + \lambda)^{K_n\sigma - n} e^{-|\alpha|(1+\lambda)^\sigma}$.

The normalized generalized gamma CRM is the only normalized random measure of Gibbs type (see Lijoi et al. 2008, Proposition 2).

Example 14.58 (Normalized extended gamma CRM) An example of a non-homogeneous CRM is given by the extended gamma process defined in Example J.15. This has intensity measure given by $s^{-1}e^{-s/b(x)}\alpha(dx)\,ds$, for a positive measurable function b and a σ-finite measure α with $\int \log(1 + b)\,d\alpha < \infty$. The corresponding normalized measure is called an *extended gamma CRM*.

Corollary 14.59 (Predictive distribution) *If $X_1, X_2, \ldots \mid P$ are a random sample from a normalized random measure P with intensity $\nu(dx, ds) = \rho(ds\mid x)\,\alpha(dx)$ satisfying the conditions of Theorem 14.56, then*

$$X_{n+1}\mid X_1, \ldots, X_n \sim \sum_{j=1}^{K_n} p_j(X_1, \ldots, X_n)\delta_{\tilde{X}_j} + p_{K_n+1}(X_1, \ldots, X_n\mid x)\,\bar{\alpha},$$

for, with $\tau_n(\lambda\mid x) = \int s^n e^{-\lambda s}\,\rho(ds\mid x)$,

$$p_j(X_1, \ldots, X_n) = \frac{\int_0^\infty \lambda^n e^{-\psi(\lambda)}\tau_{N_{j,n}+1}(\lambda\mid \tilde{X}_i) \prod_{i=1, i\neq j}^{K_n} \tau_{N_{i,n}}(\lambda\mid \tilde{X}_i)\,d\lambda}{n\int_0^\infty \lambda^{n-1}e^{-\psi(\lambda)} \prod_{i=1}^{K_n} \tau_{N_{i,n}}(\lambda\mid \tilde{X}_i)\,d\lambda},$$

$$p_{K_n+1}(X_1, \ldots, X_n\mid x) = \frac{\int_0^\infty \lambda^n e^{-\psi(\lambda)}\tau_1(\lambda\mid x) \prod_{i=1}^{K_n} \tau_{N_{i,n}}(\lambda\mid \tilde{X}_i)\,d\lambda}{n\int_0^\infty \lambda^{n-1}e^{-\psi(\lambda)} \prod_{i=1}^{K_n} \tau_{N_{i,n}}(\lambda\mid \tilde{X}_i)\,d\lambda}.$$

Proof The predictive distribution is the mean $\mathrm{E}\big[P\mid X_1, \ldots, X_n\big]$ of the posterior distribution of P. By Theorem 14.56 the posterior distribution of P is a mixture of normalized completely random measures, and hence the posterior mean is a mixture of means of normalized completely random measures. For the latter means, we apply the general formula given in Lemma 14.60, with the intensity measure equal to the one given in Theorem 14.56. The Laplace exponent of the latter measure is $\psi_{\Phi|\lambda, D_n}(\theta) = \int\int (1-e^{-\theta s})e^{-\lambda s}\rho(ds\mid x)\,\alpha(dx) = \psi(\theta + \lambda) - \psi(\lambda)$, for ψ the Laplace exponent of the prior intensity measure, while for $k = 0, 1$,

$$\int s^k e^{-\theta s}\,d\nu^d_{\Phi|\lambda, D_n}(\{\tilde{X}_i\}, ds) = \frac{\tau_{N_{i,n}+k}(\theta + \lambda\mid \tilde{X}_i)}{\tau_{N_{i,n}}(\lambda\mid \tilde{X}_i)}.$$

We substitute these expressions in the formula of Lemma 14.60, taking the integration variable there equal to θ, multiply with the density of λ from Theorem 14.56, which is proportional to $\lambda^{n-1}e^{-\psi(\lambda)}\prod_i \tau_{N_{i,n}}(\lambda\mid X_i)$, and integrate over λ. In the resulting double integral with respect to $(\lambda, \theta) > 0$, we make the change of variables $\theta + \lambda = u$, next integrate over $\lambda \in (0, u)$, giving the factor u^n/n. The sum over the fixed jump points then gives the formulas for p_j and $j \leq K_n$, while the continuous part gives the formula for p_{K_n+1}, after an application of Fubini's theorem. ∎

Lemma 14.60 (Mean) *The mean measure of the normalized completely random measure $P = \Phi/\Phi(\mathfrak{X})$ with intensity measure ν and strictly positive fixed jumps $\Phi(\{a_j\})$ at the points (a_j) is given by, for $\psi(\lambda) = \int\int(1-e^{\lambda s})\,\nu^c(dx, ds)$,*

$$\mathrm{E}\Big[\frac{\Phi(A)}{\Phi(\mathfrak{X})}\Big] = \int e^{-\psi(\lambda)} \int_A\int s e^{-\lambda s}\,\nu^c(dx, ds) \prod_j \int e^{-\lambda s}\,\nu^d(\{a_j\}, ds)\, d\lambda$$
$$+ \sum_{j:a_j\in A} \int e^{-\psi(\lambda)} \int s e^{-\lambda s}\,\nu^d(\{a_j\}, ds) \prod_{i\neq j} \int e^{-\lambda s}\,\nu^d(\{a_i\}, ds)\, d\lambda.$$

Proof Applying the identity $y^{-1} = \int e^{-\lambda y}\,d\lambda$ to $y = \Phi(\mathfrak{X})$ and Fubini's theorem, we can write the left side in the form $\int \mathrm{E}\big[\Phi(A)e^{-\lambda\Phi(\mathfrak{X})}\big]\,d\lambda$. Next we decompose $\Phi(\mathfrak{X}) = \Phi(A) + \Phi(A^c)$ and use the independence of the two variables on the right to further rewrite the left side of the lemma as $\int \mathrm{E}\big[\Phi(A)e^{-\lambda\Phi(A)}\big]\mathrm{E}\big[e^{-\lambda\Phi(A^c)}\big]\,d\lambda$. By combining (J.7) and (J.10), we have

$$\mathrm{E}\big[e^{-\lambda\Phi(A^c)}\big] = e^{-\int_{A^c}\int(1-e^{-\lambda s})\,\nu^c(dx,ds)} \prod_{j:a_j\in A^c} \int e^{-\lambda s}\,\nu^d(\{a_j\}, ds).$$

Applying equations (J.7) and (J.10) a second time, we also find

$$\mathrm{E}\big[\Phi(A)e^{-\lambda\Phi(A)}\big] = -\frac{d}{d\lambda}\Big[e^{-\int_A\int(1-e^{-\lambda s})\,\nu^c(dx,ds)} \prod_{j:a_j\in A} \int e^{-\lambda s}\,\nu^d(\{a_j\}, ds)\Big].$$

We substitute these expressions in the integral, first for a set A without fixed jumps, and next for $A = \{a_j\}$, and simplify the resulting expression to the one given in the lemma. □

The structure of the predictive distribution is similar to that obtained in species sampling models, in Lemma 14.11. A difference is that the probabilities of redrawing a preceding observation or generating a new value depend on the *values* of the preceding observations, next to the partition they generate. This is due to the fact that for a general normalized completely random measure (when the intensity measure is not homogeneous) the value of an observation and the probability that it is drawn are confounded.

Even though a NCRM is in general not a species sampling model, a random sample $X_1, X_2, \ldots$ from an NCRM is still exchangeable and hence generates an infinite exchangeable partition. The following theorem shows that its EPPF has exactly the same form as that of a special Poisson-Kingman process, given in Theorem 14.44.

Theorem 14.61 (EPPF) *The EPPF of the exchangeable partition generated by a random sample from an NCRM with intensity measure ν without fixed atoms and Laplace exponent $\psi(\lambda) = \int\int(1-e^{-\lambda s})\,\nu(dx, ds)$ is given by, for $n = \sum_{j=1}^k n_j$,*

$$p(n_1, \ldots, n_k) = \frac{(-1)^{n-k}}{\Gamma(n)} \int \lambda^{n-1} e^{-\psi(\lambda)} \prod_{j=1}^{k} \psi^{(n_j)}(\lambda)\, d\lambda.$$

Proof Because ν does not have fixed atoms, the corresponding (extended) Poisson process has at most one point on every half line $\{x\} \times (0, \infty)$. Sampling an observation X from a NCRM $\Phi/\Phi(\mathfrak{X})$ given Φ is therefore equivalent to choosing one of the points (x_j, s_j) from the realization of Φ. The pattern of ties in a random sample $X_1, X_2, \ldots$ arises by the observations X_i choosing the same or different points. The probability of choosing a point (x_j, s_j) is equal to $s_j / \sum_k s_k$, for $\sum_k s_k$ equal to the realization of $\Phi(\mathfrak{X})$; it does not depend on the value of x_j. It follows that the mechanism of forming ties depends on the weights s_j only, which form a Poisson process on $(0, \infty)$ with intensity measure the marginal measure of ν given by $\rho_2(D) = \nu(\mathfrak{X} \times D)$. The Laplace exponent of ρ_2 is equal to $\int (1 - e^{-\lambda s})\, \rho_2(ds) = \int\int (1 - e^{-\lambda s})\, \nu(dx, ds) = \psi(\lambda)$. The theorem is a corollary of Theorem 14.44. ☐

If we write $\kappa_n(\lambda) = \int\int s^n e^{-\lambda s}\, \nu(dx, ds)$, then the theorem shows that given λ the EPPF is essentially of the product form $\prod_{j=1}^{k} \kappa_{n_j}(\lambda)$, and hence the exchangeable partition generated by a normalized completely random measure is conditionally Gibbs, by Lemma 14.21. By reweighting we can write, for any positive probability density g,

$$p(n_1, \ldots, n_k) = \int p(n_1, \ldots, n_k |\, \lambda)\, g(\lambda)\, d\lambda, \qquad p(n_1, \ldots, n_k |\, \lambda) = \frac{\lambda^{n-1} e^{-\psi(\lambda)}}{\Gamma(n) g(\lambda)} \prod_{j=1}^{k} \kappa_{n_j}(\lambda).$$

The Gibbs form is convenient, as general results on Gibbs processes become available, conditional on λ.

In a species sampling model the distinct values $\tilde{X}_1, \tilde{X}_2, \ldots$ can be viewed as generated independently from the center measure G and attached to the partitioning sets as the common values of the observations (see Theorem 14.13). For general normalized random measures the partitioning and the distinct values are dependent. Their joint distribution is given in the following lemma, a version of which was used in the proof of Theorem 14.44.

Lemma 14.62 *The distinct values $\tilde{X}_1, \tilde{X}_2, \ldots$ and partitions $\mathcal{P}_n$ generated by a random sample from a normalized completely random measure with intensity measure $\nu(dx, ds) = \rho(ds|\, x)\, \alpha(dx)$ without fixed atoms and Laplace exponent $\psi(\lambda) = \int\int (1 - e^{-\lambda s})\, \nu(dx, ds)$ satisfy, for every measurable sets $B_1, \ldots, B_k$ and partition $A_1, \ldots, A_k$ of $\{1, 2, \ldots, n\}$, with $\tau_n(\lambda |\, x) = \int s^n e^{-\lambda s}\, \rho(ds |\, x)$,*

$$\mathrm{P}(\tilde{X}_1 \in B_1, \ldots, \tilde{X}_k \in B_k, \mathcal{P}_n = \{A_1, \ldots, A_k\})$$
$$= \int_{B_1} \cdots \int_{B_k} \int_0^\infty \frac{\lambda^{n-1}}{\Gamma(n)} e^{-\psi(\lambda)} \prod_{j=1}^{k} \tau_{n_j}(\lambda |\, x_j)\, d\lambda\, \alpha(dx_1) \cdots \alpha(dx_k).$$

Proof For notational convenience we give the proof in the case that $k = 2$ only. The partition structure $\mathcal{P}_n = \{A_1, A_2\}$ expresses that the observations $X_1, \ldots, X_n$ are obtained in a particular order in two groups, where two different points (x_1, s_1) and (x_2, s_2) are chosen from the point process N on $\mathfrak{X} \times (0, \infty)$ that generates the NCRM, by two particular observations (say with indices $\min A_1$ and $\min A_2$), setting these equal to x_1 and x_2, and all other observations choose either (x_1, s_1) or (x_2, s_2), depending on whether their index is in A_1 or in A_2. Given N the observations are independent and choose a point (x, s) with probability s/S, for $S = \int\int s\, N(dx, ds)$ the "total weight" in N. It follows that, for $(n_1, n_2) = (|A_1|, |A_2|)$,

$$\mathrm{P}(\tilde{X}_1 \in B_1, \tilde{X}_2 \in B_2, \mathcal{P}_n = \{A_1, A_2\} \mid N)$$
$$= \int_{B_1}\int \Big(\frac{s_1}{S}\Big)^{n_1-1} \int_{B_2}\int \Big(\frac{s_2}{S}\Big)^{n_2-1} \frac{s_2}{S}\,(N - \{(x_1, s_1)\})(dx_2, ds_2)\,\frac{s_1}{S}\,N(dx_1, ds_1).$$

Here $N - \{(x_1, s_1)\}$ is the point process N with the point (x_1, s_1) removed; this arises because the point (x_2, s_2) must be different from the point (x_1, s_1). The two factors $(s_i/S)^{n_i-1}$ are the (conditional) probabilities that $n_i - 1$ points choose the point (x_i, s_i), and $(s_1/S)\,N(dx_1, ds_1)$ and $(s_2/S)\,(N - \{(x_1, s_1)\})(dx_2, ds_2)$ are the law of the first distinct point, and the conditional law of the second distinct point given that it is different from the first, respectively.

The probability in the lemma is the expectation over the preceding display relative to N. We next proceed by exchanging the order of expectation and integration twice, with the help of formula (J.1), which is based on Palm theory for the Poisson process N. Application of the formula on $N(dx_1, ds_1)$, with the other, inner integral treated as a fixed function of N, gives that the expectation of the preceding display is equal to

$$\int_{B_1}\int \mathrm{E}\Big(\frac{s_1}{S+s_1}\Big)^{n_1} \int_{B_2}\int \Big(\frac{s_2}{S+s_1}\Big)^{n_2} N(dx_2, ds_2)\,\nu(dx_1, ds_1).$$

To see this, note that (J.1) demands to replace N by the process $N \cup \{(x_1, s_1)\}$, thus changing $S = \int\int s\, N(dx, ds)$ into $S + s_1$, and $N - \{(x_1, s_1)\}$ into N. A second application of formula (J.1), now on $N(dx_2, ds_2)$, allows to rewrite this further as

$$\int_{B_1}\int \int_{B_2}\int \mathrm{E}\Big(\frac{s_1}{S+s_1+s_2}\Big)^{n_1}\Big(\frac{s_2}{S+s_1+s_2}\Big)^{n_2} \nu(dx_2, ds_2)\,\nu(dx_1, ds_1).$$

The remaining expectation refers only to the variable S, which is the total weight of the Poisson process with intensity measure $\nu(\mathfrak{X}, ds)$ on $(0, \infty)$ and has density f with Laplace transform $e^{-\psi(\lambda)}$. Thus the preceding display is equal to

$$\int_{B_1}\int \int_{B_2}\int \int \Big(\frac{s_1}{t+s_1+s_2}\Big)^{n_1}\Big(\frac{s_2}{t+s_1+s_2}\Big)^{n_2} f(t)\,dt\,\nu(dx_2, ds_2)\,\nu(dx_1, ds_1).$$

To arrive at the expression given by the lemma we express the factor $(t + s_1 + s_2)^{-n}$, as $\Gamma(n)^{-1}\int_0^\infty \lambda^{n-1} e^{-\lambda(t+s_1+s_2)}\,d\lambda$, rearrange by several applications of Fubini's theorem, and use that $\int e^{-\lambda t} f(t)\,dt = e^{-\psi(\lambda)}$. ☐

14.8 Relations between Classes of Discrete Random Probability Measures

Figure 14.5 pictures the relations between the various classes of discrete random probability measures developed in the preceding sections. Except for the non-homogeneous normalized

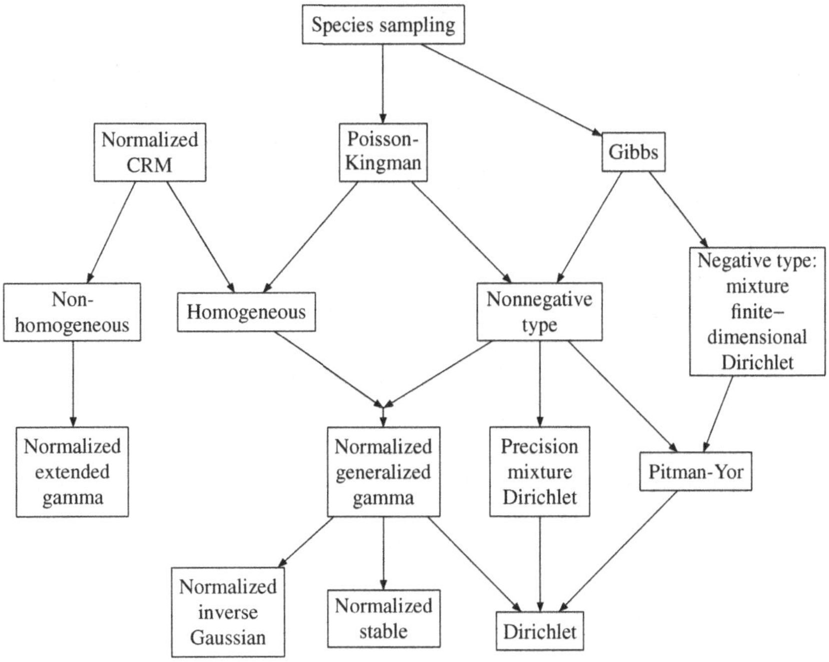

Figure 14.5 Relations between classes of random probability measures. An arrow indicates that a target is a special (or limiting) case of the origin of the arrow.

completely random measures, all classes are species sampling models, and characterized by the distribution of an infinite probability vector. In a Poisson-Kingman process this distribution is induced by normalizing the points of a Poisson process in $(0, \infty)$ by its sum. This distribution may be conditioned on the sum, or mixed over the sum, using either the intrinsic distribution of the sum or an arbitrary distribution. Mixtures with the intrinsic distribution give exactly the same class of random distributions as the homogeneous normalized completely random measures. Mixtures with an arbitrary distribution and the Poisson process equal to a σ-stable process or generalized gamma process give all Gibbs processes of type $0 < \sigma < 1$. Gibbs processes are precisely those species sampling processes for which the exchangeable partition probability function factorizes (and the probability of discovering a new species depends only on the sample size and the number of distinct elements). A Gibbs process of type $\sigma = 0$ is a mixture of Dirichlet DP(MG)-distributions over M, while negative type Gibbs processes are mixtures of finite-dimensional Dirichlet Dir(m; $-\sigma, \ldots, -\sigma$)-distributions over m. Within the Gibbs processes the Pitman-Yor processes are prominent, and contain the Dirichlet process as a special case. They can be obtained as Poisson-Kingman processes with σ-stable intensity and mixing distribution a polynomially tilted positive σ-stable distribution. All Pitman-Yor processes admit a stick-breaking representation with independent beta variables. Normalized completely random measures with generalized gamma intensity form the intersection of Gibbs processes and normalized completely random measures. This class parameterized by (σ, τ) contains the Dirichlet process in the limit $\sigma \to 0$, the normalized inverse-Gaussian process for $\sigma = 1/2$, and the normalized σ-stable process in the limit $\tau \to 0$.

14.9 Dependent Random Discrete Distributions

Many applications, especially in spatial statistics, involve a collection of distributions P_z on a sample space $\mathfrak{X}$, indexed by a parameter z belonging to some "covariate" space $\mathfrak{Z}$. A typical example is that P_z is the conditional distribution of a variable X given a covariate $Z = z$; *density regression* then focuses on estimating the conditional density of X given Z. A useful prior distribution on the P_z should not treat these measures separately, but view them as related quantities. As an extreme case all P_z could be taken equal, and a prior distribution be imposed on the common measure, but such total dependence is usually too restrictive. More typical would be to incorporate some form of continuity of the map $z \mapsto P_z$ into the prior specification.

One possibility is to model the variables x and z simultaneously, for instance through a stochastic process $(\xi(x, z): (x, z) \in \mathfrak{X} \times \mathfrak{Z})$ or a set of basis functions on the product space $\mathfrak{X} \times \mathfrak{Z}$. For instance, a prior distribution on the conditional density of X given Z can be constructed from a Gaussian process ξ and a link function Ψ through the relation

$$p(x|\, z) = \frac{\Phi(\xi(x, z))}{\int \Phi(\xi(y, z))\, dy}.$$

Alternatively, a random series in (x, z) may replace the Gaussian process. Continuity in z is then expressed in the stochastic process or basis functions, where it may be sensible to treat the arguments x and z asymmetrically.

In this section we focus on priors of the form

$$P_z = \sum_{j=1}^{\infty} W_j(z)\delta_{\theta_j(z)}, \tag{14.25}$$

for $\mathfrak{Z}$-indexed stochastic processes of "weights" $(W_j(z): z \in \mathfrak{Z})$ and "locations" $(\theta_j(z): z \in \mathfrak{Z})$. The processes θ_j may be i.i.d. copies of some $\mathfrak{X}$-valued process; a Gaussian process may be convenient; an infinite-dimensional exponential family, or a random series are alternatives. The weights W_j could be constructed using stick breaking:

$$W_j(z) = V_j(z) \prod_{l=1}^{j-1} (1 - V_l(z)). \tag{14.26}$$

The processes $(V_j(z): z \in \mathfrak{Z})$ can be taken independent (or even i.i.d.), and should then satisfy $\sum_{j=1}^{\infty} \log \mathrm{E}\big[1 - V_j(z)\big] = -\infty$ (cf. Lemma 3.4) to ensure that the P_z are proper probability distributions (and in addition a regularity condition in z such as separability to ensure that all P_z are simultaneously proper almost surely).

A process of random measures $(P_z: z \in \mathfrak{Z})$ defined by the combination of (14.25) and (14.26) is called a *dependent stick-breaking process* (DSBP).

In a dependent stick-breaking process closeness of $V_j(z)$ and $V_j(z')$, and $\theta_j(z)$ and $\theta_j(z')$, for $j \in \mathbb{N}$ and $z, z' \in \mathfrak{Z}$, implies closeness of P_z and $P_{z'}$. One of the processes V_j or θ_j may be chosen constant in z without losing too much flexibiity. The extent of sharing of atoms across different locations might be controlled by an auxiliary variable.

An advantage of stick-breaking weights is that the series can be truncated to a finite number of terms, thus effectively reducing computations to finitely many variables. For posterior computation, the Metropolis-Hastings algorithm is applicable, but more specialized algorithms, such as a block-Gibbs sampler (cf. Ishwaran and James 2001), often work better.

Measurability issues may be resolved by the result that a map of two arguments that is continuous in the first argument and measurable in the second, is measurable provided the spaces involved are nice, such as Polish. In the present case P_z is a measurable random element in $\mathfrak{M}$, for every z, and $z \mapsto P_z$ is weakly continuous if the processes V_j and θ_j have weakly continuous sample paths, since $x \mapsto \delta_x$ is weakly continuous.

In view of the success of the Dirichlet process as a prior distribution, it is reasonable to make each marginal prior P_z a Dirichlet process. By the stick-breaking representation of a Dirichlet process, the specifications $\theta_j(z) \stackrel{\text{iid}}{\sim} G_z$ and $V_j(z) \stackrel{\text{iid}}{\sim} \text{Be}(1, M_z)$, for $j = 1, 2, \ldots$ and a probability measure G_z on $\mathfrak{X}$ and some $M_z > 0$, ensure that $P_z \sim \text{DP}(M_z G_z)$, for every fixed $z \in \mathfrak{Z}$. A process satisfying the latter requirement is referred to as a *dependent Dirichlet process* (DDP). We discuss some specific constructions later in this section.

Instead of Dirac measures $\delta_{\theta_j}(z)$ in (14.25) one might use general random probability distributions. Choosing these independent of z leads to priors of the form

$$P_z = \sum_{j=1}^{\infty} W_j(z) G_j. \tag{14.27}$$

The measures G_j might be i.i.d. from a prior on the set of probability distributions, such as a Dirichlet process.

The density regression problem requires a prior on functions rather than on measures, and an extra smoothing step is necessary to employ a DSBP or a DDP. This can be written in a hierarchical structure with latent variables as

$$Y_i \mid X_i, Z_i \stackrel{\text{iid}}{\sim} \psi(\cdot, X_i, \varphi), \qquad X_i \mid Z_i \stackrel{\text{ind}}{\sim} P_{Z_i}, \qquad i = 1, 2, \ldots,$$

where ψ is a kernel with a (possibly additional) parameter φ, and $\{P_z, z \in \mathfrak{Z}\}$ may follow a DDP or DSBP.

14.9.1 Kernel Stick-Breaking Process

The *kernel stick-breaking process* (KSBP) is the combination of (14.27) with $G_j \stackrel{\text{iid}}{\sim} \text{DP}(MG)$ and stick-breaking weights (14.26) with relative stick lengths of the form $V_j(z) = U_j K(z, \Gamma_j)$, for random sequences (U_j) and (Γ_j), and a given "kernel function" $K: \mathfrak{Z} \times \mathfrak{Z} \to [0, 1]$. In the special case $M = 0$ we understand $\text{DP}(MG)$ to generate pointmasses $G_j = \delta_{\theta_j}$, for $\theta_j \stackrel{\text{iid}}{\sim} G$, and then (14.27) reduces to (14.25), with θ_j free of z.

It is sensible to choose the kernel K to have large values near the diagonal, so that $K(z, \gamma)$ and $K(z', \gamma)$ are similar if z and z' are close. For instance, K could be a decreasing function of the distance between its arguments. Typical choices are $U_j \stackrel{\text{ind}}{\sim} \text{Be}(a_j, b_j)$ and, independently, $\Gamma_j \stackrel{\text{iid}}{\sim} H$ for some distribution H, for $j = 1, 2, \ldots$. If $\text{E}\big[K(z, \Gamma)\big] > 0$ and $a_j/(a_j + b_j)$ remains bounded away from 0, then $\sum_{j=1}^{\infty} \log \text{E}[1 - V_j(z)] = -\infty$, and the construction marginally defines proper random probability distributions.

Since the G_j have expectation G and are independent of the weights, it follows that $\mathrm{E}(P_z|\,U_1, \Gamma_1, U_2, \Gamma_2, \ldots) = G$, for all $z \in \mathfrak{Z}$, as in Subsection 3.4.2. Similarly the (conditional) covariance between the measures at two values z, z' can be computed as

$$\mathrm{cov}\,(P_z(A), P_{z'}(A)|\,U_1, \Gamma_1, U_2, \Gamma_2, \ldots) = \sum_{j=1}^{\infty} W_j(z) W_j(z') \frac{G(A)G(A^c)}{M+1}.$$

Unconditional second moments can also be found easily, and particularly attractive formulas are found when the parameters a_j and b_j do not change with j.

Suppose that, given covariates (Z_i), we generated observations $X_i|\,Z_i \overset{\text{ind}}{\sim} P_{Z_i}$, for $i = 1, \ldots, n$. Since the realizations G_j from the Dirichlet distribution are discrete, with different supports for different j, two observations X_i and X_j can be equal only if they are sampled from the same component G_j. In that case they are as a sample of size 2 from a Dirichlet distribution, they are equal with probability $1/(M+1)$, by (4.13). Thus the probability of pairwise clustering $\mathrm{P}(X_i = X_j|\,Z_i, Z_j)$, in the KSBP model with i.i.d. vectors $(U_j, \Gamma_j) \sim (U, \Gamma)$ can be computed as

$$\begin{aligned}
\mathrm{P}(X_i = X_j|\,Z_i, Z_j) &= \sum_{k=1}^{\infty} \mathrm{E}\big[W_k(Z_i) W_k(Z_j)\big] \frac{1}{M+1} \\
&= \frac{\mathrm{E}\big[U^2 K(Z_i, \Gamma) K(Z_j, \Gamma)\big]}{M+1} \sum_{k=1}^{\infty} \Big(\mathrm{E}\big[(1 - UK(Z_i, \Gamma))(1 - UK(Z_j, \Gamma))\big]\Big)^{k-1} \\
&= \frac{\mathrm{E}\big[U^2 K(Z_i, \Gamma) K(Z_j, \Gamma)\big]}{(M+1)(\mathrm{E}\big[UK(Z_i, \Gamma)\big] + \mathrm{E}\big[UK(Z_j, \Gamma)\big] - \mathrm{E}\big[U^2 K(Z_i, \Gamma) K(Z_j, \Gamma)\big]}.
\end{aligned}$$

For the important special case that $M = 0$, and $U \sim \mathrm{Be}(a, b)$ independent of Γ, this reduces to

$$\mathrm{P}(X_i = X_j|\,Z_i, Z_j) = \frac{\frac{a+1}{a+b+1} \mathrm{E}\big[K(Z_i, \Gamma) K(Z_j, \Gamma)\big]}{\mathrm{E}\big[K(Z_i, \Gamma)\big] + \mathrm{E}\big[K(Z_j, \Gamma)\big] - \frac{a+1}{a+b+1} \mathrm{E}\big[K(Z_i, \Gamma) K(Z_j, \Gamma)\big]}. \tag{14.28}$$

14.9.2 Local Dirichlet Process

Let $d\colon \mathfrak{Z} \times \mathfrak{Z}' \to \mathbb{R}^+$ be some function that "connects" the points of two measurable spaces $\mathfrak{Z}$ and $\mathfrak{Z}'$, and for given $\epsilon > 0$ let $N_\epsilon(z) = \{z' \in \mathfrak{Z}' \colon d(z, z') < \epsilon\}$ be the points in $\mathfrak{Z}'$ connected to a given $z \in \mathfrak{Z}$. These "neighborhoods" are used to couple the points z in a random fashion as follows. We generate a sample $\Gamma_1, \Gamma_2, \ldots \overset{\text{iid}}{\sim} H$, for a given probability measure H on $\mathfrak{Z}'$, and determine for every $z \in \mathfrak{Z}$ the set of variables Γ_j that fall into the neighborhood $N_\epsilon(z)$. Coupling arises as z with overlapping neighborhoods will share many Γ_j.

Assume that $H(N_\epsilon(z)) > 0$ for all $z \in \mathfrak{Z}$, so that each neighborhood $N_\epsilon(z)$ receives infinitely many Γ_j. For given z let $I_1(z, \epsilon) < I_2(z, \epsilon) < \cdots$ be the (random) indices such that $\Gamma_j \in N_\epsilon(z)$, in increasing order. Next for random samples $V_j \overset{\text{iid}}{\sim} \mathrm{Be}(1, M)$ and $\bar{\theta}_j \overset{\text{iid}}{\sim} G$ define stick lengths and support points by

$$W_j(z) = V_{I_j(z,\epsilon)} \prod_{l<j} (1 - V_{I_l(z,\epsilon)}), \qquad \theta_j(z) = \bar{\theta}_{I_j(z,\epsilon)}, \qquad j = 1, 2, \ldots$$

The resulting process (14.25) is called the *local Dirichlet process* (LDP).

Since for a given z, the variables $V_{I_l(z,\epsilon)}$ used for breaking the sticks are a random sample from the Be$(1, M)$-distribution, each process P_z is a DP(MG)-process, for any z. Thus a local Dirichlet process is a dependent Dirichlet process (an "LDP" is a "DDP"). The measures P_z are dependent through the auxiliary sequence Γ_j. If two points z and z' are close, then their sets of indices $(I_j(z, \epsilon))$ and $(I_j(z', \epsilon))$ will have a large intersection, and hence the stick breaks of P_z and $P_{z'}$ are more similar. On the other hand, measures P_z with disjoint sequences of indices are independent.

The parameter ϵ is a tuning parameter, and may be equipped with a prior.

The stick-breaking variables $(V_{I_j(z,\epsilon)})$ attached to z can be augmented to the sequence $V_j(z) = V_j \mathbb{1}\{\Gamma_j \in N_\epsilon(z)\}$, for $j = 1, 2, \ldots$, which has zero values at indices j between the values of the $I_j(z, \epsilon)$ and the values $V_{I_j(z,\epsilon)}$ inserted at their "correct" positions. The corresponding sequence of weights $\tilde{W}_j(z) = V_j(z) \prod_{l<j}(1 - V_l(z))$ has zeros at the same set of "in-between" indices and the values $W_j(z)$ at the indices $I_j(z, \epsilon)$. This shows that $P_z = \sum_{j=1}^\infty \tilde{W}_j(z)\delta_{\bar{\theta}_j}$ is an alternative representation of the local Dirichlet process P_z and hence the LDP is a special case of the KSBP with $U_j = V_j$ and the kernel $K(z, \Gamma) = \mathbb{1}\{\gamma \in N_\epsilon(z)\}$. Conequently, the clustering probability can be obtained from (14.28). It can be written in the form

$$\mathrm{P}(X_i = X_j \mid Z_i, Z_j) = \frac{2P_{Z_i,Z_j}}{(1 + P_{Z_i,Z_j})M + 2},$$

where $P_{z,z'} = H(N_\epsilon(z) \cap N_\epsilon(z'))/H(N_\epsilon(z) \cup N_\epsilon(z'))$.

14.9.3 Probit Stick-Breaking Process

A *probit stick-breaking process* is a dependent stick-breaking process (14.25) with stick-breaking weights (14.26)

$$V_j(z) = \Phi(\alpha_j + f_j(z)), \qquad \alpha_j \stackrel{\text{iid}}{\sim} \mathrm{Nor}(\mu, 1), \qquad f_j \stackrel{\text{iid}}{\sim} H, \qquad j = 1, 2, \ldots,$$

where Φ is the standard normal cumulative distribution function. The idea behind using the probit transformation is that it is easier to propose sensible models for (dependent) general real variables than for variables restricted to $[0, 1]$. For example, a Gaussian process prior may be used for the functions f_j. A potential disadvantage is that the $V_j(z)$ are not beta distributed, and hence the resulting process will not be a dependent Dirichlet process.

Conditions for the weights $W_j(z)$ to sum to one can be given in various ways. If the functions f_j are forced to be nonnegative, then $V_j(z) \geq \Phi(\alpha_j)$ by the monotonicity of Φ, so that $\sum_{j=1}^\infty \log \mathrm{E}[1 - V_j(z)] \leq \sum_{j=1}^\infty \log \mathrm{E}[1 - \Phi(\alpha_j)] = -\infty$ because the α_j are i.i.d. and the expectation is strictly smaller than 1.

14.9.4 Ordering Dependent Stick-Breaking Processes

A random discrete distribution $\sum_{j=1}^\infty W_j\delta_{\theta_j}$ is distributionally invariant under permutations of its terms. If for each given z the terms are permuted by a permutation of $\mathbb{N}$ that depends on z, then the marginal distributions remain the same, and at the same time dependence across z arises. In particular, if the original discrete discrete distribution is a Dirichlet process, then the result is a dependent Dirichlet process. This process is called an *order based dependent Dirichlet process* (πDDP). In practice the permutation may be limited to a finite set $\{1, \ldots, N\}$, and this is clearly sufficient if the series is truncated at a finite number of terms.

Random permutations $\{\mathfrak{p}(z): z \in \mathfrak{Z}\}$ may be generated with the help of a point process $(T_1, T_2, \ldots)$ in $\mathfrak{Z}$. Assume that a given neighborhood $U(z)$ of z contains only finitely many points of this process, and given a metric on $\mathfrak{Z}$ define $\mathfrak{p}_1(z), \mathfrak{p}_2(z), \ldots$ by the relations

$$d(z, T_{\mathfrak{p}_z(1)}) < d(z, T_{\mathfrak{p}_z(2)}) < \cdots < d(z, T_{\mathfrak{p}_z(N_z)}).$$

Specifically, one can consider a Poisson process with absolutely continuous intensity measure. For an expression for the correlation between $P_z(A)$ and $P_{z'}(A)$, see Griffin and Steel (2006).

14.9.5 Nested Dirichlet Processes

For α a (atomless) finite measure on a (Polish) sample space $(\mathfrak{X}, \mathscr{X})$, the specification $Q \sim \mathrm{DP}(M\mathrm{DP}(\alpha))$ defines a random measure Q *on the space of measures* on $(\mathfrak{X}, \mathscr{X})$; equivalently Q is a random element in $\mathfrak{M}(\mathfrak{M}(\mathfrak{X}))$. The stick-breaking representation is helpful to understand this complicated structure: it takes the form $Q = \sum_{j=1}^{\infty} \pi_j \delta_{G_j}$, where $G_1, G_2, \ldots \overset{\text{iid}}{\sim} \mathrm{DP}(\alpha)$ is a sample of *measures* on $(\mathfrak{X}, \mathscr{X})$. This structure is called a *nested Dirichlet process* (NDP).

A sample $F_1, F_2, \ldots | Q \overset{\text{iid}}{\sim} Q$ are again measures on $(\mathfrak{X}, \mathscr{X})$, which cluster by their dependence through Q, and thus might be a reasonable model for the distributions of multiple samples of observations in $\mathfrak{X}$ that share common features. The clustering of the F_k is the usual one in the Dirichlet process, and is clearest from the stick-breaking representation: every F_k will be one of the measures G_j in the realization of $Q = \sum_{j=1}^{\infty} \pi_j \delta_{G_j}$, and the F_k cluster in groups of equal distributions.

As $F_1, F_2, \ldots$ are a sample from a Dirichlet process, they are subject to the predictive formulas of Section 4.1.4, where $F_1, F_2, \ldots$ take the place of the observations $X_1, X_2, \ldots$ in that section. In particular, the marginal distribution of every F_k is equal to the center measure $\mathrm{DP}(\alpha)$, and F_2 is equal to F_1 or a new draw from $\mathrm{DP}(\alpha)$ with probabilities $1/(M+1)$ and $M/(M+1)$, respectively. By exchangeability it follows that, for every $k \neq k'$ and measurable set B,

$$\mathrm{P}(F_k = F_{k'}) = \frac{1}{M+1}, \quad \mathrm{cov}\,(F_k(B), F_{k'}(B)) = \frac{\mathrm{var}\,(F_k(B))}{M+1} = \frac{\bar{\alpha}(B)\bar{\alpha}(B^c)}{(M+1)(|\alpha|+1)}.$$

In particular, the correlation $\mathrm{corr}\,(F_k(B), F_{k'}(B)) = 1/(M+1)$ is free of B. The first equality in the display shows that the nested Dirichlet process is substantially different from the hierarchical Dirichlet process, introduced in Example 5.12, for which $\mathrm{P}(F_k = F_{k'}) = 0$, as only atoms are shared but not weights.

Samples $X_{k,i} | F_k, Q \overset{\text{iid}}{\sim} F_k$ from the distributions F_k cluster as well. As each F_k is marginally from a $\mathrm{DP}(\alpha)$-distribution, two different ($i \neq i'$) observations $X_{k,i}$ and $X_{k,i'}$ from the same sample are equal with probability $\mathrm{P}(X_{k,i} = X_{k,i'}) = 1/(|\alpha|+1)$, as for an ordinary Dirichlet process. On the other hand, two different observations $X_{k,i}$ and $X_{k',i'}$ from different samples (thus $k \neq k'$ and $i \neq i'$) are equal with probability

$$P(X_{k,i} = X_{k',i'}) = \frac{1}{M+1} \times \frac{1}{|\alpha|+1}.$$

This follows, because the distributions $G_k \sim \mathrm{DP}(\alpha)$ will have disjoint (countable) supports, and hence so will F_k and $F_{k'}$) unless they cluster on the same G_k.

Thus observations from the same sample are more likely to be equal than observations from different samples, but clustering across groups occurs with positive probability. This feature is shared with the hierarchical Dirichlet process.

14.10 The Indian Buffet Process

The Chinese restaurant process (see Section 14.1.2) induces clustering through the values of a single variable. In some situations individuals may be marked by multiple, potentially even infinitely many, features. Some features (or feature values) may be more popular than others, and be shared by more individuals. The Indian buffet process (IBP) is a stochastic process that models this phenomenon. Like the Chinese restaurant process, presence of a feature in more subjects enhances the probability that a future subject possesses the same feature, and thus builds up a clustering effect. However, feature sharing in an Indian buffet process occurs separately for each feature. In that sense, the process is a factorial analog of the Chinese restaurant process.

The Indian buffet process assumes that feature values are binary, with a 1 or 0 interpreted as possessing or not possessing the feature, and allows infinitely many features. Subjects are then characterized by an infinite row of 0s and 1s, and data can be represented in a binary matrix, with rows corresponding to subjects and columns corresponding to features. It will be assumed that every subject can possess only a finite number of features, so that the binary matrix has finite row sums. Thus the Indian buffet process specifies a probability distribution on sparse infinite binary matrices.

The Indian buffet process will be defined as a limit in distribution of $(n \times K)$-matrices $Z^K = ((Z_{ik}))$, corresponding to n subjects and K features, with n fixed and $K \to \infty$. For given K and parameter $M > 0$ the distribution of the matrix Z^K is specified as

$$Z_{ik} \mid \pi_1, \ldots, \pi_k \stackrel{\text{ind}}{\sim} \mathrm{Bin}(1, \pi_k), \qquad \pi_k \stackrel{\text{iid}}{\sim} \mathrm{Be}(M/K, 1).$$

Thus π_k is the probability that the kth feature is present in a subject, and given these probabilities all variables Z_{ik} are mutually independent. The probability of a particular $(n \times K)$ matrix (or "configuration") z with m_k values 1 and $n - m_k$ values 0 in particular locations in the kth column is then

$$P(Z^K = z) = \prod_{k=1}^{K} \int \pi_k^{m_k} (1-\pi_k)^{n-m_k} \frac{M}{K} \pi_k^{M/K-1}\, d\pi_k = \prod_{k=1}^{K} \frac{M\Gamma(m_k + M/K)\Gamma(n - m_k + 1)}{K\Gamma(n + 1 + M/K)}.$$

The K columns of the matrix Z^K are i.i.d. and every column is an exchangeable n-vector, which makes the matrix "exchangeable in both rows and columns." The expected number of 1s per column is equal to $n(M/K)/(1+M/K)$, giving an expected total number of 1s equal to $nM/(1 + M/K)$. For $K \to \infty$ this stabilizes to nM, so that sparsity is ensured. This is

01000110001011000010		11111110000000000000
10000100010001010001		11000001111000000000
01000100101000001001	left-ordering	10110001000100000000
01001001001001000010	→	01111000000011000000
00010100001001000100		11100000000000110000

Figure 14.6 Example of a matrix showing $K = 20$ features scored on $n = 5$ subjects (left) and the corresponding left-ordered matrix (right).

the rationale behind the choice M/K for the parameter of the beta distribution for the π_k — this parameter must decrease at the rate K^{-1} to balance the effect of the growing number of columns if the total number of 1s must remain finite.

The ordering of the features is considered irrelevant, and hence matrices that are identical up to a permutation of columns will be identified. It is convenient to choose a representative of every equivalence class of matrices thus created, defined by the *left-ordering procedure*. This consists of first permuting columns so that all 1s appear before all 0s in the first row. This forms two groups of columns, defined by their value in the first row, which are separately permuted by the same procedure, but now taking account of 1s and 0s in the second row. Next there are four groups of columns, defined by their values in the first two rows, and the procedure continues with four separate permutations based on the third row. The procedure continues sequentially until all rows have been operated on. (The final result can also be described as an ordering of the columns in terms of the value of the column viewed as a binary number.) Figure 14.6 shows an example of a feature matrix and the left-ordered representative of its equivalence class.

We define a probability distribution on the set of left-ordered matrices by assigning every such matrix the sum of all probabilities received by some member of its equivalence class. Since the columns of the matrix Z^K are exchangeable, all members of an equivalence class have the same probability, and this sum is simply equal to the cardinality of the equivalence class times $\mathrm{P}(Z^K = z)$. If the vector (K_h), for $h = 0, \ldots, 2^{n-1}$, gives the number of times that every of the 2^n possible columns (in $\{0, 1\}^n$) occurs in a matrix, then the cardinality of its equivalence class is equal to $K!/\prod_{h=0}^{2^n-1} K_h!$, being the number of permutations of the columns, divided by the number of permutations that yield the same left-ordered matrix. Thus with $[Z^K]$ denoting a left-ordered equivalence class, we assign to a configuration $[z]$ characterized by (K_h) and (m_k) the probability

$$\mathrm{P}([Z^K] = [z]) = \frac{K!}{\prod_{h=0}^{2^n-1} K_h!} \prod_{k=1}^{K} \frac{M\Gamma(m_k + M/K)\Gamma(n - m_k + 1)}{K\Gamma(n + 1 + M/K)}. \tag{14.29}$$

The numbers m_k enter this expression as an unordered set $\{m_1, \ldots, m_K\}$, but in their original interpretation they refer to the number of 1s in specific columns. In particular, after appropriate permutation every m_k refers to a column that belongs to one of the groups of columns whose cardinalities are counted by the K_h, for $h = 0, \ldots, 2^n - 1$. The values of the m_k as a group can for a given configuration be derived from the numbers K_h, which give not only the number of 1s but also their positions in the columns. Thus we can view the preceding probability as a probability distribution on the set of vectors (K_h). These have length 2^n, and nonnegative integer-valued coordinates with sum $\sum_h K_h = K$. Each vector corresponds to a left-ordered $(n \times K)$-matrix. For convenience of notation let $h = 0$ refer to the zero column, so that K_0 is the number of columns with only zero coordinates (i.e.

columns with $m_k = 0$). We shall take the limit as $K \to \infty$ with the numbers $(K_h)_{h>0}$ of nonzero columns fixed; hence $K_0 \to \infty$ and the number of columns k with $m_k > 0$ is also fixed.

Lemma 14.63 *In the preceding setting we have that* $K - \mathrm{E}K_0 = \sum_{i=1}^n M/i + O(1/K)$, *as* $K \to \infty$ *for fixed* n, *and hence* $\sum_{h>0} K_h$ *is bounded in probability. Furthermore, for fixed values* $(K_h)_{h>0}$, *expression* (14.29) *converges to, as* $K \to \infty$ *for fixed* n,

$$\mathrm{P}([Z] = [z]) := \frac{M^{\sum_{h>0} K_h}}{\prod_{h>0} K_h!} e^{-M \sum_{i=1}^n i^{-1}} \prod_{k:m_k>0} \frac{(n - m_k)!(m_k - 1)!}{n!}.$$

This expression gives a probability density function of a variable $[Z]$ *with values in the set of left-ordered binary matrices with* n *rows and infinitely many columns. The corresponding distribution is exchangeable in the rows of the matrix.*

Proof The expected value $\mathrm{E}K_0$ is K times the probability that a given column of the $(n \times K)$ matrix Z^K has only zero coordinates. This latter probability is equal to $\mathrm{E}(1 - \pi_k)^n$, for $\pi_k \sim \mathrm{Be}(M/K, 1)$. Since $\Gamma(1 + \alpha) = \alpha\Gamma(\alpha)$, for $\alpha > 0$, this can be computed to be

$$\frac{\Gamma(n+1)\Gamma(1 + M/K)}{\Gamma(n + 1 + M/K)} = \Big[\Big(1 + \tfrac{M}{Kn}\Big)\Big(1 + \tfrac{M}{K(n-1)}\Big) \times \cdots \times \Big(1 + \tfrac{M}{K}\Big)\Big]^{-1}.$$

The right side can be seen to be $1 - (M/K) \sum_{i=1}^n 1/i + O(1/K^2)$, from which the assertion on $K - \mathrm{E}K_0$ follows. Since $\sum_{h>0} K_h = K - K_0$, it follows that the sequence of nonnegative variables $\sum_{h>0} K_h$ has a bounded mean value and hence is bounded in probability.

To prove the convergence of (14.29), we first note that for any configuration (K_h) with $\sum_{h>0} K_h = K - K_0$ fixed, we have that $K!/(K_0! K^{K-K_0}) \to 1$, as $K \to \infty$. Furthermore, we have the identities, where $\Gamma(0)$ is interpreted as 1,

$$\prod_{k=1}^K \frac{\Gamma(m_k + M/K)}{\Gamma(m_k)} = \Gamma\Big(\tfrac{M}{K}\Big)^{K_0} \prod_{k:m_k>1} \Big[\Big(1 + \tfrac{M}{K(m_k-1)}\Big) \times \cdots \times \Big(1 + \tfrac{M}{K}\Big)\Big]\Big[\tfrac{M}{K}\Gamma\Big(\tfrac{M}{K}\Big)\Big]^{K-K_0},$$

$$\prod_{k=1}^K \frac{\Gamma(n + 1 + M/K)}{\Gamma(n+1)} = \Big[\Big(1 + \tfrac{M}{Kn}\Big)\Big(1 + \tfrac{M}{K(n-1)}\Big) \times \cdots \times \Big(1 + \tfrac{M}{K}\Big)\Big]^K \Big[\tfrac{M}{K}\Gamma\Big(\tfrac{M}{K}\Big)\Big]^K.$$

When taking the quotient of these two expressions, the factors $\Gamma(M/K)$ cancel. Furthermore, for fixed $(K_h)_{h>0}$ the product on the right of the first equation has a bounded number of terms, each of which tends to 1, so that the product tends to 1. The first term in the second equation can be seen to converge to $e^{M \sum_{i=1}^n 1/i}$. The convergence of (14.29) follows by substituting these findings in the quotient of the latter expression and its claimed limit, and simplify the result.

By construction, the function $z \mapsto p_K(z)$ on the right side of (14.29) is a probability density function on the set of left-ordered $(n \times K)$-matrices z. We can view p_k also as a probability density function on the set $\mathfrak{K}$ of all vectors $(K_h)_{h>0}$ of length $2^n - 1$ with coordinates in the nonnegative integers with $\sum_{h>0} K_h < \infty$. The first part of the proof shows that it gives probability at least $1 - \epsilon$ to the finite subset of all such vectors with $\sum_{h>0} K_h \le L$, for a sufficiently large large constant L. Thus the sequence p_K is uniformly tight for the discrete topology on $\mathfrak{K}$. Then its pointwise limit must be a probability density function on $\mathfrak{K}$. □

Definition 14.64 (Indian buffet process) The (one-parameter) *Indian buffet process* (IBP) with parameter M is the probability distribution on the set of left-ordered binary matrices of dimension $(n \times \infty)$ given in Lemma 14.63, and defined for $n \in \mathbb{N}$.

Like the Chinese restaurant process, the Indian buffet process can be described by a culinary metaphor. Imagine that customers enter an Indian restaurant that serves a lunch buffet with an unlimited number of dishes. The first customer arriving in the restaurant chooses a Poisson(M) number of dishes from the buffet. The second customer chooses every dish tasted by the first customer independently and randomly with probability $1/2$ and also independently chooses an additional Poisson($M/2$) number of new dishes. The probability $1/2$ and parameter $M/2$ are chosen so that marginally the number of dishes selected by the second customer is also Poisson(M). In a similar manner, the ith customer chooses every previously tasted dish k with probability $m_{k,i-1}/i$, where $m_{k,i-1}$ is the number of previous customers (among $1, \ldots, i-1$) who tasted dish k, and also chooses an additional Poisson(M/i) number of so-far untasted dishes. After continuing up to n customers, we can form a binary matrix with the customers as rows and the columns as dishes, with 1s at coordinates (i, k) such that customer i tasted dish k.

Lemma 14.65 *The left-ordering of the binary matrix constructed in the preceding paragraph is a realization of the Indian buffet process.*

Proof This may be proved by writing down the explicit probability of obtaining a matrix configuration with given $(K_h)_{h>0}$, and verifying that this agrees with the expression given in Lemma 14.63. □

Conditional distributions in an Indian buffet process are easy to compute. These distributions can be used to implement a Gibbs sampling procedure when an Indian buffet prior is applied on latent features, similar as for a Dirichlet mixture process. If $Z_{-i,k}$ stands for the kth column of Z with the ith coordinate deleted and $m_{-i,k}$ is the corresponding column sum, then $\mathrm{P}(Z_{ik} = 1 | Z_{-i,k}) = m_{-i,k}/n$. This can be easily derived by taking the limit as $K \to \infty$ of the corresponding probability in the finite version of the model with K features. Then $Z_{-i,k}$ is a vector of Bernoulli variables Z_{jk}, for $j \neq i$, with success probability $\pi_k \sim \mathrm{Be}(M/K, 1)$, and the corresponding probability is $\mathrm{E}(\pi_k | Z_{-i,k}) = (m_{-i,k} + M/K)/(n + M/K)$, since the posterior distribution of π_k given $Z_{-i,k}$ is $\mathrm{Be}(M/K + m_{-i,k}, n - m_{-i,k})$.

Problem 14.19 describes a type of stick-breaking representation of the Indian buffet process.

A different way to construct the Indian buffet process is through the de Finetti representation of the distribution of the rows of the binary matrix, in the same way as the Chinese restaurant process arises from a random sample of observations from the Dirichlet process.

In the context of the Indian buffet process the observations (the rows of the matrix) are random variables with values in $\{0, 1\}^\infty$, and the mixing distribution is a measure on the set of distributions of such infinite vectors.

It is convenient to describe the construction in a continuous time framework, where the 1s are placed at random locations in $(0, \infty)$, and the remaining times are interpreted as 0s. Since we are interested in the pattern of sharing of 1s, the additional zeros do not affect the interpretation. Then the two distributions involved in the de Finetti representation are the Bernoulli process for the observations and the beta process for the mixing distribution. The beta process is described in Section 13.3 and Definition 13.3. It is a family of stochastic processes with independent increments on the time set $[0, \infty)$, parameterized by a pair (c, Λ) of a positive function c and a cumulative hazard function Λ;[13] the function c will here be restricted to be constant; the function Λ is the mean function of the process and will be important here only through its total mass, which will be assumed finite. The Bernoulli process is another process with independent increments, parameterized by a cumulative hazard function H, and defined as follows.

Definition 14.66 (Bernoulli process) A *Bernoulli process* with parameter H is an independent increment process on $[0, \infty)$ with intensity measure given by $\nu(dx, ds) = H(dx)\delta_1(ds)$, where H is a cumulative hazard function and δ_1 is the Dirac measure at 1.

The definition uses the language of completely random measures, as described in Appendix J, but the Bernoulli process can also be described directly as follows. In the case that H is a discrete measure the Bernoulli process can be identified with an infinite sequence of independent Bernoulli variables h_k, a variable $h_k \sim \text{Bin}(1, H\{x_k\})$ for every atom x_k of H. The sample paths of the Bernoulli process are then zero at time zero and jump up by 1 at every x_k such that $h_k = 1$.[14] If $H = H^c + H^d$ consists of a continuous and a discrete part, then the Bernoulli process is the sum $Z = Z^c + Z^d$ of two independent processes, where Z^c is a Poisson process with mean measure $\text{E}[Z^c(t)] = H^c(t)$ and $Z^d(t) = \sum_k h_k \mathbb{1}\{x_k \leq t\}$ is as described previously (with $H = H^d$). While as an independent increment process the Bernoulli process is understood to increase by 1 at all jump points, below we view the process as a measure and identify it with its jumps, making the sample paths take values 0 and 1 rather than increasing functions.

Given a constant $c > 0$ and an atomless cumulative hazard function Λ, consider the model

$$Z_1, Z_2, \ldots \mid H \stackrel{\text{iid}}{\sim} \text{Bernoulli process}(H), \qquad H \sim \text{Beta process}(c, \Lambda). \tag{14.30}$$

Because the beta process produces a discrete cumulative hazard function H with probability one, the realizations of the Bernoulli processes Z_i can be identified with infinite sequences of 0s and 1s (where the order does not matter as long it is consistent across i). Thus for every n we can form the binary $(n \times \infty)$-matrix with rows $Z_1, \ldots, Z_n$. We show below that for $c = 1$ and Λ of total mass M, the left-ordered version of this matrix is distributed

[13] The cumulative distribution function of a measure on $(0, \infty)$ that is finite on finite intervals and has atoms strictly smaller than one; it may be identified with such a measure.

[14] The total number of jumps on finite intervals is finite, as it has expectation $\sum_{x_k \leq t} H\{x_k\} < \infty$, for every t.

as the Indian buffet process with parameter M. For other values of c we obtain different distributions, each with exchangeable rows by construction.

Definition 14.67 (Two-parameter Indian buffet process) The *two-parameter Indian buffet process* with parameters c and M is the distribution of the left-ordering of the $(n\times\infty)$ matrix with rows the processes $Z_1, \ldots, Z_n$ defined in (14.30) (or rather the values of their jumps at a countable set containing all atoms of H). Here $c > 0$ and M is the total mass of Λ.

To connect this definition to the earlier one-parameter Indian buffet process, we derive the conditional distribution of a new row Z_{n+1} given the preceding ones. The following result is a process analog of the familiar binomial-beta conjugacy.

Lemma 14.68 *The posterior distribution of H given $Z_1, \ldots, Z_n$ in the model* (14.30) *is a beta process with parameters $c + n$ and $c\Lambda/(c + n) + \sum_{i=1}^n Z_i/(c + n)$. Furthermore, marginally every Z_i is a Bernoulli process with parameter Λ.*

Proof It suffices to give the proof for $n = 1$, because the general result can next be obtained by posterior updating. A proof can be based on a discrete approximation to the beta process, and calculations of the posterior distribution in the finite-dimensional case. Alternatively we can base a proof on the representations of the processes through point processes, as follows.

The beta process H can be identified with a point process $N = N^c + N^d$ on $(0, \infty) \times (0, 1)$, where N^c is a Poisson process with intensity measure $\nu^c(dx, ds) = cs^{-1}(1 - s)^{c-1}\, d\Lambda^c(x)\, ds$, and N^d consists of points (x, s), with x restricted to the atoms of Λ and $s \sim \mathrm{Be}(c\Lambda\{x\}, c(1 - \Lambda\{x\}))$, independently across atoms, and independently of N^c (see Section 13.3).

Given N (or H), the Bernoulli process Z_1 jumps (always by 1) with probability s at every point x with $(x, s) \in N$. In other words, the process Z_1 jumps at every x from the set of points (x, s) obtained from *thinning* N with probabilities s.

This thinning transforms the Poisson process N^c into a Poisson process M^c of intensity $s\nu^c(dx, ds) = c(1-s)^{c-1}\, d\Lambda^c(x)\, ds$. Given M^c, the process N^c is distributed as the union of M^c and a Poisson process K generated by thinning N^c with the complementary probabilities $1 - s$, hence with intensity measure $(1 - s)\nu^c(dx, ds) = cs^{-1}(1 - s)^c\, d\Lambda^c(x)\, ds$, which is the intensity measure of a beta process with parameters $(c + 1, c\Lambda^c/(c + 1))$. Observing Z_1 is less informative than observing M^c, as Z_1 only (but fully) reveals the horizontal locations x of the points (x, s) in M^c. Since the locations x and heights s of the points (x, s) in M^c can be realized by first spreading the locations x according to a Poisson process with intensity measure Λ^c and next generating the heights s independently from the $\mathrm{Be}(1, c)$-distribution, the conditional distribution of M^c given Z_1 is to reconstruct the heights independently from this same beta distribution. Since $\mathrm{Be}(1, c) = \mathrm{Be}((c+1)\Delta, (c+1)(1-\Delta))$, for $\Delta = 1/(c+1)$, it follows that the points in $M^c|\, Z_1$ can be described as the fixed atoms of the beta process with parameters $(c+1, Z_1^c/(c+1))$, for Z_1^c the jumps of Z_1 that do not belong to the support of Λ^d.

A point (x, s) of N^d leads to a jump in Z_1 with probability s, and this will sit at an atom of Λ^d. If x is such an atom and also a jump point of Z_1, then the conditional density of s is proportional to s times the $\mathrm{Be}(c\Lambda\{x\}, c(1 - \Lambda\{x\}))$-density, while if x is an atom of Λ^d and *not* a jump point of Z_1, then the conditional density of s is proportional to $1 - s$ times the latter density. This leads to $\mathrm{Be}(c\Lambda\{x\} + 1, c(1 - \Lambda\{x\}))$- and $\mathrm{Be}(c\Lambda\{x\}, c(1 - \Lambda\{x\}) + 1)$-distributions for s in the two cases, which are consistent with fixed atoms of a beta process with parameters $c+1$ and discrete cumulative hazard component $c\Lambda^d/(c+1) + Z_1^d/(c+1)$, for Z_1^d the jumps of Z_1 that are also atoms of Λ^d.

The jumps of the process Z_1 occur at locations x obtained from the points (x, s) in N by thinning these points with probability s. The points (x, s) left from the Poisson process N^c form a Poisson process with intensity $s\nu^c(dx, ds) = c(1 - s)^{c-1}\, d\Lambda^c(x)\, ds$, and project to a Poisson process with intensity $\Lambda^c(dx)$ on the first coordinate. The points (x, s) left from N^d are located at atoms of Λ^d, and an atom remains with probability s, where $s \sim \mathrm{Be}(c\Lambda\{x\}, c(1 - \Lambda\{x\}))$, whence with marginal probability $\mathrm{E}[s] = \Lambda\{x\}$. □

Now specialize to the situation where $c > 0$ and Λ is an atomless measure of total mass M. Then the Bernoulli processes Z_i are marginally Poisson processes with intensity measure Λ, by the second assertion of Lemma 14.68. In particular, their total numbers of jumps are $\mathrm{Poi}(M)$-distributed.

Because given H the Z_i are i.i.d. Bernoulli processes with intensity H, the predictive distribution of Z_{n+1} given $Z_1, \ldots, Z_n$ is a Bernoulli process given an intensity process generated from the conditional distribution of H given $Z_1, \ldots, Z_n$. By the first assertion Lemma 14.68, this is a beta process with updated parameters, whence by the last assertion of the lemma Z_{n+1} given $Z_1, \ldots, Z_n$ is a Bernoulli process with the updated cumulative hazard function. The latter function can alternatively be written as

$$\frac{c}{c+n}\Lambda + \sum_{k=1}^{\infty} \frac{m_{k,n}}{c+n}\delta_{x_k},$$

where $m_{k,n}$ is the number of processes $Z_1, \ldots, Z_n$ that have a jump at atom x_k (and the sequence $x_1, x_2, \ldots$ must contain all jump points of all Z_i). Thus Z_{n+1} given $Z_1, \ldots, Z_n$ is a Bernoulli process with both a continuous component and a discrete component. By the definition of a Bernoulli process the continuous component is the Poisson process with intensity measure $c\Lambda/(c + n)$, while the discrete component is the process that jumps with probability equal to $m_{k,n}/(c + n)$ at every location x_k, with jump size 1. Because Λ is atomless by assumption, the events of the continuous component will with probability 1 not occur at any of the points x_k, and their number will be $\mathrm{Poi}(cM/(c + n))$. For $c = 1$ the probability $m_{k,n}/(1 + n)$ of a jump at an "old location" and the distribution $\mathrm{Poi}(M/(1 + n))$ of the number of jumps at a "new location" agree exactly with the dish-sharing probability and the distribution of the number of new dishes in the Indian buffet process with parameter M. Since Z_1 starts with the same number of dishes and all conditionals are as in the Indian buffet process, the two-parameter Indian buffet process with parameter $c = 1$ is equal to the one-parameter process of Definition 14.64.

The new number of dishes tasted by the ith customer arises from the Poisson part of $Z_{i+1} \mid Z_1, \ldots, Z_i$, and hence is Poisson distributed with mean $cM/(c + i)$. Therefore the expected total number of dishes tasted by at least one of the first n customers in a two-parameter Indian buffet process can be calculated as $Mc\sum_{i=0}^{n-1}(c + i)^{-1}$. The logarithmic growth as $n \to \infty$ for fixed c is reminiscent of the growth of the number of clusters in the Chinese restaurant process (see Proposition 4.8). The parameter c can be interpreted as a measure of concentration. If $c \to 0$, then the expected total number of dishes tasted tends to M, and hence (on the average) all customers share the same dishes (maximum concentration), whereas if $c \to \infty$, the expected number of dishes tasted tends to Mn, meaning that (on the average) customers are not sharing at all (minimum concentration).

14.11 Historical Notes

Exchangeable random partitions were first studied by Kingman (1978, 1982) and later by Aldous (1985), Perman et al. (1992) and Pitman (1995, 2006). Kingman's representation for exchangeable partitions, Theorem 14.7, appeared in Kingman (1978); the short proof presented here was taken from Aldous (1985). Lemma 14.8 appeared explicitly in Lee et al. (2013). Partial exchangeability and Theorem 14.13 and Corollary 14.14 are due to Pitman (1995). Theorem 14.18 was obtained by Pitman (1996b). The important role of size-biased permutations is pointed out in Perman et al. (1992) and Pitman (1995, 1996a). The name *Chinese restaurant process* was apparently coined by Dubins and Pitman (see Aldous (1985) and Pitman (2006)). The sampling scheme was already known in Bayesian Nonparametrics as the *Blackwell-MacQueen urn scheme* or *generalized Pólya urn scheme*, and appeared in Blackwell and MacQueen (1973). In population genetics Hoppe (1984) introduced a different sampling scheme, which also leads to the EPPF induced by sampling from a Dirichlet process. The Chinese restaurant franchise process and the hierarchical Dirichlet process were studied by Teh et al. (2006). Proposition 14.9 characterizing the Dirichlet process as a special species sampling process is due to Gnedin and Pitman (2006); the proof given here is taken from Lee et al. (2013). The abstract posterior consistency result for species sampling process given by Theorem 14.19 was obtained for the case $\alpha_n = 1$ and $\beta_n = 0$ by Jang et al. (2010), by a longer proof. The present generalization covers the case of Gibbs processes, which were treated separately by De Blasi et al. (2013). Gibbs processes were introduced and systematically studied by Gnedin and Pitman (2006), although some of the results were anticipated in Pitman (2003). The main example, the two-parameter model, goes back to at least Perman et al. (1992) and Pitman (1995). They were used in Bayesian nonparametrics by Lijoi et al. (2007b,a). Theorems 14.26 and 14.27 and Lemma 14.28 are due to De Blasi et al. (2013). De Blasi et al. (2015) gave a review of the topic. The Pitman-Yor process appears explicitly in Perman et al. (1992), and Pitman (1995) who refer to Perman's 1990 thesis for the stick-breaking representation. This extended the characterization by McCloskey (1965) that the Dirichlet process is the only discrete random probability measure for which size-biased random permutation of random atoms admits a stick-breaking representation in terms of a collection of *i.i.d.* random variables. Pitman (1995) studied its partition structure, and Pitman and Yor (1997) explored connections with stable processes and gave an alternative construction via the Radon-Nikodym derivative of a σ-stable CRM. Proposition 14.34 is due to Pitman and Yor (1997), Theorem 14.37 is due to Pitman (1996b), Theorem 14.33 is due to Perman et al. (1992). Proposition 14.35 is due to Vershik et al. (2001). The distribution of linear functionals of a Pitman-Yor process was characterized in terms of generalized Cauchy-Stieltjes transform by James et al. (2008). Theorem 14.38 and Lemma 14.39 were obtained by James (2008). Poisson-Kingman processes were introduced and studied in Pitman (2003), following Perman et al. (1992). The latter reference obtains various distributional properties, including the characterization through residual allocation (or stick breaking) given in Theorem 14.49. Various applications of the gamma Poisson-Kingman process can be found in the monograph Feng (2010) and the references therein.

The EPPF and PPF of a special Poisson-Kingman process was also obtained by Regazzini et al. (2003) by using the normalized CRM approach. Gnedin and Pitman (2006) characterized all Gibbs type processes with type parameter $\sigma \in (0, 1)$ precisely as Poisson-Kingman processes based on σ-stable CRM. The *σ-diversity* for a Poisson-Kingman process based on a σ-stable or generalized gamma CRM was derived by Pitman (2003), based on a characterization of the diversity in terms of the weight sequence by Karlin (1967). Posterior inference about the σ-diversity with applications in Bayesian nonprametrics are discussed in Favaro et al. (2009) and Favaro et al. (2012a). Normalized completely random measures were introduced in the context of prior distributions on probability measures on the real line in Regazzini et al. (2003). The posterior updating rule was obtained in James et al. (2009). Results on the partition structure were also obtained in this paper, but follow from the case of Poisson-Kingman processes obtained in earlier work by Pitman (2003). The general stick-breaking representation Theorem 14.49 of homogeneous NCRM (or Poisson-Kingman processes) was given in Perman et al. (1992), and repeated in Pitman (2003), who also discussed the partitioning structure. It was specialized to the normalized inverse-Gaussian process as in Proposition 14.54 in Lijoi et al. (2005a). The generalized gamma NCRM was constructed as a Poisson-Kingman process in Pitman (2003) (apparently following a 1995 preprint), and as an NCRM in Lijoi and Prünster (2003); also see Cerquetti (2007). This class was characterized as the intersection of NCRM processes and Gibbs processes in Lijoi et al. (2008) and Cerquetti (2008). Bayesian analysis with such a prior distribution appeared in Lijoi et al. (2007b), extending the special case of the normalized inverse-Gaussian in Lijoi et al. (2005a). To date the normalized inverse-Gaussian and the Dirichlet process are the only processes with explicit descriptions of marginal distributions of sets in a partition. The generalized Dirichlet process discussed in Lijoi et al. (2005b) is another example for which many closed form expressions can be obtained, notwithstanding the fact that it falls outside the class of generalized gamma NCRM, and is not Gibbs; see Problem 14.12. Regazzini et al. (2003) and James et al. (2010) studied distributions of linear functionals of an NCRM process. Nieto-Barajas et al. (2004) obtained distribution of the mean functional of both prior and posterior process of NCRM processes. A review of many results on distribution of linear functionals of various type of processes is given by Lijoi and Prünster (2009). Thc first contribution on dependent discrete random probability distributions was made by Cifarelli and Regazzini (1978), who proposed a nonparametric prior for partially exchangeable arrays, defined as a mixture of Dirichlet processes. MacEachern (1999) developed the idea in a modern context and introduced the idea of dependent stick breaking. The concept was quickly followed up and led to various constructions of dependent processes, and applications in different contexts such as density regression. The order based dependent Dirichlet process appeared in Griffin and Steel (2006). The kernel stick-breaking process appeared in Dunson and Park (2008), the probit stick-breaking process in Rodríguez and Dunson (2011) and the local Dirichlet process in Chung and Dunson (2011). The nested Dirichlet process, which is similar in spirit to the hierarchical Dirichlet process, was introduced by Rodríguez et al. (2008). The one-parameter Indian buffet process was introduced by Griffiths and Ghahramani (2006). Thibaux and Jordan (2007) generalized it to the two-parameter situation and obtained the de Finetti representation through the beta process. The stick-breaking representation of the the Indian buffet process was obtained by Teh et al. (2007).

Problems

14.1 (Kingman's formula for EPPF) Show that the EPPF of a proper species sampling model can be written in the form

$$p(n_1,\ldots,n_k)=\sum_{1\le i_1\neq\cdots\neq i_k<\infty}\mathrm{E}\prod_{j=1}^{k}W_{i_j}^{n_j}.$$

14.2 (De Blasi et al. 2013, Gnedin 2010) Suppose that P follows a Gibbs prior distribution with type-parameter $\sigma=-1$ and mixing distribution in (14.19) given by $\pi(j)=\gamma(1-\gamma)^{[j-1]}/j!$, where $0<\gamma<1$. Show that $V_{n+1,K_n+1}/V_{n,K_n}\to 1$ a.s. if K_n is the number of distinct values in a random sample of size n from a true distribution P_0, and conclude that the posterior distribution of P concentrates near the center measure G a.s. [Thus the posterior is *totally inconsistent*: the data have asymptotically no influence and the posterior concentrates near the prior mean.]

14.3 (De Blasi et al. 2013) Suppose that P follows a Gibbs distribution with type-parameter $\sigma=-1$ with mixing distribution in (14.19) the geometric distribution with parameter $\gamma\in(0,1)$. Show that $V_{n+1,n+1}/V_{n,n}\to(2-\gamma-2\sqrt{1-\gamma})/\gamma$ a.s., and conclude that the posterior distribution based on a random sample from an atomless true distribution P_0 is inconsistent.

14.4 (Kingman 1993) Let $W_{1:n}\le\cdots\le W_{n:n}$ be the order statistics of a $\mathrm{Dir}(n;\alpha_{1,n},\ldots,\alpha_{n,n})$-distributed random vector. Assume that $\max\{\alpha_{n,j}: 1\le j\le n\}\to 0$ and $\sum_{j=1}^n\alpha_{n,j}\to M\in(0,\infty)$ as $n\to\infty$. Show that $(W_{n:n},W_{n-1:n},\ldots,W_{n-k:n})\rightsquigarrow(P_1,P_2,\ldots,P_k)$ for any fixed k, where $(P_1,P_2,\ldots)$ follows a one-parameter Poisson-Dirichlet distribution with parameter M.

14.5 (Kingman 1993) If $(P_1,P_2,\ldots)$ follows a one-parameter Poisson-Dirichlet distribution with parameter M, show that $-\log P_k\sim k/M$ a.s. as $k\to\infty$.

14.6 Show that Proposition 14.5 reduces to Ewens's sampling formula when applied to a Dirichlet process partition.

14.7 (Perman et al. 1992) If $\varepsilon_i\overset{\text{iid}}{\sim}\mathrm{Exp}(1)$, then show that the reordering $\tilde w_1,\tilde w_2,\ldots$ of a probability vector $w_1,w_2,\ldots$ such that the corresponding variables ε_j/w_j are in increasing order, is size-biased.

14.8 Reinterpret equation (14.23) in Theorem 14.49 to show that the sequence of variables $T_0:=Y, T_1:=\sum_{j\ge2}\tilde Y_j, T_3:=\sum_{j\ge3}\tilde Y_j,\ldots$ is a Markov chain with stationary transition density.

14.9 Show that the Pitman-Yor process can be obtained as a $\mathrm{Ga}(M/\sigma,1)$-mixture of generalized gamma processes $\rho_{\sigma,\xi}$, as claimed in Example 14.48(iv). [Hint: By Example 14.47 the Pitman-Yor distribution is a mixture of the $\mathrm{PK}(\rho_\sigma|\,t)$-distributions with respect to the density $h(t)=\sigma\Gamma(M)/\Gamma(M/\sigma)f_\sigma(t)$. Because conditional Poisson-Kingman distributions are invariant under exponential tilting of the intensity measure, $\mathrm{PK}(\rho_\sigma|\,t)=\mathrm{PK}(\rho_{\sigma,\xi}|\,t)$, and hence it suffices to show that the mixture $\int f_{\sigma,\xi}(t)g(\xi)\,d\mathrm{Ga}(M/\sigma,1)(\xi)$. of the densities $f_{\sigma,\xi}$ associated with $\rho_{\sigma,\xi}$ is equal to h. Consider Laplace transforms and write $t^{-M}=\int\mu^{M-1}e^{-\mu t}\,d\mu/\Gamma(M)$.]

14.10 (Perman et al. 1992) Show that for a Poisson-Kingman process, with the random variables $Y, V_1, V_2, \ldots$ appearing in the stick-breaking representation given by Theorem 14.49, the following assertions are equivalent:

(a) P follows a Dirichlet process;
(b) Y and V_1 are independent;
(c) $Y \sim \mathrm{Ga}(M, \lambda)$ and independently $V_1 \sim \mathrm{Be}(1, M)$, for some $M, \lambda > 0$;
(d) $Y, V_1, V_2, \ldots$ are independent;
(e) $Y \sim \mathrm{Ga}(M, \lambda)$ and $V_1, V_2, \ldots \overset{\text{iid}}{\sim} \mathrm{Be}(1, M)$, for some $M, \lambda > 0$.

14.11 (Perman et al. 1992) Show that for a Poisson-Kingman process, with the random variables $Y, V_1, V_2, \ldots$ appearing in the stick-breaking representation given by Theorem 14.49, the following assertions are equivalent:

(a) $Y - \tilde{Y}_1$ and V_1 are independent;
(b) Y is a positive σ-stable random variable.

If either and hence both conditions are satisfied, then $Y_n, V_1, \ldots, V_n$ are independent for all n and $V_i \overset{\text{ind}}{\sim} \mathrm{Be}(1 - \sigma, M + i\sigma), i = 1, 2, \ldots$

14.12 (Generalized Dirichlet distribution) A random probability vector $p \in \mathbb{S}_k$ has a *generalized Dirichlet distribution* with parameters $(a, b) \in \mathbb{R}^{k-1} \times \mathbb{R}^{k-1}$ if the density is

$$\prod_{i=1}^{k-1} \frac{\Gamma(a_i + b_i)}{\Gamma(a_i)\Gamma(b_i)} \Big(\prod_{i=1}^{k-1} p_i^{a_i - 1}\Big) p_k^{b_{k-1}-1} \Big[\prod_{i=1}^{k-2} \Big(1 - \sum_{j=1}^{i} p_j\Big)^{b_i - (a_{i+1} + b_{i+1})}\Big].$$

Show that if $V_i \overset{\text{ind}}{\sim} \mathrm{Be}(a_i, b_i)$, for $i = 1, \ldots, k-1$, and $V_k = 1$, then the stick-breaking weights $(W_1, \ldots, W_k)$ obtained by $W_i = V_i \prod_{j=1}^{i-1}(1 - V_j)$, $i = 1, \ldots, k$, follow the generalized Dirichlet distribution with parameters $((a_1, a_2, \ldots, a_{k-1}), (b_1, b_2, \ldots, b_{k-1}))$.

14.13 (Lijoi et al. 2005a) Let P follow the normalized inverse-Gaussian process with parameter α, where $\int_0^\infty \sqrt{x}\, d\alpha(x) < \infty$. Show that the cumulative distribution function of $L = \int x\, dP(x)$ satisfies

$$\begin{aligned} \mathrm{P}(L \leq s) = \frac{1}{2} - \frac{e^{|\alpha|}}{\pi} \int_0^\infty t^{-1} \exp\Big[-\int (1 + 4t^2(x - s)^2)^{1/4} \\ \cos\Big(\tfrac{1}{2} \arctan(2t(x - s))\Big)\, d\alpha(x)\Big] \times \sin\Big[-\int [1 + 4t^2(x - s)^2]^{1/4} \\ \sin\Big(\frac{1}{2} \arctan(2t(x - s))\Big)\, d\alpha(x)\Big]\, dt. \end{aligned}$$

14.14 (Lijoi et al. 2005a) Let P follow the normalized inverse-Gaussian process with parameter α, where α is atomless and has support in $[c, \infty)$, for $c \geq -\infty$. Show that the posterior density of $L = \int x\, dP(x)$ based on $(X_1, \ldots, X_n) |\, P \overset{\text{iid}}{\sim} P$ is given by $\pi^{-1} I_{c+}^{n-1} \mathrm{Im}[\psi]$ if n is odd and $-\pi^{-1} I_{c+}^{n-1} \mathrm{Re}[\psi]$ if n is even, where I_{c+}^{m} is the fractional integral (11.1) and

$$\psi(s) = \frac{1}{Q}(n-1)!\,2^{n-1}|\alpha|^{-(2n-(2+K))} \int_0^\infty t^{n-1} e^{-\int \sqrt{1-2it(x-s)}\,d\alpha(x)}$$
$$\prod_{j=1}^{K} (1 - 2it(\tilde{X}_j - s))^{-N_{j,n}+1/2}\,dt,$$

for $Q = \sum_{r=0}^{n-1} \binom{n-1}{r}(-1)^r |\alpha|^{-2r} \Gamma(K + 2 + 2r - 2n; |\alpha|)$.

14.15 Let $(X, Y) \sim P$, where P follows a stick-breaking process with reference measure $\alpha = \alpha_1 \alpha_{2|1}$. Let P_1 be the marginal of $X \sim P$ and $P_{2|1}(\cdot|\, X)$ be the regular conditional distribution of Y given X. Show that if α_1 is atomless, P_1 has a stick-breaking representation. If $\alpha_{2|1}(\cdot|\, x)$ is also atomless for all x, then show that $P_{2|1}(\cdot|\, X)$ is δ_Y with $Y \sim \alpha_{2|1}(\cdot|\, X)$.

14.16 Consider density estimation with a general kernel in the setting of Theorem 7.15, but the mixing distribution follows a stick-breaking process with independent stick-breaking variables $V_1, V_2, \ldots$ Formulate an analogous consistency theorem.

14.17 (Lijoi et al. 2008) Show that the intersection of the class of Gibbs processes and normalized (homogeneous) CRM is the class of normalized generalized gamma processes.

14.18 (James et al. 2006) Show that if the posterior distribution for an NCRM under i.i.d. sampling is again an NCRM, then the resulting prior must be a Dirichlet process.

14.19 (IBP stick breaking) Let $\pi_{(1),K} \geq \pi_{(2),K} \geq \cdots \geq \pi_{(K),K}$ be the ordered values of a sample $\pi_1, \pi_2, \ldots \pi_K \stackrel{\text{iid}}{\sim} \text{Be}(M/K, 1)$. Then $\pi_{(k),K} = \prod_{i=1}^{k} \tau_{i,K}$, for $\tau_{i,K} := \pi_{(i),K}/\pi_{(i-1),K}$ and $\pi_{(0),K}$ interpreted as 1. Show that $\tau_{i,K} \stackrel{\text{iid}}{\sim} \text{Be}(M, 1)$, for $i = 1, \ldots, K$.

Appendix A

Space of Probability Measures

The space of all probability measures on a given sample space can be equipped with multiple topologies and metrics and corresponding measurable structures. In this appendix we study foremost the weak topology and its Borel σ-field, and next turn attention to a number of different metrics.

Throughout $\mathfrak{M}$ stands for the set of all probability measures on a "sample space" $(\mathfrak{X}, \mathscr{X})$.

A.1 Borel Sigma Field

The *Borel σ-field* $\mathscr{X}$ on a topological space $\mathfrak{X}$ is defined as the smallest σ-field making all open subsets of $\mathfrak{X}$ measurable. To avoid complications, we restrict to separable metric spaces $(\mathfrak{X}, d)$, unless otherwise specified, and often to a *Polish space*, a space with a topology that is generated by a metric that renders it complete and separable. The following result gives alternative characterizations of $\mathscr{X}$; its proof can be found in standard texts.

Theorem A.1 (Borel sigma field) *The Borel σ-field $\mathscr{X}$ on a separable metric space $\mathfrak{X}$ is also*

(i) *the smallest σ-field containing a generator $\mathscr{G}$ for the topology on $\mathfrak{X}$;*
(ii) *the smallest σ-field containing containing all balls of rational radius with center located in a countable dense subset of $\mathfrak{X}$;*
(iii) *the smallest σ-field making all continuous, real-valued functions on $\mathfrak{X}$ measurable.*

A *Borel isomorphism* between two measurable spaces is a one-to-one, onto, bimeasurable map; that is, the map and its inverse are both measurable. A measurable space is called a standard Borel space if it is Borel isomorphic to a Polish space. The cardinality of such a set is at most c, the cardinality of $\mathbb{R}$. A remarkable theorem in descriptive set theory says that any uncountable standard Borel space is Borel isomorphic to $\mathbb{R}$ (or equivalently the unit interval); see Theorem 3.3.13 of Srivastava (1998).

A.2 Weak Topology

The *weak topology* on $\mathfrak{M}$ is defined by the convergence of all nets $\int \psi\, P_\alpha \to \int \psi\, dP$, for ψ ranging over $\mathfrak{C}_b(\mathfrak{X})$, the space of all bounded continuous real-valued functions on $\mathfrak{X}$. Equivalently, it is the weakest topology that makes all maps $P \mapsto \int \psi\, dP$ continuous.

We denote the weak topology on $\mathfrak{M}$ by $\mathscr{W}$ and the weak convergence of a net P_α on $\mathfrak{M}$ to a $P \in \mathfrak{M}$ by $P_\alpha \rightsquigarrow P$.

The *portmanteau theorem* characterizes weak convergence in several different ways.

Theorem A.2 (Portmanteau) *The following statements are equivalent for any net P_α in $\mathfrak{M}$ and $P \in \mathfrak{M}$:*

(i) *P_α converges weakly to P;*
(ii) *$\int \psi\, dP_\alpha \to \int \psi\, dP$ for all bounded uniformly continuous $\psi: \mathfrak{X} \to \mathbb{R}$;*
(iii) *$\int \psi\, dP_\alpha \to \int \psi\, dP$ for all bounded continuous $\psi: \mathfrak{X} \to \mathbb{R}$ with compact support;*
(iv) *$\int \psi\, dP_\alpha \to \int \psi\, dP$ for all bounded Lipschitz continuous $\psi: \mathfrak{X} \to \mathbb{R}$;*
(v) *$\limsup_\alpha P_\alpha(F) \le P(F)$ for every closed subset F;*
(vi) *$\liminf_\alpha P_\alpha(G) \ge P(G)$ for every open subset G;*
(vii) *$\lim_\alpha P_\alpha(A) = P(A)$ for every Borel subset A with $P(\partial A) = 0$*[1] *where $\partial A = \bar{A} \cap \bar{A^c}$ stands for the topological boundary of A.*

If $\mathfrak{X}$ is a Euclidean space, then these statements are also equivalent to, for F_α and F the distribution functions of P_α and P,

(viii) *$F_\alpha(x) \to F(x)$ at all x such that $P(\{y: y_i \le x_i \text{ for all } i,\ y_i = x_i \text{ for some } i\}) = 0$.*

A base for the neighborhood system of the weak topology at a given $P_0 \in \mathfrak{M}$ is given by open sets of the form $\{P: |\int \psi_i\, dP - \int \psi_i\, dP_0| < \epsilon, i = 1, \ldots, k\}$, where $0 \le \psi_i \le 1$, $i = 1, \ldots, k$, are continuous functions, $k \in \mathbb{N}$ and $\epsilon > 0$. It follows that a subbase for the neighborhood system is given by $\{P: \int \psi\, dP < \int \psi\, dP_0 + \epsilon\}$, where $0 \le \psi \le 1$ is a continuous function. These neighborhoods are convex, and hence the weak topology is locally convex, that is, the weak topology admits a convex base for the neighborhood system.

The following theorem shows that the weak topology has simple and desirable topological properties.

Theorem A.3 (Weak topology) *The weak topology $\mathscr{W}$ on the set $\mathfrak{M}$ of Borel measures on a separable metric space $\mathfrak{X}$ is metrizable and separable. Furthermore,*

(i) *$\mathfrak{M}$ is complete if and only if $\mathfrak{X}$ is complete.*
(ii) *$\mathfrak{M}$ is Polish if and only if $\mathfrak{X}$ is Polish.*
(iii) *$\mathfrak{M}$ is compact if and only if $\mathfrak{X}$ is compact.*

The set of measures $\sum_{i=1}^k w_i \delta_{x_i}$ for $(w_i) \in (\mathbb{Q}^+)^k$, with $\sum_{i=1}^k w_i = 1$ and $k \in \mathbb{N}$, and (x_i) ranging over a countable dense subset of $\mathfrak{X}$ is dense in $\mathfrak{M}$.

Two metrics inducing $\mathscr{W}$ are the *Lévy-Prohorov distance* and *bounded Lipschitz distance*, given by

$$d_L(P, Q) = \inf\Big\{\epsilon > 0: P(A) < Q(A^\epsilon) + \epsilon,\ Q(A) < P(A^\epsilon) + \epsilon\Big\}, \tag{A.1}$$

$$d_{BL}(P, Q) = \sup_{\|f\|_{\mathfrak{C}^1} \le 1} \Big|\int f\, dP - \int f\, dQ\Big|, \tag{A.2}$$

[1] Such a set is known as a P-continuity set.

where $A^\epsilon = \{y: d(x, y) < \epsilon \text{ for some } x \in A\}$ and the supremum is over all functions $f: \mathfrak{X} \to [-1, 1]$ with $|f(x) - f(y)| \leq d(x, y)$ for all $x, y \in \mathfrak{X}$. The bounded Lipschitz distance can also be defined on the set of all signed Borel measures, and then is induced by the *bounded Lipschitz norm* $\|P\| = \sup\{|\int f\, dP|: \|f\|_{\mathfrak{C}^1} \leq 1\}$. Under this norm the signed Borel measures form a Banach space, with the set of all bounded Lipschitz functions as its dual space. This justifies the name *weak topology*.[2]

Prohorov's theorem characterizes the weakly compact subsets of $\mathfrak{M}$. A subset $\Gamma \subset \mathfrak{M}$ is called *tight* if, given any $\epsilon > 0$, there exists a compact subset K_ϵ of $\mathfrak{X}$ such that $P(K_\epsilon) \geq 1 - \epsilon$ for every $P \in \Gamma$.

Theorem A.4 (Prohorov) *If $\mathfrak{X}$ is Polish, then $\Gamma \subset \mathfrak{M}$ is pre-compact if and only if Γ is tight.*

Let $\mathscr{M}$ be the Borel σ-field on $\mathfrak{M}$ generated by the weak topology $\mathscr{W}$. If $\mathfrak{X}$ is Polish, then this is also the σ-field generated by the evaluation maps or "projections" $P \mapsto P(A)$.

Proposition A.5 (Borel sigma field weak topology) *If $\mathfrak{X}$ is Polish, and $\mathscr{X}_0$ is a generator of $\mathscr{X}$, then the Borel σ-field $\mathscr{M}$ on $\mathfrak{M}$ for the weak topology is also*

(i) *the smallest σ-field on $\mathfrak{M}$ making all maps $P \mapsto P(A)$ measurable, for $A \in \mathscr{X}$;*
(ii) *the smallest σ-field on $\mathfrak{M}$ making all maps $P \mapsto P(A)$ measurable, for $A \in \mathscr{X}_0$;*
(iii) *the smallest σ-field on $\mathfrak{M}$ making all maps $P \mapsto \int \psi\, dP$ measurable, for $\psi \in \mathfrak{C}_b(\mathfrak{X})$.*

Consequently, a finite measure on $(\mathfrak{M}, \mathscr{M})$ is completely determined by the set of distributions induced under the maps (a) or (b) or (c) given by

(a) $P \mapsto (P(A_1), \ldots, P(A_k))$, *for $A_1, \ldots, A_k \in \mathscr{X}_0$ and $k \in \mathbb{N}$;*
(b) $P \mapsto (P(A_1), \ldots, P(A_k))$, *for every partition $A_1, \ldots, A_k$ of $\mathfrak{X}$ in sets in $\mathscr{X}$ and $k \in \mathbb{N}$;*
(c) $P \mapsto \int \psi\, dP$, *for $\psi \in \mathfrak{C}_b(\mathfrak{X})$.*

Proof The maps $P \mapsto \int \psi\, dP$ in (iii) are $\mathscr{M}$-measurable, because weakly continuous, and hence generate a σ-field smaller than the weak Borel σ-field $\mathscr{M}$. Conversely, these maps generate the weak topology, so that finite intersections of sets of the form $\{P: \int \psi\, dP < c\}$ are a basis for the weakly open sets. Because the weak topology is Polish, every open set is a countable union of these basis sets, and hence is contained in the σ-field generated by the maps in (iii). Then so is $\mathscr{M}$, and the proof of (iii) is complete.

For (i) we first show that the map $P \mapsto P(C)$ is $\mathscr{M}$-measurable for every closed subset C of $\mathfrak{X}$. Because the set $C_n = \{y: d(x, y) < n^{-1} \text{ for some } x \in C\}$ is open, for $n \in \mathbb{N}$, by Urysohn's lemma there exists a continuous function $0 \leq \psi_n \leq 1$ such that $\psi_n(x) = 1$ for all $x \in C$ and $\psi_n(x) = 0$ for all $x \notin C_n$. Thus $\mathbb{1}\{x \in C\} \leq \psi_n(x) \leq \mathbb{1}\{x \in C_n\}$, and hence $P(C) \leq \int \psi_n\, dP \leq P(C_n) \to P(C)$, for any $P \in \mathfrak{M}$. Thus $P \mapsto P(C)$ is the limit of the sequence of $P \mapsto \int \psi_n\, dP$, and hence is $\mathscr{M}$-measurable by (iii).

[2] However, this name was used before this justification was found, and some authors prefer weak*-convergence over weak convergence, considering $\mathfrak{M}$ as embedded in the dual space of $\mathfrak{C}_b(\mathfrak{X})$.

Now the collection of sets $\{A \in \mathscr{X}: P \mapsto P(A) \text{ is measurable}\}$ can be checked to be a Λ-system.[3] As it contains the Π-system[4] of all closed sets, it must be equal to $\mathscr{X}$, by Dynkin's Π-Λ theorem. Thus *all* evaluation maps are $\mathscr{M}$-measurable.

Conversely, let $\bar{\mathscr{M}}$ be the smallest σ-field on $\mathfrak{M}$ making all maps $\{P \mapsto P(A),\ A \in \mathscr{X}\}$ measurable. For every continuous function $\psi: \mathfrak{X} \to \mathbb{R}$ there exists a sequence of simple functions with $0 \le \psi_n \uparrow \psi$. Clearly $P \mapsto \int \psi_n \, dP$ is $\bar{\mathscr{M}}$-measurable for every n, and so $P \mapsto \int \psi \, dP = \lim_{n\to\infty} \int \psi_n \, dP$ is also $\bar{\mathscr{M}}$-measurable. This implies that $\mathscr{M} \subset \bar{\mathscr{M}}$, in view of (iii).

The equivalence of (i) and (ii) follows from standard measure theoretic arguments using the good sets principle.

It remains to prove the assertions (a)–(c). The induced distribution of $(P(A_1), \ldots, P(A_k))$ determines the finite measure on the σ-algebra generated by the maps $P \mapsto P(A_i)$, for $i = 1, \ldots, k$. Thus the collection of all induced distributions of this type determines the measure on the union of all these σ-fields, which is a field, that generates $\mathscr{M}$ by (ii). Assertion (a) follows because fields are measure-determining. Next assertion (b) follows, because the coordinates of vectors in (a) can be written as finite sums over measures of partitions as in (b). The proof of (c) is similar to the proof of (a), after we first note that the set of *univariate* distributions of maps of the type $\int \sum_i a_i \psi_i \, dP$ for a given finite set $\psi_1, \ldots, \psi_k$ determines the joint distribution of the vector $(\int \psi_1 \, dP, \ldots, \int \psi_k \, dP)$. ☐

By Proposition A.5 a measurable function $P: (\Omega, \mathscr{A}) \to (\mathfrak{M}, \mathscr{M})$ from some measurable space $(\Omega, \mathscr{A})$ in $(\mathfrak{M}, \mathscr{M})$ is a Markov kernel from Ω to $\mathfrak{X}$: a map $P: \Omega \times \mathscr{X} \to [0, 1]$ such that, for every $A \in \mathscr{X}$, the map $\omega \mapsto P(\omega, A)$ is measurable and, for every $\omega \in \Omega$, the map $A \mapsto P(\omega, A)$ is a probability measure on $\mathfrak{X}$. If $(\Omega, \mathscr{A})$ is equipped with a probability measure, then the induced probability measure of the random measure P is called a *random probability distribution* on $(\mathfrak{M}, \mathscr{M})$.

The *expectation measure* $\mu(A) = \int P(A) \, d\Pi(P)$ of a given random probability distribution Π on $(\mathfrak{M}, \mathscr{M})$ is a well-defined probability measure on $(\mathfrak{X}, \mathscr{X})$. The tightness of a family of probability measures is equivalent to the tightness of the corresponding collection of expectation measures.

Theorem A.6 *A family of distributions* $\{\Pi_\lambda: \lambda \in \Lambda\}$ *on* $(\mathfrak{M}, \mathscr{M})$ *is tight if and only if the corresponding family* $\{\mu_\lambda: \lambda \in \Lambda\}$ *of expectation measures on* $(\mathfrak{X}, \mathscr{X})$ *is tight.*

Proof Fix $\epsilon > 0$. If $\{\Pi_\lambda: \lambda \in \Lambda\}$ is tight, there exists a compact set $\Gamma \subset \mathfrak{M}$ such that $\Pi_\lambda(\Gamma^c) < \epsilon/2$ for all $\lambda \in \Lambda$. By Prohorov's theorem, there exists a compact set $K \subset \mathfrak{X}$ such that $P(K^c) < \epsilon/2$ for all $P \in \Gamma$. Then, for all $\lambda \in \Lambda$,

$$\mu_\lambda(K^c) = \left(\int_{\Gamma^c} + \int_{\Gamma}\right) P(K^c) \, d\Pi_\lambda(P) \le \Pi_\lambda(\Gamma^c) + \sup_{P \in \Gamma} P(K^c) < \epsilon.$$

Conversely, given $\epsilon > 0$ and $m \in \mathbb{N}$, let K_m be a compact subset of $\mathfrak{X}$ such that $\mu_\lambda(K_m^c) < 6\epsilon\pi^{-2}m^{-3}$ for all $\lambda \in \Lambda$. Define closed sets $\Gamma_m = \{P \in \mathfrak{M}: P(K_m^c) \le m^{-1}\}$, and set

[3] A class of sets closed under countable disjoint union and proper differencing.
[4] A class of sets closed under finite intersection.

$\Gamma = \cap_{m=1}^{\infty} \Gamma_m$. If $\eta > 0$ is given and $m > \eta^{-1}$, then $P(K_m^c) < \eta$ for every $P \in \Gamma$. Therefore Γ is a tight family, and closed. By Prohorov's theorem it is compact. Now for all $\lambda \in \Lambda$,

$$\Pi_\lambda(\Gamma^c) \leq \sum_{m=1}^{\infty} \Pi_\lambda(\Gamma_m^c) \leq \sum_{m=1}^{\infty} m\mu_\lambda(K_m^c) \leq 6\epsilon\pi^{-2} \sum_{m=1}^{\infty} m^{-2} = \epsilon.$$

Here the second inequality follows, because $\Pi_\lambda(P: P(K_m^c) > m^{-1}) \leq m\mathrm{E}P(K_m^c)$. □

Many interesting subsets of $\mathfrak{M}$ or maps on this space are $\mathscr{M}$-measurable. The following results give some examples.

Proposition A.7 *The following sets are contained in* $\mathscr{M}$*:*

(i) *The set* $\{P \in \mathfrak{M}: P\{x: P(\{x\}) > 0\} = 1\}$ *of all discrete probability measures.*
(ii) *The set* $\{P \in \mathfrak{M}: P\{x: P(\{x\}) > 0\} = 0\}$ *of all atomless probability measures.*

Proof The set $\mathcal{E}$ as defined in Lemma A.8 (ii) is measurable in the product σ-field and hence the map $P \mapsto \int \mathbb{1}\{(P, x) \in \mathcal{E}\}\, dP(x) = P\{x: P(\{x\}) > 0\}$ is measurable by Lemma A.8 (i). The sets in (i) and (ii) are the sets where this map is 1 and 0, respectively. □

Lemma A.8 (i) *If* $f: \mathfrak{M} \times \mathfrak{X} \to \mathbb{R}$ *is* $\mathscr{M} \otimes \mathscr{X}$*-measurable and integrable, then the map* $P \mapsto \int f(P, x)\, dP(x)$ *is* $\mathscr{M}$*-measurable.*
(ii) *The set* $\mathcal{E} = \{(P, x) \in \mathfrak{M} \times \mathfrak{X}: P(\{x\}) > 0\}$ *is* $\mathscr{M} \otimes \mathscr{X}$*-measurable.*

Proof (i). For $f = \mathbb{1}_{\mathcal{U}}$ for a set $\mathcal{U}$ in $\mathscr{M} \otimes \mathscr{X}$, we have $\int f(P, x)\, dP(x) = P(\mathcal{U}_P)$, for $\mathcal{U}_P := \{x: (P, x) \in \mathcal{U}\}$ the P-cross section of $\mathcal{U}$. It can be checked that $\mathscr{G} = \{\mathcal{U} \in \mathscr{M} \otimes \mathscr{X}: P \mapsto P(\mathcal{U}_P)$ is $\mathscr{M}$-measurable$\}$ is a Λ-class. It contains the Π-class of all product type sets $\mathcal{U} = C \times A$, with $C \in \mathscr{M}$, $A \in \mathscr{X}$, since for a set of this type $P(\mathcal{U}_P) = P(A)\mathbb{1}\{P \in C\}$ is a measurable function of P, by Proposition A.5. By Dynkin's theorem $\mathscr{G}$ is equal to the product σ-field $\mathscr{M} \otimes \mathscr{X}$.

It follows that the lemma is true for every f of the form $f = \mathbb{1}_{\mathcal{U}}$, for $\mathcal{U}$ in $\mathscr{M} \otimes \mathscr{X}$. By routine arguments the result extends to simple functions, nonnegative measurable functions, and finally to all jointly measurable functions.

(ii). It suffices to show that $(P, x) \mapsto P(\{x\})$ is measurable. The class of sets $\{F \in \mathscr{X} \otimes \mathscr{X}: (P, x) \mapsto P(F_x)$ is measurable$\}$ can be checked to be a Λ-class of subsets of $\mathfrak{X} \times \mathfrak{X}$. It contains the Π-system of all measurable rectangles of $\mathfrak{X} \times \mathfrak{X}$, which generates the product σ-field. Hence by Dynkin's theorem this class is equal to $\mathscr{X} \otimes \mathscr{X}$. The result follows, because the diagonal $\{(x, x): x \in \mathfrak{X}\}$, is a product measurable subset. □

Proposition A.9 *The map* $\psi: \mathfrak{M}(\mathbb{R}) \times \mathbb{R} \to \mathfrak{M}(\mathbb{R})$ *defined by* $\psi(P, \theta) = P(\cdot - \theta)$ *is continuous. Furthermore, the restriction of* ψ *to* $\mathfrak{F} \times \mathbb{R}$ *for* $\mathfrak{F} \subset \mathfrak{M}(\mathbb{R})$ *a set of probability measures with a unique* q*th quantile at a given number* a *is one-to-one, and has an inverse* ψ^{-1} *which is also continuous. In particular, the map* $P(\cdot - \theta) \mapsto \theta$ *is continuous.*

Proof For $\theta_n \to \theta$ and a bounded uniformly continuous function $\phi: \mathbb{R} \to \mathbb{R}$, the sequence $\phi(x + \theta_n)$ converges uniformly in x to $\phi(x + \theta)$. Hence if $P_n \rightsquigarrow P$, then

$$\int \phi(x)\,dP_n(x-\theta_n) = \int \phi(x+\theta_n)\,dP_n(x) \to \int \phi(x+\theta)\,dP(x) = \int \phi(x)\,dP(x-\theta).$$

This proves the continuity of ψ.

If $P_n(\cdot - \theta_n) \rightsquigarrow P(\cdot - \theta)$ and every P_n and P have the same unique qth quantile, then the unique qth quantile $a+\theta_n$ of $P_n(\cdot - \theta_n)$ converges to the unique qth quantile $a+\theta$ of $P(\cdot - \theta)$. Thus $\theta_n \to \theta$, and for any uniformly continuous function ϕ,

$$\int \phi\,dP_n = \int \phi(x-\theta_n)\,dP_n(x-\theta_n) \to \int \phi(x-\theta)\,dP(x-\theta) = \int \phi\,dP.$$

We conclude that also $P_n \rightsquigarrow P$, whence ψ^{-1} is continuous. □

A.3 Other Topologies

Next to the weak topology some other topologies on $\mathfrak{M}$ are of interest. The *total variation distance* is defined as

$$d_{TV}(P,Q) = \|P-Q\|_{TV} = \sup_{A\in\mathscr{X}} |P(A)-Q(A)|. \tag{A.3}$$

Convergence $P_n \to P$ in the corresponding topology is equivalent to the uniform convergence of $P_n(A)$ to $P(A)$, for A ranging over all Borel sets in $\mathfrak{X}$. The supremum in the definition of d_{TV} does not change if it is restricted to a field that generates $\mathscr{X}$, as the probability of any set can be approximately closely by the probability of a set from the field. In particular, it can be restricted to a countable collection. It may also be checked that

$$d_{TV}(P,Q) = \sup\Bigl\{\Bigl|\int \psi\,dP - \int \psi\,dQ\Bigr|:\ \psi\colon\mathfrak{X}\to[0,1],\text{ measurable}\Bigr\}. \tag{A.4}$$

The space $\mathfrak{M}$ is complete under the total variation metric, but it is separable only if $\mathfrak{X}$ is countable, since $d_{TV}(\delta_x,\delta_y)=1$ for all $x\neq y$.

This nonseparability causes that the Borel σ-field relative to d_{TV}, generated by all d_{TV}-open subsets, is not equal to the ball σ-field on $\mathfrak{M}$, generated by all open d_{TV}-balls. In fact, in general the d_{TV}-ball σ-field is strictly smaller than $\mathscr{M}$, which in turn is strictly smaller than the d_{TV}-Borel σ-field (Problems A.2 and A.3). These properties make the total variation topology on the whole of $\mathfrak{M}$ somewhat clumsy. However, when restricted to subsets of $\mathfrak{M}$ the total variation topology may be separable, and the Borel σ-field more manageable. In particular, this is true for the set of all probability measures that are absolutely continuous with respect to a given measure ν.

Proposition A.10 (Domination) *The subset $\mathfrak{M}_0 \subset \mathfrak{M}$ of all probability measures on a standard Borel space that are absolutely continuous relative to a given σ-finite measure is complete and separable relative to the total variation metric. The trace of the d_{TV}-Borel σ-field on $\mathfrak{M}_0$ coincides with the trace of $\mathscr{M}$.*

Proof The first assertion follows, because $\mathfrak{M}_0$ is isomorphic to $\mathbb{L}_1(\nu)$, for ν the dominating measure, which is well known to be complete and separable. See Appendix B for details.

Because the supremum in the definition of d_{TV} can be restricted to a countable set, and every set $\{P\colon|P(A)-Q(A)|<\epsilon\}$ is contained in $\mathfrak{M}$ by Proposition A.5, every open total variation ball is contained in $\mathscr{M}$. In a separable metric space any open set is a countable union of open balls. Hence every (relatively) d_{TV}-open set in $\mathfrak{M}_0$ is contained in $\mathscr{M} \cap \mathfrak{M}_0$. □

For a Euclidean space $\mathfrak{X} = \mathbb{R}^k$ another topology on $\mathfrak{M}$ is given by the uniform convergence of the corresponding cumulative distribution functions (c.d.f.) $F(x) = P((-\infty, x])$, metrized by the familiar *Kolmogorov-Smirnov distance*

$$d_{KS}(P, Q) = \sup_{x \in \mathbb{R}} |P((-\infty, x]) - Q((-\infty, x])|. \tag{A.5}$$

This topology is intermediate between the weak and total variation topologies. Like the total variation distance, d_{KS} makes $\mathfrak{M}$ complete, but not separable (as $d_{KS}(\delta_x, \delta_y) = 1$ for all $x \neq y$, as before). Its main appeal is its importance in the Glivenko-Cantelli theorem. On the subspace of continuous probability measures the Kolmogorov-Smirnov and weak topologies coincide. This is a consequence of *Pólya's theorem.*

Proposition A.11 (Pólya) *Any sequence of distribution functions that converge weakly to a continuous distribution function converges also in the Kolmogorov-Smirnov metric. Consequently any d_{KS}-open neighborhood of an atomless $P_0 \in \mathfrak{M}(\mathbb{R}^k)$ contains a weak neighborhood of P_0.*

A fourth topology, also intermediate between the weak and total variation topologies, is the *topology of setwise convergence*: the weakest topology such that the maps $A \mapsto P(A)$, for $A \in \mathscr{X}$ are continuous. In this topology a net P_α converges to a limit P if and only if $P_\alpha(A) \to P(A)$, for all $A \in \mathscr{X}$. This topology has the inconvenience of being not metrizable (see Problem A.4). It is inherited from the product topology on $[0, 1]^{\mathscr{X}}$ after embedding $\mathfrak{M}$ through the map $P \mapsto (P(A): A \in \mathscr{X})$.[5] A second inconvenience is that $\mathfrak{M}$ is not measurable as a subset of $[0, 1]^{\mathscr{X}}$ relative to the product σ-field, since every product measurable subset is determined by only countably many coordinates. These disadvantages disappear by restricting to a countable collection $\{A_i: i \in \mathbb{N}\}$ of sets. The pointwise convergence $P_\alpha(A_i) \to P(A_i)$ for every $i \in \mathbb{N}$ is equivalent to convergence for the semimetric

$$d(P, Q) = \sum_{i=1}^{\infty} 2^{-i} |P(A_i) - Q(A_i)|.$$

For a sufficiently rich collection of sets, this will be stronger than weak convergence. For instance, all intervals in $\mathfrak{X} = \mathbb{R}^k$ with rational endpoints, or all finite unions of balls around points in a countable dense set of radii from a countable dense set in a general space $\mathfrak{X}$.

A.4 Support

Intuitively, the *support* of a measure on a measurable space $(\mathfrak{X}, \mathscr{X})$ is the smallest subset outside which no set gets positive measure. Unfortunately, such a smallest subset may not exist if $\mathfrak{X}$ is uncountable. For a topological space $\mathfrak{X}$ with its Borel σ-field a more fruitful notion of support of a measure μ is

[5] According to the *Vitale-Hahn-Sacks theorem,* $\mathfrak{M}$ is sequentially closed as a subset of $[0, 1]^{\mathscr{X}}$.

$$\text{supp}(\mu) = \Big(\bigcup_{\substack{U \text{ open},\\ \mu(U)=0}} U\Big)^c = \{x \in \mathfrak{X}: \mu(U) > 0 \text{ for every open } U \ni x\}. \tag{A.6}$$

Thus the support is the residual closed set that remains after removing all open sets of measure zero. For a probability measure P, the definition can be rewritten as

$$\text{supp}(P) = \bigcap_{\substack{C \text{ closed},\\ P(C)=1}} C = \{x \in \mathfrak{X}: P(U) > 0 \text{ for every open } U \ni x\}. \tag{A.7}$$

Clearly supp(P) is closed and no proper closed subset of it can have full probability content. For separable metric spaces the support itself also has probability one, and hence can be characterized as the *smallest closed subset of probability one*, or equivalently, as the complement of the largest open subset of probability zero. (This is true more generally if there exists a separable closed subset with probability one. For general topological spaces the support may fail to have full probability, and may even be empty.) In particular, the support of a prior distribution Π on the space $(\mathfrak{M}, \mathscr{M})$ of Borel probability measures on a separable metric space with the weak topology is given by

$$\text{supp}(\Pi) = \bigcap_{\substack{\mathcal{C} \text{ weakly closed},\\ \Pi(\mathcal{C})=1}} \mathcal{C} = \{P \in \mathfrak{M}: \Pi(\mathcal{U}) > 0 \text{ for every open } \mathcal{U} \ni P\}. \tag{A.8}$$

A.5 Historical Notes

The results in this appendix are all well known, albeit not easy to find in one place. General references include Parthasarathy (2005), Billingsley (1968), van der Vaart and Wellner (1996) and Dudley (2002). In particular for a proof of the Portmanteau theorem, see Parthasarathy (2005), pages 40–42, or van der Vaart (1998), Lemma 2.2; for Prohorov's theorem, see Parthasarathy (2005), Theorem 6.7, or van der Vaart and Wellner (1996), Theorem 1.3.8; for metrizability and separability, see van der Vaart and Wellner (1996), Chapter 1.12, or Parthasarathy (2005), pages 39–52.

Problems

A.1 Show that $x \mapsto \delta_x$ is a homeomorphism between $\mathfrak{X}$ and the space of all degenerate probability measures under the weak topology. In particular, the correspondence is a Borel isomorphism.

A.2 Show that the trace of the σ-field generated by all total variation open sets on the space of all degenerate probability measures is the power set, and hence has cardinality bigger than $\mathbb{R}$. As $\mathfrak{M}$ is a Polish space under the weak topology, its cardinality equals that of $\mathbb{R}$. Conclude that the Borel σ-field corresponding to the total variation topology is strictly bigger than $\mathscr{M}$.

A.3 Show that the σ-field generated by all total variation open balls is strictly smaller than $\mathscr{M}$.

A.4 Consider the topology of setwise convergence on $\mathfrak{M}$.

(a) Find an uncountable collection of disjoint open sets under the topology. Thus $\mathfrak{M}$ is not separable under the topology induced by the setwise convergence.
(b) Show that a continuous probability measure P_0 does not have a countable base of its neighborhood system. Thus $\mathfrak{M}$ is neither separable nor metrizable.

A.5 Kuratowski's theorem implies that the inverse of a Borel measurable one-to-one map from standard Borel space onto a metrizable space is also measurable (see Corollary 4.5.5 of Srivastava 1998). Use this to give another proof of the fact that the strong and the weak topologies on the space of absolutely continuous probability measures with respect to a σ-finite measure ν generate the same σ-field.

Appendix B

Space of Probability Densities

There are many useful distances on a set of density functions, some of a statistical origin. A number of "divergences" lack symmetry, but similarly measure discrepancies between pairs of densities. In this chapter we review many distances, divergences and their relations, with particular attention for distances on probability densities.

Throughout the chapter $(\mathfrak{X}, \mathscr{X})$ is a measurable space equipped with σ-finite measure ν. A "density" is a measurable function $p: \mathfrak{X} \to \mathbb{R}$, typically nonnegative and integrable relative to ν. We denote a measurable function $p: \mathfrak{X} \to \mathbb{R}$ by a lowercase letter, and the induced measure $A \mapsto \int_A p\, d\nu$ by the corresponding upper case letter P.

B.1 Distances and Divergences

As probability densities are nonnegative and integrate to one, the γth power of their γth root is well defined and integrable (with integral one), and hence the distance $r_\gamma(p, q) = (\int |p^{1/\gamma} - q^{1/\gamma}|^\gamma\, d\nu)^{(1/\gamma)\wedge 1}$ is well defined for any $\gamma > 0$. The special cases $\gamma = 1$ and $\gamma = 2$ are the $\mathbb{L}_1$-*distance* (equivalent with the *total variation distance*) and *Hellinger distance*, given by

$$\|p - q\|_1 = \int |p - q|\, d\nu = 2 - 2\int p \wedge q\, d\nu = 2\|P - Q\|_{TV}, \tag{B.1}$$

$$d_H(p, q) = \Big(\int (\sqrt{p} - \sqrt{q})^2\, d\nu\Big)^{1/2}.$$

The total variation distance was defined in (A.3) as a supremum over sets; in terms of densities this supremum can be seen to be attained at the set $\{x: p(x) > q(x)\}$. All distances r_γ are invariant with respect to changing the dominating measure. For the $\mathbb{L}_1$-distance this is clear from its expression as $2\|P - Q\|_{TV}$. A similar, symbolic expression for the Hellinger distance, which also eliminates the dominating measure, is $(\int (\sqrt{dP} - \sqrt{dQ})^2)^{1/2}$. On the set of probability densities the two distances are maximally 2 and $\sqrt{2}$, respectively; the maximum values are attained by pairs of probability densities with disjoint regions of positivity.

For densities whose rth power is integrable, the $\mathbb{L}_r$-*distance* is defined as

$$\|p - q\|_r = \Big(\int |p - q|^r\, d\nu\Big)^{(1/r)\wedge 1}.$$

The case $r = 2$ is often used in classical density estimation, because the corresponding risks admit explicit expressions. A drawback of these distances is, that they are not invariant under the choice of a dominating measure.

The limiting case $r = \infty$ is the *essential supremum* of $|p - q|$ relative to ν. This is bounded above by the *supremum distance* (or *uniform distance*)

$$\|p - q\|_\infty = \sup_{x \in \mathfrak{X}} |p(x) - q(x)|.$$

Even if this is too strong in some applications, it can be useful as a technical device, especially when calculating the size of a space. Estimates of entropies (see Appendix C) relative to the supremum distance are readily available for several function spaces, and carry over to other distances, by domination (on bounded domains) or under other restrictions (e.g. for normal mixtures). All metrics r_γ for $\gamma \geq 1$ and $\mathbb{L}_r$ for $r \geq 1$ lead to locally convex topologies.

A fundamental quantity, both in non-Bayesian and Bayesian statistics, is the *Kullback-Leibler divergence* (or KL divergence)

$$K(p; q) = \int p \log(p/q)\, d\nu.$$

Here, $\log(p/q)$ is understood to be ∞ if $q = 0 < p$, meaning that $K(p; q) = \infty$ if $P(q = 0) > 0$. (Because $\log_-(p/q) \leq q/p - 1$, and $p(q/p - 1)$ is integrable, the integral $K(p; q)$ is always well defined.) The importance of the KL divergence stems from the fact that the likelihood ratio $\prod_{i=1}^n (q/p)(X_i)$ under i.i.d. sampling from p roughly behaves like $e^{-nK(p;q)}$. Like the total variation and Hellinger distances, the Kullback-Leibler divergence is not dependent on the choice of the dominating measure. Furthermore, it is nonnegative, and $K(p; q) = 0$ if and only if $P = Q$. However it is not a metric – it lacks both symmetry and transitivity. The asymmetry is sometimes rectified by considering the *symmetrized KL divergence* $K_S(p, q) = K(p; q) \wedge K(q; p)$, especially in model selection problems and in the context of reference priors.

Higher-order KL divergences and their centered versions are defined by, for $k > 1$,

$$V_k(p; q) = \int p \big|\log(p/q)\big|^k\, d\nu,$$

$$V_{k,0}(p; q) = \int p \big|\log(p/q) - K(p; q)\big|^k\, d\nu. \tag{B.2}$$

The most important case is $k = 2$. We refer to them as *Kullback-Leibler variations.*

The following lemmas describe relationships between these distances and discrepancies. The $\mathbb{L}_1$- and Hellinger distances induces the same topology, which is stronger than the weak topology, and called the *strong topology* or *norm topology*. The norm and weak topologies agree on norm-compact classes of densities. Furthermore, they induce the same σ-field on any class of densities, by Proposition A.10.

Lemma B.1 *For any pair of probability densities p, q,*

(i) $\|p - q\|_1 \leq d_H(p, q)\sqrt{4 - d_H^2(p, q)} \leq 2d_H(p, q)$.

(ii) $d_H^2(p, q) \leq \|p - q\|_1$.

(iii) $\|p-q\|_1^2 \le 2K(p;q)$ *(Kemperman's inequality or Pinsker's inequality).*
(iv) $d_H^2(p,q) \le K(p;q)$.
(v) $d_H(p,q) \le \|(\sqrt{p}+\sqrt{q})^{-1}\|_\infty \|p-q\|_2$.
(vi) $\|p-q\|_2 \le \|\sqrt{p}+\sqrt{q}\|_\infty d_H(p,q)$.
(vii) $\|p-q\|_r \le \|p-q\|_\infty \nu(\mathfrak{X})^{1/r}$, *for* $r \ge 1$.

Proof For (i) we factorize $|p-q| = |\sqrt{p}-\sqrt{q}|\,|\sqrt{p}+\sqrt{q}|$, and next use the Cauchy-Schwarz inequality to obtain the bound $\|p-q\|_1 \le d_H(p,q)\|\sqrt{p}+\sqrt{q}\|_2$. Finally we compute $\|\sqrt{p}+\sqrt{q}\|_2^2 = 2+2\int\sqrt{p}\sqrt{q}\,d\nu = 4-d_H^2(p,q)$.

Assertion (ii) follows from the inequality $|\sqrt{p}-\sqrt{q}|^2 \le |p-q|$, for all $p,q\ge 0$.

(iii). Apply the inequality $|x-1| \le ((4+2x)/3)^{1/2}(x\log x - x + 1)^{1/2}$, which is valid for $x \ge 0$, followed by the Cauchy-Schwarz inequality, to $\int |p/q-1|\,dQ$.

(iv). Since $\log x \le 2(\sqrt{x}-1)$, for $x\ge 0$, we have $K(p;q) = -2\int p\log(\sqrt{q}/\sqrt{p})\,d\nu \ge -2\int p(\sqrt{q/p}-1)\,d\nu$. This can be rewritten as $2-2\int\sqrt{pq}\,d\nu = d_H^2(p,q)$.

Assertions (v) and (vi) follow from the factorization as (i); (vii) is immediate from the definitions. ∎

Inequality (iv) cannot be reversed in general, as the Hellinger distance is bounded by $\sqrt{2}$ and the Kullback-Leibler divergence can be infinite. However, if the quotient p/q is bounded (or suitably integrable), then the square Hellinger distance and Kullback-Leibler convergence are (almost) comparable.

Lemma B.2 *For every $b>0$, there exists a constant $\epsilon_b>0$ such that for all probability densities p and densities q with $0<d_H^2(p,q)<\epsilon_b P(p/q)^b$,*

$$K(p;q) \lesssim d_H^2(p,q)\Big(1+\frac{1}{b}\log_- d_H(p,q)+\frac{1}{b}\log_+ P\Big(\frac{p}{q}\Big)^b\Big)+1-Q(\mathfrak{X}),$$

$$V_2(p;q) \lesssim d_H^2(p,q)\Big(1+\frac{1}{b}\log_- d_H(p,q)+\frac{1}{b}\log_+ P\Big(\frac{p}{q}\Big)^b\Big)^2.$$

Furthermore, for every pair of probability densities p and q and any $0<\epsilon<0.4$,

$$K(p;q) \le d_H^2(p,q)(1+2\log_-\epsilon)+2P\Big[\Big(\log\frac{p}{q}\Big)\mathbb{1}\{q/p\le\epsilon\}\Big],$$

$$V_2(p;q) \le d_H^2(p,q)(12+2\log_-^2\epsilon)+8P\Big[\Big(\log\frac{p}{q}\Big)^2\mathbb{1}\{q/p\le\epsilon\}\Big].$$

Consequently, for every pair of probability densities p and q,

$$K(p;q) \lesssim d_H^2(p,q)\Big(1+\log\Big\|\frac{p}{q}\Big\|_\infty\Big) \le 2d_H^2(p,q)\Big\|\frac{p}{q}\Big\|_\infty,$$

$$V_2(p;q) \lesssim d_H^2(p,q)\Big(1+\log\Big\|\frac{p}{q}\Big\|_\infty\Big)^2 \le 2d_H^2(p,q)\Big\|\frac{p}{q}\Big\|_\infty.$$

Proof The function $r\colon(0,\infty)\to\mathbb{R}$ defined implicitly by $\log x = 2(\sqrt{x}-1)-r(x)(\sqrt{x}-1)^2$ possesses the following properties:

(i) r is nonnegative and decreasing.
(ii) $r(x)\sim\log_- x$ as $x\downarrow 0$ and $r(x)\le 2\log_- x$ for $x\in[0,0.4]$.

(iii) For every $b > 0$ there exists $\epsilon'_b > 0$ such that $x \mapsto x^b r(x)$ is increasing on $[0, \epsilon'_b]$.

By these properties and the identity $d_H^2(p,q) = -2P(\sqrt{q/p} - 1) + 1 - Q(\mathfrak{X})$,

$$K(p;q) + Q(\mathfrak{X}) - 1 = d_H^2(p,q) + P\Big[r\Big(\frac{q}{p}\Big)\Big(\sqrt{\frac{q}{p}} - 1\Big)^2\Big]$$
$$\le d_H^2(p,q) + r(\epsilon) d_H^2(p,q) + P\Big[r\Big(\frac{q}{p}\Big)\mathbb{1}_{q/p\le\epsilon}\Big],$$
$$\le d_H^2(p,q) + 2(\log_- \epsilon) d_H^2(p,q) + 2\epsilon^b \log_- \epsilon\, P\Big(\frac{p}{q}\Big)^b,$$

for $\epsilon \le \epsilon'_b \wedge 0.4$. The first inequality now follows by choosing $\epsilon^b = d_H^2(p,q)/P(p/q)^b$ and $\epsilon_b = (\epsilon'_b \wedge 0.4)^b$. The first inequality of the second pair is the intermediate second result in the display.

For the proof of the second inequality, we first use the inequality $\log x \le 2(\sqrt{x} - 1)$ to see that

$$P\Big[\Big(\log \frac{p}{q}\Big)^2 \mathbb{1}_{q/p\ge 1}\Big] \le 4P\Big(\sqrt{\frac{q}{p}} - 1\Big)^2 = 4d_H^2(p,q).$$

Next, for $\epsilon \le \epsilon'_{b/2}$, in view of the third property of r,

$$P\Big[\Big(\log \frac{p}{q}\Big)^2 \mathbb{1}_{q/p\le 1}\Big] \le 8P\Big(\sqrt{\frac{q}{p}} - 1\Big)^2 + 2P\Big[r^2\Big(\frac{q}{p}\Big)\Big(\sqrt{\frac{q}{p}} - 1\Big)^4 \mathbb{1}_{q/p\le 1}\Big]$$
$$\le 8d_H^2(p,q) + 2r^2(\epsilon) d_H^2(p,q) + 2\epsilon^b r^2(\epsilon) P\Big(\frac{p}{q}\Big)^b.$$

With $\epsilon^b = d_H^2(p,q)/P(p/q)^b$ and $\epsilon_b \le (0.4 \wedge \epsilon'_{b/2})^b$, this can be bounded by the right side of the second inequality, by property (ii). The second inequality in the second pair is again the intermediate result.

For $b \ge 1$ we can choose $\epsilon'_b = 1$ in property (iii). Furthermore, if we replace the inequality in (ii) by $r(x) \le 2 + 2\log_- x$, which is valid for $x \in [0,1]$, then we can choose $\epsilon_b = 1$ later in the proof. This leads to a multiple of the bounds as obtained before, which are then valid for every $b \ge 2$ and any probability densities p and q with $d_H^2(p,q) \le P(p/q)^b$. Here $P(p/q)^b = Q(p/q)^{b+1} \ge (Q(p/q))^{b+1} \ge 1$ for $b > 1$, by Jensen's inequality. Thus the bounds are true for every sufficiently large b and every p and q with $d_H^2(p,q) \le 1$. Now as $b \uparrow \infty$, we have $b^{-1}\log_- d_H(p,q) \to 0$ and $(P(p/q)^b)^{1/b} \to \|p/q\|_\infty$. □

Lemma B.3 *For any pair of probability densities p and q, and $k \in \mathbb{N}$, $k \ge 2$,*

$$P\Big[e^{|\log(q/p)|} - 1 - |\log(q/p)|\Big] \le 2d_H^2(p,q)\Big\|\frac{p}{q}\Big\|_\infty, \tag{B.3}$$
$$V_k(p;q) \le k!2d_H^2(p,q)\Big\|\frac{p}{q}\Big\|_\infty. \tag{B.4}$$

Proof For every $c \ge 0$ and $x \ge -c$, we have the inequality $e^{|x|} - 1 - |x| \le 2e^c(e^{x/2} - 1)^2$. If $c = \log\|p/q\|_\infty$, then $\log(q/p) \ge -c$ and hence the integrand on the left side of (B.3) is bounded above by $2e^c(e^{\frac{1}{2}\log(q/p)} - 1)^2 = 2e^c(\sqrt{q/p} - 1)^2$. The result now follows by integration.

The second inequality follows by expanding the exponential in the left side of the first inequality. □

The Kullback-Leibler divergence may not be continuous in its arguments, but is always lower semicontinuous in its second argument.

Lemma B.4 *If p, q_n, q are probability densities such that $\|q_n - q\|_1 \to 0$ as $n \to \infty$, then* $\liminf_{n\to\infty} K(p; q_n) \geq K(p; q)$.

Proof If $X_n = q_n/p$ and $X = q/p$, then $X_n \to X$ in P-probability and in mean by assumption. We can write $K(p; q_n)$ as the sum of $\mathrm{E}(\log X_n)\mathbb{1}_{X_n>1}$ and $\mathrm{E}(\log X_n)\mathbb{1}_{X_n\leq 1}$. Because $0 \leq (\log x)\mathbb{1}_{x>1} \leq x$, the sequence $(\log X_n)\mathbb{1}_{X_n>1}$ is dominated by $|X_n|$, and hence is uniformly integrable. Because $x \mapsto (\log x)\mathbb{1}_{x>0}$ is continuous, $\mathrm{E}(\log X_n)\mathbb{1}_{X_n>1} \to \mathrm{E}(\log X)\mathbb{1}_{X>1}$. Because the variables $(\log X_n)\mathbb{1}_{X_n<1}$ are nonpositive, we can apply Fatou's lemma to see that $\limsup \mathrm{E}(\log X_n)\mathbb{1}_{X_n<1} \leq \mathrm{E}(\log X)\mathbb{1}_{X<1}$. □

B.2 Hellinger Transform, Affinity and Other Divergences

Other measures of "divergence" between densities are the *Hellinger transform*, *α-divergence* and *Renyi divergence*, given by

$$\rho_\alpha(p; q) = \int p^\alpha q^{1-\alpha}\, d\nu, \tag{B.5}$$

$$D_\alpha(p; q) = 1 - \rho_\alpha(p; q), \tag{B.6}$$

$$R_\alpha(p; q) = -\log \rho_\alpha(p; q). \tag{B.7}$$

The Hellinger transform at $\alpha = 1/2$ is also called *affinity*. For $0 \leq \alpha \leq 1$ and integrable functions p, q, the Hellinger transform is finite (and bounded by one for probability densities), by Hölder's inequality, but this is not guaranteed for $\alpha > 1$. The Hellinger transform for $\alpha = -1$ is equal to twice the *χ^2-divergence* $\int (p-q)^2/q\, d\nu$.

Lemma B.5 *For a probability density p and a density q, the function $\alpha \mapsto \rho_\alpha(p; q)$ is convex on* $[0, 1]$ *with limits $Q(p > 0)$ and $P(q > 0)$ as $\alpha \downarrow 0$ or $\alpha \uparrow 1$, respectively, and derivatives from the right and left equal to $-K(q\mathbb{1}_{p>0}; p)$ and $K(p\mathbb{1}_{q>0}; q)$ at $\alpha = 0$ and $\alpha = 1$, respectively. Furthermore, for $0 < \alpha < 1$ and probability densities p, q,*

(i) $2D_{1/2}(p; q) = d_H^2(p, q) = 2 - 2\rho_{1/2}(p, q)$.
(ii) $\|p - q\|_1^2 \leq 4(1 - \rho_{1/2}^2(p, q))$
(iii) $\rho_{1/2}(p, q) \geq 1 - \|p - q\|_1/2$.
(iv) $\min(\alpha, 1-\alpha) d_H^2(p, q) \leq D_\alpha(p; q)$.
(v) $D_\alpha(p; q) \leq d_H^2(p, q)$.
(vi) $D_\alpha(p; q) \leq R_\alpha(p; q) \leq D_\alpha(p; q)/\rho_\alpha(p; q)$.

Proof The function $\alpha \mapsto e^{\alpha y}$ is convex on $(0, 1)$ for all $y \in [-\infty, \infty)$, implying the convexity of $\alpha \mapsto \rho_\alpha(q; p) = P(q/p)^\alpha$ on $(0, 1)$. The function $\alpha \mapsto y^\alpha = e^{\alpha \log y}$ is continuous on [0, 1] for any $y > 0$, is decreasing for $y < 1$, increasing for $y > 1$ and constant for $y = 1$. By the monotone convergence theorem, as $\alpha \downarrow 0$,

$$Q\Big[\Big(\frac{p}{q}\Big)^{\alpha} \mathbb{1}_{0<p<q}\Big] \uparrow Q\Big[\Big(\frac{p}{q}\Big)^{0} \mathbb{1}_{0<p<q}\Big] = Q(0 < p < q). \tag{B.8}$$

By the dominated convergence theorem, with the functions $(p/q)^{\alpha}\mathbb{1}_{p\geq q}$ for $\alpha \leq \frac{1}{2}$ dominated by $(p/q)^{1/2}\mathbb{1}_{p\geq q}$, we have, as $\alpha \to 0$,

$$Q\Big[\Big(\frac{p}{q}\Big)^{\alpha} \mathbb{1}_{p\geq q}\Big] \to Q\Big[\Big(\frac{p}{q}\Big)^{0} \mathbb{1}_{p\geq q}\Big] = Q(p \geq q). \tag{B.9}$$

Together (B.8) and (B.9) show that $\rho_\alpha(q; p) = Q(p/q)^{\alpha} \to Q(p > 0)$, as $\alpha \downarrow 0$.

By the convexity of the function $\alpha \mapsto e^{\alpha y}$, the map $\alpha \mapsto f_\alpha(y) = (e^{\alpha y} - 1)/\alpha$ decreases as $\alpha \downarrow 0$, to $(d/d\alpha)|_{\alpha=0}e^{\alpha y} = y$, for every y. For $y \leq 0$ we have $f_\alpha(y) \leq 0$, while for $y \geq 0$, by Taylor's formula,

$$f_\alpha(y) \leq \sup_{0<\alpha'\leq\alpha} y e^{\alpha' y} \leq y e^{\alpha y} \leq \epsilon^{-1} e^{(\alpha+\epsilon)y}.$$

Hence we conclude that $f_\alpha(y) \leq 0 \vee \epsilon^{-1} e^{(\alpha+\epsilon)y}\mathbb{1}_{y\geq 0}$. Consequently $\alpha^{-1}(e^{\alpha \log(p/q)} - 1)$ decreases to $\log(p/q)$ as $\alpha \downarrow 0$ and is bounded above by $0 \vee \epsilon^{-1}(p/q)^{2\epsilon}\mathbb{1}_{p\geq q}$ for small $\alpha > 0$, which is Q-integrable for $\epsilon < \frac{1}{2}$. We conclude that

$$\frac{1}{\alpha}(\rho_\alpha(p; q) - \rho_0(p; q)) = \frac{1}{\alpha} Q\Big[((p/q)^{\alpha} - 1)\mathbb{1}_{p>0}\Big] \downarrow Q\Big[\log(p/q)\mathbb{1}_{p>0}\Big],$$

as $\alpha \downarrow 0$, by the monotone convergence theorem.

Assertion (i) is clear by expanding the square in the integrand of d_H^2. Next (ii) and (iii) are rewrites of the first (not the second!) inequality in (i) and of (ii) of Lemma B.1, respectively.

(iv). By convexity $\rho_\alpha(p; q) \leq (1 - 2\alpha)\rho_0(p; q) + 2\alpha\rho_{1/2}(p; q)$, for $\alpha \leq 1/2$, which can be further bounded by $1 - 2\alpha + 2\alpha\rho_{1/2}(p; q)$. This can be rearranged to obtain (iv) for $\alpha \leq 1/2$. The case $\alpha > 1/2$ follows similarly or by symmetry.

(v). By convexity $\rho_{1/2}(p; q) \leq (1/2)\rho_\alpha(p; q) + (1/2)\rho_{1-\alpha}(p; q)$, for $\alpha \leq 1/2$, which can be further bounded by $(1/2)\rho_\alpha(p; q) + 1/2$.

(vi). This follows by applying the inequalities $1 - x \leq -\log x \leq x^{-1} - 1$, valid for $x > 0$, to $x = \rho_\alpha(p; q)$. □

If $Q \ll P$, then $K(q\mathbb{1}_{p>0}; p) = K(q; p)$, and hence the lemma shows that the right derivative at $\alpha = 0$ of the Hellinger transform is equal to $-K(q; p)$. Similarly if $P \ll Q$, then the left derivative at $\alpha = 1$ is $K(p; q)$. Without absolute continuity minus/plus these derivatives are not equal to the Kullback-Leibler divergences, and may also be negative, even when both P and Q are probability measures.

The following two lemmas quantify the remainder if $\rho_\alpha(p; q)$ is approximated by its Taylor expansion at $\alpha = 0$, in a "misspecified setting", when the "true measure" P_0 can be different from P.

Lemma B.6 *There exists a universal constant C such that for any probability measure P_0 and all finite measures $P, Q, Q_1, \ldots, Q_m$ and constants $0 < \alpha \leq 1$, $\lambda_i \geq 0$ with $\sum_{i=1}^m \lambda_i = 1$,*

$$\Big|1 - P_0\Big(\frac{q}{p}\Big)^{\alpha} - \alpha P_0 \log\frac{p}{q}\Big| \le \alpha^2 C P_0\Big[\log^2\frac{p}{q}\Big(\Big(\frac{q}{p}\Big)^{\alpha}\mathbb{1}\{q > p\} + \mathbb{1}\{q \le p\}\Big)\Big],$$

$$\Big|1 - P_0\Big(\frac{\sum_{i=1}^m \lambda_i q_i}{p}\Big)^{\alpha} - \alpha P_0 \log\frac{p}{\sum_{i=1}^m \lambda_i q_i}\Big| \le 2\alpha^2 C \sum_{i=1}^m \lambda_i P_0\Big[\log^2\frac{q_i}{p}\Big\{\Big(\frac{q_i}{p}\Big)^2 + 1\Big\}\Big].$$

Proof The function R defined by $R(x) = (e^x - 1 - x)/(x^2 e^x)$ if $x > 0$, $(e^x - 1 - x)/x^2$ if $x < 0$ and $R(0) = \frac{1}{2}$ is uniformly bounded on $\mathbb{R}$ by a constant $C < \infty$. The first inequality follows by applying this with $x = \alpha \log(q/p)$.

Taking $q = \sum_{i=1}^m \lambda_i q_i$ in the first inequality, we see that for the proof of the second it suffices to bound

$$P_0\Big[\Big(\log\frac{\sum_{i=1}^m \lambda_i q_i}{p}\Big)^2\Big\{\Big(\frac{\sum_{i=1}^m \lambda_i q_i}{p}\Big)^{\alpha}\mathbb{1}\{\sum_{i=1}^m \lambda_i q_i > p\} + \mathbb{1}\{\sum_{i=1}^m \lambda_i q_i \le p\}\Big\}\Big].$$

Replacing α by 2 makes the expression larger. Next we bound the two terms corresponding to the decomposition by indicators separately.

If $\sum_{i=1}^m \lambda_i q_i > p$, then we first bound, using convexity of $x \mapsto x \log x$,

$$\Big(\log\frac{\sum_{i=1}^m \lambda_i q_i}{p}\Big)\Big(\frac{\sum_{i=1}^m \lambda_i q_i}{p}\Big) \le \sum_{i=1}^m \lambda_i \Big(\log\frac{q_i}{p}\Big)\Big(\frac{q_i}{p}\Big). \tag{B.10}$$

Since the left side is positive, squaring preserves the inequality, where on the right side the square can be lowered within the sum, by Jensen's inequality.

If $\sum_{i=1}^m \lambda_i q_i < p$, then concavity of the logarithm gives

$$0 < -\log\Big(\sum_{i=1}^m \lambda_i q_i/p\Big) \le -\sum_{i=1}^m \lambda_i \log(q_i/p).$$

As before, we square this and next lower the square into the sum by Jensen's inequality. □

Lemma B.7 *There exists a universal constant C such for any probability measure P_0 and any finite measures $P, Q, Q_1, \ldots, Q_m$ and any $0 < \alpha \le 1$, $\lambda_1, \ldots, \lambda_m \ge 0$ with $\sum_{i=1}^m \lambda_i = 1$ and $0 < \alpha \le 1$,*

$$\Big|1 - P_0\Big(\frac{q}{p}\Big)^{\alpha} - \alpha P_0 \log\frac{p}{q}\Big| \le \alpha^2 C P_0\Big[\Big(\sqrt{\frac{q}{p}} - 1\Big)^2\mathbb{1}\{q > p\} + \log^2\frac{p}{q}\mathbb{1}\{q \le p\}\Big],$$

$$\Big|1 - P_0\Big(\frac{\sum_{i=1}^m \lambda_i q_i}{p}\Big)^{\alpha} - \alpha P_0 \log\frac{p}{\sum_{i=1}^m \lambda_i q_i}\Big| \le 2\alpha^2 C \sum_{i=1}^m \lambda_i P_0\Big[\Big(\sqrt{\frac{q_i}{p}} - 1\Big)^2 + \log^2\frac{q_i}{p}\Big].$$

Proof The function R defined by $R(x) = (e^x - 1 - x)/\alpha^2(e^{x/2\alpha} - 1)^2$ for $x \ge 0$ and $R(x) = (e^x - 1 - x)/x^2$ for $x \le 0$ is uniformly bounded on $\mathbb{R}$ by a constant C, independent of $\alpha \in (0, 1]$. Now the proof follows by proceeding as in the proof of Lemma B.6 and using the convexity of $x \mapsto |\sqrt{x} - 1|^2$ on $[0, \infty)$. □

B.3 Product Densities

Given densities p_i relative to σ-finite measures ν_i on measurable spaces $(\mathfrak{X}_i, \mathscr{X}_i)$, let $\otimes_{i=1}^n p_i$ denote the density of the product measure. Some distances are friendlier than other with product densities. For instance, the following lemma bounds the rate of growth of the $\mathbb{L}_1$-distance between products of n identical densities as n, whereas for the Hellinger distance the rate is $\sqrt{n}$.

Lemma B.8 *For any probability densities $p_1, \ldots, p_n$ and $q_1, \ldots, q_n$,*

(i) $\| \otimes_{i=1}^n p_i - \otimes_{i=1}^n q_i \|_1 \le \sum_{i=1}^n \|p_i - q_i\|_1$.
(ii) $\rho_\alpha(\otimes_{i=1}^n p_i; \otimes_{i=1}^n q_i) = \prod_{i=1}^n \rho_\alpha(p_i; q_i)$.
(iii) $d_H^2(\otimes_{i=1}^n p_i, \otimes_{i=1}^n q_i) \le \sum_{i=1}^n d_H^2(p_i, q_i)$.
(iv) $\rho_\alpha(\otimes_{i=1}^n p_i; \otimes_{i=1}^n q_i) \le \prod_{i=1}^n [1 - \beta d_H^2(p_i, q_i)]$, *for* $\beta = \alpha \wedge (1-\alpha)$.
(v) $K(\otimes_{i=1}^n p_i; \otimes_{i=1}^n q_i) = \sum_{i=1}^n K(p_i; q_i)$.
(vi) $V_{2,0}\left(\otimes_{i=1}^n p_i; \otimes_{i=1}^n q_i\right) = \sum_{i=1}^n V_{2,0}(p_i; q_i)$.
(vii) $V_{k,0}\left(\otimes_{i=1}^n p_i; \otimes_{i=1}^n q_i\right) \le d_k n^{k/2-1} \sum_{i=1}^n V_{k,0}(p_i; q_i)$, *for a constant d_k that depends on $k \ge 2$ only.*

Proof For (iii), we rewrite the Hellinger distance in terms of the affinity, and use the inequality $\prod_i (1 - x_i) \ge 1 - \sum_i x_i$, which is valid for any numbers $x_i \in [0, 1]$, with $1 - x_i$ the affinities. For (iv) we employ Lemma B.5(iv).

The left side of (vii) is $\mathrm{E}|\sum_{i=1}^n (Y_i - \mathrm{E}(Y_i))|^k$, for $Y_i = \log(p_i/q_i)$, where the expectation is taken with respect to the density $\otimes_{i=1}^n p_i$. The Marcinkiewicz-Zygmund inequality (see Lemma K.4) implies that this is bounded by $d_k n^{k/2-1} \sum_{i=1}^n \mathrm{E}|Y_i - \mathrm{E}(Y_i)|^k$, which is the right side of (vii). The proofs of the other assertions are easier. □

Corollary B.9 (Kakutani's criterion) *Two infinite product probability measures $\prod_{i=1}^\infty P_i$ and $\prod_{i=1}^\infty Q_i$ are either mutually absolutely continuous or mutually singular, depending on whether $\sum_{i=1}^\infty d_H^2(P_i, Q_i)$ converges or diverges; equivalently on whether $\prod_{i=1}^\infty \rho_{1/2}(P_i; Q_i)$ is positive or zero. In particular, P^∞ and Q^∞ are mutually singular if $P \ne Q$.*

B.4 Discretization

Given a measurable partition $\{\mathfrak{X}_1, \ldots, \mathfrak{X}_k\}$ of $\mathfrak{X}$ in sets of finite ν-measure, and a density p, consider the density p^* defined by

$$p^*(x) = \sum_{j=1}^k \frac{P(\mathfrak{X}_j)}{\nu(\mathfrak{X}_j)} \mathbb{1}\{x \in \mathfrak{X}_j\}. \tag{B.11}$$

The density p^* is constant on each of the partitioning sets, with level equal to the average of p over the set. The following lemma measures the "distance" between p and its "locally uniform discretization" p^* in terms of the *variation* of p or $\log p$ over the partition. The variation of a function $p\colon \mathfrak{X} \to \mathbb{R}$ is defined by

$$\Delta p = \max_{j=1,\ldots,k} \sup_{x,y\in\mathfrak{X}_j} |p(x) - p(y)|. \tag{B.12}$$

Lemma B.10 *For any nonnegative density p and finite, measurable partition $\{\mathfrak{X}_1, \ldots, \mathfrak{X}_k\}$ in sets of finite measure,*

(i) $\|p - p^*\|_1 \le \nu(\mathfrak{X})\|p - p^*\|_\infty \le \nu(\mathfrak{X})\,\Delta p$,
(ii) $K(p; p^*) \le \Delta \log p$,
(iii) $\|p\|_\infty^{-1}\Delta p \le \Delta \log p \le \|p^{-1}\|_\infty \Delta p$ *if p is bounded away from zero and infinity.*

Furthermore, if ν is a probability measure and $\int p \log p\, d\nu < \infty$, then $\int p \log p^* d\nu \to K(p; \nu)$ as $k \to \infty$.

Proof The proofs of (i) and (iii) are straightforward. For the proof of (ii), by concavity of the logarithm,

$$K(p; p^*) = \sum_{j=1}^k \int_{\mathfrak{X}_j} p \log \frac{p}{\int_{\mathfrak{X}_j} p\, d\nu/\nu(\mathfrak{X}_j)}\, d\nu \le \sum_{j=1}^k \int_{\mathfrak{X}_j}\int_{\mathfrak{X}_j} p(x) \log \frac{p(x)}{p(y)} \frac{d\nu(y)}{\nu(\mathfrak{X}_j)}\, d\nu(x).$$

Here $\log(p(x)/p(y))$ can be bounded uniformly by $\Delta \log p$, after which the remaining integral can be evaluated as one.

To prove the final assertion, write $\int \log(p/p^*) p_0\, d\nu = K(p; \nu) - \int (\log p^*)\, p^*\, d\nu$. Let $\mathscr{T}_k = \sigma\langle \mathfrak{X}_1, \ldots, \mathfrak{X}_k\rangle$. By the convexity of the function $x \mapsto x \log x$ and Jensen's inequality, $p^* \log p^* \le \mathrm{E}_\nu(p \log p|\, \mathscr{T}_k)$, which is ν-uniformly integrable, as $p \log p$ is integrable by assumption. Because also $p^* \log p^* \ge -e^{-1}$ it follows that the sequence $p^* \log p^*$, as k varies, is ν-uniformly integrable, whence $\int p \log p^*\, d\nu = \int p^* \log p^*\, d\nu \to K(p; \nu)$. ☐

B.5 Information Loss

Kullback-Leibler information and negative Hellinger affinities are measures of statistical separation. They decrease if statistical information is lost by mapping or randomization.

Lemma B.11 *If $\tilde p$ and $\tilde q$ are the densities of a measurable function $T(X, U)$ of a variable $U \sim$ Unif[0, 1] and an independent variable X with densities p or q, respectively, then*

$$\begin{aligned} &K(\tilde p; \tilde q) \le K(p; q), \\ &\rho_\alpha(\tilde p; \tilde q) \ge \rho_\alpha(p; q), \qquad 0 < \alpha < 1, \\ &d_H(\tilde p, \tilde q) \le d_H(p, q), \\ &\|\tilde p - \tilde q\|_1 \le \|p - q\|_1. \end{aligned}$$

Proof Because all four quantities are independent of the dominating measure ν, we may without loss of generality assume that ν is a probability measure. Then $\mathrm{P}_p(T \in A) = \mathrm{E}_\nu \mathbb{1}\{T \in A\} p(X) = \mathrm{E}_\nu \mathbb{1}\{T \in A\} \mathrm{E}_\nu(p(X)|\, T)$, and hence $\tilde p(T) = \mathrm{E}_\nu(p(X)|\, T)$ is a density of $\tilde p$ relative to $\nu \circ T^{-1}$. The analogous formula for $\tilde q$, convexity of the map $(u, v) \mapsto$

$u\log(u/v)$ on $[0,\infty)$ and Jensen's inequality give

$$K(\tilde p;\tilde q)=\mathrm{E}_{\nu\circ T^{-1}}\tilde p(T)\log\frac{\tilde p}{\tilde q}(T)\le \mathrm{E}_{\nu\circ T^{-1}}\mathrm{E}_\nu\Big[p(X)\log\frac{p}{q}(X)|\,T\Big]=K(p;q).$$

The assertion for the affinity, Hellinger distance and total variation follow similarly, now using the concavity of $(u,v)\mapsto u^\alpha v^{1-\alpha}$, or the convexity of the maps $(u,v)\mapsto|\sqrt{u}-\sqrt{v}|^2$ or $(u,v)\mapsto|u-v|$. (Alternatively, the Hellinger distance can be treated by writing its square as $(1-\rho_{1/2})/2$.) ☐

Example B.12 The mixture densities $\int p_\theta\,dG$ and $\int q_\theta\,dG$ obtained from two families of (jointly measurable) densities p_θ and q_θ on a measurable space indexed by a common parameter θ and a given probability distribution G satisfy

$$d_H^2\Big(\int p_\theta\,dG(\theta),\int q_\theta\,dG(\theta)\Big)\le\int d_H^2(p_\theta,q_\theta)\,dG(\theta). \qquad \text{(B.13)}$$

In particular, for any three Lebesgue densities p,q and ϕ on $\mathbb{R}^d$,

$$d_H(\phi*p,\phi*q)\le d_H(p,q). \qquad \text{(B.14)}$$

Similar inequalities are valid for the Kullback-Leibler divergence, the total variation distance and (the reverse for) the Hellinger transform.

To see this, consider the model in which $Z|\,\theta$ has density p_θ or q_θ, and $\theta\sim G$. Then Z is distributed according to the mixture density and (Z,θ) is distributed according to $p=G\otimes p_\theta$ or $q=G\otimes q_\theta$, respectively. Hence by Lemma B.11 with $T(Z,\theta,U)=Z$, the Hellinger distance between the mixtures is bounded by the Hellinger distance between the measures p and q, which has square $\int\int(\sqrt{p_\theta}-\sqrt{q_\theta})^2\,d\nu\,dG(\theta)=\int d_H^2(p_\theta,q_\theta)\,dG(\theta)$.

B.6 Signed Kullback-Leibler and Comparisons

Define the *positive part Kullback-Leibler divergence* and *negative part Kullback-Leibler divergence* respectively by $K^+(p;q)=P\log_+(p/q)$ and $K^-(p;q)=P\log_-(p/q)$, and define positive and negative Kullback-Leibler variations $V^\pm(p;q)=P(\log_\pm(p/q))^2$, $V_0^\pm(p;q)=P(\log(p/q)-K(p;q))_\pm^2$, higher-order variations $V_k^\pm(p;q)=P(\log_\pm(p/q))^k$, $V_{k,0}^\pm(p;q)=P(\log(p/q)-K(p;q))_\pm^k$, similarly. Note that $K=K^+-K^-$ and $V=V^++V^-$.

Lemma B.13 *For any two densities p and q,*

$$K^-(p;q)\le\tfrac12\|p-q\|_1\le\sqrt{\tfrac12K(p;q)}, \qquad \text{(B.15)}$$

$$K^+(p;q)\le\tfrac12\|p-q\|_1+K(p;q)\le K(p;q)+\sqrt{\tfrac12K(p;q)}, \qquad \text{(B.16)}$$

$$V^-(p;q)\le 4d_H^2(p,q)\le 4K(p;q). \qquad \text{(B.17)}$$

Proof Using $\log x\le x-1$ and (B.1), we obtain

$$K^-(p;q)=\int_{q>p}p\log(q/p)\,d\nu\le\int_{q>p}(q-p)\,d\nu=\|p-q\|_1/2.$$

The first part of relation (B.16) follows because $K^+ = K + K^-$. The second parts of (B.15) and (B.16) follow from Kemperman's inequality connecting the KL divergence and the $\mathbb{L}_1$-distance. The first inequality on V^- follows using the inequality $(\log_- x)^2 \le 4(\sqrt{x}-1)^2$; the second is part (iv) of Lemma B.7. ☐

The following results bound the Kullback-Leibler divergence and variation in terms of the same objects evaluated at a third density.

Lemma B.14 *For any probability densities p, q, r,*

$$K(p;q) \le \log\Big\|\frac{p}{r}\Big\|_\infty + \Big\|\frac{p}{r}\Big\|_\infty\Big(K(r;q) + \sqrt{\tfrac{1}{2}K(r;q)}\Big),$$

$$V^+(p;q) \le 2\Big\|\frac{p}{r}\Big\|_\infty \log^2\Big\|\frac{p}{r}\Big\|_\infty + 2\Big\|\frac{p}{r}\Big\|_\infty V^+(r;q).$$

Proof If $p \le Cr$, then $K(p;q)$ is bounded by $P\log(Cr/q) = \log C + P\log(r/q)$. We bound $\log(r/q)$ by its positive part, and next P by CR, and finally apply (B.16). The inequality for V^+ follows similarly. ☐

Lemma B.15 *For any probability densities p, q and r,*

$$K(p;q) \le K(p;r) + 2d_H(r,q)\Big\|\frac{p}{q}\Big\|_\infty^{1/2},$$

$$V_2(p;q) \le 4V_2(p;r) + 16d_H^2(p,r) + 16d_H^2(r,q)\Big\|\frac{p}{q}\Big\|_\infty + 16d_H^2(p,q).$$

Here p/q is read as 0 if $p = 0$ and as ∞ if $q = 0 < p$.

Proof Writing $P\log(p/q) = P\log(p/r) + P\log(r/q)$ and using $\log x \le 2(\sqrt{x}-1)$ and the Cauchy-Schwarz inequality, we have

$$P\log\frac{r}{q} \le 2\int\frac{p}{\sqrt{q}}(\sqrt{r}-\sqrt{q}) \le 2\Big\|\frac{p}{q}\Big\|_\infty^{1/2}\int\sqrt{p}(\sqrt{r}-\sqrt{q})\,d\lambda \le 2\Big\|\frac{p}{q}\Big\|_\infty^{1/2} d_H(r,q).$$

By the relations $\log_+ x \le 2|\sqrt{x}-1|$ and $\log_- x = \log_+(1/x) \le 2|\sqrt{1/x}-1|$, we have for any probability densities p, q, r,

$$P\log_+^2\frac{r}{q} \le 4P\Big(\sqrt{\frac{r}{q}}-1\Big)^2 \le 4\Big\|\frac{p}{q}\Big\|_\infty d_H^2(r,q),$$

$$P\log_-^2\frac{p}{q} \le 4P\Big(\sqrt{\frac{q}{p}}-1\Big)^2 = 4d_H^2(p,q).$$

Since $|\log(p/q)| \le \log(p/r) + \log_-(p/r) + \log_+(r/q) + \log_-(p/q)$ the second relation follows from the triangle inequality for the $\mathbb{L}_2(P)$-norm. ☐

Problems

B.1 (Normal distribution) Let f_μ be the density of the Nor(μ, σ^2)-distribution, for given $\sigma > 0$. Show that

(i) $\|f_\mu - f_\nu\|_1 \le \sqrt{2/\pi}|\mu - \nu|/\sigma$,
(ii) $\rho_{1/2}(f_\mu, f_\nu) = e^{-(\mu-\nu)^2/(8\sigma^2)}$,
(iii) $d_H(f_\mu, f_\nu) \le |\mu - \nu|/(2\sigma)$,
(iv) $K(f_\mu; f_\nu) = (\mu - \nu)^2/(2\sigma^2)$,
(v) $V_{2,0}(f_\mu; f_\nu) = (\mu - \nu)^2/\sigma^2$.

Find the corresponding expressions for multivariate normal densities.

B.2 (Bernoulli distribution) Let p_θ stand for the density of the $\mathrm{Bin}(1, \theta)$-distribution. Show that

(i) $\|p_\mu - p_\nu\|_1 = 2|\mu - \nu|$,
(ii) $\rho_{1/2}(p_\mu, p_\nu) = \sqrt{\mu\nu} + \sqrt{1-\mu}\sqrt{1-\nu}$,
(iii) $d_H^2(p_\mu, p_\nu) = (\sqrt{\mu} - \sqrt{\nu})^2 + (\sqrt{1-\mu} - \sqrt{1-\nu})^2$,
(iv) $K(p_\mu; p_\nu) = \mu \log(\mu/\nu) + (1-\mu)\log((1-\mu)/(1-\nu))$,
(v) $V_2(p_\mu; p_\nu) = \mu \log^2(\mu/\nu) + (1-\mu)\log^2((1-\mu)/(1-\nu))$.

Show that the latter three quantities are bounded above by a multiple of $|\mu - \nu|^2$ if μ and ν are bounded away from 0 and 1.

B.3 The condition of boundedness of likelihood ratio in Lemma B.3 can be relaxed to an integrability condition, but then the bound will not be equally sharp. Show that for any pair of probability densities p and p_0 such that $P_0(p_0/p) < \infty$,

$$P_0\left(e^{|\log(p/p_0)|} - 1 - |\log(p/p_0)|\right) \le 4d_H^2(p, p_0)\left(1 + \Phi^{-1}(d_H^2(p, p_0))\right),$$

for $\Phi^{-1}(\epsilon) = \sup\{M: \Phi(M) \ge \epsilon\}$ inverse of $\Phi(M) = M^{-1}P_0(p_0/p)\mathbb{1}\{p_0/p \ge M\}$.

B.4 Let P and Q be probability measures and define $K^*(P; Q) = P[\log(p/q)\mathbb{1}\{q > 0\}]$. Show that $K^*(P^n; Q^n) = K^*(P; Q)(P\{q > 0\})^{n-1}$.

B.5 Let $\mathfrak{F}$ be a class of densities and $f_0 \in \mathfrak{F}$ such that for every x the map $f \mapsto f(x)$ is continuous on $\mathfrak{F}$. Show that $f \mapsto K(f_0; f)$ is a measurable (possibly extended) real-valued functional. In particular, show that Kullback-Leibler neighborhood $\{f: K(f_0; f) < \epsilon\}$ is a Borel measurable subset of $\mathfrak{F}$.

B.6 By developing the exponentials in their power series, show that $(e^x - 1 - x)/(e^{x/2} - 1)^2$ and $\alpha^{-1}(e^x - 1)/(e^{\alpha x} - 1)$ are bounded independently of α.

B.7 Show that $K^-(p; q) \le \|p - q\|_1 - d_H^2(p; q)$.

B.8 Using the convexity of $x \mapsto x(\log_+ x)^2$, show that the positive part Kullback-Leibler variation satisfies a information loss inequality analogous to Lemma B.11.

Appendix C

Packing, Covering, Bracketing and Entropy Numbers

Covering and bracketing numbers provide ways to measure the complexity of a metric space, or a set of functions. In this appendix we give their definitions, and a number of examples.

C.1 Definitions

A subset S of a semimetric space (T, d) is said to be *ϵ-dispersed* if $d(s, s') \geq \epsilon$ for all $s, s' \in S$ with $s \neq s'$. The maximum cardinality of an ϵ-dispersed subset of T is known as the *ϵ-packing number* of T and is denoted by $D(\epsilon, T, d)$.

A set S is called an *ϵ-net* for T if for every $t \in T$, there exists $s \in S$ such that $d(s, t) < \epsilon$, or equivalently T is covered by the collection of balls of radius ϵ around the points in S. The minimal cardinality of an ϵ-net is known as the *ϵ-covering number* of T and is denoted by $N(\epsilon, T, d)$. If T is itself a subset of a larger semimetric space, then one may allow the points of the ϵ-net not to belong to T. In general this results in a smaller covering number, but as every ball of radius ϵ around a point not in T is contained in a ball of radius 2ϵ around a point of T (unless it does not intersect T) covering numbers at 2ϵ that are restricted to centers in T will not be bigger.

Because a maximal ϵ-dispersed set is necessarily an ϵ-net, and $\epsilon/2$-balls centered at points in an ϵ-dispersed set are disjoint, we have the inequalities, for every $\epsilon > 0$,

$$N(\epsilon, T, d) \leq D(\epsilon, T, d) \leq N(\epsilon/2, T, d). \tag{C.1}$$

This means that packing and covering numbers can be used interchangeably if their order of magnitude and not exact constants are important.

For two given functions $u, l: \mathfrak{X} \to \mathbb{R}$ on a set $\mathfrak{X}$ with $l \leq u$, the *bracket* $[l, u]$ is defined as the set of all functions $f: \mathfrak{X} \to \mathbb{R}$ such that $l \leq f \leq u$ everywhere. For a semimetric d that is compatible with pointwise partial ordering in the sense that $d(l, u) = \sup\{d(f, g): f, g \in [l, u]\}$, the bracket is said to be of size ϵ if $d(l, u) < \epsilon$. For a given set T of functions the minimal number of ϵ-brackets needed to cover T is called the *ϵ-bracketing number* of T, and is denoted by $N_{[\,]}(\epsilon, T, d)$. In this definition the boundary functions l, u of the brackets are restricted to a given function space that contains T, and the bracketing numbers may depend on this space.

Because a bracket of size ϵ is contained in the ball of radius ϵ around its lower bracket, it follows that, for every $\epsilon > 0$,[1]

[1] Typically this can be slightly improved, with 2ϵ rather than ϵ in the right side, by considering balls around the mid function $(l + u)/2$, but this requires a stronger condition on the metric.

$$N(\epsilon, T, d) \le N_{[\,]}(\epsilon, T, d). \tag{C.2}$$

In general, there is no inequality in the reverse direction, except when d is the uniform distance, in which case the left side coincides with the expression on the right for ϵ replaced by 2ϵ.

The logarithms of the packing (or covering) and bracketing numbers are called the (metric) *entropy* and the *bracketing entropy*, respectively.

All these numbers grow to infinity as ϵ decays to zero, unless T is a finite set, in which case the both packing and covering numbers are bounded by the cardinality of the sets. The rate of growth qualitatively measures the size of the space, and is fundamental in empirical process theory and Bayesian nonparametrics.

In some applications the lower bounds of the brackets are not needed. It is therefore useful to define *upper bracketing numbers* $N_{]}(\epsilon, T, d)$ as the minimal number of functions $u_1, \ldots, u_m$ such that for every $p \in T$, there exist a function u_i such that both $p \le u_i$ and $d(u_i, p) < \epsilon$. The upper bracketing numbers are clearly smaller than the corresponding bracketing numbers.

When more than one semimetric is relevant, the corresponding packing and covering numbers may be related. If d_1 and d_2 are semimetrics such that $d_1(t, t') \le H(d_2(t, t'))$ for all $t, t' \in T$, for some strictly increasing continuous function $H\colon [0, \infty) \to [0, \infty)$ with $H(0) = 0$, then $N(\epsilon, T, d_1) \le N(H^{-1}(\epsilon), T, d_2)$. Similar conclusions hold for packing and bracketing numbers. In particular, if $\mathcal{F}$ is a set of probability densities, then $N(\epsilon, \mathcal{F}, d_H) \le N(\epsilon^2, \mathcal{F}, \|\cdot\|_1)$.

C.2 Examples

In this section we present examples of bounds for covering and bracketing numbers, first for subsets of Euclidean spaces and next for the most important spaces.

Proposition C.1 (Unit simplex) *For the norm $\|x\|_1 = \sum_i |x_i|$ on the m-dimensional unit simplex $\mathbb{S}_m$, for $0 < \epsilon \le 1$,*

$$D(\epsilon, \mathbb{S}_m, \|\cdot\|_1) \le \left(\frac{5}{\epsilon}\right)^{m-1}.$$

Proof For $x \in \mathbb{S}_m$, let x^* denote the vector of its first $m-1$ coordinates. Then x^* belongs to the set $\mathbb{D}_{m-1} := \{(y_1, \ldots, y_{m-1})\colon y_i \ge 0, \sum_{i=1}^{m-1} y_i \le 1\}$. The correspondence $x \mapsto x^*$ is one-to-one and $\|x_1 - x_2\|_1 \le 2\|x_1^* - x_2^*\|_1$. Let $x_1, \ldots, x_N \in \mathbb{S}_m$ such that $\|x_i - x_j\|_1 > \epsilon$, for $i, j = 1, \ldots, N$, $i \ne j$. Then $\|x_i^* - x_j^*\|_1 > \epsilon/2$, for $i, j = 1, \ldots, N$, $i \ne j$, and so that the ℓ_1-balls in $\mathbb{R}^{m-1}$ of radius $\epsilon/4$ centered at x_i^*, are disjoint. The union of these balls is clearly contained in the set

$$\left\{(y_1, \ldots, y_{m-1})\colon \sum_{i=1}^{m-1} |y_i| \le (1 + \epsilon/4)\right\}. \tag{C.3}$$

Let V_{m-1} be the volume of the unit ℓ_1-ball in $\mathbb{R}^{m-1}$. Then the volume of the set in (C.3) is $(1 + \epsilon/4)^{m-1} V_{m-1} \le (5/4)^{m-1} V_{m-1}$, while the volume of each $\epsilon/4$-ball around a point x_i^* is $(\epsilon/4)^{m-1} V_{m-1}$. A comparison of volumes gives the desired bound. ☐

Proposition C.2 (Euclidean space) *For* $\|x\|_p = (\sum_i |x_i|^p)^{1/p}$ *and* $p \geq 1$, *for any* M *and* $0 < \epsilon < M$,

$$D(\epsilon, \{x \in \mathbb{R}^m: \|x\|_p \leq M\}, \|\cdot\|_p) \leq \left(\frac{3M}{\epsilon}\right)^m.$$

The proof of Proposition C.2 is similar to that of Proposition C.1. The lemma, shows, in particular, that the *local entropy*

$$\log D(\epsilon, \{x \in \mathbb{R}^m: \|x\|_p \leq k\epsilon\}, \|\cdot\|_p)$$

is bounded by $m\log(3k)$, independently of ϵ, for any fixed k. Thus this quantity behaves essentially as the dimension of the space.

Recall the variation Δp of a function p over a given partition, defined in (B.12).

Corollary C.3 *Given a measurable partition* $\{\mathfrak{X}_1, \ldots, \mathfrak{X}_m\}$ *of a probability space* $(\mathfrak{X}, \mathscr{X}, \nu)$, *let* $\mathcal{P}_1$ *and* $\mathcal{P}_2$ *be the classes of all probability densities* p *with* $\Delta p < \epsilon$ *and* $\Delta \log p < \epsilon$, *respectively. Then* $N(2\epsilon, \mathcal{P}_1, \|\cdot\|_1) \leq (5/\epsilon)^m$ *and* $N(\epsilon + \sqrt{2\epsilon}, \mathcal{P}_2, \|\cdot\|_1) \leq (5/\epsilon)^m$.

Proof For a given probability density p let p^* be its projection on the set of probability densities that are constant on each partitioning set, as in (B.11). By Lemma B.10 $\|p - p^*\|_1 \leq \Delta p \leq \epsilon$, for every $p \in \mathcal{P}_1$, and $\|p - p^*\|_1^2 \leq 2K(p; p^*) < 2\epsilon$, for every $p \in \mathcal{P}_2$. Furthermore, for any pair of discretizations, $\|p^* - q^*\|_1 = \sum_{j=1}^m |P(\mathfrak{X}_j) - Q(\mathfrak{X}_j)|$, so that the ϵ-covering number of the set of discretizations is bounded by $(5/\epsilon)^m$, by Proposition C.1. □

The preceding bounds show that entropy numbers of sets in Euclidean spaces grow logarithmically. For infinite-dimensional spaces the growth is much faster, as is illustrated by the following examples.

Definition C.4 (Hölder space) The *Hölder space* $\mathfrak{C}^\alpha(\mathfrak{X})$ is the class of all functions $f: \mathfrak{X} \to \mathbb{R}$ with domain a bounded, convex subset $\mathfrak{X} \subset \mathbb{R}^m$ such that $\|f\|_{\mathfrak{C}^\alpha} < \infty$, where the *Hölder norm* of order α is defined as

$$\|f\|_{\mathfrak{C}^\alpha} = \max_{k:|k|\leq \underline{\alpha}} \sup_{x\in\mathcal{X}} |D^k f(x)| + \max_{k:|k|=\underline{\alpha}} \sup_{x,y\in\mathcal{X}:x\neq y} \frac{|D^k f(x) - D^k f(y)|}{\|x-y\|^{\alpha-\underline{\alpha}}}.$$

Here $\underline{\alpha}$ is the biggest integer strictly smaller than α, and for a vector $k = (k_1, \ldots, k_m)$ of integers with sum $|k|$, D^k is the differential operator

$$D^k = \frac{\partial^{|k|}}{\partial x_1^{k_1} \cdots \partial x_m^{k_m}}.$$

Proposition C.5 (Hölder space) *There exists a constant* K *depending only on the dimension* m *and smoothness* α *such that, with* $\mathfrak{X}^* = \{y: \|y - x\| \leq 1\}$,

$$\log N(\epsilon, \{f \in \mathfrak{C}^\alpha: \|f\|_{\mathfrak{C}^\alpha} \leq M\}, \|\cdot\|_\infty) \leq K \text{vol}(\mathfrak{X}^*) \left(\frac{M}{\epsilon}\right)^{m/\alpha}.$$

Definition C.6 (Sobolev space) The *Sobolev space* $\mathfrak{W}^\alpha(\mathfrak{X})$ of order $\alpha \in \mathbb{N}$ for an interval $\mathfrak{X} \subset \mathbb{R}$ is the class of all functions $f \in \mathbb{L}_2(\mathfrak{X})$ that possess an absolutely continuous $(\alpha - 1)$th derivative whose (weak) derivative $f^{(\alpha)}$ is contained in $\mathbb{L}_2(\mathfrak{X})$; the space is equipped with the norm $\|f\|_{2,2,\alpha} = \|f\|_2 + \|f^{(\alpha)}\|_2$. The Sobolev space $\mathfrak{W}^\alpha(\mathbb{R}^m)$ of order $\alpha > 0$ is the class of all functions $f \in \mathbb{L}_2(\mathbb{R}^m)$ with Fourier transform $\hat{f}$ satisfying $v_\alpha(f)^2 := \int |\lambda|^{2\alpha} |\hat{f}(\lambda)|^2 \, d\lambda < \infty$; the space is equipped with the norm $\|f\|_{2,2,\alpha} = \|f\|_2 + v_\alpha(f)$.

The preceding definition is restricted to functions of "integral" smoothness $\alpha \in \mathbb{N}$ with domain an interval in the real line or functions of general smoothness $\alpha > 0$ with domain a full Euclidean space. The relation between the two cases is that for $\alpha \in \mathbb{N}$ the function $\lambda \mapsto (i\lambda)^\alpha \hat{f}(\lambda)$ is the Fourier transform of the (weak) αth derivative $f^{(\alpha)}$ of a function $f: \mathbb{R} \to \mathbb{R}$ and hence $v_\alpha(f) = \|f^{(\alpha)}\|_2$. The Fourier transform is not entirely natural for functions not defined on the full Euclidecan space, which makes definitions of Sobolev spaces for general domains and smoothness a technical matter. One possibility suggests itself by the fact that in both cases the Sobolev space is known to be equivalent to the Besov space $\mathfrak{B}^\alpha_{2,2}(\mathfrak{X})$, defined in Definition E.8. Hence we may define a Sobolev space $\mathfrak{W}^\alpha(\mathfrak{X})$ in general as the corresponding Besov space. This identification suggested the notation $\|\cdot\|_{2,2,\alpha}$ for the norm. It is known that functions defined on a sufficiently regular domain (such as an interval) that belong to a given Besov space extend to a function on the full Euclidean domain of the same Besov norm. Thus it is also reasonable to define the Sobolev space $\mathfrak{W}^\alpha(\mathfrak{X})$ on a domain $\mathfrak{X} \subset \mathbb{R}^m$ in general as the set of functions $f: \mathfrak{X} \to \mathbb{R}$ that possess an extension belonging to $\mathfrak{W}^\alpha(\mathbb{R}^d)$, equipped with the norm of a "minimal" extension. Another possibility arise for *periodic* functions, which we briefly indicate below.

The Sobolev spaces of functions on a compact domain (identified with the corresponding Besov space) are compact in $\mathbb{L}_r(\mathbb{R})$ if the smoothness level is high enough: $\alpha > m(1/2 - 1/r)_+$. The entropy is not bigger than the entropy of the smaller Hölder spaces and hence the following proposition generalizes and improves Proposition C.5.

Proposition C.7 (Sobolev space) *For $\alpha > m(1/2 - 1/r)_+$ and $r \in (0, \infty]$ there exists a constant K that depends only on α and r such that*

$$\log N(\epsilon, \{f \in \mathfrak{W}^\alpha[0,1]^m : \|f\|_{2,2,\alpha} \le M\}, \mathbb{L}_r([0,1]^m)) \le K\Big(\frac{M}{\epsilon}\Big)^{m/\alpha}.$$

Proposition C.8 (Monotone functions) *The collection $\mathcal{F}$ of monotone functions $f: \mathfrak{X} \to [-M, M]$ on an interval $\mathfrak{X} \subset \mathbb{R}$ satisfies, for $\|\cdot\|_{r,Q}$ the $\mathbb{L}_r(Q)$ norm relative to a probability measure Q, any $r \ge 1$ and a constant K that depends on r only,*

$$\log N_{[\,]}(\epsilon, \mathcal{F}, \|\cdot\|_{r,Q}) \le K\frac{M}{\epsilon}.$$

Proposition C.9 (Analytic functions) *The class $\mathfrak{A}_A[0,1]^m$ of all functions $f: [0,1]^m \to \mathbb{R}$ that can be extended to an analytic function on the set $G = \{z \in \mathbb{C}^m : \|z - [0,1]^m\|_\infty < A\}$ with $\sup_{z \in G} |f(z)| \le 1$ satisfies, for $\epsilon < 1/2$ and a constant c that depends on m only,*

$$\log N(\epsilon, \mathfrak{A}_A[0,1]^m, \|\cdot\|_\infty) \le c\Big(\frac{1}{A}\Big)^m \Big(\log\frac{1}{\epsilon}\Big)^{1+m}.$$

A function $f \in \mathbb{L}_2[0, 2\pi]$ may be identified with its sequence of Fourier coefficients $(f_j) \in \ell_2$. The αth derivative of f has Fourier coefficients $((ij)^\alpha f_j)$. This suggests to think of the ℓ_2-norm of the sequence $(j^\alpha f_j)$ as a *Sobolev norm* of order α. The following proposition gives the entropy of the corresponding Sobolev sequence space relative to the ℓ_2-norm.

Proposition C.10 (Sobolev sequence) *For $\|\theta\|_2 = (\sum_{i=1}^\infty \theta_i^2)^{1/2}$ the norm of ℓ_2 and $\alpha > 0$, for all $\epsilon > 0$,*

$$\log D\Big(\epsilon, \{\theta \in \ell_2: \sum_{i=1}^\infty i^{2\alpha}\theta_i^2 \le B^2\}, \|\cdot\|_2\Big) \le \log(4(2e)^{2\alpha})\Big(\frac{3B}{\epsilon}\Big)^{1/\alpha}.$$

C.3 Historical Notes

Metric entropy was introduced by Kolmogorov and Tihomirov (1961) and subsequently developed by many authors. The results mentioned here are only a few examples from the literature. Proofs can be found at many places, including Edmunds and Triebel (1996), Dudley (1984, 2014), van der Vaart and Wellner (1996, 2017) and Giné and Nickl (2015). For instance, for Propositions C.5, see van der Vaart and Wellner (1996), pages 155–156; for Proposition C.8 see pages 159–162 in the same reference; for Proposition C.7, see van der Vaart and Wellner (2017) or Edmunds and Triebel (1996), page 105; for Proposition C.9, see Kolmogorov and Tihomirov (1961) or van der Vaart and Wellner (2017); for Proposition C.10 see e.g. Belitser and Ghosal (2003). These references also give many other results and a more extensive bibliography.

Problems

C.1 (Wu and Ghosal 2010) Consider a class $\mathcal{F}_{a,\Sigma}$ of mixtures of multivariate normal density given by $p(x) = p_{F,\Sigma} = \int \phi_d(x; \theta, \Sigma)\, dF(\theta)$, where Σ is a fixed $d \times d$ nonsingular matrix and $F(\|\theta\|_\infty \le a) = 1$. Show that

$$\log N(2\epsilon, \mathcal{F}_{a,\Sigma}, \|\cdot\|_1) \le \left(\sqrt{\frac{8d}{\pi \det(\Sigma)}}\frac{a}{\epsilon} + 1\right)[1 + \log(1 + \epsilon^{-1})].$$

Now let $\mathcal{F}^M_{a,h,\epsilon}$ stand for the class of all normal mixtures $p_{F,\Sigma}$, where $F(\|\theta\|_\infty > a) \le \epsilon$, $h \le \mathrm{eig}_1(\Sigma) \le \mathrm{eig}_d(\Sigma) \le M$, and $a > \sqrt{d}M\epsilon^{-1/2}$. Then

$$\log N(4\epsilon, \mathcal{F}^M_{a,h,\epsilon}, \|\cdot\|_1) \le \left(\sqrt{\frac{8d}{\pi}}\frac{2a}{h^{d/2}\epsilon} + 1\right)[1 + \log(1 + \epsilon^{-1})].$$

C.2 (Separability of Hölder spaces) Show that the functions $f_b: [0, 1] \to \mathbb{R}$ given by $f_b(x) = |x - b|^\alpha$, for $b \in (0, 1)$ and $\alpha \in (0, 1]$, satisfy $\sup_{x \ne y} |(f_b - f_c)(x) - (f_b - f_c)(y)|/|x - y|^\alpha \ge 2\alpha|b - c|^{\alpha-1}$. Conclude that $\mathfrak{C}^\alpha[0, 1]$ is not separable under its norm. [Hint: for $b \le c$ consider the increase of $f_b - f_c$ over the interval $[b, c]$.]

Appendix D

Hypothesis Tests

This appendix presents theory on the construction and existence of exponentially powerful tests. The focus is towards applications to proofs of consistency and contraction rates of posterior distributions. The appendix starts with general theory, and next focuses on a number of different statistical models.

D.1 Minimax Risk

Let P be a probability measure and let $\mathcal{Q}$ be a collection of finite measures on a measurable space $(\mathfrak{X}, \mathscr{X})$. The *minimax risk for testing* P versus $\mathcal{Q}$, weighted by positive numbers a and b, is defined by

$$\pi(P, \mathcal{Q}) = \inf_{\phi}\Bigl(aP\phi + b \sup_{Q \in \mathcal{Q}} Q(1 - \phi)\Bigr). \tag{D.1}$$

The infimum is taken over all *tests*, i.e. measurable functions $\phi\colon \mathfrak{X} \to [0, 1]$. The problem is to give a manageable bound on this risk, or equivalently on its two components, the probabilities of *errors of the first kind* $P\phi$ and *of the second kind* $Q(1 - \phi)$. Consideration of the symmetric case $a = b$ and probability measures Q suffices for most applications, but considering weights and general finite measures is useful and not more difficult. For simplicity, we assume throughout the section that P and $\mathcal{Q}$ are dominated by a σ-finite measure μ, and denote by p and q the densities of the measures P and Q.

The *Hellinger transform* $\rho_\alpha(p; q)$ is defined in (B.5) for pairs of densities. We also write $\rho_\alpha(P; Q)$ and for ease of notation

$$\rho_\alpha(P; \mathcal{Q}) = \sup\{\rho_\alpha(P; Q)\colon Q \in \mathcal{Q}\}. \tag{D.2}$$

Proposition D.1 (Minimax risk) *If P and $\mathcal{Q}$ are dominated, then the infimum in* (D.1) *is attained, and for every* $0 < \alpha < 1$,

$$\pi(P, \mathcal{Q}) = \sup_{Q \in \mathrm{conv}(\mathcal{Q})} \frac{1}{2}(a\|p\|_1 + b\|q\|_1 - \|ap - bq\|_1) \le a^\alpha b^{1-\alpha} \rho_\alpha(P; \mathrm{conv}(\mathcal{Q})).$$

Proof The set of test-functions ϕ can be identified with the nonnegative functions in the unit ball Φ of $\mathbb{L}_\infty(\mathfrak{X}, \mathscr{X}, \mu)$, which is dual to $\mathbb{L}_1(\mathfrak{X}, \mathscr{X}, \mu)$, since μ is σ-finite. The set Φ is compact and Hausdorff with respect to the weak*-topology, by the Banach-Alaoglu theorem (cf. Theorem 3.15 of Rudin 1973) and weak*-closure of the set of positive functions. Because the map $(\phi, Q) \mapsto aP\phi + bQ(1 - \phi)$ from $\mathbb{L}_\infty(\mathfrak{X}, \mathscr{X}, \mu) \times \mathbb{L}_1(\mathfrak{X}, \mathscr{X}, \mu)$ to $\mathbb{R}$ is convex and weak*-continuous in ϕ and linear in Q, the infimum over ϕ is attained, and the minimax theorem, Theorem L.5, gives

$$\inf_{\phi\in\Phi}\sup_{Q\in\mathrm{conv}(\mathcal{Q})}(aP\phi+bQ(1-\phi))=\sup_{Q\in\mathrm{conv}(\mathcal{Q})}\inf_{\phi\in\Phi}(aP\phi+bQ(1-\phi)).$$

The expression on the left side is the minimax testing risk $\pi(P,\mathcal{Q})$. The infimum on the right side is attained for $\phi=\mathbb{1}\{ap<bq\}$, and the minimal value $aP(ap<bq)+bQ(ap\geq bq)=b\|q\|_1-\int(ap-bq)^-\,d\mu$ can be rewritten as in the equality in the lemma.

For the inequality, we write

$$aP(ap<bq)+bQ(ap\geq bq)=a\int_{ap<bq}p\,d\mu+b\int_{ap\geq bq}q\,d\mu,$$

and bound p in the first integral by $p^\alpha(bq/a)^{1-\alpha}$, and q in the second integral by $(ap/b)^\alpha q^{1-\alpha}$. □

In particular, for probability measures P and Q and $a=b=1$, the minimax risk can be written in the form

$$\pi(P,\mathcal{Q})=1-\tfrac{1}{2}\|P-\mathrm{conv}(\mathcal{Q})\|_1=1-d_{TV}(P,\mathrm{conv}(\mathcal{Q})). \tag{D.3}$$

This exact expression is often difficult to handle. In Section D.4 we shall see that the further bound using the Hellinger transform is easy to manipulate for product measures.

The Hellinger distance is closely related to testing through its link to the affinity (see Lemma B.5). Because $-\log\rho_{1/2}=R_{1/2}\geq d_H^2/2$, the preceding display shows that convex alternatives $\mathcal{Q}$ with $d_H(P,\mathcal{Q})>\epsilon$ can be tested with errors bounded by $e^{-\epsilon^2/2}$. That the Hellinger distance is a Hilbert space norm also makes it manageable for direct constructions of tests. In the following lemma, this is exploited to construct a likelihood ratio test $\phi=\mathbb{1}\{\bar q/\bar p>c^{-2}\}$ between Hellinger balls around two given densities p and q based on a "least-favorable" pair $\bar p$ and $\bar q$ of densities. This lemma will be the basis of constructing tests for non-identically distributed observations and Markov chains, later on.

Lemma D.2 (Basic Hellinger testing) *Given arbitrary probability densities p and q, there exist probability densities $\bar p$ and $\bar q$ such that for, any probability density r,*

$$R\sqrt{\frac{\bar q}{\bar p}}\leq 1-\frac{1}{6}d_H^2(p,q)+d_H^2(p,r),\qquad R\sqrt{\frac{\bar p}{\bar q}}\leq 1-\frac{1}{6}d_H^2(p,q)+d_H^2(q,r).$$

Proof The linear subspace $\{a\sqrt{p}+b\sqrt{q}: a,b\in\mathbb{R}\}$ of $\mathbb{L}_2(\nu)$ is isometric to $\mathbb{R}^2$ equipped with the inner product $\langle(a,b),(a',b')\rangle_\omega=(a,b)V_\omega(a',b')^\top$, for V_ω the nonnegative-definite matrix, with $\omega\in[0,\pi/2]$ such that $\rho_{1/2}(p;q)=\cos\omega$,

$$V_\omega=\begin{pmatrix}1&\cos\omega\\ \cos\omega&1\end{pmatrix}=L^\top L,\qquad L=\begin{pmatrix}1&\cos\omega\\ 0&\sin\omega\end{pmatrix}.$$

The collection of linear combinations $a\sqrt{p}+b\sqrt{q}$ with $\mathbb{L}_2(\nu)$-norm equal to 1 is represented by the vectors (a,b) such that $\langle(a,b),(a,b)\rangle_\omega=a^2+b^2+2ab\cos\omega=1$. In terms of the Choleski decomposition $V_\omega = L^\top L$ as given in the display this ellipsoid can also be parameterized as

$$\begin{pmatrix} a \\ b \end{pmatrix} = L^{-1}\begin{pmatrix} \cos t \\ \sin t \end{pmatrix} = \frac{1}{\sin\omega}\begin{pmatrix} \sin(\omega - t) \\ \sin t \end{pmatrix}, \qquad 0 \le t \le 2\pi.$$

Denote the element $a\sqrt{p} + b\sqrt{q} \in \mathbb{L}_2(\nu)$ corresponding to t by $g(t)$. Then $g(0) = \sqrt{p}$ and $g(\omega) = \sqrt{q}$, and the inner product $\int g(t_1)g(t_2)\,d\nu$, which is equal to the $\langle\cdot,\cdot\rangle_\omega$ inner product between the corresponding vectors (a_1, b_1) and (a_2, b_2), is given by the Euclidean inner product between the vectors $(\cos t_1, \sin t_1)$ and $(\cos t_2, \sin t_2)$, which is $\cos t_1 \cos t_2 + \sin t_1 \sin t_2 = \cos(t_1 - t_2)$.

We define $\bar p$ and $\bar q$ by their roots, through $\sqrt{\bar p} = g(\omega/3)$ and $\sqrt{\bar q} = g(2\omega/3)$, respectively. Then

$$\frac{\sqrt{\bar q}}{\sqrt{\bar p}} = \frac{g(2\omega/3)}{g(\omega/3)} = \frac{\sin(\omega/3)\sqrt{p} + \sin(2\omega/3)\sqrt{q}}{\sin(2\omega/3)\sqrt{p} + \sin(\omega/3)\sqrt{q}} \le \frac{\sin(2\omega/3)}{\sin(\omega/3)} = 2\cos(\omega/3).$$

The root $\sqrt{r}$ of a general element $r \in \mathbb{L}_2(\mu)$ can be decomposed as its projection onto $\mathrm{lin}(\sqrt{p}, \sqrt{q})$ and an orthogonal part. Because its projection integrates to at most 1, it corresponds to a point inside the ellipsoid and can be represented as $\theta g(\gamma)$, for some $\theta \in [0, 1]$ and $\gamma \in [0, 2\pi]$. It follows that $\int \sqrt{r} g(t)\,d\nu = \theta\cos(\gamma - t)$, for any $t \in [0, 2\pi]$.

Now

$$R\sqrt{\frac{\bar q}{\bar p}} = \int \sqrt{\frac{\bar q}{\bar p}}(\sqrt{r} - \sqrt{\bar p})^2\,d\nu + 2\int \sqrt{\bar q}\sqrt{r}\,d\nu - \int \sqrt{\bar q}\sqrt{\bar p}\,d\nu.$$

Bounding the first term on the right by $2\cos(\omega/3)d_H^2(r, \bar p)$, using $d_H^2 = 2 - 2\rho_{1/2}$, and expressing the inner products in their angles as found previously, we see that this is bounded above by

$$\begin{aligned} &2\cos(\omega/3)(2 - 2\theta\cos(\gamma - \omega/3)) + 2\theta\cos(\gamma - 2\omega/3) - \cos(\omega/3) \\ &\quad = 3\cos(\omega/3) - 2\theta\cos\gamma \le 3\Big[\tfrac{8}{9} + \tfrac{1}{9}\cos\omega\Big] - 2\theta\cos\gamma. \end{aligned}$$

Finally we substitute $\cos\omega = \int \sqrt{p}\sqrt{q}\,d\nu = 1 - d_H^2(p, q)/2$, and $\theta\cos\gamma = \int \sqrt{p}\sqrt{r}\,d\nu = 1 - d_H^2(p, r)/2$. This concludes the proof of the first inequality; the second follows by symmetry. □

D.2 Composite Alternatives

Proposition D.1 shows the importance of the convex hull of the alternatives $\mathcal{Q}$. Not the separation of $\mathcal{Q}$ from the null hypothesis, but the separation of its convex hull drives the error probabilities. Unfortunately the complement $\{Q: d(P, Q) > \epsilon\}$ of a ball around P is not convex, and hence even though it is separated from P by distance ϵ, the existence of good tests, even consistent ones in an asymptotic set-up, is not guaranteed for this alternative in general. This is true even for d the total variation distance!

To handle a nonconvex alternative using a metric structure, such an alternative may be covered by convex sets, and corresponding tests combined into a single overall test. The power will then depend on the number of sets needed in a cover, and their separation from the null hypothesis. In the following basic lemma we make this precise in a form suited to posterior analysis. The following definition is motivated by the fact that convex alternatives at Hellinger distance ϵ from the null hypothesis can be tested with error probabilities bounded by $e^{-\epsilon^2/2}$ (see the remarks following (D.3), or Lemma D.2).

Given a semimetric d on a collection of measures that contains null and alternative hypotheses (or an arbitrary function $Q \mapsto d(P, Q)$ from $\mathcal{Q}$ to $[0, \infty)$), and for given positive constants c and K, and every $\epsilon > 0$, define the *covering number for testing* $N_t(\epsilon, \mathcal{Q}, d)$ as the minimal number of sets $\mathcal{Q}_1, \ldots, \mathcal{Q}_N$ in a partition of $\{Q \in \mathcal{Q}: \epsilon < d(P, Q) < 2\epsilon\}$ such that for every partitioning set $\mathcal{Q}_l$ there exists a test ψ_l with

$$P\psi_l \leq c\, e^{-K\epsilon^2}, \qquad \sup_{Q \in \mathcal{Q}_l} Q(1 - \psi_l) \leq c^{-1} e^{-K\epsilon^2}. \tag{D.4}$$

Lemma D.3 *If $N_t(\epsilon, \mathcal{Q}, d) \leq N(\epsilon)$ for every $\epsilon > \epsilon_0 \geq 0$ and some nonincreasing function $N: (0, \infty) \to (0, \infty)$, then for every $\epsilon > \epsilon_0$ there exists a test ϕ such that, for all $j \in \mathbb{N}$,*

$$P\phi \leq c\, N(\epsilon) \frac{e^{-K\epsilon^2}}{1 - e^{-K\epsilon^2}}, \qquad \sup_{Q \in \mathcal{Q}: d(P,Q) > j\epsilon} Q(1 - \phi) \leq c^{-1} e^{-K\epsilon^2 j^2}.$$

Proof For a given $j \in \mathbb{N}$, choose a minimal partition of $\mathcal{Q}_j := \{Q \in \mathcal{Q}: j\epsilon < d(P, Q) < 2j\epsilon\}$ as in the definition of $N_t(j\epsilon, \mathcal{Q}_j, d)$, and let $\phi_{j,l}$ be the corresponding tests. This gives $N_t(j\epsilon, \mathcal{Q}_j, d) \leq N(j\epsilon) \leq N(\epsilon)$ tests for every j. Let ϕ be the supremum of the countably many tests obtained in this way, when j ranges over $\mathbb{N}$. Then

$$P\phi \leq \sum_{j=1}^{\infty} \sum_{l} c e^{-Kj^2\epsilon^2} \leq c \sum_{j=1}^{\infty} N(\epsilon) e^{-Kj^2\epsilon^2} \leq cN(\epsilon) \frac{e^{-K\epsilon^2}}{1 - e^{-K\epsilon^2}}$$

and, for every $j \in \mathbb{N}$,

$$\sup_{Q \in \cup_{i>j} \mathcal{Q}_i} Q(1 - \phi) \leq \sup_{i>j} c^{-1} e^{-Ki^2\epsilon^2} \leq c^{-1} e^{-Kj^2\epsilon^2},$$

since for every $Q \in \mathcal{Q}_i$ there exists a test $\phi_{i,l}$ with $\phi \geq \phi_{i,l}$ that satisfies $Q(1 - \phi_{i,l}) \leq c^{-1} e^{-Ki^2\epsilon^2}$ by construction. ☐

In view of Proposition D.1 (applied with $a = c^{-1}$ and $b = c$) the cover $\mathcal{Q}_1, \ldots, \mathcal{Q}_N$ of the alternative $\mathcal{Q}$ in the definition of the testing number $N_t(\epsilon, \mathcal{Q}, d)$ can consist of any sets with

$$\inf_{0<\alpha<1} \left[c^{1-2\alpha} \rho_\alpha(P; \operatorname{conv}(\mathcal{Q}_l))\right] \leq e^{-K\epsilon^2}. \tag{D.5}$$

If restricted to probability measures the Hellinger transform satisfies $-\log \rho_{1/2} = R_{1/2} \geq d_H^2/2$, by Lemma B.5 (vi) and (i), and hence the left side of the display is bounded above by $\exp(-d_H^2(P, \operatorname{conv}(\mathcal{Q}_l)/2)$ (use $\alpha = 1/2$). Consequently *convex* sets $\mathcal{Q}_l$ with $d_H(P, \mathcal{Q}_l) \geq \sqrt{2K}\epsilon$ will do.[1]

[1] Because the log affinity responds better to operations such as forming product measures, (D.5) is better left in terms of the affinity than weakened to a bound in terms of the square Hellinger distance.

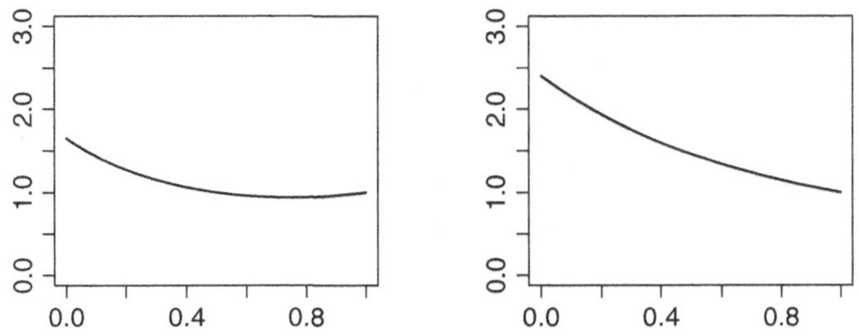

Figure D.1 The Hellinger transforms $\alpha \mapsto \rho_\alpha(p; q)$, for $P = \text{Nor}(0, 2)$ and Q the measure defined by $q = (d\text{Nor}(1, 1)/d\text{Nor}(0, 1))\, p$ (left) and $q = (d\text{Nor}(1, 1)/d\text{Nor}(3/2, 1))\, p$ (right). Intercepts with the vertical axis at the left and right of the graphs equal $\|q\|$ and $\|p\| = 1$, respectively. The slope at 1 equals $K(p; q)$, and has positive and negative in the two cases. In the left side the hypotheses p and q are testable versus each other, but in the right side they are not. (The measure $P^* = N(0, 1)$ is the point in the model minimizing $K(P; P^*)$ over the model $\{\text{Nor}(\theta, 1): \theta \in \mathbb{R}\}$; the slope at 1 is also positive if $N(1, 1)$ is replaced by $\text{Nor}(\theta, 1)$ for arbitrary θ. The measure $\text{Nor}(3/2, 1)$ is also in the model, but not the projection, which implies the existence of θ (such as $\theta = 1$) yielding a graph as on the right.)

By the same lemma, there is no point in using a Hellinger transform ρ_α for a value of α different from $1/2$ if the hypotheses involve only *probability* measures, but a different value $\alpha \in (0, 1)$ can be useful if the measures Q possess total mass bigger than 1. While the Hellinger transform $\rho_\alpha(p; q)$ of two *probability* densities $p \neq q$ is strictly smaller than 1 for any $0 < \alpha < 1$, it may assume values above 1 if q is a general density. For equivalent p and q the Hellinger transform $\alpha \mapsto \rho_\alpha(p; q)$ is a convex function, which takes the values $\|q\|_1$ at $\alpha = 0$ and $\|p\|_1 = 1$ at $\alpha = 1$. If $\|q\|_1 > 1 = \|p\|_1$, then it assumes values strictly smaller than 1 if and only if its left derivative at $\alpha = 1$, which is the Kullback-Leibler divergence $K(p; q)$, is strictly positive (see Lemma B.5). Figure D.1 illustrates that the Hellinger transform may be smaller than one for some α (left panel), or may be bigger than one for any $\alpha \in (0, 1)$. In the second case, the bound on the minimax testing risk given by Proposition D.1 is the useless bound 1.

If $K(p; q) > 0$, then a value of α close to 1 will be appropriate for constructing a test, even if q is not a probability density. By Lemma B.7 (with $p = p_0$), for any set of densities $\mathcal{Q}$,

$$\sup_{q \in \text{conv}(\mathcal{Q})} |1 - \rho_{1-\alpha}(p; q) - \alpha K(p; q)| \lesssim \alpha^2 \sup_{q \in \mathcal{Q}} \Big[d_H^2(p, q) + V_2(p; q) \Big].$$

If the right side can be controlled appropriately this means that the preceding argument can be made uniform in q, yielding that $\rho_\alpha(p; \mathcal{Q}) < 1$ for α sufficiently close to 1. The second inequality of the lemma allows this to extend to the convex hull of $\mathcal{Q}$.

We summarize the preceding in the following theorem.

Theorem D.4 *If there exists a nonincreasing function $N: (0, \infty) \to (0, \infty)$ such that for every $\epsilon > \epsilon_0 \geq 0$ the set $\{Q \in \mathcal{Q}: d(P, Q) < 2\epsilon\}$ can be covered with $N(\epsilon)$ sets $\mathcal{Q}_l$ satisfying* (D.5), *then for every $\epsilon > \epsilon_0$ there exist a test ϕ such that, for all $j \in \mathbb{N}$,*

$$P\phi \le c\,N(\epsilon)\frac{e^{-K\epsilon^2}}{1-e^{-K\epsilon^2}}, \qquad \sup_{Q\in\mathcal{Q}:d(P,Q)>j\epsilon} Q(1-\phi) \le c^{-1}e^{-K\epsilon^2 j^2}.$$

Proof For $a^{-1} = c = b$, condition (D.5) says that $\inf_{0<\alpha<1}\left[a^\alpha b^{1-\alpha}\rho_\alpha(p, \mathrm{conv}(\mathcal{Q}_l))\right]$ is bounded above by $e^{-K\epsilon^2}$. Hence by Proposition D.1, there exists a test ψ_l satisfying (D.4). The theorem therefore follows from Lemma D.3. □

D.3 Testing and Metric Entropy

If the sets $\mathcal{Q}_l$ are constructed as balls relative to a metric, then the covering number for testing can be bounded by ordinary metric entropy. The following *basic testing assumption*, which will be verified for a number of statistical setups in the next sections, is instrumental. Suppose that d and e are semimetrics such that for universal constants $K > 0$ and $\xi \in (0, 1)$, there exists for every $\epsilon > 0$ and every $Q \in \mathcal{Q}$ with $d(P, Q) > \epsilon$ a test ϕ with

$$P\phi \le c\,e^{-K\epsilon^2}, \qquad \sup_{R\in\mathcal{Q}:e(R,Q)<\xi\epsilon} R(1-\phi) \le c^{-1}\,e^{-K\epsilon^2}. \tag{D.6}$$

In the common case that $d = e$, this requires that the null hypothesis P can be tested against any ball with errors that are exponential in minus a constant times the square distance of the ball to P, as illustrated in Figure D.2. It follows readily from the definitions that in this situation

$$N_t(\epsilon, \mathcal{Q}, d) \le N(\xi\epsilon, \{Q \in \mathcal{Q}: d(P, Q) < 2\epsilon\}, e), \qquad \epsilon > 0.$$

The quantity on the right is known as the *local covering number*, and its logarithm as the *Le Cam dimension* (at level ϵ) of $\mathcal{Q}$. Clearly the local covering number is upper bounded by the ordinary covering numbers $N(\xi\epsilon, \mathcal{Q}, e)$. In infinite-dimensional situations these often have the same order of magnitude as $\epsilon \downarrow 0$, and hence nothing is lost by this simplification. For finite-dimensional models, local covering numbers are typically of smaller order. The difference is also felt in asymptotics where the dimension of the model tends to infinity.

The Le Cam dimension $\log N(\epsilon, \{\theta \in \mathbb{R}^k: \|\theta\| < 2\epsilon\}, \|\cdot\|_2)$ of Euclidean space is bounded by a multiple of its dimension k, in view of Proposition C.2, uniformly in ϵ. This is one motivation for the term "dimension."

The following result is a consequence of the aggregation of basic tests and is the basis for nearly all theorems on posterior consistency and rates of contraction.

Theorem D.5 (Basic testing) *If the basic testing assumption* (D.6) *holds for arbitrary semimetrics d and e and $N(\xi\epsilon, \{Q \in \mathcal{Q}: d(P, Q) < 2\epsilon\}, e) \le N(\epsilon)$ for every $\epsilon > \epsilon_0 \ge 0$*

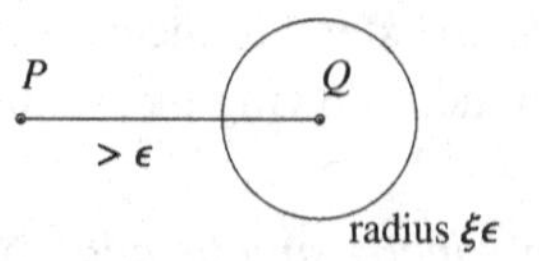

Figure D.2 Assumption (D.6) assumes existence of a test of a simple null hypothesis P against a ball of radius $\xi\epsilon$ with center Q located at distance ϵ from P.

and some nonincreasing function $N: (0, \infty) \to (0, \infty)$, *then for every* $\epsilon > \epsilon_0$ *there exist a test* ϕ *such that, for all* $j \in \mathbb{N}$,

$$P\phi \le c\,N(\epsilon)\frac{e^{-K\epsilon^2}}{1 - e^{-K\epsilon^2}}, \qquad \sup_{Q\in\mathcal{Q}: d(P,Q) > j\epsilon} Q(1 - \phi) \le c^{-1} e^{-K\epsilon^2 j^2}.$$

D.4 Product Measures

For $i = 1, \ldots, n$, let P_i and $\mathcal{Q}_i$ be a probability measure and a set of probability measures on an arbitrary measurable space $(\mathfrak{X}_i, \mathscr{X}_i)$, and consider testing the product $\otimes_i P_i$ versus the set $\otimes_i \mathcal{Q}_i$ of products $\otimes_i Q_i$ with Q_i ranging over $\mathcal{Q}_i$.

The Hellinger transform factorizes on pairs of product measures. For set-valued arguments it is defined as a supremum over the convex hull. Then it does not factorize, but as shown in the following lemma, it is still "sub-multiplicative."

Lemma D.6 *For any* $0 < \alpha < 1$ *and any measures* P_i *and any sets* $\mathcal{Q}_i$ *of finite measures* $\rho_\alpha(\otimes_i P_i; \mathrm{conv}(\otimes_i \mathcal{Q}_i)) \le \prod_i \rho_\alpha(P_i; \mathrm{conv}(\mathcal{Q}_i))$.

Proof If suffices to give the proof for $n = 2$; the general case follows by repetition. Any measure $Q \in \mathrm{conv}(\mathcal{Q}_1 \times \mathcal{Q}_2)$ can be represented by a density of the form $q(x, y) = \sum_j \kappa_j q_{1j}(x) q_{2j}(y)$, for nonnegative constants κ_j with $\sum_j \kappa_j = 1$, and q_{ij} densities of measures belong to $\mathcal{Q}_i$. Then $\rho_\alpha(p_1 \times p_2; q)$ can be written in the form

$$\int p_1(x)^\alpha \Bigl(\sum_j \kappa_j q_{1j}(x)\Bigr)^{1-\alpha} \Bigl[\int p_2(y)^\alpha \Bigl(\frac{\sum_j \kappa_j q_{1j}(x) q_{2j}(y)}{\sum_j \kappa_j q_{1j}(x)}\Bigr)^{1-\alpha} d\mu_2(y)\Bigr]\, d\mu_1(x).$$

(If $\sum_j \kappa_j q_{1j}(x) = 0$, the quotient in the inner integral is interpreted as 0.) The inner integral is bounded by $\rho_\alpha(P_2; \mathrm{conv}(\mathcal{Q}_2))$ for every fixed $x \in \mathfrak{X}$. After substitution of this upper bound the remaining integral is bounded by $\rho_\alpha(P_1; \mathrm{conv}(\mathcal{Q}_1))$. □

Combining Lemma D.6 and Proposition D.1 (with $\alpha = 1/2$), we obtain the following corollary.

Corollary D.7 *For any probability measures* P_i *and sets of densities* $\mathcal{Q}_i$ *and* $a, b > 0$ *there exists a test* ϕ *such that*

$$a(\otimes_{i=1}^n P_i)\phi + b \sup_{q_i \in \mathcal{Q}_i} (\otimes_{i=1}^n Q_i)(1 - \phi) \le \sqrt{ab} \prod_{i=1}^n \rho_{1/2}(p_i; \mathrm{conv}(\mathcal{Q}_i)).$$

In particular, for identically distributed observations and null and alternative hypotheses P and $\mathcal{Q}$ there exists a test ϕ with (since $\rho_{1/2} \le e^{-d_H^2/2}$)

$$aP^n\phi + b \sup_{Q\in\mathcal{Q}} Q^n(1 - \phi) \le \sqrt{ab}\, e^{-nd_H^2(p, \mathrm{conv}(\mathcal{Q}))/2}.$$

This shows that (D.6) is satisfied (with $c = 1$) for collections of product measures equipped with $d = e$ the Hellinger distance on the marginal distributions, with the constant K equal to $n(1-\xi)^2/2$. In the following theorem we record the more general fact that it is valid for any metric that generates convex balls and is bounded above by a multiple of the Hellinger distance. Examples are the total variation distance, and the $\mathbb{L}_2$-distance if the densities are uniformly bounded.

Proposition D.8 *Suppose that d is a semimetric that generates convex balls and satisfies $d(p,q) \le d_H(p,q)$ for every q. Then for every $c>0$, $n\in\mathbb{N}$ and $\epsilon>0$ and probability densities p and q with $d(p,q)>\epsilon$, there exists a test ϕ such that*

$$P^n\phi \le c\,e^{-n\epsilon^2/8}, \qquad \sup_{r:d(r,q)<\epsilon/2} R^n(1-\phi) \le c^{-1}\,e^{-n\epsilon^2/8}.$$

Proof If $d(p,q)>\epsilon$, then the d-ball B of radius $\epsilon/2$ around q has d-distance and hence Hellinger distance at least $\epsilon/2$ to p. Consequently $-\log\rho_{1/2}(p;B)\ge d_H^2(p,B)/2\ge \epsilon^2/8$, by Lemma B.5(vi) and (i). Because B is convex, the result follows from the display preceding the statement of the lemma, applied with $\sqrt{b/a}=c$. ☐

By the same arguments the general result (D.6) can be verified relative to the Hellinger distance on general product measures, with possibly different components. However, in this situation a more natural distance is the *root average square Hellinger metric*

$$d_{n,H}\Big(\bigotimes_{i=1}^n p_i, \bigotimes_{i=1}^n q_i\Big) = \sqrt{\frac{1}{n}\sum_{i=1}^n d_H^2(p_i,q_i)}.$$

By Lemma B.8 (iv) this is an upper bound on the Hellinger distance on the product measures, but the two distances are not equivalent. The following lemma nevertheless guarantees the existence of tests as in (D.6).

Proposition D.9 *For every $c>0$, $n\in\mathbb{N}$ and $\epsilon>0$ and probability densities p_i and q_i with $d_{n,H}(\otimes_{i=1}^n p_i, \otimes_{i=1}^n q_i)>\epsilon$, there exists a test ϕ such that*

$$\Big(\bigotimes_{i=1}^n P_i\Big)\phi \le c\,e^{-n\epsilon^2/8}, \qquad \sup_{\otimes_{i=1}^n r_i: d_{n,H}(\otimes_{i=1}^n r_i, \otimes_{i=1}^n q_i)<\epsilon/5} \Big(\bigotimes_{i=1}^n R_i\Big)(1-\phi) \le c^{-1}\,e^{-n\epsilon^2/8}.$$

Proof Let $\phi = \mathbb{1}\{\otimes_{i=1}^n \sqrt{\bar q_i/\bar p_i} > c^{-1}\}$, for $\bar p_1,\ldots,\bar p_n$ and $\bar q_1,\ldots,\bar q_n$ the densities attached to $p_1,\ldots,p_n$ and $q_1,\ldots,q_n$ in Lemma D.2. By Markov's inequality

$$\Big(\bigotimes_{i=1}^n P_i\Big)\phi \le c\prod_{i=1}^n P_i\sqrt{\frac{\bar q_i}{\bar p_i}} \le c\prod_{i=1}^n \big(1-\tfrac16 d_H^2(p_i,q_i)\big) \le c e^{-n d_{n,H}^2(\otimes p_i,\otimes q_i)/6},$$

$$\begin{aligned}\Big(\bigotimes_{i=1}^n R_i\Big)(1-\phi) &\le c^{-1}\prod_{i=1}^n R_i\sqrt{\frac{\bar p_i}{\bar q_i}} \le c^{-1}\prod_{i=1}^n \big(1-\tfrac16 d_H^2(p_i,q_i)+d_H^2(r_i,q_i)\big)\\ &\le c^{-1} e^{-n d_{n,H}^2(\otimes p_i,\otimes q_i)/6 + n d_{n,H}^2(\otimes r_i,\otimes q_i)}.\end{aligned}$$

For $d_{n,H}(\otimes p_i,\otimes q_i)>\epsilon$ and $d_{n,H}(\otimes r_i,\otimes q_i)<\epsilon/\sqrt{24}$, the right sides are bounded as in the lemma (as $1/6-1/24=1/8$). ☐

Corollary D.7 is in terms of the affinity or Hellinger distance between the null hypothesis and the *convex hull* of the alternative hypothesis. Analytic computations on this hull may be avoided by constructing tests of the hypotheses. Tests with uniform error probabilities automatically refer to the convex hull, and may be suggested by statistical reasoning. In view of the characterization of the minimax testing risk in terms of the $\mathbb{L}_1$-norm (see Theorem D.1) and the relation between this norm and the affinity (see Lemma B.5(ii)) these two approaches are equivalent. The next lemma makes this explicit.

Lemma D.10 (Aggregating tests) *If for each $i = 1, \ldots, n$, there exists a test ϕ_i such that*

$$P_i\phi_i \leq \alpha_i \leq \gamma_i \leq \inf_{Q_i \in \mathcal{Q}_i} Q_i\phi_i, \tag{D.7}$$

then for $\epsilon^2 = n^{-1}\sum_{i=1}^n (\gamma_i - \alpha_i)^2$ there exist a test ϕ such that $(\otimes_{i=1}^n P_i)\phi < e^{-n\epsilon^2/2}$ and $(\otimes_{i=1}^n Q_i)(1-\phi) < e^{-n\epsilon^2/2}$, for every $Q_i \in \mathcal{Q}_i$.

Proof The condition implies that $\|p_i - q_i\|_1/2 = d_{TV}(P_i, Q_i) \geq \gamma_i - \alpha_i$, for every $Q_i \in \mathrm{conv}(\mathcal{Q}_i)$. By Lemma B.5(ii), $\rho_{1/2}(p_i; q_i)^2 \leq 1 - \|p_i - q_i\|_1^2/4 \leq 1 - (\gamma_i - \alpha_i)^2$, for any $q_i \in \mathrm{conv}(\mathcal{Q}_i)$. Next apply Lemma D.6 to see that $\rho_{1/2}(\otimes_i P_i; Q)^2 \leq e^{-n\epsilon^2}$ for any $Q \in \mathrm{conv}(\otimes_i \mathcal{Q}_i)$, and finally apply Proposition D.1 (with $a = b = 1$ and $\alpha = 1/2$). □

In the case of i.i.d. observations there is nothing special about exponentially small error probabilities: if a fixed set $\mathcal{Q}$ can be uniformly *consistently* tested versus P, then it can automatically be tested with exponentially small error probabilities.

Lemma D.11 (Fixed alternative) *If there exist tests ψ_n such that, for a given probability measure P and a set $\mathcal{Q}$ of probability measures, $P^n\psi_n \to 0$ and $\sup_{Q\in\mathcal{Q}} Q^n(1-\psi_n) \to 0$, then there exist tests ϕ_n and a constant $K > 0$ such that*

$$P^n\phi_n \leq e^{-Kn}, \qquad \sup_{Q\in\mathcal{Q}} Q^n(1-\phi_n) \leq e^{-Kn}.$$

Proof For any $0 < \alpha < \gamma < 1$, there exists n_0 such that $P^{n_0}\psi_{n_0} < \alpha < \gamma < Q^{n_0}\psi_{n_0}$. We apply Lemma D.10 with P_i and Q_i of the lemma equal to the present P^{n_0} and Q^{n_0}. For a given n we can construct $\lfloor n/n_0 \rfloor \asymp n$ blocks of size n_0, and obtain a test with error probabilities that are exponential in minus $\lfloor n/n_0 \rfloor (\gamma - \alpha)^2/2$. □

D.5 Markov Chains

Let p be a (Markov) transition density $(x, y) \mapsto p(y|\, x)$ and $\mathcal{Q}$ a collection of Markov transition densities $(x, y) \mapsto q(y|\, x)$ from a given sample space $(\mathfrak{X}, \mathscr{X})$ into itself, relative to a dominating measure ν. Consider testing the distribution of a Markov chain $X_0, X_1, \ldots, X_n$ that evolves according to either p or some $q \in \mathcal{Q}$. Let an initial value X_0 be distributed according to a measure P_0 or Q_0, which may or may not be stationary distributions to the transition kernels p and q. Denote the laws of $(X_1, \ldots, X_n)$ under the null and alternative hypotheses by $P^{(n)}$ and $Q^{(n)}$.

The first result is a generalization of Lemma D.6. For given x let $\rho_\alpha(p; q|\,x)$ denote the Hellinger affinity between the probability densities $p(\cdot|\,x)$ and $q(\cdot|\,x)$.

Lemma D.12 *For any $0 < \alpha < 1$ and any transition density p and class $\mathcal{Q}$ of transition densities,*

$$\rho_\alpha(P^{(n)}; \operatorname{conv}(\mathcal{Q}^{(n)})) \le \Big(\sup_{x\in\mathfrak{X}} \sup_{q\in\operatorname{conv}(\mathcal{Q})} \rho_\alpha(p; q|\,x)\Big)^n.$$

Proof The proof can evolve as the proof of Lemma D.6, where we peel off observations in the order $X_n, X_{n-1}, \ldots, X_0$, each time bounding the integral

$$\int p^\alpha(x_i|\,x_{i-1})\Big(\frac{\sum_j \kappa_j q_j^{(i-1)}(x_1,\ldots,x_{i-1}) q_j(x_i|\,x_{i-1})}{\sum_j \kappa_j q_j^{(i-1)}(x_1,\ldots,x_{i-1})}\Big)^{1-\alpha}\,d\nu(x_i)$$

by the supremum over the convex hull of the transition densities $y \mapsto q(y|\,x_{i-1})$, and next the supremum over x_{i-1}, of the Hellinger affinity, yielding the n terms in the product on the right side. ☐

Combining Lemma D.12 and Proposition D.1, we can obtain the analog of Corollary D.7. This may next be translated into the following testing statement, which uses the *supremum Hellinger distance*, defined by

$$d_{H,\infty}(p,q) = \sup_{x\in\mathfrak{X}} d_H(p(\cdot|\,x), q(\cdot|\,x)).$$

Corollary D.13 *For any transition density p and convex set of transition densities $\mathcal{Q}$ with $\inf_{x\in\mathfrak{X}}\inf_{q\in\mathcal{Q}} d_H^2(p(\cdot|\,x), q(\cdot|\,x)) \ge \epsilon$, any initial distributions P_0 and Q_0, and any $a, b > 0$, there exists a test ϕ such that*

$$aP^{(n)}\phi + b\sup_{q\in\mathcal{Q}} Q^{(n)}(1-\phi) \le \sqrt{ab}\,\epsilon^{-n\epsilon^2/2}.$$

Furthermore, if $\mathcal{Q}$ is not convex, then there still exists a test satisfying the preceding display with the right-hand side replaced by $N(\epsilon/4, \mathcal{Q}, d_{H,\infty})e^{-n\epsilon^2/2}$.

The supremum Hellinger distance may be too strong, and the condition that the infimum of the Hellinger distances is bounded away from zero unworkable. An alternative is a *weighted Hellinger distance* of the form, for some measure μ,

$$d_{H,\mu}(q_1,q_2) = \Big(\int\!\!\int\Big(\sqrt{q_1(y|\,x)} - \sqrt{q_2(y|\,x)}\Big)^2\,d\nu(y)\,d\mu(x)\Big)^{1/2}. \tag{D.8}$$

We can use these distances for μ bounding the transition probabilities of the Markov chain as follows. For $n \in \mathbb{N}$ let $(x, A) \mapsto Q^n(A|\,x)$ be the n-step transition kernel, given recursively by

$$Q^1(A|\,x) = \int_A q(y|\,x)\,d\nu(y), \qquad Q^{n+1}(A|\,x) = \int Q^n(A|\,y)\,Q(dy|\,x).$$

Then assume that there exist some $k, l \in \mathbb{N}$ and measures $\underline{\mu}, \overline{\mu}$ such that for every element of $\{P\} \cup \mathcal{Q}$, every $x \in \mathfrak{X}$ and every $A \in \mathscr{X}$,

$$\underline{\mu}(A) \leq \frac{1}{k}\sum_{j=1}^{k} Q^j(A|\,x), \qquad Q^l(A|\,x) \leq \overline{\mu}(A). \tag{D.9}$$

This condition requires that the transitions out of the possible states x occur with a certain uniformity in the initial state, captured by the measures $\underline{\mu}$ and $\overline{\mu}$.

The following lemma, due to Birgé (1983a,b), shows that tests satisfying (D.6) exist for the semimetrics d_n and e_n equal to the weighted Hellinger distances $d_{H,\underline{\mu}}$ and $d_{H,\overline{\mu}}$, respectively.

Proposition D.14 *There exist a constant K depending only on (k, l) such that for every transition kernels p and q with $d_{H,\underline{\mu}}(p, q) > \epsilon$ there exist tests ϕ such that, for every $n \in \mathbb{N}$,*

$$P^{(n)}\phi \leq e^{-Kn\epsilon^2}, \qquad \sup_{r: d_{H,\overline{\mu}}(r,q) \leq \epsilon/5} R^{(n)}(1-\phi) \leq e^{-Kn\epsilon^2}.$$

Proof For $m = k + l$ partition the n observations in $N = \lfloor n/m \rfloor$ blocks of m consecutive observations, and a remaining set of observations, which are discarded. Let $I_1, \ldots, I_N$ be independent random variables, with I_j uniformly distributed on the set of the k last indices of the jth block: $\{(j-1)m+l+1, (j-1)m+l+2, \ldots, (j-1)m+l+k\}$. Then for $\bar{p}(\cdot|\,x)$ and $\bar{q}(\cdot|\,x)$ the densities attached to $p(\cdot|\,x)$ and $q(\cdot|\,x)$ in Lemma D.2, for every given x, define $\phi = \mathbb{1}\{\sum_{j=1}^N \log(\bar{q}/\bar{p})(X_{I_j}|\,X_{I_{j-1}}) > -\log(2c)\}$.

By Markov's inequality

$$P^{(n)}\phi \leq cP^{(n)}\Bigl[\prod_{j=1}^{N}\sqrt{\frac{\bar{q}}{\bar{p}}}(X_{I_j}|\,X_{I_j-1})\Bigr], \quad R^{(n)}(1-\phi) \leq c^{-1}R^{(n)}\Bigl[\prod_{j=1}^{N}\sqrt{\frac{\bar{p}}{\bar{q}}}(X_{I_j}|\,X_{I_j-1})\Bigr].$$

We evaluate these expressions by peeling off the terms of the products one-by-one from the right, each time conditioning on the preceding observations. For the error probability of the second kind, the peeling step proceeds as

$$\begin{aligned}
&\mathrm{E}_R\Bigl(\sqrt{\frac{\bar{p}}{\bar{q}}}(X_{I_j}|\,X_{I_j-1})|\,X_1, \ldots, X_{(j-1)m}\Bigr) \\
&= \frac{1}{k}\sum_{i=1}^{k}\int\!\!\int \sqrt{\frac{\bar{p}}{\bar{q}}}(y|\,x)\,r(y|\,x)\,d\nu(x)\,R^{l+i-1}(dx|\,X_{(j-1)m}) \\
&\leq \frac{1}{k}\sum_{i=1}^{k}\Bigl(1 - \tfrac{1}{6}d_H^2(p(\cdot|\,x), q(\cdot|\,x)) + d_H^2(r(\cdot|\,x), q(\cdot|\,x))\Bigr)\,R^{l+i-1}(dx|\,X_{(j-1)m}),
\end{aligned}$$

by Lemma D.2. By the Chapman-Kolmogorov equations and the bounds (D.9) on the transition kernels $R^{l+i}(A|\,x) = \int R^l(A|\,y)\,R^i(dy|\,x) \leq \overline{\mu}(A)$, for every $i \geq 0$; and also

$$k^{-1}\sum_{i=1}^{k} R^{l+i-1}(A|\,x) = k^{-1}\sum_{i=1}^{k}\int R^i(A|\,y)\,R^{l-1}(dy|\,x) \geq \underline{\mu}(A).$$

Therefore, the previous display is bounded above by $1 - \frac{1}{6}d_{H,\mu}^2(p,q) + d_{H,\overline{\mu}}^2(r,q) < 1 - \epsilon^2/8$, for $d_{H,\mu}(p,q) > \epsilon$ and $d_{H,\overline{\mu}}(r,q) < \epsilon/\sqrt{24}$. Peeling off all N terms in this manner, we obtain the upper bound $(1-\epsilon^2/8)^N$, which is bounded as desired.

The probability of an error of the first kind can be handled similarly. □

D.6 Gaussian Time Series

Let $P_f^{(n)}$ denote the distribution of $(X_1,\ldots,X_n)$ for a stationary Gaussian time series $(X_t: t\in\mathbb{Z})$ with mean zero and spectral density f. For $k\in\mathbb{Z}$ let $\gamma_f(h) = \int_{-\pi}^{\pi} e^{ih\lambda} f(\lambda)\,d\lambda$ define the corresponding autocovariance function.

The following proposition, whose proof can be found in Birgé (1983a,b), shows that (D.6) is satisfied under some restrictions, for d equal to the $\mathbb{L}_2[0,\pi]$-metric and e equal to the uniform metric.

Proposition D.15 *Let $\mathcal{F}$ be a set of measurable functions $f:[0,\pi]\to[0,\infty)$ such that $\|\log f\|_\infty \le M$ and $\sum_{h\in\mathbb{Z}} |h|\gamma_f^2(h) \le N$ for all $f\in\mathcal{F}$. Then there exist constants ξ and K depending only on M and N such that for every $\epsilon \ge n^{-1/2}$ and every $f_0, f_1\in\mathcal{F}$ with $\|f_1 - f_0\|_2 \ge \epsilon$,*

$$P_{f_0}^{(n)}\phi_n \le e^{-Kn\epsilon^2}, \qquad \sup_{f\in\mathcal{F}:\|f-f_1\|_\infty\le\xi\epsilon} P_f^{(n)}(1-\phi_n) \le e^{-Kn\epsilon^2}.$$

D.7 Gaussian White Noise

For $\theta\in\Theta\subset\mathbb{L}_2[0,1]$, let $P_\theta^{(n)}$ be the distribution on $\mathfrak{C}[0,1]$ of the stochastic process $X^{(n)} = (X_t^{(n)}: 0\le t\le 1)$ defined structurally relative to a standard Brownian motion W as

$$X_t^{(n)} = \int_0^t \theta(s)\,ds + \frac{1}{\sqrt{n}} W_t.$$

An equivalent experiment is obtained by expanding $dX^{(n)}$ on an arbitrary orthonormal basis $e_1, e_2, \ldots$ of $\mathbb{L}_2[0,1]$, giving the random vector $X_n = (X_{n,1}, X_{n,2},\ldots)$, for $X_{n,i} = \int e_i(t)\,dX_t^{(n)}$. The vector X_n is sufficient in the experiment consisting of observing $X^{(n)}$. Its coordinates are independent and normally distributed with means the coefficients $\theta_i := \int e_i(t)\theta(t)\,dt$ of θ relative to the basis and variance $1/n$.

Lemma D.16 *For any $\theta_0,\theta_1\in\Theta$ with $\|\theta-\theta_1\|\ge\epsilon$ the test $\phi_n = \mathbb{1}\{2\langle\theta_1-\theta_0, X^{(n)}\rangle > \|\theta_1\|^2 - \|\theta_0\|^2\}$ satisfies*

$$P_{\theta_0}^{(n)}\phi_n \le 1 - \Phi(\sqrt{n}\epsilon/2) \le e^{-n\epsilon^2/8},$$

$$\sup_{\theta:\|\theta-\theta_1\|<\epsilon/4} P_\theta^{(n)}(1-\phi_n) \le 1 - \Phi(\sqrt{n}\epsilon/4) \le e^{-n\epsilon^2/32}.$$

Proof The test ϕ_n rejects the null hypothesis for positive values of the statistic $T_n = \langle\theta_1-\theta_0, dX^{(n)}\rangle - \frac{1}{2}\|\theta_1\|^2 + \frac{1}{2}\|\theta_0\|^2$. Under $P_\theta^{(n)}$ this possesses a normal distribution with mean $\langle\theta_1-\theta_0, \theta-\theta_1\rangle + \frac{1}{2}\|\theta_1-\theta_0\|^2$ and variance $\|\theta_1-\theta_0\|^2/n$. Under $P_{\theta_0}^{(n)}$ the mean is

$-\frac{1}{2}\|\theta_0 - \theta_1\|^2$, giving $P_{\theta_0}^{(n)}(T_n < 0) \leq \Phi(-\sqrt{n}\|\theta_1 - \theta_0\|/2)$, whereas by the Cauchy-Schwarz inequality under $P_\theta^{(n)}$ for $\|\theta - \theta_1\| \leq \|\theta_1 - \theta_0\|/4$, the mean is greater than $\|\theta_0 - \theta_1\|^2/4$, giving $P_\theta^{(n)}(T_n < 0) \leq 1 - \Phi(\sqrt{n}\|\theta_1 - \theta_0\|/4)$. The second inequalities follow by Lemma K.6. □

D.8 Historical Notes

Proposition D.1 is due to Le Cam (1986). The form of Lemma D.2 presented here was personally communicated by Birgé, and is modified from Birgé (1979) and Birgé (1983b). The combining technique of Lemma D.3 is by Le Cam (1986). The generalization to non-probability measures for the use in misspecified models presented in Theorem D.4 is due to Kleijn and van der Vaart (2006). Lemma D.6 and Proposition D.6 are due to Le Cam (1986). Lemma D.9 is due to Birgé (1979) and Birgé (1983b). Lemma D.11 was observed by Le Cam (1986). Proposition D.14 is due to Birgé (1983b) and Proposition D.15 is obtained from Birgé (1983a).

Problems

D.1 (Ghosal et al. 1999b) Use Hoeffding's inequality to show that for every pair P_0 and P_1 of probability measures there exist tests ϕ_n such that (D.6) holds for the product measures P_0^n and P^n with $d = e$ the total variation distance d_{TV} on the marginal distributions. Identify the resulting constants.

D.2 Extend Lemma D.11 by weakening the assumption to the following condition: For some $m \in \mathbb{N}$, there exists a test $\phi = \phi(X_1, \ldots, X_m)$ such that

$$\sup_{P \in \mathcal{P}_0} P^m\phi + \sup_{P \in \mathcal{P}_1} P^m(1 - \phi) < 1.$$

Appendix E

Polynomials, Splines and Wavelets

This appendix gathers results on three classes of functions used to approximate other functions: Bernstein polynomials, splines and wavelets. We focus on properties that are used throughout the main text.

E.1 Polynomials

Polynomials are widely used because of their simplicity and approximation ability. The classical Weierstrass theorem asserts that every continuous function on a compact interval in the real line can be uniformly approximated by polynomials. To approximate a general function, the degree of the polynomials must tend to infinity, where the rate of approximation as a function of the degree may be arbitrarily slow. However, the uniform distance between a function f in the Hölder space $\mathfrak{C}^\alpha[0,1]$ and the closest polynomial of degree k is of the order $k^{-\alpha}$.

Proposition E.1 (Polynomial approximation) *There exists a constant D that depends only on α such that for every $f \in \mathfrak{C}^\alpha[0,1]$ and $k \in \mathbb{N}$ there exists a polynomial P of degree k such that $\|f - P\|_\infty \leq Dk^{-\alpha}\|f\|_{\mathfrak{C}^\alpha}$.*

A related result is that a periodic function in $\mathfrak{C}^\alpha[0,1]$ can be approximated with the same order of accuracy $k^{-\alpha}$ by trigonometric polynomials (Jackson 1912; these are polynomials in sine and cosine functions; a Fourier series is an example of such a polynomial, but gives the approximation only up to logarithmic factor). The relation between the two settings arises by considering ordinary polynomials in $z = e^{it}$, which are trigonometric polynomials as functions of t.

The remainder of this section is concerned with Bernstein polynomials, which are a special type of polynomial that can be obtained in a constructive manner from a given function. Even though their accuracy of approximation is suboptimal, they are interesting for retaining properties such as monotonicity or positiveness.

For a continuous function $F: (0,1] \to \mathbb{R}$, the associated *Bernstein polynomial* is defined as

$$B(x; k, F) = \sum_{j=0}^{k} F\Big(\frac{j}{k}\Big)\binom{k}{j} x^j (1-x)^{k-j}. \tag{E.1}$$

As $k \to \infty$ the functions $x \mapsto B(x; k, F)$ converge uniformly to F. This can be derived from the law of large numbers by the representation

$$B(x; k, F) = \mathrm{E}\Big[F\Big(\frac{J}{k}\Big)\Big], \qquad J \sim \mathrm{Bin}(k, x).$$

The result follows since $J/k \to x$ in probability, uniformly in x. The representation also shows that if $F(x)$ takes values in an interval $[c_1, c_2]$, so does $B(x; k, F)$.

The derivative of $B(x; k, F)$ in $(0, 1)$ is

$$b(x; k, F) = \sum_{j=1}^{k} \Big(F\Big(\frac{j}{k}\Big) - F\Big(\frac{j-1}{k}\Big)\Big) \,\mathrm{be}(x; j, k - j + 1), \tag{E.2}$$

where $\mathrm{be}(x; a, b)$ stands for the beta density $\mathrm{be}(x; a, b) = \Gamma(a + b)/(\Gamma(a)\Gamma(b))\, x^{a-1}(1 - x)^{b-1}$. In particular, the derivative is positive if F is monotone. Further, if F is a cumulative distribution function on $[0, 1]$, then $B(1; k, F) = 1$, so that $B(x; k, F)$ is also a cumulative distribution function on $[0, 1]$. These two shape-preserving properties of Bernstein polynomials are attractive for the purpose of approximating a function, as natural constraints can be maintained. Also observe that if F is a probability distribution on $[0, 1]$ with $F(0) = 0$, then $B(x; k, F)$ has a density representable as a mixture of beta densities $\mathrm{be}(x; j, k - j + 1)$, $j = 1, \ldots, k$.

More generally, given a weight sequence $w = (w_j: 1 \le j \le k)$, $k \in \mathbb{N}$, we define a *Bernstein polynomial density corresponding to* w by

$$b(x; k, w) := \sum_{j=1}^{k} w_j \mathrm{be}(x; j, k - j + 1) = k \sum_{j=1}^{k} w_j \binom{k-1}{j-1} x^{j-1}(1 - x)^{k-j}, \tag{E.3}$$

If w is a probability vector, then p is a probability density function. Note the abuse of the notation $b(\cdot; k, \cdot)$ applied to both a function and a weight sequence in its third argument.

Lemma E.2 *For every $w \in \mathbb{R}^k$, $\|b(\cdot; k, w)\|_\infty \le k\|w\|_\infty$ and*

$$\|b(\cdot; k, w) - b(\cdot; k, w')\|_1 \le \|w - w'\|_1.$$

Proof In view of (E.3) the norm $\|b(\cdot; k, w)\|_\infty$ is bounded by

$$k \max_j |w_j| \sup_x \sum_{i=0}^{k-1} \binom{k-1}{i} x^i (1 - x)^{k-1-i}.$$

This is equal to $k\|w\|_\infty$ by the binomial formula. The second relation follows by integrating the first middle expression in (E.3) after taking absolute difference at w and w'. □

The following result describes the accuracy of approximation of Bernstein polynomials at α-smooth densities in the sense of Definition C.4. Recall that the Lipschitz constant of order $\alpha \in (0, 1]$ of a continuous function f is given by $L_\alpha(f) = \sup\{|f(x_1) - f(x_2)|/|x_1 - x_2|^\alpha : x_1 \ne x_2\}$. In the lemmas below the target function f need not be a probability density.

Lemma E.3 *For $f \in \mathfrak{C}^\alpha[0, 1]$ and $\alpha \in (0, 2]$, define $M_\alpha(f)$ to be $L_\alpha(f)$ if $\alpha \le 1$ and to be $L_{\alpha-1}(f') + \|f'\|_\infty$ if $1 < \alpha \le 2$. Then $\|f - b(\cdot; k, F)\|_\infty \le 3M_\alpha(f)k^{-\alpha/2}$ for any $\alpha \in (0, 2]$, where F is a primitive function of f.*

Proof By (E.3), with J a variable with the Bin$(k-1, x)$-distribution,

$$b(x;k,F) = k\mathrm{E}\Big[F\Big(\frac{J+1}{k}\Big) - F\Big(\frac{J}{k}\Big)\Big]. \tag{E.4}$$

First consider the case that $\alpha \in (0,1]$. By the mean value theorem $F((y+1)/k) - F(y/k) = k^{-1}f(\xi_y)$, for some $\xi_y \in [y/k, (y+1)/k]$. Consequently, by (E.4),

$$\begin{aligned}|f(x) - b(x;k,F)| &= |f(x) - \mathrm{E}f(\xi_J)| \le \mathrm{E}|f(x) - f(\xi_J)| \\ &\le L_\alpha(f)\mathrm{E}|x-\xi_J|^\alpha \le L_\alpha(f)(\mathrm{E}|x-\xi_J|)^\alpha,\end{aligned} \tag{E.5}$$

by Jensen's or Hölder's inequality. Since $|\xi_J - J/k| \le k^{-1}$ and $\mathrm{E}(J/k) = x - x/k$,

$$\mathrm{E}|x-\xi_J| \le \frac{1}{k} + \mathrm{E}\Big|x - \frac{J}{k}\Big| \le \frac{1+x}{k} + \mathrm{E}\Big|\mathrm{E}\frac{J}{k} - \frac{J}{k}\Big| \le \frac{2}{k} + \sqrt{\mathrm{var}\,\frac{J}{k}} \le \frac{2}{k} + \frac{1}{2\sqrt{k}}.$$

The assertion of the lemma follows upon inserting this bound on the right side of (E.5).

If $\alpha \in (1,2]$, we can use Taylor's theorem with integral remainder to obtain

$$k\mathrm{E}\Big[F\Big(\frac{J+1}{k}\Big) - F\Big(\frac{J}{k}\Big)\Big] = \mathrm{E}f\Big(\frac{J}{k}\Big) + \frac{1}{k}\int_0^1 f'\Big(\frac{J}{k} + \frac{s}{k}\Big)(1-s)\,ds. \tag{E.6}$$

The second term on the right is bounded above by $(2k)^{-1}\|f'\|_\infty$. By the mean value theorem there exists for every t and h some $\xi \in [0,1]$ such that

$$\big|f(t+h) - f(t) - hf'(t)\big| = \big|hf'(t+\xi h) - hf'(t)\big| \le |h|L_{\alpha-1}(f')|\xi h|^{\alpha-1} \le L_{\alpha-1}(f')|h|^\alpha,$$

since f' is $(\alpha-1)$-smooth. Substituting $t = (k-1)x/k$ and $h = J/k - (k-1)x/k$, we find

$$\Big|f\Big(\frac{J}{k}\Big) - f\Big(\frac{k-1}{k}x\Big) - \Big(\frac{J}{k} - \frac{k-1}{k}x\Big)f'\Big(\frac{k-1}{k}x\Big)\Big| \le L_{\alpha-1}(f')\Big|\frac{J}{k} - \frac{k-1}{k}x\Big|^\alpha. \tag{E.7}$$

Hölder's inequality gives

$$\mathrm{E}\Big|\frac{J}{k} - \frac{k-1}{k}x\Big|^\alpha \le \Big(\mathrm{E}\Big|\frac{J}{k} - \frac{k-1}{k}x\Big|^2\Big)^{\alpha/2} = \Big(\mathrm{var}\,\frac{J}{k}\Big)^{\alpha/2} \le \Big(\frac{1}{4k}\Big)^{\alpha/2}. \tag{E.8}$$

Combining (E.4), (E.6), (E.7) and (E.8), and using that $\mathrm{E}(J/k) = (k-1)x/k$, we see that $\|f - b(\cdot;k,F)\|_\infty$ is bounded by

$$\begin{aligned}&\sup_{0\le x\le 1}\Big|f(x) - f\Big(\frac{(k-1)x}{k}\Big)\Big| + \sup_{0\le x\le 1}\Big|f\Big(\frac{(k-1)x}{k}\Big) - \mathrm{E}f\Big(\frac{J}{k}\Big)\Big| + \frac{1}{2k}\|f'\|_\infty \\ &\le L_1(f)\frac{1}{k} + L_{\alpha-1}(f')\Big(\frac{1}{4k}\Big)^{\alpha/2} + \frac{1}{2k}\|f'\|_\infty.\end{aligned}$$

For $\alpha \le 2$ the middle term dominates in order. □

The approximation error of Bernstein polynomials does not compare well with other approximation techniques in terms of dimension. Splines, wavelets or general polynomials based on k terms achieve an approximation error of the order $k^{-\alpha}$ for α-smooth densities. This may be thought of as a price paid for the shape-preserving property of Bernstein polynomials. Their use of many similar and hence redundant terms translates into higher complexity, and may reduce the speed of posterior convergence, depending on the prior. This may be avoided by clumping together terms, thus reducing the dimension of the approximating linear combination of the basis without sacrificing the quality of approximation.

For $k = l^2$, $l \in \mathbb{N}$ and $w \in \mathbb{S}_l$, define the *coarsened Bernstein polynomial* as

$$\tilde{b}(x; l^2, w) = \sum_{i=1}^{l} w_i \, l^{-1} \sum_{j=(i-1)l+1}^{il} \mathrm{be}(x; j, k+1-j). \tag{E.9}$$

If $w_j = F((j-1)/l, j/l])$, we shall also write $\tilde{b}(x; l^2, w)$ as $\tilde{b}(x; l^2, F)$. Like the classical Bernstein polynomials, these functions are mixtures of beta-densities $\mathrm{be}(x; j, k+1-j)$, for $j = 1, \ldots, k$. However they are linear combinations of only l fixed functions, equal to the averages of the beta-densities in blocks of l consecutive elements. Although these functions have lower complexity, their accuracy of approximation is comparable with Bernstein polynomials of order k, as the following result shows.

Lemma E.4 *For $\alpha \in (0, 1]$*[1] *and $f \in \mathfrak{C}^\alpha[0, 1]$ we have $\|f - \tilde{b}(\cdot; k, F)\|_\infty \leq 6L_\alpha(f)k^{-\alpha/2}$, where $F(x) = \int_0^x f(u)\,du$.*

Proof Define a function ϕ_k by $\phi_k(y) = il$ if $(i-1)l \leq y < il$, for $i = 1, \ldots, l = \sqrt{k}$. It is easy to verify that, for $J \sim \mathrm{Bin}(k-1, x)$,

$$\tilde{b}(x; k, F) = l\,\mathrm{E}\Big[F\Big(\frac{\phi_k(J)}{k}\Big) - F\Big(\frac{\phi_k(J)-l}{k}\Big)\Big]. \tag{E.10}$$

In view of Lemma E.3 it suffices to show that $\|b(\cdot; k, F) - \tilde{b}(\cdot; k, F)\|_\infty \leq 3L_\alpha(f)k^{-\alpha/2}$. By the definition of ϕ_k it follows that $|y - (\phi_k(y) - l)| \leq l$ for every y, whence

$$\mathrm{E}\Big|\frac{J}{k} - \frac{\phi_k(J)-l}{k}\Big|^\alpha \leq k^{-\alpha/2}. \tag{E.11}$$

By Taylor's theorem with integral remainder,

$$k\Big[F\Big(\frac{J+1}{k}\Big) - F\Big(\frac{J}{k}\Big)\Big]f = \int_0^1 f\Big(\frac{J}{k} + \frac{s}{k}\Big)\,ds,$$
$$l\Big[F\Big(\frac{\phi_k(J)}{k}\Big) - F\Big(\frac{\phi_k(J)-l}{k}\Big)\Big] = \int_0^1 f\Big(\frac{\phi_k(J)-l}{k} + \frac{s}{l}\Big)\,ds.$$

By (E.4) and (E.10), for all $x \in [0, 1]$,

$$\begin{aligned}
\Big|b(x; k, F) - \tilde{b}(x; k, F)\Big| &\leq \int_0^1 \mathrm{E}\Big|f\Big(\frac{\phi_k(J)-l}{k} + \frac{s}{l}\Big) - f\Big(\frac{J}{k} + \frac{s}{k}\Big)\Big|\,ds \\
&\leq L_\alpha(f)\int_0^1 \mathrm{E}\Big|\frac{\phi_k(J)-l}{k} + \frac{s}{l} - \frac{J}{k} - \frac{s}{k}\Big|^\alpha ds \\
&\leq L_\alpha(f)\int_0^1 \Big[\Big|\frac{s}{k} + \frac{s}{l}\Big|^\alpha + \mathrm{E}\Big|\frac{J}{k} - \frac{\phi_k(J)-l}{k}\Big|^\alpha\Big]\,ds \leq 3k^{-\alpha/2},
\end{aligned}$$

where the last inequality holds because of (E.11). ∎

[1] For $\alpha \in (1, 2]$, the result is false. Indeed, coarsening may be counterproductive in this case; see Problem E.3.

E.2 Splines

A *spline* function on an interval $[a, b)$ in $\mathbb{R}$ is a piecewise polynomial function with a given level of global smoothness. More precisely, given $K+1$ *knots* $a = t_0 < t_1 < \cdots < t_K = b$, a function $f\colon [a, b] \to \mathbb{R}$ is a spline of *order* q if the restriction $f|_{[t_{k-1},t_k]}$ of f to any subinterval is a polynomial of degree at most $q-1$ and $f \in \mathfrak{C}^{q-2}(a, b]$ (provided that $q \geq 2$).

Splines form a linear subspace of the space of all *piecewise polynomials*, which are the functions f such that every restriction $f|_{[t_{k-1},t_k]}$ is a polynomial. A piecewise polynomial of order q is described by q free coefficients in every interval, and hence the linear space of all piecewise polynomials of order K has dimension Kq. Splines are restricted to have equal left and right derivatives of orders $0, 1, \ldots, q-2$ at every interior knot point, leading to $q-1$ linear constraints at each of the $K-1$ interior knot points $t_1, \ldots, t_{K-1}$, and hence to $(K-1)(q-1)$ constraints in total. Thus the dimension of the spline space of order q is $J = Kq - (K-1)(q-1) = q + K - 1$.

Splines occur naturally as solutions in interpolation problems, by their ability to approximate smooth functions. In statistics they appear naturally as the solution of the penalized least square regression problem:

$$\operatorname*{argmin}_f \left\{ \sum_{i=1}^{n} (Y_i - f(X_i))^2 + \lambda \int f^{(m)}(x)^2\, dx \right\}.$$

The solution can be shown to be a spline of order $2m-1$ with knots at the observations (and the begin and end of the domain); it is known as a *smoothing spline*. The common choice $m = 2$ leads to *cubic splines*.

A convenient basis for the space of splines is the set of *B-splines* $B_{0,q}, \ldots, B_{J-1,q}$. For the interval $[0, 1]$ these can be defined recursively, as follows. For $q = 1$ the functions $B_{0,1}, \ldots, B_{K-1,1}$ are simply indicator functions $\mathbb{1}\{t_j \leq x < t_{j+1}\}$, for $j = 0, 1, \ldots, K-1$, having discontinuities at interior knot points $t_1, \ldots, t_{K-1}$. For $q = 2$, the basis functions are *tent functions*

$$B_{j,2}(x) = \begin{cases} \frac{x-t_j}{t_{j+1}-t_j}, & x \in [t_j, t_{j+1}), \\ \frac{t_{j+2}-x}{t_{j+2}-t_{j+1}}, & x \in [t_{j+1}, t_{j+2}), \\ 0, & \text{otherwise}, \end{cases} \qquad j = -1, 0, \ldots, K-1.$$

Here by convention $t_{-1} = t_0 = 0$ and $t_K = t_{K+1} = 1$. These functions are continuous, but not differentiable at the interior knot points. For a general q, define an extended knot sequence, with 0 and 1 are deliberately repeated q times, as

$$\overbrace{0, 0, \ldots, 0}^{q \text{ times}}, t_1, \ldots, t_{K-1}, \overbrace{1, 1, \ldots, 1}^{q \text{ times}}.$$

Then B-splines of order q can be written in terms of lower order ones as

$$B_{j,q}(x) = \frac{x - t_j}{t_{j+q-1} - t_j} B_{j,q-1}(x) + \frac{t_{j+q} - x}{t_{j+q} - t_{j+1}} B_{j+1,q-1}(x), \qquad j = -q+1, \ldots, K-1.$$

Figure E.1 gives a visible impression of these functions.

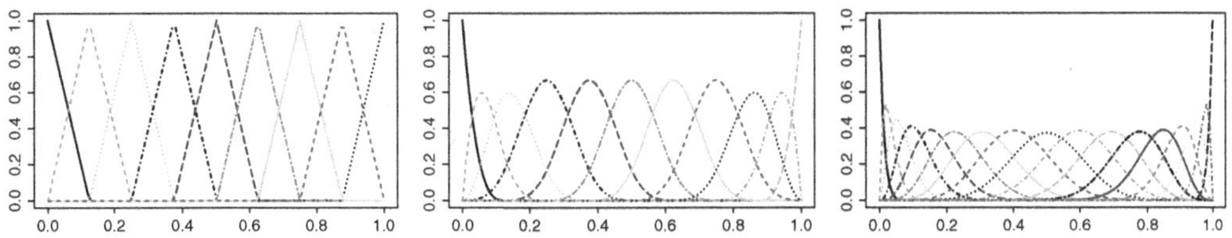

Figure E.1 B-spline basis functions of order 2, 4 and 10, for the knots at $0, 1/10, 2/10, \ldots, 1$ (i.e. $q = 2, 4, 10$ and $K = 10$).

From now on we assume that the order of the splines has been fixed at some value q. This will be dropped from the notation and the B-spline basis denoted by $B_1, B_2, \ldots, B_J$. Furthermore, for $\theta \in \mathbb{R}^J$ we write $\theta^\top B$ for the linear combination $\sum_{j=1}^J \theta_j B_j$.

For the half open unit interval $[0, 1)$, the uniform knot sequence k/K, for $k = 0, 1, \ldots, K$, is typically sufficient, but most of the following results are true more generally for "quasi-uniform" knot sequences, meaning that the ratio of the largest and smallest spacing between the knots is bounded by a universal constant. The following properties of B-splines are relevant for theoretical studies:

(i) $B_j \geq 0$, $j = 1, \ldots, J$,
(ii) $\sum_{j=1}^J B_j = 1$,
(iii) B_j is supported inside an interval of length q/K,
(iv) at most q functions B_j are nonzero at any given x.
(v) the integrals are given by

$$\int_0^1 B_j(x)\,dx = \begin{cases} j/(q(J-q+1)), & j = 1, 2, \ldots, q-1, \\ 1/(J-q+1), & j = q, q+1, \ldots, J-q+1, \\ (J-j+1)/(q(J-q+1)), & j = J-q+2, \ldots, J. \end{cases} \quad \text{(E.12)}$$

The first two properties express that the basis elements form a partition of unity, and the third and fourth properties mean that their supports are close to being disjoint if K is very large relative to q. The integrals of the B-spline functions located in the middle of the interval are identical as these functions are location shifts of each other. Only the most extreme B-spline functions have different integrals.

The following result shows that spline functions have excellent approximation properties for smooth functions.

Lemma E.5 *Suppose $q \geq \alpha > 0$. There exists a constant C depending only on q and α such that for every $f \in \mathfrak{C}^\alpha[0, 1]$ there exists $\theta \in \mathbb{R}^J$ with $\|\theta\|_\infty < \|f\|_{\mathfrak{C}^\alpha}$ such that*

$$\|\theta^\top B - f\|_\infty \leq C J^{-\alpha} \|f\|_{\mathfrak{C}^\alpha}.$$

Furthermore,

(a) *If f is strictly positive, for large J the vector θ in (a) can be chosen to have strictly positive coordinates.*[2]

[2] The condition that f is strictly positive is crucial. The best approximation $\theta^\top B$ with nonnegative coefficients θ of a nonnegative function f may have approximation error only $O(J^{-1})$ no matter how smooth f is; see De Boor and Daniel (1974).

(b) *If* $0 < f < 1$, *for large* J *the coordinates of the vector* θ *can be chosen between* 0 *and* 1.

(c) *If* f *is a probability density, then for large* J *there exists* $\theta \in \mathbb{S}_J$ *such that, for* $B_j^* = B_j / \int_0^1 B_j(x)\,dx$,

$$\|f - \theta^\top B^*\|_\infty \leq C J^{-\alpha} \|f\|_{\mathfrak{C}^\alpha}. \tag{E.13}$$

Proof The first part is well known in approximation theory and can be found in de Boor (1978), page 170.

For the proof of assertion (a), suppose that $f \geq \epsilon > 0$ pointwise. By Corollaries 4 and 6 in Chapter 11 of de Boor (1978), for each θ_i, there exists a constant C_1 that depends only on q such that $|\theta_j - c| \leq C_1 \sup_{x \in [t_{j+1}, t_{j+q-1}]} |f(x) - c|$, for any choice of the constant c. Choose c equal to the minimum of f over the interval $[t_{j+1}, t_{j+q-1}]$. If this minimum is attained at t^*, then $|f(x) - c| \leq C_2 |x - t^*|^{\min(\alpha,1)} \leq C_2 (q/J)^{\min(\alpha,1)}$, for every $x \in [t_{j+1}, t_{j+q-1}]$ and some constant $C_2 > 0$, since $f \in \mathfrak{C}^\alpha$. Then for $J > q(C_1 C_2/\epsilon)^{\max(1/\alpha,1)}$, we have $\theta_j > c - C_1 (q/J)^{\min(\alpha,1)} \geq 0$.

Part (b) follows from the preceding proof of (a) when also applied to $1 - f > 0$.

For part (c), we apply (b) to obtain existence of $\eta_1 \in (0,\infty)^J$ such that $\|f - \eta_1^\top B\| \lesssim J^{-\alpha}$. If $\eta_{2,i} = \eta_{1,i} \int_0^1 B_j(x)\,dx$, for $j = 1, \ldots, J$, then $\|f - \eta_2^\top B^*\|_\infty \lesssim J^{-\alpha}$, and in particular $\|\eta_2^\top B\|$ is bounded. By integration it follows that $|1 - \|\eta_2\|_1| = |1 - \sum_{j=1}^J \eta_{2,j}| \lesssim J^{-\alpha}$. Finally, if $\theta = \eta_2 / \|\eta_2\|_1$, then $\|f - \theta^\top B^*\|_\infty \leq \|f - \eta_2^\top B^*\|_\infty + \|\eta_2^\top B^*\|_\infty |1 - (\|\eta_2\|_1)^{-1}| \lesssim J^{-\alpha}$. □

Lemma E.6 *For any* $\theta \in \mathbb{R}^J$, *with* $\|\cdot\|_2$ *denoting the Euclidean and the* $\mathbb{L}_2[0,1]$*-norm,*

$$\|\theta\|_\infty \lesssim \|\theta^\top B\|_\infty \leq \|\theta\|_\infty, \qquad \|\theta\|_2 \lesssim \sqrt{J}\, \|\theta^\top B\|_2 \lesssim \|\theta\|_2.$$

Proof The first inequality is proved by de Boor (1978), page 156, Corollary XI.3). The second is immediate from the fact that the B-spline basis forms a partition of unity.

Let I_j be the interval $[(j-q)/K \vee 0, j/K \wedge 1]$. By Equation (2) on page 155 of de Boor (1978), we have

$$\sum_j \theta_j^2 \lesssim \sum_i \|\theta^\top B_{|I_j}\|_\infty^2 \lesssim \sum_j K \|\theta^\top B_{|I_j}\|_2^2.$$

The last inequality follows, because $\theta^\top B_{|I_j}$ consists of at most q polynomial pieces, each on an interval of length K^{-1}, and the supremum norm of a polynomial of order q on an interval of length L is bounded by $L^{-1/2}$ times the $\mathbb{L}_2$-norm, up to a constant depending on q. To see the third, observe that the squared $\mathbb{L}_2[0,1]$-norm of the polynomial $x \mapsto \sum_{j=0}^{q-1} \alpha_j x^j$ on $[0,1]$ be the quadratic form $\alpha^\top \mathrm{E}(U_q U_q^T)\alpha$ for $U_q = (1, U, \ldots, U^{q-1})$ and U a uniform $[0,1]$ variable. The second moment matrix $\mathrm{E}(U_q U_q^T)$ is nonsingular and hence the quadratic form is bounded below by a multiple of $\|\alpha\|^2$. This yields the third inequality.

By property (iii) of the B-spline basis, at most q elements $B_j(x)$ are nonzero for every given x, say for $j \in J(x)$. Therefore,

$$(\theta^\top B(x))^2 = \Big(\sum_{j\in J(x)} \theta_j B_j(x)\Big)^2 \leq \sum_{j\in J(x)} \theta_j^2 B_j^2(x)\, q, \tag{E.14}$$

by the Cauchy-Schwarz inequality. Since each B_j is supported on an interval of length proportional to $K^{-1} \asymp J^{-1}$ and takes its values in $[0, 1]$, its $\mathbb{L}_2[0, 1]$-norm is of the order $J^{-1/2}$. Combined with (E.14), this yields $\int_0^1 (\theta^\top B(x))^2\, dx \lesssim q\|\theta\|_2^2/J$, leading to the fourth inequality. ☐

Smooth functions on the multi-dimensional domain $[0, 1]^k$ may be approximated using a basis formed by tensor products of univariate splines:

$$B_{j_1\cdots j_k}(x_1, \ldots, x_k) = \prod_{l=1}^k B_{j_l}(x_l), \qquad 1 \leq j_l \leq J_l, \text{ for } l = 1, \ldots, k.$$

This collection of tensor-products inherits several useful properties of univariate B-splines, such as nonnegativity, adding to unity, being supported within a cube with sides of length q/K, and at most q^k many tensor-product B-splines being nonzero at any given point. The approximation properties also carry over.

For $\alpha = (\alpha_1, \ldots, \alpha_k) \in \mathbb{N}^k$, define the *anisotropic Hölder class* $\mathfrak{C}^\alpha([0, 1]^k)$ to be the collection of all functions that are α_l times partially differentiable in their lth coordinate with partial derivative satisfying

$$\Big\| \frac{\partial^{\sum_{l=1}^k r_l} f}{\partial x_1^{r_1} \cdots \partial x_k^{r_k}} \Big\|_\infty < \infty, \qquad 0 \leq r_l \leq \alpha_l, \ \text{ for } l = 1, \ldots, k. \tag{E.15}$$

Lemma E.7 *There exists a constant $C > 0$ that depends only on q such that for every $f \in \mathfrak{C}^\alpha([0, 1]^k)$ there exists $\theta = (\theta_{j_1\cdots j_k}: 1 \leq j_l \leq J_l), \in \mathbb{R}^{\prod_{l=1}^k J_l}$ with*

$$\|f - \theta^\top B\|_\infty \leq C \sum_{l=1}^k J_l^{-\alpha_l} \Big\| \frac{\partial^{\alpha_l} f}{\partial x_l^{\alpha_l}} \Big\|_\infty.$$

Furthermore,

(a) *If f is strictly positive, then every component of θ can be chosen to be positive.*
(b) *If $0 < f < 1$, then every component of θ can be taken to lie between 0 and 1.*
(c) *If f is a probability density function, the approximation order is maintained when replacing the tensor product splines by their normalized versions $B^*_{j_1\cdots, j_k}$ and restricting θ to $\mathbb{S}_{\prod_{l=1}^k J_l}$.*

If $\alpha_1 = \cdots = \alpha_k$, then these assertions extend to non-integer values of α and the isotropic Hölder class of order α (see Definition C.4).

Proof The first assertion is established in Theorem 12.7 in Schumaker (2007).

Assertion (a) can be proved using the arguments given in the proof of part (a) of Lemma E.5, once we show that for any $c > 0$,

$$|\theta_{j_1\cdots j_k} - c| \leq C_1 \max_l \sup_{x_l \in [t_{i+1,l}, t_{i+q-1,l}]} |f(x) - c|.$$

This requires bounding the absolute values of the coefficients using the values of the target function, which can be accomplished by using a dual-basis – a class of linear functionals which recovers values of the coefficients in a spline expansion from the values of the function; see Schumaker (2007). Clearly a dual basis for the multivariate B-splines is formed by tensor products of univariate dual bases and these can be chosen to be uniformly bounded; see Theorem 4.41 of Schumaker (2007). This bounds the maximum value of coefficients of spline approximations by a constant multiple of the $\mathfrak{L}_\infty$-norm of the target function. This gives the desired bound.

Proofs of parts (b) and (c) can follow the same arguments used to derive the respective parts of Lemma E.5. Finally, for the isotropic case, the approximation property is not restricted to the case of integer smoothness; see Schumaker (2007). The rest can be completed by repeating the same arguments. □

E.3 Wavelets

A drawback of polynomials and trigonometric functions is that they are globally defined, so that precise behavior of an approximation at a given point automatically has consequences for the whole domain. Spline functions overcome this drawback by partitioning the interval, and permit local modeling through the B-spline basis, but they lack the ease of an orthonormal basis. Wavelets provide both local modeling and orthogonality, and have excellent approximation properties relative to a wide range of norms. They come as a double-indexed basis ψ_{jk}, where j corresponds to a "resolution level," with bigger j focusing on smaller details, and k corresponds to location.

For a given function $g: \mathbb{R} \to \mathbb{R}$ and $j, k \in \mathbb{Z}$, let g_{jk} denote the function

$$g_{jk}(x) = 2^{j/2} g(2^j x - k).$$

Thus j gives a binary scaling (or "dilation") and k shifts the function. The leading scaling factor $2^{j/2}$ is chosen so that every g_{jk} has the same $\mathbb{L}_2(\mathbb{R})$-norm as g. For a given resolution j one needs 2^j functions g_{jk} to shift the function over a unit length interval.

The surprising fact is that there exist compactly supported functions $\psi: \mathbb{R} \to \mathbb{R}$, called *wavelet* or *mother wavelet*, such that $\{\psi_{jk}: j, k \in \mathbb{Z}\}$ are an orthonormal basis of $\mathbb{L}_2(\mathbb{R})$, and, moreover, the projection of a function on the subset of functions with resolution up to a given level is an excellent approximation to the function. Furthermore, the contributions of the resolution levels j up to a given level can be expressed as an infinite series of orthonormal translates of another compactly supported function $\phi: \mathbb{R} \to \mathbb{R}$, called *scaling function* or *father wavelet*. Thus a wavelet expansion of a given function $f \in \mathbb{L}_2(\mathbb{R})$ (starting at level $j = 0$) takes the form

$$f(x) = \sum_{k\in\mathbb{Z}} \alpha_k \phi_{0k}(x) + \sum_{j=0}^{\infty} \sum_{k\in\mathbb{Z}} \beta_{jk} \psi_{jk}(x). \tag{E.16}$$

The functions ϕ_{0k}, ψ_{jk}, for $k \in \mathbb{Z}$ and $j = 0, 1, \ldots$ form an orthonormal basis of $\mathbb{L}_2(\mathbb{R})$.

These functions are said to give a *multi-resolution* decomposition. If V_0 is the space spanned by the functions ϕ_{0k}, for $k \in \mathbb{Z}$, and W_j is the space spanned by ψ_{jk}, for $k \in \mathbb{Z}$, then $V_0, W_0, W_1, \ldots$ are orthogonal and the spaces $V_j = V_0 \oplus W_0 \oplus \cdots \oplus W_j$ are a nested sequence $V_0 \subset V_1 \subset \cdots$ whose span is $\mathbb{L}_2(\mathbb{R})$. Rather than at $j = 0$ as in the preceding display, the expansion can be started at an arbitrary $j > 0$, corresponding to the decomposition $\mathbb{L}_2(\mathbb{R}) = V_j \oplus W_j \oplus W_{j+1} \oplus \cdots$. The dilations ϕ_{jk}, for $k \in \mathbb{Z}$, of the father wavelet are an orthonormal basis of V_j, for every j. Therefore the expansion (E.16) remains valid if the ϕ_{0k} are replaced by the ϕ_{Jk} (with adapted coefficients α_k) and simultaneously the sum over j is started at $j = J$ instead of at $j = 0$.

By orthonormality the coefficients in (E.16) can be calculated as the inner products $\alpha_k = \int f(x)\phi_{0k}(x)\,dx$ and $\beta_{jk} = \int f(x)\psi_{jk}(x)\,dx$, for $k \in \mathbb{Z}$ and $j = 0, 1, \ldots$ Furthermore, the $\mathbb{L}_2$-norm of the function f is simply the ℓ_2-norm of the set of all coefficients in the expansion. For sufficiently regular wavelets, $\mathbb{L}_r$-norms for $r \neq 2$ can also be related to ℓ_r-norms of the coefficients, but less perfectly. Within a resolution level there is equivalence up to constants: for $1 \leq r \leq \infty$ and $j = 0, 1, \ldots$,

$$\Bigl\|\sum_k \alpha_k\phi_{0k}\Bigr\|_r \asymp \Bigl(\sum_k |\alpha_k|^r\Bigr)^{1/r},$$

$$\Bigl\|\sum_k \beta_{jk}\psi_{jk}\Bigr\|_r \asymp 2^{j(1/2-1/r)}\Bigl(\sum_k |\beta_{jk}|^r\Bigr)^{1/r}. \tag{E.17}$$

The factors $2^{j(1/2-1/r)}$ may seem surprising, but they simply arise because the wavelets ψ_{jk} have been scaled so that they have the same $\mathbb{L}_2$-norm; the factor $2^{j(1/2-1/r)}$ corrects this for the $\mathbb{L}_r$-norm. In general the equivalence in the preceding display does not extend to the full representation (E.16), but together with the triangle inequality they can be used to obtain at least an upper bound on the $\mathbb{L}_r$-norm of a function f in terms of its wavelet coefficients.

An approximation to f is obtained by truncating the series in j at a given resolution level J. In fact, this gives the orthogonal projection onto the span of the functions ϕ_{0k}, ψ_{jk}, for $k \in \mathbb{Z}$ and $j \leq J$. The accuracy of this approximation is usually described in terms of Besov spaces, which are defined as follows.

For a function f and number h the first difference is defined by $\Delta_h f(x) = f(x+h) - f(x)$; higher-order differences are defined by repetition as $\Delta_h^{r+1} f = \Delta_h\Delta_h^r f$, for $r \in \mathbb{N}$, where $\Delta_h^0 f = f$. The modulus of order r of f is defined as

$$\omega_{r,p}(f;t) = \sup_{h:0<h<t} \|\Delta_h^r f\|_p.$$

If f is defined on an interval in $\mathbb{R}$, then the rth difference $\Delta_h^r f(x)$ is only defined when both x and $x + rh$ are contained in the interval. The $\mathbb{L}_p$-norm $\|\Delta_h^r f\|_p$ is then understood to be relative to the Lebesgue measure on the corresponding reduced domain.

Definition E.8 (Besov space) For $p, q \geq 1$ and $\alpha > 0$ the *Besov* (p, q, α) *seminorm* of a measurable function $f: \mathfrak{X} \to \mathbb{R}$ on a bounded or unbounded interval $\mathfrak{X} \subset \mathbb{R}$ is defined as

$$\|f\|_{p,q,\alpha} = \begin{cases} \|f\|_p + \bigl[\int_0^\infty (h^{-\alpha}\omega_{\bar\alpha,p}(f;h))^q\,h^{-1}\,dh\bigr]^{1/q}, & 1 \leq q < \infty, \\ \|f\|_p + \sup_{0<h<1} h^{-s}\omega_{\bar\alpha,p}(f;h), & q = \infty. \end{cases} \tag{E.18}$$

Here $\bar{\alpha}$ may be any integer that is strictly bigger than α. The corresponding *Besov space* $\mathfrak{B}^\alpha_{p,q}(\mathfrak{X})$ is defined as the set of all functions $f: \mathfrak{X} \to \mathbb{R}$ with $\|f\|_{p,q,\alpha} < \infty$.

This definition is intimidating, by referring to three parameters, difference operators of order bigger than α and a weighted integral. The parameter α is the level of smoothness of the functions; the pair (p, q) determines the way in which smoothness is measured. The norm becomes more stringent (the norm up to equivalence bigger; the Besov space smaller) if p or α are increased, or if q is decreased.

In the important special case that $p = q = \infty$, the Besov space with parameter α contains the Hölder space of order α (see Definition C.4), and it reduces to this space if $\alpha \notin \mathbb{N}$. Thus the Besov scale may be viewed as a refinement of the Hölder scale.

Another important special case is obtained for $p = q = 2$, when the Besov space reduces to the *Sobolev space* $\mathfrak{W}^\alpha(\mathfrak{X})$. For $\alpha \in \mathbb{N}$ this is set of functions whose αth derivative (in the distributional sense) is contained in $\mathbb{L}_2(\mathfrak{X})$; the norm $\|f\|_2 + \|f^{(\alpha)}\|_2$ is equivalent to the Besov norm. For noninteger values of α, the Sobolev space $\mathfrak{W}^\alpha(\mathbb{R})$ is defined as the set of all functions $f \in \mathbb{L}_2(\mathbb{R})$ with Fourier transform $\hat{f}$ satisfying $\int (1 + |\lambda|^{2\alpha})|\hat{f}(\lambda)|^2\, d\lambda < \infty$; the Sobolev norm is the square root of the latter integral and is equivalent to the Besov norm.

Besov spaces can also be described through wavelet expansions of the form (E.16). If the wavelets are sufficiently regular ("higher than α"; we skip the details), then the Besov norm of a function $f \in \mathfrak{B}_{p,q,\alpha}(\mathbb{R})$ is equivalent to a norm on the coefficients in its wavelet expansion, as follows:

$$\|f\|_{p,q,\alpha} \asymp \Big(\sum_{k\in\mathbb{Z}} |\alpha_k|^p\Big)^{1/p} + \Big[\sum_{j=0}^{\infty} 2^{j(\alpha+1/2-1/p)q}\Big(\sum_{k\in\mathbb{Z}} |\beta_{jk}|^p\Big)^{q/p}\Big]^{1/q}, \qquad q < \infty,$$

$$\|f\|_{p,\infty,\alpha} \asymp \Big(\sum_{k\in\mathbb{Z}} |\alpha_k|^p\Big)^{1/p} + \sup_{j\geq 0}\Big[2^{j(\alpha+1/2-1/p)}\Big(\sum_{k\in\mathbb{Z}} |\beta_{jk}|^p\Big)^{1/p}\Big].$$

Thus a bound on the Besov norms implies a bound on the coefficients of its wavelet expansion, which in turn gives control on the approximation error.

Proposition E.9 *Let $f_{\leq J}$ be the orthogonal projection of the function f on the linear span up to resolution level J of a suitable wavelet basis of regularity higher than α. For any $1 \leq p, q \leq \infty$ and for $f \in \mathfrak{B}^\alpha_{p,q}(\mathbb{R})$ there exists a sequence $c(f) \in \ell_q$ with $\|c(f)\|_q \lesssim \|f\|_{p,q,\alpha}$ such that $\|f - f_{\leq J}\|_p \leq 2^{-J\alpha} c_J(f)$, for every J. In particular $\|f - f_{\leq J}\|_p \lesssim 2^{-J\alpha}\|f\|_{p,\infty,\alpha}$, for every J.*

Proof If $\|\beta_{j\cdot}\|_p$ is the ℓ_p-norm of the sequence $(\beta_{jk})_{k=1}^\infty$ of coefficients in the wavelet expansion of f, then $\|f - f_{\leq J}\|_p \lesssim \sum_{j>J} \|\beta_{j\cdot}\|_p 2^{j(1/2-1/p)}$, by the triangle inequality and (E.17). Furthermore, for f contained in the Besov space the ℓ_q-norm of the sequence $v_j = \|\beta_{j\cdot}\|_p 2^{j(\alpha+1/2-1/p)}$ is bounded above by $\|f\|_{p,q,\alpha}$. This readily gives the result in the case $q = \infty$. For finite q, we have $\sum_{j>J} \|\beta_{j\cdot}\|_p 2^{j(1/2-1/p)} = 2^{-J\alpha} c_J$, for $c_J = \sum_{j>J} v_j 2^{-(j-J)\alpha}$. By Hölder's inequality, with q' the conjugate of q,

$$|c_J|^q \le \sum_{j>J} |v_j|^q 2^{-(j-J)\alpha q/2} \Big(\sum_{j>J} 2^{-(j-J)\alpha q'/2}\Big)^{q/q'}.$$

The second sum on the right is bounded by a constant, and hence $\sum_J |c_J|^q \le \sum_J \sum_{j>J} |v_j|^q 2^{-(j-J)\alpha q/2}$, which can be seen to be bounded by a multiple of $\sum_j |v_j|^q$ with the help of Fubini's theorem. □

The Haar basis is a simple example of a wavelet basis. Its father and mother wavelet are $\phi = \mathbb{1}_{[0,1]}$ and $\psi = -\frac{1}{2}\mathbb{1}_{[0,\frac{1}{2})} + \frac{1}{2}\mathbb{1}_{[\frac{1}{2},1)}$, and the functions up to resolution level J are the functions that are constant on the dyadic intervals $(k2^{-J}, (k+1)2^{-J})$. Unfortunately, the discontinuity of these functions makes the Haar basis unsuitable for approximating functions of smoothness α higher than 1.

There are several popular wavelet bases of higher regularity, the most popular being *Daubechies wavelets*. No closed-form expressions of these functions exist, but many mathematical or statistical software packages contain a wavelet toolbox that allows numerical calculations.

Because a wavelet basis is defined through translation, its application to functions that are defined on a subinterval of $\mathbb{R}$ is awkward. One way around is to extend a given function to the full domain $\mathbb{R}$, and apply the preceding results to the extension. It can be shown that any function in a Besov space $\mathfrak{B}^\alpha_{p,q}[a,b]$ possesses such an extension of equal Besov norm, whence the Besov norm on a reduced interval can still be used to control approximation. If the interval $[a,b]$ is compact, then for each resolution level the support of only finitely many (of the order 2^j) of the translates ψ_{jk} will intersect the interval and hence the expansion can be reduced to finite sums in k.

Another solution is to adapt a given wavelet basis near the boundaries of the interval, so that orthonormality is retained within the interval. One construction (Cohen et al. 1993) gives for any smoothness level $\alpha > 0$ a wavelet basis $\{\psi_{jk}: 0 \le k \le 2^j - 1, j = 0, 1, \ldots\}$ for the interval $[0,1]$ with the following properties:

- $\mathrm{diam}(\mathrm{supp}(\psi_{jk})) \lesssim 2^{-j}$;
- $\|\psi_{jk}\|_\infty \lesssim 2^{j/2}$;
- $\#\{k': \mathrm{supp}(\psi_{j'k'}) \cap \mathrm{supp}(\psi_{jk}) \ne \varnothing\}$ is bounded by a universal constant, for any j and $j' \le j$;
- $\#\{k': \mathrm{supp}(\psi_{j'k'}) \cap \mathrm{supp}(\psi_{jk}) \ne \varnothing\} \lesssim 2^{j'-j}$, for any j and $j' > j$;
- $\sum_{k=0}^{2^j-1} \|\psi_{jk}\|_\infty \lesssim 2^{j/2}$;
- $\int \psi_{jk}(x)\,dx = 0$, for all sufficiently large j;
- $g \in \mathfrak{B}^\alpha_{p,q}[0,1]$ if and only if $|\int g\psi_{jk}(x)\,dx| \lesssim 2^{-j(\alpha+1/2)}$.

E.4 Historical Notes

The results on polynomials and splines are classical; the results on wavelets date back to the 1980s and 90s. The books de Boor (1978), Cohen et al. (1993), DeVore and Lorentz (1993), Daubechies (1992), Schumaker (2007) and Zygmund (2002) contain most results and many references. For introductions from a statistical perspective, we refer to Härdle et al. (1998) and Giné and Nickl (2015).

Problems

E.1 Give a simpler proof of Lemma E.3 for the case $\alpha = 2$.

E.2 Let k be an odd integer. Show that the sum of the standard deviations of $\mathrm{Be}(\cdot; j, k + 1 - j)$, $j = 1, \ldots, k$, is $O(\sqrt{k})$. Hence for $k \to \infty$, there is increasing overlap in the terms of the mixture, especially for terms corresponding to j close to $k/2$. This motivates the idea behind Lemma E.4.

E.3 Let $f(x) = F'(x) = 2x$. Show that if k is a perfect square,

$$\tilde{b}(1; k, F) = \frac{1}{\sqrt{k}}\Big[F(1) - F\Big(\frac{\sqrt{k}-1}{\sqrt{k}}\Big)\Big]\beta(1; k, 1) = \sqrt{k}\Big[F(1) - F\Big(\frac{k-1}{k}\Big)\Big]$$
$$= 2 - \frac{1}{\sqrt{k}}.$$

Thus the approximation error of the coarsened Bernstein polynomial is equal to $|f(1) - \tilde{b}(1; k, F)| = k^{-1/2}$, bigger than the approximation k^{-1} of the ordinary Bernstein polynomial.

Appendix F

Elements of Empirical Processes

This appendix introduces the empirical measure and the empirical process, and states the most important results on these objects.

The *empirical measure* of the random elements $X_1, \ldots, X_n$ in a sample space $(\mathfrak{X}, \mathscr{X})$ is the random measure given by

$$\mathbb{P}_n(C) = \frac{1}{n}\sum_{i=1}^{n}\delta_{X_i}(C) = \frac{1}{n}\#\{1 \leq i \leq n: X_i \in C\}.$$

For a given class $\mathcal{F}$ of measurable functions $f: \mathfrak{X} \to \mathbb{R}$, we also identify the empirical measure with the a map $f \mapsto \mathbb{P}_n(f)$. The corresponding *empirical process* is defined as

$$\mathbb{G}_n(f) = \sqrt{n}(\mathbb{P}_n(f) - P[f]) = \frac{1}{\sqrt{n}}\sum_{i=1}^{n}(f(X_i) - P[f]).$$

Here it is assumed that $X_1, \ldots, X_n$ have a common marginal distribution P, and that $P[f]$ exists for every $f \in \mathcal{F}$.

If $X_1, \ldots, X_n$ are i.i.d., then $\mathbb{P}_n(f) \to P[f]$ a.s. by the strong law of large numbers for every f with $P[|f|] < \infty$; if also $P[f^2] < \infty$, then also $\mathbb{G}_n(f) \rightsquigarrow \mathrm{Nor}(0, P[(f - Pf)^2])$ by the central limit theorem. A class $\mathcal{F}$ for which these theorems are true "uniformly in $\mathcal{F}$" is called Glivenko-Cantelli or Donsker, respectively.

More precisely, a class $\mathcal{F}$ of functions is called *Glivenko-Cantelli* if[1]

$$\sup_{f\in\mathcal{F}} |\mathbb{P}_n f - Pf| \to 0, \qquad \text{a.s.}$$

The classical Glivenko-Cantelli theorem is the special case that $X_1, \ldots, X_n$ are real-valued, and $\mathcal{F}$ is the class of all indicator functions of cells $(-\infty, t] \subset \mathbb{R}$.

For the Donsker property it is assumed that the sample paths $f \mapsto \mathbb{G}_n(f)$ are uniformly bounded, and the empirical process is viewed as a map into the metric space $\mathfrak{L}_\infty(\mathcal{F})$ of all uniformly bounded functions $z: \mathcal{F} \to \mathbb{R}$, metrized by the uniform norm. Then $\mathcal{F}$ is defined to be *Donsker* if $\mathbb{G}_n$ tends in distribution in $\mathfrak{L}_\infty(\mathcal{F})$ to a tight, Borel measurable random element in $\mathfrak{L}_\infty(\mathcal{F})$.[2] The limit $\mathbb{G}$ in the Donsker theorem can be identified

[1] In the case that the suprema S_n are not measurable, the almost sure convergence is interpreted in the sense $\mathrm{P}^*\big(\max_{m\geq n} |S_m| > \epsilon\big) \to 0$, for every $\epsilon > 0$, where the asterisk denotes outer probability.

[2] If the empirical process is not Borel measurable, convergence in distribution is understood as convergence of the *outer* expectations $\mathrm{E}^*\psi(\mathbb{G}_n)$ of bounded, measurable functions $\psi: \mathfrak{L}_\infty(\mathcal{F}) \to \mathbb{R}$. cf. Section 1.3 of van der Vaart and Wellner (1996) for details.

from its "marginals" $(\mathbb{G}(f_1), \ldots, \mathbb{G}(f_k))$, which are the weak limits of the random vectors $(\mathbb{G}_n(f_1), \ldots, \mathbb{G}_n(f_k))$. By the multivariate central limit theorem $(\mathbb{G}(f_1), \ldots, \mathbb{G}(f_k))$ must be normally distributed with mean vector 0 and covariance kernel

$$\operatorname{cov}(\mathbb{G}(f_i), \mathbb{G}(f_j)) = P[f_i f_j] - P[f_i]P[f_j].$$

Thus $\mathbb{G}$ is a Gaussian process; it is known as a *P-Brownian bridge*. The classical Brownian bridge is the special case where P is the uniform measure on the unit interval.

Sufficient conditions for the Glivenko-Cantelli and Donsker properties can be given in terms of the entropy of the class $\mathcal{F}$ and integrability of its envelope function. A measurable function $F: \mathfrak{X} \to \mathbb{R}$ is called an *envelope function* for $\mathcal{F}$ if $|f(x)| \le F(x)$ for all $f \in \mathcal{F}$ and $x \in \mathfrak{X}$.

The entropy conditions use either $\mathbb{L}_r(P)$-bracketing numbers, or $\mathbb{L}_r(Q)$-covering numbers, for Q ranging over discrete measures. In the latter case it is also necessary to make some measurability assumptions. A class of functions $\mathcal{F}$ is called *P-measurable* if the supremum $\sup_{f\in\mathcal{F}} |\sum_{i=1}^n \xi_i f(X_i)|$ is a measurable variable for every $(\xi_1, \ldots, \xi_n) \in \{0, 1\}^n$. Write $N(\epsilon, \mathcal{F}, \mathbb{L}_r)$ for the *uniform covering number* of a class $\mathcal{F}$:

$$N(\epsilon, \mathcal{F}, \mathbb{L}_r) = \sup_Q N(\epsilon \|F\|_{Q,r}, \mathcal{F}, \mathbb{L}_r(Q)),$$

where the supremum is taken over all finitely discrete probability measures $Q \in \mathfrak{M}(\mathfrak{X})$. For proofs of the following theorems, see e.g. van der Vaart and Wellner (1996), Sections 2.4, 2.5 and 2.6.

Theorem F.1 (Glivenko-Cantelli) *A P-measurable class $\mathcal{F} \subset \mathbb{L}_1(P)$ is Glivenko-Cantelli if and only if $P\|f - Pf\|_{\mathcal{F}} < \infty$ and $n^{-1} \log N(\epsilon, \{f 1\!\!1_{F \le M}: f \in \mathcal{F}\}, \mathbb{L}_1(\mathbb{P}_n)) \to_p 0$ for every $\epsilon > 0$ and $M \in (0, \infty)$. Either one the following two conditions is sufficient:*

(i) $N_{[\,]}(\epsilon, \mathcal{F}, \mathbb{L}_1(P)) < \infty$, *for every* $\epsilon > 0$.
(ii) $N(\epsilon, \mathcal{F}, \mathbb{L}_1) < \infty$ *for every* $\epsilon > 0$, $PF < \infty$ *and* $\mathcal{F}$ *is P-measurable.*

Theorem F.2 (Donsker) *Either one of the two conditions is sufficient for $\mathcal{F}$ to be Donsker:*

(i) $\int_0^\infty \sqrt{\log N_{[\,]}(\epsilon, \mathcal{F}, \mathbb{L}_2(P))}\, d\epsilon < \infty$.
(ii) $\int_0^\infty \sqrt{\log N(\epsilon, \mathcal{F}, \mathbb{L}_2)}\, d\epsilon < \infty$, $PF^2 < \infty$ *and* $(\mathcal{F} - \mathcal{F})^2$ *and* $\mathcal{F}_\delta := \{f - g: f, g \in \mathcal{F}, \|f - g\|_{P,2} < \delta\}$ *are P-measurable for every* $\delta > 0$.

Another mode of convergence than that provided by the Glivenko-Cantelli theorem is convergence in distribution. In this case the empirical measure is viewed as a random element with values in the space of probability measures $(\mathfrak{M}, \mathscr{M})$ on $(\mathfrak{X}, \mathscr{X})$ equipped with the weak topology. On a Polish sample space the convergence is true without any additional condition.

Proposition F.3 (Weak convergence) *If $(\mathfrak{X}, \mathscr{X})$ is a Polish space with the Borel σ-field, then the empirical distribution $\mathbb{P}_n$ of a random sample from P converges weakly to P, almost surely.*

Proof The Borel measurability of P_n follows from Proposition A.5.

The weak convergence $P_n \rightsquigarrow P$ of a general sequence of measures P_n can be described by the convergence $P_n[f] \to P[f]$ of expectations, for every f in a suitable countable class of bounded, continuous functions. (See e.g. van der Vaart and Wellner 1996, Theorem 1.12.2.) For the empirical measure and a single function this convergence follows from the strong law of large numbers, and hence the proposition follows by countably many applications of the strong law. □

The rate of convergence of maximum likelihood estimators can be characterized through the bracketing entropy integral relative to the Hellinger distance; see Wong and Shen (1995) or van der Vaart and Wellner (1996, 2017), Chapter 3.4 for a proof, and other results of a similar nature.

Theorem F.4 (MLE) *If $\hat{p}_n$ is the maximizer of $p \mapsto \prod_{i=1}^n p(X_i)$ over a collection $\mathcal{P}_n$ of probability densities, for a random sample $X_1 \ldots, X_n \stackrel{iid}{\sim} p_0 \in \mathcal{P}_n$, then $d_H(\hat{p}_n, p_0) = O_{P_0}(\epsilon_n)$ for any ϵ_n satisfying*

$$\int_0^{\epsilon_n} \sqrt{\log N_{[\,]}(u, \mathcal{P}_n, d_H)}\, du \lesssim \sqrt{n}\epsilon_n^2. \tag{F.1}$$

Appendix G

Finite-Dimensional Dirichlet Distribution

This Dirichlet distribution on the finite-dimensional unit simplex is the basis for the definition of the Dirichlet process, and therefore of central importance to Bayesian nonparametric statistics. This chapter states many of its properties and characterizations.

A random vector $X = (X_1, \ldots, X_k)$ with values in the k-dimensional unit simplex $\mathbb{S}_k := \{(s_1, \ldots, s_k): s_j \geq 0, \sum_{j=1}^k s_j = 1\}$ is said to possess a *Dirichlet distribution* with parameters $k \in \mathbb{N}$ and $\alpha_1, \ldots, \alpha_k > 0$ if it has density proportional to $x_1^{\alpha_1 - 1} \cdots x_k^{\alpha_k - 1}$ with respect to the Lebesgue measure on $\mathbb{S}_k$.

The unit simplex $\mathbb{S}_k$ is a subset of a $(k-1)$-dimensional affine space, and so "its Lebesgue measure" is to be understood to be $(k-1)$-dimensional Lebesgue measure appropriately mapped to $\mathbb{S}_k$. The normalizing constant of the Dirichlet density depends on the precise construction. Alternatively, the vector X may be described through the vector $(X_1, \ldots, X_{k-1})$ of its first $k-1$ coordinates, as the last coordinate satisfies $X_k = 1 - \sum_{i=1}^{k-1} X_i$. This vector has a density with respect to the usual $(k-1)$-dimensional Lebesgue measure on the set $\mathbb{D}_{k-1} = \{(x_1, \ldots, x_{k-1}): \min_i x_i \geq 0, \sum_{i=1}^{k-1} x_i \leq 1\}$. The inverse of the normalizing constant is given by the *Dirichlet form*

$$\int_0^1 \int_0^{1-x_1} \cdots \int_0^{1-x_1-\cdots-x_{k-2}} x_1^{\alpha_1-1} x_2^{\alpha_2-1} \cdots x_{k-1}^{\alpha_{k-1}-1} \times (1 - x_1 - \cdots - x_{k-1})^{\alpha_k - 1}\, dx_{k-1} \cdots dx_2\, dx_1. \tag{G.1}$$

The Dirichlet distribution takes its name from this integral, which by successive integrations and scalings to beta integrals can be evaluated to $\Gamma(\alpha_1) \cdots \Gamma(\alpha_k) / \Gamma(\alpha_1 + \cdots + \alpha_k)$.

Definition G.1 (Dirichlet distribution) The *Dirichlet distribution* $\mathrm{Dir}(k; \alpha)$ with parameters $k \in \mathbb{N} \setminus \{1\}$ and $\alpha = (\alpha_1, \ldots, \alpha_k) > 0$ is the distribution of a vector $(X_1, \ldots, X_k)$ such that $\sum_{i=1}^k X_i = 1$ and such that $(X_1, \ldots, X_{k-1})$ has density

$$\frac{\Gamma(\alpha_1 + \cdots + \alpha_k)}{\Gamma(\alpha_1) \cdots \Gamma(\alpha_k)} x_1^{\alpha_1} x_2^{\alpha_2 - 1} \cdots x_{k-1}^{\alpha_{k-1}-1} (1 - x_1 - \cdots - x_{k-1})^{\alpha_k - 1}, \quad x \in \mathbb{D}_{k-1}. \tag{G.2}$$

The Dirichlet distribution with parameters k and $\alpha \geq 0$, where $\alpha_i = 0$ for $i \in I \subsetneq \{1, \ldots, k\}$, is the distribution of the vector $(X_1, \ldots, X_k)$ such that $X_i = 0$ for $i \in I$ and such that $(X_i: i \notin I)$ possesses a lower-dimensional Dirichlet distribution, given by a density of the form (G.2).

For $k = 2$, the vector (X_1, X_2) is completely described by a single coordinate, where $X_1 \sim \mathrm{Be}(\alpha_1, \alpha_2)$ and $X_2 = 1 - X_1 \sim \mathrm{Be}(\alpha_2, \alpha_1)$. Thus the Dirichlet distribution is a multivariate generalization of the beta distribution. The $\mathrm{Dir}(k; 1, \ldots, 1)$-distribution is the uniform distribution on $\mathbb{S}_k$.

Throughout the section we write $|\alpha|$ for $\sum_{i=1}^k \alpha_i$.

There are several handy structural characterizations of the Dirichlet distribution.

Proposition G.2 (Representations) *For random variables $Y_1, \ldots, Y_k$ and $Y = \sum_{i=1}^k Y_i$,*

(i) *If $Y_i \overset{ind}{\sim} \mathrm{Ga}(\alpha_i, 1)$, then $(Y_1, \ldots, Y_k)/Y \sim \mathrm{Dir}(k; \alpha_1, \ldots, \alpha_k)$, and is independent of Y.*
(ii) *If $Y_i \overset{ind}{\sim} \mathrm{Be}(\alpha_i, 1)$, then $((Y_1, \ldots, Y_k)| \, Y = 1) \sim \mathrm{Dir}(k; \alpha_1, \ldots, \alpha_k)$.*
(iii) *If $Y_i \overset{ind}{\sim} \mathrm{Exp}(\alpha_i)$, then $((e^{-Y_1}, \ldots, e^{-Y_k})| \sum_{i=1}^k e^{-Y_i} = 1) \sim \mathrm{Dir}(k; \alpha_1, \ldots, \alpha_k)$.*

Proof We may assume that all α_i are positive.

(i). The Jacobian of the inverse of the coordinate transformation

$$(y_1, \ldots, y_k) \mapsto (y_1/y, \ldots, y_{k-1}/y, y) =: (x_1, \ldots, x_{k-1}, y)$$

is given by y^{k-1}. The density of the $\mathrm{Ga}(\alpha_i, 1)$-distribution is proportional to $e^{-y_i} y_i^{\alpha_i - 1}$. Therefore the joint density of $(Y_1/Y, \ldots, Y_{k-1}/Y, Y)$ is proportional to

$$e^{-y} y^{|\alpha|-1} x_1^{\alpha_1 - 1} \cdots x_{k-1}^{\alpha_{k-1}-1} (1 - x_1 - \cdots - x_{k-1})^{\alpha_k - 1}.$$

This factorizes into a Dirichlet density of dimension $k - 1$ and the $\mathrm{Ga}(|\alpha|, 1)$-density of Y.

(ii). The Jacobian of the transformation $(y_1, \ldots, y_k) \mapsto (y_1, \ldots, y_{k-1}, y)$, for $y = \sum_{i=1}^k y_i$, is equal to 1. The $\mathrm{Be}(\alpha_i, 1)$-density is proportional to $y_i^{\alpha_i - 1}$. Therefore the joint density of $(Y_1, \ldots, Y_{k-1}, Y)$ is proportional to $y_1^{\alpha_1 - 1} \cdots y_{k-1}^{\alpha_{k-1}-1} (y - y_1 - \cdots - y_{k-1})^{\alpha_k - 1}$, for $0 < y_i < 1$ for $i = 1, \ldots, k - 1$ and $0 < y - \sum_{i=1}^{k-1} y_i < 1$. The conditional density of $(Y_1, \ldots, Y_{k-1})$ given $Y = 1$ is proportional to this expression with y taken equal to 1, i.e. a Dirichlet density (G.2).

(iii). This follows from (ii), since $e^{-Y_i} \overset{\text{ind}}{\sim} \mathrm{Be}(\alpha_i, 1)$, $i = 1, \ldots, k$. □

The gamma representation in Proposition G.2(i) leads to several important properties. The first result states that the conditional distribution of a collection of coordinates given their sum inherits the Dirichlet structure and is independent of the total mass that is conditioned.

Proposition G.3 (Aggregation) *If $X \sim \mathrm{Dir}(k; \alpha_1, \ldots, \alpha_k)$ and $Z_j = \sum_{i \in I_j} X_i$ for a given partition $I_1, \ldots, I_m$ of $\{1, \ldots, k\}$, then*

(i) *$(Z_1, \ldots, Z_m) \sim \mathrm{Dir}(m; \beta_1, \ldots, \beta_m)$, where $\beta_j = \sum_{i \in I_j} \alpha_i$, for $j = 1, \ldots, m$.*
(ii) *$(X_i/Z_j : i \in I_j) \overset{ind}{\sim} \mathrm{Dir}(\#I_j; \alpha_i, i \in I_j)$, for $j = 1, \ldots, m$.*
(iii) *$(Z_1, \ldots, Z_m)$ and $(X_i/Z_j : i \in I_j)$, $j = 1, \ldots, m$, are independent.*

Conversely, if X is a random vector such that (i)–(iii) hold, for a given partition $I_1, \ldots, I_m$ and $Z_j = \sum_{i \in I_j} X_i$, then $X \sim \mathrm{Dir}(k; \alpha_1, \ldots, \alpha_k)$.

Proof In terms of the gamma representation $X_i = Y_i/Y$ of Proposition G.2 we have

$$Z_j = \frac{\sum_{i\in I_j} Y_i}{Y}, \quad \text{and} \quad \frac{X_i}{Z_j} = \frac{Y_i}{\sum_{i\in I_j} Y_i}.$$

Because $W_j := \sum_{i\in I_j} Y_i \overset{\text{ind}}{\sim} \text{Ga}(\beta_j, 1)$ for $j = 1, \ldots, m$, and $\sum_j W_j = Y$, the Dirichlet distributions in (i) and (ii) are immediate from Proposition G.2. The independence in (ii) is immediate from the independence of the groups $(Y_i : i \in I_j)$, for $j = 1, \ldots, m$. By Proposition G.2 W_j is independent of $(Y_i/W_j : i \in I_j)$, for every j, whence by the independence of the groups the variables W_j, $(Y_i/W_j : i \in I_j)$, for $j = 1, \ldots, m$, are jointly independent. Then (iii) follows, because $(X_i/Z_j : i \in I_j, j = 1, \ldots, m)$ is a function of $(Y_i/W_j : i \in I_j, j = 1, \ldots, m)$ and $(Z_1, \ldots, Z_m)$ is a function of $(W_j : j = 1, \ldots, m)$.

The converse also follows from the gamma representation. □

Corollary G.4 (Moments) *If* $X \sim \text{Dir}(k; \alpha_1, \ldots, \alpha_k)$, *then* $X_i \sim \text{Be}(\alpha_i, |\alpha| - \alpha_i)$. *In particular,* $\text{E}(X_i) = \alpha_i/|\alpha|$ *and* $\text{var}(X_i) = \alpha_i(|\alpha| - \alpha_i)/(|\alpha|^2(|\alpha| + 1))$. *Furthermore,* $\text{cov}(X_i, X_j) = -\alpha_i\alpha_j/(|\alpha|^2(|\alpha| + 1))$ *and, with* $r = r_1 + \cdots + r_k$,

$$\text{E}(X_1^{r_1} \cdots X_k^{r_k}) = \frac{\Gamma(\alpha_1 + r_1) \cdots \Gamma(\alpha_k + r_k)}{\Gamma(\alpha_1) \cdots \Gamma(\alpha_k)} \times \frac{\Gamma(|\alpha|)}{\Gamma(|\alpha| + r)}. \tag{G.3}$$

In particular, if $r_1, \ldots, r_k \in \mathbb{N}$, *then the expression in* (G.3) *is equal to* $\alpha_1^{[r_1]} \cdots \alpha_k^{[r_k]}/|\alpha|^{[r]}$, *where* $x^{[m]} = x(x + 1) \cdots (x + m - 1)$, $m \in \mathbb{N}$, *stands for the ascending factorial.*

Proof The first assertion follows from Proposition G.3 by taking $m = 2$, $I_i = \{i\}$, $I_2 = I \setminus \{i\}$, for $I = \{1, \ldots, k\}$. Next the expressions for expectation and variance follow by the properties of the beta distribution.

For the second assertion, we take $m = 2$, $I_1 = \{i, j\}$ and $I_2 = I \setminus I_1$ in Proposition G.3 to see that $X_i + X_j \sim \text{Be}(\alpha_i + \alpha_j, |\alpha| - \alpha_i - \alpha_j)$. This gives $\text{var}(X_i + X_j) = (\alpha_i + \alpha_j)(|\alpha| - \alpha_i - \alpha_j)/(|\alpha|^2(|\alpha| + 1))$, and allows to obtain the expression for the covariance from the identity $2\,\text{cov}(X_i, X_j) = \text{var}(X_i + X_j) - \text{var}(X_i) - \text{var}(X_j)$.

For the derivation of (G.3), observe that the mixed moment is the ratio of two Dirichlet forms (G.1) with parameters $(\alpha_1 + r_1, \ldots, \alpha_k + r_k)$ and $(\alpha_1, \ldots, \alpha_k)$. □

The *stick-breaking* representation of a vector $(X_1, \ldots, X_k)$ is given by

$$X_j = V_j \prod_{i<j}(1 - V_i), \qquad V_j = \frac{X_j}{1 - \sum_{i<j} X_i}, \qquad j = 1, \ldots, k - 1.$$

(If the vector takes its values in the unit simplex, then X_k is redundant, but one may define $V_k = 1$ to make the first equation true also for $j = k$.) The stick lengths V_j of a Dirichlet vector possess a simple characterization.

Corollary G.5 (Stick-breaking) *A vector* $(X_1, \ldots, X_k)$ *is* $\text{Dir}(k; \alpha_1, \ldots, \alpha_k)$*-distributed if and only if the stick lengths* $V_1, \ldots, V_{k-1}$ *are independent with* $V_j \sim \text{Be}(\alpha_j, \sum_{i>j} \alpha_i)$, *for* $j = 1, \ldots, k - 1$.

Proof Since the map between the vector $(X_1, \ldots, X_{k-1})$ and the stick lengths is a bimeasurable bijection, it suffices to derive the distribution of the stick lengths from the Dirichlet distribution of X. By (iii) of Proposition G.3 applied with the partition $I_1 = \{1\}$ and $I_2 = \{2, \ldots, k\}$ the variable $V_1 = X_1$ and vector $(X_2, \ldots, X_k)/(1 - X_1)$ are independent, by (i) the first is $\mathrm{Be}(\alpha_1, \sum_{i>1} \alpha_i)$-distributed and by (ii) the second possesses a $\mathrm{Dir}(k - 1; \alpha_2, \ldots, \alpha_k)$-distribution. We repeat this argument on the vector $(X_2, \ldots, X_k)/(1-X_1)$, giving that $V_2 = X_2/\sum_{i>2} X_i$ is independent of $(X_3, \ldots, X_k)/(1-X_1 - X_2)$ (where the factor $1 - X_1$ has cancelled out upon computing the quotients), and this variable and vector possess $\mathrm{Be}(\alpha_2, \sum_{i>2} \alpha_i)$- and $\mathrm{Dir}(k - 2; \alpha_3, \ldots, \alpha_k)$-distributions. Further repetitions lead to $V_1, \ldots, V_{k-2}$ being independent beta variables independent of $(X_{k-1}, X_k)/(1 - \sum_{i<k-1} X_i)$, from which we extract the distribution of its first coordinate V_{k-1}. □

Dirichlet distributions of fixed dimension depend continuously on their parameter vector α. If the sum of the parameters tends to zero or infinity, then the weak limit points are discrete, and supported on the vertices $e_1, \ldots, e_k$ of the unit simplex or at a single point, respectively.

Proposition G.6 (Limits) (i) *If* $\alpha \to \beta \in [0, \infty)^k$, $\beta \neq 0$, *then* $\mathrm{Dir}(k; \alpha) \rightsquigarrow \mathrm{Dir}(k; \beta)$.
(ii) *If* $|\alpha| \to 0$ *such that* $\alpha/|\alpha| \to \rho$, *then* $\mathrm{Dir}(k; \alpha) \rightsquigarrow \sum_{i=1}^k \rho_i \delta_{e_i}$.
(iii) *If* $|\alpha| \to \infty$ *such that* $\alpha/|\alpha| \to \rho$, *then* $\mathrm{Dir}(k; \alpha) \rightsquigarrow \delta_\rho$.

Proof For convenience of notation let $(X_1, \ldots, X_k) \sim \mathrm{Dir}(k; \alpha)$.

(i). If all coordinates of β are positive, then the result follows by the (pointwise) convergence of densities, since the gamma function is continuous. If some coordinate β_i is zero, but $\beta = \lim \alpha > 0$, then $\mathrm{E}X_i = \alpha_i/|\alpha| \to 0$, and hence X_i tends to zero in distribution. After eliminating these indices, convergence of the vector of the remaining coordinates follows by convergence of densities.

(ii). From the form of the beta densities of the coordinates it follows that $\mathrm{P}(\epsilon < X_i < 1 - \epsilon) \to 0$ for every $\epsilon > 0$. Therefore every weak limit point $(Y_1, \ldots, Y_k)$ has $Y_i \in \{0, 1\}$ a.s. Because $\sum_{i=1}^k Y_i = 1$, it follows that $(Y_1, \ldots, Y_k) \in \{e_1, \ldots, e_k\}$. Also by convergence of moments $\mathrm{E}(Y_1, \ldots, Y_k) = \rho$.

(iii). By assumption $\mathrm{E}X_i = \alpha_i/|\alpha| \to \rho_i$, while $\mathrm{var}\, X_i \to 0$ for every i. □

We may include the limit distributions in (ii) and (iii) into an extended family of *generalized Dirichlet distributions* $\overline{\mathrm{Dir}}(k; M, \rho)$ parameterized by three parameters, and defined to be the ordinary Dirichlet distribution $\mathrm{Dir}(k; M\rho)$ if $M \in (0, \infty)$ and $\rho \in [0, \infty)^k \setminus \{0\}$, to be $\sum_{i=1}^k \rho_i \delta_{e_i}$ if $M = 0$, and to be δ_ρ if $M = \infty$. This family is then closed under weak convergence, and is characterized by conditional independence properties in the following proposition. The parameter ρ is the mean vector, while M is a shape parameter (which is identifiable provided we do not allow $\rho = e_i$ if $M = 0$).

Let $(X_1, \ldots, X_n)$ be a vector of nonnegative variables X_i with $\sum_{i=1}^n X_i \leq 1$, and write $S_j = \sum_{i=1}^j X_j$ and $S_I = \sum_{i \in I} X_i$. The vector $(X_i : i \in I)$ is said to be *neutral* in $(X_1, \ldots, X_n)$ if it is independent of the vector $(X_j/(1 - S_I) : j \notin I)$. The vector $(X_1, \ldots, X_n)$ is said to be *completely neutral* if the variables $X_1, X_2/(1 - S_1), X_3/(1 - S_2), \ldots, X_n/(1 - S_{n-1})$ are mutually independent. To give a meaning to the independence conditions if the variables $S_I, S_1, \ldots, S_{n-1}$ can assume the value one, each quotient A/B should be replaced by a variable Y such that $A = BY$ a.s. and the independence condition made to refer to the Ys. An equivalent definition of complete neutrality is that there exist independent variables $Y_1, \ldots, Y_n$ such that $X_1 = Y_1, X_2 = (1 - Y_1)Y_2, X_3 = (1 - Y_1)(1 - Y_2)Y_3, \ldots$

If $X_1, \ldots, X_n$ are probabilities of sets in a partition $\{B_1, \ldots, B_n\}$ of a measurable space, then complete neutrality expresses a stick-breaking mechanism to distribute the total mass. The variable X_1 is the mass of the first set B_1, the variable X_2 is the part of the remaining probability assigned to B_2, etc., and all these conditional probabilities are independent. A bit in contradiction to its name, "complete neutrality" is dependent on the ordering of the variables. The masses may well be distributed in complete neutrality in one direction, but not in other "directions."

The following proposition only considers the symmetric case.

Proposition G.7 (Characterizations) *Let $(X_1, \ldots, X_k)$ be a random vector that takes values in $\mathbb{S}_k$ for $k \geq 3$ and be such that none of its coordinates vanishes a.s. Then the following statements are equivalent:*

(i) *X_i is neutral in $(X_1, \ldots, X_{k-1})$, for every $i = 1, \ldots, k-1$.*
(ii) *$(X_j: j \neq i, k)$ is neutral in $(X_1, \ldots, X_{k-1})$, for every $i = 1, \ldots, k-1$.*
(iii) *$(X_1, \ldots, X_{k-1})$ is completely neutral and X_{k-1} is neutral in $(X_1, \ldots, X_{k-1})$.*
(iv) *$(X_1, \ldots, X_k)$ possesses a Dirichlet distribution or a weak limit of Dirichlet distributions.*

Proof The equivalence of (i), (ii) and (iv) is the main result of Fabius (1964). The equivalence of (iii) and (iv) is Theorem 2 of James and Mosimann (1980). □

An important reason why the Dirichlet distribution is important in Bayesian inference is its conjugacy with respect to the multinomial likelihood.

Proposition G.8 (Conjugacy) *If $p \sim$ Dir$(k; \alpha)$ and $N|\, p \sim$ M$_k(n; p)$, then $p|\, N \sim$ Dir$(k; \alpha + N)$.*

Proof If some coordinate α_i of α is zero, then the corresponding coordinate p_i of p is zero with probability one, and hence so is the coordinate N_i of N. After removing these coordinates we can work with densities. The product of the Dirichlet density and the multinomial likelihood is proportional to

$$p_1^{\alpha_1 - 1} \cdots p_k^{\alpha_k - 1} \times p_1^{N_1} \cdots p_k^{N_k} = p_1^{\alpha_1 + N_1 - 1} \cdots p_k^{\alpha_k + N_k - 1}.$$

This is proportional to the density of the Dir$(k; \alpha_1 + N_1, \ldots, \alpha_k + N_k)$-distribution. □

The following three results give a mixture decomposition and two regenerative properties with respect to taking certain random convex combinations.

Proposition G.9 *For $k \in \mathbb{N}$ and any α with $|\alpha| > 0$,*

$$\sum_{i=1}^{k} \frac{\alpha_i}{|\alpha|} \mathrm{Dir}(k; \alpha_1, \ldots \alpha_{i-1}, \alpha_i + 1, \alpha_{i+1}, \ldots, \alpha_k) = \mathrm{Dir}(k; \alpha_1, \ldots, \alpha_k). \tag{G.4}$$

Proof If $p \sim \mathrm{Dir}(k; \alpha)$ and $N|\, p \sim \mathrm{MN}_k(1; p)$, then $\mathrm{P}(N = i) = \alpha_i/|\alpha|$, and $p|\,\{N = i\} \sim \mathrm{Dir}(k; \alpha + e_i)$, where e_i is the ith unit vector by Proposition G.8. The assertion now follows from decomposing the marginal distribution of p over its conditionals, given N. □

Proposition G.10 *If $X \sim \mathrm{Dir}(k; \alpha)$, $Y \sim \mathrm{Dir}(k; \beta)$, and $V \sim \mathrm{Be}(|\alpha|, |\beta|)$ are independent random vectors, then $VX + (1 - V)Y \sim \mathrm{Dir}(k; \alpha + \beta)$. In particular, if $X \sim \mathrm{Dir}(k; \alpha)$ and $V \sim \mathrm{Be}(|\alpha|, |\beta|)$, then $VX + (1 - V)e_i \sim \mathrm{Dir}(k; \alpha + \beta e_i)$, where e_i is the ith unit vector in $\mathbb{R}^k$, and $i \in \{1, \ldots, k\}$.*

Proof Let $W_i \overset{\text{ind}}{\sim} \mathrm{Ga}(\alpha_i, 1)$, $R_i \overset{\text{ind}}{\sim} \mathrm{Ga}(\beta_i, 1)$, $i = 1, \ldots, k$, be independent collections and represent $X_i = W_i/W$, $Y_i = R_i/R$, $V = W/(W + R)$, where $W = \sum_{i=1}^k W_i$, $R = \sum_{i=1}^k R_i$. Then the ith component of the convex combination $VX + (1 - V)Y$ is given by $(W_i + R_i)/(W + R)$, and hence the proof of the first assertion follows from the gamma representation, Proposition G.2(i).

The second assertion follows from the first by viewing $Y = e_i$ as a random vector distributed as $\mathrm{Dir}(k; \beta e_i)$. □

Proposition G.11 *If $p \sim \mathrm{Dir}(k; \alpha)$, $N \sim \mathrm{MN}_k(1; \alpha)$ and $V \sim \mathrm{Be}(1, |\alpha|)$ are independent, then $VN + (1 - V)p \sim \mathrm{Dir}(k; \alpha)$.*

Proof If $Y_0, Y_1, \ldots, Y_k \overset{\text{ind}}{\sim} \mathrm{Ga}(\alpha_i, 1)$, $i = 0, 1, \ldots, k$, where $\alpha_0 = 1$, then the vector (Y_0, Y) for $Y = \sum_{i=1}^k Y_i$ is independent of $p := (Y_1/Y, \ldots, Y_k/Y) \sim \mathrm{Dir}(k, \alpha_1, \ldots, \alpha_k)$ and $V = Y_0/(Y_0 + Y) \sim \mathrm{Be}(1, |\alpha|)$, by Propositions G.3 and G.2. Furthermore

$$(V, (1 - V)p) = (Y_0, Y_1, \ldots, Y_k)/(Y_0 + Y) \sim \mathrm{Dir}(k + 1; 1, \alpha).$$

Next merging the 0th cell with the ith, for any $i = 1, \ldots, k$, we obtain, again from Proposition G.3, that

$$(Ve_i + (1 - V)p) \sim \mathrm{Dir}(k; \alpha + e_i), \qquad i = 1, \ldots, k. \tag{G.5}$$

The distribution of the left side with e_i replaced by N is the mixture at the left-hand side of (G.4). □

Lemma G.12 *If $V_a \sim \mathrm{Be}(a, a)$, then as $a \to \infty$,*

(i) $\mathrm{P}(|2V_a - 1| > x) \le 2\pi^{-1/2} a^{1/2} e^{-x^2 a}$.
(ii) $a\,\mathrm{E} \log(2V_a) = O(1)$.
(iii) $\sqrt{a}\,\mathrm{E}|\log(2V_a)| \to \pi^{-1/2}$ *and* $a\,\mathrm{E}(\log(2V_a))^2 \to 1/2$.

Proof (i). Because the integrand is decreasing on $(1/2, 1]$, the integral $\int_{1/2+x/2}^1 v^{a-1}(1 - v)^{a-1}\,dv$ is bounded by $(1/2 + x/2)^{a-1}(1/2 - x/2)^{a-1}(1 - 1/2 - x/2) \le 2^{-2a+1}(1 -$

$x^2)^a$. Combine this with the inequality $B(a,a) \geq \pi^{1/2}2^{-2a+1}/\sqrt{a}$, which follows from the *duplication property* of the gamma function $\Gamma(2a) = \pi^{-1/2}2^{2a-1}\Gamma(a)\Gamma(a+\frac{1}{2})$ and the estimate $\Gamma(a+\frac{1}{2}) \leq \Gamma(a)\sqrt{a}$.

(ii) The moment $\mathrm{E}V_a^{-m}$ can be expressed as the ratio of two beta functions, and thus be seen to tend to 2^m, as $a \to \infty$. Because $|\log v| \lesssim v^{-1}$ for $v < 1/4$, we see that $\mathrm{E}|\log(2V_a)|\mathbb{1}_{V_a<1/4} \leq (\mathrm{E}(V_a^{-1}))^{1/2}\mathrm{P}(V_a \leq 1/4)^{1/2}$, which tends to zero exponentially fast by (i). Since $|\log(2x) - 2(x-1/2)| \lesssim (x-1/2)^2$ for $x > 1/4$, we find that

$$\begin{aligned}|\mathrm{E}(\log(2V_a))| &= |\mathrm{E}(\log(2V_a) - 2(V_a - 1/2))| \\ &\lesssim o(a^{-1}) + \mathrm{E}(V_a - 1/2)^2\mathbb{1}\{V_a > 1/4\} = O(a^{-1}).\end{aligned}$$

(iii) By Proposition G.2 and Corollary G.4 we can represent V_a as $Y_a/(Y_a+Z_a)$ for Y_a, Z_a i.i.d. $\mathrm{Ga}(a,1)$-variables, whence

$$\log(2V_a) = -\log\Big(1 + \frac{Z_a/a - Y_a/a}{2Y_a/a}\Big).$$

By the central limit theorem, $\sqrt{a}(Y_a/a - 1)$ and $\sqrt{a}(Z_a/a-1)$ converge jointly in distribution to independent standard normal variables. From this it can be shown by the delta-method that $\sqrt{a}\log(2Y_a)$ tends in distribution to an $\mathrm{Nor}(0,2)$-distribution. If it can be shown that all absolute moments of these variables are bounded, then all moments converge to the corresponding $\mathrm{Nor}(0,2)$-moment, from which the assertion follows. Now by the same argument as under (ii) $\mathrm{E}|\sqrt{a}\log(2V_a)|^m\mathbb{1}\{V < 1/4\} \lesssim a^{m/2}\sqrt{\mathrm{E}(V_a)}\mathrm{P}(V_a \leq 1/4)^{1/2}$ is exponentially small, while $\mathrm{E}|\sqrt{a}\log(2V_a)|^m\mathbb{1}\{V > 1/4\} \leq \mathrm{E}|\sqrt{a}(V_a - 1/2)|^m = O(1)$, in view of (i). ☐

The following lemma plays a fundamental role in estimating prior probabilities of neighborhoods in relation to Dirichlet process and Dirichlet process mixture priors.

Lemma G.13 (Prior mass) *If $(X_1,\ldots,X_k) \sim \mathrm{Dir}(k;\alpha_1,\ldots,\alpha_k)$, where $A\epsilon^b \leq \alpha_j \leq M$, $Mk\epsilon \leq 1$, and $\sum_{j=1}^k \alpha_j = m$ for some constants A, ϵ, b, M and $M \geq m$, then there exist positive constants c and C depending only on A, M, m and b such that for any point $(x_1,\ldots,x_k)$ in the k-simplex $\mathbb{S}_k$,*

$$P\Big(\sum_{i=1}^k |X_i - x_i| \leq 2\epsilon,\ \min_{1\leq i\leq k} X_i > \frac{\epsilon^2}{2}\Big) \geq Ce^{-ck\log_-\epsilon}.$$

Proof First assume that $M = 1$, so that $\epsilon < k^{-1}$. There is at least one index i with $x_i \geq k^{-1}$; by relabeling, we can assume that $i = k$, and then $\sum_{i=1}^{k-1} x_i = 1 - x_k \leq (k-1)/k$. If $(y_1,\ldots,y_k)$ is contained in the k-simplex and $|y_i - x_i| \leq \epsilon^2$ for $i = 1,\ldots,k-1$, then

$$\sum_{i=1}^{k-1} y_i \leq \sum_{i=1}^{k-1} x_i + (k-1)\epsilon^2 \leq (k-1)(k^{-1} + \epsilon^2) \leq 1 - \epsilon^2 < 1.$$

Furthermore, $\sum_{i=1}^k |y_i - x_i| \leq 2\sum_{i=1}^{k-1}|y_i - x_i| \leq 2\epsilon^2(k-1) \leq 2\epsilon$ and $y_k > \epsilon^2 > \epsilon^2/2$ in view of the preceding display. Therefore the probability on the left side of the lemma is bounded below by

$$\mathrm{P}\Big(\max_{1\le i\le k-1}|X_i - x_i| \le \epsilon^2\Big) \ge \frac{\Gamma(m)}{\prod_{i=1}^k \Gamma(\alpha_i)} \prod_{i=1}^{k-1} \int_{\max((x_i-\epsilon^2),0)}^{\min((x_i+\epsilon^2),1)} 1\, dy_i,$$

because $\prod_{i=1}^k y_i^{\alpha_i - 1} \ge 1$ for every $(y_1, \ldots, y_k) \in \mathbb{S}_k$, as $\alpha_i - 1 \le 0$ by assumption. Since each interval of integration contains an interval of at least length ϵ^2, and $\alpha\Gamma(\alpha) = \Gamma(\alpha + 1) \le 1$ for $0 < \alpha \le 1$, the last display is bounded from below by

$$\Gamma(m)\epsilon^{2(k-1)} \prod_{i=1}^k \alpha_i \ge \Gamma(m)\epsilon^{2(k-1)}(A\epsilon^b)^k \ge Ce^{-ck\log_- \epsilon}.$$

This concludes the proof in the case that $M = 1$.

We may assume without loss of generality that a "general" M is an integer, and represent the jth component of the Dirichlet vector as block sums of $(X_{j,m}: m = 1, \ldots, M)$, where the whole collection $\{X_{j,m}:\ j = 1, \ldots, k,\ m = 1, \ldots, M\}$ is Dirichlet distributed with parameters 1 and $\alpha_{j,m} = \alpha_j/M$, $j = 1, \ldots, k, m = 1, \ldots, M$, and it satisfies the conditions of the lemma with $M = 1$ and k replaced by Mk. Clearly, the event on the left side of the lemma contains

$$\Big\{\sum_{j=1}^k \sum_{m=1}^M |X_{j,m} - \frac{x_j}{M}| \le 2\epsilon, \min_{1\le j\le k-1, 1\le m\le M} X_{j,m} > \epsilon^2/2\Big\}.$$

The result now follows from the special case. □

Problems

G.1 Show that if $\alpha \ne \beta$ are nonnegative k-vectors, then $\mathrm{Dir}(k; \alpha) \ne \mathrm{Dir}(k; \beta)$, unless both α and β are degenerate at the same component.

G.2 Prove the converse part of Proposition G.3.

G.3 If $Y \sim \mathrm{Be}(a, b)$, show that $\mathrm{E}(\log Y) = \Psi(a) - \Psi(a + b)$, where $\Psi(x) = \frac{d}{dx}\log\Gamma(x)$ is the digamma function.

G.4 If $U \sim \mathrm{Be}(a_0, b_0)$, $V \sim \mathrm{Be}(a_1, b_1)$, $W \sim \mathrm{Be}(a_2, b_2)$ are independently distributed, and $a_0 = a_1 + b_1$, $b_0 = a_2 + b_2$, then $(UV, U(1 - V), (1 - U)W, (1 - U)(1 - W)) \sim \mathrm{Dir}(4; a_1, b_1, a_2, b_2)$. In particular, $UV \sim \mathrm{Be}(a_1, b_0 + b_1)$.

G.5 (Connor and Mosimann 1969) For nonnegative random variables $X_1, \ldots, X_n$ which satisfy $\sum_{i=1}^n X_i \le 1$, the following assertions are equivalent:

(a) $(X_1, \ldots, X_j)$ is neutral in $(X_1, \ldots, X_n)$, for every $j = 1, \ldots, n$.

(b) There exist independent random variables $Y_1, \ldots, Y_n$ such that $X_j = Y_j(1 - Y_1)\cdots(1 - Y_{j-1})$, for every $j = 1, \ldots, n$.

If either (a) or (b) is true, then the vector $(X_1, \ldots, X_n)$ is called *completely neutral*. [Hint: Condition (a) means the existence of a random vector $W_{>j}$ independent of $(X_1, \ldots, X_j)$ such that $(X_{j+1}, \ldots, X_n) = W_{>j}(1 - S_j)$, for every j. The variables $Y_1, Y_2, \ldots$ are the consecutive relative lengths in the stick-breaking algorithm, $1 - S_j = (1 - Y_1)\cdots(1 - Y_j)$ are the remaining stick lengths, and $W_{>j}$ are relative lengths of all the remaining sticks after j cuts. Given the Ys we can define

$W_{>j,k} = (1 - Y_{j+1}) \cdots (1 - Y_{j+k-1}) Y_{j+k}$, for $k = 1, \dots, n - j$, which is a function of the $(Y_{j+1}, \dots, Y_n)$ and hence is independent of $(X_1, \dots, X_j)$, which is a function of $(Y_1, \dots, Y_j)$. Given the Ws, we can take $Y_1 = X_1$, $Y_2 = W_{>1,1}$, $Y_3 = W_{>2,1}$, etc., which can be seen to be independent by arguing that Y_j is independent of $(Y_1, \dots, Y_j)$ for every j.]

Appendix H

Inverse-Gaussian Distribution

The normalized inverse-Gaussian distribution is a probability distribution on the unit simplex, and can be viewed as an analog of the finite-dimensional Dirichlet distribution. It arises from normalizing a vector of independent inverse-Gaussian variables, which replace the gamma variables that would yield the Dirichlet distribution.

Recall that the *inverse-Gaussian distribution* $\mathrm{IGau}(\alpha, \gamma)$ with shape parameter $\alpha > 0$ and scale parameter $\gamma > 0$ is the probability distribution on $(0, \infty)$ with density

$$y \mapsto \frac{\alpha e^{\alpha\gamma}}{\sqrt{2\pi}} \frac{1}{y^{3/2}} e^{-\frac{1}{2}(\alpha^2/y + \gamma^2 y)}.$$

The $\mathrm{IGau}(0, \gamma)$ distribution is understood to be the distribution degenerate at 0. (The distribution takes its name from the distribution of the time until a Brownian motion $B = (B_t : t \geq 0)$ with linear drift reaches a given level: the hitting time $\inf\{t > 0: B_t + t = \alpha\}$ possesses the $\mathrm{IGau}(\alpha, 1)$-distribution. This hitting time can be considered to be "inverse" to the (Gaussian) distribution of this process itself.)

Definition H.1 (Normalized inverse-Gaussian distribution) Given $Y_i \overset{\mathrm{ind}}{\sim} \mathrm{IGau}(\alpha_i, 1)$ and $Y = \sum_{i=1}^k Y_i$, the vector $(Y_1, \ldots, Y_k)/Y$ is said to possess the *normalized inverse-Gaussian distribution* $\mathrm{NIGau}(k; \alpha)$ with parameters $k \in \mathbb{N}$ and $\alpha = (\alpha_1, \ldots, \alpha_k) \geq 0$.

As the second parameter γ of an inverse-Gaussian distribution is a scale parameter, and one degree of freedom is lost by normalization, taking this parameter in the distribution of the variables Y_i in the preceding definition to be 1 is not a loss of generality.

The following propositions give elementary properties of inverse-Gaussian distributions; these resemble properties of the Dirichlet distribution.

Proposition H.2 (Aggregation)

(i) *If* $Y_i \overset{ind}{\sim} \mathrm{IGau}(\alpha_i, \gamma)$, *then* $\sum_i Y_i \sim \mathrm{IGau}(\sum_i \alpha_i, \gamma)$.

(ii) *If* $(W_1, \ldots, W_k) \sim \mathrm{NIGau}(k; \alpha_1, \ldots, \alpha_k)$ *and* $Z_j = \sum_{i \in I_j} W_i$ *for a given partition* $I_1, \ldots, I_m$ *of* $\{1, \ldots, k\}$, *then* $(Z_1, \ldots, Z_m) \sim \mathrm{NIGau}(m; \beta_1, \ldots, \beta_m)$, *where* $\beta_j = \sum_{i \in I_j} \alpha_i$, *for* $j = 1, \ldots, m$.

Proposition H.3 (Density) *If $\alpha_i > 0$ for $i = 1, \dots, k$, then the vector $(W_1, \dots, W_k) \sim \mathrm{NIGau}(k; \alpha_1, \dots, \alpha_k)$ possesses a probability density function, given by*

$$w = (w_1, \dots, w_k) \mapsto \frac{e^{\sum_{i=1}^k \alpha_i} \prod_{i=1}^k \alpha_i}{2^{k/2-1}\pi^{k/2}} \prod_{i=1}^k \frac{1}{w_i^{3/2}} \frac{K_{-k/2}\Big(\big(\sum_{i=1}^k \alpha_i^2/w_i\big)^{1/2}\Big)}{\big(\sum_{i=1}^k \alpha_i^2/w_i\big)^{k/4}}, \qquad w \in \mathbb{S}_k,$$

where K_m is the modified Bessel function of the third kind *of order m (see Watson (1995)). In particular, the marginal density of W_1 is*

$$w \mapsto \frac{e^{\alpha_1+\alpha_2}\alpha_1\alpha_2}{\pi} \frac{1}{w^{3/2}(1-w)^{3/2}} \frac{K_{-1}\Big(\big(\alpha_1^2/w + \alpha_2^2/(1-w)\big)^{1/2}\Big)}{\big(\alpha_1^2/w + \alpha_2^2/(1-w)\big)^{1/2}}, \qquad w \in (0, 1).$$

Proposition H.4 (Laplace transform) *The Laplace transform of $Y \sim \mathrm{IGau}(\alpha, \gamma)$ is given by $\mathrm{E}(e^{-\lambda Y}) = \exp\big[-\alpha(\sqrt{2\lambda + \gamma^2} - \gamma)\big]$, for $\lambda \geq 0$.*

Proposition H.5 (Mean) *If $(W_1, \dots, W_k) \sim \mathrm{NIGau}(k; \alpha_1, \dots, \alpha_k)$, then, $\mathrm{E}(W_i) = \alpha_i / \sum_{j=1}^k \alpha_j$.*

Proof Write $W_i = Y_i/Y$, where $Y_j \overset{\text{ind}}{\sim} \mathrm{IGau}(\alpha_i, 1)$ and $Y = \sum_{j=1}^k Y_j$. Also let $Y_{-i} = \sum_{j \neq i} Y_j$ and $|\alpha| = \sum_{j=1}^k \alpha_j$. Then

$$\begin{aligned}\mathrm{E}(W_i) &= \int_0^\infty \mathrm{E}(Y_i e^{-uY})\, du = \int_0^\infty \mathrm{E}(Y_i e^{-uY_i})\mathrm{E}(e^{-uY_{-i}})\, du \\ &= -\int_0^\infty \frac{d}{du}\mathrm{E}(e^{-uY_i})\, \mathrm{E}(e^{-uY_{-i}})\, du.\end{aligned}$$

Now $Y_{-i} \sim \mathrm{IGau}(|\alpha| - \alpha_i, \gamma)$, by Proposition H.2. Therefore, by Proposition H.4 the preceding display is equal to

$$-\int_0^\infty \frac{d}{du} e^{-\alpha_i(\sqrt{2u+1}-1)}\, e^{-(|\alpha|-\alpha_i)(\sqrt{2u+1}-1)}\, du = \alpha_i \int_0^\infty (2u+1)^{-1/2} e^{-|\alpha|(\sqrt{2u+1}-1)} du.$$

We conclude the proof by the change of variables $y = \sqrt{2u+1} - 1$. ☐

The "generalized" inverse-Gaussian distribution replaces the power $y^{-3/2}$ in the density of an inverse-Gaussian distribution by a general one.

Definition H.6 (Generalized inverse-Gaussian distribution) The *generalized inverse-Gaussian distribution* $\mathrm{GIGau}(a, b, p)$ with parameters $\alpha > 0$, $\gamma > 0$ and $p \in \mathbb{R}$ is the probability distribution on $(0, \infty)$ with density

$$y \mapsto \frac{(\alpha/\gamma)^p}{2K_p(\alpha\gamma)} y^{p-1} e^{-\frac{1}{2}(\alpha^2/y + \gamma^2 y)},$$

where K_p is the modified Bessel function of the third kind.

Appendix I

Gaussian Processes

This appendix gives background on Gaussian processes and Gaussian random elements in Banach spaces. The focus is on properties that are important for their use as priors, as in Chapter 11.

For basic definitions and examples of Gaussian processes see Section 11.1.

I.1 Sample Path Properties

Sample path properties of a Gaussian process $W = (W_t: t \in T)$, such as boundedness, continuity or differentiability, are (of course) determined by the covariance kernel $K(s,t) = \mathrm{cov}(W_s, W_t)$ of the process. In this section we discuss continuity for an abstract index set T, and next differentiability for processes indexed by Euclidean space.

Although a precise characterization of continuity and boundedness in terms of majorizing measures is available (see e.g. Talagrand 1987, 1992), the following result in terms of the entropy of the index set with respect to the *intrinsic semimetric* ρ is a good compromise between generality and simplicity. The square of this metric is

$$\rho^2(s,t) = \mathrm{var}(W_s - W_t) = K(t,t) + K(s,s) - 2K(s,t).$$

Proposition I.1 (Continuity) *If $(W_t: t \in T)$ is a separable, mean-zero Gaussian process with intrinsic metric ρ, then for any $\delta > 0$,*

$$\mathrm{E}\Big[\sup_{\rho(s,t)\le\delta} |W_s - W_t|\Big] \lesssim \int_0^\delta \sqrt{\log D(\epsilon, T, \rho)}\, d\epsilon. \tag{I.1}$$

Furthermore, for $J(\delta)$ the integral on the right side,

$$\mathrm{E} \sup_{\rho(s,t)<\delta} \frac{|W_s - W_t|}{J(\rho(s,t))} < \infty.$$

Consequently, $|W_s - W_t| = O(J(\rho(s,t)))$, uniformly in (s,t) with $\rho(s,t) \to 0$, almost surely.

Proof The display is known as *Dudley's bound*, motivated from Dudley (1967); see for instance van der Vaart and Wellner (1996), Corollary 2.2.8 for a proof. The second inequality is equivalent to the finiteness of the supremum of the absolute values of the Gaussian process $(s,t) \mapsto (W_s - W_t)/J(\rho(s,t))$, which was also established in Dudley (1967), and is stronger than (I.1). □

Proposition I.2 (Modulus) *If $(W_t : t \in T)$ is a mean-zero Gaussian process indexed by a compact set $T \subset \mathbb{R}^d$ with $\mathrm{E}\|W_s - W_t\|^2 \le \|s-t\|^{2\alpha}$, for all $s, t \in T$ and some $\alpha \in (0, 1]$, then W possesses a version with continuous sample paths such that $|W_s - W_t| = O(\|s - t\|^\alpha \log_- \|s-t\|)$, uniformly in (s,t) with $\|s-t\| \to 0$, almost surely.*

Proof Because the intrinsic metric ρ of W is by assumption bounded above by the Euclidean metric to the power α, the entropy of a compact interval is bounded above by a multiple of $\epsilon^{-d/\alpha}$. Hence the entropy integral is of the order $\int_0^\delta \sqrt{\log_- \epsilon}\, d\epsilon \sim \delta\sqrt{\log_- \delta}$, as $\delta \to 0$. As the process is continuous in probability, there exists a separable version of the process, by general results on stochastic processes. The result follows by applying the preceding proposition to such a version. □

For a multi-index $j = (j_1, \ldots, j_d)$ of nonnegative integers , let D^j denote the mixed partial derivative operator $\partial^{|j|}/\partial t_1^{j_1} \cdots \partial t_d^{j_d}$. Furthermore, for a function K of two arguments, let $D_s^i D_t^j K(s,t)$ denote the function obtained by differentiating the function i times with respect to s and j times with respect to t.

Proposition I.3 (Differentiability) *If the partial derivatives $(s,t) \mapsto D_s^j D_t^j K(s,t)$ of order j of the covariance kernel K of the mean-zero Gaussian process $(W_t : t \in T)$, for T an interval in $\mathbb{R}^d$, exist and are Lipschitz continuous of order $\alpha > 0$, then W possesses a version whose sample paths are partially differentiable up to order j with jth order derivative that is Lipschitz of order a, for any $a < \alpha$. Furthermore, the derivative process $D^j W$ is Gaussian with covariance function* $\mathrm{cov}(D^j W_s, D^j W_t) = D_s^j D_t^j K(s,t)$.

Proof First consider the case that $d = 1$ and $j = 1$ and set $Y_h^t = W_{t+h} - W_t$, for given t and h. By linearity of the covariance and continuous differentiability of K, for every s, t, g, h,

$$\begin{aligned}\mathrm{cov}(Y_g^s, Y_h^t) &= K(s+g, t+h) - K(s+g, t) - K(s, t+h) + K(s,t)\\ &= \int_s^{s+g}\int_t^{t+h} D_u D_v K(u,v)\, dv\, du.\end{aligned}$$

Then $\mathrm{var}(Y_h^t/h - Y_{h'}^t/h') = \mathrm{var}(Y_h^t/h) + \mathrm{var}(Y_{h'}^t/h') - 2\,\mathrm{cov}(Y_h^t/h, Y_{h'}^t/h') \to 0$ as $h, h' \to 0$, as the two variances and the covariance on the right all tend to $D_s D_t K(s,t)_{|s=t}$, by continuity of $D_s D_t K$. Thus Y_h^t/h is a Cauchy net and converges in $\mathbb{L}_2$ to a limit; denote this by $\dot W_t$. The latter process is mean-zero Gaussian with covariance function $\mathrm{cov}(\dot W_s, \dot W_t) = \lim \mathrm{cov}(Y_h^s/h, Y_h^t/h) = D_s D_t K(s,t)$, by the preceding display. By assumption the latter function is Lipschitz of order α (or of order 1 if $j > 1$), and hence the process $\dot W_t$ possesses a version that is Lipschitz of order $a < \alpha$, by Proposition I.2. We claim that this version $\dot W_t$ is also a pathwise derivative of $t \mapsto W_t$. Indeed, for arbitrary $t_0 \in T$ the process $V_t := W_{t_0} + \int_{t_0}^t \dot W_s\, ds$ is pathwise continuously differentiable, and can be seen to be also differentiable in $\mathbb{L}_2$, with derivative $\dot W_t$. Thus the process $W_t - V_t$ possesses $\mathbb{L}_2$-derivative equal to 0 and vanishes at t_0. This implies that the function $t \mapsto \mathrm{E}[(V_t - W_t)H]$ is continuously differentiable, for any square-integrable H, with derivative $\mathrm{E}[0H] = 0$, and hence is zero by the mean value theorem. Thus $V_t = W_t$ almost surely.

We can proceed to $j > 1$ by induction.

For $d > 1$ we can apply the argument coordinatewise, keeping the other coordinates fixed. This shows that the process is partially differentiable with continuous partial derivatives. Hence it is totally differentiable. □

Proposition I.4 (Stationary processes) *Suppose that $(W_t: t \in \mathbb{R}^d)$ is a mean-zero stationary Gaussian process with spectral measure μ.*

(i) *If $\int \|\lambda\|^{2\alpha d}\, d\mu(\lambda) < \infty$, then W has a version whose sample paths are partially differentiable up to order the biggest integer k strictly smaller than α with partial derivatives of order k that are Lipschitz of order $a - k$, for any $a < \alpha$.*
(ii) *If $\int e^{c\|\lambda\|}\, d\mu(\lambda) < \infty$ for some $c > 0$, then W has a version with analytic sample paths.*

Proof By the dominated convergence theorem, for any multi-indices j and h with $\sum_l (j_l + h_l) \le 2k$,

$$D_s^j D_t^h K(s,t) = \int (i\lambda)^j (-i\lambda)^h e^{i\langle (s-t),\lambda\rangle}\, d\mu(\lambda).$$

Furthermore, the right-hand side is Lipschitz in (s,t) of order $\alpha - k$. Thus (i) follows from Proposition I.3.

For the proof of (ii) we note that the function $K(s,t) = \int e^{i\langle s-\bar{t},\lambda\rangle}\, d\mu(\lambda)$ is now also well defined for complex-valued s, t with absolute imaginary parts smaller than $c/2$ (and $\bar{t}$ the complex conjugate of t). The function is conjugate-symmetric and nonnegative-definite and hence defines a covariance function of a (complex-valued) stochastic process $(W_t: t \in T)$, indexed by $T = \{t \in \mathbb{C}: |\operatorname{Im} t| < c/2\}$. By an extension of Proposition I.3 this can be seen to have sample paths with continuous partial derivatives, which satisfy the Cauchy-Riemann equations, and hence the process is differentiable on its complex domain. □

I.2 Processes and Random Elements

If the sample paths of a Gaussian stochastic process $W = (W_t: t \in T)$ belong to a Banach space $\mathbb{B}$ of functions $z: T \to \mathbb{R}$, then W can be viewed as a map from the underlying probability space into $\mathbb{B}$. If this map is Borel measurable, then W induces a distribution on the Borel σ-field of $\mathbb{B}$. In Definition 11.2 W is defined to be a *Gaussian random element* if $b^*(W)$ is normally distributed, for every element b^* of the dual space $\mathbb{B}^*$. In this section we link the two definitions of being Gaussian, as a process and as a Borel measurable map.

We start with considering Borel measurability of a stochastic process.

Lemma I.5 *If the sample paths $t \mapsto W_t$ of the stochastic process $W = (W_t: t \in T)$ belong to a separable subset of a Banach space and the norm $\|W - w\|$ is a random variable for every w in the subset, then W is a Borel measurable map in this space.*

Proof The condition implies that $\{W \in B\}$ is measurable for every open or closed ball B in the given subset $\mathbb{B}_0$ in which W takes its values. Thus W is measurable in the σ-field generated in $\mathbb{B}_0$ by these balls. As $\mathbb{B}_0$ is separable by assumption, its Borel σ-field is equal to the ball σ-field (see e.g. Chapter 1.7 in van der Vaart and Wellner 1996). Thus W is a

Borel measurable map in $\mathbb{B}_0$. As the trace of the Borel σ-field in the encompassing Banach space is the Borel σ-field in $\mathbb{B}_0$, this is then also true for W as a map in $\mathbb{B}$. □

The lemma applies for instance to the space of continuous functions on a compact metric space, where the supremum norm can be seen to be measurable by reducing the supremum to a countable, dense set. The assumption of the lemma that the sample paths belong to a *separable* subset is not harmless. For instance, the Hölder norm of a stochastic process with sample paths in a Hölder space $\mathfrak{C}^\alpha[0,1]$ is easily seen to be measurable, but the Hölder space itself is not separable relative to its norm. The sample paths must be contained in a smaller set for the lemma to apply; for instance, a Hölder space of higher order $\beta > \alpha$ would do.

If a Gaussian stochastic process is a Borel measurable map in a Banach function space, then it is still not clear that it also fulfills the requirement of a Gaussian random element that every variable $b^*(W)$ is Gaussian. The following lemma addresses this for the Banach space of bounded functions.

Lemma I.6 *If the stochastic process $W = (W_t : t \in T)$ is a Borel measurable random element in a separable subset of the Banach space $\mathfrak{L}_\infty(T)$ equipped with the supremum norm, then W is a Gaussian random element in this space.*

Proof Every coordinate projection π_t, defined by $\pi_t z = z(t)$, is an element of the dual space $\mathfrak{L}_\infty(T)^*$, and so are linear combinations of coordinate projections. The assumption that W is Gaussian implies that $b^*(W)$ is Gaussian for every such linear combination $b^* = \sum_i \alpha_i \pi_{s_i}$. We shall show that a general b^* is a pointwise limit of linear combinations of coordinate projections, at least on a set in which W takes its values. Then Gaussianity of $b^*(W)$ follows from the fact an almost sure limit of Gaussian variables is Gaussian.

We may assume without loss of generality that the separable subset of $\mathfrak{L}_\infty(T)$ in which W takes its values is complete. Then W is automatically a tight random element, and it is known that there exists a semimetric ρ on T under which T is totally bounded and such that W takes its values in the subspace $\mathfrak{UC}(T,\rho)$ of functions $f : T \to \mathbb{R}$ that are uniformly continuous relative to ρ (e.g. van der Vaart and Wellner 1996, Lemma 1.5.9). Thus we may assume without loss of generality that W takes its values in $\mathfrak{UC}(T,\rho)$ for such a semimetric ρ.

Only the restriction to $\mathfrak{UC}(T,\rho)$ of a given element of $\mathfrak{L}_\infty(T)^*$ is now relevant, and this is contained in $\mathfrak{UC}(T,\rho)^*$. By the Riesz representation theorem an arbitrary element of $\mathfrak{UC}(T,\rho)^*$ is a map $f \mapsto \int \bar{f}(t)\, d\mu(t)$ for a signed Borel measure μ on the completion $\bar{T}$ of T and $\bar{f} : \bar{T} \to \mathbb{R}$ the continuous extension of f. Because T is totally bounded we can write it for each $m \in \mathbb{N}$ as a finite union of sets of diameter smaller than $1/m$. If we define μ_m as the measure obtained by concentrating the masses of μ on the partitioning sets in a fixed, single point in the partitioning set, then $\int \bar{f}\, d\mu_m \to \int \bar{f}\, d\mu$ as $m \to \infty$, for each $f \in \mathfrak{UC}(T,\rho)$. The map $f \mapsto \int \bar{f}\, d\mu_m$ is a linear combination of coordinate projections. It follows that for any $b^* \in \mathfrak{UC}(T,\rho)^*$ there exists a sequence b_m^* of linear combinations of coordinate projections that converges pointwise on $\mathfrak{UC}(T,\rho)$ to b^*. □

The preceding proof is based on approximating an arbitrary element of the dual space $\mathbb{B}^*$ pointwise by a sequence of linear combinations of coordinate projections. Thus a sufficient general condition is that the linear span of the coordinate projections $z \mapsto z(t)$ form a weak-* dense subset of the dual space. However, as the proof shows weak-* approximation need only hold on a subset of $\mathbb{B}$ where W takes its values.

Consider the special example of the Hölder space $\mathfrak{C}^\alpha([0,1]^d)$. A simple sufficient condition is that the sample paths of the process are smoother than α.

Lemma I.7 *A Gaussian process $W = (W_t: t \in [0,1]^d)$ with sample paths in $\mathfrak{C}^\beta([0,1]^d)$ is a Gaussian random element in $\mathfrak{C}^\alpha([0,1]^d)$, for $\alpha < \beta$.*

Proof For $\alpha < \beta$ the space $\mathfrak{C}^\beta([0,1]^d)$ is a separable subset of $\mathfrak{C}^\alpha([0,1]^d)$ for the $\mathfrak{C}^\alpha$-norm, and the norm $\|W - w\|_{\mathfrak{C}^\alpha}$ can be seen to be a random variable for every $w \in \mathfrak{C}^\beta([0,1]^d)$ by writing it as a countable supremum. Therefore W is Borel measurable by Lemma I.5. To show that it is Gaussian, for simplicity restrict to the subspace $\mathfrak{C}_0^\alpha[0,1]^d$ of functions that vanish at 0. A function f in this space can be identified with the function $v_f: [0,1]^{2d} \setminus D \to \mathbb{R}$, for $D = \{(t,t): t \in [0,1]^d\}$, defined by

$$v_f(s,t) = \frac{f(s) - f(t)}{|s-t|^\alpha}.$$

In fact this identification gives an isometry of $\mathfrak{C}_0^\alpha([0,1]^d)$ onto the set V of all functions v_f equipped with the uniform norm. Since every sample of W is in $\mathfrak{C}^\beta([0,1]^d)$, the image v_W takes its values in the subspace $V_0 \subset V$ of functions that can be continuously extended to a function in $\mathfrak{C}([0,1]^{2d})$ that vanishes on D. The restriction of $b^* \in V^*$ to V_0 is a continuous linear map on V_0 that can be extended to an element of the dual space of $\mathfrak{C}([0,1]^{2d})$, by the Hahn-Banach theorem. By the Riesz representation theorem it is representable by a finite Borel measure μ on $[0,1]^{2d}$. Putting it all together we see that $b^*(W) = \int (W(s) - W(t))/|s-t|^\alpha \, d\mu(s,t)$. We now approximate μ by a sequence of discrete measures and finish as in the proof of Lemma I.6. □

I.3 Probability Bounds

The distribution of a Gaussian process or random element is tightly concentrated around its mean. *Borell's inequality*, Proposition 11.17, is an expression of this fact. The following inequality is often also called Borell's inequality, or *Borell-Sudakov-Tsirelson inequality*, but concerns the concentration of the norm of the process. The inequality applies both to random elements in a Banach space and to stochastic processes.

For W a mean-zero Gaussian element in a separable Banach space, let $M(W)$ be the median of $\|W\|$, and

$$\sigma^2(W) = \sup_{b^* \in \mathbb{B}^*: \|b^*\| = 1} \mathrm{E}\big[b^*(W)^2\big]. \tag{I.2}$$

Alternatively, for $W = (W_t: t \in T)$ a separable mean-zero Gaussian process, assume that $\|W\| := \sup_t |W_t|$ is finite almost surely, and let $M(W)$ be the median of $\|W\|$, and

$$\sigma^2(W) = \sup_t \mathrm{var}(W_t).$$

Proposition I.8 (Borell-Sudakov-Tsirelson) *Let W be a mean-zero Gaussian random element in a Banach space or a mean-zero separable Gaussian process such that $\|W\| < \infty$ almost surely. Then $\sigma(W) \le M(W)/\Phi^{-1}(3/4)$, and for any $x > 0$,*

$$\mathrm{P}\Big(\big|\|W\| - M(W)\big| > x\Big) \le 2\big[1 - \Phi(x/\sigma(W))\big],$$
$$\mathrm{P}\Big(\big|\|W\| - \mathrm{E}\|W\|\big| > x\Big) \le 2e^{-x^2/(2\sigma^2(W))}.$$

Proof For W a Gaussian process, see for instance van der Vaart and Wellner (1996), Proposition A.2.1. The upper bound on $\sigma(W)$ follows from the facts that $X_t/\operatorname{sd}(X_t) \sim \mathrm{Nor}(0, 1)$ and $\mathrm{P}(|X_t| \le M(X)) \ge 1/2$, for every t. The Banach space version can be reduced to the stochastic process version of the process $(b^*(W)\colon \|b^*\| = 1)$. The bound on the upper tail $\mathrm{P}(\|W\| - M(W) > x)$ is also a consequence of Borell's inequality, Proposition 11.17, upon taking $\epsilon = M(W)$ and $M = x/\sigma(W)$. □

The remarkable fact is that these bounds are independent of the size or complexity of the index set T or the Banach space. For any index set the tails of the distribution of $\|W\|$ away from its median or mean are smaller than Gaussian. On the other hand, the size of this median or mean, and hence the location of the distribution of $\|W\|$, strongly depends on the complexity of T or the Banach space. Proposition I.1 gives an upper bound in terms of the metric entropy of T.

Somewhat remarkable too is that the variance of the Gaussian tail bound is equal to $\sigma^2(W)$, which is the maximal variance of a variable W_t or $b^*(W)$. Thus the bound is no worse than the Gaussian tail bound for a single (worse case) variable in the supremum. By integrating $2x$ times the second bound we see that $\operatorname{var}[\|W\|] \le 4\sigma^2(W)$, so that the variance of the norm is bounded by a multiple of the maximal variance.

By integrating the first inequality it is seen that $|\mathrm{E}\|W\| - M(W)| \le \sigma(W)\sqrt{2/\pi}$, which shows that the mean is finite as soon as $\|W\|$ is a finite random variable. By integrating px^{p-1} times the first inequality we see that all moments of $\|W\|$ are finite, and can be bounded in terms of the first two moments, in the same way as for a univariate Gaussian variable.

I.4 Reproducing Kernel Hilbert Space

In Section 11.2 the *reproducing kernel Hilbert space* (RKHS) $\mathbb{H}$ attached to a Gaussian variable is defined both for Gaussian processes and for Gaussian random elements in a separable Banach space. In this section we give further background.

The definition of the Banach space RKHS uses the *Pettis integral* of a random element X with values in a Banach space $\mathbb{B}$. This is defined as the unique element $\mu \in \mathbb{B}$ such that $b^*(\mu) = \mathrm{E}[b^*(X)]$ for every $b^* \in \mathbb{B}^*$; it is denoted by $\mu = \mathrm{E}(X)$. The following lemma gives a condition for existence of this expectation.

Lemma I.9 *If X is a Borel measurable map in a separable Banach space $\mathbb{B}$ with $\mathrm{E}\|X\| < \infty$, then there exists an element $\mu \in \mathbb{B}$ such that $b^*(\mu) = \mathrm{E}[b^*(X)]$, for every $b^* \in \mathbb{B}^*$.*

Proof As a random element X taking values in a complete, separable metric space is automatically tight (cf. Parthasarathy 2005, or van der Vaart and Wellner 1996, 1.3.2), for any $n \in \mathbb{N}$ there exists a compact set K such that $\mathrm{E}[\|X\| \mathbb{1}_{\{X \notin K\}}] < n^{-1}$. Partition K in finitely many sets $B_1, \ldots, B_k$ of diameter less than n^{-1}. For increasing n, these partitions can be chosen successive refinements without loss of generality. Let $X_n = \sum_{i=1}^k b_i \mathbb{1}_{\{X \in B_i\}}$ for b_i an arbitrary point in B_i. Then $\mathrm{E}(X_n) := \sum_{i=1}^k b_i \mathrm{P}(X \in B_i)$ satisfies $b^*(\mathrm{E}(X_n)) = \mathrm{E}(b^*(X_n))$ for all $b^* \in \mathbb{B}^*$. Also $\|\mathrm{E}(X_n) - \mathrm{E}(X_m)\| = \sup_{\|b^*\|=1} |\mathrm{E}b^*(X_n - X_m)| \leq \mathrm{E}\|X_n - X_m\| \to 0$ as $n, m \to \infty$. Thus $\mathrm{E}(X_n)$ is a Cauchy sequence in $\mathbb{B}$, and hence converges to some limit μ. Because $\mathrm{E}\|X_n - X\| < 2n^{-1}$, we have that $b^*(\mu) = \lim_{n\to\infty} b^*(\mathrm{E}X_n) = \lim_{n\to\infty} \mathrm{E}[b^*(X_n)] = \mathrm{E}[b^*(X)]$, for every $b^* \in \mathbb{B}^*$. □

For a mean-zero Gaussian random element W in a Banach space the RKHS is defined as the completion of the set of Pettis integrals $Sb^* = \mathrm{E}\big[b^*(W)W\big]$ with respect to the norm $\|Sb^*\|_{\mathbb{H}} = \mathrm{sd}\big[b^*(W)\big]$. By the Hahn-Banach theorem and the Cauchy-Schwarz inequality,

$$\begin{aligned}\|Sb^*\| &= \sup_{b_2^* \in \mathbb{B}^*: \|b_2^*\|=1} |b_2^*(Sb^*)| = \sup_{b_2^* \in \mathbb{B}^*: \|b_2^*\|=1} |\mathrm{E}[b_2^*(W) b^*(W)]| \\ &\leq \sigma(W)\, \mathrm{sd}[b^*(W)] = \sigma(W)\, \|Sb^*\|_{\mathbb{H}}. \end{aligned} \tag{I.3}$$

Thus the RKHS-norm $\|\cdot\|_{\mathbb{H}}$ on $S\mathbb{B}^*$ is stronger than the original norm $\|\cdot\|$, so that a $\|\cdot\|_{\mathbb{H}}$-Cauchy sequence in $S\mathbb{B}^* \subset \mathbb{B}$ is a $\|\cdot\|$-Cauchy sequence in $\mathbb{B}$. Consequently, the RKHS $\mathbb{H}$, which is the completion of $S\mathbb{B}^*$ under the RKHS norm, can be identified with a subset of $\mathbb{B}$. In terms of the unit balls $\mathbb{B}_1$ and $\mathbb{H}_1$ of $\mathbb{B}$ and $\mathbb{H}$ the preceding display can be written

$$\mathbb{H}_1 \subset \sigma(W)\mathbb{B}_1. \tag{I.4}$$

In other words, the norm of the embedding $\iota: \mathbb{H} \to \mathbb{B}$ is bounded by $\sigma(W)$.

Lemma I.10 *The map $S: \mathbb{B}^* \to \mathbb{H}$ is weak*-continuous.*

Proof The unit ball $\mathbb{B}_1^*$ of the dual space is weak*-metrizable (cf. Rudin 1973, 3.16). Hence the restricted map $S: \mathbb{B}_1^* \to \mathbb{H}$ is weak*- continuous if and only if weak*-convergence of a sequence b_n^* in $\mathbb{B}_1^*$ to an element b^* implies that $Sb_n^* \to Sb^*$ in $\mathbb{H}$. As the weak*-convergence $b_n^* \to b^*$ is the same as the pointwise convergence on $\mathbb{B}$, it implies that $(b_n^* - b^*)(W) \to 0$ a.s., and hence $(b_n^* - b^*)(W) \rightsquigarrow 0$. Since the latter variables are mean-zero Gaussian, the convergence is equivalent to the convergence of the variances $\|Sb_n^* - Sb^*\|_{\mathbb{H}}^2 = \mathrm{var}\,((b_n^* - b)(W)) \to 0$.

This concludes the proof that the restriction of S to the unit ball $\mathbb{B}_1^*$ is continuous. A weak*-converging net b_n^* in $\mathbb{B}^*$ is necessarily bounded in norm, by the Banach-Steinhaus theorem (Rudin 1973, 2.5), and hence is contained in a multiple of the unit ball. The continuity of the restriction then shows that $Sb_n^* \to Sb^*$, which concludes the proof. □

Corollary I.11 *If $\mathbb{B}_0^*$ is a weak*-dense subset of $\mathbb{B}^*$, then $\mathbb{H}$ is the completion of $S\mathbb{B}_0^*$.*

By the definitions $\langle Sb^*, S\underline{b}^* \rangle_{\mathbb{H}} = \mathrm{E}\big[b^*(W)\underline{b}^*(W)\big] = b^*(S\underline{b}^*)$, for any $b^*, \underline{b}^* \in \mathbb{B}^*$. By continuity of the inner product this extends to the *reproducing formula*: for every $h \in \mathbb{H}$ and

$b^* \in \mathbb{B}^*$,

$$\langle Sb^*, h\rangle_{\mathbb{H}} = b^*(h). \tag{I.5}$$

Read from right to left this expresses that the restriction of an element b^* of the dual space of $\mathbb{B}$ to the RKHS can be represented as an inner product.

Just as for stochastic processes there is an alternative representation of the RKHS for Banach space valued random elements through *first chaos*. In the present setting the latter is defined as the closed linear span of the variables $b^*(W)$ in $\mathbb{L}_2(\Omega, \mathscr{U}, P)$, for $(\Omega, \mathscr{U}, P)$ the probability space on which the Gaussian process is defined. The elements Sb^* of the RKHS can be written $Sb^* = \mathrm{E}[HW]$ for $H = b^*(W)$, and the RKHS-norm of Sb^* is by definition the $\mathbb{L}_2(\Omega, \mathscr{U}, P)$-norm of this H. This immediately implies the following lemma. Note that $\mathrm{E}[HW]$ is well defined as a Pettis integral for every $H \in \mathbb{L}_2(\Omega, \mathscr{U}, P)$, by Lemma I.9.

Lemma I.12 *The RKHS of the random element W is the set of Pettis integrals* $\mathrm{E}[HW]$ *for H ranging over the closed linear span of the variables $b^*(W)$ in $\mathbb{L}_2(\Omega, \mathscr{U}, P)$ with inner product* $\langle \mathrm{E}[H_1 W], \mathrm{E}[H_2 W]\rangle_{\mathbb{H}} = \mathrm{E}[H_1 H_2]$.

It is useful to decompose the map $S: \mathbb{B}^* \to \mathbb{B}$ as $S = A^*A$ for $A^*: \mathbb{L}_2(\Omega, \mathscr{U}, P) \to \mathbb{B}$ and $A: \mathbb{B}^* \to \mathbb{L}_2(\Omega, \mathscr{U}, P)$ given by

$$A^*H = \mathrm{E}[HW], \qquad Ab^* = b^*(W).$$

It may be checked that the operators A and A^* are indeed adjoints, after identifying $\mathbb{B}$ with a subset of its second dual space $\mathbb{B}^*$ under the canonical embedding (Rudin 1973, 3.15 and 4.5), as the notation suggests. By the preceding lemma the RKHS is the image of the first chaos space under A^*. Because $\mathrm{Range}(A)^\perp = \mathrm{Null}(A^*)$, the full range $A^*(\mathbb{L}_2(\Omega, \mathscr{U}, P))$ is not bigger than the image of the first chaos, but $A^*: \mathbb{L}_2(\Omega, \mathscr{U}, P) \to \mathbb{H}$ is an isometry if restricted to the first chaos space.

Recall that an operator is compact if it maps bounded sets into precompact sets, or, equivalently, maps bounded sequences into sequences that possess a converging subsequence.

Lemma I.13 *The maps $A^*: \mathbb{L}_2(\Omega, \mathscr{A}, \mathrm{P}) \to \mathbb{B}$ and $A: \mathbb{B}^* \to \mathbb{L}_2(\Omega, \mathscr{A}, \mathrm{P})$ and $S: \mathbb{B}^* \to \mathbb{B}$ are compact, and the unit ball $\mathbb{H}_1$ of the RKHS is precompact in $\mathbb{B}$.*

Proof An operator is compact if and only if its adjoint is compact, and a composition with a compact operator is compact (see Rudin 1973, 4.19). Fix some sequence b_n^* in the unit ball $\mathbb{B}_1^*$. As the unit ball is weak*- compact by the Banach-Alaoglu theorem (see Rudin 1973, 4.3(c)), there exists a subsequence along which $b_{n_j}^*$ converges pointwise on $\mathbb{B}^*$ to a limit b^*. Thus $b_{n_j}^*(W) \to b^*(W)$ a.s., and hence $\mathrm{var}\,(b_{n_j}^*(W) - b^*(W)) \to 0$. This shows that the operator A is compact.

The final assertion of the lemma follows from the fact that $\mathbb{H}_1 = A^*\mathbb{U}_1$, for $\mathbb{U}_1$ the unit ball of $\mathbb{L}_2(\Omega, \mathscr{A}, \mathrm{P})$, and hence is precompact by the compactness of A^*. □

Example I.14 (Hilbert space) The *covariance operator* of a mean-zero Gaussian random element W taking values in a Hilbert space $(\mathbb{B}, \langle\cdot,\cdot\rangle)$ is the continuous, linear, nonnegative, self-adjoint operator $S: \mathbb{B} \to \mathbb{B}$ satisfying

$$\mathrm{E}\big[\langle W, b_1\rangle\langle W, b_2\rangle\big] = \langle b_1, Sb_2\rangle, \qquad b_1, b_2 \in \mathbb{B}.$$

The RKHS of W is given by Range$(S^{1/2})$ equipped with the norm $\|S^{1/2}b\|_{\mathbb{H}} = \|b\|$. Here $S^{1/2}$ is the square root of S: the positive, self-adjoint operator $S^{1/2}: \mathbb{B} \to \mathbb{B}$ such that $S^{1/2}S^{1/2} = S$.

To prove this, observe first that the covariance operator S coincides with the operator S defined by the Pettis integral $Sb^* = \mathrm{E}\big[b^*(W)W\big]$ under the natural identification of $\mathbb{B}^*$ with $\mathbb{B}$ (where $b \in \mathbb{B}$ corresponds to the element $b_1 \mapsto \langle b, b_1\rangle$ of $\mathbb{B}^*$). Hence the RKHS is the completion of $\{Sb: b \in \mathbb{B}\}$ under the norm $\|Sb\|_{\mathbb{H}} = \mathrm{sd}[\langle W, b\rangle] = \|S^{1/2}b\|$. This is the same as the completion of $\{S^{1/2}c: c \in S^{1/2}\mathbb{B}\}$ under the norm $\|S^{1/2}c\|_{\mathbb{H}} = \|c\|$. The latter set is already complete, so that the completion operation is superfluous.

Example I.15 ($\mathbb{L}_2$-space) A measurable stochastic process W with $\int_0^1 W_s^2\,ds < \infty$, for every sample path, is a random element in $\mathbb{L}_2[0,1]$. The dual space of $\mathbb{L}_2[0,1]$ consists of the maps $g \mapsto \int g(s)f(s)\,ds$ for f ranging over $\mathbb{L}_2[0,1]$. By Fubini's theorem, with K the covariance function of W,

$$Sf(t) = \mathrm{E}\Big[W_t \int_0^1 W_s f(s)\,ds\Big] = \int_0^1 K(s,t)f(s)\,ds.$$

Thus S coincides with the kernel operator with kernel K. The RKHS is the completion of the range of K under the inner product $\langle Sf, Sg\rangle_{\mathbb{H}} = \int_0^1\int_0^1 K(s,t)f(s)g(t)\,dsdt$.

RKHS under Transformation

The image of a Gaussian random element under a continuous, linear map is again a Gaussian random element. If the map is also one-to-one, then the RKHS is transformed in parallel.

Lemma I.16 *Let W be a mean-zero Gaussian random element in $\mathbb{B}$ with RKHS $\mathbb{H}$. Let $T: \mathbb{B} \to \underline{\mathbb{B}}$ be a one-to-one, continuous, linear map from $\mathbb{B}$ into another Banach space $\underline{\mathbb{B}}$. Then the RKHS of the Gaussian random element TW in $\underline{\mathbb{B}}$ is given by $T\mathbb{H}$ and $T: \mathbb{H} \to \underline{\mathbb{H}}$ is a Hilbert space isometry.*

Proof Let $T^*: \underline{\mathbb{B}}^* \to \mathbb{B}^*$ be the adjoint of T, so that $(T^*\underline{b}^*)(b) = \underline{b}^*(Tb)$, for every $\underline{b}^* \in \underline{\mathbb{B}}^*$ and $b \in \mathbb{B}$. The RKHS $\underline{\mathbb{H}}$ of TW is the completion of $\{\underline{S}\underline{b}^*: \underline{b}^* \in \underline{\mathbb{B}}^*\}$, where

$$\underline{S}\underline{b}^* = \mathrm{E}\big[(TW)\underline{b}^*(TW)\big] = T\Big(\mathrm{E}\big[W\underline{b}^*(TW)\big]\Big) = T\Big(\mathrm{E}\big[W(T^*\underline{b}^*)(W)\big]\Big) = TST^*\underline{b}^*.$$

Furthermore, the inner product in $\underline{\mathbb{H}}$ is given by

$$\langle \underline{S}\underline{b}_1^*, \underline{S}\underline{b}_2^*\rangle_{\underline{\mathbb{H}}} = \mathrm{E}\big[\underline{b}_1^*(TW)\underline{b}_2^*(TW)\big] = \mathrm{E}\big[(T^*\underline{b}_1^*W)(T^*\underline{b}_2^*W)\big] = \langle ST^*\underline{b}_1^*, ST^*\underline{b}_2^*\rangle_{\mathbb{H}}.$$

Hence it follows that $\underline{S}\underline{b}^* = T(ST^*\underline{b}^*)$, and $\|\underline{S}\underline{b}^*\|_{\underline{\mathbb{H}}} = \|ST^*\underline{b}^*\|_{\mathbb{H}}$. Thus the linear map $T: ST^*\underline{\mathbb{B}}^* \subset \mathbb{H} \to \underline{S}\underline{\mathbb{B}}^* \subset \underline{\mathbb{H}}$ is an isometry for the RKHS-norms. It extends by continuity to a linear map from the completion $\mathbb{H}_0$ of $ST^*\underline{\mathbb{B}}^*$ in $\mathbb{H}$ to $\underline{\mathbb{H}}$. Because T is continuous for the norms of $\mathbb{B}$ and $\underline{\mathbb{B}}$ and the RKHS-norms are stronger, this extension agrees with T. Since $\mathbb{H}_0$ and $\underline{\mathbb{H}}$ are, by definition, the completions of $ST^*\underline{\mathbb{B}}^*$ and $\underline{S}\underline{\mathbb{B}}^*$, we have that $T: \mathbb{H}_0 \to \underline{\mathbb{H}}$ is an isometry onto $\underline{\mathbb{H}}$. It remains to show that $\mathbb{H}_0 = \mathbb{H}$.

Because T is one-to-one, its range $T^*(\mathbb{B}^*)$ is weak*- dense in $\mathbb{B}^*$; see Rudin (1973), Corollary 4.12. In view of Lemma I.10, the map $S: \mathbb{B}^* \to \mathbb{H}$ is continuous relative to the weak*- and RKHS topologies. Thus $S(T^*\mathbb{B}^*)$ is dense in $S\mathbb{B}^*$ for the RKHS-norm of $\mathbb{H}$ and hence is dense in $\mathbb{H}$. Thus $\mathbb{H}_0 = \mathbb{H}$. □

RKHS Relative to Different Norms

A stochastic process W can often be viewed as a map into several Banach spaces. For instance, a process indexed by the unit interval with continuous sample paths is a Borel measurable map in $\mathfrak{C}[0, 1]$, as well as in $\mathbb{L}_2[0, 1]$. A process with continuously differentiable sample paths is a map in $\mathfrak{C}[0, 1]$, in addition to being a map in $\mathfrak{C}^\alpha[0, 1]$, for $\alpha < 1$. The RKHS obtained from using a weaker Banach space norm (corresponding to a continuous embedding in a bigger Banach space) is typically the same. One could say that RKHS is *intrinsic* to the process.

Lemma I.17 *Let $(\mathbb{B}, \|\cdot\|)$ be a separable Banach space and let $\|\cdot\|'$ be a norm on $\mathbb{B}$ with $\|b\|' \leq \|b\|$. Then the RKHS of a Gaussian random element W in $(\mathbb{B}, \|\cdot\|)$ is the same as the RKHS of W viewed as a random element in the completion of $\mathbb{B}$ under $\|\cdot\|'$.*

Proof Let $\mathbb{B}'$ be the completion of $\mathbb{B}$ relative to $\|\cdot\|'$. The assumptions imply that the identity map $\iota: (\mathbb{B}, \|\cdot\|) \to (\mathbb{B}', \|\cdot\|')$ is continuous, linear and one-to-one. Hence the proposition follows from Lemma I.16. □

RKHS under Independent Sum

If a given Gaussian prior misses certain desirable "directions" in its RKHS, then these can be filled in by adding independent Gaussian components in these directions.

Recall that a closed linear subspace $\mathbb{B}_0 \subset \mathbb{B}$ of a Banach space $\mathbb{B}$ is *complemented* if there exists a closed linear subspace $\mathbb{B}_1$ with $\mathbb{B} = \mathbb{B}_0 + \mathbb{B}_1$ and $\mathbb{B}_0 \cap \mathbb{B}_1 = \{0\}$. All closed subspaces of a Hilbert space are complemented, but in a general Banach space this is not the case. However, finite-dimensional subspaces and subspaces that are the full space up to a finite-dimensional space are complemented in every Banach space, as a consequence of the Hahn-Banach theorem.

Lemma I.18 *Let V and W be independent mean-zero Gaussian random elements taking values in subspaces $\mathbb{B}^V$ and $\mathbb{B}^W$ of a separable Banach space $\mathbb{B}$ with RKHSs $\mathbb{H}^V$ and $\mathbb{H}^W$ respectively. Assume that $\mathbb{B}^V \cap \mathbb{B}^W = \{0\}$ and that $\mathbb{B}^V$ is complemented in $\mathbb{B}$ by a subspace that contains $\mathbb{B}^W$. Then the RKHS of $V + W$ is given by the direct sum $\mathbb{H}^V \oplus \mathbb{H}^W$ and the RKHS norms satisfy $\|h^V + h^W\|^2_{\mathbb{H}^{V+W}} = \|h^V\|^2_{\mathbb{H}^V} + \|h^W\|^2_{\mathbb{H}^W}$.*

Proof By the independence of V and W, for any $b^* \in \mathbb{B}^*$,

$$S^{V+W}b^* = \mathrm{E}\big[b^*(V+W)(V+W)\big] = S^V b^* + S^W b^*.$$

The assumptions that $\mathbb{B}^V \cap \mathbb{B}^W = \{0\}$ and $\mathbb{B}^V$ is complemented by a subspace that contains $\mathbb{B}^W$ imply that there exists a continuous linear map $\Pi: \mathbb{B} \to \mathbb{B}^V$ such that $\Pi b = b$ if $b \in \mathbb{B}^V$ and $\Pi b = 0$ if $b \in \mathbb{B}^W$. (This is a consequence of the Hahn-Banach Theorem [cf.

Theorem 3.2 of Rudin 1973].) Then $\Pi V = V$ and $(I-\Pi)W = W$ and $(I-\Pi)V = \Pi W = 0$ a.s., which can be seen to imply, for every $b^* \in \mathbb{B}^*$,

$$S^V b^* = S^V(b^*\Pi), \qquad S^W b^* = S^W(b^*(I-\Pi)), \qquad S^V(b^*(I-\Pi)) = S^W(b^*\Pi) = 0.$$

Given $b_1^*, b_2^* \in \mathbb{B}^*$ the map $b^* = b_1^*\Pi + b_2^*(I-\Pi)$ is also an element of $\mathbb{B}^*$, and $S^{V+W}b^* = S^V b_1^* + S^W b_2^*$. It is also seen that $\|S^{V+W}b^*\|^2_{\mathbb{H}^{V+W}} = \mathrm{var}[b^*(V+W)] = \mathrm{var}[(b_1^*(V) + b_2^*(W)]$ is the sum of $\|S^V b^*\|^2_{\mathbb{H}^V}$ and $\|S^W b^*\|^2_{\mathbb{H}^W}$. □

The assumption that $\mathbb{B}^V \cap \mathbb{B}^W = \{0\}$ can be interpreted as requiring "linear independence" rather than some form of orthogonality of the supports of V and W. The lemma shows that stochastic independence of V and W translates the linear independence into orthogonality in the RKHS of $V + W$.

The lemma assumes trivial intersection of the *supports* of the variables V and W, not merely trivial intersection of linear subspaces containing the ranges of the variables. As the RKHS is independent of the norm in view of Lemma I.17, the closure may be taken with respect to the strongest possible norm, to make the supports as small as possible and enhance the possibility of a trivial intersection.

The assumption of trivial intersection cannot be removed, as the following example shows.

Example I.19 Let $Z_{i1}, Z_{i2} \overset{\text{iid}}{\sim} \mathrm{Nor}(0,1)$, for $i = 1, 2, \ldots$, and define two independent Gaussian processes by the series $V_j = \sum_{i=1}^\infty \mu_{ij} Z_{ij} \psi_i$, for $j = 1, 2$, where $\{\psi_i\}$ is a basis in some Banach space. Then

$$V_1 + V_2 = \sum_{i=1}^\infty (\mu_{i1} Z_{i1} + \mu_{i2} Z_{i2})\psi_i = \sum_{i=1}^\infty \mu_i Z_i \psi_i,$$

for $\mu_i^2 = \mu_{i1}^2 + \mu_{i2}^2$ and $Z_i \overset{\text{iid}}{\sim} \mathrm{Nor}(0,1)$, for $i = 1, 2, \ldots$. As shown in Section I.6, the RKHS of $V_1 + V_2$ consists of the series $\sum_{i=1}^\infty w_i \psi_i$ with $\sum_{i=1}^\infty (w_i^2/\mu_i^2) < \infty$. The RKHSs of V_1 and V_2 are characterized similarly, and in general the RKHS of $V_1 + V_2$ is not an orthogonal sum of the latter RKHSs. In fact, the RKHS of $V_1 + V_2$ depends crucially on the order at which μ_i tends to as $i \to \infty$, and this is determined by the larger of μ_{i1} and μ_{i2}. If $\mu_{i1}/\mu_{i2} \to 0$, then the RKHS of $V_1 + V_2$ essentially coincides with the RKHS of V_2, but the distributions of V_1 and V_2 are orthogonal in that case.

I.5 Absolute Continuity

For a mean-zero Gaussian random element W in a separable Banach space $\mathbb{B}$ defined on a probability space $(\Omega, \mathscr{U}, P)$ and $\mathbb{H}$ its RKHS, define a map U by

$$U(Sb^*) = b^*(W), \qquad b^* \in \mathbb{B}^*. \tag{I.6}$$

By the definition of the RKHS the map $S\mathbb{B}^*: \mathbb{H} \to \mathbb{L}_2(\Omega, \mathscr{U}, P)$ is an isometry. Let $U: \mathbb{H} \to \mathbb{L}_2(\Omega, \mathscr{U}, P)$ be its extension to the full RKHS.

Proposition I.20 *If W is a mean-zero Gaussian random element in a separable Banach space and h is an element of its RKHS, then the distributions P^{W+h} and P^W of $W+h$ and W on $\mathbb{B}$ are equivalent with Radon-Nikodym density*

$$\frac{dP^{W+h}}{dP^W}(W) = e^{Uh-\frac{1}{2}\|h\|_{\mathbb{H}}^2}, \qquad \text{a.s.}$$

Proof Because Uh is in the closed linear span of the mean-zero Gaussian variables $b^*(W)$, it is itself mean-zero Gaussian variable. By the isometry property of U its variance is equal to $\text{var}[Uh] = \|h\|_{\mathbb{H}}^2$, and hence

$$dQ = e^{Uh-\frac{1}{2}\|h\|_{\mathbb{H}}^2}\, dP$$

defines a probability measure on $(\Omega, \mathscr{U})$. For any pair $b_1^*, b_2^* \in \mathbb{B}^*$ the joint distribution of the random vector $(USb_1^*, USb_2^*) = (b_1^*W, b_2^*W)$ under P is bivariate normal with mean zero and covariance matrix $((\langle Sb_i^*, Sb_j^*\rangle_{\mathbb{H}}))_{i,j=1,2}$. By taking limits we see that for every $h \in \mathbb{H}$ the joint distribution of the vector (b_1^*W, Uh) is bivariate normal with mean zero and covariance matrix Σ with $\Sigma_{1,1} = \|Sb_1^*\|_{\mathbb{H}}^2$, $\Sigma_{1,2} = \langle Sb_1^*, h\rangle_{\mathbb{H}}$ and $\Sigma_{2,2} = \|h\|_{\mathbb{H}}^2$. Thus

$$\begin{aligned} \mathrm{E}_Q e^{ib_1^*(W)} &= \mathrm{E}e^{ib_1^*(W)}e^{Uh-\frac{1}{2}\|h\|_{\mathbb{H}}^2} = e^{\frac{1}{2}(i,1)\Sigma(i,1)^\top}e^{-\frac{1}{2}\|h\|_{\mathbb{H}}^2} = e^{-\frac{1}{2}\Sigma_{1,1}+i\Sigma_{1,2}} \\ &= \mathrm{E}e^{ib_1^*W+i\langle Sb_1^*,h\rangle_{\mathbb{H}}}. \end{aligned}$$

By the reproducing formula (I.5) we have $\langle Sb_1^*, h\rangle_{\mathbb{H}} = b_1^*(h)$, whence the right side is equal to $\mathrm{E}e^{ib_1^*(W+h)}$. From uniqueness of characteristic functions we conclude that the distribution of $b^*(W)$ under Q is the same as the distribution of $b^*(W+h)$ under P, for every $b^* \in \mathbb{B}^*$. This implies that the distribution of $W+h$ under P is the same as the distribution of W under Q, i.e. $P(W+h \in B) = \mathrm{E}_Q\mathbb{1}_B(W) = \mathrm{E}\mathbb{1}_B(W)(dQ/dP)$, which shows that dQ/dP is the desired Radon-Nikodym derivative. ☐

The preceding lemma requires that the shift h is contained in the RKHS. If this is not the case, then the two Gaussian measures are orthogonal and hence there is no density (see e.g. van der Vaart and van Zanten 2008b, Lemma 3.3 for a proof).

I.6 Series Representation

Consider a covariance kernel K of a Gaussian process $W = (W_t: t \in T)$ of the form, for given $\lambda_1, \lambda_2, \ldots > 0$ and arbitrary functions $\phi_j: T \to \mathbb{R}$,

$$K(s,t) = \sum_{j=1}^{\infty} \lambda_j \phi_j(s)\phi_j(t). \tag{I.7}$$

The series is assumed to converge pointwise on $T \times T$; the index set T may be arbitrary. The following theorem characterizes the stochastic process RKHS, under the condition that the functions ϕ_j are *infinitely linearly independent* in the following sense: if $\sum_{j=1}^\infty w_j\phi_j(t) = 0$ for every $t \in T$ and some sequence w_j with $\sum_{j=1}^\infty w_j^2\lambda_j^{-1} < \infty$, then $w_j = 0$ for every $j \in \mathbb{N}$.

Theorem I.21 (RKHS series) *If the covariance function K of a mean-zero Gaussian stochastic process $W = (W_t: t \in T)$ can be represented as in (I.7) for positive numbers λ_j and infinitely linearly independent functions $\phi_j: T \to \mathbb{R}$ such that $\sum_{j=1}^{\infty} \lambda_j \phi_j^2(t) < \infty$ for every $t \in T$, then the RKHS of W consists of all functions of the form $\sum_{j=1}^{\infty} w_j \phi_j$ with $\sum_{j=1}^{\infty} w_j^2 \lambda_j^{-1} < \infty$ with inner product*

$$\Big\langle \sum_{j=1}^{\infty} v_j \phi_j, \sum_{j=1}^{\infty} w_j \phi_j \Big\rangle_{\mathbb{H}} = \sum_{j=1}^{\infty} \frac{v_j w_j}{\lambda_j}.$$

Proof By the Cauchy-Schwarz inequality and the assumption that $\sum_{j=1}^{\infty} \lambda_j \phi_j^2(t) < \infty$ for every $t \in T$, the series (I.7) converges absolutely for every $(s, t) \in T \times T$. The same is true for every series $f_w(t) := \sum_{j=1}^{\infty} w_j \phi_j(t)$ with coefficients such that $(w_j/\lambda_j^{-1/2}) \in \ell_2$. Thus $f_w: T \to \mathbb{R}$ defines a function, which by the assumption of linear independence of the basis functions corresponds to a unique sequence of coefficients (w_j). The inner product $\langle f_v, f_w \rangle_{\mathbb{H}}$ as given in the theorem gives an isometry between the functions and the sequences $(w_j/\lambda_j^{-1/2}) \in \ell_2$ and hence the set of all functions f_w is a Hilbert space H under this inner product.

It suffices to show that H coincides with the RKHS. By (I.7) for every $s \in T$ the function $t \mapsto K(s, t)$ is contained in H with coefficients $w_j = \lambda_j \phi_j(s)$. Furthermore, for $s, t \in T$,

$$\langle K(s, \cdot), K(t, \cdot) \rangle_{\mathbb{H}} = \sum_{k=1}^{\infty} \frac{\lambda_k \phi_k(s) \lambda_k \phi_k(t)}{\lambda_k} = K(s, t).$$

By Definition 11.12 and the following discussion, this shows that on the linear span of the functions $K(s, \cdot)$ the given inner product indeed coincides with the RKHS inner product. The RKHS is the completion of this linear span by definition, whence it suffices to show that H is not bigger than the RKHS. For $t \in T$ and any $f_w \in H$,

$$\langle f_w, K(t, \cdot) \rangle_{\mathbb{H}} = \Big\langle \sum_{j=1}^{\infty} w_j \phi_j, \sum_{j=1}^{\infty} \lambda_j \phi_j(t) \phi_j \Big\rangle_{\mathbb{H}} = \sum_{j=1}^{\infty} \frac{w_j \lambda_j \phi_j(t)}{\lambda_j} = f_w(t).$$

If $f_w \in H$ with $f_w \perp \mathbb{H}$, then in particular $f_w \perp K(t, \cdot)$ for every $t \in T$. Then the preceding reproducing formula show that $f_w(t) = 0$, for all t. Hence H is equal to the RKHS. □

Series expansions of the type (I.7) are not unique, and some may be more useful than others. They may arise as an eigenvalue expansion of the operator corresponding to the covariance function. However, this is not a requirement of the proposition, which applies to arbitrary functions ϕ_j.

Example I.22 (Eigen expansion) Suppose that $(T, \mathscr{T}, \nu)$ is a measure space and $K: T \times T \to \mathbb{R}$ is a covariance kernel such that $\iint K^2(s, t)\, d\nu(s)\, d\nu(t) < \infty$. Then the integral operator $K: \mathbb{L}_2(T, \mathscr{T}, \nu) \to \mathbb{L}_2(T, \mathscr{T}, \nu)$ defined by

$$Kf(t) = \int f(s)\, K(s, t)\, d\nu(t)$$

is compact and positive self-adjoint. Then there exists a sequence of eigenvalues $\lambda_j \downarrow 0$ and an orthonormal system of eigenfunctions $\phi_j \in \mathbb{L}_2(T, \mathscr{T}, \nu)$ (thus $K\phi_j = \lambda_j \phi_j$ for every $j \in \mathbb{N}$) such that (I.7) holds, except that the series converges in $\mathbb{L}_2(T \times T, \mathscr{T} \times \mathscr{T}, \nu \times \nu)$. The series $\sum_j w_j \phi_j$ now converges in $\mathbb{L}_2(T, \mathscr{T}, \nu)$ for any sequence (w_j) in ℓ_2. By the orthonormality of the functions ϕ_j, they are certainly linearly independent.

If the series (I.7) also converges pointwise on $T \times T$, then in particular $K(t,t) = \sum_j \lambda_j \phi_j^2(t) < \infty$ for all $t \in T$ and Theorem I.21 shows that the RKHS is the set of all functions $\sum_k w_k \phi_k$ for sequences (w_j) such that $(w_j/\lambda_j^{-1/2}) \in \ell_2$.

If the kernel is suitably regular, then we can apply the preceding with many choices of measure ν, leading to different eigenfunction expansions.

If the series (I.7) does not converge pointwise, then the preceding theorem does not apply. However, by Example I.15 the RKHS can be characterized as the range of the operator K with square norm $\|Kf\|_{\mathbb{H}}^2 = \int (Kf) f\, d\nu$. Since $f = \sum_j f_j \phi_j$ for $(f_j) \in \ell_2$ and $Kf = \sum_j \lambda_j f_j \phi_j$ and $\int (Kf) f\, d\nu = \sum_j \lambda_j f_j^2$, this shows that the analogous result is still true (make the substitution $\lambda_j f_j = w_j$).

Consider the stochastic process of the form, for a sequence of numbers μ_j, i.i.d. standard normal variables (Z_j) and suitable functions ϕ_j,

$$W_t = \sum_{j=1}^{\infty} \mu_j Z_j \phi_j(t).$$

If this series converges in $\mathbb{L}_2(\Omega, \mathscr{U}, P)$ for every t, then (I.7) holds with $\lambda_j = \mu_j^2$. The stochastic process RKHS then takes the form given by the preceding proposition.

The following theorem gives a Banach space version of this result. Say that a sequence (h_j) of elements of a separable Banach space is *linearly independent* over ℓ_2 if $\sum_{j=1}^{\infty} w_j h_j = 0$ for for some $w \in \ell_2$, where the convergence of the series is in $\mathbb{B}$, implies that $w = 0$.

Theorem I.23 (RKHS series) *If for a given sequence (h_j) of ℓ_2-linearly independent elements of a separable Banach space $\mathbb{B}$ and a sequence (Z_j) of i.i.d. standard normal variables the series $W = \sum_{j=1}^{\infty} Z_j h_j$ converges almost surely in $\mathbb{B}$, then the RKHS of W as a map in $\mathbb{B}$ is the set of all elements $\sum_{i=1}^{\infty} w_j h_j$ for $w \in \ell_2$ with squared norm $\|\sum_j w_j h_j\|_{\mathbb{H}}^2 = \sum_j w_j^2$.*

Proof The almost sure convergence of the series $W = \sum_j Z_j h_j$ in $\mathbb{B}$ implies the almost sure convergence of the series $b^*(W) = \sum_j Z_j b^*(h_j)$ in $\mathbb{R}$, for any $b^* \in \mathbb{B}^*$. Because the partial sums of the last series are mean-zero Gaussian, the series converges also in $\mathbb{L}_2(\Omega, \mathscr{U}, P)$. Hence for any $b^*, \underline{b}^* \in \mathbb{B}^*$,

$$\mathrm{E}\big[b^*(W)\underline{b}^*(W)\big] = \mathrm{E}\Big[\sum_{j=1}^{\infty} Z_j b^*(h_j) \sum_{j=1}^{\infty} Z_j \underline{b}^*(h_j)\Big] = \sum_{j=1}^{\infty} b^*(h_j)\underline{b}^*(h_j).$$

In particular, the sequence $(b^*(h_j))_{j=1}^\infty$ is contained in ℓ_2 for every $b^* \in \mathbb{B}^*$, with square norm $\mathrm{E}\big[b^*(W)^2\big]$.

For $w \in \ell_2$ and natural numbers $m < n$, by the Hahn-Banach theorem and the Cauchy-Schwarz inequality,

$$\Big\|\sum_{m<j\le n} w_j h_j\Big\|^2 = \sup_{\|b^*\|\le 1}\Big\|\sum_{m<j\le n} w_j b^*(h_j)\Big\|^2 \le \sum_{m<j\le n} w_j^2 \sup_{\|b^*\|\le 1}\sum_{m<j\le n}(b^*(h_j))^2.$$

As $m, n \to \infty$ the first factor on the far right tends to zero, since $w \in \ell_2$. By the first paragraph the second factor is bounded by $\sup_{\|b^*\|\le 1}\mathrm{E}\big[b^*(W)^2\big] \le \mathrm{E}\|W\|^2$. Hence the partial sums of the series $\sum_j w_j h_j$ form a Cauchy sequence in $\mathbb{B}$, whence the infinite series converges.

Because the sequence $w_j = b^*(h_j)$ is contained in ℓ_2, the series $\sum_j b^*(h_j)h_j$ converges in $\mathbb{B}$, and hence $\underline{b}^*(\sum_j b^*(h_j)h_j) = \sum_j b^*(h_j)\underline{b}^*(h_j) = \mathrm{E}\big[b^*(W)\underline{b}^*W\big]$, for any $\underline{b}^* \in \mathbb{B}^*$. This shows that $Sb^* = \sum_j (b^*h_j)h_j$ and hence the RKHS is not bigger than the space as claimed.

The space would be smaller than claimed if there existed $w \in \ell_2$ that is not in the closure of the linear span of the elements $b^*(h_j)$ of ℓ_2 when b^* ranges over $\mathbb{B}^*$. We can take this w without loss of generality orthogonal to the latter collection, i.e. $\sum_j w_j b^*(h_j) = 0$ for every $b^* \in \mathbb{B}^*$. This is equivalent to $\sum_j w_j h_j = 0$, but this is excluded for any $w \neq 0$ by the assumption of linear independence of the h_j. □

The sequence (h_j) in the preceding theorem may consist of arbitrary elements of the Banach space, only restricted by linear independence over ℓ_2 and the convergence of the random sequence $\sum_j Z_j h_j$. The theorem shows that when combined in a series with i.i.d. standard normal coefficients, then this sequence turns into an *orthonormal* basis of the RKHS.

From the proof it can be seen that the linear independence is necessary; see Problem I.2.

Example I.24 (Polynomials) For $Z_0, \ldots, Z_k$ i.i.d. standard normal variables consider the polynomial process $t \mapsto \sum_{j=0}^k Z_j t^j/j!$ viewed as a map in (for instance) $\mathfrak{C}[0,1]$. The RKHS of this process is equal to the set of kth degree polynomials $P_a(t) = \sum_{j=0}^k a_j t^j/j!$ with square norm $\|P_a\|_{\mathbb{H}}^2 = \sum_{j=0}^k a_j^2$. In other words, the kth degree polynomials P with square norm $\|P\|_{\mathbb{H}}^2 = \sum_{j=0}^k P^{(j)}(0)^2$.

The following theorem shows that, conversely, any Gaussian random element W in a separable Banach space can be expanded in a series $W = \sum_{j=1}^\infty Z_j h_j$, for i.i.d. standard normal variables Z_j and any orthonormal basis (h_j) of its RKHS, where the series converges in the norm of the Banach space. Because we can rewrite this expansion as $W = \sum_j \|h_j\| Z_j \tilde{h}_j$, where $\tilde{h}_j = h_j/\|h_j\|$ is a sequence of norm one, the corresponding "eigenvalues" λ_j are in this case the square norms $\|h_j\|^2$.

To formulate the theorem, recall the isometry $U: \mathbb{H} \to \mathbb{L}_2(\Omega, \mathscr{U}, P)$ defined by $U(Sb^*) = b^*(W)$ in (I.6).

Theorem I.25 (Series representation) *Let (h_j) be a complete orthonormal system in the RKHS of a Borel measurable, mean-zero Gaussian random element W in a separable Banach space $\mathbb{B}$. Then $Uh_1, Uh_2, \ldots$ is an i.i.d. sequence of standard normal variables and $W = \sum_{i=1}^{\infty}(Uh_j)h_j$, where the series converges in the norm of $\mathbb{B}$, almost surely.*

Proof It is immediate from the definitions of U and the RKHS that $U: \mathbb{H} \to \mathbb{L}_2(\Omega, U, P)$ is an isometry. Because U maps the subspace $S\mathbb{B}^* \subset \mathbb{H}$ into the Gaussian process $b^*(W)$ indexed by $b^* \in \mathbb{B}^*$, it maps the completion $\mathbb{H}$ of $S\mathbb{B}^*$ into the completion of the linear span of this process in $\mathbb{L}_2(\Omega, \mathscr{U}, P)$, which consists of normally distributed variables. Because U retains inner products, it follows that $Uh_1, Uh_2, \ldots$ is a sequence of i.i.d. standard normal variables.

If $W_n = \sum_{j=1}^{n}(Uh_j)h_j$, then $\mathrm{E}(W_n | Uh_1, \ldots, Uh_m) = W_m$, for every $m \leq n$, in a Banach space sense. Convergence of the infinite series follows by a martingale convergence theorem for Banach space valued variables; see Ledoux and Talagrand (1991), Proposition 3.6. ☐

I.7 Support and Concentration

In this section we provide proofs of the key concentration lemmas Lemma 11.18 and 11.19.

In Chapter 11 the *concentration function* of the Gaussian random element W is defined by

$$\varphi_w(\epsilon) = \inf_{h \in \mathbb{H}: \|h-w\| \leq \epsilon} \tfrac{1}{2}\|h\|_{\mathbb{H}}^2 - \log \mathrm{P}(\|W\| < \epsilon).$$

Lemma I.26 *For any w in $\mathbb{B}$ the concentration function $\epsilon \mapsto \varphi_w(\epsilon)$ of a nondegenerate mean-zero Gaussian random element in a separable Banach space is strictly decreasing and convex on $(0, \infty)$, and hence continuous.*

Proof The centered small ball exponent φ_0 is convex by Lemma 1.1 of Gaenssler et al. (2007). The decentering function (the infimum) is also convex, as a consequence of the convexity of the norms of $\mathbb{B}$ and $\mathbb{H}$.

The decentering function is clearly nondecreasing. We show that the centered small ball exponent is strictly decreasing by showing that $\mathrm{P}(\epsilon < \|W\| < \epsilon') > 0$, whenever $0 < \epsilon < \epsilon'$. Since W is nondegenerate, the RKHS $\mathbb{H}$ contains some nonzero element h, which we can scale so that $\epsilon < \|h\| < \epsilon'$. Then a ball of sufficiently small radius centered at h is contained in $\{b \in \mathbb{B}: \epsilon < \|b\| < \epsilon'\}$ and has positive probability, as the RKHS belongs to the support of W. ☐

Lemma I.27 *For every h in the RKHS of mean-zero Gaussian random element W in a separable Banach space $\mathbb{B}$ and every Borel measurable set $C \subset \mathbb{B}$ with $C = -C$,*

$$\mathrm{P}(W - h \in C) \geq e^{-\frac{1}{2}\|h\|_{\mathbb{H}}^2}\mathrm{P}(W \in C).$$

Proof Since $W =_d -W$ and $C = -C$ we have $\mathrm{P}(W + h \in C) = \mathrm{P}(-W + h \in -C) = \mathrm{P}(W - h \in C)$. By Lemma I.20,

$$\mathrm{P}(W + h \in C) = \mathrm{E}[\mathbb{1}_C(W + h)] = \mathrm{E}[e^{Uh - \frac{1}{2}\|h\|_{\mathbb{H}}^2}\mathbb{1}_C(W)].$$

Since the left side remains the same if h is replaced $-h$,

$$\mathrm{P}(W - h \in C) = \tfrac{1}{2}\mathrm{E}[e^{Uh - \frac{1}{2}\|h\|_{\mathbb{H}}^2}\mathbb{1}_C(W)] + \tfrac{1}{2}\mathrm{E}[e^{U(-h) - \frac{1}{2}\|-h\|_{\mathbb{H}}^2}\mathbb{1}_C(W)].$$

The result follows since $(e^x + e^{-x})/2 \geq 1$, for every x. ☐

The lemma with $C = \epsilon\mathbb{B}_1$ relates the decentered small ball probability $\mathrm{P}(\|W - h\| < \epsilon)$ to the corresponding centered small ball probability, for $h \in \mathbb{H}_1$. The following lemma extends to a ball around a general element of $\mathbb{B}$.

Lemma I.28 *For every w in the closure of the RKHS of a mean-zero Gaussian random element W in a separable Banach space $\mathbb{B}$, and every $\epsilon > 0$,*

$$\varphi_w(\epsilon) \leq -\log \mathrm{P}(\|W - w\| < \epsilon) \leq \varphi_w(\epsilon/2).$$

Proof If $h \in \mathbb{H}$ is such that $\|h - w\| \leq \epsilon$, then $\|W - w\| \leq \epsilon + \|W - h\|$ by the triangle inequality and hence

$$\mathrm{P}(\|W - w\| < 2\epsilon) \geq \mathrm{P}(\|W - h\| < \epsilon) \geq e^{-\|h\|_{\mathbb{H}}^2/2}\mathrm{P}(\|W\| < \epsilon),$$

by Lemma I.27. Taking the negative logarithm and optimizing over $h \in \mathbb{H}$, we obtain the upper bound of the lemma.

The set $B_\epsilon = \{h \in \mathbb{H}: \|h - w\| \leq \epsilon\}$ is convex, and closed in $\mathbb{H}$, because the RKHS norm is stronger than the Banach space norm. Thus the convex map $h \mapsto \|h\|_{\mathbb{H}}^2$ attains a minimum on B_ϵ at some point h_ϵ. Because $(1 - \lambda)h_\epsilon + \lambda h \in B_\epsilon$ for every $h \in B_\epsilon$ and $0 \leq \lambda \leq 1$, we have $\|(1 - \lambda)h_\epsilon + \lambda h\|_{\mathbb{H}}^2 \geq \|h_\epsilon\|_{\mathbb{H}}^2$, whence $2\lambda\langle h - h_\epsilon, h_\epsilon\rangle_{\mathbb{H}} + \lambda^2\|h - h_\epsilon\|_{\mathbb{H}}^2 \geq 0$. Since $0 \leq \lambda \leq 1$ can be arbitrary, this gives $\langle h - h_\epsilon, h_\epsilon\rangle_{\mathbb{H}} \geq 0$, or

$$\langle h, h_\epsilon\rangle_{\mathbb{H}} \geq \|h_\epsilon\|_{\mathbb{H}}^2 \text{ for every } h \in B_\epsilon.$$

By Theorem I.25 the Gaussian element can be represented as $W = \sum_{i=1}^\infty (Uh_j)h_j$, where the convergence is in the norm of $\mathbb{B}$, almost surely, and where $\{h_1, h_2, \ldots\}$ is a complete orthonormal system for $\mathbb{H}$. The variable $W_m := \sum_{i=1}^m (Uh_j)h_j$ takes its values in $\mathbb{H}$, and for any $g \in \mathbb{H}$ satisfies $\|W_m - g - w\| < \epsilon$ for sufficiently large m, a.s. on the event $\|W - g - w\| < \epsilon$. In other words, $W_m - g \in B_\epsilon$ eventually a.s. on the event $\|W - g - w\| < \epsilon$, and hence by the preceding display $\langle W_m - g, h_\epsilon\rangle_{\mathbb{H}} \geq \|h_\epsilon\|_{\mathbb{H}}^2$ for all sufficiently large m a.s. Since $\langle W_m, h_\epsilon\rangle_{\mathbb{H}} = \sum_{i=1}^m (Uh_j)\langle h_j, h_\epsilon\rangle_{\mathbb{H}} = U\sum_{i=1}^m h_j\langle h_j, h_\epsilon\rangle_{\mathbb{H}}$, the sequence $\langle W_m, h_\epsilon\rangle_{\mathbb{H}}$ tends to Uh_ϵ as $m \to \infty$, in $\mathbb{L}_2(\Omega, \mathscr{U}, P)$ and hence also almost surely along a subsequence. We conclude that $Uh_\epsilon - \langle g, h_\epsilon\rangle_{\mathbb{H}} \geq \|h_\epsilon\|_{\mathbb{H}}^2$ a.s. on the event $\|W - g - w\| < \epsilon$. For $g = -h_\epsilon$, this gives that $Uh_\epsilon \geq 0$ a.s. on the event $\|W + h_\epsilon - w\| < \epsilon$. By Lemma I.20,

$$\begin{aligned}\mathrm{P}(W \in w + \epsilon\mathbb{B}_1) &= \mathrm{P}(W - h_\epsilon \in w - h_\epsilon + \epsilon\mathbb{B}_1)\\ &= \mathrm{E}\big[e^{-Uh_\epsilon - \frac{1}{2}\|h_\epsilon\|_{\mathbb{H}}^2}\mathbb{1}\{W \in w - h_\epsilon + \epsilon\mathbb{B}_1\}\big].\end{aligned}$$

Since $Uh_\epsilon \geq 0$ on the event $W \in w - h_\epsilon + \epsilon\mathbb{B}_1$, the exponential is bounded above by $e^{-\frac{1}{2}\|h_\epsilon\|_{\mathbb{H}}^2}$. Furthermore, the probability $\mathrm{E}[\mathbb{1}\{W \in w - h_\epsilon + \epsilon\mathbb{B}_1\}]$ is bounded above by $\mathrm{P}(W \in \epsilon\mathbb{B}_1)$, by Anderson's lemma. □

By Lemma I.27 a ball of radius ϵ around a point h in the unit ball $\mathbb{H}_1$ of the RKHS contains mass at least $e^{-1/2}\mathrm{P}(\|W\| < \epsilon) = e^{-1/2}e^{-\varphi_0(\epsilon)}$. One can place $D(2\epsilon, \mathbb{H}_1, \|\cdot\|)$ points in $\mathbb{H}_1$ so that their surrounding balls of radius ϵ are disjoint. The law of total probability then gives that $1 \geq D(2\epsilon, \mathbb{H}_1, \|\cdot\|)e^{-1/2}e^{-\varphi_0(\epsilon)}$, or $\varphi_0(\epsilon) \geq \log D(2\epsilon, \mathbb{H}_1, \|\cdot\|) - 1/2$. The following two results refine this estimate and also give a bound in the other direction. The first lemma roughly shows that

$$\varphi_0(2\epsilon) \lesssim \log N\Big(\frac{\epsilon}{\sqrt{2\varphi_0(\epsilon)}}, \mathbb{H}_1, \|\cdot\|\Big) \lesssim \varphi_0(\epsilon),$$

but this is true only if the modulus and entropy are sufficiently regular functions.

Lemma I.29 *Let $f\colon (0,\infty) \to (0,\infty)$ be regularly varying*[1] *at zero. Then*

(i) $\log N(\epsilon/\sqrt{2\varphi_0(\epsilon)}, \mathbb{H}_1, \|\cdot\|) \gtrsim \varphi_0(2\epsilon)$;
(ii) *if* $\varphi_0(\epsilon) \lesssim f(\epsilon)$, *then* $\log N(\epsilon/\sqrt{f(\epsilon)}, \mathbb{H}_1, \|\cdot\|) \lesssim f(\epsilon)$;
(iii) *if* $\log N(\epsilon, \mathbb{H}_1, \|\cdot\|) \gtrsim f(\epsilon)$, *then* $\varphi_0(\epsilon) \gtrsim f(\epsilon/\sqrt{\varphi_0(\epsilon)})$;
(iv) *if* $\log N(\epsilon, \mathbb{H}_1, \|\cdot\|) \lesssim f(\epsilon)$, *then* $\varphi_0(2\epsilon) \lesssim f(\epsilon/\sqrt{\varphi_0(\epsilon)})$.

Lemma I.30 *For $\alpha > 0$ and $\beta \in \mathbb{R}$, as $\epsilon \downarrow 0$, $\varphi_0(\epsilon) \asymp \epsilon^{-\alpha}(\log_- \epsilon)^\beta$ if and only if* $\log N(\epsilon, \mathbb{H}_1, \|\cdot\|) \asymp \epsilon^{-2\alpha/(2+\alpha)}(\log_- \epsilon)^{2\beta/(2+\alpha)}$.

Proofs See Kuelbs and Li (1993) and Li and Linde (1998). □

The following result shows that the concentration function of the sum of independent Gaussian random elements can be estimated from the concentration functions of the components.

Lemma I.31 *Let $W = \sum_{i=1}^N W_i$ be the sum of finitely many independent Gaussian random elements in a separable Banach space $(\mathbb{B}, \|\cdot\|)$ with concentration functions φ_{i,w_i} for given $w_i in \mathbb{B}$. Then, the concentration function φ_w of W around $w = \sum_{i=1}^N w_i$ satisfies*

$$\varphi_w(N\epsilon) \leq \sum_{i=1}^N \varphi_{i,w_i}(\epsilon/2).$$

Proof By the independence of the W_is, $\mathrm{P}(\|W - w\| < N\epsilon) \geq \prod_{i=1}^N \mathrm{P}(\|W_i - w_i\| < \epsilon)$. Next two applications of Lemma I.28 lead to the desired bound. □

[1] A function $f(x)$ is called regularly varying at zero if $\lim_{t\to 0} f(tx)/f(t) = x^\alpha$ for some $\alpha > 0$. Examples include polynomial functions.

I.8 Historical Notes

The main part of this appendix is based on van der Vaart and van Zanten (2008b), which reviews literature on reproducing kernel Hilbert spaces. Overviews of Gaussian process theory are given by Li and Shao (2001) and Lifshits (2012). Key original references include Kuelbs and Li (1993); Ledoux and Talagrand (1991); Borell (1975); Li and Linde (1998); Kuelbs et al. (1994); Kuelbs and Li (1993). The proof of Proposition I.20 is taken from Proposition 2.1 in de Acosta (1983). For a version of the proposition for processes, see van der Vaart and van Zanten (2008b), Lemma 3.1. Lemma I.26 is taken from Castillo (2008).

Problems

I.1 In Example I.14, show that the square root $S^{1/2}$ of the operator S can be described as having the same eigenfunctions as S with eigenvalues the square roots of the eigenvalues of S.

I.2 (van der Vaart and van Zanten 2008b) Show that the ℓ_2-linear independence assumption in Theorem I.23 cannot be dropped. [Hint: If $\{h_j\}$ is not ℓ_2-linearly independent, then the RKHS is $\overline{\text{lin}}\{\sum_{i=1}^{\infty} w_i h_i : w \in \overline{\text{lin}}\{b^*(h_i) : b^* \in \mathbb{B}^*\}\}$ with squared norm $\sum_{i=1}^{\infty} w_i^2$. Taking these linear combinations for all $w \in \ell_2$ gives the same set, but the ℓ_2-norm should be computed for a projected w.]

Appendix J

Completely Random Measures

Completely random measures are measures whose values on disjoint sets are independent random variables. They arise as priors, or building blocks for priors, in many Bayesian nonparametric applications. In this appendix we first discuss the special case of measures that assume only (nonnegative) integer values and are Poisson distributed. Next we generalize to general completely random measures. Finally we specialize to the case that the sample space is the positive half line, when the measures can be described as processes with independent increments through their cumulative distribution functions.

Throughout the chapter $(\Omega, \mathscr{A}, P)$ is a probability space, rich enough to support countably many independent nontrivial random variables, and $(\mathfrak{X}, \mathscr{X})$ is a Polish space with its Borel σ-field. Furthermore $\mathfrak{M}_\infty = \mathfrak{M}_\infty(\mathfrak{X}, \mathscr{X})$ is the set of all measures on $(\mathfrak{X}, \mathscr{X})$, and $\mathscr{M}_\infty$ is the σ-field on $\mathfrak{M}_\infty$ generated by the evaluation maps $\mu \mapsto \mu(A)$, for $A \in \mathscr{X}$.

J.1 Poisson Random Measures

A "Poisson random subset" is a model for randomly spreading out points in space. It can be viewed as a random subset, a stochastic process, or a random measure with values in the integers, depending on the aspect one focuses on.

Definition J.1 (Poisson random subset) A *Poisson random subset* (*PRS*) of $\mathfrak{X}$ is a map Π from Ω into the collection of subsets of $\mathfrak{X}$ of at most countably many elements such that $N(A) := \operatorname{card}(\Pi \cap A)$ is a random variable for every $A \in \mathscr{X}$ and $N(A_i) \overset{\text{ind}}{\sim} \operatorname{Poi}(\mu(A_i))$, for every finite collection of disjoint sets $A_1, \ldots, A_k \in \mathscr{X}$[1] and a measure μ on $(\mathfrak{X}, \mathscr{X})$, called the *intensity measure*. The stochastic process $N = (N(A): A \in \mathscr{X})$ is called a *Poisson process* on $\mathfrak{X}$ with intensity measure μ. The corresponding counting measure on the points Π is called a *Poisson random measure*.

It is included in the definition that the total number of points $N(\mathfrak{X})$ is also a Poisson variable and hence it is either infinite or finite almost surely. That μ is a measure is actually implied by the other requirements (see Problem J.2). It is the mean measure of N, and also called the *compensator* of the Poisson process. Because $N(\{x\}) \leq 1$ by construction and a nontrivial Poisson variable has support $\mathbb{N}$, it follows that $\mu(\{x\}) = 0$ for every x, i.e. μ is atomless. In the case that $\mathfrak{X} = \mathbb{R}^d$ and μ is absolutely continuous with respect to the

[1] By convention, Poi(0) and Poi(∞) are the measures that are degenerate at 0 and ∞, respectively.

Lebesgue measure, the density function of μ is called the *intensity function* (and is often denoted by λ).

By the additivity of N, the joint distribution of $(N(A_1), \dots, N(A_k))$ for sets $A_1, \dots, A_k$ that are not necessarily disjoint can be computed from the distribution of the vector of counts $(N(B_1), \dots, N(B_n))$, in the collection $\{B_1, \dots, B_n\}$ of all intersections of sets from $\{A_1, \dots, A_k\}$. In particular, the distribution of a PRS is completely determined by its intensity measure. In view of the bilinearity of the covariance it also follows that $\operatorname{cov}(N(A_1), N(A_2)) = \mu(A_1 \cap A_2)$, for every pair of measurable sets.

The following properties of a PRS will be important. Their proofs, which are not too difficult, can be found in Kingman (1993).

(i) *Disjointness:* If Π_1, Π_2 are independent PRS on $\mathfrak{X}$ with intensity measures μ_1 and μ_2 respectively, then $\Pi_1 \cap \Pi_2 \cap A = \varnothing$ with probability one, for any $A \in \mathscr{A}$ with $\mu_1(A), \mu_2(A) < \infty$.

(ii) *Superposition:* If $\Pi_1, \Pi_2, \dots$ are independent PRS on $\mathfrak{X}$ with finite intensity measures $\mu_1, \mu_2, \dots$ respectively, then $\Pi = \cup_{i=1}^\infty \Pi_i$ is a PRS on $\mathfrak{X}$ with intensity measure $\mu = \sum_{i=1}^\infty \mu_i$.

(iii) *Restriction:* If $\mathfrak{X}' \in \mathscr{X}$ and Π is a PRS on $\mathfrak{X}$, then $\Pi' = \Pi \cap \mathfrak{X}'$ is a PRS on $\mathfrak{X}'$ with intensity measure μ' given by e $\mu'(A) = \mu(A \cap \mathfrak{X}')$.

(iv) *Transformation:* If Π is a PRS on $\mathfrak{X}$ and $\psi\colon \mathfrak{X} \to \mathfrak{Y}$ is a measurable map into another measurable space $(\mathfrak{Y}, \mathscr{Y})$ such that the induced measure $\mu\psi^{-1}$ is atomless, then $\psi(\Pi)$ is a PRS on $\mathfrak{Y}$ with intensity measure $\mu\psi^{-1}$.

(v) *Conditioning:* If Π is a PRS on $\mathfrak{X}$ and $\mathfrak{X}' \in \mathscr{X}$ satisfies $0 < \mu(\mathfrak{X}') < \infty$, then

$$(N(A_1 \cap \mathfrak{X}'), \dots, N(A_k \cap \mathfrak{X}') | \, N(\mathfrak{X}') = n) \sim \mathrm{MN}_k(n; \bar{\mu}(A_1), \dots, \bar{\mu}(A_k)),$$

for any measurable partition $A_1, \dots, A_k$ of $\mathfrak{X}$, where $\bar{\mu}(A) = \mu(A \cap \mathfrak{X}')/\mu(\mathfrak{X}')$.

(vi) *Thinning:* If Π is a PRS on $\mathfrak{X}$ with σ-finite intensity measure μ and $f\colon \mathfrak{X} \to [0,1]$ is a measurable function, then the random set of points obtained by deleting every point x of Π independently with probability $f(x)$, is a PRS with intensity measures $A \mapsto \int_A f\, d\mu$.

The process $\Pi' = \Pi \cap \mathfrak{X}'$ with counting measure $N'(A) = \mathrm{card}(\Pi \cap \mathfrak{X}' \cap A)$, as in (v), conditioned on the total number of points in $\mathfrak{X}'$ being n is a suitable model for spreading out a fixed finite number of points randomly in a space. It will be called a *multinomial random set* with parameters n and $\bar{\mu}$. The special case corresponding to $n = 1$ is equivalent to a single random variable on $\mathfrak{X}$ with distribution $\bar{\mu}$.

The existence of a PRS can be proved constructively by using these properties in a converse direction. Let μ be a measure on $(\mathfrak{X}, \mathscr{X})$ such that there exist finite measures μ_n, $n \in \mathbb{N}$, with $\mu = \sum_{n=1}^\infty \mu_n$. (This certainly includes every σ-finite measure μ.) Let $\bar{\mu}_n = \mu_n/\mu_n(\mathfrak{X})$ be the probability measures obtained by normalizing μ_n, and on $(\Omega, \mathscr{A}, P)$ construct mutually independent random variables $N_n \sim \mathrm{Poi}(\mu_n(\mathfrak{X}))$ and $X_{n,r} \overset{\text{iid}}{\sim} \bar{\mu}_n$, for $r, n \in \mathbb{N}$. Then define $\Pi_n = \{X_{n,1}, \dots, X_{n,N_n}\}$, for $n \in \mathbb{N}$, and

$$\Pi = \bigcup_{n=1}^\infty \Pi_n = \bigcup_{n=1}^\infty \{X_{n,1}, \dots, X_{n,N_n}\}.$$

Clearly, $\Pi_n | N_n = m$ is a multinomial random set with parameters m and $\bar{\mu}_n$, whence Π is a PRS on $\mathfrak{X}$ with intensity measure μ, by the properties (ii) and the converse of (v).

Interpreting N as a (random) counting measure that puts weight 1 at every point in Π, we can write $\int f\,dN = \sum_{x\in\Pi} f(x)$, for a given measurable function $f: \mathfrak{X} \to \mathbb{R}$. The distribution of this random variable can be easily characterized by its Laplace transform. As a function of f (and with $\theta = 1$) the left side of the following lemma is called the *Laplace functional* of N.

Lemma J.2 *If N is a Poisson random measure with intensity measure μ and $f: \mathfrak{X} \to \mathbb{R}$ is nonnegative or $\int f\,dN$ is finite almost surely, then, for $\theta > 0$,*

$$\mathrm{E}\Big[e^{-\theta \int f\,dN}\Big] = \exp\Big[-\int (1 - e^{-\theta f(x)})\,d\mu(x)\Big].$$

Proof For an indicator function $f(x) = \mathbb{1}_A(x)$ the formula follows from the formula for the Laplace transform of the Poisson variable $N(A)$. It extends to linear combinations of indicators of disjoint sets, by the independence of the corresponding Poisson variables. A nonnegative function f can be approximated from below by such simple functions; the formula is preserved under such approximation by the monotone convergence theorem. If $\int f\,dN < \infty$ almost surely, then it is the difference of the variables $\int f^+\,dN$ and $\int f^-\,dN$, which are independent, because f^+ and f^- have disjoint supports. Both left and right sides factorize over this decomposition. ☐

J.1.1 Palm Theory

The *Campbell measure* of a Poisson process N is the measure on $(\mathfrak{X} \times \mathfrak{M}_\infty, \mathscr{X} \otimes \mathscr{M}_\infty)$ determined by $C(A \times B) = \mathrm{E}\big[N(A)\mathbb{1}\{N \in B\}\big]$, for $A \in \mathscr{X}$ and $B \in \mathscr{M}_\infty$. Here the expectation is relative to N, which we view as a random element in $\mathfrak{M}_\infty$ by identifying it with the counting measure on its point set. By monotone approximation, for every nonnegative (or suitably integrable) measurable function $g: \mathfrak{X} \times \mathfrak{M}_\infty \to \mathbb{R}$,

$$\int_{\mathfrak{X}\times\mathfrak{M}_\infty} g(x,\xi)\,C(dx,d\xi) = \mathrm{E}\int g(x,N)\,N(dx) = \int_{\mathfrak{M}_\infty}\int_{\mathfrak{X}} g(x,N)\,N(dx)\,\mathcal{P}(dN),$$

where $\mathcal{P}$ is the law of N on $(\mathfrak{M}_\infty, \mathscr{M}_\infty)$ (and we abuse notation by employing the symbol N also as a dummy variable for integration). This shows that the measure $N(dx)\,\mathcal{P}(dN)$ is a disintegration of C. Since $C(A \times \mathfrak{M}_\infty) = \mathrm{E}\big[N(A)\big]$, the marginal distribution of C on its first coordinate is the intensity measure μ of N. Thus the disintegration of C with its two coordinates swapped takes the form

$$\mathrm{E}\int g(x,N)\,N(dx) = \int_{\mathfrak{X}}\int_{\mathfrak{M}_\infty} g(x,N)\,\mathcal{P}_x(dN)\,\mu(dx).$$

The measure $\mathcal{P}_x$ in the right side is known as the *Palm measure* of N at x. By definition of *disintegration* every $\mathcal{P}_x$ is a measure on $(\mathfrak{M}_\infty, \mathscr{M}_\infty)$ and the map $x \mapsto \mathcal{P}_x(B)$ is measurable for every $B \in \mathscr{M}_\infty$.

For a more detailed treatment of these concepts and the (not so difficult) proof of the following proposition, see Chapter 13 of Daley and Vere-Jones (2008).

Proposition J.3 (Palm measure) *The Palm measure at x of the Poisson random measure N is the counting measure on the points $N \cup \{x\}$.*

The proposition is equivalent to the assertion that, for every nonnegative measurable function $g\colon \mathfrak{X} \times \mathfrak{M}_\infty \to \mathbb{R}$,

$$\mathrm{E}\int g(x, N)\, N(dx) = \int \mathrm{E}g(x, N \cup \{x\})\, \mu(dx). \tag{J.1}$$

(Both E-signs mean expectation relative to N.) The formula is often read as saying that "a Poisson process conditioned to have a point at x (referring to $\mu(dx)$ at the far right) is distributed as before but with the point at x added." In the simpler case that the function g is free of N the formula reduces to *Campbell's theorem.* Various extensions (to other random measures or processes, higher-order versions) carry additional names, including Mecke, Hardy and Palm.

One might interpret the definition of the Campbell measure C as a recipe for generating a pair (x, N) in $\mathfrak{X} \times \mathfrak{M}_\infty$ by first generating N and next given N generating x from N viewed as a measure on $\mathfrak{X}$. Because C will typically be infinite, the notion of "generating" does not carry immediate meaning. Ignoring this difficulty, we might interpret the Palm measures as a recipe for generating the same point (x, N) by first generating x from μ and next N from $\mathcal{P}_x$. It is possible to introduce and interpret Palm measures for general random measures in this way. Proposition J.3 shows that for a Poisson random measure the second step comes down to generating a realization of the original process N and adding the point x to it.

In a similar spirit, but thinking of the law of N as a prior on $\mathfrak{M}_\infty$ and x as an observation from N, we can also think of the Palm measure as the posterior distribution given the data. Consideration of multiple data points could be implemented through an extension to multivariate Campbell measures of the form $C_n(A_1 \times \cdots \times A_n \times B) = \mathrm{E}\big[\prod_{i=1}^n N(A_i)\mathbb{1}\{N \in B\}\big]$.

One application of Proposition J.3 is to the distribution of a Poisson process when a point is removed. Let $w\colon \mathfrak{X} \to \mathbb{R}^+$ be a given weight function such that $\int w(x)\, N(dx) < \infty$ almost surely, and suppose that given N a point Y_1 from N is chosen with probability proportional to $w(Y_1)$, i.e. for every measurable set A,

$$\mathrm{P}(Y_1 \in A \mid N) = \frac{\int_A w(x)\, N(dx)}{\int w(x)\, dN(x)}.$$

Let $N_1 = N - \{Y_1\}$ be the process N with the chosen point removed, and let $T = \int w(x)\, N(dx)$ and $T_1 = \int w(x)\, N_1(dx)$ be the total weights of the points in N and N_1, respectively (so that $T = T_1 + w(Y_1)$).

Lemma J.4 *If N is a Poisson process with intensity measure μ, then for $\mathcal{P}$ the distribution of N on $(\mathfrak{M}_\infty, \mathscr{M}_\infty)$,*

(i) *(Y_1, N_1) has density $(y, \xi) \mapsto w(y)/(\int w(x)\, \xi(dx) + w(y))$ relative to the measure $\mu \times \mathcal{P}$;*
(ii) *T_1 has density $t \mapsto \int w(x)/(t + w(x))\, d\mu(x)$ relative to the distribution of T;*
(iii) *Y_1 and N_1 are conditionally independent given T_1;*
(iv) *$N_1 \mid T_1 = t \sim N \mid T = t$, for almost every t, $[P^T]$.*

Proof (i). The definition of Y_1 implies that $\mathrm{E}\big[f(Y_1)|\, N\big] = \int f(x)w(x)\, dN(dx)/T$, for every bounded, measurable function $f\colon \mathfrak{X} \to \mathbb{R}$. We apply this with the function $y \mapsto f(y)g(N - \{y\})$, for given N and given bounded, measurable functions $f\colon \mathfrak{X} \to \mathbb{R}$ and $g\colon \mathfrak{M}_\infty \to \mathbb{R}$, to find that

$$\mathrm{E}\big[f(Y_1)g(N_1)\big] = \mathrm{E}\,\mathrm{E}\big[f(Y_1)g(N - \{Y_1\})|\, N\big] = \mathrm{E}\Big[\frac{\int f(x)g(N - \{x\})w(x)\, N(dx)}{\int w(y)\, dN(y)}\Big],$$

where the expectation on the right is with respect to N. By the definition of the Palm measures, the right side can also be written as

$$\iint \frac{f(x)g(\xi - \{x\})w(x)}{\int w(y)\,\xi(dy)}\, \mathcal{P}_x(d\xi)\,\mu(dx) = \iint \frac{f(x)g(N)w(x)}{\int w(y)\, N(dy) + w(x)}\, \mathcal{P}(dN)\,\mu(dx).$$

The last step follows because the Palm measure $\mathcal{P}_x$ is the distribution of $N \cup \{x\}$, by Proposition J.3. Assertion (i) follows, since it has been established that the left side of the preceding display is equal to the right side of the last display for every bounded, measurable functions f and g.

(ii), (iii) and (iv). All three assertions are consequences of the fact that the density of (Y_1, N_1) in (i) relative to the product dominating measure $\mathcal{P} \times \mu$ depends on (Y_1, T_1) only. For a precise proof, let (Y', N') be distributed according to $\mu \times \mathcal{P}$ and let $T' = \int w(x)\, N'(dx)$. Then, by (i), for every bounded, measurable functions g and h,

$$\begin{aligned}\mathrm{E}\big[g(N_1)h(Y_1, T_1)\big] &= \mathrm{E}\Big[g(N')h(Y', T')\frac{w(Y')}{T' + w(Y')}\Big]\\ &= \mathrm{E}\Big[\mathrm{E}(g(N')|\, Y', T')h(Y', T')\frac{w(Y')}{T' + w(Y')}\Big].\end{aligned}$$

Taking g equal to 1, we can conclude from the equation that (Y_1, T_1) has density $(y, t) \mapsto w(y)/(t + w(y))$ relative to the distribution of (Y', T'). Since the law of (Y', T') is a product distribution, the marginal distribution of T_1 has density relative to the distribution of T' equal to the integral of the joint density of (Y_1, T_1) relative to the density of Y'. Since $T' =_d T \sim \mu$, this gives (ii). If we set $k(Y', T') = \mathrm{E}(g(N')|\, Y', T')$, then the far right side of the display is equal to the expectation of $k(Y', T')h(Y', T')$ times the density of (Y_1, T_1) relative to (Y', T'), whence it is equal to $\mathrm{E}[k(Y_1, T_1)h(Y_1, T_1)]$. Since this is true for every bounded, measurable h, it follows that $k(Y_1, T_1) = \mathrm{E}[g(N_1)|\, Y_1, T_1]$. By the independence of Y' and N', the function k actually depends only on its second argument, and hence $\mathrm{E}[g(N_1)|\, Y_1, T_1]$ is free of Y_1, proving (iii). Moreover, taken at a value $T_1 = t$ the conditional expectation is equal to $k(Y_1, t) = \mathrm{E}[g(N')|\, T' = t]$, which is identical to $\mathrm{E}[g(N)|\, T = t]$, since $N =_d N'$. This proves (iv). ☐

J.2 Completely Random Measures

A Poisson process N takes its values in the integers $\{0, 1, \ldots, \infty\}$, a restriction that is too rigid for a general prior. This constraint is released in the following definition of a general completely random measure.

Definition J.5 (Completely random measure) A measurable map $\Phi: \Omega \to (\mathfrak{M}_\infty, \mathscr{M}_\infty)$ is a *completely random measure* (*CRM*) on $\mathfrak{X}$ if the random variables $\Phi(A_1), \ldots, \Phi(A_k)$ are mutually independent, for any disjoint sets $A_1, \ldots, A_k \in \mathscr{X}$.

As in the case of a Poisson process the joint distribution of $(\Phi(A_1), \ldots, \Phi(A_k))$ for arbitrary measurable subsets $A_1, \ldots, A_k$ of $\mathfrak{X}$ can be obtained through additivity of Φ from its distribution over the partition generated by these sets. By the assumed independence this is fixed by the distributions of the univariate random variables $\Phi(A)$. Thus the distribution of a completely random measure Φ on $\mathfrak{X}$ is completely determined by the marginal distributions of the variables $\Phi(A)$, for $A \in \mathscr{X}$.

The Laplace transform $\theta \mapsto \mathrm{E}e^{-\theta\Phi(A)}$ of such a variable is defined and finite (at least) for $\theta > 0$. The *cumulant measures* of Φ are defined as minus the logarithms of these Laplace transforms:

$$\lambda_\theta(A) = -\log \mathrm{E}(e^{-\theta\Phi(A)}).$$

From the countable additivity of Φ and independence over disjoint sets, it can be checked that the cumulant measures are indeed measures on $(\mathfrak{X}, \mathscr{X})$. By uniqueness of Laplace transforms (Feller 1971, Theorem XIII.1.1) they collectively, for $\theta > 0$, determine the distribution of any $\Phi(A)$, and hence the full distribution of Φ. It can also be seen that $\lambda_\theta(A) = 0$ if and only if $\Phi(A) = 0$ a.s.; and $\lambda_\theta(A) = \infty$ if and only if $\Phi(A) = \infty$ a.s. In particular $\lambda_\theta(A)$ is zero or nonzero finite or infinite for some $\theta > 0$ if and only if the same is true for *every* $\theta > 0$.

We shall assume that λ_θ is σ-finite for some (and hence by the preceding observation all) $\theta > 0$. Then it has at most countably many atoms and the sample space can be partitioned as $\mathfrak{X} = \{a_1, a_2, \ldots\} \cup \cup_j \mathfrak{X}_j$, where $a_1, a_2, \ldots$ are the atoms and the $\mathfrak{X}_j$ are measurable sets with $0 < \lambda_\theta(\mathfrak{X}_j) < \infty$. The set of atoms is the same for every θ, and the sets $\mathfrak{X}_j$ can be chosen independent of θ as well. By the definition of a CRM the random measures in the corresponding decomposition

$$\Phi(\cdot) = \sum_j \Phi(\{a_j\})\delta_{a_j}(\cdot) + \sum_j \Phi(\cdot \cap \mathfrak{X}_j) \tag{J.2}$$

are jointly independent. Because $\Phi(\{a\})$ is a non-degenerate variable if and only if a is an atom of λ_θ (and otherwise is identically zero), $a_1, a_2, \ldots$ are called *fixed atoms* of Φ. Every of the processes $\Phi(\cdot \cap \mathfrak{X}_j)$ is a CRM without fixed atoms that is neither identically zero nor identically infinite; its λ_θ-measure is the restriction of λ_θ to $\mathfrak{X}_j$. Thus to construct a CRM we may construct countably many independent CRMs with disjoint, finite and atomless λ_θ-measures and countably many independent variables $\Phi(\{a_j\})$, and add them together as in (J.2).

It turns out that apart from a deterministic component an arbitrary CRM can be described as a purely atomic measure with atoms of random weights at random locations in $\mathfrak{X}$. Furthermore, the "non-fixed atoms" are placed at the points of a PRS, and also the pairs (x, s) of a location x and a weight s are given by a PRS, on the product space $\mathfrak{X} \times (0, \infty]$. That the pairs (x, s) are spread according to a PRS means that they are "spatially independent" in the product space, but location x and weight s are not independent in general.

The representation is in "in law" (or "weak") in that the proposition shows how a CRM with a given distribution can be constructed on a suitable probability space.

Proposition J.6 *Any CRM Φ with σ-finite cumulant measures λ_θ can be represented uniquely as $\Phi = \sum_j \Phi(\{a_j\})\delta_{a_j} + \beta + \Psi$, for fixed points $a_1, a_2, \ldots$ in $\mathfrak{X}$, independent nonnegative random variables $\Phi(\{a_1\}), \Phi(\{a_2\}), \ldots$, a deterministic σ-finite Borel measure β on $\mathfrak{X}$, and a CRM Ψ of the form*

$$\Psi(A) = \sum_{(x,s)\in\Pi^c, x\in A} s = \iint \mathbb{1}_A(x) s\, N^c(dx, ds), \tag{J.3}$$

for Π^c and N^c a PRS on $\mathfrak{X} \times (0, \infty]$, independent of $\Phi(\{a_1\}), \Phi(\{a_2\}), \ldots$, with intensity measure ν^c such that $\nu^c(\{x\} \times (0, \infty]) = 0$ for every $x \in \mathfrak{X}$, and $\int_{\mathfrak{X}_j} \int (s \wedge 1)\, \nu^c(dx, ds) < \infty$, for every set $\mathfrak{X}_j$ in a countable measurable partition of $\mathfrak{X}$. The cumulant measures of Ψ take the form

$$-\log \mathrm{E}(e^{-\theta\Psi(A)}) = \int_A \int_{(0,\infty]} (1 - e^{-\theta s})\, \nu^c(dx, ds). \tag{J.4}$$

Conversely, every $a_1, a_2, \ldots$, $\Phi(\{a_1\}), \Phi(\{a_2\}), \ldots$, β and N^c and ν^c with the given properties define a CRM with σ-finite cumulant measures λ_θ.

Proof As shown in (J.2) the fixed atoms of the CRM can be separated off and the CRM can be decomposed over subsets $\mathfrak{X}_j$, of finite λ_θ measure. Thus it is not a loss of generality to assume that the λ_θ are nontrivial, finite, atomless measures. In this case the distribution of $\Phi(A)$ (on $[0, \infty]$) is necessarily *infinitely divisible*: for every $n \in \mathbb{N}$ it can be written as the convolution of the distributions of n independent variables that are asymptotically negligible as $n \to \infty$. Indeed, $\Phi(A) = \sum_{j=1}^n \Phi(A_{n,j})$ for any measurable partition $A = \cup_{j=1}^n A_{n,j}$; this can be chosen so that $\lambda_1(A_{n,j}) = \lambda_1(A)/n$ if λ_1 is atomless, in which case, for any $\epsilon > 0$ as $n \to \infty$,

$$\max_{1\le j\le n} \mathrm{P}(\Phi(A_{n,j}) \ge \epsilon) = \max_{1\le j\le n} \mathrm{P}(1 - e^{-\Phi(A_{n,j})} \ge 1 - e^{-\epsilon}) \le \frac{1 - e^{-\lambda_1(A)/n}}{1 - e^{-\epsilon}} \to 0.$$

Since the variable $\Phi(A)$ is infinitely divisible, its negative log-Laplace transform admits the *Lévy-Khinchine representation* (Kallenberg 1986, Corollary 15.8, or Feller 1971, Theorem XIII.7.2[2]), for all $\theta > 0$,

$$\lambda_\theta(A) = -\log \mathrm{E}(e^{-\theta\Phi(A)}) = \theta\beta(A) + \int_{(0,\infty]} (1 - e^{-\theta s})\, \gamma(A, ds). \tag{J.5}$$

Here $\beta(A)$ is a nonnegative constant, the *deterministic component* of $\Phi(A)$, and $\gamma(A, \cdot)$ is a measure on $(0, \infty]$ with the property $\int_{(0,\infty]} (s \wedge 1)\, \gamma(A, ds) < \infty$, the *Lévy measure* of $\Phi(A)$. The characteristics $\beta(A)$ and $\gamma(A, \cdot)$ are uniquely determined by the distribution of $\Phi(A)$. From the σ-additivity of λ_θ it follows that $\sum_i \beta(A_i)$ and $\sum_i \gamma(A_i, \cdot)$ can be

[2] Unlike in these references, the present $\Phi(A)$ can be infinite with positive probability. This is accommodated by allowing $\gamma(A, \cdot)$ to have an atom at ∞: $\mathrm{P}\big(\Phi(A) < \infty\big) = e^{-\gamma(A, \{\infty\})}$, as follows by letting $\theta \downarrow 0$ in the formula. The references incorporate the term $\theta\beta(A)$ in the integral by allowing an atom at 0 in the (redefined) Lévy measure.

used as the number and measure in the right side of (J.5) to represent $\lambda_\theta(\cup_i A_i)$, while the number 0 and the zero measure represent $\lambda_\theta(\varnothing)$ in its left side. From the uniqueness of the representation we conclude that the maps $A \mapsto \beta(A)$ and $A \mapsto \gamma(A, B)$, for a fixed measurable $B \subset (0, \infty]$, are atomless measures.

First assume that the measure μ defined by $\mu(A) = \gamma(A, (0, \infty])$ is σ-finite. Then the measures μ_s defined by $\mu_s(A) = \gamma(A, (0, s])$ are right-continuous and increasing in $s > 0$ and $\mu_s \ll \mu$. By the Radon-Nikodym theorem there exists for every s a measurable function $F(\cdot, s): \mathfrak{X} \to [0, \infty)$ such that $\mu_s(A) = \int_A F(x, s)\, d\mu(x)$ for every A. By the monotonicity in s these functions automatically satisfy $F(x, s) \le F(x, t)$ for μ-almost every x whenever $s \le t$. This and right continuity can be used to show that the functions $\inf_{s\in\mathbb{Q}, s>t} F(x, s)$ also represent the measures μ_t. As the latter functions are right-continuous and monotone in t for *every* x, it is not a loss of generality to assume that the original function F possesses this property. Then $F(x, \cdot)$ is a Lebesgue-Stieltjes distribution function, and we can define a measure ν^c by $d\nu^c(x, s) = F(x, ds)\, d\mu(x)$. By construction this satisfies $\nu^c(A \times B) = \gamma(A, B)$.

If μ is not σ-finite, define measures $\bar\mu_k(A) = \gamma(A, (1/(k+1), 1/k])$, for $k \in \mathbb{N}$. Since $\gamma(A, (\epsilon, \infty]) < \infty$ for all A with $\lambda_1(A) < \infty$ and $\epsilon > 0$, by the condition $\int_{(0,\infty]} (s \wedge 1)\, \gamma(A, ds) < \infty$, and λ_1 is finite by assumption, it follows that $\bar\mu_k$ is σ-finite for every k. Now define $\nu^c = \sum_k \nu_k$ for the measures ν_k defined by $d\nu_k(x, s) = F_k(x, ds)\, d\mu_k(x)$, where F_k is constructed from $\bar\mu_k$ as previously F from μ. Again $\nu(A \times B) = \gamma(A, B)$, and hence $\int_A \int (s \wedge 1)\, \nu^c(dx, ds) = \int (s \wedge 1)\, \gamma(A, ds)$, which is finite, also for $A = \mathfrak{X}$.

For disjoint sets $A_1, \ldots, A_k$ the variables $\Psi(A_1), \ldots, \Psi(A_k)$ obtained from the process (J.3) are mutually independent, as they are functions of the independent Poisson processes $\{N^*(A_i \times (0, s]): s > 0\}$, for $i = 1, \ldots, k$. Adding the deterministic component β does not change this independence. Therefore $\Phi = \beta + \Psi$ is a CRM with characteristics β and γ as soon as (J.5) holds. To verify this we first validate formula (J.4) by applying Lemma J.2 to the Poisson process N^c on $\mathfrak{X} \times [0, \infty)$ and the function $f(x, s) = \mathbb{1}_A(x)s$. Next we note that $\int_A \int_0^\infty f(s)\, \nu^c(dx, ds) = \int_0^\infty f(s)\, \gamma(A, ds)$ for the functions $f(s) = 1 - e^{-\theta s}$, as $\nu^c(A \times B) = \gamma(A, B)$ by construction.

Taking $A = \{x\}$ in (J.4) gives that $\int_{\{x\}} \int (1 - e^{-\theta s})\, \nu^c(dx, ds) = \lambda_\theta(\{x\}) = 0$ and hence $\nu^c(\{x\} \times (0, \infty]) = 0$. ∎

The measure ν^c in the proposition, the intensity measure of the Poisson process N^c associated with the CRM Φ, is by an abuse of terminology also referred to as the *intensity measure* of the CRM Φ.

Combination of the Poisson representation and Lemma J.2 gives for the integral $\int f\, d\Psi$ of a nonnegative measurable function $f: \mathfrak{X} \to \mathbb{R}$ with respect to the CRM Ψ in (J.3) (or if f is integrable with respect to Ψ a.s.) as

$$\log \mathrm{E}(e^{-\theta \int f\, d\Psi}) = -\iint (1 - e^{-\theta s f(x)})\, \nu^c(dx, ds). \tag{J.6}$$

The exponent of the left side viewed as a function of f (and with $\theta = 1$) is called the *Laplace functional* of Ψ. Expanding the exponential in the right side and integrating term by term gives the cumulant generating function of $\int f\, d\Psi$, for Ψ as in (J.3), as

$$\log \mathrm{E}(e^{-\theta \int f\,d\Psi}) = \sum_{j=1}^{\infty} \frac{(-\theta)^j}{j!} \iint_{\mathfrak{X}\times\mathbb{R}^+} s^j f^j(x)\, \nu^c(dx, ds). \tag{J.7}$$

The jth cumulant of $\int f\,d\Psi$ is the coefficient of $(-\theta)^j/j!$ in the right side (provided the integrals exist). In particular,

$$\mathrm{E}\Big(\int f\,d\Psi\Big) = \iint_{\mathfrak{X}\times\mathbb{R}^+} sf(x)\, \nu^c(dx, ds), \tag{J.8}$$

$$\mathrm{var}\Big(\int f\,d\Psi\Big) = \iint_{\mathfrak{X}\times\mathbb{R}^+} s^2 f^2(x)\, \nu^c(dx, ds). \tag{J.9}$$

These formulas are valid for CRMs without fixed atoms and without deterministic component. Adding a fixed component β shifts the mean by $\int f\,d\beta$, but does not affect the higher cumulants. Adding a fixed atom at a simply adds the cumulant of the variable $f(a)\Phi(\{a\})$.

The preceding construction exhibits the CRM without fixed atoms as a purely atomic random measure, with weights of size s at points x, for (x, s) the points in the associated PRS. Since finite-dimensional distributions do not determine path properties, it is not immediate that *any* version of the CRM is purely atomic, but this is nevertheless true. For a gamma process on a Polish space, this was established by Blackwell (1973), and for general CRMs by Kingman (1975).

The counting process $A \mapsto N^c(A \times (0,\infty])$ giving the "marginal distribution of the locations" will often be infinite, but is a PRS with intensity measure $A \mapsto \nu^c(A \times (0,\infty])$. The equations $\nu^c(\{x\} \times (0,\infty]) = 0$ show that the latter is atomless, so that with probability one the Poisson process N^c has at most one point (x, s) on every half line $\{x\} \times (0,\infty]$.[3] This is natural, as per given location x only one weight s need be specified.

The fixed atoms are special in two ways: their locations $a_1, a_2, \ldots$ are deterministic; and their weights need not be "Poisson" distributed. It is possible to represent the fixed atoms through the random measure N^d on $\mathfrak{X} \times (0,\infty]$ given by

$$N^d = \sum_j \delta_{a_j, \Phi(\{a_j\})}.$$

Then $\int_A \int s\, N^d(dx, ds)$ is identical to the first sum on the right side of (J.2) evaluated at A, and the representation of a CRM Φ as in Proposition J.6 can be written in the form

$$\Phi(A) = \beta(A) + \iint \mathbb{1}_A(x) s\, N(dx, ds), \qquad N = N^c + N^d.$$

The sum N is a CRM on $\mathfrak{X} \times (0,\infty]$, but it is not a Poisson random measure. For instance, the variable $N(\{a_j\} \times A)$, which is almost surely equal to $\mathbb{1}\{\Phi(\{a_j\}) \in A\}$, is not Poisson, but Bernoulli distributed. Jacod and Shiryaev (2003) (in the case that $\mathfrak{X} = \mathbb{R}^+$ and $(0,\infty]$ is replaced by a general space E) call N an *extended Poisson random measure*. The mean measure of N is given by

$$\nu(A) := \mathrm{E}N(A) = \nu^c(A) + \nu^d(A), \qquad \nu^d(A \times B) := \sum_j \mathbb{1}_A(a_j) \mathrm{P}(\Phi(\{a_j\}) \in B).$$

[3] For a finite measure ν^c this interpretation is clear from representing Π^c as $N_n \sim \mathrm{Poi}(|\nu^c|)$ points (x, s) drawn i.i.d. from $\bar{\nu}^c$; the general case follows by decomposition.

As $\nu^c(\{x\} \times (0, \infty]) = 0$ for every x, the measures ν^c and ν^d are concentrated off and on the set $\cup_j \{a_j\} \times (0, \infty]$, respectively, and hence are orthogonal. Their sum $\nu = \nu^c + \nu^d$ is called the *intensity measure* of the CRM Φ, and together with the deterministic part β uniquely identifies its distribution. Because the variable $\Phi(\{a_j\})$ may be zero with positive probability, the measure $\nu^d(\{a_j\} \times \cdot) = \mathrm{P}(\Phi(\{a_j\}) \in \cdot)$ may be a subprobability measure on $(0, \infty]$; the "missing mass" satisfies $1 - \nu^d(\{a_j\} \times (0, \infty]) = \mathrm{P}(\Phi(\{a_j\}) = 0)$. We can compute[4]

$$\mathrm{E}e^{-\theta\Phi(\{a_j\})} = 1 - \nu^d(\{a_j\}, (0, \infty]]) + \int_{(0,\infty]} e^{-\theta s}\, \nu^d(\{a_j\}, ds). \tag{J.10}$$

This may be combined with (J.4) to obtain an expression for the cumulant measure of a general CRM in terms of the intensity measure ν and deterministic component β, but because of the conceptually different roles played by the fixed atoms and continuous part, the resulting formula may be obscure rather than helpful. In particular note that replacing ν^c by ν in the right sight of (J.7) does not give a valid formula for the cumulant generating function of $\int f\, d\Psi$ for a general CRM.

By expressing the mean and variance of the variables $f(a_j)\Phi(\{a_j\})$ in ν^d (for instance by differentiating the preceding display) and some subsequent manipulation, we can extend formulas (J.8) and (J.9) for the mean and variance of $\int f\, d\Phi$ for a CRM without fixed atoms to a general CRM Φ, with intensity measure[5] $\nu = \nu^c + \nu^d$

$$\mathrm{E}\Big(\int f\, d\Psi\Big) = \iint_{\mathfrak{X}\times\mathbb{R}^+} s f(x)\, \nu(dx, ds), \tag{J.11}$$

$$\mathrm{var}\Big(\int f\, d\Psi\Big) = \iint_{\mathfrak{X}\times\mathbb{R}^+} s^2 f^2(x)\, \nu(dx, ds) - \sum_x \Big(\int s f(x)\, \nu(\{x\}, ds)\Big)^2. \tag{J.12}$$

Example J.7 (Gamma process) Perhaps the most important example of a CRM is the *gamma process*. In this case $\Phi(A)$ is gamma distributed with parameters $\alpha(A)$ and 1, for a given atomless σ-finite measure α on $\mathfrak{X}$. The corresponding characteristics are $\beta = 0$ and $\nu(dx, ds) = s^{-1}e^{-s}\, ds\, d\alpha(x)$. Indeed, with these choices equation (J.7) reduces to

$$\begin{aligned}\log \mathrm{E}\left(e^{-\theta \int f\, d\Phi}\right) &= \sum_{j=1}^{\infty} \frac{(-\theta)^j}{j!} \int_0^{\infty} s^{j-1} e^{-s}\, ds \int f^j(x)\, d\alpha(x) \\ &= -\int \log\left(1 + \theta f(x)\right) d\alpha(x).\end{aligned} \tag{J.13}$$

In the special case $f = \mathbb{1}_A$ this reduces to $-\alpha(A)\log(1+\theta)$, which is indeed the cumulant generating function of a gamma distribution with parameters $\alpha(A)$ and 1.

It is often convenient to disintegrate the intensity measure as $\nu(dx, ds) = \rho_x(ds)\, \alpha(dx)$, where α is a measure on $\mathfrak{X}$ and ρ_x a transition kernel on $\mathfrak{X} \times \mathbb{R}^+$. The more interesting processes have $\rho_x(\mathbb{R}^+) = \infty$, for all x. The measure α refers to the distribution of the locations of the masses, and ρ_x to the size of a mass at x. If ρ_x does not depend on x, then the

[4] The notation $\int f(s)\, \nu(\{a\}, ds)$ means the integral of f relative to the measure $B \mapsto \nu(\{a\} \times B)$.

[5] Note that $\nu(\{x\}, ds) = \nu^d(\{x\}, ds)$, but ν in the other terms cannot be replaced by ν^c.

locations and sizes are stochastically independent. The CRM is then called *homogeneous*. (This same word is also employed to refer to uniformity in space or time, for instance (only) if ρ_x is the Lebesgue measure.)

Example J.8 (Product intensities) If $\nu(dx, ds) = \rho(ds)\,\alpha(dx)$ for a finite measure α, then it is no loss of generality to assume that α is a probability measure. Also assume that α is atomless. Below we show that the CRM can then be generated as $\Psi(A) = \sum_{i\in I} S_i \mathbb{1}\{X_i \in A\}$, for a Poisson random set $(S_i\colon i \in I)$ with intensity measure ρ on $(0, \infty)$, and an independent random sample $X_1, X_2, \ldots \overset{\text{iid}}{\sim} \alpha$.

If $\rho(0, \infty) < \infty$, then the sum (over I) will have finitely many terms (namely a Poisson number), but typically the sum will be infinite. The requirement $\int (s \wedge 1)\, \rho(ds) < \infty$, resulting from Proposition J.6, shows that $\rho([\epsilon, \infty)) < \infty$ for every $\epsilon > 0$, so that the number of jumps bigger than ϵ is finite almost surely, and hence the jumps will cluster at zero only. The requirement also shows that $\int_0^1 s\, \rho(ds) < \infty$ so that the sum of the "small" jumps has finite expectation and hence the sum of all jumps is finite almost surely.

To prove the claim that Ψ has the given representation it suffices to verify that the point process defined by $N(A) := \#(i\colon (X_i, S_i) \in A)$ is a Poisson process with intensity ν. For product sets $A = C \times D$, the counts $N(A)$ arise as the thinning of the process $\#(i\colon S_i \in D)$ by removing points i with $X_i \notin C$, and hence $N(C \times D)$ is a Poisson process in D for every fixed C. Together with independence across disjoint C, this identifies N as a Poisson process, since the maps $M \mapsto M(C \times D)$ generate the same σ-field as the maps $A \mapsto M(A)$ (cf. Proposition A.5 together with a truncation argument to accommodate infinite random measures).

Example J.9 (Simulation) The preceding example extends to more general intensities $\nu(dx, ds) = \rho_x(ds)\,\alpha(dx)$ for a probability measure α, not necessarily of product form. We generate a sequence $X_1, X_2, \ldots \overset{\text{iid}}{\sim} \alpha$ of locations as before, but define the corresponding weights as $S_i = L_{X_i}^{-1}(V_i)$, for a standard, homogeneous Poisson process $V_1, V_2, \ldots$ on $(0, \infty)$ and $L_x(s) = \rho_x((s, \infty))$. The CRM is obtained as $\Psi(A) = \sum_i S_i \mathbb{1}\{X_i \in A\}$, as before.

To see that this works it suffices again to show that the point process of the points (X_i, S_i) is a Poisson process with intensity measure ν. Arguing as in the preceding example, we see that the points (X_i, V_i) form a Poisson process with intensity measure $ds\,\alpha(dx)$. The transformed points (X_i, S_i) are obtained by applying the deterministic transformation $(x, v) \mapsto (x, L_x^{-1}(v))$ to the points of this process and hence still form a Poisson process. By the definition of L_x, for every measurable set A and $0 \le u < v$,

$$\begin{aligned}
\mathrm{E}\#(X_i \in A, u < S_i \le v) &= \sum_{i=1}^{\infty} \int_A \mathrm{E}\mathbb{1}\{L_x(u) < V_i \le L_x(v)\}\, d\alpha(x) \\
&= \int_A (L_x(v) - L_x(u))\, d\alpha(x) = \nu(A \times (u, v]).
\end{aligned}$$

Thus the transformed points possess intensity measure equal to ν.

J.3 Completely Random Measures on the Positive Half Line

CRMs on $\mathfrak{X} = \mathbb{R}^+$ are especially important in survival analysis. If they are finite on finite intervals, then they can be described by their Lebesgue-Stieltjes distribution function, for which we shall use the (unusual) notation X, so that $X(t) = \Phi((0,t])$. The sample paths $(X(t): t \in \mathbb{R}^+)$ of this stochastic process are nondecreasing right-continuous functions with independent increments over disjoint intervals. Thus X is an *independent increment process* (or *IIP*) with nonnegative increments. If the increments are stationary, then X is also a "Lévy process" and a "subordinator." In general, a *Lévy process* has cadlag sample paths and stationary independent increments, and a *subordinator* is a Lévy process with nonnegative increments. Several examples are given below. A Brownian motion is also an example of a Lévy process, but its increments are not nonnegative. A general Lévy process is the sum of an affine linear function, a multiple of Brownian motion, and a pure jump process. For CRMs the Brownian motion is absent and the pure jump processes have nonnegative jumps, and hence sample paths of bounded variation.

A CRM on $\mathfrak{X} = \mathbb{R}^+$ corresponds to a jump measure N on $\mathbb{R}^+ \times (0,\infty)$. Its distribution function can be represented in terms of $N = N^d + N^c$ as

$$\begin{aligned} X(t) &= \beta(t) + \int_{(0,t]} \int_{(0,\infty)} s\, N(dx, ds) \\ &= \beta(t) + \sum_{j: a_j \le t} \Delta X(a_j) + \int_{(0,t]} \int_{(0,\infty)} s\, N^c(dx, ds), \end{aligned}$$

where the variables $\Delta X(a_j)$ give the fixed atoms of the CRM, but are referred to as *fixed jumps* in this context. The fixed jumps are independent nonnegative random variables independent of the PRS N^c, the *fixed jump times* a_j are arbitrary positive numbers, and β is a cumulative distribution function on $\mathbb{R}^+$. If it is understood that $\Delta X(a)$ is the zero variable if $a \notin \{a_1, a_2, \ldots\}$, then the sum in the far right side of the display can also be extended to range over all $a \le t$. The mean measure ν of N can be decomposed as $\nu = \nu^d + \nu^c$ for ν^c the intensity measure of the PRS N^c, and ν^d describing the distribution of the fixed jumps: it concentrates on $\cup_j \{a_j\} \times (0,\infty)$ and $\nu^d(\{a_j\} \times A) = \mathrm{P}(\Delta X(a_j) \in A)$, for $A \subset (0,\infty)$ and every j. The random measure N^c is a PRS, while N is an *extended Poisson random measure*, in the terminology of Jacod and Shiryaev (2003). By formulas (J.11) and (J.12) the mean and variance of $X(t)$ are given in terms of ν by

$$\begin{aligned} \mathrm{E}\big[X(t)\big] &= \beta(t) + \int_{(0,t]} \int s\, \nu(dx, ds), \\ \mathrm{var}\big[X(t)\big] &= \int_{(0,t]} \int s^2\, \nu(dx, ds) - \sum_{x: x \le t} \Big(\int s\, \nu(\{x\}, ds)\Big)^2. \end{aligned}$$

Here $\nu(\{x\}, \cdot) = \nu^d(\{x\}, \cdot)$ is nonzero only at fixed jump times x, and hence the second term in the variance formula disappears if the process is without fixed jumps.

Because its sample paths are nondecreasing, the process X is trivially a *submartingale*, provided it is integrable. By the independence of increments the process $X(t) - \mathrm{E}\big[X(t)\big]$ is a martingale, whence $t \mapsto \mathrm{E}\big[X(t)\big]$ is the *compensator* of X.

Example J.10 (Time change) For a given process X with independent, nonnegative increments and a deterministic cadlag, nondecreasing function A, the process defined by $Y(t) = X(A(t))$ also possesses independent, nonnegative increments. This is called a *time change*. In the representation of X through a point process N^c on $\mathbb{R}^+ \times (0, \infty]$, the transformation simply moves the points, giving $(0, t] \times D \mapsto N^c((0, A(t)] \times D)$ as the point process representing Y, with intensity measure $(0, t] \times D \mapsto \nu^c((0, A(t)] \times D)$, for ν^c the intensity measure of X. Fixed atoms of X are moved similarly to fixed atoms of Y.

Example J.11 (Integration) Another operation that preserves independent, nonnegative increments is the formation of the process $Y(t) = \int_{(0,t]} b\,dX$, for a deterministic, nonnegative function b. Because an increment $Y(t_2) - Y(t_1)$ of the latter process depends only on the variables $\{X(u) - X(t_1): t_1 < u \leq t_2\}$, the increments of Y are independent if those of X are independent. If X does not have fixed jumps, then the Laplace transform of the process Y can be computed using (J.6): if X and Y also denote the corresponding CRMs, then $\int f\,dY = \int fb\,dX$ and hence the formula gives, for ν the intensity measure of X,

$$-\log \mathrm{E}\left(e^{-\theta \int f\,dY}\right) = \iint (1 - e^{-\theta s f(x) b(x)})\,\nu(dx, ds).$$

In order to identify the intensity measure ν^* of Y the right side must be written in the form $\iint (1 - e^{-\theta s f(x)})\,\nu^*(dx, ds)$. This gives

$$\nu^*(A) = \iint \mathbb{1}_A(x, sb(x))\,\nu(dx, ds). \tag{J.14}$$

Example J.15 gives a concrete illustration.

Example J.12 (Compound Poisson process) A *standard Poisson process* on $\mathbb{R}^+$ is a subordinator. Its intensity measure is given by $\nu(dt, ds) = dt\,\delta_1(ds)$, for δ_1 the Dirac measure at 1. The Dirac measure arises as all jumps have size 1.

In a *compound Poisson process* the jumps occur at the times of a standard Poisson process, but their sizes are independently generated from a fixed probability distribution ρ, on $(0, \infty)$ to obtain nonnegative increments. The corresponding intensity measure is $\nu(dt, ds) = dt\,\rho(ds)$.

These processes are said to have "finite activity," as only finitely many jumps occur in bounded time intervals.

Example J.13 (σ-stable process) The *σ-stable subordinator*, for $0 < \sigma < 1$, corresponds to the intensity measure

$$\nu(dt, ds) = \frac{\sigma}{\Gamma(1-\sigma)} s^{-1-\sigma}\,dt\,ds.$$

In this case the distribution of $X(t)$ is positive σ-stable: the sum of n independent copies of $X(t)$ is distributed like $n^{1/\sigma} X(t)$, for any n.

Example J.14 (Gamma process) The marginal distribution of the increments $X(t) - X(s)$ of the process resulting from the intensity measure $\nu(dt, ds) = a s^{-1} e^{-bs}\,dt\,ds$ are $\mathrm{Ga}(at -$

$as, b)$, as noted in Example J.7 when $b = 1$. The process X is called a *(time-homogeneous) gamma process* with parameters (a, b); if $b = 1$ it is called a *standard gamma process*.

The more flexible process obtained by allowing a to depend on t, with intensity measure $\nu(dt, ds) = a(t)s^{-1}e^{-bs}\, dt\, ds$, is called the *time-inhomogeneous gamma process*. This process can also be defined for a general increasing, cadlag process A, not necessarily absolutely continuous, by the intensity measure

$$\nu(dt, ds) = s^{-1}e^{-bs}\, dA(t)\, ds.$$

The process can be constructed by a *time change* from a standard gamma process Y as $X(t) = Y(A(t))$. This shows that the marginal distribution of $X(t)$ is $\mathrm{Ga}(A(t), b)$.

The behavior of $X(t)$ as $t \to \infty$ can be controlled by the corresponding behavior of $A(t)$. If A increases to infinity (i.e. $A(\infty) = \infty$), then $X(t) \to \infty$ a.s., whereas in the other case $X(\infty) := \lim_{t\to\infty} X(t)$ is a real-valued random variable distributed as $\mathrm{Ga}(A(\infty), 1)$.

Example J.15 (Extended gamma process) If X is a gamma process with intensity measure $s^{-1}e^{-s}\, dA(t)\, ds$ and b is a given positive, measurable function, then the process $Y(t) = \int_{(0,t]} b\, dX$ is an independent increment process with intensity measure

$$\nu^c(dt, ds) = s^{-1}e^{-s/b(t)}\, dA(t)\, ds. \tag{J.15}$$

This follows from the general formula (J.14), and the change of variables $(t, sb(t)) \to (t, u)$ in the integral. Because $Y(t) = \int \mathbb{1}_{(0,t]} b\, dX$, the cumulant generating function of $Y(t)$ is equal to $-\int_{(0,t]} \log(1 + \theta b)\, dA$, by formula (J.13). It follows that $\mathrm{E}[Y(t)] = \int_{(0,t]} b\, dA$ and $\mathrm{var}[Y(t)] = \int_{(0,t]} b^2\, dA$.

The process Y is known as an *extended gamma process*. When b is constant, the process reduces to an ordinary gamma process. This is the only case in which the extended gamma process is homogeneous.

Example J.16 (Beta process) The *standard beta process* with parameters $a, b > 0$ has intensity measure

$$\nu(dt, ds) = B(a, b)^{-1}s^{a-2}(1 - s)^{b-1}\mathbb{1}_{(0,1)}(s)\, dt\, ds.$$

If $a > 1$, then the function $s \mapsto s^{a-2}(1 - s)^{b-1}$ is integrable, and one could construct the process by generating jump sizes from the beta distribution to be placed at jump locations forming a Poisson process, as in Example J.8 but with the roles of locations and sizes swapped. If $a \leq 1$ the sizes follow a "beta distribution with infinite weight near zero."

Example J.17 (Generalized gamma process) The *generalized gamma process* is the independent increment process with intensity measure, for $0 < \sigma < 1$ and $\tau \geq 0$,

$$\nu(dt, ds) = \frac{\sigma}{\Gamma(1 - \sigma)} s^{-(1+\sigma)}e^{-\tau s} dA(t)\, ds.$$

For $\tau = 0$ this reduces to the σ-stable process, whereas letting $\sigma \to 0$ takes it to the homogeneous gamma process. The *inverse-Gaussian process* is obtained for $\sigma = 1/2$.

Two CRMs with a similar intensity measure are close. The following proposition gives a simple criterion for the convergence of random cumulative distribution functions in terms of their intensity measures. The convergence is in the *Skorohod space* $\mathfrak{D}[0,\infty)$ relative to the Skorohod topology on compacta (see Billingsley 1968 or Pollard 1984, Chapter VI). The use of the latter topology, which is weaker than the topology of uniform convergence on compacta, is natural as the jumps of the processes occur at random times, and are not restricted to a given countable set. For simplicity we restrict to the situation that the limiting CRM possesses a continuous intensity measure.[6]

Proposition J.18 (Convergence of CRMs) *If X_n are processes with nonnegative, independent increments with intensity measures ν_n such that $(s\wedge 1)\,\nu_n(dx,ds) \rightsquigarrow (s\wedge 1)\,\nu(dx,ds)$ on $[0,t]\times\mathbb{R}^+$, for every $t>0$,[7] for an intensity measure ν with zero fixed jump component ν^d, then $X_n \rightsquigarrow X$ in $\mathfrak{D}[0,\infty)$ relative to the Skorohod topology on compacta, for X the process with nonnegative, independent increments with intensity measure ν.*

Proof By combining (J.6) and (J.10) we see that, for a given continuous function $f\colon\mathbb{R}^+\to[0,1]$ with compact support, and with $R(x)=\log(1-x)+x$,

$$\begin{aligned}-\log \mathrm{E}[e^{-\int f\,dX_n}] &= \int\!\!\int (1-e^{-f(x)s})\,\nu_n^c(dx,ds) - \sum_x \log\Big(1-\int (1-e^{-f(x)s})\,\nu_n^d(\{x\},ds)\Big)\\ &= \int\!\!\int (1-e^{-f(x)s})\,\nu_n(dx,ds) - \sum_x R\Big(\int (1-e^{-f(x)s})\,\nu_n^d(\{x\},ds)\Big).\end{aligned}$$

Since the function $(x,s)\mapsto 1-e^{-f(x)s}$ is continuous, nonnegative and bounded above by $s\wedge 1$, the first term on the far right tends to the same expression, but with ν_n replaced by ν. If it can be shown that the last term on the right tends to zero, then it follows that the Laplace transforms of the variables $\int f\,dX_n$ tend to the Laplace transform of $\int f\,dX$, which implies their convergence in distribution (e.g. Chung 2001, Theorem 6.6.3) and next marginal convergence of X_n.

In view of the assumed continuity of the first marginal of ν, we also have that the maps $t\mapsto L_n(t){:=}\int_{(0,t]}\int(1-e^{-f(x)s})\,\nu_n(dx,ds)$ converge pointwise to the map L defined in the same way from ν, and then by Pólya's theorem also uniformly on compacta. The integrals $\int(1-e^{-f(x)s})\,\nu_n^d(\{x\},ds)$ are the jumps of the maps L_n, and hence converge uniformly to the jumps of L, which are zero. This allows to complete the argument of the preceding paragraph.

To verify that the sequence X_n is uniformly tight in $\mathfrak{D}[0,\infty)$, assume first that the intensity measures ν_n concentrate on $\mathbb{R}^+\times[0,K]$, for some K. Then the function $(x,s)\mapsto s$ is bounded above by a multiple of $s\wedge 1$, whence, for $t>0$,

[6] For the general case, and also for stronger versions of the proposition, see Theorems VII.3.13 and VII.3.4 in Jacod and Shiryaev (2003). In the case of a CRM with $\mathrm{E}[X(t)]<\infty$, the characteristic triple of X is $(B,C,\nu){:=}(\mathrm{E}\tilde X(t),0,\nu)$, where $\tilde X_t$ is X_t with the big jumps removed, and the modified second characteristic $\tilde C$ is $\int_0^t\int h^2(s)\,\nu(dx,ds)-\sum_{x:x\le t}(\int h(s)\,\nu(\{x\},ds))^2$, where h is the truncation function.

[7] This means $\int_0^t\int g(x,s)(s\wedge 1)\,\nu_n(dx,ds)\to\int_0^t\int g(x,s)(s\wedge 1)\,\nu(dx,ds)$ for every continuous bounded function $g\colon[0,t]\times(0,\infty)\to\mathbb{R}$.

$$\mathrm{E}[X_n(t)] = \int_{(0,t]}\int s\,\nu_n(dx,ds) \to \int_{(0,t]}\int s\,\nu(dx,ds) = \mathrm{E}[X(t)].$$

Because the limit function is continuous by assumption, this convergence is automatically uniform. It follows that $\mathrm{E}|X_n(v) - X_n(u)|$ converges uniformly in (u, v) belonging to compacta to $\mathrm{E}|X(v) - X(u)|$, which can be made uniformly (arbitrarily) small by restricting u, v to $|u - v| < \delta$, for a small δ. This implies that *Aldous condition* for uniform tightness is satisfied (see Pollard 1984, Theorem VI.16 and Example VI.18).

Finally the weak convergence of the measures $(s \wedge 1)\,\nu_n(dx, ds)$ implies their uniform tightness, so that $\sup_n \nu_n([0,t]\times(K,\infty)) \to 0$ as $K \to \infty$. The number of non-fixed jumps of size bigger than K in X_n is a Poisson variable with mean $\nu_n^c([0,t]\times(K,\infty))$ and hence is zero with probability tending to one, as $K \to \infty$. Similarly, the probability that at least one fixed jump height exceeds K is bounded above by $\nu_n^d([0,t] \times (K,\infty))$ and tends to zero. This means that $\inf_n \mathrm{P}(X_n = X_n^K) \to 1$, for X_n^K the process defined by removing from X_n the jumps of heights bigger than K. By the preceding paragraph $X_n^K \rightsquigarrow X^K$, as $n \to \infty$, for every fixed K, and $X^K \rightsquigarrow X$ as $K \to \infty$. We can choose a suitable $K_n \to \infty$ such that both $\mathrm{P}(X_n = X_n^{K_n}) \to 1$ and $X_n^{K_n} \rightsquigarrow X$ to complete the proof. □

The sample paths of most of the preceding examples of nondecreasing, independent increment processes are delicate, as they increase by countably many jumps in any interval. This complicates a description of their rates of growth. In the following result this is given for the standard gamma process (for a proof see Fristedt and Pruitt 1971)). Similar results can be obtained for other subordinators (see Section III.4 of Bertoin 1996).

Theorem J.19 *Let X be the standard gamma process on* $[0,\infty)$ *and* $h\colon (0,\infty) \to (0,\infty)$.

(i) *If h is strictly increasing, then either* $\mathrm{P}(\liminf_{t\to 0} X(t)/h(t) = 0) = 1$ *or* $\mathrm{P}(\liminf_{t\to 0} X(t)/h(t) = \infty) = 1$.

(ii) *Let h be strictly increasing and convex. Then*

$$\mathrm{P}(\limsup_{t\to 0} X(t)/h(t) = 0) = \begin{cases} 1, & \text{if } \int_0^1 \int_{h(t)}^{\infty} e^{-x} x^{-1} dx\, dt < \infty, \\ 0, & \text{if } \int_0^1 \int_{h(t)}^{\infty} e^{-x} x^{-1} dx\, dt = \infty. \end{cases}$$

(iii) *The variable* $\liminf_{t\to 0} X(t) \exp[r \log|\log t|/t]$ *is equal to* 0 *a.s. if* $r < 1$ *and equal to* ∞ *a.s. if* $r > 1$.

J.4 Historical notes

Kingman (1967, 1975) developed the theory of completely random measures on general sample spaces. For general point processes (treated within the context of random measures) and their historical development, also see Daley and Vere-Jones (2003, 2008). For the special case of Lévy processes, which were developed earlier, see Bertoin (1996).

Lemma J.4 is based on Section 4 of Perman et al. (1992).

Problems

J.1 Using the time change technique of Lévy processes, state and prove a version of Theorem J.19 for a non-homogeneous gamma processes.

J.2 (Kingman 1993) Show that the mean μ of a Poisson process is a measure. Further show that μ is atomless. [Hint: Use $N(\{x\}) \leq 1$ for all $x \in \mathfrak{X}$.]

J.3 (Kingman 1993) Show that $\mathrm{cov}\,(N(A_1), N(A_2)) = \mu(A_1 \cap A_2)$ for a Poisson process N.

J.4 (Kingman 1993) Let $Z = Z(\Pi)$ be a random variable depending on a PRS Π on $\mathfrak{X}$ with finite mean measure μ. Derive the following compounding formula:

$$\mathrm{E}(Z) = \sum_{n=0}^{\infty} \mathrm{E}_n(Z) e^{-\mu(\mathfrak{X})} \mu(\mathfrak{X})^n / n!,$$

where $\mathrm{E}_n(Z) = \mathrm{E}[Z(\Pi_n)]$, $\Pi_0 = \emptyset$, and Π_n is a multinomial random set with parameters n and $P := \mu/\mu(\mathfrak{X})$.

J.5 If $(X(t): t \geq 0)$ is an independent increment process with cadlag sample paths, then it defines a random distribution function and hence a random measure Φ. The independence of the increments imply that the variables $\Phi(A_1), \ldots, \Phi(A_k)$ are independent whenever $A_1, \ldots, A_k$ are disjoint intervals. Show that these variables are independent for any disjoint Borel sets $A_1, \ldots, A_k$, and $k \in \mathbb{N}$, so that Φ is also a CRM according to the definition. [Hint: use the monotone class theorem and the "good sets principle" in succession on $A_1, \ldots, A_k$.]

Appendix K

Inequalities and Estimates

This appendix collects some results for ease of reference.

The classical inequalities named after Hoeffding and Bernstein give exponential bounds on the (tail) probabilities of deviation from their means of sums of independent variables. The first gives "sub-Gaussian" tails, but applies to variables with finite range only. The second gives mixed sub-Gaussian and sub-exponential tails, under exponential moments bounds, and involves the variance of the variables. We include only the proof of Hoeffding's inequality; a proof of Bernstein's inequality can be found in Bennett (1962), pages 37–38.

Theorem K.1 (Hoeffding) *For any independent random variables $X_1, \ldots, X_n$ such that $a_i \le X_i \le b_i$, $i = 1, \ldots, n$, and any $t > 0$,*

$$\mathrm{P}\Big(\textstyle\sum_{i=1}^n (X_i - \mathrm{E}(X_i)) \ge t\Big) \le e^{-2t^2/\sum_{i=1}^n (b_i - a_i)^2}$$

$$\mathrm{P}\Big(\Big|\textstyle\sum_{i=1}^n (X_i - \mathrm{E}(X_i))\Big| \ge t\Big) \le 2e^{-2t^2/\sum_{i=1}^n (b_i - a_i)^2}.$$

Proof The second inequality follows from the first applied to both the variables X_i and $-X_i$, and Boole's inequality. To prove the first we apply Markov's inequality with the function e^{hx}, for $h > 0$ to be chosen later, to obtain

$$\mathrm{P}\Big(\sum_{i=1}^n (X_i - \mathrm{E}(X_i)) \ge t\Big) \le e^{-hnt} \prod_{i=1}^n \mathrm{E}(e^{h(X_i - \mathrm{E}(X_i))}).$$

By convexity of the exponential function $e^{hX} \le ((b-X)e^{ha} + (X-a)e^{hb})/(b-a)$ whenever $a \le X \le b$, whence, by taking expectation,

$$\mathrm{E}(e^{hX}) \le e^{ha}\frac{b - \mathrm{E}(X)}{b-a} + e^{hb}\frac{\mathrm{E}(X) - a}{b-a} = e^{g(\xi)},$$

where $g(\xi) = \log(1 - p + pe^{\xi}) - p\xi$, for $\xi = (b-a)h$ and $p = (\mathrm{E}(X) - a)/(b-a)$.

Now $g(0) = 0$, $g'(0) = 0$ and $g''(\xi) = (1-p)pe^{\xi}/(1 - p + pe^{\xi})^2 \le \frac{1}{4}$ for all ξ, so that a second order Taylor's expansion gives $g(\xi) \le \xi^2/8$. Combining this with the preceding displays, we obtain, for any $h > 0$,

$$\mathrm{P}\Big(\sum_{i=1}^n (X_i - \mathrm{E}(X_i)) \ge t\Big) \le \exp\Big(-hnt + h^2 \sum_{i=1}^n (b_i - a_i)^2\Big).$$

The result follows upon choosing $h = 4nt/\sum_{i=1}^n (b_i - a_i)^2$. □

Theorem K.2 (Bernstein) *For independent random variables* $X_1, \ldots, X_n$ *with mean* 0 *and* $\mathrm{E}|X_i|^m \leq m!M^{m-2}v_i/2$, *for some* $M, v_1, \ldots, v_n > 0$, *all* $i = 1, \ldots, n$ *and all* $m \in \mathbb{N}$, *for any* $t > 0$,

$$\mathrm{P}\Bigl(\Bigl|\sum_{i=1}^n X_i\Bigr| \geq t\Bigr) \leq 2e^{-t^2/(Mt+\sum_{i=1}^n v_i)}.$$

In particular, this is true with $M = 2K$ *if* $K^2\mathrm{E}(e^{|X_i|/K} - 1 - |X_i|/K) \leq v_i/4$.

Lemma K.3 *If* $\psi: [0, \infty) \to \mathbb{R}$ *is convex and nondecreasing, then* $\mathrm{E}\psi(|X - \mathrm{E}X|) \leq \mathrm{E}\psi(2|X|)$ *for every random variable* X.

Proof The map $y \mapsto \psi(|y|)$ is convex on $\mathbb{R}$. If X' is an independent copy of X, then the left side is equal to $\mathrm{E}\psi(|X - \mathrm{E}X'|) \leq \mathrm{E}\psi(|X - X'|)$, by Jensen's inequality. Next bound $|X - X'| \leq |X| + |X'|$ and use the monotonicity and convexity of ψ again to bound the expectation by $\mathrm{E}(\frac{1}{2}\psi(2|X|) + \frac{1}{2}\psi(2|X'|))$, which is the right side. ☐

The variance of a sum of independent variables is the sum of the variances of the individual variables. The following inequality bounds the order of a general central moment of a sum of variables.

Lemma K.4 (Marcinkiewicz-Zygmund) *For any independent, mean-zero, random variables* $X_1, \ldots, X_n$, *any* $p \geq 2$, *and a constant* C_p *that depends on* p *only,*

$$\mathrm{E}\Bigl|\sum_{i=1}^n X_i|^p \leq C_p\mathrm{E}\Bigl(\sum_{i=1}^n X_i^2\Bigr)^{p/2} \leq C_p\, n^{p/2-1}\sum_{i=1}^n \mathrm{E}|X_i|^p.$$

The following lemma, due to Ibragimov (1962), extends this (for $p = 2$) to dependent random variables. The *α-mixing coefficients* of a sequence of random variables X_n are defined by, with the supremum taken over all measurable sets A, B in the sample space,

$$\alpha_k = \sup_{m\in\mathbb{N}} \sup_{A,B} |\mathrm{P}(X_m \in A, X_{m+k} \in B) - \mathrm{P}(X_m \in A)\mathrm{P}(X_{m+k} \in B)|. \tag{K.1}$$

Lemma K.5 *If* $X_0, X_1, \ldots$ *is a sequence of random variables with α-mixing coefficients* $\{\alpha_k\}$, *then for any* $s > 2$,

$$\mathrm{var}\Bigl(\sum_{i=1}^n f(X_i)\Bigr) \leq 4n\sum_{k=0}^\infty \alpha_k^{1-2/s} \sup_i (\mathrm{E}|f(X_i)|^s)^{2/s}. \tag{K.2}$$

Lemma K.6 *For* Φ *the standard normal cumulative distribution function and* ϕ *its density,*

(i) $(x^{-1} - x^{-3})\phi(x) \leq 1 - \Phi(x) \leq \min(x^{-1}\phi(x), \frac{1}{2}e^{-x^2/2})$, *for* $x > 0$.
(ii) $\Phi^{-1}(u) \geq -\sqrt{2\log_- u}$, *for* $0 < u < 1$.
(iii) $\Phi^{-1}(u) \leq -\frac{1}{2}\sqrt{\log_- u}$, *for* $0 < u < 1/2$.
(iv) $\Phi(\sqrt{2x} + \Phi^{-1}(e^{-x})) \geq \frac{1}{2}$, *for* $x > 0$.

Proof Inequalities (i) follow by one or two partial integrations of $\int_x^\infty \phi(s)\,ds$, observing that $\phi'(x) = -x\phi(x)$, and deleting the integral term, which takes different signs in the two cases. For (ii)–(iii) it suffices to show that $\Phi(-\sqrt{2\log_- u}) \leq u$, for $0 < u < 1/2$, and $\Phi(-\sqrt{\log_- u}/2) \geq u$, for $0 < u < 1/4$. □

Lemma K.7 *For any $q \geq 0$, $t \geq -2q$, $u > 0$ and $v \geq 0$, as $N \to \infty$,*

$$\sup_{\sum_{i=1}^\infty i^{2q}\xi_i^2 \leq 1} \sum_{i=1}^\infty \frac{\xi_i^2 i^{-t}}{(1+Ni^{-u})^v} \asymp N^{-((t+2q)/u)\wedge v}.$$

Moreover, for every fixed ξ with $\sum_{i=1}^\infty i^{2q}\xi_i^2 < \infty$, as $N \to \infty$,

$$N^{((t+2q)/u)\wedge v} \sum_{i=1}^\infty \frac{\xi_i^2 i^{-t}}{(1+Ni^{-u})^v} \to \begin{cases} 0, & \text{if } (t+2q)/u < v, \\ \|\xi\|^2_{2,2,(uv-t)/2}, & \text{if } (t+2q)/u \geq v. \end{cases}$$

The last assertion remains true if the sum is limited to the terms $i \leq cN^{1/u}$, for any $c > 0$.

Lemma K.8 *For any $t, v \geq 0$, $u > 0$, and (ξ_i) such that $|\xi_i| = i^{-q-1/2}S(i)$ for $q > -t/2$ and a slowly varying function $S\colon (0,\infty) \to (0,\infty)$, as $N \to \infty$,*

$$\sum_i \frac{\xi_i^2 i^{-t}}{(1+Ni^{-u})^v} \asymp \begin{cases} N^{-(t+2q)/u} S^2(N^{1/u}), & \text{if } (t+2q)/u < v, \\ N^{-v}\sum_{i\leq N^{1/u}} S^2(i)/i, & \text{if } (t+2q)/u = v, \\ N^{-v}, & \text{if } (t+2q)/u > v. \end{cases}$$

Moreover, for every $c > 0$, the sum on the left is asymptotically equivalent to the same sum restricted to the terms $i \leq cN^{1/u}$ if and only if $(t+2q)/u \geq v$.

Proofs See Knapik et al. (2011), Lemmas 8.1 and 8.2. □

The following result is useful in lower bounding the measure of the set where the values of one function are separated from those of another function, if the two functions are not very close in the $\mathbb{L}_1$-sense. Such an estimate is useful to construct test functions given positive $\mathbb{L}_1$-separation.

Lemma K.9 *Let ν be a finite measure on $\mathfrak{X}$ and let f and g be nonnegative, measurable functions such that $\int |f-g|\,d\nu > \eta$.*

(i) *If $f, g \leq M$, then $\nu\{|f-g| > \epsilon\} \geq (\eta - \nu(\mathfrak{X})\epsilon)/M$.*
(ii) *If $g \leq M$ and $\int f\,d\nu \leq \int g\,d\nu + \eta'$, then $\nu\{f < g - \epsilon\} \geq (\eta - \eta' - 2\nu(\mathfrak{X})\epsilon)/(2M)$.*

Proof For (i) we decompose $\|f-g\|_1$ as $\int_{|f-g|>\epsilon} |f-g|\,d\nu + \int_{|f-g|\leq\epsilon} |f-g|\,d\nu$, which is bounded above by $M\nu\{|f-g| > \epsilon\} + \epsilon\nu(\mathfrak{X})$. The estimate follows by rearranging terms.

For (ii), using the identity $|x| = 2x^+ - x$, we decompose $\|f-g\|_1$ as

$$2\int_{f\leq g} (g-f)\,d\nu - \int (g-f)\,d\nu \leq 2\int_{f<g-\epsilon} (g-f)\,d\nu + 2\int_{g-\epsilon\leq f\leq g} (g-f)\,d\nu + \eta'.$$

This can be further bounded by $2M\nu\{f < g-\epsilon\} + 2\epsilon\nu(\mathfrak{X}) + \eta'$. The estimate again follows by rearranging terms. □

In the following two lemmas $(z, B) \mapsto \Psi(B|\, z)$ is a Markov kernel between measurable spaces $(\mathfrak{Z}, \mathscr{Z})$ and $(\mathfrak{X}, \mathscr{X})$, and for a probability distribution F on $(\mathfrak{Z}, \mathscr{Z})$, we write P_F for the mixture distribution given by $P_F(A) = \int \psi(A|\, z)\, dF(z)$.

Lemma K.10 *If F^* is the renormalized restriction of a probability measure F on $(\mathfrak{Z}, \mathscr{Z})$ to a set $\mathfrak{Z}_0$ of positive F-measure (i.e. $F^*(A) = F(A \cap \mathfrak{Z}_0)/F(\mathfrak{Z}_0)$ for all A), then $\|P_{F^*} - P_F\|_{TV} \leq F(\mathfrak{Z}_0^c)$.*

Proof For any set B the difference $P_{F^*}(B) - P_F(B)$ can be decomposed as $\int_{\mathfrak{Z}_0} (\Psi(A|\, z) - 1/2)\, dF(z) F(\mathfrak{Z}_0^c)/F(\mathfrak{Z}_0) - \int_{\mathfrak{Z}_0^c} (\Psi(A|\, z) - 1/2)\, dF(z)$. Applying the triangle inequality and next bounding $|\Psi(A|\, z) - 1/2|$ by $1/2$ we see that this is bounded by $F(\mathfrak{Z}_0^c)$. □

Lemma K.11 *Suppose that $\mathfrak{Z}$ and $\mathfrak{X}$ are Polish spaces, and*

(i) *the map $F \mapsto P_F$ is one-to-one and continuous for the weak topology,*
(ii) *for every $\epsilon > 0$ and compact set $C \subset \mathfrak{X}$, there exists a compact set $D \subset \mathfrak{Z}$ with $\Psi(C|\, z) < \epsilon$ for every $z \notin D$.*

Then $P_{F_n} \rightsquigarrow P_F$ implies that $F_n \rightsquigarrow F$. The first condition is satisfied in particular if the mixing distribution is identifiable and the Markov kernel admits a transition density $(x, z) \mapsto \psi(x|\, z)$ that is bounded and continuous in z, for every given x.

Proof If p_{F_n} converges weakly, then it is tight by Prohorov's theorem. Thus for every $\epsilon > 0$ there exists a compact set C with $\int \Psi(C|\, z)\, dF_n(z) = P_{F_n}(C) > 1 - \epsilon$, for every n. By assumption there exists a compact set D with $\int_{D^c} \Psi(C|\, z)\, dF_n(z) < \epsilon$, whence $1 - 2\epsilon < \int_D \Psi(C|\, z)\, dF_n(z) < F_n(D)$. We conclude that the sequence F_n is tight, and hence every subsequence has a weakly converging subsequence, by Prohorov's theorem.

If F_n converges to F^* along a subsequence, then $P_{F_n} \rightsquigarrow P_{F^*}$, by the assumed continuity. If $P_{F_n} \rightsquigarrow P_F$, then we conclude that $P_{F^*} = P_F$ and hence $F = F^*$. It follows that F is the only weak limit point of the sequence F_n. □

The following result, known as Anderson's lemma, compares the probability of a symmetric convex set and its shift for symmetric unimodal distributions. For a proof and more general results, see Ibragimov and Has'minskiĭ (1981), page 155.

Lemma K.12 (Anderson) *Let X be an $\mathbb{R}^d$-valued random variable with a density p such that the set $\{x: p(x) \geq u\}$ is convex and symmetric for every $u \geq 0$. Then, for every $y \in \mathbb{R}^d$,*

(i) $\mathrm{P}(X + y \in A) \leq \mathrm{P}(X \in A)$, *for any convex, symmetric set A.*
(ii) $\mathrm{E}[l(X+y)] \geq \mathrm{E}[l(X)]$, *for every function $l: \mathbb{R}^d \to [0, \infty)$ such that the set $\{x: l(x) < u\}$ is convex and symmetric for all $u > 0$.*

In particular, the inequalities hold for any normally distributed random vector X with mean zero.

Lemma K.13 *The volume of the unit ball with respect to the Euclidean distance in* $\mathbb{R}^J$ *satisfies, as* $J \to \infty$,

$$\text{vol}(x \in \mathbb{R}^J: \|x\| \le 1) = \frac{\pi^{J/2}}{\Gamma(J/2+1)} = \frac{(2\pi e)^{J/2}}{\sqrt{\pi J}} J^{-J/2}(1+o(1)).$$

Proof The exact formula follows by recursive integration; the asymptotic approximation by Stirling's approximation $\Gamma(m+1) = \sqrt{2\pi} m^{m+1/2} e^{-m}(1+o(1))$ as $m \to \infty$. □

The following result is useful in bounding prior probabilities related to a normal prior.

Lemma K.14 *If* $W_i \stackrel{ind}{\sim} \text{Nor}(\xi_i, \sigma_i^2)$, $i = 1, \ldots, N$, *then, for* $\underline{\sigma}_N = \min_{1 \le i \le N} \sigma_i$,

$$\text{P}\Big(\sum_{i=1}^N W_i^2 \le \delta^2\Big) \ge \frac{\underline{\sigma}_N^N}{\prod_{i=1}^N (\sqrt{2}\sigma_i)} e^{-\sum_{i=1}^N \xi_i^2/\sigma_i^2} \text{P}\Big(\sum_{i=1}^N V_i^2 \le \frac{2\delta^2}{\underline{\sigma}_N^2}\Big),$$

where $V_1, \ldots, V_N$ *are independent standard normal random variables. In particular, if* $\sigma_i = i^{-d}$ *for some* $d > 0$, *then the leading factor on the right is further bounded below by* $(e^d\sqrt{2})^{-N}$.

Proof Using the inequality $(a+b)^2 \le 2(a^2+b^2)$, we see that the density $\phi_{\xi,\sigma}$ of the univariate normal distribution satisfies, for every $\xi \in \mathbb{R}$, $\sigma > 0$ and $0 < \tau \le \sigma/\sqrt{2}$,

$$\phi_{\xi,\sigma}(w) \ge \frac{\tau}{\sigma} e^{-\xi^2/\sigma^2} \phi_{0,\tau}(w).$$

Therefore, the probability on the left of the lemma is bounded below by

$$\prod_i (\underline{\sigma}_N/\sqrt{2}/\sigma_i) e^{-\sum_i \xi_i^2/\sigma_i^2} \text{P}\Big(\sum_i \tilde{V}_i^2 \le \delta^2\Big),$$

for $\tilde{V}_1, \ldots, V_N \stackrel{\text{iid}}{\sim} \text{Nor}(0, \underline{\sigma}_N^2/2)$. This is identical to the right side.

The final assertion follows with the help of the inequality $N! \ge e^{-N} N^N$. □

Appendix L

Miscellaneous Results

This appendix collects results for ease of reference.

The following result concludes that given any probability distribution there exists a discrete distribution with at most $N+1$ support points that has the same (generalized) moments up to degree N.

Lemma L.1 (Caratheodory's lemma) *Let K be a compact metric space and let $\psi_1, \ldots, \psi_N$ be continuous functions from K to $\mathbb{R}$. Then for any probability measure P on K, there exists a discrete probability measure Q on K with at most $N+1$ support points such that $\int \psi_j \, dQ = \int \psi_j \, dP$ for all $j = 1, \ldots, N$.*

Proof The set $C = \{(\psi_1(x), \ldots, \psi_N(x)) : x \in K\}$ is compact in $\mathbb{R}^N$, whence its convex hull $\operatorname{conv}(C)$ is also compact. Then for every probability measure P on K,

$$v_0 = \Bigl(\int \psi_1 \, dP, \ldots, \int \psi_N \, dP\Bigr)^\top \in \operatorname{conv}(C) \tag{L.1}$$

(see, e.g., Lemma 3 on page 74 of Ferguson 1967). Since $\operatorname{conv}(C) \subset \mathbb{R}^N$, any element of $\operatorname{conv}(C)$ may be written as a convex combination of at most $N+1$ elements of C (see, e.g. Rudin 1973, page 73). Thus there exist $\lambda_1, \ldots, \lambda_{N+1} \geq 0$, $\sum_{j=1}^{N+1} \lambda_j = 1, x_1 \ldots, x_{N+1} \in K$ such that

$$v_0 = \sum_{j=1}^{N+1} \lambda_j (\psi_1(x_j), \ldots, \psi_N(x_j))^\top = \Bigl(\int \psi_1 \, dQ, \ldots, \int \psi_N \, dQ\Bigr)^\top.$$

Thus $Q = \sum_{j=1}^{N+1} \lambda_j \delta_{x_j}$ satisfies the requirements. □

Lemma L.2 *Let X and Δ be random elements on the same probability space with values in a normed linear space $\mathfrak{L}$ and the interval $[-1, 1]$, respectively, and such that $\mathrm{P}(|\Delta| = 1) < 1$. Then the distribution of any random element Y in $\mathfrak{L}$ that is independent of (X, Δ) and satisfies the distributional equation $Y =_d X + \Delta Y$ is unique.*

Proof Let (Δ_n, X_n) be a sequence of i.i.d. copies of (Δ, X), and for two solutions Y and Y' that are independent of this sequence, set $Y_1 = Y$, $Y_1' = Y'$, and recursively define $Y_{n+1} = X_n + \Delta_n Y_n$, $Y_{n+1}' = X_n + \Delta_n Y_n'$, for $n \in \mathbb{N}$. Then every Y_n is distributed as Y and every Y_n' is distributed as Y', because each of them satisfies the distributional equation. Also

$$\|Y'_{n+1} - Y_{n+1}\| = |\Delta_n| \, \|Y'_n - Y_n\| = \prod_{i=1}^n |\Delta_i| \, \|Y'_1 - Y_1\| \to 0$$

with probability 1, since Δ's are i.i.d. and $P(|\Delta| = 1) < 1$. This forces the distributions of Y and Y' to agree. □

Lemma L.3 *The following pairs of measures are mutually absolutely continuous:*

(i) $\otimes_{i=1}^\infty \mathrm{Nor}(0, \sigma^2)$ *and* $\otimes_{i=1}^\infty \mathrm{Nor}(\theta_i, \sigma^2)$, *for any* $(\theta_1, \theta_2, \ldots) \in \ell_2$ *and any* $\sigma > 0$.
(ii) $\otimes_{i=1}^\infty \mathrm{Nor}(0, \sigma_i^2)$ *and* $\otimes_{i=1}^\infty \mathrm{Nor}(0, \tau_i^2)$, *whenever* $\sum_{i=1}^\infty (\tau_i/\sigma_i - 1)^2 < \infty$.

A particular case of (ii) is $\sigma_i^2 = \sigma^2$ *and* $\tau_i^2 = \sigma^2 + s_i$ *for* $\sum_i |s_i| < \infty$.

Proof By Kakutani's criterion (Corollary B.9) it suffices to verify that the product of the affinities of the terms of the products is positive. Now

$$\rho_{1/2}(\mathrm{Nor}(\mu_1, \sigma_1^2), \mathrm{Nor}(\mu_2, \sigma_2^2)) = \left[1 - \frac{(\sigma_1 - \sigma_2)^2}{\sigma_1^2 + \sigma_2^2}\right]^{1/2} \exp\left\{-\frac{(\mu_1 - \mu_2)^2}{4(\sigma_1^2 + \sigma_2^2)}\right\}. \quad \text{(L.2)}$$

Thus (i) is immediate. For (ii) we note that a product of the form $\prod_{i=1}^\infty (1 - a_i)$, with all $a_i > 0$, converges to a positive limit if and only if $\sum_{i=1}^\infty a_i$ converges. This leads to the condition $\sum_i (\tau_i/\sigma_i - 1)^2/(\tau_i^2/\sigma_i^2 + 1) < \infty$. This can be satisfied only if $\tau_i/\sigma_i \to 1$, and hence reduces to the condition as given. For the final assertion note that $\sqrt{1 + s_i/\sigma} - 1$ is bounded above and below by a multiple of s_i/σ_i if s_i/σ_i lies in a neighborhood of 1. □

Lemma L.4 *For* $P_\theta = \otimes_{i=1}^\infty \mathrm{Nor}(\theta_i, \sigma^2)$, *for* $\theta \in \ell_2$,

(i) $\dfrac{dP_\theta}{dP_{\theta_0}}(Y) = \exp\Big[\sum_{i=1}^\infty (Y_i - \theta_{i0})(\theta_i - \theta_{i0})/\sigma^2 - \tfrac{1}{2}\|\theta - \theta_0\|^2/\sigma^2\Big]$, *where the series converges almost surely and in second mean and is normally distributed.*
(ii) $K(P_{\theta_0}; P_\theta) = \frac{1}{2}\|\theta - \theta_0\|^2/\sigma^2$.
(iii) $V_{2,0}(P_{\theta_0}; P_\theta) = \|\theta - \theta_0\|^2/\sigma^2$.
(iv) $d_H^2(P_{\theta_0}; P_\theta) = 2(1 - \exp\big[-\frac{1}{8}\|\theta - \theta_0\|^2/\sigma^2\big])$.

Proof The density in (i) can be obtained as the limit as $N \to \infty$ of the densities of the finite product measures $\otimes_{i=1}^N \mathrm{Nor}(\theta_i, \sigma^2)$, by the martingale convergence theorem. The convergence of the series also follows by the martingale convergence theorem (note that $\mathrm{E}\sum_{i=1}^N (Y_i - \theta_{i0})^2(\theta_i - \theta_{i0})^2 \le \sigma^2\|\theta - \theta_0\|^2 < \infty$, for every N), or by direct calculation and the Itô-Nisio theorem. The remaining assertions follows by calculations on the Radon-Nikodym density in (i). □

Theorem L.5 (Minimax theorem) *Let* T *be a compact, convex set of a locally convex topological vector space (for example a normed linear space) and* S *be a convex subset of a linear space. Let* $f: T \times S \to \mathbb{R}$ *be a function such that*

(i) $t \mapsto f(t, s)$ *is continuous and concave for all* $s \in S$;
(ii) $s \mapsto f(t, s)$ *is convex for all* $t \in T$.

Then

$$\inf_{s\in S}\sup_{t\in T} f(t,s) = \sup_{t\in T}\inf_{s\in S} f(t,s). \tag{L.3}$$

Furthermore, the supremum on the right side is attained at some $t_0 \in T$.

Proof See Strasser (1985), pages 239–241. □

Lemma L.6 *If* μ *is a signed measure on* $\mathbb{R}^d$ *with total variation measure* $|\mu|$ *such that* $\int e^{\delta\|\lambda\|}\, d|\mu|(\lambda) < \infty$, *for some* $\delta > 0$, *then the set of all polynomials is dense in* $\mathbb{L}_2(\mu)$.

Proof The following extends the proof for $d = 1$ given in Parthasarathy (2005). For $\psi \in \mathbb{L}_2(\mu)$ that is orthogonal to all polynomials, we can define a complex measure ν by $\nu(B) = \int_B \overline{\psi(\lambda)}\, \mu(d\lambda)$, for $\overline{\psi}$ the complex conjugate of ψ. It suffices to show that $\nu = 0$.

Under the given integrability condition the map $z \mapsto \int e^{\langle\lambda,z\rangle}\, d\nu(\lambda)$ is analytic on the strip $\Omega = \{z \in \mathbb{C}^d: \max\{|\operatorname{Re} z_1|, \ldots, |\operatorname{Re} z_d|\} < \delta/\sqrt{d}\}$, as (repeated) differentiation under the integral sign can be justified by the dominated convergence theorem. Again, by the dominated convergence theorem, if $z \in \mathbb{R}^d$,

$$\int e^{\langle\lambda,z\rangle}\, d\nu(\lambda) = \int \sum_{n=0}^{\infty} \frac{\langle\lambda, z\rangle^n}{n!}\, d\nu(\lambda) = \sum_{n=0}^{\infty} \int \frac{\langle\lambda, z\rangle^n}{n!}\, \overline{\psi}(\lambda)\, d\mu(\lambda) = 0,$$

because ψ is orthogonal to all polynomials. Thus $z \mapsto \int e^{\langle\lambda,z\rangle}\, d\nu(\lambda)$ vanishes on $\{z \in \Omega: \operatorname{Im} z = 0\}$, which contains a nontrivial interval in $\mathbb{R}$ for every coordinate. By repeated analytic continuation, this implies that the function vanishes on Ω. In particular, the Fourier transform $t \mapsto \int e^{i\langle\lambda,t\rangle}\, d\nu(\lambda)$ of ν vanishes on $\mathbb{R}^d$, whence ν is the zero-measure. □

Lemma L.7 *Let* P *and* μ *be probability measures on a measurable space* $(\Omega, \mathscr{A})$ *with restrictions* P_m *and* μ_m *to the elements of an increasing sequence of* σ*-fields* $\mathscr{A}_1 \subset \mathscr{A}_2 \subset \cdots \subset \mathscr{A}$ *that generates* $\mathscr{A}$. *If* $P_m \ll \mu_m$ *for every* m, *then* $dP_m/d\mu_m \to dP^a/d\mu$, *almost surely* $[\mu]$, *for* P^a *the part of* P *that is absolutely continuous with respect to* μ. *Furthermore, this convergence is also in* $\mathbb{L}_1(\mu)$ *if and only if* $P \ll \mu$.

Proof If $P \ll \mu$, then $dP_m/d\mu_m$ is the conditional expectation $\mathrm{E}_\mu(dP/d\mu|\,\mathscr{A}_m)$, and hence a uniformly integrable μ-martingale that converges to $dP/d\mu$, both in $\mathbb{L}_1$ and almost surely. General P and μ are both absolutely continuous with respect to $Q = (P+\mu)/2$, and hence satisfy both $p_m := dP_m/dQ_m \to dP/dQ$ and $r_m := d\mu_m/dQ_m \to d\mu/dQ$. It follows that $dP_m/d\mu_m = p_m/r_m \to (dP/dQ)/(d\mu/dQ)$, Q-almost surely. The right side can be identified as $dP^a/d\mu$. If $dP_m/d\mu_m$ converges also in $\mathbb{L}_1(\mu)$, then it is a uniformly integrable martingale with expected value 1, and hence its limit $dP^a/d\mu$ integrates to 1. This implies that $P^a = P$. □

Theorem L.8 *Let* $I_n(\lambda) = (2\pi n)^{-1}|\sum_{t=1}^n X_t e^{-it\lambda}|^2$ *be the the periodogram of a stationary Gaussian time series* $(X_t: t = 0, 1, \ldots)$ *with mean zero, autocovariances* $\gamma_r = \mathrm{E}(X_t X_{t+r})$, *and spectral density* $f(\lambda) = (2\pi)^{-1}\sum_{r=0}^{\infty} \gamma_r e^{-ir\lambda}$. *If* $\sum_{r=0}^{\infty} r^\alpha |\gamma_r| < \infty$ *for some* $\alpha > 0$ *and* $f(\lambda) > 0$ *for all* $\lambda \in [-\pi, \pi]$, *then the joint distributions of*

$(I_n(\lambda_j): j = 1, \ldots, \lfloor n/2 \rfloor)$ and of a vector of independent exponential random variables with means $f(\lambda_j)$, for $j = 1, \ldots, \lfloor n/2 \rfloor$, at the frequencies $\lambda_j = 2\pi j/n$, $j = 1, \ldots, \lfloor n/2 \rfloor$, are mutually contiguous.[1,2]

For a proof, see Choudhuri et al. (2004b).

Problems

L.1 Suppose that X and Y are identically distributed random variables defined on the same probability space with $Y \geq X$ almost surely, then show that $X = Y$ almost surely. [For all t we have $\mathrm{P}(Y > t \geq X) = \mathrm{P}(Y > t) - \mathrm{P}(X > t)$.]

L.2 (Gurland 1948) Inversion formula: Show that the cumulative distribution function F of a real-valued random variable can be expressed in its characteristic function ϕ as

$$F(x) + F(x-) = 1 - \frac{2}{\pi} \lim_{\epsilon \downarrow 0, T \uparrow \infty} \int_\epsilon^T \frac{\mathrm{Im}(e^{-ixt}\phi(t)}{t}) \, dt.$$

[Hint: Because we can shift to the variable $-x$, it suffices to prove the identity for $x = 0$. Because $h(T) := \int_0^T (\sin t)/t \, dt \to \pi/2$, as $T \to \infty$, we have $\int_\epsilon^T \sin(tx)/t \, dt \to \pi \operatorname{sign}(x)/2$, as $\epsilon \downarrow 0$ and $T \uparrow \infty$, for every $x \in \mathbb{R}$ (where $\operatorname{sign}(x)$ is $-1, 0, 1$ if $x < 0, = 0, > 0$ respectively). Because, moreover, this expression is equal to $h(Tx) - h(\epsilon x)$ and hence is uniformly bounded (as h is), the dominated convergence theorem gives that $\int \int_\epsilon^T \sin(tx)/t \, dt \, dF(x) \to \pi \int \operatorname{sign}(x) \, dF(x)/2 = \pi(1 - F(0) - F(0-))/2$. Finally, by Fubini's theorem we can exchange the order of integration in the double integral on the left, as the function $(t, x) \mapsto \sin(tx)/t$ is uniformly bounded on $t > \epsilon$. Finally $\int \sin(tx) \, dF(x) = \mathrm{Im}\, \phi(t)$.]

L.3 (Skorokhod 1965, Lo 1983) Let $\{X(t), t \in [a, b]\}$ be a separable, continuous time Markov process with separator S.[3] Let $\mathscr{F}_t = \sigma\langle X(s), s \leq t\rangle$. If for $\epsilon > 0$, $0 \leq \alpha < 1$

$$\sup_{t \in [a,b] \cap S} \mathrm{P}\{|X(b) - X(t)| > \epsilon | \mathscr{F}_t\} \leq \alpha \text{ a.s.},$$

then show that

$$\mathrm{P}\{\sup_{t \in [a,b]} \mathrm{P}\{|X(t) - X(a)| > 2\epsilon\} \leq \frac{1}{1-\alpha} \mathrm{P}\{|X(b) - X(a)| > \epsilon\}.$$

L.4 (Lo 1983) Let $\{X_n(t): t \in [0, 1]\}$ be a sequence of continuous time Markov processes taking values in the Skorohod space $\mathfrak{D}[0, 1]$ of cadlag functions. Assume that

(a) $(X_n(t_1), \ldots, X_n(t_k)) \rightsquigarrow (X(t_1), \ldots, X(t_k))$, for all $t_1, \ldots, t_k \in [0, 1]$, $k \in \mathbb{N}$, where X is a mean-zero Gaussian process taking values in $\mathfrak{D}[0, 1]$;

[1] The likelihood of the exponential variables evaluated with the observables $I_n(\lambda_j)$ substituted for these variables is called the Whittle likelihood for f, after Whittle (1957).

[2] Two sequences of probability measures P_n and Q_n on measurable spaces $(\Omega_n, \mathscr{A}_n)$, for $n \in \mathbb{N}$, are said to be mutually contiguous if $P_n(A_n) \to 0$ if and only if $Q_n(A_n) \to 0$, for any $A_n \in \mathscr{A}_n$; see Le Cam (1986) or van der Vaart (1998) for details.

[3] This means that if Y is another separable Markov process such that $X(s) = Y(s)$ a.s. for all $s \in S$, then $\mathrm{P}\{X(t) = Y(t) \text{ for all } t \in [a, b]\} = 1$.

(b) for all $\epsilon > 0$, there exists $\alpha \in (0, 1)$, $n_0 \in \mathbb{N}$, $0 < \delta < 1$, such that for all $n \geq n_0$, $\delta \leq \delta_0$,

$$\sup_{k \leq \lfloor \delta^{-1} \rfloor} \sup_{t \in (k\delta, (k+1)\delta] \cap [0,1]} \mathrm{P}\{|X_n((k+1)\delta) - X_n(t)| > \epsilon | \mathscr{F}_{t,n}\} \leq \alpha \text{ a.s.},$$

where $\mathscr{F}_{t,n} = \sigma\langle X_n(s), s \leq t \rangle$;

(c) $\mathrm{E}|X(t) - X(s)|^2 \leq G(t) - G(s)$, $s \leq t$, for some continuous, increasing function $G: [0, 1] \to (0, \infty)$ with $G(0) = 0$.

Show that $X_n \rightsquigarrow X$.

Appendix M

Elements of Markov Chain Monte Carlo

Because analytic formulas are typically intractable, a posterior distribution is often "calculated" through stochastic simulation. Posterior characteristics are approximated by the corresponding sample characteristics from a sample from the posterior distribution. In a Markov Chain Monte Carlo (MCMC) algorithm the sample is a Markov chain whose stationary distribution is the posterior distribution. Typically it is not possible to start the chain from the stationary distribution, but the chain will settle down to approximate stationarity after a "burn-in-period." The initial sample values are then discarded and averages over the remaining values are used as estimates of posterior characteristics, or a histogram of these values as an approximation to the posterior density. For an efficient algorithm the burn-in period is short, and the consecutive sample values are not too dependent and "explore" the range of the posterior distribution. A Markov chain with properties is said to be *mixing*, or to mix well.

In general it is difficult to assess when a chain has "converged" to its stationary state. Typically one checks whether the sample paths of consecutive draws, known as trace plots, mix well over the target range. It is also common practice to start the chain from different initial states, and check if and when the resulting chains have nearly reached similar behavior.

From the many MCMC techniques we briefly discuss Gibbs sampling, the Metropolis-Hastings algorithm, slice sampling and approximate Bayesian computation, which are based on general principles and widely used. In a Bayesian analysis these techniques are often combined with *data augmentation*, which is to introduce latent (unobservable) variables with the purpose to create a "complete likelihood" from which simulation is easy. After simulating a joint sample of the variables of interest and the latent variables, the latter are next discarded.

Gibbs Sampling

The *Gibbs sampler* simulates a vector $(X_1, \ldots, X_k)$ from the conditional distributions of its coordinates X_i given the other variables $(X_j\colon j \neq i)$. The latter lower-dimensional distributions are often more tractable. The algorithm is made possible by the fact that a joint distribution is indeed determined by the set of its conditional distributions, by the *Hammersley-Clifford theorem.*

For each i let $x_i \mapsto p_i(x_i \mid x_j, j \neq i)$ be the conditional density of X_i, given $(X_j\colon j \neq i)$. It is assumed that it is easy to sample from these densities.

Then Gibbs sampling for generating samples from the joint density of $(X_1, \ldots, X_k)$ consists of consecutively updating the coordinates, as follows:

(a) Start with arbitrary values $x_1, \ldots, x_k$.
(b) Given a current value $(x_1, \ldots, x_k)$, draw a sample

$$\begin{aligned} x_1' &\sim p_1(\cdot|\, x_2, \ldots, x_k),\\ x_2' &\sim p_2(\cdot|\, x_1', x_3, \ldots, x_k),\\ &\vdots\\ x_k' &\sim p_k(\cdot|\, x_1', \ldots, x_{k-1}'). \end{aligned}$$

(c) Replace $(x_1, \ldots, x_k)$ by $(x_1', \ldots, x_k')$ and repeat step (b).
(d) Continue steps (b)-(c) until convergence and collect sample values after burn-in.

The coordinates x_i in this scheme may be univariate variables or may be vectors themselves. In the second case the algorithm is known as the *block Gibbs sampler*.

It may be shown that the joint distribution of $(X_1, \ldots, X_k)$ is indeed the stationary distribution of the chain, and under mild conditions the chain converges to stationarity at a geometric rate.

Metropolis-Hastings Algorithm

The *Metropolis-Hastings algorithm* is a universally applicable method to simulate a sample from a density whose values can be computed up to a norming constant. It is suitable for Bayesian analysis, as a posterior density is proportional to the product of the likelihood and the prior density. The algorithm avoids computing the norming constant, which is the integral of the product and may be difficult to calculate, especially for high-dimensional parameters.

The method works with a "proposal density" $y \mapsto q(y|\, x)$, which is the density of a Markov kernel from the sample space into itself. The efficiency of the algorithm strongly depends on a proper choice of this density, although for mere "convergence" it can be almost arbitrary. It is assumed that it is easy to simulate a value of y from this density, given x, and that the values $q(y|\, x)$ can be computed next to the values $p(x)$, up to norming constants.

Let $x \mapsto p(x)$ be the density of the random variable (or vector) X that must be simulated. The Metropolis-Hastings algorithm is described as follows:

(a) Start with an arbitrary value x.
(b) Given x, simulate $y \sim q(\cdot|\, x)$ and compute $r = q(x|\, y)p(y)/(q(y|\, x)p(x))$.
(c) "Accept" y with probability $\min(r, 1)$ and "reject" it with probability $1 - \min(r, 1)$; in the second case set y equal to x.
(d) Replace x by y and repeat step (b).
(c) Continue steps (b)–(c) until convergence and collect sample values after burn-in.

It can be checked that p is a stationary density for the resulting Markov chain and under mild conditions the Metropolis-Hastings algorithm converges to stationarity. Two popular special cases are: the "random walk," which has $q(y|\, x) = g(y-x)$ for a symmetric density g, and the "independent proposal," which has $q(y|\, x) = g(y)$. The quotient r simplifies to $p(y)/p(x)$ and $g(x)p(y)/(g(y)p(x))$ in the two cases. A proposal density with thick tails (such as a t-density) often allows the chain to move better over the whole state space, and thus makes the algorithm more efficient.

The Metropolis-Hastings algorithm may be used within a Gibbs loop to sample from a conditional density. The resulting procedure is called "Metropolis step within Gibbs sampler."

Slice Sampling

The *slice sampler* performs Gibbs sampling after introducing auxiliary uniform variables.

In its simplest form it assumes that the values $p(x)$ of the target density can be computed up to a norming constant and that it is possible to determine the sections $\{x: p(x) > u\}$ of the area under the graph of this density, for $u > 0$.

For f a function that is proportional to p the algorithm performs the steps:

(a) Start with arbitrary values x, u.
(b) Given a current value (x, u), draw

$$\begin{aligned} x' &\sim \text{Unif}(\{t: f(t) > u\}), \\ u' &\sim \text{Unif}((0, f(x'))). \end{aligned}$$

(c) Replace (x, u) by (x', u') and repeat step (b).
(d) Continue steps (b)-(c) until convergence and collect sample values x' after burn-in.

Here $\text{Unif}(\mathfrak{X})$ is the uniform distribution on $\mathfrak{X}$.

The algorithm is a Gibbs sampler applied to the joint density proportional to $(x, u) \mapsto \mathbb{1}\{u < f(x)\}$, which gives the uniform distribution on the area $\{(x, u): 0 \le u < f(x)\}$ under the graph of f. For $f \propto p$, the first marginal density of this joint density is indeed p, and the conditional densities are uniform on the horizontal and vertical sections under the graph.

The algorithm can be extended to a density $p(x)$ that is proportional to the product $x \mapsto \prod_{i=1}^k f_i(x)$ of several functions $f_1, \ldots, f_k$. Then introduce k slice variables $U_1, \ldots, U_k$ and apply the Gibbs sampler to the density proportional to $(x, u_1, \ldots, u_k) \mapsto \mathbb{1}\{u_1 < f_1(x), \ldots, u_k < f_k(x)\}$.

(a) Start with arbitrary values $x, u_1, \ldots, u_k$.
(b) Given a current value $(x, u_1, \ldots, u_k)$, draw

$$\begin{aligned} x' &\sim \text{Unif}(\{t: f_1(t) > u_1, \ldots, f_k(t) > u_k\}), \\ u_1' &\sim \text{Unif}((0, f_1(x'))), \\ &\vdots \\ u_k' &\sim \text{Unif}((0, f_k(x'))). \end{aligned}$$

(c) Replace $(x, u_1, \ldots, u_k)$ by $(x', u_1', \ldots, u_k')$ and repeat step (b).
(d) Continue steps (b)–(c) until convergence and collect sample values x' after burn-in.

Approximate Bayesian Computation

Approximate Bayesian computation or *ABC* is an algorithm to simulate from the *conditional* distribution of a variable X given $Z = z$ in a situation that good samplers for the marginal of X and of the distribution of Z given X are available. It is used when it is not possible to calculate the values of the relevant conditional density $x \mapsto p(x|\,z)$, so that a Metropolis-Hastings algorithm is impossible.

The algorithm differs from the preceding algorithms in that it addresses the simulation from a distribution conditioned on an exact value z of another variable. In the Bayesian setup the latter variable would be the observed data. Whereas the other algorithms will be set up with p being the conditional density, and the data are fixed and suppressed in all further steps, the ABC algorithm also generates new data.

The algorithm is easy and exact if Z possesses a discrete distribution. Let p be the marginal density of x and $z \mapsto p_2(\cdot|\, x)$ the conditional density of Z given $X = x$. The value z is given.

(a) Simulate $x \sim p$.
(b) Simulate $z' \sim p_2(\cdot|\, x)$.
(c) If $z' = z$ output x; otherwise return to step (a).
(d) Collect the value x. Continue steps (a)-(c) as often as desired to collect further values.

Thus we generate from the joint distribution of (X, Z) until the second coordinate is equal to the given z. It can be seen that the value x obtained at that time is indeed sampled from its conditional distribution.

The algorithm will finish if the support of Z is finite, but may never end otherwise. To remedy this approximate Bayesian calculation replaces step (c) by:

(c) If $d(z', z) < \epsilon$ output x; otherwise return to step (a).

Here d is some distance and ϵ some tolerance. The accuracy and efficiency of the algorithm strongly depend on the choice of these quantities.

References

Albert, J., and S. Chib (1993). Bayesian analysis of binary and polychotomous response data. *J. Amer. Statist. Assoc. 88*(422), 669–679.

Aldous, D. (1985). Exchangeability and related topics. In *École d'été de probabilités de Saint-Flour, XIII—1983*, Volume 1117 of *Lecture Notes in Math.*, pp. 1–198. Berlin: Springer.

Amewou-Atisso, M., S. Ghosal, J. Ghosh, and R. Ramamoorthi (2003). Posterior consistency for semi-parametric regression problems. *Bernoulli 9*(2), 291–312.

Andersen, P., O. Borgan, R. Gill, and N. Keiding (1993). *Statistical Models Based on Counting Processes*. Springer Series in Statistics. New York: Springer-Verlag.

Antoniak, C. (1974). Mixtures of Dirichlet processes with applications to Bayesian nonparametric problems. *Ann. Statist. 2*, 1152–1174.

Arratia, R., L. Goldstein, and L. Gordon (1990). Poisson approximation and the Chen–Stein method. *Statist. Sci. 5*(4), 403–434. With comments and a rejoinder by the authors.

Barrios, E., A. Lijoi, L. Nieto-Barajas, and I. Pruenster (2013). Modeling with normalized random measure mixture models. *Stat. Sci. 277*, 313–334.

Barron, A. (1986). Discussion of "on the consistency of Bayes estimates" by P. Diaconis and D. Freedman. *Ann. Statist. 14*(1), 26–30.

Barron, A. (1999). Information-theoretic characterization of Bayes performance and the choice of priors in parametric and nonparametric problems. In *Bayesian Statistics, 6 (Alcoceber, 1998)*, pp. 27–52. New York: Oxford Univ. Press.

Barron, A., M. Schervish, and L. Wasserman (1999). The consistency of posterior distributions in nonparametric problems. *Ann. Statist. 27*(2), 536–561.

Bauer, H. (2001). *Measure and Integration Theory*, Volume 26 of *de Gruyter Studies in Mathematics*. Berlin: Walter de Gruyter & Co. Translated from the German by Robert B. Burckel.

Beal, M., and Z. Ghahramani (2003). The variational Bayesian EM algorithm for incomplete data: with application to scoring graphical model structures. In *Bayesian Statistics, 7 (Tenerife, 2002)*, pp. 453–463. New York: Oxford Univ. Press.

Begun, J., W. Hall, W.-M. Huang, and J. Wellner (1983). Information and asymptotic efficiency in parametric–nonparametric models. *Ann. Statist. 11*(2), 432–452.

Belitser, E., and S. Ghosal (2003). Adaptive Bayesian inference on the mean of an infinite-dimensional normal distribution. *Ann. Statist. 31*(2), 536–559. Dedicated to the memory of Herbert E. Robbins.

Bennett, B. (1962). On multivariate sign tests. *J. Roy. Statist. Soc. Ser. B 24*, 159–161.

Berger, J., and A. Guglielmi (2001). Bayesian and conditional frequentist testing of a parametric model versus nonparametric alternatives. *J. Amer. Statist. Assoc. 96*(453), 174–184.

Bernardo, J., and A. Smith (1994). *Bayesian Theory*. Wiley Series in Probability and Mathematical Statistics: Probability and Mathematical Statistics. Chichester: John Wiley & Sons Ltd.

Bertoin, J. (1996). *Lévy Processes*, Volume 121 of *Cambridge Tracts in Mathematics*. Cambridge: Cambridge University Press.

Bhattacharya, A., and D. Dunson (2010). Nonparametric Bayesian density estimation on manifolds with applications to planar shapes. *Biometrika 97*(4), 851–865.

Bickel, P., C. Klaassen, Y. Ritov, and J. Wellner (1998). *Efficient and Adaptive Estimation for Semiparametric Models*. New York: Springer-Verlag. Reprint of the 1993 original.

Bickel, P., and B. Kleijn (2012). The semiparametric Bernstein–von Mises theorem. *Ann. Statist. 40*(1), 206–237.

Bickel, P., and M. Rosenblatt (1973). On some global measures of the deviations of density function estimates. *Ann. Statist. 1*, 1071–1095.

Billingsley, P. (1968). *Convergence of Probability Measures*. New York: John Wiley & Sons Inc.

Billingsley, P. (1979). *Probability and Measure*. New York–Chichester–Brisbane: John Wiley & Sons. Wiley Series in Probability and Mathematical Statistics.

Birgé, L. (1979). Sur les tests de deux ensembles convexes compacts de probabilités. *C. R. Acad. Sci. Paris Sér. A-B 289*(3), A237–A240.

Birgé, L. (1983a). Approximation dans les espaces métriques et théorie de l'estimation. *Z. Wahrsch. Verw. Gebiete 65*(2), 181–237.

Birgé, L. (1983b). Robust testing for independent nonidentically distributed variables and Markov chains. In *Specifying Statistical Models (Louvain-la-Neuve, 1981)*, Volume 16 of *Lecture Notes in Statist.*, pp. 134–162. New York: Springer.

Blackwell, D. (1973). Discreteness of Ferguson selections. *Ann. Statist. 1*, 356–358.

Blackwell, D., and L. Dubins (1962). Merging of opinions with increasing information. *Ann. Math. Statist. 33*, 882–886.

Blackwell, D., and J. MacQueen (1973). Ferguson distributions via Pólya urn schemes. *Ann. Statist. 1*, 353–355.

Blei, D., and M. Jordan (2006). Variational inference for Dirichlet process mixtures. *Bayesian Anal. 1*(1), 121–143 (electronic).

Borell, C. (1975). The Brunn–Minkowski inequality in Gauss space. *Invent. Math. 30*(2), 207–216.

Breiman, L., L. Le Cam, and L. Schwartz (1964). Consistent estimates and zero-one sets. *Ann. Math. Statist. 35*, 157–161.

Brown, L., and M. Low (1996). Asymptotic equivalence of nonparametric regression and white noise. *Ann. Statist. 24*(6), 2384–2398.

Brunner, L. (1992). Bayesian nonparametric methods for data from a unimodal density. *Statist. Probab. Lett. 14*(3), 195–199.

Brunner, L., and A. Lo (1989). Bayes methods for a symmetric unimodal density and its mode. *Ann. Statist. 17*(4), 1550–1566.

Carter, C., and R. Kohn (1997). Semiparametric Bayesian inference for time series with mixed spectra. *J. Roy. Statist. Soc. Ser. B 59*(1), 255–268.

Castillo, I. (2008). Lower bounds for posterior rates with Gaussian process priors. *Electron. J. Stat. 2*, 1281–1299.

Castillo, I. (2012a). Semiparametric Bernstein–von Mises theorem and bias, illustrated with Gaussian process priors. *Sankhya A 74*(2), 194–221.

Castillo, I. (2012b). A semiparametric Bernstein–von Mises theorem for Gaussian process priors. *Probab. Theory Related Fields 152*(1–2), 53–99.

Castillo, I. (2014). On Bayesian supremum norm contraction rates. *Ann. Statist. 42*(5), 2058–2091.

Castillo, I., and R. Nickl (2013). Nonparametric Bernstein–von Mises theorems in Gaussian white noise. *Ann. Statist. 41*(4), 1999–2028.

Castillo, I., and R. Nickl (2014). On the Bernstein–von Mises phenomenon for nonparametric Bayes procedures. *Ann. Statist. 42*(5), 1941–1969.

Castillo, I., and J. Rousseau (2015). A general Bernstein–von Mises theorem in semiparametric models. *Ann. Statist. 43*(6), 2353–2383.

Catoni, O. (2004). *Statistical Learning Theory and Stochastic Optimization*, Volume 1851 of *Lecture Notes in Mathematics*. Berlin: Springer-Verlag. Lecture notes from the 31st Summer School on Probability Theory held in Saint-Flour, July 8–25, 2001.

Cerquetti, A. (2007). A note on Bayesian nonparametric priors derived from exponentially tilted Poisson–Kingman models. *Statist. Probab. Lett. 77*(18), 1705–1711.

Cerquetti, A. (2008). On a Gibbs characterization of normalized generalized gamma processes. *Statist. Probab. Lett. 78*(18), 3123–3128.

Chen, L. (1975). Poisson approximation for dependent trials. *Ann. Probab. 3*(3), 534–545.

Chen, L., L. Goldstein, and Q.-M. Shao (2011). *Normal Approximation by Stein's Method*. Probability and Its Applications. New York, Heidelberg: Springer.
Choi, T., and R. Ramamoorthi (2008). Remarks on consistency of posterior distributions. In *Pushing the Limits of Contemporary Statistics: Contributions in Honor of Jayanta K. Ghosh*, Volume 3 of *Inst. Math. Stat. Collect.*, pp. 170–186. Beachwood, OH: Inst. Math. Statist.
Choi, T., and M. Schervish (2007). On posterior consistency in nonparametric regression problems. *J. Multivariate Anal. 98*(10), 1969–1987.
Choudhuri, N. (1998). Bayesian bootstrap credible sets for multidimensional mean functional. *Ann. Statist. 26*(6), 2104–2127.
Choudhuri, N., S. Ghosal, and A. Roy (2004a). Bayesian estimation of the spectral density of a time series. *J. Amer. Statist. Assoc. 99*(468), 1050–1059.
Choudhuri, N., S. Ghosal, and A. Roy (2004b). Contiguity of the Whittle measure for a Gaussian time series. *Biometrika 91*(1), 211–218.
Choudhuri, N., S. Ghosal, and A. Roy (2007). Nonparametric binary regression using a Gaussian process prior. *Stat. Methodol. 4*(2), 227–243.
Chung, K.-L. (2001). *A Course in Probability Theory* (Third ed.). San Diego, CA: Academic Press Inc.
Chung, Y., and D. Dunson (2011). The local Dirichlet process. *Ann. Inst. Statist. Math. 63*(1), 59–80.
Cifarelli, D., and E. Regazzini (1978). Problemi statistici non parametrici in condizioni di scambiabilita parziale e impiego di medie associative. *Quad. Istit. Mat. Finanz. Univ. Torino 3*, 1–13.
Cifarelli, D., and E. Regazzini (1990). Distribution functions of means of a Dirichlet process. *Ann. Statist. 18*(1), 429–442.
Cohen, A., I. Daubechies, and P. Vial (1993). Wavelets on the interval and fast wavelet transforms. *Applied and Computational Harmonic Analysis 1*(1), 54–81.
Connor, R., and J. Mosimann (1969). Concepts of independence for proportions with a generalization of the Dirichlet distribution. *J. Amer. Statist. Assoc. 64*, 194–206.
Coram, M., and S. Lalley (2006). Consistency of Bayes estimators of a binary regression function. *Ann. Statist. 34*(3), 1233–1269.
Cox, D. D. (1993). An analysis of Bayesian inference for nonparametric regression. *Ann. Statist. 21*(2), 903–923.
Cox, D. R. (1972). Regression models and life-tables. *J. Roy. Statist. Soc. Ser. B 34*, 187–220. With discussion by F. Downton, R. Peto, D. Bartholomew, D. Lindley, P. Glassborow, D. Barton, S. Howard, B. Benjamin, J. Gart, L. Meshalkin, A. Kagan, M. Zelen, R. Barlow, J. Kalbfleisch, R. Prentice and N. Breslow, and a reply by D. R. Cox.
Cseke, B., and T. Heskes (2011). Approximate marginals in latent Gaussian models. *J. Mach. Learn. Res. 12*, 417–454.
Csörgő, M., and P. Révész (1975). A new method to prove Strassen type laws of invariance principle. I, II. *Z. Wahrscheinlichkeitstheorie und Verw. Gebiete 31*, 255–259.
Csörgő, M., and P. Révész (1981). *Strong Approximations in Probability and Statistics*. Probability and Mathematical Statistics. New York: Academic Press Inc.
Dalal, S. (1979). Dirichlet invariant processes and applications to nonparametric estimation of symmetric distribution functions. *Stochastic Process. Appl. 9*(1), 99–107.
Daley, D. J., and D. Vere-Jones (2003). *An Introduction to the Theory of Point Processes*. Vol. I (Second ed.). Probability and Its Applications (New York). Springer-Verlag, New York. Elementary theory and methods.
Daley, D. J., and D. Vere-Jones (2008). *An Introduction to the Theory of Point Processes*. Vol. II (Second ed.). Probability and Its Applications (New York). New York: Springer. General theory and structure.
Damien, P., P. Laud, and A. Smith (1995). Approximate random variate generation from infinitely divisible distributions with applications to Bayesian inference. *J. Roy. Statist. Soc., Ser. B. 57*(3), 547–563.
Damien, P., P. Laud, and A. Smith (1996). Implementation of Bayesian non-parametric inference based on beta processes. *Scand. J. Statist. 23*(1), 27–36.
Dass, S., and J. Lee (2004). A note on the consistency of Bayes factors for testing point null versus nonparametric alternatives. *J. Statist. Plann. Inference 119*(1), 143–152.

Daubechies, I. (1992). *Ten Lectures on Wavelets*, Volume 61 of *CBMS-NSF Regional Conference Series in Applied Mathematics*. Philadelphia, PA: Society for Industrial and Applied Mathematics (SIAM).

de Acosta, A. (1983). Small deviations in the functional central limit theorem with applications to functional laws of the iterated logarithm. *Ann. Probab. 11*(1), 78–101.

De Blasi, P., S. Favaro, A. Lijoi, R. Mena, I. Prünster, and M. Ruggiero (2015). Are Gibbs-type priors the most natural generalization of the Dirichlet process? *IEEE Trans. Pattern Anal. Machine Intell. 37*, 212–229.

De Blasi, P., A. Lijoi, and I. Prünster (2013). An asymptotic analysis of a class of discrete nonparametric priors. *Statistica Sinica 23*(3), 1299–1321.

De Blasi, P., G. Peccati, and I. Prünster (2009). Asymptotics for posterior hazards. *Ann. Statist. 37*(4), 1906–1945.

de Boor, C. (1978). *A Practical Guide to Splines*, Volume 27 of *Applied Mathematical Sciences*. New York: Springer-Verlag.

de Boor, C., and J. Daniel (1974). Splines with nonnegative B-spline coefficients. *Mathematics of Computation 28*(126), 565–568.

de Finetti, B. (1937). La prévision: ses lois logiques, ses sources subjectives. *Ann. Inst. H. Poincaré 7*(1), 1–68.

de Jonge, R., and H. van Zanten (2010). Adaptive nonparametric Bayesian inference using location-scale mixture priors. *Ann. Statist. 38*(6), 3300–3320.

DeVore, R. A., and G. G. Lorentz (1993). *Constructive Approximation*, Volume 303 of *Grundlehren der Mathematischen Wissenschaften [Fundamental Principles of Mathematical Sciences]*. Berlin: Springer-Verlag.

Dey, D., P. Müller, and D. Sinha (1998). *Practical Nonparametric and Semiparametric Bayesian Statistics*. Springer Heidelberg.

Dey, J., R. Erickson, and R. Ramamoorthi (2003). Some aspects of neutral to right priors. *Int. Stat. Rev. 71*(2), 383–401.

Diaconis, P., and D. Freedman (1986a). On inconsistent Bayes estimates of location. *Ann. Statist. 14*(1), 68–87.

Diaconis, P., and D. Freedman (1986b). On the consistency of Bayes estimates. *Ann. Statist. 14*(1), 1–67. With a discussion and a rejoinder by the authors.

Diaconis, P., and D. Freedman (1993). Nonparametric binary regression: a Bayesian approach. *Ann. Statist. 21*(4), 2108–2137.

Diaconis, P., and J. Kemperman (1996). Some new tools for Dirichlet priors. In *Bayesian Statistics, 5 (Alicante, 1994)*, Oxford Sci. Publ., pp. 97–106. New York: Oxford Univ. Press.

Doksum, K. (1974). Tail-free and neutral random probabilities and their posterior distributions. *Ann. Probab. 2*, 183–201.

Doksum, K., and A. Lo (1990). Consistent and robust Bayes procedures for location based on partial information. *Ann. Statist. 18*(1), 443–453.

Doob, J. (1949). Application of the theory of martingales. In *Le Calcul des probabilités et ses applications*, Colloques Internationaux du Centre National de la Recherche Scientifique, no. 13, pp. 23–27. Paris: Centre National de la Recherche Scientifique.

Doss, H. (1985a). Bayesian nonparametric estimation of the median. I. Computation of the estimates. *Ann. Statist. 13*(4), 1432–1444.

Doss, H. (1985b). Bayesian nonparametric estimation of the median. II. Asymptotic properties of the estimates. *Ann. Statist. 13*(4), 1445–1464.

Doss, H., and T. Sellke (1982). The tails of probabilities chosen from a Dirichlet prior. *Ann. Statist. 10*(4), 1302–1305.

Drăghici, L., and R. Ramamoorthi (2000). A note on the absolute continuity and singularity of Pólya tree priors and posteriors. *Scand. J. Statist. 27*(2), 299–303.

Drăghici, L., and R. Ramamoorthi (2003). Consistency of Dykstra–Laud priors. *Sankhyā 65*(2), 464–481.

Dubins, L., and D. Freedman (1963). Random distribution functions. *Bull. Amer. Math. Soc. 69*, 548–551.

Dudley, R. (1967). The sizes of compact subsets of Hilbert space and continuity of Gaussian processes. *J. Functional Analysis 1*, 290–330.

Dudley, R. (2002). *Real Analysis and Probability*, Volume 74 of *Cambridge Studies in Advanced Mathematics*. Cambridge: Cambridge University Press. Revised reprint of the 1989 original.

Dudley, R. M. (1984). A course on empirical processes. In *École d'été de probabilités de Saint-Flour, XII—1982*, Volume 1097 of *Lecture Notes in Math.*, pp. 1–142. Berlin: Springer.

Dudley, R. M. (2014). *Uniform Central Limit Theorems* (Second ed.), Volume 142 of *Cambridge Studies in Advanced Mathematics*. New York: Cambridge University Press.

Dunson, D., and J.-H. Park (2008). Kernel stick-breaking processes. *Biometrika 95*(2), 307–323.

Dykstra, R., and P. Laud (1981). A Bayesian nonparametric approach to reliability. *Ann. Statist. 9*(2), 356–367.

Edmunds, D., and H. Triebel (1996). *Function Spaces, Entropy Numbers, Differential Operators*, Volume 120 of *Cambridge Tracts in Mathematics*. Cambridge: Cambridge University Press.

Efroĭmovich, S., and M. Pinsker (1984). A self-training algorithm for nonparametric filtering. *Avtomat. i Telemekh.* (11), 58–65.

Eichelsbacher, P., and A. Ganesh (2002). Moderate deviations for Bayes posteriors. *Scand. J. Statist. 29*(1), 153–167.

Escobar, M. (1994). Estimating normal means with a Dirichlet process prior. *J. Amer. Statist. Assoc. 89*(425), 268–277.

Escobar, M., and M. West (1995). Bayesian density estimation and inference using mixtures. *J. Amer. Statist. Assoc. 90*(430), 577–588.

Fabius, J. (1964). Asymptotic behavior of Bayes' estimates. *Ann. Math. Statist. 35*, 846–856.

Favaro, S., A. Lijoi, R. Mena, and I. Prünster (2009). Bayesian non-parametric inference for species variety with a two-parameter Poisson–Dirichlet process prior. *J. Roy. Stat. Soc. Ser. B 71*(5), 993–1008.

Favaro, S., A. Lijoi, and I. Prünster (2012a). Asymptotics for a Bayesian nonparametric estimator of species variety. *Bernoulli 18*(4), 1267–1283.

Favaro, S., A. Lijoi, and I. Prünster (2012b). On the stick-breaking representation of normalized inverse Gaussian priors. *Biometrika 99*(3), 663–674.

Favaro, S., and Y. Teh (2013). MCMC for normalized random measure mixture models. *Stat. Sci 28*, 335–350.

Feller, W. (1968). *An Introduction to Probability Theory and Its Applications.* Vol. I. (Third ed.). New York: John Wiley & Sons Inc.

Feller, W. (1971). *An Introduction to Probability Theory and Its Applications.* Vol. II. (Second ed.). New York: John Wiley & Sons Inc.

Feng, S. (2010). *The Poisson–Dirichlet Distribution and Related Topics: Models and Asymptotic Behaviors.* Probability and its Applications. New York, Heidelberg: Springer.

Ferguson, T. (1967). *Mathematical Statistics: A Decision Theoretic Approach.* Probability and Mathematical Statistics, Vol. 1. New York: Academic Press.

Ferguson, T. (1973). A Bayesian analysis of some nonparametric problems. *Ann. Statist. 1*, 209–230.

Ferguson, T. (1974). Prior distributions on spaces of probability measures. *Ann. Statist. 2*, 615–629.

Ferguson, T. (1983). Bayesian density estimation by mixtures of normal distributions. In *Recent Advances in Statistics*, pp. 287–302. New York: Academic Press.

Ferguson, T., and M. Klass (1972). A representation of independent increment processes without Gaussian components. *Ann. Math. Statist. 43*, 1634–1643.

Ferguson, T., and E. Phadia (1979). Bayesian nonparametric estimation based on censored data. *Ann. Statist. 7*(1), 163–186.

Freedman, D. (1963). On the asymptotic behavior of Bayes' estimates in the discrete case. *Ann. Math. Statist. 34*, 1386–1403.

Freedman, D. (1965). On the asymptotic behavior of Bayes estimates in the discrete case. II. *Ann. Math. Statist. 36*, 454–456.

Freedman, D. (1999). On the Bernstein–von Mises theorem with infinite-dimensional parameters. *Ann. Statist. 27*(4), 1119–1140.

Freedman, D., and P. Diaconis (1982). On inconsistent M-estimators. *Ann. Statist. 10*(2), 454–461.

Fristedt, B., and W. Pruitt (1971). Lower functions for increasing random walks and subordinators. *Z. Wahrsch. Verw. Gebiete 18*, 167–182.

Gaenssler, P., P. Molnár, and D. Rost (2007). On continuity and strict increase of the CDF for the sup-functional of a Gaussian process with applications to statistics. *Results Math. 51*(1-2), 51–60.

Ganesh, A., and N. O'Connell (2000). A large-deviation principle for Dirichlet posteriors. *Bernoulli 6*(6), 1021–1034.

Gangopadhyay, A., B. Mallick, and D. Denison (1999). Estimation of spectral density of a stationary time series via an asymptotic representation of the periodogram. *J. Statist. Plann. Inference 75*(2), 281–290.

Gasparini, M. (1996). Bayesian density estimation via Dirichlet density processes. *J. Nonparametr. Statist. 6*(4), 355–366.

Gelfand, A., and A. Kottas (2002). A computational approach for full nonparametric Bayesian inference under Dirichlet process mixture models. *J. Comput. Graph. Statist. 11*(2), 289–305.

Ghorai, J., and H. Rubin (1982). Bayes risk consistency of nonparametric Bayes density estimates. *Austral. J. Statist. 24*(1), 51–66.

Ghosal, S. (2001). Convergence rates for density estimation with Bernstein polynomials. *Ann. Statist. 29*(5), 1264–1280.

Ghosal, S., J. Ghosh, and R. Ramamoorthi (1997). Non-informative priors via sieves and packing numbers. In *Advances in Statistical Decision Theory and Applications*, Stat. Ind. Technol., pp. 119–132. Boston, MA: Birkhäuser Boston.

Ghosal, S., J. Ghosh, and R. Ramamoorthi (1999a). Consistent semiparametric Bayesian inference about a location parameter. *J. Statist. Plann. Inference 77*(2), 181–193.

Ghosal, S., J. Ghosh, and R. Ramamoorthi (1999b). Posterior consistency of Dirichlet mixtures in density estimation. *Ann. Statist. 27*(1), 143–158.

Ghosal, S., J. Ghosh, and T. Samanta (1995). On convergence of posterior distributions. *Ann. Statist. 23*(6), 2145–2152.

Ghosal, S., J. Ghosh, and A. van der Vaart (2000). Convergence rates of posterior distributions. *Ann. Statist. 28*(2), 500–531.

Ghosal, S., J. Lember, and A. van der Vaart (2003). On Bayesian adaptation. In *Proceedings of the Eighth Vilnius Conference on Probability Theory and Mathematical Statistics, Part II (2002)*, Volume 79, pp. 165–175.

Ghosal, S., J. Lember, and A. van der Vaart (2008). Nonparametric Bayesian model selection and averaging. *Electron. J. Stat. 2*, 63–89.

Ghosal, S., and A. Roy (2006). Posterior consistency of Gaussian process prior for nonparametric binary regression. *Ann. Statist. 34*(5), 2413–2429.

Ghosal, S., A. Roy, and Y. Tang (2008). Posterior consistency of Dirichlet mixtures of beta densities in estimating positive false discovery rates. In *Beyond Parametrics in Interdisciplinary Research: Festschrift in Honor of Professor Pranab K. Sen*, Volume 1 of *Inst. Math. Stat. Collect.*, pp. 105–115. Beachwood, OH: Inst. Math. Statist.

Ghosal, S., and Y. Tang (2006). Bayesian consistency for Markov processes. *Sankhyā 68*(2), 227–239.

Ghosal, S., and A. van der Vaart (2001). Entropies and rates of convergence for maximum likelihood and Bayes estimation for mixtures of normal densities. *Ann. Statist. 29*(5), 1233–1263.

Ghosal, S., and A. van der Vaart (2003). Discussion of "new tools for consistency in Bayesian nonparametrics" by Gabriella Salinetti. In *Bayesian Statistics, 7 (Tenerife, 2002)*, pp. 382. New York: Oxford Univ. Press.

Ghosal, S., and A. van der Vaart (2007a). Convergence rates of posterior distributions for non-i.i.d. observations. *Ann. Statist. 35*(1), 192–223.

Ghosal, S., and A. van der Vaart (2007b). Posterior convergence rates of Dirichlet mixtures at smooth densities. *Ann. Statist. 35*(2), 697–723.

Ghosh, J., S. Ghosal, and T. Samanta (1994). Stability and convergence of the posterior in non-regular problems. In *Statistical Decision Theory and Related Topics, V (West Lafayette, IN, 1992)*, pp. 183–199. New York: Springer.

Ghosh, J., and R. Ramamoorthi (1995). Consistency of Bayesian inference for survival analysis with or without censoring. In *Analysis of Censored Data (Pune, 1994/1995)*, Volume 27 of *IMS Lecture Notes Monogr. Ser.*, pp. 95–103. Hayward, CA: Inst. Math. Statist.

Ghosh, J., and R. Ramamoorthi (2003). *Bayesian Nonparametrics*. Springer Series in Statistics. New York: Springer-Verlag.

Ghosh, J., and S. Tokdar (2006). Convergence and consistency of Newton's algorithm for estimating mixing distribution. In *Frontiers in Statistics*, pp. 429–443. London: Imp. Coll. Press.

Gilks, W., A. Thomas, and D. Spiegelhalter (1994). A language and program for complex Bayesian modelling. *The Statistician*, 169–177.

Gill, R., and S. Johansen (1990). A survey of product-integration with a view toward application in survival analysis. *Ann. Statist. 18*(4), 1501–1555.

Giné, E., and R. Nickl (2015). *Mathematical Foundations of Infinite-Dimensional Statistical Models*. Cambridge University Press.

Gnedin, A. (2010). A species sampling model with finitely many types. *Electron. Commun. Probab. 15*, 79–88.

Gnedin, A., and J. Pitman (2006). Exchangeable Gibbs partitions and Stirling triangles. *Journal of Mathematical Sciences 138*(3), 5674–5685.

Grenander, U. (1981). *Abstract Inference*. New York: John Wiley & Sons Inc. Wiley Series in Probability and Mathematical Statistics.

Griffin, J., and M. Steel (2006). Order-based dependent Dirichlet processes. *J. Amer. Statist. Assoc. 101*(473), 179–194.

Griffiths, T., and Z. Ghahramani (2006). Infinite latent feature models and the indian buffet process. Volume 18 of *Advances in Neural Information Processing Systems*. Cambridge, MA: MIT Press.

Gu, J., and S. Ghosal (2008). Strong approximations for resample quantile processes and application to ROC methodology. *J. Nonparametr. Stat. 20*(3), 229–240.

Gurland, J. (1948). Inversion formulae for the distribution of ratios. *Ann. Math. Statist. 19*, 228–237.

Hansen, B., and J. Pitman (2000). Prediction rules for exchangeable sequences related to species sampling. *Statist. Probab. Lett. 46*(3), 251–256.

Hanson, T., and W. Johnson (2002). Modeling regression error with a mixture of Pólya trees. *J. Amer. Statist. Assoc. 97*(460), 1020–1033.

Härdle, W., G. Kerkyacharian, A. Tsybakov, and D. Picard (1998). *Wavelets, Approximation and Statistical Applications*. New York: Springer-Verlag.

Helson, H. (1983). *Harmonic Analysis*. Reading, MA: Addison-Wesley Publishing Company Advanced Book Program.

Hesthaven, J., S. Gottlieb, and D. Gottlieb (2007). *Spectral Methods for Time Dependent Problems*. Cambridge, UK: Cambridge University Press.

Hjort, N. (1990). Nonparametric Bayes estimators based on beta processes in models for life history data. *Ann. Statist. 18*(3), 1259–1294.

Hjort, N. (1996). Bayesian approaches to non- and semiparametric density estimation. In *Bayesian Statistics, 5 (Alicante, 1994)*, Oxford Sci. Publ., pp. 223–253. New York: Oxford University Press.

Hoppe, F. (1984). Pólya-like urns and the Ewens' sampling formula. *J. Math. Biol. 20*(1), 91–94.

Huang, T.-M. (2004). Convergence rates for posterior distributions and adaptive estimation. *Ann. Statist. 32*(4), 1556–1593.

Ibragimov, I. (1962). On stationary Gaussian processes with a strong mixing property. *Dokl. Akad. Nauk SSSR 147*, 1282–1284.

Ibragimov, I., and R. Has'minskiĭ (1981). *Statistical Estimation: Asymptotic Theory*, Volume 16 of *Applications of Mathematics*. New York: Springer-Verlag. Translated from the Russian by Samuel Kotz.

Ibragimov, I., and R. Has'minskiĭ (1982). An estimate of the density of a distribution belonging to a class of entire functions. *Teor. Veroyatnost. i Primenen. 27*(3), 514–524.

Ishwaran, H., and L. James (2001). Gibbs sampling methods for stick-breaking priors. *J. Amer. Statist. Assoc. 96*(453), 161–173.

Ishwaran, H., and M. Zarepour (2002). Dirichlet prior sieves in finite normal mixtures. *Statist. Sinica 12*(3), 941–963.

Jackson, D. (1912). On approximation by trigonometric sums and polynomials. *Trans. Amer. Math. Soc. 13*(4), 491–515.

Jacobsen, M. (1989). Existence and unicity of MLEs in discrete exponential family distributions. *Scand. J. Statist. 16*(4), 335–349.

Jacod, J., and A. Shiryaev (2003). *Limit Theorems for Stochastic Processes* (Second ed.), Volume 288 of *Grundlehren der Mathematischen Wissenschaften [Fundamental Principles of Mathematical Sciences]*. Berlin: Springer-Verlag.

James, I., and J. Mosimann (1980). A new characterization of the Dirichlet distribution through neutrality. *Ann. Statist. 8*(1), 183–189.

James, L. (2005). Bayesian Poisson process partition calculus with an application to Bayesian Lévy moving averages. *Ann. Statist. 33*(4), 1771–1799.

James, L. (2008). Large sample asymptotics for the two-parameter Poisson-Dirichlet process. In *Pushing the Limits of Contemporary Statistics: Contributions in Honor of Jayanta K. Ghosh*, Volume 3 of *Inst. Math. Stat. Collect.*, pp. 187–199. Beachwood, OH: Institute of Mathematical Statistics.

James, L., A. Lijoi, and I. Prünster (2006). Conjugacy as a distinctive feature of the Dirichlet process. *Scand. J. Statist. 33*(1), 105–120.

James, L., A. Lijoi, and I. Prünster (2008). Distributions of linear functionals of two parameter Poisson-Dirichlet random measures. *Ann. Appl. Probab. 18*(2), 521–551.

James, L., A. Lijoi, and I. Prünster (2009). Posterior analysis for normalized random measures with independent increments. *Scand. J. Stat. 36*(1), 76–97.

James, L., A. Lijoi, and I. Prünster (2010). On the posterior distribution of classes of random means. *Bernoulli 16*(1), 155–180.

Jang, G., J. Lee, and S. Lee (2010). Posterior consistency of species sampling priors. *Statist. Sinica 20*(2), 581–593.

Jara, A. (2007). Applied Bayesian non-and semi-parametric inference using dppackage. *R News 7*, 17–26.

Jara, A., T. Hanson, F. Quintana, P. Müller, and G. Rosner (2015). Package dppackage.

Jiang, W. (2007). Bayesian variable selection for high dimensional generalized linear models: convergence rates of the fitted densities. *Ann. Statist. 35*(4), 1487–1511.

Kalbfleisch, J. (1978). Non-parametric Bayesian analysis of survival time data. *J. Roy. Statist. Soc. Ser. B 40*(2), 214–221.

Kallenberg, O. (1986). *Random Measures* (Fourth ed.). Berlin: Akademie-Verlag.

Karlin, S. (1967). Central limit theorems for certain infinite urn schemes. *J. Math. Mech. 17*, 373–401.

Kim, Y. (1999). Nonparametric Bayesian estimators for counting processes. *Ann. Statist. 27*(2), 562–588.

Kim, Y. (2001). Mixtures of beta processes priors for right censored survival data. *J. Korean Statist. Soc. 30*(1), 127–138.

Kim, Y. (2006). The Bernstein–von Mises theorem for the proportional hazard model. *Ann. Statist. 34*(4), 1678–1700.

Kim, Y., and J. Lee (2001). On posterior consistency of survival models. *Ann. Statist. 29*(3), 666–686.

Kim, Y., and J. Lee (2003a). Bayesian analysis of proportional hazard models. *Ann. Statist. 31*(2), 493–511. Dedicated to the memory of Herbert E. Robbins.

Kim, Y., and J. Lee (2003b). Bayesian bootstrap for proportional hazards models. *Ann. Statist. 31*(6), 1905–1922.

Kim, Y., and J. Lee (2004). A Bernstein–von Mises theorem in the nonparametric right-censoring model. *Ann. Statist. 32*(4), 1492–1512.

Kimeldorf, G., and G. Wahba (1970). A correspondence between Bayesian estimation on stochastic processes and smoothing by splines. *Ann. Math. Statist. 41*, 495–502.

Kingman, J. (1967). Completely random measures. *Pacific J. Math. 21*, 59–78.

Kingman, J. (1975). Random discrete distribution. *J. Roy. Statist. Soc. Ser. B 37*, 1–22. With a discussion by S. J. Taylor, A. G. Hawkes, A. M. Walker, D. R. Cox, A. F. M. Smith, B. M. Hill, P. J. Burville, T. Leonard, and a reply by the author.

Kingman, J. (1978). The representation of partition structures. *J. London Math. Soc. (2) 18*(2), 374–380.

Kingman, J. (1982). The coalescent. *Stochastic Process. Appl. 13*(3), 235–248.

Kingman, J. (1993). *Poisson Processes*, Volume 3 of *Oxford Studies in Probability*. New York: The Clarendon Press Oxford University Press. Oxford Science Publications.

Kleijn, B., and A. van der Vaart (2006). Misspecification in infinite-dimensional Bayesian statistics. *Ann. Statist. 34*(2), 837–877.

Knapik, B. T., B. T. Szabó, A. W. van der Vaart, and J. H. van Zanten (2016). Bayes procedures for adaptive inference in inverse problems for the white noise model. *Probab. Theory Related Fields 164*(3-4), 771–813.

Knapik, B., A. van der Vaart, and H. van Zanten (2011). Bayesian inverse problems with Gaussian priors. *Ann. Statist. 39*(5), 2626–2657.

Kolmogorov, A., and V. Tihomirov (1961). ε-entropy and ε-capacity of sets in functional space. *Amer. Math. Soc. Transl. (2) 17*, 277–364.

Komlós, J., P. Major, and G. Tusnády (1975). An approximation of partial sums of independent RVs and the sample DF. I. *Z. Wahrscheinlichkeitstheorie und Verw. Gebiete 32*, 111–131.

Korwar, R., and M. Hollander (1973). Contributions to the theory of Dirichlet processes. *Ann. Probab. 1*, 705–711.

Kraft, C. (1964). A class of distribution function processes which have derivatives. *J. Appl. Probab. 1*, 385–388.

Kruijer, W., J. Rousseau, and A. van der Vaart (2010). Adaptive Bayesian density estimation with location-scale mixtures. *Electron. J. Stat. 4*, 1225–1257.

Kruijer, W., and A. van der Vaart (2008). Posterior convergence rates for Dirichlet mixtures of beta densities. *J. Statist. Plann. Inference 138*(7), 1981–1992.

Kruijer, W., and A. van der Vaart (2013). Analyzing posteriors by the information inequality. In *From Probability to Statistics and Back: High-Dimensional Models and Processes – A Festschrift in Honor of Jon A. Wellner*. IMS.

Kuelbs, J., and W. Li (1993). Metric entropy and the small ball problem for Gaussian measures. *J. Funct. Anal. 116*(1), 133–157.

Kuelbs, J., W. Li, and W. Linde (1994). The Gaussian measure of shifted balls. *Probab. Theory Related Fields 98*(2), 143–162.

Kuo, L. (1986). Computations of mixtures of Dirichlet processes. *SIAM J. Sci. Statist. Comput. 7*(1), 60–71.

Lavine, M. (1992). Some aspects of Pólya tree distributions for statistical modelling. *Ann. Statist. 20*(3), 1222–1235.

Lavine, M. (1994). More aspects of Pólya tree distributions for statistical modelling. *Ann. Statist. 22*(3), 1161–1176.

Le Cam, L. (1953). On some asymptotic properties of maximum likelihood estimates and related Bayes' estimates. *Univ. California Publ. Statist. 1*, 277–329.

Le Cam, L. (1986). *Asymptotic Methods in Statistical Decision Theory*. Springer Series in Statistics. New York: Springer-Verlag.

Leahu, H. (2011). On the Bernstein–von Mises phenomenon in the Gaussian white noise model. *Electron. J. Stat. 5*, 373–404.

Ledoux, M., and M. Talagrand (1991). *Probability in Banach spaces*, Volume 23. Berlin: Springer-Verlag.

Lee, J. (2007). Sampling methods of neutral to the right processes. *J. Comput. Graph. Statist. 16*(3), 656–671.

Lee, J., and Y. Kim (2004). A new algorithm to generate beta processes. *Comput. Statist. Data Anal. 47*(3), 441–453.

Lee, J., F. Quintana, P. Müller, and L. Trippa (2013). Defining predictive probability functions for species sampling models. *Stat. Sci. 28*(2), 209–222.

Lember, J., and A. van der Vaart (2007). On universal Bayesian adaptation. *Statist. Decisions 25*(2), 127–152.

Lenk, P. (1988). The logistic normal distribution for Bayesian, nonparametric, predictive densities. *J. Amer. Statist. Assoc. 83*(402), 509–516.

Leonard, T. (1978). Density estimation, stochastic processes and prior information. *J. Roy. Statist. Soc. Ser. B 40*(2), 113–146. With discussion.

Li, W., and W. Linde (1998). Existence of small ball constants for fractional Brownian motions. *C. R. Acad. Sci. Paris Sér. I Math. 326*(11), 1329–1334.

Li, W., and Q.-M. Shao (2001). Gaussian processes: inequalities, small ball probabilities and applications. In *Stochastic Processes: Theory and Methods*, Volume 19 of *Handbook of Statist.*, pp. 533–597. Amsterdam: North-Holland.

Lifshits, M. (2012). *Lectures on Gaussian Processes*. Springer Briefs in Mathematics. Heidelberg: Springer.

Lijoi, A., R. Mena, and I. Prünster (2005a). Bayesian nonparametric analysis for a generalized Dirichlet process prior. *Stat. Inference Stoch. Process. 8*(3), 283–309.

Lijoi, A., R. Mena, and I. Prünster (2005b). Hierarchical mixture modeling with normalized inverse-Gaussian priors. *J. Amer. Statist. Assoc. 100*(472), 1278–1291.

Lijoi, A., R. Mena, and I. Prünster (2007a). Bayesian nonparametric estimation of the probability of discovering new species. *Biometrika 94*(4), 769–786.

Lijoi, A., R. Mena, and I. Prünster (2007b). Controlling the reinforcement in Bayesian non-parametric mixture models. *J. Roy. Stat. Soc. Ser. B, Stat. Methodol. 69*(4), 715–740.

Lijoi, A., and I. Prünster (2003). On a normalized random measure with independent increments relevant to Bayesian nonparametric inference. In *Proceedings of the 13th European Young Statisticians Meeting*, pp. 123–134. Bernoulli Society.

Lijoi, A., and I. Prünster (2009). Distributional properties of means of random probability measures. *Stat. Surv. 3*, 47–95.

Lijoi, A., I. Prünster, and S. Walker (2008). Bayesian nonparametric estimators derived from conditional Gibbs structures. *Ann. Appl. Probab. 18*(4), 1519–1547.

Lijoi, A., and E. Regazzini (2004). Means of a Dirichlet process and multiple hypergeometric functions. *Ann. Probab. 32*(2), 1469–1495.

Lindsay, B. (1995). *Mixture Models: Theory, Geometry and Applications.* In *NSF-CBMS Regional Conference Series in Probability and Statistics*, pp. i–163. Institute of Mathematical Statistics.

Liseo, B., D. Marinucci, and L. Petrella (2001). Bayesian semiparametric inference on long-range dependence. *Biometrika 88*(4), 1089–1104.

Lo, A. (1983). Weak convergence for Dirichlet processes. *Sankhyā Ser. A 45*(1), 105–111.

Lo, A. (1984). On a class of Bayesian nonparametric estimates. I. Density estimates. *Ann. Statist. 12*(1), 351–357.

Lo, A. (1986). A remark on the limiting posterior distribution of the multiparameter Dirichlet process. *Sankhyā Ser. A 48*(2), 247–249.

Lo, A. (1987). A large sample study of the Bayesian bootstrap. *Ann. Statist. 15*(1), 360–375.

Lo, A. (1991). A characterization of the Dirichlet process. *Statist. Probab. Lett. 12*(3), 185–187.

Lo, A. (1993). A Bayesian bootstrap for censored data. *Ann. Statist. 21*(1), 100–123.

MacEachern, S. (1994). Estimating normal means with a conjugate style Dirichlet process prior. *Comm. Statist. Simulation Comput. 23*(3), 727–741.

MacEachern, S. (1999). Dependent nonparametric processes. In *ASA Proceedings of the Section on Bayesian Statistical Science*, pp. 50–55. American Statistical Association, pp. 50–55, Alexandria, VA.

MacEachern, S., and P. Müller (1998). Estimating mixture of dirichlet process models. *J. Comp. Graph. Statist. 7*(2), 223–238.

Mandelbrot, B., and J. Van Ness (1968). Fractional Brownian motions, fractional noises and applications. *SIAM Rev. 10*, 422–437.

Mauldin, R., W. Sudderth, and S. Williams (1992). Pólya trees and random distributions. *Ann. Statist. 20*(3), 1203–1221.

McCloskey, J. (1965). *A Model for the Distribution of Individuals by Species in an Environment.* ProQuest LLC, Ann Arbor, MI. Thesis (Ph.D.)–Michigan State University.

McVinish, R., J. Rousseau, and K. Mengersen (2009). Bayesian goodness of fit testing with mixtures of triangular distributions. *Scand. J. Stat. 36*(2), 337–354.

Meyn, S., and R. Tweedie (1993). *Markov Chains and Stochastic Stability.* Communications and Control Engineering Series. London: Springer-Verlag London Ltd.

Micchelli, C. (1980). A constructive approach to Kergin interpolation in $\mathbb{R}^k$: multivariate B-splines and Lagrange interpolation. *Rocky Mountain J. Math. 10*(3), 485–497.

Muirhead, R. (1982). *Aspects of Multivariate Statistical Theory.* New York: John Wiley & Sons Inc. Wiley Series in Probability and Mathematical Statistics.

Muliere, P., and L. Tardella (1998). Approximating distributions of random functionals of Ferguson-Dirichlet priors. *Canad. J. Statist. 26*(2), 283–297.

Müller, P., F. Quintana, A. Jara, and T. Hanson (2015). *Bayesian Nonparametric Data Analysis*. Cambridge: Springer.

Murphy, S., and A. van der Vaart (2000). On profile likelihood. *J. Amer. Statist. Assoc. 95*(450), 449–485. With comments and a rejoinder by the authors.

Neal, R. (2000). Markov chain sampling methods for Dirichlet process mixture models. *J. Comput. Graph. Statist. 9*(2), 249–265.

Newton, M. (2002). On a nonparametric recursive estimator of the mixing distribution. *Sankhyā Ser. A 64*(2), 306–322. Selected articles from San Antonio Conference in honour of C. R. Rao (San Antonio, TX, 2000).

Newton, M., F. Quintana, and Y. Zhang (1998). Nonparametric Bayes methods using predictive updating. In *Practical Nonparametric and Semiparametric Bayesian Statistics*, Volume 133 of *Lecture Notes in Statistics*, pp. 45–61. New York: Springer.

Newton, M., and Y. Zhang (1999). A recursive algorithm for nonparametric analysis with missing data. *Biometrika 86*(1), 15–26.

Nieto-Barajas, L., I. Prünster, and S. Walker (2004). Normalized random measures driven by increasing additive processes. *Ann. Statist. 32*(6), 2343–2360.

Nussbaum, M. (1996). Asymptotic equivalence of density estimation and Gaussian white noise. *Ann. Statist. 24*(6), 2399–2430.

Panzar, L., and H. van Zanten (2009). Nonparametric Bayesian inference for ergodic diffusions. *J. Statist. Plann. Inference 139*(12), 4193–4199.

Parthasarathy, K. (2005). *Probability Measures on Metric Spaces*. AMS Chelsea Publishing, Providence, RI. Reprint of the 1967 original.

Peccati, G., and I. Prünster (2008). Linear and quadratic functionals of random hazard rates: an asymptotic analysis. *Ann. Appl. Probab. 18*(5), 1910–1943.

Perman, M., J. Pitman, and M. Yor (1992). Size-biased sampling of Poisson point processes and excursions. *Probab. Theory Related Fields 92*(1), 21–39.

Petrone, S. (1999a). Bayesian density estimation using Bernstein polynomials. *Canad. J. Statist. 27*(1), 105–126.

Petrone, S. (1999b). Random Bernstein polynomials. *Scand. J. Statist. 26*(3), 373–393.

Petrone, S., and P. Veronese (2010). Feller operators and mixture priors in Bayesian nonparametrics. *Statist. Sinica 20*(1), 379–404.

Petrone, S., and L. Wasserman (2002). Consistency of Bernstein polynomial posteriors. *J. Roy. Stat. Soc. Ser. B, Stat. Methodol. 64*(1), 79–100.

Pinsker, M. (1980). Optimal filtration of square-integrable signals in Gaussian noise. *Problems Inform. Transmission 16*.

Pitman, J. (1995). Exchangeable and partially exchangeable random partitions. *Probab. Theory Related Fields 102*(2), 145–158.

Pitman, J. (1996a). Random discrete distributions invariant under size-biased permutation. *Adv. in Appl. Probab. 28*(2), 525–539.

Pitman, J. (1996b). Some developments of the Blackwell-MacQueen urn scheme. In *Statistics, Probability and Game Theory*, Volume 30 of *IMS Lecture Notes Monogr. Ser.*, pp. 245–267. Hayward, CA: Institute of Mathematical Statistics.

Pitman, J. (2003). Poisson-Kingman partitions. In *Statistics and Science: a Festschrift for Terry Speed*, Volume 40 of *IMS Lecture Notes Monogr. Ser.*, pp. 1–34. Beachwood, OH: Institute of Mathematical Statistics.

Pitman, J. (2006). *Combinatorial Stochastic Processes*, Volume 1875 of *Lecture Notes in Mathematics*. Berlin: Springer-Verlag. Lectures from the 32nd Summer School on Probability Theory held in Saint-Flour, July 7–24, 2002. With a foreword by Jean Picard.

Pitman, J., and M. Yor (1997). The two-parameter Poisson-Dirichlet distribution derived from a stable subordinator. *Ann. Probab. 25*(2), 855–900.

Pollard, D. (1984). *Convergence of Stochastic Processes*. Springer Series in Statistics. New York: Springer-Verlag.

Præstgaard, J., and J. Wellner (1993). Exchangeably weighted bootstraps of the general empirical process. *Ann. Probab. 21*(4), 2053–2086.

Prünster, I. (2012). Proof of conjugacy of neutral to the right processes for censored data.

Ramamoorthi, R., L. Drăghici, and J. Dey (2002). Characterizations of tailfree and neutral to the right priors. In *Advances on theoretical and methodological aspects of probability and statistics (Hamilton, ON, 1998)*, pp. 305–316. London: Taylor & Francis.

Rasmussen, C., and C. Williams (2006). *Gaussian Processes for Machine Learning*. Adaptive Computation and Machine Learning. Cambridge, MA: MIT Press.

Ray, K. (2014). Adaptive Bernstein–von Mises theorems in Gaussian white noise. *arXiv:1407.3397*.

Regazzini, E., A. Guglielmi, and G. Di Nunno (2002). Theory and numerical analysis for exact distributions of functionals of a Dirichlet process. *Ann. Statist. 30*(5), 1376–1411.

Regazzini, E., A. Lijoi, and I. Prünster (2003). Distributional results for means of normalized random measures with independent increments. *Ann. Statist. 31*(2), 560–585. Dedicated to the memory of Herbert E. Robbins.

Rivoirard, V., and J. Rousseau (2012). Bernstein–von Mises theorem for linear functionals of the density. *Ann. Statist. 40*(3), 1489–1523.

Rodríguez, A., and D. Dunson (2011). Nonparametric Bayesian models through probit stick-breaking processes. *Bayesian Anal. 6*(1), 145–177.

Rodríguez, A., D. Dunson, and A. Gelfand (2008). The nested Dirichlet process. *J. Amer. Statist. Assoc. 103*(483), 1131–1144.

Rousseau, J. (2010). Rates of convergence for the posterior distributions of mixtures of betas and adaptive nonparametric estimation of the density. *Ann. Statist. 38*(1), 146–180.

Roy, A., S. Ghosal, and W. Rosenberger (2009). Convergence properties of sequential Bayesian D-optimal designs. *J. Statist. Plann. Inference 139*(2), 425–440.

Rubin, D. (1981). The Bayesian bootstrap. *Ann. Statist. 9*(1), 130–134.

Rudin, W. (1973). *Functional Analysis*. New York: McGraw-Hill Book Co. McGraw-Hill Series in Higher Mathematics.

Rue, H., S. Martino, and N. Chopin (2009). Approximate Bayesian inference for latent Gaussian models by using integrated nested Laplace approximations. *J. Roy. Stat. Soc. Ser. B Stat. Methodol. 71*(2), 319–392.

Salinetti, G. (2003). New tools for consistency in Bayesian nonparametrics. In *Bayesian Statistics, 7 (Tenerife, 2002)*, pp. 369–384. New York: Oxford Univ. Press. With discussions by J. K. Ghosh, S. Ghosal and A. van der Vaart, and a reply by the author.

Schervish, M. (1995). *Theory of Statistics*. Springer Series in Statistics. New York: Springer-Verlag.

Schumaker, L. (2007). *Spline Functions: Basic Theory* (Third ed.). Cambridge Mathematical Library. Cambridge: Cambridge University Press.

Schwartz, L. (1965). On Bayes procedures. *Z. Wahrsch. Verw. Gebiete 4*, 10–26.

Schwarz, G. (1978). Estimating the dimension of a model. *Ann. Statist. 6*(2), 461–464.

Scricciolo, C. (2006). Convergence rates for Bayesian density estimation of infinite-dimensional exponential families. *Ann. Statist. 34*(6), 2897–2920.

Scricciolo, C. (2007). On rates of convergence for Bayesian density estimation. *Scand. J. Statist. 34*(3), 626–642.

Scricciolo, C. (2014). Adaptive Bayesian density estimation in L^p-metrics with Pitman-Yor or normalized inverse-Gaussian process kernel mixtures. *Bayesian Anal. 9*(2), 475–520.

Sethuraman, J. (1994). A constructive definition of Dirichlet priors. *Statist. Sinica 4*(2), 639–650.

Sethuraman, J., and R. Tiwari (1982). Convergence of Dirichlet measures and the interpretation of their parameter. In *Statistical Decision Theory and Related Topics, III, Vol. 2 (West Lafayette, Ind., 1981)*, pp. 305–315. New York: Academic Press.

Shen, W., and S. Ghosal (2015). Adaptive Bayesian procedures using random series prior. *Scand. J. Statist. 42*, 1194–1213.

Shen, W., S. Tokdar, and S. Ghosal (2013). Adaptive Bayesian multivariate density estimation with Dirichlet mixtures. *Biometrika 100*(3), 623–640.

Shen, X., and L. Wasserman (2001). Rates of convergence of posterior distributions. *Ann. Statist. 29*(3), 687–714.

Shively, T., R. Kohn, and S. Wood (1999). Variable selection and function estimation in additive nonparametric regression using a data-based prior. *J. Amer. Statist. Assoc. 94*(447), 777–806. With comments and a rejoinder by the authors.

Skorohod, A. (1974). *Integration in Hilbert space*. New York: Springer-Verlag. Translated from the Russian by Kenneth Wickwire, Ergebnisse der Mathematik und ihrer Grenzgebiete, Band 79.

Sniekers, S., and A. van der Vaart (2015a). Adaptive Bayesian credible sets in regression with a Gaussian process prior. *Electron. J. Stat. 9*(2), 2475–2527.

Sniekers, S., and A. van der Vaart (2015b). Credible sets in the fixed design model with Brownian motion prior. *J. Statist. Plann. Inference 166*, 78–86.

Srivastava, S. (1998). *A Course on Borel Sets*, Volume 180 of *Graduate Texts in Mathematics*. New York: Springer-Verlag.

Stein, E. (1970). *Singular Integrals and Differentiability Properties of Functions*. Princeton Mathematical Series, No. 30. Princeton, N.J.: Princeton University Press.

Stone, C. (1990). Large-sample inference for log-spline models. *Ann. Statist. 18*(2), 717–741.

Stone, C. (1994). The use of polynomial splines and their tensor products in multivariate function estimation. *Ann. Statist. 22*(1), 118–184. With discussion by A. Buja and T. Hastie and a rejoinder by the author.

Strasser, H. (1985). *Mathematical Theory of Statistics*, Volume 7 of *de Gruyter Studies in Mathematics*. Berlin: Walter de Gruyter & Co. Statistical experiments and asymptotic decision theory.

Susarla, V., and J. Van Ryzin (1976). Nonparametric Bayesian estimation of survival curves from incomplete observations. *J. Amer. Statist. Assoc. 71*(356), 897–902.

Szabó, B., A. van der Vaart, and H. van Zanten (2013). Empirical Bayes scaling of Gaussian priors in the white noise model. *Electron. J. Statist. 7*, 991–1018.

Szabó, B., A. W. van der Vaart, and J. H. van Zanten (2015). Frequentist coverage of adaptive nonparametric Bayesian credible sets. *Ann. Statist. 43*(4), 1391–1428.

Talagrand, M. (1987). Regularity of Gaussian processes. *Acta Math. 159*(1-2), 99–149.

Talagrand, M. (1992). A simple proof of the majorizing measure theorem. *Geom. Funct. Anal. 2*(1), 118–125.

Tang, Y., and S. Ghosal (2007a). A consistent nonparametric Bayesian procedure for estimating autoregressive conditional densities. *Comput. Statist. Data Anal. 51*(9), 4424–4437.

Tang, Y., and S. Ghosal (2007b). Posterior consistency of Dirichlet mixtures for estimating a transition density. *J. Statist. Plann. Inference 137*(6), 1711–1726.

Teh, Y., D. Görür, and Z. Ghahramani (2007). Stick-breaking construction for the Indian buffet process. In *International Conference on Artificial Intelligence and Statistics*, pp. 556–563.

Teh, Y., M. Jordan, M. Beal, and D. Blei (2006). Hierarchical Dirichlet processes. *J. Amer. Statist. Assoc. 101*(476), 1566–1581.

Teicher, H. (1961). Identifiability of mixtures. *Ann. Math. Statist. 32*, 244–248.

Thibaux, R., and M. Jordan (2007). Hierarchical beta processes and the Indian buffet process. In *International Conference on Artificial Intelligence and Statistics*, pp. 564–571.

Tierney, L., and J. Kadane (1986). Accurate approximations for posterior moments and marginal densities. *J. Amer. Statist. Assoc. 81*(393), 82–86.

Tokdar, S. (2006). Posterior consistency of Dirichlet location-scale mixture of normals in density estimation and regression. *Sankhyā 68*(1), 90–110.

Tokdar, S. (2007). Towards a faster implementation of density estimation with logistic Gaussian process priors. *J. Comput. Graph. Statist. 16*(3), 633–655.

Tokdar, S., and J. Ghosh (2007). Posterior consistency of logistic Gaussian process priors in density estimation. *J. Statist. Plann. Inference 137*(1), 34–42.

Tokdar, S., R. Martin, and J. Ghosh (2009). Consistency of a recursive estimate of mixing distributions. *Ann. Statist. 37*(5A), 2502–2522.

Tsiatis, A. (1981). A large sample study of Cox's regression model. *Ann. Statist. 9*(1), 93–108.

van der Meulen, F., A. van der Vaart, and H. van Zanten (2006). Convergence rates of posterior distributions for Brownian semimartingale models. *Bernoulli 12*(5), 863–888.

van der Vaart, A. (1991). On differentiable functionals. *Ann. Statist. 19*(1), 178–204.

van der Vaart, A. (1998). *Asymptotic Statistics*, Volume 3 of *Cambridge Series in Statistical and Probabilistic Mathematics*. Cambridge: Cambridge University Press.

van der Vaart, A. (2010). Bayesian regularization. In *Proceedings of the International Congress of Mathematicians. Volume IV*, pp. 2370–2385. Hindustan Book Agency, New Delhi.

van der Vaart, A., and H. van Zanten (2007). Bayesian inference with rescaled Gaussian process priors. *Electron. J. Stat. 1*, 433–448 (electronic).

van der Vaart, A., and H. van Zanten (2008a). Rates of contraction of posterior distributions based on Gaussian process priors. *Ann. Statist. 36*(3), 1435–1463.

van der Vaart, A., and H. van Zanten (2008b). Reproducing kernel Hilbert spaces of Gaussian priors. In *Pushing the Limits of Contemporary Statistics: Contributions in Honor of Jayanta K. Ghosh*, Volume 3 of *Inst. Math. Stat. Collect.*, pp. 200–222. Beachwood, OH: Institute of Mathematical Statistics.

van der Vaart, A., and H. van Zanten (2009). Adaptive Bayesian estimation using a Gaussian random field with inverse gamma bandwidth. *Ann. Statist. 37*(5B), 2655–2675.

van der Vaart, A., and H. van Zanten (2011). Information rates of nonparametric Gaussian process methods. *J. Mach. Learn. Res. 12*, 2095–2119.

van der Vaart, A., and J. Wellner (1996). *Weak Convergence and Empirical Processes*. Springer Series in Statistics. New York: Springer-Verlag. With applications to statistics.

van der Vaart, A., and J. Wellner (2017). *Weak Convergence and Empirical Processes* (Second ed.). Springer Series in Statistics. New York: Springer-Verlag. With applications to statistics.

van Waaij, J., and H. van Zanten (2016). Gaussian process methods for one-dimensional diffusions: optimal rates and adaptation. *Electron. J. Statist. 10*(1), 628–645.

Verdinelli, I., and L. Wasserman (1998). Bayesian goodness-of-fit testing using infinite-dimensional exponential families. *Ann. Statist. 26*(4), 1215–1241.

Vershik, A., M. Yor, and N. Tsilevich (2001). The Markov-Kreĭn identity and the quasi-invariance of the gamma process. *Zap. Nauchn. Sem. S.-Peterburg. Otdel. Mat. Inst. Steklov. (POMI) 283*(Teor. Predst. Din. Sist. Komb. i Algoritm. Metody. 6), 21–36, 258.

Wahba, G. (1978). Improper priors, spline smoothing and the problem of guarding against model errors in regression. *J. Roy. Statist. Soc. Ser. B 40*(3), 364–372.

Walker, S. (2003). On sufficient conditions for Bayesian consistency. *Biometrika 90*(2), 482–488.

Walker, S. (2004). New approaches to Bayesian consistency. *Ann. Statist. 32*(5), 2028–2043.

Walker, S. (2007). Sampling the Dirichlet mixture model with slices. *Comm. Statist. Simulation Comput. 36*(1-3), 45–54.

Walker, S., and N. Hjort (2001). On Bayesian consistency. *J. Roy. Statist. Soc. Ser. B 63*(4), 811–821.

Walker, S., A. Lijoi, and I. Prünster (2005). Data tracking and the understanding of Bayesian consistency. *Biometrika 92*(4), 765–778.

Walker, S., A. Lijoi, and I. Prünster (2007). On rates of convergence for posterior distributions in infinite-dimensional models. *Ann. Statist. 35*(2), 738–746.

Walker, S., and P. Muliere (1997). Beta-Stacy processes and a generalization of the Pólya-urn scheme. *Ann. Statist. 25*(4), 1762–1780.

Watson, G. (1995). *A treatise on the theory of Bessel functions*. Cambridge: Cambridge University Press.

Whitney, H. (1934). Analytic extensions of differentiable functions defined in closed sets. *Trans. Amer. Math. Soc. 36*(1), 63–89.

Whittle, P. (1957). Curve and periodogram smoothing. *J. Roy. Statist. Soc. Ser. B. 19*, 38–47 (discussion 47–63).

Williamson, R. (1956). Multiply monotone functions and their Laplace transforms. *Duke Math. J. 23*, 189–207.

Wolpert, R., and K. Ickstadt (1998). Simulation of Lévy random fields. In *Practical Nonparametric and Semiparametric Bayesian Statistics*, pp. 227–242. New York: Springer.

Wong, W., and X. Shen (1995). Probability inequalities for likelihood ratios and convergence rates of sieve MLEs. *Ann. Statist. 23*(2), 339–362.

Wood, S., and R. Kohn (1998). A Bayesian approach to robust nonparametric binary regression. *J. Amer. Statist. Assoc. 93*, 203–213.

Wu, Y., and S. Ghosal (2008a). Kullback-Leibler property of kernel mixture priors in Bayesian density estimation. *Electron. J. Stat. 2*, 298–331.

Wu, Y., and S. Ghosal (2008b). Posterior consistency for some semi-parametric problems. *Sankhyā, Ser. A 70*(2), 267–313.

Wu, Y., and S. Ghosal (2010). The L_1-consistency of Dirichlet mixtures in multivariate Bayesian density estimation. *J. Multivariate Anal. 101*(10), 2411–2419.

Xing, Y. (2008). On adaptive Bayesian inference. *Electron. J. Stat. 2*, 848–862.

Xing, Y. (2010). Rates of posterior convergence for iid observations. *Comm. Statist. Theory Methods 39*(19), 3389–3398.

Xing, Y., and B. Ranneby (2009). Sufficient conditions for Bayesian consistency. *J. Statist. Plann. Inference 139*(7), 2479–2489.

Yamato, H. (1984). Expectations of functions of samples from distributions chosen from Dirichlet processes. *Rep. Fac. Sci. Kagoshima Univ. Math. Phys. Chem.* (17), 1–8.

Yang, Y., and A. Barron (1999). Information-theoretic determination of minimax rates of convergence. *Ann. Statist. 27*(5), 1564–1599.

Yang, Y., and S. T. Tokdar (2015). Minimax-optimal nonparametric regression in high dimensions. *Ann. Statist. 43*(2), 652–674.

Zhang, T. (2006). From ϵ-entropy to KL-entropy: analysis of minimum information complexity density estimation. *Ann. Statist. 34*(5), 2180–2210.

Zhao, L. (2000). Bayesian aspects of some nonparametric problems. *Ann. Statist. 28*(2), 532–552.

Zygmund, A. (2002). *Trigonometric series. Vol. I, II* (Third ed.). Cambridge Mathematical Library. Cambridge University Press, Cambridge. With a foreword by Robert A. Fefferman.
peacoke

Author Index

Subject Index

Printed in the United States
By Bookmasters